Differential-Algebraic Equations Forum

The series *Differential-Algebraic Equations Forum* is concerned with analytical, algebraic, control theoretic and numerical aspects of differential algebraic equations, as well as their applications in science and engineering. It is aimed to contain survey and mathematically rigorous articles, research monographs and textbooks. Proposals are assigned to a Publishing Editor, who recommends publication on the basis of a detailed and careful evaluation by at least two referees. The appraisals will be based on the substance and quality of the exposition.

More information about this series at http://www.springer.com/series/11221

Ulrich Oberst · Martin Scheicher ·
Ingrid Scheicher

Linear Time-Invariant Systems, Behaviors and Modules

Springer

Ulrich Oberst
Department of Mathematics
University of Innsbruck
Innsbruck, Austria

Martin Scheicher

Ingrid Scheicher

ISSN 2199-7497 ISSN 2199-840X (electronic)
Differential-Algebraic Equations Forum
ISBN 978-3-030-43935-4 ISBN 978-3-030-43936-1 (eBook)
https://doi.org/10.1007/978-3-030-43936-1

This Springer imprint is published by the registered company Springer Nature Switzerland AG
The registered company address is: Gewerbestrasse 11, 6330 Cham, Switzerland

Contents

About the Author

Ulrich Oberst is Professor Emeritus at the University of Innsbruck, Austria. Earlier, he was professor of mathematics at the same university from 1972–2009. He completed his Ph.D. in Mathematics from the University of Munich, Germany, in 1965, under the guidance of Prof. F. Kasch who, in particular, taught injective co-generators—a notion from homological algebra which plays an important part in the present book. He has visited a number of universities in several countries on invitations, including Pennsylvania State University, USA, University of California San Diego, USA, University of Chicago, USA, University of Florida, USA, and University of Rome II, Italy. His areas of interest include algebra and applied mathematics with special emphasis on constructive methods, in particular multidimensional linear systems theory and discrete Gelfand, Fourier and Hartley transforms. A father of two and grandfather of six, Prof. Oberst is happily married to Karla. He is also an experienced piano player and has participated in two chamber music ensembles, playing piano trios, quartets and quintets with, in particular, string players from the Tyrolean Symphony Orchestra Innsbruck.

Chapter 1
Introduction

In Sect. 2.1, we explain the basic objects and goals of LTI (linear time-invariant) systems theory for the continuous-time standard case over the complex field in the behavioral language. In Sect. 2.9, we point out the other LTI theories treated in this book, in particular, the discrete-time theory over an arbitrary field and the module-behavior duality enabling the simultaneous treatment of all these cases. In Sect. 2.11, we mention various other systems theories that are not discussed in this book and are often more general and difficult and less developed.

All notions, models, and theories of this book from systems theory and electrical engineering have been taken from the engineering literature, mainly from the textbooks [1–12]. Further textbooks on the subjects of this book are [13–18]. Many different mathematical subjects have been used in the original literature and in the cited books and are also applied, but often in a different form, in the present book. Among these fields are linear and polynomial algebra, module theory, differential equations, topology, convolution, distributions, Fourier series and transform, Laplace transform, cf. [2, XVII].

The goal of this book is

1. to derive the systems theoretic and electrical engineering results, mainly taken from or inspired by the quoted textbooks, by partly new mathematical methods, in particular, by module-behavior duality.
2. to give complete and exact proofs of all results from item 1 and also of all used mathematical results that go beyond the first two years of university mathematics. The latter will be recalled, but without proofs.
3. to accompany all important results by algorithms that can be implemented in all computer algebra systems, for instance, in MAPLE, and to demonstrate such implementations in several nontrivial examples that are mainly exposed in the later application Chaps. 8, 10 and 12. The examples use and demonstrate many of the results and algorithms of the preceding chapters. The algorithms require, of course, an understanding of the meaning of the quantities that appear in them,

U. Oberst et al., *Linear Time-Invariant Systems, Behaviors and Modules*, Differential-Algebraic Equations Forum,
https://doi.org/10.1007/978-3-030-43936-1_1

but not of all mathematical details of their derivations. Application of the algorithms to various problems of the quoted textbooks will further demonstrate their applicability and usefulness.

Except that mentioned in 2 we do not assume any previous knowledge, in particular of systems and control theory, in contrast to several of the cited books. Like in most of these the exposition of the necessary mathematics requires substantial space in this book. This applies, in particular, to module-behavior duality, quotient rings and modules and simplified versions of the last four analysis subjects from above. Of course, the study of this material can be omitted if the reader knows it.

The book can be studied by everyone who is interested in the treated subjects, cf. the Contents and Chap. 2, and has the prerequisites mentioned in 2. In addition, a certain so-called mathematical maturity, i.e., an experience with rigorous proofs and algorithms, is desirable. Knowledge of physics or engineering is not required, but of course an advantage. Electrical and translational mechanical networks are described from scratch, but other parts of mechatronics like electromechanical systems are not touched at all. For these, we refer to the books on mechatronics, for instance, [19–21]. From our study of the quoted engineering textbooks and from our experience as mathematicians, we conclude that readers with either a mathematical or an engineering background will have no problems with the presupposed analysis, in particular, with a higher order ordinary linear differential equation with constant coefficients and with elementary complex variables. The Laplace transform, in particular of the Dirac distribution and its derivatives, is assumed as standard knowledge of engineering students in most cited engineering textbooks, whereas it is not discussed at all in the regular curriculum of the first two years in mathematics. It is a difficult subject, cf. [22, Chap. VIII], [12, Sect. 12.3.4], and is, therefore, fully developed in this book by a new rigorous method that resembles Heaviside's unproven original operational calculus. In contrast, the algebraic prerequisites are standard knowledge of mathematicians, but, as far as we can infer from the quoted textbooks, not of engineers. So readers with this background will have to study some algebra that is recalled in the book, but without proofs, mainly basic definitions and results on (Noetherian) commutative rings and modules, in particular Hom and exactness, principal ideal domains, and the Smith form of polynomial matrices.

We refer to the cited books for the history of systems theory, for much larger bibliographies than in the present book that also list numerous original papers, for various introductions to the methods and results up to 1970 and the validity of the used models and their technical boundaries. Very many outstanding and well-known scientists contributed to the field, and their previous ideas and important work are, of course, also the basis of this book. Due to their large number we can only mention some of them. We refer to their homepages for bibliographical data. We do not quote the mathematical details of the original papers since ours are different in general. Expert systems theorists will, of course, recognize how we adapted the ideas of our predecessors to our framework. In general, we do not discuss this transformation process. For many results of the book, we point, however, to corresponding results from the cited books, and the reader can thus compare the results and methods of these books with ours.

The *Contents* list the discussed subjects of the book. Chapter 2 is a detailed comment on the content and a self-contained survey over larger and, according to the cited textbooks and the engineering community, highly significant parts of linear time-invariant systems theory and electrical engineering and the decisive equations of these fields, on the basis of mathematical knowledge of two university years. In this chapter, we present the most important methods and results of the book. We state the results and refer to the sections or theorems where they are discussed, and also to corresponding results in the cited books. Of course, the chapter contains no proofs and does not assume any knowledge from other chapters. Many advanced notions, methods, and results have to be explained. We have done this in a mathematical language that is known after two years of studies in mathematics. All additional notions are introduced in the chapter. As the title *Survey* of Chap. 2 indicates its content will be discussed in detail in the later chapters and is, of course, not presupposed in these. So a potential reader need not read Chap. 2 to understand the following chapters. However, we recommend this.

Most of the book's results are constructive and accompanied by partly new algorithms, but the latter are not exposed in Chap. 2. They can be carried out with all computer algebra systems. The most important tools are the computations of the Smith form of polynomial matrices and of the complex roots and complex partial fraction decomposition of rational polynomials. Over the base fields of rational and Gaussian numbers as in all practical cases and over finite fields the Smith form is given precisely. Over the real or complex numbers, problems with numerical computations may arise, but are not discussed in this book. As important applications we discuss *electrical and translational mechanical networks*. From an application point of view, the following sections of this book are the most important ones:

4. Sections 8.3 and 8.4 on *electrical and mechanical networks*, cf. their survey in Sect. 2.7. They furnish comprehensive tools for the analysis and synthesis of these networks, but we do not treat the vast field of synthesis of networks with specific properties. Theorems 8.3.11, 8.3.18, 8.3.21, 8.3.23, 8.3.32, 8.3.40, and 8.3.43 are the main results. Examples 8.3.17, 8.3.25, 8.3.33, 8.4.6, and 8.4.7 demonstrate the algorithms and their implementation.
5. Section 10.2, cf. its survey in Sect. 2.8, on the construction and parametrization, for a stabilizable plant, of all *stabilizing feedback compensators that perform the tasks of tracking and disturbance rejection, and on their robustness. The main construction resp. robustness results are Theorems* 10.2.8, 10.2.11, 10.2.17 *resp. Theorems* 10.2.33, 10.2.47, 10.2.50. Examples 10.1.17, 10.1.22, 10.1.23, and 10.2.12 demonstrate the algorithms and their implementation.
6. In Sect. 12, we compute state space realizations of input/output behaviors by means of Gröbner bases, cf. Sect. 2.2. This method gives more general and more constructive results than in the literature. Examples 12.3.11, 12.4.9, 12.4.11, and 12.4.15 demonstrate the algorithms and their implementation.
7. In Chap. 13, we extend the standard *fractional calculus* considerably and solve complicated linear systems of generalized fractional integral/differential equations constructively. Theorem 13.1.3 is the main result and Example 13.1.7 gives simple, but instructive examples.

Compared to the existing literature and especially to the quoted textbooks essential and, in our opinion, of course, favorable modifications are carried out in the following subjects:

8. Behavior-module duality instead of time-frequency domain duality. In the latter, the transformation from the time domain to the frequency domain is, in general, connected with a loss of information. This is avoided by the categorical module-behavior duality. In particular, the fine properties of autonomous and noncontrollable behaviors can thus be studied.
9. Behavior isomorphisms instead of system equivalences. The categorical module-behavior duality enables this and simplifies the study of system equivalences.
10. Algebraic definition of the rational transfer matrix without the impulse response and without the Laplace transform. All standard properties of the transfer matrix hold and are proven.
11. Stability theory by means of the characteristic variety and of quotient modules.
12. (Periodic) distributions, Laplace transform and Fourier series and construction of transfer operators (input/output maps) without impulse responses and without integral operators. In its simplest and most important form, the inverse Laplace transform describes the bijection of the predefined rational transfer matrices onto their (possibly distributional) impulse responses.
13. The input/output representation of an electrical or mechanical network by means of the simple Gauß algorithm instead of the usual tree-cotree graph theoretical methods and its study by means of the transfer matrix and operator.
14. The construction of stabilizing compensators and the study of their robustness by means of quotient signal modules.
15. State space realizations by means of Buchberger's Gröbner basis algorithm. This method gives more precise results than usual and is fully constructive since this algorithm is implemented in all computer algebra systems.
16. Generalized fractional calculus and behaviors via vector space-behavior duality and constructive solution of multivariable linear systems of generalized fractional integral/differential equations.

For the study of electrical networks in Sect. 8.3, most results of Chaps. 3–8 are needed except those of Chap. 7, Sect. 8.1 and those on state space behaviors and on Rosenbrock equations in Sects. 4.1.2–4.1.3, 6.3.2–6.3.5. In Chap. 10, Chaps. 7 resp. 9 on (feedback) interconnections resp. on stability via quotient modules are essential additional tools. For the application of the results to state space systems, their previous study is, of course, required.

We use the standard notations $\mathbb{N}$ ($\mathbb{Z}$, $\mathbb{Q}$, $\mathbb{R}$, $\mathbb{C}$) for the natural numbers (integers, rational numbers, real numbers, complex numbers). The real resp. imaginary part of a complex number z is denoted by $\Re(z)$ resp. $\Im(z)$. The number of elements of a finite set S is $\sharp(S)$. Other more specific notations are listed in the index of the book.

Acknowledgements

1. We thank Christian Bargetz for critical reading of Sects. 10.2.4, 10.2.5.
2. We thank two anonymous reviewers for their reviews, suggestions, and considerable work for our book.
3. We thank the editors for having accepted the book for the DAE series.

References

1. W.A. Wolovich, *Linear Multivariable Systems*. Applied Mathematical Sciences, vol. 11 (Springer, New York, 1974)
2. T. Kailath, *Linear Systems* (Prentice-Hall, Englewood Cliffs, 1980)
3. C.T. Chen, *Linear System Theory and Design* (Harcourt Brace College Publishers, Fort Worth, 1984)
4. F.M. Callier, C.A. Desoer, *Multivariable Feedback Systems* (Springer Texts in Electrical Engineering. Springer, New York, 1982)
5. F.M. Callier, C.A. Desoer, *Linear System Theory* (Springer, New York, 1991)
6. M. Vidyasagar, *Control System Synthesis. A Factorization Approach*, vol. 7. MIT Press Series in Signal Processing, Optimization, and Control (MIT Press, Cambridge, 1985)
7. R. Unbehauen, *Elektrische Netzwerke. Eine Einführung in die Analyse* (Springer, Berlin, 1987)
8. A.I.G. Vardulakis, *Linear Multivariable Control. Algebraic Analysis and Synthesis Methods* (John Wiley & Sons, Chichester, 1991)
9. M. Albach, *Grundlagen der Elektrotechnik 2. Periodische und nichtperiodische Signalformen* (Pearson Studium, München, 2005)
10. L.P. Schmidt, G. Schaller, S. Martius, *Grundlagen der Elektrotechnik 3. Netzwerke* (Pearson Studium, München, 2006)
11. P.J. Antsaklis, A.N. Michel, *Linear Systems*, 2nd edn. (Birkhäuser, Boston, 2006)
12. H. Bourlès, *Linear Systems* (ISTE, London, 2010)
13. P.A. Fuhrmann, *Linear Systems and Operators in Hilbert Space* (McGraw-Hill, New York, 1981)
14. E.D. Sontag, *Mathematical Control Theory* (Springer, New York, 1990)
15. K. Zhou, J.C. Doyle, K. Glover, *Robust and Optimal Control* (Prentice-Hall, 1996)
16. D. Hinrichsen, A.J. Pritchard, *Mathematical Systems Theory I. Modelling, State Space Analysis, Stability and Robustness*, vol. 48 *Texts in Applied Mathematics* (Springer, Berlin, 2005)
17. P.A. Fuhrmann, U. Helmke, *The Mathematics of Networks of Linear Systems* (Springer International Publishing, 2015)
18. D. Hinrichsen, A.J. Pritchard with the cooperation of F. Colonius, T. Damm, A. Ilchmann, B. Jacob, F. Wirth, *Mathematical Systems Theory II. Control, Observation, Realization, and Feedback* (Springer)
19. R.G. Ballas, G. Pfeifer, R. Werthschützky, *Elektromechanische Systeme in der Mikrotechnik und Mechatronik* (Springer, Berlin, 2009)
20. K. Jantschek, *Mechatronic Systems Design. Methods, Models, Concepts* (Springer, Berlin, 2012)
21. D.C. Karnopp, D.L. Margolis, R.C. Rosenberg, *System Dynamics. Modeling, Simulation, and Control of Mechatronic Systems* (Wiley, Hoboken, 2012)
22. L. Schwartz. *Théorie des distributions* (Publications de l'Institut de Mathématique de l'Université de Strasbourg, No. IX-X. Hermann, Paris, nouvelle édition, entiérement corrigée, refondue et augmentée edition, 1966)

Chapter 2
A Survey of the Book's Content

2.1 Modules and Behaviors

In this chapter we explain the problems and results of LTI systems theory in the continuous-time case over the complex field $\mathbb{C}$. The theory for the real field $\mathbb{R}$, that is predominant in the engineering literature, is amply treated in the book. A complex polynomial in the indeterminate s has the form $f = \sum_{\mu=0}^{d} f_\mu s^\mu =: \sum_{\mu=0}^{\infty} f_\mu s^\mu,\ d \in \mathbb{N}$, where the f_μ belong to $\mathbb{C}$ and are zero for $\mu > d$. The letter s for the indeterminate comes from the Laplace transform where s denotes a complex number with sufficiently large real part. It also reminds of the shift operator in discrete-time systems theory. The set $\mathbb{C}[s]$ of all polynomials with the standard addition and multiplication is a principal ideal domain, and its quotient field $\mathbb{C}(s)$ consists of the rational functions $h(s) = f(s)g(s)^{-1},\ f, g \in \mathbb{C}[s],\ g \neq 0$.

Let $\mathcal{F}$ be a vector space of complex-valued functions $y(t)$, $t \in \mathbb{R}$, on the real line $\mathbb{R}$. In systems theory and electrical engineering $\mathbb{R}$ resp. t are interpreted as the *time axis resp. a time instant*, and the function y is called a *signal*. The basic equations for the considered theories are differential and require that $\mathcal{F}$ is closed under differentiation, i.e., $y \in \mathcal{F}$ implies $s \circ y := \mathrm{d}y/\mathrm{d}t \in \mathcal{F}$. The prototypical space with this property is the space $\mathrm{C}^\infty := \mathrm{C}^\infty(\mathbb{R}, \mathbb{C})$ of smooth complex-valued functions. This signal space is, however, too restricted for engineering applications, since these require piecewise continuous signals with jumps, for instance to describe the switching of electrical networks. The smallest space that contains these signals and is closed under differentiation is the space $\mathrm{C}^{-\infty} := \mathrm{C}^{-\infty}(\mathbb{R}, \mathbb{C})$ of *distributions of finite order* that consists of all derivatives of (piecewise) continuous signals, cf. Sects. 2.3 and 8.2 for a detailed treatment. The derivative $\mathrm{d}/\mathrm{d}t : \mathrm{C}^{-\infty} \to \mathrm{C}^{-\infty}$ is defined as a $\mathbb{C}$-linear derivation such that $\mathrm{d}y/\mathrm{d}t$ coincides with the standard derivative y' for continuously differentiable functions $y\ (\in \mathrm{C}^1)$. In particular, $\mathrm{C}^{-\infty}$ contains Dirac's δ-distribution

$$\delta := \mathrm{d}Y/\mathrm{d}t = \mathrm{d}^2y/\mathrm{d}t^2,\ Y(t) := \begin{cases} 1 & \text{if } t \geq 0 \\ 0 & \text{if } t < 0 \end{cases},\ y(t) := \begin{cases} t & \text{if } t \geq 0 \\ 0 & \text{if } t < 0 \end{cases}. \tag{2.1}$$

U. Oberst et al., *Linear Time-Invariant Systems, Behaviors and Modules*, Differential-Algebraic Equations Forum,
https://doi.org/10.1007/978-3-030-43936-1_2

Here Y is *Heaviside's step function* and δ is interpreted as an impulse at $t = 0$, cf. (2.38). We define the *scalar multiplication*,

$$f \circ y := \sum_{\mu=0}^{\infty} f_\mu y^{(\mu)},\ y^{(\mu)} := \mathrm{d}^\mu y/\mathrm{d}t^\mu,\ f = \sum_\mu f_\mu s^\mu \in \mathbb{C}[s],\ y \in \mathcal{F}, \tag{2.2}$$

that makes $\mathcal{F}$ a $\mathbb{C}[s]$-module, i.e., addition and scalar multiplication satisfy the associative, commutative, and distributive laws like a vector space. So f acts on y as differential operator. The column space $\mathcal{F}^l,\ l \in \mathbb{N}$, is also a $\mathbb{C}[s]$-module with the componentwise structure. Consider a polynomial $k \times l$-matrix

$$R = \left(R_{\alpha\beta}\right)_{1\le\alpha\le k,\ 1\le\beta\le l} \in \mathbb{C}[s]^{k\times l},\ k, l \in \mathbb{N},\ R_{\alpha\beta} = \sum_\mu R_{\alpha\beta,\mu} s^\mu \in \mathbb{C}[s]. \tag{2.3}$$

For a column vector $w = (w_1, \ldots, w_l)^\top \in \mathcal{F}^l$ we define

$$\begin{aligned} &R \circ w \in \mathcal{F}^k,\ (R \circ w)_\alpha := \sum_{\beta=1}^{l} R_{\alpha\beta} \circ w_\beta = \sum_{\beta=1}^{l} \sum_{\mu\in\mathbb{N}} R_{\alpha\beta,\mu} w_\beta^{(\mu)},\ \text{and then} \\ &\mathcal{B} := \left\{w \in \mathcal{F}^l;\ R \circ w = 0\right\} \\ &\quad = \left\{ w \in \mathcal{F}^l;\ \forall \alpha = 1, \ldots, k : \sum_{\beta=1}^{l} \sum_{\mu\in\mathbb{N}} R_{\alpha\beta,\mu} w_\beta^{(\mu)} = 0 \right\}. \end{aligned} \tag{2.4}$$

The equation $R \circ w = x$ with given right side $x \in \mathcal{F}^k$ represents an *inhomogeneous* (x arbitrary) resp. *homogeneous* ($x = 0$) *implicit system of linear differential equations with constant coefficients* $R_{\alpha\beta,\mu}$. The solution set $\mathcal{B}$ is a $\mathbb{C}[s]$-submodule of $\mathcal{F}^l$, i.e., closed under addition and scalar multiplication. Its elements are the *trajectories* of $\mathcal{B}$. According to Willems [1], these solution modules $\mathcal{B}$ are called *behaviors* in systems theory. In *Algebraic Analysis*, i.e., the algebraic theory of linear PDEs (partial differential equations), they were already extensively studied by Ehrenpreis, Malgrange, Palamodov [2–5] in the beginning 1960s, both for distributional and for smooth signals. This theory was applied to multidimensional systems theory in [6]. The present book describes, in particular, the much simpler one-dimensional version of this theory. *One-dimensional* resp. *multidimensional* systems or behaviors are described by linear systems of *ordinary* resp. of *partial* differential or difference equations with constant coefficients.

Equation systems $R \circ w = x$ and their solution modules $\mathcal{B}$ occur naturally when large systems are composed of many components that are described by basic and simple linear differential equations with constant coefficients. Such systems arise from physics and engineering, economics, biology, etc., also from more general nonlinear systems by linearization. Our models have been taken from the cited books. Prototypical examples are *electrical and mechanical networks* that will be studied in detail in Sects. 8.3 and 8.4. The primary interest of an engineer is the behavior $\mathcal{B}$ and its trajectories that can be measured, controlled, etc. and show *how the system behaves*, and hence the chosen terminology.

The matrix R gives rise to its row submodule

$$U := \mathbb{C}[s]^{1\times k}R := \sum_{\alpha=1}^{k}\mathbb{C}[s]R_{\alpha-} := \left\{\sum_{\alpha=1}^{k} f_\alpha R_{\alpha-};\ f_\alpha \in \mathbb{C}[s]\right\} \subseteq \mathbb{C}[s]^{1\times l},$$
$$R_{\alpha-} := (R_{\alpha 1},\dots,R_{\alpha l}) \in \mathbb{C}[s]^{1\times l}, \tag{2.5}$$

of the free module $\mathbb{C}[s]^{1\times l}$ of l-dimensional rows. The latter has the standard $\mathbb{C}[s]$-basis

$$\delta_\beta := (0,\dots,0,\overset{\beta}{1},0,\dots,0),\ \beta = 1,\dots,l,\ \text{with}$$
$$\xi = (\xi_1,\dots,\xi_l) = \sum_{\beta=1}^{l}\xi_\beta\delta_\beta \in \mathbb{C}[s]^{1\times l}. \tag{2.6}$$

The α-th row resp. β-th column of the matrix R are denoted by $R_{\alpha-} = R_{\alpha,-}$ resp. by $R_{-\beta} = R_{-,\beta}$. The submodule U, in turn, induces the finitely generated factor module

$$M := \mathbb{C}[s]^{1\times l}/U := \left\{\overline{\xi} := \xi + U;\ \xi \in \mathbb{C}[s]^{1\times l}\right\} = \sum_{\beta=1}^{l}\mathbb{C}[s]\overline{\delta_\beta}\ \text{with}$$
$$\overline{\xi} + \overline{\eta} := \overline{\xi+\eta},\ f\overline{\xi} := \overline{f\xi},\ \xi,\eta \in \mathbb{C}[s]^{1\times l},\ f \in \mathbb{C}[s], \tag{2.7}$$

and the distinguished list of generators $\overline{\delta_\beta}$ that satisfy the relations

$$0 = \overline{R_{\alpha-}} = \sum_{\beta=1}^{l} R_{\alpha\beta}\overline{\delta_\beta},\ \alpha = 1,\dots,k. \tag{2.8}$$

It was a simple, but important observation of Malgrange in 1962 that the map

$$\mathrm{sol}_{\mathcal{F}}(M) := \mathrm{Hom}_{\mathbb{C}[s]}(M,\mathcal{F}) \overset{\cong}{\to} \mathcal{B},\ \phi \mapsto w,\ w_\beta := \phi(\overline{\delta_\beta}), \tag{2.9}$$

is well-defined and a $\mathbb{C}[s]$-isomorphism, where $\mathrm{Hom}_{\mathbb{C}[s]}(M_1, M_2)$ denotes the $\mathbb{C}[s]$-module of all $\mathbb{C}[s]$-linear maps from a $\mathbb{C}[s]$-module M_1 into another one M_2. The isomorphism (2.9) is the first link between modules and behaviors. Equation (2.9) also was an early explicit appearance of solution modules that were later called behaviors by Willems.

For many important signal modules $\mathcal{F}$ there is a one-one correspondence between $\mathcal{B}$, U, and M. This makes the old and well-established theory of polynomial matrices and finitely generated polynomial modules available for systems theory. The same algebraic theory was used by Kalman in the 1960s to derive his *state space theory*

[7], by Rosenbrock [8] and Wolovich [9] in the 1970s for the *polynomial matrix models* or *differential operator representations* and also by Willems in his *theory of behaviors* [1], [10]. Indeed, there is no approach to LTI systems theory without univariate polynomial and rational matrices. In the so-called *frequency domain* the latter appear as rational Laplace transforms. In this book the frequency domain is replaced by the *algebraic domain* of finitely generated polynomial modules. *All* important algorithms of LTI systems theory in engineering, in systems theory, in the quoted, and in the present book rest on algorithms from univariate polynomial algebra or, in the case of state space theory, also from linear algebra over a field. This explains why *algebra* plays such a dominant part in LTI systems theory. However, *analysis* is also an essential ingredient of the theory, of the quoted books and also here. Sections 8.2 and 10.2.3–10.2.5 introduce and discuss, with complete and exact proofs, indispensable notions like distributions, in particular periodic ones, Laplace transform, convolution, Fourier series, and integral and normed linear spaces. Lebesgue's theory, i.e., measure, integral, and convolution, is not needed or used in this book.

We isolate two properties of $\mathcal{F}$ that imply the one-one correspondence $M \leftrightarrow U \leftrightarrow \mathcal{B}$. Since the rows of R generate U it is obvious that

$$\mathcal{B} = \left\{w \in \mathcal{F}^l;\ R \circ w = 0\right\} = U^\perp := \left\{w \in \mathcal{F}^l;\ \forall \xi \in U : \xi \circ w = 0\right\}, \quad (2.10)$$

i.e., $\mathcal{B}$ depends on U and M, but not on the special generating matrix R. The number $p := \operatorname{rank}(R)$ is the rank of R as matrix with entries in the field $\mathbb{C}(s)$. Since $\mathbb{C}[s]$ is a principal ideal domain, the $\mathbb{C}[s]$-module U is free of dimension p, i.e., has a basis of this length. In other words, there is a matrix $\widetilde{R} \in \mathbb{C}[s]^{p\times l}$ of

$$\operatorname{rank}(\widetilde{R}) = \operatorname{rank}(R) = p \text{ such that } U = \mathbb{C}[s]^{1\times k} R = \mathbb{C}[s]^{1\times p}\widetilde{R} = \oplus_{\alpha=1}^{p}\mathbb{C}[s]\widetilde{R}_{\alpha-}.$$

In the sequel we may and do, therefore, assume that $R = \widetilde{R}$, i.e., that $p = k$ and that the p rows of R are linearly independent and thus a $\mathbb{C}[s]$-basis of U.

Since R has rank p, there are various choices of p linearly independent columns of R. After such a choice and a possible permutation of the columns of R and the components of w we may assume that R, w, and $\mathcal{B}$ have the form

$$\begin{aligned}
&R = (P, -Q) \in \mathbb{C}[s]^{p\times(p+m)},\ m := l - p,\ \operatorname{rank}(P) = p \text{ or } \det(P) \neq 0,\\
&w = \left(\begin{smallmatrix} y\\ u\end{smallmatrix}\right) \in \mathcal{F}^{p+m},\ \mathcal{B} = \left\{\left(\begin{smallmatrix} y\\ u\end{smallmatrix}\right) \in \mathcal{F}^{p+m};\ P \circ y = Q \circ u\right\}\\
&\Longrightarrow H := P^{-1}Q \in \mathbb{C}(s)^{p\times m}.
\end{aligned} \quad (2.11)$$

Such a decomposition of R and $\mathcal{B}$ is called an *IO (input/output) decomposition* or *structure* with $u \in \mathcal{F}^m$ as *input* and $y \in \mathcal{F}^p$ as *output*, and $\mathcal{B}$ with this structure is called an *IO behavior*. The number m is called the *rank* of M and of $\mathcal{B}$. In engineering the input u is also called the *external excitation* or *cause* and y the *response, reaction*, or *effect*. This interpretation and language is appropriate only if the input u is *free*, i.e., if *each* $u \in \mathcal{F}^m$ gives rise to an output $y \in \mathcal{F}^p$, i.e., a solution of $P \circ y = Q \circ u$. A (signal) module $\mathcal{F}$ is called *injective* if this holds, i.e., if all equations $P \circ y = Q \circ u$

with given $u \in \mathcal{F}^m$, $(P, -Q) \in \mathbb{C}[s]^{p\times(p+m)}$ and $\operatorname{rank}(P) = p$ have a solution y. Since $\mathbb{C}[s]$ is a principal ideal domain, it suffices that this holds for $p = m = 1$. In particular, every $\mathbb{C}(s)$-vector space is an injective $\mathbb{C}[s]$-module. We will study injectivity in detail in Sect. 3.2. *In the rest of Chap. 2 we assume that* ${}_{\mathbb{C}[s]}\mathcal{F}$ *is injective.*

The following three important signal modules are injective, cf. Results 3.2.12, 8.2.28 and 5.3.10:

$$\begin{aligned} &\mathcal{F} := \mathrm{C}^{-\infty} = \mathrm{C}^{-\infty}(\mathbb{R}, \mathbb{C}) \supset \mathrm{C}^{\infty} = \mathrm{C}^{\infty}(\mathbb{R}, \mathbb{C}) \supset \mathrm{t}(\mathcal{F}) \\ &\mathrm{t}(\mathcal{F}) = \mathrm{t}(\mathrm{C}^{-\infty}) = \mathrm{t}(\mathrm{C}^{\infty}) = \bigoplus_{\lambda\in\mathbb{C}} \mathbb{C}[t]e^{\lambda t} = \bigoplus_{\lambda\in\mathbb{C}}\bigoplus_{k\in\mathbb{N}} \mathbb{C}t^k e^{\lambda t}. \end{aligned} \tag{2.12}$$

The module $\mathrm{t}(\mathcal{F})$ is the *torsion submodule* of $\mathcal{F}$ of all signals y that satisfy a differential equation $f \circ y = 0,\ 0 \neq f \in \mathbb{C}[s]$. It consists of the *polynomial-exponential* functions that are (finite) $\mathbb{C}$-linear combinations of functions $t^k e^{\lambda t}$, cf. Sect. 5.3.3. The following inclusions hold:

$$\begin{aligned} \mathrm{C}^{-\infty} &\supset \mathrm{C}^{0,\mathrm{pc}} := \mathrm{C}^{0,\mathrm{pc}}(\mathbb{R}, \mathbb{C}) := \{u : \mathbb{R} \to \mathbb{C} \text{ piecewise continuous}\} \\ &\supset \mathrm{C}^{0} := \mathrm{C}^{0}(\mathbb{R}, \mathbb{C}) := \{u : \mathbb{R} \to \mathbb{C} \text{ continuous}\}\,. \end{aligned} \tag{2.13}$$

The $\mathbb{C}[s]$-submodule $\mathrm{C}_{+}^{-\infty}$ of distributions *with left bounded support* is given by the derivatives of continuous functions with such support, i.e.,

$$\begin{aligned} &\mathrm{C}_{+}^{-\infty} := \mathrm{C}^{-\infty}(\mathbb{R}, \mathbb{C})_{+} := \bigcup_{n\geq 0} s^n \circ \mathrm{C}_{+}^{0} \text{ where} \\ &\mathrm{C}_{+}^{0,(\mathrm{pc})} := \mathrm{C}^{0,(\mathrm{pc})}(\mathbb{R}, \mathbb{C})_{+} := \left\{u \in \mathrm{C}^{0,(\mathrm{pc})};\ \exists t_0 \forall t \leq t_0 : u(t) = 0\right\}. \end{aligned} \tag{2.14}$$

For an obvious reason the signals in $\mathrm{C}_{+}^{-\infty}$ are called *initially-at-rest*. Since all technical systems start at some time t_0, mostly chosen as $t_0 = 0$, these signals are important. The δ-distribution according to (2.1) is obviously contained in $\mathrm{C}_{+}^{-\infty}$. The $\mathbb{C}[s]$-module $\mathrm{C}_{+}^{-\infty}$ is a $\mathbb{C}(s)$-vector space by an, necessarily unique, extension of the $\mathbb{C}[s]$-scalar multiplication, cf. Theorem 8.2.37. If u is continuous and zero for $t \leq t_0$ and $0 \neq f \in \mathbb{C}[s]$, $d := \deg_s(f) :=$ degree of f, then $y := f^{-1} \circ u$ is the unique, d times continuously differentiable solution

$$y \in \mathrm{C}^{d}(\mathbb{R}, \mathbb{C}) \text{ of } f \circ y = u \text{ with } y^{(\mu)}(t_0) = 0,\ \mu = 0, \dots, d-1, \Longrightarrow y|_{(-\infty, t_0]} = 0.$$

A very important $\mathbb{C}(s)$-subspace of $\mathrm{C}_{+}^{-\infty}$ and thus $\mathbb{C}[s]$-injective is

$$\mathcal{F}_2 := \mathbb{C}[s] \circ \delta \oplus \mathrm{t}(\mathcal{F})Y,\ \mathbb{C}[s] \circ \delta = \oplus_{k\in\mathbb{N}}\mathbb{C}\delta^{(k)},\ \delta^{(k)} := s^k \circ \delta = \mathrm{d}^k\delta/\mathrm{d}t^k. \tag{2.15}$$

Signals αY, $\alpha \in \mathrm{t}(\mathcal{F})$, occur if a polynomial-exponential signal α is started at $t = 0$. So $\mathcal{F}_2$ consists of sums of such signals and $\mathbb{C}$-linear combinations of the derivatives of the Dirac distribution δ. *All these injective signal modules will be studied in Sect.* 8.2.4.

The next property of $\mathcal{F}$ ensures that $\mathcal{B} = U^{\perp}$ contains as much information as U. The behavior $\mathcal{B} = U^{\perp} \subseteq \mathcal{F}^l$ induces its orthogonal submodule

$$U^{\perp\perp} = \mathcal{B}^{\perp} := \left\{\xi \in \mathbb{C}[s]^{1\times l};\ \xi \circ \mathcal{B} = 0\right\} \supseteq U. \tag{2.16}$$

The trivial case $\mathcal{F} = 0$ and $U^{\perp\perp} = \mathbb{C}[s]^{1\times l}$ shows that $U^{\perp\perp} = U$ need not hold. The injective signal module $\mathcal{F}$ is called a *cogenerator*, cf. Sect. 3.3, if $U^{\perp\perp} = U$ holds for all submodules $U \subseteq \mathbb{C}[s]^{1\times l}$, $l \in \mathbb{N}$, i.e., if U is determined by $\mathcal{B}$. This condition obviously implies and is indeed equivalent to the equivalences

$$\mathcal{B} = U^{\perp} = 0 \Longleftrightarrow U = \mathbb{C}[s]^{1\times l} \Longleftrightarrow M = 0. \tag{2.17}$$

The modules $\mathcal{F} = \mathrm{C}^{-\infty},\ \mathrm{C}^{\infty},\ \mathrm{t}(\mathcal{F})$ are injective cogenerators. A $\mathbb{C}(s)$-vector space, for instance $\mathcal{F}_2$, is never a $\mathbb{C}[s]$-cogenerator. The direct sum module

$$\mathcal{F}_4 := \mathcal{F}_2 \bigoplus \mathrm{t}(\mathcal{F}) = \mathbb{C}[s] \circ \delta \bigoplus \left(\oplus_{\lambda\in\mathbb{C}}\mathbb{C}[t]e^{\lambda t}\right) Y \bigoplus \oplus_{\lambda\in\mathbb{C}}\mathbb{C}[t]e^{\lambda t}, \tag{2.18}$$

however, is injective and contains the cogenerator $\mathrm{t}(\mathcal{F})$, and is thus an injective cogenerator too. It consists of sums of signals in $\mathcal{F}_2$ and of polynomial-exponential signals. All signals in $\mathcal{F}_4$ are described by finitely many complex numbers and are especially suitable for computation. In *electrical engineering* signals in $\mathcal{F}_4$ and piecewise continuous periodic signals, see Sect. 2.4, are used almost exclusively. Less important signal modules $\mathcal{F}_1$ and $\mathcal{F}_3$ will be introduced in Sect. 8.2.4.

In the rest of this chapter we assume an injective cogenerator signal module $\mathcal{F}$ and describe further important consequences of this assumption. With $R, U, M, \mathcal{B}$ from above the modules U resp. M are called the *equation resp. the system module* of $\mathcal{B}$.

For the behavior $\mathcal{B}$ from (2.4) and an arbitrary matrix $T \in \mathbb{C}[s]^{l_2\times l}$ the injectivity of $\mathcal{F}$ implies that also the image $T \circ \mathcal{B}$ is a behavior, cf. Theorem 3.2.20. An important case of this is Willems' *elimination of latent variables* [10, Chap. 6]. Assume, more generally, two behaviors

$$\mathcal{B}_i = U_i^{\perp} \subseteq \mathcal{F}^{l_i},\ U_i \subseteq \mathbb{C}[s]^{1\times l_i},\ M_i := \mathbb{C}[s]^{1\times l_i}/U_i,\ i = 1, 2. \tag{2.19}$$

A $\mathbb{C}$-linear map $\phi : \mathcal{B}_1 \to \mathcal{B}_2$ is called a *behavior morphism* if there is a matrix $T \in \mathbb{C}[s]^{l_2\times l_1}$ such that $\phi(w_1) = T \circ w_1$ for all $w_1 \in \mathcal{B}_1$, i.e., that ϕ is a differential operator. The set $\mathrm{Hom}(\mathcal{B}_1, \mathcal{B}_2)$ of all these morphisms is a proper (cf. Example 3.3.17) $\mathbb{C}[s]$-submodule of $\mathrm{Hom}_{\mathbb{C}[s]}(\mathcal{B}_1, \mathcal{B}_2)$. The injective cogenerator property of $\mathcal{F}$ implies the canonical $\mathbb{C}[s]$-isomorphism, cf. Theorem 3.3.18,

$$\begin{aligned}&\mathrm{Hom}_{\mathbb{C}[s]}(M_2, M_1) \cong \mathrm{Hom}(\mathcal{B}_1, \mathcal{B}_2),\ F \leftrightarrow \phi,\ T \in \mathbb{C}[s]^{l_2\times l_1},\\ &F(\xi_2 + U_2) = \xi_2 T + U_1,\ \phi(w_1) = T \circ w_1,\ \xi_2 \in \mathbb{C}[s]^{1\times l_2},\ w_1 \in \mathcal{B}_1.\end{aligned} \tag{2.20}$$

This isomorphism implies that the bijective correspondence $M \leftrightarrow \mathcal{B}$ is very strong. It is called a *categorical duality* and is discussed in Sect. 3.3.3. In particular, ϕ is injective (surjective, bijective) if and only if F is surjective (injective, bijective). In this book behavioral isomorphisms $\phi : \mathcal{B}_1 \xrightarrow{\cong} \mathcal{B}_2$ and the dual isomorphisms F replace the various *system equivalences* in the literature, for instance Rosenbrock's and Fuhrmann's, cf. [11, pp. 561–566], [12, Sects. 2.2, 2.3]. If ϕ is injective, the implication $\phi(w_1) = \phi(\widetilde{w}_1) \Longrightarrow w_1 = \widetilde{w}_1$ suggested the language that w_1 *is observable from* $\phi(w_1)$. If there is a surjective $\phi : \mathcal{F}^{l_1} \to \mathcal{B}_2$, $\mathcal{B}_1 := 0^\perp = \mathcal{F}^{l_1}$, the behavior $\mathcal{B}_2$ is called *controllable* and ϕ is an *image representation* of $\mathcal{B}_2$. The term *controllable* is justified by Kalman's Theorem 4.3.10 and Willems' Theorem 4.3.4. If in this case $\phi(w_1) = w_2$ Pommaret calls w_1 a *potential* of w_2, a terminology suggested by an analogue for partial differential equations. The surjection ϕ implies the injection $F : M_2 \to M_1 = \mathbb{C}[s]^{1\times l_1}/0 = \mathbb{C}[s]^{1\times l_1}$ and thus that M_2 as submodule of a free $\mathbb{C}[s]$-module is itself free of dimension $m_2 := \operatorname{rank}(\mathcal{B}_2)$. Hence there is even a bijective image representation $\mathcal{F}^{m_2} \cong \mathcal{B}_2$. Thus a behavior $\mathcal{B}$ is controllable if and only if its module M is free. Observability and controllability are studied in Chap. 4. The main application of controllability in this book is for the construction and parametrization of *stabilizing compensators* in Chap. 10. Observability is a necessary and sufficient condition for the construction and parametrization of *functional observers* in Chap. 11.

LTI systems theory has three primary tasks and goals, cf. [13, Sect. 1–1]:

(i) *Modeling*: The theory of this book applies if a real-world system can be described (approximately) by equations $R \circ w = x$ as in (2.4). Our models have been taken from the cited books.
(ii) *Analysis, both qualitative and quantitative*, i.e., to determine the properties of a given $\mathcal{B}$ by means of the properties of R, U, and M and to compute numerical solutions.
(iii) *Synthesis or design*, i.e., to construct a behavior $\mathcal{B}$ with chosen properties, mainly of its transfer matrix H and its transfer operator, see Sect. 2.2.

In this book, synthesis is mainly treated in Chap. 10 where we discuss the construction of *stabilizing compensators* with special properties, mainly *tracking and disturbance rejection*. Kalman's realization theorem of proper transfer matrices, see (2.28) below, is also a synthesis result, cf. [13, Sect. 6.1]. The analysis, but not the synthesis of electrical and mechanical networks, for instance of filters, is treated in Sect. 8.3.

2.2 The Transfer Matrix and Transfer Operator

The assumptions of the preceding section are in force.

The IO behavior (2.11) implies the behavior isomorphism

$$\mathcal{B}^0 := \{y \in \mathcal{F}^p;\ P \circ y = 0\} \cong \mathcal{B} \bigcap (\mathcal{F}^p \times \{0\}) = \{\binom{y}{u} \in \mathcal{B};\ u = 0\},\ y \mapsto \binom{y}{0},$$
$$\Longrightarrow \mathrm{Hom}_{\mathbb{C}[s]}(M^0, \mathcal{F}) \cong \mathcal{B}^0 \text{ with } M^0 := \mathbb{C}[s]^{1\times p}/U^0,\ U^0 := \mathbb{C}[s]^{1\times p}P. \tag{2.21}$$

A behavior (2.4) or (2.11) is called *autonomous* if it has no free components u or if the following equivalent properties hold, cf. Sect. 4.2:

$$\begin{aligned} &\mathrm{rank}(\mathcal{B}) = m = 0 \Longleftrightarrow \mathrm{rank}(R) = p = l \Longleftrightarrow P = R \Longleftrightarrow \mathcal{B}^0 = \mathcal{B} \\ &\Longleftrightarrow M = \mathrm{t}(M) \Longleftrightarrow \mathcal{B} = \mathrm{t}(\mathcal{B}) \Longleftrightarrow \dim_{\mathbb{C}}(M) < \infty \Longleftrightarrow \dim_{\mathbb{C}}(\mathcal{B}) < \infty \\ &\Longrightarrow \dim_{\mathbb{C}}(M) = \dim_{\mathbb{C}}(\mathcal{B}) \text{ and } \mathcal{B} \subset \mathrm{t}(\mathcal{F})^p = \oplus_{\lambda\in\mathbb{C}}\mathbb{C}[t]^p e^{\lambda t}, \end{aligned} \tag{2.22}$$

where $\dim_{\mathbb{C}}(V)$ denotes the $\mathbb{C}$-dimension of a $\mathbb{C}$-vector space. Since $P \in \mathbb{C}[s]^{p\times p}$ and $\mathrm{rank}(P) = p$ the behavior $\mathcal{B}^0 = \{y \in \mathcal{F}^p;\ P \circ y = 0\}$ is autonomous and called the *autonomous or zero-input part* of $\mathcal{B}$. It depends on $\mathcal{B}$ and its IO structure, but not on the special choice of the defining matrices. Its dimension is

$$n := \dim_{\mathbb{C}}(\mathcal{B}^0) = \dim_{\mathbb{C}}(M^0) = \deg_s(\det(P)), \tag{2.23}$$

cf. Theorem 4.2.14. If y_1, y_2 are two outputs to the same input u, then $P \circ (y_2 - y_1) = Q \circ u - Q \circ u = 0$ implies $y_2 - y_1 \in \mathcal{B}^0$ or $y_2 = y_1 + z,\ z \in \mathcal{B}^0$.

The rational matrix $H := P^{-1}Q$ from (2.11) depends on U or $\mathcal{B}$ and the chosen IO decomposition, but again not on the special choice of the matrix $R = (P, -Q)$, cf. Theorem and Definition 6.2.2. It is called the *transfer matrix* of $\mathcal{B}$ and, for $p = m = 1$, the *transfer function* , and $\mathcal{B}$ is called an *IO realization* of H. A given rational matrix $H \in \mathbb{C}(s)^{p\times m}$ trivially admits various representations $H = P^{-1}Q$ with $(P, -Q) \in \mathbb{C}[s]^{p\times(p+m)}$ and $\mathrm{rank}(P) = p$, for instance, $H = (f\,\mathrm{id}_p)^{-1}(fH)$ where f is a common denominator of the entries of H. Hence there are many IO behavior realizations of H, but only one *controllable realization*, cf. Corollary 6.2.3, that furnishes the, essentially unique, so-called *left coprime factorization* $H = P^{-1}Q$ of H.

Recall that $C_+^{-\infty}$ is a $\mathbb{C}(s)$-vector space, and hence

$$H\circ : \left(C_+^{-\infty}\right)^m \to \left(C_+^{-\infty}\right)^p,\ u \mapsto H \circ u = \left(\sum_{\mu=1}^{m} H_{\nu\mu} \circ u_\mu\right)_{\nu=1,\dots,p}, \tag{2.24}$$

is defined. This is the *transfer operator or IO map* induced by H. Note that the equation $P \circ y = Q \circ u$ does not define a linear map $u \mapsto y$ since, for general $u \in$

$\left(\mathbb{C}^{-\infty}\right)^m$, y always exists, but is unique only up to a summand in $\mathcal{B}^0$. The map, cf. Theorem and Definition 8.2.41,

$$\begin{pmatrix} H \\ \mathrm{id}_m \end{pmatrix} \circ : \ \left(\mathbb{C}_+^{-\infty}\right)^m \cong \mathcal{B} \bigcap \left(\mathbb{C}_+^{-\infty}\right)^{p+m}, \ u \mapsto \begin{pmatrix} H \circ u \\ u \end{pmatrix}, \tag{2.25}$$

is a $\mathbb{C}(s)$-isomorphism and shows that the transfer matrix determines and is determined by the *initially-at-rest part* of the IO behavior $\mathcal{B}$. If in this situation y is any other output to u, then

$$y_{ss} := H \circ u \text{ resp. } z := y - y_{ss} \in \mathcal{B}^0 \tag{2.26}$$

are often called the *steady or stationary state* resp. the *transient* of y. This language is appropriate only if $\mathcal{B}^0$ is *asymptotically stable*, i.e.,

$$\lim_{t \to \infty} z(t) = 0, \ z \in \mathcal{B}^0, \quad \text{(cf. Sect. 2.6)} \tag{2.27}$$

so that y and $y_{ss} = H \circ u$ can be identified for large t, written as $y \approx y_{ss}$. Mainly in electrical engineering the linear equation

$$y \approx y_{ss} = H \circ u \text{ or } y_\nu \approx y_{ss,\nu} = \sum_{\mu=1}^{m} H_{\nu\mu} \circ u_\mu, \ \nu = 1, \ldots, p,$$

then establishes the *superposition principle*, experimentally due to Helmholtz: The partial effects $H_{\nu\mu} \circ u_\mu$ of the different input components u_μ are added (=superposed) to form the total effect of all input components on the output component $y_\nu \approx y_{ss,\nu}$. Notice that this principle does not apply to arbitrary equations $P \circ y = Q \circ u$. In the engineering literature [14, 15] the superposition principle is essentially used and experimentally or heuristically proved, but, in general, not with all necessary mathematical details.

Any IO behavior admits a *state space representation* as follows, cf. [7], Theorem and Definition 6.3.8 and Chap. 12: There are matrices

$$\begin{aligned} & A \in \mathbb{C}^{n \times n}, \ B \in \mathbb{C}^{n \times m}, \ C \in \mathbb{C}^{p \times n} \text{ and } D \in \mathbb{C}[s]^{p \times m} \text{ such that} \\ & \begin{pmatrix} C & D \\ 0 & \mathrm{id}_m \end{pmatrix} \circ : \ \mathcal{B}_s := \left\{ \begin{pmatrix} x \\ u \end{pmatrix} \in \mathcal{F}^{n+m}; \ s \circ x = Ax + Bu \right\} \\ & = \left\{ \begin{pmatrix} x \\ u \end{pmatrix} \in \mathcal{F}^{n+m}; \ (s \, \mathrm{id}_n - A) \circ x = Bu \right\} \cong \mathcal{B}, \ \begin{pmatrix} x \\ u \end{pmatrix} \mapsto \begin{pmatrix} Cx + D \circ u \\ u \end{pmatrix}, \end{aligned} \tag{2.28}$$

is a behavior isomorphism. The matrices A, B, C are unique up to similarity, and D is unique. This means that if two quadrupels (A_i, B_i, C_i, D_i), $i = 1, 2$, of dimensions n_i, $i = 1, 2$, satisfy (2.28) then $n_1 = n_2 =: n$, and there is an invertible matrix

$$T \in \mathrm{Gl}_n(\mathbb{C}) \text{ such that } (A_2, B_2, C_2, D_2) = (TA_1T^{-1}, TB_1, C_1T^{-1}, D_1). \tag{2.29}$$

The behavior $\mathcal{B}_s$ is an IO behavior since the characteristic polynomial $\chi_A := \det(s\,\mathrm{id}_n - A)$ of A has degree n and is nonzero. The transfer matrices of $\mathcal{B}_s$ resp. $\mathcal{B}$, cf. Theorem and Definition 6.3.1, are

$$H_s = (s\,\mathrm{id}_n - A)^{-1}B \text{ resp. } H = P^{-1}Q = D + CH_s = D + C(s\,\mathrm{id}_n - A)^{-1}B. \tag{2.30}$$

If u is a piecewise continuous input, the vector x is continuous, and x and y have the standard form

$$\begin{aligned} x(t) &= e^{(t-t_0)A}x(t_0) + \int_{t_0}^{t} e^{(t-\tau)A}Bu(\tau)\mathrm{d}\tau, \quad t, t_0 \in \mathbb{R}, \\ y(t) &= D \circ u + Ce^{(t-t_0)A}x(t_0) + C\int_{t_0}^{t} e^{(t-\tau)A}Bu(\tau)\mathrm{d}\tau. \end{aligned} \tag{2.31}$$

The outputs x of $\mathcal{B}_s$ and y of $\mathcal{B}$ for $t \geq t_0$ are thus determined by the input $u|_{[t_0,\infty)}$ for $t \geq t_0$ and the initial vector $x(t_0)$ at $t = t_0$. Therefore $x \in \mathcal{F}^n$ is called the *state* of $\mathcal{B}_s$ and of $\mathcal{B}$ and $x(t_0) \in \mathbb{C}^n$ the state at time t_0. The isomorphism (2.28) is called a *state space representation or realization* of $\mathcal{B}$ and of H. Its existence is a slight variant of Kalman's famous *realization theorem.* The injectivity of (2.28) defines the *observability* of the equations $s \circ x = Ax + Bu$ and $y = Cx + D \circ u$. The isomorphism (2.28) is used, in particular, to (i) *simulate*, for $D \in \mathbb{C}^{p\times m}$, the trajectories of $\mathcal{B}$ by those of $\mathcal{B}_s$ and (ii) to derive the properties of the general IO behavior $\mathcal{B}$ from those of the state behavior $\mathcal{B}_s$, for instance, in [16, pp. 560–], [13, Chap. 6] and [17, Chap. 7, p. 283]. In this book these applications do not play a dominant role. As in electrical engineering the most important results on IO behaviors, for instance of electrical and mechanical networks, with the equations $P \circ y = Q \circ u$ will be derived directly form $(P, -Q)$ and not from the state space representation (2.28) of the IO behavior. The algorithmic computation of A, B, C, D is difficult, cf. [11, Chap. 6], [13, Chap. 6]. We compute four state space realizations of an IO behavior, usually called the *observability, observer, controllability resp. controller* realization, by means of the *Gröbner basis algorithm* in Chap. 12. These realizations depend only on the behavior, its IO structure, and a chosen term order for the Gröbner theory, and are, therefore, called *canonical.* They give rise to the *observability* and *controllability indices* in connection with (2.28), cf. Theorem 12.3.2, [11, Sect. 6.4.6]. The ensuing algorithms are stronger and more general than those of [11, Sects. 6.4, 7.1] and are directly implementable and are demonstrated in Examples 12.3.11, 12.4.9, 12.4.11, and 12.4.15. The most important consequence of the observer realization is the so-called *pole shifting* algorithm, cf. Theorem 12.4.12, Corollary 12.4.14: If observability holds, i.e., if the map (2.28) is injective, and $f \in \mathbb{C}[s]$ is any monic polynomial of degree n, then the algorithm furnishes a matrix $L \in \mathbb{C}^{n\times p}$ such that $\det(s\,\mathrm{id}_n - (A - LC)) = f$.

If D is a nonconstant polynomial matrix, the vector $D \circ u$ may be distributional, for instance, $s \circ Y = \delta$. Kalman avoided this in the following fashion: If $h = fg^{-1}$ is rational, one defines the s-degree of h as $\deg(h) := \deg_s(h) :=$

$\deg_s(f) - \deg_s(g)$, for instance, $\deg(s^{-1}) = -1$, $\deg(0) := -\infty$. Then h is called *proper* resp. *strictly proper* if $\deg(h) \le 0$ resp. $\deg(h) \le -1$. Euclidean division of f by g furnishes a unique decomposition $h = h_{\mathrm{pol}} + h_{\mathrm{spr}}$ into a polynomial h_{pol} and a strictly proper h_{spr}, for instance, $(s^2+1)(s+1)^{-1} = (s-1) + 2(s+1)^{-1}$. Then h is proper (strictly proper) if and only if $h_{\mathrm{pol}} \in \mathbb{C}$ ($h_{\mathrm{pol}} = 0$). The degree of $H = \left(H_{\nu\mu}\right)_{\nu,\mu} \in \mathbb{C}(s)^{p\times m}$ is $\deg(H) := \deg_s(H) := \max_{\mu,\nu} \deg_s(H_{\nu\mu})$. The "proper"-language and the decomposition $H = H_{\mathrm{pol}} + H_{\mathrm{spr}}$ are extended to matrices componentwise. Cramer's rule implies that $(s\,\mathrm{id}_n - A)^{-1}$ is strictly proper. Hence so is H_s, and $H = D + C(s\,\mathrm{id}_n - A)^{-1}B$ is the decomposition $H = H_{\mathrm{pol}} + H_{\mathrm{spr}}$. We infer that $D \in \mathbb{C}^{p\times m}$ if and only if H is proper. Equation (2.31) and its term $D \circ u$ then imply that *H is proper if and only if for all trajectories $\binom{y}{u} \in \mathcal{B}$ the piecewise continuity of u implies that of y, or, equivalently, if $u \in \left(C_+^{0,\mathrm{pc}}\right)^m$ implies $H \circ u \in \left(C_+^{0,\mathrm{pc}}\right)^p$*, cf. Theorem 6.3.1. Thus a proper transfer matrix induces the transfer operator $H\circ : \left(C_+^{0,\mathrm{pc}}\right)^m \to \left(C_+^{0,\mathrm{pc}}\right)^p$. The property of the input u can be chosen whereas that of y is determined by the behavior. Components of the behavior, that are distributions and not piecewise continuous, generally imply destruction or malfunctioning of a real system that is modeled by the behavior. In electrical engineering they say that the network burns out or saturates. Therefore, behaviors with nonproper transfer matrix have to be redesigned, for instance by choosing a different IO structure and ensuing transfer matrix, cf. [18, Sect. 2.5.3].

For proper H the behavior $\mathcal{B}_s$, the operator $\left(\begin{smallmatrix} C & D \\ 0 & \mathrm{id}_m \end{smallmatrix}\right)$ and thus H can be realized by interconnection of elementary building blocks, cf. Sect. 7.2. In this context one talks about the *synthesis and simulation* of H by means of $s \circ x = Ax + Bu,\ y = Cx + Du$.

Rosenbrock's equations generalize Kalman's state space equations in the form

$$\begin{aligned} &A \circ x = B \circ u,\ \ y = C \circ x + D \circ u \text{ with} \\ &A \in \mathbb{C}[s]^{n\times n},\ \mathrm{rank}(A) = n,\ \ B \in \mathbb{C}[s]^{n\times m},\ \ C \in \mathbb{C}[s]^{p\times n},\ \ D \in \mathbb{C}[s]^{p\times m}, \end{aligned} \tag{2.32}$$

cf. Theorem 6.3.1. These equations give rise to the behaviors

$$\begin{aligned} &\mathcal{B}_1 := \left\{\binom{x}{u} \in \mathcal{F}^{n+m};\ A \circ x = B \circ u\right\},\ \ H_1 := A^{-1}B, \\ &\mathcal{B}_2 := \left(\begin{smallmatrix} C & D \\ 0 & \mathrm{id}_m \end{smallmatrix}\right) \circ \mathcal{B}_1 = \left\{\binom{y}{u} \in \mathcal{F}^{p+m};\ P \circ y = Q \circ u\right\} \text{ where} \\ &\quad (P, -Q) \in \mathbb{C}[s]^{p\times(p+m)},\ \mathrm{rank}(P) = p,\ H_2 := P^{-1}Q = D + CA^{-1}B. \end{aligned} \tag{2.33}$$

It is obvious that $\mathcal{B}_1$ is an IO behavior with input u and transfer matrix H_1 and that $\mathcal{B}_2$ is a behavior as an image of $\mathcal{B}_1$. It turns out that $\mathcal{B}_2$ is also an IO behavior with input u and the indicated transfer matrix, and that the matrix $(P, -Q)$ can be computed from A, B, C, D. Here x is called the *pseudo-state*. In Willems' language the behavior $\mathcal{B}_2$ is obtained by eliminating the *latent variable* x

from $\left\{\begin{pmatrix} x \\ y \\ u \end{pmatrix} \in \mathcal{F}^{n+p+m};\ A \circ x = B \circ u,\ y = C \circ x + D \circ u\right\} \cong \mathcal{B}_1,\ \begin{pmatrix} x \\ y \\ u \end{pmatrix} \mapsto \binom{x}{u}$. Rosenbrock equations are the basic equations in [8, 9, 12, 19] and are intensively studied in [11, 13, 16, 18]. They appear at various places in this book, but are not predominant.

For the discussion below we also need the set of poles of H. Let $\mathrm{V}_{\mathbb{C}}(g) \subset \mathbb{C}$ denote the finite set of *roots or zeros* of a nonzero polynomial g. If $h = fg^{-1}$ is a rational function with coprime f and g, i.e., with greatest common divisor $\gcd(f, g) = 1$, we define the *set of poles resp. the domain* of h by $\mathrm{pole}(h) := \mathrm{V}_{\mathbb{C}}(g)$ resp. $\mathrm{dom}(h) := \mathbb{C} \setminus \mathrm{pole}(h)$. For $\lambda \in \mathrm{dom}(h)$ the value $h(\lambda) := f(\lambda)g(\lambda)^{-1} \in \mathbb{C}$ is defined, sometimes $h(\lambda) := \infty$ for $\lambda \in \mathrm{pole}(h)$ is used. The set of poles of the rational matrix H is $\mathrm{pole}(H) := \bigcup_{\beta,\alpha} \mathrm{pole}(H_{\beta\alpha})$, its complement is $\mathrm{dom}(H)$. For all $\lambda \in \mathrm{dom}(H)$ the matrix $H(\lambda) \in \mathbb{C}^{p\times m}$ is defined. The set $\mathrm{pole}(H)$ plays an important part in stability theory, see Sect. 2.6.

2.3 Distributions of Finite Order, Impulse Response and Laplace Transform

We discuss the choice of the function space $\mathcal{F}$ again. The basic equations of LTI systems theory are the differential systems $P \circ y = Q \circ u$ from (2.11) where u, y have components in $\mathcal{F}$. The entries of P, Q are polynomials of arbitrarily high degree. This leads to the requirement that $\mathcal{F}$ be closed under differentiation or a $\mathbb{C}[s]$-module. Input signals $u = \alpha Y$, $\alpha \in \mathrm{t}(\mathcal{F})^m$, with a jump at $t = 0$ play an important part in electrical engineering, but have, in general, no derivative at $t = 0$ in the standard sense. This suggests to introduce a larger space $\mathcal{F}$ that includes all these and all continuous signals and their derivatives. This is similar to the extensions $\mathbb{N} \subset \mathbb{Z} \subset \mathbb{Q} \subset \mathbb{R} \subset \mathbb{C}$. The famous solution of this problem is Schwartz' distribution theory [20, 21] and space $\mathcal{D}'$ of distributions. Let $\mathrm{C}_0^\infty := \mathrm{C}_0^\infty(\mathbb{R}, \mathbb{C})\ (\subset \mathrm{C}^\infty)$ denote the space of smooth functions φ with compact support, i.e., with $\varphi(t) = 0,\ |t| \geq r$, for some $r \geq 0$. With a suitable topology this is a topological vector space. Then $\mathcal{D}' \subset \mathrm{Hom}_{\mathbb{C}}(\mathrm{C}_0^\infty, \mathbb{C})$ is the space of *continuous* $\mathbb{C}$-linear functions from C_0^∞ to $\mathbb{C}$. The space $\mathrm{C}^{0,\mathrm{pc}}$ is embedded into $\mathcal{D}'$ via the monomorphism

$$\mathrm{C}^{0,\mathrm{pc}} \to \mathcal{D}',\ u \mapsto (\varphi \mapsto u(\varphi)),\ u(\varphi) := \int_{-\infty}^{\infty} u(t)\varphi(t)\mathrm{d}t \Longrightarrow \mathrm{C}^{0,\mathrm{pc}} \underset{\text{identification}}{\subset} \mathcal{D}'. \tag{2.34}$$

Schwartz' theory in [20] is difficult, and so is Hörmander's very elegant form of it [21]. Since these theories do not belong to the mathematical knowledge of the first two university years, neither in mathematics nor in engineering, we proceed with a less elegant, but simpler method, cf. Sect. 8.2. We do not discuss the vector space topologies and replace (2.34) by the $\mathbb{C}$-monomorphism

$$\mathrm{C}^{0,\mathrm{pc}} \to \mathcal{D}^* := \mathrm{Hom}_{\mathbb{C}}(\mathrm{C}_0^\infty, \mathbb{C}),\ u \mapsto (\varphi \mapsto u(\varphi)),\ u(\varphi) := \int_{-\infty}^{\infty} u(t)\varphi(t)\mathrm{d}t. \tag{2.35}$$

Again we identify $\mathrm{C}^{0,\mathrm{pc}} \subset \mathcal{D}^*$ via $u = (\varphi \mapsto u(\varphi))$. We make $\mathcal{D}^*$ a $\mathbb{C}[s]$-module by means of

$$(s \circ u)(\varphi) := u\left(-(s \circ \varphi)\right),\ u \in \mathcal{D}^*,\ \varphi \in \mathrm{C}_0^\infty. \tag{2.36}$$

The $-$ sign is chosen in order that $s \circ u = u'$ for a function $u \in \mathrm{C}^1$. In particular,

$$\begin{aligned}\delta := \mathrm{d}Y/\mathrm{d}t,\ \delta(\varphi) &= -\int_{-\infty}^{\infty} Y(t)\varphi'(t)\mathrm{d}t \\ &= -\int_0^{\infty} \varphi'(t)\mathrm{d}t = -(\varphi(\infty) - \varphi(0)) = \varphi(0).\end{aligned} \tag{2.37}$$

Let $u_n \geq 0$ be a sequence of continuous functions with $u_n(t) = 0$ for $|t| \geq n^{-1}$ and $\int_{-1/n}^{1/n} u_n(t)\mathrm{d}t = 1$ and hence $\lim_{n\to\infty} u_n(0) = \infty$. Then

$$\delta(\varphi) = \varphi(0) = \lim_{n\to\infty} u_n(\varphi), \tag{2.38}$$

cf. Theorem 8.2.10, and this suggested to call δ an *impulse at time* $t = 0$. Such an impulse is, of course, a mathematical idealization, as are all distributions that are not functions. Due to (2.38) a real system can be destroyed if certain components are distributions. This has to be avoided by a better design. In many engineering books δ is suggestively introduced as $\delta \geq 0$, $\delta(t) := 0$ for $t \neq 0$ and $\int_{-\infty}^{\infty} \delta(t)\varphi(t)\mathrm{d}t := \varphi(0)$. The distribution theory gives this definition a well-defined sense. Higher order derivatives of δ are needed, but are not introduced in the quoted textbooks. Distributions cannot be avoided by omitting the discontinuities of the signals [19, Sect. 3.2.1], for instance $s \circ Y = \delta$, but $s \circ Y|_{\mathbb{R}\setminus\{0\}} = 0$. If u is a Lebesgue absolutely integrable function on $\mathbb{R}$, trivially $\int_{0-}^{\infty} u(t)\mathrm{d}t = \int_{0+}^{\infty} u(t)\mathrm{d}t$. In particular, the Laplace transforms $\mathcal{L}_+$ and $\mathcal{L}_-$, defined by $\mathcal{L}_+(u)(s) = \int_{0+}^{\infty} u(t)e^{-st}\mathrm{d}t$ and $\mathcal{L}_-(u)(s) = \int_{0-}^{\infty} u(t)e^{-st}\mathrm{d}t$ for $s \in \mathbb{C}$ with $\Re(s) \geq 0$ coincide for such an u. These two integrals can differ only if u is a distribution with support in $[0, \infty)$ and the integral is properly redefined by means of distribution theory, cf. [11, Sect. 1.2], [18, pp. 381, 395].

The space $\mathcal{D}^*$ contains many elements that are of no analytic interest. Therefore, we only consider the $\mathbb{C}[s]$-submodule of $\mathcal{D}^*$, generated by C^0, i.e.,

$$\mathrm{C}^{-\infty} := \mathrm{C}^{-\infty}(\mathbb{R}, \mathbb{C}) = \bigcup_{n=0}^{\infty} s^n \circ \mathrm{C}^0 \subset \mathcal{D}' \subset \mathcal{D}^*. \tag{2.39}$$

This is the, now well-defined, $\mathbb{C}[s]$-module of all derivatives of all continuous functions, and is called the space of *distributions of finite order* [21, Theorem 4.4.7]. Many properties of $\mathcal{F} := \mathrm{C}^{-\infty}$ are first introduced for $\mathcal{D}^*$ by purely algebraic means and then carried over to $\mathrm{C}^{-\infty}$. We emphasize that this algebraic introduction of distributions works only in dimension one, i.e., for functions of one variable t. The $\mathbb{C}(s)$-vector spaces $\mathrm{C}_+^{-\infty}$ and $\mathcal{F}_2$ and the injective cogenerator $\mathcal{F}_4$ follow according to (2.14), (2.15), (2.18).

Since $\mathrm{C}_+^{-\infty}$ is a $\mathbb{C}(s)$-vector space the map $\mathbb{C}(s) \to \mathrm{C}_+^{-\infty},\ H \mapsto h := H \circ \delta$, is injective. Since δ is interpreted as an impulse, h is called the *impulse response* of H. The *partial fraction decomposition* of H, cf. Sect. 5.5.2, then implies the $\mathbb{C}(s)$-isomorphism, cf. Theorem and Definition 8.2.47,

$$\begin{gathered}\mathcal{L}^{-1}: \mathbb{C}(s) \cong \mathcal{F}_2 = \mathbb{C}[s] \circ \delta \bigoplus \left(\oplus_{\lambda \in \mathbb{C}} \mathbb{C}[t] e^{\lambda t}\right) Y,\ H \mapsto H \circ \delta, \text{ with} \\ \forall k \geq 0: \ \mathcal{L}^{-1}(s^k) = \delta^{(k)},\ \forall k \geq 1,\ \lambda \in \mathbb{C}: \ \mathcal{L}^{-1}((s-\lambda)^{-k}) = \tfrac{t^{k-1}}{(k-1)!} e^{\lambda t} Y.\end{gathered} \tag{2.40}$$

The map $\mathcal{L}^{-1}$ (an image $\mathcal{L}^{-1}(H)$) is called the *inverse Laplace transform* (of H), and its inverse $\mathcal{L}$ (image $\mathcal{L}(h)$) the *Laplace transform* (of h). There results the bijective correspondence

$$\mathcal{F}_2 \ni h = \mathcal{L}^{-1}(H) = H \circ \delta \longleftrightarrow H = \mathcal{L}(h) \in \mathbb{C}(s). \tag{2.41}$$

The constructive form of $\mathcal{L}$ follows directly from (2.40). In *electrical engineering* tables are in use to compute $\mathcal{L}^{-1}(H)$ and $\mathcal{L}(h)$ in special cases, cf. [22, pp. 253–256]. The constructive partial fraction decomposition furnishes these computations for all H and h. There are the additional equivalences

$$\begin{gathered}h := H \circ \delta = \alpha Y,\ \alpha \in \mathrm{t}(\mathcal{F}) \Longleftrightarrow H = \mathcal{L}(h) \text{ strictly proper or } \deg_s(H) \leq -1, \\ h = H \circ \delta \text{ continuous} \Longleftrightarrow sH \text{ strictly proper or } \deg_s(H) \leq -2 \Longleftrightarrow \alpha(0) = 0.\end{gathered} \tag{2.42}$$

The maps $\mathcal{L}$ and $\mathcal{L}^{-1}$ are extended to matrices componentwise such that (2.40) and (2.42) hold likewise for matrices. Assume the IO behavior from (2.11) with its transfer matrix H and its impulse response $h := H \circ \delta = \mathcal{L}^{-1}(H),\ \mathcal{L}(h) = H$. Notice that we derived the transfer matrix H of an IO behavior by module-behavior duality, whereas in almost all engineering books H is *defined* by the equation $H = \mathcal{L}(h)$. This definition depends on the special matrices, defining the behavior, and requires, of course, that $\mathcal{L}$ and h have been defined before. For this approach, that is mostly not carried out with all exact mathematical details, we refer to the quoted textbooks, especially [14, Sect. 6.6] in electrical engineering, and to (2.48)–(2.52) below.

Consider any input signal $u = H_2 \circ \delta = \mathcal{L}^{-1}(H_2),\ H_2 \in \mathbb{C}(s)^m$. These signals, in particular $u = H_2 \circ \delta = \alpha Y,\ \alpha \in \mathrm{t}(\mathcal{F})^m$, for strictly proper H_2, are by far the most important ones with left bounded support in electrical engineering, cf. [22, Sect. 5.2]. They occur if a electrical network is switched on at time $t = 0$. Then all outputs of $\mathcal{B}$ to the input u have the form, cf. Theorem 8.2.53,

$$\begin{gathered}y = y_{ss} + z,\ y_{ss} := HH_2 \circ \delta = \mathcal{L}^{-1}(HH_2),\ z \in \mathcal{B}^0 \subset \mathrm{t}(\mathcal{F})^p \\ \Longrightarrow P \circ y_{ss} = PHH_2 \circ \delta = QH_2 \circ \delta = Q \circ (H_2 \circ \delta) = Q \circ u, \\ P\mathcal{L}(y_{ss}) = PHH_2 = QH_2 = Q\mathcal{L}(u),\ \mathcal{L}(y_{ss}) = H\mathcal{L}(u).\end{gathered} \tag{2.43}$$

Here y_{ss} can be easily computed with the partial fraction decomposition of HH_2. If $\mathcal{B}$ is asymptotically stable, cf. Sect. 2.6, y_{ss} resp. z are again called the *steady or stationary state resp. the transient* of y.

In Theorem 8.2.53 we assume that HH_2 is strictly proper with $HH_2 \circ \delta = \beta Y$, $\beta \in \mathrm{t}(\mathcal{F})^p$, to compute the unique solution of the *initial value problem*, cf. [14, Sect. 6.6],

$$\begin{aligned}
&P \circ y = Q \circ (H_2 \circ \delta) \text{ with given } y_k^{(\mu)}(0+) := \lim_{t \to 0, t>0} y_k^{(\mu)}(t) \in \mathbb{C} \text{ for}\\
&(k,\mu) \in \Gamma^{\mathrm{ob}} := \left\{(k,\mu);\ 1 \le k \le p,\ 0 \le \mu \le d^{\mathrm{ob}}(k) - 1\right\} \text{ as}\\
&y = \beta Y + C^{\mathrm{ob}} e^{tA^{\mathrm{ob}}} \left(y_k^{(\mu)}(0+) - \beta_k^{(\mu)}(0)\right)_{(k,\mu)\in\Gamma^{\mathrm{ob}}} \quad \text{where}\\
&A^{\mathrm{ob}} \in F^{\Gamma^{\mathrm{ob}} \times \Gamma^{\mathrm{ob}}},\ C^{\mathrm{ob}} \in F^{p \times \Gamma^{\mathrm{ob}}},\ C^{\mathrm{ob}}_{i,(k,\mu)} = \delta_{i,k}\delta_{0,\mu} \text{ if } d^{\mathrm{ob}}(i) > 0.
\end{aligned} \tag{2.44}$$

The *observability indices* $d^{\mathrm{ob}}(k) \geq 0$, $k = 1, \ldots, p$, and the matrices A^{ob}, C^{ob} of the canonical observability realization of $\mathcal{B}$ are introduced and computed in Sect. 12.3. For instance [11, p. 11, Example 1.2-1.],

$$y' + 2y = \delta \text{ or } (s+2) \circ y = 1 \circ \delta,\ y(0+) := 2,\ (s+2)^{-1} \circ \delta = e^{-2t}Y,\ y = e^{-2t}(Y+1).$$

According to (2.44) the transient $z = y - \beta Y$ can be determined precisely if and only if the initial values $y_k^{(\mu)}(0+)$ are known precisely from exact measurements. Measuring devices for high-order derivatives of signals do not exist in general. Hence, in general, the transient is not known precisely, but the matrix $C^{\mathrm{ob}} e^{tA^{\mathrm{ob}}}$ in (2.44) determines the general form of its decay. This remark applies to most transients discussed in this book. For low-order derivatives such measuring devices exist, for instance, speedometers and accelerometers, and only for such cases examples can be found in the quoted textbooks.

For most practical signals u, (2.43) is the best method to compute $H \circ u$. For certain *proofs*, however, the representation as *convolution* is needed, cf. Sect. 8.2.5. The convolution of two continuous functions u_1 and u_2 with $u_i(t) = 0$ for $t \le t_i$, $i = 1, 2$, is the continuous function $(u_1 * u_2)(t) := \int_{-\infty}^{\infty} u_1(t-\tau)u_2(\tau)d\tau$. This integral is indeed a finite Riemann integral, no Lebesgue theory is needed or used. The convolution is commutative and associative, and is uniquely extended to all derivatives of continuous functions with left bounded support, i.e., to the convolution product $\mathrm{C}_+^{-\infty} \times \mathrm{C}_+^{-\infty} \to \mathrm{C}_+^{-\infty}$, $(u_1, u_2) \mapsto u_1 * u_2$. This makes $\mathrm{C}_+^{-\infty}$ a commutative $\mathbb{C}$-algebra with the 1-element δ, i.e., $\delta * u = u$, and the rule $H \circ (u_1 * u_2) = (H \circ u_1) * u_2$, $H \in \mathbb{C}(s)$. As usual, the convolution is extended to matrices componentwise. We infer

$$\begin{aligned}
&H \circ u = H \circ (\delta * u) = (H \circ \delta) * u = h * u,\ h := H \circ \delta, \text{ where}\\
&H = P^{-1}Q \in \mathbb{C}(s)^{p\times m},\ h \in \mathcal{F}_2^{p\times m} \subset \left(\mathrm{C}_+^{-\infty}\right)^{p\times m},\ u \in \left(\mathrm{C}_+^{-\infty}\right)^m.
\end{aligned} \tag{2.45}$$

Due to $P \circ h = Q \circ \delta$ and $P \circ (h * u) = Q \circ u$ the impulse response matrix $h = H \circ \delta$ is also called the *fundamental solution* of $P \circ y = Q \circ u$ [21, p. 80]. If

$$H = H_{\text{pol}} + H_{\text{spr}},\ H_{\text{pol}} = \sum_{k=0}^{d} H_k s^k \in \mathbb{C}[s]^{p\times m},\ H_{\text{spr}} \circ \delta = \alpha Y,\ u \in \left(\mathrm{C}_+^{0,\text{pc}}\right)^m$$

$$\text{with } H_k \in \mathbb{C}^{p\times m},\ \alpha \in \mathrm{t}(\mathcal{F})^{p\times m},\ \text{then}$$

$$H \circ u = H_{\text{pol}} \circ u + H_{\text{spr}} \circ u = \sum_{k=0}^{d} H_k s^k \circ u + \int_{-\infty}^{t} \alpha(t-\tau)u(\tau)d\tau \tag{2.46}$$

where the integral is Riemann, finite and continuous in t, cf. [19, p. 95]. If $H_{\text{pol}} = H_0$ or H is proper, then $H \circ u = H_0 u + H_{\text{spr}} \circ u \in \left(\mathrm{C}_+^{0,\text{pc}}\right)^p$. The equation $H_1 H_2 \circ \delta = H_1 H_2 \circ (\delta * \delta) = (H_1 \circ \delta) * (H_2 \circ \delta)$ implies that $\mathcal{F}_2 = \mathbb{C}[s] \circ \delta \oplus \mathrm{t}(\mathcal{F})Y$ is a subalgebra of $\mathrm{C}_+^{-\infty}$ and that the Laplace transform and its inverse are algebra isomorphisms. Since $\mathbb{C}(s)$ is a field, so is $\mathcal{F}_2$.

By reduction to the case

$$H_2 = (s-\lambda)^{-k},\ \lambda \in \mathbb{C},\ k \geq 1,\ H_2 \circ \delta = \tfrac{t^{k-1}}{(k-1)!} e^{\lambda t} Y$$

one shows

$$H_2(s) = \mathcal{L}(H_2 \circ \delta)(s) = \mathcal{L}(\alpha Y)(s) = \int_0^\infty \alpha(t)e^{-st}\mathrm{d}t \text{ for } s \in \{z \in \mathbb{C};\ \Re(z) > \sigma\}$$
$$\text{if } \deg_s(H_2) \leq -1,\ H_2 \circ \delta = \alpha Y,\ \alpha \in \mathrm{t}(\mathcal{F})^m,\ \sigma \geq \max\{\Re(\lambda);\ \lambda \in \mathrm{pole}(H_2)\}. \tag{2.47}$$

This is the standard engineering definition of the Laplace transform of αY and suggested to extend the definition of $\mathcal{L}$ to more general distributions in the following fashion, cf. Theorem 8.2.87 in Sect. 8.2.7 and Theorem 10.2.51: A function $u \in \mathrm{C}_+^{0,\text{pc}}$ is called *Laplace transformable* if there is $\sigma > 0$ such that $|u(t)|e^{-\sigma t},\ t \in \mathbb{R}$, is bounded. Then the function $u(t)e^{-st}$ for $s \in \mathbb{C}$, $\Re(s) > \sigma$, is absolutely integrable on $\mathbb{R}$ and

$$\mathcal{L}(u)(s) := \int_{-\infty}^{\infty} u(t)e^{-st}\mathrm{d}t,\ s \in \{z \in \mathbb{C};\ \Re(z) > \sigma\}, \tag{2.48}$$

is a holomorphic function of s in the open half-plane $\{z \in \mathbb{C};\ \Re(z) > \sigma\}$. The higher derivatives $s^n \circ v$, $n \geq 0$, of Laplace transformable functions v are called Laplace transformable distributions. They form the subset $\mathfrak{A}_+ \subset \mathrm{C}_+^{-\infty}$. For such an $u = s^n \circ v \in \mathfrak{A}_+$ one defines the *Laplace transform* $\mathcal{L}(u)$ of u as the holomorphic function $\mathcal{L}(u)(s) := s^n \mathcal{L}(v)(s)$, $\Re(s) > \sigma$. As usual $\mathcal{L}$ is extended to matrices componentwise. Then $\mathfrak{A}_+$ and $\mathcal{L}$ have the following properties: The set $\mathfrak{A}_+$ is a $\mathbb{C}(s)$-subspace of $\mathrm{C}_+^{-\infty}$ and $\mathcal{L}$ is $\mathbb{C}(s)$-linear on $\mathfrak{A}_+$, i.e.,

$$\mathcal{L}(H \circ u)(s) = H(s)\mathcal{L}(u)(s),\ H \in \mathbb{C}(s),\ u \in \mathfrak{A}_+,\ s \in \{z \in \mathbb{C};\ \Re(z) > \sigma\}, \tag{2.49}$$

for some $\sigma > 0$, depending on H and u. The $\mathbb{C}(s)$-subspace $\mathcal{F}_2 = \mathbb{C}[s] \circ \delta \bigoplus \oplus_{\lambda\in\mathbb{C}} \mathbb{C}[t]e^{\lambda t} \subset \mathrm{C}_+^{-\infty}$ is contained in $\mathfrak{A}_+$ and $\mathcal{L}$ extends $\mathcal{L}$ from Theorem 8.2.47, i.e., $\mathcal{L}(H \circ \delta) = H$. The map $\mathcal{L}$ is injective, i.e., $\mathcal{L}(u) = 0$ implies $u = 0$. If $u \in \mathrm{C}_+^0$ is Laplace transformable with $\sigma > 0$, if $\rho > \sigma$ and if $\mathcal{L}(u)(\rho + j\omega)$ is absolutely integrable as function of ω, then the following inversion formula holds, cf. [11, Sect. 1.2, (2), (4)], [14, (6.114)], [23, p. 485, (12)], [18, p. 398, (12.56)]:

$$u(t) = (2\pi j)^{-1} \int_{\rho-j\infty}^{\rho+j\infty} \mathcal{L}(u)(s)e^{st}ds, \; j := \sqrt{-1}. \tag{2.50}$$

Finally $\mathfrak{A}_+$ satisfies the *exchange theorem*, i.e., $\mathfrak{A}_+$ is a unital subalgebra of the convolution algebra $\left(\mathrm{C}_+^{-\infty}, *\right)$ with one-element δ, $\mathcal{L}(\delta) = 1$ and $\mathcal{L}$ is multiplicative on $\mathfrak{A}_+$, i.e.,

$$\mathcal{L}(u_1 * u_2)(s) = \mathcal{L}(u_1)(s)\mathcal{L}(u_2)(s), \; u_1, u_2 \in \mathfrak{A}_+, \; s \in \{z \in \mathbb{C}; \; \Re(z) > \sigma\} \tag{2.51}$$

for some $\sigma > 0$, depending on u_1 and u_2. If, in particular, $u \in \mathfrak{A}_+^m \subset \left(\mathrm{C}_+^{-\infty}\right)^m$ is a Laplace transformable input of the IO behavior $\mathcal{B}$ from (2.11), then the unique output $y := H \circ u \in \left(\mathrm{C}_+^{-\infty}\right)^p$ with $P \circ y = Q \circ u$ is also Laplace transformable and

$$\mathcal{L}(y)(s) = H(s)\mathcal{L}(u)(s), \; P(s)\mathcal{L}(y)(s) = Q(s)\mathcal{L}(u)(s), \; s \in \{z \in \mathbb{C}; \; \Re(z) > \sigma\}, \tag{2.52}$$

for some $\sigma > 0$. The latter equation holds without the usually required *zero initial conditions* [11, p. 551], [19, p. 94, (21)], [12, p. 55], [18, Theorem 24 on p. 37] or *relaxedness assumptions* [13, p. 82]. Equations (2.49)–(2.52) are not proven in detail in the quoted textbooks, but are essential for the *definition* of the transfer matrix H. Our proof in Theorem 8.2.89 is short and elementary and, in particular, does not use the Fourier transform of temperate distributions, cf. [18, Sect. 12.3.4]. In the present book the Laplace transform on $\mathcal{F}_2$ from (2.40) and from Theorem 8.2.47 suffices for all considered applications.

For a Laplace transformable distribution u with support in $[0, \infty)$ its Laplace transform, as already mentioned, is often *defined* [11, p. 10], [23, (5) on p. 482], [18, Sect. 12.3.4, (12.48)] as

$$\mathcal{L}(u)(s) = \int_{0-}^{\infty} u(t)e^{-st}dt.$$

Unless u is a Laplace transformable function, this expression and its further use require a precise distributional explanation. For a smooth function u the equation $(uY)' = u'Y + u(0)\delta$ implies $\mathcal{L}(u'Y) = s\mathcal{L}(uY) - u(0)$ [18, p. 396]. It is often written as $\mathcal{L}(u') = s\mathcal{L}(u) - u(0)$ [23, p. 185], [16, p. 155] and then seems to contradict the rule $\mathcal{L}(s \circ u) = s\mathcal{L}(u)$.

Many authors, e.g., [10, Sect. 2.3.2], use the space $\mathrm{L}^1_{\mathrm{loc}}(\mathbb{R}, \mathbb{C})$ as the basic signal space and a different notion of *weak* solution of a differential equation. In our opinion these are inappropriate for the following reasons: This space is a factor space $\mathrm{L}^1_{\mathrm{loc}} = \mathcal{L}^1_{\mathrm{loc}}/\mathcal{L}_0$. A function w in $\mathcal{L}^1_{\mathrm{loc}}$ is a Lebesgue measurable function whose Lebesgue integrals $\int_a^b |w(t)|\mathrm{d}t$, $a, b \in \mathbb{R}, a < b$, are finite whereas a function in $\mathcal{L}_0$ is measurable and zero almost everywhere. An element of $\mathrm{L}^1_{\mathrm{loc}}$ is a residue class $\overline{w} := w + \mathcal{L}_0$, $w \in \mathcal{L}^1_{\mathrm{loc}}$, and hence $\overline{w}(t)$, $t \in \mathbb{R}$, is not defined, i.e., $\overline{w}$ has no functional values. In contrast to piecewise continuous signals such signals can neither be measured nor generated, a basic requirement for signals in electrical engineering. If $\mathcal{L}^1_{\mathrm{loc}}$ instead of $\mathrm{L}^1_{\mathrm{loc}}$ is used, then the basic implication $\left(\int_a^b |w(t)|\mathrm{d}t = 0 \Longrightarrow w|_{[a,b]} = 0\right)$ does not hold. For $w_1, w_2 \in \mathrm{L}^1_{\mathrm{loc}}$ the following implications hold, cf. [20, Theorem III, p. 54]:

$$
\begin{aligned}
&s \circ w_1 = w_2 \text{ in } \mathrm{C}^{-\infty} \Longleftrightarrow \forall \varphi \in \mathrm{C}_0^\infty : -\int_{-\infty}^{\infty} w_1(x)\varphi'(x)\mathrm{d}x = \int_{-\infty}^{\infty} w_2(x)\varphi(x)\mathrm{d}x \\
&\Longrightarrow \exists c \in \mathbb{C} \text{ with } w_1 = \int_0^t w_2(x)\mathrm{d}x + c \in \mathrm{L}^1_{\mathrm{loc}}.
\end{aligned}
\tag{2.53}
$$

Moreover w_1 is then continuous and its usual derivative $w_1'(t)$ exists almost everywhere and coincides with w_2 in $\mathrm{L}^1_{\mathrm{loc}}$. The converse implication in the second line holds if w_1 is differentiable almost everywhere in the usual sense and $w_2 = w_1' \in \mathrm{L}^1_{\mathrm{loc}}$, but not in general. So a *weak* solution of $\mathrm{d}w_1/\mathrm{d}t = w_2$ according to [10, Def. 2.3.7], i.e., $w_1 = \int_0^t w_2(x)\mathrm{d}x + c \in \mathrm{L}^1_{\mathrm{loc}}$, does *not* imply $s \circ w_1 = w_2$ in $\mathrm{C}^{-\infty}$. Also all piecewise continuous functions belong to $\mathrm{L}^1_{\mathrm{loc}}$, for instance Y, but their derivatives like $\delta = s \circ Y$ do not.

2.4 Periodic Signals and Fourier Series

Another important class of signals are the periodic ones, and Fourier series are an essential technical tool for these, cf. [22, Chap. 3] and Sect. 8.2.8. We assume the IO behavior from (2.11) with transfer matrix H. Let $T > 0$ and $\omega := 2\pi T^{-1}$. A piecewise continuous signal u is called T-periodic if $u(t) = u(t + T)$ for all $t \in \mathbb{R}$, the *sinusoidal* or *harmonic* functions $e^{j\mu\omega t}$, $\mu \in \mathbb{Z}$, $j := \sqrt{-1}$, being the standard examples. Let $\mathcal{P}^0$ ($\mathcal{P}^{0,\mathrm{pc}}$) be the space of (piecewise) continuous, T-periodic signals. The space $\mathcal{P}^{0,\mathrm{pc}}$ has the *inner product* $\langle u_1, u_2\rangle := T^{-1}\int_0^T \overline{u_1(t)}u_2(t)\mathrm{d}t$ and the *induced norm* $\|u\|_2 := \langle u, u\rangle^{1/2}$ with $\|1\|_2 = 1$. The $e^{j\mu\omega t}$, $\mu \in \mathbb{Z}$, form the standard orthonormal family of functions. For $u \in \mathcal{P}^{0,\mathrm{pc}}$ one defines the *sequence of Fourier coefficients*

$$
\begin{aligned}
&\mathbb{F}(u) \in \mathbb{C}^{\mathbb{Z}} \text{ by } \mathbb{F}(u)(\mu) := \langle e^{j\mu\omega t}, u\rangle = T^{-1}\int_0^T e^{-j\mu\omega t}u(t)\mathrm{d}t. \text{ Then} \\
&u = \sum_{\mu\in\mathbb{Z}} \mathbb{F}(u)(\mu)e^{j\mu\omega t}, \text{ i.e., } \lim_{N\to\infty} \|u - \sum_{\mu=-N}^{N} \mathbb{F}(u)(\mu)e^{j\mu\omega t}\|_2 = 0.
\end{aligned}
\tag{2.54}
$$

The map $\mathbb{F}$ becomes a bijective transformation in the following fashion, cf. [20, Sect. VII.1]. Like $C^{-\infty}$ we define the subspace $\mathcal{P}^{-\infty}\left(\subset C^{-\infty}\right)$ of periodic distributions as the space of derivatives $s^n \circ u,\ u \in \mathcal{P}^0,\ n \geq 0$. Again, no topological vector spaces are used. We define the sequence space $\mathfrak{s}^{-\infty} \subset \mathbb{C}^{\mathbb{Z}}$ of all sequences $\widehat{u} \in \mathbb{C}^{\mathbb{Z}}$ that *grow at most polynomially*, i.e., for which there are $M > 0$ and $k \in \mathbb{Z}$ such that $|\widehat{u}(\mu)| \leq M(1+\mu^2)^k$ for all $\mu \in \mathbb{Z}$. Let $\mathbb{C}(s)_{\text{per}}$ denote the subalgebra of $\mathbb{C}(s)$ of all rational functions H without poles in $\mathbb{Z}j\omega$, i.e., for which $H(j\mu\omega) \in \mathbb{C}$ is defined for all $\mu \in \mathbb{Z}$. The space $\mathfrak{s}^{-\infty}$ becomes a $\mathbb{C}(s)_{\text{per}}$-module with the scalar multiplication $H \circ \widehat{u}$, defined by

$$(H \circ \widehat{u})(\mu) := H(j\mu\omega)\widehat{u}(\mu),\ \ H \in \mathbb{C}(s)_{\text{per}},\ \widehat{u} \in \mathfrak{s}^{-\infty},\ \mu \in \mathbb{Z}. \tag{2.55}$$

With these data the map $\mathbb{F}$ can be uniquely extended to a $\mathbb{C}(s)_{\text{per}}$-isomorphism

$$\mathbb{F}:\ \mathcal{P}^{-\infty} \cong \mathfrak{s}^{-\infty},\ \text{(cf. Theorem 8.2.99)}\,. \tag{2.56}$$

With respect to a suitable topology on $\mathcal{P}^{-\infty}$ [20, (VII, 1; 3)] that we, however, do not discuss, one obtains the convergent series $u = \sum_{\mu\in\mathbb{Z}} \mathbb{F}(u)(\mu)e^{j\mu\omega t}$ for $u \in \mathcal{P}^{-\infty}$. If $u \in \mathcal{P}^{0,\text{pc}}$ and if $\sum_{\mu\in\mathbb{Z}} |\mathbb{F}(u)(\mu)| < \infty$, then $u = \sum_{\mu\in\mathbb{Z}} \mathbb{F}(u)(\mu)e^{j\mu\omega t}$ is uniformly convergent and thus continuous.

As usual, $\mathbb{F}$ is extended to matrices componentwise. Assume the IO behavior from (2.11) and $H \in \mathbb{C}(s)_{\text{per}}^{p\times m}$. Then

$$H\circ:\ \left(\mathcal{P}^{-\infty}\right)^m \to \left(\mathcal{P}^{-\infty}\right)^p,\ u \mapsto H \circ u = \mathbb{F}^{-1}\left(H \circ \mathbb{F}(u)\right), \tag{2.57}$$

is another well-defined *transfer operator* such that for all $u \in \left(\mathcal{P}^{-\infty}\right)^m$ the trajectory $\left(\begin{smallmatrix} H\circ u \\ u \end{smallmatrix}\right)$ is periodic and belongs to $\mathcal{B}$. If, in addition, $P^{-1} \in \mathbb{C}(s)_{\text{per}}^{p\times p}$ or $V_{\mathbb{C}}(\det(P)) \subset \mathbb{C}\setminus\mathbb{Z}j\omega$, then H induces the $\mathbb{C}(s)_{\text{per}}$-isomorphism, cf. (2.25),

$$\left(\mathcal{P}^{-\infty}\right)^m \cong \mathcal{B}\bigcap\left(\mathcal{P}^{-\infty}\right)^{p+m},\ u \mapsto \left(\begin{smallmatrix} H\circ u \\ u \end{smallmatrix}\right). \tag{2.58}$$

The obvious equations $C_+^0 \cap \mathcal{P}^0 = 0$ and $C_+^{-\infty} \cap \mathcal{P}^{-\infty} = 0$ show that the maps $H\circ$ from (2.24) and (2.57) are independent of each other, but both satisfy $P \circ (H \circ u) = Q \circ u$, i.e., $\left(\begin{smallmatrix} H\circ u \\ u \end{smallmatrix}\right) \in \mathcal{B}$.

Assume additionally that $H = H_0 + H_{\text{spr}}$ is proper, i.e., $H_0 \in \mathbb{C}^{p\times m}$ and $u \in \left(\mathcal{P}^{0,\text{pc}}\right)^m$. Then the output signal $y_{ss} := H \circ u$ is, cf. Theorem 8.2.102,

$$y_{ss} = H \circ u = \sum_{\mu\in\mathbb{Z}} H(j\mu\omega)\mathbb{F}(u)(\mu)e^{j\mu\omega t} = H_0 u + \sum_{\mu\in\mathbb{Z}} H_{\text{spr}}(j\mu\omega)\mathbb{F}(u)(\mu)e^{j\mu\omega t}, \tag{2.59}$$

where the second sum $\sum_\mu$ is uniformly convergent and thus continuous. Equation 2.59 is the most general form of the *superposition principle* for periodic signals and an important tool for the analysis of electrical networks [14, Sect. 5.5.2],

[22, Sect. 3.3.1]. If u is a *sinusoidal or harmonic* input of the simple form $u = u(0)e^{j\omega t}$, $u(0) \in \mathbb{C}^m$, $\omega > 0$, then (2.59) simplifies to

$$y_{ss} = H \circ u = H(j\omega)u = H(j\omega)u(0)e^{j\omega t}. \tag{2.60}$$

If $(P, -Q) \in \mathbb{R}[s]^{p\times(p+m)}$ and thus $H \in \mathbb{R}(s)^{p\times m}$ are real, we obtain

$$\begin{aligned}
\Re(u) &= \Re(u(0))\cos(\omega t) - \Im(u(0))\sin(\omega t),\\
\Im(u) &= \Im(u(0))\cos(\omega t) + \Re(u(0))\sin(\omega t),\\
\Re(y_{ss}) &= H \circ \Re(u) = \Re(H \circ u) = \Re\left((H(j\omega)u(0))e^{j\omega t}\right)\\
&= (\Re(H(j\omega))\Re(u(0)) - \Im(H(j\omega))\Im(u(0)))\cos(\omega t)\\
&- (\Re(H(j\omega))\Im(u(0)) + \Im(H(j\omega))\Re(u(0)))\sin(\omega t).
\end{aligned} \tag{2.61}$$

The simplicity of (2.60) compared to (2.61) suggested the *complex method for real harmonic voltages and currents* in electrical engineering.

Notice again that for $\binom{y}{u} \in \left(\mathcal{P}^{0,\mathrm{pc}}\right)^{p+m}$ the basic equation $P \circ y = Q \circ u$ makes, in general, no sense without the space $\mathcal{P}^{-\infty}$ of periodic distributions that contains all derivatives of functions in $\mathcal{P}^{0,\mathrm{pc}}$. If $\mathcal{B}$ is asymptotically stable, cf. (2.102), and y is any output to u, i.e., solves $P \circ y = Q \circ u$, then y_{ss} resp. $y - y_{ss} \in \mathcal{B}^0$ are again called the *steady or stationary state resp. the transient of* y.
All practical signals in $\mathcal{P}^{0,\mathrm{pc}}$ are derived from polynomial-exponential functions, for instance, the T-periodic signals

$$u_1(t) = 2T^{-1}t,\ -T/2 \le t < T/2,\ \text{or } u_2(t) := \begin{cases} 2T^{-1}t & \text{if } 0 \le t \le T/2\\ 2T^{-1}(T-t) & \text{if } T/2 \le t \le T \end{cases}. \tag{2.62}$$

For these signals, (2.59) can be constructively improved, cf. Theorems 8.2.107 and 8.2.110.

2.5 Generalized Fractional Calculus and Behaviors

With the same methods as for the general Laplace transform in Sect. 2.3 we study *fractional or symbolic calculus*, cf. [20, Sect. VI, 5], [Wikipedia; https://en.wikipedia.org/wiki/Fractional_calculus, September 4, 2019], [24] and *fractional behaviors* in the last Chap. 13 , but we do not discuss the role of these in applications that are comprehensively treated in [24]. A suitable vector space-behavior duality is again the key to compute the trajectories of fractional behaviors, cf. Theorem 13.1.3. The partial fraction decomposition of rational matrices enables the constructive solution of very general linear systems of fractional integral/differential equations.

If f is a piecewise continuous, complex-valued function on the open interval $(0, \infty)$ we extend this to a function $f_{\mathbb{R}} : \mathbb{R} \to \mathbb{C}$ by

$$\forall t > 0 : \ f_{\mathbb{R}}(t) := f(t), \ \forall t \leq 0 : \ f_{\mathbb{R}}(t) := 0. \tag{2.63}$$

Obviously $f_{\mathbb{R}}$ is piecewise continuous on $\mathbb{R} \setminus \{0\}$. If $f(0+) := \lim_{t \to 0,\, t>0} f(t)$ exists, then $f_{\mathbb{R}}$ has the jump $f_{\mathbb{R}}(0+) - f_{\mathbb{R}}(0-) = f(0+)$ at $t = 0$ and is also piecewise continuous. If for some $a > 0$ the Riemann integral

$$\int_{-a}^{a} |f_{\mathbb{R}}(t)|\mathrm{d}t = \int_{0}^{a} |f(t)|\mathrm{d}t := \lim_{\epsilon \to 0,\, \epsilon > 0} \int_{\epsilon}^{a} |f(t)|\mathrm{d}t < \infty$$

is finite, then $f_{\mathbb{R}}$ is called *locally integrable*. The corresponding distribution is defined by

$$f_{\mathbb{R}}(\varphi) := \int_{-\infty}^{\infty} f_{\mathbb{R}}(t)\varphi(t)\mathrm{d}t = \lim_{\epsilon \to 0,\, \epsilon > 0} \int_{\epsilon}^{\infty} f(t)\varphi(t)\mathrm{d}t, \ \varphi \in \mathrm{C}_0^{\infty}, \Longrightarrow \forall u \in \mathrm{C}_+^{0,\mathrm{pc}} :$$
$$(f_{\mathbb{R}} * u)(t) = \int_{-\infty}^{t} f(t-x)u(x)\mathrm{d}x = \lim_{\epsilon \to 0, \epsilon > 0} \int_{-\infty}^{t-\epsilon} f(t-x)u(x)\mathrm{d}x. \tag{2.64}$$

If $f_{\mathbb{R}}$ is locally integrable and $u \in \mathrm{C}_+^{0,\mathrm{pc}}$, then $f_{\mathbb{R}} * u$ is continuous, i.e., $f_{\mathbb{R}} * u \in \mathrm{C}_+^0$. Since $f_{\mathbb{R}} = (s \circ Y) * f_{\mathbb{R}} = s \circ (Y * f_{\mathbb{R}})$ is the derivative of the continuous function $Y * f_{\mathbb{R}}$, $f_{\mathbb{R}}$ is a distribution of finite order and belongs to $\mathrm{C}_+^{-\infty}$.

Let $\Gamma(m)$, $m \in \mathbb{C}$, denote the meromorphic *Gamma* function with $\Gamma(m+1) = m!$ for $m \in \mathbb{N}$. According to [20, Sect. VI, 5] one defines the *fractional integral operators*

$$\forall m \in \mathbb{C} : \ I^m : \mathrm{C}_+^{-\infty} \to \mathrm{C}_+^{-\infty}, \ y \mapsto Y_m * y, \ \text{where}$$
$$Y_m := \begin{cases} \left(\Gamma(m)^{-1} t^{m-1}\right)_{\mathbb{R}} \in \mathrm{C}_+^0 & \text{if } \Re(m) - 1 > 0 \\ s^k \circ Y_{m+k} \in \mathrm{C}_+^{-\infty} & \text{if } k \in \mathbb{N} \text{ and } \Re(m) + k - 1 > 0 \end{cases}. \tag{2.65}$$

Especially this implies

$$\forall m \in \mathbb{Z} : \ Y_m = s^{-m} \circ \delta, \ \forall m \in \mathbb{N} : \ Y_{-m} = \delta^{(m)}, \ Y_0 = \delta, \ Y_1 = Y,$$
$$\forall m \in \mathbb{C} \forall u \in \mathrm{C}_+^{0,\mathrm{pc}} : \ I^m u = s^k \circ \Gamma(m+k)^{-1} \int_{-\infty}^{t} (t-x)^{m+k-1} u(x)\mathrm{d}x. \tag{2.66}$$

The general definition is independent of the choice of $k \in \mathbb{N}$ with $\Re(m) + k - 1 > 0$. We note that $0 < \Re(m) + k \leq 1$ would suffice, but then Y_{m+k} is only locally integrable, but not continuous on $\mathbb{R}$. The equations

$$Y_1 = s^{-1} \circ \delta = Y \text{ and } I^1 u = Y_1 * u = Y * u = \int_{-\infty}^{t} u(t)\mathrm{d}t, \ u \in \mathrm{C}_+^{0,\mathrm{pc}}, \tag{2.67}$$

suggested the notation Y_m for a generalized Heaviside function and to call I^m an integral operator.

The convolution equation $Y_m * Y_n = Y_{m+n}, \ m, n \in \mathbb{C}$, holds. The *fractional differential operator* is defined as $D^m := I^{-m} := Y_{-m}*$. If arbitrary $m \in \mathbb{C}$ are admitted, every fractional integral operator $I^m = D^{-m}$ can be interpreted as a differential one, and vice versa. The equation

$$\begin{aligned} &\forall m \in \mathbb{C} \setminus (-\mathbb{N}) \forall \epsilon > 0 \forall \varphi \in \mathbb{C}_0^\infty \text{ with } \varphi|_{[0,\epsilon]} = 0: \\ &Y_m(\varphi) = \Gamma(m)^{-1} \int_\epsilon^\infty t^{m-1} \varphi(t) \mathrm{d}t \end{aligned} \tag{2.68}$$

suggested to call, for $m \in \mathbb{C}$ with $\Re(m) \leq 0$, the distribution Y_m the *finite part* of the function $\left(\Gamma(m)^{-1} t^{m-1}\right)_{\mathbb{R}}$ [20, (II,2;26)].

The distribution $Y_m, \ m \in \mathbb{C}$, is Laplace transformable, and indeed

$$\begin{aligned} &\mathcal{L}(Y_m)(s) = s^{-m} := e^{-m \ln(s)}, \ \Re(s) > 0, \text{ where} \\ &s = |s| e^{j\alpha}, \ |s| > 0, \ -\pi/2 < \alpha < \pi/2, \ \ln(s) = \ln(|s|) + j\alpha, \ j = \sqrt{-1}. \end{aligned} \tag{2.69}$$

Let $\mu > 0$ be a fixed positive real number. Then one usually calls I^μ an integral operator and $D^\mu = I^{-\mu}$ a differential operator. The convolution equations $Y_{m\mu} * Y_{n\mu} = Y_{(m+n)\mu}$ imply that

$$\bigoplus_{m \in \mathbb{Z}} \mathbb{C} Y_{m\mu} \subset \left(\mathrm{C}_+^{-\infty}, *\right), \ Y_{\mu m} = Y_\mu^m, \ Y_\mu^{-1} = Y_{-\mu}, \tag{2.70}$$

is a subalgebra of $\mathrm{C}_+^{-\infty}$ with respect to the convolution multiplication. Let $\mathbb{C}[s, s^{-1}] = \bigoplus_{m \in \mathbb{Z}} \mathbb{C} s^m$ with $s^m s^n = s^{m+n}$ denote the principal ideal domain of *Laurent polynomials*. Then the map

$$\mathbb{C}[s, s^{-1}] \to \bigoplus_{m \in \mathbb{Z}} \mathbb{C} Y_{m\mu}, \ H = \sum_{m \in \mathbb{Z}} a_m s^m \mapsto H(Y_\mu) := \sum_{m \in \mathbb{Z}} a_m Y_\mu^m, \ Y_\mu^m = Y_{m\mu}, \tag{2.71}$$

is an algebra isomorphism. Hence $\mathrm{C}_+^{-\infty}$ is a $\mathbb{C}[s, s^{-1}]$-module with the scalar multiplication

$$\begin{aligned} &H \circ_\mu y = H(Y_\mu) * y, \ H \in \mathbb{C}[s, s^{-1}], \ y \in \mathrm{C}_+^{-\infty} \\ &\Longrightarrow H(Y_\mu) * y = \sum_{m \in \mathbb{Z}} a_m (I^\mu)^m y = \sum_{m \in \mathbb{Z}} a_{-m} (D^\mu)^m y. \end{aligned} \tag{2.72}$$

An equation $H \circ_\mu y = u$ with given $H \in \mathbb{C}[s, s^{-1}]$ and $u \in \mathrm{C}_+^{-\infty}$ is called an *inhomogeneous (μ)-fractional integral/differential equation* or, shorter, fractional differential equation.

We are now going to extend these equations considerably. Let $\mathbb{C}\langle\langle s\rangle\rangle$ denote the field of convergent Laurent series at 0 with at most a pole at 0. This is given by

$$\mathbb{C}\langle\langle s\rangle\rangle = \mathbb{C}[s^{-1}]\bigoplus\mathbb{C}\langle s\rangle_+,\ \mathbb{C}\langle s\rangle_+ := \left\{\sum_{m=1}^{\infty} a_m s^m;\ a_m \in \mathbb{C},\ \limsup_m \sqrt[m]{|a_m|} < \infty\right\}. \tag{2.73}$$

The series $\sum_{m=1}^{\infty} a_m s^m$ is a (locally at 0) convergent power series with constant term 0 and the convergence radius $\rho := \left(\limsup_m \sqrt[m]{|a_m|}\right)^{-1} > 0$, hence holomorphic in the disc $\{s \in \mathbb{C};\ |s| < \rho\}$. We write

$$H = H_- + H_+,\ H_- := \sum_{m=0}^{\infty} a_{-m} s^{-m} \in \mathbb{C}[s^{-1}],\ H_+ := \sum_{m=1}^{\infty} a_m s^m \in \mathbb{C}\langle s\rangle_+. \tag{2.74}$$

Almost all a_{-m} for $m \in \mathbb{N}$ are 0. We are now going to define $H(Y_\mu) := H_-(Y_\mu) + H_+(Y_\mu)$ where, of course, $H_-(Y_\mu) = \sum_{m=0}^{\infty} a_{-m} Y_{-m\mu}$. The infinite sum $H_+(Y_\mu) := \sum_{m=1}^{\infty} a_m Y_{m\mu}$ is not defined a priori in the distribution space $C_+^{-\infty}$ and hence we proceed as follows. We define

$$\begin{aligned} &\widehat{H_+}_\mu(z) := \sum_{m=0}^{\infty} a_{m+1}\Gamma((m+1)\mu)^{-1} z^m,\ z \in \mathbb{C}, \\ &H_+(Y_\mu) := \left(t^{\mu-1}\widehat{H_+}_\mu(t^\mu)\right)_{\mathbb{R}},\ H(Y_\mu) := H_-(Y_\mu) + H_+(Y_\mu), \end{aligned} \tag{2.75}$$

where $\widehat{H_+}_\mu(z)$ is an everywhere convergent power series and an entire holomorphic function on $\mathbb{C}$. This holds since the $\Gamma((m+1)\mu)$ grow very fast like factorials, due to Stirling's formula for the Γ-function. Hence $\widehat{H_+}_\mu(t^\mu)$ is continuous on $[0, \infty)$ and $H_+(Y_\mu) := \left(t^{\mu-1}\widehat{H_+}_\mu(t^\mu)\right)_{\mathbb{R}}$ is locally integrable on $\mathbb{R}$ since $\mu - 1 > -1$. In particular, $H_+(Y_\mu) * u,\ u \in C_+^{0,\mathrm{pc}}$, is continuous. We show

$$\begin{aligned} &\sum_{m\in\mathbb{Z}} a_m Y_{m\mu} := \sum_{m=1}^{\infty} a_{-m} Y_{-m\mu} + \lim_{N\to\infty}\sum_{m=1}^{N} a_m Y_{m\mu} = H(Y_\mu),\ \text{i.e.,} \\ &\quad \forall \varphi \in C_0^\infty : \sum_{m\in\mathbb{Z}} a_m Y_{m\mu}(\varphi) \\ &\qquad := \sum_{m=1}^{\infty} a_{-m} Y_{-m\mu}(\varphi) + \lim_{N\to\infty}\sum_{m=1}^{N} a_m Y_{m\mu}(\varphi) = H(Y_\mu)(\varphi) \end{aligned} \tag{2.76}$$

and

$$\forall H_1, H_2 \in \mathbb{C}\langle\langle s\rangle\rangle : (H_1 + / \cdot H_2)(Y_\mu) = H_1(Y_\mu) + / * H_2(Y_\mu). \tag{2.77}$$

Therefore the map

$$\mathbb{C}\langle\langle s\rangle\rangle \cong \mathcal{F}_{2,\mu} := \left\{H(Y_\mu);\ H \in \mathbb{C}\langle\langle s\rangle\rangle\right\},\ H \mapsto H(Y_\mu), \tag{2.78}$$

is a field isomorphism where $\mathcal{F}_{2,\mu}$ is a large subfield of $\left(\mathrm{C}_+^{-\infty}, *\right)$. This, in turn, implies that $\mathrm{C}_+^{-\infty}$ is a $\mathbb{C}\langle\langle s\rangle\rangle$-vector space with the scalar multiplication

$$H \circ_\mu u = H(Y_\mu) * u,\ H \in \mathbb{C}\langle\langle s\rangle\rangle,\ u \in \mathrm{C}_+^{-\infty},\ H \circ_\mu \delta = H(Y_\mu). \tag{2.79}$$

Then $H \circ_\mu \delta = H(Y_\mu)$ is called the *μ-impulse response* of H. As usual the action $\circ_\mu$ is extended to matrices and vectors. In particular, we consider linear systems

$$\begin{aligned}
&P \circ_\mu y = P(Y_\mu) * y = Q \circ_\mu u = Q(Y_\mu) * u \text{ where } u \in \left(\mathrm{C}_+^{-\infty}\right)^m,\ y \in \left(\mathrm{C}_+^{-\infty}\right)^p,\\
&(P, -Q) \in \mathbb{C}\langle\langle s\rangle\rangle^{p\times(p+m)},\ \operatorname{rank}(P) = p \text{ or } \det(P) \neq 0,\ H := P^{-1}Q\\
&\Longrightarrow y = H \circ_\mu u = H(Y_\mu) * u \Longrightarrow\\
&\left(\mathrm{C}_+^{-\infty}\right)^m \underset{\mathbb{C}\langle\langle s\rangle\rangle}{\cong} \mathcal{B} := \left\{\begin{pmatrix} y \\ u \end{pmatrix} \in \left(\mathrm{C}_+^{-\infty}\right)^{p+m}\ P \circ_\mu y = Q \circ_\mu u\right\},\ u \mapsto \begin{pmatrix} H\circ_\mu u \\ u \end{pmatrix}.
\end{aligned} \tag{2.80}$$

The solution $\mathbb{C}\langle\langle s\rangle\rangle$-vector space $\mathcal{B}$ is called a *generalized fractional IO behavior*. By definition its trajectories have left bounded support like the signals in connection with the Laplace transform in this book and like the often used signals in electrical engineering. Initial conditions are neither needed nor used in our approach. If, in particular, the input u is of the general form $u = H_2 \circ_\mu \delta = H_2(Y_\mu)$ where $H_2 \in \mathbb{C}\langle\langle s\rangle\rangle^m$ then

$$y := HH_2 \circ_\mu \delta = (HH_2)(Y_\mu) \text{ solves } P(Y_\mu) * y = Q(Y_\mu) * H_2(Y_\mu) \tag{2.81}$$

uniquely and $y = (HH_2)(Y_\mu)$ can be explicitly computed. Standard multivariable μ-fractional integral/differential systems are the special case where $(P, -Q) \in \mathbb{C}[s, s^{-1}]^{p\times(p+m)}$. For instance, consider the binomial power series

$$\begin{aligned}
&H := (1-\lambda s)^{-k} = \sum_{m=0}^{\infty} \binom{-k}{m} \lambda^m s^m,\ 0 \neq \lambda \in \mathbb{C},\ k \geq 1,\\
&\Longrightarrow H = \sum_{m=0}^{\infty} \binom{m+k-1}{k-1} \lambda^m s^m = 1 + H_+,\\
&\widehat{H_+}_\mu(z) = \sum_{m=0}^{\infty} \binom{m+k}{k-1} \lambda^{m+1} \Gamma((m+1)\mu)^{-1} z^m,\\
&(1-\lambda s)^{-k}(Y_\mu) = (\delta - \lambda Y_\mu)^{-k} = \delta + {}_\lambda Z_\mu^{(k)},\ {}_\lambda Z_\mu^{(k)} := \left(t^{\mu-1}\widehat{H_+}_\mu(t^\mu)\right)_{\mathbb{R}}.
\end{aligned} \tag{2.82}$$

Note that for $0 < \mu < 1$ the function ${}_\lambda Z_\mu^{(k)}$ is locally integrable, but not piecewise continuous at 0.

Let, more generally, $H \in \mathbb{C}(s)$ be an arbitrary rational function with its partial fraction decomposition, cf. Sect. 5.5.2,

$$\begin{aligned} H &= \sum_{m\in\mathbb{Z}} a_m s^m + \sum_{0\neq\lambda\in\mathrm{pole}(H)} \sum_{k=1}^{m_\lambda} a_{\lambda,k}(s-\lambda)^{-k} \\ &= \sum_{m\in\mathbb{Z}} a_m s^m + \sum_{0\neq\lambda\in\mathrm{pole}(H)} \sum_{k=1}^{m_\lambda} a_{\lambda,k}(-\lambda)^{-k}(1-\lambda^{-1}s)^{-k} \text{ with} \\ & a_m, a_{\lambda,k} \in \mathbb{C},\ 1 \le m_\lambda \in \mathbb{N},\ a_{\lambda,m_\lambda} \neq 0,\ a_m = 0 \text{ for almost all } m. \end{aligned} \tag{2.83}$$

Then

$$\begin{aligned} H(Y_\mu) &= \sum_{m\in\mathbb{Z}} a_m Y_{m\mu} + \sum_{0\neq\lambda\in\mathrm{pole}(H)} \sum_{k=1}^{m_\lambda} a_{\lambda,k}(-\lambda)^{-k}(1-\lambda^{-1}s)^{-k}(Y_\mu) \\ &= \sum_{m\in\mathbb{Z}} a_m Y_{m\mu} + \sum_{0\neq\lambda\in\mathrm{pole}(H)} \sum_{k=1}^{m_\lambda} a_{\lambda,k}(-\lambda)^{-k}\left(\delta + {}_{\lambda^{-1}}Z_\mu^{(k)}\right). \end{aligned} \tag{2.84}$$

For $H \in \mathbb{C}\langle\langle s\rangle\rangle$ the map $H(Y_\mu)* = H\circ_\mu$ induces a map $H(Y_\mu)* : \mathrm{C}_+^{0,\mathrm{pc}} \to \mathrm{C}_+^{0,\mathrm{pc}}$ if and only if 0 is not a pole of H, i.e., if H is a locally convergent power series or $a_m = 0$ for $m < 0$. The distribution $H(Y_\mu)$ is even a locally integrable function on $\mathbb{R}$ if and only if in addition $H(0) = a_0 = 0$.

For a rational function $H \in \mathbb{C}(s)$ the Laplace transform of $H(Y_\mu)$ is

$$\mathcal{L}(H(Y_\mu)) = H(s^{-\mu}),\ H \in \mathbb{C}(s). \tag{2.85}$$

In our approach to fractional differential equations this result is not applied. We do not know an analogue of (2.85) for general convergent Laurent series.

In practice only rational exponents $\mu \in \mathbb{Q}$ are considered. Assume a positive rational number

$$\begin{aligned} &\mu = r/m,\ r, m \in \mathbb{N},\ r, m > 0 \\ &\Longrightarrow H(Y_\mu) = H(s^r)(Y_{1/m}),\ H(s)\circ_\mu u = H(s^r)\circ_{1/m} u. \end{aligned} \tag{2.86}$$

For finitely many positive rational numbers μ_i with their least common denominator $m \in \mathbb{N}$ this implies

$$\mu_i = r_i/m,\quad r_i > 0,\ H(Y_{\mu_i}) = H(s^{r_i})(Y_{1/m}),\ H(s)\circ_{\mu_i} u = H(s^{r_i})\circ_{1/m} u. \tag{2.87}$$

Hence finitely many different operators $I^{\mu_i} = Y_{\mu_i}*$ with positive rational indices μ_i and their differential counterparts $D^{\mu_i} = Y_{-\mu_i}*$ can be treated with the single vector space $\left({}_{\mathbb{C}\langle\langle s\rangle\rangle}\mathrm{C}_+^{-\infty}, \circ_{1/m}\right)$. Note that

$$
\begin{aligned}
&s \circ \delta = s^{-1} \circ_1 \delta = \delta' = Y_{-1} \\
&\Longrightarrow \forall H \in \mathbb{C}(s): \; H(s) \circ \delta = H(s^{-1}) \circ_1 \delta = H(s^{-1})(Y_1), \; Y_1 = Y.
\end{aligned}
\tag{2.88}
$$

For a nonrational convergent Laurent series $H(s)$ the function $H(s^{-1})$ does not belong to $\mathbb{C}\langle\langle s\rangle\rangle$ and $H(s^{-1})(Y_1)$ is not defined.

The preceding theory can be reformulated as a theory for the vector space $\left({}_{\mathbb{C}\langle\langle s^{1/\infty}\rangle\rangle}\mathrm{C}_{+}^{-\infty}, \circ_1\right)$ where

$$
\mathbb{C}\langle\langle s^{1/\infty}\rangle\rangle := \bigcup_{m=1}^{\infty} \mathbb{C}\langle\langle s^{1/m}\rangle\rangle
\tag{2.89}
$$

is the *Puiseux field* of convergent *Puiseux series* that is the algebraic closure of the field $\mathbb{C}\langle\langle s\rangle\rangle$ of convergent Laurent series, cf. [Wikipedia; https://en.wikipedia.org/wiki/Puiseux series, 8 September 8 2019], [25, Sect. 3.1] and Chap. 13, and where the scalar multiplication is given by

$$
\begin{aligned}
&H(s^{1/m}) \circ_1 u := H(s) \circ_{1/m} u \text{ where} \\
&H \in \mathbb{C}\langle\langle s\rangle\rangle, \; H(s^{1/m}) \in \mathbb{C}\langle\langle s^{1/m}\rangle\rangle \subset \mathbb{C}\langle\langle s^{1/\infty}\rangle\rangle, \; u \in \mathrm{C}_{+}^{-\infty}.
\end{aligned}
\tag{2.90}
$$

A different application of $\mathbb{C}\langle\langle s^{1/\infty}\rangle\rangle$ for LTV-(linear time-varying) systems was described in [25].

Finally we treat a simple example that already demonstrates the power of our method for the explicit solution of fractional differential equations. Consider

$$
\begin{aligned}
&(D^{1/2} - \lambda_1)y = e^{\lambda_2 t}Y, \; \lambda_1 \neq 0, \; \lambda_2 \neq 0, \; \lambda_3^2 = \lambda_2 \neq \lambda_1^2, \; \mu := 1/2, \text{ or} \\
&(s^{-1} - \lambda_1) \circ_{1/2} y = (s - \lambda_2)^{-1} \circ \delta = (s^{-1} - \lambda_2)^{-1} \circ_1 \delta = (s^{-2} - \lambda_2)^{-1} \circ_{1/2} \delta.
\end{aligned}
\tag{2.91}
$$

This fractional differential equation has the unique solution

$$
\begin{aligned}
y &= (s^{-1} - \lambda_1)^{-1}(s^{-2} - \lambda_2)^{-1} \circ_{1/2} \delta \\
&= \left(s^3(1 - \lambda_1 s)^{-1}(1 - \lambda_3 s)^{-1}(1 + \lambda_3 s)^{-1}\right)(Y_{1/2}).
\end{aligned}
\tag{2.92}
$$

The partial fraction decomposition is

$$
\begin{aligned}
&s^3(1 - \lambda_1 s)^{-1}(1 - \lambda_3 s)^{-1}(1 + \lambda_3 s)^{-1} \\
&\qquad = a + b(1 - \lambda_1 s)^{-1} + c(1 - \lambda_3 s)^{-1} + d(1 + \lambda_3 s)^{-1} \text{ with} \\
&a = (\lambda_1 \lambda_2)^{-1}, \; b = \left(\lambda_1(\lambda_1^2 - \lambda_2)\right)^{-1}, \; c = (2\lambda_2(\lambda_3 - \lambda_1))^{-1}, \\
&d = -\,(2\lambda_2(\lambda_3 + \lambda_1))^{-1} \in \mathbb{C}, \; a + b + c + d \underset{s=0}{=} 0,
\end{aligned}
$$

and furnishes the continuous solution, cf. (2.84),

$$\begin{aligned} y &= a\delta + b\left(\delta + {}_{\lambda_1}Z_{1/2}^{(1)}\right) + c\left(\delta + {}_{\lambda_3}Z_{1/2}^{(1)}\right) + d\left(\delta + {}_{(-\lambda_3)}Z_{1/2}^{(1)}\right) \\ &= b\,{}_{\lambda_1}Z_{1/2}^{(1)} + c\,{}_{\lambda_3}Z_{1/2}^{(1)} + d\,{}_{(-\lambda_3)}Z_{1/2}^{(1)} \end{aligned} \tag{2.93}$$

with the locally integrable functions ${}_{\lambda}Z_{\mu}^{(k)}$ on $\mathbb{R}$.

2.6 Stability

A basic requirement for an IO behavior $\mathcal{B}$ from (2.11) with transfer matrix H is its *stability*. We study this by means of the *Chinese Remainder Theorem* (CRT) in Chap. 5, cf. [26]. An important consequence of the latter is the *primary direct sum decomposition* of the torsion submodule $\mathrm{t}(M)$ of any $\mathbb{C}[s]$-module M, cf. Theorem 5.3.2, viz.,

$$\begin{aligned} \mathrm{t}(M) &:= \{y \in M;\ \exists 0 \neq f \in \mathbb{C}[s]:\ f \circ y = 0\} \\ &= \bigoplus_{\lambda\in\mathbb{C}} M_\lambda \ni y = \sum_{\lambda} y_\lambda,\ M_\lambda := \left\{y \in M;\ \exists k \in \mathbb{N} \text{ with } (s-\lambda)^k \circ y = 0\right\}. \end{aligned} \tag{2.94}$$

The CRT yields the y_λ from y constructively. In particular, one gets

$$\mathrm{t}\left(\mathrm{C}^{-\infty}\right)_\lambda = \mathrm{t}\left(\mathrm{C}^{\infty}\right)_\lambda = \mathbb{C}[t]e^{\lambda t} \quad \text{(Sect. 5.3.3)}. \tag{2.95}$$

The autonomous part $\mathcal{B}^0$ of $\mathcal{B}$ thus admits the primary decomposition $\mathcal{B}^0 = \bigoplus_{\lambda\in\mathbb{C}} \mathcal{B}^0_\lambda$. If $\mathcal{B}^0_\lambda$ is nonzero, the number λ is called a *characteristic value* or *pole* of $\mathcal{B}^0$ and of $\mathcal{B}$. Since $\dim_{\mathbb{C}}(\mathcal{B}^0) < \infty$, this occurs only for finitely many λ, and indeed, cf. Sect. 5.4.1,

$$\begin{aligned} \mathrm{char}(\mathcal{B}^0) &:= \left\{\lambda \in \mathbb{C};\ \mathcal{B}^0_\lambda \neq 0\right\} = \mathrm{V}_{\mathbb{C}}(\det(P)) \\ &= \{\lambda \in \mathbb{C};\ \mathrm{rank}(P(\lambda)) < p = \mathrm{rank}(P)\} \\ \Longrightarrow \mathcal{B}^0 &= \bigoplus_{\lambda\in\mathrm{char}(\mathcal{B}^0)} \mathcal{B}^0_\lambda,\ \mathcal{B}^0_\lambda = \mathcal{B}^0 \bigcap \mathbb{C}[t]^p e^{\lambda t} \subset \mathrm{t}(\mathcal{F})^p. \end{aligned} \tag{2.96}$$

The set $\mathrm{char}(\mathcal{B}^0)$ is called the *characteristic variety* of $\mathcal{B}^0$, a term originally from *Algebraic Analysis* [2, 4, 5]. The trajectories in $\mathcal{B}^0_\lambda$, $\lambda \in \mathrm{char}(\mathcal{B}^0)$, are called the *$\lambda$-modes of $\mathcal{B}^0$*. The finite dimension $\mathrm{l}_\lambda(\mathcal{B}^0) := \mathrm{mult}(\mathcal{B}^0_\lambda) := \dim_{\mathbb{C}}\left(\mathcal{B}^0_\lambda\right)$ is called the *λ-length or λ-multiplicity* of $\mathcal{B}^0$ and gives rise to the infinite vector

$$\begin{aligned} &\mathrm{l}(\mathcal{B}^0) = \left(\mathrm{l}_\lambda(\mathcal{B}^0)\right)_{\lambda\in\mathbb{C}} \in \mathbb{N}^{(\mathbb{C})} \\ &:= \left\{\mu = (\mu(\lambda))_{\lambda\in\mathbb{C}} \in \mathbb{N}^{\mathbb{C}};\ \mathrm{supp}(\mu) := \{\lambda \in \mathbb{C};\ \mu(\lambda) \neq 0\} \text{ finite}\right\}. \end{aligned} \tag{2.97}$$

In Theorem 5.4.11 we compute a $\mathbb{C}$-basis of the space $\mathcal{B}^0_\lambda$of λ-modes, and hence the length $\mathrm{l}_\lambda(\mathcal{B}^0)$ and a basis of $\mathcal{B}^0$. In the literature [18, Sects. 7.2, 13.4.2], [12, Sect. 2.5] the elements $\mu \in \mathbb{N}^{(\mathbb{C})}$ are often written as

$$\mathrm{supp}(\mu) = \{\lambda_1, \ldots, \lambda_r\},\ \mu = \left\{\underbrace{\lambda_1, \ldots, \lambda_1}_{\mu(\lambda_1)}, \ldots, \underbrace{\lambda_r, \ldots, \lambda_r}_{\mu(\lambda_r)}\right\}. \tag{2.98}$$

The set $\mathrm{supp}(\mu)$ with the multiplicities $\mu(\lambda_i)$ is called a finite *valued subset of* $\mathbb{C}$. Notice that these μ can be added in $\mathbb{N}^{(\mathbb{C})}$ and subtracted in $\mathbb{Z}^{(\mathbb{C})}$, as is done in [12, Theorem 2.62] without detailed explanation. One writes $\mu \uplus \nu := \mu + \nu$. The multiplicities play an important part in connection with poles of *Rosenbrock equations*, cf. Sect. 6.3.5.

The state space representation (2.28) implies

$$\begin{aligned} \mathcal{B}^0_s &= \left\{x = e^{tA}x(0);\, x(0) \in \mathbb{C}^n\right\} \cong \mathcal{B}^0 \text{ and} \\ \mathrm{char}(\mathcal{B}^0) &= \mathrm{char}(\mathcal{B}^0_s) = \mathrm{V}_\mathbb{C}(\det(s\,\mathrm{id}_n - A)) = \mathrm{spec}(A), \end{aligned} \tag{2.99}$$

where $\mathrm{spec}(A)$ is the *spectrum* or *set of eigenvalues* of A. The primary components $(\mathcal{B}^0_s)_\lambda$ are related to the Jordan decomposition of A. This, in turn, is given by the *primary or Jordan decomposition* of $\mathbb{C}^n = \bigoplus_{\lambda \in \mathrm{spec}(A)}(\mathbb{C}^n)_\lambda$ into the *generalized eigenspaces* $(\mathbb{C}^n)_\lambda$ of A where $\mathbb{C}^n$ is a $\mathbb{C}[s]$ -module via $s \circ x = Ax$, cf. Sect. 5.5.1. In contrast to $\mathrm{char}(\mathcal{B}^0)$ the *characteristic variety of* $\mathcal{B}$ is

$$\begin{aligned} &\mathrm{char}(\mathcal{B}) := \{\lambda \in \mathbb{C};\ \mathrm{rank}(P(\lambda), -Q(\lambda)) < p = \mathrm{rank}(P, -Q)\} \\ &\Longrightarrow \mathrm{char}(\mathcal{B}^0) = \mathrm{char}(\mathcal{B}) \cup \mathrm{pole}(H),\ \text{cf. Theorems 5.2.7, 5.2.9.} \end{aligned} \tag{2.100}$$

The behavior $\mathcal{B}$ is controllable, i.e., its module is free, if and only if $\mathrm{char}(\mathcal{B}) = \emptyset$. Therefore the elements of $\mathrm{pole}(H)$ resp. of $\mathrm{char}(\mathcal{B})$ are called the *controllable* resp. *uncontrollable* poles of $\mathcal{B}$. Note that $\mathrm{pole}(H) \cap \mathrm{char}(\mathcal{B}) \neq \emptyset$ may occur.

The $\mathbb{C}[s]$-module of *asymptotically stable polynomial-exponential signals* is

$$\mathcal{F}_- := \left\{y \in \mathrm{t}(\mathcal{F});\ \lim_{t\to\infty} y(t) = 0\right\} = \bigoplus_{\lambda \in \mathbb{C}_-} \mathbb{C}[t]e^{\lambda t},\ \mathbb{C}_- := \{\lambda \in \mathbb{C};\ \Re(\lambda) < 0\}, \tag{2.101}$$

cf. Theorem 5.4.16. The behaviors $\mathcal{B}$ and $\mathcal{B}^0$ are called *asymptotically stable if and only if*

$$\forall z \in \mathcal{B}^0:\ \lim_{t\to\infty} z(t) = 0 \Longleftrightarrow \mathcal{B}^0 \subset \mathcal{F}^p_- \Longleftrightarrow \mathrm{char}(\mathcal{B}^0) \subset \mathbb{C}_-. \tag{2.102}$$

Recall the decompositions $y = y_{ss} + z,\ z \in \mathcal{B}^0$, into the *steady or stationary state* y_{ss} and the *transient* z from (2.26), (2.43) and (2.59). If $\mathcal{B}^0$ is asymptotically stable and hence $\lim_{t\to\infty}(y - y_{ss})(t) = 0$, y and y_{ss} can often be identified in practical situations. This suggested the *steady (stationary) state, transient* terminology, that is also used, but not justified without the asymptotic stability of $\mathcal{B}^0$. It would be appropriate to talk of *one* instead of *the* steady state y_{ss} of y, but all satisfy $\lim_{t\to\infty}(y - y_{ss})(t) = 0$. The *external stability* of $\mathcal{B}$ is a property of its transfer operator $H\circ$. We assume that H is proper, and obtain the operator $H\circ : \left(\mathrm{C}_+^{0,\mathrm{pc}}\right)^m \to \left(\mathrm{C}_+^{0,\mathrm{pc}}\right)^p$. We need the normed signal spaces L^q and L_+^q, $1 \le q \le \infty$, defined by

$$\begin{aligned}
\mathrm{L}^q &:= \left\{u \in \mathrm{C}^{0,\mathrm{pc}};\ \|u\|_q := \left(\int_{-\infty}^{\infty} |u(t)|^q \mathrm{d}t\right)^{1/q} < \infty\right\},\ q < \infty,\\
\mathrm{L}^\infty &:= \left\{u \in \mathrm{C}^{0,\mathrm{pc}};\ \|u\|_\infty := \sup_{t\in\mathbb{R}} |u(t)| < \infty\right\},\ \mathrm{L}_+^q := \mathrm{L}^q \cap \mathrm{C}_+^{0,\mathrm{pc}},\ q \le \infty,
\end{aligned} \tag{2.103}$$

with the norms $\| - \|_q$. The completions $\mathfrak{L}^q$ of L^q for $q = 1, 2, \infty$ are Banach spaces and used in Sect. 10.2.4 in connection with the *robustness of stabilizing compensators*. As usual, we also consider matrices with entries in these L_+^q. Let

$$\begin{aligned}
&H = H_0 + H_{\mathrm{spr}},\ H_0 \in \mathbb{C}^{p\times m},\ h_{\mathrm{spr}} := H_{\mathrm{spr}} \circ \delta \in \left(\mathrm{C}_+^{0,\mathrm{pc}}\right)^{p\times m}\\
&\Longrightarrow \forall u \in \left(\mathrm{C}_+^{0,\mathrm{pc}}\right)^m:\ H \circ u = H_0 u + H_{\mathrm{spr}} \circ u = H_0 u + h_{\mathrm{spr}} * u.
\end{aligned} \tag{2.104}$$

External stability of the behavior is then characterized by the following equivalent properties, cf. Theorem 8.2.83, Corollary 8.2.84:

$$\begin{aligned}
&(i)\ \ \mathrm{pole}(H) \subset \mathbb{C}_-,\\
&(ii)\ \ h_{\mathrm{spr}} \in \left(\mathrm{L}_+^1\right)^{p\times m},\\
&(iii)\ \ \text{for } q = 1 \text{ or } q = \infty:\ H \circ \left(\mathrm{L}_+^q\right)^m \subseteq \left(\mathrm{L}_+^q\right)^p,\\
&(iv)\ \ \forall q,\ 1 \le q \le \infty:\ H \circ \left(\mathrm{L}_+^q\right)^m \subseteq \left(\mathrm{L}_+^q\right)^p.
\end{aligned} \tag{2.105}$$

Moreover the operator $H\circ$ is continuous in the $\| - \|_q$-norms, i.e., the output $H \circ u$ depends continuously on the input u. The condition (iii) for $q = \infty$ is called *BIBO (bounded input/bounded output) stability*. Since $\mathrm{pole}(H) \subseteq \mathrm{char}(\mathcal{B}^0)$ we infer that asymptotic stability implies external stability.

2.7 Electrical and Mechanical Networks

The theory of *electrical networks* is both a very important source and application of systems theoretic methods, and is the only applied field that is discussed in detail in this book, cf. Sects. 3.1.3 and 8.3. According to [14, 15, 22] the following methods and results are fundamental or even the most basic results of electrical engineering.

We derive them with the systems theory of this book that is very suitable for exact mathematical derivations in this field. Several of our equations are more general than those of the cited books. Mechanical networks are then treated via the *electrical-mechanical Firestone analogy*, cf. Sect. 8.4. Examples 8.3.17, 8.3.25, 8.3.34, and 8.4.6 show how the theorems and algorithms are applied. We discuss *translational mechanical networks*, but not *rotational* ones. We refer to the books [27–29] on *mechatronics* where networks of additional energy domains and their interconnections are discussed. The mathematics of the present book is also useful for these extensions. We do not discuss the vast design part of electrical and mechanical engineering, i.e., the construction of an electrical network and not just of an arbitrary IO behavior with prescribed transfer matrix. For the latter Kalman's realization theorem solves the problem, cf. (2.28) and Chap. 12.

In electrical and mechanical engineering IO behaviors, i.e., with a decomposition of the trajectories into input and output components, are far more important than general behaviors, for instance, for steady-state and superposition principle considerations. This is in contrast to Willems' general philosophy.

Electrical networks give rise to behaviors of a special form. We use the real base field $\mathbb{R}$ and the real versions $\mathcal{F} := \mathcal{F}_{\mathbb{R}}$ of the injective cogenerator function modules, for instance, $\mathcal{F}_{\mathbb{R}} := \mathrm{C}^{\infty}(\mathbb{R}, \mathbb{R})$ or $\mathcal{F}_{\mathbb{R}} := \mathrm{C}^{-\infty}(\mathbb{R}, \mathbb{R})$, that are later precisely explained. The notion of a *network* refers to a *connected directed graph* (V, K), consisting of a finite set V of size $m := \sharp(V)$ of *nodes* or *vertices* and a finite set K of size $n := \sharp(K)$ of *branches, edges* or *arrows* with two maps dom, cod : $K \to V$ (domain, codomain), written as $k : v := \mathrm{dom}(k) \to w := \mathrm{cod}(k)$. Then k is called a *directed* branch from the node v to the node w. The connectedness means that for arbitrary $v, w \in V$ there is a path along edges from v to w. In a real electrical network a branch $k : v \to w$ is realized by a wire (short circuit), a voltage or current source or a passive electrical element with two terminals. The nodes represent the points where the wires or terminals of the different electrical elements are connected. The trajectories of the network behavior are of the form

$$\binom{U}{I} \in \mathcal{F}^{K \uplus K},\ U := (U_k)_{k \in K} \in \mathcal{F}^K,\ I := (I_k)_{k \in K} \in \mathcal{F}^K, \tag{2.106}$$

where U_k is the *voltage or potential difference* between v and w and I_k is the *current* through k from v to w. The set K is decomposed as $K = K_p \uplus K_s$ where K_p resp. K_s contain the *one-port resp. the source branches*. Along $k \in K_p$ there is an electrical device with *two terminals*, called *two-pole or one-port*, described by an equation

$$P_k \circ U_k = Q_k \circ I_k,\ P_k, Q_k \in \mathbb{R}[s],\ P_k \neq 0,\ Q_k \neq 0. \tag{2.107}$$

The prototypical one-port branches are the ideal *resistance, capacitance, inductance* (R, C, L) branches with the simple equations

$$U_k = R_k I_k,\ C_k s \circ U_k = I_k,\ U_k = L_k s \circ I_k,\ R_k, C_k, L_k > 0. \tag{2.108}$$

Voltage resp. current source branches $k \in K_s$ are characterized by given U_k resp. I_k that are supplied to the network from outside, whereas the corresponding I_k resp. U_k are determined by the network. The network without the voltage or current sources $U_k,\ I_k,\ k \in K_s$, is called *passive* and often studied. However, we always include the sources into the considerations.

The U_k of the network satisfy *Kirchhoff's circuit* or *voltage law* (KVL) and the currents I_k *Kirchhoff's node* or *current law* (KCL) that will be specified in Sect. 3.1.3. With these data the behavior of the network is

$$\mathcal{B} := \left\{ \begin{pmatrix} U \\ I \end{pmatrix} \in \mathcal{F}_{\mathbb{R}}^{K \uplus K};\ \begin{cases} (i)\text{(KVL) and (KCL) are satisfied} \\ (ii)\ \forall k \in K_p:\ P_k \circ U_k = Q_k \circ I_k \end{cases} \right\}. \tag{2.109}$$

There are networks with different equations (ii), for instance, with *ideal transformers* or *gyrators* or *controlled voltage* or *current sources*. These do not change the mathematics essentially, cf. Corollaries 8.3.14 and 8.3.22. Let

$$V_s := \{\mathrm{dom}(k), \mathrm{cod}(k);\ k \in K_s\},\ m_s := \sharp(V_s),\ n_s := \sharp(K_s),\ m_s \le 2n_s. \tag{2.110}$$

The nodes in V_s are called the *terminals or poles* of $\mathcal{B}$ and represent the connection with the outside, and $\mathcal{B}$ is called an m_s*-pole*. If $m_s = 2n_s$, i.e., if the $\mathrm{dom}(k), \mathrm{cod}(k),\ k \in K_s$, are pairwise distinct, $\mathcal{B}$ is called an n_s*-port*, and each $k : \mathrm{dom}(k) \to \mathrm{cod}(k),\ k \in K_s$, is called a *port*. New source branches between existing nodes can be added to K_s, but change the network and its behavior. In the engineering literature more special graphs are usually employed.

Usually the study of $\mathcal{B}$ begins with the *node-potential*, *mesh-current*, or *state space* method, based on graph theory, to derive the consequences of the Kirchhoff laws, cf. [14, Sect. 3.1.4], [22, Chap. 3]. These methods are, however, only special cases of the *Gauß algorithm* for the solution of linear systems over a field, and we can and do, therefore, proceed with a much simpler method. Indeed, let $A = (A_{vk})_{v\in V, k\in K} \in \mathbb{R}^{V\times K}$ denote the *incidence matrix* of (V, K), defined by

$$A_{vk} = A(v,k) = \begin{cases} 1 & \text{if } v = \mathrm{dom}(k) \neq \mathrm{cod}(k) \\ -1 & \text{if } v = \mathrm{cod}(k) \neq \mathrm{dom}(k) \\ 0 & \text{otherwise} \end{cases}. \tag{2.111}$$

The connectedness of (V, K) implies $\mathrm{rank}(A) = m - 1,\ m = \sharp(V)$. Elementary row operations and column permutations on A furnish the echelon form

$$\begin{aligned} &XA = \begin{matrix} r \\ m-r \end{matrix}\overset{\begin{matrix} K_1 & K_2 \end{matrix}}{\begin{pmatrix} \mathrm{id}_r & M \\ 0 & 0 \end{pmatrix}} \in F^{(r+(m-r))\times(K_1\uplus K_2)} \text{ where} \\ &\sharp(K_1) = r = m-1,\ M \in \mathbb{R}^{K_1\times K_2},\ X \in \mathrm{Gl}_m(F), \\ &\Longrightarrow A|_{K_2} = A|_{K_1}M,\ A|_{K_i} = (A_{vk})_{v\in V, k\in K_i} \in \mathbb{R}^{V\times K_i},\ i = 1,2. \end{aligned} \tag{2.112}$$

It implies that the columns $A_{-,k_1} = A(-, k_1)$, $k_1 \in K_1$, are an $\mathbb{R}$-basis of the column space $A\mathbb{R}^K := \sum_{k\in K} A_{-k}\mathbb{R} \subseteq \mathbb{R}^V$ of A, and that the linear relations $A_{-k_2} = \sum_{k_1\in K_1} A_{-k_1} M_{k_1 k_2}$, $k_2 \in K_2$, hold. Notice that the Gaußalgorithm and therefore the decomposition $K = K_1 \uplus K_2$ and the matrix M are not unique. This variability is essential to ensure $K_s \subseteq K_1$ or $K_s \subseteq K_2$ under suitable conditions and to derive a suitable state space representation of $\mathcal{B}$, cf. (2.123) and Theorem 8.3.32. In the electrical engineering literature [14, 15] the branches in K_1 resp. K_2 are usually obtained as *tree* resp. *cotree (tree complement, link)* branches.

It turns out, cf. Theorem 8.3.5, that the Kirchhoff voltage resp. current law is equivalent to the equation

$$\begin{aligned} &U_{K_2} = M^\top U_{K_1} \text{ resp. } I_{K_1} = -M I_{K_2} \text{ with} \\ &U_{K_i} := (U_k)_{k\in K_i} \in \mathcal{F}^{K_i},\ I_{K_i} := (I_k)_{k\in K_i} \in \mathcal{F}^{K_i}. \end{aligned} \tag{2.113}$$

With $K_{1,s} := K_1 \cap K_s$ etc., the decompositions

$$K = K_1 \uplus K_2 = K_s \uplus K_p \text{ imply } K = K_{1,s} \uplus K_{2,s} \uplus K_{1,p} \uplus K_{2,p}. \tag{2.114}$$

Define $u := \begin{pmatrix} I_{K_{2,s}} \\ U_{K_{1,s}} \end{pmatrix} \in \mathcal{F}^{K_{2,s}\uplus K_{1,s}} = \mathcal{F}^{K_s} = \mathcal{F}^{n_s}$, and let y be the subvector of $\binom{U}{I}$ that contains all components except those of u, hence $\binom{U}{I} = \binom{y}{u}$ (up to the order of the components). Then the network behavior $\mathcal{B}$ can be written as

$$\begin{aligned} &\mathcal{B} := \left\{\binom{U}{I} = \binom{y}{u} \in \mathcal{F}^{K\uplus K} = \mathcal{F}^{(2n-n_s)+n_s};\ P \circ y = Q \circ u\right\} \text{ with} \\ &(P, -Q) \in \mathbb{R}[s]^{(2n-n_s)\times((2n-n_s)+n_s)},\ n := \sharp(K),\ n_s := \sharp(K_s), \end{aligned} \tag{2.115}$$

where P and Q are easily derived from M and the P_k, Q_k, cf. (8.221). Under a weak constructive condition, that is satisfied generically or almost always, $\mathcal{B}$ is an IO behavior, cf. Theorem 8.3.11, with input u and transfer matrix $H := P^{-1}Q$. Assume this. It follows the reasonable result that the source currents I_{k_2}, $k_2 \in K_{2,s}$ and source voltages U_{k_1}, $k_1 \in K_{1,s}$, can be chosen as input and give rise to all other branch voltages U_k, $k \in K \setminus K_{1,s}$, and branch currents I_k, $k \in K \setminus K_{2,s}$. If $\mathcal{B}$ is not an IO behavior, it has to be redesigned. The characteristic variety $\mathrm{char}(\mathcal{B}^0) = \mathrm{V}_{\mathbb{C}}(\det(P))$ of $\mathcal{B}^0 := \{y;\ P \circ y = 0\}$ can be easily determined. The steady-state and superposition principle considerations concerning $\mathcal{B}$ in electrical engineering are valid if and only if the IO and asymptotic stability condition $\mathrm{V}_{\mathbb{C}}(\det(P)) \subset \mathbb{C}_-$ holds. The latter condition is often ignored, since it requires P and is hard to formulate in the usual engineering language. If it is satisfied and u has left bounded support or is periodic and y is any output to u with $P \circ y = Q \circ u$, then $y_{ss} := H \circ u$ can be identified with y (for $t \to \infty$, $y \approx y_{ss}$), and its components are all steady-state branch voltages U_k, $k \in K \setminus K_{1,s}$, and branch currents I_k, $k \in K \setminus K_{2,s}$. If, additionally, H is proper and u is piecewise continuous, so is y. This is a predominant result for the analysis of electrical networks with an arbitrary number of source branches.

If $\mathcal{B}$ is an IO behavior and an n_s-port, i.e., with $2n_s$ terminals, the current which flows into the network at one terminal of a port coincides with that which flows out of it at the other terminal of the same port. This so-called *port condition* is always satisfied *for an IO n_s-port* and need not be required as is often done in the engineering literature, cf. Corollary 8.3.12, [Wikipedia; https://en.wikipedia.org/wiki/Port (circuit theory), 24 May 2018], [15, Sect. 6.1].

Let $y_s := \begin{pmatrix} I_{K_{1,s}} \\ U_{K_{2,s}} \end{pmatrix} \in \mathcal{F}^{K_{1,s} \uplus K_{2,s}} = \mathcal{F}^{K_s} = \mathcal{F}^{n_s}$ denote the vector of complementary source currents and voltages to those of u, and define the real projection matrix C_s such that $C_s y = y_s$. Then Eq. (2.32) and Theorem 8.3.18 constructively furnish a new IO behavior

$$\begin{aligned}
&\mathcal{B}_s := \begin{pmatrix} C_s & 0 \\ 0 & \mathrm{id}_{K_s} \end{pmatrix} \mathcal{B} = \left\{ \begin{pmatrix} C_s y \\ u \end{pmatrix};\ \begin{pmatrix} y \\ u \end{pmatrix} \in \mathcal{B} \right\} = \left\{ \begin{pmatrix} y_s \\ u \end{pmatrix} \in \mathcal{F}^{K_s \uplus K_s};\ P_s \circ y_s = Q_s \circ u \right\} \\
&\text{with } (P_s, -Q_s) \in \mathbb{R}[s]^{K_s \times (K_s \uplus K_s)} = \mathbb{R}[s]^{n_s \times (n_s + n_s)}, \\
&\qquad \mathrm{rank}(P_s) = \mathrm{rank}(P_s, -Q_s) = n_s = \sharp(K_s),\ H_s := P_s^{-1} Q_s.
\end{aligned} \tag{2.116}$$

Obviously $\mathcal{B}_s$ implies the differential system $P_s \circ y_s = Q_s \circ u$ for the source voltages and currents and eliminates the one-port voltages and currents of the interior of the network. It is sometimes called a *black box* with the terminals as connection to the outside. Again (2.43) and (2.59) are applicable to inputs u and steady-state outputs $H_s \circ u$ under the specified conditions. Define

$$\begin{aligned}
&U_s := (U_k)_{k \in K_s} \in \mathcal{F}^{K_s},\ I_s := (I_k)_{k \in K_s} \in \mathcal{F}^{K_s},\ R_s := (P_s, -Q_s) \\
&\Longrightarrow w_s := \begin{pmatrix} y_s \\ u \end{pmatrix} \left(= \begin{pmatrix} U_s \\ I_s \end{pmatrix} \text{ up to the order of the components} \right), \\
&\qquad \mathcal{B}_s = \left\{ w_s \in \mathcal{F}^{K_s \uplus K_s} = \mathcal{F}^{2n_s};\ R_s \circ w_s = 0 \right\},\ \mathrm{rank}(R_s) = n_s = \sharp(K_s).
\end{aligned} \tag{2.117}$$

Any n_s $\mathbb{R}(s)$-linearly independent columns of R_s give rise to a new IO structure of $\mathcal{B}_s$ and a new IO behavior $\widetilde{\mathcal{B}_s}$ with input $\widetilde{u} \in \mathcal{F}^{n_s}$. After the standard column and component permutations this has the form

$$\widetilde{\mathcal{B}_s} = \left\{ \widetilde{w}_s = \begin{pmatrix} \widetilde{y}_s \\ \widetilde{u} \end{pmatrix} \in \mathcal{F}^{n_s + n_s};\ \widetilde{P}_s \circ \widetilde{y}_s = \widetilde{Q}_s \circ \widetilde{u} \right\},\ \mathrm{rank}(\widetilde{P}_s) = n_s,\ \widetilde{H}_s = \widetilde{P}_s^{-1} \widetilde{Q}_s. \tag{2.118}$$

Notice that $(\widetilde{P}_s, -\widetilde{Q}_s)$ resp. $\widetilde{w}_s = \begin{pmatrix} \widetilde{y}_s \\ \widetilde{u} \end{pmatrix}$ coincide with $R_s = (P_s, -Q_s)$ resp. $w_s := \begin{pmatrix} y_s \\ u \end{pmatrix}$ up to the order of the columns resp. components, and can thus be trivially computed. Also $\mathrm{rank}(\widetilde{P}_s) = n_s$ can be easily tested. Assume this in the sequel. Then $\widetilde{\mathcal{B}_s}$ and also the original network behavior $\mathcal{B}$ are IO behaviors with input $\widetilde{u}$. Thus $\mathcal{B}$ too can be written as, cf. Theorem 8.3.21,

$$\begin{aligned}
&\widetilde{\mathcal{B}} = \left\{ \widetilde{w} = \begin{pmatrix} \widetilde{y} \\ \widetilde{u} \end{pmatrix} \in \mathcal{F}^{(2n - n_s) + n_s};\ \widetilde{P} \circ \widetilde{y} = \widetilde{Q} \circ \widetilde{u} \right\} \text{ with} \\
&(\widetilde{P}, -\widetilde{Q}) \in \mathbb{R}[s]^{(2n - n_s) \times ((2n - n_s) + n_s)},\ rank(\widetilde{P}) = 2n - n_s = n + n_p,\ \widetilde{H} = \widetilde{P}^{-1} \widetilde{Q}.
\end{aligned} \tag{2.119}$$

Again $(\widetilde{P}, -\widetilde{Q})$ resp. $\widetilde{w} = \begin{pmatrix} \widetilde{y} \\ \widetilde{u} \end{pmatrix}$ coincide with $(P, -Q)$ resp. $w = \begin{pmatrix} U \\ I \end{pmatrix} = \begin{pmatrix} y \\ u \end{pmatrix}$ up to the order of the columns resp. components and can be trivially computed, and so can be char $\left(\widetilde{\mathcal{B}}^0\right) = \mathrm{V}_{\mathbb{C}}(\det(\widetilde{P}))$. Assume asymptotic stability, i.e., char $\left(\widetilde{\mathcal{B}}^0\right) \subset \mathbb{C}_-$, so that steady-state considerations are valid.

Simple applications of $\widetilde{H}$ furnish various forms of the *Helmholtz/Thévenin and the Mayer/Norton equivalents (theorems)*, cf. Example 8.3.16, [14, Sects. 4.2.1, 4.2.2], [15, Sect. 1.4].

Choose a period $T := 2\pi\omega^{-1} > 0$ and assume that $\widetilde{H}$ and thus $\widetilde{H}_s$ are proper. Also choose a piecewise continuous T-periodic input

$$\widetilde{u} = \sum_{\mu\in\mathbb{Z}} \mathbb{F}(\widetilde{u})(\mu)e^{j\mu\omega t} \text{ and define } \widetilde{y} := \widetilde{H}\circ\widetilde{u} = \sum_{\mu\in\mathbb{Z}} \widetilde{H}(j\mu\omega)\mathbb{F}(\widetilde{u})(\mu)e^{j\mu\omega t},$$
$$\widetilde{y}_s := \widetilde{H}_s\circ\widetilde{u} = \sum_{\mu\in\mathbb{Z}} \widetilde{H}_s(j\mu\omega)\mathbb{F}(\widetilde{u})(\mu)e^{j\mu\omega t}, \tag{2.120}$$

where $\widetilde{y}$ resp. $\widetilde{y}_s$ are the steady-state outputs of $\widetilde{\mathcal{B}}$ resp. $\widetilde{\mathcal{B}}_s$ to the input $\widetilde{u}$. According to (2.59) $\widetilde{y}$ resp. $\widetilde{y}_s$ are piecewise continuous and even continuous with uniform convergence of the Fourier series if $\widetilde{H}$ is strictly proper. Notice again that (2.120) gives an explicit Fourier series for *all* steady-state voltages and currents U_k and I_k, $k \in K$, for the periodic input $\widetilde{u}$ under the condition that

$$\operatorname{rank}(P) = 2n - n_s,\ \operatorname{rank}(\widetilde{P}_s) = n_s,\ \text{char}\left(\widetilde{\mathcal{B}}^0\right)\subset\mathbb{C}_-,\ \widetilde{H}\text{ proper}, \tag{2.121}$$

cf. [14, Sect. 5.5], [22, Sect. 3.3].

Assume, in particular, that $\widetilde{u} = I_s = (I_k)_{k\in K_s}$ and hence $\widetilde{y}_s = U_s = (U_k)_{k\in K_s}$. Then the matrices $\widetilde{H}_s$ resp. $\widetilde{H}_s(j\mu\omega)$ are called the *impedance transfer matrix* resp. *impedance matrix* at the frequency $\mu\omega$ of the network. If, in contrast, $\widetilde{u} = U_s$ and thus $\widetilde{y}_s = I_s$, then $\widetilde{H}_s$ resp. $\widetilde{H}_s(j\mu\omega)$, of course, with a different $\widetilde{H}_s$, are the *admittance transfer matrix* resp. *admittance matrix*. For every choice of $\widetilde{u}$ *with* $\operatorname{rank}(\widetilde{P}_s) = n_s$ the corresponding matrices $\widetilde{H}_s$, $\widetilde{H}_s(j\mu\omega)$ get special names. For two-ports with $\binom{U_s}{I_s} \in \mathcal{F}^4$ there are obviously $\binom{4}{2} = 6$ essentially different choices of $\widetilde{u} \in \mathcal{F}^2$. If $\operatorname{rank}(\widetilde{P}_s) = 2$, such a choice gives rise to a two-port $\widetilde{\mathcal{B}}_s$. These various two-ports are intensively studied in electrical engineering, cf. [15, Chap. 6].

We next explain a special state space representation of a pure RCL-network behavior $\mathcal{B}$, cf. [30], [14, Sect. 3.4], Theorem 8.3.32 and Example 8.3.33. We use a suitable Gaußalgorithm to obtain decompositions $K = K_1 \uplus K_2 = K_s \uplus K_p$ with special properties. Let K_C resp. K_L denote the set of capacitance resp. inductance branches, and define

$$K_{1,C} := K_1\cap K_C,\ K_{2,L} := K_2\cap K_L,\ U_{1,C} := (U_k)_{k\in K_{1,C}},\ I_{2,L} := (I_k)_{k\in K_{2,L}},$$
$$\widetilde{x} := \begin{pmatrix} U_{1,C}\\ I_{2,L}\end{pmatrix},\ \widetilde{u} := \begin{pmatrix} U_{1,s}\\ I_{2,s}\end{pmatrix},\ \begin{pmatrix} U\\ I\end{pmatrix} = \begin{pmatrix} \widetilde{x}\\ \widetilde{y}\\ \widetilde{u}\end{pmatrix} \text{ (up to the order of the components)}, \tag{2.122}$$

where, by definition, $\widetilde{y}$ contains all components of $\binom{U}{I}$ that are not contained in $\widetilde{x}$ and $\widetilde{u}$. Note that the tildes have a different meaning here than in the preceding considerations. Then one can compute real matrices of suitable sizes $\widetilde{A}$, $\widetilde{C}$, $\widetilde{B}_0$, $\widetilde{B}_1$, $\widetilde{D}_0$, $\widetilde{D}_1$

and then $\widetilde{B} := \widetilde{B}_1 s + \widetilde{B}_0$, $\widetilde{D} := \widetilde{D}_1 s + \widetilde{D}_0$ such that the following map is a behavior isomorphism:

$$\begin{aligned}
&\widetilde{\mathcal{B}} := \left\{\begin{pmatrix}\widetilde{x}\\ \widetilde{u}\end{pmatrix} \in \mathcal{F}^{\bullet+n_s};\ s \circ \widetilde{x} = \widetilde{A}\widetilde{x} + \widetilde{B} \circ \widetilde{u}\right\} \cong \mathcal{B},\ \begin{pmatrix}\widetilde{x}\\ \widetilde{u}\end{pmatrix} \mapsto \begin{pmatrix}U\\ I\end{pmatrix} := \begin{pmatrix}\widetilde{x}\\ \widetilde{C}\widetilde{x}+\widetilde{D}\circ\widetilde{u}\\ \widetilde{u}\end{pmatrix},\\
&\Longrightarrow \widetilde{\mathcal{B}}^0 = \left\{\widetilde{x};\ s \circ \widetilde{x} = \widetilde{A}\widetilde{x}\right\} \cong \mathcal{B}^0,\ \widetilde{x} \mapsto \begin{pmatrix}\widetilde{x}\\ \widetilde{C}\widetilde{x}\end{pmatrix},\ \mathrm{spec}(\widetilde{A}) = \mathrm{char}(\mathcal{B}^0),\\
&\widetilde{H} = (s\,\mathrm{id} - \widetilde{A})^{-1}(\widetilde{B}_1 s + \widetilde{B}_0),\ H = \begin{pmatrix}\widetilde{H}\\ \widetilde{D}+\widetilde{C}\widetilde{H}\end{pmatrix}.
\end{aligned} \tag{2.123}$$

Since $\widetilde{B}, \widetilde{D}$ are not constant, but $\deg_s(\widetilde{B}) \le 1$, $\deg_s(\widetilde{D}) \le 1$, this is generally not a state space representation according to Kalman, but very similar conclusions can be drawn, for instance, on $\mathrm{char}(\mathcal{B}^0)$, cf. Theorem 8.3.32. The transfer matrix H is proper if and only if $\widetilde{D}_1 = 0$. Notice that the state $\widetilde{x}$ is a subvector of the trajectory $\begin{pmatrix}U\\ I\end{pmatrix}$, a rare occurrence in state space equations. If $\widetilde{B}_1 = 0$, $\widetilde{D}_1 = 0$, (2.123) is a state space representation according to Kalman, and especially well suited to simulate the trajectories of $\mathcal{B}$ by those of $\widetilde{\mathcal{B}}$, including initial conditions on $\widetilde{x}$ resp. $\begin{pmatrix}U\\ I\end{pmatrix}$.

We finally discuss several results on electrical power. The instantaneous power along k is $U_k(t)I_k(t)$ where the piecewise continuity of U_k and I_k is assumed. For distributions this product makes no sense. A famous theorem with a very simple proof is *Tellegen's*, cf. Theorem 8.3.36, that says that

$$\sum_{k\in K} U_k(t)I_k(t) = 0. \tag{2.124}$$

It is an energy preservation result for the total behavior with its energy sources from outside.

Assume that the network behavior $\mathcal{B}$ is an asymptotically stable IO behavior with input $\widetilde{u} = I_s := (I_k)_{k\in K_s}$, associated IO behavior $\widetilde{\mathcal{B}}_s$, output $U_s := (U_k)_{k\in K_s}$ and proper impedance transfer matrix $\widetilde{H}_s$. For a period $T = 2\pi\omega^{-1} > 0$ this implies the impedance matrix $Z := \widetilde{H}_s(j\mu\omega)$ at the frequency $\mu\omega$. Then

$$\widetilde{H}_s^\top = \widetilde{H}_s,\ Z^\top = Z, \tag{2.125}$$

i.e., these matrices are symmetric. This holds if the network consists of source and one-port branches only. These networks are called *reciprocal*, cf. [15, Sect. 6.3.1] for 2-ports.

We finally derive the average power that is supplied to the network that may be more complicated without symmetric $\widetilde{H}_s$, but is assumed asymptotically stable with proper transfer matrix. We assume a (real) piecewise continuous, periodic current vector $I_s = (I_k)_{k\in K_s}$ and the steady-state voltage output $U_s := \widetilde{H}_s \circ I_s$ with the Fourier series

$$I_s = \sum_{\mu\in\mathbb{Z}} \mathbb{F}(I_s)(\mu)e^{j\mu\omega t},\ U_s = \sum_{\mu\in\mathbb{Z}} \mathbb{F}(U_s)(\mu)e^{j\mu\omega t},\ \mathbb{F}(U_s)(\mu) = \widetilde{H}_s(j\mu\omega)\mathbb{F}(I_s)(\mu). \tag{2.126}$$

The condition, that I_s is real, is equivalent to $\mathbb{F}(I_s)(-\mu) = \overline{\mathbb{F}(I_s)(\mu)}$, and likewise for $\mathbb{F}(U_s)$. The average of a piecewise continuous, T-periodic function f is defined as $T^{-1}\int_0^T f(t)dt$. Then the (average) *real power* of the sources of the network is, cf. [14, Sect. 5.5.4], [22, Sect. 3.3.4], Theorem 8.3.43,

$$\begin{aligned}\mathcal{P}_r := \sum_{k\in K_s} T^{-1}\int_0^T U_k(t)I_k(t)dt &= \mathbb{F}(I_s)(0)^\top \widetilde{H}_s(0)\mathbb{F}(I_s)(0)\\ &+\sum_{\mu=1}^{\infty}\mathbb{F}(I_s)(\mu)^*\left(\widetilde{H}_s(j\mu\omega)^* + \widetilde{H}_s(j\mu\omega)\right)\mathbb{F}(I_s)(\mu),\end{aligned} \tag{2.127}$$

where $M^* := \overline{M^\top}$ denotes the Hermitian adjoint of a complex matrix. Notice that $\mathbb{F}(I_s)(0)$ and $\widetilde{H}_s(0)$ are real and $\widetilde{H}_s(j\mu\omega)^* + \widetilde{H}_s(j\mu\omega)$ is Hermitian, so the expression on the right is indeed real as it should be. In our approach the voltage U_k and the current I_k have the same direction for *all* $k \in K$, also for $k \in K_s$, whereas in the engineering literature U_k and I_k, $k \in K_s$, have the opposite direction. The consequence is that in our approach a negative instantaneous power $U_k(t)I_k(t)$, $k \in K_s$, means that energy flows from the source k to the interior of the network at time t, whereas a positive power means a flow toward the source. If $\mathcal{B}$ is a one-port, i.e., $\sharp(K_s) = 1$, an *apparent resp. reactive* power $\mathcal{P}_{app} \geq 0$ resp. $\mathcal{P}_{react}$ with $\mathcal{P}_{app}^2 = \mathcal{P}_r^2 + \mathcal{P}_{react}^2$ are defined and discussed, cf. [22, 3.3.4] and Corollary 8.3.45.

2.8 Stabilizing Compensators

The notations and assumptions of the preceding sections remain in force, we denote $\mathcal{D} := \mathbb{C}[s]$. *Chap.* 10 *is a variant of essential parts of Vidyasagar's book* [31] *and also of* [19, Chaps. 6, 7], *however with modified mathematics. We refer to* [31] *for the history of this approach. It deals with the synthesis of suitable behaviors, cf. the title of* [31]. *The chapter also owes much to Bourlès' RST-controllers* [18, Chap. 6] *and his suggestions for the paper* [32]. *Its mathematical details come from the papers* [33] *and* [34]. *The use of modules in this context is due to Quadrat* [35].

The set T of asymptotically stable polynomials $t \in \mathcal{D}$, i.e., with $\mathrm{V}_{\mathbb{C}}(t) \subset \mathbb{C}_-$, is a saturated, symmetric submonoid of $\mathbb{C}[s]$, i.e., satisfies

$$1 \in T,\ 0 \notin T,\ t_1, t_2 \in T \Longleftrightarrow t_1t_2 \in T,\ t \in T \Longleftrightarrow \overline{t} \in T. \tag{2.128}$$

The following considerations hold more generally for nonempty subsets $\Lambda_1 = \overline{\Lambda_1} \subseteq \mathbb{C}_-$ and $T = \{t \in \mathcal{D};\ \mathrm{V}_{\mathbb{C}}(t) \subseteq \Lambda_1\}$. Assume this in the sequel. For *pole placement* the set Λ_1 may be chosen finite. With today's computer algebra systems the variety $\mathrm{V}_{\mathbb{C}}(t)$ and the inclusion $\mathrm{V}_{\mathbb{C}}(t) \subseteq \Lambda_1$ can be easily determined. There are other methods (*Routh-Hurwitz criterion*) to decide $t \in T$ without computing $\mathrm{V}_{\mathbb{C}}(t)$, but these are not discussed in this book.

The monoid T gives rise to the quotient ring $\mathcal{D}_T := \left\{ft^{-1} = \frac{f}{t};\ f \in \mathcal{D}, t \in T\right\} \subseteq \mathbb{C}(s)$, which is also a principal ideal domain. Likewise, every $\mathcal{D}$-module M gives rise to the $\mathcal{D}_T$*-quotient module*

$$M_T := \left\{xt^{-1} = \tfrac{x}{t};\ x \in M, t \in T\right\} \text{ with } \tfrac{f}{t_1}\tfrac{x}{t_2} := \tfrac{fx}{t_1 t_2} \tag{2.129}$$

as scalar multiplication, and the T*-torsion submodule*

$$\begin{aligned}&\mathrm{t}_T(M) := \{x \in M;\ \exists t \in T : tx = 0\} = \ker\left(M \to M_T,\ x \mapsto \tfrac{x}{1}\right) \subseteq M\\ &\Longrightarrow (\mathrm{t}_T(M) = M \Longleftrightarrow M_T = 0).\end{aligned} \tag{2.130}$$

Note that $\mathcal{D}$ has no zero divisors and thus $\mathrm{t}_T(\mathcal{D}) = 0$ whereas $\mathrm{t}_T(M)$ may be nonzero. Hence the construction of M_T and the study of its properties in Sect. 9.1 are more difficult than those of $\mathcal{D}_T$ that are known from the construction of $\mathbb{Q}$ resp. $\mathbb{C}(s)$ from $\mathbb{Z}$ resp. $\mathcal{D} = \mathbb{C}[s]$.

Let $\mathcal{F}$ be one of the injective cogenerators $\mathrm{C}^{-\infty}$, C^{∞}, $\mathrm{t}(\mathrm{C}^{\infty}) = \oplus_{\lambda\in\mathbb{C}}\mathbb{C}[t]e^{\lambda t}$. Note that, in this section, t generally denotes a polynomial in T whereas in the formulas $e^{\lambda t}$ and $\lim_{t\to\infty}$ it denotes a time instant in $\mathbb{R}$. Then $\mathcal{F}_T$ *is an injective* $\mathcal{D}_T$*-cogenerator, cf. Theorem* 9.3.6, and thus gives rise to a theory of ${}_{\mathcal{D}_T}\mathcal{F}_T$-behaviors. Moreover $\mathcal{F}$ admits a $\mathcal{D}$-linear direct sum decomposition, cf. Theorem 9.3.2,

$$\begin{aligned}&\mathcal{F} := \mathrm{t}_T(\mathcal{F}) \oplus \mathcal{F}' \ni w = w_{tr} + w_{ss},\ \mathrm{t}_T(\mathcal{F}) = \oplus_{\lambda\in\Lambda_1}\mathbb{C}[t]e^{\lambda t} \subseteq \mathcal{F}_- = \oplus_{\lambda\in\mathbb{C}_-}\mathbb{C}[t]e^{\lambda t}\\ &\text{with } \mathcal{F}' \underset{\mathcal{D}}{\cong} \mathcal{F}_T,\ \widetilde{w} \mapsto \tfrac{\widetilde{w}}{1} \Longrightarrow \mathcal{F}' \underset{\text{identification}}{=} \mathcal{F}_T,\ \widetilde{w} = \tfrac{\widetilde{w}}{1}.\end{aligned} \tag{2.131}$$

The elements $t \in T$, $H \in \mathcal{D}_T$ resp. $y \in \mathrm{t}_T(\mathcal{F})$ *are called* T*-stable polynomials, rational functions resp. signals.* Due to $\Lambda_1 \subseteq \mathbb{C}_-$ T-stability implies asymptotic stability. The components w_{tr} resp. w_{ss} are again called the *transient* resp. the *steady state* of w for this decomposition. The existence of the direct summand $\mathcal{F}'$ depends on the nonconstructive *Lemma of Zorn*, and therefore neither $\mathcal{F}'$ nor, in general, the decomposition $w = w_{tr} + w_{ss}$ can be computed explicitly. However, in many situations the *unique existence* of $w = w_{tr} + w_{ss}$ is sufficient. Due to $\mathrm{t}_T(\mathcal{F}) \subseteq \mathcal{F}_-$ the limit $\lim_{t\to\infty} w_{tr}(t) = 0$ holds, i.e., for practical purposes w and w_{ss} can be identified in many situations.

If $H = \frac{f}{t_1} \in \mathcal{D}_T \subset \mathbb{C}(s)$ is a T-stable rational function and $\widetilde{u} = \frac{u}{t_2} \in \mathcal{F}' = \mathcal{F}_T$, then $H \circ \widetilde{u} = \frac{f \circ u}{t_1 t_2}$ is defined, whereas $H \circ u$ is not defined for each $u \in \mathcal{F}$, let alone for arbitrary $H \in \mathbb{C}(s)$. In [31, Chaps. 3, 5; (1), (2)] the meaning of Hu is not explained. Note, however, that $H \circ u$ is well defined for $H \in \mathbb{C}(s)$ and $u \in \mathrm{C}_+^{-\infty}$. Any behavior

$$\begin{aligned}&\mathcal{B} = \left\{w \in \mathcal{F}^l;\ R \circ w = 0\right\} \text{ with}\\ &R \in \mathcal{D}^{p\times l},\ \mathrm{rank}(R) = p,\ U = \mathcal{B}^\perp = \mathcal{D}^{1\times p}R,\ M = \mathcal{D}^{1\times l}/U,\end{aligned}$$

implies

$$\begin{aligned}&\mathcal{B}_T=\left\{w\in\mathcal{F}_T^l;\ R\circ w=0\right\}\cong\mathrm{Hom}_{\mathcal{D}_T}(M_T,\mathcal{F}_T)\ \text{where}\\&\quad U_T=\mathcal{D}_T^{1\times p}R,\ M_T=\mathcal{D}_T^{1\times l}/U_T,\\&\mathcal{B}=\mathrm{t}_T(\mathcal{B})\oplus\mathcal{B}_T,\ \mathrm{t}_T(\mathcal{B})=\mathcal{B}\cap\mathrm{t}_T(\mathcal{F})^l,\ \mathcal{B}_T\underset{\mathcal{F}'=\mathcal{F}_T}{=}\mathcal{B}\cap(\mathcal{F}')^l=\mathcal{B}\cap\mathcal{F}_T^l.\end{aligned}\tag{2.132}$$

In particular, we infer the equivalence

$$\mathcal{B}_T=0\Longleftrightarrow M_T=0\Longleftrightarrow\exists t\in T\ \text{with}\ tM=0\Longleftrightarrow\mathcal{B}=\mathrm{t}_T(\mathcal{B})\ (\subseteq\mathrm{t}(\mathcal{B}))\,.\tag{2.133}$$

Hence, if $\mathcal{B}_T=0$, $\mathcal{B}$ is autonomous and called T-*autonomous*. The main application is to IO behaviors

$$\begin{aligned}&\mathcal{B}=\left\{\binom{y}{u}\in\mathcal{F}^{p+m};\ P\circ y=Q\circ u\right\},\ \mathcal{B}^0=\left\{y\in\mathcal{F}^p;\ P\circ y=0\right\},\\&\mathcal{B}_T=\left\{\binom{y}{u}\in\mathcal{F}_T^p;\ P\circ y=Q\circ u\right\},\ \mathcal{B}_T^0=\left\{y\in\mathcal{F}_T^p;\ P\circ y=0\right\},\end{aligned}\tag{2.134}$$

where $(P,-Q)\in\mathcal{D}^{p\times(p+m)}$, $\mathrm{rank}(P)=p$. The *IO behavior $\mathcal{B}$ is called T-stable* if it satisfies the following equivalent conditions, cf. Theorem 9.4.2:

1. $\mathcal{B}^0$ is T-autonomous or, equivalently, $\mathcal{B}_T^0=0$ or $\det(P)\in T$ or $P\in\mathrm{Gl}_p(\mathcal{D}_T)$.
2. (i) $\mathcal{B}_T$ is controllable or M_T is $\mathcal{D}_T$-free or $\mathrm{char}(\mathcal{B})\subseteq\Lambda_1$.
   (ii) H is T-stable, i.e., $H\in\mathcal{D}_T^{p\times m}$.

Assume that $\mathcal{B}$ is T-stable. Thus, ${}_{\mathcal{D}_T}\mathcal{F}_T$ and $H=P^{-1}Q$ imply

$$\begin{aligned}\mathcal{B}_T&=\left\{\binom{y}{u}\in\mathcal{F}_T^{p+m}=(\mathcal{F}')^{p+m};\ P\circ y=Q\circ u\right\}\\&=\left\{\binom{y}{u}\in\mathcal{F}_T^{p+m}=(\mathcal{F}')^{p+m};\ y=H\circ u\right\}\\&\Longrightarrow\forall\binom{y}{u}\in\mathcal{B}\ \text{with}\ u=u_{tr}+u_{ss}:\ y=y_{tr}+y_{ss},\ y_{ss}=H\circ u_{ss}.\end{aligned}\tag{2.135}$$

Hence $H\circ u_{ss}$ is the steady state of y, cf. (2.131) and (2.135). The latter equation is the main tool to simplify equations for T-stable IO behaviors.

We have seen that properness of a transfer matrix is an important property. For T-stable IO behaviors this means

$$H\in\mathbb{C}[s]_T^{p\times m}\cap\mathbb{C}(s)_{\mathrm{pr}}^{p\times m}=\mathcal{S}^{p\times m},\ \mathcal{S}:=\mathbb{C}[s]_T\cap\mathbb{C}(s)_{\mathrm{pr}}\subset\mathbb{C}(s),\tag{2.136}$$

where $\mathbb{C}(s)_{\mathrm{pr}}$ resp. $\mathcal{S}$ is the ring of proper resp. of proper and T-stable rational functions. The computations in [31] essentially use that $\mathcal{S}$ is Euclidean. Instead, we choose $\alpha\in\Lambda_1$, define the variable $\widehat{s}:=(s-\alpha)^{-1}\in\mathbb{C}(s)$, and show in Theorem 9.5.4 that

$$\mathcal{S}=\mathbb{C}[\widehat{s}]_{\widehat{T}}\subset\mathbb{C}(s),\ \widehat{T}:=\left\{\widehat{t}:=t\widehat{s}^{\deg_s(t)}=\tfrac{t}{(s-\alpha)^{\deg_s(t)}};\ t\in T\right\},\tag{2.137}$$

where $\mathbb{C}[\widehat{s}]$ is the polynomial algebra in the variable $\widehat{s}$, cf. [36]. Computations with this quotient ring $\mathcal{S} = \mathbb{C}[\widehat{s}]_{\widehat{T}}$ of a polynomial algebra are as simple as with the polynomial algebra itself, are implemented in every *Computer Algebra* system, and are simpler than those in a general Euclidean ring, e.g., in $\mathcal{S}$. In Chap. 11 the ring $\mathcal{S}$ is used for the construction of *functional observers*.

A feedback compensator or controller $\mathcal{B}_2$ of a *plant* $\mathcal{B}_1$ is used to stabilize the plant and to steer its output into a desired direction. The mathematical model is given by two IO behaviors

$$\begin{aligned}\mathcal{B}_1 &:= \left\{\begin{pmatrix} y_1 \\ u_1 \end{pmatrix} \in \mathcal{F}^{p+m};\ P_1 \circ y_1 = Q_1 \circ u_1\right\},\ (P_1, -Q_1) \in \mathcal{D}^{p\times(p+m)},\ \operatorname{rank}(P_1) = p,\\ \mathcal{B}_2 &:= \left\{\begin{pmatrix} u_2 \\ y_2 \end{pmatrix} \in \mathcal{F}^{p+m};\ P_2 \circ y_2 = Q_2 \circ u_2\right\},\ (-Q_2, P_2) \in \mathcal{D}^{m\times(p+m)},\ \operatorname{rank}(P_2) = m.\end{aligned} \tag{2.138}$$

We use $\begin{pmatrix} u_2 \\ y_2 \end{pmatrix}$ instead of $\begin{pmatrix} y_2 \\ u_2 \end{pmatrix}$ for dimension reasons since the output (input) y_1 (u_1) of the *plant* $\mathcal{B}_1$ is assumed to have the same dimension p (m) as the input (output) u_2 (y_2) of the *compensator* $\mathcal{B}_2$. *Feedback* means to add (feed back) the output of $\mathcal{B}_1$ $(\mathcal{B}_2)$ to the input of $\mathcal{B}_2$ $(\mathcal{B}_1)$. Define $y := \begin{pmatrix} y_1 \\ y_2 \end{pmatrix}$, $u := \begin{pmatrix} u_2 \\ u_1 \end{pmatrix} \in \mathcal{F}^{p+m}$. Then the feedback interconnection of the two behaviors is the behavior

$$\begin{aligned}\mathcal{B} := \operatorname{fb}(\mathcal{B}_1, \mathcal{B}_2) &:= \left\{\begin{pmatrix} y \\ u \end{pmatrix} \in \mathcal{F}^{2(p+m)};\ \begin{pmatrix} y_1 \\ u_1+y_2 \end{pmatrix} \in \mathcal{B}_1,\ \begin{pmatrix} y_2 \\ u_2+y_1 \end{pmatrix} \in \mathcal{B}_2\right\}\\ &= \left\{\begin{pmatrix} y \\ u \end{pmatrix} \in \mathcal{F}^{2(p+m)};\ P \circ y = Q \circ u\right\} \text{ where}\\ P := \begin{pmatrix} P_1 & -Q_1 \\ -Q_2 & P_2 \end{pmatrix}&,\ Q := \begin{pmatrix} 0 & Q_1 \\ Q_2 & 0 \end{pmatrix} \in \mathcal{D}^{(p+m)\times(p+m)}.\end{aligned} \tag{2.139}$$

If $\operatorname{rank}(P) = p + m$ or $\det(P) \neq 0$, this is an IO behavior with transfer matrix $H = P^{-1}Q$. One then says that *the feedback behavior is well-posed*, and calls $H = P^{-1}Q$ the *closed-loop transfer matrix*. Assume this in the sequel.

The first goal is the T-*stabilization* of $\mathcal{B}_1$ by $\mathcal{B}_2$, i.e., the T stability, especially asymptotic stability, of $\mathcal{B} = \operatorname{fb}(\mathcal{B}_1, \mathcal{B}_2)$. This means $\det(P) \in T$ or that $\mathcal{B}_T$ is controllable and $H = P^{-1}Q \in \mathcal{D}_T^{(p+m)\times(p+m)}$. Then (2.135) is applicable to $\operatorname{fb}(\mathcal{B}_1, \mathcal{B}_2)$. For a given plant $\mathcal{B}_1$ a compensator $\mathcal{B}_2$ with well-posed and T-stable $\mathcal{B} = \operatorname{fb}(\mathcal{B}_1, \mathcal{B}_2)$ exists if and only if $\mathcal{B}_{1,T}$ is controllable or $\operatorname{char}(\mathcal{B}_1) \subseteq \Lambda_1$, and one then says that $\mathcal{B}_1$ is T-*stabilizable* and $\mathcal{B}_2$ is a T-*stabilizing compensator*, cf. Corollary 10.1.11. This is the main use of controllability in this book. Assume the T-stabilizability of $\mathcal{B}_1$ in the sequel.

In Theorem 10.1.10 we construct, for T-stabilizable $\mathcal{B}_1$, *all* controllable compensators $\mathcal{B}_2$ such that $\mathcal{B} = \operatorname{fb}(\mathcal{B}_1, \mathcal{B}_2)$ is well-posed and T-stable. These $\mathcal{B}_2$ depend on T-stable rational $m \times p$-matrices X as *parameter*, and therefore one talks about the *parametrization* of these compensators. In Theorem 10.1.20 we construct (parametrize) all compensators for which, in addition, the closed-loop transfer matrix H is proper, i.e., $H \in \mathcal{S}^{(p+m)\times(p+m)}$, cf. (2.137). Finally, in Theorem 10.1.34, all $\mathcal{B}_2$ with, in addition, proper transfer matrix, $H_2 = P_2^{-1}Q_2$, are parametrized. We do not assume that the transfer matrix $H_1 = P_1^{-1}Q_1$ of $\mathcal{B}_1$ is proper, but most real plants have this property. The properness of H_2 is important since it enables a state space representation of $\mathcal{B}_2$ according to Kalman and its construction with elementary building blocks. If also $\mathcal{B}_1$ is a state space behavior, then the preceding considerations yield many more stabilizing compensators than those that are usually constructed by means of *Luenberger state observers* connected with *state feedback*, cf. [11, p. 523], [10, Sects. 10.5–10.6], Sects. 10.1.5 and 12.5.

The preceding theory is applicable to finite $\Lambda_1 \subset \mathbb{C}_-$. For the T-stability condition $\mathrm{char}(\mathcal{B}^0) \subseteq \Lambda_1$ one says that the *poles of $\mathcal{B}^0$ have been placed into or assigned to* Λ_1, cf. Theorem 10.1.26.

As explained in the preceding sections stability is a necessary condition for almost any IO behavior. But stabilization is not a goal for itself except in special cases, for instance, to stabilize a building after an earthquake. Switching off an asymptotically stable electrical network stabilizes it, but the network cannot serve any further purpose. *Control design* in Sect. 10.2 is devoted to the choice of compensators, among those just described, that serve a useful purpose. We always assume that the feedback behavior and the T-stabilizing compensator have proper transfer matrices. The main treated design problem is *tracking and disturbance rejection*. In this case a nonzero polynomial ϕ and three signals $r, u_2 \in \mathcal{F}^p$, $u_1 \in \mathcal{F}^m$, with $\phi \circ (u_2, u_1, r) = (0, 0, 0)$ are considered where r is a given *reference signal* and u_1 resp. u_2 are unknown *disturbance signals* of the input resp. output of $\mathcal{B}_1$. These signals are, of course, not assumed T-stable, i.e., $\phi \notin T$ in general. The input of the compensator is $y_1 + u_2 - r$ where $y_1 + u_2$ is the actual disturbed output of the plant in the feedback behavior. For instance, $\phi = s\ (s^2, s^3)$ means that the signals are constant (linear, quadratic) functions of t. The design goal is to construct $\mathcal{B}_2$ such that $y_1 + u_2 - r$ is T-stable, in particular, $\lim_{t \to \infty}((y_1 + u_2) - r)(t) = 0$, i.e., the actual disturbed output signal of the plant in the feedback behavior $\mathcal{B}$ coincides, for practical purposes, with the desired reference signal. One says that in $\mathcal{B}$ *the output of the plant tracks the reference signal and rejects the disturbance signals*.

The polynomial ϕ is fixed and thus restricts the admissible unknown disturbance signals, but generically (almost) all ϕ can be chosen for a given plant. We derive a necessary and sufficient condition for the existence of such compensators $\mathcal{B}_2$ for given plant $\mathcal{B}_1$ and polynomial ϕ, and parametrize all these. The most important constructive results are Theorems 10.2.8, 10.2.11, and 10.2.17 and the algorithm in Corollary 10.2.10. In Sect. 10.2.2 we also discuss the significance of the *(transmission) zeros* of the plant's transfer matrix in this context. All matrix computations require the Smith form of polynomial matrices only.

In the literature the reference signal r and likewise the disturbance signals are often assumed, cf. [19, (17), p. 198], [13, (9–100), (9–101), p. 495], [31, Sect. 7.5], as

$$\begin{aligned}
&\widetilde{r} = H_r \circ \delta = \mathcal{L}^{-1}(H_r) = \alpha_r Y, \ \ \alpha_r \in \mathrm{t}(\mathcal{F})^p \text{ where} \\
&H_r = \mathcal{L}(\widetilde{r}) = \phi^{-1} Q_r, \ Q_r \in \mathbb{C}[s]^p, \ \deg_s(\phi) > \deg_s(Q_r) \Longrightarrow \phi \circ \alpha_r = 0.
\end{aligned} \tag{2.140}$$

Conversely, if

$$\begin{aligned}&\phi \circ r = 0, \text{ then } \phi \circ (rY) = (\phi \circ r)Y + Q_r \circ \delta = Q_r \circ \delta, \ Q_r \in \mathbb{C}[s]^p, \\ &\Longrightarrow \widetilde{r} := rY = H_r \circ \delta, \ H_r := \phi^{-1} Q_r, \ \deg_s(\phi) > \deg_s(Q_r), \ \widetilde{r}|_{[0,\infty)} = r|_{[0,\infty)}.\end{aligned} \tag{2.141}$$

For $(\widetilde{u}_2, \widetilde{u}_1, \widetilde{r})$ instead of (u_2, u_1, r) the design goal is $e = y_1 + \widetilde{u}_2 - \widetilde{r} = H_e \circ \delta$ with strictly proper and T-stable $H_e \in \mathbb{C}(s)^p$, cf. [19, p. 206, (70)], [31, p. 296, (R2)]. In Remark 10.2.6 we show that our theory furnishes such an H_e. Laplace transform techniques in the quoted books require these different reference and disturbance signals.

For the preceding data assume that $\mathcal{B}_2$ is a compensator for $\mathcal{B}_1$ and that $\widetilde{\mathcal{B}}_1 \subseteq \mathcal{F}^{p+m}$ is another IO plant. One says that $\mathcal{B}_2$ is a *robust compensator for* $\mathcal{B}_1$ if it is also a compensator (with all properties from above) for all *T-stabilizable $\widetilde{\mathcal{B}}_1$ near $\mathcal{B}_1$* or all controllable $\widetilde{\mathcal{B}}_{1,T}$ near $\mathcal{B}_{1,T}$. This requires a topology on the set of controllable IO behaviors $\widetilde{\mathcal{B}}_{1,T} \subseteq \mathcal{F}_T^{p+m}$, cf. Corollary and Definition 10.2.31. The main result is proven only for the case $\Lambda_1 = \mathbb{C}_-$ where T- and asymptotic stability coincide. Two such topologies are derived from a norm $\|H\|$ and a finer norm $\|H\|_1$ on the algebra $\mathcal{S}$ of T-stable rational functions, cf. Theorem 10.2.20 and Remark 10.2.49, given for $H = H_0 + H_{\text{spr}} \in \mathcal{S}$ with $H_0 \in \mathbb{C}$ and strictly proper H_{spr} by

$$\begin{aligned}\|H\| &:= \sup_{\omega \in \mathbb{R}} |H(j\omega)| = \sup_{\omega \in \mathbb{R}} |H_0 + H_{\text{spr}}(j\omega)| \\ \|H\|_1 &:= |H_0| + \|h\|_1, \ h := H_{\text{spr}} \circ \delta, \ \|h\|_1 := \int_0^\infty |h(t)| \mathrm{d}t, \ \|H\| \le \|H\|_1.\end{aligned} \tag{2.142}$$

These norms are naturally extended to matrix norms $\|H\|$ and $\|H\|_1$. A matrix $H = H_0 + H_{\text{spr}} \in \mathcal{S}^{p+m}$ with $h := H_{\text{spr}} \circ \delta \in (\mathrm{L}^1_+)^{p+m}$ induces the operators

$$\begin{array}{cccc} H\circ: & (\mathrm{L}^2_+)^{p+m} & \longrightarrow & (\mathrm{L}^2_+)^{p+m} \\ & \cap & & \cap \\ H\circ: & (\mathfrak{L}^2)^{p+m} & \longrightarrow & (\mathfrak{L}^2)^{p+m} \end{array} \tag{2.143}$$

and

$$\begin{array}{cccc} H\circ: & (\mathrm{L}^\infty)^{p+m} & \longrightarrow & (\mathrm{L}^\infty)^{p+m} \\ & \cap & & \cap \\ H\circ: & (\mathfrak{L}^\infty)^{p+m} & \longrightarrow & (\mathfrak{L}^\infty)^{p+m}, \end{array} \tag{2.144}$$

where $\mathfrak{L}^2$ is the Banach completion of L^2_+ and of L^2 and $\mathfrak{L}^\infty$ that of L^∞. These operators, in turn, have the finite norms, cf. Theorems 10.2.36 and 10.2.46,

$$\begin{aligned}
\|H\circ\|_2 :&= \|H\circ :\ \left(\mathfrak{L}^2\right)^{p+m} \to \left(\mathfrak{L}^2\right)^{p+m}\| \\
&= \|H\| := \sup_{\omega\in\mathbb{R}} \sigma(H(j\omega)),\ j := \sqrt{-1}, \\
\|H\circ\|_\infty :&= \|H\circ :\ \left(\mathfrak{L}^\infty\right)^{p+m} \to \left(\mathfrak{L}^\infty\right)^{p+m}\| \\
&= \|H\|_1 := \max_{i=1,\dots,p} \sum_{j=1}^{m} (|H_{0,ij}| + \|h_{ij}\|_1),
\end{aligned} \tag{2.145}$$

where $\sigma(A) = \|A\|_2$ denotes the largest singular value of a complex matrix A, cf. [37], [38, p. 107], [18, p. 518]. In the literature robust control with the topology derived from $\|H\|$ is called H_∞*-control,* and $\|H\|$ is denoted as $\|H\|_\infty$ although this norm refers to the Hilbert space $\mathfrak{L}^2$ with its norm $\|-\|_2$. We do not know of a standard terminology for robust control derived from $\|H\|_1$.

Theorem 10.2.50 is the main robustness result with two assertions:
(i) *For* $\Lambda_1 = \mathbb{C}_-$ *the constructed compensators are robust with respect to both derived topologies.*
(ii) Let $\widetilde{\mathcal{B}}_1$ be near $\mathcal{B}_1$ in one of these topologies, and let $H \in \mathcal{S}^{(p+m)\times(p+m)}$ resp. $\widetilde{H} \in \mathcal{S}^{(p+m)\times(p+m)}$ be the asymptotically stable closed-loop transfer matrices of $\mathcal{B} := \mathrm{fb}(\mathcal{B}_1, \mathcal{B}_2)$ resp. $\widetilde{\mathcal{B}} = \mathrm{fb}(\widetilde{\mathcal{B}}_1, \mathcal{B}_2)$. We show in Theorems 10.2.47 and 10.2.50 that

(a) If $\lim \widetilde{\mathcal{B}}_1 = \mathcal{B}_1$ in the $\|-\|$-topology, derived from that of $\mathcal{S}$, then

$$\lim \|\widetilde{H} - H\| = \lim \|\widetilde{H}\circ - H\circ\|_2 = 0. \tag{2.146}$$

(b) If $\lim \widetilde{\mathcal{B}}_1 = \mathcal{B}_1$ in the $\|-\|_1$-topology, derived from that of $\mathcal{S}$, then

$$\lim \|\widetilde{H} - H\|_1 = \lim \|\widetilde{H}\circ - H\circ\|_\infty = 0. \tag{2.147}$$

In words, assertion (b) ((a) analogous) reads: *If the plant* $\widetilde{\mathcal{B}}_1$ *is near* $\mathcal{B}_1$ *in the* $\|-\|_1$*-topology, then the BIBO stable transfer operator* $\widetilde{H}\circ$ *exists and is near* $H\circ$ *in the* $\|-\|_\infty$*-norm.* In other words, *the BIBO stable transfer operator* $\widetilde{H}\circ$ *depends continuously on the plant* $\widetilde{\mathcal{B}}_1$.

The robustness properties (i) and (ii) of the compensator are required due to model uncertainty, cf. [38, Chap. 9], *i.e., that for various reasons the data of the plant model deviate slightly from those of the real modeled plant.*

The details for the preceding assertions are relatively difficult, but are also completely proved. In particular, continuity properties of the *Fourier integral (transform),* denoted by the same letter as the Fourier series,

$$\mathbb{F} : \mathrm{L}^1_t \to \mathrm{C}^0_\omega,\ u \mapsto \mathbb{F}(u),\ \mathbb{F}(u)(\omega) = \int_{-\infty}^{\infty} u(t)e^{-j\omega t}\mathrm{d}t, \tag{2.148}$$

and its extensions have to be used. These, in turn, imply continuity properties of the *Laplace transform*

$$\mathcal{L} : \mathrm{L}^1_{\geq 0} := \left\{u \in \mathrm{L}^1;\ \mathrm{supp}(u) \subseteq [0, \infty)\right\} \to \mathrm{C}^0,\ u \mapsto \mathcal{L}(u), \text{ with}$$
$$\mathcal{L}(u)(s) := \int_0^\infty u(t)e^{-st}\mathrm{d}t,\ s \in \mathbb{C},\ \Re(s) \geq 0, \tag{2.149}$$

that extends that from (2.40) on $Y\mathcal{F}_-$, cf. (2.101). We do not need or use the Fourier transform of general temperate distributions [21, Theorem 7.1.10].

2.9 Further Systems Theories in This Book

The theory of this book is applicable to all situations where a principal ideal operator domain $\mathcal{D}$ and an injective cogenerator signal module ${}_{\mathcal{D}}\mathcal{F}$ are given, and where linear systems $R \circ w = x$ and behaviors $\left\{w \in \mathcal{F}^l;\ R \circ w = 0\right\}$ are of interest. In most cases $\mathcal{D}$ is a polynomial algebra $\mathcal{D} = F[s]$ over a field F, but consider ${}_{\mathcal{D}_T}\mathcal{F}_T$ from (2.131) where the operator domain is not polynomial.

The theory of the preceding sections is also valid for the real base field, the injective cogenerators being

$$_{\mathbb{R}[s]}\mathrm{C}^\infty(\mathbb{R}, \mathbb{R}) \subset {}_{\mathbb{C}[s]}\mathrm{C}^\infty(\mathbb{R}, \mathbb{C}),\ {}_{\mathbb{R}[s]}\mathrm{C}^{-\infty}(\mathbb{R}, \mathbb{R}) \subset {}_{\mathbb{C}[s]}\mathrm{C}^{-\infty}(\mathbb{R}, \mathbb{C}). \tag{2.150}$$

Since real behaviors are the real parts of complex ones, most results can be directly transferred from the complex to the real case. Few real results require additional considerations, and these are carried out in detail.
The *LTI (linear time-invariant) discrete-time behaviors* in this book for an arbitrary base field F use the injective cogenerator

$$_{F[s]}F^{\mathbb{N}} \ni w = (w(0), w(1), w(2), \ldots),\ (s \circ w)(t) = w(t+1), \tag{2.151}$$

of sequences in F. For any matrix $R = \left(R_{\alpha\beta}\right)_{\alpha,\beta} \in F[s]^{k\times l}$ with $R_{\alpha\beta} = \sum_{\mu\in\mathbb{N}} R_{\alpha\beta,\mu}s^\mu$ an inhomogeneous system has the form

$$R \circ w = x,\ w \in \left(F^{\mathbb{N}}\right)^l,\ x \in \left(F^{\mathbb{N}}\right)^k, \text{ or}$$
$$\forall t \in \mathbb{N},\ \forall \alpha = 1, \ldots, k:\ \sum_{\mu\in\mathbb{N}} \sum_{\beta=1}^{l} R_{\alpha\beta,\mu} w_\beta(t+\mu) = x_\alpha(t). \tag{2.152}$$

So the basic equations are linear systems of *difference equations*, a famous one being the *Fibonacci equation*

$$(s^2 - s - 1) \circ w = 0 \Longleftrightarrow \forall t \in \mathbb{N}:\ w(t+2) = w(t+1) + w(t)$$
$$\text{with } w(0) := 1,\ w(1) := 1, \text{ hence } w = (1, 1, 2, 3, 5, 8, \ldots) \in \mathbb{R}^{\mathbb{N}}. \tag{2.153}$$

A variant of this theory is furnished by the injective cogenerator

$$\begin{aligned} &{}_{\mathcal{D}}F^{\mathbb{Z}} \ni w = (\ldots, w(-2), w(-1), w(0), w(1), w(2), \ldots),\ (s \circ w)(t) = w(t+1),\\ &\text{where } \mathcal{D} := F[s, s^{-1}] = \oplus_{\mu \in \mathbb{Z}} F s^{\mu} = \{\text{Laurent polynomials}\}\,. \end{aligned} \tag{2.154}$$

The theory is almost the same as that for (2.151), but not discussed in this book.

Over the base fields $\mathbb{C}$ and $\mathbb{R}$ almost all results of the continuous-time theory have a discrete-time analogue, in particular, those of Chap. 10 on stabilizing compensators, with the exception of $\lim \|\widetilde{H}\circ - H\circ\|_2 = 0$ in (2.146) and $\lim \|\widetilde{H}\circ - H\circ\|_\infty = 0$ in (2.147). These analogues can be derived, but we have not done this. Most proofs for the two cases, in particular, those of Chap. 10, are carried out simultaneously for ${}_{F[s]}\mathcal{F}$-behaviors with $F = \mathbb{C}, \mathbb{R}$ and injective cogenerators $\mathcal{F} = \mathrm{C}^{-\infty}(\mathbb{R}, F)$ or $\mathcal{F} = F^{\mathbb{N}}$.

The paper [39] studies more general feedback interconnections and quotes the corresponding literature, cf. [10, Sect. 10.8.2].

2.10 Additional Results

Here we mention additional results with new derivations that are, however, not further used in the book.

In Sect. 6.3.2 we explain the connection of the behavioral and the Rosenbrock languages with that of the *French school* of Fliess, Bourlès [18] et al..

Section 8.2.9 is devoted to a short explanation of *Mikusinski's calculus* that is used as an alternative for one-dimensional distribution theory, for instance by Fliess, but not in this book.

For nonproper $H \in \mathbb{C}(s)^{p\times m}$ there are inputs $u \in (\mathrm{C}^{\infty})^m\, Y$ with a jump at $t = 0$ such that $y := H \circ u$ has *impulsive components* in $\mathbb{C}[s]^p \circ \delta$. In Sect. 8.2.10 we compute these components by a modification of Bourlès' method in [40], cf. also [12, Sect. 4.2] and [41].

In Sect. 12.5 we construct compensators for state space behaviors by means of *Luenberger state observers and state feedback*, cf. [42], [10, Sects. 10.5, 10.6]. This construction method is very special and does not furnish all possible compensators, but was historically the first.

In *model matching*, cf. Sect. 10.2.6, one constructs a compensator that realizes a given proper and T-stable transfer matrix H_{y_1,u_2} from u_2 to y_1 of the closed-loop behavior.

In Chap. 11 we construct and parametrize so-called *functional T-observers*. These observers were studied by many colleagues, in particular intensively by Fuhrmann [43], and were applied for the construction of compensators, but in Chap. 10 they are not needed or used.

2.11 Systems Theories Not Discussed in This Book

For an obvious reason the following remarks are very short in those areas that we have not studied ourselves. Of course, this is no statement whatsoever on the relative importance of the areas and of the researchers' contributions.

1. *Multidimensional systems*: The multivariate polynomial algebra $\mathbb{C}[s]$ with $s := (s_1, \ldots, s_n)$, $n > 1$, acts on $u = u(x_1, \ldots, x_n) \in \mathrm{C}^\infty(\mathbb{R}^n, \mathbb{C})$ or, more generally, on $u \in \mathcal{D}'(\mathbb{R}^n, \mathbb{C})$ (Schwartz' distributions) by partial differentiation, $s_i \circ u = \partial u/\partial x_i$, and makes these signal spaces injective cogenerators. As mentioned, the injectivity was shown by very difficult work [2–4] whose usefulness for systems theory was established in [6]. There is also an analogous theory for difference equations [6]. The corresponding behaviors are called *multidimensional*, for which Theorem 3.3.18 holds. Many authors have contributed to this field in the last decades, among them Bisiacco, Bose, Fornasini, Kaczorek, Lin, Marchesini, Owens, Pommaret, Quadrat, Robertz, Rocha, Rogers, Shankar, Valcher, Willems, Wood, Zampieri, Zerz, and also the authors of this book. In general, these authors have not considered systems with additional boundary conditions.
2. *Infinite-dimensional systems*: There is a different, very important, and vast multidimensional theory of partial differential equations with boundary conditions, cf. [44]. An outstanding author in this area was J.-L. Lions. We have not studied this field.
3. *LTV (linear time-varying) state space systems*: Many advanced results on differential and difference systems

$$\mathrm{d}x/\mathrm{d}t = A(t)x(t) + B(t)u(t),\ \ y(t) = C(t)x(t) + D(t)u(t)\ \text{(continuous time)},$$
$$x(t+1) = A(t)x(t) + B(t)u(t),\ \ y(t) = C(t)x(t) + D(t)u(t)\ \text{(discrete time)}, \tag{2.155}$$

   and surveys of the literature are contained in the books [17, 45, 46] and, partially, in the other cited textbooks, cf. [13], [23, Chap. 2], [16, Sect. 2.6]. Recently Anderson, Berger, Hill, Ilchmann, and Wirth have contributed to this area.
4. *The behavioral theory of implicit LTV differential systems (difference systems)*: This is more difficult than 3 for two reasons:
   (i) The domain $\mathcal{D}$ contains the operators $f = \sum_{\mu\in\mathbb{N}} f_\mu s^\mu$ with functions (instead of constants) $f_\mu(t)$. Again f acts on u via $f \circ u = \sum_\mu f_\mu u^{(\mu)}$. The domain $\mathcal{D}$ is noncommutative since $sf_\mu = f_\mu s + f'_\mu$ (for differential equations). Its algebraic properties depend very much on the choice of the coefficient functions.
   (ii) The choice of a suitable signal module and the proof of its injectivity is not obvious.
   We refer to the papers [25, 47–52] for various solutions and references to the literature. Schmale, Ilchmann, Mehrmann, Rocha, and Zerz have recently contributed to this field.
5. *Algebraic Analysis*: Many outstanding mathematicians have contributed to this area, i.e., the algebraic theory of *noncommutative Noetherian domains of partial differential operators with variable coefficients* [5], among them Hörmander,

Kashiwara, Malgrange, Pommaret, Sato. More recently, the French school, in particular Quadrat, Robertz, and many other researchers have developed its computational side. Corresponding behaviors, i.e., solution spaces, have not been studied from the engineering point of view, but see [53]. The area of partial differential equations is, of course, one of the largest in mathematics.

6. *Convolution behaviors*: These use the signal module $\mathcal{E} := C^\infty(\mathbb{R}, \mathbb{C})$ as in this book, but the larger commutative, but non-Noetherian operator domain $\mathcal{E}'$ of all distributions with compact support with the convolution multiplication that acts on $\mathcal{E}$ by convolution. A typical case is the *delay-differential equation*

$$\left((\delta' - \delta_1) * y\right)(t) = y'(t) - y(t-1) = u(t). \tag{2.156}$$

The functions in torsion behaviors $\mathcal{B} := \{y \in \mathcal{E};\ T * y = 0\}$, $0 \neq T \in \mathcal{E}'$, are called *mean-periodic* and were studied by many outstanding analysts. The paper [32] on convolution behaviors also discusses the relevant literature and the principal contributors, among them Schwartz, Ehrenpreis, Berenstein, Glüsing-Lürssen, Zampieri.

7. *Nonlinear systems*: These are mostly described by state space equations $x' = f(x, u), y = g(x, u)$, where $x(t) \in \mathbb{R}^n$ is the state at time t and $u(t) \in \mathbb{R}^m$ the control. Most real systems are originally nonlinear. One important solution method is *linearization*, i.e., the approximation of the nonlinear system by a linear one. See [54] for a broad discussion.
8. *Optimal control*: In context with Chap. 10 this means the choice of a stabilizing compensator among all parametrized ones that, for instance, optimizes a quadratic cost function. We refer to [38, Chaps. 14, 15], [31, Chap. 6, pp. 310–], [17, Chap. 9].
9. *Stochastic systems theory*: This is used, for instance, to replace the very restricted disturbance signals u_1, u_2 with $\phi \circ u_i = 0$ for given ϕ by wider classes of signals with special probability distributions. We refer to [18, Chap. 11] for an introduction.

References

1. J.C. Willems, From time series to linear system. I. Finite-dimensional linear time invariant systems. II. Exact modelling. Automatica J. IFAC, **22**(5 and 6), 561–580 and 675–694 (1986)
2. L. Ehrenpreis, *Fourier Analysis in Several Complex Variables* (Wiley, New York, 1970)
3. B. Malgrange, Sur les systèmes différentiels à coefficients constants. In *Les Équations aux Dérivées Partielles (Paris, 1962)* (Editions du Centre National de la Recherche Scientifique, Paris, 1963), pp. 113–122
4. V.P. Palamodov, *Linear Differential Operators With Constant Coefficients* (Springer, New York, 1970)
5. J.-E. Björk, *Rings of Differential Operators* (North-Holland Publishing, Amsterdam, 1979)
6. U. Oberst, Multidimensional constant linear systems. Acta Appl. Math. **20**(1–2), 1–175 (1990)
7. R.E. Kalman, P.L. Falb, M.A. Arbib, *Topics in Mathematical System Theory* (McGraw-Hill, New York, 1969)

8. H.H. Rosenbrock, *State-Space and Multivariable Theory* (Wiley, New York, 1970)
9. W.A. Wolovich, *Linear Multivariable Systems*. Applied Mathematical Sciences, vol. 11 (Springer, New York, 1974)
10. J.W. Polderman, J.C. Willems, *Introduction to Mathematical Systems Theory. A Behavioral Approach*, Texts in Applied Mathematics, vol. 26 (Springer, New York, 1998)
11. T. Kailath, *Linear Systems* (Prentice-Hall, Englewood Cliffs, 1980)
12. A.I.G. Vardulakis, *Linear Multivariable Control. Algebraic Analysis and Synthesis Methods* (Wiley, Chichester, 1991)
13. C.T. Chen, *Linear System Theory and Design* (Harcourt Brace College Publishers, Fort Worth, 1984)
14. R. Unbehauen, *Elektrische Netzwerke: Eine Einführung in die Analyse* (Springer, Berlin, 1987)
15. L.P. Schmidt, G. Schaller, S. Martius, *Grundlagen der Elektrotechnik 3. Netzwerke* (Pearson Studium, München, 2006)
16. P.J. Antsaklis, A.N. Michel, *Linear Systems*, 2nd edn. (Birkhäuser, Boston, 2006)
17. D. Hinrichsen, A.J. Pritchard with the cooperation of F. Colonius, T. Damm, A. Ilchmann, B. Jacob, F. Wirth. *Mathematical systems theory II. Control, Observation, Realization, and Feedback* . Springer, to appear
18. H. Bourlès, *Linear Systems* (ISTE, London, 2010)
19. F.M. Callier, C.A. Desoer, *Multivariable Feedback Systems* (Springer Texts in Electrical Engineering Springer. New York, 1982)
20. L. Schwartz. *Théorie des distributions*. Publications de l'Institut de Mathématique de l'Université de Strasbourg, No. IX-X. Hermann, Paris, nouvelle édition, entiérement corrigée, refondue et augmentée edition, 1966
21. L. Hörmander, *The Analysis of Linear Partial Differential Operators. I. Distribution Theory and Fourier Analysis*. Grundlehren der Mathematischen Wissenschaften, vol. 256 (Springer, Berlin, 1983)
22. M. Albach, *Grundlagen der Elektrotechnik 2. Periodische und nichtperiodische Signalformen* (Pearson Studium, München, 2005)
23. F.M. Callier, C.A. Desoer, *Linear System Theory* (Springer, New York, 1991)
24. A. Kochubei, Y. Luchko, *Handbook of Fractional Calculus with Applications. Vol. 1: Basic Theory, Vol. 2: Fractional Differential Equations*. (De Gruyter, Berlin, 2019)
25. H. Bourlès, B. Marinescu, U. Oberst, Weak exponential stability of linear time-varying differential behaviors. Linear Algebra Appl. **486**, 523–571 (2015)
26. U. Oberst, Anwendungen des chinesischen Restsatzes. Exp. Math. **3**, 97–148 (1985)
27. R.G. Ballas, G. Pfeifer, R. Werthschützky, *Elektromechanische Systeme in der Mikrotechnik und Mechatronik* (Springer, Berlin, 2009)
28. K. Jantschek, *Mechatronic Systems Design. Methods, Models, Concepts* (Springer, Berlin, 2012)
29. D.C. Karnopp, D.L. Margolis, R.C. Rosenberg, *System Dynamics. Modeling, Simulation, and Control of Mechatronic Systems* (Wiley, Hoboken, 2012)
30. R.W. Newcomb, *Network Theory, The State-Space Approach* (Librairie Universitaire Louvain, Leuven, 1969)
31. M. Vidyasagar, *Control System Synthesis: A Factorization Approach*, MIT Press Series in Signal Processing, Optimization, and Control 7 (MIT Press, Cambridge, 1985)
32. H. Bourlès, U. Oberst, Generalized convolution behaviors and topological algebra. Acta Appl. Math. **141**, 107–148 (2016)
33. I. Blumthaler, Functional T-observers. Linear Algebra Appl. **432**(6), 1560–1577 (2010)
34. I. Blumthaler, U. Oberst, Design, parametrization, and pole placement of stabilizing output feedback compensators via injective cogenerator quotient signal modules. Linear Algebra Appl. **436**(5), 963–1000 (2012)
35. A. Quadrat, On a generalization of the Youla-Kučera parametrization. II. The lattice approach to MIMO systems. Math. Control Signals Systems, **18**(3), 199–235 (2006)
36. L. Pernebo, An algebraic theory for the design of controllers for linear multivariable systems. I. Structure matrices and feedforward design. II. Feedback realizations and feedback design. IEEE Trans. Automat. Control **26**(1), 171–182 and 183–194 (1981)

37. C.A. Desoer, M. Vidyasagar, *Feedback Systems: Input-Output Properties* (Academic Press, Cambridge, 1975)
38. K. Zhou, J.C. Doyle, K. Glover, *Robust and Optimal Control* (Prentice-Hall, Upper Saddle River, 1996)
39. I. Blumthaler, Stabilisation and control design by partial output feedback and by partial interconnection. Int. J. Control **85**(11), 1717–1736 (2012)
40. H. Bourlès, Impulsive systems and behaviors in the theory of linear dynamical systems. Forum Math. **17**(5), 781–807 (2005)
41. C. Bargetz, Impulsive solutions of differential behaviors. Master's thesis, University of Innsbruck, 2008
42. D.G. Luenberger, Observing the state of a linear system. IEEE Trans. Mil. Electron. **8**(2), 74–80 (1964)
43. P.A. Fuhrmann, Observer theory. Linear Algebra Appl. **428**(1), 44–136 (2008)
44. R.F. Curtain, H. Zwart, *An Introduction to Infinite-Dimensional Linear Systems Theory* (Springer, New York, 1995)
45. W.J. Rugh, *Linear System Theory*, 2nd edn. (Prentice-Hall, Upper Saddle River, 1996)
46. D. Hinrichsen, A.J. Pritchard, *Mathematical Systems Theory I. Modelling, State Space Analysis, Stability and Robustness*. Texts in Applied Mathematics, vol. 48 (Springer, Berlin, 2005)
47. S. Fröhler, U. Oberst, Continuous time-varying linear systems. Syst. Control Lett. **35**(2), 97–110 (1998)
48. H. Bourlès, B. Marinescu, U. Oberst, Exponentially stable linear time-varying discrete behaviors. SIAM J. Control Optim. **53**(5), 2725–2761 (2015)
49. H. Bourlès, B. Marinescu, U. Oberst, The injectivity of the canonical signal module for multidimensional linear systems of difference equations with variable coefficients. Multidimens. Syst. Signal Process. **28**(1), 75–103 (2017)
50. H. Bourlès, U. Oberst, Robust stabilization of discrete-time periodic linear systems for tracking and disturbance rejection. Math. Control Signals Syst. **28**(3), Art. 18 (2016)
51. U. Oberst, Two invariants for weak exponential stability of linear time-varying differential behaviors. Linear Algebra Appl. **504**, 468–486 (2016)
52. U. Oberst, A constructive test for exponential stability of linear time-varying discrete-time systems. Appl. Algebra Eng. Comm. Comput. **28**(5), 437–456 (2017)
53. J.-F. Pommaret, *Partial Differential Control Theory. Vol. I. Mathematical Tools; Vol. II. Control Systems* (Kluwer, Dordrecht, 2001)
54. E.D. Sontag, *Mathematical Control Theory* (Springer, New York, 1990)

Chapter 3
The Language and Fundamental Properties of Behaviors

3.1 Behaviors

3.1.1 Definition of Behaviors

See Sect. 2.1 for a survey. A behavior is the set of solutions of a system, that in most cases is the mathematical model of a technical system as, for example, a machine or an electrical or mechanical network. These solutions are also called the trajectories of the system. In this book we consider time systems, where the trajectories are functions of a continuous- or discrete-time variable and which satisfy linear differential or difference equations with constant coefficients. Therefore, the behaviors are solution spaces of linear systems of ordinary differential or difference equations with constant coefficients. In engineering mathematics, the cases of differential and of difference equations are called the *continuous-(time)* and the *discrete-(time)* cases, respectively. There are many other types of equations, systems, and behaviors, cf. Sect. 2.11, but they are not studied in this book.

As explained in Sect. 2.1, the theory requires the choice of a *signal space* and of an *operator ring* that acts on it. In the beginning, we make the following choices. Let F be a field and let $\mathcal{D} := F[s]$ be the polynomial algebra over F in one indeterminate s. The letter $\mathcal{D}$ is chosen since $\mathcal{D}$ is used as a ring of differential or difference operators.

In continuous time one chooses the base fields $F = \mathbb{R}$ or $F = \mathbb{C}$ of real or complex numbers, respectively. As usual, we identify

$$\mathbb{C} = \mathbb{R}^2 \ni z = (a, b) = a + bj, \text{ where } j := \sqrt{-1},\ a = \Re(z) \text{ and } b = \Im(z). \tag{3.1}$$

In discrete time the field F is arbitrary, and, in particular, finite fields are of interest. If F is a field and I an arbitrary (index) set, the space $F^I := \{y\colon I \to F\}$ of all functions from I to F is an F-vector space with the componentwise addition $(y^1 + y^2)(i) = y^1(i) + y^2(i)$ and scalar multiplication $(ay)(i) := ay(i)$. Such a function y is also

U. Oberst et al., *Linear Time-Invariant Systems, Behaviors and Modules*, Differential-Algebraic Equations Forum,
https://doi.org/10.1007/978-3-030-43936-1_3

written as $y = (y(i))_{i\in I} = (y_i)_{i\in I}$ and then called a *family* of elements of F, indexed by I. The following function spaces are of special interest.

Definition 3.1.1 *Continuous time.* Let F denote the field of real or complex numbers and let

$$\mathcal{F}_{\rm cont} := {\rm C}^\infty(\mathbb{R}) := {\rm C}^\infty(\mathbb{R}, F) := \{y\colon\ \mathbb{R} \longrightarrow F,\ t \longmapsto y(t);\ \ y \text{ is smooth}\}$$

be the F-vector space of infinitely often differentiable or smooth functions. The independent variable is denoted by t since in the standard cases it is interpreted as time. A complex-valued function $y\colon t \longmapsto y(t)$ is contained in ${\rm C}^\infty(\mathbb{R}, \mathbb{C})$ if and only if its real and imaginary parts belong to ${\rm C}^\infty(\mathbb{R}, \mathbb{R})$. The space $\mathcal{F}_{\rm cont}$ is an F-vector space with the usual structure. We denote the F-linear differentiation operator by

$$\phi_{\rm cont}\colon\ {\rm C}^\infty(\mathbb{R}, F) \longrightarrow {\rm C}^\infty(\mathbb{R}, F),\ y \longmapsto \phi_{\rm cont}(y) := \frac{{\rm d}y}{{\rm d}t}.$$

Discrete time. The field F is arbitrary,

$$\mathcal{F}_{\rm dis} := F^{\mathbb{N}} = \{y = (y(t))_{t\in\mathbb{N}} = \big(y(0), y(1), \ldots\big);\ y(t) \in F\}$$

is the F-vector space of F-valued sequences and

$$\phi_{\rm dis}\colon\ F^{\mathbb{N}} \longrightarrow F^{\mathbb{N}},\ y = (y(0), y(1), \ldots) \longmapsto \phi_{\rm dis}(y) := \big(y(1), y(2), \ldots\big),$$
$$\text{i.e., } \phi_{\rm dis}(y)(t) := y(t+1),$$

is the F-linear *left shift* operator. In this case, $t \in \mathbb{N}$ is interpreted as a discrete instant of time and the sequence y is also called a *time series.* ◇

Reminder 3.1.2 If F is a commutative ring and V and W are F-modules, then $\mathrm{Hom}_F(V, W)$ denotes the F-module of F-linear maps or F-homomorphisms from V to W. In particular, $\mathrm{End}_F(V) := \mathrm{Hom}_F(V, V)$ is the F-algebra of F-endomorphisms of V. If $\phi\colon\ V \longrightarrow V$ belongs to $\mathrm{End}_F(V)$, we define a new scalar multiplication

$$F[s] \times V \longrightarrow V,\quad (f, v) \longmapsto f \circ v := f(\phi)(v) = \sum_{m\in\mathbb{N}} a_m \phi^m(v),\ f = \sum_{m\in\mathbb{N}} a_m s^m. \tag{3.2}$$

With this multiplication and the given addition on V the F-module V becomes an $F[s]$-module, i.e., addition and scalar multiplication satisfy the commutative, associative, and distributive laws and $1 \circ v = v$.

Conversely, if V is an $F[s]$-module with the scalar multiplication $f \circ v$, then V is also an F-module since $F \subset F[s]$, and the map $\phi\colon\ V \longrightarrow V$ defined by $\phi(v) := s \circ v$ belongs to $\mathrm{End}_F(V)$ and satisfies $f \circ v = f(\phi)(v)$ for $f \in F[s]$. This shows that an $F[s]$-module V is an F-module V together with an F-endomorphism $\phi\colon\ V \longrightarrow V$. ◇

We apply the preceding reminder to the data from Definition 3.1.1 and obtain the following corollary.

Corollary 3.1.3 *Let F be a field and consider a polynomial $f = a_n s^n + a_{n-1}s^{n-1} + \cdots + a_0 \in F[s]$.*

Continuous time. *The space $\mathcal{F}_{\text{cont}} = \mathrm{C}^\infty(\mathbb{R}, F)$ becomes an $F[s]$-module via*

$$f \circ y := f(\phi_{\text{cont}})(y) = a_n y^{(n)} + a_{n-1} y^{(n-1)} + \cdots + a_0 y,$$

where $y^{(i)} = \frac{\mathrm{d}^i y}{\mathrm{d}t^i}$ is the i-th derivative. In particular, $s^n \circ y = \phi_{\text{cont}}^n(y) = y^{(n)}$.
Discrete time. *The sequence space $\mathcal{F}_{\text{dis}} = F^{\mathbb{N}}$ becomes an $F[s]$-module via*

$$f \circ y := f(\phi_{\text{dis}})(y),$$

i.e., for $t \in \mathbb{N}$ we have that

$$(f \circ y)(t) = a_n y(t+n) + a_{n-1} y(t+n-1) + \cdots + a_1 y(t+1) + a_0 y(t).$$

In particular, the identity $(s^n \circ y)(t) = y(t+n)$ holds. ◇

We will treat the continuous-time and the discrete-time cases simultaneously and make the following general assumption.

Assumption 3.1.4 1. All rings $\mathcal{D}$ of operators of this book are assumed commutative. Noncommutative rings occur as endomorphism rings and especially matrix rings. In general and if not explicitly stated otherwise, we make the minimal assumption that the ring $\mathcal{D}$ is a commutative Noetherian integral domain and that $\mathcal{F}$ is a $\mathcal{D}$-module. It is satisfied in most behavioral systems theories, but not in all, cf. Sect. 2.11. Noetherianess is needed for the characterization of injective modules by the fundamental principle, cf. Sect. 3.2.2. Integral domains are necessary for the consideration of divisibility, torsion modules, and autonomous behaviors, cf. Sect. 4.2. The properties of Noetherian rings and modules will be recalled in Definition 3.2.13 and Result 3.2.14. The most important systems theoretic results of this book require that $\mathcal{D}$ is a principal ideal domain. The standard examples for this setting in this book are the $F[s]$-modules $\mathcal{F}_{\text{cont}}$ and $\mathcal{F}_{\text{dis}}$ from Definition 3.1.1 and Corollary 3.1.3.
2. However, for some preparatory results, for instance, on injective and divisible modules, general commutative rings or integral domains suffice, and we use these more general assumptions after their explicit introduction. The idea is to use the weakest assumption for establishing a theorem without making the proof more difficult. This is a standard procedure in mathematics and makes clear what is really important for a proof. It also enables the use of the results in other systems theories where only more general assumptions hold and in which the reader may be interested in further study, cf. Sect. 2.11.

◇

As explained in Chap. 2 we will later use further signal modules and rings of operators.

Under Assumption 3.1.4, consider an equation

$$f \circ y = u, \text{ where } f \in \mathcal{D} \text{ and } y, u \in \mathcal{F}. \tag{3.3}$$

This linear equation is called *homogeneous* if u is zero and *inhomogeneous* if $u \neq 0$. As usual, u is called the *right-hand side* of the equation. For given f and u a *solution* y of the equation is an $y \in \mathcal{F}$ which satisfies $f \circ y = u$. In systems theory u and y are called the *input* and the *output*, respectively.

The solution space of the homogeneous equation is the *annihilator*

$$\operatorname{ann}_{\mathcal{F}}(f) := \{y \in \mathcal{F};\ f \circ y = 0\} \tag{3.4}$$

of f in $\mathcal{F}$. Since $\mathcal{D}$ is commutative, it is a $\mathcal{D}$-submodule of $\mathcal{F}$. If $y_1 \in \mathcal{F}$ is a solution of the inhomogeneous equation, then the set of all solutions of the inhomogeneous equation is the *affine* submodule

$$\{y \in \mathcal{F};\ f \circ y = u\} = y_1 + \operatorname{ann}_{\mathcal{F}}(f) \tag{3.5}$$

since

$$\begin{aligned} f \circ y = u &\Longleftrightarrow f \circ y = f \circ y_1 \Longleftrightarrow f \circ (y - y_1) = 0 \Longleftrightarrow y - y_1 \in \operatorname{ann}_{\mathcal{F}}(f) \\ &\Longleftrightarrow y \in y_1 + \operatorname{ann}_{\mathcal{F}}(f). \end{aligned}$$

In general, it is possible that the homogeneous equation (3.3) has only few solutions and that the inhomogeneous equation has none. However, the differential and difference equations discussed in this book have always "sufficiently many" solutions as the following theorem shows.

Theorem 3.1.5 *We use the data from Definition 3.1.1 with* $\mathcal{D} := F[s]$ *and* $\mathcal{F} := \mathcal{F}_{\text{cont}}$ *or* $\mathcal{F} := \mathcal{F}_{dis}$. *Let* $f = s^n + a_{n-1}s^{n-1} + \cdots + a_0 \in F[s] \setminus \{0\}$ *be a nonzero polynomial. Without loss of generality, we assume that it is monic, i.e., that its leading coefficient* $\operatorname{lc}(f)$ *is one. Let* $u \in \mathcal{F}$ *be an arbitrary input and* $x := (x_0, \ldots, x_{n-1}) \in F^n$ *an arbitrary* initial vector.

Continuous time. *The* initial value problem or Cauchy problem

$$\begin{gathered} y^{(n)} + a_{n-1}y^{(n-1)} + \cdots + a_0 y = f \circ y = u \\ \text{with the initial condition } y^{(i)}(0) = x_i \text{ for } i = 0, \ldots, n-1 \end{gathered} \tag{3.6}$$

has a unique solution $y \in C^{\infty}(\mathbb{R}, F)$. *In particular, the map*

$$\operatorname{ann}_{\mathcal{F}}(f) = \left\{y \in C^{\infty}(\mathbb{R}, F);\ y^{(n)} + a_{n-1}y^{(n-1)} + \cdots + a_0 y = 0\right\} \longrightarrow F^n$$
$$y \longmapsto \left(y(0), \ldots, y^{(n-1)}(0)\right) \tag{3.7}$$

is an F*-isomorphism, i.e., a bijective* F*-linear map.*

Discrete time. *The difference equation*

$$(f \circ y)(t) = y(t+n) + a_{n-1}y(t+n-1) + \cdots + a_0 y(t) = u(t) \text{ for } t \in \mathbb{N}$$
$$\text{with the initial condition } y(i) = x_i \text{ for } i = 0, \ldots, n-1 \tag{3.8}$$

has a unique solution $y \in F^{\mathbb{N}}$. *In particular, the map*

$$\left\{y \in F^{\mathbb{N}};\ \forall t \in \mathbb{N}:\ y(t+n) + a_{n-1}y(t+n-1) + \cdots + a_0 y(t) = 0\right\} \longrightarrow F^n$$
$$y \longmapsto \left(y(0), \ldots, y(n-1)\right) \tag{3.9}$$

is an F*-isomorphism.*

Proof *Continuous time.* The unique solvability of (3.6) is a standard result in the theory of differential equations. For a proof see Theorem 3.2.23. Obviously, the map (3.7) is F-linear. Its bijectivity follows from the unique solvability of (3.6) for $u = 0$. *Discrete time.* We solve (3.8) for $y(t+n)$ and obtain with

$$y(t+n) = -a_{n-1}y(t+n-1) - \cdots - a_0 y(t) + u(t)$$

a recursion formula for the values of $y(t)$ for $t \geq n$ if $y(0), \ldots, y(n-1)$ and $\left(u(t)\right)_{t\in\mathbb{N}}$ are known. This implies that (3.8) is solvable. The uniqueness of the solution follows as in the continuous case. □

Example 3.1.6 For $F := \mathbb{R}$ and $\mathcal{F} := \mathbb{R}^{\mathbb{N}}$ consider the *Fibonacci* equation

$$(s^2 - s - 1) \circ y = 0 \quad \text{or} \quad y(t+2) = y(t+1) + y(t)$$

with initial values $y(0) = y(1) = 1$. Its unique solution is the sequence $y = (1, 1, 2, 3, 5, 8, 13, \ldots)$. In Sect. 5.2, we will give closed-form solutions for homogeneous difference equations $f \circ y = 0$. ◇

Example 3.1.7 (*Stabilization and tracking*) We explain a typical problem of systems theory, viz., *stabilization and tracking*, by the simplest example. We consider the continuous-time case over the groundfield $\mathbb{R}$ of real numbers and the equationbreak $(s - \lambda) \circ y = u$ with $\lambda > 0$, input u, and output y in the real continuous-time case, i.e.,

$$y' = \lambda y + u, \quad \text{with } y, u \in C^{\infty}(\mathbb{R}, \mathbb{R}), \text{ parameter } \lambda > 0,$$
$$\text{initial value } y(0) \neq 0, \text{ and solution } y(t) = e^{\lambda t}y(0) + \int_0^t e^{\lambda(t-\tau)}u(\tau)\mathrm{d}\tau. \tag{3.10}$$

The solution of its homogeneous part $y' = \lambda y$ is $e^{\lambda t}y(0)$ and *unstable* since $\lim_{t\to\infty}|y(t)| = \infty$ if $y(0) \neq 0$. In addition we assume a *reference signal* $r \in \mathbb{R}$.

The engineering task is to choose an input u such that the output $y(t)$ *tracks* the reference r, i.e., $\lim_{t\to\infty} y(t) = r$. There are two solution methods, namely, *open-* and *closed-loop controls*. *Open-loop control*: By replacing y by $y - r$, we assume w.l.o.g. that $r = 0$. We make the *ansatz* $u := e^{\kappa t}u(0)$, $\kappa \neq \lambda$, which leads to the solution

$$y(t) = \big(y(0) - (\kappa - \lambda)^{-1}u(0)\big)\, e^{\lambda t} + (\kappa - \lambda)^{-1}e^{\kappa t}u(0).$$

We choose a real number $\kappa < 0$ and $u(0) := (\kappa - \lambda)y(0)$ and obtain the asymptotically stable solution $y(t) = (\kappa - \lambda)^{-1}e^{\kappa t}y(0)$.

This choice, however, is not structurally stable. If the initial value is only slightly different, for instance, due to unprecise measurements of $y(0)$, the constant $y(0) - (\kappa - \lambda)^{-1}u(0)$ in front of $e^{\lambda t}$ is not zero, and therefore the corresponding solution again satisfies $\lim_{t\to\infty}|y(t)| = \infty$. Since measurements are never absolutely precise, the preceding stabilization and tracking method is not suitable from an engineering point of view.

Feedback or closed-loop control: The term *closed loop* will be explained later in Chap. 10 when interconnections of systems have been defined. We choose κ such that

$$\begin{aligned}
&\lambda + \kappa < 0 \text{ and } u := \kappa\left(y - \kappa^{-1}(\lambda + \kappa)r\right)\\
&\Longrightarrow y' = \lambda y + u = (\lambda + \kappa)(y - r)\\
&\Longrightarrow (y - r)' = y' = (\lambda + \kappa)(y - r) \Longrightarrow y(t) - r = e^{(\lambda+\kappa)t}(y(0) - r)\\
&\Longrightarrow \lim_{t\to\infty} y(t) = r.
\end{aligned} \tag{3.11}$$

For obvious reasons, the difference $e := y - r$ is called the *error signal* and

$$u := \kappa\left(y - \kappa^{-1}(\lambda + \kappa)r\right) \tag{3.12}$$

is the *feedback term* of $y' = \lambda y + u$ because one *feeds the output* y *back to the input*. In $e' = y' = (\lambda + \kappa)e$, a positive error at time causes a decrease of the error in the future and, similarly, a negative error induces an increase of the error. Since $\lambda + \kappa < 0$, the error finally converges to zero.

The equation $y' = (\lambda + \kappa)(y - r)$ is the *closed-loop equation*. Its homogeneous part $y' = (\lambda + \kappa)y$ is called *stable* since it has the stable solution $y(t) = e^{(\lambda+\kappa)t}y(0)$. If λ varies, but κ remains fixed, the stability and tracking properties are preserved as long as $\lambda + \kappa < 0$. This property is called *robustness* of the closed loop.

The constructive solution of the general *robust tracking and disturbance rejection* problem is a main result of Chap. 10 and of this book. ◊

We generalize Eqs. (3.3) and (3.4) to linear *systems of equations*. Let k and l be natural numbers. We employ the $\mathcal{D}$-module $\mathcal{D}^{1\times l}$ of all row vectors $\xi = (\xi_1, \ldots, \xi_l)$ with components in $\mathcal{D}$, which is equipped with the usual componentwise addition and scalar multiplication. The module $\mathcal{D}^{1\times l}$ is free, i.e., it has a basis, e.g., the standard basis $(\delta_1, \dots, \delta_l)$, where

$$\begin{aligned}&\delta_j = (0\dots,0,\overset{j}{1},0,\dots,0),\ j = 1,\dots,l,\\ &\text{with the basis representation } \xi = (\xi_1,\dots,\xi_l) = \sum_{j=1}^{l} \xi_j\delta_j \in \mathcal{D}^{1\times l}.\end{aligned} \tag{3.13}$$

We also use the free $\mathcal{D}$-module $\mathcal{D}^{k\times l}$ of all $k \times l$-matrices with coefficients in $\mathcal{D}$ with the usual componentwise addition and scalar multiplication of matrices. If $k = l$, then $\mathcal{D}^{l\times l}$ is a $\mathcal{D}$-algebra. Since

$$\mathcal{D}^{k\times l} \overset{\cong}{\longleftrightarrow} \mathrm{Hom}_{\mathcal{D}}(\mathcal{D}^{1\times k}, \mathcal{D}^{1\times l}),\quad R \longmapsto \cdot R := (\xi \mapsto \xi R),$$

is a standard $\mathcal{D}$-isomorphism, we usually identify

$$\mathcal{D}^{k\times l} = \mathrm{Hom}_{\mathcal{D}}(\mathcal{D}^{1\times k}, \mathcal{D}^{1\times l}),\ R = \cdot R. \tag{3.14}$$

Finally, we employ the $\mathcal{D}$-module $\mathcal{F}^l := \mathcal{F}^{l\times 1}$ of all *column* signal vectors $y = \begin{pmatrix} y_1 \\ \vdots \\ y_l \end{pmatrix}$ with components y_j in $\mathcal{F}$ and the usual componentwise addition and scalar multiplication.

Definition and Corollary 3.1.8 For matrices $R = (R_{ij})_{i=1,\dots,k;\, j=1,\dots,l} \in \mathcal{D}^{k\times l}$ and signal vectors $w \in \mathcal{F}^l$, we define the scalar multiplication

$$R \circ w := \Big(\sum_{j=1}^{l} R_{ij} \circ w_j\Big)_{i=1,\dots,k} = (R_{i-} \circ w)_{i=1,\dots,k} \in \mathcal{F}^k.$$

This multiplication is associative, i.e., it satisfies

$$R^1 \circ (R^2 \circ w) = (R^1 R^2) \circ w \quad \text{for matrices } R^i \text{ of suitable sizes,}$$

and $\mathcal{D}$-bilinear since $\mathcal{D}$ is commutative. In particular, the map

$$\begin{aligned}\mathcal{D}^{k\times l} = \mathrm{Hom}_{\mathcal{D}}(\mathcal{D}^{1\times k}, \mathcal{D}^{1\times l}) &\longrightarrow \mathrm{Hom}_{\mathcal{D}}(\mathcal{F}^l, \mathcal{F}^k)\\ R = \cdot R &\longmapsto R\circ := (w \mapsto R \circ w)\end{aligned}$$

is $\mathcal{D}$-linear. If $u \in \mathcal{F}^k$ is a signal vector, the equation $R \circ y = u$ is called a *linear system* of equations with right-hand side u. A solution $y \in \mathcal{F}^l$ of this system is a vector which satisfies the system. The linear system is homogeneous if $u = 0$ and inhomogeneous if $u \neq 0$. ◊

Corollary and Definition 3.1.9 *Let F be a field and let $R \in F[s]^{k\times l}$ be a matrix. For the entries R_{ij} of R, we write*

$$R_{ij} := \sum_{\mu\in\mathbb{N}} R_{ij,\mu}s^{\mu} \in F[s] \text{ with } R_{ij,\mu} \in F,$$

where only finitely many of the $R_{ij,\mu}$ are nonzero. The advantage of this notation is that it is not necessary to specify the degrees of the polynomials.

Consider the data from Definition 3.1.1 and Corollary 3.1.3 in the continuous and discrete cases.

Continuous time. *In this case the equation $R \circ y = u$ represents the implicit linear system of ordinary differential equations with constant coefficients*

$$\sum_{j=1}^{l} R_{ij} \circ y_j = \sum_{j=1}^{l}\sum_{\mu\in\mathbb{N}} R_{ij,\mu}y_j^{(\mu)} = u_i, \quad i = 1, \dots, k.$$

Implicit *systems of differential equations are usually not discussed in the standard courses on differential equations.*

Discrete time. *Here $R \circ y = u$ is a shorthand notation for the implicit linear system of ordinary difference equations with constant coefficients*

$$\Big(\sum_{j=1}^{l} R_{ij} \circ y_j\Big)(t) = \sum_{j=1}^{l}\sum_{\mu\in\mathbb{N}} R_{ij,\mu}y_j(t+\mu) = u_i(t) \text{ for } i = 1, \dots, k \text{ and } t \in \mathbb{N}.$$

◇

Example 3.1.10 In the situation of the preceding corollary, let $A \in F^{n\times n}$ be a matrix with coefficients in F and let $R := s\,\mathrm{id}_n - A \in F[s]^{n\times n}$, where id_n denotes the $n \times n$-identity matrix. For a vector $x \in \mathcal{F}^n$, we get the equivalence

$$R \circ x = 0 \Longleftrightarrow s \circ x = Ax.$$

In continuous time this equation has the form $x' = Ax$ and the unique solution $x(t) = e^{tA}x(0)$, $t \in \mathbb{R}$, for given initial condition $x(0)$. In discrete time it has the form $x(t+1) = Ax(t)$ and the obvious unique solution $x(t) = A^t x(0)$ for the initial vector $x(0)$. The equation $s \circ x = Ax$ is fundamental in context with Kalman's *state space representations* and these will be discussed in detail later. ◇

Definition and Lemma 3.1.11 (Behavior) *Let $R \in \mathcal{D}^{k\times l}$ be a matrix. The solution space*

$$\mathcal{B} := \ker(R\circ : \mathcal{F}^l \longrightarrow \mathcal{F}^k) = \{w \in \mathcal{F}^l;\ R \circ w = 0\} \subset \mathcal{F}^l$$

is a $\mathcal{D}$-submodule of $\mathcal{F}^l$ and it is called a behavior *or, more precisely, a ${}_{\mathcal{D}}\mathcal{F}$-behavior.*

Let $u \in \mathcal{F}^k$. If the vector $w^1 \in \mathcal{F}^l$ is a solution of $R \circ w = u$, then the set of all solutions of the inhomogeneous equation $R \circ w = u$ is the affine submodule

$$\{w \in \mathcal{F}^l;\ R \circ w = u\} = w^1 + \mathcal{B}.$$

Proof Since $R\circ\colon \mathcal{F}^l \longrightarrow \mathcal{F}^k$ is $\mathcal{D}$-linear, its kernel $\mathcal{B}$ is a $\mathcal{D}$-submodule of $\mathcal{F}^l$. The second assertion follows as in the proof of (3.5). □

3.1.2 Examples from Mechanics

Example 3.1.12 (*The linearized pendulum*) This pendulum is a standard example from mechanics. We derive the equations directly from Newton's laws and not by means of Lagrangian or Hamilton mechanics which, in general, are not known to students after two years. We only use the laws

1. force = mass × acceleration and
2. action = reaction.

The pendulum is described by the data in Fig. 3.1. The mechanical system is situated in the Euclidean plane $\mathbb{R}^2$ with the standard inner product $\langle -, - \rangle$ and the standard orthonormal basis (e_1, e_2). For the angle φ we consider the rotated orthonormal basis

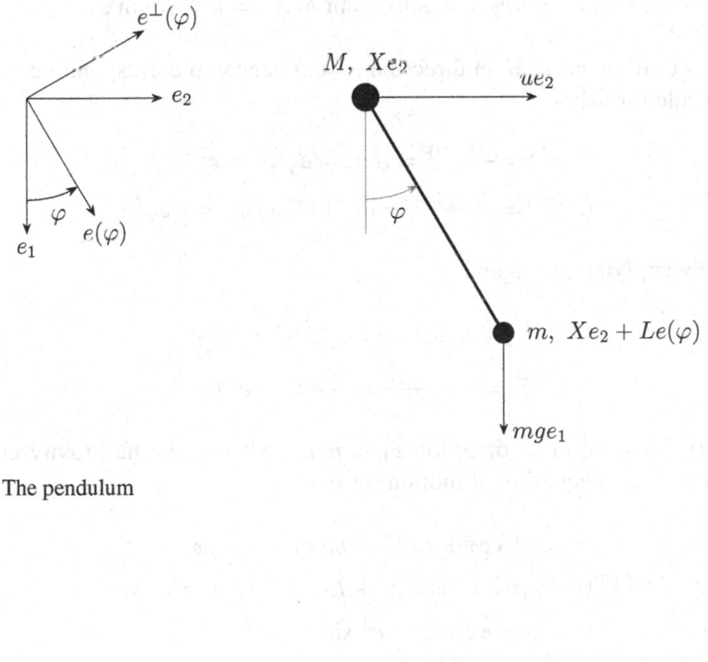

Fig. 3.1 The pendulum

$$(e(\varphi), e^{\perp}(\varphi)) := (e_1, e_2)D(\varphi), \ D(\varphi) := \begin{pmatrix} \cos\varphi & -\sin\varphi \\ \sin\varphi & \cos\varphi \end{pmatrix}$$
$$(e_1, e_2) = (e(\varphi), e^{\perp}(\varphi))D(-\varphi)$$
$$e := e(\varphi) = e_1 \cos\varphi + e_2 \sin\varphi, \ e^{\perp} := e^{\perp}(\varphi) = -e_1 \sin\varphi + e_2 \cos\varphi$$
$$e_1 = e\cos\varphi - e^{\perp} \sin\varphi, \ e_2 = e\sin\varphi + e^{\perp}\cos\varphi.$$

A simple computation shows

$$\tfrac{de}{d\varphi} = e^{\perp}, \ \tfrac{de^{\perp}}{d\varphi} = -e.$$

A mass M moves along the e_2-axis, for instance on a cart, its position at time t is $Xe_2 := X(t)e_2$. A mass m is connected with the mass M by an ideal, weightless, stiff rod of length L. At time t the angle and the position of m are $\varphi := \varphi(t)$ resp. $X(t)e_2 + Le(\varphi(t))$. A force $u(t)e_2 = ue_2$ acts on M along the e_2 -axis by which one wants to control the angle φ. Thus the state of the system at time t is completely characterized by the vector $w(t) := (X(t), \varphi(t))^{\top}$. Another force $Ke(\varphi)$ with first unknown K acts upon the mass M in direction e of the rod, whereas the inverse force $-Ke(\varphi)$ acts upon m in the inverse direction (actio = reactio). This force keeps the two masses at the same constant distance L. Gravity acts on m in direction e_1. As usual let $a' := da/dt$ denote the first derivative with respect to the time t. The velocity resp. acceleration of the moving mass M are $X'e_2$ resp. $X''e_2$. The component of the force Ke in direction e_2 is $\langle Ke, e_2\rangle e_2 = K\sin\varphi e_2$. Hence Newton's equation of motion for mass M is

$$MX''e_2 = ue_2 + K\sin\varphi e_2 \text{ or } MX'' = u + K\sin\varphi. \tag{3.15}$$

There is no movement of M in direction e_1 and hence no corresponding equation. The chain rule furnishes

$$e' := \tfrac{de(\varphi(t))}{dt} = de(\varphi)/d\varphi\varphi' = e^{\perp}\varphi'$$
$$(e^{\perp})' := \tfrac{de^{\perp}(\varphi(t))}{dt} = de^{\perp}(\varphi)/d\varphi\varphi' = -e\varphi'$$

and then, by applying this again,

$$e'' := \tfrac{d^2e(\varphi(t))}{dt^2} = -e(\varphi')^2 + e^{\perp}\varphi''$$
$$(e^{\perp})'' = \tfrac{d^2e^{\perp}(\varphi(t))}{dt^2} = -e\varphi'' - e^{\perp}(\varphi')^2.$$

The gravity force on m in direction e_1 is mge_1 where g is the gravity constant. Therefore Newton's equation of motion for m is

$$m(Xe_2 + Le)'' = mge_1 - Ke \text{ or}$$
$$\big(X''(e\sin\varphi + e^{\perp}\cos\varphi) + L(-(\varphi')^2 e + \varphi'' e^{\perp})\big) =$$
$$g(e\cos\varphi - e^{\perp}\sin\varphi) - \tfrac{K}{m}e.$$

Comparison of the coordinates with respect to the basis $(e, e^{\perp})$ and inclusion of Eq. (3.15) furnishes three equations

$$\begin{aligned} X'' \sin\varphi - L(\varphi')^2 &= g\cos\varphi - \tfrac{K}{m} \\ X'' \cos\varphi + L\varphi'' &= -g\sin\varphi \\ MX'' &= u + K\sin\varphi. \end{aligned} \tag{3.16}$$

By solving the first equation for K and by substituting this into the third equation we obtain the implicit, second-order nonlinear differential system

$$\begin{aligned} X'' \cos\varphi + L\varphi'' &= -g\sin\varphi \\ MX'' &= u + (mg\cos\varphi + mL(\varphi')^2 - mX''\sin\varphi)\sin\varphi. \end{aligned} \tag{3.17}$$

Remark 3.1.13 (*Linearization*) Such systems are hard to solve explicitly. Therefore it is customary to *linearize* them as first approximation. This is by no means possible for all *implicit nonlinear* differential systems. However, the system (3.17) can be easily solved for the highest, i.e., here, the second, derivatives—we omit the details—and can be written as $w'' = F(w, w', u),\ w := (X, \varphi)^{\top}$, or, with $v := \binom{v_1}{v_2} := \binom{w}{w'}$, as the explicit first-order differential system

$$\binom{v_1}{v_2}' = G(v, u) := \binom{v_2}{F(v,u)}. \tag{3.18}$$

The vector fields F and $G\colon \mathbb{R}^5 \to \mathbb{R}^4$ are analytic near $(v^0, u^0) := (0, 0)$ and satisfy $F(v^0, u^0) = 0,\ G(v^0, u^0) = 0$. Therefore the nonlinear differential system has the stationary solution $v^0 = \binom{w}{w'} := \binom{0}{0}$ for $u = u^0$. With $v^1 := v - v^0$ and $u^1 := u - u^0$ the linear part

$$\partial G/\partial v(v^0, u^0)v^1 + \partial G/\partial u(v^0, u^0)u^1 \text{ of } G(v, u) = G(v^0 + v^1, u^0 + u^1)$$

is a good approximation of the latter for (v, u) near (v^0, u^0), i.e., here for small v and u, and therefore one considers the linear system

$$(v^1)' = \partial G/\partial v(v^0, u^0)v^1 + \partial G/\partial u(v^0, u^0)u^1 \tag{3.19}$$

as a good approximation of $v' = G(v)$. For the original system (3.17) this signifies that all second and higher powers of the components of $w = (X, \varphi)^{\top}$ and w' are omitted from the equations. This linearization procedure furnishes

$$\begin{gathered} X'' + L\varphi'' = -g\varphi,\ \ MX'' = u + mg\varphi \text{ or} \\ \begin{pmatrix} s^2 & Ls^2+g & 0 \\ Ms^2 & -mg & -1 \end{pmatrix} \circ \begin{pmatrix} X \\ \varphi \\ u \end{pmatrix} = \binom{0}{0}. \end{gathered} \tag{3.20}$$

From these equations one can easily eliminate X to obtain the differential equation

$$\varphi'' + \tfrac{(M+m)g}{ML}\varphi = -\tfrac{1}{ML}u \tag{3.21}$$

which describes the direct connection between u and φ. Later in Sect. 3.2.3 we will systematically study *elimination of latent variables* according to J. C. Willems. ◇

Linearization of (3.17) in the neighborhood of $\varphi = \pi$ with $\psi := \varphi - \pi$ and ψ near zero, i.e., where the pendulum is in the unstable upright position and a positive ψ means its falling to the left, furnishes the equation

$$\psi'' - \tfrac{(M+m)g}{ML}\psi = \tfrac{1}{ML}u. \tag{3.22}$$

In Sect. 8.4 we will treat *translational mechanical networks.* ◇

3.1.3 Electrical Networks

Electrical networks are typical and important systems and will be studied in detail in Sect. 8.3. We discuss their mathematical model already here. The term network already suggests the connection with graphs.

Definition 3.1.14 A finite (directed) *graph* or *diagram scheme* Γ is given by the following data:

1. a finite set V whose elements v are called the *vertices* or *nodes* of the graph,
2. a finite set K of *branches*, *edges*, or *arrows*, and
3. two maps $\mathrm{dom}, \mathrm{cod}\colon K \to V,\ k \mapsto \mathrm{dom}(k), \mathrm{cod}(k)$. The vertex $v := \mathrm{dom}(k)$ of an edge $k \in K$ is called the *initial node*, *domain*, or *source* of the branch, whereas $w := \mathrm{cod}(k)$ is its *final node*, *codomain*, or *sink*.

The usual graphical representations are

$$k\colon v = \mathrm{dom}(k) \longrightarrow w = \mathrm{cod}(k) \text{ or } \overset{v}{\bullet} \overset{k}{\longrightarrow} \overset{w}{\bullet}.$$

We shortly write $\Gamma = (V, K)$ where we assume that the maps dom resp. cod are clear from the context. ◇

Loops $k\colon v \longrightarrow v$ and *multiple arrows* $k_1, k_2, \ldots, k_m\colon v \longrightarrow w$ with the same domain and codomain are permitted. Such general graphs are also applied in [1, Introduction, item 6] for electrical networks and were already used in [2, §8, pp. 160/161] for arbitrary system interconnections. Special graphs are those where K is a subset of $V \times V$ and the maps dom resp. cod are the restrictions to K of the first resp. the second projection from $V \times V$ to V.

In the sequel we assume that a finite graph $\Gamma = (V, K)$ is given.

Definition 3.1.15 (*Chain groups*) The finite-dimensional free abelian groups

$$\mathrm{C}_0(\Gamma) := \mathrm{C}_0 := \mathbb{Z}^V = \oplus_{v\in V}\mathbb{Z}\delta_v$$
$$\mathrm{C}_1(\Gamma) := \mathrm{C}_1 := \mathbb{Z}^K = \oplus_{k\in K}\mathbb{Z}\delta_k$$

with their standard bases $\delta_v := (0, \dots, 0, \overset{v}{1}, 0 \dots, 0),\ v \in V$, and $\delta_k,\ k \in K$, are called the groups of 0-chains resp. of 1-chains of the graph. It is customary to identify

$$V \subset \mathrm{C}_0,\ v = \delta_v, \text{ and } K \subset \mathrm{C}_1,\ k = \delta_k.$$

The 0-chain $d(k) := \mathrm{dom}(k) - \mathrm{cod}(k) \in \mathrm{C}_0$ is called the (oriented) boundary of $k \in K$. This map is linearly extended to the *boundary operator*

$$d := C_1 \to \mathrm{C}_0,\ c = \sum_{k\in K} c_k k \mapsto d(c) := \sum_k c_k d(k)$$
$$\text{with } d(c) = \sum_{v\in V}\left(\sum_{k:\, v\to\bullet} c_k - \sum_{k:\, \bullet\to v} c_k\right) v.$$

The matrix $A = (A(v,k))_{(v,k)\in V\times K} \in \mathbb{Z}^{V\times K},\ A_{vk} := A(v,k)$, of d for the standard bases $(k)_{k\in K} \in \left(\mathbb{Z}^K\right)^{1\times K}$ and $(v)_{v\in V} \in \left(\mathbb{Z}^V\right)^{1\times V}$ is called the *incidence matrix* of (V, K). It has the form

$$(k)_{k\in K} = (v)_{v\in V}A,\ d(k) = \sum_{v\in V} vA_{vk} \text{ with}$$
$$A_{vk} = \begin{cases} 1 & \text{if } v = \mathrm{dom}(k) \neq \mathrm{cod}(k) \\ -1 & \text{if } v = \mathrm{cod}(k) \neq \mathrm{dom}(k) \\ 0 & \text{otherwise} \end{cases},\quad d(c) = Ac,\ d(c)_v = \sum_{k\in K} A_{vk}c_k. \tag{3.23}$$

Remark 3.1.16 Here and later in the book we consider matrix groups $S^{V\times K}$ where S is an arbitrary abelian group and V and K are arbitrary finite sets. Addition of matrices is the usual componentwise one. If S is a commutative ring, V, K, L are three finite sets and $M \in S^{V\times K},\ N \in S^{K\times L}$, then the product $MN \in S^{V\times L}$ is defined as usual, i.e., by $(MN)_{vl} := \sum_{k\in K} M_{vk}N_{kl},\ v \in V, l \in L.$ ◊

A 1-chain $c = \sum_{k\in K} c_k k$ is called a 1-*cycle* or *closed* if

$$d(c) = 0 \text{ or } Ac = 0 \text{ or } \sum_{k:\, v\to\bullet} c_k = \sum_{k:\, \bullet\to v} c_k \text{ for all } v \in V.$$

The free subgroup

$$\mathrm{Z}_1 := \mathrm{Z}_1(\Gamma) := \ker(d : \mathrm{C}_1 \to \mathrm{C}_0) = \left\{c \in \mathbb{Z}^K;\ Ac = 0\right\}$$
$$= \left\{c = \sum_k c_k k \in \mathrm{C}_1;\ \forall v \in V :\ \sum_{k:\, v \to \bullet} c_k - \sum_{k:\, \bullet \to v} c_k = 0\right\}$$

is the *group of* 1-*cycles*. Its $\mathbb{Z}$-dimension is the *first Betti number* of (V, K). The chain

$$-k = -\delta_k \in \mathrm{C}_1,\ k\colon v \longrightarrow w \in K,$$

is called the *inverse branch* of the branch k with

$$\mathrm{dom}(-k) := \mathrm{cod}(k),\ \mathrm{cod}(-k) := \mathrm{dom}(k),\ -k\colon \mathrm{cod}(k) \longrightarrow \mathrm{dom}(k).$$

◇

Definition 3.1.17 A *directed path* $\omega = (k_1, \dots, k_n)\colon v \longrightarrow w$ from node v to node w is given by

$$\omega\colon \overset{v=v_0}{\bullet} \xrightarrow{k_1} \overset{v_1}{\bullet} \xrightarrow{k_2} \cdots \xrightarrow{k_{n-1}} \overset{v_{n-1}}{\bullet} \xrightarrow{k_n} \overset{v_n=w}{\bullet}$$
$$\text{where } k_i \in K \uplus (-K),\ i = 1, \dots, n \geq 0.$$
$$\text{Then } c(\omega) := k_1 + \cdots + k_n$$

is the associated 1-chain with boundary $d(c(\omega)) = v - w$. Again $\mathrm{dom}(\omega) := v$ resp. $\mathrm{cod}(\omega) := w$ are called the initial resp. final node of the path. For each node $v \in V$ and $n = 0$ there is the empty path (v) from v to v. The path is *closed* if $c(\omega)$ is a 1-cycle, i.e., $v = w$. It is a *circuit* if moreover $n > 0$ and the vertices $v_0 = v_n, v_1, \dots, v_{n-1}$ are pairwise distinct, and then $c(\omega) = k_1 + \cdots + k_n$ is also called a circuit.

The graph is called *connected* if it is not empty and if every two vertices can be connected by a path. ◇

Lemma 3.1.18 *The group* Z_1 *of* 1*-cycles is generated by its circuits. More precisely, every* 1*-cycle is a sum of circuits.*

Proof Let $c = \sum_{k \in K} c_k k \in \mathrm{Z}_1$ and $n := \sum_{k \in K} |c_k|$. We write

$$c_k k = \begin{cases} k + \cdots + k & \text{if } c_k > 0 \\ (-k) + \cdots + (-k) & \text{if } c_k < 0 \end{cases} \Longrightarrow c = \sum_{i=1}^n k_i,\ k_i \in K \uplus (-K),$$
$$\Longrightarrow 0 \underset{c \in \mathrm{Z}_1}{=} d(c) = \sum_{i=1}^n \mathrm{dom}(k_i) - \sum_{i=1}^n \mathrm{cod}(k_i) \Longrightarrow \sum_{i=1}^n \mathrm{dom}(k_i) = \sum_{i=1}^n \mathrm{cod}(k_i). \tag{3.24}$$

We show that c is a sum of circuits by induction on n. The case $n = 1$ is obvious since $c = k$ is a loop, i.e., $\mathrm{dom}(k) = \mathrm{cod}(k)$. Let $n > 1, l_1 := k_1, v_0 := \mathrm{dom}(k_1)$ and $v_1 := \mathrm{cod}(k_1) \neq v_0$. Again by induction on ν assume that a directed path

$$v_0 \xrightarrow{l_1} v_1 \longrightarrow \cdots \xrightarrow{l_\nu} v_\nu,\ \{l_1, \ldots, l_\nu\} \subseteq \{k_1, \ldots, k_n\},\ v_\nu = \mathrm{cod}(l_\nu),$$

with pairwise different v_λ has been constructed. Then also the l_λ are pairwise different. But

$$\begin{aligned} &l_\nu \in \{k_1, \ldots, k_n\},\ v_\nu = \mathrm{cod}(l_\nu) \underset{(3.24)}{\Longrightarrow} \exists l_{\nu+1} \in \{k_1, \ldots, k_n\} \text{ with} \\ &v_\nu = \mathrm{dom}(l_{\nu+1}),\ v_{\nu+1} := \mathrm{cod}(l_{\nu+1}). \end{aligned}$$

If $v_{\nu+1} \notin \{v_0, \ldots, v_\nu\}$ the path is extended by $k_{\nu+1} : v_\nu \to v_{\nu+1}$. If $v_{\nu+1} = v_\mu,\ \mu \leq \nu$, the path

$$\omega : v_\mu \xrightarrow{l_{\mu+1}} v_{\mu+1} \longrightarrow \cdots \xrightarrow{l_{\nu+1}} v_{\nu+1} = v_\mu$$

is a circuit. Since V is finite an equation $v_\mu = v_{\nu+1},\ \mu \leq \nu$, occurs after finitely many steps. Define

$$c(\omega) = \sum_{\lambda=\mu+1}^{\nu+1} l_\lambda \in \mathrm{Z}_1,\ c' := c - c(\omega) \in \mathrm{Z}_1 .$$

Then c' is a sum of at most $n - 1$ edges and, by induction, a sum $c' = \sum_{\omega'} c(\omega')$ of circuits ω'. But then $c = c' + c(\omega) = \sum_{\omega'} c(\omega') + c(\omega)$ has the same property. □

Lemma and Definition 3.1.19 *Define the surjection*

$$\epsilon : \mathrm{C}_0 = \mathbb{Z}^V \to \mathbb{Z},\ v \mapsto 1,\ \eta = \sum_{v \in V} \eta_v v \mapsto \epsilon(\eta) = \sum_{v \in V} \eta_v. \tag{3.25}$$

If (V, K) *is connected then* $\ker(\epsilon) = \mathrm{im}(d) = A\mathbb{Z}^K$.

Proof (i) The inclusion $\mathrm{im}(d) \subseteq \ker(\epsilon)$ follows from

$$d(k) = \mathrm{dom}(k) - \mathrm{cod}(k) \Longrightarrow \epsilon(d(k)) = 0 \Longrightarrow \epsilon(\mathrm{im}(d)) = 0 \Longrightarrow \mathrm{im}(d) \subseteq \ker(\epsilon).$$

(ii) Assume $\eta = \sum_{v \in V} \eta_v v \in \ker(\epsilon)$, i.e., $0 = \epsilon(\eta) = \sum_v \eta_v = 0$. Choose an arbitrary $v_0 \in V$. Since the graph is connected each $v \in V$ admits a directed path $\pi_v : v \to v_0$ and its associated 1-chain $c_v \in \mathbb{Z}^K$ with boundary $d(c_v) = v - v_0$, hence $v = v_0 + d(c_v)$. Then

$$
\begin{aligned}
\eta &= \sum_v \eta_v v = \sum_v \eta_v v_0 + \sum_v \eta_v d(c_v) \\
&= \left(\sum_v \eta_v\right) v_0 + d\left(\sum_v \eta_v c_v\right) = d\left(\sum_v \eta_v c_v\right) \in \mathrm{im}(d) \\
&\Longrightarrow \ker(\epsilon) \subseteq \mathrm{im}(d) \Longrightarrow \ker(\epsilon) = \mathrm{im}(d).
\end{aligned}
$$

□

Definition 3.1.20 (*Electrical network, RLC network*) An electrical network or RLC circuit is mathematically given by the following data and equations:
(1) **Data**.
(i) a connected finite graph (V, K).
(ii) for each edge $k \in K$ two, most often real-valued, but sometimes complex-valued functions $U_k(t)$ and $I_k(t)$ of the continuous-time variable $t \in \mathbb{R}$.
Interpretation: Let $k\colon v \longrightarrow w$ be a branch of K. The function $I_k(t)$ is the *branch current* in direction k which flows from v to w at time t and $U_k(t)$ is the *branch voltage* or *potential difference* between the vertices v and w. For $k \in K$ we define $U_{-k}(t) := -U_k(t)$ and $I_{-k}(t) := -I_k(t)$, i.e., the change of orientation of a branch changes the sign of the voltage and the current along the branch. We assume that the functions are sufficiently differentiable which is always the case for smooth functions or later for distributions. In the beginning we assume that the functions belong to $C^\infty(\mathbb{R})$.
(iii) The graph is *valued*: Each branch has the additional attribute or property to be an *external source*, *resistor*, *capacitor*, or *inductor*. The source branches k are those where energy is supplied to the network from outside. Mathematically this signifies that either $U_k(t)$ or $I_k(t)$ are given. The resistor (capacitor, inductor) branches are characterized by positive numbers R_k (C_k, L_k) which are called the *resistance* (*capacitance*, *inductance*) of the branch. The standard symbols for these different attributes are depicted in Fig. 3.2.

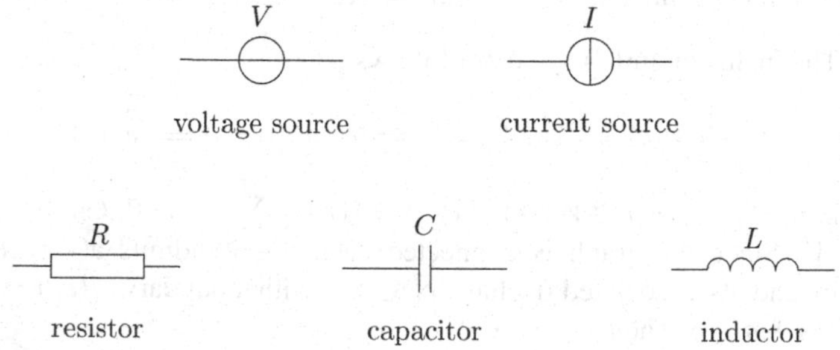

Fig. 3.2 The symbols of some standard components of electrical networks

(2) **Equations**

(i) *Kirchhoff's current or node law (KCL)*: For each vertex v the total current which flows into v equals that which flows out of v, i.e.,

$$\sum_{k:\ \bullet\to v} I_k(t) - \sum_{k:\ v\to\bullet} I_k(t) = 0 \text{ for all } v \in V. \tag{3.26}$$

(ii) *Kirchhoff's voltage or circuit law (KVL)*: The voltages or potential differences along any circuit of the graph sum up to zero, i.e., for all $t \in \mathbb{R}$

$$\sum_{i=1}^{n} U_{k_i}(t) = 0 \text{ where } \omega = (k_1, \ldots, k_n),\ k_i \in K \uplus (-K), \tag{3.27}$$

is an arbitrary circuit of the graph.
(iii) *Ohm's law*: If k is a resistor branch, there is a *resistance* $R_k > 0$ with

$$U_k = R_k I_k. \tag{3.28}$$

(iv) If k is a capacitor branch, there is a *capacitance* $C_k > 0$ with

$$I_k = C_k \overset{\bullet}{U}_k = C_k dU_k/dt = C_k(s \circ U_k). \tag{3.29}$$

Interpretation: The capacitor in branch k is a storage element for the electrical charge. Its total charge $Q_k(t)$ is proportional to the voltage between its two electrodes and the proportionality factor is its capacity (capacitance) C_k, i.e., $Q_k = C_k U_k$. Since the current I_k along k is the rate of change $\overset{\bullet}{Q}_k$ of the charge we get $I_k = C_k \overset{\bullet}{U}_k$ indeed.
(v) If k is an inductor branch, there is an *inductance* $L_k > 0$ with

$$U_k = L_k \overset{\bullet}{I}_k = L_k dI_k/dt = L_k(s \circ I_k). \tag{3.30}$$

Interpretation: The changing current induces a magnetic field and this, in turn, *self-induces* an electrical field and a voltage along k. ◊

Remarks 3.1.21 (1) In electrodynamics the preceding equations are consequences of the Maxwell equations.
(2) There are other types of networks, for instance, with *ideal transformers and gyrators* or with *controlled voltage or current* branches. In Sect. 8.3 these will be shortly discussed. ◊

In the sequel we assume that an RLC circuit according to Definition 3.1.20 is given. From Lemma 3.1.18 we infer the following.

Corollary 3.1.22 *If $c = \sum_{k\in K} c_k k \in Z_1(K)$ is a 1-cycle of the graph the equality $\sum_k c_k U_k = 0$ holds. This generalizes Kirchhoff's voltage law (KVL).*

Proof We extend the voltage vector $U = (U_k)_{k\in K} \in \mathrm{C}^\infty(\mathbb{R})^K$ linearly to the $\mathbb{Z}$-linear map

$$U : \mathrm{C}_1 \to \mathrm{C}^\infty(\mathbb{R}),\ c = \sum_{k\in K} c_k k \mapsto U(c) := \sum_k c_k U_k.$$

Such a map is called a 1-*cochain* with values in $\mathrm{C}^\infty(\mathbb{R})$. Kirchhoff's voltage law says that $U(c(\omega)) = U_{k_1} + \cdots + U_{k_n} = 0$ if $\omega = (k_1, \ldots, k_n)$ is a circuit in Γ and $c(\omega) := k_1 + \cdots + k_n$ is the associated circuit chain. By Lemma 3.1.18 these circuit chains generate Z_1 as $\mathbb{Z}$-module and since $c \mapsto U(c)$ is linear this latter map is 0 on Z_1 and this is the assertion. □

Definition 3.1.23 With the network from Definition 3.1.20 and the signal space $\mathcal{F} := \mathrm{C}^\infty(\mathbb{R})$ we define the behavior of the network. Let

$$U = (U_k)_{k\in K} \in \mathcal{F}^K,\ I = (I_k)_{k\in K} \in \mathcal{F}^K,\ y := \begin{pmatrix} U \\ I \end{pmatrix} \in \mathcal{F}^{2\sharp(K)} := \mathcal{F}^{K \uplus K}.$$

Then

$$\mathcal{B} := \left\{ w = \begin{pmatrix} U \\ I \end{pmatrix} \in \mathcal{F}^{2\sharp(K)};\ \text{The equations from 3.1.20 are satisfied} \right\}$$

is a behavior since all equations are homogeneous linear equations with coefficients in $\mathbb{R}[s]$. Kirchhoff's two laws are even $\mathbb{Z}$-linear equations. Notice that

$$U_k = R_k I_k \iff (1, -R_k) \circ \begin{pmatrix} U_k \\ I_k \end{pmatrix} = 0 \text{ (resistor branch)}$$
$$I_k = C_k \dot{U}_k \iff (-C_k s, 1) \circ \begin{pmatrix} U_k \\ I_k \end{pmatrix} = 0 \text{ (capacitor branch)}$$
$$U_k = L_k \dot{I}_k \iff (1, -L_k s) \circ \begin{pmatrix} U_k \\ I_k \end{pmatrix} = 0 \text{ (inductor branch)}.$$

The voltage and currents of the source branches appear in Kirchhoff's laws only. Differential equations for them will later be computed from the behavior by *elimination*. ◇

Example 3.1.24 Consider the *series connection* of a voltage source, a resistor, a capacitor, and an inductor according to Fig. 3.3 with the directed branches as indicated.

Kirchhoff's laws yield

$$I_R = I_L = I_C := I,\ I_s = -I \text{ and } -U_s + U_R + U_L + U_C = 0.$$

Therefore we only use the component I in the equations. The branch laws are

$$U_R = RI,\ U_L = L(s \circ I),\ I = C(s \circ U_C).$$

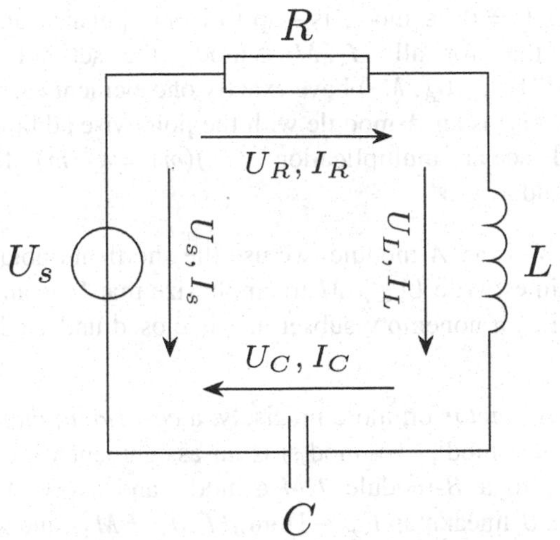

Fig. 3.3 The series connection of a voltage source, a resistor, an inductor, and a capacitor

The behavioral matrix equation for the network is

$$\widetilde{R} \circ w = 0 \text{ with } w := \begin{pmatrix} U_s \\ U_R \\ U_L \\ U_C \\ I \end{pmatrix} \text{ and } \widetilde{R} := \begin{pmatrix} -1 & 1 & 1 & 1 & 0 \\ 0 & 1 & 0 & 0 & -R \\ 0 & 0 & 1 & 0 & -Ls \\ 0 & 0 & 0 & Cs & -1 \end{pmatrix}. \qquad \diamond$$

3.1.4 Categories of Modules and Exact Sequences

We introduce resp. recall categories of modules over commutative rings, functors between these categories, exact sequences, and the universal property of the factor module. Together, these concepts form the language we use in this book to investigate behaviors with algebraic means. For a deeper treatment of category theory, see, e.g., [3–5].

In this subsection, let A and B be commutative rings.

Definition 3.1.25 We denote the *category of A-modules* by mod_A. Its *objects* are the A-modules. For each pair of A-modules M_1 and M_2 the set of *morphisms* or *arrows* from M_1 to M_2 is the set $\mathrm{Hom}_A(M_1, M_2)$ of A-linear maps (or homomorphisms) from M_1 to M_2. The *composition gf of morphisms* $M_1 \xrightarrow{f} M_2 \xrightarrow{g} M_3$ is the composition of maps, i.e., $(gf)(m_1) = g\big(f(m_1)\big)$ for $m_1 \in M_1$. For every A-module M, the *identity morphism* id_M is the identity function on M, i.e., $\mathrm{id}_M(m) = m$ for $m \in M$. The *associativity law* $h(gf) = (hg)f$ is obviously satisfied.

The *zero module* $0 = 0_A \in \text{mod}_A$ is—up to isomorphism—uniquely characterized by the fact that for all $M_1, M_2 \in \text{mod}_A$ the sets of homomorphisms $\text{Hom}_A(M_1, 0_A)$ and $\text{Hom}_A(0_A, M_2)$ have exactly one element each. The set of morphisms $\text{Hom}_A(M_1, M_2)$ is an A-module with the pointwise addition $(f+g)(m) = f(m) + g(m)$ and scalar multiplication $(af)(m) = af(m)$ for $f, g \in \text{Hom}_A(M_1, M_2)$, $a \in A$, and $m \in M_1$. ◊

To express that M is an A-module, we use the shorthand notations $M \in \text{mod}_A$ or ${}_AM$. We sometimes write $U \leq_A M$ to emphasize that U is an A-submodule of the A-module M, i.e., a nonempty subset that is closed under addition and scalar multiplication.

Definition 3.1.26 A *functor* or, more precisely, a *covariant functor* F from mod_A to mod_B (in short, $F\colon \text{mod}_A \longrightarrow \text{mod}_B$) is an assignment which maps every A-module $M \in \text{mod}_A$ to a B-module $FM \in \text{mod}_B$ and every A-linear map $\varphi \in \text{Hom}_A(M_1, M_2)$ to a B-linear map $F\varphi \in \text{Hom}_B(FM_1, FM_2)$, and which is structure-preserving, i.e., it satisfies $F\,\text{id}_M = \text{id}_{FM}$ for every $M \in \text{mod}_A$ and $F(\psi\varphi) = (F\psi)(F\varphi)$ for all A-homomorphisms $M_1 \xrightarrow{\varphi} M_2 \xrightarrow{\psi} M_3$.

A *contravariant functor* $F\colon \text{mod}_A^{\text{op}} \longrightarrow \text{mod}_B$ has the same properties as a covariant functor with the exception that it *reverses the arrows*. This means that an A-homomorphism $\varphi\colon M_1 \longrightarrow M_2$ is mapped to a B-homomorphism $F\varphi\colon FM_2 \longrightarrow FM_1$, and, consequently, the composition condition is $F(\psi\varphi) = (F\varphi)(F\psi)$ for all A-homomorphisms $M_1 \xrightarrow{\varphi} M_2 \xrightarrow{\psi} M_3$.

The superscript "op" in mod_A^{op} denotes the dual category of mod_A, which is the category where all the arrows are reversed and the associativity law is adapted accordingly. For the purpose of this book we use this notation only to distinguish between a covariant functor $F\colon \text{mod}_A \longrightarrow \text{mod}_B$ and a contravariant one $F\colon \text{mod}_A^{\text{op}} \longrightarrow \text{mod}_B$. ◊

Every (covariant or contravariant) functor maps isomorphisms to isomorphisms. In this book we will only consider *additive functors* that, by definition, satisfy $F(\varphi_1 + \varphi_2) = F\varphi_1 + F\varphi_2$ for all $M_1, M_2 \in \text{mod}_A$ and all $\varphi_1, \varphi_2 \in \text{Hom}_A(M_1, M_2)$. A direct consequence of this is that $F0_A = 0_B$, i.e., F maps the zero object of mod_A to the zero object of mod_B up to isomorphism.

Examples 3.1.27 Let $N \in \text{mod}_A$ be fixed. The assignment

$$\begin{array}{cccccc}
\text{Hom}_A(N,-)\colon & \text{mod}_A & \longrightarrow & \text{mod}_A & & \\
& M_1 & & \text{Hom}_A(N, M_1) & \ni & \alpha \\
& \downarrow \varphi & \longmapsto & \downarrow \text{Hom}_A(N,\varphi) & & \downarrow \\
& M_2 & & \text{Hom}_A(N, M_2) & \ni & \text{Hom}_A(N,\varphi)\alpha := \varphi\alpha
\end{array}$$

where $\text{Hom}_A(N, \varphi)\alpha = \varphi\alpha\colon N \xrightarrow{\alpha} M_1 \xrightarrow{\varphi} M_2$ is a covariant functor and it is called the *covariant Hom-functor* $\text{Hom}_A(N, -)$.

In contrast, the assignment

$$
\begin{array}{ccccc}
\mathrm{Hom}_A(-,N)\colon \ \mathrm{mod}_A^{\mathrm{op}} & \longrightarrow & \mathrm{mod}_A & & \\
M_1 & & \mathrm{Hom}_A(M_1,N) & \ni & \mathrm{Hom}_A(\varphi,N)\alpha := \alpha\varphi \\
\downarrow \varphi & \longmapsto & \uparrow \mathrm{Hom}_A(\varphi,N) & & \uparrow \\
M_2 & & \mathrm{Hom}_A(M_2,N) & \ni & \alpha
\end{array}
\tag{3.31}
$$

where $\mathrm{Hom}_A(\varphi, N)\alpha = \alpha\varphi\colon M_1 \xrightarrow{\varphi} M_2 \xrightarrow{\alpha} N$ is a contravariant functor and it is called the *contravariant Hom-functor* $\mathrm{Hom}_A(-, N)$.

Both Hom-functors F are A-linear, i.e., satisfy

$$
F(\varphi_1 + \varphi_2) = (F\varphi_1) + (F\varphi_2) \quad \text{and} \quad F(a\varphi_1) = a(F\varphi_1)
$$
$$
\text{for } \varphi_1, \varphi_2 \in \mathrm{Hom}_A(M_1, M_2) \text{ and } a \in A.
$$

The most important functor in this book is the contravariant Hom-functor $\mathrm{D} := \mathrm{Hom}_{\mathcal{D}}(-, \mathcal{F})$ where $\mathcal{D}$ is the ring of operators and $\mathcal{F}$ is the signal module. In Chaps. 9–11 we will also encounter the covariant quotient functor $(-)_T$. ◊

Definition 3.1.28 A sequence of A-modules and A-homomorphisms

$$
M_1 \xrightarrow{\varphi_1} M_2 \xrightarrow{\varphi_2} M_3 \tag{3.32}
$$

is a *complex* if $\varphi_2\varphi_1 = 0$ or, in other words,

$$
\mathrm{im}(\varphi_1) = \varphi_1(M_1) \subseteq \ker(\varphi_2) = \{m_2 \in M_2;\ \varphi_2(m_2) = 0\}.
$$

The sequence (3.32) is *exact* if $\mathrm{im}(\varphi_1) = \ker(\varphi_2)$.

More generally, a possibly infinite sequence

$$
M_*\colon \quad \cdots \longrightarrow M_{i+1} \xrightarrow{d_{i+1}} M_i \xrightarrow{d_i} M_{i-1} \longrightarrow \cdots, \quad i \in \mathbb{Z}, \tag{3.33}
$$

is a complex if and only if all three-member subsequences have this property, i.e., if and only if for all $i \in \mathbb{Z}$ the identity $\mathrm{im}(d_{i+1}) \subseteq \ker(d_i)$ holds. Similarly, the sequence M_* is exact if and only if $\mathrm{im}(d_{i+1}) = \ker(d_i)$ holds for all $i \in \mathbb{Z}$. ◊

Lemma 3.1.29 *A homomorphism $\varphi \in \mathrm{Hom}_A(M_1, M_2)$ is* injective *or a* monomorphism *if and only if the sequence $0 \longrightarrow M_1 \xrightarrow{\varphi} M_2$ is exact. It is* surjective *(onto) or an* epimorphism *if and only if the sequence $M_1 \xrightarrow{\varphi} M_2 \longrightarrow 0$ is exact. It is* bijective *or an* isomorphism *if and only if the sequence $0 \longrightarrow M_1 \xrightarrow{\varphi} M_2 \longrightarrow 0$ is exact. A module M is isomorphic to the zero module if and only if the sequence $0 \longrightarrow M \longrightarrow 0$ is exact.* ◊

Definition 3.1.30 A functor F is *exact* if it transforms exact sequences into exact sequences, i.e., if every exact sequence $M_1 \overset{\varphi_1}{\longrightarrow} M_2 \overset{\varphi_2}{\longrightarrow} M_3$ is mapped to the exact sequence $FM_1 \overset{F\varphi_1}{\longrightarrow} FM_2 \overset{F\varphi_2}{\longrightarrow} FM_3$ if F is covariant, and to the exact sequence $FM_1 \overset{F\varphi_1}{\longleftarrow} FM_2 \overset{F\varphi_2}{\longleftarrow} FM_3$ if F is contravariant.

A covariant functor F is *left exact* if every exact sequence

$$0 \longrightarrow M_1 \overset{\varphi_1}{\to} M_2 \overset{\varphi_2}{\to} M_3$$

is mapped to the exact sequence

$$F0 = 0 \longrightarrow FM_1 \overset{F\varphi_1}{\longrightarrow} FM_2 \overset{F\varphi_2}{\longrightarrow} FM_3,$$

whereas a contravariant functor F is *left exact* if every exact sequence

$$0 \longleftarrow M_3 \overset{\varphi_2}{\leftarrow} M_2 \overset{\varphi_1}{\leftarrow} M_1$$

is mapped to the exact sequence

$$0 \longrightarrow FM_3 \overset{F\varphi_2}{\longrightarrow} FM_2 \overset{F\varphi_1}{\longrightarrow} FM_1.$$

As a guideline, the zero has to be at the beginning of the sequence after applying the functor. ◊

The following generalization of the homomorphism theorem will be often used.

Theorem 3.1.31 *Let M be an A-module and let U be a submodule of M. The canonical map*

$$\mathrm{can}\colon M \longrightarrow M/U, \quad x \longmapsto \overline{x} := x + U,$$

has the following properties:

1. $\mathrm{can}(U) = 0$, *i.e.,* $U \subseteq \ker(\mathrm{can})$. *(Even* $U = \ker(\mathrm{can})$ *holds).*
2. *For every A-module N and for every linear map $\varphi\colon M \longrightarrow N$ with $\varphi(U) = 0$ there is a unique induced linear map $\varphi_{\mathrm{ind}}\colon M/U \longrightarrow N$ such that $\varphi_{\mathrm{ind}}\,\mathrm{can} = \varphi$, namely,*

$$\varphi_{\mathrm{ind}}\colon M/U \longrightarrow N, \quad \overline{x} = x + U \longmapsto \varphi_{\mathrm{ind}}(\overline{x}) = \varphi(x). \tag{3.34}$$

The condition $\varphi(U) = 0$ implies that $\varphi_{\mathrm{ind}}(\overline{x}) = \varphi(x)$ is independent of the choice of the representative $x \in x + U$, and hence the map φ_{ind} is well defined. *The assignment $\varphi \longmapsto \varphi_{\mathrm{ind}}$ is A-linear. In other words, property 2 signifies that for every A-module N there is the A-isomorphism*

$$\begin{aligned}\operatorname{Hom}_A(\mathrm{can}, N)\colon \operatorname{Hom}_A(M/U, N) &\overset{\cong}{\longleftrightarrow} \{\varphi \in \operatorname{Hom}_A(M, N);\ \varphi(U) = 0\},\\ \psi \longmapsto \psi\,\mathrm{can},\ &\varphi_{\mathrm{ind}} \longleftarrow\!\shortmid\ \varphi.\end{aligned} \tag{3.35}$$

These two properties characterize the factor module M/U or, more exactly, the pair $(M/U, \mathrm{can})$ uniquely up to isomorphism, and are, therefore, called the universal property *of the factor module.*

If $U = \ker(\varphi)$, then φ_{ind} is a monomorphism. If φ is an epimorphism, then φ_{ind} is an epimorphism too. From these two facts combined, we obtain the homomorphism theorem $\varphi_{\mathrm{ind}}\colon M/\ker(\varphi) \cong \operatorname{im}(\varphi)$. ◇

Lemma 3.1.32 *For every $N \in \mathrm{mod}_A$, both the covariant Hom-functor* $\operatorname{Hom}_A(N, -)$ *and the contravariant one* $\operatorname{Hom}_A(-, N)$ *are left exact.*

Proof For $\operatorname{Hom}_A(N, -)$ this follows from the definition, and for $\operatorname{Hom}_A(-, N)$ from Theorem 3.1.31. □

In Sect. 3.2.1, we need the following result on contravariant functors.

Lemma 3.1.33 *Let $F\colon \mathrm{mod}_A^{\mathrm{op}} \longrightarrow \mathrm{mod}_B$ be a contravariant functor. The following statements are equivalent:*

1. *F is exact.*
2. *F is left exact and maps monomorphisms to epimorphisms.*

Proof 1. $\Longrightarrow$ 2.: An exact functor is obviously left exact. If $\varphi \in \operatorname{Hom}_A(M_1, M_2)$ is a monomorphism, then the sequence $0 \longrightarrow M_1 \xrightarrow{\varphi} M_2$ is exact. The contravariant exact functor F maps it to the exact sequence $FM_2 \xrightarrow{F\varphi} FM_1 \longrightarrow 0$, i.e., $F\varphi$ is an epimorphism.
2. $\Longrightarrow$ 1.: Let $M_1 \xrightarrow{\varphi_1} M_2 \xrightarrow{\varphi_2} M_3$ be an exact sequence. We have to show that $FM_1 \xleftarrow{F\varphi_1} FM_2 \xleftarrow{F\varphi_2} FM_3$ is exact, i.e., that $\operatorname{im}(F\varphi_2) = \ker(F\varphi_1)$.

Consider the cokernel $\operatorname{cok}(\varphi_1) := M_2/\operatorname{im}(\varphi_1)$ with the canonical epimorphism $\mathrm{can}\colon M_2 \longrightarrow \operatorname{cok}(\varphi_1)$. The sequence

$$M_1 \xrightarrow{\varphi_1} M_2 \xrightarrow{\mathrm{can}} \operatorname{cok}(\varphi_1) \longrightarrow 0$$

is exact. Application of the left exact functor F gives the exact sequence

$$0 \longrightarrow F\big(\operatorname{cok}(\varphi_1)\big) \xrightarrow{F\,\mathrm{can}} FM_2 \xrightarrow{F\varphi_1} FM_1,$$

in particular, $\operatorname{im}(F\,\mathrm{can}) = \ker(F\varphi_1)$.

From $\operatorname{im}(\varphi_1) = \ker(\varphi_2)$ follows that $\operatorname{cok}(\varphi_1) = M_2/\operatorname{im}(\varphi_1) = M_2/\ker(\varphi_2)$. By the homomorphism theorem φ_2 induces the monomorphism

$$\varphi_{2,\mathrm{ind}}\colon \operatorname{cok}(\varphi_1) = M_2/\ker(\varphi_2) \longrightarrow M_3 \text{ with } \varphi_{2,\mathrm{ind}}\,\mathrm{can} = \varphi_2,$$

hence $(F\,\mathrm{can})(F\varphi_{2,\mathrm{ind}}) = F\varphi_2$. By assumption $F\varphi_{2,\mathrm{ind}}$ is surjective. We conclude $\mathrm{im}(F\varphi_2) = \mathrm{im}(F\,\mathrm{can}) = \ker(F\varphi_1)$ as asserted. □

3.1.5 Algebraization of Behaviors

The results of Sect. 3.1.4 *are now applied to a behavior in the sense of Definition* 3.1.11.

In the beginning $\mathcal{F}$ denotes an arbitrary module over a commutative integral domain $\mathcal{D}$ with the scalar multiplication $\circ$. Let δ_j, $j = 1, \dots, l$, denote the standard basis on $\mathcal{D}^{1\times l}$.

For every l and $\mathcal{F}$, there is the canonical $\mathcal{D}$-isomorphism

$$\mathrm{can}_l\colon \mathrm{Hom}_{\mathcal{D}}(\mathcal{D}^{1\times l}, \mathcal{F}) \cong \mathcal{F}^l, \quad \psi \longmapsto w = \begin{pmatrix} w_1 \\ \vdots \\ w_l \end{pmatrix}, \text{ where } \psi(\delta_j) := w_j, \tag{3.36}$$

since a linear map is uniquely given by the images of the basis vectors.

Lemma 3.1.34 *For $R \in \mathcal{D}^{k\times l}$ and $\cdot R\colon \mathcal{D}^{1\times k} \longrightarrow \mathcal{D}^{1\times l}$, $\xi \longmapsto \xi R$, the diagram*

$$\begin{array}{ccc} \mathrm{Hom}_{\mathcal{D}}(\mathcal{D}^{1\times l}, \mathcal{F}) & \xrightarrow{\mathrm{Hom}_{\mathcal{D}}(\cdot R, \mathcal{F})} & \mathrm{Hom}_{\mathcal{D}}(\mathcal{D}^{1\times k}, \mathcal{F}) \\ \cong \downarrow \mathrm{can}_l & & \cong \downarrow \mathrm{can}_k \\ \mathcal{F}^l & \xrightarrow{R\circ} & \mathcal{F}^k \end{array} \tag{3.37}$$

is commutative, i.e., if $\psi \in \mathrm{Hom}_{\mathcal{D}}(\mathcal{D}^{1\times l}, \mathcal{F})$, then

$$\mathrm{can}_k\big(\psi(\cdot R)\big) = (R\circ)\,\mathrm{can}_l(\psi).$$

Proof For $w := \mathrm{can}_l(\psi)$ and $v := \mathrm{can}_k(\psi(\cdot R))$ we have to show that $R \circ w = v$. But

$$v_i = \big(\psi(\cdot R)\big)(\delta_i) = \psi(\delta_i R) = \psi(R_{i-}) = \psi\Big(\sum_{j=1}^{l} R_{ij}\delta_j\Big)$$

$$= \sum_{j=1}^{l} R_{ij} \circ \psi(\delta_j) = \sum_{j=1}^{l} R_{ij} \circ w_j = R_{i-} \circ w = (R \circ w)_i.$$

□

Definition 3.1.35 Let $R \in \mathcal{D}^{k\times l}$ be a matrix. The matrix R gives rise to its row submodule $U := \mathrm{im}(\cdot R\colon \mathcal{D}^{1\times k} \to \mathcal{D}^{1\times l}) = \mathcal{D}^{1\times k}R = \sum_{i=1}^{k} \mathcal{D}R_{i-}$,

the factor module $M := \mathcal{D}^{1\times l}/U = \sum_{j=1}^{l} \mathcal{D}\overline{\delta_j} \ni \overline{(\xi_1,\ldots,\xi_l)} = \sum_{j=1}^{l} \xi_j\overline{\delta_j}$,

the canonical map $\mathrm{can}\colon \mathcal{D}^{1\times l} \longrightarrow M,\quad \xi \longmapsto \overline{\xi}$,

and the $\mathcal{F}$-behavior $\mathcal{B} := \{w \in \mathcal{F}^l;\ R \circ w = 0\}$.

◊

Theorem 3.1.36 (Algebraization of behaviors) *For the data from Definition 3.1.35, the canonical maps*

$$\mathrm{Hom}_{\mathcal{D}}(M,\mathcal{F}) \cong \mathcal{B},\quad \varphi \longleftrightarrow w = \begin{pmatrix} w_1 \\ \vdots \\ w_l \end{pmatrix},$$
$$w_j = \varphi(\overline{\delta_j}),\quad \varphi(\overline{\xi}) = \sum_{j=1}^{l} \xi_j \circ w_j = \xi \circ w, \tag{3.38}$$

are $\mathcal{D}$-isomorphisms and inverse to each other. We write

$$\mathrm{can}_M\colon \mathrm{Hom}_{\mathcal{D}}(M,\mathcal{F}) \cong \mathcal{B};\quad \varphi = (\overline{\xi} \mapsto \xi \circ w) \longmapsto w,\quad w_j := \varphi(\overline{\delta_j}).$$

Proof The commutative diagram (3.37) with the canonical vertical isomorphisms induces the isomorphism of the kernels of the horizontal maps, i.e.,

$$\ker\big(\mathrm{Hom}(\cdot R,\mathcal{F})\big) = \{\psi\colon \mathcal{D}^{1\times l} \to \mathcal{F};\ \psi(\cdot R) = 0\} \xrightarrow{\cong} \ker(R\circ\colon \mathcal{F}^l \to \mathcal{F}^k) = \mathcal{B}$$
$$\psi \longmapsto \mathrm{can}_l(\psi) = \begin{pmatrix} \psi(\delta_1) \\ \vdots \\ \psi(\delta_l) \end{pmatrix}. \tag{3.39}$$

But $\psi(\cdot R) = 0$ is equivalent to $\psi\big(\mathrm{im}(\cdot R)\big) = \psi(U) = 0$, hence

$$\ker\big(\mathrm{Hom}_{\mathcal{D}}(\cdot R,\mathcal{F})\big) = \{\psi \in \mathrm{Hom}_{\mathcal{D}}(\mathcal{D}^{1\times l},\mathcal{F});\ \psi(U) = 0\}. \tag{3.40}$$

From (3.35), we infer the isomorphism

$$\mathrm{Hom}_{\mathcal{D}}(M,\mathcal{F}) = \mathrm{Hom}_{\mathcal{D}}(\mathcal{D}^{1\times l}/U,\mathcal{F}) \overset{\cong}{\longleftrightarrow} \{\psi \in \mathrm{Hom}_{\mathcal{D}}(\mathcal{D}^{1\times l},\mathcal{F});\ \psi(U) = 0\},$$
$$\varphi = \psi_{\mathrm{ind}} \longleftrightarrow \psi,\quad \varphi(\overline{\xi}) = \psi(\xi). \tag{3.41}$$

From the isomorphism (3.39), the identity (3.40), and the isomorphism (3.41) we obtain the asserted isomorphism (3.38). □

3.2 The Fundamental Principle and Elimination

That the fundamental principle holds for a signal module means that the obvious necessary conditions for the solvability of an inhomogeneous implicit linear system, cf. Motivation and Definition 3.2.1, *are also sufficient. It is an important observation that a module satisfies the fundamental principle if and only if it is injective* (*Theorem* 3.2.16). *The latter notion from homological algebra is, therefore, the first object of study in this section. We are going to show in Corollary* 3.2.12 *that the standard signal modules* ${}_{F[s]}\mathrm{C}^\infty(\mathbb{R}, F)$ *for* $F = \mathbb{R}, \mathbb{C}$ *and* ${}_{F[s]}F^{\mathbb{N}}$ *from Sect.* 3.1 *are injective. The fundamental principle is essential for the computation of images of behaviors* (*Sect.* 3.2.3). *We apply this to matrix models and state space equations in Sects.* 3.2.4 *and* 3.2.5.

3.2.1 Injective and Divisible Modules

Assumption 3.1.4 is in force, i.e., $\mathcal{D}$ is a commutative Noetherian integral domain and $\mathcal{F}$ is a $\mathcal{D}$-module if not stated otherwise.

Motivation and Definition 3.2.1 Consider a linear system

$$R \circ y = u \quad \text{with } R \in \mathcal{D}^{k\times l},\ y \in \mathcal{F}^l,\ u \in \mathcal{F}^k, \text{ and } \xi \in \mathcal{D}^{1\times k}. \tag{3.42}$$

If $\xi R = \sum_{i=1}^k \xi_i R_{i-} = 0$ then $\xi \circ u = \xi \circ (R \circ y) = (\xi R) \circ y = 0 \circ y = 0$. A vector ξ with the linear dependence relation $\xi R = 0$ is called a *relation* or a *syzygy* of the rows of R. The solvability of $R \circ y = u$ for given right side u thus implies the linear relations

$$\xi \circ u = \sum_{i=1}^{k} \xi_i \circ u_i = 0 \quad \text{for all linear relations } \xi \text{ of the rows of } R. \tag{3.43}$$

The relations (3.43) are called the necessary *compatibility conditions* for the solvability of $R \circ y = u$. In the context of rings of differential operators, one also speaks of the necessary *integrability conditions*.

More generally, a matrix R^1 with

$$R^1 R = 0, \quad \text{i.e.,} \quad (R^1 R)_{i-} = R^1_{i-} R = 0 \text{ for all rows of } R^1$$

has the property that all its rows are syzygies of R. The matrix R^1 is called a *left annihilator* of R, and R is called a *right annihilator* of R^1. Again, the solvability of $R \circ y = u$ implies the necessary compatibility condition $R^1 \circ u = 0$. The *fundamental principle* characterizes those modules $\mathcal{F}$ where these necessary compatibility conditions are also sufficient for solvability, viz., the *injective* modules. ◇

Recall from Theorem 3.1.5 that in the standard cases each equation $f \circ y = u$ in $\mathcal{F}$ with nonzero $f \in \mathcal{D}$ has a solution y for arbitrary u. Since $f \neq 0$ and $\mathcal{D}$ is a domain, the set of relations of (the rows of) f is $\{0\} \subseteq \mathcal{D}$, and this means that the compatibility conditions are satisfied for arbitrary $u \in \mathcal{F}$. Therefore, we have shown in Theorem 3.1.5 that in the standard cases, the necessary compatibility conditions are sufficient if the system consists of one equation in one unknown.

Definition 3.2.2 Let $\mathcal{D}$ be an integral domain. A $\mathcal{D}$-module $\mathcal{F}$ is called *divisible* if for every nonzero $f \in \mathcal{D}$ and every $u \in \mathcal{F}$ the equation $f \circ y = u$ has a solution $y \in \mathcal{F}$ or, in other words, if all maps

$$f\circ\colon \mathcal{F} \longrightarrow \mathcal{F},\ y \longmapsto f \circ y, \quad \text{for } 0 \neq f \in \mathcal{D},$$

are surjective. ◊

By Theorem 3.1.5, the modules ${}_{F[s]}\mathrm{C}^{\infty}(\mathbb{R}, F)$ for $F = \mathbb{R}, \mathbb{C}$ and ${}_{F[s]}F^{\mathbb{N}}$ are divisible.

We reformulate divisibility in order to point out its connection with the notion of an injective module. Let $\mathrm{D} := \mathrm{Hom}_{\mathcal{D}}(-, \mathcal{F})$ denote the contravariant Hom-functor from (3.31) with respect to $\mathcal{F}$. We call D the *duality functor*. Let $f \in \mathcal{D}$ be nonzero and consider the principal ideal $\mathfrak{a} := \mathcal{D}f \subset \mathcal{D}$ as well as the linear injection $\mathrm{inj}\colon \mathfrak{a} \xrightarrow{\subseteq} \mathcal{D}$. The injection gives rise to the restriction map

$$\mathrm{Dinj}\colon \mathrm{D}\mathcal{D} \longrightarrow \mathrm{D}\mathfrak{a}, \quad \psi \longmapsto \psi\,\mathrm{inj} = \psi|_{\mathfrak{a}}. \tag{3.44}$$

Lemma 3.2.3 *The following assertions are equivalent for an integral domain $\mathcal{D}$ and a module $\mathcal{F}$:*

1. *The $\mathcal{D}$-module $\mathcal{F}$ is divisible.*
2. *For every principal ideal $\mathfrak{a} = \mathcal{D}f$ of $\mathcal{D}$, the map* Dinj *from* (3.44) *is surjective, i.e., for every $\varphi \in \mathrm{Hom}_{\mathcal{D}}(\mathfrak{a}, \mathcal{F})$ there is a linear map*

$$\psi\colon \mathcal{D} \longrightarrow \mathcal{F} \text{ such that } \varphi(a) = \psi(a) \text{ for all } a \in \mathfrak{a}.$$

In other words, the map φ can be linearly extended to all of $\mathcal{D}$. Since $\mathcal{D}$ is generated by 1, *the map ψ can always be described as $\psi(g) = g \circ \psi(1) = g \circ y$ for $y := \psi(1)$ and $g \in \mathcal{D}$.*

Proof 1. $\Longrightarrow$ 2.: For $\mathfrak{a} = 0$, the assertion is obviously true. So assume that $\mathfrak{a} = \mathcal{D}f \neq 0$. Since $\mathcal{F}$ is divisible, there is $y \in \mathcal{F}$ such that $u := \varphi(f) = f \circ y$. For all $a = gf \in \mathfrak{a} = \mathcal{D}f$ this implies

$$\varphi(a) = \varphi(gf) = g \circ \varphi(f) = g \circ (f \circ y) = (gf) \circ y = a \circ y,$$

i.e., the map $\psi\colon g \longmapsto g \circ y$ is a linear extension of φ.
2. $\Longrightarrow$ 1.: Let $0 \neq f \in \mathcal{D}$ and let u be any element of $\mathcal{F}$. Since $\mathcal{D}$ is an integral

domain, the ideal $\mathfrak{a} := \mathcal{D}f$ is a free $\mathcal{D}$-module with the basis f, and therefore there is a unique linear map $\varphi \in \mathrm{Hom}_{\mathcal{D}}(\mathfrak{a}, \mathcal{F})$ with $\varphi(f) = u$, namely,

$$\varphi\colon \mathfrak{a} = \mathcal{D}f \longrightarrow \mathcal{F},\ \ gf \longmapsto g \circ u.$$

Let ψ be a linear extension of φ to $\mathcal{D}$ and define $y := \psi(1)$. Then

$$u = \varphi(f) = \psi(f) = \psi(f1) = f \circ \psi(1) = f \circ y. \qquad \square$$

Corollary 3.2.4 *Let $\mathcal{D}$ be a principal ideal domain, e.g., the polynomial algebra $F[s]$ in one indeterminate over a field. Then the following conditions are equivalent:*

1. *The $\mathcal{D}$-module $\mathcal{F}$ is divisible.*
2. *For every ideal $\mathfrak{a}$ of $\mathcal{D}$ the map* D(inj) *from* (3.44) *is surjective, i.e., for every $\varphi \in \mathrm{Hom}_{\mathcal{D}}(\mathfrak{a}, \mathcal{F})$ there is a linear extension*

$$\psi\colon \mathcal{D} \longrightarrow \mathcal{F},\ \ 1 \longmapsto y,\ \ \textit{such that } \varphi(a) = \psi(a) = a \circ y \textit{ for } a \in \mathfrak{a}.$$

$\diamond$

For an arbitrary ring $\mathcal{D}$, the second condition of the preceding corollary characterizes the injectivity of the $\mathcal{D}$-module $\mathcal{F}$ according to *Baer's criterion* (see Theorem 3.2.9).

Definition 3.2.5 A $\mathcal{D}$-module $\mathcal{F}$ over an arbitrary commutative ring $\mathcal{D}$ is *injective* if the duality functor $\mathrm{D} := \mathrm{Hom}_{\mathcal{D}}(-, \mathcal{F})$ is exact, i.e., transforms exact sequences into exact sequences. $\diamond$

Before proving Baer's criterion, we characterize the exactness of the duality functor in the following lemma.

Lemma 3.2.6 *For a module $\mathcal{F} \in \mathrm{mod}_{\mathcal{D}}$, the following properties are equivalent:*

1. *The module $\mathcal{F}$ is injective, i.e., the functor* $\mathrm{D} := \mathrm{Hom}_{\mathcal{D}}(-, \mathcal{F})$ *is exact.*
2. *For every $\mathcal{D}$-module M and every submodule $M' \leq_{\mathcal{D}} M$ with the injection* $\mathrm{inj}\colon M' \longrightarrow M$*, the map*

$$\mathrm{D}\mathrm{inj}\colon \mathrm{D}M = \mathrm{Hom}_{\mathcal{D}}(M, \mathcal{F}) \longrightarrow \mathrm{D}M',\ \ \psi \longmapsto \varphi := \psi\,\mathrm{inj} = \psi|_{M'}$$

is surjective. In other words, every linear map $\varphi\colon M' \longrightarrow \mathcal{F}$ can be extended to a linear map $\psi\colon M \longrightarrow \mathcal{F}$ on the module M with $\psi|_{M'} = \varphi$.

Proof 1. $\Longrightarrow$ 2.: The injection is a monomorphism. By assumption, the functor D is exact, thus, by Lemma 3.1.33, the map Dinj is an epimorphism.
2. $\Longrightarrow$ 1.: The contravariant Hom-functor $\mathrm{D} = \mathrm{Hom}_{\mathcal{D}}(-, \mathcal{F})$ is left exact for every module $\mathcal{F}$, see Lemma 3.1.32. We will show that under the assumption of item 2,

it maps monomorphisms to epimorphisms. Then Lemma 3.1.33 implies that it is an exact functor.

Let $\varphi\colon M_1 \longrightarrow M_2$ be a monomorphism. The map φ can be factorized as

$$\varphi = \mathrm{inj}\,\varphi_{\mathrm{ind}}\colon M_1 \overset{\varphi_{\mathrm{ind}}}{\longrightarrow} \mathrm{im}(\varphi) \overset{\mathrm{inj}}{\longrightarrow} M_2,$$

where the induced map $\varphi_{\mathrm{ind}}\colon M_1 \longrightarrow \mathrm{im}(\varphi),\ x \longmapsto \varphi(x)$, is an isomorphism. All functors map isomorphisms to isomorphisms. Thus, $\mathrm{D}\varphi_{\mathrm{ind}}$ is an isomorphism too. Since Dinj is surjective by assumption, the map $\mathrm{D}\varphi = \mathrm{D}(\mathrm{inj}\,\varphi_{\mathrm{ind}}) = (\mathrm{D}\varphi_{\mathrm{ind}})(\mathrm{Dinj})$ is surjective. □

We need Zorn's lemma at some instances in this book. The first one is the proof of Baer's criterion.

Result 3.2.7 (Zorn's lemma) *A nonempty partially ordered set I contains a maximal element if every chain in I, i.e., every nonempty strictly ordered subset J of I, has an upper bound in I.* ◇

In general, the union of submodules, for instance, the union of two different lines containing 0 in the real plane, is not a submodule.

Lemma 3.2.8 *If I is a nonempty set of submodules of the $\mathcal{D}$-module M and upward directed, then its union $U' := \bigcup_{U\in I} U$ is also a submodule of M. The condition that I is upward directed signifies that for two modules U_1 and U_2 in I there is a $U_3 \in I$ such that $U_1 + U_2 \subseteq U_3$. In particular, this is satisfied if I is strictly ordered.* ◇

Theorem 3.2.9 (Baer's criterion) *For a module $\mathcal{F}$ over a commutative ring $\mathcal{D}$, the following two assertions are equivalent:*

1. *The module $\mathcal{F}$ is injective, i.e., the duality functor $\mathrm{D} = \mathrm{Hom}_{\mathcal{D}}(-, \mathcal{F})$ is exact.*
2. *For every ideal $\mathfrak{a}$ of $\mathcal{D}$ every linear map $\varphi\colon \mathfrak{a} \longrightarrow \mathcal{F}$ can be linearly extended to $\psi\colon \mathcal{D} \longrightarrow \mathcal{F}$ with $\psi|_{\mathfrak{a}} = \varphi$, i.e., there is $y \in \mathcal{F}$ such that $\varphi(a) = a \circ y$ for all $a \in \mathfrak{a}$.*

The theorem holds also for noncommutative rings, which, however, are never used in this book.

Proof 1. $\Longrightarrow$ 2.: This implication follows directly from Lemma 3.2.6 with $M' := \mathfrak{a} \subseteq M := \mathcal{D}$.
2. $\Longrightarrow$ 1.: We prove the second condition of Lemma 3.2.6. Let $M' \leq_{\mathcal{D}} M$ be a submodule of M and let $\varphi\colon M' \longrightarrow \mathcal{F}$ a linear map. We have to show that φ can be extended to a linear map $\psi\colon M \longrightarrow \mathcal{F}$ with $\psi|_{M'} = \varphi$. The proof is nonconstructive and uses Zorn's lemma (see Result 3.2.7) to obtain a maximal possible extension.

For this purpose we consider the ordered set of all linear extensions of $\varphi\colon M' \longrightarrow \mathcal{F}$ to submodules of M, more precisely,

$$I := \{(U, \psi);\ M' \leq_{\mathcal{D}} U \leq_{\mathcal{D}} M,\ \psi\colon U \longrightarrow \mathcal{F} \text{ with } \psi|_{M'} = \varphi\}.$$

The set I is partially ordered by

$$(U_1, \psi_1) \leq (U_2, \psi_2) :\Longleftrightarrow U_1 \subseteq U_2 \text{ and } \psi_2|_{U_1} = \psi_1.$$

We apply Zorn's lemma to this ordered set I. It is not empty since it contains (M', φ). Let $J \subseteq I$ be a nonempty strictly ordered subset of I. The union $U_0 := \bigcup_{(U,\psi)\in J} U$ contains M' and, since J is strictly ordered, it is a submodule of M according to Lemma 3.2.8. Define the map

$$\psi_0 \colon U_0 = \bigcup_{(U,\psi)\in J} U \longrightarrow \mathcal{F} \text{ by } \psi_0(u_0) := \psi_1(u_0) \text{ if } u_0 \in U_1,\ (U_1, \psi_1) \in J.$$

This definition is independent of the choice of (U_1, ψ_1). Indeed, if

$$(U_1, \psi_1),\ (U_2, \psi_2) \in J \text{ with } u_0 \in U_1 \cap U_2,$$

we can always assume without loss of generality that $(U_1, \psi_1) \leq (U_2, \psi_2)$ since J is strictly ordered. Hence, we obtain $\psi_1 = \psi_2|_{U_1}$ and thus $\psi_1(u_0) = \psi_2(u_0) = \psi_0(u_0)$.

The $\mathcal{D}$-linearity of ψ_0 is clear. By construction, ψ_0 extends all ψ for $(U, \psi) \in J$ which, in turn, extend $\varphi\colon M' \longrightarrow \mathcal{F}$ by definition of I. Therefore, (U_0, ψ_0) is contained in I and an upper bound of J.

Since this holds for every chain J in I, Zorn's lemma implies that the set I contains a maximal element. We denote this maximal element by $(\widetilde{U}, \widetilde{\psi})$. We have to show that $\widetilde{U} = M$. Then, by construction, the map $\widetilde{\psi}\colon M = \widetilde{U} \longrightarrow \mathcal{F}$ is the desired extension of φ to the module M.

Assume that $\widetilde{U} \subsetneq M$, let $\widehat{u} \in M \setminus \widetilde{U}$ and define $\widehat{U} := \widetilde{U} + \mathcal{D}\widehat{u} \supsetneq \widetilde{U}$. Notice that the sum $\widehat{U} = \widetilde{U} + \mathcal{D}\widehat{u}$ is not direct in general. We are going to extend $\widetilde{\psi}$ to $\widehat{\psi}\colon \widehat{U} \longrightarrow \mathcal{F}$ in contradiction to the maximality of $(\widetilde{U}, \widetilde{\psi})$. First, we give a motivation for the form of $\widehat{\psi}$. Assume that the extension

$$\begin{aligned}\widehat{\psi}\colon \widehat{U} = \widetilde{U} + \mathcal{D}\widehat{u} &\longrightarrow \mathcal{F},\\ x = \widetilde{u} + d\widehat{u} &\longmapsto \widehat{\psi}(x) = \widehat{\psi}(\widetilde{u}) + d\widehat{\psi}(\widehat{u}) = \widetilde{\psi}(\widetilde{u}) + d\widehat{\psi}(\widehat{u}),\end{aligned}$$

of $\widetilde{\psi}$ exists. If, in particular, $a\widehat{u} \in \widetilde{U} \cap \mathcal{D}\widehat{u}$, then $a\widehat{\psi}(\widehat{u}) = \widehat{\psi}(a\widehat{u}) = \widetilde{\psi}(a\widehat{u})$.

We use this observation for constructing an extension $\widehat{\psi}$ of $\widetilde{\psi}$. Namely, we consider the equivalence class $\overline{\widehat{u}} = \widehat{u} + \widetilde{U} \in M/\widetilde{U}$ and define the ideal

$$\mathfrak{a} := \operatorname{ann}_{\mathcal{D}}(\overline{\widehat{u}}) := \{a \in \mathcal{D};\ a\overline{\widehat{u}} = \overline{0}\} = \{a \in \mathcal{D};\ a\widehat{u} \in \widetilde{U}\} \leq \mathcal{D}$$

as well as the linear map

$$\chi\colon \mathfrak{a} \longrightarrow \mathcal{F},\quad a \longmapsto \widetilde{\psi}(a\widehat{u}).$$

Using the assumption in item 2 of the theorem, there is a linear extension $\gamma\colon \mathcal{D} \longrightarrow \mathcal{F}$ of χ, i.e., $\gamma|_{\mathfrak{a}} = \chi$. With this, we define

$$\widehat{\psi}\colon \widehat{U} = \widetilde{U} + \mathcal{D}\widehat{u} \longrightarrow \mathcal{F},$$
$$x = \widetilde{u} + d\widehat{u} \longmapsto \widehat{\psi}(x) := \widetilde{\psi}(\widetilde{u}) + \gamma(d).$$

We have to show that the map $\widehat{\psi}$ is well defined, i.e., that it is independent of the non-unique representation $x = \widetilde{u} + d\widehat{u}$ of x. Assume that

$$x = \widetilde{u}_1 + d_1\widehat{u} = \widetilde{u}_2 + d_2\widehat{u}.$$

Then $\widetilde{u}_1 - \widetilde{u}_2 = (d_2 - d_1)\widehat{u} \in \widetilde{U}$. Therefore, $d_2 - d_1 \in \mathfrak{a}$ and we have that

$$\widetilde{\psi}(\widetilde{u}_1 - \widetilde{u}_2) = \widetilde{\psi}\big((d_2 - d_1)\widehat{u}\big) = \chi(d_2 - d_1) = \gamma(d_2 - d_1).$$

Hence, $\widetilde{\psi}(\widetilde{u}_1) + \gamma(d_1) = \widetilde{\psi}(\widetilde{u}_2) + \gamma(d_2)$ and, consequently, the map $\widehat{\psi}$ is well defined. Clearly, $\widehat{\psi}$ is linear and an extension of $\widetilde{\psi}$ and thus also of φ. We conclude that $(\widehat{U}, \widehat{\psi}) \in I$ and $(\widetilde{U}, \widetilde{\psi}) \leq (\widehat{U}, \widehat{\psi})$.

But this contradicts the maximality of $(\widetilde{U}, \widetilde{\psi})$ in I. Therefore, $\widetilde{U} = M$ and φ has the extension $\widetilde{\psi}$ to M. □

Corollary 3.2.10 *Every vector space $\mathcal{F}$ over a field K is an injective K-module. If, in particular, M' is a subspace of a K-space M, then every linear function $\varphi\colon M' \longrightarrow \mathcal{F}$ can be linearly extended to M. For $\mathcal{F} = K$ this is the Hahn–Banach theorem for discrete vector spaces.*

Proof Since K has only the trivial ideals K and 0, condition 2 of the preceding theorem is clearly satisfied. □

Notice that the vector space M in the preceding corollary is not assumed finite dimensional. For finite-dimensional vector spaces the result is proven by extension of finite bases of subspaces—a standard result which is shown in every course on linear algebra.

Corollary 3.2.11 *For a module $\mathcal{F}$ over a principal ideal domain $\mathcal{D}$ the following properties are equivalent:*

1. *The module $\mathcal{F}$ is injective.*
2. *The module $\mathcal{F}$ is divisible.*
3. *For every ideal $\mathfrak{a}$ of $\mathcal{D}$ and every linear map $\varphi\colon \mathfrak{a} \longrightarrow \mathcal{F}$ there is $y \in \mathcal{F}$ with $\varphi(a) = a \circ y$ for all $a \in \mathfrak{a}$.*

In particular, this applies to the polynomial ring $\mathcal{D} = F[s]$ in one indeterminate over a field F and to the ring $\mathbb{Z}$ of integers.

Proof $1. \underset{\text{Theorem. 3.2.9}}{\Longleftrightarrow} 3. \underset{\text{Corollary 3.2.4}}{\Longleftrightarrow} 2.$ □

Corollary 3.2.12 *Let $F = \mathbb{R}$ or $F = \mathbb{C}$. In the continuous case, the $F[s]$-module ${}_{F[s]}\mathrm{C}^\infty(\mathbb{R}, F)$ is divisible according to Theorem 3.1.5 and thus injective by Corollary 3.2.11. Likewise, for arbitrary fields F, in the discrete case the $F[s]$-module ${}_{F[s]}F^{\mathbb{N}}$ is also injective.* ◊

3.2.2 *The Fundamental Principle*

In this section we reformulate the injectivity as the fundamental principle and show how the latter can be used constructively, cf. Sect. 2.1.

We need the concepts of Noetherian rings and modules and recall their definitions as well as their basic properties, cf. [6, Ch. X].

Definition 3.2.13 (Cf. [6, Ch. VI]) Let $\mathcal{D}$ be a commutative ring. A $\mathcal{D}$-module M is *Noetherian* if it satisfies the following equivalent conditions:

1. *Ascending chain condition*: Every ascending chain
$$U_0 \subseteq U_1 \subseteq U_2 \subseteq \cdots \subseteq M$$
of submodules U_n, $n \in \mathbb{N}$, of M becomes stationary, i.e., there exists an index n_0 such that $U_n = U_{n_0}$ for all $n \geq n_0$.
2. Every *properly* ascending chain
$$U_0 \subsetneq U_1 \subsetneq U_2 \subsetneq \cdots \subseteq M$$
of submodules is finite.
3. *Maximality condition*: Every nonempty set I of submodules of M contains a submodule U which is maximal in I. (In general, this is not a maximal submodule of M.)
4. Every submodule of M is finitely generated.

The commutative ring $\mathcal{D}$ is *Noetherian* if it has this property as module over itself, i.e., if the conditions of item 1 hold for ideals of $\mathcal{D}$. ◊

Result 3.2.14 *1. Condition 4 of Definition 3.2.13 implies that every principal ideal domain is Noetherian.*
2. *Every finitely generated module M over a commutative Noetherian ring $\mathcal{D}$ is Noetherian, i.e., all its submodules are also finitely generated. In particular, this holds for the finite-dimensional free module $\mathcal{D}^{1\times l}$ with its finite standard basis $\delta_1, \ldots, \delta_l$.*
3. Hilbert's Basissatz *[6, Thm. 4.1 and Cor. 4.2 on pp. 186–187]: The polynomial ring $A[s_1, \ldots, s_r]$ in finitely many indeterminates s_ρ over a commutative Noetherian ring A is also Noetherian.* ◊

For the fundamental principle, we assume that $\mathcal{D}$ is a commutative Noetherian ring, not necessarily a domain.

Consider a matrix $R^0 \in \mathcal{D}^{l\times m}$ as the linear map $\cdot R^0\colon \mathcal{D}^{1\times l} \longrightarrow \mathcal{D}^{1\times m}$ and its kernel $\ker(\cdot R^0)$. From Motivation and Definition 3.2.1 we know that the vectors in this kernel are the linear relations or the syzygies of the rows of R^0, and therefore this kernel is called the syzygy module or module of relations. According to 2. resp. 4. of Results 3.2.14 resp. 3.2.13 the syzygy module is finitely generated. Let $R^1 \in \mathcal{D}^{k\times l}$ be a matrix whose rows generate this module, i.e.,

$$\mathcal{D}^{1\times k}R^1 = \sum_{i=1}^{k} \mathcal{D}R^1_{i-} = \ker(\cdot R^0\colon \mathcal{D}^{1\times l} \longrightarrow \mathcal{D}^{1\times m}). \tag{3.45}$$

In other words, the sequence

$$\mathcal{D}^{1\times k} \xrightarrow{\cdot R^1} \mathcal{D}^{1\times l} \xrightarrow{\cdot R^0} \mathcal{D}^{1\times m} \tag{3.46}$$

is exact. In particular, the matrix R^1 satisfies $R^1R^0 = 0$, i.e., it is a left annihilator of R^0.

Definition 3.2.15 If two matrices R^0 and R^1 define an exact sequence (3.46), then R^1 is called a *universal left annihilator* of R^0. ◊

If R^1 is a left annihilator of R^0, its rows are contained in the syzygy module $\ker(\cdot R^0)$. If R^1 is even universal, every syzygy ξ of R^0, i.e., with $\xi R^0 = 0$, has the form $\xi = \eta R^1 = \sum_{i=1}^{k} \eta_i R^1_{i-}$ for some $\eta \in \mathcal{D}^{1\times k}$, and hence it is a linear combination of the generating syzygies R^1_{i-}. Insofar, the rows of R^1 are universal syzygies.

Theorem 3.2.16 (Fundamental principle) *Let $\mathcal{F}$ be a module over a commutative Noetherian ring $\mathcal{D}$. Then the following assertions are equivalent:*

1. *The module $\mathcal{F}$ is injective.*
2. *For every matrix $R^0 \in \mathcal{D}^{l\times m}$ and every universal left annihilator $R^1 \in \mathcal{D}^{k\times l}$ of R^0, the system $R^0 \circ y = u \in \mathcal{F}^l$ has a solution $y \in \mathcal{F}^m$ if and only if $R^1 \circ u = 0$. In other words, the necessary compatibility condition $R^1 \circ u = 0$ for the solvability of $R^0 \circ y = u$ is also sufficient.*

If these conditions hold, we say that the fundamental principle *is valid for $\mathcal{F}$. This terminology is due to Ehrenpreis (1961) in context with linear systems of partial differential equations with constant coefficients.*

Proof 1. $\Longrightarrow$ 2.: The injectivity of $\mathcal{F}$ signifies that the duality functor $\mathrm{D} := \mathrm{Hom}_{\mathcal{D}}(-, \mathcal{F})$ preserves exact sequences. We apply D to the exact sequence (3.46) and obtain the diagram

$$\begin{array}{ccccc} \mathrm{D}\mathcal{D}^{1\times m} & \xrightarrow{\mathrm{D}(\cdot R^0)} & \mathrm{D}\mathcal{D}^{1\times l} & \xrightarrow{\mathrm{D}(\cdot R^1)} & \mathrm{D}\mathcal{D}^{1\times k} \\ \cong\downarrow \mathrm{can}_m & & \cong\downarrow \mathrm{can}_l & & \cong\downarrow \mathrm{can}_k \\ \mathcal{F}^m & \xrightarrow{R^0\circ} & \mathcal{F}^l & \xrightarrow{R^1\circ} & \mathcal{F}^k \end{array} \tag{3.47}$$

The exactness of D implies the exactness of the upper row of this diagram, whereas Lemma 3.1.34 implies its commutativity. Therefore, also the lower row is exact, i.e.,

$$R^0 \circ \mathcal{F}^m = \{u \in \mathcal{F}^l;\ R^1 \circ u = 0\}.$$

But this signifies that there is y with $u = R^0 \circ y$ if and only if $R^1 \circ u = 0$.
2. $\Longrightarrow$ 1.: We use Baer's criterion 3.2.9 to show that $\mathcal{F}$ is injective. Let $\mathfrak{a}$ be an ideal of the Noetherian ring $\mathcal{D}$ with the generators $a_1, \ldots, a_l$ and let $\varphi\colon \mathfrak{a} \longrightarrow \mathcal{F}$ be a linear map. Define $R^0 := \begin{pmatrix} a_1 \\ \vdots \\ a_l \end{pmatrix} \in \mathcal{D}^{l\times 1}$ and let $R^1 \in \mathcal{D}^{k\times l}$ be a universal left annihilator of R^0. Let $u := \begin{pmatrix} \varphi(a_1) \\ \vdots \\ \varphi(a_l) \end{pmatrix} = \varphi(R^0) \in \mathcal{F}^l$. Clearly,

$$R^1 \circ u = R^1 \circ \varphi(R^0) = \varphi(R^1 R^0) = \varphi(0) = 0$$

holds. By condition 2 of the theorem, this relation implies the existence of $y \in \mathcal{F}$ with $u = R^0 \circ y$, i.e., $\varphi(a_i) = u_i = a_i \circ y$ for $i = 1, \ldots, l$. Since the a_i generate the ideal $\mathfrak{a}$, the preceding equation implies

$$\varphi(a) = a \circ y \text{ for all } a \in \mathfrak{a}.$$

This holds for every ideal $\mathfrak{a}$ of $\mathcal{D}$. Therefore, Baer's criterion (condition 2 of Theorem 3.2.9) is satisfied and $\mathcal{F}$ is injective. □

In Theorem 3.2.18, we describe an algorithm for the computation of universal left annihilators over principal ideal domain. This allows to use the fundamental principle constructively. For this we need the Smith form of matrices $R \in \mathcal{D}^{k\times l}$ over principal ideal domain $\mathcal{D}$ and recall its definition and its basic properties, cf. [7, §XV.108], [6, §XV.2]. In Corollary 3.2.19 we extend the Smith form to matrices over the quotient field $K := \left\{\frac{f}{g};\ f, g \in \mathcal{D},\ g \neq 0\right\}$.

A *unit* of $\mathcal{D}$ is a multiplicatively invertible element of $\mathcal{D}$. We denote the group of units of $\mathcal{D}$ by $\mathrm{U}(\mathcal{D})$. For systems theory the standard case is the polynomial ring $\mathcal{D} := F[s]$ over a field F with $\mathrm{U}(F[s]) = F \setminus \{0\}$. The simplest case is given by the ring $\mathcal{D} = \mathbb{Z}$ of integers with $\mathrm{U}(\mathbb{Z}) = \{+1, -1\}$.

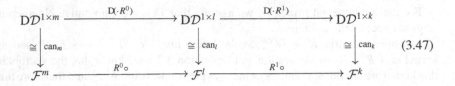

If $f, g \in \mathcal{D}$, then f *divides* g or is a *divisor* of g , in short $f \mid g$ or $f \underset{\mathcal{D}}{\mid} g$, if and only if $\mathcal{D}g \subseteq \mathcal{D}f$, i.e., if there is $d \in \mathcal{D}$ with $g = df$. Similarly, f is *associated* with g, in short $f \sim g$, if and only if $\mathcal{D}f = \mathcal{D}g$, i.e., if there is $u \in \mathrm{U}(\mathcal{D})$ with $g = uf$. The *greatest common divisor* $\gcd(f, g)$ of f and g is defined by the equation $\mathcal{D}\gcd(f,g) = \mathcal{D}f + \mathcal{D}g$, and the *least common multiple* $\mathrm{lcm}(f,g)$ by $\mathcal{D}\,\mathrm{lcm}(f,g) = \mathcal{D}f \cap \mathcal{D}g$. As usual the gcd and the lcm are unique up to association, i.e., up to units in $\mathcal{D}$. They satisfy the rules

$$\gcd(hf, hg) = h\gcd(f,g) \text{ and } \mathrm{lcm}(hf, hg) = h\,\mathrm{lcm}(f,g) \text{ for } f, g, h \in \mathcal{D}.$$

We will use the following notation. Let $k, l, r \in \mathbb{N}$ with $r \le \min(k,l)$ and let $e_1, \dots, e_r$ be elements of $\mathcal{D}$. We define the diagonal matrices

$$\mathrm{diag}(e_1, \dots, e_r) := \begin{pmatrix} e_1 & & 0 \\ & \ddots & \\ 0 & & e_r \end{pmatrix} \in \mathcal{D}^{r\times r} \text{ and}$$
$$\begin{pmatrix} \mathrm{diag}(e_1,\dots,e_r) & 0 \\ 0 & 0 \end{pmatrix} \in \mathcal{D}^{k\times l} = \mathcal{D}^{(r+(k-r))\times(r+(l-r))}.$$

The numbers k and l do not appear explicitly in the notation, but are inferred from the context.

Two matrices R and S are called *equivalent* if there are invertible matrices X, Y such that $S = XRY$. If $R \in \mathcal{D}^{k\times l}$ and $i \le \min(k,l)$, the greatest common divisor d_i of all $i \times i$ minors, i.e., of the determinants of $i \times i$-submatrices of R, is called the i-th *determinantal divisor* of R. It is unique up to association. Equivalent matrices have the same determinantal divisors (up to association). The rank of R over the quotient field K of $\mathcal{D}$, denoted as $\mathrm{rank}(R)$, satisfies

$$\mathrm{rank}(R) = \max\{i \le \min(k,l);\ d_i \neq 0\}.$$

Result 3.2.17 (Smith form) *Let $\mathcal{D}$ be a principal ideal domain and $R \in \mathcal{D}^{k\times l}$.*

1. *The matrix R is equivalent to a $k \times l$ diagonal matrix $S = \begin{pmatrix} \mathrm{diag}(e_1,\dots,e_r,0) & 0 \\ 0 & 0 \end{pmatrix}$ such that the e_i are nonzero and satisfy the divisibility relations*

$$e_1|e_2|\cdots|e_r \neq 0, \text{ in other words, } \mathcal{D}e_1 \supseteq \mathcal{D}e_2 \supseteq \cdots \supseteq \mathcal{D}e_r \supsetneq 0, \text{ hence}$$
$$\exists X \in \mathrm{Gl}_k(\mathcal{D}),\ Y \in \mathrm{Gl}_l(\mathcal{D}):\ XRY = \begin{pmatrix} \mathrm{diag}(e_1,\dots,e_r,0) & 0 \\ 0 & 0 \end{pmatrix}.$$

   In general, the matrices R and S are not similar.
2. *The number r is the rank of R and the elements $e_1, \dots, e_r$ are unique up to association, i.e., the principal ideals $\mathcal{D}e_i$ are unique. Let d_i, $i \le \min(k,l)$, be the determinantal divisors of R and hence of S. Then*

$$d_1 = e_1,\ d_2 = e_1e_2,\ \dots,\ d_r = e_1e_2\cdots e_r, \text{ i.e., } e_i = \frac{d_i}{d_{i-1}} \text{ for } i = 2, \dots, r.$$

According to the seminal and worldwide influential book [7, §XV.8], originally from 1933, the elements e_i are called the elementary divisors *of the matrix R and of the row module $\mathcal{D}^{1\times k}R$. In the engineering literature they are also called* invariant factors. *The matrix S is called the* Smith form *of R.*

For the rings $\mathcal{D} = F[s]$ resp. $\mathcal{D} = \mathbb{Z}$ one requires that the elementary divisors are monic polynomials, i.e., have leading coefficient 1, *resp. are positive. Then the Smith forms over $F[s]$ and $\mathbb{Z}$ are unique.*

3. *Two $k \times l$-matrices are equivalent if and only if they have the same Smith form.* ◊

All Smith forms of this book are derived from that for the polynomial algebra $F[s]$ which is implemented in all computer algebra systems for $F = \mathbb{Q}, \mathbb{R}, \mathbb{C}$.

Theorem 3.2.18 (Construction of universal left annihilators) *Let $\mathcal{D}$ be a principal ideal domain. Let $R \in \mathcal{D}^{k\times l}$ be a matrix, $S = XRY$ its Smith form as in Result 3.2.17 and $r :=$ rank(R) its rank. We write $X \in \mathrm{Gl}_k(\mathcal{D})$ in block form as $X = \begin{pmatrix} X^1 \\ X^2 \end{pmatrix} \in \mathcal{D}^{(r+(k-r))\times k}$. Then X^2 is a universal left annihilator of R, i.e., the sequence $0 \longrightarrow \mathcal{D}^{1\times(k-r)} \xrightarrow{\cdot X^2} \mathcal{D}^{1\times k} \xrightarrow{\cdot R} \mathcal{D}^{1\times l}$ is exact. If $r = k$, i.e., if the rows of R are linearly independent, the map $\cdot R$ is injective and $X^2 = 0$.*

Proof Since the rows of the invertible matrix are linearly independent, the matrix X^2 has rank $k - r$ and the map $\cdot X^2$ is indeed a monomorphism. We have to show that

$$\mathcal{D}^{1\times(k-r)}X^2 = \{\xi \in \mathcal{D}^{1\times k};\ \xi R = 0\}.$$

Let $\xi \in \mathcal{D}^{1\times k}$. Since $X \in \mathrm{Gl}_k(\mathcal{D})$, its inverse X^{-1} exists and has entries in $\mathcal{D}$. Therefore, we define

$$(\eta^1, \eta^2) := \eta := \xi X^{-1} \in \mathcal{D}^{1\times(r+(k-r))}.$$

We write $S = \begin{pmatrix} S^1 & 0 \\ 0 & 0 \end{pmatrix} \in \mathcal{D}^{(r+(k-r))\times(r+(l-r))}$ with $S^1 = \mathrm{diag}(e_1, \ldots, e_r)$. Then

$$\begin{aligned}
\xi R = 0 &\underset{Y\in\mathrm{Gl}_l(\mathcal{D})}{\iff} \xi X^{-1}XRY = 0 \\
&\underset{XRY=S}{\iff} \eta S = (\eta^1, \eta^2)\begin{pmatrix} S^1 & 0 \\ 0 & 0 \end{pmatrix} = (\eta^1 S^1, 0) = 0 \\
&\iff \eta^1 S^1 = 0,\ \text{i.e., } \forall i \in \{1, \ldots, r\}:\ \eta_i e_i = 0 \\
&\underset{e_i\neq 0 \text{ for } i=1,\ldots,r}{\iff} \eta^1 = 0 \iff \eta = (0, \eta^2) \in \mathcal{D}^{1\times(r+(k-r))} \\
&\iff \xi = \eta X = (0, \eta^2)\begin{pmatrix} X^1 \\ X^2 \end{pmatrix} = \eta^2 X^2 \in \mathcal{D}^{1\times(k-r)}X^2
\end{aligned}$$

as asserted. □

The following extension of the Smith form from Result 3.2.17 to matrices over the quotient field $K^{k\times l}$ is used at several instances later on. We start by extending the divisibility theory to K. Let $\mathfrak{a} \subseteq K$ be a *$\mathcal{D}$-fractional ideal*, i.e., a $\mathcal{D}$-finitely generated submodule of ${}_{\mathcal{D}}K$. Assume $\mathfrak{a} = \sum_{i=1}^{n} \mathcal{D}h_i,\ h_i \in \mathfrak{a}$. We write $h_i = a_i b^{-1} \in K$ with $a_i \in \mathcal{D}$ and a common denominator $0 \neq b \in \mathcal{D}$ for $i = 1, \ldots, n$. Let $a := \gcd(a_1, \ldots, a_n)$. Then

$$\mathfrak{a} = \sum_{i=1}^{n} \mathcal{D}a_i b^{-1} = b^{-1}\left(\sum_{i=1}^{n} \mathcal{D}a_i\right) = b^{-1}\mathcal{D}a = \mathcal{D}ab^{-1}.$$

Hence every fractional ideal is a free module of dimension one. This permits to transfer $\mathcal{D}$-divisibility from $\mathcal{D}$ to K.

If $f, g \in K$, then f *divides* g *with respect to* $\mathcal{D}$, in short, $f \underset{\mathcal{D}}{|} g$, if and only if $\mathcal{D}g \subseteq \mathcal{D}f$, i.e., if and only if there is $d \in \mathcal{D}$ with $g = df$. Notice that for nonzero f the equation $g = (gf^{-1})f$ always holds in the field K, but gf^{-1} does not belong to $\mathcal{D}$ in general. Similarly, f is *associated* with g, in short $f \underset{\mathcal{D}}{\sim} g$, if and only if $\mathcal{D}f = \mathcal{D}g$ or $g = uf$ for some $u \in \mathrm{U}(\mathcal{D})$. The *greatest common divisor* $\gcd(f, g)$ of $f, g \in K$ is again defined by the equation $\mathcal{D}\gcd(f, g) = \mathcal{D}f + \mathcal{D}g$ and the *least common multiple* $\mathrm{lcm}(f, g)$ by $\mathcal{D}\,\mathrm{lcm}(f, g) = \mathcal{D}f \cap \mathcal{D}g$. As usual, the gcd and the lcm are unique up to association. They satisfy the rules

$$\gcd(hf, hg) = h\gcd(f, g) \quad \text{and} \quad \mathrm{lcm}(hf, hg) = h\,\mathrm{lcm}(f, g) \quad \text{for } f, g, h \in K.$$

Corollary 3.2.19 (Smith form or Smith/McMillan form)

1. *We assume a matrix* $R \in K^{k\times l}$. *Let*

$$\begin{aligned} &R_{ij} = \frac{a_{ij}}{b_{ij}} \text{ with } a_{ij}, b_{ij} \in \mathcal{D} \text{ and } \gcd(a_{ij}, b_{ij}) = 1. \text{ Then} \\ &\{d' \in \mathcal{D};\ d'R \in \mathcal{D}^{k\times l}\} = \mathcal{D}d \text{ with } d := \mathrm{lcm}_{i,j}(b_{ij}),\ dR \in \mathcal{D}^{k\times l}. \end{aligned} \tag{3.48}$$

The Smith form of dR is

$$\begin{aligned} &X(dR)Y = S' = \begin{pmatrix} \mathrm{diag}(e'_1,\ldots,e'_r) & 0 \\ 0 & 0 \end{pmatrix} \in \mathcal{D}^{k\times l}, \\ &\text{where } r = \mathrm{rank}(R),\ X \in \mathrm{Gl}_k(\mathcal{D}),\ Y \in \mathrm{Gl}_l(\mathcal{D}), \text{ and } e'_1 \mid \cdots \mid e'_r \neq 0 \text{ in } \mathcal{D}, \end{aligned}$$

hence

$$\begin{aligned} XRY &= \begin{pmatrix} \mathrm{diag}(e_1,\ldots,e_r) & 0 \\ 0 & 0 \end{pmatrix} =: S \in K^{k\times l} \\ \text{with } e_1 &:= \tfrac{e'_1}{d} \underset{\mathcal{D}}{|} e_2 := \tfrac{e'_2}{d} \underset{\mathcal{D}}{|} \cdots \underset{\mathcal{D}}{|} e_r := \tfrac{e'_r}{d} \neq 0 \text{ in } K. \end{aligned} \tag{3.49}$$

Again, for $i = 1, \ldots, r$, *the product* $d_i = e_1 \cdots e_i$ *is the i-th determinantal divisor of R, and the* e_i *are thus unique up to association. In other words, the fractional ideals* $\mathcal{D}e_i$ *are unique. The matrix S is called the* Smith form *or* Smith/McMillan form *of R with respect to* $\mathcal{D}$, *and the* e_i *are its* elementary divisors. *This definition of the Smith form and the elementary divisors of* $R \in K^{k\times l}$ *depend on* $\mathcal{D}$. *A different subring* $\mathcal{D}' \subseteq K$ *with* $\mathrm{quot}(\mathcal{D}') = K$ *may lead to a different Smith form.*

2. *If* $e_i := \frac{a_i}{b_i}$ *with* $\gcd(a_i, b_i) = 1$ *for* $i = 1, \ldots, r$, *then*

$$b_r \mid b_{r-1} \mid \cdots \mid b_1 \ \ \textit{and} \ \ a_1 \mid a_2 \cdots \mid a_r, \ \textit{or} \\ \mathcal{D}b_1 \subseteq \cdots \subseteq \mathcal{D}b_r \ \ \textit{and} \ \ \mathcal{D}a_1 \supseteq \cdots \supseteq \mathcal{D}a_r. \tag{3.50}$$

Proof The assertions of item 1 are already proven in the text of the corollary. As for the second item, $e_i \underset{\mathcal{D}}{\mid} e_{i+1}$ signifies that there is a $d_i \in \mathcal{D}$ with $e_{i+1} = d_i e_i$. Thus, $a_{i+1}b_i = d_i a_i b_{i+1}$. Now we use that $\gcd(a_i, b_i) = \gcd(a_{i+1}, b_{i+1}) = 1$ and obtain that $a_i \mid a_{i+1}$ and $b_{i+1} \mid b_i$. □

3.2.3 Images of Behaviors and Elimination

In this section, we use the fundamental principle to show that the image of a behavior under a matrix is again a behavior. An important application of this is to the elimination of latent variables according to Willems [8, Thm. 6.2.6].

We assume that $\mathcal{D}$ is a Noetherian domain and that $\mathcal{F}$ is an injective $\mathcal{D}$-module. The standard continuous and discrete cases with $\mathcal{D} = F[s]$ over a field F and $\mathcal{F}_{\text{cont}} := {}_{F[s]}\mathrm{C}^\infty(\mathbb{R}, F)$ for $F = \mathbb{R}, \mathbb{C}$, and $\mathcal{F}_{\text{dis}} := {}_{F[s]}F^{\mathbb{N}}$ from systems theory satisfy these assumptions. Consider an $\mathcal{F}$-behavior

$$\mathcal{B}_1 = \{w_1 \in \mathcal{F}^{l_1};\ R_1 \circ w_1 = 0\} \text{ with } R_1 \in \mathcal{D}^{k_1 \times l_1} \ \text{ and } P \in \mathcal{D}^{l_2 \times l_1}. \tag{3.51}$$

Since $\mathcal{B}_1$ is a $\mathcal{D}$-submodule of $\mathcal{F}^{l_1}$ and $P\circ\colon \mathcal{F}^{l_1} \longrightarrow \mathcal{F}^{l_2}$ is $\mathcal{D}$-linear, the image

$$\mathcal{B}_2 := P \circ \mathcal{B}_1 = \{w_2 \in \mathcal{F}^{l_2};\ \exists w_1 \in \mathcal{B}_1 \text{ with } P \circ w_1 = w_2\}$$

is a $\mathcal{D}$-submodule of $\mathcal{F}^{l_2}$. We are going to show that $\mathcal{B}_2$ is even a behavior, i.e., it is of the form $\mathcal{B}_2 = \{w_2 \in \mathcal{F}^{l_2};\ R_2 \circ w_2 = 0\}$ for some $R_2 \in \mathcal{D}^{k_2 \times l_2}$. If R_2 exists, the equation $R_1 \circ w_1 = 0$ implies $(R_2 P) \circ w_1 = R_2 \circ (P \circ w_1) = 0$. In particular, this is true if there is a matrix X such that $R_2 P = X R_1$ or $(-X, R_2)\binom{R_1}{P} = 0$. This suggests to consider a universal left annihilator of $\binom{R_1}{P}$ for the construction of R_2 and the behavior $\mathcal{B}_2 := \{w_2 \in \mathcal{F}^{l_2};\ R_2 \circ w_2 = 0\}$.

Theorem 3.2.20 *With the data from* (3.51) *let* $(-X, R_2) \in \mathcal{D}^{k_2 \times (k_1 + l_2)}$ *be a universal left annihilator of* $\binom{R_1}{P} \in \mathcal{D}^{(k_1 + l_2) \times l_1}$. *Then*

$$\mathcal{B}_2 := \{w_2 \in \mathcal{F}^{l_2};\ R_2 \circ w_2 = 0\} = P \circ \mathcal{B}_1.$$

Proof $\mathcal{B}_2 \supseteq P \circ \mathcal{B}_1$: The left annihilator property implies

$$(-X, R_2)\binom{R_1}{P} = -X R_1 + R_2 P = 0 \Longrightarrow X R_1 = R_2 P \\ \Longrightarrow R_2 \circ (P \circ \mathcal{B}_1) = R_2 P \circ \mathcal{B}_1 = X R_1 \circ \mathcal{B}_1 = X \circ (R_1 \circ \mathcal{B}_1) = X \circ 0 = 0 \\ \Longrightarrow P \circ \mathcal{B}_1 \subseteq \mathcal{B}_2.$$

$\mathcal{B}_2 \subseteq P \circ \mathcal{B}_1$: Assume $w_2 \in \mathcal{B}_2$ or $R_2 \circ w_2 = 0$. Then $u := \binom{0}{w_2} \in \mathcal{F}^{k_1+l_2}$ is a solution of $(-X, R_2) \circ u = (-X, R_2) \circ \binom{0}{w_2} = R_2 \circ w_2 = 0$. The fundamental principle and the universality of the left annihilator $(-X, R_2)$ imply the existence of $w_1 \in \mathcal{F}^{l_1}$ such that

$$u = \binom{0}{w_2} = \binom{R_1}{P} \circ w_1 = \binom{R_1 \circ w_1}{P \circ w_1}$$
$$\Longrightarrow R_1 \circ w_1 = 0 \text{ or } w_1 \in \mathcal{B}_1 \text{ and } P \circ w_1 = w_2.$$

□

The first application of the preceding theorem is that to the *elimination of latent variables* as introduced by Willems [8]. Again, a behavior

$$\mathcal{B} = \{w \in \mathcal{F}^l;\ R \circ w = 0\},\ R \in \mathcal{D}^{k \times l}, \tag{3.52}$$

is given. Additionally, the components w_i of w are partitioned into two disjoint sets, namely, the so-called *manifest variables* and the *latent variables*. In this model the manifest variables are those in which the investigator is principally interested.

In daily life the manifest variables of a system or machine (for instance, a car, computer, TV set, cell phone, etc.) are significant for the *user* of the machine. In contrast, the latent variables are mainly important for its *constructor* who uses them to model the system and to analyze its internal working, and are, therefore, more numerous than the manifest ones in general. In electrical networks, the voltage and currents of the (external) source branches can, for instance, be chosen as manifest variables, whereas the voltages and currents of the RLC branches are the latent ones, cf. Sect. 8.3. The goal is to find a description of the system in terms of the manifest variables only and to eliminate the latent ones.

After a possible permutation of the components of w and with $l = l^1 + l^2$, we can write

$$\begin{aligned} &w = \binom{w^1}{w^2} \in \mathcal{F}^l = \mathcal{F}^{l^1+l^2},\ R = (R^1, R^2) \in \mathcal{D}^{k \times l} = \mathcal{D}^{k \times (l^1+l^2)}, \text{ and} \\ &\mathcal{B} = \left\{\binom{w^1}{w^2} \in \mathcal{F}^{l^1+l^2};\ (R^1, R^2) \circ \binom{w^1}{w^2} = R^1 \circ w^1 + R^2 \circ w^2 = 0\right\}, \end{aligned} \tag{3.53}$$

where w^1 is the vector of the latent variables and w^2 that of the manifest ones. According to the interpretation above the number l^1 of latent variables is, in general, much bigger than the number l^2 of manifest variables. The image $\mathcal{B}^{\text{man}} := P \circ \mathcal{B}$ of $\mathcal{B}$ under the projection

$$\begin{aligned} &P\circ := (0, \text{id}_{l_2})\circ\colon \mathcal{F}^{l^1+l^2} \longrightarrow \mathcal{F}^{l^2},\ \binom{w_1}{w_2} \longmapsto w_2, \text{ i.e.,} \\ &\mathcal{B}^{\text{man}} = \left\{w^2 \in \mathcal{F}^{l^2};\ \exists w^1 \in \mathcal{F}^{l^1} \text{ with } R^1 \circ w^1 + R^2 \circ w^2 = 0\right\} \end{aligned}$$

is a behavior according to Theorem 3.2.20 and is called the *manifest behavior* of $\mathcal{B}$ with respect to the chosen decomposition of the variables.

To motivate the assertion of the next Theorem 3.2.21 assume that $R^1 \circ w^1 + R^2 \circ w^2 = 0$ and that X is a left annihilator of R^1. With this, the equation $XR^1 = 0$ implies that $XR^1 \circ w^1 + XR^2 \circ w^2 = XR^2 \circ w^2 = 0$, and hence

$$\mathcal{B}^{\mathrm{man}} \subseteq \{w^2 \in \mathcal{F}^{l^2};\ XR^2 \circ w^2 = 0\}.$$

This inclusion suggests to consider a *universal* left annihilator X of R^1.

Theorem 3.2.21 (Elimination of latent variables) *[8, Thm. 6.2.6, p. 206]. We assume the behavior* (3.52) *with the chosen manifest-latent decomposition* (3.53). *Let* $X \in \mathcal{D}^{k^2 \times k}$ *be a universal left annihilator of* R^1 *and define* $R^{\mathrm{man}} := XR^2$. *Then*

$$\mathcal{B}^{\mathrm{man}} = (0, \mathrm{id}_{l^2}) \circ \mathcal{B} = \{w^2 \in \mathcal{F}^{l^2};\ R^{\mathrm{man}} \circ w^2 = 0\}.$$

Proof We use Theorem 3.2.20 and consider the matrix $\begin{pmatrix} R \\ P \end{pmatrix} = \begin{pmatrix} R^1 & R^2 \\ 0 & \mathrm{id}_{l^2} \end{pmatrix}$. Then

$$(X, -XR_2)\begin{pmatrix} R^1 & R^2 \\ 0 & \mathrm{id}_{l^2} \end{pmatrix} = (XR^1, XR^2 - XR^2) = (0, 0) = 0,$$

i.e., $(X, -XR_2) \in \mathcal{D}^{k^2 \times (k+l^2)}$ is a left annihilator of $\begin{pmatrix} R \\ P \end{pmatrix}$. To show that it is universal, let $(\xi, \eta) \in \mathcal{D}^{1 \times (k+l^2)}$ be any linear relation of the rows of $\begin{pmatrix} R \\ P \end{pmatrix}$, i.e.,

$$(\xi, \eta)\begin{pmatrix} R \\ P \end{pmatrix} = (\xi, \eta)\begin{pmatrix} R^1 & R^2 \\ 0 & \mathrm{id}_{l^2} \end{pmatrix} = (\xi R^1, \xi R^2 + \eta) = (0, 0) \Longrightarrow \xi R^1 = 0,\ \xi R^2 + \eta = 0.$$

Since X is a universal left annihilator of R^1, there is $\zeta \in \mathcal{D}^{1 \times k^2}$ with $\xi = \zeta X$, hence $\eta = -\xi R^2 = -\zeta X R^2$ and $(\xi, \eta) = \zeta(X, -XR^2) \in \mathcal{D}^{1 \times k^2}(X, -XR^2)$. This signifies that $(X, -XR^2)$ is universal as asserted.

This and Theorem 3.2.20 imply that $-XR^2 \circ w^2 = 0$ and then, of course, also $XR^2 \circ w^2 = 0$ are defining equations for $\mathcal{B}^{\mathrm{man}}$. □

3.2.4 Rosenbrock Equations or Matrix Models

Another important application of Theorem 3.2.20 is to *Rosenbrock systems* or so-called *matrix models*. In the literature, these models are defined for the polynomial algebra $\mathcal{D} = F[s]$ over a field F and the standard signal spaces, and are then called *polynomial matrix models* or *polynomial matrix descriptions* or *differential operator representations*, cf. [9, 10]. Usually, these models are defined by their equations only, but we also define the associated behaviors.

Let $\mathcal{D}$ be a Noetherian integral domain and let $\mathcal{F}$ be an injective module. These conditions are satisfied in all standard cases. A matrix model is given by four matrices

$$A \in \mathcal{D}^{n \times n},\ B \in \mathcal{D}^{n \times m},\ C \in \mathcal{D}^{p \times n},\ D \in \mathcal{D}^{p \times m} \tag{3.54}$$

with $\operatorname{rank}(A) = n$, i.e., $\det(A) \neq 0$, and the two *Rosenbrock equations*

$$A \circ x = B \circ u \text{ and } y = C \circ x + D \circ u, \tag{3.55}$$

where $x \in \mathcal{F}^n$ is the *pseudo-state*, $u \in \mathcal{F}^m$ is the *input*, and $y \in \mathcal{F}^p$ is the *output* of the system. *Input/output (IO) decompositions* of arbitrary behaviors will be discussed in Chap. 6.

Equation (3.55) gives rise to two behaviors

$$\begin{aligned}
\mathcal{B}_1 &:= \left\{ \left(\begin{smallmatrix} x \\ u \end{smallmatrix}\right) \in \mathcal{F}^{n+m};\ A \circ x = B \circ u, \text{i.e., } (A, -B) \circ \left(\begin{smallmatrix} x \\ u \end{smallmatrix}\right) = 0 \right\} \text{ and} \\
\mathcal{B}_2 &:= \left(\begin{smallmatrix} C & D \\ 0 & \mathrm{id}_m \end{smallmatrix}\right) \circ \mathcal{B}_1 \\
&= \left\{ \left(\begin{smallmatrix} y \\ u \end{smallmatrix}\right) \in \mathcal{F}^{p+m};\ \exists x \in \mathcal{F}^n \text{ with } A \circ x = B \circ u \text{ and } y = C \circ x + D \circ u \right\}.
\end{aligned} \tag{3.56}$$

The second image behavior $\mathcal{B}_2$ describes the relation between input and output.

Theorem 3.2.22 (Elimination of the pseudo-state) *Assume the Rosenbrock system* (3.55) *with the matrices from* (3.54). *Let* $(-Y, P) \in \mathcal{D}^{k \times (n+p)}$ *be a universal left annihilator of* $\left(\begin{smallmatrix} A \\ C \end{smallmatrix}\right)$. *Then the behavior* $\mathcal{B}_2$ *from* (3.56) *has the form*

$$\begin{aligned}
&\mathcal{B}_2 = \left\{ \left(\begin{smallmatrix} y \\ u \end{smallmatrix}\right) \in \mathcal{F}^{p+m};\ P \circ y = Q \circ u \right\} \ \textit{with}\ Q := YB + PD, \\
&\Longrightarrow \mathcal{B}_2^0 = C \circ \mathcal{B}_1^0 \ \textit{with}\ \mathcal{B}_1^0 := \left\{ x \in \mathcal{F}^n;\ A \circ x = 0 \right\},\ \mathcal{B}_2^0 := \left\{ y \in \mathcal{F}^p;\ P \circ y = 0 \right\}.
\end{aligned}$$

Proof $\subseteq$: Let $\left(\begin{smallmatrix} y \\ u \end{smallmatrix}\right) \in \mathcal{B}_2$ and let $x \in \mathcal{F}^n$ be such that the equations (3.55) are satisfied. We multiply the first equation of (3.55) with P and the second one with Y from the left and obtain

$$YA \circ x = YB \circ u \quad \text{and} \quad P \circ y = PC \circ x + PD \circ u.$$

Since $(-Y, P) \left(\begin{smallmatrix} A \\ C \end{smallmatrix}\right) = 0$ or $YA = PC$, we conclude that

$$P \circ y = YA \circ x + PD \circ u = YB \circ u + PD \circ u = (YB + PD) \circ u = Q \circ u.$$

$\supseteq$: The opposite inclusion can be proven by applying Theorem 3.2.20, but instead we derive it directly from the fundamental principle. Let $\left(\begin{smallmatrix} y \\ u \end{smallmatrix}\right)$ be such that the equation $P \circ y = Q \circ u = (YB + PD) \circ u$ holds. Then

$$0 = P \circ y - (YB + PD) \circ u = (-Y, P) \left(\begin{smallmatrix} B \circ u \\ y - D \circ u \end{smallmatrix}\right).$$

Since $(-Y, P)$ is a universal left annihilator of $\left(\begin{smallmatrix} A \\ C \end{smallmatrix}\right)$, the fundamental principle implies the existence of $x \in \mathcal{F}^n$ with

$$\left(\begin{smallmatrix} B \circ u \\ y - D \circ u \end{smallmatrix}\right) = \left(\begin{smallmatrix} A \\ C \end{smallmatrix}\right) \circ x, \quad \text{i.e.,} \quad A \circ x = B \circ u \text{ and } y = C \circ x + D \circ u.$$

Hence, $\binom{x}{u} \in \mathcal{B}_1$ and $\binom{y}{u} = \begin{pmatrix} C & D \\ 0 & \mathrm{id}_m \end{pmatrix} \circ \binom{x}{u} \in \mathcal{B}_2$. □

3.2.5 Kalman Systems or State Space Equations

The state space systems were introduced as basic objects of systems theory by Kalman in the 1960s, cf. [11]. They are defined for the polynomial ring $\mathcal{D} := F[s]$ in one indeterminate over a field and any injective module ${}_{F[s]}\mathcal{F}$, in particular, for $\mathcal{F}_{\mathrm{cont}} = \mathrm{C}^\infty(\mathbb{R}, F)$ with $F = \mathbb{R}$ or $F = \mathbb{C}$, and for $\mathcal{F}_{\mathrm{dis}} = F^{\mathbb{N}}$ over an arbitrary field F. They are given by four matrices

$$A \in F^{n\times n},\ B \in F^{n\times m},\ C \in F^{p\times n},\ D \in F[s]^{p\times m} \tag{3.57}$$

and the equations

$$s \circ x = Ax + Bu \text{ and } y = Cx + D \circ u, \tag{3.58}$$

where $x \in \mathcal{F}^n$ ($u \in \mathcal{F}^m,\ y \in \mathcal{F}^p$) are the *state (input, output)*. In contrast to the Rosenbrock systems of Sect. 3.2.4, the matrices A, B, and C have entries in the base field F. Kalman required $D \in F^{p\times m}$ too, but later researchers like Rosenbrock [9] and Wolovich [10] made the more general assumption $D \in F[s]^{p\times m}$. We use the notation

$$ay = a \circ y \quad \text{for } a \in F \subseteq F[s] \text{ and } y \in \mathcal{F}.$$

The equations (3.58) are transformed into the Rosenbrock equations of (3.55) (note the different meaning of A in (3.55) and(3.59)) by

$$(s\,\mathrm{id}_n - A) \circ x = B \circ u = Bu \quad \text{and} \quad y = C \circ x + D \circ u = Cx + D \circ u. \tag{3.59}$$

The associated behaviors $\mathcal{B}_1$ and $\mathcal{B}_2$ according to (3.56) are

$$\begin{aligned} \mathcal{B}_1 &:= \left\{\binom{x}{u} \in \mathcal{F}^{n+m};\ s \circ x = Ax + Bu\right\} \text{ and} \\ \mathcal{B}_2 &:= \begin{pmatrix} C & D \\ 0 & \mathrm{id}_m \end{pmatrix} \circ \mathcal{B}_1 \\ &= \left\{\binom{y}{u} \in \mathcal{F}^{p+m};\ \exists x \in \mathcal{F}^n \text{ with } s \circ x = Ax + Bu \text{ and } y = Cx + D \circ u\right\}. \end{aligned} \tag{3.60}$$

From Theorem 3.2.22 we infer

$$\begin{aligned} &\mathcal{B}_2 = \left\{\binom{y}{u} \in \mathcal{F}^{p+m};\ P \circ y = Q \circ u\right\},\ (-Y, P)\begin{pmatrix} s\,\mathrm{id}_n - A \\ C \end{pmatrix},\ Q = YB + PD, \\ &\Longrightarrow C \cdot \mathcal{B}_1^0 = \mathcal{B}_2^0,\ \mathcal{B}_1^0 = \left\{x \in \mathcal{F}^n;\ s \circ x = Ax\right\},\ \mathcal{B}_2^0 = \left\{y \in \mathcal{F}^p;\ P \circ y = 0\right\}, \end{aligned} \tag{3.61}$$

where $(-Y, P) \in F[s]^{p\times(n+p)}$ is a universal left annihilator of $\begin{pmatrix} s\,\mathrm{id}_n - A \\ C \end{pmatrix}$.

The state space behaviors in the standard cases have two big advantages, a mathematical one and an engineering one: Their solutions can be easily computed, see the next theorem, and they can be realized as interconnections of simple parts. We will discuss the second property later in Sect. 7.2.

Theorem 3.2.23 *1.* Continuous time. *For $F = \mathbb{R}$ or $F = \mathbb{C}$ and a continuous function $u \in C^0(\mathbb{R}, F)^m$ the explicit differential system*

$$\begin{aligned} x' = \frac{\mathrm{d}x}{\mathrm{d}t} &= s \circ x = Ax + Bu \\ y &= Cx + Du \end{aligned} \quad \text{with the initial condition } x(0) \in F^n \qquad (3.62)$$

has the unique continuously differentiable solution

$$\begin{aligned} x(t) &= e^{tA}x(0) + \int_0^t e^{(t-\tau)A}Bu(\tau)\,\mathrm{d}\tau = e^{tA}\left(x(0) + \int_0^t e^{-\tau A}Bu(\tau)\,\mathrm{d}\tau\right), \\ y(t) &= Du(t) + Ce^{tA}x(0) + \int_0^t Ce^{(t-\tau)A}Bu(\tau)\,\mathrm{d}\tau. \end{aligned} \qquad (3.63)$$

2. Discrete time. *For an arbitrary field F and the sequence space*

$$F^{\mathbb{N}} = \{a = (a(t))_{t\in\mathbb{N}};\ a\colon \mathbb{N} \to F,\ t \mapsto a(t)\},$$

the difference system

$$\begin{aligned} x(t+1) &= (s \circ x)(t) = Ax(t) + Bu(t) \\ y(t) &= Cx(t) + Du(t) \end{aligned} \quad \text{with the initial condition } x(0) \in F^n \qquad (3.64)$$

has the unique solution

$$\begin{aligned} x(t) &= A^t x(0) + \sum_{k=0}^{t-1} A^{t-1-k}Bu(k), \\ y(t) &= Du(t) + Cx(t) = Du(t) + CA^t x(0) + \sum_{k=0}^{t-1} CA^{t-1-k}Bu(k). \end{aligned} \qquad (3.65)$$

This theorem is the only unique existence result for differential and difference equations that is used in this book, in contrast to most of the cited books. This is enabled by the restriction to the LTI case.

Proof 1. We use the standard equation $\left(e^{tA}\right)' = Ae^{tA} = e^{tA}A$. Differentiation of the equation in the first line of (3.63) gives $x' = Ax + Bu$.
Assume that x_1 and x_2 are two solutions of $x' = Ax + Bu$ with $x_1(0) = x_2(0)$ and define $x := x_1 - x_2$ and $y := e^{-tA}x$. Then

$$x' = Ax,\ x(0) = 0 \Longrightarrow y' = 0,\ y(0) = 0 \Longrightarrow y = 0,\ x = 0,\ x_1 = x_2.$$

2. This follows from (3.64) by a simple induction. □

The time $t = 0$ is interpreted as the *present*. The solution formula signifies that if the state $x(0)$ at the *present time* 0 and the *future input* $u(t)$ for $t \geq 0$ are known, the future output $y(t), t \geq 0$, is uniquely determined. The *past input* $u(t), t < 0$, enters the computation only via the state $x(0)$. In engineering, one considers $x(0)$ as the *memory* of the system at time 0. This property also justifies referring to x as the system's *state*, a terminology known from physics.

3.3 Duality and the Category of Systems

In this section we prove the one-one correspondence between the equation module $U := \mathcal{D}^{1\times k}R,\ R \in \mathcal{D}^{k\times l}$ *and its behavior* $\mathcal{B} := U^{\perp} = \left\{w \in \mathcal{F}^{l};\ R \circ w = 0\right\}$ *in case* $\mathcal{F}$ *is a cogenerator, and the ensuing duality* $M := \mathcal{D}^{1\times l}/U \longleftrightarrow \mathcal{B}$, *cf. Sect.* 2.1. *The algebraic notion of a cogenerator has thus to be studied.*

3.3.1 *Cogenerators*

Let $\mathcal{D}$ be an arbitrary commutative ring and $\mathcal{F}$ an injective $\mathcal{D}$-module with the associated contravariant duality functor $\mathrm{D} := \mathrm{Hom}_{\mathcal{D}}(-, \mathcal{F})$.

Motivation 3.3.1 Let $\mathcal{D} := F[s]$ be the polynomial ring over a field F and let $\mathcal{F}$ be one of the standard function modules in systems theory from Corollary 3.1.3. For $0 \neq f \in \mathcal{D}$ and $M := \mathcal{D}/\mathcal{D}f$, Theorem 3.1.36 implies the isomorphism

$$\mathrm{D}M \cong \mathcal{B} := \mathrm{ann}_{\mathcal{F}}(f) = \{y \in \mathcal{F};\ f \circ y = 0\}. \tag{3.66}$$

From Equations (3.7) and (3.9) and the isomorphism

$$F^{\deg(f)} \xrightarrow{\cong} M = \mathcal{D}/\mathcal{D}f,\quad \left(a_0, \ldots, a_{\deg(f)-1}\right) \longmapsto \overline{\sum_{i=0}^{\deg(f)-1} a_i s^i},$$

we infer the equality

$$\dim_F(M) = \deg(f) = \dim_F(\mathcal{B}),\ \text{hence}\ \left(M \neq 0 \Longleftrightarrow \mathcal{B} \neq 0 \Longleftrightarrow \deg(f) > 0\right). \tag{3.67}$$

In Theorem 3.3.5,2., this will be generalized to arbitrary M and $\mathcal{B}$. ◇

Before we define and characterize cogenerators, we introduce the *Gelfand map*. For a $\mathcal{D}$-module M this is defined as the $\mathcal{D}$-linear map

$$\rho_M := (\alpha)_{\alpha\in \mathrm{D}M} : M \longrightarrow \mathrm{D}^2M = \mathrm{Hom}_{\mathcal{D}}\big(\mathrm{Hom}_{\mathcal{D}}(M,\mathcal{F}),\mathcal{F}\big) \subseteq \mathcal{F}^{\mathrm{D}M}$$
$$x \longmapsto \rho_M(x) := (\alpha(x))_{\alpha\in \mathrm{D}M} = \big(\alpha \mapsto \rho_M(x)(\alpha) := \alpha(x)\big). \tag{3.68}$$

In general, ρ_M is neither injective nor surjective.

Definition and Lemma 3.3.2 (Cogenerator) *Let $\mathcal{D}$ be a commutative ring and $\mathcal{F}$ an arbitrary $\mathcal{D}$-module (not necessarily injective). The following statements are equivalent:*

1. *If a $\mathcal{D}$-linear map $\varphi\colon M_1 \longrightarrow M_2$ is nonzero, then so is $\mathrm{D}\varphi$.*
2. *For every $\mathcal{D}$-module M, the linear maps in $\mathrm{D}M$ separate the elements of M. This means that for all nonzero $x \in M$ there is $\alpha \in \mathrm{D}M$ with $\alpha(x) \neq 0$.*
3. *For every $\mathcal{D}$-module M the map $\rho_M = (\alpha)_{\alpha\in \mathrm{D}M}$ is a monomorphism.*

If $\mathcal{F}$ has these equivalent properties, it is called a cogenerator.

Proof $1 \Longrightarrow 2.$: Let $0 \neq x \in M$. We define the nonzero map

$$\varphi\colon \mathcal{D} \longrightarrow M,\quad d \longmapsto dx.$$

By assumption also $\mathrm{D}\varphi \neq 0$ is nonzero and hence there is an $\alpha \in \mathrm{D}M$ with $0 \neq (\mathrm{D}\varphi)(\alpha) = \alpha\varphi$. For $y \in \mathcal{D}$ with $(\alpha\varphi)(y) \neq 0$ this implies

$$0 \neq (\alpha\varphi)(y) = (\alpha\varphi)(y\cdot 1) = y(\alpha\varphi)(1) = y\alpha\big(\varphi(1)\big) = y\alpha(x) \Longrightarrow \alpha(x) \neq 0.$$

$2. \Longrightarrow 1.$: If $\varphi\colon M_1 \longrightarrow M_2$ is nonzero, there is an $x_1 \in M_1$ with $\varphi(x_1) \neq 0$. Then 2 implies the existence of $\alpha \in \mathrm{D}M_2$ with $0 \neq \alpha\big(\varphi(x_1)\big) = (\alpha\varphi)(x_1)$. Hence, $(\mathrm{D}\varphi)(\alpha) = \alpha\varphi \neq 0 \Longrightarrow \mathrm{D}\varphi \neq 0$.
$2. \Longleftrightarrow 3.$: This follows from $\rho_M(x) = (\alpha(x))_{\alpha\in \mathrm{D}M}$ for $x \in M$. □

In Theorem 3.3.5 we will assume that the $\mathcal{D}$-module $\mathcal{F}$ is injective, and we will obtain stronger equivalent conditions for the cogenerator property. For this, we need some preparations.

Reminder 3.3.3 1. A module S is called *simple* or *irreducible* if it is not zero and if it has exactly the two submodules 0 and S.
2. Let M be a $\mathcal{D}$-module. A submodule $M' \leq_{\mathcal{D}} M$ is called *maximal* if it is maximal in the set of all proper submodules of M. This means that $M' \lneqq_{\mathcal{D}} M$ and if U is a submodule with $M' \leq_{\mathcal{D}} U \lneqq_{\mathcal{D}} M$, then $M' = U$.
3. Let ${}_{\mathcal{D}}M$ be a module and let $U \leq_{\mathcal{D}} M$ with the canonical map $\mathrm{can}\colon M \longrightarrow M/U$. The submodules of M/U are in one-to-one correspondence with those submodules of M which contain U. More precisely, the maps

$$\begin{aligned} \{U';\ U \leq_{\mathcal{D}} U' \leq_{\mathcal{D}} M\} &\longleftrightarrow \{V';\ V' \leq_{\mathcal{D}} M/U\},\\ U' &\longmapsto \mathrm{can}(U') = U'/U,\\ \mathrm{can}^{-1}(V') &\longleftarrow V', \end{aligned} \tag{3.69}$$

are order isomorphisms, i.e., they are bijective, inverse to each other, and order-preserving (i.e., inclusion-preserving). In particular, the left side has exactly two submodules if and only if the right side does. In other words, U is a maximal submodule of M if and only if M/U is simple. An ideal $\mathfrak{m}$ of $\mathcal{D}$ is maximal if and only if $\mathcal{D}/\mathfrak{m}$ is simple as $\mathcal{D}$-module or if and only if $\mathcal{D}/\mathfrak{m}$ is a field. ◇

Lemma 3.3.4 *Every proper submodule U_0 of a finitely generated $\mathcal{D}$-module M is contained in a maximal submodule.*

Proof The proof uses Zorn's lemma (Result 3.2.7). Let I be the ordered set $I := \{U;\ U_0 \leq_{\mathcal{D}} U \lneq_{\mathcal{D}} M\}$ with the inclusion order. Obviously, $U_0 \in I$, i.e., I is nonempty. Let $J \subseteq I$ be a chain in I. According to Lemma 3.2.8, the union $U' := \bigcup_{U \in J} U$ is a submodule of M. Furthermore, U' contains U_0 since all U in J do.

We show that U' is a proper submodule of M. Assume that $U' = M$ and, since M is finitely generated, let $x_1, \ldots, x_n \in M$ with $M = \mathcal{D}x_1 + \cdots + \mathcal{D}x_n$. Each x_i is contained in U' and therefore in some $U_i \in J$. Since J is strictly ordered, there is a largest module among the U_i, say U_k, and this contains all x_i. Hence,

$$M = \mathcal{D}x_1 + \cdots + \mathcal{D}x_n \subseteq U_k \subsetneq M,$$

a contradiction. Therefore, $U' \lneq_{\mathcal{D}} M$.

Summarizing, we have shown that $U' \in I$ is an upper bound of J in I. Thus, every chain J in I has an upper bound in I. By Zorn's lemma, I contains a submodule U'' which is maximal in I, and this is obviously a maximal submodule of M which contains U_0. □

Theorem 3.3.5 characterizes the cogenerator property for *injective* signal modules. Condition 2 is the generalization of (3.67) to arbitrary M. Condition 4 will be an essential tool in the rest of the book.

Theorem 3.3.5 *For a commutative ring $\mathcal{D}$ and an injective $\mathcal{D}$-module ${}_{\mathcal{D}}\mathcal{F}$ the following statements are equivalent:*

1. *The module $\mathcal{F}$ is a cogenerator.*
2. *Every nonzero module ${}_{\mathcal{D}}M$ satisfies* $\mathrm{D}M \neq 0$.
3. *Every simple module ${}_{\mathcal{D}}S$ satisfies* $\mathrm{D}S \neq 0$. *There is even a monomorphism $\alpha\colon S \longrightarrow \mathcal{F}$, i.e., $\mathcal{F}$ contains S up to isomorphism.*
4. *The functor* D*preserves and reflects exactness, i.e., a sequence*

$$M_*\colon M_1 \xrightarrow{\varphi_1} M_2 \xrightarrow{\varphi_2} M_3 \tag{3.70}$$

is exact if and only if

$$\mathrm{D}M_*\colon \mathrm{D}M_1 \xleftarrow{\mathrm{D}\varphi_1} \mathrm{D}M_2 \xleftarrow{\mathrm{D}\varphi_2} \mathrm{D}M_3 \tag{3.71}$$

is exact.

In particular, a linear map φ is a monomorphism, epimorphism, or isomorphism if and only if $\mathrm{D}\varphi$ *is an epimorphism, monomorphism, or isomorphism, respectively.*

Proof 1. $\Longrightarrow$ 2.: If $M \neq 0$ then $\varphi := \mathrm{id}_M \neq 0$. Therefore, by 1., $0 \neq \mathrm{D}\mathrm{id}_M = \mathrm{id}_{\mathrm{D}M}$, i.e., $\mathrm{D}M \neq 0$.
2. $\Longrightarrow$ 1.: For nonzero $\varphi \in \mathrm{Hom}_{\mathcal{D}}(M_1, M_2)$ consider the factorization

$$\varphi = \mathrm{inj}\,\psi\colon M_1 \xrightarrow[\text{epi}]{\psi} M := \mathrm{im}(\psi) \xrightarrow[\text{mono}]{\text{inj}} M_2,\ \psi(x) := \varphi(x),\ M \neq 0. \tag{3.72}$$

Since D is contravariant and exact, we obtain $\mathrm{D}\varphi = (\mathrm{D}\psi)(\mathrm{D}\,\mathrm{inj})$ with the monomorphism $\mathrm{D}\psi$ and epimorphism $\mathrm{D}\,\mathrm{inj}$. Since M and hence, by 2, also $\mathrm{D}M$ are nonzero, we infer $\mathrm{D}\varphi \neq 0$ as required.
2. $\Longrightarrow$ 3.: A simple module is nonzero by assumption, and therefore property 3 is a special case of property 2.
If $\alpha\colon S \longrightarrow \mathcal{F}$ is a nonzero homomorphism, its kernel is properly contained in S and therefore zero because S is simple. Hence, α is injective.
3. $\Longrightarrow$ 2.: Let M be nonzero and $0 \neq x \in M$. According to Lemma 3.3.4 the nonzero finitely generated submodule $U := \mathcal{D}x$ of M has a maximal submodule V. Then $S := U/V$ is simple and $\mathrm{D}S$ is nonzero by 3. We apply the contravariant exact functor D to $S = U/V \xleftarrow[\text{epi}]{\text{can}} U \xrightarrow[\text{mono}]{\text{inj}} M$ and obtain $\mathrm{D}S \xrightarrow[\text{mono}]{\mathrm{Dcan}} \mathrm{D}U \xleftarrow[\text{epi}]{\mathrm{Dinj}} \mathrm{D}M$. Since $\mathrm{D}S$ is nonzero, Dcan is injective and D inj surjective, this implies $\mathrm{D}M \neq 0$. 1., 2. $\Longrightarrow$ 4.: The implication ((3.70) exact $\Longrightarrow$ (3.71) exact) follows from the assumed injectivity of $\mathcal{F}$. We show the other implication in several steps.

(a) Let $\varphi \in \mathrm{Hom}_{\mathcal{D}}(M_1, M_2)$ and assume that $\mathrm{D}\varphi$ is surjective. We show that φ is injective, i.e., that $\ker(\varphi) = 0$. The sequence

$$0 \longrightarrow \ker(\varphi) \overset{\text{inj}}{\to} M_1 \overset{\varphi}{\to} M_2$$

is exact. Since D is an exact functor, also

$$\mathrm{D}M_2 \overset{\mathrm{D}\varphi}{\to} \mathrm{D}M_1 \overset{\mathrm{Dinj}}{\longrightarrow} \mathrm{Dker}(\varphi) \longrightarrow 0$$

is exact. By assumption, $\mathrm{D}\varphi$ is surjective, and therefore $\mathrm{Dker}(\varphi) = 0$. By 2, we conclude that $\ker(\varphi) = 0$.

(b) Assume that (3.71) is exact. In particular, it is a complex and hence $\mathrm{D}(\varphi_2\varphi_1) = (\mathrm{D}\varphi_1)(\mathrm{D}\varphi_2) = 0$. By 1, we infer that $\varphi_2\varphi_1 = 0$; in other words, (3.70) is a complex too.

(c) Assume that (3.71) is exact. We show that $\mathrm{im}(\varphi_1) = \ker(\varphi_2)$ holds. Define $N := \mathrm{cok}(\varphi_1) = M_2/\mathrm{im}(\varphi_1)$. The sequence

$$M_1 \overset{\varphi_1}{\to} M_2 \overset{\text{can}}{\longrightarrow} N \longrightarrow 0 \tag{3.73}$$

is exact and, since D is an exact functor, also

$$0 \longrightarrow \mathrm{D}N \xrightarrow{\mathrm{Dcan}} \mathrm{D}M_2 \xrightarrow{\mathrm{D}\varphi_1} \mathrm{D}M_1 \tag{3.74}$$

is exact. From (b) we infer that $\mathrm{im}(\varphi_1) \subseteq \ker(\varphi_2)$. Because of the universal property of the factor module (Theorem 3.1.31), the induced map

$$\varphi_{2,\mathrm{ind}} : N \longrightarrow M_3, \quad \overline{x_2} \longmapsto \varphi_2(x_2)$$

is well defined and satisfies $\varphi_2 = \varphi_{2,\mathrm{ind}}\,\mathrm{can}$. Application of D implies

$$\mathrm{D}\varphi_2 = (\mathrm{Dcan})(\mathrm{D}\varphi_{2,\mathrm{ind}}). \tag{3.75}$$

We infer that

$$\begin{aligned}(\mathrm{Dcan})(\mathrm{D}N) &= \mathrm{im}(\mathrm{Dcan}) \overset{(3.74)\text{ exact}}{=\!=\!=} \ker(\mathrm{D}\varphi_1) \overset{(3.71)\text{ exact}}{=\!=\!=} \mathrm{im}(\mathrm{D}\varphi_2) \\ &\overset{(3.75)}{=\!=} \mathrm{im}\left((\mathrm{Dcan})(\mathrm{D}\varphi_{2,\mathrm{ind}})\right) = (\mathrm{Dcan})\left(\mathrm{im}(\mathrm{D}\varphi_{2,\mathrm{ind}})\right).\end{aligned}$$

Since can is an epimorphism and thus Dcan a monomorphism, this equation implies $\mathrm{D}N = \mathrm{im}\left(\mathrm{D}\varphi_{2,\mathrm{ind}}\right)$ and hence that $\mathrm{D}\varphi_{2,\mathrm{ind}}$ is surjective and, by (a), that $\varphi_{2,\mathrm{ind}}$ is injective. This, in turn, implies

$$\mathrm{im}(\varphi_1) \overset{(3.73)\text{ exact}}{=\!=\!=} \ker(\mathrm{can}) \overset{\varphi_{2,\mathrm{ind}}\text{ injective}}{=\!=\!=} \ker(\varphi_{2,\mathrm{ind}}\,\mathrm{can}) = \ker(\varphi_2),$$

i.e., that (3.70) is exact.

4. $\Longrightarrow$ 2.: We prove the negation of 2, i.e., that $\mathrm{D}M = 0$ implies $M = 0$. For this, we use that a module N is zero if and only if $0 \longrightarrow N \longrightarrow 0$ is exact. Assume that $\mathrm{D}M = 0$, i.e., that $0 = \mathrm{D}0 \longrightarrow \mathrm{D}M \longrightarrow \mathrm{D}0 = 0$ is exact. Due to condition 4, the sequence $0 \longrightarrow M \longrightarrow 0$ is exact too, hence $M = 0$. □

Lemma 3.3.6 *Let $S \in \mathrm{mod}_{\mathcal{D}}$ be a simple module and let $0 \neq x \in S$. Then $S = \mathcal{D}x \cong \mathcal{D}/\mathfrak{m}$, where $\mathfrak{m} = \mathrm{ann}_{\mathcal{D}}(x) = \mathrm{ann}_{\mathcal{D}}(S) = \{d \in \mathcal{D};\ dS = 0\}$ is a maximal ideal.*

Proof It is easily seen that a simple module S is cyclic and therefore generated by any nonzero element $x \in S$. This means that the map $\mathcal{D} \longrightarrow S$, $d \longmapsto dx$ is surjective and, by the homomorphism theorem,

$$\mathcal{D}/\,\mathrm{ann}_{\mathcal{D}}(x) \xrightarrow{\cong} S, \quad \overline{d} \longmapsto dx,$$

is an isomorphism. By item 3 of Reminder 3.3.3, the ideal $\mathrm{ann}_{\mathcal{D}}(x)$ is maximal. Furthermore, $S = \mathcal{D}x$ implies that $\mathrm{ann}_{\mathcal{D}}(x) = \mathrm{ann}_{\mathcal{D}}(S)$. □

Theorem 3.3.7 *The standard function modules $\mathrm{C}^\infty(\mathbb{R}, F)$, $F = \mathbb{R}, \mathbb{C}$, and $F^{\mathbb{N}}$, F a field, over $F[s]$ from Corollary 3.1.3 are injective cogenerators.*

Proof The injectivity was shown in Corollary 3.2.12.

We check the cogenerator property by means of condition 3 of Theorem 3.3.5. Let S be a simple module. By Lemma 3.3.6, we can assume that $S = F[s]/\mathfrak{m}$, where $\mathfrak{m}$ is a maximal ideal. Since $F[s]$ is a principal ideal domain, $\mathfrak{m}$ has the form $\mathfrak{m} = F[s]f$, with irreducible $f \in F[s]$. With (3.66) and (3.67), we conclude the desired inequality

$$\mathrm{D}S \cong \{y \in \mathcal{F};\ f \circ y = 0\} \neq 0. \qquad \square$$

Corollary 3.3.8 *Every nonzero vector space $\mathcal{F}$ over a field K is an injective cogenerator in the category* mod_K *of K-vector spaces.*

Proof The injectivity of ${}_K\mathcal{F}$ was shown in Corollary 3.2.10.

Again, we use condition 3 of Theorem 3.3.5 to show that ${}_K\mathcal{F}$ is a cogenerator. Let $S = K/\mathfrak{m}$ be simple, i.e., with maximal $\mathfrak{m}$. The proper ideal $\mathfrak{m}$ of the field K is zero. Then the map $S = K \longrightarrow \mathcal{F},\ \ 1 \longmapsto y,\ \ a \longmapsto ay,\ 0 \neq y \in \mathcal{F}$ is nonzero in $\mathrm{D}K = \mathrm{D}S$. $\square$

3.3.2 Orthogonal Submodules

As explained in Sect. 3.2.2, the injectivity of the function modules is mainly used by means of the fundamental principle. Their cogenerator property, in turn, is mostly applied as a duality between the behaviors $\mathcal{B} = \ker R\circ$ and the equation modules $U = \mathcal{D}^{1\times k}R \subseteq \mathcal{D}^{1\times l}$.

We first list some facts on bilinear forms. For an arbitrary $\mathcal{D}$-module $\mathcal{F}$ and any $l > 0$, Corollary and Definition 3.1.8 implies the $\mathcal{D}$-bilinear map

$$\mathcal{D}^{1\times l} \times \mathcal{F}^l \longrightarrow \mathcal{F},\quad (\xi, y) \longmapsto \xi \circ y = \sum_{i=1}^{l} \xi_i \circ y_i. \tag{3.76}$$

If M is any $\mathcal{D}$-module, let $\mathbb{P}(M)$ denote the *projective geometry* of M, i.e., the ordered set of all submodules of M. The bilinear map (3.76) induces the two maps

$$\begin{gathered}\mathbb{P}(\mathcal{D}^{1\times l}) \rightleftarrows \mathbb{P}(\mathcal{F}^l),\quad U \longmapsto U^\perp,\quad \mathcal{B}^\perp \longleftarrow \mathcal{B},\\ \text{with } U^\perp := \{w \in \mathcal{F}^l;\ U \circ w = 0\} \text{ and } \mathcal{B}^\perp := \{\xi \in \mathcal{D}^{1\times l};\ \xi \circ \mathcal{B} = 0\},\end{gathered} \tag{3.77}$$

where $U^\perp$ and $\mathcal{B}^\perp$ are called the submodules *orthogonal* to U and $\mathcal{B}$, respectively, with respect to the given bilinear form. The two maps are order reversing, i.e., $U_1 \subseteq U_2$ implies that $U_1^\perp \supseteq U_2^\perp$ and $\mathcal{B}_1 \subseteq \mathcal{B}_2$ implies that $\mathcal{B}_1^\perp \supseteq \mathcal{B}_2^\perp$. By definition, $U \circ U^\perp = 0$ and $\mathcal{B}^\perp \circ \mathcal{B} = 0$, and hence the maps from (3.77) have the additional properties

$$U \subseteq U^{\perp\perp} \text{ and } \mathcal{B} \subseteq \mathcal{B}^{\perp\perp}. \tag{3.78}$$

Order reversing maps with these additional properties are called a *Galois correspondence*. A direct consequence of the defining properties are the identities

$$U^\perp = U^{\perp\perp\perp} \text{ and } \mathcal{B}^\perp = \mathcal{B}^{\perp\perp\perp}$$

since $U^\perp \subseteq (U^\perp)^{\perp\perp} = U^{\perp\perp\perp}$, and $U \subseteq U^{\perp\perp}$ implies $U^\perp \supseteq U^{\perp\perp\perp}$, and similarly for $\mathcal{B}$. For a matrix $R \in \mathcal{D}^{k\times l}$ and its row module $U := \mathcal{D}^{1\times k}R$ the associated behavior is

$$\begin{aligned}\mathcal{B} &:= \ker(R\circ\colon \mathcal{F}^l \longrightarrow \mathcal{F}^k) = \{w \in \mathcal{F}^l;\ R \circ w = 0\}\\ &= \{w \in \mathcal{F}^l;\ \forall \eta \in \mathcal{D}^{1\times k}\colon\ \eta R \circ w = 0\}\\ &= \{w \in \mathcal{F}^l;\ \forall \xi \in U = \mathcal{D}^{1\times k}R\colon\ \xi \circ w = 0\} = U^\perp.\end{aligned} \tag{3.79}$$

Hence every behavior is an orthogonal submodule and satisfies

$$\mathcal{B} = U^\perp = U^{\perp\perp\perp} = \mathcal{B}^{\perp\perp}. \tag{3.80}$$

The analogous equality $U = U^{\perp\perp}$ holds for a cogenerator $\mathcal{F}$ only.

Theorem 3.3.9 *Let $\mathcal{F}$ be a module over a commutative ring $\mathcal{D}$ and let $M = \mathcal{D}^{1\times l}/U$ be a finitely generated $\mathcal{D}$-module, $l \geq 0$. Then*

1. $\ker(\rho_M\colon M \longrightarrow \mathrm{D}^2M) = U^{\perp\perp}/U$, *where ρ_M is the Gelfand map, which is defined by $\rho_M(x)(\alpha) := \alpha(x)$ for $x \in M$ and $\alpha \in \mathrm{D}M$.*
2. *(Cf. [8, Thm. 3.6.2]) If $\mathcal{F}$ is a cogenerator, then*

$$U = U^{\perp\perp} \text{ for all } U \leq_{\mathcal{D}} \mathcal{D}^{1\times l}.$$

If, in particular, $U = \mathcal{D}^{1\times k}R$, $R \in \mathcal{D}^{k\times l}$, is the row space of a matrix and if $\mathcal{B} := U^\perp = \{w \in \mathcal{F}^l;\ R \circ w = 0\} \cong \mathrm{D}M$ is the associated behavior, then

$$\mathcal{D}^{1\times k}R = U = \mathcal{B}^\perp = \{\xi \in \mathcal{D}^{1\times l};\ \xi \circ \mathcal{B} = 0\}.$$

In words: The behavior as the solution module of $R \circ w = 0$ determines the row module U uniquely, whereas the matrix R itself is not unique. Also, a vector $\xi \in \mathcal{D}^{1\times l}$ is an equation for $\mathcal{B}$, i.e., $\xi \circ \mathcal{B} = 0$, if and only if it is a linear combination of the rows of any matrix R with $U = \mathcal{D}^{1\times k}R$.
3. *If $\mathcal{F}$ is injective and $\mathfrak{m} = \mathfrak{m}^{\perp\perp}$ for all maximal ideals of $\mathcal{D}$, then $\mathcal{F}$ is a cogenerator.*

Proof 1. Consider the canonical isomorphism

$$\mathrm{D}\mathcal{D}^{1\times l} = \mathrm{Hom}_{\mathcal{D}}(\mathcal{D}^{1\times l}, \mathcal{F}) \cong \mathcal{F}^l,\quad \beta \longleftrightarrow w,\quad \beta(\delta_i) = w_i.$$

For $\xi = \sum_{i=1}^{l} \xi_i\delta_i \in \mathcal{D}^{1\times l}$, this implies $\beta(\xi) = \xi \circ w$. As in Theorem 3.1.36, this isomorphism induces the isomorphism

$$\begin{aligned} \mathrm{D}M = \mathrm{D}(\mathcal{D}^{1\times l}/U) &\cong \{\beta \in \mathrm{D}\mathcal{D}^{1\times l};\ \beta(U)=0\} \cong U^{\perp} = \{w \in \mathcal{F}^l;\ U \circ w = 0\},\\ \alpha \leftrightarrow &\qquad \beta \qquad \leftrightarrow w,\\ \alpha(\overline{\xi}) = &\qquad \beta(\xi) \qquad = \xi \circ w. \end{aligned}$$

For the kernel of the Gelfand map ρ_M, this implies

$$\begin{aligned} \overline{\xi} \in \ker(\rho_M) &\Longleftrightarrow \forall \alpha \in \mathrm{D}M:\ \rho_M(\overline{\xi})(\alpha) = \alpha(\overline{\xi}) = 0\\ &\Longleftrightarrow \forall w \in U^{\perp}:\ \xi \circ w = 0 \Longleftrightarrow \xi \in U^{\perp\perp} \Longleftrightarrow \overline{\xi} \in U^{\perp\perp}/U. \end{aligned}$$

Hence, $\ker(\rho_M) = U^{\perp\perp}/U$.

2. If $\mathcal{F}$ is a cogenerator, the Gelfand maps ρ_M are monomorphisms by Definition and Lemma 3.3.2, and hence $U = U^{\perp\perp}$ for all $U \le \mathcal{D}^{1\times l}$.

3. Let $\mathcal{F}$ be injective. According to Theorem 3.3.5,3., we show $\mathrm{D}S \neq 0$ for every simple module $S = \mathcal{D}/\mathfrak{m}$ with maximal $\mathfrak{m}$. From the assumption in 3 we infer $0 = \mathfrak{m}^{\perp\perp}/\mathfrak{m} = \ker(\rho_S) = \ker\left((\alpha)_{\alpha\in \mathrm{D}S}\right)$. Let $0 \neq x \in S = \mathcal{D}/\mathfrak{m}$ and thus $\rho_S(x) \neq 0$. Hence there is some $\alpha \in \mathrm{D}S$ with $0 \neq \rho_S(x)(\alpha) = \alpha(x)$ and thus $0 \neq \alpha \in \mathrm{D}S$. According to Theorem 3.3.5,3., $\mathcal{F}$ is a cogenerator. □

In the next theorem we assume for simplicity that the ring $\mathcal{D}$ is Noetherian. This implies that every submodule U of $\mathcal{D}^{1\times l}$ is finitely generated and of the form $U = \mathcal{D}^{1\times k}R$, $R \in \mathcal{D}^{k\times l}$. It gives rise to the associated behavior $\mathcal{B}$ with

$$\mathrm{D}(\mathcal{D}^{1\times l}/U) \cong U^{\perp} = \mathcal{B} := \{w \in \mathcal{F}^l;\ R \circ w = 0\}.$$

If $U_i = \mathcal{D}^{1\times k^i} R_i \le_{\mathcal{D}} \mathcal{D}^{1\times l}$ with matrices $R_i \in \mathcal{D}^{k_i\times l}$, $i = 1, 2$, then

$$U_1 = \mathcal{D}^{1\times k_1} R_1 \subseteq \mathcal{D}^{1\times k_2} R_2 = U_2 \Longleftrightarrow \exists X \in \mathcal{D}^{k_1\times k_2} \text{ with } R_1 = XR_2. \tag{3.81}$$

Theorem 3.3.10 *Let $\mathcal{F}$ be an injective cogenerator over a commutative Noetherian ring $\mathcal{D}$. Consider two matrices $R_i \in \mathcal{D}^{k_i\times l}$, $i = 1, 2$, their row submodules $U_i := \mathcal{D}^{1\times k_i} R_i$ and their associated behaviors $\mathcal{B}_i := U_i^{\perp} = \{w \in \mathcal{F}^l;\ R_i \circ w = 0\}$. Then the following holds:*

1. *$U_1 \subseteq U_2 \Longleftrightarrow \mathcal{B}_2 \subseteq \mathcal{B}_1$. This equivalence together with* (3.81) *is crucial whenever one wants to check the inclusion of behaviors. In Theorem 3.3.12, we will give an algorithm for this purpose for principal ideal domains $\mathcal{D}$.*
2. *$\mathcal{B}_1 \cap \mathcal{B}_2 = (U_1 + U_2)^{\perp}$. Hence, the intersection of two behaviors is again a behavior. This holds for arbitrary modules ${}_{\mathcal{D}}\mathcal{F}$.*
3. *$\mathcal{B}_1 + \mathcal{B}_2 = (U_1 \cap U_2)^{\perp}$. Hence, the sum of two behaviors is again a behavior.*

Proof 1. Since $(-)^{\perp}$ is order reversing, the inclusion $U_1 \subseteq U_2$ implies $\mathcal{B}_2 = U_2^{\perp} \subseteq \mathcal{B}_1 = U_1^{\perp}$. Likewise, the latter inclusion and item 2 of Theorem 3.3.9 imply

$$U_1 = U_1^{\perp\perp} = \mathcal{B}_1^{\perp} \subseteq \mathcal{B}_2^{\perp} = U_2^{\perp\perp} = U_2.$$

2. Since $U_i \subseteq U_1 + U_2$ for $i = 1, 2$, the following equations are obvious:

$$\begin{aligned}(U_1 + U_2)^{\perp} &= \{w \in \mathcal{F}^l;\ (U_1 + U_2) \circ w = 0\} \\ &= \{w \in \mathcal{F}^l;\ U_1 \circ w = 0\} \cap \{w \in \mathcal{F}^l;\ U_2 \circ w = 0\} = U_1^{\perp} \cap U_2^{\perp}.\end{aligned}$$

3. Consider the map

$$(\mathrm{id}_l, \mathrm{id}_l)\circ\colon \mathcal{B}_1 \times \mathcal{B}_2 \longrightarrow \mathcal{F}^l,\quad \begin{pmatrix} w_1 \\ w_2 \end{pmatrix} \longmapsto (\mathrm{id}_l, \mathrm{id}_l) \circ \begin{pmatrix} w_1 \\ w_2 \end{pmatrix} = w_1 + w_2.$$

Since the Cartesian product

$$\mathcal{B}_1 \times \mathcal{B}_2 = \left\{\begin{pmatrix} w_1 \\ w_2 \end{pmatrix} \in \mathcal{F}^{l_1+l_2};\ \begin{pmatrix} R_1 & 0 \\ 0 & R_2 \end{pmatrix} \circ \begin{pmatrix} w_1 \\ w_2 \end{pmatrix} = 0\right\},$$

is obviously a behavior, the image $\mathrm{im}\left((\mathrm{id}_l\ \mathrm{id}_l) \circ\right) = \mathcal{B}_1 + \mathcal{B}_2$ is also a behavior according to Theorem 3.2.20. By (3.80) this implies that $(\mathcal{B}_1 + \mathcal{B}_2)^{\perp\perp} = \mathcal{B}_1 + \mathcal{B}_2$. As in 2 one shows $(\mathcal{B}_1 + \mathcal{B}_2)^{\perp} = \mathcal{B}_1^{\perp} \cap \mathcal{B}_2^{\perp}$ and finally

$$\begin{aligned}\mathcal{B}_1 + \mathcal{B}_2 &= (\mathcal{B}_1 + \mathcal{B}_2)^{\perp\perp} = \left((\mathcal{B}_1 + \mathcal{B}_2)^{\perp}\right)^{\perp} = (\mathcal{B}_1^{\perp} \cap \mathcal{B}_2^{\perp})^{\perp} \\ &= (U_1^{\perp\perp} \cap U_2^{\perp\perp})^{\perp} \overset{\text{Theorem 3.3.9, 2.}}{=\!=\!=} (U_1 \cap U_2)^{\perp}.\end{aligned}$$ □

Definition 3.3.11 Let $\mathcal{F}$ be an injective cogenerator over the commutative Noetherian ring $\mathcal{D}$, let $R \in \mathcal{D}^{k\times l}$ be a matrix, and let

$$U := \mathcal{D}^{1\times k} R \le \mathcal{D}^{1\times l},\ \text{and}\ \mathcal{B} := U^{\perp} = \{w \in \mathcal{F}^l;\ R \circ w = 0\} \subseteq \mathcal{F}^l.$$

We call $U = \mathcal{B}^{\perp}$ the *module of equations* and $\mathrm{M}(\mathcal{B}) = \mathcal{D}^{1\times l}/\mathcal{B}^{\perp}$ the *system module* or *module of observables* of the behavior $\mathcal{B}$. The latter terminology was introduced by Pommaret and will be justified in Example 3.3.20. ◊

Next we show how to decide the situations of Theorem 3.3.10 algorithmically in the standard cases. We assume that $\mathcal{D}$ is a principal ideal domain, for instance the polynomial ring $F[s]$ over a field F.

First we describe the solution set of a system of linear equations $\xi R = \eta$ with given $R \in \mathcal{D}^{k\times l}$ and $\eta \in \mathcal{D}^{1\times l}$, and unknown $\xi \in \mathcal{D}^{1\times k}$. Let $URV = S = \begin{pmatrix} \mathrm{diag}(e_1,\ldots,e_r) & 0 \\ 0 & 0 \end{pmatrix}$ be the Smith form of R with $U \in \mathrm{Gl}_k(\mathcal{D})$ and $V \in \mathrm{Gl}_l(\mathcal{D})$. Since U and V are invertible over $\mathcal{D}$ the equivalences

$$\begin{aligned}&\xi R = \eta \Longleftrightarrow \xi RV = \eta V =: \eta' \Longleftrightarrow \underbrace{\xi U^{-1}}_{=:\xi'}\underbrace{URV}_{=S} = \eta',\ \text{i.e.,}\ \xi' S = \eta' \\ &\Longleftrightarrow \forall i = 1, \ldots, r\colon\ \xi'_i e_i = \eta V_{-i}\ \text{and}\ \forall i = r+1, \ldots, l\colon\ 0 = \eta V_{-i}\end{aligned}$$

hold. This implies that $\xi R = \eta$ has a solution if and only if

$$\forall i = 1, \ldots, r\colon\ e_i | \eta V_{-i} \quad \text{and} \quad \forall i = r+1, \ldots, l\colon\ \eta V_{-i} = 0.$$

If this is the case, then $\xi' = \left(e_1^{-1}\eta V_{-1}, \ldots, e_r^{-1}\eta V_{-r}, 0, \ldots, 0\right) \in \mathcal{D}^{1\times(r+(k-r))}$ solves $\xi' S = \eta V$ and $\xi = \xi' U$ solves $\xi R = \eta$.

The set of all solutions is $\xi + \{\zeta;\ \zeta R = 0\}$. We decompose $U = \left(\begin{smallmatrix} U^1 \\ U^2 \end{smallmatrix}\right) \in \mathcal{D}^{(r+(k-r))\times r}$. As we have seen in Theorem 3.2.18, the matrix U^2 is a universal left annihilator of R, and hence its rows $U_{(r+1)-}, \ldots, U_{k-}$ form a basis of the solution space of the homogeneous system of equations $\zeta R = 0$.

Theorem 3.3.12 *Assume that $\mathcal{D}$ is a principal ideal domain, the data from Theorem 3.3.10 and those just introduced. Note that U, U^1, U^2 are matrices whereas U_1, U_2 are submodules.*

1. *Let $UR_2V = \left(\begin{smallmatrix} \operatorname{diag}(e_1,\ldots,e_r) & 0 \\ 0 & 0 \end{smallmatrix}\right)$ be the Smith form of R_2. Let $R_1V := (A, B) \in \mathcal{D}^{k_1\times(r+(l-r))}$. Then $U_1 \subseteq U_2$, i.e., there is a matrix $X \in \mathcal{D}^{k_1\times k_2}$ with $R_1 = XR_2$, if and only if*

$$B = 0 \ \text{and} \ \forall i = 1, \ldots, k_1\ \forall j = 1, \ldots, r\colon\ e_j | A_{ij}.$$

If this holds and

$$Y := (e_1^{-1}A_{-1}, \ldots, e_r^{-1}A_{-r}, 0, \ldots, 0) \in \mathcal{D}^{k_1\times(r+(k_2-r))}, \text{ then}$$
$$R_1 = XR_2 \text{ with } X := YU.$$

All solutions $X' \in \mathcal{D}^{k_1\times k_2}$ of $X'R_2 = 0$ and $\widetilde{X}$ of $R_1 = \widetilde{X}R_2$ are given by the isomorphism

$$\mathcal{D}^{k_1\times(k_2-r)} \cong \left\{X' \in \mathcal{D}^{k_1\times k_2};\ X'R_2 = 0\right\},\ X'' \mapsto X''U^2, \text{ and by}$$
$$\left\{\widetilde{X} \in \mathcal{D}^{k_1\times k_2};\ R_1 = \widetilde{X}R_2\right\} = X + \left\{X' \in \mathcal{D}^{k_1\times k_2};\ X'R_2 = 0\right\}$$

2. *The identity $U_1 + U_2 = \mathcal{D}^{1\times(k_1+k_2)}\left(\begin{smallmatrix} R_1 \\ R_2 \end{smallmatrix}\right)$ holds.*
3. *For the computation of $U_1 \cap U_2$ define the matrix*

$$\widetilde{R} := \begin{pmatrix} R_1 & 0 \\ 0 & R_2 \\ -\operatorname{id}_l & -\operatorname{id}_l \end{pmatrix} \in \mathcal{D}^{\widetilde{k}\times(l+l)},\ \widetilde{k} := k_1 + k_2 + l,\ \widetilde{r} = \operatorname{rank}(\widetilde{R}),$$

and let $\widetilde{U}\widetilde{R}\widetilde{V} = \widetilde{S} = \left(\begin{smallmatrix} \operatorname{diag}(\widetilde{e}_1,\ldots,\widetilde{e}_r) & 0 \\ 0 & 0 \end{smallmatrix}\right)$ be the Smith form of $\widetilde{R}$. Decompose

$$\widetilde{U} = \begin{pmatrix} U^{11} & U^{12} & U^{13} \\ U^{21} & U^{22} & U^{23} \end{pmatrix} \in \mathcal{D}^{(\widetilde{r}+(\widetilde{k}-\widetilde{r}))\times(k_1+k_2+l)}.$$

Then the rows of U^{23} form a generating system of $U_1 \cap U_2$.

Proof 1. The matrix equation $XR_2 = R_1$ are several equations of the form $\xi R_2 = \eta$ stacked together. The assertions follow from the derivations which precede the theorem.
2. Obvious.
3. By Theorem 3.2.18, the matrix (U^{21}, U^{22}, U^{23}) is a universal left annihilator of $\widetilde{R}$. We infer

$$\begin{aligned}
\eta \in U_1 \cap U_2 &\Longleftrightarrow \exists \xi_1 \in \mathcal{D}^{1\times k_1}\ \exists \xi_2 \in \mathcal{D}^{1\times k_2} \text{ with } \eta = \xi_1 R_1 = \xi_2 R_2 \\
&\Longleftrightarrow \exists \xi_1 \in \mathcal{D}^{1\times k_1}\ \exists \xi_2 \in \mathcal{D}^{1\times k_2} \text{ with } (\xi_1, \xi_2, \eta)\widetilde{R} = 0 \\
&\Longleftrightarrow \exists \xi_1,\ \xi_2,\ \zeta \text{ with } (\xi_1, \xi_2, \eta) = \zeta(U^{21}, U^{22}, U^{23}) \\
&\Longleftrightarrow \exists \zeta \text{ with } \eta = \zeta U^{23} \Longrightarrow U_1 \cap U_2 = \mathcal{D}^{1\times(\widetilde{k}-\widetilde{r})}U^{23}. \qquad \square
\end{aligned}$$

3.3.3 Behavior Morphisms

We introduce the abelian category of systems or behaviors and show that there is a categorical duality between the category of systems and the category of finitely generated modules with distinguished sets of generators. In particular, we characterize the morphisms as well as the mono-, epi-, and isomorphisms of the category of systems.

We assume that $\mathcal{D}$ is a commutative Noetherian domain and that $\mathcal{F}$ is an injective cogenerator over $\mathcal{D}$. According to Theorem 3.3.7 the standard function modules from Corollary 3.1.3 satisfy this assumption.

Naturally, the *objects* of the category of $\mathcal{F}$-behaviors shall be the behaviors $\mathcal{B} \subseteq \mathcal{F}^l$ for $l \in \mathbb{N}$. By definition, every behavior $\mathcal{B}$ comes equipped with the $\mathcal{D}$-linear embedding $\text{inj}\colon \mathcal{B} \longrightarrow \mathcal{F}^l$. Moreover, $\mathcal{B}$ gives rise to the modules $\mathcal{B}^\perp \leq_{\mathcal{D}} \mathcal{D}^{1\times l}$ and $\mathrm{M}(\mathcal{B}) := \mathcal{D}^{1\times l}/\mathcal{B}^\perp = \sum_{i=1}^{l} \mathcal{D}\overline{\delta_i}$. Thus, the finitely generated module $\mathrm{M}(\mathcal{B})$ has the distinguished set $\overline{\delta_i}$, $i = 1, \ldots, l$, of generators. In Theorem 3.1.36, we have shown that there is the canonical isomorphism

$$\begin{aligned}
\mathrm{can}_{\mathcal{B}}\colon \mathrm{Hom}_{\mathcal{D}}(\mathrm{M}(\mathcal{B}), \mathcal{F}) &\overset{\cong}{\longleftrightarrow} \mathcal{B}, \\
\alpha &\longleftrightarrow w, \quad \text{where } \alpha(\overline{\delta_i}) = w_i, \text{ i.e., } \alpha(\overline{\xi}) = \xi \circ w.
\end{aligned}$$

Assume, conversely, that ${}_{\mathcal{D}}M$ is finitely generated with a distinguished list $x_1, \ldots, x_l$ of generators. The epimorphism $\varphi\colon \mathcal{D}^{1\times l} \longrightarrow M$, $\xi \longmapsto \sum_{i=1}^{l} \xi_i x_i$, induces the *submodule of relations or syzygies* of the x_i, namely,

$$U := \ker(\varphi) = \left\{\xi \in \mathcal{D}^{1\times l};\ \sum_{i=1}^{l} \xi_i x_i = 0\right\},$$

the behavior $\mathcal{B} := U^{\perp}$ with $U = \mathcal{B}^{\perp}$, and the isomorphism

$$\varphi_{\text{ind}} \colon \mathrm{M}(\mathcal{B}) = \mathcal{D}^{1\times l}/U \overset{\cong}{\longleftrightarrow} M, \quad \overline{\delta_i} \longmapsto x_i, \quad \overline{\xi} \longmapsto \sum_{i=1}^{l} \xi_i x_i. \tag{3.82}$$

Hence a behavior can be uniquely defined by a finitely generated $\mathcal{D}$-module with distinguished generators. By this, we have derived the relation between the objects of the category of systems and the objects of the category of finitely generated modules with distinguished sets of generators.

We define the system morphisms via the morphisms in the category of finitely generated modules with distinguished sets of generators. First, we have to characterize the morphisms in the latter category (Theorem 3.3.14). The following lemma is a preparation for this.

Lemma 3.3.13 *Let $X_i \leq_{\mathcal{D}} Y_i$, $i = 1, 2$ be submodules and let* $\text{can}_i \colon Y_i \longrightarrow Y_i/X_i$ *denote the canonical maps. Then the following holds.*

1. *Let $\psi \in \mathrm{Hom}_{\mathcal{D}}(Y_2, Y_1)$. There exists a $\mathcal{D}$-homomorphism $\psi_{\text{ind}} \colon Y_2/X_2 \longrightarrow Y_1/X_1$ such that the diagram*

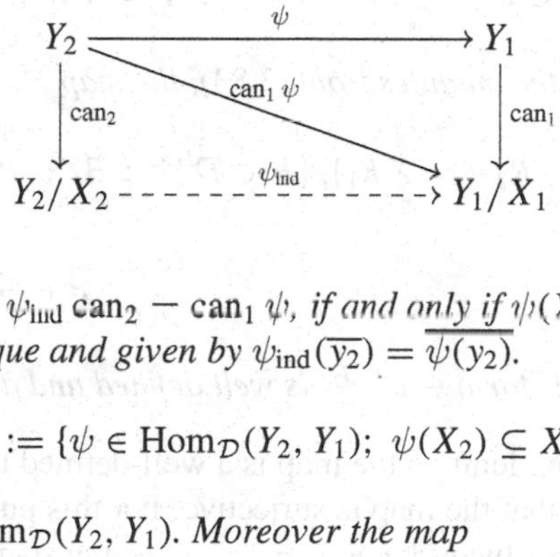

commutes, i.e., with $\psi_{\text{ind}}\,\text{can}_2 = \text{can}_1\,\psi$, if and only if $\psi(X_2) \subseteq X_1$.
If ψ_{ind} exists it is unique and given by $\psi_{\text{ind}}(\overline{y_2}) = \overline{\psi(y_2)}$.

2. *The set*

$$\mathcal{H} := \{\psi \in \mathrm{Hom}_{\mathcal{D}}(Y_2, Y_1);\ \psi(X_2) \subseteq X_1\}$$

is a submodule of $\mathrm{Hom}_{\mathcal{D}}(Y_2, Y_1)$. *Moreover the map*

$$\text{can} \colon \mathcal{H} \longrightarrow \mathrm{Hom}_{\mathcal{D}}(Y_2/X_2, Y_1/X_1), \quad \psi \longmapsto \psi_{\text{ind}},$$

is $\mathcal{D}$-linear and induces a monomorphism

$$\text{can}_{\text{ind}} \colon \mathcal{H}/\{\psi \in \mathrm{Hom}_{\mathcal{D}}(Y_2, Y_1);\ \psi(Y_2) \subseteq X_1\} \longrightarrow \mathrm{Hom}_{\mathcal{D}}(Y_2/X_2, Y_1/X_1)$$
$$\overline{\psi} \longmapsto \psi_{\text{ind}}.$$

Proof 1. According to the universal property of the factor module (Theorem 3.1.31), the map $\text{can}_1\,\psi$ admits the factorization $\text{can}_1\,\psi = \psi_{\text{ind}}\,\text{can}_2$, i.e., $\psi_{\text{ind}}(\overline{y_2}) = (\text{can}_1\,\psi)(y_2) = \overline{\psi(y_2)}$, if and only if $(\text{can}_1\,\psi)(X_2) = 0$, i.e., $\psi(X_2) \subseteq X_1$, and ψ_{ind} is the only map with this property.

2. It is easy to see that $\mathcal{H}$ is a submodule and that the map $\psi \longmapsto \psi_{\text{ind}}$ is linear. Moreover, ψ is in the kernel of this map if and only if $\psi_{\text{ind}} = 0$, and this is equivalent to $\psi_{\text{ind}}\,\text{can}_2 = \text{can}_1\,\psi = 0$, i.e., to $\psi(Y_2) \subseteq X_1$ The rest follows from the homomorphism theorem. □

We apply the preceding lemma to the modules

$$Y_i := \mathcal{D}^{1\times l_i}, \quad X_i := U_i = \mathcal{D}^{1\times k_i} R_i \subseteq \mathcal{D}^{1\times l_i}, \quad \text{and } M_i = \mathcal{D}^{1\times l_i}/U_i, \\ \text{where } R_i \in \mathcal{D}^{k_i\times l_i},\ i = 1, 2, \tag{3.83}$$

and identify

$$\mathcal{D}^{l_2\times l_1} = \text{Hom}_{\mathcal{D}}(\mathcal{D}^{1\times l_2}, \mathcal{D}^{1\times l_1}), \quad P = \cdot P.$$

The $\mathcal{D}$-module $\mathcal{H}$ and its submodule ker(can) are given as

$$\begin{aligned} \mathcal{H} &= \{P \in \mathcal{D}^{l_2\times l_1};\ U_2 P = \mathcal{D}^{1\times k_2} R_2 P \subseteq U_1 = \mathcal{D}^{1\times k_1} R_1\} \\ &= \{P \in \mathcal{D}^{l_2\times l_1};\ \exists X \in \mathcal{D}^{k_2\times k_1}:\ R_2 P = X R_1\} \quad \text{and} \\ \ker(\text{can}) &= \{P \in \mathcal{D}^{l_2\times l_1};\ \mathcal{D}^{1\times l_2} P \subseteq \mathcal{D}^{1\times k_1} R_1\} \\ &= \{P \in \mathcal{D}^{l_2\times l_1};\ \exists Y \in \mathcal{D}^{l_2\times k_1}:\ P = Y R_1\}. \end{aligned} \tag{3.84}$$

Theorem 3.3.14 *For the modules from* (3.83)*, the map*

$$\begin{aligned} \{P \in \mathcal{D}^{l_2\times l_1};\ \exists X:\ R_2 P = X R_1\}/\{P \in \mathcal{D}^{l_2\times l_1};\ \exists Y:\ P = Y R_1\} \\ \longrightarrow \text{Hom}_{\mathcal{D}}(M_2, M_1), \\ \overline{P} \longmapsto (\cdot P)_{\text{ind}}, \end{aligned} \tag{3.85}$$

where $(\cdot P)_{\text{ind}}(\overline{\eta}) = \overline{\eta P}$ *for* $\eta \in \mathcal{D}^{1\times l_2}$*, is well defined and an isomorphism.*

Proof By the preceding lemma, the map is a well-defined monomorphism.

It remains to show that the map is surjective. For this purpose, let $\varphi\colon M_2 \to M_1$ be a linear map. Let, as always be δ_i, $i = 1, \ldots, l_2$, the standard basis of $\mathcal{D}^{1\times l_2}$. For each such i we choose a row $P_{i-} \in \mathcal{D}^{1\times l_1}$ such that $\varphi(\overline{\delta_i}) = \overline{P_{i-}}$ and form the matrix $P \in \mathcal{D}^{l_2\times l_1}$ with these rows. Then, $\varphi(\overline{\delta_i}) = \overline{P_{i-}} = \overline{\delta_i P}$, hence also $\varphi(\overline{\eta}) = \overline{\eta P}$ for arbitrary $\eta \in \mathcal{D}^{1\times l_2}$. □

The preceding theorem characterizes the $\mathcal{D}$-modules $\text{Hom}_{\mathcal{D}}(M_2, M_1)$ for modules $M_i = \mathcal{D}^{1\times l_i}/\mathcal{D}^{1\times k_i} R_i$, $R_i \in \mathcal{D}^{k_i\times l_i}$, $i = 1, 2$. Let $P \in \mathcal{D}^{l_2\times l_1}$ such that $(\cdot P)_{\text{ind}} \in \text{Hom}_{\mathcal{D}}(M_2, M_1)$ as in Theorem 3.3.14. We consider the behaviors $\mathcal{B}_i = \{w_i \in \mathcal{F}^{l_i};\ R_i \circ w_i = 0\}$.

Lemma 3.3.15 *For these data* $P \circ \mathcal{B}_1 \subseteq \mathcal{B}_2$ *holds, i.e., the map* $P\circ\colon \mathcal{B}_1 \longrightarrow \mathcal{B}_2$, $w_1 \longmapsto P \circ w_1$ *is well defined. Furthermore, the diagram*

$$
\begin{array}{ccccc}
 & \alpha_1 & \longmapsto & \alpha_1(\cdot P)_{\mathrm{ind}} & \\
\alpha_1 & DM_1 = \mathrm{Hom}_{\mathcal{D}}(M_1, \mathcal{F}) & \xrightarrow{D(\cdot P)_{\mathrm{ind}}} & DM_2 = \mathrm{Hom}_{\mathcal{D}}(M_2, \mathcal{F}) & \alpha_2 \\
\updownarrow & \cong \updownarrow & & \cong \updownarrow & \updownarrow \\
w_1 & \mathcal{B}_1 & \xrightarrow{P\circ} & \mathcal{B}_2 & w_2 \\
 & w_1 & \longmapsto & P \circ w_1 &
\end{array}
$$

commutes. The vertical isomorphisms are those from Theorem 3.1.36, i.e., $\alpha_i(\overline{\xi}) = \xi \circ w_i$ *for* $\xi \in \mathcal{D}^{1\times l_i}$, $i = 1, 2$.

Proof According to Theorem 3.3.14 the matrix P satisfies $R_2 P = X R_1$ for some matrix X. For $w_1 \in \mathcal{B}_1$ this implies

$$0 = X \circ \underbrace{(R_1 \circ w_1)}_{=0} = X R_1 \circ w_1 = R_2 P \circ w_1 = R_2 \circ (P \circ w_1) \Longrightarrow P \circ w_1 \in \mathcal{B}_2.$$

Let $\alpha_1 \in DM_1$ and $w_1 \in \mathcal{B}_1$ with $\alpha_1(\overline{\xi_1}) = \xi_1 \circ w_1$. Define $\alpha_2 := \alpha_1(\cdot P)_{\mathrm{ind}}$. For all $\xi_2 \in \mathcal{D}^{1\times l_2}$ this implies

$$\xi_2 \circ w_2 = \alpha_2(\overline{\xi_2}) = \alpha_1(\overline{\xi_2 P}) = \xi_2 P \circ w_1 = \xi_2 \circ (P \circ w_1) \Longrightarrow w_2 = P \circ w_1.$$

□

Definition and Corollary 3.3.16 We define the *category of systems* or *category of behaviors*, denoted by $\mathrm{syst}_{\mathcal{D}\mathcal{F}}$, as follows. The *objects* are the behaviors $\mathcal{B} \subseteq \mathcal{F}^l$, $l \in \mathbb{N}$. The *morphisms* from a behavior $\mathcal{B}_1 \subseteq \mathcal{F}^{l_1}$ to a behavior $\mathcal{B}_2 \subseteq \mathcal{F}^{l_2}$ are the elements of the set

$$\mathrm{Hom}(\mathcal{B}_1, \mathcal{B}_2) := \{\phi \in \mathrm{Hom}_{\mathcal{D}}(\mathcal{B}_1, \mathcal{B}_2);\ \exists P \in \mathcal{D}^{l_2\times l_1} : \\ P \circ \mathcal{B}_1 \subseteq \mathcal{B}_2 \text{ and } \forall w_1 \in \mathcal{B}_1 :\ \phi(w_1) = P \circ w_1\}.$$

The set $\mathrm{Hom}(\mathcal{B}_1, \mathcal{B}_2)$ of *system morphisms* or *behavior morphisms* is a $\mathcal{D}$-submodule of $\mathrm{Hom}_{\mathcal{D}}(\mathcal{B}_1, \mathcal{B}_2)$, as is easily seen. The *composition* in the category of systems is the composition of maps. ◊

In the continuous-time and discrete-time standard cases the behavior morphisms are defined by matrices of differential and difference operators, respectively.

Example 3.3.17 The translation map $T : \mathbb{C}^\infty \to \mathbb{C}^\infty$, $u(t) \mapsto u(t-1)$, is $\mathbb{C}[s]$-linear since

$$T(s \circ u) = s \circ T(u),\ (T(s \circ u))(t) = u'(t-1) = d\,(u(t-1))\,/dt = (s \circ T(u))(t).$$

Obviously the map T is not a differential operator. Hence, in general,

$$\mathrm{Hom}(\mathcal{B}_1, \mathcal{B}_2) \subsetneq \mathrm{Hom}_{\mathbb{C}[s]}(\mathcal{B}_1, \mathcal{B}_2).$$

◊

The next theorem is the main duality result of this book and of fundamental importance in the sequel.

Theorem 3.3.18 (Duality theorem) *Let $\mathcal{F}$ be an injective cogenerator over a commutative Noetherian ring $\mathcal{D}$. Let $\mathcal{B}_i \subseteq \mathcal{F}^{l_i}$, $i = 1, 2, 3$, be behaviors with modules $\mathrm{M}(\mathcal{B}_i) = \mathcal{D}^{1\times l_i}/\mathcal{B}_i^{\perp}$. The following statements hold:*

1. *The map*

$$\mathrm{Hom}_{\mathcal{D}}\left(\mathrm{M}(\mathcal{B}_2), \mathrm{M}(\mathcal{B}_1)\right) \overset{\cong}{\longleftrightarrow} \mathrm{Hom}(\mathcal{B}_1, \mathcal{B}_2), \quad (\cdot P)_{\mathrm{ind}} \longmapsto P\circ, \tag{3.86}$$

is a $\mathcal{D}$-isomorphism.

2. *A sequence of behaviors and behavior morphisms*

$$\mathcal{B}_1 \overset{P_1\circ}{\longrightarrow} \mathcal{B}_2 \overset{P_2\circ}{\longrightarrow} \mathcal{B}_3$$

with $P_i \circ \mathcal{B}_i \subseteq \mathcal{B}_{i+1}$ is exact if and only if the corresponding module sequence

$$\mathrm{M}(\mathcal{B}_1) \overset{(\cdot P_1)_{\mathrm{ind}}}{\longleftarrow} \mathrm{M}(\mathcal{B}_2) \overset{(\circ P_2)_{\mathrm{ind}}}{\longleftarrow} \mathrm{M}(\mathcal{B}_3)$$

is exact.

In colloquial terms: Behaviors and the morphisms between them correspond one-to-one to finitely generated $\mathcal{D}$-modules with a list of generators and their linear maps. The correspondence preserves and reflects exact sequences. The term duality *instead of* equivalence *is due to the fact that arrows (maps) are reversed.*

Proof 1. We have shown in Lemma 3.3.15 that the map is well defined, and it is obviously $\mathcal{D}$-linear.

To show that it is injective, assume that $P\circ\colon \mathcal{B}_1 \longrightarrow \mathcal{B}_2$ is zero. From the commutative diagram of Lemma 3.3.15 we infer that $\mathrm{D}(\cdot P)_{\mathrm{ind}} = 0$. The cogenerator property of $\mathcal{F}$ and Theorem 3.3.5 imply that $(\cdot P)_{\mathrm{ind}} = 0$. Hence, the map (3.86) is injective.

To show that it is surjective, let $P \in \mathcal{D}^{l_2\times l_1}$ with $P \circ \mathcal{B}_1 \subseteq \mathcal{B}_2$. Let $R_i \in \mathcal{D}^{k_i\times l_i}$ be such that $\mathcal{B}_i = (\mathcal{D}^{1\times k_i} R_i)^{\perp}$ for $i = 1, 2$. Let $(-X_1, R_2') \in \mathcal{D}^{k_2'\times(k_1+l_2)}$ be a universal left annihilator of $\binom{R_1}{P}$. We infer from Theorem 3.2.20 that $P \circ \mathcal{B}_1 = \ker(R_2'\circ)$. By duality,

$$\mathcal{D}^{1\times k_2} R_2 = \mathcal{B}_2^{\perp} \subseteq (P \circ \mathcal{B}_1)^{\perp} = \ker(R_2'\circ)^{\perp} = \mathcal{D}^{1\times k_2'} R_2',$$

i.e., there is a matrix $X_2 \in \mathcal{D}^{k_2\times k_2'}$ such that $R_2 = X_2 R_2'$. The left annihilator property implies that $X_1 R_1 = R_2' P$ and thus $R_2 P = X_2 R_2' P = X_2 X_1 R_1 = X R_1$, where $X := X_2 X_1$. According to Theorem 3.3.14, this is the condition that $(\cdot P)_{\mathrm{ind}} \in \mathrm{Hom}_{\mathcal{D}}\left(\mathrm{M}(\mathcal{B}_2), \mathrm{M}(\mathcal{B}_1)\right)$.

2. This assertion is a direct consequence of item 4 of Theorem 3.3.5 and the commutativity of the diagram in Lemma 3.3.15. □

In categorical language, the preceding theorem says that the functor

$$\begin{array}{ccc} \mathrm{syst}_{\mathcal{DF}}^{\mathrm{op}} & \longrightarrow & \left\{\begin{array}{c}\text{finitely generated } \mathcal{D}\text{-modules}\\ \text{with a distinguished set of generators}\end{array}\right\} \\ \mathcal{B}_1 & & \mathrm{M}(\mathcal{B}_1) \\ \downarrow P\circ & \longmapsto & \uparrow (\cdot P)_{\mathrm{ind}} \\ \mathcal{B}_2 & & \mathrm{M}(\mathcal{B}_2) \end{array} \tag{3.87}$$

is a *contravariant equivalence*, in other words, a *duality*, between the category $\mathrm{syst}_{\mathcal{DF}}$ of embedded behaviors and the category of finitely generated $\mathcal{D}$-modules with a distinguished set of generators.

Corollary and Definition 3.3.19 *The module $\mathcal{F}$ is a behavior and* $\mathrm{M}(\mathcal{F}) = \mathcal{D}$. *For every behavior $\mathcal{B} \subseteq \mathcal{F}^l$, the map*

$$\mathrm{M}(\mathcal{B}) = \mathcal{D}^{1\times l}/\mathcal{B}^{\perp} \xrightarrow{\cong} \mathrm{Hom}(\mathcal{B}, \mathcal{F}), \quad \overline{\xi} \longmapsto \xi\circ,$$
$$\text{where } \xi \circ w = \xi_1 \circ w_1 + \cdots + \xi_l \circ w_l \text{ for } w \in \mathcal{B},$$

is a $\mathcal{D}$-isomorphism. According to Pommaret the elements of $\mathrm{M}(\mathcal{B}) \cong \mathrm{Hom}(\mathcal{B}, \mathcal{F})$ *are called the* observables *of $\mathcal{B}$.*

Proof For $0 := U \leq_{\mathcal{D}} \mathcal{D}^{1\times 1}$ we get $0^{\perp} = \mathcal{F}$, hence $\mathcal{F}$ is a behavior and $\mathrm{M}(\mathcal{F}) = \mathcal{D}/0 = \mathcal{D}$. Theorem 3.3.18 implies the asserted isomorphism

$$\begin{array}{ccccc} \mathrm{M}(\mathcal{B}) & \stackrel{\cong}{\longleftrightarrow} & \mathrm{Hom}_{\mathcal{D}}\left(\mathcal{D}^{1\times 1}, \mathrm{M}(\mathcal{B})\right) & \stackrel{\cong}{\longleftrightarrow} & \mathrm{Hom}(\mathcal{B}, \mathcal{F}^1), \\ \overline{\xi} & \longleftrightarrow & (\cdot\xi)_{\mathrm{ind}} = \left(d \mapsto d\overline{\xi}\right) & \longleftrightarrow & \xi\circ = \left(w \mapsto \xi \circ w\right). \end{array}$$

□

In the standard cases an observable $\phi : \mathcal{B} \to \mathcal{F}$ assigns a trajectory $w \in \mathcal{B}$ a signal $\phi(w)$ with values $\phi(w)(t) \in F$ that can be observed, hence the name. This term comes originally from physics. We give a standard example from classical mechanics.

Example 3.3.20 (*Observables in physics*) Let $\mathcal{F} := \mathrm{C}^{\infty}(\mathbb{R}, \mathbb{R})$. We consider the example of a particle or point mass m which moves on a curve $x \in \mathcal{F}^3$, i.e., $x(t) \in \mathbb{R}^3$ for $t \in \mathbb{R}$, with the velocity vector $v := \frac{\mathrm{d}x}{\mathrm{d}t}$, and is driven by a C^{∞}-*force vector field* $F : \mathbb{R}^3 \longrightarrow \mathbb{R}^3$. In general, the vector field F, for instance, the gravitational field, has singular points, but this is of no significance for this example and can be remedied by omitting the singular points. Newton's law is $\frac{\mathrm{d}^2x}{\mathrm{d}t^2}(t) = F(x(t))$, and this determines the movement of the point mass uniquely, once its initial position and velocity are given.

We assume that the vector field has a *potential*, i.e., that there is a C^∞-function $V\colon \mathbb{R}^3 \longrightarrow \mathbb{R}$ such that $F = -\operatorname{grad}(V)$. In Cartesian coordinates x_i, $i = 1, 2, 3$, this means

$$F_i(x) = -\frac{\partial V}{\partial x_i}(x) \text{ for } x \in \mathbb{R}^3 \text{ and } i = 1, 2, 3, \text{ where } F = \begin{pmatrix} F_1 \\ F_2 \\ F_3 \end{pmatrix}.$$

The minus sign is just a convention. Such a potential V exists if and only if $\operatorname{curl}(F) = 0$. We define the solution set

$$\mathcal{B} := \left\{x \in \mathcal{F}^3;\ \tfrac{d^2 x}{dt^2} = F(x) = -\operatorname{grad}(V)(x)\right\}.$$

This is, of course, not a linear behavior, but a nonlinear one in general. We use the Euclidean norm $\|-\|$ on $\mathbb{R}^3$ to define the nonlinear observable

$$\begin{aligned} E\colon \mathcal{B} &\longrightarrow \mathcal{F}, \\ x \longmapsto E(x) &= \tfrac{1}{2} m \left\| \tfrac{dx}{dt} \right\|^2 + V(x) \\ &= \tfrac{1}{2} m \left(\left(\tfrac{dx_1}{dt}\right)^2 + \left(\tfrac{dx_2}{dt}\right)^2 + \left(\tfrac{dx_3}{dt}\right)^2 \right) + V(x). \end{aligned}$$

The value $E(x)(t)$ is the (total) energy of the particle at time t if it moves on the trajectory $x \in \mathcal{B}$. The summands $\frac{1}{2} m \left\| \frac{dx}{dt} \right\|^2$ and $V(x)$ are called the *kinetic* and the *potential* energy of the particle.

In phase space with the trajectories $w = \binom{x}{v}$, $v := dx/dt$, the behavior is

$$\mathcal{B}_{\text{phase}} = \left\{ w = \begin{pmatrix} x \\ v \end{pmatrix} \in \mathcal{F}^6;\ \frac{d}{dt}\begin{pmatrix} x \\ v \end{pmatrix} = \begin{pmatrix} v \\ -\operatorname{grad}(V)(x) \end{pmatrix} \right\},$$

and the energy is the nonlinear observable

$$\begin{aligned} &E_{\text{phase}}\colon \mathcal{B}_{\text{phase}} \longrightarrow \mathcal{F}, \\ &\quad \tbinom{x}{v} \longmapsto E_{\text{phase}}(x, v) = \tfrac{1}{2} m \|v\|^2 + V(x) = \tfrac{1}{2} m \left(v_1^2 + v_2^2 + v_3^2\right) + V(x), \text{ with} \\ &\frac{dE}{dt} = m \sum_{i=1}^{3} \frac{dx_i}{dt} \frac{d^2 x_i}{dt^2} + \sum_{i=1}^{3} \frac{dx_i}{dt} \frac{\partial V}{\partial x_i}(x) = 0, \text{ since } m \frac{d^2 x_i}{dt^2} = F_i(x) = -\partial V(x). \end{aligned}$$

Thus the total energy along a trajectory of $\mathcal{B}$ or $\mathcal{B}_{\text{phase}}$ is time-invariant. This result describes the conservation of the total energy. ◇

Corollary 3.3.21 *In the situation of Theorem 3.2.22 the behavior epimorphisms* $\begin{pmatrix} C & D \\ 0 & \mathrm{id}_m \end{pmatrix} \circ : \mathcal{B}_1 \to \mathcal{B}_2$ *and* $C\circ : \mathcal{B}_1^0 \to \mathcal{B}_2^0$ *induce the monomorphisms*

$$\left(\cdot\begin{pmatrix} C & D \\ 0 & \mathrm{id}_m \end{pmatrix}\right)_{\mathrm{ind}} : M(\mathcal{B}_2) \to M(\mathcal{B}_1),\ (\cdot C)_{\mathrm{ind}} : M(\mathcal{B}_2^0) \to M(\mathcal{B}_1^0) \text{ where}$$
$$M(\mathcal{B}_1) = \mathcal{D}^{1\times(n+m)}/\mathcal{D}^{1\times n}(A, -B),\ M(\mathcal{B}_2) = \mathcal{D}^{1\times(p+m)}/\mathcal{D}^{1\times k}(P, -Q)$$
$$M(\mathcal{B}_1^0) = \mathcal{D}^{1\times n}/\mathcal{D}^{1\times n}A,\ M(\mathcal{B}_2^0) = \mathcal{D}^{1\times p}/\mathcal{D}^{1\times k}P.$$

This corollary will be applied in Sect. 6.3.5, for instance. ◇

Next we characterize the system mono-, epi-, and isomorphisms.

Theorem 3.3.22 *Let $R_i \in \mathcal{D}^{k_i \times l_i}$, $i = 1, 2$, and let*

$$P\circ\colon \mathcal{B}_1 = \ker(R_1\circ) \longrightarrow \mathcal{B}_2 = \ker(R_2\circ)$$

be a behavior morphism, in particular, $P \circ \mathcal{B}_1 \subseteq \mathcal{B}_2$. Let $(-X_1, R_2') \in \mathcal{D}^{k_2'\times(k_1+l_2)}$ be a universal left annihilator of $\begin{pmatrix} R_1 \\ P \end{pmatrix} \in \mathcal{D}^{(k_1+l_2)\times l_1}$, hence $P \circ \mathcal{B}_1 = \ker(R_2'\circ)$ by Theorem 3.2.20.

1. *The morphism $P\circ\colon \mathcal{B}_1 \longrightarrow \mathcal{B}_2$ is injective or a behavior monomorphism if and only if the matrix $\begin{pmatrix} R_1 \\ P \end{pmatrix} \in \mathcal{D}^{(k_1+l_2)\times l_1}$ has a left inverse $(Y, Z) \in \mathcal{D}^{l_1\times(k_1+l_2)}$, that is, a matrix which satisfies*

$$(Y, Z)\begin{pmatrix} R_1 \\ P \end{pmatrix} = YR_1 + ZP = \mathrm{id}_{l_1}. \tag{3.88}$$

In the engineering literature, a left invertible matrix $\begin{pmatrix} R_1 \\ P \end{pmatrix}$ is called right prime *and its components R_1 and P with the same number of columns are called* right coprime. *We do not use the primeness terminology, but will always talk of left or right invertibility of a matrix.*

2. *The matrix P induces $P \circ \mathcal{B}_1 = \mathcal{B}_2$ if and only if $\mathcal{D}^{1\times k_2'}R_2' = \mathcal{D}^{1\times k_2}R_2$ or, in other terms, there are matrices X, X' with $R_2' = XR_2,\ R_2 = X'R_2'$.*
3. *The morphism $P\circ\colon \mathcal{B}_1 \longrightarrow \mathcal{B}_2$ is bijective or a behavior isomorphism if and only both conditions 1 and 2 are satisfied.*

Proof 1. The following equivalences hold:

$$P\circ : \mathcal{B}_1 \to \mathcal{B}_2 \text{ is injective} \iff P\circ : \mathcal{B}_1 \to \mathcal{F}^{l_2} \text{ is injective}$$
$$\iff \left(\mathcal{D}^{1\times(k_1+l_2)}\begin{pmatrix} R_1 \\ P \end{pmatrix}\right)^{\perp} = \ker\left(\begin{pmatrix} R_1 \\ P \end{pmatrix}\circ\right) = \ker\left(P\circ : \mathcal{B}_1 \to \mathcal{F}^{l_2}\right) = 0$$
$$\iff \mathcal{D}^{1\times(k_1+l_2)}\begin{pmatrix} R_1 \\ P \end{pmatrix} = \mathcal{D}^{1\times l_1}$$
$$\iff \exists (Y, Z) \in \mathcal{D}^{l_1\times(k_1+l_2)} \text{ with } YR_1 + ZP = (Y, Z)\begin{pmatrix} R_1 \\ P \end{pmatrix} = \mathrm{id}_{l_1}.$$

2. $P \circ \mathcal{B}_1 = \left(\mathcal{D}^{1\times k_2'}R_2'\right)^{\perp},\ \mathcal{B}_2 = \left(\mathcal{D}^{1\times k_2}R_2\right)^{\perp}$. □

For a principal ideal domain $\mathcal{D}$ the preceding theorem is constructive. Its item 2, i.e., $P \circ \mathcal{B}_1 = \mathcal{B}_2$, can be checked algorithmically by Theorem 3.3.12, 1. Item 1 requires the construction of left inverses.

Theorem 3.3.23 (Left invertibility) *Let $\mathcal{D}$ be a principal ideal domain. Let $R \in \mathcal{D}^{k\times l}$ be a matrix with* $\operatorname{rank}(R) = r$ *and Smith form* $URV = S = \begin{pmatrix} \operatorname{diag}(e_1,\dots,e_r) & 0 \\ 0 & 0 \end{pmatrix}$. *The following properties are equivalent:*

1. *The matrix R is left invertible, i.e., there is a matrix $X \in \mathcal{D}^{l\times k}$ with $XR = \mathrm{id}_l$.*
2. $\operatorname{rank}(R) = l$ *and* $e_l = 1$.

These conditions imply that all elementary divisors e_i, $i = 1, \dots, l$, are equal to one (up to association) or, in other terms, that $S = \begin{pmatrix} \mathrm{id}_l \\ 0 \end{pmatrix}$.
Then $X := V(\mathrm{id}_l, 0)U$ is one left inverse of R, and all left inverses of R are obtained in the form

$$X' = X + Y, \text{ where } Y \in \mathcal{D}^{l\times k} \text{ with } YR = 0.$$

In the engineering literature a left invertible matrix R is called right prime.

Proof The isomorphism $\cdot V : \mathcal{D}^{1\times l} \xrightarrow{\cong} \mathcal{D}^{1\times l}$, $\xi \longmapsto \xi V$, induces the isomorphism

$$\mathcal{D}^{1\times k}R \cong \mathcal{D}^{1\times k}RV = \mathcal{D}^{1\times k}U^{-1}URV = \mathcal{D}^{1\times k}S = \oplus_{i=1}^{r}\mathcal{D}e_i\delta_i. \tag{3.89}$$

The following equivalences hold:

$$\begin{aligned} \exists X \in \mathcal{D}^{l\times k}:\ XR = \mathrm{id}_l &\Longleftrightarrow \mathcal{D}^{1\times k}R = \mathcal{D}^{1\times l} \\ &\overset{(3.89)}{\Longleftrightarrow} \mathcal{D}^{1\times k}S = \oplus_{i=1}^{r}\mathcal{D}e_i\delta_i = \mathcal{D}^{1\times l} = \oplus_{i=1}^{l}\mathcal{D}\delta_i \\ &\Longleftrightarrow r = l \text{ and } e_1 = \dots = e_l = 1 \Longleftrightarrow S = \begin{pmatrix} \mathrm{id}_l \\ 0 \end{pmatrix}. \end{aligned}$$

If these conditions are satisfied, we infer from $\mathrm{id}_l = (\mathrm{id}_l, 0)\begin{pmatrix} \mathrm{id}_l \\ 0 \end{pmatrix} = (\mathrm{id}_l, 0)S$ that $\mathrm{id}_l = (\mathrm{id}_l, 0)URV$ and, consequently, $\mathrm{id}_l = V(\mathrm{id}_l, 0)UR$. Hence, $X := V(\mathrm{id}_l, 0)U$ is a left inverse of R. Moreover,

$$X'R = \mathrm{id}_l = XR \Longleftrightarrow (X' - X)R = 0 \Longleftrightarrow X' = X + Y \text{ with } YR = 0. \qquad \square$$

The following analogue of the preceding theorem for right invertibility will later be used in Theorem 4.2.12 for a characterization of controllable systems.

Corollary 3.3.24 (Right invertibility) *We use the same data as in Theorem 3.3.23. The following properties are equivalent:*

1. *The matrix R is right invertible, i.e., there is a matrix Y with $RY = \mathrm{id}_k$.*
2. $\operatorname{rank}(R) = k$ *and* $e_k = 1$.

These conditions imply that all elementary divisors e_i, $i = 1, \dots, k$, are equal to 1 *(up to association) or, in other terms, that $S = (\mathrm{id}_k, 0)$.*

Then $Y := V\begin{pmatrix} \mathrm{id}_k \\ 0 \end{pmatrix}U$ is one right inverse of R, and all right inverses of R are obtained in the form

$$Y' = Y + X, \text{ where } X \in \mathcal{D}^{l\times k} \text{ with } RX = 0.$$

In the engineering literature a right invertible matrix R is called left prime.

Proof The assertions follow by applying the preceding theorem to

$$V^{\top}R^{\top}U^{\top} = S^{\top} = \begin{pmatrix} \operatorname{diag}(e_1,\ldots,e_r) & 0 \\ 0 & 0 \end{pmatrix}.$$

This equation implies that the rank and the elementary divisors of a matrix R and those of its transpose $R^{\top}$ coincide. □

References

1. J.C. Willems, From time series to linear system. I. Finite-dimensional linear time invariant systems. II. Exact modelling. *Automatica J. IFAC* **22**(5 and 6), 561–580 and 675–694 (1986)
2. U. Oberst, Multidimensional constant linear systems. Acta Appl. Math. **20**(1–2), 1–175 (1990)
3. S. Mac Lane, Categories for the working mathematician, in *Graduate Texts in Mathematics*, 2nd edn., vol. 5 (Springer, New York, 1998)
4. S.I. Gelfand, Y.I. Manin, *Methods of Homological Algebra*, 2nd edn. (Springer, Berlin, 2003)
5. J. Adámek, H. Herrlich, G.E. Strecker, *Abstract and Concrete Categories: The Joy of Cats* (Wiley, New York, 1990)
6. S. Lang, Algebra, in *Graduate Texts in Mathematics*, 3rd edn., vol. 211 (Springer, New York, 2002)
7. B.L. van der Waerden, *Modern Algebra* (Springer, 2003)
8. J.W. Polderman, J.C. Willems, Introduction to mathematical systems theory. A behavioral approach, in *Texts in Applied Mathematics*, vol. 26 (Springer, New York, 1998)
9. H.H. Rosenbrock, *State-Space and Multivariable Theory* (Wiley, New York, 1970)
10. W.A. Wolovich, Linear multivariable systems, in *Applied Mathematical Sciences*, vol. 11 (Springer, New York, 1974)
11. R.E. Kalman, P.L. Falb, M.A. Arbib, *Topics in Mathematical System Theory* (McGraw-Hill, New York, 1969)

Chapter 4
Observability, Autonomy, and Controllability of Behaviors

In this chapter we study the three fundamental system properties *observability, autonomy, and controllability*. We characterize these algebraically and in terms of the behavior's trajectories, and derive Kalman's famous results on state space systems as special cases. Both controllability and autonomy of a behavior can be described algebraically via the torsion module of the system module.

4.1 Observability of Behaviors

4.1.1 Observability

We assume that $\mathcal{D}$ is a commutative Noetherian integral domain and that ${}_{\mathcal{D}}\mathcal{F}$ is an injective cogenerator. This is satisfied in all standard cases.

Definition and Lemma 4.1.1 (Observability, *[1, §5.3]*) *Assume behaviors* $\mathcal{B}_i = \ker(R_i\circ)$, $R_i \in \mathcal{D}^{k_i\times l_i}$, $i = 1, 2$, *and a behavior morphism* $P\circ\colon \mathcal{B}_1 \longrightarrow \mathcal{B}_2$. *Let* $w_1 \in \mathcal{B}_1$ *and* $w_2 := P \circ w_1 \in \mathcal{B}_2$. *If* $P\circ$ *is a monomorphism, then* w_1 *is uniquely determined by* w_2 *and called* observable *or* reconstructible *from* w_2. *It can be computed as follows. By Theorem 3.3.22, there is a left inverse* (Y, Z) *of* $\binom{R_1}{P}$, *and the identity* $w_1 = Z \circ w_2$ *holds.*

Proof The assertion follows since

$$w_1 = \mathrm{id}_{l_1} \circ w_1 = (YR_1 + ZP) \circ w_1 = Y \circ \underbrace{R_1 \circ w_1}_{=0} + Z \circ (P \circ w_1) = Z \circ w_2. \qquad \square$$

U. Oberst et al., *Linear Time-Invariant Systems, Behaviors and Modules*, Differential-Algebraic Equations Forum,
https://doi.org/10.1007/978-3-030-43936-1_4

For the next corollary we assume that $\mathcal{D} = F[s]$ and that $\mathcal{F}$ is one of the standard function modules from Corollary 3.1.3. The polynomial matrix modules $F[s]^{k\times l}$ are $F[s]$-modules and contain $F^{k\times l}$. The direct sum

$$F[s]^{k\times l} = \bigoplus_{m=0}^{\infty} F^{k\times l} s^m \ni R = \sum_{m=0}^{\infty} R_m s^m \text{ with } R_m \in F^{k\times l}, \tag{4.1}$$

where, of course, $R_m = 0$ for almost all m, signifies that a polynomial matrix $R \in F[s]^{k\times l}$ can be written uniquely as the matrix polynomial $R = \sum_{m=0}^{\infty} R_m s^m$. The maximal m with $R_m \neq 0$ is again called the *degree* $\deg_s(R)$ of R.

Corollary 4.1.2 *Assume the continuous-time or discrete-time standard case, i.e., let $\mathcal{D} := F[s]$ and let $\mathcal{F}_{\text{cont}} := C^\infty(\mathbb{R}, F)$ with $F := \mathbb{R}, \mathbb{C}$ or let $\mathcal{F}_{\text{dis}} := F^{\mathbb{N}}$ for a field F, respectively. Let $P\circ\colon \mathcal{B}_1 \longrightarrow \mathcal{B}_2$ be a system monomorphism and let $Z \in F[s]^{l_1\times l_2}$ be as in Definition and Lemma 4.1.1. We write*

$$Z = \sum_{m=0}^{d} Z_m s^m \in F[s]^{l_1\times l_2}, \quad \text{with } Z_d \neq 0,\ d := \deg_s(Z).$$

Let $w_i \in \mathcal{B}_i$, $i = 1, 2$, with $w_2 = P \circ w_1$ and $w_1 = Z \circ w_2$. Then

$$\begin{aligned} w_1(t) &= \sum_{m=0}^{d} Z_m w_2^{(m)}(t) \text{ for } t \in \mathbb{R} && \text{in continuous time} \\ w_1(t) &= \sum_{m=0}^{d} Z_m w_2(t+m) \text{ for } t \in \mathbb{N} && \text{in discrete time.} \end{aligned} \tag{4.2}$$

Hence, $w_1(t)$ can be computed F-linearly from the vectors

$$\begin{aligned} &w_2(t), \ldots, w_2^{(d)}(t) \text{ in continuous time} \\ &w_2(t), \ldots, w_2(t+d) \text{ in discrete time.} \end{aligned}$$ ◇

Remark 4.1.3 In continuous time with $\mathcal{D} = F[s]$ and $\mathcal{F} = C^\infty(\mathbb{R}, F)$, $F = \mathbb{R}, \mathbb{C}$, the computation of w_1 from w_2 according to the preceding corollary requires *differentiators* for the computation of the higher derivatives of w_2. Such devices, for example, speedometers or accelerometers, are difficult to build for large d. In discrete time (4.2) means that one can compute $w_1(t)$ only at the time instant $t + d$ in the future, but not at the present time t. In Chap. 11 we will describe so-called *observers* that compute an approximation of w_1 on the basis of the values $w_2(t)$ only. ◇

4.1.2 Observable Rosenbrock Equations

Let $\mathcal{F}$ be an injective cogenerator over a commutative Noetherian ring $\mathcal{D}$. We consider the Rosenbrock data from Sect. 3.2.4 and from Corollary 3.3.21, i.e.,

$$\begin{aligned}
&A \circ x = B \circ u, \quad y = C \circ x + D \circ u\\
&\mathcal{B}_1 := \left\{ \left(\begin{smallmatrix} x \\ u \end{smallmatrix}\right) \in \mathcal{F}^{n+m};\ A \circ x = B \circ u \right\},\\
&\mathcal{B}_2 := \left(\begin{smallmatrix} C & D \\ 0 & \mathrm{id}_m \end{smallmatrix}\right) \circ \mathcal{B}_1 = \left\{ \left(\begin{smallmatrix} y \\ u \end{smallmatrix}\right) \in \mathcal{F}^{p+m};\ P \circ y = Q \circ u \right\} \qquad (4.3)\\
&\mathcal{B}_1^0 := \left\{ x \in \mathcal{F}^n;\ A \circ x = 0 \right\},\ \mathcal{B}_2^0 := \left\{ y \in \mathcal{F}^p;\ P \circ x = 0 \right\},\\
&(-Y, P) \in \mathcal{D}^{k \times (n+p)},\ (-Y, P) \left(\begin{smallmatrix} A \\ C \end{smallmatrix}\right) = 0 \text{ and } Q = YB + PD \text{ where}
\end{aligned}$$

$(-Y, P)$ is a universal left annihilator of $\left(\begin{smallmatrix} A \\ C \end{smallmatrix}\right)$. In the next theorem we define and characterize observable Rosenbrock systems.

Theorem and Definition 4.1.4 ([2, Definition 5.3.7 on p. 156]) *For the data from* (4.3) *the following properties are equivalent:*

1. *The epimorphism* $\left(\begin{smallmatrix} C & D \\ 0 & \mathrm{id}_m \end{smallmatrix}\right) \circ : \mathcal{B}_1 \longrightarrow \mathcal{B}_2$ *is an isomorphism, i.e., this map is also injective.*
2. *The epimorphism* $C\circ : \mathcal{B}_1^0 \longrightarrow \mathcal{B}_2^0$ *is an isomorphism.*
3. *The matrix* $\left(\begin{smallmatrix} A \\ C \end{smallmatrix}\right)$ *has a left inverse* $(X, Z) \in \mathcal{D}^{n \times (n+p)}$, *i.e.,* $XA + ZC = \mathrm{id}_n$.

If these conditions are satisfied the inverse maps of 1 resp. 2 are

$$\left(\begin{smallmatrix} Z & XB-ZD \\ 0 & \mathrm{id}_m \end{smallmatrix}\right) : \mathcal{B}_2 \xrightarrow{\cong} \mathcal{B}_1,\ Z : \mathcal{B}_2^0 \xrightarrow{\cong} \mathcal{B}_1^0. \qquad (4.4)$$

The dual module isomorphisms of the inverse isomorphisms $C\circ$ *and* $Z\circ$ *are*

$$M_1^0 := \mathcal{D}^{1\times n}/\mathcal{D}^{1\times n}A \cong M_2^0 := \mathcal{D}^{1\times p}/\mathcal{D}^{1\times k}P,\ \overline{\eta} \leftrightarrow \overline{\zeta},\ \eta = \zeta C,\ \zeta = \eta Z. \quad (4.5)$$

Under these circumstances the Rosenbrock equations (4.3) *are called* observable.

Proof 1. $\Longrightarrow$ 2.: This follows at once for $u = 0$.
2. $\Longleftrightarrow$ 3.: Apply Theorem 3.3.22, 1., to $\left(\begin{smallmatrix} A \\ C \end{smallmatrix}\right)$ instead of $\left(\begin{smallmatrix} R_1 \\ P \end{smallmatrix}\right)$.
3. $\Longrightarrow$ 1.: Let $\left(\begin{smallmatrix} x \\ u \end{smallmatrix}\right) \in \mathcal{B}_1$ and $y := C \circ x + D \circ u$. Then

$$\begin{aligned}
x = \mathrm{id}_n \circ x = (XA + ZC) \circ x &= X \circ \underbrace{A \circ x}_{=B\circ u} + Z \circ \underbrace{C \circ x}_{=y - D\circ u}\\
&= (XB - ZD) \circ u + Z \circ y \Longrightarrow \left(\begin{smallmatrix} x \\ u \end{smallmatrix}\right) = \left(\begin{smallmatrix} Z & XB-ZD \\ 0 & \mathrm{id}_m \end{smallmatrix}\right) \left(\begin{smallmatrix} y \\ u \end{smallmatrix}\right).
\end{aligned}$$

This proves (4.4) and (4.5) and 1. The equation $x = (XB - ZD) \circ u + Z \circ y$ shows that x can be computed from y and u. This justifies the *observability* terminology. □

If $\mathcal{D}$ is a principal ideal domain, for instance, $\mathcal{D} = F[s]$, Theorem 3.3.23 enables us to check the left invertibility of $\left(\begin{smallmatrix} A \\ C \end{smallmatrix}\right)$ and to construct a left inverse.

Example 4.1.5 Assume $n = p = m = 1$ in the preceding theorem. Then the system is observable if and only if the matrix $\binom{A}{C} \in \mathcal{D}^{2\times 1}$ has a left inverse. Here, this signifies that $\mathcal{D}A + \mathcal{D}C = \mathcal{D}$ or, in other words, that the elements $A, C \in \mathcal{D}$ are coprime. If $\mathcal{D}$ is a principal ideal domain, for instance, if $\mathcal{D} = F[s]$, this signifies that the greatest common divisor $\gcd(A, C)$ of A and C is 1.
Consider, for instance,

$$s \circ x = B \circ u, \quad y = (s^2 + 1) \circ x + D \circ u.$$

This system is observable because $(-s, 1)\binom{s}{s^2+1} = 1$, hence

$$x = -s \circ s \circ x + (s^2 + 1) \circ x = -s \circ B \circ u + (y - D \circ u) = y - (sB + D) \circ u.$$

◇

4.1.3 Observable State Space Equations

The characterization of observable state space equations was one of Kalman's many significant contributions to systems theory. For this we specialize the preceding section to the state space systems from Sect. 3.2.5*, i.e., to the polynomial ring* $\mathcal{D} = F[s]$ *over a field* F*, an injective* $F[s]$*-cogenerator* $\mathcal{F}$*, for instance one of those from Corollary* 3.1.3*, and to equations*

$$\begin{aligned} & s \circ x = Ax + Bu, \quad y = Cx + D \circ u \text{ with} \\ & A \in F^{n\times n}, \ B \in F^{n\times m}, \ C \in F^{p\times n}, D \in F[s]^{p\times m}, \\ & x \in \mathcal{F}^n, \ y \in \mathcal{F}^p, \ \text{and } u \in \mathcal{F}^m. \end{aligned} \tag{4.6}$$

Kalman characterized observability by the equation $n = \operatorname{rank}(\mathfrak{O})$ *where* $\mathfrak{O} \in F^{np\times n}$ *is the observability matrix, see below. In Theorem* 4.1.9 *we will describe the connection of his theory with Theorem* 4.1.4*, and in Theorem* 4.3.8 *the dual theorem for controllability.*

Equations (4.6) and (4.3) induce the behaviors

$$\begin{aligned} \mathcal{B}_1 &:= \left\{\binom{x}{u} \in \mathcal{F}^{n+m};\ s \circ x = Ax + Bu\right\} \\ &= \left\{\binom{x}{u} \in \mathcal{F}^{n+m};\ (s\,\mathrm{id}_n - A) \circ x = Bu\right\}, \\ \mathcal{B}_2 &:= \begin{pmatrix} C & D \\ 0 & \mathrm{id}_m \end{pmatrix} \circ \mathcal{B}_1 = \left\{\binom{y}{u} \in \mathcal{F}^{p+m};\ P \circ y = Q \circ u\right\} \text{ where} \\ & (-Y, P) \in F[s]^{p\times(n+p)},\ (-Y, P)\binom{s\,\mathrm{id}_n - A}{C} = 0,\ Q = YB + PD, \end{aligned} \tag{4.7}$$

and $(-Y, P)$ is a universal left annihilator of $\binom{s\,\mathrm{id}_n - A}{C}$. Since $F[s]$ is a principal ideal domain the number of rows of $(-Y, P)$ can be and is chosen as $(n + p) - \operatorname{rank}\binom{s\,\mathrm{id}_n - A}{C} = (n + p) - n = p$.

We first study the module $F[s]^{1\times n}/F[s]^{1\times n}(s\,\mathrm{id}_n - A)$ that in (4.5) appeared as $\mathcal{D}^{1\times n}/\mathcal{D}^{1\times n}A$ (with a different meaning of A). According to Reminder 3.1.2, the space $F^{1\times n}$ becomes an $F[s]$-module with scalar product $\circ_A$ via the endomorphism

$$\begin{aligned}&\Phi\colon F^{1\times n} \longrightarrow F^{1\times n},\quad \xi \longmapsto \xi A,\\ &\text{i.e., } s\circ_A \xi := \Phi(\xi) = \xi A \text{ and } f\circ_A \xi = \xi f(A) \text{ for } f\in F[s].\end{aligned} \tag{4.8}$$

There is the $F[s]$-epimorphism

$$\begin{aligned}&\varphi\colon F[s]^{1\times n} \longrightarrow F^{1\times n},\quad \delta_i \longmapsto \delta_i,\\ &(\eta_1,\ldots,\eta_n) = \sum_{i=1}^n \eta_i\delta_i \longmapsto \sum_{i=1}^n \eta_i\circ_A \delta_i = \sum_{i=1}^n \delta_i\eta_i(A) = \sum_{i=1}^n \eta_i(A)_{i-}.\end{aligned} \tag{4.9}$$

Lemma 4.1.6 *The matrix $A\in F^{n\times n}$ gives rise to its* characteristic matrix $s\,\mathrm{id}_n - A \in F[s]^{n\times n}$ *and to the row space $U := F[s]^{1\times n}(s\,\mathrm{id}_n - A)$. With these data and φ from (4.9) we get* $\ker(\varphi) = U$. *Via the homomorphism theorem the epimorphism φ induces the isomorphism*

$$\begin{aligned}&\varphi_{\mathrm{ind}}: F[s]^{1\times n}/U \xrightarrow{\cong} F^{1\times n}, \overline{\delta_i} \mapsto \delta_i,\ \overline{\eta} \leftrightarrow \xi,\ \textit{where}\\ &\xi := \sum_{i=1}^n \eta_i(A)_{i-},\ \xi_j = \sum_{i=1}^n \eta_i(A)_{ij},\ \overline{\eta} = \overline{\xi}.\end{aligned} \tag{4.10}$$

Proof The residue classes $\overline{\delta_i}\in F[s]^{1\times n}/U,\ i=1,\ldots,n$ are $F[s]$-generators of $F[s]^{1\times n}/U$. Since the rows

$$(s\,\mathrm{id}_n - A)_{i-} = s\delta_i - A_{i-} = s\delta_i - \sum_{j=1}^n A_{ij}\delta_j,\quad i=1,\ldots,n, \tag{4.11}$$

are $F[s]$-generators of U and

$$\varphi\Big(s\delta_i - \sum_{j=1}^n A_{ij}\delta_j\Big) = s\circ_A \delta_i - \sum_{j=1}^n A_{ij}\delta_j = \delta_i A - A_{i-} = A_{i-} - A_{i-} = 0,$$

we get $\varphi(U)=0$. By Theorem 3.1.31 φ induces the $F[s]$-epimorphism

$$\varphi_{\mathrm{ind}}\colon F[s]^{1\times n}/U \longrightarrow F^{1\times n},\quad \overline{\delta_i} \longmapsto \delta_i.$$

The surjectivity of φ_{ind} implies $\dim_F(F[s]^{1\times n}/U) \geq \dim_F(F^{1\times n}) = n$. Moreover we infer from (4.11) that $s\overline{\delta_i} = \sum_{j=1}^n A_{ij}\overline{\delta_j}$, and hence $s\left(\sum_{i=1}^n F\overline{\delta_i}\right) \subseteq \sum_{j=1}^n F\overline{\delta_j}$. This means that $\sum_{j=1}^n F\overline{\delta_j}$ is closed under scalar multiplication with polynomials in $F[s]$ and thus an $F[s]$-submodule of $F[s]^{1\times n}/U$. Since the $\overline{\delta_i}$ are $F[s]$-generators of this module, this implies $\sum_{j=1}^{n} F\overline{\delta_j} = F[s]^{1\times n}/U$ and, consequently, $\dim_F(F[s]^{1\times n}/U) \le n$ and $\dim_F(F[s]^{1\times n}/U) = n$. This means that φ_{ind} is indeed an isomorphism and $\ker(\varphi) = U$ as asserted. □

Let

$$f_A = s^d + \cdots \in F[s] \quad \text{with} \quad d := \deg_s(f_A)$$

be the *minimal polynomial* of A, i.e., the unique monic polynomial with

$$F[s]f_A = \{f \in F[s];\ f(A) = 0\} = \ker\left(F[s] \to F^{n\times n};\ f \mapsto f(A)\right). \tag{4.12}$$

The homomorphism theorem implies the algebra isomorphism

$$F[s]/F[s]f_A = \bigoplus_{i=0}^{d-1} F\overline{s^i} \xrightarrow{\cong} F[A] = \sum_{i=0}^{\infty} FA^i,\ \overline{s^i} \mapsto A^i,\ \overline{f} \mapsto f(A), \text{ hence}$$

$$F[A] = \sum_{i=0}^{\infty} FA^i = \bigoplus_{i=0}^{d-1} FA^i = \sum_{i=0}^{t-1} FA^i \quad \text{for all } t \ge d. \tag{4.13}$$

For any matrix $C \in F^{p\times n}$ we consider the submodule of the $F[s]$-module $F^{1\times n}$, generated by the rows of C, namely, $\sum_{j=1}^{p} F[s] \circ_A C_{j-} \subseteq F^{1\times n}$. Due to (4.13) for any $t \ge d$ we get

$$\begin{aligned}\sum_{j=1}^{p} F[s] \circ_A C_{j-} &= \sum_{j=1}^{p}\sum_{i=0}^{\infty} Fs^i \circ_A C_{j-} = \sum_{j=1}^{p}\sum_{i=0}^{\infty} FC_{j-}A^i = \sum_{i=0}^{\infty} F^{1\times p}CA^i \\ &= \sum_{i=0}^{t-1} F^{1\times p}CA^i = F^{1\times tp}\begin{pmatrix} C \\ CA \\ \vdots \\ CA^{t-1}\end{pmatrix}.\end{aligned} \tag{4.14}$$

The *characteristic polynomial* of A is defined as

$$\chi_A := \det(s\,\mathrm{id}_n - A) = s^n + \cdots \in F[s] \quad \text{with} \quad n = \deg_s(\chi_A).$$

The Cayley-Hamilton theorem [3, Theorem 3.1 on p. 561] says that $\chi_A(A) = 0$. It implies $\chi_A \in F[s]f_A$, $f_A | \chi_A$ and $d \le n$. In particular, the equality

$$\sum_{j=1}^{p} F[s] \circ_A C_{j-} = F^{1\times np}\begin{pmatrix} C \\ CA \\ \vdots \\ CA^{n-1}\end{pmatrix} \tag{4.15}$$

holds. The representation (4.15) is preferred to (4.14) since n is given, whereas d has to be computed.

Definition 4.1.7 The matrix

$$\mathfrak{O} := \begin{pmatrix} C \\ CA \\ \vdots \\ CA^{n-1} \end{pmatrix} \in F^{np\times n}$$

is called the *observability matrix* of the equations (4.6). ◊

The algebraic properties of the observability matrix are summarized in Corollary 4.1.8, and its significance for systems theory is exhibited in Theorem 4.1.9.
In context with state space equations it is customary to employ the nondegenerate bilinear form

$$F^{1\times n} \times F^n \longrightarrow F, \quad (\xi, \eta) \longmapsto \xi\eta = \xi_1\eta_1 + \cdots + \xi_n\eta_n. \tag{4.16}$$

This is the special case of the bilinear form $\mathcal{D}^{1\times n} \times \mathcal{F}^n \longrightarrow \mathcal{F}$, $(\xi, w) \longmapsto \xi \circ w$, from (3.76) for the injective F-cogenerator F, see Corollary 3.3.8. We apply Theorems 3.3.9 and 3.3.10. Every F-subspace $U = F^{1\times p}X \subseteq F^{1\times n}$, $X \in F^{p\times n}$, of dimension $\operatorname{rank}(X)$ and likewise every subspace V of F^n gives rise to its *orthogonal subspace*

$$\begin{aligned} U^\perp &= \{\eta \in F^n;\ U\eta = 0\} = \left\{\eta \in F^n;\ X\eta = 0\right\} = \ker(X\cdot) \text{ and} \\ V^\perp &= \{\xi \in F^{1\times n};\ \xi V = 0\} \text{ with} \\ U^{\perp\perp} &= U,\ \dim_F(U) = \operatorname{rank}(X),\ \dim_F(U^\perp) = n - \operatorname{rank}(X). \end{aligned} \tag{4.17}$$

This implies the *duality theorem of projective geometry*, namely, that the maps

$$\mathbb{P}(F^{1\times n}) \rightleftarrows \mathbb{P}(F^n), \quad U \longmapsto U^\perp, \quad V^\perp \longleftarrow V, \tag{4.18}$$

are order reversing bijections, mutually inverse to each other, and map intersections resp. sums of subspaces into sums resp. intersections.
Similarly to (4.8) we consider the column space F^n as an $F[s]$-module via

$$\begin{aligned} &s\ {}_A{\circ}\ \eta := A\eta \text{ and thus } f\ {}_A{\circ}\ \eta = f(A)\eta \quad \text{for } \eta \in F^n \text{ and } f \in F[s] \\ &\Longrightarrow (s \circ_A \xi)\eta = \xi A\eta = \xi(s\ {}_A{\circ}\ \eta),\ \left(YF^p\right)^\perp = \ker(\cdot Y),\ Y \in F^{n\times p}. \end{aligned} \tag{4.19}$$

Corollary 4.1.8 *For all $t \geq \deg_s(f_A)$, especially for all $t \geq n$, the equalities*

$$\sum_{j=1}^{p} F[s] \circ_A C_{j-} = \sum_{i=0}^{\infty} F^{1\times p}CA^i = \sum_{i=0}^{t-1} F^{1\times p}CA^i = F^{1\times tp}\begin{pmatrix} C \\ CA \\ \vdots \\ CA^{t-1} \end{pmatrix} = F^{1\times tp}\mathfrak{O}$$

hold. This space is the least $F[s]$-submodule of $(F^{1\times n}, \circ_A)$ containing the row space $F^{1\times p}C$. Application of $(-)^\perp$ implies

$$\left(\sum_{j=1}^{p} F[s] \circ_A C_{j-}\right)^{\perp} = \left(\sum_{i=0}^{t-1} F^{1\times p} C A^i\right)^{\perp} = (F^{1\times np}\mathfrak{O})^{\perp}$$
$$= \bigcap_{i=0}^{t-1} \ker(CA^i \cdot) = \ker(\mathfrak{O}\cdot) \subseteq F^n.$$

This is the largest $F[s]$-submodule of $(F^n, {}_A\circ)$ which is contained in $\ker(C\cdot)$. *The preceding identities imply the dimension formulas*

$$\dim_F\left(\sum_{j=1}^{p} F[s] \circ_A C_{j-}\right) = \operatorname{rank}(\mathfrak{O}) \quad \textit{and}$$
$$\dim_F\left(\bigcap_{i=0}^{\infty} \ker(CA^i \cdot)\right) = n - \operatorname{rank}(\mathfrak{O}).$$

Proof The assertions follow directly from (4.17)–(4.18). □

Next we characterize observable state space equations.

Theorem and Definition 4.1.9 *Compare [2, Theorem 3.5.26 on p. 75, Cor. 5.3.6 on p. 156]. For the equations from* (4.6) *and* (4.7) *the following properties are equivalent:*

1. *The epimorphism* $\begin{pmatrix} C & D \\ 0 & \mathrm{id}_m \end{pmatrix} \circ : \mathcal{B}_1 \longrightarrow \mathcal{B}_2$ *is an isomorphism.*
2. *The matrix* $\begin{pmatrix} s\,\mathrm{id}_n - A \\ C \end{pmatrix}$ *has a left inverse* $(X, Z) \in \mathcal{D}^{n\times(n+p)}$, *i.e., the identity* $X(s\,\mathrm{id}_n - A) + ZC = \mathrm{id}_n$ *holds.*
3. *The map*
$$C\cdot : \mathcal{B}_1^0 = \{x \in \mathcal{F}^n;\ s \circ x = Ax\} \longrightarrow \mathcal{B}_2^0 = \{y \in \mathcal{F}^p;\ P \circ y = 0\},\ x \mapsto Cx,$$
is an isomorphism. With Z from 2 its inverse is $Z\circ :\ \mathcal{B}_2^0 \to \mathcal{B}_1^0,\ y \mapsto Z \circ y$.
4. *The map*
$$(\cdot C)_{\mathrm{ind}} : M_2^0 = F[s]^{1\times p}/F[s]^{1\times p}P$$
$$\longrightarrow M_1^0 = F[s]^{1\times n}/F[s]^{1\times n}(s\,\mathrm{id}_n - A),\ \overline{\zeta} \mapsto \overline{\zeta C},$$
is an isomorphism. Its inverse is $(\cdot Z)_{\mathrm{ind}} :\ M_1^0 \to M_2^0,\ \overline{\eta} \mapsto \overline{\eta Z}$, *cf. 2.*
5. $F^{1\times n} = \sum_{j=1}^{p} F[s] \circ_A C_{j-} = F^{1\times pn}\mathfrak{O}$ *or* $\operatorname{rank}(\mathfrak{O}) = n$.
6. $\ker(\mathfrak{O}\cdot) = \left(\sum_{j=1}^{p} F[s] \circ_A C_{j-}\right)^{\perp} = 0.$

Equations (4.6) *that satisfy the equivalent conditions of this theorem are called* observable.

Proof The equivalences 1. $\iff$ 2. $\iff$ 3. $\iff$ 4. have been shown in Theorem 4.1.4. The equivalence 5. $\iff$ 6. follows from Corollary 4.1.8. It remains

$$2. \iff F[s]^{1\times n} = F[s]^{1\times(n+p)}\begin{pmatrix} s\,\mathrm{id}_n - A \\ C\end{pmatrix} = F[s]^{1\times n}(s\,\mathrm{id}_n - A) + F[s]^{1\times p}C$$

$$\iff F[s]^{1\times n} = F[s]^{1\times n}(s\,\mathrm{id}_n - A) + \sum_{j=1}^{p} F[s]C_{j-}$$

$$\iff F[s]^{1\times n}/F[s]^{1\times n}(s\,\mathrm{id}_n - A) = \sum_{j=1}^{p} F[s]\overline{C_{j-}}$$

$$\underset{\text{Lemma 4.1.6}}{\iff} F^{1\times n} = \sum_{j=1}^{p} F[s]\circ_A C_{j-} \iff 5. \qquad \square$$

Corollary 4.1.10 *Consider the observable equations* (4.6) *and their consequences from Theorem 4.1.9, in particular,* $(X, Z)\begin{pmatrix} s\,\mathrm{id}_n - A \\ C\end{pmatrix} = \mathrm{id}_n$. *The isomorphisms* (4.10), $(\cdot C)_{\mathrm{ind}}$ *and* $(\cdot Z)_{\mathrm{ind}}$ *induce the isomorphisms*

$$F^{1\times n} \cong F[s]^{1\times n}/F[s]^{1\times n}(s\,\mathrm{id}_n - A) \cong F[s]^{1\times p}/F[s]^{1\times p}P,\ \xi \leftrightarrow \overline{\eta} \leftrightarrow \overline{\zeta},\ \textit{where}$$

$$\xi_j = \sum_{i=1}^{n}\eta_i(A)_{ij} = \sum_{i=1}^{n}\sum_{k=1}^{p}\zeta_k(A)_{ij}C_{ki},\ \eta = \zeta C,\ \zeta = \eta Z. \tag{4.20}$$

$\Diamond$

Finally, we consider state space systems that are not necessarily observable, and construct their observable factor system. We define

$$\begin{aligned} &\mathcal{B} := \left\{\begin{pmatrix} x \\ u\end{pmatrix} \in \mathcal{F}^{n+m};\ s\circ x = Ax + Bu\right\} \text{ and} \\ &\mathcal{B}^0 := \left\{x \in \mathcal{F}^n;\ s\circ x = Ax\right\},\ \text{hence} \\ &\ker(C\circ\colon \mathcal{B}^0 \to \mathcal{F}^p) = \ker\left(\begin{pmatrix} s\,\mathrm{id}_n - A \\ C\end{pmatrix}\circ\right) \\ &\qquad \overset{\cong}{\longrightarrow} \ker\left(\begin{pmatrix} C & D \\ 0 & \mathrm{id}_m\end{pmatrix}\circ\colon \mathcal{B} \to \mathcal{F}^{p+m}\right),\ x \longmapsto \begin{pmatrix} x \\ 0\end{pmatrix}. \end{aligned} \tag{4.21}$$

Define

$$n_1 := \dim_F\left(\sum_{j=1}^{p} F[s]\circ_A C_{j-}\right) = \mathrm{rank}(\mathfrak{O}) \text{ and}$$

$$n_2 := n - n_1 = \dim_F\left(\ker(\mathfrak{O}\cdot\colon F^n \to F^{np})\right).$$

Thus, there is a matrix $V_2 \in F^{n\times n_2}$ with $V_2F^{n_2} = \ker(\mathfrak{O}\cdot)$ and $\mathrm{rank}(V_2) = \dim_F\left(\ker(\mathfrak{O}\cdot)\right) = n_2$. We extend this matrix to an $n\times(n_1+n_2)$- matrix

$$V = (V_1, V_2) \in \mathrm{Gl}_n(F) \text{ with } \begin{pmatrix} V_1^{-1} \\ V_2^{-1}\end{pmatrix} := V^{-1} \in F^{(n_1+n_2)\times n}. \tag{4.22}$$

By Corollary 4.1.8 $\ker(\mathfrak{O}\cdot)$ is an $F[s]$-submodule of $\ker(C\cdot) \subseteq F^n$ and hence $s \circ_A \ker(\mathfrak{O}\cdot) \subseteq \ker(\mathfrak{O}\cdot) = V_2 F^{n_2}$. Therefore there are matrices $A_{22} \in F^{n_2 \times n_2}$, $A_{11} \in F^{n_1 \times n_1}$, $A_{21} \in F^{n_2 \times n_1}$ such that

$$AV_2 = s \circ_A V_2 = V_2 A_{22},\ AV = A(V_1, V_2) = V \begin{pmatrix} A_{11} & 0 \\ A_{21} & A_{22} \end{pmatrix}. \tag{4.23}$$

Since $\mathfrak{O} V_2 = 0$ implies $CV_2 = 0$, we obtain $CV = (CV_1, 0) \in F^{p \times (n_1+n_2)}$.
As motivation for the next theorem we observe the equivalence of the following equations:

$$\begin{aligned}
&\begin{cases} s \circ x = Ax + Bu \\ y = Cx + D \circ u \end{cases} \overset{V^{-1}\cdot}{\iff} \begin{cases} s \circ V^{-1}x = V^{-1}AV(V^{-1}x) + V^{-1}Bu \\ y = CV(V^{-1}x) + D \circ u \end{cases} \\
&\overset{\binom{x_1}{x_2} := V^{-1}x}{\iff} \begin{cases} s \circ \binom{x_1}{x_2} = \begin{pmatrix} A_{11} & 0 \\ A_{21} & A_{22} \end{pmatrix} \binom{x_1}{x_2} + \begin{pmatrix} V_1^{-1}B \\ V_2^{-1}B \end{pmatrix} u \\ y = (CV_1, 0) \binom{x_1}{x_2} + D \circ u \\ \quad = CV_1 x_1 + D \circ u \end{cases} \\
&\iff \begin{cases} s \circ x_1 = A_{11}x_1 + V_1^{-1}Bu \\ y = CV_1 x_1 + D \circ u \\ s \circ x_2 = A_{22}x_2 + (A_{21}x_1 + V_2^{-1}Bu). \end{cases}
\end{aligned} \tag{4.24}$$

These equations suggest to introduce the new state space equations

$$s \circ x_1 = A_{11}x_1 + V_1^{-1}Bu, \quad y = CV_1 x_1 + D \circ u \tag{4.25}$$

and their behavior

$$\mathcal{B}_{\text{obs}} := \left\{ \binom{x_1}{u};\ s \circ x_1 = A_{11}x_1 + V_1^{-1}Bu \right\}.$$

Theorem 4.1.11 (Cf. [4, Theorem 6.2.2 on p. 362], [1, §5.4])

1. *The system* (4.25) *is observable, i.e.,* $\begin{pmatrix} CV_1 & D \\ 0 & \mathrm{id}_m \end{pmatrix} \circ : \mathcal{B}_{\text{obs}} \longrightarrow \mathcal{F}^{p+m}$ *is a monomorphism.*
2. *The morphism*

$$\begin{pmatrix} V_1^{-1} & 0 \\ 0 & \mathrm{id}_m \end{pmatrix} \cdot : \mathcal{B} \longrightarrow \mathcal{B}_{\text{obs}},\quad \binom{x}{u} \longmapsto \binom{V_1^{-1}x}{u}, \tag{4.26}$$

is surjective and

$$\begin{pmatrix} CV_1 & D \\ 0 & \mathrm{id}_m \end{pmatrix} \begin{pmatrix} V_1^{-1} & 0 \\ 0 & \mathrm{id}_m \end{pmatrix} \circ\Big|_{\mathcal{B}} = \begin{pmatrix} C & D \\ 0 & \mathrm{id}_m \end{pmatrix} \circ|_{\mathcal{B}}. \tag{4.27}$$

Therefore,

$$\begin{pmatrix} CV_1 & D \\ 0 & \mathrm{id}_m \end{pmatrix} \circ : \mathcal{B}_{\mathrm{obs}} \xrightarrow{\cong} \begin{pmatrix} C & D \\ 0 & \mathrm{id}_m \end{pmatrix} \circ \mathcal{B}$$
$$= \left\{ \begin{pmatrix} y \\ u \end{pmatrix} \in \mathcal{F}^{p+m};\ \exists x : \begin{array}{r} s \circ x = Ax + Bu \\ y = Cx + D \circ u \end{array} \right\} \tag{4.28}$$
$$\begin{pmatrix} x_1 \\ u \end{pmatrix} \longmapsto \begin{pmatrix} CV_1x_1 + D\circ u \\ u \end{pmatrix}$$

is an isomorphism and $\ker(C \circ |_{\mathcal{B}^0}) = \ker(V_1^{-1}|_{\mathcal{B}^0})$. *Thus,* (4.28) *is an* observable *state space representation of the image behavior* $\begin{pmatrix} C & D \\ 0 & \mathrm{id}_m \end{pmatrix} \circ \mathcal{B}$ *and called the* observable factor *of* (4.6). *It is unique up to similarity, as will be shown in Theorem 6.3.8.*

Proof 1. We show that the enlarged observability matrix of the system (4.25), namely, $\begin{pmatrix} CV_1 \\ \vdots \\ CV_1A_{11}^{n-1} \end{pmatrix} \in F^{np \times n_1}$, has rank n_1. Then the system is observable by Theorem and Definition 4.1.9 and Eq. (4.14). The equations $V^{-1}AV = \begin{pmatrix} A_{11} & 0 \\ A_{21} & A_{22} \end{pmatrix}$ and $CV = (CV_1, 0)$ imply, for all $i \in \mathbb{N}$,

$$V^{-1}A^iV = \begin{pmatrix} A_{11}^i & 0 \\ * & * \end{pmatrix},\ CA^iV = CVV^{-1}A^iV = (CV_1, 0)\begin{pmatrix} A_{11}^i & 0 \\ * & * \end{pmatrix} = (CV_1A_{11}^i, 0),$$

$$\Longrightarrow \mathfrak{O}V = \begin{pmatrix} CV \\ \vdots \\ CA^{n-1}V \end{pmatrix} = \begin{pmatrix} CV_1 & 0 \\ \vdots & \vdots \\ CV_1A_{11}^{n-1} & 0 \end{pmatrix}$$

$$\underset{V \in \mathrm{Gl}_n(F)}{\Longrightarrow} n_1 = \mathrm{rank}(\mathfrak{O}) = \mathrm{rank}(\mathfrak{O}V) = \mathrm{rank}\begin{pmatrix} CV_1 \\ \vdots \\ CV_1A_{11}^{n-1} \end{pmatrix}.$$

2. The motivational equations (4.21) imply that (4.26) is well defined, i.e., that $\begin{pmatrix} V_1^{-1} & 0 \\ 0 & \mathrm{id}_m \end{pmatrix} \circ$ maps $\mathcal{B}$ into $\mathcal{B}_{\mathrm{obs}}$. Equation (4.24) implies (4.27).
It remains to show that (4.26) is surjective. For this, let $\begin{pmatrix} x_1 \\ u \end{pmatrix} \in \mathcal{B}_{\mathrm{obs}}$. Since $\chi_{A_{22}} = \det(s\,\mathrm{id}_{n_2} - A_{22}) \neq 0$, the map

$$\cdot(s\,\mathrm{id}_{n_2} - A_{22}) : F[s]^{1 \times n_2} \longrightarrow F[s]^{1 \times n_2}$$

is injective and, by duality, $(s\,\mathrm{id}_{n_2} - A_{22})\circ : \mathcal{F}^{n_2} \longrightarrow \mathcal{F}^{n_2}$ is surjective. In particular, for arbitrary u there is a solution x_2 of the last equation in (4.24), namely, of $(s\,\mathrm{id}_{n_2} - A_{22}) \circ x_2 = A_{21}x_1 + V_2^{-1}Bu$. Define $x := V\begin{pmatrix} x_1 \\ x_2 \end{pmatrix}$ and thus $\begin{pmatrix} x_1 \\ x_2 \end{pmatrix} = \begin{pmatrix} V_1^{-1} \\ V_2^{-1} \end{pmatrix} x$ and $x_1 = V_1^{-1}x$. Equation (4.24) implies $\begin{pmatrix} x \\ u \end{pmatrix} \in \mathcal{B}$ and

$$\begin{pmatrix} x_1 \\ u \end{pmatrix} = \begin{pmatrix} V_1^{-1} & 0 \\ 0 & \mathrm{id}_m \end{pmatrix}\begin{pmatrix} x \\ u \end{pmatrix} \in \begin{pmatrix} V_1^{-1} & 0 \\ 0 & \mathrm{id}_m \end{pmatrix} \circ \mathcal{B} \Longrightarrow \mathcal{B}_{\mathrm{obs}} = \begin{pmatrix} V_1^{-1} & 0 \\ 0 & \mathrm{id}_m \end{pmatrix} \circ \mathcal{B}.$$

That (4.28) is well defined and an isomorphism follows directly from 1, the epimorphism (4.26) and (4.27). □

4.2 Definition and Algebraic Characterization of Controllable and Autonomous Behaviors

In this section we will define and characterize controllability and autonomy of a behavior algebraically, and in Sect. 4.3, *for the standard signal modules, by means of the behavior's trajectories. These two behavior properties are opposites: The trajectories of an autonomous behavior are determined by their initial values, whereas those of a controllable behavior can be steered to a desired trajectory or, in case of state space systems, to a desired state.*

We assume that $\mathcal{D}$ is a Noetherian integral domain with quotient field $K = \mathrm{quot}(\mathcal{D})$ and that ${}_{\mathcal{D}}\mathcal{F}$ is an injective cogenerator.
As before we consider a matrix $R \in \mathcal{D}^{k\times l}$ and the induced objects

$$\begin{aligned} &U := \mathcal{D}^{1\times k}R \subseteq \mathcal{D}^{1\times l},\ M := \mathcal{D}^{1\times l}/U,\ \mathcal{B} := U^{\perp} = \ker(R\circ) \subseteq \mathcal{F}^{l} \\ &\Longrightarrow U = \mathcal{B}^{\perp} \text{ and } M = \mathrm{M}(\mathcal{B}) = \mathcal{D}^{1\times l}/\mathcal{B}^{\perp}. \end{aligned} \tag{4.29}$$

Definition 4.2.1 An element $x \in M$ is called *free* if it is linearly independent, i.e., if the map $\mathcal{D} \longrightarrow M,\ \ d \longmapsto dx$, is injective. If it is linearly dependent or the preceding map is not injective, the element x is also called *bound*. ◊

The subset

$$\mathrm{t}(M) := \{x \in M;\ \exists d \in \mathcal{D},\ d \neq 0,\ \text{with } dx = 0\} \tag{4.30}$$

of all linearly dependent elements of M is a submodule of M. It is called the *torsion (sub)module* of M. The module M is called *torsionfree* if $\mathrm{t}(M) = 0$. The factor module $M/\mathrm{t}(M)$ is always torsionfree.

Lemma 4.2.2 *1. A linear map $\varphi\colon M_1 \longrightarrow M_2$ induces linear maps*

$$\begin{aligned} &\varphi|_{\mathrm{t}(M_1)} : \mathrm{t}(M_1) \longrightarrow \mathrm{t}(M_2),\ x_1 \longmapsto \varphi(x_1),\ \textit{and} \\ &\varphi_{\mathrm{ind}} : M_1/\mathrm{t}(M_1) \longrightarrow M_2/\mathrm{t}(M_2),\ \overline{x_1} = x_1 + \mathrm{t}(M_1) \longmapsto \overline{\varphi(x_1)}. \end{aligned}$$

In other words, the assignments

$$M \longmapsto \mathrm{t}(M),\ \varphi \longmapsto \varphi|_{\mathrm{t}(M)},\ \textit{and } M \longmapsto M/\mathrm{t}(M),\ \varphi \longmapsto \varphi_{\mathrm{ind}},$$

are additive functors on the category of $\mathcal{D}$-modules. In particular, if φ is an isomorphism then so are $\varphi|_{\mathrm{t}(M)}$ and φ_{ind}.

2. *If φ is injective, then $\varphi^{-1}(\mathrm{t}(M_2)) = \mathrm{t}(M_1)$ or, equivalently, $\mathrm{t}(M_2) \cap \varphi(M_1) = \varphi\varphi^{-1}(\mathrm{t}(M_2)) = \varphi(\mathrm{t}(M_1))$. Moreover φ_{ind} is injective too.*

3. *The functor* $M \longmapsto \mathrm{t}(M)$ *is left exact, i.e., if*

$$0 \longrightarrow M_1 \xrightarrow{\varphi} M_2 \xrightarrow{\psi} M_3$$

is exact, then so is

$$0 \longrightarrow \mathrm{t}(M_1) \xrightarrow{\varphi|_{\mathrm{t}(M_1)}} \mathrm{t}(M_2) \xrightarrow{\psi|_{\mathrm{t}(M_2)}} \mathrm{t}(M_3).$$

In general, the surjectivity of $\psi\colon\ M_2 \longrightarrow M_3$ *does not imply that of* $\psi|_{\mathrm{t}(M_2)}\colon\ \mathrm{t}(M_2) \longrightarrow \mathrm{t}(M_3)$. *If, for instance,* $M_2 = \mathcal{D}$, $\mathfrak{a} \lneq \mathcal{D}$ *is a proper nonzero ideal and* $M_3 = \mathcal{D}/\mathfrak{a}$, *then* $\mathrm{t}(\mathcal{D}) = 0$ *and* $\mathrm{t}(\mathcal{D}/\mathfrak{a}) = \mathcal{D}/\mathfrak{a} \neq 0$.

4. *If* $M_1 = \mathrm{t}(M_1)$ *is a torsion module and if*

$$0 \longrightarrow M_1 \xrightarrow{\varphi} M_2 \xrightarrow{\psi} M_3 \longrightarrow 0 \qquad (4.31)$$

is exact, then so is

$$0 \longrightarrow \mathrm{t}(M_1) \xrightarrow{\varphi|_{\mathrm{t}(M_1)}} \mathrm{t}(M_2) \xrightarrow{\psi|_{\mathrm{t}(M_2)}} \mathrm{t}(M_3) \longrightarrow 0. \qquad (4.32)$$

Proof 1. Assume $x_1 \in \mathrm{t}(M_1)$ and $d \neq 0$ with $dx_1 = 0$. Then also $d\varphi(x_1) = \varphi(dx_1) = 0$, i.e., $\varphi(x_1) \in \mathrm{t}(M_2)$, and hence $\varphi(\mathrm{t}(M_1)) \subseteq \mathrm{t}(M_2)$. This inclusion and Lemma 3.3.13 imply the existence of φ_{ind}.
2. The inclusion $\varphi(\mathrm{t}(M_1)) \subseteq \mathrm{t}(M_2)$ implies $\mathrm{t}(M_1) \subseteq \varphi^{-1}(\mathrm{t}(M_2))$. For the converse, let $x_1 \in \varphi^{-1}(\mathrm{t}(M_2))$. Then $\varphi(x_1) \in \mathrm{t}(M_2)$ and there exists $d \neq 0$ with $0 = d\varphi(x_1) = \varphi(dx_1)$. Since φ is injective, $dx_1 = 0$ and $x_1 \in \mathrm{t}(M_1)$ follow.
Finally, the kernel of φ_{ind} is

$$\ker(\varphi_{\mathrm{ind}}) = \varphi^{-1}(\mathrm{t}(M_2))/\,\mathrm{t}(M_1) = \mathrm{t}(M_1)/\,\mathrm{t}(M_1) = 0.$$

3. Since φ is injective, obviously, $\varphi|_{\mathrm{t}(M_1)}$ is injective too. Moreover,

$$\varphi(\mathrm{t}(M_1)) \overset{2.}{=} \mathrm{t}(M_2) \cap \varphi(M_1) = \mathrm{t}(M_2) \cap \ker(\psi) = \ker\left(\psi|_{\mathrm{t}(M_2)}\right).$$

4. We have still to show $\psi(\mathrm{t}(M_2)) = \mathrm{t}(M_3)$ only. For $x_3 \in \mathrm{t}(M_3)$ there is $d_3 \neq 0$ with $d_3x_3 = 0$. Since ψ is surjective, $x_3 = \psi(x_2)$ for some $x_2 \in M_2$. We infer

$$\begin{aligned}
&\psi(d_3x_2) = d_3x_3 = 0 \Longrightarrow d_3x_2 \in \ker(\psi) = \mathrm{im}(\varphi)\\
&\Longrightarrow \exists x_1 \in M_1 = \mathrm{t}(M_1) \text{ with } d_3x_2 = \varphi(x_1)\\
&\Longrightarrow \exists d_1 \neq 0 \text{ with } d_1x_1 = 0 \text{ and } d_1d_3x_2 = \varphi(d_1x_1) = \varphi(0) = 0\\
&\underset{d_1d_3\neq 0}{\Longrightarrow} x_2 \in \mathrm{t}(M_2) \text{ with } \psi(x_2) = x_3 \Longrightarrow \psi(\mathrm{t}(M_2)) = \mathrm{t}(M_3).
\end{aligned}$$

□

According to Corollary and Definition 3.3.19, we have the $\mathcal{D}$-isomorphisms

$$\begin{aligned} \mathrm{M}(\mathcal{B}) = \mathcal{D}^{1\times l}/\mathcal{B}^{\perp} &\overset{\cong}{\longleftrightarrow} \mathrm{Hom}_{\mathcal{D}}(\mathcal{D}, \mathrm{M}(\mathcal{B})) &\overset{\cong}{\longleftrightarrow} \mathrm{Hom}(\mathcal{B}, \mathcal{F}), \\ \overline{\xi} &\longleftrightarrow \quad (\cdot\xi)_{\mathrm{ind}} &\longleftrightarrow \xi\circ, \quad \xi \in \mathcal{D}^{1\times l}. \end{aligned} \tag{4.33}$$

The elements of $M(\mathcal{B})$ and $\mathrm{Hom}(\mathcal{B}, \mathcal{F})$ are the observables of $\mathcal{B}$.

Corollary 4.2.3 *An observable $\xi\circ\colon \mathcal{B} \longrightarrow \mathcal{F}$, $\overline{\xi} \in M(\mathcal{B})$, is free if and only if $\xi\circ$ is surjective or $\xi \circ \mathcal{B} = \mathcal{F}$. Hence, ξ is bound if and only if*

$$\xi \circ \mathcal{B} \subsetneq \mathcal{F} \Longleftrightarrow \exists d \neq 0 \text{ with } d\overline{\xi} = 0 \Longleftrightarrow \exists d \neq 0 \text{ with } d \circ (\xi \circ \mathcal{B}) = d\xi \circ \mathcal{B} = 0.$$

Proof By duality, $\xi\circ\colon \mathrm{D}M(\mathcal{B}) = \mathcal{B} \to \mathrm{D}\mathcal{D} = \mathcal{F}$ is surjective if and only if $\mathcal{D} \to M(\mathcal{B})$, $d \mapsto d\overline{\xi}$, is injective. □

Definition 4.2.4 The behavior $\mathcal{B}$ is called *controllable* if its nonzero observables are free or, in other terms, if the module $\mathrm{M}(\mathcal{B})$ is torsionfree.
The system $\mathcal{B}$ is called *autonomous* if all its observables are bound or $\mathrm{M}(\mathcal{B})$ is a torsion module. ◊

Lemma 4.2.5 *1. A subbehavior and an image of an autonomous behavior are again autonomous.*
2. An image of a controllable behavior is controllable.

Proof If $P \circ \mathcal{B}$ is an image of $\mathcal{B}$, the epimorphism $P\circ\colon \mathcal{B} \longrightarrow P \circ \mathcal{B}$ induces the dual monomorphism $(\cdot P)_{\mathrm{ind}}\colon \mathrm{M}(P \circ \mathcal{B}) \longrightarrow \mathrm{M}(\mathcal{B})$, i.e., $\mathrm{M}(P \circ \mathcal{B}) \subseteq \mathrm{M}(\mathcal{B})$ up to isomorphism. If $\mathcal{B}$ is autonomous, the module $\mathrm{M}(\mathcal{B})$ is a torsion module, and so is its submodule $M(P \circ \mathcal{B})$, and hence $P \circ \mathcal{B}$ is also autonomous. Similarly, if $\mathcal{B}$ is controllable, then $\mathrm{M}(\mathcal{B})$ is torsion free, and so is its submodule $\mathrm{M}(P \circ \mathcal{B})$. Hence, $P \circ \mathcal{B}$ is controllable.

If $\mathcal{B}_1$ is a subbehavior of the autonomous behavior $\mathcal{B}$, the injection $\mathrm{inj}\colon \mathcal{B}_1 \longrightarrow \mathcal{B}$ induces the dual surjection $\mathrm{inj}_{\mathrm{ind}}\colon \mathrm{M}(\mathcal{B}) \longrightarrow \mathrm{M}(\mathcal{B}_1)$. Since $M(\mathcal{B})$ is a torsion module, so is its image $M(\mathcal{B}_1)$, and hence $\mathcal{B}_1$ is autonomous too. □

The annihilator of a $\mathcal{D}$-module N is the ideal $\mathrm{ann}_{\mathcal{D}}(N) = \{d \in \mathcal{D};\ dN = 0\}$.

Lemma 4.2.6 *1.* $\mathrm{ann}_{\mathcal{D}}\left(\mathrm{M}(\mathcal{B})\right) = \mathrm{ann}_{\mathcal{D}}(\mathcal{B})$.
2. $\mathrm{M}(\mathcal{B})$ *is a torsion module if and only if* $\mathrm{ann}_{\mathcal{D}}(M(\mathcal{B})) = \mathrm{ann}_{\mathcal{D}}(\mathcal{B}) \neq 0$.

Proof 1. Let $M := M(\mathcal{B})$. By duality, $d\cdot : M \to M,\ x \mapsto dx$, is zero if and only if $d\circ = \mathrm{D}(d\cdot) : \mathcal{B} \to \mathcal{B}$ is zero, hence $d \in \mathrm{ann}_{\mathcal{D}}(M) \Longleftrightarrow d \in \mathrm{ann}_{\mathcal{D}}(\mathcal{B})$.
2. (a) If $d \neq 0$ and $dM = 0$ then $dx = 0$ for all $x \in M$ and hence $M = \mathrm{t}(M)$.
(b) We use $M = \sum_{i=1}^{l} \mathcal{D}x_i,\ x_i := \overline{\delta_i}$. Assume $M = \mathrm{t}(M)$. Therefore

$$\forall i = 1, \ldots, l \exists d_i \neq 0 \text{ with } d_i x_i = 0 \Longrightarrow d := \prod_{i=1}^{l} d_i \neq 0 \text{ and } \forall i : \ dx_i = 0$$

$$\Longrightarrow dM = \sum_{i=1}^{l} \mathcal{D} dx_i = 0 \Longrightarrow d \in \operatorname{ann}_{\mathcal{D}}(M). \qquad \square$$

Corollary 4.2.7 *The behavior $\mathcal{B}$ is autonomous if and only if there is a nonzero $d \in \mathcal{D}$ with $d \circ \mathcal{B} = 0$, i.e., if and only if all trajectories $w \in \mathcal{B}$ satisfy the same equation $d \circ w = 0$ for some nonzero d.* ◊

4.2.1 The Decomposition of Finitely Generated Modules and Behaviors over Principal Ideal Domains

We assume that $\mathcal{D}$ is principal ideal domain. We will relate the torsion module $\mathrm{t}(M)$ of $M = \mathcal{D}^{1\times l}/\mathcal{D}^{1\times k} R$ *as well as the module $M/\mathrm{t}(M)$ with the data from the Smith form of R.*

For a matrix $R \in \mathcal{D}^{k\times l}$ consider its Smith form, cf. Result 3.2.17:

$$S := URV = \begin{pmatrix} \Delta & 0 \\ 0 & 0 \end{pmatrix} \text{ with } p := \operatorname{rank}(R),$$
$$\Delta := \operatorname{diag}(e_1, \ldots, e_p), \quad e_1 | \cdots | e_p \neq 0, \quad U \in \mathrm{Gl}_k(\mathcal{D}), \text{ and } \quad V \in \mathrm{Gl}_l(\mathcal{D}). \tag{4.34}$$

In addition, we define $e_j := 0$ for $j = p+1, \ldots, l$. The row module of S is

$$\mathcal{D}^{1\times k} S = \bigoplus_{j=1}^{p} \mathcal{D} e_j \delta_j = \bigoplus_{j=1}^{l} \mathcal{D} e_j \delta_j, \text{ hence}$$

$$\mathcal{D}^{1\times l}/\mathcal{D}^{1\times k} S = \left(\bigoplus_{j=1}^{l} \mathcal{D}\delta_j\right) / \left(\bigoplus_{j=1}^{l} \mathcal{D} e_j \delta_j\right) \overset{\cong}{\longleftrightarrow} \prod_{j=1}^{l} \mathcal{D}/\mathcal{D} e_j,$$
$$\overline{(\eta_1, \ldots, \eta_l)} \longleftrightarrow (\overline{\eta_1}, \ldots, \overline{\eta_l}), \text{ where}$$
$$\overline{\eta} := \eta + \mathcal{D}^{1\times k} S \text{ for } \eta \in \mathcal{D}^{1\times k} \quad \text{and} \quad \overline{\eta_j} := \eta_j + \mathcal{D} e_j \text{ for } \eta_j \in \mathcal{D}.$$

In the sequel we will identify these modules, i.e.,

$$\mathcal{D}^{1\times l}/\mathcal{D}^{1\times k} S = \prod_{j=1}^{l} \mathcal{D}/\mathcal{D} e_j = \prod_{j=1}^{p} \mathcal{D}/\mathcal{D} e_j \times \mathcal{D}^{1\times m}, \ m := l - p, \tag{4.35}$$
$$\overline{(\eta_1, \ldots, \eta_l)} = (\overline{\eta_1}, \ldots, \overline{\eta_l}) = (\overline{\eta_1}, \ldots, \overline{\eta_p}, \eta_{p+1}, \ldots, \eta_l).$$

Then

$$\mathcal{D}^{1\times l}/\mathcal{D}^{1\times k}S = \bigoplus_{j=1}^{l} \mathcal{D}\overline{\delta_j} \quad \text{and} \quad \operatorname{ann}_{\mathcal{D}}(\overline{\delta_j}) = \mathcal{D}e_j. \tag{4.36}$$

The behavior $\ker(S\circ)$ is

$$\begin{aligned}\ker(S\circ\colon \mathcal{F}^l \to \mathcal{F}^k) &= (\mathcal{D}^{1\times k}S)^{\perp} = \{v \in \mathcal{F}^l;\ \forall j = 1,\dots,l\colon\ e_j \circ v_j = 0\}\\ &= \prod_{j=1}^{l} \operatorname{ann}_{\mathcal{F}}(e_j) = \prod_{j=1}^{p} \operatorname{ann}_{\mathcal{F}}(e_j) \times \mathcal{F}^m.\end{aligned} \tag{4.37}$$

We decompose U, V, and V^{-1} in block form as

$$\begin{aligned} U &= \begin{pmatrix} U_I \\ U_{II} \end{pmatrix} \in \mathcal{D}^{(p+(k-p))\times k},\ V = (V_I, V_{II}) \in \mathcal{D}^{l\times(p+m)},\ \text{and}\\ V^{-1} &= \begin{pmatrix} V_I^{-1} \\ V_{II}^{-1} \end{pmatrix} = \begin{pmatrix} (V^{-1})_I \\ (V^{-1})_{II} \end{pmatrix} \in \mathcal{D}^{(p+m)\times l}.\end{aligned} \tag{4.38}$$

Theorem and Definition 4.2.8 *1. The invertible matrix $V \in \mathrm{Gl}_l(\mathcal{D})$ induces the $\mathcal{D}$-isomorphism $\cdot V : \mathcal{D}^{1\times k}R \cong \mathcal{D}^{1\times k}S = \bigoplus_{j=1}^{p} \mathcal{D}e_j\delta_j$, hence*

$$\mathcal{D}^{1\times k}R = \bigoplus_{j=1}^{p} \mathcal{D}e_j V_{j-}^{-1} \ \textit{and}$$

$$\begin{aligned} \mathrm{M}(\mathcal{B}) = \mathcal{D}^{1\times l}/\mathcal{D}^{1\times k}R &\overset{\cong}{\longleftrightarrow} \mathcal{D}^{1\times l}/\mathcal{D}^{1\times k}S = \prod_{j=1}^{l} \mathcal{D}/\mathcal{D}e_j,\\ \overline{\xi} = \xi + \mathcal{D}^{1\times k}R &\longmapsto \overline{\xi V} = (\cdot V)_{\mathrm{ind}}(\overline{\xi}),\ \overline{\eta V^{-1}} \longleftarrow \overline{\eta},\end{aligned} \tag{4.39}$$

with the inverse map $\left((\cdot V)_{\mathrm{ind}}\right)^{-1} = (\cdot V^{-1})_{\mathrm{ind}}$. Thus,

$$\mathrm{M}(\mathcal{B}) = \bigoplus_{j=1}^{l} \mathcal{D}\overline{V_{j-}^{-1}} \quad \textit{and} \quad \operatorname{ann}_{\mathcal{D}}(\overline{V_{j-}^{-1}}) = \mathcal{D}e_j. \tag{4.40}$$

2. By duality, V also induces the mutually inverse behavior isomorphisms

$$\mathcal{B} \overset{\cong}{\longleftrightarrow} \prod_{j=1}^{l} \operatorname{ann}_{\mathcal{F}}(e_j),\ w = V \circ v \leftrightarrow v = V^{-1} \circ w. \tag{4.41}$$

3. *The equation* $U_I RV = (\Delta, 0)$ *implies*

$$\mathcal{D}^{1\times k}R = \mathcal{D}^{1\times p}U_I R,\ \operatorname{rank}(U_I R) = p,\ \mathcal{B} = \ker(R\circ) = \ker(U_I R\circ).$$

Therefore, by replacing $R \in \mathcal{D}^{k\times l}$ *by* $U_I R \in \mathcal{D}^{p\times l}$, *one may assume that the rows of the defining matrix of a behavior are linearly independent or that R has* $k = \operatorname{rank}(R) = p$ *rows. This assumption will often be made.*
4. *The elementary divisors* e_i *depend on the submodule* $\mathcal{D}^{1\times k}R$ *only and are called the* elementary divisors *of* $\mathcal{D}^{1\times k}R \subseteq \mathcal{D}^{1\times l}$.

Proof 1. We have $\mathcal{D}^{1\times k}RV = \mathcal{D}^{1\times k}U^{-1}URV = \mathcal{D}^{1\times k}S$, hence $\mathcal{D}^{1\times k}R = \mathcal{D}^{1\times k}SV^{-1}$. This and Lemma 3.3.13 imply the induced maps

$$(\cdot V)_{\text{ind}}\colon \mathcal{D}^{1\times l}/\mathcal{D}^{1\times k}R \longrightarrow \mathcal{D}^{1\times l}/\mathcal{D}^{1\times k}S,\quad \overline{\xi} \longmapsto \overline{\xi V} \quad \text{and}$$
$$(\cdot V^{-1})_{\text{ind}}\colon \mathcal{D}^{1\times l}/\mathcal{D}^{1\times k}S \longrightarrow \mathcal{D}^{1\times l}/\mathcal{D}^{1\times k}R,\quad \overline{\eta} \longmapsto \overline{\eta V^{-1}}.$$

They are obviously inverse of each other and therefore isomorphisms. Application of $(\cdot V^{-1})_{\text{ind}}$ to (4.36) with $(\cdot V^{-1})_{\text{ind}}(\overline{\delta_j}) = \overline{\delta_j V^{-1}} = \overline{V_{j-}^{-1}}$ furnishes the statements in (4.40).
2. This follows directly from (4.39) and by duality, cf. Theorem 3.3.18.
3. Since the last $k - p$ rows of S are zero, we get $UR = SV^{-1} = \binom{U_I R}{0}$, hence

$$\begin{aligned}\mathcal{D}^{1\times k}R &= \mathcal{D}^{1\times k}UR = \mathcal{D}^{1\times p}U_I R,\\ p &= \operatorname{rank}(R) = \operatorname{rank}(U_I R), \quad \text{and}\\ \ker(R\circ) &= (\mathcal{D}^{1\times k}R)^{\perp} = (\mathcal{D}^{1\times k}U_I R)^{\perp} = \ker(U_I R\circ).\end{aligned}$$

4. Assume $\mathcal{D}^{1\times k}R = \mathcal{D}^{1\times k'}R'$, hence $\operatorname{rank}(R') = \operatorname{rank}(R) = p$. Let $U'R'V' = \begin{pmatrix}\Delta' & 0\\ 0 & 0\end{pmatrix}$ with $\Delta' \in \mathcal{D}^{p\times p}$ be the Smith form of R' and $U' = \begin{pmatrix}U'_I\\ U'_{II}\end{pmatrix} \in \mathcal{D}^{(p+(k'-p))\times k'}$. With this, we obtain $\mathcal{D}^{1\times k}R = \mathcal{D}^{1\times p}U_I R = \mathcal{D}^{1\times p}U'_I R'$ and the Smith forms $(U_I R)V = (\Delta, 0)$ and $(U'_I R')V' = (\Delta', 0)$.
Moreover there are matrices $X, X' \in \mathcal{D}^{p\times p}$ with $U'_I R' = XU_I R$ and $U_I R = X'U'_I R'$. The linear independence of the rows of $U_I R$ and $U'_I R'$ implies $XX' = \mathrm{id}_p$ and hence $X \in \mathrm{Gl}_p(\mathcal{D})$. Therefore the matrices $U_I R$ and $U'_I R'$ are equivalent and have the same Smith form $(\Delta, 0) = (\Delta', 0)$, $\Delta = \Delta'$. □

Theorem 4.2.9 *1. The torsion module of* $M := \mathrm{M}(\mathcal{B})$ *is*

$$\mathrm{t}(M) = \bigoplus_{j=1}^{p} \mathcal{D}\overline{V_{j-}^{-1}} \cong \prod_{j=1}^{p} \mathcal{D}/\mathcal{D}e_j$$

and has the annihilator

$$\operatorname{ann}_{\mathcal{D}}\left(\mathrm{t}(M)\right) = \bigcap_{j=1}^{p} \mathcal{D}e_j = \mathcal{D}e_p.$$

In particular, the ideal $\mathcal{D}e_p$ is an invariant of $\mathrm{t}(M)$, thus of M and of $\mathcal{B}$.

2. *The torsion submodule is constructively given as*

$$\mathrm{t}(M) = \mathcal{D}^{1\times p}V_I^{-1}/\mathcal{D}^{1\times k}R \subseteq M = \mathcal{D}^{1\times l}/\mathcal{D}^{1\times k}R,$$

where $V_I^{-1} \in \mathcal{D}^{p\times l}$ is the matrix from (4.38).

3. *The module*

$$N := \bigoplus_{j=p+1}^{l} \mathcal{D}\overline{V_{j-}^{-1}} \cong \bigoplus_{j=p+1}^{l} \mathcal{D}\delta_j = \mathcal{D}^{1\times m}$$

is free and has the basis $\overline{V_{j-}^{-1}}$, $j = p+1, \ldots, l$ of length $m = l - p$.
In particular, M can be written as the direct sum $M = \mathrm{t}(M) \oplus N$ and the isomorphisms $M/\mathrm{t}(M) \cong N \cong \mathcal{D}^{1\times m}$ hold.

Proof 1. Theorem and Definition 4.2.8, 1, implies

$$\operatorname{ann}_{\mathcal{D}}\left(\overline{V_{j-}^{-1}}\right) = \operatorname{ann}_{\mathcal{D}}(\overline{\delta_j}) = \mathcal{D}e_j = \begin{cases} \neq 0 & \text{if } j \leq p \\ = 0 & \text{if } j > p \end{cases}, \text{ hence}$$

$$\mathrm{t}\left(\mathcal{D}\overline{V_{j-}^{-1}}\right) = \begin{cases} \mathcal{D}\overline{V_{j-}^{-1}} & \text{if } j \leq p \\ 0 & \text{if } j > p \end{cases}.$$

With the obvious equation $\mathrm{t}(M_1 \oplus \cdots \oplus M_l) = \mathrm{t}(M_1) \oplus \cdots \oplus \mathrm{t}(M_l)$ for a direct sum of modules we conclude

$$\mathrm{t}(M) = \mathrm{t}\left(\bigoplus_{j=1}^{l} \mathcal{D}\overline{V_{j-}^{-1}}\right) = \bigoplus_{j=1}^{l} \mathrm{t}\left(\mathcal{D}\overline{V_{j-}^{-1}}\right) = \bigoplus_{j=1}^{p} \mathcal{D}\overline{V_{j-}^{-1}}.$$

2. $$\mathcal{D}^{1\times k}R = \underbrace{\mathcal{D}^{1\times k}U^{-1}}_{=\mathcal{D}^{1\times k}}\underbrace{URV}_{=S}V^{-1} = \mathcal{D}^{1\times k}SV^{-1} = \bigoplus_{j=1}^{p} \mathcal{D}e_jV_{j-}^{-1} \subseteq \bigoplus_{j=1}^{p} \mathcal{D}V_{j-}^{-1}$$

$$\Longrightarrow \mathrm{t}(M) \underset{1.}{=} \bigoplus_{j=1}^{p} \mathcal{D}\overline{V_{j-}^{-1}} = \left(\bigoplus_{j=1}^{p} \mathcal{D}V_{j-}^{-1}\right)/\mathcal{D}^{1\times k}R = \mathcal{D}^{1\times p}V_I^{-1}/\mathcal{D}^{1\times k}R.$$

3. The isomorphism (4.39) induces the isomorphism

$$N := \bigoplus_{j=p+1}^{l} \mathcal{D}\overline{V_{j-}^{-1}} \overset{\cong}{\longleftrightarrow} \bigoplus_{j=p+1}^{l} \mathcal{D}\delta_j = \prod_{j=p+1}^{l} \mathcal{D}/\mathcal{D}e_j = \mathcal{D}^{1\times m}.$$

Hence N is free and

$$M = \bigoplus_{j=1}^{l} \mathcal{D}\overline{V_{j-}^{-1}} = \bigoplus_{j=1}^{p} \mathcal{D}\overline{V_{j-}^{-1}} \oplus \bigoplus_{j=p+1}^{l} \mathcal{D}\overline{V_{j-}^{-1}} = \mathrm{t}(M) \oplus N$$
$$\Longrightarrow M/\mathrm{t}(M) \cong N \cong \mathcal{D}^{1\times m}.$$

□

Remark 4.2.10 *Invariant factors.* In Theorem 4.2.9, we have seen that e_p is an invariant of M, up to association. But, indeed, all e_i which are nonunits are invariants of M. To concretize this, let $q \le p$ be such that

$$\mathcal{D} = \mathcal{D}e_1 = \cdots = \mathcal{D}e_q \supsetneq \mathcal{D}e_{q+1} \supseteq \cdots \supseteq \mathcal{D}e_p \supsetneq \mathcal{D}e_{p+1} = \cdots = \mathcal{D}e_l = 0,$$

i.e., the e_j which are units are those with $j \in \{1, \ldots, q\}$. Let $n := l - q$ and define $a_j := e_{j+q}$ for $j = 1, \ldots, n$. Then $\mathcal{D} \supsetneq \mathcal{D}a_1 \supseteq \cdots \supseteq \mathcal{D}a_n$ and

$$M \cong \prod_{j=q+1}^{l} \mathcal{D}/\mathcal{D}e_j \cong \prod_{j=1}^{n} \mathcal{D}/\mathcal{D}a_j, \tag{4.42}$$

i.e., M is a direct product of nonzero cyclic modules. The number n of nonzero factors $\mathcal{D}/\mathcal{D}a_j$ of M and the ideals $\mathcal{D}a_j$ are uniquely determined by (the isomorphism class of) M. The elements a_j are unique up to association and are called the *invariant factors* of M. This implies that two finitely generated modules are isomorphic if and only if they have the same invariant factors, whence the terminology.

The proof of this uniqueness result, which can be found, e.g., in [3, Theorem 7.8 on p. 153], is more difficult than the proof of the uniqueness of $m = \dim_{\mathcal{D}}(M/\mathrm{t}(M))$ and of e_p with $\mathrm{ann}_{\mathcal{D}}(\mathrm{t}(M)) = \mathcal{D}e_p$ which we gave in items 3 and 1 of Theorem 4.2.9. The uniqueness of the invariant factors is not a simple consequence of that of the elementary divisors e_j of R since R itself is not an invariant of the isomorphism class of M.

However, the converse implication, namely, that the invariant factors a_j, $j = 1, \ldots, n$, determine the number m and the elementary divisors $e_1, \ldots, e_p$, is simple. Indeed, Eq. (4.42) implies

$$\begin{aligned}
m &= \dim_{\mathcal{D}}(M/\mathrm{t}(M)) = \sharp\{j \le n;\ a_j = 0\}\\
q &= l - n,\\
e_1 &= \cdots = e_q = 1 \text{ (up to association), and}\\
e_j &= a_{j-q} \text{ for } j = q+1, \ldots, p.
\end{aligned}$$

It is obvious that neither the number l of generators of $M = \sum_{j=1}^{l} \mathcal{D}\overline{\delta_j}$ nor the rank p of R are invariants of M.

Similarity invariants. Consider a field F, a matrix $A \in F^{n\times n}$, the polynomial algebra $\mathcal{D} := F[s]$, and the characteristic matrix $s\,\mathrm{id}_n - A \in F[s]^{n\times n}$ of rank n. We use the $F[s]$-isomorphism

$$F[s]^{1\times n}/F[s]^{1\times n}(s\,\mathrm{id}_n - A) \overset{\cong}{\longrightarrow} (F^{1\times n}, \circ_A), \quad \overline{\delta_j} \longmapsto \delta_j,$$

from Lemma 4.1.6.
The characteristic matrix $s\,\mathrm{id}_n - A$ has rank n and therefore n nonzero elementary divisors $e_1 | \cdots | e_n$ which give rise to the isomorphism

$$(F^{1\times n}, \circ_A) \cong F[s]^{1\times n}/F[s]^{1\times n}(s\,\mathrm{id}_n - A) \cong \prod_{j=1}^{n} F[s]/F[s]e_j.$$

According to item 1 above, the nonunit elementary divisors e_j are the invariant factors of this $F[s]$-module. In this context, the elementary divisors are called *similarity invariants*, and this terminology will be justified in the following theorem. ◊

Theorem and Definition 4.2.11 *For two matrices $A_1, A_2 \in F^{n\times n}$, the following properties are equivalent:*

1. *A_1 and A_2 are similar.*
2. *The characteristic matrices $s\,\mathrm{id}_n - A_1$ and $s\,\mathrm{id}_n - A_2$ are similar.*
3. *The characteristic matrices $s\,\mathrm{id}_n - A_1$ and $s\,\mathrm{id}_n - A_2$ are equivalent, i.e., they have the same elementary divisors $e_1 \mid e_2 \mid \cdots \mid e_n \neq 0$.*

For an obvious reason, these e_i are called the similarity invariants *of A_i.*

Proof 1. $\Longrightarrow$ 2. $\Longrightarrow$ 3. Obvious.
3. $\Longrightarrow$ 1. According to the preceding remarks, we obtain an $F[s]$-isomorphism

$$\cdot U : (F^{1\times n}, \circ_{A_1}) \overset{\cong}{\longrightarrow} (F^{1\times n}, \circ_{A_2}), \quad \xi \longmapsto \xi U,$$

with some matrix $U \in \mathrm{Gl}_n(F)$. The $F[s]$-linearity signifies that

$$\xi A_1 U = (s \circ_{A_1} \xi)U = s \circ_{A_2} (\xi U) = \xi U A_2 \quad \text{for all } \xi \in F^{1\times n}.$$

Hence, $A_1 U = U A_2$ or$A_2 = U^{-1} A_1 U$, i.e., A_1 and A_2 are similar. □

4.2.2 The Algebraic Characterization of Controllability and Autonomy

In the following theorem we characterize controllability and autonomy of $\mathcal{B} =$ $\ker(R\circ)$ *from* (4.29) *by means of Theorem 4.2.9. For $\mathcal{D} = F[s]$, $F = \mathbb{R}, \mathbb{C}$, we will complement this in Theorem 5.4.4 by conditions on the characteristic variety.*

Theorem 4.2.12 *1. The following properties are equivalent:*

(a) *The behavior $\mathcal{B}$ is autonomous; in other words, M is a torsion module.*
(b) $p = rank(R) = l$, *i.e., the columns of R are linearly independent. This is equivalent to* $\mathrm{rank}(M) = m = l - p = 0$.

2. *The following properties are equivalent:*

(a) *The behavior $\mathcal{B}$ is controllable; in other words, M is torsionfree.*
(b) *The module M is free, i.e., has a basis.*
(c) $e_p = 1$.
(d) *If, in addition, $k = p = \mathrm{rank}(R)$: The matrix R is right invertible.*

If these properties are satisfied, then the map

$$(\cdot V_{II})_{\mathrm{ind}}\colon \mathrm{M}(\mathcal{B}) \longrightarrow \mathcal{D}^{1\times m}, \quad \overline{\xi} \longmapsto \xi V_{II},$$

is an isomorphism and its inverse is

$$(\cdot V_{II}^{-1})_{\mathrm{ind}}\colon \mathcal{D}^{1\times m} \longrightarrow \mathrm{M}(\mathcal{B}), \quad \eta_{II} = (\eta_{p+1}, \ldots, \eta_l) \mapsto \overline{\eta_{II} V_{II}^{-1}}. \tag{4.43}$$

Proof 1. We use the direct sum decomposition $M = \mathrm{t}(M) \oplus N$ with $N \cong \mathcal{D}^{1\times m}$ from item 3 of Theorem 4.2.9. Then

$$M = \mathrm{t}(M) \Longleftrightarrow N = 0 \Longleftrightarrow l - p = m = 0 \Longleftrightarrow \mathrm{rank}(R) = l.$$

2. (a) $\Longrightarrow$ (b). Again, we use the decomposition $M = \mathrm{t}(M) \oplus N$. If $\mathrm{t}(M) = 0$, then $M = N$ is free.
(b) $\Longrightarrow$ (a). If M is free, then $M \cong \mathcal{D}^{1\times q}$ for some $q \in \mathbb{N}$, and thus

$$\mathrm{t}(M) \cong \mathrm{t}(\mathcal{D}^{1\times q}) = \mathrm{t}(\mathcal{D})^{1\times q} = 0^{1\times q} = 0.$$

(b) $\Longleftrightarrow$ (c). The isomorphism $\mathrm{t}(M) \cong \prod_{j=}^{p} \mathcal{D}/\mathcal{D}e_j$ from Theorem 4.2.9, 1, implies that

$$\begin{aligned} \mathrm{t}(M) = 0 &\Longleftrightarrow \forall j = 1, \ldots, p\colon\ \mathcal{D}/\mathcal{D}e_j = 0, \text{ i.e., } \mathcal{D}e_j = \mathcal{D} \\ &\Longleftrightarrow \forall j = 1, \ldots, p\colon\ e_j = 1 \text{ (up to association)} \\ &\overset{e_j \mid e_p}{\Longleftrightarrow} e_p = 1. \end{aligned}$$

(c) $\Longleftrightarrow$ (d). We have shown this in Corollary 3.3.24.

If the conditions (a)–(d) are satisfied, the isomorphism (4.39) induces the mutually inverse isomorphisms

$$\begin{array}{ccccc} M(\mathcal{B}) & \overset{\cong}{\longleftrightarrow} & \prod_{j=1}^{l} \mathcal{D}/\mathcal{D}e_j & = \{0\} \times \prod_{j=p+1}^{l} \mathcal{D}/\mathcal{D}e_j & \overset{\cong}{\longleftrightarrow} \mathcal{D}^{1\times m}, \\ \overline{\xi} \longmapsto & & \overline{\xi V} & = \overline{(\xi V_I, \xi V_{II})} & \longmapsto \xi V_{II}, \\ \overline{\eta_{II} V_{II}^{-1}} = \overline{(0, \eta_{II}) V^{-1}} \longleftarrow & & & \overline{(0, \eta_{II})} & \longleftarrow \eta_{II}, \end{array}$$

where $V^{-1} = \begin{pmatrix} V_I^{-1} \\ V_{II}^{-1} \end{pmatrix} \in \mathcal{D}^{(p+m)\times l}$. □

Corollary and Definition 4.2.13 *Cf. [1, Theorem 6.6.1 on p. 229]. The following properties are equivalent:*

1. *The behavior is controllable.*
2. *There is a behavior epimorphism $P\circ : \mathcal{F}^{\widetilde{m}} \to \mathcal{B}$ or, in other terms, an exact behavior sequence $\mathcal{F}^{\widetilde{m}} \xrightarrow{P\circ} \mathcal{F}^l \xrightarrow{R\circ} \mathcal{F}^k$.*
3. *There is a behavior isomorphism $P\circ : \mathcal{F}^m \cong \mathcal{B}$.*

If these properties are satisfied, then, with the data of the preceding theorem,

$$V_{II}\circ : \mathcal{F}^m \longrightarrow \mathcal{B} \ \text{ and } \ V_{II}^{-1}\circ : \mathcal{B} \longrightarrow \mathcal{F}^m \tag{4.44}$$

are mutually inverse behavior isomorphisms.
According to Willems [1, §6.6] $P\circ$ from 2 (3) is called a (bijective) image representation *of $\mathcal{B}$ for obvious reasons. For $w = P \circ u \in \mathcal{B}$, $u \in \mathcal{F}^{\widetilde{m}}$, Pommaret calls u a* potential *of the* field *w. This terminology comes from multidimensional systems theory and partial differential equations.*

Proof Recall that $\mathcal{F}^{\widetilde{m}}$ is a behavior and that $\mathrm{M}(\mathcal{F}^{\widetilde{m}}) = \mathcal{D}^{1\times\widetilde{m}}$ is free, hence $\mathcal{F}^{\widetilde{m}}$ is a controllable behavior. We use the duality theorem (Theorem 3.3.18).

3. $\Longrightarrow$ 2. Obvious.
2. $\Longrightarrow$ 1. Lemma 4.2.5 shows that the image $P \circ \mathcal{F}^{\widetilde{m}} = \mathcal{B}$ of the controllable behavior $\mathcal{F}^{\widetilde{m}}$ is controllable too.
1. $\Longrightarrow$ 3. Since $M(\mathcal{B})$ is free, the inverse isomorphisms in (4.43) and then their duals in (4.44) hold. This implies 3 with $P := V_{II}$. □

Theorem and Definition 4.2.14 *Assume the data of the standard cases, i.e., $\mathcal{D} = F[s]$ and $\mathcal{F} = \mathrm{C}^\infty(\mathbb{R}, F)$, $F = \mathbb{R}$ or $F = \mathbb{C}$, in continuous time, and $\mathcal{F} = F^{\mathbb{N}}$, F a field, in discrete time.*

1. *The unique monic polynomial $f = e_p$ with $\mathrm{ann}_{F[s]}\left(\mathrm{t}(M)\right) = F[s]f$ also satisfies*

$$F[s]f = \ker\Big(F[s] \to \mathrm{End}_F\left(\mathrm{t}(M)\right),\ g \mapsto (x \mapsto gx)\Big),$$

and is, therefore, called the minimal polynomial *of $\mathrm{t}(M)$, M, or $\mathcal{B}$.*
For $A \in F^{n\times n}$ the minimal polynomial of A according to (4.12) coincides with that of the torsion module $F[s]^{1\times n}/F[s]^{1\times n}(s\,\mathrm{id}_n - A) \cong (F^{1\times n}, \circ_A)$ where $s \circ_A \xi = \xi A$.

2. *The following properties are equivalent:*

   (a) $\mathcal{B}$ *is autonomous, i.e.,* $p = l$.
   (b) $\dim_F(M) < \infty$.
   (c) $\dim_F(\mathcal{B}) < \infty$.

   If these properties are satisfied, then $\dim_F(M) = \dim_F(\mathcal{B}) = \sum_{j=1}^{l} \deg_s(e_j)$.

Proof 1. Obvious.

2. Let $0 \neq g \in F[s]$ and $n := \deg_s(g)$. Recall the $F[s]$-isomorphism $\operatorname{ann}_{\mathcal{F}}(g) \cong F^n$ from Theorem 3.1.5. Because of this,

$$n = \deg_s(g) = \dim_F\left(F[s]/F[s]g\right) = \dim_F\left(\operatorname{ann}_{\mathcal{F}}(g)\right).$$

On the other hand, both $F[s]$ and $\mathcal{F}$ are infinite dimensional over F. We have the isomorphisms

$$M \cong \prod_{j=1}^{p} F[s]/F[s]e_j \times F[s]^{1\times m} \quad \text{and} \quad \mathcal{B} \cong \prod_{j=1}^{p} \operatorname{ann}_{\mathcal{F}}(e_j) \times \mathcal{F}^m$$

from (4.39) and (4.41). Hence,

$$\begin{aligned}\mathcal{B} \text{ is autonomous} &\overset{\text{Theorem 4.2.12}}{\Longleftrightarrow} l = p \Longleftrightarrow m = l - p = 0\\ &\Longleftrightarrow \dim_F(M) < \infty \Longleftrightarrow \dim_F(\mathcal{B}) < \infty.\end{aligned}$$

If this is the case, then we infer that

$$\begin{aligned}\dim_F(M) &= \sum_{j=1}^{p} \dim_F\left(F[s]/F[s]e_j\right) = \sum_{j=1}^{p} \deg_s(e_j)\\ &= \sum_{j=1}^{p} \dim_F\left(\operatorname{ann}_{\mathcal{F}}(e_j)\right) = \dim_F(\mathcal{B}).\end{aligned}$$

□

Theorem 4.2.15 (The largest controllable subbehavior) *Compare [1, Theorem 5.2.14 on p.161]. Assume that* $\mathcal{D}$ *is a Noetherian integral domain with quotient field* $K = \operatorname{quot}(\mathcal{D})$ *and the data* R, U, M, *and* $\mathcal{B}$ *from* (4.29), *as well as the vector space* $KU = K^{1\times k}U$. *Define*

$$\begin{aligned}&U_{\text{cont}} := \mathcal{D}^{1\times l} \cap KU = \mathcal{D}^{1\times l} \cap K^{1\times k}R \supseteq U, \; M_{\text{cont}} := \mathcal{D}^{1\times l}/U_{\text{cont}},\\ &\mathcal{B}_{\text{cont}} := U_{\text{cont}}^{\perp} \subseteq \mathcal{B} = U^{\perp}.\end{aligned} \tag{4.45}$$

1. *The torsion module* $\mathrm{t}(M)$ *admits the representation*

$$\mathrm{t}(M) = \ker\left(\mathcal{D}^{1\times l}/U \longrightarrow K^{1\times l}/KU, \; \overline{\xi} \longmapsto \overline{\xi}\right) = U_{\text{cont}}/U.$$

2. *The behavior* $\mathcal{B}_{\rm cont}$ *is the largest controllable subbehavior of* $\mathcal{B}$ *and*

$$M_{\rm cont} = \mathcal{D}^{1\times l}/U_{\rm cont} \cong \left(\mathcal{D}^{1\times l}/U\right)/\left(U_{\rm cont}/U\right) = M/\mathrm{t}(M). \tag{4.46}$$

Proof 1. The equality $\ker\left(\mathcal{D}^{1\times l}/U \to K^{1\times l}/KU,\ \overline{\xi}\mapsto\overline{\xi}\right) = \left(\mathcal{D}^{1\times l}\cap KU\right)/U$ is obvious.

$\subseteq$. Let $\overline{\xi}\in \mathrm{t}(M)$ and let $0\neq d\in\mathcal{D}$ with $0 = d\overline{\xi} = \overline{d\xi}$. Then $d\xi\in U$ and $\xi\in \mathcal{D}^{1\times l}\cap KU$.

$\supseteq$. Let $\xi = \frac{a}{d}\eta\in \mathcal{D}^{1\times l}\cap KU$ with $\eta\in U$. Then $d\xi = a\eta\in U$, i.e., $d\overline{\xi} = 0$ or $\overline{\xi}\in\mathrm{t}(M)$.

2. The isomorphism (4.46) follows from the isomorphism theorem. The behavior $\mathcal{B}_{\rm cont}$ is controllable since its module $M_{\rm cont}\cong M/\mathrm{t}(M)$ is torsionfree.
To show that $\mathcal{B}_{\rm cont}$ is the largest controllable subbehavior of $\mathcal{B}$, assume that $\mathcal{B}'\subseteq\mathcal{B}$ is any controllable subbehavior. Then $U = \mathcal{B}^{\perp}\subseteq U' := \mathcal{B}'^{\perp}$ and $\mathcal{D}^{1\times l}/U'$ is torsionfree. We show $\mathcal{B}'\subseteq\mathcal{B}_{\rm cont}$ or, equivalently, $U_{\rm cont}\subseteq U'$.
If $\xi\in U_{\rm cont} = \mathcal{D}^{1\times l}\cap KU$, then there is a nonzero $d\in\mathcal{D}$ with $d\xi\in U\subseteq U'$, i.e., $0 = d\overline{\xi}\in\mathcal{D}^{1\times l}/U'$. The torsionfreeness of $\mathcal{D}^{1\times l}/U'$ implies $0 = \overline{\xi}\in\mathcal{D}^{1\times l}/U'$, hence $\xi\in U'$ and $U_{\rm cont}\subseteq U'$. □

In the next theorem we assume that $\mathcal{D}$ is a principal ideal domain. We use the behavior $\mathcal{B} = (\mathcal{D}^{1\times k}R)^{\perp}$ with $\mathrm{rank}(R) = p$ from (4.29), the Smith form $\left(\begin{smallmatrix}\Delta & 0\\ 0 & 0\end{smallmatrix}\right) = URV$ of R from (4.34), and the decompositions

$$U = \begin{pmatrix}U_I\\ U_{II}\end{pmatrix}\in\mathcal{D}^{(p+(k-p))\times p},\qquad V = (V_I, V_{II})\in\mathcal{D}^{l\times(p+(l-p))},$$
$$U^{-1} = (U_I^{-1}, U_{II}^{-1})\in\mathcal{D}^{k\times(p+(k-p))},\quad\text{and}\quad V^{-1} = \begin{pmatrix}V_I^{-1}\\ V_{II}^{-1}\end{pmatrix}\in\mathcal{D}^{(p+(l-p))\times l} \tag{4.47}$$

from (4.38) and define

$$\begin{aligned}&X_{\rm cont} := U_I^{-1}\Delta\in\mathcal{D}^{k\times p},\quad R_{\rm cont} := V_I^{-1}\in\mathcal{D}^{p\times l}\\ &\Longrightarrow R = U_I^{-1}\Delta V_I^{-1} = X_{\rm cont}R_{\rm cont},\ \mathrm{rank}(X_{\rm cont}) = \mathrm{rank}(R_{\rm cont}) = p.\end{aligned} \tag{4.48}$$

Theorem 4.2.9, 2, and Theorem 4.2.15, 1, imply

$$\begin{aligned}&\mathrm{t}(M) = \mathcal{D}^{1\times p}V_I^{-1}/\mathcal{D}^{1\times k}R = U_{\rm cont}/\mathcal{D}^{1\times k}R \text{ and hence}\\ &U_{\rm cont} = \mathcal{D}^{1\times p}R_{\rm cont} = \mathcal{D}^{1\times p}V_I^{-1},\ \mathcal{B}_{\rm cont} = U_{\rm cont}^{\perp} = \left\{w\in\mathcal{F}^l;\ V_I^{-1}\circ w = 0\right\}.\end{aligned} \tag{4.49}$$

This furnishes an algorithm for the computation of $\mathcal{B}_{\rm cont}\subseteq\mathcal{B}$.

Theorem 4.2.16 *With the data from* (4.47)–(4.49) *the following holds.*

1. *The following map is well defined and an isomorphism:*

$$(\cdot V_I^{-1})_{\rm ind}:\ \mathcal{D}^{1\times p}/\mathcal{D}^{1\times k}X_{\rm cont}\xrightarrow{\cong}\mathrm{t}(M) = \mathcal{D}^{1\times p}V_I^{-1}/\mathcal{D}^{1\times k}R. \tag{4.50}$$

2. *The following sequence of modules is exact:*

$$
\begin{aligned}
&0 \longrightarrow M_{\text{uncont}} \xrightarrow{(\cdot R_{\text{cont}})_{\text{ind}}} M \xrightarrow{\text{can}} M_{\text{cont}} \longrightarrow 0, \quad \textit{where}\\
&M_{\text{uncont}} := \mathcal{D}^{1\times p}/\mathcal{D}^{1\times k}X_{\text{cont}},\ M = \mathcal{D}^{1\times l}/\mathcal{D}^{1\times k}R,\\
&M_{\text{cont}} := M/\operatorname{t}(M) = \mathcal{D}^{1\times l}/\mathcal{D}^{1\times p}R_{\text{cont}} = \mathcal{D}^{1\times l}/\mathcal{D}^{1\times p}V_I^{-1}.
\end{aligned}
\tag{4.51}
$$

It implies the dual exact behavior sequence

$$
\begin{aligned}
&0 \longleftarrow \mathcal{B}_{\text{uncont}} \xleftarrow{R_{\text{cont}}\circ} \mathcal{B} \xleftarrow{\supseteq} \mathcal{B}_{\text{cont}} \longleftarrow 0, \quad \textit{where}\\
&\mathcal{B}_{\text{uncont}} = \{y \in \mathcal{F}^p;\ X_{\text{cont}} \circ y = 0\},\ \mathcal{B} = \{w \in \mathcal{F}^l;\ R \circ w = 0\},\\
&\mathcal{B}_{\text{cont}} = \{w \in \mathcal{F}^l;\ R_{\text{cont}} \circ w = 0\} = \{w \in \mathcal{F}^l;\ V_I^{-1} \circ w = 0\}.\\
&\Longrightarrow \mathcal{B}/\mathcal{B}_{\text{cont}} \xrightarrow{\cong} \mathcal{B}_{\text{uncont}},\quad w + \mathcal{B}_{\text{cont}} \longmapsto R_{\text{cont}} \circ w = V_I^{-1} \circ w.
\end{aligned}
\tag{4.52}
$$

The behavior $\mathcal{B}_{\text{uncont}}$ is autonomous with $M(\mathcal{B}_{\text{uncont}}) \cong \operatorname{t}(M)$, and is called the uncontrollable factor *of $\mathcal{B}$. It is uniquely determined by M or $\mathcal{B}$ up to isomorphism.*

Proof 1. The kernel of the linear surjection

$$
\begin{aligned}
\operatorname{can}(\cdot R_{\text{cont}}):\ \mathcal{D}^{1\times p} &\xrightarrow{\cdot R_{\text{cont}}} \mathcal{D}^{1\times p}R_{\text{cont}} &&\xrightarrow{\text{can}} \operatorname{t}(M) = \mathcal{D}^{1\times p}R_{\text{cont}}/\mathcal{D}^{1\times k}R,\\
\xi &\longmapsto \quad \xi R_{\text{cont}} &&\longmapsto \xi R_{\text{cont}} + \mathcal{D}^{1\times k}R,
\end{aligned}
$$

is $\mathcal{D}^{1\times k}X_{\text{cont}}$, since for $\xi \in \mathcal{D}^{1\times p}$ the equivalences

$$
\begin{aligned}
\xi \in \ker\left(\operatorname{can}(\cdot R_{\text{cont}})\right) &\Longleftrightarrow \exists \eta \in \mathcal{D}^{1\times k}:\ \xi R_{\text{cont}} = \eta R = \eta X_{\text{cont}} R_{\text{cont}}\\
&\overset{\operatorname{rank}(R_{\text{cont}})=p}{\Longleftrightarrow} \xi = \eta X_{\text{cont}} \in \mathcal{D}^{1\times p}X_{\text{cont}}
\end{aligned}
$$

hold. The homomorphism theorem then implies the isomorphism (4.50).
2. The equation $\operatorname{im}((\cdot R_{\text{cont}})_{\text{ind}}) = \operatorname{t}(M) = \mathcal{D}^{1\times p}R_{\text{cont}}/\mathcal{D}^{1\times k}R = \ker(\text{can})$ and the injectivity of $(\cdot R_{\text{cont}})_{\text{ind}}$ follow from 1. This implies the exactness of (4.51) and, by duality, that of (4.52). □

In Theorem 4.2.18 we derive the standard connection between split exact sequences and direct sum decompositions. We will use this result for the controllable/autonomous decomposition in Corollary and Definition 4.2.19 as well as in Chaps. 10 and 11.

Lemma 4.2.17 *Let $\mathcal{D}$ be a commutative ring, let M be a $\mathcal{D}$-module, and let $V \leq_{\mathcal{D}} M$ be a submodule. A* projection *is an endomorphism $p \in \operatorname{End}_{\mathcal{D}}(M) = \operatorname{Hom}_{\mathcal{D}}(M, M)$ with $p^2 = pp = p$. The maps*

$$\{W \leq_{\mathcal{D}} DM;\ V \oplus W = M\} \rightleftarrows \{p \in \mathrm{End}_{\mathcal{D}}(M) \text{ projection with } \mathrm{im}(p) = V\},$$
$$W \longmapsto \big(x \mapsto v, \text{ where } x = v + w \in M = V \oplus W\big),$$
$$\ker(p) \longleftarrow p,$$

are well defined, bijective, and inverses of each other. In other words, a projection p *with* $\mathrm{im}(p) = V$ *induces a direct sum decomposition*

$$M = \mathrm{im}(p) \oplus \ker(p) = V \oplus \ker(p) \ni m = p(m) + (m - p(m)),$$

and all direct sum decompositions can be obtained in this way.

Proof For $m \in M$ we have that $p\big(m - p(m)\big) = p(m) - p\big(p(m)\big) = p(m) - p(m) = 0$, and thus $m - p(m) \in \ker(p)$. The rest of the proof is easy. □

We write $V \leq_{\mathcal{D}} M$ instead of $V \subseteq M$ to emphasize that V is a $\mathcal{D}$-*submodule* and not only a *subset* of M.

Theorem and Definition 4.2.18 *Let* $\mathcal{D}$ *be a commutative ring and let*

$$0 \longrightarrow M' \xrightarrow{f_1} M \xrightarrow{g_1} M'' \longrightarrow 0 \tag{4.53}$$

be an exact sequence of $\mathcal{D}$*-modules, in particular,* $\mathrm{im}(f_1) = \ker(g_1)$.

1. *The following assertions are equivalent:*
   (a) *The submodule* $\mathrm{im}(f_1) = \ker(g_1)$ *is a direct summand of* M.
   (b) *The monomorphism* f_1 *has a linear retraction* $g_2\colon M \longrightarrow M'$, *i.e.,* $g_2 f_1 = \mathrm{id}_{M'}$.
   (c) *The epimorphism* g_1 *has a linear section* $f_2\colon M'' \longrightarrow M$, *i.e.,* $g_1 f_2 = \mathrm{id}_{M''}$.
   (d) *For every module* N, *the map*

$$\mathrm{Hom}(f_1, N)\colon \mathrm{Hom}_{\mathcal{D}}(M, N) \longrightarrow \mathrm{Hom}_{\mathcal{D}}(M', N),\quad k \longmapsto kf_1,$$

   is surjective.

   If these conditions are satisfied the sequence (4.53) *is called* split exact *or* splitting.
2. *The following maps are bijective:*

$$\begin{array}{ccc} \{W \leq_{\mathcal{D}} M;\ M = f_1(M') \oplus W\} & \ni & W \\ \updownarrow & & \updownarrow \\ \{g_2 \in \mathrm{Hom}_{\mathcal{D}}(M, M');\ g_2 f_1 = \mathrm{id}_{M'}\} & \ni & g_2 \\ \updownarrow & & \updownarrow \\ \{f_2 \in \mathrm{Hom}_{\mathcal{D}}(M'', M);\ g_1 f_2 = \mathrm{id}_{M''}\} & \ni & f_2 \end{array} \tag{4.54}$$

   where $W = \ker(g_2) = \mathrm{im}(f_2)$, *the map* g_2 *is the unique linear retraction of* f_1 *with* $\ker(g_2) = W$, *and the map* f_2 *is the unique section of* g_1 *with* $\mathrm{im}(f_2) = W$.

Furthermore, the sequence

$$0 \longleftarrow M' \xleftarrow{g_2} M \xleftarrow{f_2} M'' \longleftarrow 0$$

is exact and $\mathrm{id}_M = f_1 g_2 + f_2 g_1$.

3. *If 1. holds and if f_2^0 and g_2^0 are related by (4.54), then the sets from (4.54) are also in one-one correspondence with* $\mathrm{Hom}_{\mathcal{D}}(M'', M')$ *via*

$$h \in \mathrm{Hom}_{\mathcal{D}}(M'', M'),\ f_2 = f_2^0 + f_1 h,\ g_2 = g_2^0 - h g_1. \qquad (4.55)$$

Proof 1.(a) $\Longrightarrow$ (b). Let $W \leq_{\mathcal{D}} M$ with $M = \mathrm{im}(f_1) \oplus W$. Every $m \in M$ can be written uniquely as $m = m_1 + m_2 \in M = \mathrm{im}(f_1) \oplus W$, and, since f_1 is injective, there is a unique $m' \in M'$ with $f_1(m') = m_1$. Define $g_2 \colon M \longrightarrow M'$ by $g_2(m) := m'$. This map g_2 is a well-defined linear retraction of f_1 and obviously the unique one with $\ker(g_2) = W$.

(a) $\Longrightarrow$ (c). Let $W \leq_{\mathcal{D}} M$ with $M = \ker(g_1) \oplus W$. Since g_1 is surjective, we have the isomorphism

$$\varphi\colon W \xrightarrow{\cong} M/\ker(g_1) \xrightarrow{\cong} M'',\quad w \longmapsto w + \ker(g_1) \longmapsto g_1(w).$$

Let $f_2 := \mathrm{inj}\,\varphi^{-1}\colon M'' \xrightarrow{\varphi^{-1}} W \xrightarrow{\subseteq} M$. By construction, this map satisfies $g_1 f_2 = \mathrm{id}_{M''}$, i.e., it is a section of g_1 and indeed the unique one with $\mathrm{im}(f_2) = W$.

(b) $\Longrightarrow$ (a). The map $p := f_1 g_2$ is a projection since $g_2 f_1 = \mathrm{id}_{M'}$, and, consequently, $p^2 = f_1(g_2 f_1) g_2 = f_1 g_2 = p$. By Lemma 4.2.17, this induces the decomposition $M = \mathrm{im}(p) \oplus \ker(p)$, and, since g_2 is surjective, we have that $\mathrm{im}(p) = \mathrm{im}(f_1 g_2) = \mathrm{im}(f_1)$.

(c) $\Longrightarrow$ (a). The map $q := f_2 g_1$ is a projection since $g_1 f_2 = \mathrm{id}_{M''}$, and, consequently, $q^2 = f_2 g_1 f_2 g_1 = f_2 g_1 = q$. By Lemma 4.2.17, this induces the decomposition $M = \ker(q) \oplus \mathrm{im}(q)$, and, since f_2 is injective, we have that $\ker(q) = \ker(f_2 g_1) = \ker(g_1)$.

(d) $\Longrightarrow$ (b). For $h \in \mathrm{Hom}_{\mathcal{D}}(M', N)$ we get

$$h = h\,\mathrm{id}_{M'} = h g_2 f_1 = \mathrm{Hom}(f_1, N)(h g_2) \in \mathrm{im}\left(\mathrm{Hom}(f_1, N)\right).$$

(b) $\Longrightarrow$ (d). For $N := M'$ and $\mathrm{id}_{M'} \in \mathrm{Hom}_{\mathcal{D}}(M', M')$ there is a $g_2 \in \mathrm{Hom}_{\mathcal{D}}(M, M')$ with $\mathrm{id}_{M'} = \mathrm{Hom}(f_1, M')(g_2) = g_2 f_1$.

2. The bijections (4.54) are derived in the proof of 1. and imply the exact sequence. To show the identity $\mathrm{id}_M = f_1 g_2 + f_2 g_1$, let $m \in M$. We use that $M = \mathrm{im}(f_1) \oplus \mathrm{im}(f_2) \ni m = f_1(m') + f_2(m'')$ for some $m' \in M'$ and $m'' \in M''$. The equations $\mathrm{im}(f_2) = \ker(g_2)$ and $g_2 f_1 = \mathrm{id}_{M'}$ imply

$$g_2(m) = g_2\big(f_1(m')\big) + g_2\big(f_2(m'')\big) = g_2\big(f_1(m')\big) = m'.$$

Similarly, $\operatorname{im}(f_1) = \ker(g_1)$ and $g_1 f_2 = \mathrm{id}_{M''}$ imply that $g_1(m) = g_1\big(f_2(m'')\big) = m''$. By putting everything together, we conclude that

$$m = f_1(m') + f_2(m'') = (f_1 g_2)(m) + (f_2 g_1)(m).$$

3. Consider a pair (f_2, g_2) from (4.54). From

$$g_1(f_2 - f_2^0) = g_1 f_2 - g_1 f_2^0 = \mathrm{id}_{M''} - \mathrm{id}_{M''} = 0,$$

we infer that $\operatorname{im}(f_2 - f_2^0) \subseteq \ker(g_1) = \operatorname{im}(f_1) \cong M'$. Therefore the map $f_2 - f_2^0$ admits the unique factorization $f_2 - f_2^0 = f_1 h$, where

$$h\colon M'' \longrightarrow M', \quad m'' \longmapsto h(m'') := f_1^{-1}\big((f_2 - f_2^0)(m'')\big) \Longrightarrow f_2 = f_2^0 + f_1 h.$$

From $\mathrm{id}_M = f_1 g_2 + f_2 g_1$ and $\mathrm{id}_M = f_1 g_2^0 + f_2^0 g_1$, we infer the identity

$$\begin{aligned} f_1 g_2 &= \mathrm{id}_M - f_2 g_1 = \mathrm{id}_M - f_2^0 g_1 - f_1 h g_1 = \mathrm{id}_M - (\mathrm{id}_M - f_1 g_2^0) - f_1 h g_1 \\ &= f_1(g_2^0 - h g_1) \underset{f_1 \text{ injective}}{\Longrightarrow} g_2 = g_2^0 - h_1. \end{aligned}$$

□

Corollary and Definition 4.2.19 *Let $\mathcal{D}$ be a principal ideal domain and assume the data from Theorem 4.2.16. Since M_{cont} is free,* can *has a section σ (with* can $\sigma =$ $\mathrm{id}_{M_{\text{cont}}}$*) and the exact sequence* (4.50)*, namely,*

$$0 \longrightarrow M_{\text{uncont}} \xrightarrow{(\cdot R_{\text{cont}})_{\text{ind}}} M \xrightarrow{\text{can}} M_{\text{cont}} = M/\operatorname{t}(M) \longrightarrow 0,$$

splits. Dually, there is a behavior retraction $\rho\colon \mathcal{B} \longrightarrow \mathcal{B}_{\text{cont}}$ with $\rho|_{\mathcal{B}_{\text{cont}}} = \mathrm{id}_{\mathcal{B}_{\text{cont}}}$. By Theorem 4.2.18, this implies that

$$\begin{aligned} &M = \operatorname{t}(M) \oplus \sigma(M_{\text{cont}}), \ M_{\text{cont}} \cong \sigma(M_{\text{cont}}) \ \textit{free}, \\ &\mathcal{B} = \mathcal{B}_{\text{cont}} \oplus \mathcal{B}', \quad \textit{with} \quad \mathcal{B}' = \ker(\rho) \cong \mathcal{B}/\mathcal{B}_{\text{cont}} \cong \mathcal{B}_{\text{uncont}} \ \textit{autonomous}. \end{aligned}$$

This direct decomposition is called a controllable/autonomous decomposition *of $\mathcal{B}$. The autonomous direct summand $\mathcal{B}'$ is unique up to isomorphism.* ◇

4.3 The Trajectories of Controllable Behaviors

We investigate controllable behaviors in the standard cases and connect, according to Willems and to Kalman, controllability with the ability to steer the system in finite time to a desired trajectory or, for state space systems, to a desired state. These properties suggested the term *controllable*. We also derive two of Kalman's famous results for state space systems $\mathcal{B}$, namely, the characterization of the controllability of $\mathcal{B}$ by means of the full row rank of the controllability matrix and the construction of a state space representation for $\mathcal{B}_{\text{cont}}$. We repeat our remark from the end of Sect. 2.1 that in this book the main property of controllable behaviors is the freeness of the system module and is used for the construction of stabilizing compensators in Chap. 10. The examples in that chapter discuss controllability, in particular, cf. Examples 8.3.17, 8.3.20 (8.287), 10.1.23, 8.4.6. The famous results of Kalman and Willems are not essentially applied in this book, and therefore we do not present examples here.

4.3.1 Controllability via Steering of Trajectories

Theorem 4.3.1 (Controllable discrete-time behaviors) *We consider the discrete-time case as in Corollary 3.1.3 with a field F, the ring $\mathcal{D} := F[s]$ of operators, and the signal space $\mathcal{F} := F^{\mathbb{N}}$. Let $\mathcal{B} = \ker(R\circ)$ be a behavior as in (4.29) and assume that it is controllable. Then there is $d \in \mathbb{N}$ such that*

$$\forall w_1, w_2 \in \mathcal{B}\ \forall t_1 \in \mathbb{N}\ \exists w \in \mathcal{B} \text{ with } w(t) = \begin{cases} w_1(t) & \text{for } t \le t_1, \\ w_2(t) & \text{for } t \ge t_2 := t_1 + d + 1. \end{cases}$$

In other words, one can steer the system from the trajectory w_1 to the trajectory w_2 in the time interval from t_1 to t_2. In the proof w is constructed.

Proof From Corollary and Definition 4.2.13, we obtain the image representation $V_{II}\circ\colon \mathcal{F}^m \xrightarrow{\cong} \mathcal{B}$ with inverse $V_{II}^{-1}\circ\colon \mathcal{B} \xrightarrow{\cong} \mathcal{F}^m$. We define

$$z_j := V_{II}^{-1} \circ w_j, \quad \text{for } j = 1, 2,$$

hence $w_j = V_{II} \circ z_j$. We write

$$P := V_{II} = \sum_{i=0}^{d} P_i s^i \in F[s]^{l\times m}, \quad \text{with } P_i \in F^{l\times m}.$$

For every $t \in \mathbb{N}$, we get

$$w_j(t) = (P \circ z_j)(t) = \sum_{i=0}^{d} P_i (s^i \circ z_j)(t) = \sum_{i=0}^{d} P_i z_j(t+i). \tag{4.56}$$

We define $z \in \mathcal{F}^m$ by

$$z(t) := \begin{cases} z_1(t) & \text{for } t \le t_1 + d, \\ z_2(t) & \text{for } t \ge t_2 = t_1 + d + 1, \end{cases}$$

and $w := P \circ z \in P \circ \mathcal{F}^m = \mathcal{B}$. Let

$$t \le t_1 \text{ and } i \le d \Longrightarrow t + i \le t_1 + d \Longrightarrow z(t+i) = z_1(t+i)$$
$$\Longrightarrow w_1(t) = \sum_{i=0}^{d} P_i z_1(t+i) = \sum_{i=0}^{d} P_i z(t+i) = w(t) \quad \text{for } t \le t_1.$$

If, on the other hand, $t \ge t_2$ and $i \ge 0 \Longrightarrow t + i \ge t_2$, then

$$w_2(t) = \sum_{i=0}^{d} P_i z_2(t+i) = \sum_{i=0}^{d} P_i z(t+i) = w(t).$$

□

The following example shows that the preceding property does not characterize controllability. This is due to $\mathbb{N}$, and not $\mathbb{Z}$, being the time axis here. For the time domain $\mathbb{Z}$ controllability is characterized by the property of the preceding theorem. This is also true for the continuous-time domain $\mathbb{R}$, as will be shown in Theorem 4.3.4.

Example 4.3.2 Let $n > 0$ and consider the behavior

$$\mathcal{B} := \ker(s^n \circ) = \{w \in F^{\mathbb{N}};\ w(t) = 0 \text{ for } t \ge n\} = \bigoplus_{i=0}^{n-1} F\delta_i$$
$$\subseteq F^{(\mathbb{N})} = \{w \in F^{\mathbb{N}};\ w(t) = 0 \text{ for almost all } t\}.$$

The module of $\mathcal{B}$ is the torsion module $\mathrm{M}(\mathcal{B}) = F[s]/F[s]s^n$, and hence this behavior is nonzero and autonomous, but not controllable. But any two trajectories w_1, w_2 are zero for $t \ge n$. Hence, $w := w_1$ itself trivially satisfies the property $w_2(t) = w(t)$ for $t \ge t_1 + n$ of the preceding theorem. ◊

For the proof of the analogue of Theorem 4.3.1 in the continuous-time case we need a preparation on C^∞-functions. The standard example for a smooth function which is not a power series, i.e., not represented by its Taylor series, is the function $\varphi_1 \in \mathrm{C}^\infty(\mathbb{R}, \mathbb{R})$ given by

$$\varphi_1(t) := \begin{cases} e^{-\frac{1}{t}} = \frac{1}{e^{1/t}} & \text{if } t > 0 \\ 0, & \text{if } t \le 0, \end{cases} \tag{4.57}$$

and this function satisfies $\varphi^{(m)}(0) = \frac{\mathrm{d}^m \varphi}{\mathrm{d}t^m}(0) = 0$ for all $m \in \mathbb{N}$. We also consider the mirrored function

$$\varphi_{-1}(t) := \varphi_1(-t) = \begin{cases} = 0 & \text{if } t \ge 0, \\ > 0 & \text{if } t < 0. \end{cases} \tag{4.58}$$

Both functions are smooth and bounded by 1. Finally, for $\epsilon > 0$, we define the C^∞-function

$$\varphi_{0,\epsilon}(t) := \varphi_1\left(1 - \left(\tfrac{t}{\epsilon}\right)^2\right) = \begin{cases} > 0 & \text{if } |t| < \epsilon, \\ = 0 & \text{if } |t| \geq \epsilon, \end{cases} \tag{4.59}$$

which has compact support. The functions φ_j are nonnegative and their sum

$$\varphi := \varphi_{-1} + \varphi_{0,\epsilon} + \varphi_1 \in C^\infty(\mathbb{R}, \mathbb{R}) \tag{4.60}$$

is positive, and hence we obtain the partition of unity $1 = \frac{\varphi_{-1}}{\varphi} + \frac{\varphi_{0,\epsilon}}{\varphi} + \frac{\varphi_1}{\varphi}$. The function $\psi_\epsilon := \frac{\varphi_{-1}}{\varphi} + \frac{\varphi_{0,\epsilon}}{\varphi}$ is smooth and satisfies

$$\begin{cases} \psi_\epsilon(t) = 1 & \text{for } t \leq 0, \\ 0 < \psi_\epsilon(t) < 1 & \text{for } 0 < t < \epsilon, \\ \psi_\epsilon(t) = 0 & \text{for } \epsilon \leq t, \end{cases} \tag{4.61}$$

as is easily checked.

In the following corollary we show how to connect signals in $\mathcal{F}^m$ smoothly, which is decisive for the continuous analogue of Theorem 4.3.1. In Corollary 4.3.3 and in Theorem 4.3.4, we consider the continuous-time standard case $\mathcal{D} = F[s]$ and $\mathcal{F} = C^\infty(\mathbb{R}, F)$ with $F = \mathbb{R}$ or $F = \mathbb{C}$.

Corollary 4.3.3 *Let $z_1, z_2 \in \mathcal{F}^m$ be two vector functions, let $t_1, t_2 \in \mathbb{R}$ with $t_1 < t_2$, and define $\epsilon := t_2 - t_1$. Then the function*

$$z(t) := \psi_\epsilon(t - t_1)z_1(t) + \psi_\epsilon(t_2 - t)z_2(t) \tag{4.62}$$

is smooth and satisfies

$$z(t) = \begin{cases} z_1(t) & \text{for } t \leq t_1, \\ z_2(t) & \text{for } t \geq t_2. \end{cases}$$

Proof For $t \leq t_1$, we have $t - t_1 \leq 0$ and $t_2 - t \geq t_2 - t_1 = \epsilon$, hence $\psi_\epsilon(t - t_1) = 1$ as well as $\psi_\epsilon(t_2 - t) = 0$, and thus $z(t) = z_1(t)$.
Similarly, for $t \geq t_2$, we have $t - t_1 \geq t_2 - t_1 = \epsilon$ and $t_2 - t \leq 0$, hence $\psi_\epsilon(t - t_1) = 0$ as well as $\psi_\epsilon(t_2 - t) = 1$, and thus $z(t) = z_2(t)$. □

Theorem 4.3.4 (Controllable continuous-time behaviors) *Compare [1, Definition 5.2.2 on p. 157]. We consider the continuous-time case and the behavior $\mathcal{B} = \ker(R\circ)$ from (4.29). The following properties of $\mathcal{B}$ are equivalent:*

1. *$\mathcal{B}$ is controllable.*
2. *For all pairs w_1, w_2 of trajectories in $\mathcal{B}$ and for all $t_1, t_2 \in \mathbb{R}$ with $t_1 < t_2$, there is a trajectory $w \in \mathcal{B}$ such that*

$$w(t) = \begin{cases} w_1(t) & \textit{for } t \le t_1, \\ w_2(t) & \textit{for } t \ge t_2. \end{cases}$$

In other words, one can steer the system from the trajectory w_1 *to the trajectory* w_2 *in the time interval from* t_1 *to* t_2.

Proof 1. $\Longrightarrow$ 2. Since the behavior $\mathcal{B}$ is controllable, there are mutually inverse behavior isomorphisms

$$V_{II}\circ\colon \mathcal{F}^m \overset{\cong}{\longrightarrow} \mathcal{B} \text{ and } V_{II}^{-1}\circ\colon \mathcal{B} \overset{\cong}{\longrightarrow} \mathcal{F}^m$$

from Corollary 4.2.13. As in Theorem 4.3.1 we define

$$P := V_{II} = \sum_{i=0}^{d} P_i s^i \in F[s]^{l\times m}, \; P_i \in F^{l\times m}.$$

Let $z_j = V_{II}^{-1} \circ w_j$, hence $w_j = P \circ z_j = \sum_{i=0}^d P_i s^i \circ z_j = \sum_{i=0}^d P_i z_j^{(i)}$. With $\epsilon := t_2 - t_1$, we use the function $z \in \mathcal{F}^m$ from Corollary 4.3.3 that satisfies

$$z|_{(-\infty,t_1]} = z_1|_{(-\infty,t_1]} \text{ and } z|_{[t_2,\infty)} = z_2|_{[t_2,\infty)}. \tag{4.63}$$

Using the fact that if two smooth functions coincide on an interval of positive length, then so do all their derivatives, we conclude that the same equalities (4.63) hold for all derivatives of the functions z_1, z_2, and z. The trajectory $w := P \circ z \in P \circ \mathcal{F}^m = \mathcal{B}$ then satisfies

$$w(t) = \sum_{i=0}^{d} P_i z^{(i)}(t) = \begin{cases} \sum_{i=0}^d P_i z_1^{(i)}(t) = w_1(t) & \text{for } t \le t_1, \\ \sum_{i=0}^d P_i z_2^{(i)}(t) = w_2(t) & \text{for } t \ge t_2. \end{cases}$$

2. $\Longrightarrow$ 1. We show that condition 2 is false for a noncontrollable behavior $\mathcal{B}$. Assume this, i.e., that $\mathrm{M}(\mathcal{B})$ is not torsionfree. This and the isomorphism from Corollary and Definition 3.3.19,

$$\mathrm{M}(\mathcal{B}) = \mathcal{D}^{1\times l}/\mathcal{B}^{\perp} \overset{\cong}{\longleftrightarrow} \mathrm{Hom}(\mathcal{B}, \mathcal{F}), \; \overline{\xi} \longleftrightarrow (\xi\circ\colon \mathcal{B} \to \mathcal{F}),$$

imply that there are a nonzero observable $\xi\colon \mathcal{B} \longrightarrow \mathcal{F}$ and a nonzero polynomial $f \in F[s]$ such that $f\overline{\xi} = 0$, hence $\xi \circ \mathcal{B} \neq 0$, but $f \circ (\xi \circ \mathcal{B}) = 0$. Choose a trajectory $w_1 \in \mathcal{B}$ with $\xi \circ w_1 \neq 0$ and $t_1 \in \mathbb{R}$ with $(\xi \circ w_1)(t_1) \neq 0$. Moreover let $t_1 < t_2$ and $w_2 := 0 \in \mathcal{F}^l$. If 2 holds, there is a connecting trajectory $w \in \mathcal{B}$ such that

$$w(t) = \begin{cases} w_1(t) & \text{for } t \le t_1, \\ w_2(t) = 0 & \text{for } t \ge t_2. \end{cases}$$

Application of $\xi\circ\colon \mathcal{B} \longrightarrow \mathcal{F}$ yields $\alpha := \xi \circ w \in \xi \circ \mathcal{B}$ and

$$\alpha(t) = (\xi \circ w)(t) = \begin{cases} (\xi \circ w_1)(t) & \text{for } t \leq t_1, \\ 0 & \text{for } t \geq t_2. \end{cases}$$

But $f \circ \alpha = f \circ (\xi \circ w) \in f \circ (\xi \circ \mathcal{B}) = 0$ and thus $f \circ \alpha = 0$. So α satisfies the nontrivial differential equation $f \circ \alpha = 0$ and is a polynomial-exponential function and, in particular, analytic. From $\alpha|_{[t_2,\infty)} = 0$ we infer $\alpha = 0$ in contradiction to $\alpha(t_1) = (\xi \circ w)(t_1) \neq 0$, hence 2 does not hold. □

4.3.2 Controllable Rosenbrock and State Space Equations

In the next theorem we characterize controllable Rosenbrock systems, cf. Sect. 3.2.4, over a principal ideal domain $\mathcal{D}$. By definition Rosenbrock equations

$$A \circ x = B \circ u, \quad \text{with } A \in \mathcal{D}^{n\times n},\ \det(A) \neq 0, \text{ and } B \in \mathcal{D}^{n\times m}, \tag{4.64}$$

are controllable if the behavior

$$\mathcal{B} := \left\{ \binom{x}{u} \in \mathcal{F}^{n+m};\ A \circ x = B \circ u, \text{ i.e., } (A, -B) \circ \binom{x}{u} = 0 \right\}$$

is controllable. Since controllability for these equations concerns the input and the pseudo-state only, we omitted the equations for the output in (4.64).

For $\mathcal{D} = F[s]$ with $F = \mathbb{R}$ or $F = \mathbb{C}$ we will amend the following characterization of controllable Rosenbrock equations in Corollary 5.4.6.

Theorem 4.3.5 (Controllable Rosenbrock equations, cf. [2, Definition 5.3.7 on p. 156]) *We assume that $\mathcal{D}$ is a principal ideal domain. For the Rosenbrock equations from* (4.64) *the following properties are equivalent:*

1. *The Rosenbrock equations* (4.64) *are controllable, i.e., the behavior $\mathcal{B}$ is controllable.*
2. *The elementary divisor e_n of $(A, -B)$ is* 1 *and hence all elementary divisors are equal to* 1.
3. *The matrix $(A, -B) \in \mathcal{D}^{n\times(n+m)}$ or, equivalently, the matrix (A, B) has a right inverse, i.e., there are matrices X and Y with $AX + BY = \mathrm{id}_n$.*

Proof The condition $\det(A) \neq 0$ implies the rank condition $(A, -B) = n$, i.e., the linear independence of the rows of $(A, -B)$. Hence there are n nonzero elementary divisors $e_1 | \cdots | e_n \neq 0$. The equivalent conditions of the present theorem are exactly those of Theorem 4.2.12,2., (a) $\Longleftrightarrow$ (c) $\Longleftrightarrow$ (d). □

Example 4.3.6 Let $n = m = 1$ in the preceding theorem. The system $A \circ x = B \circ u$ is controllable if and only if (A, B) has a right inverse, in other words, if and only if A and B are coprime, i.e., $\mathcal{D} = \mathcal{D}A + \mathcal{D}B$ or $\gcd(A, B) = 1$. Thus, the system $s \circ x = (s+1) \circ u$ is controllable, whereas $s \circ x = s \circ u$ is not. ◊

Next we will characterize the controllability of state space equations

$$
\begin{aligned}
s \circ x &= Ax + Bu, \ A \in F^{n\times n}, \ B \in F^{n\times m}, \\
\mathcal{B} &:= \left\{ \binom{x}{u} \in \mathcal{F}^{n+m};\ s \circ x = Ax + Bu \right\}.
\end{aligned}
\tag{4.65}
$$

Dually to $\mathfrak{O}$ from Definition 4.1.7 we define the *controllability matrix*

$$
\mathfrak{C} := (B, AB, \ldots, A^{n-1}B) \in F^{n\times nm}. \tag{4.66}
$$

As in (4.19) and (4.8) we consider F^n and $F^{1\times n}$ as $F[s]$-modules via

$$
s\ {}_A\!\circ \eta = A\eta \quad \text{and} \quad s \circ_A \xi = \xi A \quad \text{for } \eta \in F^n \text{ and } \xi \in F^{1\times n}
$$

and apply the results from Sect. 4.1.3. Obviously, the maps

$$
\begin{aligned}
(F^n, {}_A\circ) &\overset{\cong}{\longrightarrow} (F^{1\times n}, \circ_{A^\top}), \quad \eta \longmapsto \eta^\top, \quad \text{and} \\
(F^{1\times n}, \circ_A) &\overset{\cong}{\longrightarrow} (F^n, {}_{A^\top}\circ), \quad \xi \longmapsto \xi^\top,
\end{aligned}
\tag{4.67}
$$

are $F[s]$-isomorphisms. The following corollary is the analogue to Corollary 4.1.8.

Corollary 4.3.7 *Let f_A be the minimal polynomial of A. For all $t \geq \deg_s(f_A)$, especially for all $t \geq n$, the equalities*

$$
\sum_{j=1}^{m} F[s]\ {}_A\!\circ B_{-j} = \sum_{i=0}^{\infty} A^i B F^m = (B, AB, \ldots, A^{t-1}B)F^{tm} = \mathfrak{C}F^{nm} \tag{4.68}
$$

hold. This space is the least $F[s]$-submodule of $(F^n, {}_A\circ)$ containing the column space BF^m. For the associated orthogonal subspaces this implies

$$
\begin{aligned}
\left(\sum_{j=1}^{m} F[s]\ {}_A\!\circ B_{-j}\right)^{\perp} &= \left(\sum_{i=0}^{\infty} A^i B F^m\right)^{\perp} = (\mathfrak{C}F^{nm})^{\perp} \\
&= \bigcap_{i=0}^{\infty} \ker(\cdot A^i B) = \ker(\cdot \mathfrak{C}) \subseteq F^{1\times n}.
\end{aligned}
$$

This is the largest $F[s]$-submodule of $(F^{1\times n}, \circ_A)$ contained in $\ker(\cdot B)$.
The preceding identities imply the dimension formulas

$$\dim_F\left(\sum_{j=1}^{m} F[s]\,{}_A\circ B_{-j}\right) = \operatorname{rank}(\mathfrak{C}) \quad \textit{and}$$

$$\dim_F\left(\bigcap_{i=0}^{\infty} \ker(\cdot A^i B)\right) = n - \operatorname{rank}(\mathfrak{C}).$$

Proof This follows from (4.18) as in Corollary 4.1.8. □

Theorem 4.3.8 *Let $\mathcal{D} = F[s]$ be the polynomial ring over a field F. By definition, the state space equations* (4.65) *are* controllable *if and only if the behavior*

$$\mathcal{B} := \left\{\binom{x}{u} \in \mathcal{F}^{n+m};\ s \circ x = Ax + Bu,\ \text{i.e., } (s\,\mathrm{id}_n - A, -B) \circ \binom{x}{u}\right\} \tag{4.69}$$

is controllable. Let $e_1, \dots, e_n$ with $e_1|\cdots|e_n \neq 0$ denote the n elementary divisors of $(s\,\mathrm{id}_n - A, -B)$. The following properties are equivalent:

1. *The behavior $\mathcal{B}$ is controllable.*
2. *$e_n = 1$, hence $e_1 = \cdots = e_n = 1$.*
3. *The matrix $(s\,\mathrm{id}_n - A, -B) \in F[s]^{n\times(n+m)}$ or, equivalently, $(s\,\mathrm{id}_n - A, B)$ has a right inverse, i.e., there are matrices X and Y with $(s\,\mathrm{id}_n - A)X + BY = \mathrm{id}_n$.*
4. *$F^n = \sum_{j=1}^m F[s]\,{}_A\circ B_{-j} = \mathfrak{C}F^{nm}$ or* rank$(\mathfrak{C}) = n$.
5. $\ker(\cdot\mathfrak{C}) = \left(\sum_{j=1}^m F[s]\,{}_A\circ B_{-j}\right)^\perp = 0.$

Proof 1. $\Longleftrightarrow$ 2. $\Longleftrightarrow$ 3. Theorem 4.3.5.
3. $\Longleftrightarrow$ 4. The following chain of equivalences holds:

$$\begin{aligned}
&(s\,\mathrm{id}_n - A, B) \text{ has a right inverse}\\
&\underset{\text{transposition}}{\Longleftrightarrow} \begin{pmatrix} s\,\mathrm{id}_n - A^\top \\ B^\top \end{pmatrix} \text{ has a left inverse}\\
&\underset{\text{Theorem 4.1.9}}{\Longleftrightarrow} F^{1\times n} = \sum_{j=1}^m F[s] \circ_{A^\top} (B^\top)_{j-}\\
&\underset{(4.67)}{\Longleftrightarrow} F^n = \sum_{j=1}^m F[s]\,{}_A\circ B_{-j}.
\end{aligned}$$

4. $\Longleftrightarrow$ 5. Corollary 4.3.7. □

For $F = \mathbb{R}$ or $F = \mathbb{C}$ Corollary 5.4.7 will supplement the preceding theorem. Next we characterize controllability by *state controllability*.

Theorem 4.3.9 (Discrete time) *Consider the discrete-time state space system* (4.65) *over an arbitrary field F and the signal space $\mathcal{F} = F^{\mathbb{N}}$. Let $d := \deg_s(f_A)\ (\leq n)$ be the degree of the minimal polynomial of A. The following properties of are equivalent:*

1. *The behavior $\mathcal{B}$ from* (4.65) *is controllable, cf. Theorem 4.3.8.*
2. *There is a time instant $t_2 \in \mathbb{N}$ such that for arbitrary state vectors $v_1, v_2 \in F^n$ one can find an input sequence $u \in \mathcal{F}^m$ with the property that the unique trajectory $x \in \mathcal{F}^n$ with $s \circ x = Ax + Bu$ and $x(0) = v_1$ satisfies $x(t_2) = v_2$. In other words, one can steer the system from any initial state v_1 to any final state v_2 in the time interval from 0 to t_2. A state space system with this property is called* state controllable.

If these properties are satisfied one may choose any $t_2 \geq d$ in 2.

Proof By 3.65, the state $x(t)$ at time $t \in \mathbb{N}$ is given by

$$x(t) = A^t v_1 + \sum_{i=0}^{t-1} A^i Bu(t-1-i) = A^t v_1 + (B, AB, \ldots, A^{t-1}B) \begin{pmatrix} u(t-1) \\ u(t-2) \\ \vdots \\ u(0) \end{pmatrix}. \tag{4.70}$$

1. $\Longrightarrow$ 2.: Choose any $t_2 \geq d$. The controllability implies

$$F^n \underset{\text{Theorem 4.3.8}}{=} (B, AB, \ldots, A^{t_2-1}B) F^{t_2 m}.$$

Hence there is a vector $\begin{pmatrix} u(t_2-1) \\ \vdots \\ u(0) \end{pmatrix} \in F^{t_2 m}$ with

$$v_2 - A^{t_2} v_1 = (B, \ldots, A^{t_2-1}B) \begin{pmatrix} u(t_2-1) \\ \vdots \\ u(0) \end{pmatrix}. \tag{4.71}$$

We extend this vector to $u \in (F^{\mathbb{N}})^m$ arbitrarily, i.e., by arbitrary $u(t),\ t \geq t_2$. Let $x \in (F^{\mathbb{N}})^n$ be the state with initial state $x(0) = v_1$ and input u according to (4.70). Equations (4.70) and (4.71) imply $x(t_2) = v_2$.
2. $\Longrightarrow$ 1.: Let $v_1 := 0$ and $v_2 \in F^n$ be arbitrary. By assumption there is an input $u \in \mathcal{F}^m$ such that the sequence x with $s \circ x = Ax + Bu$ and $x(0) = v_1 = 0$ satisfies

$$\begin{aligned} v_2 = x(t_2) &= A^{t_2} 0 + (B, \ldots, A^{t_2-1}B) \begin{pmatrix} u(t_2-1) \\ \vdots \\ u(0) \end{pmatrix} \\ &\in (B, \ldots, A^{t_2-1}B) F^{t_2 m} \subseteq \mathfrak{C} F^{nm} \Longrightarrow F^n = \mathfrak{C} F^{nm}. \end{aligned}$$

By Theorem 4.3.8 the system is controllable. □

Theorem 4.3.10 (Continuous time, cf. [2, Theorem 3.5.3 on p. 66, Cor. 5.3.6 on p. 156], [1, Theorem 5.2.22 on p. 166, Theorem 5.2.27 on p. 170]) *Consider the continuous-time state space system $\mathcal{B}$ (4.65) over the field $F = \mathbb{R}$ or $F = \mathbb{C}$ and the signal space $\mathcal{F} = \mathrm{C}^\infty(\mathbb{R}, F)$. The following properties are equivalent:*

1. *The behavior $\mathcal{B}$ is controllable, cf. Theorem 4.3.8.*
2. *For every $t_2 > 0$ and arbitrary state vectors $v_1, v_2 \in F^n$ there is an input $u \in \mathcal{F}^m$ with the property that the unique trajectory $x \in \mathcal{F}^n$ with $x' = s \circ x = Ax + Bu$*

and initial state $x(0) = v_1$ satisfies $x(t_2) = v_2$. In other words, one can steer the system from any initial state v_1 to any final state v_2 in any time interval $[0, t_2],\ t_2 > 0$.

Proof By (3.63) the unique solution of the Cauchy problem $x' = Ax + Bu,\ x(0) = v_1$, is

$$x(t) = e^{tA}\left(v_1 + \int_0^t e^{-\tau A}u(\tau)\,\mathrm{d}\tau\right). \tag{4.72}$$

1. $\Longrightarrow$ 2. Let $x_1(t) := e^{tA}v_1$ and $x_2(t) := e^{(t-t_2)A}v_2$ for $t \in \mathbb{R}$, hence $x_1(0) = v_1$ and $x_2(t_2) = v_2$. The trajectories $w_j := \binom{x_j}{0} \in \mathcal{F}^{n+m},\ j = 1, 2$, belong to $\mathcal{B}$. Since $\mathcal{B}$ is controllable by assumption, Theorem 4.3.4 furnishes a trajectory $w = \binom{x}{u} \in \mathcal{B}$ such that $w|_{(-\infty,0]} = w_1|_{(-\infty,0]}$ and $w|_{[t_2,\infty)} = w_2|_{[t_2,\infty)}$ and hence $x(0) = x_1(0) = v_1$ and $x(t_2) = x_2(t_2) = v_2$.

non-1. $\Longrightarrow$ non-2. Since $\mathcal{B}$ is not controllable, we have $\operatorname{rank}(\mathfrak{C}) < n$ and $\ker(\cdot\mathfrak{C}) \neq 0$. Let $\xi \in F^{1\times n} \setminus \{0\}$ with $\xi\mathfrak{C} = 0$ and define the hyperplane $H := \{\eta \in F^n;\ \xi\eta = 0\} \subsetneq F^n$. For $v_1 := 0$ and any $u \in \mathcal{F}^m$ the unique solution x of $x' = Ax + Bu$ with $x(0) = v_1 = 0$ is

$$x = e^{tA}\left(v_1 + \int_0^t e^{-\tau A}u(\tau)\,\mathrm{d}\tau\right) = \int_0^t e^{(t-\tau)A}u(\tau)\,\mathrm{d}\tau \in \mathcal{F}^n.$$

From $\mathfrak{C}F^{nm} = \sum_{i=0}^{\infty} A^i B F^m$ and $\xi\mathfrak{C} = 0$, we infer that $\xi A^k B = 0$ for all $k \in \mathbb{N}$. Therefore,

$$\xi x(t) = \xi \int_0^t e^{(t-\tau)A} B\,\mathrm{d}\tau = \int_0^t \sum_{k=0}^{\infty} \frac{(t-\tau)^k}{k!} \underbrace{\xi A^k B}_{=0}\,\mathrm{d}\tau = 0,$$

i.e., $x(t) \in H$ for all $t \in \mathbb{R}$. Thus, any $v_2 \in F^n \setminus H \neq \emptyset$ cannot be reached from $v_1 = 0$. Hence 2 does not hold. □

In analogy to the observable factor of $\mathcal{B}$ from Theorem 4.1.11 we finally construct an observable state space representation for the controllable part $\mathcal{B}_{\text{cont}}$ of $\mathcal{B}$ from (4.65). The notations are those of Theorem 4.3.8.

Let $n_1 := \operatorname{rank}(\mathfrak{C})$ and $n_2 := n - n_1$. The Gauss algorithm applied to the matrix $(\mathfrak{C}, \mathrm{id}_n)$ furnishes a basis of $\mathfrak{C}F^{nm}$ and its prolongation to a basis of F^n or, in other terms, a matrix

$$V = (V_1, V_2) \in F^{n\times(n_1+n_2)} \text{ with } V \in \mathrm{Gl}_n(F) \text{ and } \mathfrak{C}F^{nm} = V_1 F^{n_1}. \tag{4.73}$$

Since, by Corollary 4.3.7, $\mathfrak{C}F^{nm} = V_1 F^{n_1}$ is an $F[s]$-submodule of F^n we get $s\ {}_A\!\circ (V_1)_{-j} = A(V_1)_{-j} \in V_1 F^{n_1}$ for every column $(V_1)_{-j}$ of V_1. Hence there are matrices $A_{11} \in F^{n_1\times n_1},\ A_{12} \in F^{n_1\times n_2},\ A_{22} \in F^{n_2\times n_2}$ such that

$$\begin{aligned} AV_1 = V_1A_{11},\ AV = A(V_1, V_2) = (V_1, V_2)\begin{pmatrix} A_{11} & A_{12} \\ 0 & A_{22}\end{pmatrix}, \\ \Longrightarrow \begin{pmatrix} A_{11} & A_{12} \\ 0 & A_{22}\end{pmatrix} = V^{-1}AV. \end{aligned} \tag{4.74}$$

Due to $BF^m \subseteq \mathfrak{C}F^{nm}$ there is also $B_1 \in F^{n_1 \times m}$ with

$$B = V_1B_1 \Longrightarrow B = (V_1, V_2)\begin{pmatrix} B_1 \\ 0\end{pmatrix} = V\begin{pmatrix} B_1 \\ 0\end{pmatrix} \Longrightarrow \begin{pmatrix} B_1 \\ 0\end{pmatrix} = V^{-1}B. \tag{4.75}$$

With these data we introduce the new state equations and behavior

$$s \circ x_1 = A_{11}x_1 + B_1u \text{ and } \mathcal{B}_1 := \left\{\begin{pmatrix} x_1 \\ u\end{pmatrix} \in \mathcal{F}^{n_1+m};\ s \circ x_1 = A_{11}x_1 + B_1u\right\}. \tag{4.76}$$

Theorem 4.3.11 (Controllable part of a state space behavior, cf. [1, §5.4 on p. 184]) *We assume the state space behavior $\mathcal{B}$ from* (4.65) *and the data from* (4.73)–(4.76). *The behavior $\mathcal{B}_1$ is controllable, and the map*

$$\begin{pmatrix} V_1 & 0 \\ 0 & \mathrm{id}_m\end{pmatrix}\cdot : \mathcal{B}_1 \xrightarrow{\cong} \mathcal{B}_{\mathrm{cont}},\ \begin{pmatrix} x_1 \\ u\end{pmatrix} \longmapsto \begin{pmatrix} V_1x_1 \\ u\end{pmatrix}, \tag{4.77}$$

is well defined and an isomorphism. In other words, $\mathcal{B}_1$ is an observable state space realization of the largest controllable subbehavior $\mathcal{B}_{\mathrm{cont}}$ of $\mathcal{B}$.

Proof (i) We show first that $\mathcal{B}_1$ is controllable. Since $V = (V_1, V_2) \in \mathrm{Gl}_n(F)$, the map $V_1\cdot : F^{n_1} \longrightarrow F^n$ is a monomorphism of vector spaces. Since $s\ {}_{A_{11}}\circ\zeta = A_{11}\zeta$ is mapped to $V_1A_{11}\zeta = AV_1\zeta = s\ {}_A\circ(V_1\zeta)$, it is even $F[s]$ linear, and thus we have the $F[s]$-monomorphism

$$V_1\cdot : (F^{n_1}, {}_{A_{11}}\circ) \longrightarrow (F^n, {}_A\circ),\quad \zeta \longmapsto V_1\zeta.$$

Since $B = V_1B_1$ and thus $B_{-j} = V_1(B_1)_{-j}$, the map induces the isomorphism

$$\begin{aligned} &\sum_{j=1}^m F[s]\ {}_{A_{11}}\circ(B_1)_{-j} \cong \sum_{j=1}^m F[s]\ {}_A\circ B_{-j},\ \text{ hence } \dim_F\left(\sum_{j=1}^m F[s]\ {}_{A_{11}}\circ(B_1)_{-j}\right) \\ &= \dim_F\left(\sum_{j=1}^m F[s]\ {}_A\circ B_{-j}\right) \overset{\text{Cor. 4.3.7}}{=\!=\!=} \mathrm{rank}(\mathfrak{C}) = n_1. \end{aligned}$$

Thus $\mathcal{B}_1$ is controllable by Theorem 4.3.8.
The maps $V_1\cdot : \mathcal{F}^{n_1} \to \mathcal{F}^n$ and $\begin{pmatrix} V_1 & 0 \\ 0 & \mathrm{id}_m\end{pmatrix}\cdot$ are injective. It remains to show $\begin{pmatrix} V_1 & 0 \\ 0 & \mathrm{id}_m\end{pmatrix}\cdot \mathcal{B}_1 = \mathcal{B}_{\mathrm{cont}}$.

$\subseteq$. Due to (4.74)–(4.76) the equation $s \circ x_1 = A_{11}x_1 + B_1u$ implies

$$s \circ V_1x_1 = V_1A_{11}x_1 + V_1B_1u = A(V_1x_1) + Bu \Longrightarrow \begin{pmatrix} V_1 & 0 \\ 0 & \mathrm{id}_m\end{pmatrix}\cdot \mathcal{B}_1 \subseteq \mathcal{B}.$$

By Lemma 4.2.5 the image of the controllable behavior $\mathcal{B}_1$ is controllable too, and thus $\left(\begin{smallmatrix} V_1 & 0 \\ 0 & \mathrm{id}_m \end{smallmatrix}\right) \cdot \mathcal{B}_1 \subseteq \mathcal{B}_{\text{cont}}$.

$\supseteq$. Decompose $V^{-1} = \begin{pmatrix} V_1^{-1} \\ V_2^{-1} \end{pmatrix} \in F^{(n_1+n_2)\times n}$ and write $\left(\begin{smallmatrix} x_1 \\ x_2 \end{smallmatrix}\right) := V^{-1}x = \begin{pmatrix} V_1^{-1}x \\ V_2^{-1}x \end{pmatrix}$ for $x \in \mathcal{F}^n$. Then the equivalences

$$\begin{aligned} & s \circ x = Ax + Bu \Longleftrightarrow s \circ V^{-1}x = V^{-1}AV(V^{-1}x) + V^{-1}Bu \\ & \Longleftrightarrow s \circ \left(\begin{smallmatrix} x_1 \\ x_2 \end{smallmatrix}\right) = \left(\begin{smallmatrix} A_{11} & A_{12} \\ 0 & A_{22} \end{smallmatrix}\right)\left(\begin{smallmatrix} x_1 \\ x_2 \end{smallmatrix}\right) + \left(\begin{smallmatrix} B_1 \\ 0 \end{smallmatrix}\right) u \\ & \Longleftrightarrow \begin{cases} s \circ x_1 = A_{11}x_1 + A_{12}x_2 + B_1 u \\ s \circ x_2 = A_{22}x_2 \end{cases} \Longleftrightarrow \begin{cases} x_1 \in \mathcal{B}_1 \\ x_2 \in \mathcal{B}_2 \end{cases} \end{aligned} \tag{4.78}$$

hold, where $\mathcal{B}_2 := \{x_2 \in \mathcal{F}^{n_2};\ \ s \circ x_2 = A_{22}x_2\}$ is autonomous by Theorem 4.2.12. The morphism $\mathcal{B}_{\text{cont}} \longrightarrow \mathcal{B}_2,\ \left(\begin{smallmatrix} x \\ u \end{smallmatrix}\right) \longmapsto x_2 = V_2^{-1}x$, has an image that is both controllable and autonomous by Lemma 4.2.5 and thus zero, i.e., $x_2 = V_2^{-1}x = 0$ for all $\left(\begin{smallmatrix} x \\ u \end{smallmatrix}\right) \in \mathcal{B}_{\text{cont}}$. We conclude

$$\begin{aligned} & \forall \left(\begin{smallmatrix} x \\ u \end{smallmatrix}\right) \in \mathcal{B}_{\text{cont}} \subseteq \mathcal{B} \text{ and } \left(\begin{smallmatrix} x_1 \\ x_2 \end{smallmatrix}\right) := V^{-1}x :\ x_2 = 0,\ x = V_1 x_1 \\ & \underset{(4.78)}{\Longrightarrow} s \circ x_1 = A_{11}x_1 + B_1 u \text{ or } \left(\begin{smallmatrix} x_1 \\ u \end{smallmatrix}\right) \in \mathcal{B}_1 \Longrightarrow \left(\begin{smallmatrix} x \\ u \end{smallmatrix}\right) = \left(\begin{smallmatrix} V_1 x_1 \\ u \end{smallmatrix}\right) \in \left(\begin{smallmatrix} V_1 & 0 \\ 0 & \mathrm{id}_m \end{smallmatrix}\right) \cdot \mathcal{B}_1 \\ & \Longrightarrow \mathcal{B}_{\text{cont}} \subseteq \left(\begin{smallmatrix} V_1 & 0 \\ 0 & \mathrm{id}_m \end{smallmatrix}\right) \cdot \mathcal{B}_1. \end{aligned}$$

□

References

1. J.W. Polderman, J.C. Willems, *Introduction to Mathematical Systems Theory. A Behavioral Approach*. Texts in Applied Mathematics, vol. 26 (Springer, New York, 1998)
2. W.A. Wolovich, *Linear Multivariable Systems*. Applied Mathematical Sciences, vol. 11 (Springer, New York, 1974)
3. S. Lang, *Algebra*. Graduate Texts in Mathematics, vol. 211, 3rd edn (Springer, New York, 2002)
4. T. Kailath, *Linear Systems* (Prentice-Hall, Englewood Cliffs, 1980)

Chapter 5
Applications of the Chinese Remainder Theorem

The *Chinese remainder theorem*, in short, CRT, is ubiquitous in applied and engineering mathematics, and especially in systems theory. It applies to coprime ideals in a not necessarily commutative ring. However, in one-dimensional systems theory as in this book, it is needed only for principal ideal domains like the polynomial ring $F[s]$ over a field F.

We recall the standard theory in Sect. 5.1 and prove the Chinese remainder theorem for modules. We apply this to fundamental systems of single differential and difference equations in Sect. 5.2 and to the primary decomposition of torsion modules and of autonomous behaviors in Sect. 5.3. In Sects. 5.4 we apply this, in particular, to autonomous behaviors $\mathcal{B}^0$ in the continuous- or discrete-time standard cases. We discuss the characteristic variety or set of characteristic values or poles of $\mathcal{B}^0$, the multiplicity of the poles, the computation of a fundamental system or basis of $\mathcal{B}^0$, and the stability of $\mathcal{B}^0$. We conclude the chapter with Sect. 5.5, where we present the Jordan decomposition for matrices and the partial fraction decomposition for rational functions in one indeterminate as further applications of the Chinese remainder theorem. This chapter is based on [1].

If $f(s)$ is a polynomial we write $\deg(f) = \deg_s(f)$ for its degree.

5.1 The Chinese Remainder Theorem for Modules

We assume that $\mathcal{D}$ is a principal ideal domain with its quotient field K, its group $\mathrm{U}(\mathcal{D})$ of units, and a representative system $\mathcal{P}$ of prime elements in $\mathcal{D}$ (up to association). Then every nonzero element in K has the unique prime factor decomposition

$$d = u \prod_{p \in \mathcal{P}} p^{m(p)} \text{ with } u \in \mathrm{U}(\mathcal{D}) \text{ and } m = \big(m(p)\big)_{p \in \mathcal{P}} \in \mathbb{Z}^{(\mathcal{P})}. \tag{5.1}$$

U. Oberst et al., *Linear Time-Invariant Systems, Behaviors and Modules*, Differential-Algebraic Equations Forum,
https://doi.org/10.1007/978-3-030-43936-1_5

For almost all $p \in \mathcal{P}$ the exponent $m(p)$ is zero. It is called the p-order $\mathrm{ord}_p(d) := m(p)$ of d.

For the standard case $\mathcal{D} = F[s]$, we have

$$\begin{aligned} \mathrm{U}(F[s]) = F \setminus \{0\} \quad \text{and} \\ \mathcal{P} = \{f = s^n + \cdots \in F[s];\ f \text{ monic and irreducible}\}, \end{aligned} \tag{5.2}$$

where a polynomial f is called *monic* if its leading coefficient $\mathrm{lc}(f)$ is 1. All linear polynomials $s - \lambda$, $\lambda \in F$, belong to $\mathcal{P}$. For $F = \mathbb{R}$ and $F = \mathbb{C}$ we obtain

$$\begin{aligned} \mathcal{P}_{\mathbb{C}} &= \{s - \lambda \in \mathbb{C}[s];\ \lambda \in \mathbb{C}\} \quad \text{and} \\ \mathcal{P}_{\mathbb{R}} &= \mathcal{P}_{\mathbb{R},1} \uplus \mathcal{P}_{\mathbb{R},2}, \\ \text{where} \quad \mathcal{P}_{\mathbb{R},1} &:= \{s - \lambda \in \mathbb{R}[s];\ \lambda \in \mathbb{R}\} \quad \text{and} \\ \mathcal{P}_{\mathbb{R},2} &:= \{s^2 + as + b \in \mathbb{R}[s];\ a, b \in \mathbb{R},\ a^2 - 4b < 0\}. \end{aligned} \tag{5.3}$$

For elements $f, g \in \mathcal{D}$ the greatest common divisor $\gcd(f, g)$ and the least common multiple $\mathrm{lcm}(f, g)$ exist and are unique up to association. In $F[s]$ they are assumed monic and thus unique and computed by the Euclidean algorithm. They satisfy

$$\mathcal{D}\gcd(f, g) = \mathcal{D}f + \mathcal{D}g,\ \mathcal{D}\,\mathrm{lcm}(f, g) = \mathcal{D}f \cap \mathcal{D}g,\ fg = \gcd(f, g)\,\mathrm{lcm}(f, g).$$

Lemma and Definition 5.1.1 *For* $f, g \in \mathcal{D}$ *the property* $\gcd(f, g) = 1$, *in other words,* $\mathcal{D}f + \mathcal{D}g = \mathcal{D}$ *is equivalent to* $\overline{f} \in \mathrm{U}(\mathcal{D}/\mathcal{D}g)$ *and to* $\overline{g} \in \mathrm{U}(\mathcal{D}/\mathcal{D}f)$. *If these conditions hold, then* f *and* g *are called* coprime. *If* $f_1, g_1 \in \mathcal{D}$ *are such that* $f_1 f + g_1 g = 1$, *then* $(\overline{f})^{-1} = \overline{f_1} \in \mathrm{U}(\mathcal{D}/\mathcal{D}g)$. ◊

For the Chinese remainder theorem we assume that $f_1, \ldots, f_r$ are nonzero elements in $\mathcal{D}$ and write $f := f_1 \cdots f_r$ for their product. The canonical maps $\mathrm{can}\colon \mathcal{D}/\mathcal{D}f \longrightarrow \mathcal{D}/\mathcal{D}f_i$ induce the *diagonal map*

$$\begin{aligned} \Delta\colon\ \mathcal{D}/\mathcal{D}f &\longrightarrow \prod_{i=1}^{r} \mathcal{D}/\mathcal{D}f_i, \\ \overline{g} = g + \mathcal{D}f &\longmapsto \Delta(\overline{g}) = (\overline{g}, \ldots, \overline{g}) = (g + \mathcal{D}f_1, \ldots, g + \mathcal{D}f_r), \end{aligned} \tag{5.4}$$

which is a ring homomorphism. Its kernel is

$$\ker(\Delta) = \left(\bigcap_{i=1}^{r} \mathcal{D}f_i\right)/\mathcal{D}f = \mathcal{D}\,\mathrm{lcm}(f_1, \ldots, f_r)/\mathcal{D}f_1 \cdots f_r.$$

Result 5.1.2 (Chinese remainder theorem (CRT), see [2, Theorem 2.1 and Corollary 2.2 on pp. 94–95]) *We use the data we have just introduced.*

1. *The following conditions are equivalent:*

   (a) *The elements f_i are pairwise coprime, i.e.,* $\gcd(f_i, f_j) = 1$ *for* $i \neq j$.
   (b) *For arbitrary* $g_1, \ldots, g_r$, *the system* $g \equiv g_i \mod f_i$, $i = 1, \ldots, r$, *of congruences has a solution* g.
   (c) *The diagonal map* (5.4) *is an isomorphism.*

2. *If the equivalent conditions of 1 are satisfied, the elements* f_i *and* $\frac{f}{f_i}$ *are also coprime for all* i. *Hence, for all* $i = 1, \ldots, r$, *there are representations*

$$1 = f_{i,1}f_i + f_{i,2}\frac{f}{f_i} = a_i + b_i, \quad \text{where } b_i := f_{i,2}\frac{f}{f_i} \text{ and } a_i := f_{i,1}f_i. \tag{5.5}$$

For $\mathcal{D} = F[s]$ *they are computed by means of the Euclidean algorithm.*
Then $g := \sum_{i=1}^{r} g_i b_i$ *is a solution of the congruence system in 1(b), the set of all solutions is* $g + \mathcal{D}f$, *and the inverse of* Δ *is given by*

$$\Delta^{-1}(\overline{g_1}, \ldots, \overline{g_r}) = \overline{\sum_{i=1}^{r} g_i b_i}. \tag{5.6}$$

◇

Equation (5.6) essentially proves the CRT constructively.

Corollary and Definition 5.1.3 *The elements* $e_i := \overline{b_i} \in \mathcal{D}/\mathcal{D}f$ *are* complete orthogonal idempotents, *i.e., they satisfy the equations*

$$\begin{aligned} e_i^2 &= e_i \quad \text{for } i = 1, \ldots, r && \textit{(idempotency)},\\ e_i e_j &= 0 \quad \text{for } i \neq j && \textit{(orthogonality), and}\\ \sum_{i=1}^{r} e_i &= \overline{1} = 1_{\mathcal{D}/\mathcal{D}f} && \textit{(completeness)}. \end{aligned}$$

The first two conditions can be summed up as $e_i e_j = \delta_{i,j} e_i$ *for* $i, j \in \{1, \ldots, r\}$.

Proof Define $\delta_j := (0, \ldots, 0, \overset{i}{1}, 0, \ldots, 0) \in \prod_{i=1}^{r} \mathcal{D}/\mathcal{D}f_i$ for $j = 1, \ldots, r$. Then $e_i = \Delta^{-1}(\delta_i)$. The δ_i are obviously complete orthogonal idempotents. Since Δ^{-1} is a ring isomorphism, the three conditions of the assertion are preserved and show that the e_i are complete orthogonal idempotents too. □

In the sequel we assume the conditions and the data of the CRT (Result 5.1.2). We are going to apply the CRT to the decomposition of modules. So let M be a $\mathcal{D}$-module.

If $fM = 0$ for some $f \in \mathcal{D}$, the module M can be considered as a $\mathcal{D}/\mathcal{D}f$-module by means of the well-defined scalar multiplication

$$\overline{g}x := gx, \quad \text{where } g \in \mathcal{D},\ \overline{g} = g + \mathcal{D}f \in \mathcal{D}/\mathcal{D}f, \text{ and } x \in M. \tag{5.7}$$

For $g \in \mathcal{D}$ we use the *annihilator submodule*

$$\mathrm{ann}_M(g) := \ker\big(g\colon M \to M,\ x \mapsto gx\big) = \{x \in M;\ gx = 0\} \subseteq M. \quad (5.8)$$

Since $g\,\mathrm{ann}_M(g) = 0$, this annihilator can be and is considered as a $\mathcal{D}/\mathcal{D}g$-module. Conversely, for any $\mathcal{D}$-module N we use the *annihilator ideal*

$$\mathrm{ann}_{\mathcal{D}}(N) := \{g \in \mathcal{D};\ gN = 0\} \subseteq \mathcal{D}. \quad (5.9)$$

Theorem 5.1.4 (CRT for modules)

1. *Assume the data of Result 5.1.2 and let M be a $\mathcal{D}$-module. Then*

$$\begin{aligned}\mathrm{ann}_M(f) = \bigoplus_{i=1}^r b_i\,\mathrm{ann}_M(f) \ni x = \sum_{i=1}^r b_i x \quad \textit{and}\\ b_i\,\mathrm{ann}_M(f) = \mathrm{ann}_M(f_i) \quad \textit{for } i = 1,\dots,r.\end{aligned} \quad (5.10)$$

2. *The surjective projection*

$$\varphi_i\colon\ \mathrm{ann}_M(f) \longrightarrow b_i\,\mathrm{ann}_M(f) = \mathrm{ann}_M(f_i), \quad x \longmapsto b_i x,$$

induces the isomorphism

$$\varphi_{i,\mathrm{ind}}\colon\ \mathrm{ann}_M(f)/f_i\,\mathrm{ann}_M(f) \xrightarrow{\cong} b_i\,\mathrm{ann}_M(f), \quad \overline{x} = x + f_i\,\mathrm{ann}_M(f) \longmapsto b_i x. \quad (5.11)$$

3. *The equation $g\,\mathrm{ann}_M(g) = 0$, i.e., $\mathcal{D}g \subseteq \mathrm{ann}_{\mathcal{D}}(\mathrm{ann}_M(g))$, holds for all $g \in \mathcal{D}$. If, moreover, $\mathrm{ann}_{\mathcal{D}}(\mathrm{ann}_M(f)) = \mathcal{D}f$, then also $\mathrm{ann}_{\mathcal{D}}\big(\mathrm{ann}_M(f_i)\big) = \mathcal{D}f_i$ for $i = 1,\dots,r$.*

Proof 1. The module $U := \mathrm{ann}_M(f)$ satisfies $fU = 0$, and is thus a $\mathcal{D}/\mathcal{D}f$-module according to (5.7). By Corollary 5.1.3, the ring $\mathcal{D}/\mathcal{D}f$ has the complete orthogonal idempotents $\overline{b_i}$, $i = 1,\dots,r$. Hence,

$$U = \overline{1}U = \left(\sum_{i=1}^r \overline{b_i}\right)U = \sum_{i=1}^r \overline{b_i}U = \sum_{i=1}^r b_i U,$$

where the last equality follows from (5.7). Assume

$$0 = \sum_{i=1}^r b_i x_i = \sum_{i=1}^r \overline{b_i} x_i \Longrightarrow \forall j = 1,\dots,r:$$

$$0 = \sum_{i=1}^r \overline{b_j}\,\overline{b_i} x_i = \sum_{i=1}^r \delta_{j,i}\overline{b_i} x_i = \overline{b_j} x_j = b_j x_j \Longrightarrow U = \bigoplus_{i=1}^r b_i U.$$

Next we show that $b_i U = \operatorname{ann}_M(f_i)$ for $i = 1, \ldots, r$.

$\subseteq$. The expression $b_i = f_{i,2}\frac{f}{f_i}$ implies that $f \mid f_i b_i$, and therefore $f_i b_i U \subseteq fU = 0$. Thus, $b_i U \subseteq \operatorname{ann}_M(f_i)$.

$\supseteq$. If, conversely, $x \in \operatorname{ann}_M(f_i)$, i.e., $f_i x = 0$, then also $0 = f_{i,1} f_i x = a_i x$. Thus, $x = 1x = (a_i + b_i)x = b_i x \in b_i U$, and we conclude that $\operatorname{ann}_M(f_i) \subseteq b_i U$.

2. From $f \operatorname{ann}_M(f) = 0$ we infer that

$$\varphi_i(f_i \operatorname{ann}_M(f)) = b_i f_i \operatorname{ann}_M(f) \overset{(5.5)}{=\!=} f_{i,2} f \operatorname{ann}_M(f) = 0.$$

Hence $f_i \operatorname{ann}_M(f) \subseteq \ker(\varphi_i)$ and thus $\varphi_{i,\text{ind}}$ is well defined and surjective. To show its injectivity, let $x \in \operatorname{ann}_M(f)$ with $0 = \varphi_{i,\text{ind}}(\overline{x}) = b_i x$. Hence,

$$x = a_i x + b_i x = a_i x = f_{i,1} f_i x = f_i(f_{i,1} x).$$

Since $f_{i,1} x \in f_{i,1} \operatorname{ann}_M(f) \subseteq \operatorname{ann}_M(f)$, we infer $x \in f_i \operatorname{ann}_M(f)$and $\overline{x} = 0$.
3. Finally we assume $\operatorname{ann}_{\mathcal{D}}(U) = \mathcal{D}f$. For $i = 1, \ldots, r$ let $g_i \in \mathcal{D}$ be such that

$$\mathcal{D}g_i = \operatorname{ann}_{\mathcal{D}}(\operatorname{ann}_M(f_i)) = \operatorname{ann}_{\mathcal{D}}(b_i U).$$

Hence, $\mathcal{D}f_i \subseteq \mathcal{D}g_i$, i.e., $f_i = \xi_i g_i$ for some ξ_i, and $g_i b_i U = 0$. With $g := g_1 \cdots g_r$, we obtain $\mathcal{D}f \subseteq \mathcal{D}g$, as well as $g b_i U = 0$. Thus, $gU = \sum_{i=1}^r g b_i U = 0$, and consequently $g \in \operatorname{ann}_{\mathcal{D}}(U) = \mathcal{D}f$ and $g = \eta f$ for some $\eta \in \mathcal{D}$. We conclude that

$$g = \eta f = \eta f_1 \cdots f_r = \eta(\xi_1 g_1) \cdots (\xi_r g_r) = \eta \xi_1 \cdots \xi_r g_1 \cdots g_r = \eta \xi_1 \cdots \xi_r g.$$

Cancellation of g gives $1 = \eta \xi_1 \cdots \xi_r$, i.e., η as well as all the ξ_i, $i = 1, \ldots, r$, are invertible. This implies that $\mathcal{D}f_i = \mathcal{D}g_i = \operatorname{ann}_{\mathcal{D}}(b_i U)$ for all i. □

The following technical lemma is often useful.

Lemma 5.1.5 *Let $f, g \in \mathcal{D}$ be coprime elements with a representation $1 = f_1 f + g_1 g$ and let M be a $\mathcal{D}$-module. Then*

$$\operatorname{ann}_M(f) \overset{\cong}{\rightleftarrows} \operatorname{ann}_M(f), \quad x \longmapsto gx, \quad g_1 x_1 \longleftarrow\!\shortmid x_1.$$

Proof If $fx = 0$, then $fgx = 0$. Equally, if $fx_1 = 0$, then $fg_1 x_1 = 0$. Thus both maps are well defined. Since $x = 1x = (f_1 f + g_1 g)x = g_1(gx)$ and, equally $x_1 = g_1 g x_1 = g(g_1 x_1)$, they are inverse to each other. □

In the following corollary we apply the CRT to a monic polynomial in $F[s]$ which splits in the field F, i.e., which is a product of linear factors. That f is monic is assumed for notational simplicity only. So assume

$$f = \prod_{i=1}^{r}(s-\lambda_i)^{m(i)},\ f_i := (s-\lambda_i)^{m(i)},\ \tfrac{f}{f_i} = \prod_{j\neq i}(s-\lambda_j)^{m(j)},$$
$$\Longrightarrow \gcd(f_i, \tfrac{f}{f_i}) = 1 \Longrightarrow \exists f_{i,1}, f_{i,2} \text{ with } 1 = f_{i,1}f_i + f_{i,2}\tfrac{f}{f_i} = a_i + b_i. \tag{5.12}$$

Corollary 5.1.6 *If $f \in F[s]$ splits according to* (5.12) *and M is an $F[s]$-module, then*

$$\operatorname{ann}_M(f) = \bigoplus_{i=1}^{r} \operatorname{ann}_M\left((s-\lambda_i)^{m(i)}\right) \quad \text{and}$$
$$\operatorname{ann}_M\left((s-\lambda_i)^{m(i)}\right) = \left\{x \in M;\ (s-\lambda_i)^{m(i)}x = 0\right\} = b_i \operatorname{ann}_M(f).$$

If, moreover, $\operatorname{ann}_{F[s]}\left(\operatorname{ann}_M(f)\right) = F[s]f$, *then also*

$$\operatorname{ann}_{F[s]}\left(\operatorname{ann}_M\left((s-\lambda_i)^{m(i)}\right)\right) = F[s](s-\lambda_i)^{m(i)} \quad \text{for } i = 1, \ldots, r.$$

Proof This follows directly from (5.12) and Theorem 5.1.4. □

5.2 Fundamental Systems of Single Equations

In the next theorems we apply the CRT to the computation of fundamental systems of solutions of differential and difference equations for the standard injective $F[s]$-cogenerators $\mathcal{F}$ from Theorem 3.3.7. *Here we focus on systems which are given by a single homogeneous equation in one unknown. In Sect.* 5.4.1, *specifically in Theorem* 5.4.11 *and Corollary* 5.4.14, *we will generalize these results to autonomous systems of several equations in several unknowns. We treat the continuous-time and the discrete-time settings separately. The first observation, however, holds in both cases.*

Let $\mathcal{D}$ be a commutative ring, $g \in \mathcal{D}$ and ${}_{\mathcal{D}}\mathcal{F}$ a cogenerator. Then $\operatorname{ann}_{\mathcal{F}}(g) = \{y \in \mathcal{F};\ g \circ y = 0\} = (\mathcal{D}g)^{\perp}$, and thus, by Theorem 3.3.9,

$$\mathcal{D}g = (\mathcal{D}g)^{\perp\perp} = \{d \in \mathcal{D};\ d \circ \operatorname{ann}_{\mathcal{F}}(g) = 0\} = \operatorname{ann}_{\mathcal{D}}\left(\operatorname{ann}_{\mathcal{F}}(g)\right).$$

In particular, if $f = f_1 \cdots f_r \in \mathcal{D}$ is a product of elements of $\mathcal{D}$, then

$$\mathcal{D}f = \operatorname{ann}_{\mathcal{D}}\left(\operatorname{ann}_{\mathcal{F}}(f)\right) \quad \text{and}$$
$$\mathcal{D}f_i = \operatorname{ann}_{\mathcal{D}}\left(\operatorname{ann}_{\mathcal{F}}(f_i)\right) \quad \text{for } i = 1, \ldots, r.$$

5.2.1 The Continuous-Time Case

First we treat the continuous-time case with the field $\mathbb{C}$ and the $\mathbb{C}[s]$-module

$$(\mathcal{F}, \circ) := \mathrm{C}^{\infty}(\mathbb{R}, \mathbb{C}) \quad \text{with } s \circ y = \tfrac{\mathrm{d}y}{\mathrm{d}t}.$$

Let $f \in \mathcal{D}$ and let $f = \prod_{i=1}^{r}(s-\lambda_i)^{m(i)}$ be its decomposition into linear factors. By Corollary 5.1.6, we obtain the direct decomposition

$$\operatorname{ann}_{\mathcal{F}}(f) = \{y \in \mathcal{F};\ f \circ y = 0\} = \bigoplus_{i=1}^{r} \operatorname{ann}_{\mathcal{F}}\left((s-\lambda_i)^{m(i)}\right)$$

of the behavior of f. In order to find a $\mathbb{C}$-basis of the behavior $\operatorname{ann}_{\mathcal{F}}(f)$, we will investigate behaviors $\operatorname{ann}_{\mathcal{F}}\left((s-\lambda)^m\right)$ for $\lambda \in \mathbb{C}$ and $m \in \mathbb{N}$ first.
For this we fix $\lambda \in \mathbb{C}$ and consider the semi-linear isomorphism

$$\mathcal{F} \rightleftarrows \mathcal{F}, \quad y \longmapsto ye^{\lambda t}, \quad ze^{-\lambda t} \longleftarrow z. \tag{5.13}$$

Semi-linearity means that it is linear up to a ring isomorphism. Here we have

$$(s-\lambda) \circ \left(ye^{\lambda t}\right) = (s \circ y)e^{\lambda t} \quad \text{and thus} \quad f(s-\lambda) \circ \left(ye^{\lambda t}\right) = \left(f(s) \circ y\right)e^{\lambda t}.$$

Thus the ring isomorphism is

$$\mathbb{C}[s] \longrightarrow \mathbb{C}[s], \quad s \longmapsto s-\lambda, \quad f(s) \longmapsto f(s-\lambda).$$

Corollary 5.2.1 *For $\lambda \in \mathbb{C}$, $f \in \mathbb{C}[s]$, and $\mathcal{F} = \mathrm{C}^{\infty}(\mathbb{R}, \mathbb{C})$ the isomorphism* (5.13) *induces the semi-linear isomorphism*

$$\operatorname{ann}_{\mathcal{F}}(f) \overset{\cong}{\longleftrightarrow} \operatorname{ann}_{\mathcal{F}}(f(s-\lambda)), \quad y \longmapsto ye^{\lambda t}, \quad ze^{-\lambda t} \longleftarrow z. \tag{5.14}$$

In particular, this implies the isomorphism $\operatorname{ann}_{\mathcal{F}}(s^m) \cong \operatorname{ann}_{\mathcal{F}}\left((s-\lambda)^m\right)$. ◇

For $k \in \mathbb{N}$ the identity

$$s \circ \tfrac{t^k}{k!} = \begin{cases} \tfrac{t^{k-1}}{(k-1)!} & \text{if } k > 0, \\ 0 & \text{if } k = 0, \end{cases}$$

holds. Using the isomorphism (5.13), we infer, for $k \in \mathbb{N}$ and $\lambda \in \mathbb{C}$,

$$(s-\lambda) \circ \left(\tfrac{t^k}{k!}e^{\lambda t}\right) = \begin{cases} \tfrac{t^{k-1}}{(k-1)!}e^{\lambda t} & \text{if } k > 0, \\ 0 & \text{if } k = 0. \end{cases}$$

To obtain the simple equation $(s-\lambda)\circ e_{\lambda,k}=e_{\lambda,k-1},\ k\in\mathbb{N},\ k>0,\ \lambda\in\mathbb{C}$, we define

$$e_{\lambda,k}(t):=\tfrac{t^k}{k!}e^{\lambda t} \text{ and conclude } (s-\lambda)^l\circ e_{\lambda,k}=\begin{cases} e_{\lambda,k-l} & \text{if } l\le k,\\ 0 & \text{if } l>k.\end{cases} \tag{5.15}$$

Theorem 5.2.2 (Compare [3, Theorem 3.2.5 on p. 69]) *For the polynomial* $f=\prod_{i=1}^r(s-\lambda_i)^{m(i)}\in\mathbb{C}[s]$, *the* b_i *from* (5.12), *and the module* $\mathcal{F}:=\mathrm{C}^\infty(\mathbb{R},\mathbb{C})$ *the behavior* $\mathcal{B}:=\mathrm{ann}_{\mathcal{F}}(f)=\{y\in\mathcal{F};\ f\circ y=0\}$ *admits the direct decomposition*

$$\mathrm{ann}_{\mathcal{F}}(f)=\bigoplus_{i=1}^r\mathrm{ann}_{\mathcal{F}}\left((s-\lambda_i)^{m(i)}\right) \text{ where}$$

$$\mathrm{ann}_{\mathcal{F}}\left((s-\lambda_i)^{m(i)}\right)=\bigoplus_{k=0}^{m(i)-1}\mathbb{C}e_{\lambda_i,k}=\mathbb{C}[t]_{\le m(i)-1}e^{\lambda_i t} \text{ and}$$

$$\mathbb{C}[t]_{\le m-1}:=\bigoplus_{k=0}^{m-1}\mathbb{C}\tfrac{t^k}{k!}=\bigoplus_{k=0}^{m-1}\mathbb{C}t^k=\mathrm{ann}_{\mathcal{F}}(s^m).$$

Therefore the functions $e_{\lambda_i,k},\ i=1,\dots,r,\ k=0,\dots,m(i)-1$, *are a* fundamental system, *i.e., a basis, of the solution space of the differential equation* $f\circ y=0$. *With* b_i *from* (5.12) *the decomposition of a solution* $y\in\mathcal{B}$ *is given by*

$$y=\sum_{i=1}^r b_i\circ y=\sum_{i=1}^r b_i\left(\tfrac{\mathrm{d}}{\mathrm{d}t}\right)(y)\in\mathcal{B}=\bigoplus_{i=1}^r\mathrm{ann}_{\mathcal{F}}\left((s-\lambda_i)^{m(i)}\right).$$

Up to the last equation, this is a standard result on differential equations, but, in general, not proven by means of the CRT.

Proof Corollary 5.1.6 yields the desired decomposition

$$\mathrm{ann}_{\mathcal{F}}(f)=\bigoplus_{i=1}^r\mathrm{ann}_{\mathcal{F}}\left((s-\lambda_i)^{m(i)}\right).$$

According to (5.15) (with $\lambda=0$), the space $\mathrm{ann}_{\mathcal{F}}(s^m)$ contains the m linearly independent polynomials $\frac{t^k}{k!}$. By (3.7), its dimension is m, and hence these polynomials are a basis and $\mathrm{ann}_{\mathcal{F}}(s^m)=\bigoplus_{k=0}^{m-1}\mathbb{C}\frac{t^k}{k!}=\mathbb{C}[t]_{\le m-1}$. The isomorphism (5.14) then implies $\mathrm{ann}_{\mathcal{F}}\left((s-\lambda_i)^{m(i)}\right)=\bigoplus_{k=0}^{m(i)-1}\mathbb{C}\frac{t^k}{k!}e^{\lambda_i t}$, and thus

$$\mathrm{ann}_{\mathcal{F}}(f)=\bigoplus_{i=1}^r\bigoplus_{k=0}^{m(i)-1}\mathbb{C}\tfrac{t^k}{k!}e^{\lambda_i t}.$$ □

The preceding theorem also implies the corresponding result for real differential equations.

Corollary 5.2.3 (Compare [3, Theorem 3.2.5 on p. 69]) *Let $f \in \mathbb{R}[s]$ be an irreducible polynomial in $\mathbb{R}[s]$, let $m \geq 0$ and $\mathcal{F} := \mathrm{C}^\infty(\mathbb{R}, \mathbb{R})$. We consider two cases of the behavior*

$$\mathcal{B}_\mathbb{R} := \mathrm{ann}_\mathcal{F}(f^m) = \left\{y \in \mathrm{C}^\infty(\mathbb{R}, \mathbb{R});\ f^m \circ y = 0, \text{i.e., } f^m\left(\tfrac{\mathrm{d}}{\mathrm{d}t}\right)(y) = 0\right\}.$$

1. *If $f = s - \lambda$ for some $\lambda \in \mathbb{R}$, then $\mathcal{B}_\mathbb{R} = \bigoplus_{k=0}^{m-1} \mathbb{R}e_{\lambda,k}$, where $e_{\lambda,k} = \frac{t^k}{k!}e^{\lambda t}$.*
2. *If $f = (s - \lambda)(s - \overline{\lambda})$ has two complex, nonreal roots λ and $\overline{\lambda}$, we assume $\lambda = \alpha + j\omega,\ j := \sqrt{-1}$, with $\Im(\lambda) = \omega > 0$. Then*

$$\begin{aligned}\mathcal{B}_\mathbb{R} &= \bigoplus_{k=0}^{m-1} \mathbb{R}\tfrac{t^k}{k!}e^{\alpha t}\cos(\omega t) \oplus \bigoplus_{k=0}^{m-1} \mathbb{R}\tfrac{t^k}{k!}e^{\alpha t}\sin(\omega t)\\ &= \mathbb{R}[t]_{\leq m-1}e^{\alpha t}\big(\mathbb{R}\cos(\omega t) \oplus \mathbb{R}\sin(\omega t)\big).\end{aligned}$$

The CRT in the form of Corollary 5.1.6 then also furnishes a fundamental system of solutions of $f \circ y = 0$ in $\mathrm{C}^\infty(\mathbb{R}, \mathbb{R})$ for arbitrary $f \in \mathbb{R}[s]$.

Proof The first case, namely, $f = s - \lambda$, is analogous to the complex situation. As for the second case $f = (s - \lambda)(s - \overline{\lambda})$ with $\lambda \notin \mathbb{R}$, the real dimension of $\mathcal{B}_\mathbb{R}$ is $\dim_\mathbb{R}(\mathcal{B}_\mathbb{R}) = \deg(f^m) = 2m$ by (3.7). By the preceding theorem the complex solution space

$$\mathcal{B}_\mathbb{C} := \{y \in \mathrm{C}^\infty(\mathbb{R}, \mathbb{C});\ f^m \circ y = 0\}$$

has dimension $\dim_\mathbb{C}(\mathcal{B}_\mathbb{C}) = 2m$ and the $\mathbb{C}$-basis $e_{\lambda,k}, e_{\overline{\lambda},k} = \overline{e_{\lambda,k}}, k = 0, \ldots, m-1$. Since f^m is a real polynomial, the real and the imaginary parts of a complex solution are real solutions and hence

$$\Re\big(e_{\lambda,k}(t)\big) = \tfrac{t^k}{k!}e^{\alpha t}\cos(\omega t) \in \mathcal{B}_\mathbb{R} \quad\text{and}\quad \Im\big(e_{\lambda,k}(t)\big) = \tfrac{t^k}{k!}e^{\alpha t}\sin(\omega t) \in \mathcal{B}_\mathbb{R} \tag{5.16}$$

for $k = 0, \ldots, m-1$. Let now $y = \overline{y} \in \mathcal{B}_\mathbb{R} \subset \mathcal{B}_\mathbb{C}$ be any real solution. It admits the complex basis representation

$$\begin{aligned}&y = \sum_{k=0}^{m-1} a_{\lambda,k}e_{\lambda,k} + \sum_{k=0}^{m-1} b_{\lambda,k}e_{\overline{\lambda},k},\ a_{\lambda,k}, b_{\lambda,k} \in \mathbb{C},\\ &\Longrightarrow y = \overline{y} = \sum_{k=0}^{m-1} a_{\lambda,k}e_{\lambda,k} + \sum_{k=0}^{m-1} b_{\lambda,k}e_{\overline{\lambda},k} = \sum_{k=0}^{m-1} \overline{a_{\lambda,k}}e_{\overline{\lambda},k} + \sum_{k=0}^{m-1} \overline{b_{\lambda,k}}e_{\lambda,k}.\end{aligned}$$

Comparison of coefficients of these basis representations yields $b_{\lambda,k} = \overline{a_{\lambda,k}}$ for $k = 0, \ldots, m-1$. Since $ae + \overline{ae} = 2\Re(ae) = 2\big(\Re(a)\Re(e) - \Im(a)\Im(e)\big)$ holds for arbitrary complex numbers $a, e \in \mathbb{C}$, we infer

$$y = \sum_{k=0}^{m-1}(a_{\lambda,k}e_{\lambda,k} + \overline{a_{\lambda,k}e_{\lambda,k}}) = \sum_{k=0}^{m-1}\big(2\Re(a_{\lambda_k})\Re(e_{\lambda_k}) - 2\Im(a_{\lambda_k})\Im(e_{\lambda,k})\big)$$
$$\in \sum_{k=0}^{m-1}\big(\mathbb{R}\Re(e_{\lambda,k}) + \mathbb{R}\Im(e_{\lambda,k})\big).$$

Therefore, the $2m = \dim_{\mathbb{R}}(\mathcal{B}_{\mathbb{R}})$ solutions from (5.16) are $\mathbb{R}$-generators of $\mathcal{B}_{\mathbb{R}}$ and thus an $\mathbb{R}$-basis of this space. □

5.2.2 *The Discrete-Time Case*

The discrete-time analogue of Theorem 5.2.2 also holds. To present this, we assume that F is an arbitrary field and that $\mathcal{F} := F^{\mathbb{N}}$ is the sequence space with the left shift action of $F[s]$. Analogous considerations apply to the injective signal module ${}_{F[s,s^{-1}]}F^{\mathbb{Z}}$, cf. (2.154).

Let $f \in F[s]$ be monic and assume that f splits over F into linear factors, i.e., we have a representation $f = \prod_{i=1}^{r}(s - \lambda_i)^{m(i)}$ as in (5.12) with $\lambda_i \in F$. By Corollary 5.1.6 we have the direct sum decomposition

$$\{y \in \mathcal{F};\ f \circ y = 0\} = \operatorname{ann}_{\mathcal{F}}(f) = \bigoplus_{i=1}^{r} \operatorname{ann}_{\mathcal{F}}\big((s - \lambda_i)^{m(i)}\big).$$

Therefore, we first describe behaviors of the form $\operatorname{ann}_{\mathcal{F}}\big((s-\lambda)^m\big)$ for $\lambda \in F$ and $m \in \mathbb{N}$. We will construct the discrete analogues of the functions $e_{\lambda,k}$ from the continuous case. In contrast to the continuous setting, we have to distinguish the cases $\lambda = 0$ and $\lambda \neq 0$. We start with the simpler case $\lambda = 0$.

Lemma and Definition 5.2.4 *For $\lambda := 0$ we define the sequences*

$$e_{0,k} := \delta_k = (0, \ldots, 0, \overset{k}{1}, 0, 0, \ldots) \in F^{\mathbb{N}},\ k \in \mathbb{N},\quad \textit{i.e., } e_{0,k}(t) = \delta_{k,t}. \tag{5.17}$$

These are the standard basis vectors of

$$F^{(\mathbb{N})} := \{y \in F^{\mathbb{N}};\ y(t) = 0 \textit{ for almost all } t\} = \bigoplus_{k=0}^{\infty} Fe_{0,k} \supset$$
$$\operatorname{ann}_{\mathcal{F}}(s^m) = \left\{y \in F^{\mathbb{N}};\ \forall t \geq m:\ y(t) = 0\right\} = \bigoplus_{k=0}^{m-1} Fe_{0,k},\ m \geq 0. \tag{5.18}$$

Proof The $e_{0,k} \in \operatorname{ann}_{\mathcal{F}}(s^m)$, $k = 0, \ldots, m-1$, are linearly independent and thus a basis of $\operatorname{ann}_{\mathcal{F}}(s^m)$ since $\dim_F(\operatorname{ann}_{\mathcal{F}}(s^m)) = m$. Moreover $F^{(\mathbb{N})} = \bigcup_{m=0}^{\infty} \operatorname{ann}_{\mathcal{F}}(s^m)$. □

Lemma and Definition 5.2.5 *For $\lambda = 1$ we define the sequences*

$$e_{1,k} \in F^{\mathbb{N}} \ \textit{ by } \ e_{1,k}(t) := \binom{t}{k} 1_F \ \textit{ for } t, k \in \mathbb{N}.$$

For $m \in \mathbb{N}$ they satisfy the equations

$$(s-1)^m \circ e_{1,k} = \begin{cases} e_{1,k-m} & \textit{if } m \leq k \\ 0 & \textit{if } m > k \end{cases}, \ \textit{and} \ \operatorname{ann}_{\mathcal{F}}\left((s-1)^m\right) = \bigoplus_{k=0}^{m-1} Fe_{1,k}. \tag{5.19}$$

Proof The recursion formula for the binomial coefficients implies

$$\big((s-1) \circ e_{1,k}\big)(t) = e_{1,k}(t+1) - e_{1,k}(t) = \binom{t+1}{k} 1_F - \binom{t}{k} 1_F = e_{1,k-1}(t).$$

The first equation from (5.19) follows by induction. For $k = 0, \ldots, m-1$ it implies $e_{1,k} \in \operatorname{ann}_{\mathcal{F}}((s-1)^m)$. Moreover,

$$\binom{t}{k} = \begin{cases} 0 & \text{for } t < k \\ 1 & \text{for } t = k, \end{cases} \Longrightarrow e_{1,k} = (0, \ldots, 0, \overset{k}{1}, *, \ldots) \in F^{\mathbb{N}}.$$

Hence the $e_{1,k}$ are F-linearly independent and an F-basis of the m-dimensional behavior $\operatorname{ann}_{\mathcal{F}}((s-1)^m)$. □

Remark 5.2.6 If F has characteristic zero, we may and do identify

$$\mathbb{Q} \subseteq F,\ \alpha = \alpha 1_F, \quad \text{and} \quad F[t] \subset F^{\mathbb{N}},\ f = (f(t))_{t \in \mathbb{N}}.$$

In particular, $e_{1,k}(t) = \binom{t}{k} = \frac{t(t-1)\cdots(t-k+1)}{k!}$ is a polynomial in t, and the preceding lemma furnishes

$$\operatorname{ann}_{\mathcal{F}}((s-1)^m) = \bigoplus_{k=0}^{m-1} Fe_{1,k} = \bigoplus_{k=0}^{m-1} Ft^k = F[t]_{\leq m-1}. \tag{5.20}$$

If F has characteristic $p > 0$, we identify

$$\mathbb{Z}/\mathbb{Z}p \subseteq F, \quad \bar{t} := t + \mathbb{Z}p = t1_F = (t+p)1_F.$$

This implies that every polynomial function $y \in F^{\mathbb{N}}$ is p-periodic, i.e., satisfies $y(t+p) = y(t)$ or $(s^p - 1) \circ y = 0$. But

$$e_{1,mp}(t) = \begin{cases} 0 & \text{if } t < mp \\ 1_F = \binom{mp}{mp} 1_F & \text{if } t = mp \end{cases} \Longrightarrow 0 = e_{1,mp}(0) \neq 1 = e_{1,mp}(mp).$$

Hence $e_{1,mp} \in F^{\mathbb{N}}$ is neither p-periodic nor a polynomial function. ◇

Let now $\lambda \in F \setminus \{0\}$ be arbitrary. We will reduce this case to the situation $\lambda = 1$. The geometric sequence

$$\underline{\lambda} := (1, \lambda, \lambda^2, \lambda^3, \ldots) \quad \text{satisfies } (s - \lambda) \circ \underline{\lambda} = 0 \tag{5.21}$$

and is obviously the discrete analogue of the exponential function.

Lemma and Definition 5.2.7 *Let $0 \neq \lambda \in F$ and $y \in \mathcal{F} = F^{\mathbb{N}}$. By $y\underline{\lambda}$, we denote the componentwise product of the two sequences, i.e., $(y\lambda)(t) = y(t)\lambda^t$.*

1. *The following identities hold:*

$$\forall m \in \mathbb{N}: \ (s-\lambda)^m \circ (y\underline{\lambda}) = \lambda^m \big((s-1)^m \circ y\big)\underline{\lambda} = \big((\lambda(s-1))^m \circ y\big)\underline{\lambda}$$
$$\Longrightarrow \forall f \in F[s]: \ f(s-\lambda) \circ (y\underline{\lambda}) = \big(f\big(\lambda(s-1)\big) \circ y\big)\underline{\lambda}.$$

2. *We define the sequences*

$$e_{\lambda,k} := \lambda^{-k} e_{1,k}\underline{\lambda} = \left(\binom{t}{k}\lambda^{t-k}\right)_{t\in\mathbb{N}}, \ k \in \mathbb{N}. \tag{5.22}$$

For $m \in \mathbb{N}$ these sequences satisfy

$$(s-\lambda)^m \circ e_{\lambda,k} = \begin{cases} e_{\lambda,k-m} & \text{if } m \leq k \\ 0 & \text{if } m > k \end{cases} \Longrightarrow \operatorname{ann}_{\mathcal{F}}\big((s-\lambda)^m\big) = \bigoplus_{k=0}^{m-1} F e_{\lambda,k}, \tag{5.23}$$

i.e., the $e_{\lambda,k}$, $k = 0, \ldots, m-1$, form an F-basis of $\operatorname{ann}_{\mathcal{F}}\big((s-\lambda)^m\big)$.

Again, as in (5.15), the $e_{\lambda,k}$ are defined such that the simple recursion equation $(s-\lambda) \circ e_{\lambda,k} = e_{\lambda,k-1}$ holds for all $\lambda \in \mathbb{C}$ and $k \geq 1$.

Proof 1. For $t \in \mathbb{N}$, we have that

$$\big((s-\lambda) \circ (y\underline{\lambda})\big)(t) = y(t+1)\lambda^{t+1} - \lambda y(t)\lambda^t = \lambda\big(y(t+1) - y(t)\big)\lambda^t$$
$$= \lambda\big((s-1) \circ y\big)(t)\underline{\lambda}(t).$$

The assertion for arbitrary $m \in \mathbb{N}$ follows by induction, and for arbitrary $f \in F[s]$ by the F-bilinearity of $\circ$.

2. Equation 5.23 follows from

$$\begin{aligned}(s-\lambda)^m \circ e_{\lambda,k} &= (s-\lambda)^m \circ (\lambda^{-k} e_{1,k}\underline{\lambda}) = \lambda^{-k}(s-\lambda)^m \circ (e_{1,k}\underline{\lambda}) \\ &\overset{1.}{=} \lambda^{-k}\lambda^m\big((s-1)^m \circ e_{1,k}\big)\underline{\lambda} \\ &\underset{\text{Lemma 5.2.5}}{=} \begin{cases}\lambda^{-(k-m)} e_{1,k-m}\underline{\lambda} = e_{\lambda,k-m} & \text{if } m \le k \\ 0 & \text{if } m > k\end{cases} \\ &\Longrightarrow \text{for } k = 0,\ldots,m-1:\ e_{\lambda,k} \in \operatorname{ann}_{\mathcal{F}}\big((s-\lambda)^m\big) \\ &\Longrightarrow \operatorname{ann}_{\mathcal{F}}\big((s-\lambda)^m\big) \underset{\text{cf.(5.19)}}{=} \bigoplus_{k=0}^{m-1} \mathbb{C} e_{\lambda,k}.\end{aligned}$$

□

With these preparations we can now describe a fundamental system of the difference equation $f \circ y = 0$.

Theorem 5.2.8 *Let F be an arbitrary field, $\mathcal{F} := F^{\mathbb{N}}$ and $f \in F[s]$ a splitting polynomial as in* (5.12)*. The behavior*

$$\mathcal{B} := (F[s]f)^{\perp} = \operatorname{ann}_{\mathcal{F}}(f) = \{y \in \mathcal{F};\ f \circ y = 0\}$$

has the direct decomposition

$$\operatorname{ann}_{\mathcal{F}}(f) = \bigoplus_{i=1}^{r} \operatorname{ann}_{\mathcal{F}}\big((s-\lambda_i)^{m(i)}\big) \ \textit{with}\ \operatorname{ann}_{\mathcal{F}}\big((s-\lambda_i)^{m(i)}\big) = \bigoplus_{k=0}^{m(i)-1} F e_{\lambda_i,k}.$$

The sequences $e_{\lambda_i,k}$, $i = 1,\ldots,r$, $k = 0,\ldots,m(i)-1$, are a fundamental system, *i.e., a basis, of the solution space $\mathcal{B}$ of the difference equation $f \circ y = 0$. With b_i from* (5.12) *the decomposition of a solution $y \in \mathcal{B}$ is*

$$y = \sum_{i=1}^{r} b_i \circ y \in \mathcal{B} = \bigoplus_{i=1}^{r} \operatorname{ann}_{\mathcal{F}}((s-\lambda_i)^{m(i)}).$$

Since the field $\mathbb{C}$ is algebraically closed, every polynomial $f \in \mathbb{C}[s]$ splits. Therefore this theorem is applicable to all $f \in \mathbb{C}[s]$.

Proof Corollary 5.1.6, Lemma and Definition 5.2.4 and Lemma and Definition 5.2.7. □

Remark 5.2.9 In Theorem 5.2.8 we do not treat the case that F is not a splitting field for the polynomial f. This case was, however, completely solved for perfect base fields F by means of Galois theory in [4, Sect. 5], even for multidimensional (multivariate) difference equations. All fields of characteristic zero and all finite fields are perfect. ◇

In the following corollary we treat the situation of a real difference equation. This is the discrete-time analogue of Corollary 5.2.3.

Corollary 5.2.10 *Let* $f := (s - \lambda)(s - \overline{\lambda}) \in \mathbb{R}[s]$ *with* $\lambda \in \mathbb{C}$ *and* $\Im(\lambda) > 0$. *We write* λ *in polar form as* $\lambda = \rho e^{j\omega}$ *with* $\rho \in (0, \infty)$ *and* $\omega \in (0, \pi)$. *Then* $\lambda^t = \rho^t e^{j\omega t} = \rho^t\big(\cos(\omega t) + j\sin(\omega t)\big)$. *For* $m \in \mathbb{N}$ *we get*

$$\begin{aligned}\mathcal{B}_{\mathbb{R}} &:= \operatorname{ann}_{\mathbb{R}^{\mathbb{N}}}(f^m) = \{y \in \mathbb{R}^{\mathbb{N}};\ f^m \circ y = 0\}\\ &= \mathbb{R}[t]_{\leq m-1}\rho^t\big(\mathbb{R}\cos(\omega t) \oplus \mathbb{R}\sin(\omega t)\big).\end{aligned}$$

Here we write $y(t)$ *for the whole function* $y : \mathbb{N} \to \mathbb{R},\ t \mapsto y(t)$.

Proof The proof is analogous to the one of Corollary 5.2.3 and uses Theorem 5.2.8 and Remark 5.2.6. □

5.3 The Primary Decomposition and Autonomy

The CRT furnishes the primary decomposition of the torsion submodule $\mathrm{t}(M)$ of every module M over the principal ideal ring $\mathcal{D}$. We derive this result in Theorem 5.3.2. In Sect. 5.3.2, we apply the primary decomposition to injective cogenerator signal modules and to behaviors, and in Sect. 5.3.3 to the standard signal modules. The constructive primary decomposition is highly important in LTI systems theory and the main tool in the treatment of the following subjects of the present chapter concerning autonomous behaviors: polynomial-exponential signals, characteristic values and variety, modal decomposition, index and multiplicity of a characteristic value, fundamental systems (bases) of solutions, stability, Jordan decomposition of complex matrices, constructive forms of A^t and e^{tA} for $A \in \mathbb{C}^{n\times n}$, constructive partial fraction decomposition of rational functions. The fine pole structure of Rosenbrock equations in Sect. 6.3.5 is essentially a statement on various multiplicities of characteristic values.

5.3.1 *The Primary Decomposition of Torsion Modules*

We have seen in the preceding section that the annihilators $\operatorname{ann}_M(f^j)$ of powers of $f \in \mathcal{D}$, especially of irreducible f, play an important role. For this reason, we study these annihilators for increasing powers j in the following lemma.

Lemma and Definition 5.3.1 *For a* $\mathcal{D}$*-module* M *and a nonzero* $f \in \mathcal{D}$, *we consider the increasing sequence of submodules of* M

$$
\begin{aligned}
0 = \operatorname{ann}_M(f^0) \subseteq \cdots \subseteq \operatorname{ann}_M(f^j) \subseteq \operatorname{ann}_M(f^{j+1}) \subseteq \cdots \\
\subseteq \operatorname{ann}_M(f^\infty) := \bigcup_{j=0}^{\infty} \operatorname{ann}_M(f^j) = \{x \in M;\ \exists j \text{ with } f^j x = 0\}.
\end{aligned}
\tag{5.24}
$$

In this sequence the equality $\operatorname{ann}_M(f^j) = \operatorname{ann}_M(f^{j+1})$ *implies* $\operatorname{ann}_M(f^{j+1}) = \operatorname{ann}_M(f^{j+2})$. *Hence, either the increasing sequence (5.24) is properly increasing or there is a least number m where it becomes stationary, i.e.,*

$$
0 = \operatorname{ann}_M(f^0) \subsetneq \cdots \subsetneq \operatorname{ann}_M(f^m) = \operatorname{ann}_M(f^{m+1}) = \cdots = \operatorname{ann}_M(f^\infty). \tag{5.25}
$$

We define the index *of the pair* (M, f) *as*

$$
\operatorname{index}(M, f) := \begin{cases} \infty & \text{in the first case,} \\ m & \text{in the second case.} \end{cases}
$$

If M is finitely generated and hence Noetherian, only the second case occurs.

Proof Assume that $\operatorname{ann}_M(f^j) = \operatorname{ann}_M(f^{j+1})$. The following implications hold:

$$
\begin{aligned}
& x \in \operatorname{ann}_M(f^{j+2}) \Longrightarrow 0 = f^{j+2}x = f^{j+1}(fx) \Longrightarrow fx \in \operatorname{ann}_M(f^{j+1}) = \operatorname{ann}_M(f^j) \\
& \Longrightarrow 0 = f^j(fx) = f^{j+1}x \Longrightarrow x \in \operatorname{ann}_M(f^{j+1}) \\
& \Longrightarrow \operatorname{ann}_M(f^{j+2}) \subseteq \operatorname{ann}_M(f^{j+1}) \Longrightarrow \operatorname{ann}_M(f^{j+2}) = \operatorname{ann}_M(f^{j+1}).
\end{aligned}
$$

□

Let $\mathcal{P}$ be a representative system of the irreducible elements of $\mathcal{D}$. For $p \in \mathcal{P}$ we define the *p-component*

$$
\mathrm{t}_p(M) := \operatorname{ann}_M(p^\infty) = \bigcup_{j=0}^{\infty} \operatorname{ann}_M(p^j) = \{x \in M;\ \exists j \text{ with } p^j x = 0\}. \tag{5.26}
$$

Theorem 5.3.2 (Primary decomposition) *The torsion module* $\mathrm{t}(M)$ *of a* $\mathcal{D}$*-module* M *has the* primary decomposition

$$
\mathrm{t}(M) = \bigoplus_{p \in \mathcal{P}} \mathrm{t}_p(M).
$$

Proof First we show the equality $\mathrm{t}(M) = \sum_{p \in \mathcal{P}} \mathrm{t}_p(M)$, then we prove that the sum is direct. The inclusion $\supseteq$ is obvious.

$\subseteq$. Let $x \in \mathrm{t}(M)$ and let $0 \neq f \in \mathcal{D}$ with $fx = 0$. Consider the unique prime factor decomposition of f, i.e.,

$$f = u\prod_{i=1}^{r} p_i^{m(i)} = f_1 \cdots f_r,\ u \in \mathrm{U}(\mathcal{D}),\ f_1 := up_1^{m(1)},\ f_i := p_i^{m(i)} \text{ for } i = 2, \ldots, r$$

with pairwise coprime f_i. The CRT for modules (Theorem 5.1.4) implies

$$x \in \mathrm{ann}_M(f) = \bigoplus_{i=1}^{r} \mathrm{ann}_M(f_i) = \bigoplus_{i=1}^{r} \mathrm{ann}_M(p_i^{m(i)}) \subseteq \sum_{i=1}^{r} \mathrm{t}_{p_i}(M).$$

To show that the sum is direct, assume that

$$0 = x_1 + \cdots + x_r \in \mathrm{t}_{p_1}(M) + \cdots + \mathrm{t}_{p_r}(M)$$

with pairwise distinct p_i. Since $x_i \in \mathrm{t}_{p_i}(M)$, there is an exponent $m(i)$ with $p_i^{m(i)} x_i = 0$. Let $f := \prod_{i=1}^{r} p_i^{m(i)}$. Again by the CRT,

$$0 = \sum_{i=1}^{r} x_i \in \mathrm{ann}_M(f) = \bigoplus_{i=1}^{r} \mathrm{ann}_M(p_i^{m(i)}) \Longrightarrow x_i = 0 \text{ for } i = 1, \ldots, r.$$

□

Corollary 5.3.3 *Let ${}_{\mathcal{D}}M$ be a torsion module and assume* $\mathrm{ann}_{\mathcal{D}}(M) \neq 0$. *This holds, in particular, if M is finitely generated, cf. Lemma 4.2.6, 2. Assume*

$$\mathrm{ann}_{\mathcal{D}}(M) = \mathcal{D}f,\ f \neq 0, \text{ and let } f = \prod_{i=1}^{r} p_i^{m(i)},\ p_i \in \mathcal{P},\ r \geq 0,\ m(i) > 0,$$

be the prime factor decomposition. For $p \in \mathcal{P}$ this implies

$$\mathrm{t}_p(M) = \begin{cases} \mathrm{ann}_M(p_i^{m(i)}) \cong M/p_i^{m(i)}M \neq 0 & \text{if } p = p_i \text{ for some } i \in \{1, \ldots, r\}, \\ 0 & \text{if } p \neq p_j \text{ for all } j \in \{1, \ldots, r\}. \end{cases}$$

Moreover the following equalities hold:

$$\mathrm{index}(M, p) = \begin{cases} m(i) & \text{if } p = p_i \text{ for some } i \in \{1, \ldots, r\}, \\ 0 & \text{if } p \neq p_j \text{ for all } j \in \{1, \ldots, r\}. \end{cases}$$

Proof Theorems 5.1.4 and 5.3.2 provide us with the direct sum decompositions

$$M = \mathrm{ann}_M(f) \underset{\text{Theorem 5.1.4}}{=} \bigoplus_{i=1}^{r} \mathrm{ann}_M(p_i^{m(i)}) \underset{\text{Theorem 5.3.2}}{=} \bigoplus_{p \in \mathcal{P}} \mathrm{t}_p(M).$$

Since $\mathrm{ann}_M(p_i^{m(i)}) \subseteq \mathrm{t}_{p_i}(M)$, the corresponding summands must be equal. Item 2. of Theorem 5.1.4 implies the isomorphism

$$M/p_i^{m(i)}M \xrightarrow{\cong} \mathrm{ann}_M(p_i^{m(i)}), \quad \overline{x} \longmapsto b_i x.$$

Moreover, $\mathrm{ann}_{\mathcal{D}}(M) = \mathcal{D}f$ implies $M = \mathrm{ann}_M(f)$ and hence $\mathrm{ann}_{\mathcal{D}}(\mathrm{ann}_M(f)) = \mathrm{ann}_{\mathcal{D}}(M) = \mathcal{D}f$. We apply item 3. of Theorem 5.1.4 to obtain

$$\forall i \in \{1, \dots, r\}\colon\ \mathrm{ann}_{\mathcal{D}}(\mathrm{ann}_M(p_i^{m(i)})) = \mathcal{D}p_i^{m(i)} \subsetneq \mathcal{D} \Longrightarrow \mathrm{ann}_M(p_i^{m(i)}) \neq 0.$$

The equality $\mathrm{t}_{p_i}(M) = \mathrm{ann}_{\mathcal{D}}(p_i^{m(i)})$ implies $k(i) := \mathrm{index}(M, p_i) \leq m(i)$ for all i. Let $g := \prod_{i=1}^r p_i^{k(i)}$. Then

$$\begin{aligned}
&0 = p_i^{k(i)} \mathrm{ann}_M(p_i^{m(i)}) = p_i^{k(i)} \mathrm{t}_{p_i}(M) \Longrightarrow gM = g\bigoplus_{i=1}^r \mathrm{t}_{p_i}(M) = 0\\
&\Longrightarrow g \in \mathrm{ann}_{\mathcal{D}}(M) = \mathcal{D}f \Longrightarrow f \mid g \Longrightarrow \forall i = 1, \dots, r:\ m(i) \leq k(i)\\
&\Longrightarrow \forall i = 1, \dots, r:\ m(i) = k(i) = \mathrm{index}(M, p_i).
\end{aligned}$$

□

Assume that M and N are arbitrary $\mathcal{D}$-modules and let $p \in \mathcal{P}$ and $j \geq 0$. Then we identify

$$\begin{aligned}
\mathrm{Hom}_{\mathcal{D}}(M/p^jM, N) &\underset{\text{identification}}{=} \{\varphi \in \mathrm{Hom}_{\mathcal{D}}(M, N);\ \varphi(p^jM) = (p^j\varphi)(M) = 0\}\\
&= \{\varphi \in \mathrm{Hom}_{\mathcal{D}}(M, N);\ p^j\varphi = 0\} = \mathrm{ann}_{\mathrm{Hom}_{\mathcal{D}}(M,N)}(p^j)\\
&\underset{\text{identification}}{=} \mathrm{Hom}_{\mathcal{D}}(M, \mathrm{ann}_N(p^j)),
\end{aligned} \tag{5.27}$$

the last equality following from

$$(p^j\varphi)(M) = \varphi(p^jM) = p^j\varphi(M), \text{ hence } \big(\varphi(M) \subseteq \mathrm{ann}_N(p^j) \Longleftrightarrow p^j\varphi = 0\big).$$

By definition, $\mathrm{t}_p(N) = \bigcup_{j\geq 0} \mathrm{ann}_N(p^j)$ holds. If M is finitely generated, then so is $\varphi(M)$. By Lemma and Definition 5.3.1, this means that $k := \mathrm{index}(\varphi(M), p)$ is finite, i.e., $\varphi(M) \subseteq \mathrm{ann}_N(p^k)$ and thus $\varphi \in \mathrm{Hom}_{\mathcal{D}}(M, \mathrm{ann}_N(p^k))$. Using this and the identifications from (5.27), we obtain for finitely generated M that

$$\mathrm{Hom}_{\mathcal{D}}(M, \mathrm{t}_p(N)) = \bigcup_{j\geq 0} \mathrm{Hom}_{\mathcal{D}}(M, \mathrm{ann}_N(p^j)) \underset{\text{ident.}}{=} \bigcup_{j\geq 0} \mathrm{Hom}_{\mathcal{D}}(M/p^jM, N). \tag{5.28}$$

Corollary 5.3.4 *Assume a finitely generated torsion module M and an arbitrary $\mathcal{D}$-module N. Let $p \in \mathcal{P}$ and $m :=$ index(M, p). Then we have the isomorphisms*

$$M/p^m M \overset{\cong}{\longleftrightarrow} \mathrm{t}_p(M) = \mathrm{ann}_M(p^m),\quad x + p^m M \longmapsto x_p,\quad x_p + p^m M \longleftarrow x_p,$$
$$\textit{where } x = \sum_{q\in\mathcal{P}} x_q \in M = \oplus_{q\in\mathcal{P}} \mathrm{t}_q(M) \textit{ and}$$
$$\mathrm{Hom}_{\mathcal{D}}(\mathrm{t}_p(M), N) \cong \mathrm{Hom}_{\mathcal{D}}(M/p^m M, N) = \mathrm{Hom}_{\mathcal{D}}(M, \mathrm{t}_p(N)) \subseteq \mathrm{Hom}_{\mathcal{D}}(M, N). \tag{5.29}$$

The following equality holds:

$$\mathrm{index}(M, p) = \min\{j \in \mathbb{N};\ p^j M = p^{j+1} M\} \tag{5.30}$$

Proof Corollary 5.3.3 furnishes $M/p^m M \cong \mathrm{t}_p(M) = \mathrm{ann}_M(p^m)$.
Let $q \in \mathcal{P} \setminus \{p\}$. Then q and p are coprime and Lemma 5.1.5 furnishes

$$\begin{aligned}
&\forall k \in \mathbb{N}:\ p\,\mathrm{ann}_N(q^k) = \mathrm{ann}_N(q^k) \Longrightarrow p\,\mathrm{t}_q(M) = \mathrm{t}_q(M)\\
&\Longrightarrow \forall j \in \mathbb{N}:\ p^j M = p^j\left(\mathrm{t}_p(M) \bigoplus \oplus_{q\in\mathcal{P}\setminus\{p\}} \mathrm{t}_q(M)\right)\\
&= p^j\,\mathrm{ann}_M(p^m) \bigoplus \oplus_{q\in\mathcal{P}\setminus\{p\}} p^j\,\mathrm{t}_q(M) = p^j\,\mathrm{ann}_M(p^m) \bigoplus \oplus_{q\in\mathcal{P}\setminus\{p\}} \mathrm{t}_q(M)\\
&\Longrightarrow \forall j \geq m:\ p^j M = p^m M = \oplus_{q\neq p} \mathrm{t}_q(M).
\end{aligned} \tag{5.31}$$

We infer (5.29) via

$$\begin{aligned}
\mathrm{Hom}_{\mathcal{D}}(\mathrm{t}_p(M), N) &\cong \mathrm{Hom}_{\mathcal{D}}(M/p^m M, N) \overset{(5.31)}{=\!=} \bigcup_{j\geq m} \mathrm{Hom}_{\mathcal{D}}(M/p^j M, N)\\
&\overset{(5.27)}{=\!=} \bigcup_{j\geq m} \mathrm{Hom}_{\mathcal{D}}(M, \mathrm{ann}_N(p^j)) = \mathrm{Hom}_{\mathcal{D}}(M, \mathrm{t}_p(N)),
\end{aligned}$$

where the isomorphism comes from Corollary 5.3.3. The last equality holds since M is finitely generated and the $\mathrm{ann}_N(p^j)$ are increasing.
In order to show (5.30), let $j \in \mathbb{N}$ with $p^j M = p^{j+1} M$. Then (5.31) implies

$$\begin{aligned}
&p^j\,\mathrm{t}_p(M) = p^{j+1}\,\mathrm{t}_p(M) \Longrightarrow p^j\,\mathrm{t}_p(M) = p^{j+m}\,\mathrm{t}_p(M) = p^{j+m}\,\mathrm{ann}_M(p^m) = 0\\
&\Longrightarrow p^j \in \mathrm{ann}_{\mathcal{D}}(\mathrm{ann}_M(p^m)) = \mathcal{D}p^m\\
&\Longrightarrow j \geq m \underset{(5.31)}{\Longrightarrow} m = \min\left\{j \in \mathbb{N};\ p^j M = p^{j+1} M\right\}.
\end{aligned}$$
□

5.3.2 *The Injectivity of* $\mathrm{t}(\mathcal{F})$ *and* $\mathcal{F}/\,\mathrm{t}(\mathcal{F})$ *and the Primary Decomposition of Behaviors*

We apply the results of Sect. 5.3.1 to signal modules. We show that if $\mathcal{F}$ is an injective signal module over a principal ideal domain $\mathcal{D}$, then so are $\mathrm{t}(\mathcal{F})$, $\mathrm{t}_p(\mathcal{F})$, and $\mathcal{F}/\,\mathrm{t}(\mathcal{F})$.

We also prove that if $\mathcal{D}$ is not a field and $\mathcal{F}$ is an injective cogenerator then so is $\mathrm{t}(\mathcal{F})$. The primary decomposition of $\mathrm{t}(\mathcal{F})$ induces the primary decomposition of autonomous behaviors and, for $p \in \mathcal{P}$, its p-length or p-multiplicity. The theorems we prove here will be applied in Sects. 6.3.6, 6.4, 8.2, and 9.3.

5.3.2.1 Products and Coproducts

We start by summarizing the most important facts on products (Cartesian products) and coproducts (direct sums). For this, let $\mathcal{D}$ be a commutative ring.

Let $\mathcal{S}$ denote the class of all sets. For an index set I, a family $(M_i)_{i\in I}$ of sets M_i, viz., a map $I \to \mathcal{S}$, $i \mapsto M_i$, gives rise to the *Cartesian product*, *direct product*, or simply *product*

$$\prod_{i\in I} M_i := \left\{(m_i)_{i\in I} \colon I \to \bigcup\nolimits_{j\in I} M_j,\ i \mapsto m_i;\ \forall i \in I\colon\ m_i \in M_i\right\}, \tag{5.32}$$

which comes equipped with the projections

$$\mathrm{proj}_j\colon \prod_{i\in I} M_i \longrightarrow M_j,\quad (m_i)_{i\in I} \longmapsto m_j,$$

for every $j \in I$. If the M_i are modules over $\mathcal{D}$, then $\prod_{i\in I} M_i$ is a $\mathcal{D}$-module with the componentwise addition and scalar multiplication, and the projections are homomorphisms. We will assume this from now on. The product—more exactly, the pair $\left(\prod_{i\in I} M_i, (\mathrm{proj}_i)_{i\in I}\right)$—satisfies the universal property that for every module X the map

$$\mathrm{Hom}_{\mathcal{D}}\Big(X, \prod_{i\in I} M_i\Big) \longrightarrow \prod_{i\in I} \mathrm{Hom}_{\mathcal{D}}(X, M_i),\quad \varphi \longmapsto (\mathrm{proj}_i\,\varphi)_{i\in I}, \tag{5.33}$$

is an isomorphism, and its inverse is

$$\big(x \mapsto (\varphi_i(x))_{i\in I}\big) \longleftarrow (\varphi_i)_{i\in I}.$$

A family of linear maps $(\varphi_i\colon M_i \longrightarrow N_i)_{i\in I}$ gives rise to the linear map

$$\prod_{i\in I}\varphi_i\colon \prod_{i\in I} M_i \longrightarrow \prod_{i\in I} N_i, \quad (m_i)_{i\in I} \longmapsto \big(\varphi_i(m_i)\big)_{i\in I}. \tag{5.34}$$

Using (5.32) together with (5.34), we can regard the product $\prod_{i\in I}$ as an additive covariant functor

$$\prod_{i\in I}\colon \operatorname{mod}_{\mathcal{D}}^{I} \longrightarrow \operatorname{mod}_{\mathcal{D}}, \quad (M_i)_{i\in I} \longmapsto \prod_{i\in I} M_i, \quad (\varphi_i)_{i\in I} \longmapsto \prod_{i\in I}\varphi_i, \tag{5.35}$$

where $\operatorname{mod}_{\mathcal{D}}^{I}$ is the product category given by the data

objects	$(M_i)_{i\in I}$,	where $M_i \in \operatorname{mod}_{\mathcal{D}}$,
morphisms	$(\varphi_i)_{i\in I}$,	where $\varphi_i \in \operatorname{Hom}_{\mathcal{D}}(M_i, N_i)$,
composition	$(\psi_i)_{i\in I}(\varphi_i)_{i\in I} = (\psi_i\varphi_i)_{i\in I}$,	and
identity morphism	$\operatorname{id}_{(M_i)_{i\in I}} = (\operatorname{id}_{M_i})_{i\in I}$.	

The functor $\prod_{i\in I}$ is covariant since it preserves the arrows, i.e.,

$$\prod_{i\in I}\operatorname{id}_{M_i} = \operatorname{id}_{\prod_{i\in I} M_i} \quad \text{and}$$

$$\Big(\prod_{i\in I}\psi_i\Big)\Big(\prod_{i\in I}\varphi_i\Big) = \prod_{i\in I}\psi_i\varphi_i \quad \text{for } M_i \xrightarrow{\varphi_i} N_i \xrightarrow{\psi_i} P_i,$$

and it is additive since

$$\prod_{i\in I}\psi_i + \prod_{i\in I}\varphi_i = \prod_{i\in I}(\psi_i + \varphi_i) \quad \text{for } \psi_i, \varphi_i\colon M_i \longrightarrow N_i.$$

It is *faithfully exact*, i.e.,

$$\begin{aligned} & M_i \xrightarrow{\ \varphi_i\ } N_i \xrightarrow{\ \psi_i\ } P_i \quad \text{is exact for all } i \in I \\ \Longleftrightarrow\ & \prod_{i\in I} M_i \xrightarrow{\prod_{i\in I}\varphi_i} \prod_{i\in I} N_i \xrightarrow{\prod_{i\in I}\psi_i} \prod_{i\in I} P_i \text{ is exact.} \end{aligned}$$

As a direct consequence, we obtain the following lemma.

Lemma 5.3.5 *The module $\prod_{i\in I} M_i$ is injective if and only if all its factors M_i are injective.*

Proof The isomorphism (5.33) and the faithful exactness of the product imply that $\prod_{i\in I} M_i$ is injective, i.e., $\operatorname{Hom}_{\mathcal{D}}\big(-, \prod_{i\in I} M_i\big)$ is an exact functor, if and only if all $\operatorname{Hom}_{\mathcal{D}}(-, M_i)$ are exact, i.e., all the M_i are injective. □

Dual to the product, the *coproduct* or *external direct sum* of a family $(M_i)_{i\in I}$ of $\mathcal{D}$-modules is the set

$$\coprod_{i\in I} M_i := \left\{(m_i)_{i\in I} \in \prod_{i\in I} M_i;\ m_i = 0 \text{ for almost all } i \in I\right\},$$

which is obviously a submodule of $\prod_{i\in I} M_i$, and equal to the product if I is finite. The coproduct comes equipped with the canonical injections

$$\mathrm{inj}_j\colon M_j \longrightarrow \coprod_{i\in I} M_i,\quad m_j \longmapsto \mathrm{inj}_j(m_j),\ \text{where}\ \left(\mathrm{inj}_j(m_j)\right)_i = \begin{cases} m_j & \text{if } i = j,\\ 0 & \text{if } i \neq j,\end{cases}.$$

The equality $(m_i)_{i\in I} = \sum_{i\in I} \mathrm{inj}_i(m_i) \in \coprod_{i\in I} M_i$ holds. This sum makes sense since only finitely many m_i are nonzero. The coproduct has the dual universal property to the product, namely, that for all $\mathcal{D}$-modules Y the map

$$\mathrm{Hom}_{\mathcal{D}}\left(\coprod_{i\in I} M_i, Y\right) \longrightarrow \prod_{i\in I} \mathrm{Hom}_{\mathcal{D}}(M_i, Y),\ \varphi \longmapsto (\varphi\,\mathrm{inj}_i)_{i\in I}, \tag{5.36}$$

is an isomorphism with the inverse $(\varphi_i)_{i\in I} \longmapsto \left((m_i)_{i\in I} \mapsto \sum_{i\in I} \varphi_i(m_i)\right)$. If $\left(\varphi_i\colon M_i \longrightarrow N_i\right)_{i\in I}$ is a family of linear maps, then $\prod_{i\in I} \varphi_i$ induces the linear map

$$\coprod_{i\in I} \varphi_i\colon \coprod_{i\in I} M_i \longrightarrow \coprod_{i\in I} N_i,\quad (m_i)_{i\in I} \longmapsto \left(\varphi_i(m_i)\right)_{i\in I}.$$

In this manner and like the product the coproduct becomes a covariant additive functor

$$\coprod_{i\in I}\colon \mathrm{mod}_{\mathcal{D}}^{I} \longrightarrow \mathrm{mod}_{\mathcal{D}},\quad (M_i)_{i\in I} \longmapsto \coprod_{i\in I} M_i,\quad (\varphi_i)_{i\in I} \longmapsto \coprod_{i\in I} \varphi_i, \tag{5.37}$$

which is again faithfully exact.

If the M_i, $i \in I$, are submodules of a fixed module M, then the injections $\mathrm{inj}_i\colon M_i \overset{\subseteq}{\longrightarrow} M$ induce the map

$$\varphi\colon \coprod_{i\in I} M_i \longrightarrow M,\quad (m_i)_{i\in I} \longmapsto \sum_{i\in I} m_i.$$

If this map is an isomorphism, i.e., if every $m \in M$ admits a unique sum representation

$$m = \sum_{i\in I} m_i \quad \text{with } m_i = 0 \text{ for almost all } i \in I,$$

then $\bigoplus_{i\in I} M_i := M$ is called the *internal direct sum* of the submodules M_i. This has already been used at various instances, for instance, in the primary decomposition $\mathrm{t}\left(\mathrm{C}^\infty(\mathbb{R},\mathbb{C})\right) = \bigoplus_{\lambda\in\mathbb{C}} \mathbb{C}[t]e^{\lambda t}$.

For an arbitrary family $(M_i)_{i\in I}$ one often identifies

$$m_j = \mathrm{inj}_j(m_j) = (0,\ldots,0,\overset{j}{m_j},0,\ldots,0) \in M_j = \mathrm{inj}_j(M_j) \subseteq \coprod_{i\in I} M_i \text{ and}$$

$$\coprod_{i\in I} M_i = \bigoplus_{i\in I} M_i \ni (m_i)_{i\in I} = \sum_{i\in I} \mathrm{inj}_i(m_i) = \sum_{i\in I} m_i.$$

This is often tacitly done in the literature without the preceding justifications. For finite I one may, therefore, identify the product and the external and internal direct sums, i.e.,

$$\prod_{i\in I} M_i = \coprod_{i\in I} M_i = \bigoplus_{i\in I} M_i \ni (m_i)_{i\in I} = \sum_{i\in I} m_i.$$

5.3.2.2 The Injectivity of $\mathrm{t}(\mathcal{F})$ and $\mathcal{F}/\mathrm{t}(\mathcal{F})$

We start with some technical results before we formulate the main theorem of this section (Theorem 5.3.10). Then we apply this theorem to the standard signal modules $\mathcal{F}$.

We assume that $\mathcal{D}$ is an integral domain with quotient field $K = \mathrm{quot}(\mathcal{D})$.

Corollary 5.3.6 *If a coproduct $\coprod_{i\in I} M_i$ is injective, then all direct summands M_i are injective too.*

Proof For $j \in I$, we use the isomorphism

$$\coprod_{i\in I} M_i \overset{\cong}{\longleftrightarrow} M_j \times \coprod_{i\in I,\, i\neq j} M_i, \quad (m_i)_{i\in I} \longleftrightarrow \left(m_j, (m_i)_{i\in I,\, i\neq j}\right).$$

The injectivity of $\coprod_{i\in I} M_i$ and Lemma 5.3.5 imply that M_j is injective. □

Remark 5.3.7 If $\mathcal{D}$ is Noetherian, the injectivity of the M_i also implies that of $\coprod_{i\in I} M_i$, as can be easily proven by means of the Baer criterion for injectivity. However, we do not need this result. ◊

Lemma 5.3.8 *An injective submodule M' of a module M is a direct summand.*

Proof Since $\mathrm{Hom}_{\mathcal{D}}(-, M')$ is exact, the injection $f_1 := \mathrm{inj}\colon M' \overset{\subseteq}{\longrightarrow} M$ induces the surjection

$$\mathrm{Hom}(f_1, M')\colon \mathrm{Hom}_{\mathcal{D}}(M, M') \longrightarrow \mathrm{Hom}_{\mathcal{D}}(M', M'), \quad g \longmapsto gf_1,$$

and thus the existence of a retraction $g_2 \in \mathrm{Hom}_{\mathcal{D}}(M, M')$ with $g_2 f_1 = \mathrm{id}_{M'}$. Item 1 of Theorem 4.2.18 implies that $M' = \mathrm{im}(f_1)$ is a direct summand. □

Theorem 5.3.9 *Let $\mathcal{D}$ be an integral domain and let $\mathcal{F}$ be a $\mathcal{D}$-module. The following properties are equivalent:*

1. *The module $\mathcal{F}$ is a K-vector space, where the scalar product $\circ_K : K \times \mathcal{F} \longrightarrow \mathcal{F}$ of the K-vector space $\mathcal{F}$ is the canonical extension of the scalar product $\circ\colon \mathcal{D} \times \mathcal{F} \longrightarrow \mathcal{F}$ of the $\mathcal{D}$-module $\mathcal{F}$.*
2. *The module $\mathcal{F}$ is injective and torsionfree.*
3. *The module $\mathcal{F}$ is divisible and torsionfree.*

Recall that $\mathcal{F}$ is torsionfree if $d \circ y = 0$ implies $d = 0$ or $y = 0$, or, in other words, if for every nonzero $b \in \mathcal{D}$ the linear map

$$\mu_b := b\circ\colon \mathcal{F} \longrightarrow \mathcal{F}, \quad y \longmapsto b \circ y, \tag{5.38}$$

is a monomorphism. Condition 3 signifies that all maps μ_b, $0 \neq b \in \mathcal{D}$, are isomorphisms.

In particular, the $\mathcal{D}$-module K is injective.

Proof 1. $\Longrightarrow$ 2. We use Baer's criterion, namely, $\mathcal{F}$ is injective if and only if

$$\forall \mathfrak{a} \leq \mathcal{D}\, \forall \varphi\colon \mathfrak{a} \longrightarrow \mathcal{F}\, \exists \psi\colon \mathcal{D} \longrightarrow \mathcal{F}\colon\ \psi|_{\mathfrak{a}} = \varphi.$$

For $\mathfrak{a} = 0$ this is trivially true. Assume a nonzero ideal $\mathfrak{a}$, $0 \neq a_1 \in \mathfrak{a}$ and $\varphi \in \mathrm{Hom}_{\mathcal{D}}(\mathfrak{a}, \mathcal{F})$. Define

$$\psi\colon \mathcal{D} \to \mathcal{F},\ d \mapsto \psi(d) := a_1^{-1} \circ \varphi(da_1),$$
$$\Longrightarrow \forall a \in \mathfrak{a} :\ \psi(a) = a_1^{-1} \circ \varphi(aa_1) = a_1^{-1} \circ a_1 \circ \varphi(a) = \varphi(a) \Longrightarrow \psi|_{\mathfrak{a}} = \varphi.$$

2. $\Longrightarrow$ 3. Every injective module is divisible, see Theorem 3.2.9 and Lemma 3.2.3.

3. $\Longrightarrow$ 1. We have to construct an extension $\circ_K$ of $\circ$. As motivation, assume that $\circ_K =: \circ$ exists. For $ab^{-1} \in K = \mathrm{quot}(\mathcal{D})$ and $y \in \mathcal{F}$ this implies

$$y = b \circ b^{-1} \circ y = \mu_b\left(b^{-1} \circ y\right) \Longrightarrow b^{-1} \circ y = \mu_b^{-1}(y) \text{ and}$$
$$ab^{-1} \circ y = a \circ b^{-1} \circ y = a \circ \mu_b^{-1}(y) = \mu_b^{-1}(a \circ y).$$

Therefore we now define $\forall ab^{-1} \in K \forall y \in \mathcal{F} :\ ab^{-1} \circ y := \mu_b^{-1}(ay)$. It is easy to check that this multiplication is well defined, makes $\mathcal{F}$ a K-vector space, and extends the given scalar multiplication. We omit these details. □

Theorem 5.3.10 *Let $\mathcal{D}$ be an integral domain and let $\mathcal{F}$ be an injective $\mathcal{D}$-module with the torsion submodule $\mathrm{t}(\mathcal{F})$.*

1. *The torsion module $\mathrm{t}(\mathcal{F})$ is injective and a direct summand of $\mathcal{F}$. The factor module $\mathcal{F}/\mathrm{t}(\mathcal{F})$ is injective and torsionfree and a K-vector space.*

2. *If $\mathcal{F}$ is an injective cogenerator and $\mathcal{D}$ is not a field (as in all interesting cases), then $\mathrm{t}(\mathcal{F})$ is also an injective cogenerator. Observe that if $\mathcal{D}$ is a field, the torsion module is zero and certainly not a cogenerator.*

Proof 1. We apply Baer's criterion (Theorem 3.2.9) to prove the injectivity of $\mathrm{t}(\mathcal{F})$. Let $\mathfrak{a} \subset \mathcal{D}$ be a nonzero ideal and let $\varphi\colon \mathfrak{a} \longrightarrow \mathrm{t}(\mathcal{F})$ be a linear map. Since $\mathrm{t}(\mathcal{F}) \subseteq \mathcal{F}$, we extend the codomain to obtain the map $\varphi_1\colon \mathfrak{a} \longrightarrow \mathcal{F}$, $a \longmapsto \varphi(a)$. Since $\mathcal{F}$ is injective, there is an extension $\psi\colon \mathcal{D} \longrightarrow \mathcal{F}$ with $\psi|_{\mathfrak{a}} = \varphi_1$, i.e., for all $a \in \mathfrak{a}$, $\psi(a) = \varphi_1(a) = \varphi(a)$ holds. To prove $\mathrm{im}(\psi) \subseteq \mathrm{t}(\mathcal{F})$, it suffices to show $\psi(1) \in \mathrm{t}(\mathcal{F})$. Let

$$
\begin{aligned}
&0 \neq a_0 \in \mathfrak{a} \Longrightarrow \varphi(a_0) \in \mathrm{t}(\mathcal{F}) \Longrightarrow \exists f \neq 0 \text{ with } f \circ \varphi(a_0) = 0\\
&\Longrightarrow (f a_0) \circ \psi(1) = f \circ \big(a_0 \circ \psi(1)\big) = f \circ \psi(a_0) = f \circ \varphi(a_0) = 0\\
&\Longrightarrow \psi(1) \in \mathrm{t}(\mathcal{F}).
\end{aligned}
$$

By Lemma 5.3.8 $\mathrm{t}(\mathcal{F})$ is a direct summand of $\mathcal{F}$, so there is a module M'' with $\mathcal{F} = \mathrm{t}(\mathcal{F}) \oplus M''$ and hence $M'' \cong \mathcal{F}/\mathrm{t}(\mathcal{F})$. Since $\mathcal{F}$ is injective, M'' and then $\mathcal{F}/\mathrm{t}(\mathcal{F})$ are injective too by Corollary 5.3.6. Since this factor module is also torsionfree, it is a K-vector space according to Theorem 5.3.9.

2. According to Theorem 3.3.5 the injective module $\mathrm{t}(\mathcal{F})$ is a cogenerator if it contains all simple modules S, up to isomorphism. Assume such an S, without loss of generality of the form $S = \mathcal{D}/\mathfrak{m}$ for some maximal ideal $\mathfrak{m}$. Since $\mathcal{D}$ is not a field, this maximal ideal is nonzero, and therefore S is a torsion module. Since $\mathcal{F}$ is an injective cogenerator, we infer $S \subseteq \mathcal{F}$, up to isomorphism, and thus $S = \mathrm{t}(S) \subseteq \mathrm{t}(\mathcal{F})$, up to isomorphism. □

Corollary 5.3.11 (Autonomous behaviors) *Let $\mathcal{F}$ be an injective cogenerator with $\mathrm{t}(\mathcal{F}) \subsetneq \mathcal{F}$. The standard function modules $F^{\mathbb{N}}$ and $\mathrm{C}^\infty(\mathbb{R}, F)$ satisfy this assumption. Then a behavior $\mathcal{B} \subseteq \mathcal{F}^l$ is autonomous if and only if $\mathcal{B} \subseteq \mathrm{t}(\mathcal{F})^l$.*

Proof $\Longrightarrow$. According to Lemma 4.2.6, we have $\mathrm{ann}_{\mathcal{D}}(\mathrm{M}(\mathcal{B})) = \mathrm{ann}_{\mathcal{D}}(\mathcal{B}) \neq 0$. This implies that $\mathcal{B}$ is a torsion module and thus $\mathcal{B} = \mathrm{t}(\mathcal{B}) \subseteq \mathrm{t}(\mathcal{F}^l) = \mathrm{t}(\mathcal{F})^l$.
$\Longleftarrow$. Assume that $\mathcal{B}$ is not autonomous. According to Corollary 4.2.3 there is a free observable $\xi \in M(\mathcal{B})$ with $\xi \circ \mathcal{B} = \mathcal{F}$. We infer the contradiction

$$
\mathcal{F} = \xi \circ \mathcal{B} \subseteq \xi \circ \mathrm{t}(\mathcal{F}^l) \overset{\text{Lemma 4.2.2}}{\subseteq} \mathrm{t}(\mathcal{F}) \subsetneq \mathcal{F}. \qquad \square
$$

Corollary 5.3.12 *Assume that $\mathcal{D}$ is a principal ideal domain, but not a field, with the representative system $\mathcal{P}$ of prime elements and that $\mathcal{F}$ is an injective cogenerator. According to Theorem 5.3.2, the primary decomposition*

$$
\mathrm{t}(\mathcal{F}) = \bigoplus_{p \in \mathcal{P}} \mathrm{t}_p(\mathcal{F}), \quad \textit{where } \mathrm{t}_p(\mathcal{F}) := \{y \in \mathcal{F};\ \exists j \textit{ with } p^j \circ y = 0\},
$$

holds. Then the module $\mathrm{t}(\mathcal{F})$ *is an injective cogenerator,* $\mathcal{F}/\mathrm{t}(\mathcal{F})$ *is a* K*-vector space, and all* p*-components* $\mathrm{t}_p(\mathcal{F})$ *are injective* $\mathcal{D}$*-modules. In particular, the fundamental principle is applicable to* $\mathcal{F}$, $\mathrm{t}(\mathcal{F})$, $\mathcal{F}/\mathrm{t}(\mathcal{F})$, *and* $\mathrm{t}_p(\mathcal{F})$.

Proof The injectivity of $\mathrm{t}_p(\mathcal{F})$ follows from Corollary 5.3.6. □

5.3.2.3 The Primary Decomposition of Autonomous Behaviors

The primary decomposition of the injective cogenerator $\mathrm{t}(\mathcal{F})$ implies that of autonomous behaviors, as we will show in the following. Assume $\mathcal{D}$ and $\mathcal{P}$ as in Corollary 5.3.12 and a finitely generated torsion module

$$M = \mathcal{D}^{1\times l}/U \quad \text{with } U := \mathcal{D}^{1\times k}R,\ R \in \mathcal{D}^{k\times l},\ \mathrm{rank}(R) = l. \tag{5.39}$$

According to Theorem 3.1.36

$$\begin{aligned} \mathrm{D}(M) = \mathrm{Hom}_{\mathcal{D}}(M,\mathcal{F}) &\overset{\cong}{\longleftrightarrow} \mathcal{B} = \{w \in \mathcal{F}^l;\ R\circ w = 0\},\\ (\xi + U \longmapsto \xi\circ w) &\longleftrightarrow \qquad w \end{aligned} \tag{5.40}$$

is an isomorphism. As in Corollary 5.3.3 Lemma 4.2.6 implies

$$\mathrm{ann}_{\mathcal{D}}(\mathcal{B}) = \mathrm{ann}_{\mathcal{D}}(M) = \mathcal{D}f \text{ for some } f = \prod_{i=1}^{r} p_i^{m(i)},\ p_i \in \mathcal{P},\ m(i) > 0. \tag{5.41}$$

Theorem and Definition 5.3.13 *For* (5.39)–(5.41) *the following holds.*

1. *The primary decomposition of* $\mathcal{B}$ *is*

$$\begin{aligned} &\mathcal{B} = \bigoplus_{p\in\mathcal{P}} \mathrm{t}_p(\mathcal{B}) \text{ with } \mathrm{t}_p(\mathcal{B}) = \begin{cases} \mathrm{ann}_{\mathcal{B}}(p_i^{m(i)}) & \text{if } \exists i \text{ with } p = p_i\\ 0 & \text{if } \forall j:\ p \neq p_j \end{cases}, \text{ and}\\ &\forall p \in \mathcal{P}:\ \mathrm{index}(\mathcal{B},p) = \mathrm{index}(M,p) = \begin{cases} m(i) & \text{if } \exists i \text{ with } p = p_i\\ 0 & \text{if } \forall j:\ p \neq p_j \end{cases}. \end{aligned} \tag{5.42}$$

2. *The trajectories in* $\mathrm{t}_p(\mathcal{B})$ *are called the* p-modes *of* $\mathcal{B}$ *and especially* λ*-modes if* $\mathcal{D} = F[s]$ *and* $p = s - \lambda$. *Nonzero* p*-modes exist if and only if there is a nonzero* $w \in \mathcal{B}$ *with* $p \circ w = 0$ *or if and only if* $p = p_i$ *for some* $i = 1, \dots, r$.
3. *If* $p \in \mathcal{P}$ *and* $m = \mathrm{index}(M, p) = \mathrm{index}(\mathcal{B}, p)$, *then*

$$\begin{aligned} \mathrm{Hom}_{\mathcal{D}}(\mathrm{t}_p(M),\mathcal{F}) &\cong \mathrm{Hom}_{\mathcal{D}}(M/p^mM,\mathcal{F}) = \mathrm{D}(M/p^mM)\\ &= \mathrm{ann}_{\mathrm{D}(M)}(p^m) = \mathrm{t}_p(\mathrm{D}(M)) = \mathrm{Hom}_{\mathcal{D}}(M,\mathrm{t}_p(\mathcal{F}))\\ &\cong \mathrm{ann}_{\mathcal{B}}(p^m) = \mathrm{t}_p(\mathcal{B}) = \mathcal{B}\cap \mathrm{t}_p(\mathcal{F})^l. \end{aligned} \tag{5.43}$$

4. *The matrix representation of* $\mathrm{t}_p(M)$ *is*

$$\begin{aligned}&\mathrm{t}_p(M) \cong \mathcal{D}^{1\times l}/\mathcal{D}^{1\times(k+l)}\begin{pmatrix} R \\ p^m \mathrm{id}_l \end{pmatrix}\\ &\Longrightarrow \mathrm{t}_p(\mathcal{B}) = \mathcal{B}\cap \mathrm{t}_p(\mathcal{F})^l = \left\{w \in \mathcal{F}^l;\ \begin{pmatrix} R \\ p^m \mathrm{id}_l \end{pmatrix} \circ w = 0\right\}.\end{aligned}$$

Proof 1. This follows from Corollary 5.3.3 applied to $\mathcal{B}$ since $\mathrm{ann}_{\mathcal{D}}(M) = \mathrm{ann}_{\mathcal{D}}(\mathcal{B}) = \mathcal{D}f \neq 0$.
2. $\mathrm{t}_p(\mathcal{B}) \neq 0 \Longleftrightarrow \mathrm{ann}_{\mathcal{B}}(p) = \{w \in \mathcal{B};\ p\circ w = 0\} \neq 0$.
3. This follows from (5.27) and Corollary 5.3.4 with $N := \mathcal{F}$. The Malgrange isomorphism $\mathrm{D}(M) \cong \mathcal{B}$ implies that $\mathrm{ann}_{\mathrm{D}(M)}(p^m) \cong \mathrm{ann}_{\mathcal{B}}(p^m)$.
4. With $U = \mathcal{D}^{1\times k}R$ and $M = \mathcal{D}^{1\times l}/U$, we get

$$\begin{aligned}\mathrm{t}_p(M) &\cong M/p^m M = \left(\mathcal{D}^{1\times l}/U\right)/p^m\left(\mathcal{D}^{1\times l}/U\right)\\ &= \left(\mathcal{D}^{1\times l}/U\right)/\left((U + p^m\mathcal{D}^{1\times l})/U\right) \cong \mathcal{D}^{1\times l}/(U + p^m\mathcal{D}^{1\times l})\\ &= \mathcal{D}^{1\times l}/\mathcal{D}^{1\times(k+l)}\begin{pmatrix} R \\ p^m \mathrm{id}_l \end{pmatrix}.\end{aligned}$$ □

We specialize the preceding results to a polynomial algebra $\mathcal{D} = F[s]$ over a field F. The set $\mathcal{P}$ consists of the monic irreducible polynomials, and we have $\dim_F(\mathcal{D}/\mathcal{D}p^j) = j\deg(p)$.

Lemma and Definition 5.3.14 *Let F be a field and $\mathcal{D} = F[s]$. Let ${}_{\mathcal{D}}M$ be a torsion module and let $p \in \mathcal{P}$. Then the degree* $\deg(p)$ *divides* $\dim_F(\mathrm{t}_p(M))$. *The quotient* $\mathrm{l}_p(M) := \mathrm{mult}(M, p) := \frac{\dim_F(\mathrm{t}_p(M))}{\deg(p)}$ *is called the p-length of M or the* multiplicity *of p in M. We define $\mathrm{l}(M) := (\mathrm{l}_p(M))_{p\in\mathcal{P}} \in \mathbb{N}^{\mathcal{P}}$. Let*

$$\mathrm{ann}_{\mathcal{D}}(M) = \mathcal{D}f,\ f \neq 0,\ \textit{and}\ f = \prod_{i=1}^{r} p_i^{m(i)},\ \ p_i \in \mathcal{P},\ m(i) > 0.$$

Then the support of $\mathrm{l}(M)$ *is the finite set*

$$\begin{aligned}&\mathrm{supp}(\mathrm{l}(M)) := \{p \in \mathcal{P};\ \mathrm{l}_p(M) \neq 0,\ \textit{i.e.},\ \mathrm{t}_p(M) \neq 0\} = \{p_1, \dots, p_r\},\ \textit{hence}\\ &\mathrm{l}(M) \in \mathbb{N}^{(\mathcal{P})} := \left\{\mu \in \mathbb{N}^{\mathcal{P}};\ \mathrm{supp}(\mu) = \{p \in \mathcal{P};\ \mu(p) \neq 0\}\ \textit{is finite}\right\} \subset \mathbb{N}^{\mathcal{P}}.\end{aligned}$$

The set $\mathbb{N}^{(\mathcal{P})}$ *is an additive monoid with the componentwise addition.*

Proof According to the first part of Remark 4.2.10, the module $\mathrm{t}_p(M)$ admits a representation $\mathrm{t}_p(M) \cong \prod_{i=1}^{r}\mathcal{D}/\mathcal{D}a_i$ with $a_i \in \mathcal{D}\setminus\{0\}$. Let $m := \mathrm{index}(M, p)$. The equation $p^m\,\mathrm{t}_p(M) = 0$ implies that $p^m\mathcal{D}/\mathcal{D}a_i = 0$ for all i, i.e., $a_i \mid p^m$. But this signifies that $a_i = p^{k(i)}$ for some $0 \leq k(i) \leq m$. Hence,

$$\dim_F(\mathrm{t}_p(M)) = \sum_{i=1}^{r} \deg(a_i) = \sum_{i=1}^{r} k(i)\deg(p)$$

$$\Longrightarrow \deg(p) \mid \dim_F(\mathrm{t}_p(M)) \text{ and } 1_p(M) = \sum_{i=1}^{r} k(i).$$

The equation for the support of $1(M)$ follows from Corollary 5.3.3. □

Remark 5.3.15 A finitely generated $\mathcal{D}$-module M is of *finite length* if it admits a *composition series* or *filtration* $M = M_0 \supsetneq M_1 \supsetneq \cdots \supsetneq M_{l-1} \supsetneq M_l = 0$, where the factors M_{i-1}/M_i, $i = 1, \ldots, l$ are *simple* modules (see Reminder 3.3.3). By the *Jordan-Hölder theorem*, the number l is independent of the choice of the M_i, and is called the *length* of M. In Lemma 5.3.14, the modules M and $\mathrm{t}_p(M)$ are F-finite dimensional and thus of finite length. The factors of $\mathrm{t}_p(M)$ are isomorphic to $\mathcal{D}/\mathcal{D}p$ and the length of $\mathrm{t}_p(M)$ is the p-length $1_p(\mathrm{t}(M)) = \frac{\dim_F(\mathrm{t}_p(M))}{\deg(p)}$. We do not need this more general notion of length. Therefore, we do not discuss it in detail and refer the reader to [2, pp. 21–22 and p. 156] for more details. ◊

Corollary 5.3.16 *In the discrete-time and continuous-time standard cases with* $\mathcal{D} = F[s]$ *and* $\mathcal{F} = F^{\mathbb{N}}$, *$F$ a field, and* $\mathcal{F} = \mathrm{C}^\infty(\mathbb{R}, F)$, $F = \mathbb{R}, \mathbb{C}$, *respectively, the equality* $\dim_F(M) = \dim_F(\mathcal{B})$ *from Theorem 4.2.14 implies*

$$1(M) = 1(\mathcal{B}) := (1_p(\mathcal{B}))_{p\in\mathcal{P}} \in \mathbb{N}^{(\mathcal{P})} \quad \textit{with } 1_p(\mathcal{B}) = \frac{\dim_F(\mathrm{t}_p(\mathcal{B}))}{\deg(p)}.$$ ◊

An element $\mu = (\mu_p)_{p\in\mathcal{P}} \in \mathbb{N}^{(\mathcal{P})}$ is called a *finite valued subset of* $\mathcal{P}$. The subset is the finite support $\mathrm{supp}(\mu) = \{p_1, \ldots, p_r\}$, and the number $\mu_{p_i} > 0$ is the *value* or *multiplicity* of p_i. In the literature these μ are written as

$$\mu = \{\overbrace{p_1, \ldots, p_1}^{\mu_{p_1}\text{ times}}, \ldots, \overbrace{p_r, \ldots, p_r}^{\mu_{p_r}\text{ times}}\}.$$

If $\mu, \nu \in \mathbb{N}^{(\mathcal{P})}$, one writes

$$\mu \uplus \nu := \mu + \nu \text{ and thus } \mathrm{supp}(\mu \uplus \nu) = \mathrm{supp}(\mu + \nu) = \mathrm{supp}(\mu) \cup \mathrm{supp}(\nu). \tag{5.44}$$

The following theorem shows the significance of the finite valued sets.

Theorem 5.3.17 *Let F be a field,* $\mathcal{D} := F[s]$ *and* $\mathcal{F}$ *an injective cogenerator.*

1. *Assume an exact sequence of finitely generated torsion modules*

$$0 \longrightarrow M_1 \longrightarrow M_2 \longrightarrow M_3 \longrightarrow 0$$

or, equivalently, the dual exact sequence of autonomous behaviors

$$0 \longleftarrow \mathcal{B}_1 \longleftarrow \mathcal{B}_2 \longleftarrow \mathcal{B}_3 \longleftarrow 0.$$

Then the identity

$$\mathrm{l}(M_2) = \mathrm{l}(M_1) + \mathrm{l}(M_3) \overset{(5.44)}{=\!=\!=} \mathrm{l}(M_1) \uplus \mathrm{l}(M_3) \tag{5.45}$$

holds. For the standard signal modules of Corollary 5.3.16 this implies

$$\mathrm{l}(\mathcal{B}_2) = \mathrm{l}(M_2) = \mathrm{l}(\mathcal{B}_1) + \mathrm{l}(\mathcal{B}_3) = \mathrm{l}(M_1) + \mathrm{l}(M_3), \ \textit{in particular,}$$
$$\mathrm{supp}(\mathrm{l}(\mathcal{B}_2)) = \mathrm{supp}(\mathrm{l}(\mathcal{B}_1)) \cup \mathrm{supp}(\mathrm{l}(\mathcal{B}_3)).$$

2. *In the situation of item 1, the following equivalences hold:*

$$\begin{aligned} &M_1 \cong M_2 \Longleftrightarrow M_3 = 0 \Longleftrightarrow \mathrm{l}(M_1) = \mathrm{l}(M_2) \Longleftrightarrow \mathrm{l}(M_3) = 0 \ \ \textit{and} \\ &M_2 \cong M_3 \Longleftrightarrow M_1 = 0 \Longleftrightarrow \mathrm{l}(M_2) = \mathrm{l}(M_3) \Longleftrightarrow \mathrm{l}(M_1) = 0. \end{aligned} \tag{5.46}$$

Proof 1. The exact sequence of torsion modules, the primary decompositions of these modules, and the exactness of the coproduct (5.37) imply the exact sequences of the p-components

$$0 \longrightarrow \mathrm{t}_p(M_1) \longrightarrow \mathrm{t}_p(M_2) \longrightarrow \mathrm{t}_p(M_3) \longrightarrow 0 \text{ and thus}$$
$$\dim_F(\mathrm{t}_p(M_2)) = \dim_F(\mathrm{t}_p(M_1)) + \dim_F(\mathrm{t}_p(M_3)) \text{ and } \mathrm{l}(M_2) = \mathrm{l}(M_1) + \mathrm{l}(M_3).$$

2. This follows directly from item 1 and from $(M = 0 \Longleftrightarrow \mathrm{l}(M) = 0)$. □

Corollary and Definition 5.3.18 *For a field F and $\mathcal{D} := F[s]$ consider*

$$\mathcal{B} = \{w \in \mathcal{F}^l;\ R \circ w = 0\},\ R \in \mathcal{D}^{k \times l},\ p := \mathrm{rank}(R),\ m := \mathrm{rank}(\mathcal{B}) = l - p.$$

Assume $m > 0$, i.e., that $\mathcal{B}$ is not *autonomous. According to Corollary and Definition 4.2.19 there are the direct decompositions*

$$\begin{aligned} &M = \mathrm{t}(M) \oplus M',\ M' \cong M_{\mathrm{cont}} = M/\mathrm{t}(M) \cong \mathcal{D}^{1 \times m} \ \textit{free}, \\ &\mathcal{B} = \mathcal{B}' \oplus \mathcal{B}_{\mathrm{cont}},\ \mathcal{B}_{\mathrm{cont}} \cong \mathcal{F}^m,\ \mathcal{B}' \cong \mathcal{B}/\mathcal{B}_{\mathrm{cont}} \cong \mathcal{B}_{\mathrm{uncont}} \ \textit{autonomous}. \end{aligned} \tag{5.47}$$

The module $\mathcal{D}^{1 \times m}$ is torsionfree and hence $\mathrm{t}_p(\mathcal{D}^{1 \times m}) = 0$ for all $p \in \mathcal{P}$. In contrast, $\mathcal{F}$ is a cogenerator and contains all simple modules $\mathcal{D}/\mathcal{D}p,\ p \in \mathcal{P}$, and hence $\mathrm{t}_p(\mathcal{F}) \neq 0$ for all $p \in \mathcal{P}$ and indeed $\dim_F(\mathrm{t}_p(\mathcal{F})) = \infty$. Equation (5.47) thus implies

$$\begin{aligned} &\mathrm{t}_p(M) = \mathrm{t}_p(\mathrm{t}(M)),\ \mathrm{t}_p(M') = 0, \\ &\mathrm{t}_p(\mathcal{B}) = \mathrm{t}_p(\mathcal{B}') \oplus \mathrm{t}_p(\mathcal{B}_{\mathrm{cont}}),\ \mathrm{t}_p(\mathcal{B}') \cong \mathrm{t}_p(\mathcal{B}_{\mathrm{uncont}}),\ \mathrm{t}_p(\mathcal{B}_{\mathrm{cont}}) \cong \mathrm{t}_p(\mathcal{F})^m. \end{aligned} \tag{5.48}$$

For nonautonomous $\mathcal{B}$ we infer $\dim_F(\mathrm{t}_p(\mathcal{B})) = \infty$ for all $p \in \mathcal{P}$. In particular, the notion of p-mode is unsuitable for nonautonomous behaviors and $\mathrm{l}_p(\mathcal{B})$ cannot be *defined by means of* $\mathrm{t}_p(\mathcal{B})$. *Therefore we define the* p-multiplicity *or* p-length *of* $\mathcal{B}$ *by means of* $\mathrm{t}_p(M)$, *i.e.,*

$$\forall p \in \mathcal{P}: \ \mathrm{l}_p(\mathcal{B}) := \mathrm{l}_p(M) := \mathrm{l}_p(\mathrm{t}(M)) := \frac{\dim_F\left(\mathrm{t}_p(M)\right)}{\deg(p)} =: \mathrm{l}_p(\mathcal{B}_{\mathrm{uncont}}). \tag{5.49}$$

◇

5.3.3 The Primary Decomposition and Autonomy for the Standard Signal Modules

We specialize the results of Sects. 5.3.1 and 5.3.2 to the standard signal modules over the polynomial algebra $F[s]$ where $F = \mathbb{R}$ or $F = \mathbb{C}$.

Corollary 5.3.19 *We treat the complex case* $F = \mathbb{C}$ *first and use the representative system* $\mathcal{P} = \{s - \lambda;\ \lambda \in \mathbb{C}\}$ *of the irreducible elements of* $\mathbb{C}[s]$. *In the continuous-time and in the discrete-time cases, the decompositions* $\mathrm{t}(\mathcal{F}) = \bigoplus_{p\in\mathcal{P}} \mathrm{t}_p(\mathcal{F})$ *are the following.*
Continuous-time case $\mathcal{F} = \mathrm{C}^\infty(\mathbb{R}, \mathbb{C})$.

$$\mathrm{t}\left(\mathrm{C}^\infty(\mathbb{R}, \mathbb{C})\right) = \bigoplus_{\lambda\in\mathbb{C}} \mathrm{t}_{s-\lambda}\left(\mathrm{C}^\infty(\mathbb{R}, \mathbb{C})\right), \ \textit{where}$$

$$\mathrm{t}_{s-\lambda}\left(\mathrm{C}^\infty(\mathbb{R}, \mathbb{C})\right) = \bigoplus_{k=0}^{\infty} \mathbb{C}e_{\lambda,k} = \mathbb{C}[t]e^{\lambda t} \ \textit{and}\ e_{\lambda,k} = \tfrac{t^k}{k!}e^{\lambda t}.$$

Discrete-time case $\mathcal{F} = \mathbb{C}^{\mathbb{N}}$.

$$\mathrm{t}(\mathbb{C}^{\mathbb{N}}) = \bigoplus_{\lambda\in\mathbb{C}} \mathrm{t}_{s-\lambda}(\mathbb{C}^{\mathbb{N}}), \ \textit{where}$$

$$\mathrm{t}_{s-\lambda}(\mathbb{C}^{\mathbb{N}}) = \bigoplus_{k=0}^{\infty} \mathbb{C}e_{\lambda,k} = \begin{cases} \mathbb{C}^{(\mathbb{N})} & \textit{if}\ \lambda = 0 \\ \mathbb{C}[t]\lambda^t & \textit{if}\ \lambda \neq 0 \end{cases} \ \textit{and}$$

$$e_{\lambda,k}(t) = \begin{cases} \delta_{k,t} & \textit{if}\ \lambda = 0 \\ \binom{t}{k}\lambda^{t-k} & \textit{if}\ \lambda \neq 0 \end{cases}.$$

In the real case $F = \mathbb{R}$ *we use the representative system*

$$\mathcal{P} = \{p_\lambda;\ \lambda \in \Lambda_{\mathbb{R}}\}, \ \textit{where}\ \lambda = \alpha + i\beta = |\lambda|e^{i\omega} \ \textit{with}\ \alpha, \beta, \omega \in \mathbb{R},$$
$$\Lambda_{\mathbb{R}} = \{\lambda \in \mathbb{C};\ \beta = \Im(\lambda) \geq 0\} = \{\lambda = |\lambda|e^{i\omega} \in \mathbb{C};\ 0 \leq \omega \leq \pi\}, \ \textit{and}$$
$$p_\lambda = \begin{cases} s - \lambda & \textit{if}\ \lambda \in \mathbb{R}, \\ (s-\lambda)(s-\overline{\lambda}) = s^2 - 2\alpha s + \alpha^2 + \beta^2 & \textit{if}\ \Im(\lambda) > 0. \end{cases}$$

The primary decompositions of the torsion modules of the standard signal spaces are the following.
Continuous-time case $\mathcal{F} = \mathrm{C}^\infty(\mathbb{R}, \mathbb{R})$.

$$\mathrm{t}(\mathrm{C}^\infty(\mathbb{R}, \mathbb{R})) = \bigoplus_{\lambda \in \Lambda_\mathbb{R}} \mathrm{t}_{p_\lambda}(\mathrm{C}^\infty(\mathbb{R}, \mathbb{R})) \quad \textit{where}$$

$$\mathrm{t}_{p_\lambda}(\mathrm{C}^\infty(\mathbb{R}, \mathbb{R})) = \begin{cases} \mathbb{R}[t]e^{\lambda t} & \textit{if } \lambda = \alpha \in \mathbb{R} \\ \mathbb{R}[t]e^{\alpha t}\left(\mathbb{R}\cos(\beta t) \oplus \mathbb{R}\sin(\beta t)\right) & \textit{if } \beta = \Im(\lambda) > 0. \end{cases}$$

Discrete-time case $\mathcal{F} = \mathbb{R}^\mathbb{N}$.

$$\mathrm{t}(\mathbb{R}^\mathbb{N}) = \bigoplus_{\lambda \in \Lambda_\mathbb{R}} t_{p_\lambda}(\mathbb{R}^\mathbb{N}), \quad \textit{where}$$

$$t_{p_\lambda}(\mathbb{R}^\mathbb{N}) = \begin{cases} \mathbb{R}^{(\mathbb{N})} & \textit{if } \lambda = 0,\ p_\lambda = s, \\ \mathbb{R}[t]\lambda^t & \textit{if } 0 \neq \lambda \in \mathbb{R},\ p_\lambda = s - \lambda\,. \\ \mathbb{R}[t]|\lambda|^t\left(\mathbb{R}\cos(\omega t) \oplus \mathbb{R}\sin(\omega t)\right) & \textit{if } \lambda = |\lambda|e^{i\omega},\ 0 < \omega < \pi \end{cases}$$

The functions in $\mathrm{t}(\mathcal{F})$ *are called* polynomial-exponential.

Proof The assertions for the complex continuous, the complex discrete, the real continuous, and the real discrete case are direct consequences of Theorem 5.2.2, Theorem 5.2.8, Corollary 5.2.3, and Corollary 5.2.10, respectively. We show this for the complex continuous case only.

For $\mathcal{F} = \mathrm{C}^\infty(\mathbb{R}, \mathbb{C})$ and $\lambda \in \mathbb{C}$ we obtain

$$\mathrm{t}_{s-\lambda}\left(\mathrm{C}^\infty(\mathbb{R}, \mathbb{C})\right) = \bigcup_{m=0}^{\infty} \mathrm{ann}_\mathcal{F}\left((s-\lambda)^m\right) \underset{\text{Theorem 5.2.2}}{=} \bigcup_{m=0}^{\infty} \bigoplus_{k=0}^{m-1} \mathbb{C}e_{\lambda,k}$$
$$= \bigcup_{m=0}^{\infty} \mathbb{C}[t]_{\leq m-1}e^{\lambda t} = \bigoplus_{k=0}^{\infty} \mathbb{C}e_{\lambda,k} = \mathbb{C}[t]e^{\lambda t}.$$

□

Corollary 5.3.20 *Let* $\mathcal{D} := \mathbb{C}[s]$ *be the complex polynomial algebra with* $\mathcal{P} = \{s - \lambda;\ \lambda \in \mathbb{C}\}$.

1. *For the injective cogenerator* $\mathcal{F} = \mathrm{C}^\infty(\mathbb{R}, \mathbb{C})$, *Corollary 5.3.19 furnishes*

$$\mathrm{t}(\mathcal{F}) = \bigoplus_{\lambda \in \mathbb{C}} \mathbb{C}[t]e^{\lambda t} \quad \textit{and} \quad \mathrm{t}_{s-\lambda}(\mathcal{F}) = \mathbb{C}[t]e^{\lambda t} \ \textit{for } \lambda \in \mathbb{C}.$$

All these modules are injective and the fundamental principle is applicable to them. The space $\mathrm{t}(\mathcal{F})$ *of polynomial-exponential functions is an injective cogenerator over* $\mathbb{C}[s]$, *and the space* $\mathrm{C}^\infty(\mathbb{R}, \mathbb{C})/\left(\bigoplus_{\lambda \in \mathbb{C}} \mathbb{C}[t]e^{\lambda t}\right)$ *is a* $\mathbb{C}(s)$*-vector space.*

2. *For the injective cogenerator $\mathcal{F} = \mathbb{C}^{\mathbb{N}}$, Corollary 5.3.19 furnishes*

$$\mathrm{t}(\mathcal{F}) = \mathbb{C}^{(\mathbb{N})} \oplus \bigoplus_{\lambda \in \mathbb{C}\setminus\{0\}} \mathbb{C}[t]\lambda^t \quad \text{and} \quad \mathrm{t}_{s-\lambda}(\mathcal{F}) = \begin{cases} \mathbb{C}^{(\mathbb{N})} & \text{if } \lambda = 0, \\ \mathbb{C}[t]\lambda^t & \text{if } \lambda \neq 0. \end{cases}$$

Again, all these modules are injective and the fundamental principle is applicable to them. The torsion module $\mathrm{t}(\mathcal{F})$ is an injective cogenerator and the space $\mathbb{C}^{\mathbb{N}}/(\mathbb{C}^{(\mathbb{N})} \oplus \bigoplus_{0\neq\lambda\in\mathbb{C}} \mathbb{C}[t]\lambda^t)$ is a $\mathbb{C}(s)$-vector space.

The counterpart of these assertions for the real polynomial algebra $\mathcal{D} = \mathbb{R}[s]$ and the injective cogenerators $\mathrm{C}^\infty(\mathbb{R}, \mathbb{R})$ and $\mathbb{R}^{\mathbb{N}}$ hold mutatis mutandis and follow also from Corollary 5.3.19. ◊

Remark 5.3.21 1. For $\mathcal{F} := \mathrm{C}^\infty(\mathbb{R}, \mathbb{C})$, the injective submodule $\mathrm{t}(\mathcal{F})$ of polynomial-exponential functions gives rise to a direct decomposition

$$\mathcal{F} = \mathcal{F}_1 \oplus \mathrm{t}(\mathcal{F}) \ni y = y_1 + y_2 \Longrightarrow \mathcal{F}_1 \cong \mathcal{F}/\,\mathrm{t}(\mathcal{F}),\ y_1 \mapsto y_1 + \mathrm{t}(\mathcal{F}).$$

The submodule $\mathcal{F}_1$ is not unique and cannot be described constructively. This is due to the fact that Baer's criterion for injectivity uses the nonconstructive Lemma of Zorn, cf. Result 3.2.7 and Theorem 3.2.9.

2. For an arbitrary field F the primary decomposition of $\mathrm{t}({}_{F[s]}F^{\mathbb{N}})$ is derived in Sect. 8.1.2. ◊

5.4 The Characteristic Variety and Stability

The characteristic variety of an autonomous system $\mathcal{B}$ over the real or complex polynomial algebra is a finite subset of the complex field which generalizes the spectrum or set of complex eigenvalues of a complex square matrix. It is of fundamental importance in engineering mathematics. In Sect. 5.4.1 *we introduce the characteristic variety and use it for deriving fundamental systems, i.e., vector space bases, of autonomous behaviors in the standard cases. In Sect.* 5.4.2 *we apply it to the characterization of stability of autonomous systems.*

5.4.1 The Characteristic Variety

Let F be the real or complex field and $\mathcal{F}$ an injective cogenerator over the polynomial algebra $\mathcal{D} := F[s]$. Let $\mathcal{B} \subseteq \mathcal{F}^l$ be any behavior with the data

$$\begin{aligned} &R \in F[s]^{k\times l},\ p := \mathrm{rank}(R),\ \mathcal{B} := \{w \in \mathcal{F}^l;\ R \circ w = 0\}, \\ &M := \mathrm{M}(\mathcal{B}) := \mathcal{D}^{1\times l}/\mathcal{D}^{1\times k}R. \end{aligned} \tag{5.50}$$

Moreover, we use the Smith form

$$S = URV = \begin{pmatrix} \operatorname{diag}(e_1, \ldots, e_p) & 0 \\ 0 & 0 \end{pmatrix}, \\ e_1 \mid \cdots \mid e_p, \quad U \in \mathrm{Gl}_k(F[s]), \quad V \in \mathrm{Gl}_l(\mathcal{D}), \tag{5.51}$$

with the uniquely determined monic elementary divisors $e_i \in F[s]$ according to Result 3.2.17. By Theorem 4.2.9 the polynomial e_p is the unique monic generator of the ideal $\operatorname{ann}_{\mathcal{D}}(\mathrm{t}(M))$ and thus uniquely determined by M or $\mathcal{B}$.

The *variety*, *vanishing set* or *zero set* of a nonzero ideal $\mathfrak{a} = F[s]f$ of $F[s]$ is the finite set of complex zeros of f, i.e.,

$$\mathrm{V}_{\mathbb{C}}(\mathfrak{a}) := \{\lambda \in \mathbb{C};\ \forall g \in \mathfrak{a}\colon\ g(\lambda) = 0\} = \mathrm{V}_{\mathbb{C}}(f) := \{\lambda \in \mathbb{C};\ f(\lambda) = 0\}. \tag{5.52}$$

Definition and Corollary 5.4.1 *The variety*

$$\operatorname{char}(M) := \operatorname{char}(\mathcal{B}) := \mathrm{V}_{\mathbb{C}}(e_p) = \mathrm{V}_{\mathbb{C}}\left(\operatorname{ann}_{\mathcal{D}}(\mathrm{t}(M))\right)$$

is called the characteristic variety *of the module M and the behavior $\mathcal{B}$. It coincides with the characteristic variety of the torsion submodule* $\mathrm{t}(M)$ *and of its associated autonomous behavior $\mathcal{B}_{\mathrm{uncont}}$ according to Theorem and Definition 4.2.16 since*

$$\operatorname{ann}_{\mathcal{D}}(\mathcal{B}_{\mathrm{uncont}}) = \operatorname{ann}_{\mathcal{D}}(\mathrm{t}(M)) = \mathcal{D}e_p.$$

The elements of $\operatorname{char}(\mathcal{B})$ *are also called the* characteristic values *or* natural (complex) frequencies *of $\mathcal{B}$.* ◇

Theorem 5.4.2 (Rank singularities) *The characteristic variety of $\mathcal{B}$ can also be described by*

$$\operatorname{char}(\mathcal{B}) = \left\{\lambda \in \mathbb{C};\ \operatorname{rank}(R(\lambda)) < \operatorname{rank}(R) = p\right\}.$$

Proof Substitution of λ for s in (5.51) gives

$$S(\lambda) = U(\lambda)R(\lambda)V(\lambda) = \begin{pmatrix} \operatorname{diag}(e_1(\lambda), \ldots, e_p(\lambda)) & 0 \\ 0 & 0 \end{pmatrix}.$$

Since U and V are invertible, their determinants are units in $F[s]$, i.e., they are nonzero constants in F. Therefore,

$$\det\left(U(\lambda)\right) = \det(U) \in F \setminus \{0\} \quad \text{and} \quad \det\left(V(\lambda)\right) = \det(V) \in F \setminus \{0\},$$

i.e., $U(\lambda)$ and $V(\lambda)$ are invertible matrices over $\mathbb{C}$. Thus we have that

$$\operatorname{rank}(R(\lambda)) = \operatorname{rank}(S(\lambda)) = \operatorname{rank}\left(\operatorname{diag}(e_1(\lambda), \ldots, e_p(\lambda))\right) = \sharp\{i;\ e_i(\lambda) \neq 0\}.$$

Since e_i divides e_p, the equation $e_i(\lambda) = 0$ implies $e_p(\lambda) = 0$. We conclude the asserted equivalence

$$\lambda \in \operatorname{char}(\mathcal{B}) \Longleftrightarrow e_p(\lambda) = 0 \Longleftrightarrow \operatorname{rank}(R(\lambda)) < p = \operatorname{rank}(R).$$

□

Corollary 5.4.3 *1. If $\mathcal{B} = \{w \in \mathcal{F}^l;\ R \circ w = 0\}$ is an autonomous behavior, i.e., if $\operatorname{rank}(R) = l$, and if $R \in F[s]^{l \times l}$ is a square matrix, then*

$$\operatorname{char}(\mathcal{B}) = \left\{\lambda \in \mathbb{C};\ \operatorname{rank}\left(R(\lambda)\right) < l\right\} = \operatorname{V}_{\mathbb{C}}\left(\det(R)\right).$$

2. The characteristic variety of an autonomous state space behavior

$$\mathcal{B} := \{x \in \mathcal{F}^n;\ s \circ x = Ax\} = \{x \in \mathcal{F}^n;\ (s\,\mathrm{id}_n - A) \circ x = 0\}, \quad \textit{is}$$
$$\operatorname{char}(\mathcal{B}) = \{\lambda \in \mathbb{C};\ \operatorname{rank}(\lambda\,\mathrm{id}_n - A) < \operatorname{rank}(s\,\mathrm{id}_n - A) = n\} = \operatorname{spec}(A),$$

where $A \in F^{n \times n}$ and $\operatorname{spec}(A)$ is the complex spectrum of A, i.e., the set of complex eigenvalues of A. Since the matrix $s\,\mathrm{id}_n - A$ and its determinant $\chi_A = \det(s\,\mathrm{id}_n - A) \in F[s]$ are called the characteristic matrix *and* characteristic polynomial, *respectively, of the matrix A, this justifies the term* characteristic variety *also in the more general situation of Definition and Corollary 5.4.1.* ◇

As a supplement to Theorem 4.2.12, we characterize controllability of a behavior by its characteristic variety.

Theorem 5.4.4 (Compare [3, Theorem 5.2.10 on p. 159])

1. *The system $\mathcal{B}$ from (5.50) is controllable if and only if $\operatorname{char}(\mathcal{B}) = \emptyset$ or, in other terms, if and only if $\operatorname{rank}(R(\lambda)) = \operatorname{rank}(R)$ for all $\lambda \in \mathbb{C}$.*
2. *If $\mathcal{B}$ is autonomous, then $\operatorname{char}(\mathcal{B}) = \emptyset$ if and only if $\mathcal{B} = 0$.*

Proof 1. The following equivalences hold:

$$\mathcal{B} \text{ is controllable} \underset{\text{Theorem 4.2.12}}{\Longleftrightarrow} e_p = 1 \Longleftrightarrow \operatorname{V}_{\mathbb{C}}(e_p) = \operatorname{char}(\mathcal{B}) = \emptyset.$$

2. An autonomous behavior is controllable if and only if it is zero. □

Remark 5.4.5 In most practical situations the matrix $R \in F[s]^{k \times l}$ has coefficients in $F = \mathbb{Q}$ or in $F = \mathbb{Q}(i) = \mathbb{Q} \oplus \mathbb{Q}i$. The Smith form in $F[s]^{k \times l}$ can be computed using rational operations only and thus without any numerical errors. In particular, the condition $e_p = 1$ and the controllability of $\mathcal{B}$ can be checked exactly and easily. Theorem 5.4.4 is, therefore, of very limited value only. This remark applies also to the following corollaries. ◇

5.4.1.1 The Characteristic Variety for Rosenbrock and State Space Equations

We amend the results on the observability and the controllability of Rosenbrock systems, cf. Theorem and Definition 4.1.4 and Theorem 4.3.5, and of state space systems, cf. Theorem and Definition 4.1.9 and Theorem and Definition 4.3.8, by means of the characteristic variety.

Corollary 5.4.6 *Consider Rosenbrock equations*

$$A \circ x = B \circ u, \quad y = C \circ x + D \circ u$$
$$\text{with } A \in F[s]^{n\times n},\ \det(A) \neq 0,\ B \in F[s]^{n\times m},\ C \in F[s]^{p\times n},\ D \in F[s]^{p\times m}.$$

Define $\mathcal{B}^0 := \{x \in \mathcal{F}^n;\ A \circ x = 0\}$, *hence, by Corollary 5.4.3,*

$$\operatorname{char}(\mathcal{B}^0) = \{\lambda \in \mathbb{C};\ \operatorname{rank}(A(\lambda)) < n\} = \mathrm{V}_{\mathbb{C}}(\det(A)).$$

1. *The following properties are equivalent:*
   (a) *The equations are observable.*
   (b) *The matrix* $\binom{A}{C}$ *has a left inverse or, equivalently, its n-th elementary divisor* e_n *is one.*
   (c) *For all* $\lambda \in \operatorname{char}(\mathcal{B}^0)$ *the condition* $\operatorname{rank}\begin{pmatrix} A(\lambda) \\ C(\lambda) \end{pmatrix} = n$ *holds or, in other terms, for* $\xi \in F^n$ *the equation* $\begin{pmatrix} A(\lambda) \\ C(\lambda) \end{pmatrix} \xi = 0$ *implies* $\xi = 0$.
2. *The following properties are equivalent:*
   (a) *The equations are controllable.*
   (b) *The matrix* $(A, -B)$ *has a right inverse or, equivalently, its n-th elementary divisor* e_n *is one.*
   (c) *For all* $\lambda \in \operatorname{char}(\mathcal{B}^0)$ *the condition* $\operatorname{rank}(A(\lambda), -B(\lambda)) = n$ *holds or, in other terms, for* $\xi \in F^{1\times n}$ *the equation* $\xi(A(\lambda), -B(\lambda)) = 0$ *implies* $\xi = 0$.

Proof The condition $\det(A) \neq 0$ implies $\operatorname{rank}(A, -B) = \operatorname{rank}\binom{A}{C} = n$ and that the n-th elementary divisor e_n of the matrices is nonzero.

1.(a) $\Longleftrightarrow$ 1.(b). Theorems 3.3.23 and 4.1.4.
1.(b) $\Longleftrightarrow$ 1.(c). As in Theorem 5.4.2, we have the equivalences

$$e_n = 1 \Longleftrightarrow \mathrm{V}_{\mathbb{C}}(e_n) = \emptyset \Longleftrightarrow \forall \lambda \in \mathbb{C}:\ \operatorname{rank}\begin{pmatrix} A(\lambda) \\ C(\lambda) \end{pmatrix} = n.$$

For $\lambda \notin \operatorname{char}(\mathcal{B}^0)$ the equation $\operatorname{rank}(A(\lambda)) = n$ implies $\operatorname{rank}\begin{pmatrix} A(\lambda) \\ C(\lambda) \end{pmatrix} = n$.
2.(a) $\Longleftrightarrow$ 2.(b). Theorem 4.3.5 and Corollary 3.3.23.
2.(b) $\Longleftrightarrow$ 2.(c). Analogous to 1.(b) $\Longleftrightarrow$ 1.(c). □

Corollary 5.4.7 (Popov-Belevitch-Hautus (PBH) test) *Consider state space equations*

$$s \circ x = Ax + Bu,\ y = Cx + D \circ u \text{ with}$$
$$A \in F^{n\times n},\ B \in F^{n\times m},\ C \in F^{p\times n},\ D \in F[s]^{p\times m}.$$

1. *The following properties are equivalent:*
   (a) *The equations are observable.*
   (b) *The* n*th elementary divisor of* $\begin{pmatrix} s\,\mathrm{id}_n - A \\ C \end{pmatrix}$ *is one, i.e., for all* $\lambda \in \mathbb{C}$ *the condition* $\operatorname{rank}\begin{pmatrix} \lambda\,\mathrm{id}_n - A \\ C \end{pmatrix} = n$ *holds.*
   (c) *For all* $\lambda \in \operatorname{spec}(A)$ *the condition* $\operatorname{rank}\begin{pmatrix} \lambda\,\mathrm{id}_n - A \\ C \end{pmatrix} = n$ *holds.*
2. *The following properties are equivalent:*
   (a) *The equations are controllable.*
   (b) *The* n*th elementary divisor of* $(s\,\mathrm{id}_n - A, -B)$ *is one, i.e., for all* $\lambda \in \mathbb{C}$ *the condition* $\operatorname{rank}(\lambda\,\mathrm{id}_n - A, -B) = n$ *holds.*
   (c) *For all* $\lambda \in \operatorname{spec}(A)$ *the condition* $\operatorname{rank}(\lambda\,\mathrm{id}_n - A, -B) = n$ *holds.*

Proof In this case, we have

$$\mathcal{B}^0 = \left\{x \in \mathcal{F}^n;\ (s\,\mathrm{id}_n - A) \circ x = 0\right\} \quad \text{and} \quad \operatorname{char}(\mathcal{B}^0) = \operatorname{spec}(A).$$

□

5.4.1.2 The Connection Between the Characteristic Variety and the Length

We finally establish the connection between the characteristic variety and the multiplicity. For $F = \mathbb{C}$ or $F = \mathbb{R}$ let $\mathcal{D} = F[s]$. Consider an autonomous behavior $\mathcal{B}$ and its dual torsion module M with the data

$$\begin{aligned} &R \in \mathcal{D}^{k\times l},\ \operatorname{rank}(R) = l,\ \mathcal{B} = \{w \in \mathcal{F}^l;\ R \circ w = 0\},\\ &M := M(\mathcal{B}) := \mathcal{D}^{1\times l}/\mathcal{D}^{1\times k}R,\\ &\operatorname{ann}_{\mathcal{D}}(M) = \operatorname{ann}_{\mathcal{D}}(\mathcal{B}) = \mathcal{D}e_l,\ \operatorname{char}(M) = \operatorname{char}(\mathcal{B}) = \mathrm{V}_{\mathbb{C}}(e_l), \end{aligned} \tag{5.53}$$

where e_l is the highest (nonzero) elementary divisor of R. The next theorem follows directly from the results of Theorem 5.3.13, Lemma and Definition 5.3.14, Remark 5.3.15, Corollary 5.3.16, and Theorem 5.3.17.

Theorem 5.4.8 *For* $F = \mathbb{C}$ *we use the representative system* $\mathcal{P} := \{s - \lambda;\ \lambda \in \mathbb{C}\}$ *of irreducible polynomials and identify* $\mathbb{N}^{(\mathbb{C})} = \mathbb{N}^{(\mathcal{P})}$. *Assume the data from* (5.53), *hence* $M = \bigoplus_{\lambda\in\mathbb{C}} \mathrm{t}_{s-\lambda}(M)$ *and* $\mathcal{B} = \bigoplus_{\lambda\in\mathbb{C}} \mathrm{t}_{s-\lambda}(\mathcal{B})$. *Then*

$$e_l = \prod_{\lambda\in\mathbb{C}} (s-\lambda)^{m(\lambda)} \quad with \quad \begin{cases} m(\lambda) > 0 & if\ \lambda \in \mathrm{char}(M) = \mathrm{char}(\mathcal{B}) \\ m(\lambda) = 0 & if\ \lambda \notin \mathrm{char}(M) = \mathrm{char}(\mathcal{B}). \end{cases}$$

For all $\lambda \in \mathbb{C}$ the cited results imply the following:

1. $\mathrm{index}(M, s-\lambda) = \mathrm{index}(\mathcal{B}, s-\lambda) = m(\lambda)$.
2. $$\begin{aligned} \mathrm{t}_{s-\lambda}(M) &= \mathrm{ann}_M((s-\lambda)^{m(\lambda)}) \\ &\cong M/(s-\lambda)^{m(\lambda)}M \cong \mathcal{D}^{1\times l}/\mathcal{D}^{1\times(k+l)}\begin{pmatrix} R \\ (s-\lambda)^{m(\lambda)}\,\mathrm{id}_l \end{pmatrix}. \end{aligned}$$
3. *The equivalences* $\mathrm{t}_{s-\lambda}(M) \neq 0 \iff m(\lambda) > 0 \iff \lambda \in \mathrm{char}(M)$ *hold.*
4. *The* $s-\lambda$*-length of* M *is* $\mathrm{l}_{s-\lambda}(M) = \mathrm{mult}(M, s-\lambda) = \dim_{\mathbb{C}}(\mathrm{t}_{s-\lambda}(M))$ *and the length vector of* M *is* $\mathrm{l}(M) = \left(\dim_{\mathbb{C}}(\mathrm{t}_{s-\lambda}(M))\right)_{\lambda\in\mathbb{C}} \in \mathbb{N}^{(\mathbb{C})}$.
   As we have seen in the proof of Lemma and Definition 5.3.14, the module $\mathrm{t}_{s-\lambda}(M)$ *has a representation* $\mathrm{t}_{s-\lambda}(M) \cong \bigoplus_{i=1}^{r} \mathcal{D}/\mathcal{D}(s-\lambda)^{k(i)}$ *with* $k(i) \leq m(\lambda)$*, and then*
   $$\mathrm{l}_{s-\lambda}(M) = \mathrm{mult}(M, s-\lambda) = \sum_{i=1}^{r} k(i). \tag{5.54}$$
   For the behavior $\mathcal{B}$ *this implies*
5. $$\begin{aligned} \mathrm{t}_{s-\lambda}(\mathcal{B}) &= \mathrm{ann}_{\mathcal{B}}((s-\lambda)^{m(\lambda)}) = \left\{w \in \mathcal{F}^l;\ \begin{pmatrix} R \\ (s-\lambda)^{m(\lambda)}\,\mathrm{id}_l \end{pmatrix} \circ w = 0\right\} \\ &\cong \mathrm{Hom}_{\mathcal{D}}(\mathrm{t}_{s-\lambda}(M), \mathcal{F}). \end{aligned}$$
6. *The following equivalences hold:*
   $$\begin{aligned} \mathrm{t}_{s-\lambda}(\mathcal{B}) \neq 0 &\iff \lambda \in \mathrm{char}(\mathcal{B}) \iff m(\lambda) = \mathrm{index}(\mathcal{B}, s-\lambda) > 0 \\ &\iff \mathrm{ann}_{\mathcal{B}}(s-\lambda) \neq 0 \iff \exists w \in \mathcal{B} \text{ with } w \neq 0,\ (s-\lambda)\circ w = 0. \end{aligned}$$
7. $\mathcal{B} = \bigoplus_{\lambda\in\mathrm{char}(\mathcal{B})} \mathrm{t}_{s-\lambda}(\mathcal{B})$.
8. *For the standard signal spaces, the identity* $\dim_{\mathbb{C}}(M) = \dim_{\mathbb{C}}(\mathcal{B})$ *implies*
   $$\mathrm{l}(\mathcal{B}) = \mathrm{l}(M) \in \mathbb{N}^{(\mathbb{C})} \quad and \quad \mathrm{l}_{s-\lambda}(\mathcal{B}) = \mathrm{mult}(\mathcal{B}, s-\lambda) = \dim_{\mathbb{C}}(\mathrm{t}_{s-\lambda}(\mathcal{B})).$$

The elements in $\mathrm{t}_{s-\lambda}(\mathcal{B})$ *are called the* λ-modes *or* modes *of* $\mathcal{B}$ *for the characteristic value or natural complex frequency* λ*, cf. [5, Sect. 2.5.2].* ◇

Corollary and Definition 5.4.9 *Let* $\mathcal{D} := \mathbb{C}[s]$*. Assume a not necessarily autonomous behavior* $\mathcal{B} = \{w \in \mathcal{F}^l;\ R\circ w = 0\}$ *with* $R \in \mathcal{D}^{k\times l}$ *and system module* $M = \mathcal{D}^{1\times l}/\mathcal{D}^{1\times k}R$*. The behavior* $\mathcal{B}$ *has the controllable subbehavior* $\mathcal{B}_{\mathrm{cont}}$ *and autonomous factor behavior* $\mathcal{B}_{\mathrm{uncont}} \cong \mathcal{B}/\mathcal{B}_{\mathrm{cont}} \cong \mathrm{D}(\mathrm{t}(M))$*. According to Corollary and Definition 5.3.18 the length vector is defined by means of* $\mathrm{t}(M)$*:*

$$\mathrm{l}(\mathcal{B}) := (\mathrm{l}_{s-\lambda}(\mathcal{B}))_{\lambda\in\mathbb{C}} := \mathrm{l}(M) := \mathrm{l}(\mathrm{t}(M)),\ \mathrm{l}_{s-\lambda}(\mathrm{t}(M)) := \dim_{\mathbb{C}}\left(\mathrm{t}_{s-\lambda}(M)\right).$$

For the real case $F = \mathbb{R}$ *and* $\mathcal{D} = \mathbb{R}[s]$*, we will determine the length* $\mathrm{l}(M)$ *in Corollary 5.4.12.* ◇

Theorem 5.4.10 (Cf. [6, Lem. 559 on p. 500]) *Assume* $F := \mathbb{C}$, $\mathcal{D} := \mathbb{C}[s]$ *and* $\mathcal{F} := \mathrm{C}^\infty(\mathbb{R}, \mathbb{C})$ *or* $\mathcal{F} := \mathbb{C}^\mathbb{N}$. *For an exact sequence*

$$0 \longrightarrow M_1 \longrightarrow M_2 \longrightarrow M_3 \longrightarrow 0$$

of finitely generated torsion $\mathcal{D}$*-modules, the identity* $\mathrm{l}(M_2) = \mathrm{l}(M_1) + \mathrm{l}(M_3)$ *holds, and this implies that* $\mathrm{char}(M_2) = \mathrm{char}(M_1) \cup \mathrm{char}(M_3)$. *Let*

$$0 \longleftarrow \mathcal{B}_1 \longleftarrow \mathcal{B}_2 \longleftarrow \mathcal{B}_3 \longleftarrow 0$$

be the dual exact sequence of the autonomous behaviors $\mathcal{B}_i \cong \mathrm{Hom}_\mathcal{D}(M_i, \mathcal{F})$ *with the characteristic varieties* $\mathrm{char}(\mathcal{B}_i) = \mathrm{char}(M_i) = \mathrm{supp}(\mathrm{l}(M_i)) = \mathrm{char}(M_i)$. *Then*

$$\mathrm{l}(\mathcal{B}_2) = \mathrm{l}(\mathcal{B}_1) + \mathrm{l}(\mathcal{B}_3), \quad \textit{where } \mathrm{l}_{s-\lambda}(\mathcal{B}_i) = \dim_\mathbb{C}(\mathrm{t}_{s-\lambda}(\mathcal{B}_i)).$$

Proof This follows directly from Theorem 5.3.17. □

5.4.1.3 Fundamental Systems of Autonomous Behaviors

Next we compute an F-basis or *fundamental system* of the autonomous behavior $\mathcal{B}$ from Theorem 5.4.8 for $\mathcal{D} = F[s]$ and $\mathcal{F} = \mathrm{C}^\infty(\mathbb{R}, F)$ or $\mathcal{F} = F^\mathbb{N}$. We start with the complex case $F = \mathbb{C}$, cf. Theorem 5.4.11, and treat the real case $F = \mathbb{R}$ afterward in Corollary 5.4.14. For the latter, we need also the analogous results of Theorem 5.4.8 and 5.4.10 for the real numbers, which we formulate in Corollaries 5.4.12 and 5.4.13. Since $\mathcal{B} = \bigoplus_{\lambda \in \mathrm{char}(\mathcal{B})} \mathrm{t}_{s-\lambda}(\mathcal{B})$, it suffices to compute a basis of

$$\begin{aligned} \mathrm{t}_{s-\lambda}(\mathcal{B}) &= \left\{w \in \mathcal{F}^l;\ R \circ w = 0,\ (s-\lambda)^{m(\lambda)} \circ w = 0\right\} \\ &= \left\{w \in \mathrm{ann}_\mathcal{F}\left((s-\lambda)^{m(\lambda)}\right)^l;\ R \circ w = 0\right\}. \end{aligned} \tag{5.55}$$

Recall from Theorem 5.2.2 and Lemma and Definition 5.2.7 that

$$\begin{aligned} &\mathrm{ann}_\mathcal{F}((s-\lambda)^{m(\lambda)}) = \bigoplus_{i=0}^{m(\lambda)-1} \mathbb{C}e_{\lambda,i}, \quad \text{where} \\ &e_{\lambda,i}(t) = \begin{cases} \frac{t^i}{i!}e^{\lambda t} & \text{if } \mathcal{F} = \mathrm{C}^\infty(\mathbb{R}, \mathbb{C}), \\ \binom{t}{i}\lambda^{t-i} & \text{if } \mathcal{F} = \mathbb{C}^\mathbb{N},\ \lambda \neq 0 \text{ with} \\ \delta_{i,t} & \text{if } \mathcal{F} = \mathbb{C}^\mathbb{N},\ \lambda = 0 \end{cases} \end{aligned} \tag{5.56}$$

$$(s-\lambda)^j \circ e_{\lambda,i} = \begin{cases} e_{\lambda,i-j} & \text{if } j \leq i, \\ 0 & \text{if } j > i. \end{cases} \tag{5.57}$$

Hence we have that $\mathrm{t}_{s-\lambda}(\mathcal{B}) \subseteq \bigoplus_{i=0}^{m(\lambda)-1} \mathbb{C}^l e_{\lambda,i}$. By (5.55) and (5.56), every mode $w \in \mathrm{t}_{s-\lambda}(\mathcal{B})$ admits a representation

$$w = \sum_{i=0}^{m(\lambda)-1} w_{\lambda,i} e_{\lambda,i} \quad \text{with } w_{\lambda,i} \in \mathbb{C}^l. \tag{5.58}$$

We write the matrix R as

$$R(s+\lambda) = \sum_{j=0}^{\infty} R_{\lambda,j} s^j \quad \text{with } R_{\lambda,j} \in \mathbb{C}^{k\times l}, \text{ hence } R(s) = \sum_{j=0}^{\infty} R_{\lambda,j}(s-\lambda)^j. \tag{5.59}$$

With the matrices $R_{\lambda,j}$ we form the block matrix

$$A_\lambda := \begin{pmatrix} R_{\lambda,0} & R_{\lambda,1} & R_{\lambda,2} & \cdots & R_{\lambda,m(\lambda)-1} \\ 0 & R_{\lambda,0} & R_{\lambda,1} & \cdots & R_{\lambda,m(\lambda)-2} \\ 0 & 0 & R_{\lambda,0} & \cdots & R_{\lambda,m(\lambda)-3} \\ \vdots & \vdots & 0 & \ddots & \vdots \\ 0 & 0 & 0 & \cdots & R_{\lambda,0} \end{pmatrix} \in \mathbb{C}^{m(\lambda)k\times m(\lambda)l}. \tag{5.60}$$

Theorem 5.4.11 (Cf. [3, Theorem 3.2.16 on p. 77]) *For the autonomous behavior $\mathcal{B}$ from Theorem 5.4.8 and the matrices A_λ, $\lambda \in \mathrm{char}(\mathcal{B})$, from (5.60), the linear map*

$$\begin{aligned} \{\xi \in \left(\mathbb{C}^l\right)^{m(\lambda)} = \mathbb{C}^{m(\lambda)l};\ A_\lambda\xi = 0\} &\longrightarrow \mathrm{t}_{s-\lambda}(\mathcal{B}) \ (\subset \mathcal{F}^l) \\ w_\lambda := \begin{pmatrix} w_{\lambda,0} \\ \vdots \\ w_{\lambda,m(\lambda)-1} \end{pmatrix} &\longmapsto \sum_{i=0}^{m(\lambda)-1} w_{\lambda,i} e_{\lambda,i} \end{aligned} \tag{5.61}$$

is an isomorphism. Hence, if $n(\lambda) := m(\lambda)l - \mathrm{rank}(A_\lambda)$ and if $w_\lambda^{(1)}, \ldots, w_\lambda^{(n(\lambda))}$ is a basis of the solution space of A_λ, then, with $\mathrm{l}_{s-\lambda}(\mathcal{B}) = \dim_{\mathbb{C}}(\mathrm{t}_{s-\lambda}(\mathcal{B}))$,

$$\mathrm{l}_{s-\lambda}(\mathcal{B}) = n(\lambda) \text{ and } w^{(\nu)} := \sum_{i=0}^{m(\lambda)-1} w_{\lambda,i}^{(\nu)} e_{\lambda,i}, \quad \nu = 1, \ldots, n(\lambda),$$

is a basis or a fundamental system *of $\mathrm{t}_{s-\lambda}(\mathcal{B})$. The union of these bases over all $\lambda \in \mathrm{char}(\mathcal{B})$ is a basis of $\mathcal{B}$ and $\mathrm{l}(\mathcal{B}) = (n(\lambda))_{\lambda\in\mathbb{C}} \in \mathbb{N}^{(\mathbb{C})}$.*

Proof For $w = \sum_{i=0}^{m(\lambda)-1} w_{\lambda,i} e_{\lambda,i} \in \bigoplus_{i=0}^{m(\lambda)-1} \mathbb{C}^l e_{\lambda,i} = \mathrm{ann}_{\mathcal{F}}\left((s-\lambda)^{m(\lambda)}\right)^l$ from (5.58) define $w_\lambda := \begin{pmatrix} w_{\lambda,0} \\ \vdots \\ w_{\lambda,m(\lambda)-1} \end{pmatrix} \in \left(\mathbb{C}^l\right)^{m(\lambda)} = \mathbb{C}^{m(\lambda)l}$. Then

$$R \circ w = \sum_{j=0}^{\infty} R_{\lambda,j}(s-\lambda)^j \circ \sum_{i=0}^{m(\lambda)-1} w_{\lambda,i} e_{\lambda,i} = \sum_{j=0}^{\infty} \sum_{i=0}^{m(\lambda)-1} R_{\lambda,j} w_{\lambda,i} (s-\lambda)^j \circ e_{\lambda,i}$$

$$\overset{(5.57)}{=\!=} \sum_{0\le j\le i\le m(\lambda)-1} R_{\lambda,j} w_{\lambda,i} e_{\lambda,i-j} \overset{q:=i-j}{=\!=} \sum_{q=0}^{m(\lambda)-1}\left(\sum_{i=q}^{m(\lambda)-1} R_{\lambda,i-q} w_{\lambda,i}\right) e_{\lambda,q}$$
$$= \sum_{q=0}^{m(\lambda)-1}\left(R_{\lambda,0} w_{\lambda,q} + \cdots + R_{\lambda,m(\lambda)-1-q} w_{\lambda,m(\lambda)-1}\right) e_{\lambda,q}.$$

The isomorphy of (5.61) follows from this equation and the equivalences

$$R \circ w = 0 \underset{e_{\lambda,q}\ \text{linearly independent}}{\Longleftrightarrow}$$
$$\forall q = 0, \ldots, m(\lambda) - 1 :\ R_{\lambda,0} w_{\lambda,q} + \cdots + R_{\lambda,m(\lambda)-1-q} w_{\lambda,m(\lambda)-1} = 0$$
$$\Longleftrightarrow A_\lambda w_\lambda = \left(R_{\lambda,0} w_{\lambda,q} + \cdots + R_{\lambda,m(\lambda)-1-q} w_{\lambda,m(\lambda)-1}\right)_{0\le q\le m(\lambda)-1} = 0.$$

□

We draw slightly different conclusions for the real case $F = \mathbb{R}$, $\mathcal{D} = \mathbb{R}[s]$, where we use the representative system of irreducible polynomials

$$\mathcal{P} = \{p_\lambda;\ \lambda \in \Lambda_{\mathbb{R}}\} \text{ where } \Lambda_{\mathbb{R}} := \{\lambda \in \mathbb{C};\ \Im(\lambda) \ge 0\} \text{ and}$$
$$p_\lambda := \begin{cases} s - \lambda & \text{if } \lambda \in \mathbb{R}, \\ (s-\lambda)(s-\overline{\lambda}) = s^2 - 2\Re(\lambda)s + |\lambda|^2 & \text{if } \Im(\lambda) > 0. \end{cases} \tag{5.62}$$

The bijection $\Lambda_{\mathbb{R}} \overset{\cong}{\longrightarrow} \mathcal{P}$, $\lambda \longmapsto p_\lambda$, allows to identify $\mathbb{N}^{(\Lambda_{\mathbb{R}})} = \mathbb{N}^{(\mathcal{P})}$.

Corollary 5.4.12 *We assume that $F = \mathbb{R}$, the data from (5.53), hence $\mathcal{B} = \bigoplus_{\lambda\in\Lambda_{\mathbb{R}}} \mathrm{t}_{p_\lambda}(\mathcal{B})$, and (5.62). Then*

$$e_l = \prod_{\lambda\in\Lambda_{\mathbb{R}}} p_\lambda^{m(\lambda)} \ \ \textit{with} \ \begin{cases} m(\lambda) > 0 & \textit{if } \lambda \in \mathrm{char}(M) \\ m(\lambda) = 0 & \textit{if } \lambda \notin \mathrm{char}(M) \end{cases}.$$

The characteristic variety is

$$\mathrm{char}(M) = \mathrm{char}(\mathcal{B}) = \mathrm{V}_{\mathbb{C}}(e_l) = \left(\mathbb{R} \cap \mathrm{char}(M)\right) \bigoplus \left\{\lambda, \overline{\lambda};\ \lambda \in \mathrm{char}(M),\ \Im(\lambda) > 0\right\}.$$

Again, the results from Theorems 5.3.13 to 5.3.17 imply that for all $\lambda \in \Lambda_{\mathbb{R}}$ the identity

$$\mathrm{index}(M, p_\lambda) = \mathrm{index}(\mathcal{B}, p_\lambda) = m(\lambda) \begin{cases} > 0 & \textit{if } \lambda \in \mathrm{char}(M), \\ = 0 & \textit{if } \lambda \notin \mathrm{char}(M) \end{cases}$$

and the isomorphisms

$$\mathrm{t}_{p_\lambda}(M) \cong M/p_\lambda^{m(\lambda)}M = \mathcal{D}^{1\times l}/\mathcal{D}^{1\times(k+l)}\begin{pmatrix} R \\ p_\lambda^{m(\lambda)}\mathrm{id}_l \end{pmatrix} \quad \textit{and}$$
$$\mathrm{t}_{p_\lambda}(\mathcal{B}) \cong \mathrm{Hom}_{\mathcal{D}}(\mathrm{t}_{p_\lambda}(M), \mathcal{F})$$

hold. The length vector of M *is*

$$\mathrm{l}(M) = \left(\mathrm{l}_{p_\lambda}(M)\right)_{\lambda\in\Lambda_\mathbb{R}} \in \mathbb{N}^{(\Lambda_\mathbb{R})} \quad \textit{with}$$
$$\mathrm{l}_{p_\lambda}(M) = \mathrm{mult}(M, p_\lambda) = \begin{cases} \dim_\mathbb{R}(\mathrm{t}_{s-\lambda}(M)) & \textit{if } \lambda \in \mathbb{R}, \\ \frac{1}{2}\dim_\mathbb{R}(\mathrm{t}_{p_\lambda}(M)) & \textit{if } \Im(\lambda) > 0. \end{cases}$$

For the signal spaces $\mathrm{C}^\infty(\mathbb{R}, \mathbb{R})$ *and* $\mathbb{R}^\mathbb{N}$ *with* $\dim_\mathbb{R}(M) = \dim_\mathbb{R}(\mathcal{B})$ *we obtain*

$$\mathrm{l}(\mathcal{B}) = \mathrm{l}(M) \in \mathbb{N}^{(\Lambda_\mathbb{R})}, \quad \textit{i.e.,} \quad \mathrm{l}_{p_\lambda}(\mathcal{B}) = \mathrm{l}_{p_\lambda}(M) \textit{ for all } \lambda \in \Lambda_\mathbb{R}. \qquad \diamondsuit$$

Corollary 5.4.13 *Under the assumptions of Corollary 5.4.12 let*

$$0 \longrightarrow M_1 \longrightarrow M_2 \longrightarrow M_3 \longrightarrow 0$$

be an exact sequence of finitely generated torsion $\mathbb{R}[s]$*-modules* M_i *with their dual autonomous behaviors* $\mathcal{B}_i$*, length vectors* $\mathrm{l}(M_i) \in \mathbb{N}^{(\Lambda_\mathbb{R})}$ *with* $\mathrm{l}_\lambda(M_i) = \frac{\dim_\mathbb{R}(\mathrm{t}_{p_\lambda}(M_i))}{\deg(p_\lambda)}$ *and characteristic varieties* $\mathrm{char}(M_i) = \mathrm{char}(\mathcal{B}_i) = \{\lambda, \overline{\lambda};\ \lambda \in \Lambda_\mathbb{R},\ \mathrm{t}_{p_\lambda}(M_i) \neq 0\}$*. Then the following equations hold:*

$$\mathrm{l}(M_2) = \mathrm{l}(M_1) + \mathrm{l}(M_3) \textit{ and } \mathrm{char}(M_2) = \mathrm{char}(M_1) \cup \mathrm{char}(M_3).$$

For the signal spaces $\mathrm{C}^\infty(\mathbb{R}, \mathbb{R})$ *and* $\mathbb{R}^\mathbb{N}$ *with* $\dim_\mathbb{R}(M_i) = \dim_\mathbb{R}(\mathcal{B}_i)$ *also* $\mathrm{l}(\mathcal{B}_2) = \mathrm{l}(\mathcal{B}_1) + \mathrm{l}(\mathcal{B}_3)$ *holds.*

Proof The proof is analogous to that of Theorem 5.4.10. □

Real signals and behaviors are most easily treated by means of *complexification*. We will use this method to compute an $\mathbb{R}$-basis for an autonomous behavior and the signal space $\mathrm{C}^\infty(\mathbb{R}, \mathbb{R})$ and $\mathbb{R}^\mathbb{N}$.

The complexification of a real vector space ${}_\mathbb{R}U$ is the complex space

$$\begin{aligned} &\mathbb{C} \otimes_\mathbb{R} U := U_\mathbb{C} := U \times U \text{ with } \mathbb{C} = \mathbb{R} \oplus j\mathbb{R} \ni a + jb,\ j := \sqrt{-1}, \\ &(a + jb)(u, v) := (au - bv, av + bu) \quad \text{for } a, b \in \mathbb{R} \text{ and } u, v \in U. \end{aligned} \tag{5.63}$$

With the identification $U = U \times 0 \ni u = (u, 0)$, we obtain $jU = 0 \times U \ni ju = j(u, 0) = (0, u)$ and $\mathbb{C} \otimes_\mathbb{R} U = U \oplus jU$. If $w = u + jv \in \mathbb{C} \otimes_\mathbb{R} U = U \oplus jU$, then

$$\overline{w} := u - jv,\ \Re(w) := u = (w + \overline{w})/2,\ \Im(w) := v = (w - \overline{w})/(2j), \tag{5.64}$$

are called the complex conjugate, the real part, and the imaginary part of w, respectively. Every $\mathbb{R}$-basis of U is also a $\mathbb{C}$-basis of $\mathbb{C}\otimes_{\mathbb{R}} U = U\oplus jU$, hence $\dim_{\mathbb{C}}(\mathbb{C}\otimes_{\mathbb{R}} U) = \dim_{\mathbb{R}}(U)$. The projection $\Re:\ \mathbb{C}\otimes_{\mathbb{R}} U \longrightarrow U,\ w\longmapsto \Re(w)$, is an $\mathbb{R}$-epimorphism. The complexification $U\longmapsto \mathbb{C}\otimes_{\mathbb{R}} U$ is an exact additive functor from the category of $\mathbb{R}$-spaces to the category of $\mathbb{C}$-spaces.

Additional algebraic structures of U are canonically transferred to the complexification. If, for instance, A is an $\mathbb{R}$-algebra and M is an A-module, then $\mathbb{C}\otimes_{\mathbb{R}} A$ is a $\mathbb{C}$-algebra and $\mathbb{C}\otimes_{\mathbb{R}} M$ is a $(\mathbb{C}\otimes_{\mathbb{R}} A)$-module. We naturally identify

$$\begin{aligned}&\mathbb{C}\otimes_{\mathbb{R}}\mathbb{R}[s]=\mathbb{C}[s]\quad\text{and}\\ &\mathbb{C}\otimes_{\mathbb{R}}\mathcal{F}_{\mathbb{R}}=\mathcal{F}_{\mathbb{C}},\quad\text{where } \mathcal{F}_F=\begin{cases}C^\infty(\mathbb{R},F) & \text{continuous-time case}\\ F^{\mathbb{N}} & \text{discrete-time case.}\end{cases}\end{aligned}\tag{5.65}$$

In the sequel we assume these signal modules $\mathcal{F}_F$. The matrix $R\in\mathbb{R}[s]^{k\times l}$ gives rise to the real and the complex behaviors

$$\mathcal{B}_F := \{w\in\mathcal{F}_F^l;\ R\circ w=0\},\quad F=\mathbb{R},\mathbb{C}.\tag{5.66}$$

The behavior $\mathcal{B}_{\mathbb{C}}$ is the complexification of $\mathcal{B}_{\mathbb{R}}$, i.e.,

$$\begin{aligned}&\mathcal{B}_{\mathbb{C}}=\mathbb{C}\otimes_{\mathbb{R}}\mathcal{B}_{\mathbb{R}}=\{w=u+jv\in\mathcal{F}_{\mathbb{C}}^l=\mathbb{C}\otimes_{\mathbb{R}}\mathcal{F}_{\mathbb{R}}^l;\ R\circ u=R\circ v=0\},\ \text{hence}\\ &\dim_{\mathbb{R}}(\mathcal{B}_{\mathbb{R}})=\dim_{\mathbb{C}}(\mathcal{B}_{\mathbb{C}}).\end{aligned}$$

For $\lambda\in\mathbb{R}\cap\operatorname{char}(\mathcal{B})$, we can compute an $\mathbb{R}$-basis of $\mathrm{t}_{s-\lambda}(\mathcal{B})$ in the same way as in the complex case from Theorem 5.4.11. It remains to compute $\mathrm{t}_{p_\lambda}(\mathcal{B})$ for

$$\begin{aligned}&\lambda=\alpha+j\beta=|\lambda|e^{j\omega}\in\operatorname{char}(\mathcal{B}),\ \text{where } \beta>0 \text{ and } 0<\omega<\pi,\quad\text{and}\\ &p_\lambda=(s-\lambda)(s-\overline{\lambda})=s^2-2\alpha s+\alpha^2+\beta^2.\end{aligned}\tag{5.67}$$

Using Theorem 5.4.11, we compute a basis of $\mathrm{t}_{s-\lambda}(\mathcal{B}_{\mathbb{C}})$

$$\begin{aligned}&w^{(\nu)}=\sum_{k=0}^{m(\lambda)-1} w_{\lambda,k}^{(\nu)}e_{\lambda,k},\ \nu=1,\dots,n(\lambda),\ \text{hence}\\ &\mathrm{t}_{s-\lambda}(\mathcal{B}_{\mathbb{C}})=\left\{w\in\mathcal{F}_{\mathbb{C}}^l;\ R\circ w=(s-\lambda)^{m(\lambda)}\circ w=0\right\}=\bigoplus_{\nu=1}^{n(\lambda)}\mathbb{C}w^{(\nu)},\end{aligned}\tag{5.68}$$

where the $w_{\lambda,k}^{(\nu)}\in\mathbb{C}^l$ are specified in Theorem 5.4.11 and

$$e_{\lambda,k}=\begin{cases}\frac{t^k}{k!}e^{\lambda t}=\frac{t^k}{k!}e^{\alpha t}e^{j\beta t} & \text{continuous-time case}\\ \binom{t}{k}\lambda^{t-k}=\binom{t}{k}|\lambda|^{t-k}e^{j\omega(t-k)} & \text{discrete-time case.}\end{cases}\tag{5.69}$$

Corollary 5.4.14 *With the data from (5.67) and (5.68), the p_λ-component $\mathrm{t}_{p_\lambda}(\mathcal{B}_\mathbb{R}) = \mathrm{ann}_{\mathcal{B}_\mathbb{R}}(p_\lambda^{m(\lambda)})$ of $\mathcal{B}_\mathbb{R}$ has the $\mathbb{R}$-basis $\Re(w^{(\nu)})$, $\Im(w^{(\nu)})$, $\nu = 1, \ldots, n(\lambda)$. The basis vectors are obtained as*

$$\Re(w^{(\nu)}) = \sum_{k=0}^{m(\lambda)-1} \Re\big(w_{\lambda,k}^{(\nu)} e_{\lambda,k}\big) = \sum_{k=0}^{m(\lambda)-1} \Re(w_{\lambda,k}^{(\nu)})\Re(e_{\lambda,k}) - \Im(w_{\lambda,k}^{(\nu)})\Im(e_{\lambda,k}),$$

$$\Im(w^{(\nu)}) = \sum_{k=0}^{m(\lambda)-1} \Im\big(w_{\lambda,k}^{(\nu)} e_{\lambda,k}\big) = \sum_{k=0}^{m(\lambda)-1} \Re(w_{\lambda,k}^{(\nu)})\Im(e_{\lambda,k}) + \Im(w_{\lambda,k}^{(\nu)})\Re(e_{\lambda,k}).$$

In the continuous-time case, we have

$$\Re(e_{\lambda,k}) = \Re\big(\tfrac{t^k}{k!} e^{\lambda t}\big) = \tfrac{t^k}{k!} e^{\alpha t}\cos(\beta t) \quad \textit{and}$$
$$\Im(e_{\lambda,k}) = \Im\big(\tfrac{t^k}{k!} e^{\lambda t}\big) = \tfrac{t^k}{k!} e^{\alpha t}\sin(\beta t), \quad \textit{where } \lambda = \alpha + j\beta,$$

and in the discrete-time case, we have

$$\Re(e_{\lambda,j}) = \Re\big(\tbinom{t}{k}\lambda^{t-k}\big) = \tbinom{t}{k}|\lambda|^{t-k}\cos(\omega(t-k)) \quad \textit{and}$$
$$\Im(e_{\lambda,k}) = \Im\big(\tbinom{t}{k}\lambda^{t-k}\big) = \tbinom{t}{k}|\lambda|^{t-k}\sin(\omega(t-k)), \quad \textit{where } \lambda = |\lambda| e^{j\omega}.$$

Proof The polynomial $p_\lambda^{m(\lambda)} = (s-\lambda)^{m(\lambda)}(s-\overline{\lambda})^{m(\lambda)}$ has the two coprime factors $(s-\lambda)^{m(\lambda)}$ and $(s-\overline{\lambda})^{m(\lambda)}$ in $\mathbb{C}[s]$. Therefore, $\mathrm{t}_{p_\lambda}(\mathcal{B}_\mathbb{R}) = \mathrm{ann}_{\mathcal{B}_\mathbb{R}}(p_\lambda^{m(\lambda)})$ and the Chinese remainder theorem imply

$$\begin{aligned}\mathbb{C}\otimes_\mathbb{R} \mathrm{t}_{p_\lambda}(\mathcal{B}_\mathbb{R}) &= \mathrm{ann}_{\mathcal{B}_\mathbb{C}}(p_\lambda^{m(\lambda)})\\ &\underset{\text{Corollary 5.1.6}}{=} \mathrm{ann}_{\mathcal{B}_\mathbb{C}}\big((s-\lambda)^{m(\lambda)}\big) \oplus \mathrm{ann}_{\mathcal{B}_\mathbb{C}}\big((s-\overline{\lambda})^{m(\lambda)}\big).\end{aligned}$$

Since R is a matrix with real entries, the complex conjugation induces the isomorphisms

$$\mathcal{B}_\mathbb{C} \xrightarrow{\cong} \mathcal{B}_\mathbb{C} \quad \text{and}$$
$$\mathrm{ann}_{\mathcal{B}_\mathbb{C}}\big((s-\lambda)^{m(\lambda)}\big) = \bigoplus_{\nu=1}^{n(\lambda)} \mathbb{C}w^{(\nu)} \xrightarrow{\cong} \mathrm{ann}_{\mathcal{B}_\mathbb{C}}\big((s-\overline{\lambda})^{m(\lambda)}\big),$$
$$w \longmapsto \overline{w}.$$

This implies

$$\mathrm{ann}_{\mathcal{B}_\mathbb{C}}\big((s-\overline{\lambda})^{m(\lambda)}\big) = \bigoplus_{\nu=1}^{n(\lambda)} \mathbb{C}\overline{w^{(\nu)}} \text{ and } \mathrm{ann}_{\mathcal{B}_\mathbb{C}}(p_\lambda^{m(\lambda)}) = \bigoplus_{\nu=1}^{n(\lambda)} \Big(\mathbb{C}w^{\nu} \oplus \mathbb{C}\overline{w^{(\nu)}}\Big).$$

With $\Re(w^{(\nu)}) = \frac{1}{2}\left(w^{(\nu)} + \overline{w^{(\nu)}}\right)$ and $\Im(w^{(\nu)}) = \frac{1}{2j}\left(w^{(\nu)} - \overline{w^{(\nu)}}\right)$ we get

$$\begin{aligned}
\mathbb{C}w^{(\nu)} \oplus \mathbb{C}\overline{w^{(\nu)}} &= \mathbb{C}\Re(w^{(\nu)}) \oplus \mathbb{C}\Im(w^{(\nu)}),\\
\operatorname{ann}_{\mathcal{B}_{\mathbb{C}}}(p_\lambda^{m(\lambda)}) &= \bigoplus_{\nu=1}^{n(\lambda)} \left(\mathbb{C}\Re(w^{(\nu)}) \oplus \mathbb{C}\Im(w^{(\nu)})\right) \text{ and}\\
\operatorname{ann}_{\mathcal{B}_{\mathbb{R}}}(p_\lambda^{m(\lambda)}) &\supseteq \bigoplus_{\nu=1}^{n(\lambda)} \left(\mathbb{R}\Re(w^{(\nu)}) \oplus \mathbb{R}\Im(w^{(\nu)})\right).
\end{aligned}$$

Since $\dim_{\mathbb{R}}\left(\operatorname{ann}_{\mathcal{B}_{\mathbb{R}}}(p_\lambda^{m(\lambda)})\right) = \dim_{\mathbb{C}}\left(\operatorname{ann}_{\mathcal{B}_{\mathbb{C}}}(p_\lambda^{m(\lambda)})\right) = 2n(\lambda)$, the last inclusion is indeed an equality, and the $\Re(w^{(\nu)})$ and $\Im(w^{(\nu)})$ for $\nu = 1, \dots, n(\lambda)$ are an $\mathbb{R}$-basis of $\operatorname{ann}_{\mathcal{B}_{\mathbb{R}}}(p_\lambda^{m(\lambda)})$. □

5.4.2 *Stability of Autonomous Behaviors*

Stability of systems is one of the most important subjects in systems theory. In this section, we treat the stability of autonomous behaviors. Stability theory concerns the behavior of trajectories $w(t)$ for $t \to \infty$. Since every real trajectory can be considered as a complex one, it suffices to consider autonomous behaviors

$$\mathcal{B} \subseteq \mathcal{F}^l, \quad \text{with} \quad \mathcal{F} = \mathcal{F}_{\mathbb{C}} = \begin{cases} C^\infty(\mathbb{R}, \mathbb{C}) & \text{continuous-time case,} \\ \mathbb{C}^{\mathbb{N}} & \text{discrete-time case.} \end{cases}$$

The properties of the trajectories of a behavior for large times t will be deduced from the way the basis functions $e_{\lambda,k}(t)$ behave for large t. We summarize these properties in the following reminder.

Reminder 5.4.15 We discuss the behavior of the functions $e_{\lambda,k}(t)$ for $t \to \infty$ and therefore assume $t \geq 0$. This means that we are interested in the future behavior of a system, but not in the past.

Continuous-time case. We observe that $|e^{\lambda t}| = e^{\Re(\lambda)t}$ and thus $|e^{\lambda t}| = 1$ if $\Re(\lambda) = 0$. The behavior of the functions $e_{\lambda,k}(t) = \frac{t^k}{k!}e^{\lambda t}$ for large t is as follows:

$$\lim_{t\to\infty} |e_{\lambda,k}(t)| = \begin{cases} 0 & \text{if } \Re(\lambda) < 0, \\ 1 & \text{if } \Re(\lambda) = 0 \text{ and } k = 0, \\ \infty & \text{if } \Re(\lambda) = 0 \text{ and } k > 0, \\ \infty & \text{if } \Re(\lambda) > 0. \end{cases} \tag{5.70}$$

Discrete-time case. We observe that the binomial coefficient $\binom{t}{k}$ is a rational polynomial in t of degree k. The behavior of the sequences

$$e_{\lambda,k}(t) = \begin{cases} \delta_{t,k} & \text{if } \lambda = 0, \\ \binom{t}{k}\lambda^{t-k} & \text{if } \lambda \neq 0 \end{cases}$$

for large t is

$$\lim_{t\to\infty} |e_{\lambda,k}(t)| = \begin{cases} 0 & \text{if } |\lambda| < 1, \\ 1 & \text{if } |\lambda| = 1, \text{ and } k = 0, \\ \infty & \text{if } |\lambda| = 1, \text{ and } k > 0. \\ \infty & \text{if } |\lambda| > 1. \end{cases} \tag{5.71}$$

◇

In the following two theorems, we discuss asymptotic stability and stability of autonomous behaviors.

Theorem and Definition 5.4.16 (Compare [3, Theorem 7.2.2 on p. 245]) *For the injective cogenerators $\mathcal{F} = \mathrm{C}^\infty(\mathbb{R}, \mathbb{C})$ and $\mathcal{F} = \mathbb{C}^\mathbb{N}$ in the continuous- and discrete-time cases, respectively, and for the autonomous $\mathcal{F}$-behavior $\mathcal{B} = \mathcal{B}_\mathbb{C}$ from Theorems 5.4.8 and 5.4.11, the following properties are equivalent:*

1. *Every trajectory $w \in \mathcal{B}$ converges to 0 for $t \to \infty$. A behavior with this property is called* asymptotically stable.
2. *Every characteristic value $\lambda \in \mathrm{char}(\mathcal{B})$ of $\mathcal{B}$ satisfies $\Re(\lambda) < 0$ in the continuous-time case, and $|\lambda| < 1$ in the discrete-time case.*

These equivalent conditions also hold for real behaviors since every real trajectory is also a complex one.

Proof 2. $\Longrightarrow$ 1. According to Theorem 5.4.11, every trajectory $w \in \mathcal{B}$ admits a representation

$$w(t) = \sum_{\lambda\in\mathrm{char}(\mathcal{B})} \sum_{k=0}^{m(\lambda)-1} w_{\lambda,k} e_{\lambda,k}(t), \quad \text{with } w_{\lambda,k} \in \mathbb{C}^l.$$

Condition 2 and Eqs. (5.70) and (5.71) imply that all $e_{\lambda,k}(t)$ converge to 0 for $t \to \infty$. Hence, so does w.

non-2. $\Longrightarrow$ non-1. Let $\lambda \in \mathrm{char}(\mathcal{B})$ be such that Condition 2 is not satisfied. According to item 6 of Theorem 5.4.8, there is a nonzero trajectory $w \in \mathrm{t}_{s-\lambda}(\mathcal{B})$ that satisfies $(s - \lambda) \circ w = 0$ and therefore $w(t) = w(0)e_{\lambda,0}(t),\ 0 \neq w(0) \in \mathbb{C}^l$. Due to (5.70) and (5.71), the functions $e_{\lambda,0}(t)$ and hence also $w(t)$ do not converge to 0 and thus condition 1 is not satisfied. □

Theorem and Definition 5.4.17 (Compare [5, p. 177], [3, Theorem 7.2.2 on p. 245], [7, Theorem 3.6 on p. 563]) *For the injective cogenerators $\mathcal{F} = \mathrm{C}^\infty(\mathbb{R}, \mathbb{C})$ or $\mathcal{F} = \mathbb{C}^\mathbb{N}$ and for the autonomous $\mathcal{F}$-behavior $\mathcal{B}$ from Theorems 5.4.8 the following properties are equivalent:*

1. *Every trajectory* $w \in \mathcal{B}$ *is bounded for* $t \to \infty$, *in other words,*

$$\sup_{t \ge 0} \max\{|w_j(t)|;\ j = 1, \dots, l\} < \infty.$$

A behavior with this property is called stable.

2. *Every characteristic value* $\lambda \in \operatorname{char}(\mathcal{B})$ *with* $m(\lambda) = \operatorname{index}(\mathcal{B}, s - \lambda)$ *satisfies*

$$\begin{cases} 3\Re(\lambda) < 0 \textit{ or } \big(\Re(\lambda) = 0 \textit{ and } \operatorname{index}(\mathcal{B}, s - \lambda) = 1\big) & \textit{(continuous time)} \\ |\lambda| < 1 \textit{ or } \big(|\lambda| = 1 \textit{ and } \operatorname{index}(\mathcal{B}, s - \lambda) = 1\big) & \textit{(discrete time).} \end{cases}$$

The same equivalence is valid for real behaviors.

Proof 2. $\Longrightarrow$ 1. The proof is the same as that of the corresponding part of the preceding theorem.

non-2. $\Longrightarrow$ non-1. Assume that $\lambda \in \operatorname{char}(\mathcal{B})$ does not satisfy the conditions of 2. We distinguish two cases.

$\Re(\lambda) > 0$ or $|\lambda| > 1$, respectively. Again, there is a nonzero trajectory in $\mathcal{B}$ with $(s - \lambda) \circ w = 0$, and therefore $w(t) = w(0)e_{\lambda,0}(t)$ with $0 \ne w(0) \in \mathbb{C}^l$. Since $e_{\lambda,0}(t)$ is not bounded for $t \to \infty$, neither is w.

$\big(\Re(\lambda) = 0$ or $|\lambda| = 1$, respectively$\big)$ and $m(\lambda) = \operatorname{index}(\mathcal{B}, s - \lambda) \ge 2$. Since

$$\begin{aligned} 0 &\subsetneq \operatorname{ann}_{\mathcal{B}}(s - \lambda) \subsetneq \cdots \subsetneq \operatorname{ann}_{\mathcal{B}}\big((s - \lambda)^{m(\lambda)-1}\big) \\ &\subsetneq \operatorname{ann}_{\mathcal{B}}\big((s - \lambda)^{m(\lambda)}\big) = \mathrm{t}_{s-\lambda}(\mathcal{B}) \end{aligned}$$

and $m(\lambda) \ge 2$, there is a trajectory $w \in \mathrm{t}_{s-\lambda}(\mathcal{B}) \setminus \operatorname{ann}_{\mathcal{B}}\big((s - \lambda)^{m(\lambda)-1}\big)$, and this trajectory can be written as $w = \sum_{k=0}^{m(\lambda)-1} w_{\lambda,k} e_{\lambda,k}$ with $w_{\lambda,k} \in \mathbb{C}^l$. The properties (5.56) imply that

$$\begin{aligned} &0 \ne (s - \lambda)^{m(\lambda)-1} \circ w = w_{\lambda,m(\lambda)-1} e_{\lambda,0}, \quad \text{and} \\ &0 \ne (s - \lambda)^{m(\lambda)-2} \circ w = w_{\lambda,m(\lambda)-2} e_{\lambda,0} + w_{\lambda,m(\lambda)-1} e_{\lambda,1} \in \mathrm{t}_{s-\lambda}(\mathcal{B}). \end{aligned}$$

From the first line, we infer that $w_{\lambda,m(\lambda)-1} \ne 0$. The trajectory $(s - \lambda)^{m(\lambda)-2} \circ w$ lies in $\mathrm{t}_{s-\lambda}(\mathcal{B})$ and thus in $\mathcal{B}$, because those are $\mathbb{C}[s]$-modules. Since $e_{\lambda,0}$ is bounded, the first summand in the second line is bounded too, whereas $e_{\lambda,1}(t)$ is not bounded and has the nonzero coefficient $w_{\lambda,m(\lambda)-1}$. This implies that $(s - \lambda)^{m(\lambda)-2} \circ w$ is a trajectory of $\mathcal{B}$ which is not bounded. Hence, Condition 1 is not satisfied. □

5.5 The Jordan Decomposition and the Partial Fraction Decomposition

In this section we prove the decompositions of the title with the Chinese remainder theorem.

5.5.1 The Jordan Decomposition for Matrices

The CRT furnishes a constructive proof for the Jordan decomposition of a matrix $A \in \mathbb{C}^{n\times n}$, *cf. Theorem and Definition* 5.5.2. *See, e.g.,* [2, Theorem 2.5 and Corollary 2.5 on pp. 558, 559] *for the usual proof of this result. In Corollaries* 5.5.3 *and* 5.5.4 *we compute the powers* $A^t,\ t \in \mathbb{N}$, *and the exponentials* $e^{tA},\ t \in \mathbb{R}$, *of the matrix, and solve the state space equations* $x(t+1) = Ax(t)$ *resp.* $x' = Ax$.

Consider a matrix $A \in F^{n\times n}$ over a field F with its characteristic polynomial $\chi_A := \det(s\,\mathrm{id}_n - A)$, its minimal polynomial f_A, and the finite-dimensional $F[s]$-module $(F^n, {}_A\circ)$ with $s\ {}_A\circ\,\eta = A\eta$. Then

$$\begin{aligned} F[s]f_A &= \{f \in F[s];\ f(A) = 0\} \\ &= \{f \in F[s];\ f(A)F^n = f\ {}_A\circ\, F^n = 0\} = \mathrm{ann}_{F[s]}(F^n, {}_A\circ), \end{aligned} \tag{5.72}$$

in particular, $f_A\ {}_A\circ\, F^n = 0$ and

$$F[s]/F[s]f_A \xrightarrow{\cong} F[A],\quad \overline{g} \longmapsto g(A). \tag{5.73}$$

The next corollary complements Theorem 4.2.11.

Corollary 5.5.1 *Let* $e_1|\cdots|e_n$ *be the* n *nonzero monic similarity invariants of* A, *i.e., the elementary divisors of the characteristic matrix* $s\,\mathrm{id}_n - A$. *Then*

$$f_A = e_n \quad \text{and} \quad \chi_A = \prod_{i=1}^{n} e_i,$$

hence, $f_A | \chi_A | f_A^n$. *In particular,* f_A *and* χ_A *have the same irreducible factors, however, with different multiplicities in general.*

Proof Since $\mathrm{rank}(s\,\mathrm{id}_n - A) = n$, the matrix has n nonzero elementary divisors $e_1, \ldots, e_n$, and the Smith form is

$$U(s\,\mathrm{id}_n - A)V = \mathrm{diag}(e_1, \ldots, e_n) \quad \text{with } U, V \in \mathrm{Gl}_n(F[s]),$$

i.e., $\det(U),\ \det(V) \in F \setminus \{0\}$. By taking determinants, we obtain

$$\det(U)\det(V)\chi_A = e_1 \cdots e_n,$$

and from this follows $\chi_A = e_1 \cdots e_n$ since χ_A and the e_i are monic. □

For the Jordan decomposition we do not assume that the field F is algebraically closed, but only that χ_A splits over F, i.e.,

$$\chi_A = \prod_{i=1}^{r}(s - \lambda_i)^{n(i)}, \; n := \deg(\chi_A) = \sum_{i=1}^{r} n(i), \tag{5.74}$$

with pairwise distinct $\lambda_i \in F$ and $n(i) > 0$. The λ_i are the r different eigenvalues of A. From $f_A | \chi_A | f_A^n$, we infer that

$$f_A = \prod_{i=1}^{r}(s - \lambda_i)^{m(i)}, \quad \text{with } 0 < m(i) \leq n(i). \tag{5.75}$$

We also use the additional polynomials from (5.12), i.e.,

$$1 = a_i + b_i = f_{i,1}(s - \lambda_i)^{n(i)} + f_{i,2}\frac{f}{(s - \lambda_i)^{n(i)}}. \tag{5.76}$$

For every $i = 1, \ldots, r$ and every $j \in \mathbb{N}$, we have

$$\begin{aligned} \operatorname{ann}_{F^n}\left((s - \lambda_i)^j\right) &= \left\{x \in F^n;\; (s - \lambda_i)^j \;{}_A\!\circ x = (A - \lambda_i \operatorname{id}_n)^j x = 0\right\} \\ &= \ker\left((A - \lambda_i \operatorname{id}_n)^j \cdot\right) \\ &\subseteq V_{\lambda_i} := \ker\left((A - \lambda_i \operatorname{id}_n)^\infty \cdot\right) := \operatorname{ann}_{F^n}\left((s - \lambda_i)^\infty\right). \end{aligned}$$

For $j = 1$ the submodule $\operatorname{ann}_{F^n}(s - \lambda_i) = \ker\left((A - \lambda_i \operatorname{id}_n)\cdot\right)$ is the eigenspace of A for the eigenvalue λ_i. Therefore $\ker\left((A - \lambda_i \operatorname{id}_n)^\infty \cdot\right)$ is called the *generalized eigenspace* of A for the eigenvalue λ_i. Since F^n has finite F-dimension, the indices $\operatorname{index}(F^n, s - \lambda_i)$ are finite.

Now we apply Corollary 5.1.6 to the module $(F^n, {}_A\circ)$ and f_A.

Theorem and Definition 5.5.2 (Jordan decomposition) *Consider a matrix* $A \in F^{n \times n}$ *with the derived data from* (5.72), (5.74), (5.75). *For every eigenvalue* λ_i *of* A, *let*

$$V_{\lambda_i} := \operatorname{ann}_{F^n}\left((s - \lambda_i)^\infty\right) = \{x \in F^n;\; \exists j \in \mathbb{N} \text{ with } (A - \lambda_i \operatorname{id}_n)^j x = 0\}$$

denote the generalized eigenspace.

1. *The identity* $\operatorname{index}(F^n, s - \lambda_i) = m(i)$ *holds for every eigenvalue* λ_i, *hence*

$$V_{\lambda_i} = \operatorname{ann}_{F^n}\left((s - \lambda_i)^\infty\right) = \operatorname{ann}_{F^n}\left((s - \lambda_i)^{m(i)}\right) = \ker\left((A - \lambda_i \operatorname{id}_n)^{m(i)} \cdot\right).$$

2. *The direct sum decomposition*

$$F^n = \bigoplus_{i=1}^{r} V_{\lambda_i} \ni x = \sum_{i=1}^{r} b_i \,{}_A\!\circ x = \sum_{i=1}^{r} b_i(A)x$$

of the $F[s]$-module $(F^n, {}_A\circ)$ into the generalized eigenspaces V_{λ_i} holds. This direct sum is called the Jordan decomposition *of $(F^n, {}_A\circ)$.*
With the complete orthogonal idempotents $\overline{b_i} \in F[s]/F[s]f_A$, we obtain the decomposition

$$\overline{s} = \sum_{i=1}^{r} \overline{sb_i} = \sum_{i=1}^{r} \left(\overline{\lambda_i b_i} + \overline{(s-\lambda_i)b_i}\right).$$

This, together with the isomorphism (5.73), *gives rise to the matrices*

$$\begin{aligned} P_i &:= b_i(A) \in F[A] \quad \textit{and} \\ N_i &:= \big((s-\lambda_i)b_i\big)(A) = (A - \lambda_i \operatorname{id}_n)P_i \in F[A] \end{aligned} \tag{5.77}$$

and the identities

$$\begin{aligned} &A = \sum_{i=1}^{r}(\lambda_i P_i + N_i), \ \sum_{i=1}^{r} P_i = \operatorname{id}_n, \ P_iP_j = \delta_{i,j}P_i, \\ &P_iN_j = \delta_{i,j}N_i, \ N_iN_j = 0 \textit{ for } i \neq j, \ N_i^{m(i)} = 0, \ \textit{but } N_i^{m(i)-1} \neq 0. \end{aligned} \tag{5.78}$$

Then $A = \sum_{i=1}^{r}(\lambda_i P_i + N_i)$ is called the Jordan decomposition *of A.*
Since $F[A]$ is commutative, all considered matrices commute with each other. The matrix $P_i := b_i(A)$ is the projection of F^n onto V_{λ_i} with respect to this decomposition. The matrix $N_i = (A - \lambda_i \operatorname{id}_n)P_i$ is nilpotent.

3. *The identity $\dim_F(V_{\lambda_i}) = n(i)$ holds for every eigenvalue λ_i. The dimensions $\dim_F\left(\ker\left((A-\lambda_i \operatorname{id}_n)\cdot\right)\right)$ resp. $n(i) = \dim_F(V_{\lambda_i})$ are called the* geometric *resp. the* algebraic multiplicity *of the eigenvalue λ_i, respectively.*

Proof Corollary 5.1.6 furnishes $F^n = \bigoplus_{i=1}^{r} \operatorname{ann}_{F^n}\left((s-\lambda_i)^{m(i)}\right)$. The equation

$$\operatorname{ann}_{F[s]}\big(\underbrace{\operatorname{ann}_{F^N}(f_A)}_{\underset{(5.72)}{=} F^n}\big) = \operatorname{ann}_{F[s]}(F^n) \overset{(5.72)}{=\!=} F[s]f_A$$

and Theorem 5.1.4, 3 imply

$$\operatorname{ann}_{F[s]}\left(\operatorname{ann}_{F^n}\left((s-\lambda_i)^{m(i)}\right)\right) = F[s](s-\lambda_i)^{m(i)}. \tag{5.79}$$

1. First we show that $k(i) := \operatorname{index}(F^n, s-\lambda_i) = m(i)$. An inequality $k(i) \leq m(i)$ implies $\operatorname{ann}_{F^n}\left((s-\lambda_i)^{k(i)}\right) = \operatorname{ann}_{F^n}\left((s-\lambda_i)^{m(i)}\right)$ and then

$$(s-\lambda_i)^{k(i)} \in \mathrm{ann}_{F[s]}\left(\mathrm{ann}_{F^n}\left((s-\lambda_i)^{k(i)}\right)\right)$$
$$= \mathrm{ann}_{F[s]}\left(\mathrm{ann}_{F^n}\left((s-\lambda_i)^{m(i)}\right)\right) \overset{(5.79)}{=} F[s](s-\lambda_i)^{m(i)} \Longrightarrow \forall i: \; m(i) \le k(i).$$

Define $g := \prod_{i=1}^{r}(s-\lambda_i)^{k(i)}$, hence $g \in F[s]f_A$. The equation $f_A \;{}_A\!\circ F^n = 0$ implies $g_A\!\circ F^n = 0$ and $\mathrm{ann}_{F^n}(g) = F^n$. We apply Corollary 5.1.6 to f_A and to g and obtain

$$\begin{aligned} F^n &= \mathrm{ann}_{F^n}(f_A) = \bigoplus_{i=1}^{r} \mathrm{ann}_{F^n}\left((s-\lambda_i)^{m(i)}\right) \\ &= \mathrm{ann}_{F^n}(g) = \bigoplus_{i=1}^{r} \mathrm{ann}_{F^n}\left((s-\lambda_i)^{k(i)}\right). \end{aligned} \tag{5.80}$$

From $m(i) \le k(i)$, we infer $\mathrm{ann}_{F^n}\left((s-\lambda_i)^{m(i)}\right) \subseteq \mathrm{ann}_{F^n}\left((s-\lambda_i)^{k(i)}\right)$, and with (5.80), even

$$\mathrm{ann}_{F^n}\left((s-\lambda_i)^{m(i)}\right) = \mathrm{ann}_{F^n}\left((s-\lambda_i)^{k(i)}\right) = V_{\lambda_i}.$$

This means that $m(i) \ge \mathrm{index}(F^n, s-\lambda_i) = k(i)$ and hence $m(i) = k(i)$.
2. Above we have already shown the direct sum decomposition $F^n = \bigoplus_{i=1}^{r} V_{\lambda_i}$. The $\overline{b_i} \in F[s]/F[s]f_A$, $i = 1,\ldots r$, are complete orthogonal idempotents. The isomorphism (5.73) implies the same property for the $P_i = b_i(A)$. The commutative algebra $F[A] \cong F[s]/F[s]f_A$ contains all $g(A)$, $g \in F[s]$, in particular, $N_i = ((s-\lambda_i)b_i)(A) = (A - \lambda_i\,\mathrm{id}_n)P_i$. We conclude that

$$P_iP_j = \delta_{ij}P_i,\; \sum_{i=1}^{r} P_i = \mathrm{id}_n,\; N_iP_j = P_jN_i = \delta_{ij}N_i,\; \text{and } \forall i \ne j:\; N_iN_j = 0.$$

We combine the isomorphism (5.73) with the CRT (Result 5.1.2) and obtain the isomorphisms

$$\begin{array}{ccccc} \prod_{i=1}^{r} F[s]/F[s](s-\lambda_i)^{m(i)} & \underset{\text{CRT}}{\overset{\cong}{\longleftrightarrow}} & F[s]/F[s]f_A & \underset{(5.73)}{\overset{\cong}{\longleftrightarrow}} & F[A], \\ (\overline{g_1},\ldots,\overline{g_r}) & \longmapsto & \sum_{i=1}^{r} \overline{b_i g_i},\;\; \overline{g} & \longmapsto & g(A). \end{array} \tag{5.81}$$

In particular, for $i = 1,\ldots,r$ and $1 \le k$, we obtain

$$\prod_{i=1}^{r} F[s]/F[s](s-\lambda_i)^{m(i)} \xrightarrow{\cong} F[A],$$
$$(0,\ldots,0,\overbrace{\overline{(s-\lambda_i)^k}}^{i},0,\ldots,0) \longmapsto (A-\lambda_i\,\mathrm{id}_n)^k b_i(A) = (A-\lambda_i\,\mathrm{id}_n)^k P_i$$
$$= (A-\lambda_i\,\mathrm{id}_n)^k P_i^k = N_i^k. \tag{5.82}$$

Finally

$$F[s]/F[s](s-\lambda_i)^{m(i)} \ni \overline{(s-\lambda_i)^k} = \begin{cases} \neq 0 & \text{if } k < m(i) \\ = 0 & \text{if } k \geq m(i) \end{cases}$$
$$\Longrightarrow N_i^k = \begin{cases} \neq 0 & \text{if } k < m(i) \\ = 0 & \text{if } k \geq m(i). \end{cases}$$

Notice that (5.82) is false for $k = 0$ due to $N_i^0 = \mathrm{id}_n$. This completes the proof of (5.78).

3. For $l(i) := \dim_F(V_{\lambda_i})$, we have to show that $l(i) = n(i)$.

Let $U_i \in F^{n\times l(i)}$ be a matrix whose columns are a basis of V_{λ_i}. Since $F^n = \bigoplus_{i=1}^r V_{\lambda_i}$, the columns of the block matrix $U := (U_1,\ldots,U_r)$ form a basis of F^n, i.e., $U \in \mathrm{Gl}_n(F)$. Since V_{λ_i} is an $F[s]$-submodule of F^n, we infer that $s\ {}_A{\circ}(U_i)_{-j} = A(U_i)_{-j} \in V_{\lambda_i} = U_i F^{l(i)}$, i.e., there is a matrix $A_i \in F^{l(i)\times l(i)}$ with $AU_i = U_i A_i$ and hence $AU = U\,\mathrm{diag}(A_1,\ldots,A_r)$. Therefore A and $\mathrm{diag}(A_1,\ldots,A_r)$ are similar and have the same characteristic polynomial $\chi_A = \prod_{i=1}^r \chi_{A_i}$. The equation $U^{-1}U = \mathrm{id}_n$ implies

$$E_i := U^{-1}U_i = (0,\ldots,0,\overset{i}{\mathrm{id}_{l(i)}},0,\ldots,0)^\top \in F^{n\times l(i)}.$$

From $V_{\lambda_i} = \ker\big((A-\lambda_i\,\mathrm{id}_n)^{m(i)}\cdot\big)$ we infer, for $i = 1,\ldots,r$, the identity $(A-\lambda_i\,\mathrm{id}_n)^{m(i)}(U_iF^{l(i)}) = 0$. Thus $(A-\lambda_i\,\mathrm{id}_n)^{m(i)}U_i = 0$ and

$$\begin{aligned} 0 &= U^{-1}(A-\lambda_i\,\mathrm{id}_n)^{m(i)}UU^{-1}U_i = (U^{-1}AU-\lambda_i\,\mathrm{id}_n)^{m(i)}E_i \\ &= \big(\mathrm{diag}(A_1,\ldots,A_r)-\lambda_i\,\mathrm{id}_n\big)^{m(i)}E_i \\ &= \mathrm{diag}\big((A_1-\lambda_i\,\mathrm{id}_{l(1)})^{m(i)},\ldots,(A_r-\lambda_i\,\mathrm{id}_{l(r)})^{m(i)}\big)(0,\ldots,\overset{i}{\mathrm{id}_{l(i)}},\ldots,0)^\top \\ &= \begin{pmatrix} 0 \\ \vdots \\ (A_i-\lambda_i\,\mathrm{id}_{l(i)})^{m(i)} \\ \vdots \\ 0 \end{pmatrix} \Longrightarrow (s-\lambda_i)^{m(i)}(A_i) = (A_i-\lambda_i\,\mathrm{id}_{l(i)})^{m(i)} = 0. \end{aligned}$$

Hence the minimal polynomial f_{A_i} of A_i divides $(s-\lambda_i)^{m(i)}$. Since, by Corollary 5.5.1, χ_{A_i} and f_{A_i} have the same irreducible factors, and since $\deg(\chi_{A_i}) = l(i)$, we conclude $\chi_{A_i} = (s-\lambda_i)^{l(i)}$ and therefore also

$$\prod_{i=1}^r (s-\lambda_i)^{n(i)} \underset{(5.74)}{=} \chi_A = \prod_{i=1}^r \chi_{A_i} = \prod_{i=1}^r (s-\lambda_i)^{l(i)} \Longrightarrow \forall i = 1,\ldots,r:\ l(i) = n(i).$$

□

Finally we use the Jordan decomposition for computing the powers A^t, $t \in \mathbb{N}$, and the matrix exponential e^{tA}, $t \in \mathbb{R}$. We apply this to autonomous discrete-time and continuous-time state space systems, cf. [3, pp. 131, 132].

Corollary 5.5.3 *Under the assumptions of the preceding theorem the powers A^t, $t \in \mathbb{N}$, of A can be computed as*

$$A^t = \sum_{i=1}^{r} \sum_{k=0}^{\min(t,m(i)-1)} e_{\lambda_i,k}(t) N_i^k P_i \underset{N_i^{m(i)}=0}{=} \sum_{i=1}^{r} \sum_{k=0}^{t} \binom{t}{k} \lambda_i^{t-k} N_i^k P_i, \text{ hence}$$

$$\forall t \geq \max\{m(i);\ i = 1, \ldots, r\} - 1:\ A^t = \sum_{i=1}^{r} \sum_{k=0}^{m(i)-1} e_{\lambda_i,k}(t) N_i^k P_i,$$

where $e_{0,k}(t) = \delta_{k,t}$ and $e_{\lambda,k}(t) = \binom{t}{k}\lambda^{t-k}$, $\lambda \neq 0$, comes from (5.17) and (5.22). The autonomous state space system

$$x(t+1) = Ax(t) \quad \text{with the initial value } x(0) = x_0 \in F^n$$

thus has the unique solution

$$x(t) = A^t x_0 = \sum_{i=1}^{r} \sum_{k=0}^{\min(t,m(i)-1)} e_{\lambda_i,k}(t) N_i^k P_i x_0 \quad \text{for } t \in \mathbb{N}.$$

Proof We use the representation $f_A = \prod_{i=1}^{r}(s-\lambda_i)^{m(i)}$ of the minimal polynomial f_A of A from (5.75) and the isomorphism (5.81).

In $F[s]/F[s](s-\lambda_i)^{m(i)}$, we have $\overline{(s-\lambda_i)}^k = 0$ for $k \geq m(i)$. For $i = 1, \ldots, r$, the binomial theorem furnishes (also for $\lambda_i = 0$)

$$\overline{s}^t = \overline{(\lambda_i + (s-\lambda_i))}^t = \sum_{k=0}^{t} \binom{t}{k} \lambda_i^{t-k} \overline{(s-\lambda_i)}^k = \sum_{k=0}^{\min(t,m(i)-1)} \binom{t}{k} \lambda_i^{t-k} \overline{(s-\lambda_i)}^k$$

$$= \sum_{k=0}^{\min(t,m(i)-1)} e_{\lambda_i,k}(t) \overline{(s-\lambda_i)}^k \in F[s]/F[s](s-\lambda_i)^{m(i)}.$$

With (5.81) we infer

$$\overline{s}^t = \sum_{i=1}^{r} \sum_{k=0}^{\min(t,m(i)-1)} e_{\lambda_i,k}(t) \overline{(s-\lambda_i)}^k\, \overline{b_i}$$

$$= \sum_{i=1}^{r} \sum_{k=0}^{\min(t,m(i)-1)} e_{\lambda_i,k}(t) \overline{(s-\lambda_i) b_i}^k\, \overline{b_i} \in F[s]/F[s] f_A.$$

We substitute A for s and get, with $b_i(A) = P_i$ and $((s - \lambda_i)b_i)(A) = N_i$,

$$A^t = \sum_{i=1}^{r} \sum_{k=0}^{\min(t,m(i)-1)} e_{\lambda_i,k}(t) N_i^k P_i \in F[A].$$

□

Corollary 5.5.4 *For $F = \mathbb{C}$, the exponential e^{tA} is given as*

$$e^{tA} = \sum_{i=1}^{r} \sum_{k=0}^{m(i)-1} e_{\lambda_i,k}(t) N_i^k P_i,$$

where $e_{\lambda,k}(t) = \frac{t^k}{k!} e^{\lambda t}$ for $t \in \mathbb{R}$. The autonomous state space system

$$x' = Ax, \quad \text{with the initial value } x(0) = x_0 \in \mathbb{C}^n$$

thus has the unique solution

$$x(t) = e^{tA} x_0 = \sum_{i=1}^{r} \sum_{k=0}^{m(i)-1} e_{\lambda_i,k}(t) N_i^k P_i x_0 \quad \text{for } t \in \mathbb{R}.$$

Proof With $N_i^k = 0$ for $k \geq m(i)$ and $A^m = \sum_{i=1}^r \sum_{k=0}^m \binom{m}{k} \lambda_i^{m-k} N_i^k P_i$ from Corollary 5.5.3 we derive

$$e^{tA} = \sum_{m=0}^{\infty} \frac{t^m}{m!} A^m = \sum_{m=0}^{\infty} \sum_{i=1}^{r} \sum_{k=0}^{m} \frac{t^m}{m!} \frac{m!}{k!(m-k)!} \lambda_i^{m-k} N_i^k P_i$$

$$\overset{j:=m-k}{=\!=\!=} \sum_{i=1}^{r} \sum_{k=0}^{\infty} \frac{t^k}{k!} N_i^k P_i \sum_{j=0}^{\infty} \frac{(\lambda_i t)^j}{j!} = \sum_{i=1}^{r} \sum_{k=0}^{m(i)-1} \frac{t^k}{k!} e^{\lambda_i t} N_i^k P_i$$

$$= \sum_{i=1}^{r} \sum_{k=0}^{m(i)-1} e_{\lambda_i,k}(t) N_i^k P_i.$$

□

5.5.2 The Partial Fraction Decomposition of Rational Functions in One Indeterminate

As the final application of the CRT and, in particular, of the primary decomposition we present the partial fraction decomposition of rational functions. In Chaps. 8 and 10 *it will be an essential tool.* Let F be a field. Consider the quotient field $F(s)$ of rational functions, i.e.,

$$F(s) = \left\{\alpha = \tfrac{g}{f};\ g, f \in F[s],\ f \neq 0\right\} \supseteq F[s].$$

The degree function for polynomials is extended to rational functions via

$$\deg := \deg_s :\ F(s) \setminus \{0\} \longrightarrow \mathbb{Z},\ \alpha = \tfrac{g}{f} \mapsto \deg(\alpha) := \deg_s(g) - \deg_s(f), \tag{5.83}$$

and $\deg(0) := -\infty$. The degree function satisfies

$$\deg(\alpha + \beta) \leq \max(\deg(\alpha), \deg(\beta)) \text{ and } \deg(\alpha\beta) = \deg(\alpha) + \deg(\beta). \tag{5.84}$$

Therefore the sets of rational functions

$$F(s)_{\mathrm{pr}} := \{\alpha \in F(s);\ \deg(\alpha) \leq 0\} \supset F(s)_{\mathrm{spr}} := \{\alpha \in F(s);\ \deg(\alpha) < 0\} \tag{5.85}$$

form a subring $F(s)_{\mathrm{pr}}$ of $F(s)$ and an ideal $F(s)_{\mathrm{spr}}$ of $F(s)_{\mathrm{pr}}$. These are called the ring $F(s)_{\mathrm{pr}}$ of *proper rational functions* and the ideal $F(s)_{\mathrm{spr}}$ of *strictly proper rational functions*.

Lemma 5.5.5 *The field $F(s)$ can be decomposed as*

$$\begin{aligned} F(s) = F[s] \oplus F(s)_{\mathrm{spr}} \ni H = \tfrac{g}{f} = H_{\mathrm{pol}} + H_{\mathrm{spr}} := h + \tfrac{r}{f}, \text{ where} \\ g = hf + r \text{ with } h, r \in F[s], \text{ and } \deg(r) < \deg(f). \end{aligned} \tag{5.86}$$

The representation $g = hf + r$ is obtained by Euclidean division.

Proof This follows directly from the Euclidean division. □

In the following fashion the primary decomposition implies the partial fraction decomposition. The ring $F[s]$ is an $F[s]$-submodule of $F(s)$. The factor module $M := F(s)/F[s]$ with elements $\overline{\alpha} := \alpha + F[s], \alpha \in F(s)$, is a torsion module since $f\overline{\tfrac{g}{f}} = \overline{g} = 0 \in F(s)/F[s]$.

Lemma 5.5.6 *We denote the set of monic irreducible polynomials in $F[s]$ by $\mathcal{P}$. The primary decomposition of $F(s)/F[s]$ is*

$$F(s)/F[s] = \bigoplus_{p \in \mathcal{P}} F[s]_p/F[s] \quad \text{where } F[s]_p := \left\{\tfrac{g}{p^k};\ g \in F[s],\ k \geq 0\right\}.$$

Proof By Theorem 5.3.2 it suffices to show $F[s]_p/F[s] = \mathrm{t}_p\left(F(s)/F[s]\right)$. The inclusion $F[s]_p/F[s] \subseteq \mathrm{t}_p\left(F(s)/F[s]\right)$ follows from $p^k\overline{\tfrac{g}{p^k}} = \overline{g} = 0 \in F(s)/F[s]$. If, conversely, $\overline{\tfrac{g}{f}} \in \mathrm{t}_p\left(F(s)/F[s]\right)$, we assume without loss of generality that $\gcd(f, g) = 1$. Let $k \in \mathbb{N}$ be such that $p^k\overline{\tfrac{g}{f}} = 0$. Then $p^k\tfrac{g}{f} = \tfrac{p^k g}{f} \in F[s]$, i.e., $f \mid p^k g$. Since g and f are coprime, we infer $f \mid p^k$. With $f_1 := \tfrac{p^k}{f} \in F[s]$ we get $\tfrac{g}{f} = \tfrac{f_1 g}{f_1 f} = \tfrac{f_1 g}{p^k} \in F[s]_p$. □

We also need the p-adic representation.

Lemma 5.5.7 *For $p \in \mathcal{P}$ every nonzero polynomial g admits a unique* p-adic representation

$$g = \sum_{j=0}^{k} d_j p^j \ \text{with}\ d_j \in F[s],\ \deg(d_j) < \deg(p),\ \text{and}\ d_k \neq 0,\ \text{where}$$

$$k \deg(p) \leq \deg(g) < (k+1)\deg(p)\ \text{or}\ k = \left[\tfrac{\deg(g)}{\deg(p)}\right] := \max\left\{y \in \mathbb{Z};\ y \leq \tfrac{\deg(g)}{\deg(p)}\right\}. \tag{5.87}$$

Proof By induction on $k = 0, 1, \dots$ we prove

$$g = g_k p^k + \sum_{j=0}^{k-1} d_j p^j,\ \deg(d_j) < \deg(p),\ g_k = g_{k+1} p + d_{k+1}.$$

The induction starts with $g =: g_0 = g_0 p^0$ and stops if $\deg(g_k) < \deg(p)$ and $d_k = g_k$. □

If $\frac{g}{p^m} \in F[s]_p \cap F(s)_{\mathrm{spr}}$ and thus $\deg(g) < \deg(p^m) = m\deg(p)$, then

$$k < m \text{ and } g = \sum_{j=0}^{m-1} d_j p^j,\ \text{where } d_j := 0 \text{ for } k+1 \leq j \leq m-1,\ \text{hence}$$
$$g p^{-m} = \sum_{j=0}^{m-1} d_j p^{j-m} \underset{k:=m-j}{=} \sum_{k=1}^{m} c_k p^{-k} \text{ with } c_k := d_{m-k}. \tag{5.88}$$

Theorem 5.5.8 (Partial fraction decomposition) *Let $\mathcal{P}$ be the set of monic irreducible polynomials in $F[s]$. Then*

$$F(s) = F[s] \oplus F(s)_{\mathrm{spr}} = F[s] \oplus \bigoplus_{p \in \mathcal{P}} \bigoplus_{j=1}^{\infty} F[s]_{\leq \deg(p)-1} p^{-j}. \tag{5.89}$$

In other terms, every rational function $\alpha \in F(s)$ has a unique representation

$$\alpha = h + \sum_{p \in \mathcal{P}} \sum_{j=1}^{\infty} \frac{c_{p,j}}{p^j},\quad \text{with } h, c_{p,j} \in F[s] \text{ and } \deg(c_{p,j}) < \deg(p), \tag{5.90}$$

where, of course, almost all $c_{p,j}$ are zero.

The algorithm requires the prime factor decomposition $f = p_1^{m(1)} \cdots p_r^{m(r)}$ of (monic) polynomials. For all algebraic number fields and all finite fields this is contained in computer algebra systems like MAPLE. Roots of complex polynomials *as needed in systems theory have to be computed numerically with possible numerical errors.*

Proof We show the existence of the representation by describing an algorithm for its computation. The uniqueness follows easily from the uniqueness of the primary decomposition of $M = F(s)/F[s]$. We omit this part of the proof.

Let $\alpha = \frac{g}{f} \in F(s)$. We first compute the greatest common divisor $\gcd(f, g)$ with the Euclidean algorithm and divide f and g by it. Hence we assume from now on that $\gcd(f, g) = 1$. Furthermore, we divide f and g by the leading coefficient of f and assume thus that f is monic.

Let $f = p_1^{m(1)} \cdots p_r^{m(r)}$ be the prime factor decomposition of f with distinct monic irreducible polynomials p_i and exponents $m(i) > 0$. Again with the Euclidean algorithm, we compute for all $i = 1, \ldots, r$ the representations

$$1 = f_{i,1} p_i^{m(i)} + f_{i,2} \frac{f}{p_i^{m(i)}} = a_i + b_i \text{ with } b_i := f_{i,2} \frac{f}{p_i^{m(i)}}. \tag{5.91}$$

Since $f\overline{\alpha} = f\frac{g}{f} = 0 \in M := F(s)/F[s]$, the CRT for modules (Theorem 5.1.4) furnishes

$$\overline{\alpha} = \sum_{i=1}^{r} \overline{b_i \alpha} \in \operatorname{ann}_M(f) = \bigoplus_{i=1}^{r} \operatorname{ann}_M\left(p_i^{m(i)}\right) \subset M = F(s)/F[s], \tag{5.92}$$

where $b_i \alpha = f_{i,2} \frac{f}{p_i^{m(i)}} \frac{g}{f} = \frac{f_{i,2} g}{p_i^{m(i)}}$.

Due to (5.92), $h_0 := \alpha - \sum_{i=1}^{r} b_i \alpha$ is a polynomial. For all $i = 1, \ldots, r$, Euclidean division furnishes polynomials h_i and g_i with

$$\begin{aligned} &f_{i,2} g = h_i p_i^{m(i)} + g_i \text{ with } \deg(g_i) < \deg\left(p_i^{m(i)}\right) \Longrightarrow b_i \alpha = \frac{f_{i,2} g}{p_i^{m(i)}} = h_i + \frac{g_i}{p_i^{m(i)}} \\ &\Longrightarrow \alpha = h_0 + \sum_{i=1}^{r} b_i \alpha = \sum_{i=0}^{r} h_i + \sum_{i=1}^{r} \frac{g_i}{p_i^{m(i)}} \end{aligned} \tag{5.93}$$

Due to $\deg(g_i) < \deg\left(p_i^{m(i)}\right)$, the p-adic representation of $g_i p_i^{-m(i)}$ from (5.88) can be uniquely written as $g_i p_i^{-m(i)} = \sum_{j=1}^{m(i)} c_{p_i,j} p_i^{-j}$ with $\deg(c_{p_i,j}) < \deg(p_i)$. We conclude that

$$\alpha = h + \sum_{i=1}^{r} \sum_{j=1}^{m(i)} c_{p_i,j} p_i^{-j} \text{ with } h = \sum_{i=0}^{r} h_i,\ c_{p_i,j} \in F[s],\ \deg(c_{p_i,j}) < \deg(p_i). \quad \square$$

Corollary 5.5.9 *The field $\mathbb{C}(s)$ of complex rational functions admits the direct sum decomposition*

$$\mathbb{C}(s) = \mathbb{C}[s] \oplus \mathbb{C}(s)_{\mathrm{spr}} = \mathbb{C}[s] \oplus \bigoplus_{\lambda \in \mathbb{C}} \bigoplus_{j=1}^{\infty} \mathbb{C}(s-\lambda)^{-j}. \tag{5.94}$$

In other terms, each rational function $\alpha \in \mathbb{C}(s)$ *has a unique representation*

$$\alpha = h + \sum_{\lambda \in \mathbb{C}} \sum_{j=1}^{\infty} \frac{c_{\lambda,j}}{(s-\lambda)^j} \text{ with } h \in \mathbb{C}[s] \text{ and } c_{\lambda,j} \in \mathbb{C}. \tag{5.95}$$

Of course, almost all $c_{\lambda,j}$ *are zero.* ◇

References

1. U. Oberst, Anwendungen des chinesischen Restsatzes. Exp. Math. **3**, 97–148 (1985)
2. S. Lang, *Algebra, Graduate Texts in Mathematics*, vol. 211, 3rd edn. (Springer, New York, 2002)
3. J.W. Polderman, J.C. Willems, *Introduction to Mathematical Systems Theory: A Behavioral Approach*, Texts in Applied Mathematics, vol. 26 (Springer, New York, 1998)
4. U. Oberst, Variations on the fundamental principle for linear systems of partial differential and difference equations with constant coefficients. AAECC **6**, 211–243 (1995)
5. T. Kailath, *Linear systems* (Prentice-Hall, Englewood Cliffs, 1980)
6. H. Bourlès, *Linear Systems* (ISTE, London, 2010)
7. P.J. Antsaklis, A.N. Michel, *Linear Systems*, 2nd edn. (Birkhäuser, Boston, 2006)

Chapter 6
Input/Output Behaviors

In this chapter we discuss behaviors whose trajectories are decomposed into an input and an output component and their associated transfer matrices that are derived by module-behavior duality. Most nonautonomous systems are of this type. Later in Sect. 8.2.4 *the transfer matrix is used to define the impulse response and the Laplace transform for continuous-time systems, whereas usually the Laplace transform is introduced analytically and applied to define the transfer matrix.*

6.1 Transfer Spaces

In this section we introduce the transfer space *of a behavior. We relate this transfer space to the behavior's controllable part.*

In Sects. 6.1 *and* 6.2 *we assume that $\mathcal{F}$ is an injective cogenerator over a Noetherian integral domain $\mathcal{D}$ with the quotient field K. This is satisfied in all standard cases. The finer theory requires a principal ideal domain $\mathcal{D}$.*

As usual, we assume an $\mathcal{F}$-system

$$\begin{aligned}
&\mathcal{B} := \{w \in \mathcal{F}^l;\ R \circ w = 0\} \quad \text{with} \\
&R \in \mathcal{D}^{k \times l},\ p := \operatorname{rank}(R),\ m := l - p, \\
&U := \mathcal{D}^{1 \times k} R,\ \text{i.e., } \mathcal{B} = U^{\perp},\ \text{and } M := \mathcal{D}^{1 \times l}/\mathcal{D}^{1 \times k} R = \mathrm{M}(\mathcal{B}), \\
&M = \sum_{j=1}^{l} \mathcal{D}\overline{\delta_j},\ \delta_j = (0, \ldots, \overset{j}{1}, \ldots, 0) \in \mathcal{D}^{1 \times l},\ \overline{\delta_j} := \delta_j + U.
\end{aligned} \tag{6.1}$$

U. Oberst et al., *Linear Time-Invariant Systems, Behaviors and Modules*, Differential-Algebraic Equations Forum,
https://doi.org/10.1007/978-3-030-43936-1_6

The rank of R is its rank as matrix with entries in K. We recall from Theorem 3.2.10 that ${}_{\mathcal{D}}K$ is an injective module and hence the functor $\mathrm{Hom}_{\mathcal{D}}(-, K)$ is exact on $\mathcal{D}$-modules.

Lemma and Definition 6.1.1 (Transfer space) *The K-solution space*

$$\mathcal{B}_K := \{\widetilde{w} \in K^l;\ R\widetilde{w} = 0\} = (KU)^{\perp} = \left(K^{1\times k}R\right)^{\perp} \tag{6.2}$$

of dimension $m = l - \mathrm{rank}(R) = l - p$ is called the transfer space *of the behavior $\mathcal{B}$, where $(KU)^{\perp}$ is the orthogonal subspace of $KU = K^{1\times k}R$ with respect to the nondegenerate bilinear form $K^{1\times l} \times K^l \longrightarrow K$, $(\xi, \eta) \longmapsto \xi\eta$, from (4.16), applied to K. The map*

$$\mathrm{can}\colon\ \mathrm{Hom}_{\mathcal{D}}(M, K) \overset{\cong}{\longleftrightarrow} \mathcal{B}_K,\quad \varphi \longleftrightarrow \widetilde{w},\quad \varphi(\overline{\delta_j}) = \widetilde{w}_j, \tag{6.3}$$

is an isomorphism. This shows that $\mathcal{B}_K$ depends on M or $\mathcal{B}$ only, but not on the choice of the matrix R. We define the rank of M and of $\mathcal{B}$ as

$$\mathrm{rank}(\mathcal{B}) := \mathrm{rank}(M) := \dim_K(\mathcal{B}_K) = l - \mathrm{rank}(R) \Longrightarrow \mathrm{rank}(U) = \mathrm{rank}(R). \tag{6.4}$$

Proof (i) The isomorphism (6.3) follows from Theorem 3.1.36.
(ii) The equation $KU = K^{1\times k}R$ implies

$$\mathrm{rank}(R) = \dim_K(KU) = \dim_K(\mathrm{Hom}_K(KU, K)).$$

Moreover there is the canonical isomorphism

$$\begin{aligned}&\mathrm{Hom}_{\mathcal{D}}(U, K) \cong \mathrm{Hom}_K(KU, K),\ \varphi = \psi|U \leftrightarrow \psi,\ \psi(\alpha u) := \alpha\varphi(u),\ \alpha \in K\\ &\Longrightarrow \mathrm{rank}(U) = \dim_K\left(\mathrm{Hom}_{\mathcal{D}}(U, K)\right) = \dim_K\left(\mathrm{Hom}_K(KU, K)\right) = \mathrm{rank}(R).\end{aligned}$$

□

Corollary 6.1.2 *Exact module resp. dual behavior sequences*

$$\begin{aligned}&0 \to M_1 \to M_2 \to M_3 \to 0 \textit{ resp. } 0 \leftarrow \mathcal{B}_1 \leftarrow \mathcal{B}_2 \leftarrow \mathcal{B}_3 \leftarrow 0 \textit{ imply}\\ &\mathrm{rank}(M_2) = \mathrm{rank}(M_1) + \mathrm{rank}(M_3),\ \mathrm{rank}(\mathcal{B}_2) = \mathrm{rank}(\mathcal{B}_1) + \mathrm{rank}(\mathcal{B}_3).\end{aligned}$$

Proof Since ${}_{\mathcal{D}}K$ is injective the functor $\mathrm{Hom}_{\mathcal{D}}(-, K)$ is exact and therefore

$$\begin{aligned}&0 \leftarrow \mathrm{Hom}_{\mathcal{D}}(M_1, K) \leftarrow \mathrm{Hom}_{\mathcal{D}}(M_2, K) \leftarrow \mathrm{Hom}_{\mathcal{D}}(M_3, K) \leftarrow 0 \text{ is exact and}\\ &\dim_K\left(\mathrm{Hom}_{\mathcal{D}}(M_2, K)\right) = \dim_K\left(\mathrm{Hom}_{\mathcal{D}}(M_1, K)\right) + \dim_K\left(\mathrm{Hom}_{\mathcal{D}}(M_3, K)\right).\end{aligned}$$

□

Corollary 6.1.3 *A behavior $\mathcal{B}$ is autonomous if and only if $\mathcal{B}_K$ is zero, i.e., if* $\operatorname{rank}(\mathcal{B}) = 0$ *or* $\operatorname{rank}(R) = l$. *This generalizes the corresponding result from Theorem 4.2.12 (where we assumed $\mathcal{D}$ to be a principal ideal domain) to the more general situation of the present paragraph.*

Proof Recall from Definition 4.2.4 that $\mathcal{B}$ is autonomous if and only if M is a torsion module.
1. If $\mathcal{B}$ is autonomous, then $\mathcal{B}_K \cong \operatorname{Hom}_{\mathcal{D}}(M, K) = 0$ since there is only the zero homomorphism from the torsion module M into the torsionfree K.
2. Conversely, if $0 = \dim(\mathcal{B}_K) = l - \operatorname{rank}(R)$, i.e., $\operatorname{rank}(R) = l$, then $K^{1\times k}R = K^{1\times l}$ and thus $K^{1\times l}/K^{1\times k}R = 0$. But this means that

$$M = \ker\left(M \longrightarrow K^{1\times l}/K^{1\times k}R\right) \underset{\text{Theorem 4.2.15, 1.}}{=} \mathrm{t}(M). \qquad \square$$

The assignment $\mathcal{B} \longmapsto \mathcal{B}_K$ can be extended to a covariant functor from the category of behaviors into that of finite-dimensional K-vector spaces. Indeed, as in Section 3.3.3, consider two behaviors $\mathcal{B}_i \subseteq \mathcal{F}^{l_i}$ with $M_i := \mathcal{D}^{1\times l_i}/\mathcal{B}_i^{\perp}$, $i = 1, 2$, and a behavior morphism $P\circ \in \operatorname{Hom}(\mathcal{B}_1, \mathcal{B}_2)$ given by a matrix $P \in \mathcal{D}^{l_2\times l_1}$. Then $(\cdot P)_{\mathrm{ind}}\colon M_2 \longrightarrow M_1$ is a well-defined linear map, and it induces the commutative diagram

$$\begin{array}{ccc}
\operatorname{Hom}_{\mathcal{D}}(M_1, K) & \xrightarrow{\operatorname{Hom}_{\mathcal{D}}\left((\cdot P)_{\mathrm{ind}}, K\right)} & \operatorname{Hom}_{\mathcal{D}}(M_2, K) \\
\downarrow{\scriptstyle \mathrm{can}_1} & & \downarrow{\scriptstyle \mathrm{can}_2} \\
\mathcal{B}_{1,K} & \xrightarrow{P\cdot} & \mathcal{B}_{2,K} \\
\widetilde{w}_1 & \longmapsto & P\widetilde{w}_1.
\end{array} \tag{6.5}$$

The commutativity of this diagram follows as in Lemma 3.3.15.

Theorem 6.1.4 (The transfer space as additive exact functor) *The assignment*

$$(-)_K\colon\ \mathrm{syst}_{\mathcal{D}\mathcal{F}} \longrightarrow \mathrm{mod}_K,\quad \begin{array}{c} \mathcal{B}_1 \\ \downarrow{\scriptstyle P\circ} \\ \mathcal{B}_2 \end{array} \longmapsto \begin{array}{c} \mathcal{B}_{1,K} \\ \downarrow{\scriptstyle P\cdot} \\ \mathcal{B}_{2,K} \end{array}$$

from the category of ${}_{\mathcal{D}}\mathcal{F}$-behaviors to the category of K-vector spaces is a covariant additive and exact functor. The latter property signifies that an exact behavior sequence

$$\mathcal{B}_1 \xrightarrow{P_1\circ} \mathcal{B}_2 \xrightarrow{P_2\circ} \mathcal{B}_3 \tag{6.6}$$

induces an exact K-vector space sequence

$$\mathcal{B}_{1,K} \xrightarrow{P_1\cdot} \mathcal{B}_{2,K} \xrightarrow{P_2\cdot} \mathcal{B}_{3,K}. \tag{6.7}$$

Proof The functor property follows at once from

$$\mathrm{id}_{\mathcal{B}} = \mathrm{id}_l \circ \longmapsto \mathrm{id}_l \cdot = \mathrm{id}_{\mathcal{B}_K} \quad \text{and}$$
$$(P_2\circ)(P_1\circ) = (P_2P_1)\circ \longmapsto (P_2P_1)\cdot = (P_2\cdot)(P_1\cdot),$$

and the additivity from

$$(P_1\circ) + (P_2\circ) = (P_1 + P_2)\circ \longmapsto (P_1 + P_2)\cdot = (P_1\cdot) + (P_2\cdot)$$

for matrices of suitable sizes. According to Theorem 3.3.18, the exactness of (6.6) implies that of the module sequence

$$M_1 \xleftarrow{(\cdot P_1)_{\mathrm{ind}}} M_2 \xleftarrow{(\cdot P_2)_{\mathrm{ind}}} M_3.$$

Application of the exact contravariant functor $\mathrm{Hom}_{\mathcal{D}}(-, K)$ to this sequence and the commutative diagram (6.5) imply the exactness of (6.7). □

Corollary 6.1.5 $\mathcal{B}_{\mathrm{cont},K} = \mathcal{B}_K$.

Proof From Theorem 4.2.15 consider the modules $M = \mathcal{D}^{1\times\ell}/U$ and $M_{\mathrm{cont}} = \mathcal{D}^{1\times\ell}/U_{\mathrm{cont}}$ with $\mathrm{t}(M) = U_{\mathrm{cont}}/U$ and the associated behaviors

$$\mathcal{B} := U^\perp \cong \mathrm{Hom}_{\mathcal{D}}(M, \mathcal{F}),\ \mathcal{B}_{\mathrm{cont}} := U_{\mathrm{cont}}^\perp \cong \mathrm{Hom}_{\mathcal{D}}(M_{\mathrm{cont}}, \mathcal{F}),$$
$$\mathrm{D}(\mathrm{t}(M)) = \mathrm{Hom}_{\mathcal{D}}(\mathrm{t}(M), \mathcal{F})$$

where $\mathrm{t}(M)$ is a torsion module and $\mathrm{D}(\mathrm{t}(M))$ the dual autonomous behavior with $\mathrm{D}(\mathrm{t}(M))_K = 0$ by Corollary 6.1.3. There result the exact module and dual behavior sequence

$$0 \to \mathrm{t}(M) \to M \to M_{\mathrm{cont}} \to 0 \text{ and } 0 \leftarrow \mathrm{D}(\mathrm{t}(M)) \leftarrow \mathcal{B} \supseteq \mathcal{B}_{\mathrm{cont}} \leftarrow 0.$$

Application of the exact functor $(-)_K$ implies the exact sequence

$$0 \leftarrow \mathrm{D}(\mathrm{t}(M))_K = 0 \leftarrow \mathcal{B}_K \supseteq \mathcal{B}_{\mathrm{cont},K} \leftarrow 0 \Longrightarrow \mathcal{B}_K = \mathcal{B}_{\mathrm{cont},K}.$$ □

Corollary 6.1.6 *Let $\mathcal{B}_1, \mathcal{B}_2 \subseteq \mathcal{F}^l$ be two behaviors. According to Theorem 3.3.10, their intersection $\mathcal{B}_1 \cap \mathcal{B}_2$ and their sum $\mathcal{B}_1 + \mathcal{B}_2$ are also behaviors. The following identities hold:*

$$(\mathcal{B}_1 \cap \mathcal{B}_2)_K = \mathcal{B}_{1,K} \cap \mathcal{B}_{2,K} \quad \textit{and} \quad (\mathcal{B}_1 + \mathcal{B}_2)_K = \mathcal{B}_{1,K} + \mathcal{B}_{2,K}.$$

Proof 1. If $\mathcal{B}_i = \{w \in \mathcal{F}^l;\ R_i \circ w = 0\}$, $i = 1, 2$, then

$$\mathcal{B}_1 \cap \mathcal{B}_2 = \left\{w \in \mathcal{F}^l;\ \begin{pmatrix} R_1 \\ R_2 \end{pmatrix} \circ w = 0\right\},$$

and hence

$$\begin{aligned}(\mathcal{B}_1 \cap \mathcal{B}_2)_K &= \left\{\widetilde{w} \in K^l;\ \left(\begin{smallmatrix} R_1 \\ R_2 \end{smallmatrix}\right)\widetilde{w} = 0\right\} \\ &= \{\widetilde{w} \in K^l;\ R_1\widetilde{w} = 0\} \cap \{\widetilde{w} \in K^l;\ R_2\widetilde{w} = 0\} = \mathcal{B}_{1,K} \cap \mathcal{B}_{2,K}.\end{aligned}$$

2. The product of $\mathcal{B}_1$ and $\mathcal{B}_2$ is

$$\mathcal{B}_1 \times \mathcal{B}_2 := \left\{\left(\begin{smallmatrix} w_1 \\ w_2 \end{smallmatrix}\right) \in \mathcal{F}^{2l} = \mathcal{F}^l \times \mathcal{F}^l;\ \left(\begin{smallmatrix} R_1 & 0 \\ 0 & R_2 \end{smallmatrix}\right) \circ \left(\begin{smallmatrix} w_1 \\ w_2 \end{smallmatrix}\right) = 0\right\}$$

and its transfer space is

$$(\mathcal{B}_1 \times \mathcal{B}_2)_K = \left\{\left(\begin{smallmatrix} \widetilde{w}_1 \\ \widetilde{w}_2 \end{smallmatrix}\right) \in K^{2l};\ \left(\begin{smallmatrix} R_1 & 0 \\ 0 & R_2 \end{smallmatrix}\right) \circ \left(\begin{smallmatrix} \widetilde{w}_1 \\ \widetilde{w}_2 \end{smallmatrix}\right) = 0\right\} = \mathcal{B}_{1,K} \times \mathcal{B}_{2,K}.$$

According to Theorem 6.1.4, the epimorphism

$$\mathcal{B}_1 \times \mathcal{B}_2 \xrightarrow{(\mathrm{id}_l,\mathrm{id}_l)\circ} \mathcal{B}_1 + \mathcal{B}_2,\quad \left(\begin{smallmatrix} w_1 \\ w_2 \end{smallmatrix}\right) \longmapsto w_1 + w_2,$$

induces the epimorphism

$$(\mathcal{B}_1 \times \mathcal{B}_2)_K = \mathcal{B}_{1,K} \times \mathcal{B}_{2,K} \xrightarrow{(\mathrm{id}_l,\mathrm{id}_l)\cdot} (\mathcal{B}_1 + \mathcal{B}_2)_K,\quad \left(\begin{smallmatrix} \widetilde{w}_1 \\ \widetilde{w}_2 \end{smallmatrix}\right) \longmapsto \widetilde{w}_1 + \widetilde{w}_2,$$

whose image is $\mathcal{B}_{1,K} + \mathcal{B}_{2,K}$, hence $(\mathcal{B}_1 + \mathcal{B}_2)_K = \mathcal{B}_{1,K} + \mathcal{B}_{2,K}$.

□

The next theorem shows that the transfer space of a behavior determines its controllable part.

Theorem 6.1.7 ([1, Theorem 8.2.7 on p. 289]) *For two behaviors $\mathcal{B}_1, \mathcal{B}_2 \subseteq \mathcal{F}^l$:*

$$\mathcal{B}_{1,\mathrm{cont}} = \mathcal{B}_{2,\mathrm{cont}} \Longleftrightarrow \mathcal{B}_{1,K} = \mathcal{B}_{2,K}.$$

Proof $\Longrightarrow$. This follows directly from Corollary 6.1.5.
$\Longleftarrow$. 1. Assume first $\mathcal{B}_1 \subseteq \mathcal{B}_2$. Consider $U_2 := \mathcal{B}_2^\perp \subseteq U_1 := \mathcal{B}_1^\perp$ and $M_i = \mathcal{D}^{1\times l}/U_i$ for $i = 1, 2$. Let $R_1 \in \mathcal{D}^{k\times l}$ be a matrix with $U_1 = \mathcal{D}^{1\times k}R_1$ and

$$V := \ker\left(\mathcal{D}^{1\times k} \xrightarrow{\cdot R_1} U_1 \xrightarrow{\mathrm{can}} U_1/U_2\right).$$

Then $(\cdot R_1)_{\mathrm{ind}}\colon\ \mathcal{D}^{1\times k}/V \longrightarrow \mathcal{D}^{1\times l}/U_2 = M_2$ is a monomorphism with image U_1/U_2. This implies the exact module sequence

$$0 \to \mathcal{D}^{1\times k}/V \xrightarrow{(\cdot R_1)_{\mathrm{ind}}} M_2 \xrightarrow{\mathrm{can}} M_1 \to 0 \text{ with } \mathrm{can}(m + U_2) := m + U_1,$$

and, by duality, the exact behavior sequence

$$0 \longleftarrow V^{\perp} \overset{R_1\circ}{\longleftarrow} \mathcal{B}_2 \overset{\text{inj}}{\leftarrow} \mathcal{B}_1 \longleftarrow 0.$$

The latter and Theorem 6.1.4 induce the exact vector space sequence

$$0 \longleftarrow (V^{\perp})_K \overset{R_1\cdot}{\longleftarrow} \mathcal{B}_{2,K} \overset{\text{inj}}{\leftarrow} \mathcal{B}_{1,K} \longleftarrow 0.$$

But $\mathcal{B}_{1,K} = \mathcal{B}_{2,K}$ and thus $V_K^{\perp} = 0$. According to Corollary 6.1.3, this means that $V^{\perp}$ is autonomous and $\mathcal{D}^{1\times k}/V \cong U_1/U_2$ is a torsion module. Thus, $U_1/U_2 \subseteq \mathrm{t}(M_2) = U_{2,\mathrm{cont}}/U_2$, i.e., $U_1 \subseteq U_{2,\mathrm{cont}}$ and $\mathcal{B}_{2,\mathrm{cont}} \subseteq \mathcal{B}_1$.

Since $\mathcal{B}_{1,\mathrm{cont}}$ is the largest controllable subbehavior of $\mathcal{B}_1$, we infer $\mathcal{B}_{2,\mathrm{cont}} \subseteq \mathcal{B}_{1,\mathrm{cont}}$. Moreover $\mathcal{B}_1 \subseteq \mathcal{B}_2$ implies $\mathcal{B}_{1,\mathrm{cont}} \subseteq \mathcal{B}_{2,\mathrm{cont}}$, hence $\mathcal{B}_{1,\mathrm{cont}} = \mathcal{B}_{2,\mathrm{cont}}$.
2. For arbitrary $\mathcal{B}_1, \mathcal{B}_2 \subseteq \mathcal{F}^l$ we define $\mathcal{B} := \mathcal{B}_1 + \mathcal{B}_2 \supseteq \mathcal{B}_1$. Corollary 6.1.6 and the assumption imply $\mathcal{B}_{1,K} = \mathcal{B}_{1,K} + \mathcal{B}_{2,K} = \mathcal{B}_K$ and, by 1, $\mathcal{B}_{1,\mathrm{cont}} = \mathcal{B}_{\mathrm{cont}}$ and $\mathcal{B}_{2,\mathrm{cont}} \subseteq \mathcal{B}_{\mathrm{cont}} = \mathcal{B}_{1,\mathrm{cont}}$. The reverse inclusion follows likewise. □

If $V \subseteq K^l$ is any subspace, a behavior $\mathcal{B} \subseteq \mathcal{F}^l$ with $\mathcal{B}_K = V$ is called a *(behavior) realization* of V.

Theorem and Definition 6.1.8 (The controllable realization of a subspace) *For every K-subspace $V \subseteq K^l$ there is a* unique controllable realization $\mathcal{B} \subseteq \mathcal{F}^l$ *of V. It is also the least realization of V and therefore called the* minimal realization. *The following proof and Theorem 4.2.15 contain an algorithm for the computation of a matrix $R_{\mathrm{cont}} \in \mathcal{D}^{p\times l}$ of rank $p := l - \dim(V)$ with*

$$\mathcal{B} = \{w \in \mathcal{F}^l;\ R_{\mathrm{cont}} \circ w = 0\}.$$

Proof *Uniqueness.* This follows directly from Theorem 6.1.7.
Existence. Every subspace $V \subseteq K^l$ is a solution space of the form

$$V = \{\widetilde{w} \in K^l;\ \widetilde{R}\widetilde{w} = 0\} \quad \text{for some matrix } \widetilde{R} \in K^{k\times l}.$$

Multiplication of $\widetilde{R}$ with a common denominator d of its entries furnishes the matrix $R := d\widetilde{R} \in \mathcal{D}^{k\times l}$ with the same solution space V and $\mathrm{rank}(R) = l - \dim(V)$. The matrix R gives rise to the behavior $\mathcal{B}' := \{w \in \mathcal{F}^l;\ R \circ w = 0\}$. Clearly, $\mathcal{B}'_K = V$, but, in general, $\mathcal{B}'$ is not controllable. However, $\mathcal{B} := \mathcal{B}'_{\mathrm{cont}}$ is controllable and, by Corollary 6.1.5, $\mathcal{B}_K = \mathcal{B}'_{\mathrm{cont},K} = \mathcal{B}'_K = V$. A matrix R_{cont} with $\mathcal{B}'_{\mathrm{cont}} = (\mathcal{D}^{1\times k_c} R_{\mathrm{cont}})^{\perp}$ can be computed by means of Theorem 4.2.15.

If $\mathcal{B}''$ is any realization of V, then so is its controllable part. We infer

$$V = \mathcal{B}'_{\mathrm{cont},K} = \mathcal{B}''_{\mathrm{cont},K} \underset{\text{Theorem 6.1.7}}{\Longrightarrow} \mathcal{B} = \mathcal{B}'_{\mathrm{cont}} = \mathcal{B}''_{\mathrm{cont}} \subseteq \mathcal{B}'',$$

hence $\mathcal{B} = \mathcal{B}'_{\mathrm{cont}}$ is the least realization. □

According to Corollary 3.3.8, the field K is an injective cogenerator in the category of K-vector spaces. Thus, it induces the *category of K-behaviors* in the same fashion as the injective cogenerator ${}_{\mathcal{D}}\mathcal{F}$ induces the category of $\mathcal{F}$-behaviors $\mathcal{B} = \{w \in \mathcal{F}^l;\ R \circ w = 0\} \subseteq \mathcal{F}^l$ with its $\mathcal{D}$-module of system morphisms

$$\mathrm{Hom}(\mathcal{B}_1, \mathcal{B}_2) := \{\phi \in \mathrm{Hom}_{\mathcal{D}}(\mathcal{B}_1, \mathcal{B}_2);\ \exists P \in \mathcal{D}^{l_2 \times l_1}\ \forall w_1 \in \mathcal{B}_1:\ \phi(w_1) = P \circ w_1\},$$

where $\mathcal{B}_i \subseteq \mathcal{F}^{l_i}$, $i = 1, 2$, are behaviors. Notice that, in general, not every $\mathcal{D}$-linear map from $\mathcal{B}_1$ to $\mathcal{B}_2$ is a system morphism.

In analogy to this the objects of the category of K-behaviors are the vector spaces

$$\widetilde{\mathcal{B}} = \{\widetilde{w} \in K^l;\ R\widetilde{w} = 0\}, \quad \text{where } R \in K^{k \times l},$$

i.e., the solution spaces of matrices. Since every subspace of K^l has this form, the K-behaviors are just the finite-dimensional K-vector spaces $\widetilde{\mathcal{B}}$ with a given inclusion $\widetilde{\mathcal{B}} \subseteq K^l$. In contrast to general behaviors all K-linear maps between these spaces are behavior morphisms, as the following corollary shows.

Corollary 6.1.9 *The identity* $\mathrm{Hom}(\widetilde{\mathcal{B}}_1, \widetilde{\mathcal{B}}_2) = \mathrm{Hom}_K(\widetilde{\mathcal{B}}_1, \widetilde{\mathcal{B}}_2)$ *holds, i.e., all K-linear maps $\phi\colon\ \widetilde{\mathcal{B}}_1 \longrightarrow \widetilde{\mathcal{B}}_2$ are morphisms of K-behaviors.*

Proof If $\mathcal{B}_i = K^{l_i}$, the standard identification

$$\mathrm{Hom}_K(K^{l_1}, K^{l_2}) \overset{\cong}{\rightleftarrows} K^{l_2 \times l_1}, \quad (\widetilde{w} \mapsto P\widetilde{w}) \longleftarrow P,$$
$$\varphi \longmapsto P \text{ with } P_{ij} = \big(\varphi(\delta_j)\big)_i,$$

shows that the assertion holds for $\widetilde{\mathcal{B}}_i = K^{l_i}$.

Since ${}_K K$ and thus also ${}_K K^{l_2}$ are K-injective, every K-linear map $\phi\colon\ \widetilde{\mathcal{B}}_1 \longrightarrow \widetilde{\mathcal{B}}_2 \subseteq K^{l_2}$ can be extended to a linear map $P\circ\colon\ K^{l_1} \longrightarrow K^{l_2}$ and thus satisfies $\phi(\widetilde{w_1}) = P\widetilde{w_1}$ for $w_1 \in \mathcal{B}_1$ as asserted. □

Corollary and Definition 6.1.10 *The category of K-behaviors as just introduced is also called the category of* transfer spaces. *Hence the functor* transfer space

$$(-)_K\colon\ \mathcal{B} = \{w \in \mathcal{F}^l;\ R \circ w = 0\} \longmapsto \mathcal{B}_K = \{\widetilde{w} \in K^l;\ R\widetilde{w} = 0\}$$
$$(P\circ\colon\ \mathcal{B}_1 \to \mathcal{B}_2) \longmapsto (P\cdot\colon\ \mathcal{B}_{1,K} \to \mathcal{B}_{2,K})$$

is one from the category of $\mathcal{F}$-behaviors into that of transfer spaces. ◊

6.2 IO Decompositions and Transfer Matrices

In Sect. 6.2.1 *we define input/output (IO) structures or decompositions of behaviors and transfer matrices with respect to them. Every behavior admits, in general, several such decompositions. We introduce the unique controllable realization of a transfer matrix. A left coprime factorization of the latter furnishes a kernel representation of this realization. In Sect.* 6.2.2 *we discuss three autonomous behaviors connected with an IO behavior. They appear in a short exact sequence and give rise to the controllable and uncontrollable poles of the behavior. As in the preceding section, we assume that $\mathcal{D}$ is a Noetherian integral domain with quotient field K and that ${}_{\mathcal{D}}\mathcal{F}$ is an injective cogenerator. The finer results require a principal ideal domain $\mathcal{D}$.*

6.2.1 IO Structures or Decompositions

In Definition 6.2.1, *we define the IO structures or decompositions algebraically and then justify the terminology in Theorem and Definition* 6.2.2.

Definition 6.2.1 (*Input/Output decomposition*) An *input/output (IO) structure* or *IO decomposition* of the behavior $\mathcal{B} = \{w \in \mathcal{F}^l;\ R \circ w = 0\}$ or its submodule $U = \mathcal{B}^\perp = \mathcal{D}^{1\times k}R$ is given by a subset $J \subseteq \{1, \ldots, l\}$ of $p = \operatorname{rank}(R)$ elements such that the columns R_{-j}, $j \in J$, are K-linearly independent. $\diamondsuit$

Let, more generally, $J \subseteq \{1, \ldots, l\}$ be any subset and let $p := \sharp(J)$ denote the number of elements of J. For notational simplicity it is customary to first apply a permutation to the columns of R such that the subset J under discussion coincides with $\{1, \ldots, p\}$. We will always do this in the sequel and use the following adapted standard notation with $m := l - p$:

$$\begin{aligned} &w =: \left(\begin{smallmatrix} y \\ u \end{smallmatrix}\right) \in \mathcal{F}^l = \mathcal{F}^{p+m},\ R =: (P, -Q) \in \mathcal{D}^{k\times(p+m)},\\ &\mathcal{B} = \left\{w = \left(\begin{smallmatrix} y \\ u \end{smallmatrix}\right) \in \mathcal{F}^l = \mathcal{F}^{p+m};\ R \circ w = 0, \text{ i.e., } P \circ y = Q \circ u\right\}, \text{ and}\\ &\mathcal{B}_K = \left\{\widetilde{w} = \left(\begin{smallmatrix} \widetilde{y} \\ \widetilde{u} \end{smallmatrix}\right) \in K^l = K^{p+m};\ R\widetilde{w} = 0, \text{ i.e., } P\widetilde{y} = Q\widetilde{u}\right\}. \end{aligned} \tag{6.8}$$

These data induce the new modules, behavior, and vector space

$$\begin{aligned} &U^0 := U\left(\begin{smallmatrix} \mathrm{id}_p \\ 0 \end{smallmatrix}\right) = \mathcal{D}^{1\times k}(P, -Q)\left(\begin{smallmatrix} \mathrm{id}_p \\ 0 \end{smallmatrix}\right) = \mathcal{D}^{1\times k}P,\ M^0 := \mathcal{D}^{1\times p}/U^0,\\ &\mathcal{B}^0 := (U^0)^\perp = \{y \in \mathcal{F}^p;\ P \circ y = 0\}, \text{ and } KU^0 := K^{1\times k}P. \end{aligned} \tag{6.9}$$

With $KU = K^{1\times k}R = K^{1\times k}(P, -Q)$ we therefore get

$$\dim_K(KU) = \operatorname{rank}(R) \quad\text{and}\quad \dim_K(KU^0) = \operatorname{rank}(P).$$

The data p and the set $J = \{1, \ldots, p\}$ form an IO decomposition of $\mathcal{B}$ if $\operatorname{rank}(P) = \operatorname{rank}(R) = p$, i.e., if $\dim(KU^0) = \dim(KU) = p$ or $\operatorname{rank}(\mathcal{B}^0) = \operatorname{rank}(\mathcal{B}) = m$.

It is obvious that the behavior sequence

$$\begin{array}{ccccccc} 0 \longrightarrow & \mathcal{B}^0 & \xrightarrow{\mathrm{inj}=\left(\begin{smallmatrix} \mathrm{id}_p \\ 0 \end{smallmatrix}\right)\circ} & \mathcal{B} & \xrightarrow{\mathrm{proj}=(0,\mathrm{id}_m)\circ} & \mathcal{F}^m, \\ & y & \longmapsto & \left(\begin{smallmatrix} y \\ 0 \end{smallmatrix}\right),\ \left(\begin{smallmatrix} y \\ u \end{smallmatrix}\right) & \longmapsto & u, \end{array} \tag{6.10}$$

is exact. This implies the exactness of the dual module sequence

$$\begin{array}{ccccccc} 0 \longleftarrow & M^0 & \xleftarrow{\text{proj}=\left(\cdot\binom{\text{id}_p}{0}\right)_{\text{ind}}} & M & \xleftarrow{\text{inj}=(\cdot(0,\text{id}_m))_{\text{ind}}} & \mathcal{D}^{1\times m}, \\ & \overline{\xi} & \longleftarrow\!\shortmid & \overline{(\xi,\eta)},\ \overline{(0,\eta)} & \longleftarrow\!\shortmid & \eta. \end{array} \tag{6.11}$$

Theorem and Definition 6.2.2 (Input/output behavior and transfer matrix, cf. [2, Sect. 5.2 on pp. 135], [3, Sect. 8.1 on pp. 551], and [1, Sect. 3.3 on pp. 80, Sect. 8.2 on pp. 282]) *For the data from* (6.8) *and* (6.9), *the following properties of the system* $\mathcal{B}$ *and the decomposition* $\{1,\dots,l\} = \{1,\dots,p\} \uplus \{p+1,\dots,l=p+m\}$ *are equivalent:*

1. *They constitute an IO decomposition of* $\mathcal{B}$, *i.e.,*

$$p = \text{rank}(P) = \text{rank}(R) = \text{rank}(P,-Q) \quad \textit{or}$$
$$p = \dim_K(KU^0) = \dim_K(KU) \quad \textit{or}$$
$$K^{1\times p} = KU^0 \cong KU.$$

   Then there is a unique matrix $H \in K^{p\times m}$ *such that*

$$PH = Q \quad \textit{and} \quad KU = K^{1\times p}(\text{id}_p, -H).$$

2. *There is a matrix* $H \in K^{p\times m}$ *such that* $KU = K^{1\times p}(\text{id}_p, -H)$. *This matrix* H *coincides with the unique matrix from item 1.*
3. *The projection* $\text{proj} = \cdot\binom{\text{id}_p}{0} : K^{1\times(p+m)} \longrightarrow K^{1\times p}$ *induces an isomorphism* $KU \cong K^{1\times p}$.
4. *The projection*

$$\text{proj} = (0, \text{id}_m)\cdot : \mathcal{B}_K \longrightarrow K^m, \quad \binom{\widetilde{y}}{\widetilde{u}} \longmapsto \widetilde{u}, \tag{6.12}$$

   is a K*-isomorphism. Its inverse is given by the graph of* $H\cdot$, *where* H *is the unique matrix from item 1, i.e., by*

$$K^m \longrightarrow \mathcal{B}_K, \quad \widetilde{u} \longmapsto \binom{H\widetilde{u}}{\widetilde{u}}, \tag{6.13}$$

   and hence, $\mathcal{B}_K = \left\{\binom{H\widetilde{u}}{\widetilde{u}};\ \widetilde{u} \in K^m\right\} = \binom{H}{\text{id}_m} K^m$.
5. *In the sequence* (6.10) *the behavior* $\mathcal{B}^0$ *is autonomous, and the projection*

$$\text{proj} = (0, \text{id}_m)\circ : \mathcal{B} \longrightarrow \mathcal{F}^m, \quad \binom{y}{u} \longmapsto u,$$

   is surjective, i.e., for every $u \in \mathcal{F}^m$ *there is a trajectory* $\binom{y}{u} \in \mathcal{B}$.

6. *In the exact sequence* (6.11) *the module* M^0 *is a torsion module, and the map*

$$(\cdot(0, \mathrm{id}_m))_{\mathrm{ind}} : \mathcal{D}^{1\times m} \longrightarrow M, \quad \eta \longmapsto \overline{(0, \eta)},$$

is a monomorphism. In particular, $\mathrm{inj}(\mathcal{D}^{1\times m})$ *is a free submodule of* M *of dimension* $m = \mathrm{rank}(M)$.

Item 5 says that the component $u \in \mathcal{F}^m$ *can be freely chosen and gives rise to a component or solution* y *with* $P \circ y = Q \circ u$ *or* $\binom{y}{u} \in \mathcal{B}$. *Therefore* u *resp.* y *are called the* input *resp. the* output *of the behavior for the chosen decomposition. Moreover* $\mathcal{B}^0$ *is autonomous without free components. Any other output* $\widehat{y}$ *to the same* u *has the form* $\widehat{y} = y + z$ *with* $z \in \mathcal{B}^0$. *The behavior* $\mathcal{B}$ *with its distinguished IO structure* $\{1, \dots, p\}$ *is called an* IO behavior *or* IO system, *and* $\mathcal{B}^0$ *is called its* autonomous part *or* zero-input part. *The number* $m = \mathrm{rank}(\mathcal{B}) = \mathrm{rank}(M)$ *of input components is the same for all IO structures of* $\mathcal{B}$ *and is therefore called the* input rank *of* $\mathcal{B}$.

Equations (6.12) *and* (6.13) *imply that* H *is uniquely determined by the transfer space* $\mathcal{B}_K$ *of* $\mathcal{B}$ *and its distinguished IO decomposition. The matrix* H *is called the* transfer matrix *of the IO system* $\mathcal{B}$ *with this structure and, conversely,* $\mathcal{B}$ *is called an* IO realization *of* H.

If the rows of R *are linearly independent, in other words, if*

$$R = (P, -Q) \in \mathcal{D}^{p\times(p+m)} \quad \textit{and} \quad p = \mathrm{rank}(R) = \mathrm{rank}(P),$$

then $P \in \mathrm{Gl}_p(K)$ *and* $H = P^{-1}Q$. *If* $\mathcal{D}$ *is a principal ideal domain, then this assumption can be made without loss of generality.*

Proof 1. The condition $\mathrm{rank}(P) = \mathrm{rank}(P, -Q) = p$ implies that the columns of Q are K-linearly dependent on those of P and thus the existence of H with $PH = Q$. The uniqueness of H follows from $\mathrm{rank}(P) = p$ with the ensuing cancellation property

$$P\widetilde{y_1} = P\widetilde{y_2} \Longrightarrow \widetilde{y_1} = \widetilde{y_2} \quad \text{for } \widetilde{y_1}, \widetilde{y_2} \in K^p.$$

1. $\Longrightarrow$ 2. From $KU^0 = K^{1\times k}P = K^{1\times p}$ and $PH = Q$ we infer

$$KU = K^{1\times k}(P, -Q) = K^{1\times k}P(\mathrm{id}_p, -H) = K^{1\times p}(\mathrm{id}_p, -H),$$

and thus 2.

2. $\Longrightarrow$ 3. Obvious.

3. $\Longrightarrow$ 1. From $KU = K^{1\times k}R = K^{1\times k}(P, -Q) \cong K^{1\times p}$ we infer that

$$\mathrm{rank}(R) = \dim_K(KU) = p \text{ and}$$
$$K^{1\times p} = \mathrm{proj}(KU) = K^{1\times k}(P, -Q)\begin{pmatrix}\mathrm{id}_p \\ 0\end{pmatrix} = K^{1\times k}P \Longrightarrow \mathrm{rank}(P) = p.$$

1. $\Longleftrightarrow$ 4. There is the obvious isomorphism

$$\begin{aligned}\{\widetilde{y} \in K^p;\ P\widetilde{y} = 0\} &\stackrel{\cong}{\longrightarrow} \ker\left(\mathrm{proj}\colon\ \mathcal{B}_K \to K^m,\ \left(\begin{smallmatrix}\widetilde{y}\\ \widetilde{u}\end{smallmatrix}\right) \mapsto \widetilde{u}\right)\\ &= \left\{\left(\begin{smallmatrix}\widetilde{y}\\ 0\end{smallmatrix}\right) \in \mathcal{B}_K,\ \text{i.e., } P\widetilde{y} = 0\right\},\\ \widetilde{y} &\longmapsto \left(\begin{smallmatrix}\widetilde{y}\\ 0\end{smallmatrix}\right).\end{aligned}$$

Thus proj is injective if and only if $\mathrm{rank}(P) = p$. With this proj is even an isomorphism if and only if $\dim(\mathcal{B}_K) = m$ or $l - \mathrm{rank}(R) = l - p$, i.e., if and only if $\mathrm{rank}(R) = p$. Under these conditions we have the equivalences

$$\left(\begin{smallmatrix}\widetilde{y}\\ \widetilde{u}\end{smallmatrix}\right) \in \mathcal{B}_K \Longleftrightarrow P\widetilde{y} = Q\widetilde{u} = PH\widetilde{u} \underset{\mathrm{rank}(P)=p}{\Longleftrightarrow} \widetilde{y} = H\widetilde{u}.$$

This signifies that $\widetilde{u} \longmapsto \left(\begin{smallmatrix}H\widetilde{u}\\ \widetilde{u}\end{smallmatrix}\right)$ is indeed the inverse of proj.

1. $\Longrightarrow$ 6. From $\mathrm{rank}(P) = p$ we infer that $0 = p - \mathrm{rank}(P) = \dim_K(\mathcal{B}^0_K)$. By Corollary 6.1.3 this signifies that $\mathcal{B}^0$ is autonomous and hence M^0 is a torsion module. Moreover the map

$$(\cdot(0, \mathrm{id}_m))_{\mathrm{ind}} : \mathcal{D}^{1\times m} \longrightarrow M, \quad \eta \longmapsto \overline{(0, \eta)},$$

is injective: If $\overline{(0, \eta)} = \overline{0} \in M = \mathcal{D}^{1\times l}/\mathcal{D}^{1\times k}(P, -Q)$, i.e., if there is a $\zeta \in \mathcal{D}^{1\times k}$ such that $(0, \eta) = \zeta(P, -Q)$, then

$$0 = \zeta P \quad \text{and} \quad \eta = -\zeta Q = -\zeta(PH) = -(\zeta P)H = 0H = 0.$$

6. $\Longrightarrow$ 1. Since M^0 is a torsion module and thus $\mathcal{B}^0$ is autonomous, we obtain $p = \mathrm{rank}(P)$ as before via Corollary 6.1.3. The monomorphism $\mathcal{D}^{1\times m} \longrightarrow M$ implies

$$m \le \mathrm{rank}(M) = l - \mathrm{rank}(R) \le l - \mathrm{rank}(P) = l - p = m,$$

hence $p = \mathrm{rank}(R) = \mathrm{rank}(P)$.

5. $\Longleftrightarrow$ 6. By duality. □

Corollary 6.2.3 (Cf. [1, Theorem 8.2.7 on p. 289])

1. *The IO structures of a submodule $U \subseteq \mathcal{D}^{1\times l}$ or its associated behavior $\mathcal{B} := U^\perp$ depend only on the vector space $KU \subseteq K^{1\times l}$ generated by U. In other words, submodules $U_1, U_2 \subseteq \mathcal{D}^{1\times l}$ or their behaviors $\mathcal{B}_i := U_i^\perp$ admit the same IO decompositions if $KU_1 = KU_2$.*
2. *For two IO systems $\mathcal{B}_1, \mathcal{B}_2 \subseteq \mathcal{F}^{p+m}$ with the same IO decomposition $\{1, \dots, p\}$ and the transfer matrices H_1, H_2 the following equivalences hold:*

$$\mathcal{B}_{1,\mathrm{cont}} = \mathcal{B}_{2,\mathrm{cont}} \Longleftrightarrow \mathcal{B}_{1,K} = \mathcal{B}_{2,K} \Longleftrightarrow H_1 = H_2.$$

Proof 1. This follows from item 2 of Theorem 6.2.2.

2. The first equivalence follows from Theorem 6.1.7, the second one from (6.13). □

Next we construct the unique controllable realization of an arbitrary matrix $H \in K^{p\times m}$.

Theorem 6.2.4 (Controllable realization of a matrix, cf. [1, Theorem 5.2.14 on p. 161]) *Let $H \in K^{p\times m}$ be a matrix and let d be a common denominator of the entries of H such that $dH \in \mathcal{D}^{p\times m}$.*

1. *If $\mathcal{B} \subseteq \mathcal{F}^{p+m}$ is an IO realization of $H \in K^{p\times m}$, then so is its largest controllable subbehavior $\mathcal{B}_{\text{cont}}$. In particular, this means that a behavior $\mathcal{B}$ and its controllable part $\mathcal{B}_{\text{cont}}$ admit the same IO decompositions and have the same transfer matrices with respect to them.*
   Moreover, $\mathcal{B} = \mathcal{B}^0 + \mathcal{B}_{\text{cont}}$, where we identify $\mathcal{B}^0 \underset{\text{ident.}}{=} \mathcal{B} \cap (\mathcal{F}^p \times \{0\}) \ni y = \binom{y}{0}$.
   In general, the intersection $\mathcal{B}^0 \cap \mathcal{B}_{\text{cont}}$ is not zero and $\mathcal{B} = \mathcal{B}^0 + \mathcal{B}_{\text{cont}}$ is not the controllable/autonomous direct decomposition from Corollary and Definition 4.2.19.
2. *The matrix H has a unique controllable IO realization $\mathcal{B}_{\text{cont}}$. It is also the least IO realization of H and therefore called the minimal realization of H. Let*

$$U_{\text{cont}} = \mathcal{B}_{\text{cont}}^{\perp} = \mathcal{D}^{1\times k_c} R_{\text{cont}} \subseteq \mathcal{D}^{1\times l} \tag{6.14}$$

   with $R_{\text{cont}} = (P_{\text{cont}}, -Q_{\text{cont}}) \in \mathcal{D}^{k_c\times(p+m)}$. Then $\operatorname{rank}(P_{\text{cont}}) = p$ and the identity $P_{\text{cont}} H = Q_{\text{cont}}$ holds. Furthermore, U_{cont} is given as

$$\begin{aligned} U_{\text{cont}} &= \ker\left(\cdot \begin{pmatrix} H \\ \mathrm{id}_m \end{pmatrix} : \mathcal{D}^{1\times(p+m)} \longrightarrow K^{1\times m}\right) \\ &= \left\{(\xi, \eta) \in \mathcal{D}^{1\times(p+m)};\ \xi H + \eta = 0\right\} \\ &= \ker\left(\cdot \begin{pmatrix} dH \\ d\,\mathrm{id}_m \end{pmatrix} : \mathcal{D}^{1\times(p+m)} \longrightarrow \mathcal{D}^{1\times m}\right) \\ &= \left\{(\xi, \eta) \in \mathcal{D}^{1\times(p+m)};\ \xi(dH) + d\eta = 0\right\}. \end{aligned} \tag{6.15}$$

   In other words, the sequences

$$\begin{aligned} &\mathcal{D}^{1\times k_c} \xrightarrow{\cdot(P_{\text{cont}}, -Q_{\text{cont}})} \mathcal{D}^{1\times(p+m)} \xrightarrow{\cdot\begin{pmatrix} H \\ \mathrm{id}_m \end{pmatrix}} K^{1\times m} \\ &\mathcal{D}^{1\times k_c} \xrightarrow{\cdot(P_{\text{cont}}, -Q_{\text{cont}})} \mathcal{D}^{1\times(p+m)} \xrightarrow{\cdot\begin{pmatrix} dH \\ d\,\mathrm{id}_m \end{pmatrix}} \mathcal{D}^{1\times m} \end{aligned} \tag{6.16}$$

   are exact, i.e., R_{cont} is a universal left annihilator of $\begin{pmatrix} H \\ \mathrm{id}_m \end{pmatrix}$, which can be computed via Theorem 3.2.18 if $\mathcal{D}$ is a principal ideal domain.
   Moreover

$$U_{\text{cont}}^0 = (U_{\text{cont}})^0 = \mathcal{D}^{1\times k_c} P_{\text{cont}} = \{\xi \in \mathcal{D}^{1\times p};\ \xi H \in \mathcal{D}^{1\times m}\}. \tag{6.17}$$

If $\mathcal{D}$ is a principal ideal domain, then we can assume without loss of generality that $k_c = p$ and $P_{\text{cont}} \in \mathrm{Gl}_p(K)$. In this case, the matrix $(P_{\text{cont}}, -Q_{\text{cont}})$ is right invertible by Theorem 4.2.12, 2, the equation $H = P_{\text{cont}}^{-1} Q_{\text{cont}}$ holds, and the map $\cdot(P_{\text{cont}}, -Q_{\text{cont}})$ from (6.16) is injective.

Proof 1. The first assertion follows directly from

$$\mathcal{B}_{\text{cont},K} = \mathcal{B}_K = \left\{ \begin{pmatrix} H\widetilde{u} \\ \widetilde{u} \end{pmatrix};\ \widetilde{u} \in K^m \right\}$$

by means of Corollary 6.1.5 and Theorem 6.2.2, 4. Let $\begin{pmatrix} y \\ u \end{pmatrix} \in \mathcal{B}$. Since u is also an input for $\mathcal{B}_{\text{cont}}$, there is an $y_{\text{cont}} \in \mathcal{F}^p$ such that $\begin{pmatrix} y_{\text{cont}} \\ u \end{pmatrix} \in \mathcal{B}_{\text{cont}} \subseteq \mathcal{B}$. Hence,

$$\begin{pmatrix} y \\ u \end{pmatrix} = \begin{pmatrix} y - y_{\text{cont}} \\ 0 \end{pmatrix} + \begin{pmatrix} y_{\text{cont}} \\ u \end{pmatrix} \in \mathcal{B}^0 + \mathcal{B}_{\text{cont}}.$$

2. *Uniqueness.* If $\mathcal{B}_1$ and $\mathcal{B}_2$ are two controllable realizations of H, then $\mathcal{B}_i = \mathcal{B}_{i,\text{cont}}$, $i = 1, 2$, and from Corollary 6.2.3, 2, we infer that $\mathcal{B}_1 = \mathcal{B}_2$.
Existence. The equations

$$H = (d\,\mathrm{id}_p)^{-1}(dH) \quad \text{and} \quad p = \mathrm{rank}(d\,\mathrm{id}_p) = \mathrm{rank}(d\,\mathrm{id}_p, -dH)$$

imply that the behavior $\mathcal{B} = U^\perp$ with $U := \mathcal{D}^{1\times p}(d\,\mathrm{id}_p, -dH)$ is an IO realization of H. By 1 its controllable part $\mathcal{B}_{\text{cont}}$ is a controllable IO realization of H.

Theorem and Definition 6.2.2, 2 implies

$$\begin{aligned} KU &= K^{1\times p}(\mathrm{id}_p, -H) = \ker\left(\cdot \begin{pmatrix} H \\ \mathrm{id}_m \end{pmatrix} :\ K^{1\times(p+m)} \longrightarrow K^{1\times m}\right) \\ &= K^{1\times p}(d\,\mathrm{id}_p, -dH) = \ker\left(\cdot \begin{pmatrix} dH \\ d\,\mathrm{id}_m \end{pmatrix} :\ K^{1\times(p+m)} \longrightarrow K^{1\times m}\right). \end{aligned}$$

From this and Theorem 4.2.15 we infer

$$\begin{aligned} U_{\text{cont}} &= \mathcal{D}^{1\times(p+m)} \cap KU = \mathcal{D}^{1\times(p+m)} \cap K^{1\times p}(\mathrm{id}_p, -H) \\ &= \ker\left(\cdot \begin{pmatrix} H \\ \mathrm{id}_m \end{pmatrix} :\ \mathcal{D}^{1\times(p+m)} \longrightarrow K^{1\times m}\right) \\ &= \mathcal{D}^{1\times(p+m)} \cap K^{1\times p}(d\,\mathrm{id}_p, -dH) \\ &= \ker\left(\cdot \begin{pmatrix} dH \\ d\,\mathrm{id}_m \end{pmatrix} :\ \mathcal{D}^{1\times(p+m)} \longrightarrow \mathcal{D}^{1\times m}\right). \end{aligned}$$

The inclusion $\mathcal{D}^{1\times k_c} P_{\text{cont}} \subseteq \{\xi \in \mathcal{D}^{1\times p};\ \xi H \in \mathcal{D}^{1\times m}\}$ follows from $P_{\text{cont}} H = Q_{\text{cont}}$. As for the converse inclusion, let $\xi \in \mathcal{D}^{1\times p}$ with $\eta := \xi H \in \mathcal{D}^{1\times m}$. Then $(\xi, -\eta)\begin{pmatrix} H \\ \mathrm{id}_m \end{pmatrix} = 0$, and, by the first exact sequence in (6.16), there is a $\zeta \in \mathcal{D}^{1\times k_c}$ with $(\xi, -\eta) = \zeta(P_{\text{cont}}, -Q_{\text{cont}})$. Therefore, $\xi = \zeta P_{\text{cont}} \in \mathcal{D}^{1\times k_c} P_{\text{cont}}$. □

In the literature, the unique controllable realization of a transfer matrix H is usually treated by means of its *coprime factorizations*.

Theorem 6.2.5 (Coprime factorizations) *We assume that $\mathcal{D}$ is a principal ideal domain and that $H \in K^{p\times m}$.*

1. *If* $(P_{\mathrm{cont}}, -Q_{\mathrm{cont}}) \in \mathcal{D}^{p\times(p+m)}$ *with* $\mathcal{D}^{1\times p}P_{\mathrm{cont}} = \{\xi \in \mathcal{D}^{1\times p};\ \xi H \in \mathcal{D}^{1\times m}\}$ *is a matrix of the controllable realization of* $H = P_{\mathrm{cont}}^{-1}Q_{\mathrm{cont}}$*, then the pair* $(P_{\mathrm{cont}}, -Q_{\mathrm{cont}})$ *is unique up to row equivalence, i.e., up to left multiplication with a matrix in* $\mathrm{Gl}_p(\mathcal{D})$*, and has a right inverse by Theorem 4.2.12, 2. Any representation*

$$H = P^{-1}Q \quad \text{with } P \in \mathcal{D}^{p\times p},\ Q \in \mathcal{D}^{p\times m},$$
$$\det(P) \neq 0 \text{ and right invertible } (P, -Q)$$

is called a left coprime factorization *of* H *in the literature [3, p. 379], [4, Definition 1.36], [5, Definition 506]. Up to row equivalence, it coincides with* $H = P_{\mathrm{cont}}^{-1}Q_{\mathrm{cont}}$*, hence the sequence*

$$0 \longrightarrow \mathcal{D}^{1\times p} \xrightarrow{\cdot(P,-Q)} \mathcal{D}^{1\times(p+m)} \xrightarrow{\cdot\binom{H}{\mathrm{id}_m}} K^{1\times m} \tag{6.18}$$

is exact and the identity $\mathcal{D}^{1\times p}P = \{\xi \in \mathcal{D}^{1\times p};\ \xi H \in \mathcal{D}^{1\times m}\}$ *holds.*

2. *A* right coprime factorization *of* H *is a representation*

$$H = ND^{-1} \quad \text{with } D \in \mathcal{D}^{m\times m},\ N \in \mathcal{D}^{p\times m},$$
$$\det(D) \neq 0 \text{ and left invertible } \binom{N}{D}.$$

It is unique up to column equivalence and characterized by the exact sequence

$$0 \to \mathcal{D}^m \xrightarrow{\binom{N}{D}\cdot} \mathcal{D}^{p+m} \xrightarrow{(\mathrm{id}_p,-H)\cdot} K^p. \tag{6.19}$$

The identity $D\mathcal{D}^m = \{\xi \in \mathcal{D}^m;\ H\xi \in \mathcal{D}^p\}$ *holds. In other words, the matrix* $(D^\top, -N^\top)$ *is the, up to row equivalence unique, matrix of the controllable IO realization of* $H^\top$ *and can be computed as universal left annihilator of* $\begin{pmatrix} H^\top \\ \mathrm{id}_p \end{pmatrix}$.

3. *If* $\mathcal{B} := \left\{\binom{y}{u} \in \mathcal{F}^{p+m};\ P \circ y = Q \circ u\right\}$ *with* $P \in \mathcal{D}^{p\times p}$, $\det(P) \neq 0$, *is any IO realization of* $H \in K^{p\times m}$*, i.e.,* $H = P^{-1}Q$*, and if* $H = ND^{-1}$ *is the unique (up to column equivalence) right coprime factorization of* H*, then the sequence*

$$0 \longrightarrow \mathcal{D}^{1\times p} \xrightarrow{\cdot(P,-Q)} \mathcal{D}^{1\times(p+m)} \xrightarrow{\cdot\binom{N}{D}} \mathcal{D}^{1\times m} \longrightarrow 0 \tag{6.20}$$

is a complex with a monomorphism $\cdot(P, -Q)$ *and an epimorphism* $\cdot\binom{N}{D}$*. It is exact if and only if* $\mathcal{B}$ *is controllable, i.e., if* $H = P^{-1}Q$ *is the unique (up to row equivalence) left coprime factorization of* H*.*

4. *Assume* $(P, -Q) \in \mathcal{D}^{p\times(p+m)}$ *with* $\mathrm{rank}(P) = p$ *and* $H := P^{-1}Q$*. Consider the Smith form*

$$U(P,-Q)V=(E,0),\ E=\operatorname{diag}(e_1,\dots,e_p),\ e_1 \underset{\mathcal{D}}{|} e_2 \underset{\mathcal{D}}{|} \cdots \underset{\mathcal{D}}{|} e_p,\ \textit{and}$$
$$\begin{pmatrix} D_2 & N_1\\ N_2 & D_1\end{pmatrix} := V,\ \begin{pmatrix} P_1 & -Q_1\\ -Q_2 & P_2\end{pmatrix} := V^{-1}\in \mathrm{Gl}_{p+m}(\mathcal{D}).\ \textit{Then} \tag{6.21}$$
$$\operatorname{rank}(P_1)=p,\ \operatorname{rank}(D_1)=m \textit{ and } H=P_1^{-1}Q_1=N_1D_1^{-1}$$

are the essentially unique coprime factorizations, and easy to compute.
If f is a common denominator of the entries of H, then $H=(f\,\mathrm{id}_p)^{-1}(fH)$ is such a representation $H=P^{-1}Q$.

Proof 1. If $H=P^{-1}Q$ is any left coprime factorization of H, the right invertibility of $(P,-Q)$ is equivalent with the controllability of the associated behavior, cf. Theorem 4.2.12, 2. Hence this behavior is the unique controllable realization of H and therefore, $(P,-Q)$ and $(P_{\mathrm{cont}},-Q_{\mathrm{cont}})$ coincide up to row equivalence. Equation (6.18) then follows from (6.16).
2. This follows from 1 by transposition, i.e., $(D^\top,-N^\top)$ is the matrix of the unique controllable realization of $H^\top$. Also, (6.19) follows from (6.18) applied to $H^\top$ by transposition and an obvious sign change.
3. The monomorphy of $\cdot(P,-Q)$ follows from $\operatorname{rank}(P,-Q)=p$, the epimorphy of $\cdot\binom{N}{D}$ from its left invertibility. The equation $H=P^{-1}Q=ND^{-1}$ is equivalent with $PN=QD$, in other words, $(P,-Q)\binom{N}{D}=0$. This signifies that (6.20) is a complex at $\mathcal{D}^{1\times(p+m)}$.

From the equation $\binom{N}{D}=\binom{H}{\mathrm{id}_m}D$ and $\det(D)\neq 0$ we infer that

$$\begin{aligned}\ker\left\{\cdot\begin{pmatrix}H\\ \mathrm{id}_m\end{pmatrix}: \mathcal{D}^{1\times(p+m)}\to K^{1\times m}\right\} &= \ker\left\{\cdot\begin{pmatrix}N\\ D\end{pmatrix}: \mathcal{D}^{1\times(p+m)}\to K^{1\times m}\right\}\\ &= \ker\left\{\cdot\begin{pmatrix}N\\ D\end{pmatrix}: \mathcal{D}^{1\times(p+m)}\to \mathcal{D}^{1\times m}\right\}.\end{aligned}$$

According to 1, the sequence (6.18) is exact, i.e., the kernel we have just computed coincides with $\mathcal{D}^{1\times p}(P,-Q)$, if and only if $(P,-Q)$ has a right inverse, i.e., if and only if $H=P^{-1}Q$ is a left coprime factorization.
4. The equation $V^{-1}V=\mathrm{id}_{p+m}$ implies that $(P_1,-Q_1)$ resp. $\binom{N_1}{D_1}$ are right resp. left invertible since $(P_1,-Q_1)\binom{D_2}{N_2}=\mathrm{id}_p$ and $(-Q_2,P_2)\binom{N_1}{D_1}=\mathrm{id}_m$. Moreover

$$\begin{aligned}&U(P,-Q)V=(E,0)\Longrightarrow U(PN_1-QD_1)=0 \underset{\operatorname{rank}(U)=p}{\Longrightarrow} PN_1=QD_1=PHD_1\\ &\underset{\operatorname{rank}(P)=p}{\Longrightarrow} N_1=HD_1\Longrightarrow \begin{pmatrix}N_1\\ D_1\end{pmatrix}=\begin{pmatrix}H\\ \mathrm{id}_m\end{pmatrix}D_1\\ &\underset{\operatorname{rank}\binom{N_1}{D_1}=m}{\Longrightarrow} \operatorname{rank}(D_1)=m\Longrightarrow H=N_1D_1^{-1}.\end{aligned}$$

Likewise

$$\begin{aligned}&(P_1,-Q_1)\begin{pmatrix}N_1\\ D_1\end{pmatrix}=0\Longrightarrow P_1HD_1=P_1N_1=Q_1D_1 \underset{\operatorname{rank}(D_1)=m}{\Longrightarrow} P_1H=Q_1\\ &\Longrightarrow (P_1,-Q_1)=P_1(\mathrm{id}_p,-H) \underset{\operatorname{rank}(P_1,-Q_1)=p}{\Longrightarrow} \operatorname{rank}(P_1)=p\Longrightarrow H=P_1^{-1}Q_1.\end{aligned}$$

Hence $H = P_1^{-1}Q_1 = N_1 D_1^{-1}$ are the essentially unique coprime factorizations. □

Left and right coprime factorizations of $H \in K^{p\times m}$ can also be read off the Smith/McMillan form of H from Corollary 3.2.19: Assume that

$$\begin{aligned} &UHV = \begin{pmatrix} \operatorname{diag}(a_1/b_1, a_2/b_2, \dots, a_r/b_r) & 0 \\ 0 & 0 \end{pmatrix},\ U \in \mathrm{Gl}_p(\mathcal{D}),\ V \in \mathrm{Gl}_m(\mathcal{D}),\ \operatorname{rank}(H) = r \\ &a_i, b_i \in \mathcal{D},\ \gcd(a_i, b_i) = 1,\ a_1|a_2|\cdots|a_r \neq 0,\ b_r|b_{r-1}|\cdots|b_1 \neq 0, \end{aligned} \tag{6.22}$$

is this form. Define

$$\begin{aligned} P'_{\text{cont}} &:= \begin{pmatrix} \operatorname{diag}(b_1, \dots, b_r) & 0 \\ 0 & \mathrm{id}_{p-r} \end{pmatrix} \in \mathcal{D}^{p\times p} \cap \mathrm{Gl}_p(K) \\ Q'_{\text{cont}} &:= \begin{pmatrix} \operatorname{diag}(a_1, \dots, a_r) & 0 \\ 0 & 0 \end{pmatrix} \in \mathcal{D}^{p\times m},\ \operatorname{rank}(Q'_{\text{cont}}) = \operatorname{rank}(H) = r \\ D' &:= \begin{pmatrix} \operatorname{diag}(b_1, \dots, b_r) & 0 \\ 0 & \mathrm{id}_{m-r} \end{pmatrix} \in \mathcal{D}^{m\times m} \cap \mathrm{Gl}_m(K) \\ N' &:= \begin{pmatrix} \operatorname{diag}(a_1, \dots, a_r) & 0 \\ 0 & 0 \end{pmatrix} = Q'_{\text{cont}} \in \mathcal{D}^{p\times m},\ \operatorname{rank}(N') = \operatorname{rank}(H) = r. \end{aligned} \tag{6.23}$$

Since $\gcd(a_i, b_i) = 1$, $i = 1, \dots, r$, it is easy to see that $(P'_{\text{cont}}, -Q'_{\text{cont}})$ has a right inverse and $\binom{N'}{D'}$ has a left inverse. Therefore

$$UHV = (P'_{\text{cont}})^{-1} Q'_{\text{cont}} = N' D'^{-1}$$

are the essentially unique left resp. right coprime factorization of UHV.

Corollary and Definition 6.2.6 *(i) With the data from* (6.22) *and* (6.23) *the essentially unique left resp. right coprime factorization of $H \in K^{p\times m}$ are*

$$\begin{aligned} &H = P_{\text{cont}}^{-1} Q_{\text{cont}} = ND^{-1} \text{ with } P_{\text{cont}} := P'_{\text{cont}} U,\ Q_{\text{cont}} := Q'_{\text{cont}} V^{-1} = N'V^{-1}, \\ &D := VD',\ N := U^{-1}N',\ Q_{\text{cont}} = UNV^{-1}. \end{aligned} \tag{6.24}$$

In particular, Q_{cont} and N are equivalent.
(ii) For the unique controllable realization $\mathcal{B}_{\text{cont}} = \left\{\binom{y}{u} \in \mathcal{F}^{p+m};\ P_{\text{cont}} \circ y = Q_{\text{cont}} \circ u\right\}$ of H this implies

$$\begin{aligned} &\mathcal{B}^0_{\text{cont}} = \left\{y \in \mathcal{F}^p;\ P_{\text{cont}} \circ y = 0\right\}, \\ &M^0_{\text{cont}} = M\left(\mathcal{B}^0_{\text{cont}}\right) = \mathcal{D}^{1\times p}/\mathcal{D}^{1\times p} P_{\text{cont}} \xrightarrow{(\cdot U^{-1})_{\text{ind}},\cong} \mathcal{D}^{1\times p}/\mathcal{D}^{1\times p} P'_{\text{cont}} = \prod_{i=1}^{r} \mathcal{D}/\mathcal{D}b_i \times \{0\}. \end{aligned} \tag{6.25}$$

(iii) Consider the real resp. complex standard situation from Theorem 5.4.8 resp. Corollary 5.4.12 with the length vector $\mathrm{l}(M^0_{\text{cont}}) = \mathrm{l}(\mathcal{B}^0_{\text{cont}})$. Let $b_i = \prod_{p\in\mathcal{P}} p^{\mu(i,p)}$ be the prime factor decomposition where almost all exponents $\mu(i, p)$ are zero. The Chinese remainder theorem implies

$$\mathcal{D}/\mathcal{D}b_i = \prod_{p,\mu(i,p)>0} \mathcal{D}/\mathcal{D}p^{\mu(i,b)} \Longrightarrow 1_p(\mathcal{D}/\mathcal{D}b_i) = \mu(i,p)$$
$$\Longrightarrow 1_p(H) := 1_p\left(\mathcal{B}^0_{\mathrm{cont}}\right) = 1_p(M^0_{\mathrm{cont}}) = \sum_{i=1}^{r} \mu(i,p). \tag{6.26}$$

We define $1(H) := \left(1_p(H)\right)_{p\in\mathcal{P}} = 1(M^0_{\mathrm{cont}}) = 1(\mathcal{B}^0_{\mathrm{cont}})$. ◇

6.2.2 The Autonomous Behaviors $\mathcal{B}^0$, $\mathcal{B}^0_{\mathrm{cont}}$ and $\mathcal{B}_{\mathrm{uncont}}$

We assume a principal ideal domain $\mathcal{D}$ *and the data from Corollary and Definition* 6.2.5 *and Theorems* 4.2.15 *and* 4.2.16.

The IO behavior $\mathcal{B}$ gives rise to the three autonomous behaviors $\mathcal{B}^0$, $\mathcal{B}^0_{\mathrm{cont}} = (\mathcal{B}_{\mathrm{cont}})^0 \subseteq \mathcal{B}^0$, and $\mathcal{B}_{\mathrm{uncont}}$. In the next theorem we discuss their relationship. As usual, we use the matrices resp. modules

$$\begin{aligned}
&R = (P, -Q),\ R_{\mathrm{cont}} = (P_{\mathrm{cont}}, -Q_{\mathrm{cont}}) \in \mathcal{D}^{p\times(p+m)}\\
&U = \mathcal{D}^{1\times p}R = \mathcal{B}^\perp \subseteq U_{\mathrm{cont}} = \mathcal{D}^{1\times p}R_{\mathrm{cont}} = \mathcal{B}^\perp_{\mathrm{cont}} \text{ and}\\
&U^0 = \mathcal{D}^{1\times p}P = \mathcal{B}^{0\perp} \subseteq U^0_{\mathrm{cont}} = \mathcal{D}^{1\times p}P_{\mathrm{cont}} = (\mathcal{B}^0_{\mathrm{cont}})^\perp,
\end{aligned} \tag{6.27}$$

as well as the canonical epimorphisms

$$\begin{aligned}
M &:= \mathcal{D}^{1\times(p+m)}/U \overset{\mathrm{can}}{\longrightarrow} M_{\mathrm{cont}} := \mathcal{D}^{1\times(p+m)}/U_{\mathrm{cont}}, \text{ and}\\
M^0 &:= \mathcal{D}^{1\times p}/U^0 \overset{\mathrm{can}}{\longrightarrow} M^0_{\mathrm{cont}} := \mathcal{D}^{1\times p}/U^0_{\mathrm{cont}}.
\end{aligned} \tag{6.28}$$

The torsion module of M is $\mathrm{t}(M) = U_{\mathrm{cont}}/U$. There is a unique matrix $X_{\mathrm{cont}} \in \mathcal{D}^{p\times p}$ with

$$\begin{aligned}
&P = X_{\mathrm{cont}}P_{\mathrm{cont}}, \text{ i.e., } X_{\mathrm{cont}} = PP^{-1}_{\mathrm{cont}},\ \mathrm{rank}(X_{\mathrm{cont}}) = p\\
&\Longrightarrow R = P(\mathrm{id}_p, -H) = X_{\mathrm{cont}}P_{\mathrm{cont}}(\mathrm{id}_p, -H) = X_{\mathrm{cont}}R_{\mathrm{cont}}.
\end{aligned} \tag{6.29}$$

The data induce the exact module resp. dual behavior sequence

$$\begin{aligned}
&0 \to \quad M_{\mathrm{uncont}} := \mathcal{D}^{1\times p}/\mathcal{D}^{1\times p}X_{\mathrm{cont}} \xrightarrow{(\cdot R_{\mathrm{cont}})_{\mathrm{ind}}} M \overset{\mathrm{can}}{\longrightarrow} M_{\mathrm{cont}} \to 0 \text{ resp.}\\
&0 \leftarrow \mathcal{B}_{\mathrm{uncont}} := \{z \in \mathcal{F}^p;\ X_{\mathrm{cont}} \circ z = 0\} \xleftarrow{R_{\mathrm{cont}}\circ} \mathcal{B} \overset{\mathrm{inj}}{\leftarrow} \mathcal{B}_{\mathrm{cont}} \leftarrow 0\\
&\Longrightarrow \mathcal{B}/\mathcal{B}_{\mathrm{cont}} \cong \mathcal{B}_{\mathrm{uncont}},\ w + \mathcal{B}_{\mathrm{cont}} \mapsto R_{\mathrm{cont}} \circ w.
\end{aligned} \tag{6.30}$$

In part 2. of the next Theorem 6.2.7 we make the following standard assumption:

$$F=\mathbb{R},\mathbb{C},\ \mathcal{D}:=F[s],\ \mathcal{F}:=\begin{cases}C^{\infty}(\mathbb{R},F) & \text{continuous case}\\ F^{\mathbb{N}} & \text{discrete case}\end{cases}$$

$$\Lambda_F:=\begin{cases}\mathbb{C} & \text{if } F=\mathbb{C}\\ \Lambda_F=\{\lambda\in\mathbb{C};\ \Im(\lambda)\geq 0\} & \text{if } F=\mathbb{R}\end{cases},\quad \mathcal{P}:=\{p_\lambda;\ \lambda\in\Lambda_F\}\subset F[s]$$

$$\text{where } \forall\lambda\in\Lambda_F:\ p_\lambda:=\begin{cases}s-\lambda & \text{if } F=\mathbb{C},\ \lambda\in\Lambda_{\mathbb{C}}=\mathbb{C}\\ s-\lambda & \text{if } F=\mathbb{R},\ \lambda\in\mathbb{R}\subset\Lambda_{\mathbb{R}}\\ (s-\lambda)(s-\overline{\lambda}) & \text{if } F=\mathbb{R},\ \Im(\lambda)>0\end{cases}. \tag{6.31}$$

Theorem 6.2.7 *1. The projection* $\mathrm{proj}=\cdot\begin{pmatrix}\mathrm{id}_p\\0\end{pmatrix}:\ \mathcal{D}^{1\times(p+m)}\longrightarrow\mathcal{D}^{1\times p}$ *implies* $\mathrm{proj}(U)=U^0$ *and* $\mathrm{proj}(U_{\mathrm{cont}})=U^0_{\mathrm{cont}}$ *and hence the epimorphism*

$$\begin{array}{ccccc} M_{\mathrm{uncont}}=\mathcal{D}^{1\times p}/\mathcal{D}^{1\times p}X_{\mathrm{cont}} & \xrightarrow{(\cdot R_{\mathrm{cont}})_{\mathrm{ind}}\cong} & \mathrm{t}(M)=U_{\mathrm{cont}}/U & \xrightarrow{\left(\cdot\begin{pmatrix}\mathrm{id}_p\\0\end{pmatrix}\right)_{\mathrm{ind}}} & U^0_{\mathrm{cont}}/U^0,\\ \overline{\eta}=\eta+\mathcal{D}^{1\times p}X_{\mathrm{cont}} & \mapsto & \overline{\eta R_{\mathrm{cont}}} & \mapsto & \overline{\eta P_{\mathrm{cont}}}. \end{array} \tag{6.32}$$

This epimorphism (6.32) *is indeed an isomorphism or, in other terms, there is the commutative diagram with exact rows*

$$\begin{array}{ccccccccc} 0 & \longrightarrow & M_{\mathrm{uncont}} & \xrightarrow{(\cdot R_{\mathrm{cont}})_{\mathrm{ind}}} & M & \xrightarrow{\mathrm{can}} & M_{\mathrm{cont}} & \longrightarrow & 0\\ & & \Big\updownarrow = & & \Big\downarrow \mathrm{proj}=\left(\cdot\begin{pmatrix}\mathrm{id}_p\\0\end{pmatrix}\right)_{\mathrm{ind}} & & \Big\downarrow \mathrm{proj} & & \\ 0 & \longrightarrow & M_{\mathrm{uncont}} & \xrightarrow{(\cdot P_{\mathrm{cont}})_{\mathrm{ind}}} & M^0 & \xrightarrow{\mathrm{can}} & M^0_{\mathrm{cont}} & \longrightarrow & 0 \end{array} \tag{6.33}$$

where $M^0=\mathcal{D}^{1\times p}/U^0,\ M^0_{\mathrm{cont}}=\mathcal{D}^{1\times p}/U^0_{\mathrm{cont}},\ M_{\mathrm{uncont}}\overset{(\cdot R_{\mathrm{cont}})_{\mathrm{ind}}}{\cong}\mathrm{t}(M)$.

This diagram induces its dual commutative diagram with exact rows

$$\begin{array}{ccccccccc} 0 & \longleftarrow & \mathcal{B}_{\mathrm{uncont}} & \xleftarrow{R_{\mathrm{cont}}\circ} & \mathcal{B} & \overset{\supseteq}{\longleftarrow} & \mathcal{B}_{\mathrm{cont}} & \longleftarrow & 0\\ & & \Big\updownarrow = & & \Big\uparrow \cup\!\mathsf{I} & & \Big\uparrow \cup\!\mathsf{I} & & \\ 0 & \longleftarrow & \mathcal{B}_{\mathrm{uncont}} & \xleftarrow{P_{\mathrm{cont}}\circ} & \mathcal{B}^0 & \overset{\supseteq}{\longleftarrow} & \mathcal{B}^0_{\mathrm{cont}} & \longleftarrow & 0 \end{array} \tag{6.34}$$

where $\mathcal{B}^0/\mathcal{B}^0_{\mathrm{cont}}\overset{(P_{\mathrm{cont}}\circ)_{\mathrm{ind}}}{\cong}\mathcal{B}_{\mathrm{uncont}},\ y+\mathcal{B}^0_{\mathrm{cont}}\mapsto P_{\mathrm{cont}}\circ y,\ \mathcal{B}_{\mathrm{uncont}}\cong\mathrm{D}(\mathrm{t}(M)),$

and where we identify $y \in \mathcal{B}^0$ *with* $\binom{y}{0} \in \mathcal{B}$.
2. *Under the condition of* (6.31) *the exact sequence* (6.33) *and Theorem 5.4.8 and Corollaries 5.4.10 and 5.4.12 imply*

$$\begin{aligned}
&\mathrm{l}(M^0) = \mathrm{l}(\mathrm{t}(M)) + \mathrm{l}\left(M^0_{\mathrm{cont}}\right) = \\
&\mathrm{l}(\mathcal{B}^0) = \mathrm{l}\left(\mathcal{B}_{\mathrm{uncont}}\right) + \mathrm{l}\left(\mathcal{B}^0_{\mathrm{cont}}\right) \in \mathbb{N}^{(\Lambda_F)},\ \mathrm{l}(\mathrm{t}(M)) = \mathrm{l}\left(M_{\mathrm{uncont}}\right) \\
&\mathrm{char}(\mathcal{B}^0) = \mathrm{char}(M^0) = \mathrm{char}(\mathrm{t}(M)) \cup \mathrm{char}(M^0_{\mathrm{cont}}) = \mathrm{char}(\mathcal{B}) \cup \mathrm{char}(\mathcal{B}^0_{\mathrm{cont}}) \\
&\text{where } \forall \lambda \in \Lambda_F:\ \mathrm{l}_{p_\lambda}(M^0) = \frac{\dim_F\left(\mathrm{t}_{p_\lambda}(M^0)\right)}{\deg(p_\lambda)} = \frac{\dim_F\left(\mathrm{t}_{p_\lambda}(\mathcal{B}^0)\right)}{\deg(p_\lambda)} \text{ etc.}
\end{aligned} \tag{6.35}$$

Recall from Sect. 5.4.1 that

$$\begin{aligned}
&\mathrm{ann}_{\mathcal{D}}(M^0) = \mathcal{D}e^0_p,\ \mathrm{ann}_{\mathcal{D}}(\mathrm{t}(M)) = \mathcal{D}e_p,\ \mathrm{ann}_{\mathcal{D}}\left(M^0_{\mathrm{cont}}\right) = \mathcal{D}e^0_{p,\mathrm{cont}}, \\
&\mathrm{char}(\mathcal{B}^0) = \mathrm{V}_{\mathbb{C}}(e^0_p),\ \mathrm{char}(\mathcal{B}^0_{\mathrm{cont}}) = \mathrm{V}_{\mathbb{C}}\left(e^0_{p,\mathrm{cont}}\right) \\
&\mathrm{char}(\mathcal{B}) = \mathrm{char}(\mathcal{B}_{\mathrm{uncont}}) = \mathrm{V}_{\mathbb{C}}(e_p) = \{\lambda \in \mathbb{C};\ \mathrm{rank}(X_{\mathrm{cont}}(\lambda)) < p\}
\end{aligned} \tag{6.36}$$

where e^0_p, e_p *and* $e^0_{p,\mathrm{cont}}$ *are the highest, i.e., p-th, elementary divisors of* P, R, *and* P_{cont}, *respectively.*

Proof From (6.11) and Theorem 6.2.2, 6, we obtain the exact sequence

$$0 \longrightarrow \mathcal{D}^{1\times m} \xrightarrow{\mathrm{inj}=(\cdot(0,\mathrm{id}_m))_{\mathrm{ind}}} M \xrightarrow{\mathrm{proj}_{\mathrm{ind}}=\left(\cdot\binom{\mathrm{id}_p}{0}\right)_{\mathrm{ind}}} M^0 \longrightarrow 0.$$

The kernel $\ker(\mathrm{proj}_{\mathrm{ind}}) = \mathrm{inj}(\mathcal{D}^{1\times m})$ is a free module because inj is a monomorphism, and thus the intersection $\ker(\mathrm{proj}_{\mathrm{ind}}) \cap \mathrm{t}(M)$ is zero. But this is exactly the kernel of the map

$$\mathrm{proj}_{\mathrm{ind}}\,|_{\mathrm{t}(M)} = \left(\cdot\binom{\mathrm{id}_p}{0}\right)_{\mathrm{ind}} :\ \mathrm{t}(M) = U_{\mathrm{cont}}/U \longrightarrow U^0_{\mathrm{cont}}/U^0\ (\subseteq M^0)$$

from (6.32). Therefore, the epimorphism (6.32) is also a monomorphism.
2. This follows directly from the quoted results. □

Definition 6.2.8 (*Poles of IO behaviors*) For (6.31) the complex numbers in the finite characteristic variety $\mathrm{char}(\mathcal{B}^0)$ are called the *poles* of the IO behavior $\mathcal{B}$, the ones in $\mathrm{char}(\mathcal{B}^0_{\mathrm{cont}})$ are the *controllable poles*, and the elements of $\mathrm{char}(\mathcal{B}) = \mathrm{char}(\mathcal{B}_{\mathrm{uncont}})$ are the *uncontrollable poles* or *input decoupling zeros*. ◇

The following theorem shows that $e^0_{p,\mathrm{cont}}$ is the least common denominator of the entries of H and implies that for $\mathcal{D} = F[s]$, $F = \mathbb{R}, \mathbb{C}$, the characteristic variety of $\mathcal{B}^0_{\mathrm{cont}}$ coincides with the pole variety of H. The latter is defined as follows: If $F = \mathbb{C}$ and $h = \frac{a}{b}$ with $a, b \in \mathbb{C}[s]$, $b \neq 0$, $\gcd(a, b) = 1$, is the reduced quotient representation of a rational function $h \in \mathbb{C}(s)$ the zeros of b are called the *poles* of h. We define

$$\mathrm{pole}(h) := \mathrm{V}_{\mathbb{C}}(b). \tag{6.37}$$

The complement $\mathrm{dom}(h) := \mathbb{C} \setminus \mathrm{pole}(h)$ is called the *domain* (of definition) of h, and h can be considered as the function

$$h\colon\ \mathrm{dom}(h) \longrightarrow \mathbb{C},\ \lambda \longmapsto h(\lambda) := \tfrac{a(\lambda)}{b(\lambda)}.$$

More generally, let $H \in \mathbb{C}(s)^{p\times m}$ be a matrix and let $H_{ij} = \frac{a_{ij}}{b_{ij}}$ be the reduced quotient representations of the entries of H with their least common denominator $\mathrm{lcm}_{i,j}(b_{ij})$. We call the finite set

$$\mathrm{pole}(H) := \bigcup_{i,j} \mathrm{pole}(H_{ij}) = \bigcup_{i,j} \mathrm{V}_{\mathbb{C}}(b_{ij}) = \mathrm{V}_{\mathbb{C}}\left(\mathrm{lcm}_{i,j}(b_{ij})\right) \tag{6.38}$$

the set of *poles* of H and its complement $\mathrm{dom}(H)$ is called the domain of H. Again, we obtain the decomposition $\mathbb{C} = \mathrm{dom}(H) \uplus \mathrm{pole}(H)$ and the matrix function

$$H\colon\ \mathrm{dom}(H) \longrightarrow \mathbb{C}^{p\times m},\quad \lambda \longmapsto H(\lambda) := \left(H_{ij}(\lambda)\right)_{i,j}. \tag{6.39}$$

Theorem 6.2.9 *1. Let $H \in K^{p\times m}$ and let $H_{ij} = \frac{a_{ij}}{b_{ij}}$ with $\gcd(a_{ij}, b_{ij}) = 1$ be the unique reduced quotient representations of the entries of H. As in Theorem 6.2.7, let P_{cont} be a matrix which generates the autonomous part U^0_{cont} of the unique controllable realization of H, and let $e^0_{p,\mathrm{cont}}$ be its p-th elementary divisor. Then*

$$e^0_{p,\mathrm{cont}} = \mathrm{lcm}_{i,j}(b_{ij}) := \mathrm{lcm}\left\{b_{ij};\ i \in \{1,\dots,p\},\ j \in \{1,\dots,m\}\right\} \tag{6.40}$$

is the least common denominator of the entries of H and

$$\begin{aligned}\mathrm{ann}_{\mathcal{D}}(\mathcal{B}^0_{\mathrm{cont}}) \underset{\text{Lemma 4.2.6}}{=} \mathrm{ann}_{\mathcal{D}}(M^0_{\mathrm{cont}}) \underset{\text{Theorem 4.2.9,1.}}{=} \mathcal{D}e^0_{p,\mathrm{cont}} \\ = \{d \in \mathcal{D};\ dH \in \mathcal{D}^{p\times m}\} = \mathcal{D}\,\mathrm{lcm}_{i,j}(b_{ij}).\end{aligned} \tag{6.41}$$

For $F = \mathbb{C}, \mathbb{R}$ and $\mathcal{D} = F[s]$ we infer

$$\mathrm{pole}(H) = \mathrm{V}_{\mathbb{C}}\left(\mathrm{lcm}_{i,j}(b_{ij})\right) = \mathrm{V}_{\mathbb{C}}\left(e^0_{p,\mathrm{cont}}\right) = \mathrm{char}(\mathcal{B}^0_{\mathrm{cont}}). \tag{6.42}$$

Thus the controllable poles of $\mathcal{B}$ coincide with the poles of the transfer matrix.
2. With the data from Corollary and Definition 6.2.6 for the Smith/McMillan form of H one obtains

$$\begin{aligned}&e^0_{p,\mathrm{cont}} = b_1 \text{ and } \mathrm{pole}(H) = \mathrm{V}_{\mathbb{C}}(e^0_{p,\mathrm{cont}}) = \mathrm{V}_{\mathbb{C}}(b_1),\\ &\mathrm{l}(H) = \mathrm{l}(M^0_{\mathrm{cont}}) = \mathrm{l}(\mathcal{B}^0_{\mathrm{cont}}),\ \mathrm{l}_p(H) = \sum_{i=1}^{r} \mu(i,p),\ b_i = \prod_{p\in\mathcal{P}} p^{\mu(i,p)}.\end{aligned} \tag{6.43}$$

Proof 1. The relation $\{d \in \mathcal{D};\ dH \in \mathcal{D}^{p\times m}\} = \mathcal{D}\,\mathrm{lcm}_{i,j}(b_{i,j})$ follows from

$$dH \in \mathcal{D}^{p\times m} \Longleftrightarrow \forall i,j:\ dH_{ij} = \frac{da_{ij}}{b_{ij}} \in \mathcal{D} \Longleftrightarrow \forall i,j:\ b_{ij} \mid da_{ij}$$
$$\underset{\gcd(a_{ij},b_{ij})=1}{\Longleftrightarrow} \forall i,j:\ b_{ij} \mid d \Longleftrightarrow d \in \mathcal{D}\,\mathrm{lcm}_{i,j}(b_{ij}).$$

From (6.17) we get

$$U^0_{\mathrm{cont}} = \{\xi \in \mathcal{D}^{1\times p};\ \xi H \in \mathcal{D}^{1\times m}\} = \ker\left(\mathcal{D}^{1\times p} \to K^{1\times m}/\mathcal{D}^{1\times m},\ \xi \mapsto \overline{\xi H}\right).$$

Therefore the induced map

$$(\cdot H)_{\mathrm{ind}}:\ M^0_{\mathrm{cont}} = \mathcal{D}^{1\times p}/U^0_{\mathrm{cont}} \longrightarrow K^{1\times m}/\mathcal{D}^{1\times m},\quad \overline{\xi} \longmapsto \overline{\xi H}$$

is a monomorphism. Since M^0_{cont} is generated by the $\overline{\delta_i}$, $i = 1,\dots,p$, we obtain

$$d \in \mathrm{ann}_{\mathcal{D}}(M^0_{\mathrm{cont}}) \Longleftrightarrow \forall i = 1,\dots,p:\ 0 = d\overline{\delta_i} = \overline{d\delta_i}$$
$$\underset{(\cdot H)_{\mathrm{ind}}\ \mathrm{mono}}{\Longleftrightarrow} \forall i = 1,\dots,p:\ 0 = \overline{d\delta_i H} = \overline{dH_{i-}} \in K^{1\times m}/\mathcal{D}^{1\times m}$$
$$\Longleftrightarrow \forall i = 1,\dots,p:\ dH_{i-} \in \mathcal{D}^{1\times m} \Longleftrightarrow dH \in \mathcal{D}^{p\times m}$$
$$\Longleftrightarrow d \in \mathcal{D}\,\mathrm{lcm}_{i,j}(b_{ij}).$$

2. Since $b_r | b_{r-1} | \cdots | b_1$ the matrix $P'_{\mathrm{cont}} = \mathrm{diag}(b_1,\dots,b_r,1\dots,1)$ has the highest elementary divisor b_1. Since $P_{\mathrm{cont}} = P'_{\mathrm{cont}}U$ is equivalent to P'_{cont} we infer $b_1 = e^0_{p,\mathrm{cont}}$. □

A pole may be both controllable and uncontrollable, cf. Example 6.2.10.

Example 6.2.10 We consider a single input/single output (SISO) behavior for the standard case $F = \mathbb{C}$, $\mathcal{D} = \mathbb{C}[s]$ and $\mathcal{F} = \mathrm{C}^\infty(\mathbb{R},\mathbb{C})$, $\mathbb{C}^{\mathbb{N}}$. We choose two nonzero coprime polynomials P_{cont}, $Q_{\mathrm{cont}} \in \mathbb{C}[s]$, an arbitrary nonzero polynomial X_{cont} and define $(P, -Q) := X_{\mathrm{cont}}(P_{\mathrm{cont}}, -Q_{\mathrm{cont}})$. Then

$$\mathcal{B} := \left\{\binom{y}{u} \in \mathcal{F}^2;\ P \circ y = Q \circ u\right\}$$

is an IO behavior with transfer function $H = P^{-1}Q = P^{-1}_{\mathrm{cont}}Q_{\mathrm{cont}} \in \mathbb{C}(s)$. Its controllable part is

$$\mathcal{B}_{\mathrm{cont}} := \left\{\binom{y}{u} \in \mathcal{F}^2;\ P_{\mathrm{cont}} \circ y = Q_{\mathrm{cont}} \circ u\right\}$$

and its uncontrollable behavior is $\mathcal{B}_{\mathrm{uncont}} = \{y \in \mathcal{F};\ X_{\mathrm{cont}} \circ y = 0\}$. The sequence

$$0 \to \mathcal{B}^0_{\mathrm{cont}} = \{y \in \mathcal{F};\ P_{\mathrm{cont}} \circ y = 0\} \xrightarrow{\subseteq} \mathcal{B}^0 = \{y \in \mathcal{F};\ P \circ y = 0\}$$
$$\xrightarrow{P_{\mathrm{cont}}\circ} \mathcal{B}_{\mathrm{uncont}} = \{y \in \mathcal{F};\ X_{\mathrm{cont}} \circ y = 0\} \to 0$$

is exact and implies

$$\begin{aligned} &1(\mathcal{B}^0) = 1(\mathcal{B}^0_{\text{cont}}) + 1(\mathcal{B}_{\text{uncont}}), \ \text{char}(\mathcal{B}^0) = \text{char}(\mathcal{B}^0_{\text{cont}}) \bigcup \text{char}(\mathcal{B}_{\text{uncont}}) \\ &\text{char}(\mathcal{B}^0) = \text{V}_{\mathbb{C}}(P), \ \text{char}(\mathcal{B}^0_{\text{cont}}) = \text{pole}(H) = \text{V}_{\mathbb{C}}(P_{\text{cont}}), \\ &\text{char}(\mathcal{B}_{\text{uncont}}) = \text{char}(\mathcal{B}) = \text{V}_{\mathbb{C}}(X_{\text{cont}}). \end{aligned} \tag{6.44}$$

If $P = \prod_{\lambda \in \mathbb{C}} (s - \lambda)^{\mu_P(\lambda)}$ is the prime factor decomposition where almost all $\mu_P(\lambda)$ are zero then $1_{s-\lambda}(\mathcal{B}^0) = \mu_P(\lambda)$ and likewise for $\mathcal{B}^0_{\text{cont}}$ and $\mathcal{B}_{\text{uncont}}$. It is obvious that the varieties $\text{char}(\mathcal{B}^0_{\text{cont}}) = \text{V}_{\mathbb{C}}(P_{\text{cont}})$ and $\text{char}(\mathcal{B}_{\text{uncont}}) = \text{V}_{\mathbb{C}}(X_{\text{cont}})$ and the exponents $\mu(\lambda)$ can be chosen arbitrarily. In particular, a pole of $\mathcal{B}$ may be both controllable and uncontrollable, and $\mathcal{B}_{\text{uncont}}$ and its characteristic variety $\text{char}(\mathcal{B}) = \text{char}(\mathcal{B}_{\text{uncont}})$ do *not* determine $\text{char}(\mathcal{B}^0)$ and the dynamic behavior of $\mathcal{B}$. ◇

6.3 Rosenbrock and State Space Realizations

We derive the transfer matrix and the observable factor of a Rosenbrock system. Moreover we discuss similar state space behaviors and characterize controllable and observable state space equations. We show that observable state space realizations of a given IO behavior and controllable and observable such realizations of a given transfer matrix are similar.

Let $\mathcal{D}$ be principal ideal domain and ${}_{\mathcal{D}}\mathcal{F}$ an injective cogenerator.

6.3.1 Rosenbrock or Pseudo-State Realizations

As in Theorem 3.2.22 we consider Rosenbrock equations or matrix descriptions

$$\begin{gathered} A \circ x = B \circ u, \qquad y = C \circ x + D \circ u \\ \text{with } A \in \mathcal{D}^{n \times n}, \ B \in \mathcal{D}^{n \times m}, \ C \in \mathcal{D}^{p \times n}, \ D \in \mathcal{D}^{p \times m} \\ \text{and } \text{rank}(A) = n, \text{ i.e., } \det(A) \neq 0. \end{gathered} \tag{6.45}$$

Associated with these equations are the two behaviors

$$\begin{aligned} &\mathcal{B}_1 := \left\{ \binom{x}{u} \in \mathcal{F}^{n+m}; \ A \circ x = B \circ u \text{ or } (A, -B) \circ \binom{x}{u} = 0 \right\} \ \text{ and} \\ &\mathcal{B} := \begin{pmatrix} C & D \\ 0 & \text{id}_m \end{pmatrix} \circ \mathcal{B}_1 = \left\{ \binom{y}{u} \in \mathcal{F}^{p+m}; \ \exists x \in \mathcal{F}^n: \ (6.45) \text{ is satisfied} \right\}. \end{aligned} \tag{6.46}$$

The assumption $\det(A) \neq 0$ implies that $\mathcal{B}_1$ is an IO behavior with input u, autonomous part

$$\mathcal{B}_1^0 := \{x \in \mathcal{F}^n;\ A \circ x = 0\}, \tag{6.47}$$

and transfer matrix $H_1 = A^{-1}B$.

Theorem and Definition 6.3.1 (Rosenbrock or pseudo-state realization) *Under the preceding assumptions the behavior $\mathcal{B}$ is an IO behavior with transfer matrix $H = D + CH_1 = D + CA^{-1}B$.*

The behavior $\mathcal{B}$ is called the IO behavior of the Rosenbrock equations *and H is their transfer matrix. Conversely, Eq.* (6.45) *is called a* Rosenbrock *or* pseudo-state realization *of $\mathcal{B}$ or of H, and x is called the* pseudo-state.

Proof As in Theorem 3.2.22 we consider a universal left annihilator $(-Y, P) \in \mathcal{D}^{p\times(n+p)}$ of the matrix $\binom{A}{C} \in \mathcal{D}^{(n+p)\times n}$, hence $YA = PC$. Since $\mathcal{D}$ is a principal ideal domain we may and do assume that $(-Y, P)$ has indeed

$$\operatorname{rank}(-Y, P) = (n+p) - \operatorname{rank}\binom{A}{C} = (n+p) - n = p$$

rows. According to Theorem 3.2.22 the behavior $\mathcal{B}$ is given by

$$\mathcal{B} = \left\{\binom{y}{u} \in \mathcal{F}^{p+m};\ P \circ x = Q \circ u\right\} \text{ with } Q := YB + PD.$$

Since $\begin{pmatrix} C & D \\ 0 & \mathrm{id}_m \end{pmatrix} \circ : \mathcal{B}_1 \to \mathcal{B}$ is surjective, so is $C\circ :\ \mathcal{B}_1^0 \longrightarrow \mathcal{B}^0$ where $\mathcal{B}^0 := \{y \in \mathcal{F}^p;\ P \circ y = 0\}$. Since $\mathcal{B}_1^0$ is autonomous, so is $\mathcal{B}^0$ (Lemma 4.2.5) and hence $\operatorname{rank}(P) = p$. The equation

$$Q = PD + YB = PD + YAH_1 = PD + PCH_1 = P(D + CH_1) \text{ implies}$$
$$\operatorname{rank}(P, -Q) = \operatorname{rank}(P) = p \text{ and } PH = Q,\ H := D + CH_1.$$

This shows that $\mathcal{B}$ is an IO behavior with transfer matrix H. □

Definition 6.3.2 (*State space realization*) If in the preceding theorem $\mathcal{D} := F[s]$ and the Rosenbrock equations have the special state space form

$$s \circ x = Ax + Bu, \text{ i.e., } (s\,\mathrm{id}_n - A) \circ x = Bu \text{ and } y = Cx + D \circ u \tag{6.48}$$

with constant matrices A, B, C, and a polynomial matrix D, then (6.48) is called a *Kalman* or *state space realization* of the IO system $\mathcal{B}$ and of the transfer matrix $H = D + C(s\,\mathrm{id}_n - A)^{-1}B$. ◊

In accordance with the literature we use the term *realization* in three different meanings, namely,

1. as the IO realization of a transfer matrix H,
2. as the Rosenbrock or state space realization of an IO behavior, and
3. as the Rosenbrock or state space realization of a transfer matrix.

6.3.2 The Terminology of the French School

According to Fliess, Bourlès et al. [5, Sect. 7.1] a system is just a finitely generated module M over a polynomial algebra $\mathcal{D} = F[s]$ where F is a field. Let ${}_{\mathcal{D}}\mathcal{F}$ be an injective cogenerator. The standard case from [5, Sect. 7.1] is $F = \mathbb{R}$, $\mathcal{D} = \mathbb{R}[\partial]$, $s = \partial = \frac{\mathrm{d}}{\mathrm{d}t}$, and $\mathcal{F} = \mathrm{C}^\infty(\mathbb{R}, \mathbb{R})$ with the action $s \circ w = \frac{\mathrm{d}w}{\mathrm{d}t}$.

Consider the behavior and modules induced by a matrix $R \in \mathcal{D}^{k\times l}$, namely,

$$\begin{aligned}&\mathcal{B}_1 := \{w \in \mathcal{F}^l;\ R \circ w = 0\} = U_1^\perp,\ U_1 := \mathcal{D}^{1\times k}R = \mathcal{B}_1^\perp\\ &M_1 := \mathcal{D}^{1\times l}/U_1 = \mathrm{M}(\mathcal{B}_1) \cong \mathrm{Hom}(\mathcal{B}_1, \mathcal{F}) = \{\text{observables of } \mathcal{B}_1\} \text{ with}\\ &n := \dim(U_1) = \mathrm{rank}(R),\ m := \mathrm{rank}(M_1) = \mathrm{rank}(\mathcal{B}_1) = l - n.\end{aligned}$$

Without loss of generality we assume $k = n = \mathrm{rank}(R)$. The use of the letter w for the trajectories of $\mathcal{B}$ is due to Willems [1].

The standard basis $\delta_j \in \mathcal{D}^{1\times l}$, $j = 1, \dots, l$, induces the generating system

$$\widehat{w}_j := \overline{\delta_j} = \delta_j + U_1 \in M_1, \quad j = 1, \dots, l,$$

of M_1, i.e., $M_1 = \sum_{j=1}^l \mathcal{D}\widehat{w}_j$. In [5, p. 176], the module M_1 of observables of the behavior with its finite (re)presentation $M_1 = \mathcal{D}^{1\times l}/U_1$ is called a *system with the variables* $\widehat{w}_j$ *of rank* m. This terminology is due to Fliess. Let $\widehat{w} = (\widehat{w}_1, \dots, \widehat{w}_l)^\top \in M_1^l$ denote the column with the components $\widehat{w}_j$. In [5, p. 176], the module M_1 is written as $[\widehat{w}]_{\mathcal{D}} = M_1 = \mathcal{D}^{1\times l}\widehat{w}$.

Since for $i = 1, \dots, n$, the row $R_{i-} = \sum_{j=1}^l R_{ij}\delta_j$ belongs to U_1, the list $\widehat{w}$ of generators of $M_1 = [\widehat{w}]_{\mathcal{D}}$ satisfies the relation $R\widehat{w} = 0$, in other words,

$$\sum_{j=1}^{l} R_{ij}\widehat{w}_j = 0 \in M_1, \quad i = 1, \dots, n.$$

Thus, the column $\widehat{w}$ of *variables* of the system, that in our terminology is a list of *observables* of $\mathcal{B}_1$, and the trajectories $w \in \mathcal{B}_1 \subseteq \mathcal{F}^l$ satisfy the same relation

$$R\widehat{w} = 0 \text{ and } \quad R \circ w = 0,$$

although they belong to different modules M_1^l and $\mathcal{B}_1\ \left(\subseteq \mathcal{F}^l\right)$. The Malgrange isomorphism

$$\mathrm{Hom}_{\mathcal{D}}(M_1, \mathcal{F}) \xrightarrow{\cong} \mathcal{B}_1, \quad \varphi \longmapsto w := \varphi(\widehat{w}), \text{ i.e., } w_j = \varphi(\widehat{w}_j),$$

explains the connection of $\widehat{w}$ and w. But whereas in the standard continuous-time situation $R \circ w = R(\partial)(w) = 0$ is a differential system for a column of signals, the equation $R\widehat{w} = R(\partial)\widehat{w} = 0$, in spite of its suggestive notation, is not a differential system, but the defining family of linear relations between the generators of the module M_1, and no functions or signals are involved. This has to be kept in mind when studying [5] or papers of the French school. In the following additional explanations we will always distinguish a module element $\widehat{x} \in M_1$ from a corresponding signal $x \in \mathcal{F}^\bullet$.

In the French terminology an IO or *control system* is a triple $(M_1, \widehat{u}, \widehat{y})$ with the following specifications [5]: M_1 is a finitely generated $\mathcal{D}$-module of $\mathrm{rank}(M_1) =: m$, $\widehat{u} = (\widehat{u}_1, \dots, \widehat{u}_m)^\top \in M_1^m$ is a column of m $\mathcal{D}$-linearly independent elements of M_1, and $\widehat{y} = (\widehat{y}_1, \dots, \widehat{y}_p)^\top \in M_1^p$ is an arbitrary column of p elements of M_1. The columns $\widehat{u}$ and $\widehat{y}$ give rise to the modules

$$
\begin{aligned}
&[\widehat{u}]_{\mathcal{D}} := \mathcal{D}^{1\times m}\widehat{u} = \oplus_{j=1}^m \mathcal{D}\widehat{u}_j,\\
&[\widehat{y}]_{\mathcal{D}} := \mathcal{D}^{1\times p}\widehat{y} = \sum_{i=1}^p \mathcal{D}\widehat{y}_j,\\
&\quad M := [\widehat{y}, \widehat{u}]_{\mathcal{D}} = [\widehat{y}]_{\mathcal{D}} + [\widehat{u}]_{\mathcal{D}} = \mathcal{D}^{1\times(p+m)}\left(\begin{smallmatrix}\widehat{y}\\ \widehat{u}\end{smallmatrix}\right) = \sum_{i=1}^p \mathcal{D}\widehat{y}_i + \sum_{j=1}^m \mathcal{D}\widehat{u}_j \subseteq M_1,\\
&\quad M_1^0 := M_1/[\widehat{u}]_{\mathcal{D}},\ \mathrm{rank}(M_1^0) = \mathrm{rank}(M_1) - \mathrm{rank}([\widehat{u}]_{\mathcal{D}}) = m - m = 0,\\
&\quad M^0 := M/[\widehat{u}]_{\mathcal{D}} \subseteq M_1^0 = M_1/[\widehat{u}]_{\mathcal{D}},\ \mathrm{rank}(M^0) = 0.
\end{aligned}
\tag{6.49}
$$

The equation $\mathrm{rank}(M_1^0) = \mathrm{rank}(M^0) = 0$ signifies that M_1^0 and its submodule M^0 are torsion modules. The module $M := [\widehat{y}, \widehat{u}]_{\mathcal{D}}$ contains $[\widehat{u}]_{\mathcal{D}}$ and, conversely, every submodule of M_1 containing $[\widehat{u}]_{\mathcal{D}}$ is of the form $M := [\widehat{y}, \widehat{u}]_{\mathcal{D}}$ by the choice of a generating system $\left(\begin{smallmatrix}\widehat{y}\\ \widehat{u}\end{smallmatrix}\right) \in M_1^{p+m}$ of M. The submodules M of M_1 containing $[\widehat{u}]_{\mathcal{D}}$ are in one-one correspondence with the submodules $M^0 \subseteq M_1^0$ since

$$
\begin{aligned}
&M \underset{[\widehat{u}]_{\mathcal{D}}\subseteq M}{=} \mathrm{can}^{-1}\,\mathrm{can}(M) = \mathrm{can}^{-1}(M^0) = Y + [\widehat{u}]_{\mathcal{D}} \text{ where}\\
&\mathrm{can} : M_1 \to M_1^0,\ x \mapsto x + [\widehat{u}]_{\mathcal{D}},\ \ker(\mathrm{can}) = [\widehat{u}]_{\mathcal{D}},\ Y = \mathcal{D}^{1\times p}\widehat{y},\ \mathrm{can}(Y) = M^0.
\end{aligned}
$$

In addition we choose

$$
\widehat{x} = (\widehat{x}_1, \dots, \widehat{x}_n)^\top \in M_1^n,\ \text{such that } M_1 = [\widehat{x}, \widehat{u}]_{\mathcal{D}} = \mathcal{D}^{1\times(n+m)}\left(\begin{smallmatrix}\widehat{x}\\ \widehat{u}\end{smallmatrix}\right). \tag{6.50}
$$

Hence the column $\left(\begin{smallmatrix}\widehat{x}\\ \widehat{u}\end{smallmatrix}\right) \in M_1^{n+m}$ is a generating system of M_1 that contains the $\mathcal{D}$-linearly independent vectors $\widehat{u}_j$.

We are now going to show that the data $(M_1 = [\widehat{x}, \widehat{u}]_{\mathcal{D}}, \widehat{u}, \widehat{y})$ give rise to essentially unique Rosenbrock equations

$$
A \circ x = B \circ u,\ y = C \circ x + D \circ u.
$$

The kernel of the epimorphism

$$\mathcal{D}^{1\times(n+m)} \to M_1,\ (\xi, \eta) \mapsto (\xi, \eta)\begin{pmatrix}\widehat{x}\\ \widehat{u}\end{pmatrix},$$

is free of dimension $(n+m) - \operatorname{rank}(M_1) = (n+m) - m = n$. Hence there is a matrix $(A, -B) \in \mathcal{D}^{n\times(n+m)}$ of rank n that gives rise to the isomorphism

$$\begin{array}{llll}\mathrm{can} := \mathrm{can}_{\widehat{x},\widehat{u}}: & \mathcal{D}^{1\times(n+m)}/\mathcal{D}^{1\times n}(A, -B) & \cong & M_1\\ & \overline{(\xi, \eta)} := (\xi, \eta) + \mathcal{D}^{1\times n}(A, -B) & \mapsto & (\xi, \eta)\begin{pmatrix}\widehat{x}\\ \widehat{u}\end{pmatrix} = \xi\widehat{x} + \eta\widehat{u}\end{array}. \tag{6.51}$$

There results the commutative diagram with exact rows and vertical isomorphisms, cf. (6.11),

$$\begin{array}{ccccccccc}
0 & \longrightarrow & \mathcal{D}^{1\times m} & \xrightarrow{(\cdot(0,\ \mathrm{id}_m))_{\mathrm{ind}}} & \mathcal{D}^{1\times(n+m)}/\mathcal{D}^{1\times n}(A, -B) & \xrightarrow{\left(\cdot\begin{pmatrix}\mathrm{id}_n\\ 0\end{pmatrix}\right)_{\mathrm{ind}}} & \mathcal{D}^{1\times n}/\mathcal{D}^{1\times n}A & \longrightarrow & 0\\
 & & = \big\downarrow \mathrm{id}_m & & \cong \big\downarrow \mathrm{can} & & \cong \big\downarrow \mathrm{can} & & \\
0 & \longrightarrow & \mathcal{D}^{1\times m} & \longrightarrow & M_1 & \xrightarrow{\mathrm{can}} & M_1^0 & \longrightarrow & 0\\
 & & \eta & \longmapsto & \eta\widehat{u} & & & & \\
 & & & & (\xi\ \ \eta)\begin{pmatrix}\widehat{x}\\ \widehat{u}\end{pmatrix} = \xi\widehat{x} + \eta\widehat{u} & \longmapsto & \xi\overline{\widehat{x}} & &
\end{array} \tag{6.52}$$

where

$$\overline{\widehat{x}} = (\overline{\widehat{x}_1}, \ldots, \overline{\widehat{x}_n})^\top,\ \overline{\widehat{x}_i} = \widehat{x}_i + [\widehat{u}] \in M_1^0 = M_1/[\widehat{u}]_{\mathcal{D}}.$$

Since M_1^0 is a torsion module Theorem 6.2.2 implies that

$$\mathcal{B}_1 := \left\{\begin{pmatrix}x\\ u\end{pmatrix} \in \mathcal{F}^{n+m};\ A \circ x = B \circ u\right\} \text{ with } \mathcal{B}_1^0 = \left\{x \in \mathcal{F}^n;\ A \circ x = 0\right\} \tag{6.53}$$

is an IO behavior with modules

$$M(\mathcal{B}_1) = \mathcal{D}^{1\times(n+m)}/\mathcal{D}^{1\times n}(A, -B) \cong M_1 \text{ and } M(\mathcal{B}_1^0) = \mathcal{D}^{1\times n}/\mathcal{D}^{1\times n}A \cong M_1^0. \tag{6.54}$$

Since $\begin{pmatrix}\widehat{x}\\ \widehat{u}\end{pmatrix}$ is a generating system of M_1 there is a matrix

$$\begin{aligned}&(C, D) \in \mathcal{D}^{p\times(n+m)} \text{ such that } \widehat{y} = (C, D)\begin{pmatrix}\widehat{x}\\ \widehat{u}\end{pmatrix}\\ &\Longrightarrow M = [\widehat{y}, \widehat{u}]_{\mathcal{D}} = \mathcal{D}^{1\times(p+m)}\begin{pmatrix}C & D\\ 0 & \mathrm{id}_m\end{pmatrix}\begin{pmatrix}\widehat{x}\\ \widehat{u}\end{pmatrix}.\end{aligned} \tag{6.55}$$

By Theorem 6.3.1 there is the image IO behavior

$$\begin{aligned}&\mathcal{B} := \begin{pmatrix}C & D\\ 0 & \mathrm{id}_m\end{pmatrix} \circ \mathcal{B}_1 = \left\{\begin{pmatrix}y\\ u\end{pmatrix} \in \mathcal{F}^{p+m};\ P \circ y = Q \circ u\right\} \text{ with}\\ &\mathcal{B}^0 = C \circ \mathcal{B}_1^0 = \left\{y \in \mathcal{F}^p;\ P \circ y = 0\right\}.\end{aligned} \tag{6.56}$$

The epimorphism $\begin{pmatrix}C & D\\ 0 & \mathrm{id}_m\end{pmatrix} \circ$ induces the dual monomorphism

$$\phi = \cdot\left(\begin{pmatrix} C & D \\ 0 & \mathrm{id}_m \end{pmatrix}\right)_{\mathrm{ind}} : M(\mathcal{B}) = \mathcal{D}^{1\times(p+m)}/\mathcal{D}^{1\times p}(P, -Q) \longrightarrow M(\mathcal{B}_1). \tag{6.57}$$

The module $M(\mathcal{B})$ has the elements $\overline{(\xi,\eta)} = (\xi,\eta) + \mathcal{D}^{1\times p}(P, -Q)$ with $(\xi,\eta) \in \mathcal{D}^{1\times(p+m)}$. There results the isomorphism

$$\begin{array}{rl} \mathrm{can}_{\widehat{x},\widehat{u}}\,\phi : M(\mathcal{B}) \cong & M = \mathcal{D}^{1\times(p+m)}\begin{pmatrix} C & D \\ 0 & \mathrm{id}_m \end{pmatrix}\begin{pmatrix} \widehat{x} \\ \widehat{u} \end{pmatrix} \\ \overline{(\xi,\eta)} \mapsto & (\xi,\eta)\begin{pmatrix} C & D \\ 0 & \mathrm{id}_m \end{pmatrix}\begin{pmatrix} \widehat{x} \\ \widehat{u} \end{pmatrix} \end{array} \tag{6.58}$$

With the identification

$$M(\mathcal{B}_1) = M_1 \ni \overline{(\xi,\eta)} = (\xi,\eta) + \mathcal{D}^{1\times n}(A, -B) = (\xi,\eta)\begin{pmatrix} \widehat{x} \\ \widehat{u} \end{pmatrix} = \xi\widehat{x} + \eta\widehat{u},$$

we obtain

$$\phi = \left(\cdot\begin{pmatrix} C & D \\ 0 & \mathrm{id}_m \end{pmatrix}\right)_{\mathrm{ind}} : M(\mathcal{B}) \cong \phi(M(\mathcal{B})) = M. \tag{6.59}$$

Hence the Rosenbrock equations $A \circ x = B \circ u,\ y = C \circ x + D \circ u$ and their associated behaviors and modules according to Theorem 6.3.1 induce the data $(M_1, \widehat{y}, \widehat{u})$ and their derived modules via the isomorphisms

$$M(\mathcal{B}_1) \cong M_1,\ M(\mathcal{B}_1^0) \cong M_1^0,\ M(\mathcal{B}) \cong M,\ M(\mathcal{B}^0) \cong M^0. \tag{6.60}$$

6.3.3 The Observable Factor of Rosenbrock Equations

In Theorem 4.1.11, we constructed the observable factor of a state space behavior. The next theorem does the same job for the Rosenbrock equations. The data are those of (6.45) and (6.46). Consider the transfer matrix $H' := CA^{-1} \in K^{p\times n}$ and let $H' = C_{\mathrm{obs}}A_{\mathrm{obs}}^{-1}$ with $A_{\mathrm{obs}} \in \mathcal{D}^{n\times n}$, $C_{\mathrm{obs}} \in \mathcal{D}^{p\times n}$, be the unique (up to column equivalence) right coprime factorization of H' according to Corollary and Definition 6.2.5. Then $H'^{\top} = \left(A_{\mathrm{obs}}^{\top}\right)^{-1} C_{\mathrm{obs}}^{\top}$ is the left coprime factorization of $H'^{\top}$ and can be computed according to (6.18). These matrices are characterized by the exact sequence

$$\begin{array}{l} 0 \to \mathcal{D}^n \overset{\begin{pmatrix} A_{\mathrm{obs}} \\ -C_{\mathrm{obs}} \end{pmatrix}\cdot}{\longrightarrow} \mathcal{D}^{n+p} \overset{(H',\mathrm{id}_p)\cdot}{\longrightarrow} K^p \text{ and} \\ A_{\mathrm{obs}}\mathcal{D}^n = \left\{\xi \in \mathcal{D}^n;\ H'\xi \in \mathcal{D}^p\right\},\ \mathrm{rank}(A_{\mathrm{obs}}) = n. \end{array} \tag{6.61}$$

Moreover the matrix $\begin{pmatrix} A_{\mathrm{obs}} \\ -C_{\mathrm{obs}} \end{pmatrix}$ has a left inverse. The equation $H'A = C \in \mathcal{D}^{p\times n}$ implies that there is $X_{\mathrm{obs}} \in \mathcal{D}^{n\times n}$ with

$$\begin{array}{l} A = A_{\mathrm{obs}}X_{\mathrm{obs}} \text{ or } A_{\mathrm{obs}}^{-1}A = X_{\mathrm{obs}} \in \mathcal{D}^{n\times n},\ \mathrm{rank}(X_{\mathrm{obs}}) = n, \text{ and} \\ C = H'A = H'A_{\mathrm{obs}}X_{\mathrm{obs}} = C_{\mathrm{obs}}X_{\mathrm{obs}},\ \begin{pmatrix} A \\ C \end{pmatrix} = \begin{pmatrix} A_{\mathrm{obs}} \\ C_{\mathrm{obs}} \end{pmatrix}X_{\mathrm{obs}}. \end{array} \tag{6.62}$$

Theorem and Definition 6.3.3 (The observable factor of Rosenbrock equations) *With the Rosenbrock data of* (6.45) *and* (6.46), *and the matrices* A_{obs} *and* C_{obs} *from* (6.61) *we define the new Rosenbrock equations*

$$A_{\mathrm{obs}} \circ x_{\mathrm{obs}} = B \circ u,\ y = C_{\mathrm{obs}} \circ x_{\mathrm{obs}} + D \circ u \tag{6.63}$$

and the IO behavior

$$\mathcal{B}_{1,\mathrm{obs}} := \left\{ \begin{pmatrix} x_{\mathrm{obs}} \\ u \end{pmatrix} \in \mathcal{F}^{n+m};\ A_{\mathrm{obs}} \circ x_{\mathrm{obs}} = B \circ u \right\}.$$

Then the map

$$\begin{pmatrix} X_{\mathrm{obs}} & 0 \\ 0 & \mathrm{id}_m \end{pmatrix} \circ\colon\ \mathcal{B}_1 \longrightarrow \mathcal{B}_{1,\mathrm{obs}},\ \begin{pmatrix} x \\ u \end{pmatrix} \longmapsto \begin{pmatrix} X_{\mathrm{obs}} \circ x \\ u \end{pmatrix},$$

is a behavior epimorphism and the map

$$\begin{pmatrix} C_{\mathrm{obs}} & D \\ 0 & \mathrm{id}_m \end{pmatrix} \circ\colon\ \mathcal{B}_{1,\mathrm{obs}} \xrightarrow{\cong} \mathcal{B} = \begin{pmatrix} C & D \\ 0 & \mathrm{id}_m \end{pmatrix} \circ \mathcal{B}_1,\ \begin{pmatrix} x_{\mathrm{obs}} \\ u \end{pmatrix} \longmapsto \begin{pmatrix} C_{\mathrm{obs}} \circ x_{\mathrm{obs}} + D \circ u \\ u \end{pmatrix} \tag{6.64}$$

is a behavior isomorphism. In particular, the Rosenbrock equations (6.45) *and* (6.63) *give rise to the same IO behavior* $\begin{pmatrix} C & D \\ 0 & \mathrm{id}_m \end{pmatrix} \circ \mathcal{B}_1 = \begin{pmatrix} C_{\mathrm{obs}} & D \\ 0 & \mathrm{id}_m \end{pmatrix} \circ \mathcal{B}_{1,\mathrm{obs}}$, *and the Rosenbrock equations* (6.63) *are observable. We call* $\mathcal{B}_{1,\mathrm{obs}}$ *the* observable factor *of* $\mathcal{B}_1$ *or of the Rosenbrock equations* (6.45).

The isomorphism (6.64) *induces the isomorphism*

$$C_{\mathrm{obs}} \circ\colon\ \mathcal{B}^0_{1,\mathrm{obs}} := \{x_{\mathrm{obs}} \in \mathcal{F}^n;\ A_{\mathrm{obs}} \circ x_{\mathrm{obs}} = 0\} \xrightarrow{\cong} \mathcal{B}^0 = \{y \in \mathcal{F}^p;\ P \circ y = 0\}.$$

Proof We use $A = A_{\mathrm{obs}} X_{\mathrm{obs}}$. From $\begin{pmatrix} x \\ u \end{pmatrix} \in \mathcal{B}_1$ we infer

$$B \circ u = A \circ x = (A_{\mathrm{obs}} X_{\mathrm{obs}}) \circ x = A_{\mathrm{obs}} \circ (X_{\mathrm{obs}} \circ x),$$

hence $\begin{pmatrix} X_{\mathrm{obs}} \circ x \\ u \end{pmatrix} \in \mathcal{B}_{1,\mathrm{obs}}$ and the map $\begin{pmatrix} X_{\mathrm{obs}} & 0 \\ 0 & \mathrm{id}_m \end{pmatrix} \circ\colon\ \mathcal{B}_1 \longrightarrow \mathcal{B}_{1,\mathrm{obs}}$ is well defined. Since $\mathrm{rank}(X_{\mathrm{obs}}) = n$ the map $\cdot X_{\mathrm{obs}} : \mathcal{D}^{1\times n} \to \mathcal{D}^{1\times n}$ is injective and $X_{\mathrm{obs}} \circ : \mathcal{F}^n \to \mathcal{F}^n$ is surjective. Hence for $\begin{pmatrix} x_{\mathrm{obs}} \\ u \end{pmatrix} \in \mathcal{B}_{1,\mathrm{obs}}$ there is $x \in \mathcal{F}^n$ such that $\begin{pmatrix} X_{\mathrm{obs}} & 0 \\ 0 & \mathrm{id}_m \end{pmatrix} \circ \begin{pmatrix} x \\ u \end{pmatrix} = \begin{pmatrix} x_{\mathrm{obs}} \\ u \end{pmatrix}$. The equation

$$B \circ u = A_{\mathrm{obs}} \circ x_{\mathrm{obs}} = A_{\mathrm{obs}} \circ (X_{\mathrm{obs}} \circ x) = (A_{\mathrm{obs}} X_{\mathrm{obs}}) \circ x = A \circ x$$

shows that

$$\begin{pmatrix} x \\ u \end{pmatrix} \in \mathcal{B}_1 \Longrightarrow \begin{pmatrix} X_{\mathrm{obs}} & 0 \\ 0 & \mathrm{id}_m \end{pmatrix} \circ \mathcal{B}_1 = \mathcal{B}_{1,\mathrm{obs}}.$$

With $\begin{pmatrix} C_{\mathrm{obs}} & D \\ 0 & \mathrm{id}_m \end{pmatrix} \begin{pmatrix} X_{\mathrm{obs}} & 0 \\ 0 & \mathrm{id}_m \end{pmatrix} = \begin{pmatrix} C & D \\ 0 & \mathrm{id}_m \end{pmatrix}$, we get

$$\mathcal{B} = \begin{pmatrix} C & D \\ 0 & \mathrm{id}_m \end{pmatrix} \circ \mathcal{B}_1 = \begin{pmatrix} C_{\mathrm{obs}} & D \\ 0 & \mathrm{id}_m \end{pmatrix} \circ \left(\begin{pmatrix} X_{\mathrm{obs}} & 0 \\ 0 & \mathrm{id}_m \end{pmatrix} \circ \mathcal{B}_1 \right) = \begin{pmatrix} C_{\mathrm{obs}} & D \\ 0 & \mathrm{id}_m \end{pmatrix} \circ \mathcal{B}_{1,\mathrm{obs}}.$$

Since $\begin{pmatrix} A_{\text{obs}} \\ C_{\text{obs}} \end{pmatrix}$ has a left inverse the Rosenbrock equations (6.63) are observable and

$$\begin{pmatrix} C_{\text{obs}} & D \\ 0 & \text{id}_m \end{pmatrix} : \mathcal{B}_{1,\text{obs}} \cong \mathcal{B} = \begin{pmatrix} C_{\text{obs}} & D \\ 0 & \text{id}_m \end{pmatrix} \circ \mathcal{B}_{1,\text{obs}}$$

is an isomorphism. □

Corollary 6.3.4 *With the data from the preceding theorem, the following properties are equivalent:*

1. $\mathcal{B} = \left\{ \binom{y}{u} \in \mathcal{F}^{n+m};\ P \circ y = Q \circ u \right\}$ *is controllable.*
2. $\mathcal{B}_{1,\text{obs}}$ *is controllable.*
3. $(A_{\text{obs}}, -B)$ *is right invertible, i.e., by Corollary 3.3.24, its n-th elementary divisor is* 1.

Proof 1. $\Longleftrightarrow$ 2. This follows from the system isomorphism (6.64).
2. $\Longleftrightarrow$ 3. Theorem 4.3.5. □

Example 6.3.5 Consider nonzero polynomials $A_{\text{obs}}, C_{\text{obs}}, X_{\text{obs}}, B \in \mathbb{C}[s]$ with coprime A_{obs} and C_{obs} and the Rosenbrock equations

$$A \circ x = B \circ u,\ y = C \circ x \text{ with } \binom{A}{C} := \begin{pmatrix} A_{\text{obs}} \\ C_{\text{obs}} \end{pmatrix} X_{\text{obs}}$$
$$\mathcal{B}_1 := \left\{ \binom{x}{u} \in \mathcal{F}^2;\ A \circ x = B \circ u \right\},\ \mathcal{B} := \begin{pmatrix} C & 0 \\ 0 & 1 \end{pmatrix} \circ \mathcal{B}_1.$$

Then $\gcd(A, C) = X_{\text{obs}}$ and $(C_{\text{obs}}, -A_{\text{obs}})$ is a universal left annihilator of $\binom{A}{C}$. By Theorem 6.3.1 the behavior $\mathcal{B}$ is given as

$$\mathcal{B} = \left\{ \binom{y}{u} \in \mathcal{F}^2;\ A_{\text{obs}} \circ y = C_{\text{obs}} B \circ u \right\}. \qquad \diamond$$

We will later apply the preceding theorem to derive necessary and sufficient conditions for the controllability of series and parallel interconnections in Corollaries 7.3.1 and 7.4.1, respectively.

6.3.4 The Existence and Uniqueness of Observable State Space Realizations

In this section we show that every IO behavior admits an observable state space realization and that this is unique up to similarity. Likewise two observable and controllable state space representations of a transfer matrix H are similar. These results are among Kalman's seminal contributions to systems theory. We are going to show the uniqueness first and then the existence. In Chap. 12 *we will* construct *several* canonical *realizations by means of Gröbner bases.*

We assume the polynomial algebra $\mathcal{D} := F[s]$ over a field, an injective $F[s]$-cogenerator $\mathcal{F}$, and the data from Definition 6.3.2, namely, state space equations

$$\begin{aligned}&(s\,\mathrm{id}_n - A)\circ x = Bu, \qquad y = Cx + D\circ u\\ &\text{with } A\in F^{n\times n},\ B\in F^{n\times m},\ C\in F^{p\times n}, \text{ and } D\in F[s]^{p\times m},\end{aligned} \tag{6.65}$$

as well as the induced IO behaviors

$$\begin{aligned}\mathcal{B}_1 &:= \left\{\left(\begin{smallmatrix}x\\u\end{smallmatrix}\right)\in\mathcal{F}^{n+m};\ (s\,\mathrm{id}_n - A)\circ x = Bu\right\} \quad\text{and}\\ \mathcal{B} &:= \left(\begin{smallmatrix}C & D\\ 0 & \mathrm{id}_m\end{smallmatrix}\right)\circ\mathcal{B}_1 = \left\{\left(\begin{smallmatrix}y\\u\end{smallmatrix}\right)\in\mathcal{F}^{p+m};\ P\circ x = Q\circ u\right\}\end{aligned} \tag{6.66}$$

as in Theorem and Definition 6.3.1. The transfer matrix $H\in F(s)^{p\times m}$ of $\mathcal{B} = \left(\begin{smallmatrix}C & D\\ 0 & \mathrm{id}_m\end{smallmatrix}\right)\circ\mathcal{B}_1$ with $PH = Q$ is $H = D + C(s\,\mathrm{id}_n - A)^{-1}B$.

Lemma and Definition 6.3.6 (Similar state space equations) *Consider the state space equations*

$$\begin{aligned}s\circ x_1 &= A_1x_1 + B_1u,\ y = C_1x_1 + D\circ u\\ A_1&\in F^{n\times n},\ B_1\in F^{n\times m},\ C_1\in F^{p\times n},\ D\in F[s]^{p\times m},\end{aligned} \tag{6.67}$$

with the behaviors

$$\mathcal{B}_1 = \left\{\left(\begin{smallmatrix}x_1\\u\end{smallmatrix}\right)\in\mathcal{F}^{n+m};\ (s\,\mathrm{id}_n - A_1)\circ x_1 = B_1u\right\}\ \textit{and}\ \mathcal{B} := \left(\begin{smallmatrix}C_1 & D\\ 0 & \mathrm{id}_m\end{smallmatrix}\right)\circ\mathcal{B}_1,$$

as well as an invertible matrix $U\in \mathrm{Gl}_n(F)$. Multiplication with U from the left shows that the state x_1 satisfies the preceding equations if and only if the new state $x_2 := Ux_1$ satisfies the new equations

$$s\circ x_2 = UA_1U^{-1}x_2 + UB_1u,\ y = C_1U^{-1}x_2 + D\circ u \tag{6.68}$$

with the behaviors

$$\begin{aligned}&\mathcal{B}_2 := \left\{\left(\begin{smallmatrix}x_2\\u\end{smallmatrix}\right)\in\mathcal{F}^{n+m};\ (s\,\mathrm{id}_n - UA_1U^{-1})\circ x_2 = UB_1u\right\}\ \textit{and}\\ &\mathcal{B} = \begin{pmatrix}C_1U^{-1} & D\\ 0 & \mathrm{id}_m\end{pmatrix}\circ\mathcal{B}_2.\end{aligned}$$

These two state space systems have the same transfer matrix, namely,

$$\begin{aligned}H = D + C_1(s\,\mathrm{id} - A_1)^{-1}B_1 &= D + (C_1U^{-1})\big(U(s\,\mathrm{id} - A_1)^{-1}U^{-1}\big)(UB_1)\\ &= D + (C_1U^{-1})(s\,\mathrm{id} - UA_1U^{-1})^{-1}(UB_1),\end{aligned}$$

and the map

$$\left(\begin{smallmatrix}U & 0\\ 0 & \mathrm{id}_m\end{smallmatrix}\right)\circ\colon\ \mathcal{B}_1 \xrightarrow{\cong} \mathcal{B}_2,\quad \left(\begin{smallmatrix}x_1\\u\end{smallmatrix}\right)\longmapsto\left(\begin{smallmatrix}Ux_1\\u\end{smallmatrix}\right),$$

is a behavior isomorphism with the additional property

$$\begin{pmatrix}C_1U^{-1} & D\\ 0 & \mathrm{id}_m\end{pmatrix}\left(\begin{smallmatrix}U & 0\\ 0 & \mathrm{id}_m\end{smallmatrix}\right) = \left(\begin{smallmatrix}C_1 & D\\ 0 & \mathrm{id}_m\end{smallmatrix}\right).$$

If $\mathfrak{O}_l$ and $\mathfrak{C}_l$, $l = 1, 2$, denote the observability and controllability matrix, respectively, of (6.67) *and* (6.68), *cf. Definition 4.1.7 and Eq.* (4.66), *then the identities*

$$\mathfrak{O}_2 = \mathfrak{O}_1 U^{-1} \text{ and } \mathfrak{C}_2 = U\mathfrak{C}_1$$

hold.
The two systems (6.67) *and* (6.68) *are called* similar *state space realizations of the IO system $\mathcal{B}$ and of the transfer matrix H since the matrices A_1 and UA_1U^{-1} are similar.*

Proof The proof of the lemma is contained in the explicit formulas. □

Corollary 6.3.7 (Nonsimilar state space realizations) *Consider the IO systems $\mathcal{B}_1$ and $\mathcal{B} = \begin{pmatrix} C_1 & D \\ 0 & \mathrm{id}_m \end{pmatrix} \circ \mathcal{B}_1$ of Lemma and Definition 6.3.6.*

In Theorem 4.1.11 it was shown that the observable factor $\mathcal{B}_{1,\mathrm{obs}}$ derived from Eq. (6.67) *gives rise to the same image IO system $\mathcal{B}$ and, in particular, has the transfer matrix $H = D + C_1(s\,\mathrm{id}_n - A_1)^{-1}B_1$. However, $\mathcal{B}_1$ and $\mathcal{B}_{1,\mathrm{obs}}$ are similar if and only if their state space has the same dimension n, and this is the case if and only if $\mathcal{B}_1$ is already observable.*

In other words, a given IO behavior $\mathcal{B}$ or transfer matrix H admit essentially different state space realizations. The next theorem, however, shows that two observable *state space realizations of an IO system are always similar.* ◇

Theorem 6.3.8 (Similarity of observable state space realizations, cf. [1, Theorem 6.5.8 on p. 227] for the SISO case) *Let*

$$\mathcal{B} := \left\{ \begin{pmatrix} y \\ u \end{pmatrix} \in \mathcal{F}^{p+m};\ P \circ y = Q \circ u \right\} \text{ with } P \in F[s]^{p\times p} \text{ and } Q \in F[s]^{p\times m}$$

be an IO system over ${}_{F[s]}\mathcal{F}$ with transfer matrix H, i.e., $\mathrm{rank}(P) = p$ *and $H = P^{-1}Q$. Assume that*

$$\begin{aligned} &s \circ x_l = A_l x_l + B_l u, \quad\ y = C_l x_l + D_l \circ u \text{ with } x_l \in \mathcal{F}^{n_l}, \\ &\mathcal{B}_l := \left\{ \begin{pmatrix} x_l \\ u \end{pmatrix} \in \mathcal{F}^{n_l+m};\ (s\,\mathrm{id}_{n_l} - A_l) \circ x_l = B_l u \right\}, \quad l = 1, 2, \end{aligned} \tag{6.69}$$

are two observable state space realizations of $\mathcal{B}$, i.e., that the two maps

$$\begin{pmatrix} C_l & D_l \\ 0 & \mathrm{id}_m \end{pmatrix} \circ \colon\ \mathcal{B}_l \xrightarrow{\cong} \mathcal{B},\ \begin{pmatrix} x_l \\ u \end{pmatrix} \longmapsto \begin{pmatrix} C_l x_l + D_l \circ u \\ u \end{pmatrix}, \tag{6.70}$$

are behavior isomorphisms. Then the following holds:

1. *The two state space realizations* (6.69) *are similar, i.e., $n := n_1 = n_2$, and there is a matrix $U \in \mathrm{Gl}_n(F)$ such that*

$$A_2 = UA_1U^{-1},\quad B_2 = UB_1,\quad C_2 = C_1U^{-1},\quad \text{and } D_1 = D_2. \tag{6.71}$$

The matrix U is uniquely determined as solution of $\mathfrak{O}_2 = \mathfrak{O}_1 U^{-1}$ where $\mathfrak{O}_1$ and $\mathfrak{O}_2$ are the observability matrices of the state space systems (6.69) of rank n according to Definition 4.1.7. If $(X_2, Z_2) \in F[s]^{n\times(n+p)}$ is a left inverse of $\begin{pmatrix} s\,\mathrm{id}_n - A_2 \\ C_2 \end{pmatrix}$, computed by means of Theorem 3.3.22, then

$$U_{\alpha\beta} = \sum_{i=1}^{n} \eta_i^{(1)}(A_1)_{i\beta},\ 1 \le \alpha, \beta \le n,\ \text{where}$$
$$\delta_\alpha = (0, \dots, \overset{\alpha}{1}, \dots, 0),\ \eta^{(1)} := \delta_\alpha Z_2 C_1 = (Z_2)_{\alpha -} C_1 \in F[s]^{1\times n}. \tag{6.72}$$

The computation of U needs the powers A_1^k for $0 \le k \le \deg_s(Z_2C_1)$. The inverse U^{-1} is obtained by replacing Z_2 and C_1 by Z_1 and C_2, respectively.

2. *Let $e_1 | \cdots | e_p$ be the monic elementary divisors of P, $\mathcal{B}^0 := \{y \in \mathcal{F}^p;\ P \circ y = 0\}$ and $M^0 := M(\mathcal{B}^0) = F[s]^{1\times p}/F[s]^{1\times p}P$. Then*

$$\det(P(s)) = a \prod_{j=1}^{p} e_j \text{ and } n = \deg(\det(P(s))) = \sum_{j=1}^{p} \deg(e_j) = \dim_F(M^0) \tag{6.73}$$

with some nonzero $a \in F$. Moreover the nonconstant similarity invariants of A_1 and A_2 coincide with the nonconstant elementary divisors of P, which, in turn, are the invariant factors of M^0. In particular, the similarity invariants of the A_i are uniquely determined by $\mathcal{B}^0$ or M^0.

Proof 1. (i) We provide $F^{1\times n_l}$ with the $F[s]$-structures $s \circ_{A_l} \xi := \xi A_l$, $l = 1, 2$. Since the state equations are assumed observable the matrices $\begin{pmatrix} s\,\mathrm{id}_{n_l} - A_l \\ C_l \end{pmatrix}$ admit left inverses $(X_l, Z_l) \in F[s]^{n_l\times(n_l+p)}$ for $l = 1, 2$. From Corollary 4.1.10 we use the $F[s]$-isomorphisms

$$F^{1\times n_l} \cong M_l^0 = F[s]^{1\times n_l}/F[s]^{1\times n_l}(s\,\mathrm{id}_{n_l} - A_l)$$
$$\cong M^0 = F[s]^{1\times p}/F[s]^{1\times p}P,\ \xi^{(l)} \leftrightarrow \overline{\eta^{(l)}} \leftrightarrow \overline{\zeta^{(l)}},\ l = 1, 2,$$
$$\text{where } \xi_j^{(l)} = \sum_{i=1}^{n_l} \eta_i^{(l)}(A_l)_{ij},\ \overline{\eta^{(l)}} = \overline{\xi^{(l)}} = \overline{\zeta^{(l)}C_l},\ \overline{\zeta^{(l)}} = \overline{\xi^{(l)}Z_l} = \overline{\eta^{(l)}Z_l}$$
$$\Longrightarrow n := n_1 = n_2 = \dim_F(M^0) = \deg_s(\det(P(s))).\ \text{(cf. (6.75).)}$$

We infer the commutative diagram of $F[s]$-isomorphisms

$$
\begin{array}{ccccc}
F^{1\times n} & \xleftrightarrow{\cong} & M_2^0 & \xleftrightarrow{\cong} & M^0 \\
\downarrow \phi & & & & \updownarrow = \\
F^{1\times n} & \xleftrightarrow{\cong} & M_1^0 & \xleftrightarrow{\cong} & M^0
\end{array}
\qquad\qquad
\begin{array}{ccccc}
\xi^{(2)} & \longleftrightarrow & \overline{\eta^{(2)}} & \longleftrightarrow & \overline{\zeta} \\
\downarrow & & & & \updownarrow \\
\xi^{(1)} & \longleftrightarrow & \overline{\eta^{(1)}} & \longleftrightarrow & \overline{\zeta}
\end{array}
$$

$$
\text{where } \overline{\zeta} = \overline{\xi^{(2)} Z_2},\ \overline{\eta^{(2)}} = \overline{\zeta C_2},\ \overline{\eta^{(1)}} = \overline{\zeta C_1} = \overline{\xi^{(2)} Z_2 C_1},\ \xi^{(1)}_\beta = \sum_{i=1}^n \eta_i^{(1)} (A_1)_{i\beta}. \tag{6.74}
$$

The $F[s]$-automorphism ϕ is uniquely determined by the commutativity of the diagram. In particular, it is F-linear and thus has the form $\phi(\xi^{(2)}) = \xi^{(2)} U$ with a matrix $U \in \mathrm{Gl}_n(F)$.
(ii) The $F[s]$-linearity of ϕ implies

$$
\begin{aligned}
&\forall \xi^{(2)} \in F^{1\times n} :\ \xi^{(2)} U A_1 = s \circ_{A_1} \phi(\xi^{(2)}) = \phi(s \circ_{A_2} \xi^{(2)}) = \phi(\xi^{(2)} A_2) = \xi^{(2)} A_2 U \\
&\Longrightarrow U A_1 = A_2 U \Longrightarrow A_2 = U A_1 U^{-1}.
\end{aligned}
$$

Hence A_1 and A_2 are similar.

For $\zeta \in F^{1\times p} \subset F[s]^{1\times p}$ the corresponding vectors are

$$
\begin{aligned}
&\overline{\eta^{(l)}} = \overline{\zeta C_l},\ \zeta C_l \in F^{1\times n} \subset F[s]^{1\times n} \Longrightarrow \xi^{(l)} = \zeta C_l \\
&\Longrightarrow \zeta C_1 = \xi^{(1)} = \xi^{(2)} U = \zeta C_2 U \Longrightarrow C_1 = C_2 U \Longrightarrow C_2 = C_1 U^{-1}.
\end{aligned}
$$

(iii) The matrix U can be computed as follows. Let δ_α, $\alpha = 1, \dots, n$, denote the standard basis of $F^{1\times n}$. Then $\delta_\alpha U = U_{\alpha -}$ and $U_{\alpha\beta} = (\delta_\alpha U)_\beta$. In Eq. (6.74) choose

$$
\begin{aligned}
&\xi^{(2)} := \delta_\alpha \Longrightarrow \xi^{(1)} = \delta_\alpha U,\ \xi^{(1)}_\beta = U_{\alpha\beta},\ \overline{\zeta} = \overline{\delta_\alpha Z_2},\ \overline{\eta^{(1)}} = \overline{\zeta C_1} = \overline{\delta_\alpha Z_2 C_1} \\
&\Longrightarrow U_{\alpha\beta} = \xi^{(1)}_\beta = \sum_{i=1}^n \eta_i^{(1)} (A_1)_{i\beta},\ \eta^{(1)} = \delta_\alpha Z_2 C_1.
\end{aligned}
$$

This proves (6.72).

(iv) We use the matrix U and Lemma and Definition 6.3.6 to define the new state space equations

$$
\begin{aligned}
&s \circ x = U A_1 U^{-1} x + U B_1 u = A_2 x + U B_1 u, \\
&\qquad\qquad y = C_1 U^{-1} x + D_1 u = C_2 x + D_1 u \text{ and} \\
&\widehat{\mathcal{B}}_2 = \left\{ \left(\begin{smallmatrix} x \\ u \end{smallmatrix}\right) \in \mathcal{F}^{n+m};\ (s\,\mathrm{id}_n - A_2) \circ x = U B_1 u \right\} \\
&\Longrightarrow \left(\begin{smallmatrix} U & 0 \\ 0 & \mathrm{id}_m \end{smallmatrix}\right) : \mathcal{B}_1 \cong \widehat{\mathcal{B}}_2,\ \left(\begin{smallmatrix} C_2 & D_1 \\ 0 & \mathrm{id}_m \end{smallmatrix}\right) \circ = \left(\begin{smallmatrix} C_1 & D_1 \\ 0 & \mathrm{id}_m \end{smallmatrix}\right) \left(\begin{smallmatrix} U^{-1} & 0 \\ 0 & \mathrm{id}_m \end{smallmatrix}\right) :\ \widehat{\mathcal{B}}_2 \cong \mathcal{B} \\
&\Longrightarrow D_1 + C_2 (s\,\mathrm{id}_n - A_2)^{-1} (U B_1) = H = P^{-1} Q \\
&\qquad\qquad = D_2 + C_2 (s\,\mathrm{id}_n - A_2)^{-1} B_2 \in F(s)^{p\times m} = F[s]^{p\times m} \oplus F(s)^{p\times m}_{\mathrm{spr}}
\end{aligned}
$$

$$\Longrightarrow D_1 = D_2,\ C_2(s\,\mathrm{id}_n - A_2)^{-1}(UB_1) = C_2(s\,\mathrm{id}_n - A_2)^{-1}B_2.$$

The last equation implies $C_2(s\,\mathrm{id}_n - A_2)^{-1}B = 0,\ B := UB_1 - B_2$. We use

$$\begin{aligned} &X_2(s\,\mathrm{id}_n - A_2) + Z_2C_2 = \mathrm{id}_n \Longrightarrow X_2 + Z_2C_2(s\,\mathrm{id}_n - A_2)^{-1} = (s\,\mathrm{id}_n - A_2)^{-1} \\ &\Longrightarrow X_2B - (s\,\mathrm{id}_n - A_2)^{-1}B \\ &\quad = -Z_2C_2(s\,\mathrm{id}_n - A_2)^{-1}B = 0 = 0 + 0 \in F[s]^{n\times m} \oplus F(s)^{n\times m}_{\mathrm{spr}} \\ &\Longrightarrow X_2B = (s\,\mathrm{id}_n - A_2)^{-1}B = 0 \Longrightarrow B = 0 \Longrightarrow B_2 = UB_1. \end{aligned}$$

Thus Eq. (6.71) has been shown. The equation $\mathfrak{O}_2 = \mathfrak{O}_1U^{-1}$ follows directly. Since $\mathrm{rank}(\mathfrak{O}_1) = \mathrm{rank}(\mathfrak{O}_2) = n$ this equation determines U uniquely.

2. Let $e_1|\cdots|e_p$ denote the $p = \mathrm{rank}(P)$ elementary divisors of P and assume $e_1 = \cdots = e_s = 1$, but $\deg(e_{s+1}) > 0$. Theorem 4.2.8 yields the $F[s]$-isomorphism $M^0 \cong \prod_{j=s+1}^{p} F[s]/F[s]e_j$ and the invariant factors $e_j,\ j = s+1,\ldots,p$, of M^0. This signifies that the nonconstant elementary divisors of P coincide with the invariant factors of M^0. We also infer

$$\det(P(s)) = a\prod_{j=1}^{p} e_j,\ 0 \neq a \in F,\ \deg(\det(P(s))) = \sum_{j=1}^{p}\deg(e_j) = \dim_F(M^0). \tag{6.75}$$

By definition, the elementary divisors of $s\,\mathrm{id}_n - A_i$ are the similarity invariants of A_i. As before, those among these which are not 1 are the invariant factors of M_i^0. But the isomorphic modules M^0 and M_i^0 have the same invariant factors, see Remark 4.2.10. This proves 2. □

Remark 6.3.9 If in the preceding theorem $\mathcal{B}_2$ is given, the matrices $(P, -Q)$ that define $\mathcal{B}$ are computed as a universal left annihilator $(-Y, P) \in F[s]^{p\times(n+p)}$ of $\begin{pmatrix} s\,\mathrm{id}_n - A_2 \\ C_2 \end{pmatrix}$ and $Q = PD + YB_2$ according to Theorem 3.2.18. Then $\mathcal{B}_2$ is observable if and only if $\begin{pmatrix} s\,\mathrm{id}_n - A_2 \\ C_2 \end{pmatrix}$ has a left inverse $(X_2, Z_2) \in F[s]^{n\times(n+p)}$, and this is checked by means of Theorem 3.3.23. The matrices $(-Y, P)$ and (X_2, Z_2) are both easily derived from the Smith form of $\begin{pmatrix} s\,\mathrm{id}_n - A_2 \\ C_2 \end{pmatrix}$. ◇

In the preceding theorem we have treated observable state space realizations of a given IO behavior $\mathcal{B}$. In contrast, the following result treats the controllable and observable state space realization of a given transfer matrix H.

Theorem and Definition 6.3.10 (Characterization of minimal state space realizations. Cf. [6, Sect. 6.8 on p. 292], [3, Theorem 6.5.1 on p. 439], [1, Sect. 6.5 on p. 224])

1. *Consider state space equations*

$$s \circ x = Ax + Bu,\quad y = Cx + D\circ u,\quad \textit{with } x \in \mathcal{F}^n,\ u \in \mathcal{F}^m,\ \textit{and } y \in \mathcal{F}^p,$$

and the associated behaviors

$$\mathcal{B}_1 = \left\{\begin{pmatrix} x \\ u \end{pmatrix} \in \mathcal{F}^{n+m};\ (s\,\mathrm{id}_n - A) \circ y = Bu\right\} \quad \text{and}$$
$$\mathcal{B} = \begin{pmatrix} C & D \\ 0 & \mathrm{id}_m \end{pmatrix} \circ \mathcal{B}_1 = \left\{\begin{pmatrix} y \\ u \end{pmatrix} \in \mathcal{F}^{p+m};\ P \circ y = Q \circ u\right\}$$

with $P \in F[s]^{p \times p}$ *and* $\mathrm{rank}(P) = n$. *Hence the transfer matrix* H *of* $\mathcal{B}$ *is*

$$H = P^{-1}Q = D + C(s\,\mathrm{id}_n - A)^{-1}B.$$

The following properties of the system are equivalent:

(a) *The state space equations are controllable and observable, i.e.,* $\mathcal{B}_1$ *is controllable and*

$$\begin{pmatrix} C & D \\ 0 & \mathrm{id}_m \end{pmatrix} \circ \colon\ \mathcal{B}_1 \overset{\cong}{\to} \mathcal{B} \tag{6.76}$$

is an isomorphism.

(b) *The state dimension is*

$$n = \dim\left(F[s]^{1 \times p} / F[s]^{1 \times p} P_{\mathrm{cont}}\right) = \deg(e_1) + \cdots + \deg(e_p),$$

where $P_{\mathrm{cont}} \in F[s]^{p \times p}$ *is a matrix with*

$$F[s]^{1 \times p} P_{\mathrm{cont}} = \{\xi \in F[s]^{1 \times p};\ \xi H \in F[s]^{1 \times m}\},$$

i.e., the behavior $\left(F[s]^{1 \times p}(P_{\mathrm{cont}}, -P_{\mathrm{cont}} H)\right)^{\perp}$ *is the unique controllable realization of* H *according to Theorem 6.2.4, and the polynomials* $e_1 | \cdots | e_p$ *are the elementary divisors of* P_{cont}. *Also recall from Theorem 6.2.5 that the product* $H = P_{\mathrm{cont}}^{-1}(P_{\mathrm{cont}} H)$ *is the essentially unique left coprime factorization of* H.

Recall from (6.75) that for every square *polynomial matrix the product of its elementary divisors is its determinant, up to a nonzero factor in* F, *and hence the sum of their degrees is the determinantal degree. In particular, here we have that*

$$\deg\left(\det(P_{\mathrm{cont}})\right) = \deg(e_1) + \cdots + \deg(e_p).$$

If the equivalent conditions (a) and (b) are satisfied, the state space system is called a minimal *realization of* H, *and the nonconstant elementary divisors of* P_{cont} *coincide with the nonconstant similarity invariants of* A.

2. *Two minimal state space realizations of a transfer matrix* $H \in F(s)^{p \times m}$ *are similar. In other terms,* H *admits essentially just one minimal realization.*
3. *Consider the Smith/McMillan form of* H, *i.e.,*

$$\begin{pmatrix} \mathrm{diag}(a_1/b_1, \ldots, a_r/b_r) & 0 \\ 0 & 0 \end{pmatrix} \in F(s)^{(r+(p-r)) \times (r+(m-r))},\ a_1 | \cdots a_r \neq 0,\ b_r | \cdots | b_1 \neq 0.$$

Corollary and Definition 6.2.6 and Eq. (6.25) imply

$$M^0_{\text{cont}} \underset{F[s]}{\cong} \prod_{i=1}^{r} F[s]/F[s]b_i \text{ and}$$

$$\dim_F(\mathcal{B}^0_{\text{cont}}) = \dim_F(M^0_{\text{cont}}) = \deg\left(\det(P_{\text{cont}})\right) = \sum_{i=1}^{r} \deg(b_i).$$

This number is also called the McMillan degree *of H, cf. [3, p. 466], [7, pp. 299,396], and can be computed directly from H.*

Proof 1. $(a) \Longrightarrow (b)$. The isomorphism (6.76) and the controllability of $\mathcal{B}_1$ imply that $\mathcal{B}$ is controllable. According to Theorem 6.2.4, it is the *unique* controllable IO realization of the transfer matrix H, hence

$$\mathcal{B} = \left\{\left(\begin{smallmatrix} y \\ u \end{smallmatrix}\right) \in \mathcal{F}^{p+m};\ P_{\text{cont}} \circ y = (P_{\text{cont}} H) \circ u\right\}.$$

From (6.75) we infer

$$n = \deg(\det(P_{\text{cont}})) = \dim_F\left(F[s]^{1\times p}/F[s]^{1\times p} P_{\text{cont}}\right).$$

1. $(b) \Longrightarrow (a)$. The inclusion $\mathcal{B}_{\text{cont}} \subseteq \mathcal{B}$ implies $\mathcal{B}^0_{\text{cont}} \subseteq \mathcal{B}^0$. Moreover the map $C\circ\colon \mathcal{B}^0_1 \longrightarrow \mathcal{B}^0$ is an epimorphism. Thus we have the maps

$$\mathcal{B}^0_1 \xrightarrow{\text{epimorphism}} \mathcal{B}^0 \underset{\supseteq}{\xleftarrow{\text{monomorphism}}} \mathcal{B}^0_{\text{cont}},\quad x \longmapsto Cx = y \longleftarrow\!\shortmid y.$$

By duality, we obtain

$$M^0_1 \xleftarrow{\text{monomorphism}} M^0 \xrightarrow{\text{epimorphism}} M^0_{\text{cont}}.$$

The outer modules are finite-dimensional F-vector spaces. Condition (b) signifies that they have the same dimension $n = \dim_F(M^0_1) = \dim_F(M^0_{\text{cont}})$. This implies that these maps are isomorphisms.

The isomorphism $M^0 \cong M^0_{\text{cont}}$ furnishes $F[s]^{1\times p} P = F[s]^{1\times p} P_{\text{cont}}$, hence

$$\begin{aligned} F[s]^{1\times p}(P, -Q) &= F[s]^{1\times p} P(\mathrm{id}_p, -H) = F[s]^{1\times p} P_{\text{cont}}(\mathrm{id}_p, -H) \\ &= F[s]^{1\times p}(P_{\text{cont}}, -P_{\text{cont}} H). \end{aligned}$$

Thus $\mathcal{B} = \mathcal{B}_{\text{cont}}$ and $\mathcal{B}$ is controllable.

The isomorphism $M^0_1 \cong M^0$ and its dual isomorphism $C\circ\colon \mathcal{B}^0_1 \cong \mathcal{B}^0$ signify that the state space system is observable, i.e., that $\left(\begin{smallmatrix} C & D \\ 0 & \mathrm{id}_m \end{smallmatrix}\right)\circ\colon \mathcal{B}_1 \cong \mathcal{B}$ is an isomorphism. Since $\mathcal{B}$ is controllable, so is $\mathcal{B}_1$, i.e., the state space system is controllable according to Theorem and Definition 4.3.8.

2. By 1 a minimal state space realization of H induces the isomorphism

$$\left(\begin{smallmatrix} C & D \\ 0 & \mathrm{id}_m \end{smallmatrix}\right)\circ:\ \mathcal{B}_1 \cong \mathcal{B} = \mathcal{B}_{\mathrm{cont}} = \{\left(\begin{smallmatrix} y \\ u \end{smallmatrix}\right);\ P_{\mathrm{cont}} \circ y = (P_{\mathrm{cont}} H) \circ u\},$$

where the IO system on the right depends on H only. Hence two such realizations are observable state space realizations of the same IO system $\mathcal{B}_{\mathrm{cont}}$. By Theorem 6.3.8 they are similar. □

We finally establish the *existence* of the (essentially unique) observable state space realization of an IO behavior. The unconstructive proof is due to M. Fliess [5, Theorem 129]. In Chap. 12 we will *construct the canonical observer and observability realizations* by means of Gröbner bases. We assume an IO behavior

$$\begin{aligned} &\mathcal{B} = \left\{\left(\begin{smallmatrix} y \\ u \end{smallmatrix}\right) \in \mathcal{F}^{p+m};\ P \circ y = Q \circ u\right\} \text{ where} \\ &\quad (P, -Q) \in \mathcal{D}^{p\times(p+m)},\ \mathrm{rank}(P) = p,\ H := P^{-1}Q, \\ &U := \mathcal{D}^{1\times p}(P, -Q),\ U^0 := \mathcal{D}^{1\times p}P,\ M := \mathcal{D}^{1\times(p+m)}/U,\ M^0 := \mathcal{D}^{1\times p}/U^0. \end{aligned} \tag{6.77}$$

According to Theorem 6.2.2 we obtain the exact sequence

$$0 \longrightarrow \mathcal{D}^{1\times m} \xrightarrow{\mathrm{inj}_{\mathrm{ind}} = (\circ(0,\mathrm{id}_m))_{\mathrm{ind}}} M \xrightarrow{\mathrm{proj}_{\mathrm{ind}} = \left(\circ\left(\begin{smallmatrix} \mathrm{id}_p \\ 0 \end{smallmatrix}\right)\right)_{\mathrm{ind}}} M^0 \longrightarrow 0. \tag{6.78}$$

We use the terminology from Sect. 6.3.2. Let $\left(\begin{smallmatrix} \widetilde{y} \\ \widetilde{u} \end{smallmatrix}\right) \in M^{p+m}$ and $\widetilde{y}^0 \in (\mathrm{M}^0)^p$ denote the column vectors of residue classes of the standard bases. This implies

$$(\gamma, \eta) + U = (\gamma, \eta)\left(\begin{smallmatrix} \widetilde{y} \\ \widetilde{u} \end{smallmatrix}\right),\ \gamma + U^0 = \gamma\widetilde{y}^0,\ (\gamma, \eta) \in \mathcal{D}^{1\times(p+m)}.$$

The module M^0 is torsion and hence $n := \dim_F(M^0) < \infty$. Let $\widetilde{x}^0 \in (M^0)^n$ be an F-basis of M^0 and $\widetilde{z} \in M^n$ an inverse image of $\widetilde{x}^0$ under $\mathrm{proj}_{\mathrm{ind}}$, i.e.,

$$\begin{aligned} &\mathrm{proj}_{\mathrm{ind}}(\widetilde{z}) = \widetilde{x}^0,\ M^0 = F^{1\times n}\widetilde{x}^0 \Longrightarrow M = F^{1\times n}\widetilde{z} + \mathcal{D}^{1\times m}\widetilde{u} \\ &\Longrightarrow \exists A \in F^{n\times n},\ \widetilde{B} = \sum_{k=0}^{d} \widetilde{B}_k s^k \in \mathcal{D}^{n\times m},\ \widetilde{B}_k \in F^{n\times m},\ d := \deg_s(\widetilde{B}),\ \widetilde{B}_d \neq 0, \\ &\text{with } s\widetilde{z} = A\widetilde{z} + \widetilde{B}\widetilde{u} = A\widetilde{z} + \sum_{k=0}^{d} \widetilde{B}_k s^k \widetilde{u}. \end{aligned} \tag{6.79}$$

If $d = \deg_s(\widetilde{B}) \leq 0$ then $B := \widetilde{B} \in F^{n\times m}$. If $d > 0$ define

$$\widetilde{z}^{(d)} := \widetilde{z},\ \widetilde{B}^{(d)} := \widetilde{B},\ \widetilde{z}^{(d-1)} := \widetilde{z}^{(d)} - \widetilde{B}_d s^{d-1}\widetilde{u} \text{ and}$$
$$\widetilde{B}^{(d-1)} := \sum_{k=0}^{d-2} \widetilde{B}_k s^k + (\widetilde{B}_{d-1} + A\widetilde{B}_d)s^{d-1}$$
$$\underset{\mathrm{proj}_{\mathrm{ind}}(\widetilde{u})=0}{\Longrightarrow} \mathrm{proj}_{\mathrm{ind}}(\widetilde{z}^{(d-1)}) = \widetilde{x}^0 \text{ and}$$
$$s\widetilde{z}^{(d-1)} = A\widetilde{z}^{(d-1)} + \widetilde{B}^{(d-1)}\widetilde{u},\ \deg_s(\widetilde{B}^{(d-1)}) \le d-1.$$

By recursion from d to 0 we finally obtain

$$\begin{aligned} &\widetilde{x} := \widetilde{z}^0 \in M^n,\ B := \widetilde{B}^0 \in F^{n\times m},\ \mathrm{proj}_{\mathrm{ind}}(\widetilde{x}) = \widetilde{x}^0,\ s\widetilde{x} = A\widetilde{x} + B\widetilde{u} \\ &\Longrightarrow M = F^{1\times n}\widetilde{x} + \mathcal{D}^{1\times m}\widetilde{u},\ \exists C \in F^{p\times n},\ D \in \mathcal{D}^{p\times m} :\ \widetilde{y} = C\widetilde{x} + D\widetilde{u}. \end{aligned} \tag{6.80}$$

The matrices A, B, C, D give rise to state space equations, modules, and behaviors:

$$\begin{aligned} &s \circ x = Ax + Bu,\ y = Cx + D \circ u,\ x \in \mathcal{F}^n,\ u \in \mathcal{F}^m, \\ &U_1 := \mathcal{D}^{1\times n}(s\,\mathrm{id}_n - A, -B),\ U_1^0 := \mathcal{D}^{1\times n}(s\,\mathrm{id}_n - A), \\ &M_1 := \mathcal{D}^{1\times(n+m)}/U_1,\ M_1^0 := \mathcal{D}^{1\times n}/U_1^0 \\ &\mathcal{B}_1 := \left\{\left(\begin{smallmatrix} x \\ u \end{smallmatrix}\right) \in \mathcal{F}^{n+m};\ (s\,\mathrm{id}_n - A, -B) \circ \left(\begin{smallmatrix} x \\ u \end{smallmatrix}\right) = 0\right\} \\ &0 \longrightarrow \mathcal{D}^{1\times m} \xrightarrow{\mathrm{inj}_{\mathrm{ind}}} M_1 \xrightarrow{\mathrm{proj}_{\mathrm{ind}}} M_1^0 \longrightarrow 0 \text{ exact.} \end{aligned} \tag{6.81}$$

Moreover let $\left(\begin{smallmatrix} \widehat{x} \\ \widehat{u} \end{smallmatrix}\right) \in M_1^{n+m}$ resp $\widehat{x}^0 \in (M_1^0)^n$ denote the canonical generators induced from the standard bases. Then $\widehat{x}^0$ is an F-basis of M_1^0.

Theorem 6.3.11 *With the preceding data the map*

$$\left(\begin{smallmatrix} C & D \\ 0 & \mathrm{id}_m \end{smallmatrix}\right)\circ : \mathcal{B}_1 \to \mathcal{B},\ \left(\begin{smallmatrix} x \\ u \end{smallmatrix}\right) \mapsto \left(\begin{smallmatrix} y \\ u \end{smallmatrix}\right) := \left(\begin{smallmatrix} Cx + D\circ u \\ u \end{smallmatrix}\right),$$

is well defined and a behavior isomorphism, i.e., $\mathcal{B}_1 \cong \mathcal{B}$ *is the, by Theorem 6.3.8 unique up to similarity, observable state space realization of* $\mathcal{B}$.

Proof We establish the dual module isomorphism. The epimorphism

$$\varphi : \mathcal{D}^{1\times(n+m)} \to M = \mathcal{D}^{1\times(n+m)}\left(\begin{smallmatrix} \widetilde{x} \\ \widetilde{u} \end{smallmatrix}\right),\ (\xi, \eta) \mapsto (\xi, \eta)\left(\begin{smallmatrix} \widetilde{x} \\ \widetilde{u} \end{smallmatrix}\right),$$

satisfies

$$\varphi(U_1) = \varphi\left(\mathcal{D}^{1\times n}(s\,\mathrm{id}_n - A, -B)\right) = \mathcal{D}^{1\times n}(s\widetilde{x} - A\widetilde{x} - B\widetilde{u}) \underset{(6.80)}{=} 0.$$

Hence

$$\varphi_{\mathrm{ind}} :\ M_1 \to M,\ (\xi, \eta) + U_1 = (\xi, \eta)\left(\begin{smallmatrix} \widehat{x} \\ \widehat{u} \end{smallmatrix}\right) \mapsto (\xi, \eta)\left(\begin{smallmatrix} \widetilde{x} \\ \widetilde{u} \end{smallmatrix}\right),$$

is a well-defined epimorphism. Likewise

$$\psi_{\mathrm{ind}} : M_1^0 \to M^0,\ \xi + U_1^0 = \xi\widehat{x}^0 \mapsto \xi\widetilde{x}^0,\ \widehat{x}^0 \mapsto \widetilde{x}^0,$$

is a well-defined epimorphism. It is even an isomorphism since $\widehat{x}^0$ resp. $\widetilde{x}^0$ are F-bases of M_1^0 resp. M^0. The diagram

$$\begin{array}{ccccccccc}
0 & \longrightarrow & \mathcal{D}^{1\times m} & \xrightarrow{\mathrm{inj}_{\mathrm{ind}}} & M_1 & \xrightarrow{\mathrm{proj}_{\mathrm{ind}}} & M_1^0 & \longrightarrow & 0 \\
 & & \downarrow{\scriptstyle \mathrm{id}_m} & & \downarrow{\scriptstyle \varphi_{\mathrm{ind}}} & & \downarrow{\scriptstyle \psi_{\mathrm{ind}}} & & \\
0 & \longrightarrow & \mathcal{D}^{1\times m} & \xrightarrow{\mathrm{inj}_{\mathrm{ind}}} & M & \xrightarrow{\mathrm{proj}_{\mathrm{ind}}} & M^0 & \longrightarrow & 0
\end{array}$$

has exact rows by (6.11) and Theorem 6.2.2 and is commutative by definition of φ_{ind} and ψ_{ind}. Since id_m and ψ_{ind} are isomorphisms so is φ_{ind} as is easily proven. The defining equation $\varphi_{\mathrm{ind}}\left(\begin{smallmatrix}\widehat{x}\\ \widehat{u}\end{smallmatrix}\right) = \left(\begin{smallmatrix}\widetilde{x}\\ \widetilde{u}\end{smallmatrix}\right)$ implies

$$\begin{aligned}
&\varphi_{\mathrm{ind}}^{-1}\left(\begin{smallmatrix}\widetilde{x}\\ \widetilde{u}\end{smallmatrix}\right) = \left(\begin{smallmatrix}\widehat{x}\\ \widehat{u}\end{smallmatrix}\right)\\
\Longrightarrow\ &\varphi_{\mathrm{ind}}^{-1}((\gamma,\eta)+U) = \varphi_{\mathrm{ind}}^{-1}\left((\gamma,\eta)\left(\begin{smallmatrix}\widetilde{y}\\ \widetilde{u}\end{smallmatrix}\right)\right) \underset{(6.80)}{=} \varphi_{\mathrm{ind}}^{-1}\left((\gamma,\eta)\left(\begin{smallmatrix}C & D\\ 0 & \mathrm{id}_m\end{smallmatrix}\right)\left(\begin{smallmatrix}\widetilde{x}\\ \widetilde{u}\end{smallmatrix}\right)\right)\\
&\qquad = (\gamma,\eta)\left(\begin{smallmatrix}C & D\\ 0 & \mathrm{id}_m\end{smallmatrix}\right)\left(\begin{smallmatrix}\widehat{x}\\ \widehat{u}\end{smallmatrix}\right) = (\gamma,\eta)\left(\begin{smallmatrix}C & D\\ 0 & \mathrm{id}_m\end{smallmatrix}\right) + U_1\\
\Longrightarrow\ &\varphi_{\mathrm{ind}}^{-1}((\gamma,\eta)+U) = (\gamma,\eta)\left(\begin{smallmatrix}C & D\\ 0 & \mathrm{id}_m\end{smallmatrix}\right) + U_1 = \left(\circ\left(\begin{smallmatrix}C & D\\ 0 & \mathrm{id}_m\end{smallmatrix}\right)\right)_{\mathrm{ind}}((\gamma,\eta)+U).
\end{aligned}$$

Hence $\left(\circ\left(\begin{smallmatrix}C & D\\ 0 & \mathrm{id}_m\end{smallmatrix}\right)\right)_{\mathrm{ind}} : M \to M_1$ and $\left(\begin{smallmatrix}C & D\\ 0 & \mathrm{id}_m\end{smallmatrix}\right)\circ : \mathcal{B}_1 \to \mathcal{B}$ are well-defined isomorphisms. □

6.3.5 Characteristic Varieties and Multiplicities for Rosenbrock Equations

In Theorem 6.3.13 *we prove an analogue of Theorem* 6.2.7 *for the Rosenbrock equations* (6.45) *and* (6.46) *and then combine these two theorems to derive the fine structure of the poles of the Rosenbrock equations. In general, we assume a principal ideal domain $\mathcal{D}$ and an injective cogenerator signal module ${}_{\mathcal{D}}\mathcal{F}$. In connection with multiplicities we assume the real or complex coefficient field $F = \mathbb{C}, \mathbb{R}$, the operator algebra $\mathcal{D} = F[s]$, and the standard signal modules $\mathcal{F} := \mathcal{F}_F$ from* (6.31).

The Rosenbrock equations give rise to the exact behavior sequences

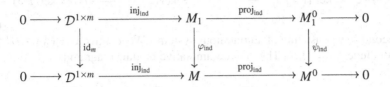

$$
\begin{array}{ccccccccc}
0 & \longrightarrow & \mathcal{C} := \ker\left(\left(\begin{smallmatrix} A \\ C \end{smallmatrix}\right)\circ\right) & \xrightarrow{\text{inj}} & \mathcal{B}_1 & \xrightarrow{\left(\begin{smallmatrix} C & D \\ 0 & \mathrm{id}_m \end{smallmatrix}\right)\circ} & \mathcal{B} & \longrightarrow & 0 \\
 & & \updownarrow = & & \uparrow \cup & & \uparrow \cup & & \\
0 & \longrightarrow & \mathcal{C} := \ker\left(\left(\begin{smallmatrix} A \\ C \end{smallmatrix}\right)\circ\right) & \xrightarrow{\text{inj}} & \mathcal{B}_1^0 = \ker(A\circ) & \xrightarrow{C\circ} & \mathcal{B}_0 = \ker(P\circ) & \longrightarrow & 0
\end{array}
\tag{6.82}
$$

the second row consisting of autonomous systems. We identify $x = \left(\begin{smallmatrix} x \\ 0 \end{smallmatrix}\right) \in \mathcal{B}_1^0 \subseteq \mathcal{B}_1$ and conclude $\text{inj} = \mathrm{id}_n \circ$. The dual commutative module diagram is

$$
\begin{array}{ccccccccc}
0 & \longleftarrow & \mathrm{M}(\mathcal{C}) = \mathcal{D}^{1\times n}/\mathcal{D}^{1\times(n+p)}\left(\begin{smallmatrix} A \\ C \end{smallmatrix}\right) & \xleftarrow{\text{can}} & M_1 & \xleftarrow{\phi} & M & \longleftarrow & 0 \\
 & & \updownarrow = & & \downarrow \text{proj} & & \downarrow \text{proj} & & \\
0 & \longleftarrow & \mathrm{M}(\mathcal{C}) = \mathcal{D}^{1\times n}/\mathcal{D}^{1\times(n+p)}\left(\begin{smallmatrix} A \\ C \end{smallmatrix}\right) & \xleftarrow{\text{can}} & M_1^0 & \xleftarrow{(\cdot C)_{\text{ind}}} & M_0 & \longleftarrow & 0
\end{array}
\tag{6.83}
$$

It has exact rows, and the maps are given by

$$
\begin{aligned}
\phi = \left(\cdot \left(\begin{smallmatrix} C & D \\ 0 & \mathrm{id}_m \end{smallmatrix}\right)\right)_{\text{ind}} :\ & M = \mathcal{D}^{1\times(p+m)}/\mathcal{D}^{1\times p}(P, -Q) \\
& \longrightarrow M_1 = \mathcal{D}^{1\times(n+m)}/\mathcal{D}^{1\times n}(A, -B) \\
\text{can} = (\cdot\, \mathrm{id}_n)_{\text{ind}} &: M_1^0 = \mathcal{D}^{1\times n}/\mathcal{D}^{1\times n}A \to \mathcal{D}^{1\times n}/\mathcal{D}^{1\times(n+p)}\left(\begin{smallmatrix} A \\ C \end{smallmatrix}\right) \\
\text{proj} = \left(\cdot \left(\begin{smallmatrix} \mathrm{id}_n \\ 0 \end{smallmatrix}\right)\right)_{\text{ind}} &: M_1 = \mathcal{D}^{1\times(n+m)}/\mathcal{D}^{1\times n}(A, -B) \to M_1^0 = \mathcal{D}^{1\times n}/\mathcal{D}^{1\times n}A.
\end{aligned}
$$

Corollary 6.3.12 *Theorem and Definition 6.3.3 implies the identity resp. isomorphism*

$$
\begin{aligned}
\mathcal{C} = \ker\left(\left(\begin{smallmatrix} A \\ C \end{smallmatrix}\right)\circ\right) &= \ker\left(X_{\text{obs}}\circ\right) \ \textit{resp.} \\
\mathcal{B}^0_{1,\text{obs}} = \ker(A_{\text{obs}}\circ) &\overset{C_{\text{obs}}\circ}{\cong} \mathcal{B}^0 = \ker\left(P\circ\right).
\end{aligned}
$$

Under the condition (6.31) *this implies*

$$
\begin{aligned}
\operatorname{char}(\mathcal{C}) &= \left\{\lambda \in \mathbb{C};\ \operatorname{rank}\left(\begin{smallmatrix} A(\lambda) \\ C(\lambda) \end{smallmatrix}\right) < n\right\} = \{\lambda \in \mathbb{C};\ \operatorname{rank}\left(X_{\text{obs}}(\lambda)\right) < n\} \\
&= \left\{\lambda \in \mathbb{C};\ \exists x = x_0 e_{\lambda,0} \in \mathcal{B}_1^0,\ 0 \neq x_0 \in \mathbb{C}^n,\ \textit{with } C \circ x = 0\right\} \\
\operatorname{char}\left(\mathcal{B}^0_{1,\text{obs}}\right) &= \{\lambda \in \mathbb{C};\ \operatorname{rank}\left(A_{\text{obs}}(\lambda)\right) < n\} \\
&= \operatorname{char}(\mathcal{B}^0) = \{\lambda \in \mathbb{C};\ \operatorname{rank}(P(\lambda)) < p\}.
\end{aligned}
\tag{6.84}
$$

Proof 1. The first equality follows from $\left(\begin{smallmatrix} A \\ C \end{smallmatrix}\right) = \left(\begin{smallmatrix} A_{\text{obs}} \\ C_{\text{obs}} \end{smallmatrix}\right) X_{\text{obs}}$ and the left invertibility of $\left(\begin{smallmatrix} A_{\text{obs}} \\ C_{\text{obs}} \end{smallmatrix}\right)$. Moreover

$$\operatorname{rank}\begin{pmatrix} A(\lambda) \\ C(\lambda) \end{pmatrix} < n \Longleftrightarrow \exists 0 \neq x_0 \in \mathbb{C}^n \text{ with } \begin{pmatrix} A(\lambda) \\ C(\lambda) \end{pmatrix} x_0 = 0,\ x := x_0 e_{\lambda,0}$$
$$\Longleftrightarrow A \circ x = A(\lambda)x_0 e_{\lambda,0} = 0,\ C \circ x = C(\lambda)x_0 e_{\lambda,0} = 0 \Longleftrightarrow x \in \mathcal{B}_1^0,\ C \circ x = 0.$$

Cf. Sect. 6.3.6 and Theorem 6.3.19 for analogous computations.
2. The second isomorphism is a consequence of $\mathcal{B}_{1,\mathrm{obs}} \cong \mathcal{B}$. □

Theorem and Definition 6.3.13 (Characteristic varieties, observability, and output decoupling zeros, cf. [4, p. 79]) *Assume the Rosenbrock equations* (6.45) *and* (6.46), *the standard signal modules from* (6.31) *and* $\mathcal{C} = \ker\left(\left(\begin{smallmatrix} A \\ C \end{smallmatrix}\right)\circ\right) = \ker(X_{\mathrm{obs}})$ *from* (6.82). *Then the identities*

$$\mathrm{l}(\mathcal{B}_1^0) = \mathrm{l}(\mathcal{C}) + \mathrm{l}(\mathcal{B}^0) \in \mathbb{N}^{(\Lambda_F)},\ \operatorname{char}(\mathcal{B}_1^0) = \operatorname{char}(\mathcal{C}) \cup \operatorname{char}(\mathcal{B}^0) \tag{6.85}$$

hold. The characteristic varieties are $\operatorname{char}(\mathcal{B}_1^0) = \{\lambda \in \mathbb{C};\ \operatorname{rank}(A(\lambda)) < n\}$ *and those from* (6.84). *The valued set or length vector* $\mathrm{l}(\mathcal{B}_1^0)$ *is given by*

$$\mathrm{l}_{p_\lambda}(\mathcal{B}_1^0) = \frac{\dim_F\left(\mathrm{t}_{p_\lambda}(\mathcal{B}_1^0)\right)}{\deg(p_\lambda)},\ \lambda \in \Lambda_F, \tag{6.86}$$

and analogous equations hold for $\mathrm{l}(\mathcal{C})$ *and* $\mathrm{l}(\mathcal{B}^0)$.

In particular, the Rosenbrock equations are observable if and only if $\left(\begin{smallmatrix} A \\ C \end{smallmatrix}\right)$ *has a left inverse or, equivalently,*

$$\mathcal{C} = 0 \Longleftrightarrow \mathcal{B}_1^0 \cong \mathcal{B}^0 \Longleftrightarrow \mathcal{B}_1 \cong \mathcal{B} \Longleftrightarrow \operatorname{char}(\mathcal{C}) = \emptyset$$
$$\Longleftrightarrow \mathrm{l}(\mathcal{C}) = 0 \Longleftrightarrow \mathrm{l}(\mathcal{B}_1^0) = \mathrm{l}(\mathcal{B}^0) \Longleftrightarrow \forall \lambda \in \mathbb{C}:\ \operatorname{rank}\begin{pmatrix} A(\lambda) \\ C(\lambda) \end{pmatrix} = n$$
$$\Longrightarrow \operatorname{char}(\mathcal{B}_1^0) = \operatorname{char}(\mathcal{B}^0),\ \operatorname{char}(\mathcal{B}_1) = \operatorname{char}(\mathcal{B}).$$

The elements of $\operatorname{char}(\mathcal{C})$ *are called the* unobservable poles *or* output decoupling zeros *of the Rosenbrock equations, those of* $\operatorname{char}(\mathcal{B}^0) = \operatorname{char}\left(\mathcal{B}_{1,\mathrm{obs}}^0\right)$ *the* observable poles.

This theorem is especially applicable to state space systems

$$s \circ x = Ax + Bu, \qquad y = Cx + D \circ u,$$

where $\mathcal{C} = \ker\left(\left(\begin{smallmatrix} s\,\mathrm{id}_n - A \\ C \end{smallmatrix}\right)\circ\right)$, *and* $\operatorname{char}(\mathcal{C}) = \left\{\lambda \in \mathbb{C};\ \operatorname{rank}\left(\begin{smallmatrix} \lambda\,\mathrm{id}_n - A \\ C \end{smallmatrix}\right) < n\right\}$.

Proof The assertions follow directly from the exact sequence (6.82) and from Theorems 5.4.8 and 5.4.10. □

Theorems 6.2.7 and 6.3.13 can be combined. The following diagram is commutative and has exact rows and columns:

$$
\begin{array}{ccccccccc}
 & & 0 & & 0 & & 0 & & \\
 & & \uparrow & & \uparrow & & \uparrow & & \\
0 & \longleftarrow & M_1/(\phi(M)+\mathrm{t}(M_1)) & \xleftarrow{\mathrm{can}} & M_1/\mathrm{t}(M_1) & \xleftarrow{\phi_{\mathrm{ind}}} & M/\mathrm{t}(M) & \longleftarrow & 0 \\
 & & \uparrow{\scriptstyle \mathrm{can}} & & \uparrow{\scriptstyle \mathrm{can}} & & \uparrow{\scriptstyle \mathrm{can}} & & \\
0 & \longleftarrow & M_1/\phi(M) & \xleftarrow{\mathrm{can}} & M_1 & \xleftarrow{\phi} & M & \longleftarrow & 0 \\
 & & \uparrow{\scriptstyle \mathrm{inj}} & & \uparrow{\scriptstyle \subseteq} & & \uparrow{\scriptstyle \subseteq} & & \\
0 & \longleftarrow & \mathrm{t}(M_1)/(\mathrm{t}(M_1)\cap\phi(M)) & \xleftarrow{\mathrm{can}} & \mathrm{t}(M_1) & \xleftarrow{\phi|_{\mathrm{t}(M)}} & \mathrm{t}(M) & \longleftarrow & 0 \\
 & & \uparrow & & \uparrow & & \uparrow & & \\
 & & 0 & & 0 & & 0 & &
\end{array}
\tag{6.87}
$$

Recall $M_1/\,\mathrm{t}(M_1) = M_{1,\mathrm{cont}}$ and $M/\,\mathrm{t}(M) = M_{\mathrm{cont}}$. That $\phi|_{\mathrm{t}(M)}$ and ϕ_{ind} are well-defined monomorphisms follows from Lemma 4.2.2. The exactness of the left column is a consequence of the isomorphism theorem. Due to the commutative exact diagrams (6.34) and (6.83) the commutative exact diagram (6.87) induces the commutative exact diagram of torsion modules

$$
\begin{array}{ccccccccc}
 & & 0 & & 0 & & 0 & & \\
 & & \uparrow & & \uparrow & & \uparrow & & \\
0 & \longleftarrow & M_1/(\phi(M)+\mathrm{t}(M_1)) & \xleftarrow{\mathrm{can}} & (M_1/\mathrm{t}(M_1))^0 & \xleftarrow{\phi_{\mathrm{ind}}} & (M/\mathrm{t}(M))^0 & \longleftarrow & 0 \\
 & & \uparrow{\scriptstyle \mathrm{can}} & & \uparrow{\scriptstyle \mathrm{can}} & & \uparrow{\scriptstyle \mathrm{can}} & & \\
0 & \longleftarrow & M_1/\phi(M) & \xleftarrow{\mathrm{can}} & M_1^0 & \xleftarrow{(\cdot C)_{\mathrm{ind}}} & M^0 & \longleftarrow & 0 \\
 & & \uparrow{\scriptstyle \mathrm{inj}} & & \uparrow{\scriptstyle \subseteq} & & \uparrow{\scriptstyle \subseteq} & & \\
0 & \longleftarrow & \mathrm{t}(M_1)/(\mathrm{t}(M_1)\cap\phi(M)) & \xleftarrow{\mathrm{can}} & \mathrm{t}(M_1) & \xleftarrow{\phi|_{\mathrm{t}(M)}} & \mathrm{t}(M) & \longleftarrow & 0 \\
 & & \uparrow & & \uparrow & & \uparrow & & \\
 & & 0 & & 0 & & 0 & &
\end{array}
\tag{6.88}
$$

The dual commutative diagram of behaviors with exact rows and columns is

$$
\begin{array}{ccccccccc}
 & & 0 & & 0 & & 0 & & \\
 & & \downarrow & & \downarrow & & \downarrow & & \\
0 & \longrightarrow & \mathcal{C}_{\mathrm{cont}} & \xrightarrow{\subseteq} & \mathcal{B}^0_{1,\mathrm{cont}} & \xrightarrow{C\circ} & \mathcal{B}^0_{\mathrm{cont}} & \longrightarrow & 0 \\
 & & \downarrow \cap & & \downarrow \cap & & \downarrow \cap & & \\
0 & \longrightarrow & \mathcal{C} & \xrightarrow{\subseteq} & \mathcal{B}^0_1 & \xrightarrow{C\circ} & \mathcal{B}^0 & \longrightarrow & 0 \\
 & & \downarrow \mathrm{can} & & \downarrow A_{\mathrm{cont}}\circ & & \downarrow P_{\mathrm{cont}}\circ & & \\
0 & \longrightarrow & \mathcal{C}/\mathcal{C}_{\mathrm{cont}} & \xrightarrow{\mathrm{D(can)}} & \mathcal{B}_{1,\mathrm{uncont}} & \xrightarrow{\mathrm{D}(\phi)} & \mathcal{B}_{\mathrm{uncont}} & \longrightarrow & 0 \\
 & & \downarrow & & \downarrow & & \downarrow & & \\
 & & 0 & & 0 & & 0 & &
\end{array}
\tag{6.89}
$$

where

$$
\begin{aligned}
&\mathcal{C}_{\mathrm{cont}} = \ker\left(\left(\begin{smallmatrix} A_{\mathrm{cont}} \\ C \end{smallmatrix}\right)\circ\right) = \mathcal{C}\bigcap \mathcal{B}^0_{1,\mathrm{cont}} \subseteq \mathcal{B}^0_{1,\mathrm{cont}} = \ker\left(A_{\mathrm{cont}}\circ\right),\\
&\mathcal{B}^0_{\mathrm{cont}} = \ker\left(P_{\mathrm{cont}}\circ\right),\ \mathcal{C} = \ker\left(\left(\begin{smallmatrix} A \\ C \end{smallmatrix}\right)\circ\right) \subseteq \mathcal{B}^0_1 = \ker\left(A\circ\right),\ \mathcal{B}^0 = \ker\left(P\circ\right)\\
&M(\mathcal{B}_{1,\mathrm{uncont}}) \cong \mathrm{t}(M_1),\ M(\mathcal{B}_{\mathrm{uncont}}) \cong \mathrm{t}(M),\\
&M(\mathcal{C}/\mathcal{C}_{\mathrm{cont}}) \cong \mathrm{t}(M_1)/(\mathrm{t}(M_1)\bigcap \phi(M)).
\end{aligned}
\tag{6.90}
$$

Recall from Theorem 5.3.17 that under the standard assumptions (6.31) an exact sequence $0 \to \mathcal{B}^{(1)} \to \mathcal{B}^{(2)} \to \mathcal{B}^{(3)} \to 0$ of autonomous behaviors implies

$$
\mathrm{l}\left(\mathcal{B}^{(2)}\right) = \mathrm{l}\left(\mathcal{B}^{(1)}\right) + \mathrm{l}\left(\mathcal{B}^{(3)}\right) \text{ and } \mathrm{char}(\mathcal{B}^{(2)}) = \mathrm{char}(\mathcal{B}^{(1)})\bigcup \mathrm{char}(\mathcal{B}^{(3)}). \tag{6.91}
$$

Theorem 6.3.14 *Assume the standard data from* (6.31).

1. *Each exact row or column of the diagram* (6.89) *gives rise to the equations from* (6.91), *in particular,*

$$
\begin{aligned}
\mathrm{l}(\mathcal{C}/\mathcal{C}_{\mathrm{cont}}) &= \mathrm{l}(\mathcal{C}) - \mathrm{l}(\mathcal{C}_{\mathrm{cont}}) = \mathrm{l}(\mathcal{B}_{1,\mathrm{uncont}}) - \mathrm{l}(\mathcal{B}_{\mathrm{uncont}})\\
\mathrm{l}(\mathcal{B}^0_{\mathrm{cont}}) &=: \mathrm{l}(H) \quad \left(\mathrm{char}(\mathcal{B}^0_{\mathrm{cont}}) = \mathrm{pole}(H)\right)\\
\mathrm{l}(\mathcal{B}^0_1) &= \mathrm{l}(H) + \left(\mathrm{l}(\mathcal{C}) + \mathrm{l}(\mathcal{B}_{1,\mathrm{uncont}}) - \mathrm{l}(\mathcal{C}/\mathcal{C}_{\mathrm{cont}})\right)\\
&= \mathrm{l}(H) + (\mathrm{l}(\mathcal{C}_{\mathrm{cont}}) + \mathrm{l}(\mathcal{B}_{\mathrm{uncont}}) + \mathrm{l}(\mathcal{C}/\mathcal{C}_{\mathrm{cont}}))\,.
\end{aligned}
\tag{6.92}
$$

Notice that the four summands of the last row of (6.92) *are the length vectors of the autonomous behaviors in the four corners of diagram* (6.89).

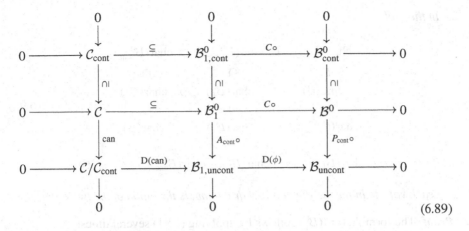

2. *In the table*

$$\begin{array}{ccccc} \mathrm{char}(\mathcal{C}_{\mathrm{cont}}) & \subset & \mathrm{char}(\mathcal{B}^0_{1,\mathrm{cont}}) & \supset & \mathrm{char}(\mathcal{B}^0_{\mathrm{cont}}) \\ \cap & & \cap & & \cap \\ \mathrm{char}(\mathcal{C}) & \subset & \mathrm{char}(\mathcal{B}^0_1) & \supset & \mathrm{char}(\mathcal{B}^0) \\ \cup & & \cup & & \cup \\ \mathrm{char}(\mathcal{C}/\mathcal{C}_{\mathrm{cont}}) & \subset & \mathrm{char}(\mathcal{B}_1) & \supset & \mathrm{char}(\mathcal{B}) \end{array} \tag{6.93}$$

$$\textit{with} \quad \mathrm{char}(\mathcal{B}_{1,\mathrm{uncont}}) = \mathrm{char}(\mathcal{B}_1), \quad \mathrm{char}(\mathcal{B}_{\mathrm{uncont}}) = \mathrm{char}(\mathcal{B})$$

each variety in the middle of a row or column is the union of the two outer ones.

Proof The formula for $\mathrm{l}(\mathcal{B}^0_1)$ follows by applying (6.91) several times:

$$\begin{aligned} \mathrm{l}(\mathcal{B}^0_1) &= \mathrm{l}(\mathcal{C}) + \mathrm{l}(\mathcal{B}^0) = \mathrm{l}(\mathcal{C}) + \mathrm{l}\left(\mathcal{B}^0_{\mathrm{cont}}\right) + \mathrm{l}\left(\mathcal{B}_{\mathrm{uncont}}\right) \\ &= \mathrm{l}(\mathcal{C}) + \mathrm{l}\left(\mathcal{B}^0_{\mathrm{cont}}\right) + \mathrm{l}\left(\mathcal{B}_{1,\mathrm{uncont}}\right) - \mathrm{l}(\mathcal{C}/\mathcal{C}_{\mathrm{cont}}) \\ &\underset{\mathrm{l}(H):=\mathrm{l}(\mathcal{B}^0_{\mathrm{cont}})}{=} \mathrm{l}(H) + \left(\mathrm{l}(\mathcal{C}) + \mathrm{l}\left(\mathcal{B}_{1,\mathrm{uncont}}\right) - \mathrm{l}(\mathcal{C}/\mathcal{C}_{\mathrm{cont}})\right) \\ &= \mathrm{l}(H) + \left(\mathrm{l}(\mathcal{C}_{\mathrm{cont}}) + \mathrm{l}\left(\mathcal{B}_{\mathrm{uncont}}\right) + \mathrm{l}(\mathcal{C}/\mathcal{C}_{\mathrm{cont}})\right). \end{aligned}$$

□

In diagram (5.93) the poles in the characteristic varieties of the

$$\left\{\begin{array}{l} \text{right column} \\ \text{left column} \\ \text{upper row} \\ \text{lower row} \end{array}\right. \quad \text{are called} \quad \left\{\begin{array}{l} \text{observable} \\ \text{unobservable} \\ \text{controllable} \\ \text{uncontrollable} \end{array}\right. \tag{6.94}$$

with respect to $\mathcal{B}_1$. The observable poles of $\mathcal{B}_1$ are the poles of $\mathcal{B}$.
Rosenbrock introduced an alternative terminology for the characteristic varieties, cf. [4, Sect. 2.4], [5, Sect. 7.2], viz.,

$$\begin{aligned} \mathrm{char}(\mathcal{C}) &=: \{\text{output decoupling zeros}\} \\ \mathrm{char}(\mathcal{C}_{\mathrm{cont}}) &=: \{\text{strict output decoupling zeros}\} \\ \mathrm{char}(\mathcal{B}_1) = \mathrm{char}(\mathcal{B}_{1,\mathrm{uncont}}) &=: \{\text{input decoupling zeros}\} \\ \mathrm{char}(\mathcal{B}) = \mathrm{char}(\mathcal{B}_{\mathrm{uncont}}) &=: \{\text{strict input decoupling zeros}\} \\ \mathrm{char}(\mathcal{C}/\mathcal{C}_{\mathrm{cont}}) &=: \{\text{input/output decoupling zeros}\}. \end{aligned} \tag{6.95}$$

The length vector

$$\mathrm{l}(\mathcal{B}^0_1) - \mathrm{l}(H) = \mathrm{l}(\mathcal{C}) + \mathrm{l}(\mathcal{B}_{1,\mathrm{uncont}}) - \mathrm{l}(\mathcal{C}/\mathcal{C}_{\mathrm{cont}}) = \mathrm{l}(\mathcal{C}_{\mathrm{cont}}) + \mathrm{l}(\mathcal{B}_{\mathrm{uncont}}) + \mathrm{l}(\mathcal{C}/\mathcal{C}_{\mathrm{cont}}) \tag{6.96}$$

is called the *length vector of hidden modes or decoupling zeros* of $\mathcal{B}_1$ [3, (25) on p. 581], [4, Remark 2.50, Theorem 2.62], [5, Definition 174, Theorem 175]. It determines those poles of $\mathcal{B}_1^0$ and their multiplicity that are not poles of the transfer matrix H. The behavior $\mathcal{B}_{\text{cont}}$ is the unique controllable realization of the transfer matrix $H = D + CA^{-1}B$. The transformation of $\mathcal{B}_1$ to $\mathcal{B}_{\text{cont}}$ can thus be interpreted in two ways:
(i) Replacement of $\mathcal{B}_1$ by controllable and observable Rosenbrock equations or by the unique controllable realization of H.
(ii) Elimination of the hidden modes of $\mathcal{B}_1$.

Remark 6.3.15 In the situation of Theorem 6.3.14 consider a direct decomposition $M_1 = X \oplus \mathrm{t}(M_1)$ where $X \cong M_1/\mathrm{t}(M_1)$ is torsionfree and thus free. In [5, Definition 174 on p. 192] the author defines the *module of hidden modes* of $\mathcal{B}_1$ as $M_2 := M_1/(X \bigcap \phi(M))$. The exact sequence

$$\begin{array}{ccccccc} 0 \to M_2 & \to & M_1/X \times M_1/\phi(M) & \to & M_1/\left(X + \phi(M)\right) \to 0 \\ \overline{x_2} & \mapsto & (\overline{x_2}, \overline{x_2}),\ (\overline{x}, \overline{y}) & \mapsto & \overline{x - y} \end{array} \tag{6.97}$$

consists of torsion modules since

$$M_1/X \cong \mathrm{t}(M_1) \text{ and } M_1/\phi(M) \cong \mathcal{D}^{1\times n}/\mathcal{D}^{1\times(n+p)}\begin{pmatrix} A \\ C \end{pmatrix}$$

are torsion modules. With (6.91) we get

$$\begin{aligned} \mathrm{l}(M_2) &= \mathrm{l}\left(M_1/X \times M_1/\phi(M)\right) - \mathrm{l}\left(M_1/\left(X + \phi(M)\right)\right) \\ &= \mathrm{l}(M_1/X) + \mathrm{l}(M_1/\phi(M)) - \mathrm{l}\left(M_1/\left(X + \phi(M)\right)\right) \\ &= \mathrm{l}(\mathcal{B}_{1,\text{uncont}}) + \mathrm{l}(\mathcal{C}) - \mathrm{l}\left(M_1/\left(X + \phi(M)\right)\right). \end{aligned} \tag{6.98}$$

Recall $\mathrm{l}\left(\mathcal{C}/\mathcal{C}_{\text{cont}}\right) = \mathrm{l}\left(\mathrm{t}(M_1)/(\mathrm{t}(M_1) \bigcap \phi(M))\right)$. On the other hand we have

$$\begin{aligned} & M_1 = X \oplus \mathrm{t}(M_1) \supseteq X + \phi(M) \Longrightarrow X + \phi(M) = X \oplus \left(\mathrm{t}(M_1) \bigcap (X + \phi(M))\right) \\ & \Longrightarrow M_1/(X + \phi(M)) = (X \oplus \mathrm{t}(M_1))/\left(X \oplus (\mathrm{t}(M_1) \bigcap (X + \phi(M)))\right) \\ & \cong \mathrm{t}(M_1)/\left(\mathrm{t}(M_1) \bigcap (X + \phi(M))\right) \\ & \Longrightarrow \mathrm{l}\left(M_1/(X + \phi(M))\right) = \mathrm{l}\left(\mathrm{t}(M_1)/\left(\mathrm{t}(M_1) \bigcap (X + \phi(M))\right)\right) \\ & \Longrightarrow \mathrm{l}(M_2) = \mathrm{l}\left(M_1/(X \bigcap \phi(M))\right) = \mathrm{l}(\mathcal{B}_{1,\text{uncont}}) + \mathrm{l}(\mathcal{C}) - \mathrm{l}(\mathcal{C}/\mathcal{C}_{\text{cont}}) \\ & + \mathrm{l}\left(\mathrm{t}(M_1)/\left(\mathrm{t}(M_1) \bigcap \phi(M)\right)\right) - \mathrm{l}\left(\mathrm{t}(M_1)/\left(\mathrm{t}(M_1) \bigcap (X + \phi(M))\right)\right). \end{aligned}$$

Recall $\mathrm{l}(\mathcal{B}_1^0) - \mathrm{l}(H) = \mathrm{l}(\mathcal{B}_{1,\text{uncont}}) + \mathrm{l}(\mathcal{C}) - \mathrm{l}(\mathcal{C}/\mathcal{C}_{\text{cont}})$ from (6.92). Moreover we use Theorem 5.3.17 several times and obtain

$$
\begin{aligned}
&\mathrm{l}\left(\mathrm{t}(M_1)/(\mathrm{t}(M_1)\bigcap\phi(M))\right)=\mathrm{l}(\mathrm{t}(M_1))-\mathrm{l}\left(\mathrm{t}(M_1)\bigcap\phi(M)\right)\\
&\mathrm{l}\left(\mathrm{t}(M_1)/(\mathrm{t}(M_1)\bigcap(X+\phi(M))\right)=\mathrm{l}(\mathrm{t}(M_1))-\mathrm{l}\left(\mathrm{t}(M_1)\bigcap(X+\phi(M))\right)\\
&\quad\Longrightarrow \mathrm{l}\left(\mathrm{t}(M_1)/(\mathrm{t}(M_1)\bigcap\phi(M))\right)-\mathrm{l}\left(\mathrm{t}(M_1)/(\mathrm{t}(M_1)\bigcap(X+\phi(M)))\right)\\
&\quad=\mathrm{l}\left(\mathrm{t}(M_1)\bigcap(X+\phi(M))/\,\mathrm{t}(M_1)\bigcap\phi(M)\right)\\
&\quad\Longrightarrow \mathrm{l}(M_2)=\mathrm{l}(\mathcal{B}_1^0)-\mathrm{l}(H)+\mathrm{l}\left((\mathrm{t}(M_1)\bigcap(X+\phi(M)))/(\mathrm{t}(M_1)\bigcap\phi(M))\right).
\end{aligned}
\tag{6.99}
$$

◊

Corollary 6.3.16 *From above we have the module of hidden modes*

$$M_2 := M_1/\,(\mathrm{t}(M_1)\cap\phi(M)) \text{ and } \mathrm{l}(\mathcal{B}_1^0)-\mathrm{l}(H)=\mathrm{l}(\mathcal{B}_{1,\mathrm{uncont}})+\mathrm{l}(\mathcal{C})-\mathrm{l}(\mathcal{C}/\mathcal{C}_{\mathrm{cont}}),$$

cf. [5, Definition 174 on p. 192]. Then

$$
\begin{aligned}
&\mathrm{l}(M_2)=\mathrm{l}(\mathcal{B}_1^0)-\mathrm{l}(H)+\mathrm{l}\left((\mathrm{t}(M_1)\bigcap(X+\phi(M)))/(\mathrm{t}(M_1)\bigcap\phi(M))\right),\ \textit{hence}\\
&\left(\mathrm{l}(M_2)=\mathrm{l}(\mathcal{B}_1^0)-\mathrm{l}(H)\iff\phi(\mathrm{t}(M))=\mathrm{t}(M_1)\bigcap\phi(M)=\mathrm{t}(M_1)\bigcap(X+\phi(M))\right).
\end{aligned}
\tag{6.100}
$$

Example 6.3.17 shows that $\mathrm{l}(M_2)=\mathrm{l}(\mathcal{B}_1^0)-\mathrm{l}(H)$ *does not hold in general, in contrast to [5, Theorem 175, Corollary 176, Lemma 560, Proposition 561].* ◊

Example 6.3.17 We construct an IO system $(M_1,\widehat{u},\widehat{y})$ according to the French terminology from Sect. 6.3.2 and identify $M=[\widehat{u},\widehat{y}]_{\mathcal{D}}=\phi(M(\mathcal{B}))$. Let

$$
\begin{aligned}
&M_1:=\mathcal{D}\times(\mathcal{D}/\mathcal{D}s(1-s))=\mathcal{D}^{1\times 2}/\mathcal{D}(0,s(1-s))\\
&\Longrightarrow \mathrm{rank}(M_1)=1,\ \mathrm{t}(M_1)=\mathcal{D}(0,\overline{1})\text{ where}\\
&\forall a\in\mathcal{D}=F[s]:\ \overline{a}:=a+\mathcal{D}s(1-s)\in\mathcal{D}/\mathcal{D}s(1-s).
\end{aligned}
$$

(i) The module $\mathcal{D}(s,\overline{1})$ is a *maximal free* submodule of M_1: Assume $\mathcal{D}(s,\overline{1})\subseteq Y$ with free Y. We have to show that $\mathcal{D}(s,\overline{1})=Y$. The dimension of Y is $1=\mathrm{rank}(M_1)$ and hence

$$Y=\mathcal{D}(a,\overline{b}),\ a,b\in\mathcal{D}\Longrightarrow\exists t\in\mathcal{D}\text{ with }(s,\overline{1})=t(a,\overline{b})\Longrightarrow s=ta,\ \overline{1}=\overline{tb}.$$

First case: t is a unit and thus $\mathcal{D}(s,\overline{1})=Y$ as asserted.
Second case: t is not a unit. Since s is irreducible this implies

$$s=ta,\ 0\neq a\in F\Longrightarrow t=a^{-1}s\Longrightarrow\overline{1}=\overline{tb}=\overline{a^{-1}sb}=\overline{s}\overline{a^{-1}b}\in\mathcal{D}/\mathcal{D}s(1-s).$$

This is a contradiction since $\overline{s}\overline{1-s}=\overline{0}\in\mathcal{D}/\mathcal{D}s(1-s)$ and thus $\overline{s}$ is a zero divisor and not invertible in $\mathcal{D}/\mathcal{D}s(1-s)$. Hence only the first case applies and $\mathcal{D}(s,\overline{1})\subset$

M_1 is maximal free.
(ii) $\mathcal{D}(s,\overline{1})\oplus \mathrm{t}(M_1)\subsetneq M_1$:

$$\mathcal{D}(s,\overline{1})\oplus \mathrm{t}(M_1)=\mathcal{D}(s,\overline{1})\oplus\mathcal{D}(0,\overline{1})$$
$$\underset{(s,\overline{1})=(s,\overline{0})+(0,\overline{1})}{=}\mathcal{D}(s,\overline{0})\oplus\mathcal{D}(0,\overline{1})\subsetneq\mathcal{D}(1,\overline{0})\oplus\mathcal{D}(0,\overline{1})=M_1.$$

The assertions (i) and (ii) contradict [5, Lemma 560 on p. 500] *where* $M_1 = Y\oplus \mathrm{t}(M_1)$ *is asserted for any maximal free submodule* Y of M_1. The reason for this contradiction is the following: Recall

$$\mathcal{D}(s,\overline{1})+\mathrm{t}(M_1)=\mathcal{D}(s,\overline{1})+\mathcal{D}(0,\overline{1})=\mathcal{D}(s,\overline{0})\oplus\mathcal{D}(0,\overline{1})$$
$$\Longrightarrow m:=(1,-\overline{1})\in M_1\setminus\left(\mathcal{D}(s,\overline{1})+\mathrm{t}(M_1)\right),$$
$$0\neq(s,\overline{1})-s(1,-\overline{1})=(0,\overline{1+s})=(1+s)(0,\overline{1})\in\mathrm{t}(M_1)$$
$$\Longrightarrow 0\neq(s,\overline{1})-s(1,-\overline{1})\in\mathrm{t}\left(\mathcal{D}(s,\overline{1})+\mathcal{D}m\right).$$

In contrast to the statement in the proof of [5, Lemma 560 on p. 500] the module $\mathcal{D}(s,\overline{1})+\mathcal{D}m$ is not torsionfree.
(iii) Define

$$\widehat{u}:=(s,\overline{1}),\ \widehat{y}:=(1,\overline{1}),\ M:=[\widehat{u},\widehat{y}]_{\mathcal{D}}=\mathcal{D}(s,\overline{1})+\mathcal{D}(1,\overline{1}),\ X:=\mathcal{D}(1,\overline{0}).$$

Then

$$X+\mathrm{t}(M_1)=\mathcal{D}(1,\overline{0})\oplus\mathcal{D}(0,\overline{1})=M_1 \text{ and}$$
$$X+M=\mathcal{D}(1,\overline{0})+\mathcal{D}(s,\overline{1})+\mathcal{D}(1,\overline{1})\underset{(0,\overline{1})=-(1,\overline{0})+(1,\overline{1})}{=}\mathcal{D}(1,\overline{0})+\mathcal{D}(0,\overline{1})=M_1.$$

(iv) $\mathrm{t}(M_1)\not\subseteq M$: We show $(0,\overline{1})\notin M$. Assume

$$(0,\overline{1})\in M=\mathcal{D}(s,\overline{1})+\mathcal{D}(1,\overline{1})\Longrightarrow\exists a,b\in\mathcal{D}\text{ with }(0,\overline{1})=a(s,\overline{1})+b(1,\overline{1})$$
$$\Longrightarrow 0=as+b,\ \overline{1}=\overline{a+b}=\overline{a-as}=\overline{(1-s)a}=\left(\overline{1-s}\right)\overline{a}.$$

This is a contradiction since $\overline{1-s}\in\mathcal{D}/\mathcal{D}s(1-s)$ is a zero divisor and not invertible in $\mathcal{D}/\mathcal{D}s(1-s)$.

With $\phi(M)\underset{\text{ident.}}{=}M$ we conclude

$$\mathrm{t}(M)=\mathrm{t}(M_1)\bigcap M\subsetneq\mathrm{t}(M_1)=\mathrm{t}(M_1)\bigcap(X+M)$$
$$\underset{\text{Corollary 6.3.16}}{\Longrightarrow}\mathrm{l}(M_2)>\mathrm{l}(\mathcal{B}_1^0)-\mathrm{l}(H).\qquad\diamond$$

The exact first column of (6.88) or dual of the first column of (6.89) is also given as

$$
\begin{aligned}
&0 \to \mathcal{D}^{1\times(n+p)}\begin{pmatrix} A_{\rm cont} \\ C \end{pmatrix} / \mathcal{D}^{1\times(n+p)}\begin{pmatrix} A \\ C \end{pmatrix} \\
&\quad \overset{\subseteq}{\to} \mathcal{D}^{1\times n} / \mathcal{D}^{1\times(n+p)}\begin{pmatrix} A \\ C \end{pmatrix} \overset{\rm can}{\longrightarrow} \mathcal{D}^{1\times n} / \mathcal{D}^{1\times(n+p)}\begin{pmatrix} A_{\rm cont} \\ C \end{pmatrix} \to 0.
\end{aligned}
\tag{6.101}
$$

The submodule $\mathcal{D}^{1\times(n+p)}\begin{pmatrix} A \\ C \end{pmatrix} = \mathcal{D}^{1\times n}A + \mathcal{D}^{1\times p}C \subseteq \mathcal{D}^{1\times n}$ is free of dimension n and therefore of the form

$$
\mathcal{D}^{1\times(n+p)}\begin{pmatrix} A \\ C \end{pmatrix} = \mathcal{D}^{1\times n}\widehat{A},\ \widehat{A} \in \mathcal{D}^{n\times n},\ \operatorname{rank}(\widehat{A}) = n, \tag{6.102}
$$

the rows of $\widehat{A}$ being a basis of the row module of $\begin{pmatrix} A \\ C \end{pmatrix}$. Likewise one obtains

$$
\begin{aligned}
&\mathcal{D}^{1\times(n+p)}\begin{pmatrix} A_{\rm cont} \\ C \end{pmatrix} = \mathcal{D}^{1\times n}A_{\rm cont} + \mathcal{D}^{1\times p}C = \mathcal{D}^{1\times n}\widehat{A_{\rm cont}} \\
&\widehat{A_{\rm cont}} \in \mathcal{D}^{n\times n},\ \operatorname{rank}(\widehat{A_{\rm cont}}) = n.
\end{aligned}
\tag{6.103}
$$

The inclusion $\mathcal{D}^{1\times n}A \subseteq \mathcal{D}^{1\times n}A_{\rm cont}$ implies

$$
\begin{aligned}
&\mathcal{D}^{1\times n}\widehat{A} \subseteq \mathcal{D}^{1\times n}\widehat{A_{\rm cont}} \Longrightarrow \exists Y \in \mathcal{D}^{n\times n} \text{ with } \widehat{A} = Y\widehat{A_{\rm cont}} \text{ and } \operatorname{rank}(Y) = n \\
&\Longrightarrow \left(\cdot\widehat{A_{\rm cont}}\right)_{\rm ind} : \mathcal{D}^{1\times n}/\mathcal{D}^{1\times n}Y \cong \mathcal{D}^{1\times n}\widehat{A_{\rm cont}}/\mathcal{D}^{1\times n}\widehat{A},\ \overline{\eta} \mapsto \overline{\eta\widehat{A_{\rm cont}}}.
\end{aligned}
\tag{6.104}
$$

Corollary 6.3.18 *By duality Eq.* (6.104) *induces the isomorphism*

$$
\begin{aligned}
&\widehat{A_{\rm cont}}\circ : \mathcal{C}/\mathcal{C}_{\rm cont} \overset{\cong}{\to} \left\{y \in \mathcal{F}^n;\ Y \circ y = 0\right\},\ x + \mathcal{C}_{\rm cont} \mapsto \widehat{A_{\rm cont}} \circ x \\
&\Longrightarrow \operatorname{char}(\mathcal{C}/\mathcal{C}_{\rm cont}) = \{\lambda \in \mathbb{C};\ \operatorname{rank}(Y(\lambda)) < n\} = \mathrm{V}_{\mathbb{C}}(\widehat{e_n}),
\end{aligned}
\tag{6.105}
$$

where $\widehat{e}_n$ is the highest elementary divisor of Y. ◇

6.3.6 The Interpretation of the Transfer Matrix as Gain Matrix in the Complex Standard Cases

In this section, we assume the standard data of complex systems theory, i.e., the ring $\mathcal{D} = \mathbb{C}[s]$ of operators and the injective $\mathbb{C}[s]$-cogenerators $\mathcal{F} = \mathrm{C}^\infty(\mathbb{R}, \mathbb{C})$ in the continuous-time case or $\mathcal{F} = \mathbb{C}^{\mathbb{N}}$ in the discrete-time case. The following considerations are mainly applied in the situation of Sect. 6.3.7 *where the complex signals are induced from real ones. For IO behaviors $\mathcal{B}$ with transfer matrix H and asymptotically stable autonomous part we investigate the asymptotic or* steady-state *output to an input*

$$
u = u(0)e^{j\omega t},\ j := \sqrt{-1},\ \omega \in \mathbb{R},\ u(0) \in \mathbb{C}^m. \tag{6.106}
$$

If $\lambda := j\omega$ resp. $\lambda := e^{j\omega}$ in the continuous-time resp. discrete-time cases is not a pole of the behavior and thus of the transfer matrix H, λ belongs to the domain of H and the asymptotic output, which is called the frequency response, is $y^{ss} = H(\lambda)u(0)e^{j\omega t}$. The entry $H(\lambda)_{\nu\mu}$ of $H(\lambda)$ *describes the frequency response from the input μ to the output ν at the frequency ω; it amplifies or dampens the input. Therefore the transfer matrix is interpreted as gain matrix.*

For $\lambda \in \mathbb{C}$ we recall from Sect. 5.2 that

$$\mathrm{ann}_{\mathcal{F}}(s-\lambda) = \mathbb{C}e_{\lambda,0} \subset \mathrm{t}_{s-\lambda}(\mathcal{F}),$$

where in the continuous-time case of $\mathcal{F} = \mathrm{C}^\infty(\mathbb{R}, \mathbb{C})$

$$e_{\lambda,0}(t) = e^{at}e^{j\omega t} = e^{at}\left(\cos(\omega t) + j\sin(\omega t)\right) \quad \text{if } \lambda = a + j\omega$$

with $a, \omega, t \in \mathbb{R}$. If $a = 0$, the function $e_{j\omega,0}(t) = \cos(\omega t) + j\sin(\omega t)$ is called a *sinusoid* for obvious reasons. It is periodic with period $T = \frac{2\pi}{\omega}$ and frequency $T^{-1} := \frac{\omega}{2\pi}$. In the literature ω is mostly called the *frequency* of the sinusoid and we will also do this. If λ is an arbitrary complex number, we will call λ the *complex frequency* of the function $e_{\lambda,0}$.

In the discrete-time case, we have

$$e_{\lambda,0}(t) = \begin{cases} \delta_{0,t} = 0^t & \text{if } \lambda = 0, \\ \lambda^t = |\lambda|^t e^{j\omega t} = |\lambda|^t(\cos(\omega t) + j\sin(\omega t)) & \text{if } \lambda = |\lambda|e^{j\omega}, \end{cases}$$

with $\omega \in \mathbb{R}$, and $t \in \mathbb{N}$. If $|\lambda| = 1$, the function $e_{e^{j\omega},0}(t) = e^{j\omega t} = \cos(\omega t) + j\sin(\omega t)$ of $t \in \mathbb{N}$ is still called a sinusoid. It is periodic of period 2π as function of ω, but periodic in t only if $\lambda = e^{\frac{2\pi j}{n}}$ is a complex n-th root of 1 for some $0 < n \in \mathbb{N}$. As in the continuous-time case, we call λ the *complex frequency* of $e_{\lambda,0}$.

In this section we consider system trajectories of the form $w(t) = w(0)e_{\lambda,0}(t) \in \mathbb{C}^l e_{\lambda,0}$ or, in other terms, $w \in \mathrm{ann}_{\mathcal{B}}(s-\lambda)$, in particular, inputs $u = u(0)e_{\lambda,0} \in \mathbb{C}^m e_{\lambda,0}$ of complex frequency λ. It turns out that these trajectories or inputs determine the behavior $\mathcal{B}$ to a large extent. Moreover, such inputs can be easily generated, for instance, by a suitable electrical network.

From (5.15) resp. (5.23) in the continuous-time resp. discrete-time cases we get $(s-\lambda) \circ e_{\lambda,0} = 0$, i.e., $s \circ e_{\lambda,0} = \lambda e_{\lambda,0}$, and therefore, $f \circ e_{\lambda,0} = f(\lambda)e_{\lambda,0}$ for arbitrary polynomials $f \in \mathbb{C}[s]$.

Theorem 6.3.19 *For $\lambda \in \mathbb{C}$ we consider $\mathbb{C}$ and thus also $\mathbb{C}^l$ as $\mathbb{C}[s]$-modules with the scalar multiplication $f \circ v := f(\lambda)v$. We assume an $\mathcal{F}$-behavior*

$$\mathcal{B} := \{w \in \mathcal{F}^l;\ R \circ w = 0\},\ R \in \mathbb{C}[s]^{k \times l},\ \mathcal{F} = \begin{cases} \mathrm{C}^\infty(\mathbb{R}, \mathbb{C}) & \text{(continuous time)} \\ \mathbb{C}^{\mathbb{N}} & \text{(discrete time)} \end{cases}.$$

1. The following mutually inverse $\mathbb{C}[s]$-isomorphisms hold:

$$\ker(R(\lambda)\cdot) := \{w_0 \in \mathbb{C}^l;\ R(\lambda)w_0 = 0\} \overset{\cong}{\rightleftarrows} \operatorname{ann}_{\mathcal{B}}(s-\lambda),\quad w_0 \longmapsto w_0 e_{\lambda,0},$$
$$w(0) \longleftarrow w.$$

2. *The space from 1 is nonzero, i.e., $\mathcal{B}$ contains a nonzero trajectory $w_0 e_{\lambda,0}$, if and only if* $\operatorname{rank}(R(\lambda)) < l$.
3. *If $\mathcal{B}$ is autonomous and thus* $\operatorname{rank}(R) = l$ *the space from 1 is nonzero if and only if* $\lambda \in \operatorname{char}(\mathcal{B})$.

Proof 1. This follows at once from $\operatorname{ann}_{\mathcal{B}}(s-\lambda) \subseteq \left(\operatorname{ann}_{\mathcal{F}}(s-\lambda)\right)^l = \mathbb{C}^l e_{\lambda,0}$ and $R \circ w(0)e_{\lambda,0} = \left(R(\lambda)w(0)\right)e_{\lambda,0}$.
2. $\dim_{\mathbb{C}}\left(\ker(R(\lambda)\cdot)\right) = l - \operatorname{rank}(R(\lambda))$.
3. $\lambda \in \operatorname{char}(\mathcal{B}) \iff \operatorname{rank}(R(\lambda)) < \operatorname{rank}(R) = l$. □

In the rest of this section, we use the IO system $\mathcal{B}$ with its characteristic varieties from Theorems 6.2.7 and 6.2.9.

Definition 6.3.20 (*Steady state, transient*) Let $\binom{y}{u}$, $\binom{\widetilde{y}}{u} \in \mathcal{B}$ be two trajectories of the IO behavior $\mathcal{B}$ with the same input $u \in \mathcal{F}^m$, hence $z := \widetilde{y} - y \in \mathcal{B}^0$, $\widetilde{y} = y + z$. Assume that $\mathcal{B}^0$ is asymptotically stable, i.e.,

$$\operatorname{char}(\mathcal{B}^0) \subset \begin{cases} \{\lambda' \in \mathbb{C};\ \Re(\lambda') < 0\} & \text{(continuous time)} \\ \{\lambda' \in \mathbb{C};\ |\lambda'| < 1\} & \text{(discrete time)} \end{cases}, \text{ or}$$

$$\mathcal{B}^0 \subseteq \left\{z \in \operatorname{t}(\mathcal{F})^m;\ \lim_{t\to\infty} z(t) = 0\right\} \Longrightarrow \lim_{t\to\infty}(\widetilde{y} - y)(t) = 0.$$

In this case y is called *a steady-state output* for the input u, and $\widetilde{y}$ and y are said to have *the same steady state*. In practical situations $\widetilde{y}(t)$ and $y(t)$ are identified for $t \geq t_0$ and suitable $t_0 \geq 0$, t_0 depending on the speed of convergence to zero of $z(t)$ for $t \to \infty$. The signal z is also called the *transient*. See Eqs. (2.26), (2.27), (2.43), and (2.59) from Chap. 2 for further examples of this language. ◇

Corollary 6.3.21 *Assume that $\lambda \in \mathbb{C}$ is not a pole of the IO behavior $\mathcal{B}$, i.e., $\lambda \notin$ $\operatorname{char}(\mathcal{B}^0)$. Then the isomorphism from Theorem 6.3.19 induces the $\mathbb{C}[s]$-isomorphisms*

$$\mathbb{C}^m \overset{\cong}{\longleftrightarrow} \ker\left((P(\lambda), -Q(\lambda))\cdot\right) \overset{\cong}{\longleftrightarrow} \operatorname{ann}_{\mathcal{B}}(s-\lambda),$$
$$u(0) \longleftrightarrow \begin{pmatrix} H(\lambda) \\ \operatorname{id}_m \end{pmatrix} u(0) = w(0) \longleftrightarrow w = w(0)e_{\lambda,0}.$$

Proof The second isomorphism is from Theorem 6.3.19. The assumption on λ means that $\operatorname{rank}(P(\lambda)) = p$ and, in particular, that $H(\lambda)$ is defined. The equations $(P(\lambda), -Q(\lambda)) = P(\lambda)(\operatorname{id}_p, -H(\lambda))$ and $w(0) = \binom{y(0)}{u(0)}$ imply the first isomorphism since

$$0 = (P(\lambda), -Q(\lambda))w(0) = P(\lambda)(\mathrm{id}_p, -H(\lambda))\begin{pmatrix} y(0) \\ u(0) \end{pmatrix}$$
$$\underset{\mathrm{rank}(P(\lambda))=p}{\iff} y(0) = H(\lambda)u(0) \iff w(0) = \begin{pmatrix} H(\lambda) \\ \mathrm{id}_m \end{pmatrix} u(0).$$ □

Theorem and Definition 6.3.22 (The interpretation of $H(\lambda)$ as gain matrix, cf. [1, Sect. 8.2]) *Let $\lambda \in \mathrm{dom}(H)$, i.e., $\lambda \notin \mathrm{pole}(H) = \mathrm{char}(\mathcal{B}^0_{\mathrm{cont}})$. For $u(0) \in \mathbb{C}^m$, we choose the input $u := u(0)e_{\lambda,0}$ and define*

$$y^{\mathrm{ss}} := H(\lambda)u = H(\lambda)u(0)e_{\lambda,0}.$$

Then the following holds:

1. *$\begin{pmatrix} y^{ss} \\ u \end{pmatrix} \in \mathcal{B}_{\mathrm{cont}} \subseteq \mathcal{B}$, i.e., y^{ss} is one possible output for the chosen input $u = u(0)e_{\lambda,0}$.*
2. *If $\mathcal{B}^0$ is asymptotically stable, then y^{ss} is a steady-state output for the input u of* complex frequency *λ, and $H(\lambda)$ is called the* gain matrix *from input to output.*

Proof 1. Since $P_{\mathrm{cont}}H = Q_{\mathrm{cont}}$ and $\lambda \notin \mathrm{pole}(H)$, we obtain

$$P_{\mathrm{cont}} \circ y^{ss} = P_{\mathrm{cont}}(\lambda)y^{ss} = P_{\mathrm{cont}}(\lambda)H(\lambda)u = Q_{\mathrm{cont}}(\lambda)u = Q_{\mathrm{cont}} \circ u,$$
$$\Longrightarrow \begin{pmatrix} y^{ss} \\ u \end{pmatrix} \in \mathcal{B}_{\mathrm{cont}}.$$

2. See Definition 6.3.20. □

Definition and Corollary 6.3.23 (*Frequency response*) Assume that $\mathcal{B}^0$ is asymptotically stable and that

$$\omega \in \mathbb{R} \text{ and } \lambda := \lambda(\omega) := \begin{cases} j\omega & \text{(continuous time)} \\ e^{j\omega} & \text{(discrete time)} \end{cases}.$$

Since $\mathcal{B}^0$ is asymptotically stable and since $\mathrm{pole}(H) \subseteq \mathrm{char}(\mathcal{B}^0)$ the number λ belongs to $\mathrm{dom}(H)$, i.e., $H(\lambda) \in \mathbb{C}^{p\times m}$ is defined. We choose $1 \le \mu \le m$, the input $u = \delta_\mu e_{\lambda,0}$, i.e., $u_{\mu'} = \delta_{\mu,\mu'}e_{\lambda,0}$ for $1 \le \mu' \le m$, and the output $y^{\mathrm{ss}} := H(\lambda)\delta_\mu e_{\lambda,0}$ and get $y^{\mathrm{ss}}_\nu = H_{\nu\mu}(\lambda)e_{\lambda,0}$ for $1 \le \nu \le p$. For any other trajectory $\begin{pmatrix} y \\ u \end{pmatrix} \in \mathcal{B}$ we get $\lim_{t\to\infty}|y_\nu(t) - y^{ss}_\nu(t)| = 0$. The equations

$$e_{\lambda,0}(t) = \begin{cases} e^{\lambda t} = e^{j\omega t} & \text{(continuous time)} \\ \lambda^t = e^{j\omega t} & \text{(discrete time)} \end{cases} \text{ and } |e_{\lambda,0}(t)| = 1 \text{ imply}$$

$$0 = \lim_{t\to\infty}\left|\frac{y_\nu(t) - y^{ss}_\nu(t)}{e_{\lambda,0}(t)}\right| = \lim_{t\to\infty}\left|\frac{y_\nu(t)}{e_{\lambda,0}(t)} - H_{\nu\mu}(\lambda)\right| \tag{6.107}$$
$$\Longrightarrow H_{\nu\mu}(\lambda) = \lim_{t\to\infty}\frac{y_\nu(t)}{e_{\lambda,0}(t)}.$$

Thus, one can determine $H_{\nu\mu}(\lambda(\omega))$ with arbitrary precision by measuring, for large t, the component $y_\nu(t)$ of any trajectory $\begin{pmatrix} y \\ u \end{pmatrix}$ of $\mathcal{B}$ for the input $u = \delta_\mu e_{\lambda,0}$. By the assumed asymptotic stability of $\mathcal{B}^0$ the influence of the initial condition $y(0)$ on $y_\nu(t)$ vanishes for large t. Notice that, in general, the last equation of (6.107) does not hold for $\begin{cases} \Re(\lambda) < 0 & \text{(continuous time)} \\ |\lambda| < 1 & \text{(discrete time)} \end{cases}$, so the choice $\lambda = j\omega$ resp. $\lambda = e^{j\omega}$ is important.

The function $\omega \mapsto H_{\nu\mu}(\lambda(\omega))$ is called the *complex frequency response* from input component u_μ to output component y_ν. It is customary to write $H_{\nu\mu}(\lambda(\omega))$ in polar form, i.e., as

$$H_{\nu\mu}(\lambda(\omega)) = |H_{\nu\mu}(\lambda(\omega))| e^{j\theta_{\nu\mu}(\omega)}, \quad -\pi \leq \theta_{\nu\mu}(\omega) \leq \pi,$$

and to call the real-valued function $\omega \mapsto |H_{\nu\mu}(\lambda(\omega))|$ the *magnitude function* and $\omega \mapsto \theta_{\nu\mu}(\omega)$ the *phase function.*

For the input $u = \delta_\mu e^{j\omega t} = (0, \ldots, 0, \overset{\mu}{e^{j\omega t}}, 0, \ldots, 0)^\top$ and steady-state output $y^{ss}(t) = \left(y_\nu^{ss}(t)\right)_{1 \leq \nu \leq p}$ this implies

$$y_\nu^{ss}(t) = |H_{\nu\mu}(\lambda(\omega))| e^{j(\omega t + \theta_{\nu\mu}(\omega))}, \quad \omega \in \mathbb{R}, \ \lambda(\omega) := \begin{cases} j\omega & \text{(continuous time)} \\ e^{j\omega} & \text{(discrete time)} \end{cases}. \tag{6.108}$$

Proof Equation (6.107) follows directly from Theorem and Definition 6.3.22. □

For a SISO system $\mathcal{B} \subseteq \mathcal{F}^2$ with transfer function $H \in \mathbb{C}(s)$ the frequency response is visualized as a *Bode plot* which consists of one graph where the magnitude function—in decibels, i.e., in a logarithmic scale—is plotted as a function of the logarithm of the frequency ω, and a second one where the phase function is plotted as a function of the logarithm of the frequency. Bode plots make only sense if the system is asymptotically stable.

To assess these properties in a graphical way, one uses the *Nyquist plot* that is the curve $\omega \mapsto H(\lambda(\omega))$ in the complex plane. In many applications the equations which define the behavior have real coefficients, so the transfer function $H \in \mathbb{R}(s)$ is a real rational function. The equation

$$\begin{aligned} &\overline{\lambda(\omega)} = \begin{cases} -j\omega = \lambda(-\omega) & \text{continuous case} \\ e^{-j\omega} = \lambda(-\omega) & \text{discrete case} \end{cases} \quad \text{implies} \\ &\overline{H(\lambda(\omega))} = H(\overline{\lambda(\omega)}) = H(\lambda(-\omega)). \end{aligned} \tag{6.109}$$

Therefore it is sufficient and customary to plot the frequency response only for nonnegative frequencies $\omega \in [0, \infty)$, since the other half of the curve can be obtained by mirroring with respect to the real axis.

For systems with multiple inputs and multiple outputs, one has to draw a Bode plot or a Nyquist plot for every combination of input and output components. The resulting multitude of diagrams makes it difficult to intuitively understand the system's properties and to design systems with certain properties, and these are the main motivations for working with these plots.

Next we generalize the preceding considerations to finite sums of sinusoids. Consider a *finite* subset $\Omega \subset \mathbb{R}$, interpreted as set of frequencies, and the $\mathbb{C}[s]$-modules

$$\mathcal{F}_\Omega := \bigoplus_{\omega\in\Omega} \mathbb{C}[t]e^{j\omega t} \supset \mathcal{F}_{\Omega,0} := \bigoplus_{\omega\in\Omega} \mathbb{C}e^{j\omega t} \ni u = \sum_{\omega\in\Omega} u_\omega e^{j\omega t} \text{ with}$$
$$f \circ u = \sum_{\omega\in\Omega} f(\lambda(\omega))u_\omega e^{j\omega t},\ f \in \mathbb{C}[s],\ \lambda(\omega) = \begin{cases} j\omega & \text{(continuous time)} \\ e^{j\omega} & \text{(discrete time)} \end{cases}. \tag{6.110}$$

We make $\mathbb{C}^\Omega$ a $\mathbb{C}[s]$-module isomorphic to $\mathcal{F}_{\Omega,0}$ via

$$f \circ (u_\omega)_{\omega\in\Omega} := (f(\lambda(\omega))u_\omega)_{\omega\in\Omega} \Longrightarrow \mathcal{F}_{\Omega,0} \underset{\mathbb{C}[s]}{\cong} \mathbb{C}^\Omega,\ \sum_{\omega\in\Omega} u_\omega e^{j\omega t} \leftrightarrow (u_\omega)_{\omega\in\Omega}. \tag{6.111}$$

With the identification $(\mathbb{C}^m)^\Omega = \left(\mathbb{C}^\Omega\right)^m$ this is a $\mathbb{C}[s]$-module too. We infer

$$\forall u = (u_\omega)_{\omega\in\Omega} \in \left(\mathbb{C}^m\right)^\Omega\ \forall Q \in \mathbb{C}[s]^{p\times m}:\ Q \circ u = (Q(\lambda(\omega))u_\omega)_{\omega\in\Omega}. \tag{6.112}$$

Corollary 6.3.24 *Let $\Omega \subset \mathbb{R}$ be finite. In the continuous-time resp. discrete-time standard case assume the IO behavior*

$$\mathcal{B} = \left\{\left(\begin{smallmatrix} y \\ u \end{smallmatrix}\right) \in \mathcal{F}^{p+m};\ P \circ y = Q \circ u\right\},\ (P, -Q) \in \mathbb{C}[s]^{p\times(p+m)},\ \operatorname{rank}(P) = p,$$

with asymptotically stable $\mathcal{B}^0 = \{y \in \mathcal{F}^p;\ P \circ y = 0\}$ and transfer matrix $H = P^{-1}Q$. There is the $\mathbb{C}[s]$-isomorphism

$$\begin{aligned} &\mathcal{B}\bigcap \mathcal{F}_{\Omega,0}^{p+m} \cong (\mathbb{C}^m)^\Omega,\ \left(\begin{smallmatrix} y^{ss} \\ u \end{smallmatrix}\right) \leftrightarrow (u_\omega)_{\omega\in\Omega},\ \text{where} \\ &u := \sum_{\omega\in\Omega} u_\omega e^{j\omega t},\ y^{ss} := \sum_{\omega\in\Omega} H(\lambda(\omega))u_\omega e^{j\omega t}. \end{aligned} \tag{6.113}$$

Moreover y^{ss} is a steady-state output to the input u, i.e.,

$$\forall u = \sum_{\omega\in\Omega} u_\omega e^{j\omega t} \in \mathcal{F}_{\Omega,0}^m \forall \left(\begin{smallmatrix} y \\ u \end{smallmatrix}\right) \in \mathcal{B}:\ \lim_{t\to\infty} (y(t) - y^{ss}(t)) = 0.$$

In this fashion implicit systems of linear differential or difference equations with constant coefficients can be studied by means of matrix equations over $\mathbb{C}$. ◇

In the continuous-time case of Corollary 6.3.24 choose a frequency $\omega > 0$ and $N > 0$ and define

$$T = 2\pi\omega^{-1},\ \Omega := \{\mu\omega;\ \mu = -N, \ldots, N\},\ u_\mu := u_{\mu\omega} \in \mathbb{C}^{m\times 1},\ y_\mu := y_{\mu\omega} \in \mathbb{C}^p. \tag{6.114}$$

In this case the sum of sinusoids $u = \sum_{\mu=-N}^{N} u_\mu e^{j\mu\omega t}$ is a *finite Fourier series* and periodic of period T. Let $\mathcal{P}(T) \subset C^0(\mathbb{R}, \mathbb{C})$ be the space of continuous functions of period T. On $\mathbb{C}^m$ we consider the standard inner product

$$\langle \xi, \eta \rangle := \xi^* \eta,\ \xi, \eta \in \mathbb{C}^m,\ \xi^* := \overline{\xi}^\top = (\overline{\xi_1}, \ldots, \overline{\xi_m}) \Longrightarrow \|\xi\|_2 := \langle \xi, \xi \rangle^{1/2}. \tag{6.115}$$

Thus $\mathbb{C}^m$ is a unitary space with the 2-norm $\|\xi\|_2$ and so is the space $(\mathbb{C}^m)^\Omega$. For periodic vector *functions* $u, v \in \mathcal{P}(T)^m$ we likewise define the inner product

$$\begin{aligned} &\langle u, v \rangle := T^{-1} \int_0^T \langle u^*(t), v(t) \rangle \mathrm{d}t = T^{-1} \int_0^T \left(\overline{u}^\top v\right)(t)\mathrm{d}t \\ &\Longrightarrow \|u\|_2 := \langle u, u \rangle^{1/2} = \left(T^{-1} \int_0^T \left(\overline{u}^\top u\right)(t)\mathrm{d}t\right)^{1/2} = \left(T^{-1} \int_0^T \|u(t)\|_2^2 \mathrm{d}t\right)^{1/2}. \end{aligned} \tag{6.116}$$

Thus $\mathcal{P}(T)^m$ is an *inner product or pre-Hilbert space* with the 2-norm $\|u\|_2$. In electrical engineering the norm $\|u\|_2$ is called the *rms (root mean square)* of the periodic vector function u, for obvious reasons.

Corollary 6.3.25 *For the finite Fourier series*

$$u = \sum_{\mu=-N}^{N} u_\mu e^{j\mu\omega t},\ v = \sum_{\mu=-N}^{N} v_\mu e^{j\mu\omega t} \in \mathcal{P}(T)^m,\ u_\mu, v_\mu \in \mathbb{C}^m,\ \mu = -N, \ldots, N,$$

$$\delta_k = (0, \ldots, \overset{k}{1}, \ldots, 0)^\top \in \mathbb{C}^m,\ v_\mu = (v_{\mu,1}, \ldots, v_{\mu,m})^\top = \sum_{k=1}^{m} v_{\mu,k}\delta_k \in \mathbb{C}^m,$$

the following equations hold:

$$\langle u, v \rangle = \sum_{\mu=0}^{N} \langle u_\mu, v_\mu \rangle,\ \|u\|_2^2 = \sum_{\mu=0}^{N} \|u_\mu\|_2^2,\ \langle \delta_k e^{j\mu\omega t}, v \rangle = v_{\mu,k}. \tag{6.117}$$

These imply the unitary isomorphism or isometry

$$\oplus_{\mu=-N}^{N} \mathbb{C}^m e^{j\mu\omega t} \cong \left(\mathbb{C}^m\right)^{2N+1},\ u = \sum_{\mu=-N}^{N} u_\mu e^{j\mu\omega t} \leftrightarrow \left(u_\mu\right)_{-N\le\mu\le N}, \tag{6.118}$$

and the orthogonal projection

$$\begin{aligned} &\Pi_N : \mathcal{P}(T)^m \to \oplus_{\mu=-N}^{N} \mathbb{C}^m e^{j\mu\omega t},\ u \mapsto \Pi_N(u) := \sum_{\mu=-N}^{N} \sum_{k=1}^{m} \langle \delta_k e^{j\mu\omega t}, u \rangle \delta_k e^{j\mu\omega t},\ \text{i.e.,} \\ &\forall v \in \oplus_{\mu=-N}^{N} \mathbb{C}^m e^{j\mu\omega t} : \langle u - \Pi_N(u), v \rangle = 0. \end{aligned} \tag{6.119}$$

Proof This follows directly from the obvious equation

$$\langle e^{j\mu\omega t}, e^{j\nu\omega t}\rangle = T^{-1}\int_0^T e^{j(\nu-\mu)\omega t}\,\mathrm{d}t = \begin{cases}1 & \text{if } \mu=\nu\\ 0 & \text{if } \mu\neq\nu\end{cases}. \qquad \square$$

Corollary 6.3.25 is needed in Sects. 6.3.7 and 8.3.

The preceding theory also permits a characterization of the poles of $\mathcal{B}$ which are not uncontrollable. Recall from Theorems 6.2.7 and 6.2.9 that

$$\operatorname{char}(\mathcal{B}^0) = \left\{\lambda\in\mathbb{C};\ \operatorname{rank}(P(\lambda)) < p\right\} \quad \text{and}$$

$$\operatorname{char}(\mathcal{B}) = \operatorname{char}(\mathcal{B}_{\text{uncont}}) = \{\lambda\in\mathbb{C};\ \operatorname{rank}\,(P(\lambda), -Q(\lambda)) < p\} \subseteq \operatorname{char}(\mathcal{B}^0).$$

Also recall that

$$\mathcal{F}_\lambda := \mathrm{t}_{s-\lambda}(\mathcal{F}) = \mathbb{C}[t]e_{\lambda,0}, \quad \text{for } \begin{cases}\lambda\in\mathbb{C} & \text{(continuous time)},\\ \lambda\in\mathbb{C}\setminus\{0\} & \text{(discrete time)}\end{cases}.$$

We omit the exceptional simple space $\mathcal{F}_0 = \mathbb{C}^{(\mathbb{N})}$ in the discrete situation because it has this different form. According to Corollary 5.3.20 every $\mathcal{F}_\lambda$ is an injective $\mathbb{C}[s]$-module. Therefore any input $u\in\mathcal{F}_\lambda^m$ admits an output $y=\widetilde{y}e_{\lambda,0}$ where $\widetilde{y}\in\mathbb{C}[t]^p$ is a polynomial vector in t. We denote its degree by $\deg_t(\widetilde{y})$.

Theorem 6.3.26 (Characterization of poles which are not uncontrollable)

1. *Assume that λ is not an uncontrollable pole of $\mathcal{B}$, i.e., that $\lambda\in\mathbb{C}\setminus\operatorname{char}(\mathcal{B})$, and $\lambda\neq 0$ in discrete time. Then the following properties are equivalent:*
   (a) *λ is a pole of $\mathcal{B}$, i.e., $\lambda\in\operatorname{char}(\mathcal{B}^0)$.*
   (b) *There is an input $u=u(0)e_{\lambda,0}\in\mathbb{C}^m e_{\lambda,0}$ such that every output $y=\widetilde{y}e_{\lambda,0}\in\mathcal{F}_\lambda^p=\mathbb{C}[t]^p e_{\lambda,0}$ with $P\circ y = Q\circ u$ satisfies $\deg_t(\widetilde{y})>0$.*
2. *Applied to the controllable behavior $\mathcal{B}_{\text{cont}}$ with $\operatorname{char}(\mathcal{B}_{\text{cont}})=\emptyset$, item 1 characterizes* all *poles of H, i.e., all elements of* $\operatorname{pole}(H)=\operatorname{char}(\mathcal{B}^0_{\text{cont}})$.

Proof The second item follows directly from the first one.

1., non-$(a)\Longrightarrow$ non-(b). Assume that $\lambda\notin\operatorname{char}(\mathcal{B}^0)\supseteq\operatorname{pole}(H)$. Then $H(\lambda)$ is defined and for every input $u=u(0)e_{\lambda,0}$, the vector $y:=H(\lambda)u=H(\lambda)u(0)e_{\lambda,0}\in\mathbb{C}^p e_{\lambda,0}$ satisfies

$$P\circ y = P(\lambda)H(\lambda)u(0)e_{\lambda,0} = Q(\lambda)e_{\lambda,0} = Q\circ\left(u(0)e_{\lambda,0}\right).$$

Since $\widetilde{y}=H(\lambda)u(0)\in\mathbb{C}^p$ is a constant vector, we have that $\deg_t\left(H(\lambda)u(0)\right)\le 0$.

1., $(a)\Longrightarrow(b)$. We have $\operatorname{rank}(P(\lambda))<p$ but, since λ is not an uncontrollable pole, $\operatorname{rank}\,(P(\lambda),-Q(\lambda))=p$. In other terms, we get $P(\lambda)\mathbb{C}^p\subsetneq\mathbb{C}^p$, but

$$(P(\lambda), -Q(\lambda))\,\mathbb{C}^{p+m} = P(\lambda)\mathbb{C}^p + Q(\lambda)\mathbb{C}^m = \mathbb{C}^p.$$

This implies that $Q(\lambda)\mathbb{C}^m \not\subseteq P(\lambda)\mathbb{C}^p$. Choose $u(0) \in \mathbb{C}^m$ with $Q(\lambda)u(0) \notin P(\lambda)\mathbb{C}^p$ and let $y = \widetilde{y}e_{\lambda,0} \in \mathcal{F}_\lambda^p = \mathbb{C}[t]^p e_{\lambda,0}$ be an arbitrary solution of $P \circ y = Q \circ \big(u(0)e_{\lambda,0}\big)$. This exists since $\mathcal{F}_\lambda$ is injective. Assume that $\deg_t(\widetilde{y}) \leq 0$, i.e., that $\widetilde{y} = y(0) \in \mathbb{C}^p$ is constant. Then

$$P(\lambda)y(0)e_{\lambda,0} = P \circ y = Q \circ \big(u(0)e_{\lambda,0}\big) = Q(\lambda)u(0)e_{\lambda,0},$$

and this implies that $Q(\lambda)u(0) = P(\lambda)y(0) \in P(\lambda)\mathbb{C}^p$ in contradiction to the choice of $Q(\lambda)u(0) \notin P(\lambda)\mathbb{C}^p$. □

Example 6.3.27 This example shows that the assumption $\lambda \notin \operatorname{char}(\mathcal{B})$ is indispensable in the preceding theorem. Indeed, consider the simple uncontrollable continuous system

$$y' = s \circ y = s \circ u = u' \quad \text{with } P = Q = s,\ H = 1,$$

and the characteristic varieties $\operatorname{char}(\mathcal{B}^0) = \operatorname{char}(\mathcal{B}) = \{0\}$. Hence, 0 is an uncontrollable pole of $\mathcal{B}$, but every input $u(0)e^{0t} = u(0)$ gives rise to the output $y = u$, i.e., $\widetilde{y} = u(0)$, with $\deg_t(\widetilde{y}) \leq 0$. ◊

6.3.7 *Complex Computations for Real Systems in Continuous Time*

The complexification of real problems is performed in many mathematical subjects and is also a standard procedure for electrical networks which are excited by sources with a fixed frequency ω or by finite Fourier series. We discussed the complexification functor already in (5.63)–(5.69).

We consider the continuous real and complex signal spaces

$$\mathcal{F}_\mathbb{R} := \mathrm{C}^\infty(\mathbb{R}, \mathbb{R}) \subset \mathcal{F}_\mathbb{C} := \mathbb{C} \otimes_\mathbb{R} \mathcal{F}_\mathbb{R} := \mathrm{C}^\infty(\mathbb{R}, \mathbb{C}) = \mathcal{F}_\mathbb{R} \oplus j\mathcal{F}_\mathbb{R},$$

the real IO system

$$\mathcal{B}_\mathbb{R} := \left\{ \left(\begin{smallmatrix} y \\ u \end{smallmatrix}\right) \in \mathcal{F}_\mathbb{R}^{p+m};\ P \circ y = Q \circ u \right\}, \quad \text{with matrices } P \in \mathbb{R}[s]^{p\times p} \text{ and } Q \in \mathbb{R}[s]^{p\times m} \text{ with } \operatorname{rank}(P) = p, \tag{6.120}$$

and the transfer matrix $H \in \mathbb{R}(s)^{p\times m}$ with $PH = Q$, as well as its complex extension

$$\mathcal{B}_\mathbb{C} := \mathbb{C} \otimes_\mathbb{C} \mathcal{B}_\mathbb{R} := \left\{ \left(\begin{smallmatrix} Y \\ U \end{smallmatrix}\right) \in \mathcal{F}_\mathbb{C}^{p+m};\ P \circ Y = Q \circ U \right\} \supset \mathcal{B}_\mathbb{R} = \mathcal{B}_\mathbb{C} \cap \mathcal{F}_\mathbb{R}^{p+m}.$$

In this section lower case resp. capital letters denote real-valued resp. complex-valued trajectories.

The following lemma is simple, but nevertheless decisive, and indeed the basis for the complex calculations in the theory of alternating currents with fixed frequency $\omega > 0$. Recall from Corollary 5.2.3 that

$$\operatorname{ann}_{\mathcal{F}_{\mathbb{R}}}(s^2+\omega^2) = \mathbb{R}\cos(\omega t) \oplus \mathbb{R}\sin(\omega t).$$

Lemma 6.3.28 *1. We consider the* $\mathbb{C}[s]$*-module* $\mathbb{C}e^{j\omega t}$, $\omega > 0$, *as* $\mathbb{R}[s]$*-module with* $s \circ z e^{j\omega t} = \frac{\mathrm{d}}{\mathrm{d}t} z e^{j\omega t} = j\omega z e^{j\omega t} = \omega z e^{j(\omega t + \frac{\pi}{2})}$, $z \in \mathbb{C}$. *The maps*

$$\begin{aligned}
\operatorname{ann}_{\mathcal{F}_{\mathbb{R}}}(s^2+\omega^2) = \mathbb{R}\cos(\omega t) \oplus \mathbb{R}\sin(\omega t) &\overset{\cong}{\rightleftarrows} \operatorname{ann}_{\mathcal{F}_{\mathbb{C}}}(s - j\omega) = \mathbb{C}e^{j\omega t}\\
a\cos(\omega t) + b\sin(\omega t) &\longmapsto (a - bj)e^{j\omega t},\\
\Re(f(t)) &\longleftarrow f(t), \text{ i.e.,}\\
\cos(\omega t) = \Re(e^{j\omega t}) &\longleftrightarrow e^{j\omega t},\\
\sin(\omega t) = \Re(-je^{j\omega t}) &\longleftrightarrow -je^{j\omega t} = e^{j(\omega t - \frac{\pi}{2})},
\end{aligned}$$

are mutually inverse $\mathbb{R}[s]$*-isomorphisms.*

2. *In polar coordinates, we write* $a + jb = |a + jb|e^{j\theta}$, $\theta \in \mathbb{R}$. *Then the conjugate is* $a - jb = |a + jb|e^{-j\theta}$. *This implies* $(a - jb)e^{j\omega t} = |a + jb|e^{j(\omega t - \theta)}$ *and*

$$a\cos(\omega t) + b\sin(\omega t) = |a + jb|\cos(\omega t - \theta).$$

The numbers $|a + jb|$ *resp.* θ *are again called the* amplitude *resp. the* phase *of the real sinusoidal signal.*

3. *For every* $l > 0$, *the isomorphism from item 1 induces the* $\mathbb{R}[s]$*-isomorphism*

$$\begin{aligned}
\mathbb{R}^l\cos(\omega t) \oplus \mathbb{R}^l\sin(\omega t) &\overset{\cong}{\rightleftarrows} \mathbb{C}^l e^{j\omega t},\\
a\cos(\omega t) + b\sin(\omega t) &\longmapsto (a - bj)e^{j\omega t} \text{ for } a, b \in \mathbb{R}^l, \text{ and}\\
\xi \circ (a\cos(\omega t) + b\sin(\omega t)) &\longmapsto \xi \circ (a - bj)e^{j\omega t}\\
&= \xi(j\omega)(a - bj)e^{j\omega t} \text{ for } \xi \in \mathbb{R}[s]^{1\times l}.
\end{aligned} \tag{6.121}$$

The advantage of the right-hand side is that the action $\xi\circ$ *of the* real differential operator $\xi \in \mathbb{R}[s]^{1\times l}$ *is simply given by multiplication with the* complex row $\xi(j\omega) \in \mathbb{C}^{1\times l}$. *The same argument applies to the multiplication with matrices of real polynomials.*

Proof The map in 1 is an $\mathbb{R}$-isomorphism since $\mathbb{C}e^{j\omega t} = \mathbb{R}e^{j\omega t} \oplus \mathbb{R}(-j)e^{j\omega t}$. It is even $\mathbb{R}[s]$-linear since

$$\begin{aligned}
s \circ \cos(\omega t) = -\omega\sin(\omega t) &\longmapsto j\omega e^{j\omega t} = s \circ e^{j\omega t} \quad \text{and}\\
s \circ \sin(\omega t) = \omega\cos(\omega t) &\longmapsto \omega e^{j\omega t} = s \circ (-je^{j\omega t}).
\end{aligned}$$

□

We combine Lemma 6.3.28 with Theorem 6.3.21 and obtain the following result.

Theorem 6.3.29 *Consider the real IO system $\mathcal{B}$ from* (6.120) *and its complexification $\mathcal{B}_{\mathbb{C}}$. Assume that $\mathcal{B}^0_{\mathbb{R}}$ is asymptotically stable. Since*

$$\operatorname{char}(\mathcal{B}^0_{\mathbb{R}}) = \left\{\lambda \in \mathbb{C};\ \operatorname{rank}\left(P(\lambda)\right) < p\right\} = \operatorname{char}(\mathcal{B}^0_{\mathbb{C}}),$$

this is equivalent to the asymptotic stability of $\mathcal{B}^0_{\mathbb{C}}$. Assume $\omega > 0$ and hence $j\omega \notin \operatorname{char}(\mathcal{B}^0_{\mathbb{C}})$. Similarly as in Theorem 6.3.19, we consider $\mathbb{C}^m$ as an $\mathbb{R}[s]$-module with the scalar multiplication $\xi \circ v = \xi(j\omega)v$. Then the following maps are inverse $\mathbb{R}[s]$-isomorphisms:

$$\begin{aligned}
&\mathcal{B}_{\mathbb{R}} \cap \left(\mathbb{R}^{p+m}\cos(\omega t) \oplus \mathbb{R}^{p+m}\sin(\omega t)\right) \overset{\cong}{\rightleftarrows} \mathcal{B}_{\mathbb{C}} \cap \mathbb{C}^{p+m}e^{j\omega t} \overset{\cong}{\rightleftarrows} \mathbb{C}^m,\\
&\qquad\qquad \binom{y}{u} \qquad\longleftrightarrow\qquad \binom{Y}{U} \qquad\longleftrightarrow U(0)\\
&\text{where } \binom{y}{u} = \binom{y_{\mathrm{r}}}{u_{\mathrm{r}}}\cos(\omega t) + \binom{y_{-\mathrm{j}}}{u_{-\mathrm{j}}}\sin(\omega t) = \begin{pmatrix}\Re(Y)\\ \Re(U)\end{pmatrix},\ U(0) = u_{\mathrm{r}} - ju_{-\mathrm{j}},\\
&\binom{Y}{U} = \left(\binom{y_{\mathrm{r}}}{u_{\mathrm{r}}} - j\binom{y_{-\mathrm{j}}}{u_{-\mathrm{j}}}\right)e^{j\omega t} = \begin{pmatrix}H(j\omega)U(0)\\ U(0)\end{pmatrix}e^{j\omega t}.
\end{aligned} \tag{6.122}$$

The asymptotic stability assumption guarantees that the computed steady states Y and y are indeed the dominant outputs for large t and the bounded inputs

$$u(t) = u_{\mathrm{r}}\cos(\omega t) + u_{-\mathrm{j}}\sin(\omega t) = \Re(U(0))\cos(\omega t) - \Im(U(0))\sin(\omega t).$$

Proof Theorem 6.3.21 and Lemma 6.3.28. □

Example 6.3.30 Let $U(0) := \delta_k$ for some $1 \le k \le m$ and $U := \delta_k e^{j\omega t}$. The corresponding real input is $\Re(U(0)e^{j\omega t}) = \delta_k\cos(\omega t)$. If $Y := H(j\omega)U$ is the corresponding steady-state output and $H_{l,k}(j\omega) = |H_{l,k}(j\omega)|e^{j\theta_{l,k}(\omega)}$ for $1 \le l \le p$, then

$$y_l := \Re(Y_l) = \Re\left(H_{l,k}(j\omega)e^{j\omega t}\right) = |H_{l,k}(j\omega)|\cos(\omega t + \theta_{l,k}(\omega)). \qquad \diamondsuit$$

For $\omega = 0$ we need a modification since

$$\begin{aligned}
&\mathbb{R} = \mathbb{R}\cos(\omega t) \oplus \mathbb{R}\sin(\omega t) = \mathbb{R}e^{j\omega t} \subsetneq \mathbb{C} = \mathbb{C}e^{j\omega t}\\
&\Longrightarrow \mathcal{B}_{\mathbb{R}} \bigcap \mathbb{R}^{p+m} = \mathcal{B}_{\mathbb{C}} \bigcap \mathbb{R}^{p+m} \cong \mathbb{R}^m,\ \begin{pmatrix}H(0)U(0)\\ U(0)\end{pmatrix} \leftrightarrow U(0).
\end{aligned} \tag{6.123}$$

Theorem 6.3.29 can be generalized to the situation of Corollary 6.3.25, i.e.,

$$\begin{aligned}
&\Omega := \{\mu\omega;\ 0 \le \mu \le N\},\ \omega > 0,\ N > 0,\ T = 2\pi\omega^{-1} \\
&\Longrightarrow \mathcal{F}_{\mathbb{R},\Omega,0} := \mathbb{R} \oplus \oplus_{\mu=1}^{N} (\mathbb{R}\cos(\mu\omega t) \oplus \mathbb{R}\sin(\mu\omega t)) \\
&\qquad \underset{\mathbb{R}[s]}{\cong} \mathcal{F}_{\Omega,0} := \mathbb{R} \oplus \oplus_{\mu=1}^{N} \mathbb{C}e^{j\mu\omega t} \\
&a_0 + \sum_{\mu=1}^{N} (a_\mu \cos(\mu\omega t) + b_\mu \sin(\mu\omega t)) \longleftrightarrow a_0 + \sum_{\mu=1}^{N} (a_\mu - jb_\mu)\, e^{j\omega t}.
\end{aligned} \tag{6.124}$$

Again the signals are *finite Fourier series*, a very important class of signals in electrical engineering.

Corollary 6.3.31 *For the asymptotically stable IO behaviors $\mathcal{B}_{\mathbb{R}}$ and $\mathcal{B}_{\mathbb{C}}$ from Theorem 6.3.29 and signal spaces $\mathcal{F}_{\mathbb{R},\Omega,0}$ and $\mathcal{F}_{\Omega,0}$ there are the $\mathbb{R}[s]$-isomorphisms*

$$\begin{aligned}
&\mathcal{B}_{\mathbb{R}} \cap \mathcal{F}_{\mathbb{R},\Omega,0}^{p+m} \cong \mathcal{B}_{\mathbb{C}} \cap \mathcal{F}_{\Omega,0}^{p+m} \cong \mathbb{R}^m \oplus (\mathbb{C}^m)^N \quad \textit{where} \\
&\begin{pmatrix} y \\ u \end{pmatrix} = \begin{pmatrix} \Re(Y) \\ \Re(U) \end{pmatrix} \longleftrightarrow \begin{pmatrix} Y \\ U \end{pmatrix} \longleftrightarrow (U_\mu)_{0 \le \mu \le N} \\
&u = a_0 + \sum_{\mu=1}^{N} (a_\mu \cos(\mu\omega t) + b_\mu \sin(\mu\omega t)),\ a_0, a_\mu, b_\mu \in \mathbb{R}^m, \\
&U = U_0 + \sum_{\mu=1}^{N} U_\mu e^{j\mu\omega t} \in \mathcal{F}_{\Omega,0}^{m},\ U_0 := a_0 \in \mathbb{R}^m,\ \forall \mu \ge 1:\ U_\mu := a_\mu - jb_\mu \in \mathbb{C}^m, \\
&Y := H(0)U_0 + \sum_{\mu=1}^{N} H(j\mu\omega)U_\omega e^{j\mu\omega t} \in \mathcal{F}_{\Omega,0}^{p}, \\
&u = \Re(U) = 2^{-1}(U + \overline{U}),\ y = \Re(Y) = 2^{-1}(Y + \overline{Y}).
\end{aligned} \tag{6.125}$$

◇

In Sect. 8.2.8 we generalize the preceding considerations to arbitrary periodic signals. In Sect. 8.3.4 we compute the *active, reactive, and apparent power* of periodic voltages and currents of arbitrary n-ports. Finite Fourier series remain important as the only ones that can be exactly realized.

6.4 BIBO Stability of Standard IO Systems

We assume the standard data of complex systems theory, i.e., the ring of operators $\mathcal{D} = \mathbb{C}[s]$ and the signal modules $\mathcal{F} = \begin{cases} C^\infty(\mathbb{R}, \mathbb{C}) & \text{(continuous time)} \\ \mathbb{C}^{\mathbb{N}} & \text{(discrete time)} \end{cases}$. We discuss BIBO stability (bounded input/bounded output stability) which, in contrast to asymptotic stability of autonomous behaviors, refers to external properties of a system. The characterization of continuous-time resp. discrete-time BIBO stable IO behaviors is *given in Theorem* 6.4.5 *resp. Theorem* 6.4.8.

We need the torsion modules

$$\mathrm{t}(\mathcal{F}) = \bigoplus_{\lambda\in\mathbb{C}} \mathcal{F}_\lambda, \text{ where}$$

$$\mathcal{F}_\lambda = \mathrm{t}_{s-\lambda}(\mathcal{F}) = \begin{cases} \mathbb{C}[t]e_{\lambda,0} = \mathbb{C}[t]e^{\lambda t} & \text{(continuous time)} \\ \begin{cases} \mathbb{C}[t]e_{\lambda,0} = \mathbb{C}[t]\lambda^t & \lambda \neq 0 \\ \mathbb{C}^{(\mathbb{N})} & \lambda = 0 \end{cases} & \text{(discrete time)} \end{cases}$$

from Corollary 5.3.19.

Definition 6.4.1 As in Theorem 5.4.17 a vector function

$$t \longmapsto u(t) = \begin{pmatrix} u_1(t) \\ \vdots \\ u_m(t) \end{pmatrix} \in \mathbb{C}^m, \quad \text{where } t \in \mathbb{R} \text{ or } t \in \mathbb{N},$$

is called *bounded* (for $t \to \infty$) if

$$\sup_{t\geq 0} \max \left\{ |u_i(t)|;\ i = 1, \ldots, m \right\} < \infty.$$

The IO behavior

$$\mathcal{B} = \left\{ \binom{y}{u} \in \mathcal{F}^{p+m};\ P \circ y = Q \circ u \right\},\ P \in \mathcal{D}^{p\times p},\ Q \in \mathcal{D}^{p\times m},\ \mathrm{rank}(P) = p, \tag{6.126}$$

is called *BIBO stable* or *bounded input/bounded output stable* if it generates bounded outputs from bounded inputs or, more precisely, if for every trajectory $\binom{y}{u} \in \mathcal{B}$, boundedness of the input u implies that of the output. ◇

In this section we will consider inputs in $\mathrm{t}(\mathcal{F})^m$ only. In Sects. 8.1.3 and 8.2.6, after the introduction of distributions and of the *transfer operator* of a transfer matrix, we will extend BIBO stability to arbitrary inputs provided that the transfer matrix is proper.

Let $\mathrm{t}(\mathcal{F})^b$ denote the subspace of $\mathrm{t}(\mathcal{F})$ of all bounded functions. We characterize this space in the continuous-time and discrete-time cases in Theorem 6.4.4. The following lemmas are decisive for this characterization. They are a generalization of the $\mathbb{C}$-linear independence of the geometric sequences λ^t (Lemma 6.4.2) and the exponential functions $e^{\lambda t}$ (Lemma 6.4.3).

Lemma 6.4.2 *Let $\Lambda \subset \mathbb{C}$ be a finite subset of the unit circle, i.e., of the complex numbers with absolute value* 1, *and consider a linear combination*

$$s(t) = \sum_{\lambda\in\Lambda} a_\lambda \lambda^t \quad \textit{with } a_\lambda \in \mathbb{C} \textit{ and } t \in \mathbb{N}.$$

If $\lim_{t\to\infty} s(t) = 0$ *then* $a_\lambda = 0$ *for all* $\lambda \in \Lambda$.

Proof We write $\Lambda = \{\lambda_1, \ldots, \lambda_n\}$. Since $(\lambda_1^t)_{t\in\mathbb{N}}$ is a sequence on the *compact* unit circle, there exists a strictly increasing map $\tau_1 \in \mathbb{N}^{\mathbb{N}}$ such that $(\lambda_1^{\tau_1(t)})_{t\in\mathbb{N}}$ is a convergent subsequence of $(\lambda_1^t)_{t\in\mathbb{N}}$. Using the same argument, we infer that there exists a strictly increasing map $\tau_2 \in \mathbb{N}^{\mathbb{N}}$ such that $(\lambda_2^{\tau_2(t)})_{t\in\mathbb{N}}$ is a convergent subsequence of $(\lambda_2^{\tau_1(t)})_{t\in\mathbb{N}}$. We continue this construction until we obtain a convergent subsequence $(\lambda_n^{\tau_n(t)})_{t\in\mathbb{N}}$ of $(\lambda_{n-1}^{\tau_{n-1}(t)})_{t\in\mathbb{N}}$. In particular, for all $\lambda = \lambda_k \in \Lambda$, $(\lambda^{\tau_n(t)})_{t\in\mathbb{N}}$ is a subsequence of the convergent sequence $(\lambda_k^{\tau_k(t)})_{t\in\mathbb{N}}$, and thus the limits $c_\lambda := \lim_{t\to\infty} \lambda^{\tau_n(t)}$, $\lambda \in \Lambda$, with $|c_\lambda| = 1$ exist. For $t' \in \mathbb{N}$ we infer

$$0 = \lim_{t\to\infty} s(t) = \lim_{t\to\infty} s(\tau_n(t) + t') = \lim_{t\to\infty} \sum_{\lambda\in\Lambda} a_\lambda \lambda^{\tau_n(t)} \lambda^{t'}$$

$$= \sum_{\lambda\in\Lambda} a_\lambda \left(\lim_{t\to\infty} \lambda^{\tau_n(t)}\right) \lambda^{t'} = \sum_{\lambda\in\Lambda} a_\lambda c_\lambda \lambda^{t'} \Longrightarrow 0 = \sum_{\lambda\in\Lambda} a_\lambda c_\lambda \left(\lambda^{t'}\right)_{t'\in\mathbb{N}}.$$

Since the sequences $e_{\lambda,0} = \left(\lambda^{t'}\right)_{t'\in\mathbb{N}} \in \mathbb{C}^{\mathbb{N}}$ are $\mathbb{C}$-linearly independent we conclude $a_\lambda c_\lambda = 0$, $\lambda \in \Lambda$, and hence $a_\lambda = 0$ due to $|c_\lambda| = 1$. □

The following lemma is the continuous-time variant of the preceding one.

Lemma 6.4.3 *Let Ω be a finite subset of $\mathbb{R}$ and consider the function*

$$s\colon \mathbb{R} \longrightarrow \mathbb{C}, \quad t \longmapsto \sum_{\omega\in\Omega} a_\omega e^{j\omega t}, \quad \textit{where } a_\omega \in \mathbb{C}.$$

If $\lim_{t\to\infty} s(t) = 0$ then $a_\omega = 0$ for all $\omega \in \Omega$.

Proof The proof is almost the same as that of the preceding lemma. One uses the linear independence of the functions $\mathbb{R} \longrightarrow \mathbb{C}$, $t \longmapsto e^{j\omega t}$, for varying $\omega \in \mathbb{R}$.

Notice that the assertion of this lemma is false if $t \in \mathbb{R}$ is replaced by $t \in \mathbb{N}$.

For example, let $\Omega := \{1, 1+2\pi\}$, $a_1 := 1$, and $a_{1+2\pi} := -1$. Since $s(t) = e^{j\cdot 1\cdot t} - e^{j(1+2\pi)t} = 0$ holds for $t \in \mathbb{N}$, the sequence $(s(t))_{t\in\mathbb{N}}$ is the zero sequence as well as a nontrivial linear combination of the $e^{j\omega t}$, $\omega \in \Omega$. □

In the following theorem we characterize the bounded polynomial-exponential functions in both standard cases.

Theorem 6.4.4 *1. In the continuous-time case with $\mathcal{F} = \mathrm{C}^\infty(\mathbb{R}, \mathbb{C})$ the space of bounded (for $t \to \infty$) polynomial-exponential functions is*

$$\mathrm{t}(\mathcal{F})^b = \bigoplus_{\Re(\lambda)<0} \mathbb{C}[t]e^{\lambda t} \oplus \bigoplus_{\Re(\lambda)=0} \mathbb{C}e^{\lambda t}.$$

2. In the discrete-time case with $\mathcal{F} = \mathbb{C}^{\mathbb{N}}$ the space of bounded polynomial-exponential sequences is

$$\mathfrak{t}(\mathcal{F})^b = \mathbb{C}^{(\mathbb{N})} \oplus \bigoplus_{0<|\lambda|<1} \mathbb{C}[t]\lambda^t \oplus \bigoplus_{|\lambda|=1} \mathbb{C}\lambda^t.$$

Proof The sums are direct by Corollary 5.3.19. The inclusions $\supseteq$ follow from (5.70) and (5.71), respectively.

We prove the converse inclusion for the continuous-time case by means of Lemma 6.4.3 and omit the proof in the discrete-time case that uses Lemma 6.4.2 instead. Assume a polynomial-exponential function u which is bounded for $t \to \infty$. The direct sum decomposition $\mathfrak{t}(\mathcal{F}) = \oplus_{\lambda\in\mathbb{C}}\mathbb{C}[t]e^{\lambda t}$ from Corollary 5.3.19 implies a representation

$$u(t) = \sum_{k\in J} c_k t^{m_k} e^{\lambda_k t},$$

where J is a finite index set, the $(m_k, \lambda_k) \in \mathbb{N} \times \mathbb{C}$ are pairwise distinct, and all coefficients c_k are nonzero. We have to show for all $k \in J$ that either

$$\Re(\lambda_k) < 0 \quad \text{or} \quad \big(\Re(\lambda_k) = 0 \text{ and } m_k = 0\big).$$

Write $\lambda_k = a_k + jb_k$ with $a_k, b_k \in \mathbb{R},\ j = \sqrt{-1}$, and define

$$\begin{aligned} a &:= \max\{a_k;\ k \in J\}, \quad & J_1 &:= \{k \in J;\ a_k = a\}, \\ m &:= \max\{m_k;\ k \in J_1\}, \quad & J_2 &:= \{k \in J_1;\ m_k = m\}. \end{aligned}$$

For every $k \in J$ the corresponding summand of u can be written as

$$c_k t^{m_k} e^{\lambda_k t} = t^m e^{at} c_k t^{m_k - m} e^{(a_k - a)t} e^{jb_k t}.$$

We distinguish two cases

$k \in J_2$. We get $c_k t^{m_k} e^{\lambda_k t} = t^m e^{at} c_k e^{jb_k t}$.
$k \notin J_2$. Here we have either $a_k - a < 0$ or $\big(a_k = a$ (i.e., $k \in J_1$) and $m_k - m < 0\big)$.
In both situations follows

$$\lim_{t\to\infty} t^{m_k - m} e^{(a_k - a)t} e^{jb_k t} = 0.$$

Summarizing, we obtain the representation

$$u(t) = t^m e^{at}\big(s_1(t) + s_2(t)\big) \ \text{ with } s_1(t) := \sum_{k\in J_2} c_k e^{jb_k t} \text{ and}$$

$$s_2(t) := \sum_{k\in J\setminus J_2} c_k t^{m_k - m} e^{(a_k - a)t} e^{jb_k t}, \ \lim_{t\to\infty} s_2(t) = 0.$$

For $k \in J_2$ the (m_k, λ_k) have the form $(m_k, \lambda_k) = (m, a + jb_k)$ and hence the $b_k,\ k \in J_2$, are pairwise distinct. Since the coefficients c_k are nonzero, Lemma 6.4.3 implies that $s_1(t)$ does not converge to 0 for $t \to \infty$. Hence,

$$\exists \epsilon > 0 \, \forall t_0 \geq 0 \, \exists t \geq t_0 \colon \; |s_1(t)| \geq \epsilon,$$

or $|s_1(t)| \geq \epsilon$ for arbitrarily large $t > 0$.

On the other hand, $s_2(t)$ converges to 0 and thus $|s_2(t)| \leq \frac{\epsilon}{2}$ for almost all $t \in \mathbb{N}$. We infer that

$$\left|s_1(t) + s_2(t)\right| \geq |s_1(t)| - |s_2(t)| \geq \epsilon - \tfrac{\epsilon}{2} = \tfrac{\epsilon}{2}$$

and thus $|u(t)| \geq t^m e^{at} \frac{\epsilon}{2}$ for arbitrarily large $t > 0$. By assumption u is bounded for large $t \in (0, \infty)$ and therefore

$$a < 0 \quad \text{or} \quad \big(a = 0 \text{ and } m = 0\big). \tag{6.127}$$

For the summands $c_k t^{m_k} e^{\lambda_k t}$ of $u(t)$, Eq. (6.127) implies that $\Re(\lambda_k) = a_k \leq 0$. Thus there are two possibilities.

$\Re(\lambda_k) = a_k < 0$. Then $c_k t^{m_k} e^{\lambda_k t} \in \bigoplus_{\Re(\lambda)<0} \mathbb{C}[t] e^{\lambda t}$.

$\Re(\lambda_k) = a_k = 0$. But then, $0 = a_k \leq a = \max\{a_{k'};\ k' \in J\} \leq 0$ because of (6.127), and thus $a = 0$ and $k \in J_1$. Also, since we are in the second case in (6.127), we have that $m_k \leq m = \max\{m_{k'};\ k' \in J_1\} = 0$ and thus $m_k = 0$. We conclude that $c_k t^{m_k} e^{\lambda_k t} \in \mathbb{C} e^{\lambda_k t}$ with $\Re(\lambda_k) = 0$.

By this, we have shown that u is of the asserted form. □

In the following theorem we characterize BIBO stability in the continuous-time case. The discrete-time counterpart of this theorem is Theorem 6.4.8.

Theorem 6.4.5 *Consider the function module $\mathcal{F} = \mathrm{C}^\infty(\mathbb{R}, \mathbb{C})$ and the IO behavior $\mathcal{B}$ from* (6.126). *The following properties of $\mathcal{B}$ are equivalent:*

1. *For every trajectory $\binom{y}{u} \in \mathcal{B}$ with bounded (for $t \to \infty$) polynomial-exponential input u, also the output y is bounded for $t \to \infty$.*
2. (a) *The autonomous part $\mathcal{B}^0 = \{y \in \mathcal{F}^p;\ P \circ y = 0\}$ is stable. By Theorem and Definition 5.4.17 this means that every $\lambda \in \mathrm{char}(\mathcal{B}^0)$ satisfies*

$$\Re(\lambda) < 0 \quad \textit{or} \quad \big(\Re(\lambda) = 0 \ \textit{ and } \ \mathrm{index}(\mathcal{B}^0, s - \lambda) = 1\big).$$

   (b) *Every pole of H satisfies $\Re(\lambda) < 0$.*

In Theorem 8.2.86 the same equivalence is shown for a proper transfer matrix H and arbitrary piecewise continuous inputs.

Proof 1. $\Longrightarrow$ 2.(*a*). Let $y \in \mathcal{B}^0$. Then $\binom{y}{0} \in \mathcal{B}$, which is BIBO stable by assumption. Since $u = 0$ is bounded, also y must be bounded. By definition (Theorem and Definition 5.4.17, item 1), this means that $\mathcal{B}^0$ is stable.

(*b*). Let $\mu \in \mathrm{pole}(H) = \mathrm{char}(\mathcal{B}^0_{\mathrm{cont}}) \subseteq \mathrm{char}(\mathcal{B}^0)$, cf. Theorem 6.3.18. By (a) $\mathcal{B}^0$ is stable and thus $\Re(\mu) \le 0$.

Assume $\Re(\mu) = 0$. Since $\mu \in \mathrm{char}(\mathcal{B}^0_{\mathrm{cont}})$ and $\mathcal{B}_{\mathrm{cont}}$ has no uncontrollable poles, Theorem 6.3.26 furnishes an input $u = u(0)e^{\mu t} \in \mathbb{C}^m e^{\mu t}$ such that every output $y = \widetilde{y}e^{\mu t} \in \mathcal{F}^p_\mu = \mathbb{C}[t]^p e^{\mu t}$ with $\binom{y}{u} \in \mathcal{B}_{\mathrm{cont}} \subseteq \mathcal{B}$ satisfies $\deg_t(\widetilde{y}) > 0$. Since $\Re(\mu) = 0$ these outputs $\widetilde{y}e^{\mu t}$ are unbounded, whereas u is bounded. This contradicts 1, hence $\Re(\mu) < 0$.

2. $\Longrightarrow$ 1. Let $u \in \mathrm{t}(\mathcal{F})^m$ be a bounded polynomial-exponential input and let $y \in \mathcal{F}^p$ be an arbitrary output with $\binom{y}{u} \in \mathcal{B}$.

By Theorem 6.4.4 the input u is a finite sum $u = \sum_{\lambda \in \Lambda} u_\lambda$, where $\Lambda \subseteq \mathbb{C}$ is finite and

$$\begin{aligned} &\left(\Re(\lambda) < 0 \text{ and } u_\lambda \in \mathcal{F}^m_\lambda = \mathbb{C}[t]^m e^{\lambda t}\right) \\ \text{or } &\left(\Re(\lambda) = 0 \text{ and } u_\lambda = u(0)e^{\lambda t} \in \mathbb{C}^m e^{\lambda t}\right). \end{aligned}$$

Without loss of generality we assume that $u_\lambda \neq 0$. We choose special outputs z_λ for each of the u_λ.

(i) $\Re(\lambda) < 0$: The module $\mathcal{F}_\lambda = \mathrm{t}_{s-\lambda}(\mathcal{F})$ is injective by Corollary 5.3.12 and

$$\mathcal{B} \cap \mathcal{F}^{p+m}_\lambda = \left\{\begin{pmatrix} y' \\ u' \end{pmatrix} \in \mathcal{F}^{p+m}_\lambda;\ P \circ y' = Q \circ u'\right\}$$

is an IO behavior. Therefore, by the fundamental principle (Theorem 3.2.16), the input $u_\lambda \in \mathcal{F}^m_\lambda$ admits an output $z_\lambda \in \mathcal{F}^p_\lambda = \mathbb{C}[t]^p e^{\lambda t}$. Since $\Re(\lambda) < 0$, this trajectory converges to zero and is thus bounded.

(ii) $\Re(\lambda) = 0$: Condition 2 (b) implies that λ is not a pole of H, i.e., $\lambda \in \mathrm{dom}(H)$. By Theorem and Definition 6.3.22 $z_\lambda := H(\lambda)u = H(\lambda)u(0)e^{\lambda t}$ is an output for u_λ and obviously bounded. Thus $z := \sum_{\lambda \in \Lambda} z_\lambda$ is bounded and $\binom{z}{u} \in \mathcal{B}$, hence

$$\begin{aligned} &\binom{y}{u}, \binom{z}{u} \in \mathcal{B} \Longrightarrow y - z \in \mathcal{B}^0 \underset{2.(a)}{\Longrightarrow} y - z \text{ bounded} \\ &\underset{\binom{z}{u}\text{ bounded}}{\Longrightarrow} \binom{y}{u} = \binom{y-z}{0} + \binom{z}{u} \text{ bounded.} \end{aligned}$$

□

Corollary and Definition 6.4.6 (Resonance) *In the situation of Theorem 6.4.5 assume that $\mathcal{B}^0$ is stable and that $\mu = j\omega$, $\omega \in \mathbb{R}$, $j = \sqrt{-1}$, is a pole of H, hence 2 (a) is satisfied, but 2 (b) is not. Then there are inputs $u = u(0)e^{j\omega t} \in \mathbb{C}^m e^{i\omega t}$ such that all corresponding outputs y with $\binom{y}{u} \in \mathcal{B}$ are unbounded. The phenomenon that bounded inputs have only unbounded outputs is called* resonance *and ω is a* resonance frequency. *If the behavior $\mathcal{B}$ describes a machine, this is a critical situation.*

Proof According to Theorem 6.3.26 there exists an input $u = u(0)e^{j\omega t} \in \mathbb{C}^m e^{j\omega t}$ such that all outputs $z = \widetilde{z}e^{j\omega t} \in \mathbb{C}[t]^p e^{i\omega t} = \mathcal{F}^p_{j\omega}$ of the controllable part $\mathcal{B}_{\mathrm{cont}}$, i.e., $\binom{z}{u} \in \mathcal{B}_{\mathrm{cont}}$, satisfy $\deg_t(\widetilde{z}) > 0$. Since $|e^{j\omega t}| = 1$ this implies that z is unbounded. Such outputs z exist because $\mathcal{F}_{j\omega}$ is injective. Since $\mathcal{B}_{\mathrm{cont}} \subseteq \mathcal{B}$, the output y of any trajectory $\binom{y}{u} \in \mathcal{B}$ satisfies $y - z \in \mathcal{B}^0$ and is bounded since $\mathcal{B}^0$ is table. Hence $y = z + (y - z)$ is unbounded like z. □

Corollary 6.4.7 *In Theorem 6.4.5 assume that $\mathcal{B}$ is controllable and hence* $\text{pole}(H) = \text{char}(\mathcal{B}^0)$. *The following properties of $\mathcal{B}$ are equivalent:*

1. *$\mathcal{B}$ is BIBO stable.*
2. *$\mathcal{B}^0$ is asymptotically stable.*

Proof The implication 2. $\Longrightarrow$ 1. is obvious.
1. $\Longrightarrow$ 2.: The controllability of $\mathcal{B}$ implies $\text{char}(\mathcal{B}^0) = \text{char}(\mathcal{B}^0_{\text{cont}}) = \text{pole}(H)$. Condition 2 (b) implies $\Re(\lambda) < 0$ for all $\lambda \in \text{char}(\mathcal{B}^0)$ and, by Theorem and Definition 5.4.16, the asymptotic stability. □

We conclude this section with the discrete-time analogue of Theorem 6.4.5 and Corollary 6.4.7.

Theorem 6.4.8 *Consider the function module $\mathcal{F} = \mathbb{C}^{\mathbb{N}}$ and an IO $\mathcal{F}$-behavior $\mathcal{B}$. The following properties of $\mathcal{B}$ are equivalent:*

1. *For every trajectory $\binom{y}{u} \in \mathcal{B}$ with bounded polynomial-exponential input u, also the output y is bounded.*
2. *(a) The autonomous part $\mathcal{B}^0$ is stable in the sense of Theorem 5.4.17, i.e., every pole $\lambda \in \text{char}(\mathcal{B}^0)$ satisfies*

$$|\lambda| < 1 \quad or \quad \big(|\lambda| = 1 \text{ and } \text{index}(\mathcal{B}^0, s - \lambda) = 1\big).$$

   (b) Every pole of H satisfies $|\lambda| < 1$.

Theorem 8.1.27 shows that this theorem holds for arbitrary bounded input sequences.

If the behavior $\mathcal{B}$ is controllable, then $\mathcal{B}$ is BIBO stable if and only if $\mathcal{B}^0$ is asymptotically stable.

Proof The proof is almost the same as in the continuous case. We omit it. □

References

1. J.W. Polderman, J.C. Willems, *Introduction to Mathematical Systems Theory: A Behavioral Approach*. Texts in Applied Mathematics, vol. 26 (Springer, New York, 1998)
2. W.A. Wolovich, *Linear Multivariable Systems*. Applied Mathematical Sciences, vol. 11 (Springer, New York, 1974)
3. T. Kailath, *Linear Systems* (Prentice-Hall, Englewood Cliffs, 1980)
4. A.I.G. Vardulakis, *Linear Multivariable Control*, Algebraic Analysis and Synthesis Methods (Wiley, Chichester, 1991)
5. H. Bourlès, *Linear Systems* (ISTE, London, 2010)
6. C.T. Chen, *Linear System Theory and Design* (Harcourt Brace College Publishers, Fort Worth, 1984)
7. P.J. Antsaklis, A.N. Michel, *Linear Systems*, 2nd edn. (Birkhäuser, Boston, 2006)

Chapter 7
Interconnections of Input/Output Behaviors

In this chapter we define interconnections or networks of IO behaviors in analogy to the electrical networks from Definition 3.1.20 and show when the interconnected behaviors are again IO behaviors in a natural way (Theorem and Definition 7.1.5). We discuss the basic types of interconnections, namely, the series or cascade connection in Sect. 7.3, the parallel connection in Sect. 7.4, and the feedback interconnection in Sect. 7.5. The latter is fundamental for the construction of stabilizing compensators and for controller design in Chap. 10. Willems introduced interconnections of general, not necessarily IO behaviors, cf. [1, Sect. 10.8.2] and Remark 7.5.5. As far as we know there are no applications of these more general interconnections in electrical and mechanical engineering, and hence we do not discuss this more general concept. In Sect. 7.2 we show that every state space system with proper transfer matrix can be constructed by interconnecting elementary building blocks. This underlines the importance of state space behaviors.

We assume a principal ideal operator domain $\mathcal{D}$ with its quotient field K and an injective cogenerator ${}_{\mathcal{D}}\mathcal{F}$ as in all standard cases.

7.1 Networks of Input/Output Behaviors

Definition 7.1.1 A *network* or an *interconnection* of IO behaviors is given by the following data:

1. A finite directed graph $\Gamma = (V, E)$ according to Definition 3.1.14. In the interesting cases the graph will be connected. If it is not connected, each of its components is treated separately.
2. For every vertex (or node) $v \in V$, a node dimension $n_v > 0$ and a signal vector $x_v \in \mathcal{F}^{n_v}$. The vector x_v is interpreted as the signal of dimension n_v at the node v of Γ.

U. Oberst et al., *Linear Time-Invariant Systems, Behaviors and Modules*, Differential-Algebraic Equations Forum,
https://doi.org/10.1007/978-3-030-43936-1_7

3. For every edge (or branch) $e\colon v \to w$ of Γ an IO behavior

$$\mathcal{B}_e := \left\{ \begin{pmatrix} y_e \\ u_e \end{pmatrix} \in \mathcal{F}^{n_w+n_v};\ P_e \circ y_e = Q_e \circ u_e \right\}$$

with transfer matrix $H_e \in K^{n_w \times n_v}$, $P_e H_e = Q_e$. Notice that the rank or input dimension of $\mathcal{B}_e$ is the node dimension n_v of the domain (or source) of e and the output dimension n_w is the node dimension of the codomain (or sink) of e.

The IO behavior $\mathcal{B}_e$ along the branch $e\colon v \to w$ is illustrated as $\overset{u_e}{\bullet} \xrightarrow{\mathcal{B}_e} \overset{y_e}{\bullet}$. It is convenient to denote the vertex and the signal in this vertex by the same name. Recall that the solution y_e of $P_e \circ y_e = Q_e \circ u_e$ is not unique in general, but it is unique if $P_e = \mathrm{id}_{n_v}$. In this case, one uses the representation $\overset{u_e}{\bullet} \xrightarrow{Q_e} \overset{y_e}{\bullet}$. The network with $P_e = Q_e = \mathrm{id}$ is represented as $\overset{u_e}{\bullet} \longrightarrow \overset{y_e}{\bullet}$. ◊

In the sequel we assume that such a network is given.

Definition 7.1.2 A node v of the network is called an *input node* if no edge has codomain v. The set of all input nodes is denoted by $I(\Gamma)$. ◊

We collect the signals of all vertices, the signals of all input nodes, and the output signals of all branches in three high-dimensional signal vectors, namely,

$$\begin{aligned} x &:= (x_v)_{v \in V} \in \mathcal{F}^{\sum_{v \in V} n_v}, \\ u &:= (x_v)_{v \in I(\Gamma)} \in \mathcal{F}^{\sum_{v \in I(\Gamma)} n_v}, \text{ and} \\ y &:= (y_e)_{e \in E} \in \mathcal{F}^{\sum_{e \in E} n_{\mathrm{cod}(e)}}. \end{aligned} \tag{7.1}$$

We use the shorthand notation $\mathcal{F}^\bullet$ for these high-dimensional signal spaces whenever their size can be deduced from the context. With these notations we define the behavior associated to the network.

Definition and Corollary 7.1.3 *We associate two behaviors with the given interconnection. The first one is*

$$\mathcal{B}_{x,y} := \left\{ \begin{pmatrix} x \\ y \end{pmatrix} \in \mathcal{F}^\bullet;\ \begin{array}{l} \forall (e\colon v \to w) \in E\colon\ \begin{pmatrix} y_e \\ x_v \end{pmatrix} \in \mathcal{B}_e \\ \forall w \in V \setminus I(\Gamma)\colon\ x_w = \sum_{e\colon v \to w} y_e \end{array} \right\}. \tag{7.2}$$

For every node v the node signal x_v is used as input for all IO behaviors $\mathcal{B}_e$ with $\mathrm{dom}(e) = v$. *The output signals of all IO behaviors $\mathcal{B}_e$ with the same codomain* $\mathrm{cod}(e) = w$ *are summed up and form the signal x_w at this node.*

Using the equations of the branch behaviors $\mathcal{B}_e$, we see that

$$\mathcal{B}_{x,y} = \left\{ \begin{pmatrix} x \\ y \end{pmatrix} \in \mathcal{F}^\bullet;\ \begin{array}{l} \forall (e\colon v \to w) \in E\colon\ (P_e, -Q_e) \circ \begin{pmatrix} y_e \\ x_v \end{pmatrix} = 0 \\ \forall w \in V \setminus I(\Gamma)\colon\ x_w - \sum_{e\colon v \to w} y_e = 0 \end{array} \right\}. \tag{7.3}$$

This representation shows that $\mathcal{B}_{x,y}$ is indeed an $\mathcal{F}$-behavior.

We use the projection $\mathrm{proj} := (\mathrm{id}_x, 0)\circ\colon \binom{x}{y} \longmapsto x$ *to define the second behavior* $\mathcal{B}_x$ *associated with a network of behaviors as*

$$\begin{aligned}\mathcal{B}_x &:= \mathrm{proj}(\mathcal{B}_{x,y}) = (\mathrm{id}_x, 0)\circ\mathcal{B}_{x,y}\\ &= \left\{x = (x_v)_{v\in V} \in \mathcal{F}^\bullet;\ \exists y = (y_e)_{e\in E}\colon\ \binom{x}{y} \in \mathcal{B}_{x,y}\right\}.\end{aligned}$$

Since $\mathcal{B}_x$ *is the image of a behavior under a behavior morphism, it is also a behavior. A matrix* R *with* $\mathcal{B}_x = \{x \in \mathcal{F}^\bullet;\ R\circ x = 0\}$ *can be constructed from* (7.3) *by means of the elimination theorem (Theorem 3.2.21). In* $\mathcal{B}_x$ *the outputs* y_e *of the different branch behaviors have been eliminated and appear only in form of the node signal* $x_w = \sum_{e:v\to w} y_e$. ◊

Since no branches end in the input nodes $w \in I(\Gamma)$, one expects the components x_w, $w \in I(\Gamma)$, as the free inputs of $\mathcal{B}_{x,y}$ or $\mathcal{B}_x$. This is often the case, but not always, as the following example shows.

Example 7.1.4 Consider the graph with two nodes 1 and 2, and two edges $e_1\colon 1 \to 2$ and $e_2\colon 2 \to 2$. Associated with the edges are the two IO systems

$$\mathcal{B}_{e_1} = \left\{\binom{y_1}{u_1} \in \mathcal{F}^2;\ y_1 = u_1\right\} \quad\text{and}\quad \mathcal{B}_{e_2} = \left\{\binom{y_2}{u_2} \in \mathcal{F}^2;\ a\circ y_2 = u_2\right\}$$

with $0 \neq a \in \mathcal{D}$. The interconnection diagram for this situation is

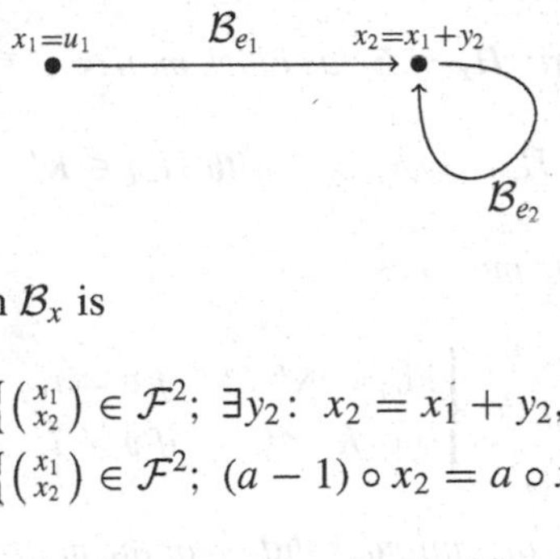

The associated system $\mathcal{B}_x$ is

$$\begin{aligned}\mathcal{B}_x &= \left\{\binom{x_1}{x_2} \in \mathcal{F}^2;\ \exists y_2\colon\ x_2 = x_1 + y_2,\ a\circ y_2 = x_2\right\}\\ &= \left\{\binom{x_1}{x_2} \in \mathcal{F}^2;\ (a-1)\circ x_2 = a\circ x_1\right\}.\end{aligned}$$

This is an IO system with input x_1 if and only if $a \neq 1$ and then $H = \frac{a}{a-1}$ is its transfer function. ◊

Some essential properties of $\mathcal{B}_{x,y}$ and $\mathcal{B}_x$ can be derived from their transfer spaces $\mathcal{B}_{x,y,K}$ and $\mathcal{B}_{x,K}$. Recall from item 4 of Theorem 6.2.2 that the transfer space of the component $\mathcal{B}_e$, $(e\colon v \to w) \in E$, of the interconnection is

$$\mathcal{B}_{e,K} = \left\{\binom{\widetilde{y}_e}{\widetilde{u}_e} \in K^{n_w+n_v};\ \widetilde{y}_e = H_e\widetilde{u}_e\right\}. \tag{7.4}$$

Theorem and Definition 7.1.5 (IO structure of interconnections)

1. *We write* $I := I(\Gamma)$. *The transfer spaces of* $\mathcal{B}_{x,y}$ *and* $\mathcal{B}_x$ *are*

$$\mathcal{B}_{x,y,K} = \left\{ \begin{pmatrix} (\widetilde{x}_v)_{v\in V} \\ (\widetilde{y}_e)_{e\in E} \end{pmatrix} \in K^\bullet;\ \begin{array}{r} \forall (e\colon v \to w) \in E\colon\ \widetilde{y}_e = H_e \widetilde{x}_v \\ \forall w \in V \setminus I\colon\ \widetilde{x}_w = \sum_{e\colon v\to w} \widetilde{y}_e \end{array} \right\} \quad \text{and}$$

$$\mathcal{B}_{x,K} = \Big\{ (\widetilde{x}_v)_v \in K^\bullet;\ \forall w \in V \setminus I : \ \widetilde{x}_w = \sum_{e\colon v\to w} H_e \widetilde{x}_v \Big\}.$$

2. *The following projection is an isomorphism:*

$$\mathrm{proj}\colon \mathcal{B}_{x,y,K} \longrightarrow \mathcal{B}_{x,K},\quad \begin{pmatrix} (\widetilde{x}_v)_v \\ (\widetilde{y}_e)_e \end{pmatrix} \longmapsto (\widetilde{x}_v)_v.$$

3. *The behaviors* $\mathcal{B}_{x,y}$ *and* $\mathcal{B}_x$ *are IO behaviors with input* $(x_i)_{i\in I}$ *if and only if for every* $\widetilde{u} = (\widetilde{u}_i)_{i\in I}$, $\widetilde{u}_i \in K^{n_i}$, *the inhomogeneous system of* K*-linear equations*

$$\widetilde{x}_w = \begin{cases} \widetilde{u}_w & \text{for } w \in I, \\ \sum_{e\colon v\to w} H_e \widetilde{x}_v & \text{for } w \in V \setminus I \end{cases} \tag{7.5}$$

has a unique solution $(\widetilde{x}_w)_{w\in V}$. *If these conditions are satisfied, we call* $\mathcal{B}_{x,y}$ *and* $\mathcal{B}_x$ *IO systems with the* canonical IO structure *and* canonical input $(x_i)_{i\in I}$. *We assume this in item 4.*

4. *We write the transfer matrix* H_x *of* $\mathcal{B}_x$ *as block matrix*

$$H_x = (H_{wi})_{w\in V\setminus I,\, i\in I} \quad \text{with } H_{wi} \in K^{n_w\times n_i}$$

and define, for $v, i \in I$, *the matrices*

$$H_{vi} := \begin{cases} \mathrm{id}_{n_i} \in K^{n_i\times n_i} & \text{if } v = i, \\ 0 \in K^{n_v\times n_i} & \text{if } v \neq i. \end{cases} \tag{7.6}$$

Then the matrices H_{wi} *are the unique solution of the matrix equations*

$$H_{wi} = \begin{cases} \mathrm{id}_{n_i} & \text{for } w = i \in I, \\ 0 & \text{for } w, i \in I,\ w \neq i, \\ \sum_{e\colon v\to w} H_e H_{vi} & \text{for } w \in V \setminus I,\ i \in I. \end{cases} \tag{7.7}$$

Proof 1. The expression for $\mathcal{B}_{x,y,K}$ follows directly from the definition (7.2) of $\mathcal{B}_{x,y}$, together with the characterization (7.4) of the transfer spaces $\mathcal{B}_{e,K}$.
Since the transfer space functor $\mathcal{B} \longmapsto \mathcal{B}_K$ is exact (Theorem 6.1.4), the projection $\mathrm{proj}\colon \mathcal{B}_{x,y} \to \mathcal{B}_x$ induces an epimorphism $\mathrm{proj}\colon \mathcal{B}_{x,y,K} \to \mathcal{B}_{x,K}$. This surjection induces the asserted expression for $\mathcal{B}_{x,K}$.

2. The surjective projection is also injective since the equations $\widetilde{y}_e = H_e\widetilde{x}_v,\ e \in E$, and $\widetilde{x}_v = 0,\ v \in V$, imply $\widetilde{y}_e = 0$ for all e.
3. The unique solvability of (7.5) signifies that the composite map

$$\mathcal{B}_{x,y,K} \xrightarrow{\cong} \mathcal{B}_{x,K} \longrightarrow K^\bullet,\quad \begin{pmatrix} (\widetilde{x}_w)_{w\in V} \\ (\widetilde{y}_e)_{e\in E} \end{pmatrix} \longmapsto (\widetilde{x}_w)_{w\in V} \longmapsto (\widetilde{x}_i)_{i\in I} = (\widetilde{u}_i)_{i\in I},$$

is bijective, and this, in turn, means that $\mathcal{B}_{x,y}$ and $\mathcal{B}_x$ are IO systems with input $(x_i)_{i\in I}$ according to Theorem 6.2.2, item 4.
4. That H_x is the transfer matrix of $\mathcal{B}_x$ means that $(\widetilde{x}_v)_{v\in V\setminus I} = H_x(\widetilde{x}_i)_{i\in I}$ or $\widetilde{x}_v = \sum_{i\in I} H_{vi}\widetilde{x}_i$ for $v \in V \setminus I$. Equation (7.6) implies the same for $v \in I$, hence

$$\forall w \in V \setminus I:\ \widetilde{x}_w \underset{(7.5)}{=} \sum_{e:\, v\to w} H_e\widetilde{x}_v = \sum_{e:\, v\to w} H_e \sum_{i\in I} H_{vi}\widetilde{x}_i = \sum_{i\in I}\sum_{e:\, v\to w} H_e H_{vi}\widetilde{x}_i$$
$$\Longrightarrow \sum_{i\in I} H_{wi}\widetilde{x}_i = \sum_{i\in I}\sum_{e:\, v\to w} H_e H_{vi}\widetilde{x}_i$$
$$\Longrightarrow \forall w \in V \setminus I \forall i \in I:\ H_{wi} = \sum_{e:\, v\to w} H_e H_{vi}.$$

□

7.2 Elementary Building Blocks

Definition 7.2.1 We define the following *elementary building blocks*. For the multipliers, integrators, and delay elements we assume $\mathcal{D} = F[s],\ F$ a field.

Disjoint sum. Consider a finite number of networks over graphs $\Gamma^j = (V^j, E^j)$, $j \in J$, with their IO behaviors $\mathcal{B}_e^j,\ e \in E^j$. The *disjoint sum* or *disjoint union* of these networks is given by the graph

$$\Gamma := (V, E) := \biguplus_{j\in J} \Gamma^j,\quad \text{i.e., } V := \biguplus_{j\in J} V^j,\ E := \biguplus_{j\in J} E^j,$$

and the branch behaviors $\mathcal{B}_e := \mathcal{B}_e^j$ for $e \in E^j \subseteq E$. There are no branches between nodes of different networks. This is the initial situation if one builds an interconnected system from its parts. The behavior $\mathcal{B}_x$ associated with this disjoint union is the direct product $\mathcal{B}_x = \prod_{j\in J} \mathcal{B}_x^j$.

Adder. The *scalar adder* is the network $\overset{\binom{u_1}{u_2}}{\bullet} \xrightarrow{\ +\ } \overset{u_1+u_2}{\bullet}$ with the IO behavior

$$+ := \left\{ \begin{pmatrix} y \\ u_1 \\ u_2 \end{pmatrix} \in \mathcal{F}^3;\ y = u_1 + u_2 \right\}.$$

The disjoint union of m scalar adders is the vector adder with the IO behavior

$$+ := \left\{ \begin{pmatrix} y \\ u_1 \\ u_2 \end{pmatrix} \in \mathcal{F}^{m+m+m};\ y = u_1 + u_2 \right\}.$$

In the engineering literature, the adder is also represented in the forms

$$\overset{u_1}{\bullet} \longrightarrow \overset{u_1+u_2}{\bullet} \longleftarrow \overset{u_2}{\bullet} \quad \text{or} \quad \overset{u_1}{\bullet} \longrightarrow \oplus \longleftarrow \overset{u_2}{\bullet}.$$

Multiplier. The scalar multiplier is the network $\overset{u}{\bullet} \xrightarrow{B} \overset{Bu}{\bullet}$, where $B \in F$. The associated IO behavior is $\mathcal{B} := \{\binom{y}{u};\ y = Bu\}$. More generally, let $B \in F^{n \times m}$ be a matrix. The graph with input nodes $u_1, \ldots, u_m$, output nodes $y_1, \ldots, y_n$, edges from u_j to y_i, and edge behaviors $\overset{u_j}{\bullet} \xrightarrow{B_{ij}} \overset{B_{ij}u_j}{\bullet}$ is

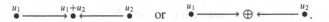

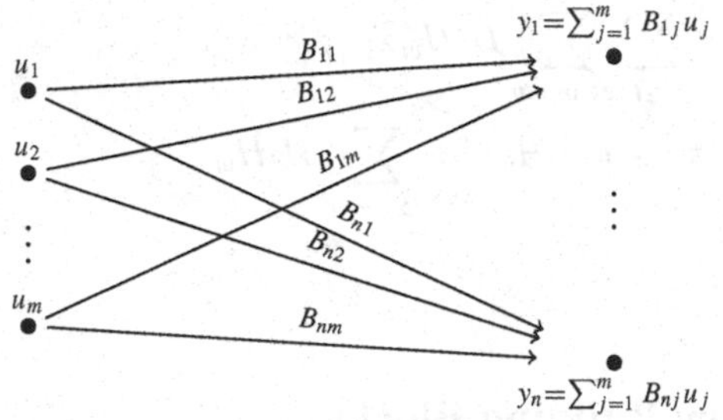

The behaviors along the edges are scalar multipliers as just introduced. With $u := (u_j)_{1 \le j \le m}$ and $y := (y_i)_{1 \le i \le n}$ the IO behavior of this network is

$$\mathcal{B} = \left\{ \binom{y}{u} \in \mathcal{F}^{n+m};\ y = Bu \right\}$$

with its canonical input u and furnishes the multiplication with the matrix B. We represent the total system as $\overset{u}{\bullet} \xrightarrow{B} \overset{y}{\bullet}$.

Integrators and delay elements. In the continuous-time case with $\mathcal{F} = C^\infty(\mathbb{R}, F)$, $F = \mathbb{R}, \mathbb{C}$, and in the discrete-time case with $\mathcal{F} = F^{\mathbb{N}}$ we consider the network with one edge $\overset{u}{\bullet} \xrightarrow{s^{-1}} \overset{y}{\bullet}$ and the IO behavior $\mathcal{B} := \left\{ \binom{y}{u} \in \mathcal{F}^2;\ s \circ y = u \right\}$. In this case it is customary to denote the behavior $\mathcal{B}$ by its transfer function s^{-1}. It has the input u and the output

$$\begin{cases} y(t) = y(t_0) + \int_{t_0}^t u(\tau)\mathrm{d}\tau,\ y(t_0) \in F & \text{(continuous time)} \\ y(t+1) = u(t),\ y(0) \in F, & \text{(discrete time).} \end{cases}$$

As with IO behaviors in general, the output y is completely determined only if both the input u and the initial condition $y(t_0) \in F$ are given, where the initial time t_0 is chosen according to the application. Then $\mathcal{B}$ is called an *integrator* in continuous time and a *delay element* in discrete time.

The disjoint union of n integrators or delay elements yields the n-dimensional counterpart $\mathcal{B} = \left\{\binom{y}{u} \in \mathcal{F}^{n+n};\ s \circ y = u\right\}$ with the diagram $\overset{u}{\bullet} \xrightarrow{s^{-1}} \overset{y}{\bullet}$. ◊

Example 7.2.2 (*State space equations*) Consider the network

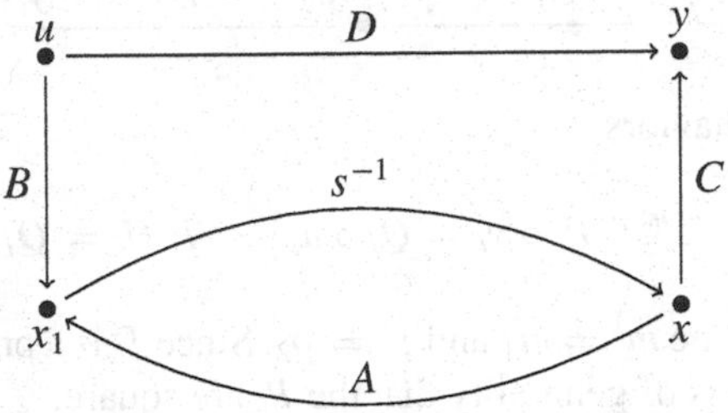

with constant matrices

$$A \in F^{n\times n}, \quad B \in F^{n\times m}, \quad C \in F^{p\times n}, \quad \text{and } D \in F^{p\times m}.$$

Its node signals are denoted by u, x, y, and x_1 as indicated. The equations of its behavior are $s \circ x = x_1$, $x_1 = Ax + Bu$, $y = Cx + Du$. By elimination of x_1 one obtains the state space equations

$$s \circ x = Ax + Bu, \quad y = Cx + Du.$$

It is one of the decisive properties of state space equations that they can be so easily realized by a simple network of elementary building blocks. ◊

Example 7.2.3 (*Rosenbrock equations*) Consider the Rosenbrock equations

$$A \circ x = B \circ u, \quad y = C \circ x + D \circ u$$

and associated IO behaviors and behavior morphism

$$\begin{pmatrix} C & D \\ 0 & \mathrm{id}_m \end{pmatrix} \circ\colon \mathcal{B}_{x,u} := \left\{\binom{x}{u} \in \mathcal{F}^{n+m};\ A \circ x = B \circ u\right\} \longrightarrow \mathcal{B}_{y,u} := \begin{pmatrix} C & D \\ 0 & \mathrm{id}_m \end{pmatrix} \circ \mathcal{B}_{x,u}.$$

The Rosenbrock equations are the equations of the network

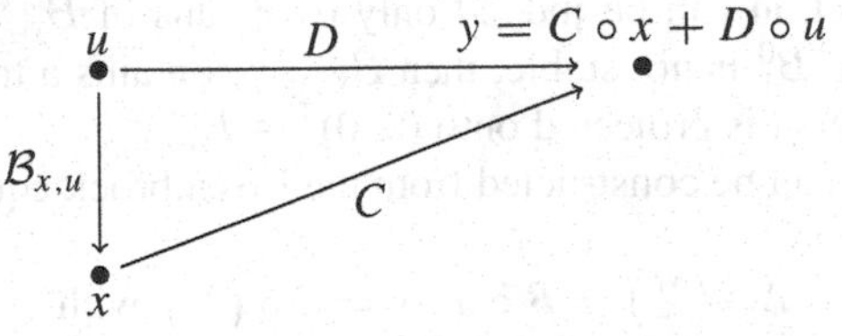

◊

7.3 Series or Cascade Connection

We consider the network with the graph

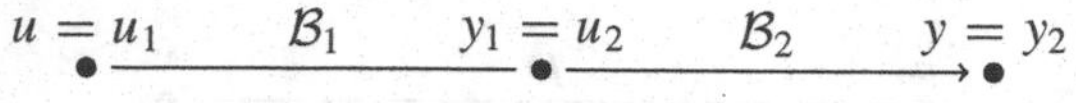

and the branch IO behaviors

$$\mathcal{B}_i := \left\{ \begin{pmatrix} y_i \\ u_i \end{pmatrix} \in \mathcal{F}^{p_i+m_i};\ P_i \circ y_i = Q_i \circ u_i \right\}, \quad P_i H_i = Q_i, \quad \text{for } i = 1, 2,$$

with $p_1 = m_2$. We define $m := m_1$ and $p := p_2$. Since $\mathcal{D}$ is a principal ideal domain, we assume without loss of generality that the P_i are square.

The behavior of the network is called the *series (inter)connection* or *cascade (inter)connection* of first $\mathcal{B}_1$ and then $\mathcal{B}_2$, and is given by the equations

$$\begin{aligned} \mathcal{B}_{y_1,y,u} &:= \left\{ \begin{pmatrix} y_1 \\ y \\ u \end{pmatrix} \in \mathcal{F}^{p_1+p+m};\ P_1 \circ y_1 = Q_1 \circ u,\ P_2 \circ y = Q_2 \circ y_1 \right\} \\ &= \left\{ \begin{pmatrix} y_1 \\ y \\ u \end{pmatrix} \in \mathcal{F}^{p_1+p+m};\ \begin{pmatrix} P_1 & 0 \\ -Q_2 & P_2 \end{pmatrix} \circ \begin{pmatrix} y_1 \\ y \end{pmatrix} = \begin{pmatrix} Q_1 \\ 0 \end{pmatrix} \circ u \right\}. \end{aligned} \tag{7.8}$$

From $\operatorname{rank}(P_1) = p_1$ and $\operatorname{rank}(P_2) = p$ we infer $\operatorname{rank}\begin{pmatrix} P_1 & 0 \\ -Q_2 & P_2 \end{pmatrix} = p_1 + p$. Hence $\mathcal{B}_{y_1,y,u}$ is an IO behavior with input u. Let H denote its transfer matrix. Then

$$\begin{pmatrix} P_1 & 0 \\ 0 & P_2 \end{pmatrix} \begin{pmatrix} \mathrm{id}_{p_1} & 0 \\ -H_2 & \mathrm{id}_{p_2} \end{pmatrix} H = \begin{pmatrix} P_1 & 0 \\ -Q_2 & P_2 \end{pmatrix} H = \begin{pmatrix} Q_1 \\ 0 \end{pmatrix} = \begin{pmatrix} P_1 & 0 \\ 0 & P_2 \end{pmatrix} \begin{pmatrix} H_1 \\ 0 \end{pmatrix}.$$

The matrix $\begin{pmatrix} P_1 & 0 \\ 0 & P_2 \end{pmatrix}$ has rank $p_1 + p_2$. Its cancellation yields the transfer matrix

$$H = \begin{pmatrix} \mathrm{id}_{p_1} & 0 \\ -H_2 & \mathrm{id}_{p_2} \end{pmatrix}^{-1} \begin{pmatrix} H_1 \\ 0 \end{pmatrix} = \begin{pmatrix} \mathrm{id}_{p_1} & 0 \\ H_2 & \mathrm{id}_{p_2} \end{pmatrix} \begin{pmatrix} H_1 \\ 0 \end{pmatrix} = \begin{pmatrix} H_1 \\ H_2 H_1 \end{pmatrix}. \tag{7.9}$$

By elimination of y_1 one obtains the IO behavior

$$\mathcal{B}_{y,u} := (0, \mathrm{id}_p, \mathrm{id}_m) \circ \mathcal{B}_{y_1,y_2,u} \tag{7.10}$$

with input u, output $y = y_2$, and the transfer matrix $H_2 H_1$. Often $\mathcal{B}_{y,u}$ is called the series or cascade connection of $\mathcal{B}_1$ and $\mathcal{B}_2$. Notice, however, that the total behavior of the series connection can be judged only by means of $\mathcal{B}_{y_1,y,u}$. If, for instance, $H_2 = Q_2 = 0$ and if $\mathcal{B}_1^0$ is not stable, then $\mathcal{B}_{y_1,y_2,u}$ contains a trajectory $(y_1, 0, 0)^\top$ with unbounded y_1 that is projected onto $(0, 0)^\top \in \mathcal{B}_{y,u}$.

The system $\mathcal{B}_{y,u}$ can be constructed from the Rosenbrock equations

$$\begin{aligned} & A \circ \begin{pmatrix} y_1 \\ y \end{pmatrix} = B \circ u,\ y = C \circ \begin{pmatrix} y_1 \\ y \end{pmatrix} \text{ with} \\ & A := \begin{pmatrix} P_1 & 0 \\ -Q_2 & P_2 \end{pmatrix},\ B := \begin{pmatrix} Q_1 \\ 0 \end{pmatrix},\ C := (0, \mathrm{id}_p). \end{aligned} \tag{7.11}$$

According to Theorem and Definition 6.3.1, $\mathcal{B}_{y,u}$ is an IO behavior

$$\mathcal{B}_{y,u} = \left\{\left(\begin{smallmatrix} y \\ u \end{smallmatrix}\right) \in \mathcal{F}^{p+m};\ P \circ y = Q \circ u\right\}$$

with input u, output y, and transfer matrix $C\left(\begin{smallmatrix} H_1 \\ H_2 H_1 \end{smallmatrix}\right) = H_2 H_1$.

The matrices P and Q are obtained by eliminating y_1 from the equations (7.11). Since $\mathcal{D}$ is a principal ideal domain, there is a universal left annihilator

$$\begin{aligned}&(-Y_1, -Y_2, P) \in \mathcal{D}^{p \times (p_1+p+p)} \text{ of } \left(\begin{smallmatrix} A \\ C \end{smallmatrix}\right) \in \mathcal{D}^{(p_1+p+p)\times(p_1+p)} \text{ since}\\ &p_1 + p + p - \operatorname{rank}\left(\begin{smallmatrix} A \\ C \end{smallmatrix}\right) = (p_1 + p + p) - (p_1 + p) = p.\end{aligned}$$

The equations of the annihilator are

$$\begin{aligned}&(Y_1 P_1 - Y_2 Q_2, Y_2 P_2) = (Y_1, Y_2)\left(\begin{smallmatrix} P_1 & 0 \\ -Q_2 & P_2 \end{smallmatrix}\right) = PC = (0, P)\\ &\Longleftrightarrow (Y_1, Y_2)\left(\begin{smallmatrix} P_1 \\ -Q_2 \end{smallmatrix}\right) = 0,\ P = Y_2 P_2.\end{aligned}$$

This means that (Y_1, Y_2) is a universal left annihilator of $\left(\begin{smallmatrix} P_1 \\ -Q_2 \end{smallmatrix}\right)$ and $P = Y_2 P_2$. According to Theorem 3.2.22 the IO behavior $\mathcal{B}_{y,u}$ is given by

$$\begin{aligned}&\mathcal{B}_{y,u} = \left\{\left(\begin{smallmatrix} y \\ u \end{smallmatrix}\right) \in \mathcal{F}^{p+m};\ P \circ y = Q \circ u\right\},\\ &P = Y_2 P_2,\ Q := (Y_1, Y_2)B = (Y_1, Y_2)\left(\begin{smallmatrix} Q_1 \\ 0 \end{smallmatrix}\right) = Y_1 Q_1.\end{aligned} \tag{7.12}$$

In the next corollary we use, for $\lambda \in \mathbb{C}$, the following equivalences:

$$\begin{aligned}&\operatorname{rank}(P_1(\lambda)) = p_1,\ \operatorname{rank}(P_2(\lambda)) = p \Longleftrightarrow P_1(\lambda) \in \mathrm{Gl}_{p_1}(F),\ P_2(\lambda) \in \mathrm{Gl}_p(F)\\ &\Longleftrightarrow \begin{pmatrix} P_1(\lambda) & 0 \\ -Q_2(\lambda) & P_2(\lambda) \end{pmatrix} \in \mathrm{Gl}_{p_1+p}(F) \Longleftrightarrow \operatorname{rank}\begin{pmatrix} P_1(\lambda) & 0 \\ -Q_2(\lambda) & P_2(\lambda) \end{pmatrix} = p_1 + p.\end{aligned}$$

These imply

$$\operatorname{rank}(P_1(\lambda)) < p_1 \text{ or } \operatorname{rank}(P_2(\lambda)) < p \Longleftrightarrow \operatorname{rank}\begin{pmatrix} P_1(\lambda) & 0 \\ -Q_2(\lambda) & P_2(\lambda) \end{pmatrix} < p_1 + p. \tag{7.13}$$

Corollary 7.3.1 *Consider the series connection IO behaviors from* (7.8)–(7.12).

1. *The series connection $\mathcal{B}_{y,u}$ is controllable if and only if $(Y_2 P_2, -Y_1 Q_1) \in \mathcal{D}^{p \times (p+m)}$ has a right inverse.*
2. *Under assumption (6.31), i.e., $F = \mathbb{C}, \mathbb{R}$, $\mathcal{D} = F[s]$ and $\mathcal{F}_F = \mathrm{C}^\infty(\mathbb{R}, F)$ with $\mathcal{F}_{\mathbb{C}} = \mathbb{C} \otimes_{\mathbb{R}} \mathcal{F}_{\mathbb{R}}$, the characteristic varieties satisfy*

$$\operatorname{char}(\mathcal{B}^0_{y,u}) \subseteq \operatorname{char}(\mathcal{B}^0_{y_1,y,u}) = \operatorname{char}(\mathcal{B}^0_1) \bigcup \operatorname{char}(\mathcal{B}^0_2).$$

If $\mathcal{B}^0_i$, $i = 1, 2$, are asymptotically stable, then so are $\mathcal{B}^0_{y_1,y,u}$ and $\mathcal{B}^0_{y,u}$.

3. *Assume (6.31), $\omega \in \mathbb{R}$ and $j\omega \notin \mathrm{char}(\mathcal{B}_1^0) \cup \mathrm{char}(\mathcal{B}_2^0)$. According to Lemma 6.3.28 there is the isomorphism*

$$\begin{array}{ccc} \mathrm{ann}_{\mathcal{F}_{\mathbb{R}}}(s^2+\omega^2) = \mathbb{R}\cos(\omega t) \oplus \mathbb{R}\sin(\omega t) & \underset{\mathbb{R}[s]}{\cong} & \mathrm{ann}_{\mathcal{F}_{\mathbb{C}}}(s-j\omega) = \mathbb{C}e^{j\omega t} \\ a\cos(\omega t) + b\sin(\omega t) = \Re\left((a-jb)e^{j\omega t}\right) & \longleftrightarrow & (a-jb)e^{j\omega t}. \end{array}$$

(a) *In the complex standard case with $F = \mathbb{C}$ there are the isomorphisms*

$$\begin{aligned} \mathcal{B}_{y_1,y,u} \bigcap \mathbb{C}^{p_1+p+m} e^{j\omega t} &= \begin{pmatrix} H_1(j\omega) \\ H_2(j\omega)H_1(j\omega) \\ \mathrm{id}_m \end{pmatrix} \mathbb{C}^m e^{j\omega t} \cong \mathbb{C}^m \\ \begin{pmatrix} H_1(j\omega) \\ H_2(j\omega)H_1(j\omega) \\ \mathrm{id}_m \end{pmatrix} u(0)e^{j\omega t} & \qquad \leftrightarrow u(0), \text{ and} \\ \mathcal{B}_{y,u} &= \begin{pmatrix} H_2(j\omega)H_1(j\omega) \\ \mathrm{id}_m \end{pmatrix} \mathbb{C}^m e^{j\omega t} \cong \mathbb{C}^m \\ \begin{pmatrix} H_2(j\omega)H_1(j\omega) \\ \mathrm{id}_m \end{pmatrix} u(0)e^{j\omega t} & \qquad \leftrightarrow u(0). \end{aligned} \tag{7.14}$$

(b) *In the real standard case with $F = \mathbb{R}$, $R \in \mathbb{R}[s]^{\bullet \times q}$ and behaviors*

$$\mathcal{B}^{\mathbb{R}} := \left\{w \in \mathcal{F}_{\mathbb{R}}^q;\ R \circ w = 0\right\} \subset \mathcal{B}^{\mathbb{C}} := \mathbb{C} \otimes_{\mathbb{R}} \mathcal{B}^{\mathbb{R}} = \left\{w \in \mathcal{F}_{\mathbb{C}}^q;\ R \circ w = 0\right\}$$

the isomorphisms in (7.14) *can be completed by*

$$\begin{array}{ccc} \mathcal{B}^{\mathbb{R}}_{y_1,y,u} \bigcap \left(\mathbb{R}^{p_1+p+m}\cos(\omega t) \oplus \mathbb{R}^{p_1+p+m}\sin(\omega t)\right) & \underset{\mathbb{R}[s]}{\cong} & \mathcal{B}^{\mathbb{C}}_{y_1,y,u} \bigcap \mathbb{C}^{p_1+p+m} e^{j\omega t} \\ a\cos(\omega t) + b\sin(\omega t) = \Re\left((a-jb)e^{j\omega t}\right) & \longleftrightarrow & (a-jb)e^{j\omega t}. \end{array} \tag{7.15}$$

4. *In an analogue for discrete time $j\omega$ is replaced by $e^{j\omega}$.*

The results 3 and 4 are useful only if $\mathcal{B}_1$ and $\mathcal{B}_2$ are asymptotically stable.

Proof 1. This follows from (7.12) and Theorem 4.2.12, item 2.
2. The characteristic variety of $\mathcal{B}^0_{y_1,y,u}$ is given by

$$\begin{aligned} \mathrm{char}\left(\mathcal{B}^0_{y_1,y,u}\right) &= \left\{\lambda \in \mathbb{C};\ \mathrm{rank}\begin{pmatrix} P_1(\lambda) & 0 \\ -Q_2(\lambda) & P_2(\lambda) \end{pmatrix} < \mathrm{rank}\begin{pmatrix} P_1 & 0 \\ -Q_2 & P_2 \end{pmatrix} = p_1 + p\right\} \\ &\underset{(7.13)}{=} \{\lambda \in \mathbb{C};\ \mathrm{rank}(P_1(\lambda)) < p_1\} \bigcup \{\lambda \in \mathbb{C};\ \mathrm{rank}(P_2(\lambda)) < p\} \\ &= \mathrm{char}(\mathcal{B}_1^0) \bigcup \mathrm{char}(\mathcal{B}_2^0). \end{aligned}$$

By definition the system morphism

$$\begin{aligned} C\circ = (0, \mathrm{id}_p)\circ : \mathcal{B}^0_{y_1,y,u} &= \left\{\begin{pmatrix} y_1 \\ y \end{pmatrix} \in \mathcal{F}^{p_1+p};\ P_1 \circ y_1 = 0,\ P_2 \circ y = Q_2 \circ y_1 = 0\right\} \\ \longrightarrow \mathcal{B}^0_{y,u} &= \left\{y \in \mathcal{F}^p;\ Y_2 P_2 \circ y = 0\right\} \end{aligned}$$

is surjective and hence $\mathrm{char}(\mathcal{B}^0_{y,u}) \subseteq \mathrm{char}(\mathcal{B}^0_{y_1,y,u})$ by Theorem 5.4.10. □

Corollary 7.3.2 and Example 7.3.3 show that the controllability of $\mathcal{B}_{y,u}$ does neither require nor follow from that of $\mathcal{B}_1$ and $\mathcal{B}_2$. We consider the SISO case $p = m = p_1 = 1$. Then (Y_1, Y_2) is given as

$$(Y_1, Y_2) = (Q_2 d^{-1}, P_1 d^{-1}) \text{ where } d := \gcd(P_1, Q_2). \tag{7.16}$$

Hence the behavior $\mathcal{B}_{y,u}$ is controllable if and only if

$$((P_1 d^{-1})P_2, -Q_1(Q_2 d^{-1})) \in \mathcal{D}^{1\times 2} \tag{7.17}$$

has a right inverse or, in other words, coprime entries.

Corollary 7.3.2 (cf. [2, Problem 2.4–6.a]) *For* $p = m = p_1 = 1$ *with* $d := \gcd(P_1, Q_2)$*, the series interconnection* $\mathcal{B}_{y,u}$ *is controllable if and only if each of* $P_1 d^{-1}$*,* P_2 *is coprime to each of* Q_1*,* $Q_2 d^{-1}$*; of course,* $P_1 d^{-1}$ *and* $Q_2 d^{-1}$ *are coprime by definition. If both* $\mathcal{B}_1$ *and* $\mathcal{B}_2$ *are controllable, i.e., if* $\gcd(P_1, Q_1) = 1$ *and* $\gcd(P_2, Q_2) = 1$*, this is equivalent to* $\gcd(P_2, Q_1) = 1$*.* ◊

Example 7.3.3 Consider nonzero elements $\widetilde{P}_1, P_2, \widetilde{Q}_1, \widetilde{Q}_2, d \in \mathcal{D}$ such that $\widetilde{P}_1, P_2$ are pairwise coprime to $\widetilde{Q}_1, \widetilde{Q}_2, d$ and d is a nonunit. Define

$$P_1 := \widetilde{P}_1 d, \; Q_1 := \widetilde{Q}_1 d, \; Q_2 := \widetilde{Q}_2 d.$$

Then $\gcd(P_1, Q_1) = d$ and $\gcd(P_1, Q_2) = d$. According to Corollary 7.3.2, the behavior $\mathcal{B}_{y,u}$ is controllable. Since d is a nonunit, P_1 and Q_1 are not coprime and thus $\mathcal{B}_1$ is not controllable.

Likewise, assume nonzero $\widetilde{P}_1, \widetilde{P}_2, d, Q_1, \widetilde{Q}_2 \in \mathcal{D}$ such that $\widetilde{P}_1, \widetilde{P}_2, d$ are pairwise coprime to $Q_1, \widetilde{Q}_2$ and d is a nonunit. Define

$$P_1 := \widetilde{P}_1 d, \; P_2 := \widetilde{P}_2 d, \; Q_2 := \widetilde{Q}_2 d.$$

Then $\gcd(P_1, Q_2) = d$ and $\gcd(P_2, Q_2) = d$. Again, the behavior $\mathcal{B}_{y,u}$ is controllable, but $\mathcal{B}_2$ is not since d is a nonunit. ◊

Next we discuss the series connection for observable Rosenbrock equations

$$A_i \circ x_i = B_i \circ u_i, \; y_i = C_i \circ x_i + D_i \circ u \text{ with } A_i \in \mathcal{D}^{n_i \times n_i}, \; \operatorname{rank}(A_i) = n_i, \; i = 1, 2, \tag{7.18}$$

with the associated behaviors

$$\begin{aligned} \widetilde{\mathcal{B}}_i &:= \left\{ \begin{pmatrix} x_i \\ u_i \end{pmatrix} \in \mathcal{F}^{n_i+m_i}; \; A_i \circ x_i = B_i \circ u_i \right\} \text{ and} \\ \mathcal{B}_i &:= \begin{pmatrix} C_i & D_i \\ 0 & \mathrm{id}_m \end{pmatrix} \circ \widetilde{\mathcal{B}}_i = \left\{ \begin{pmatrix} y_i \\ u_i \end{pmatrix} \in \mathcal{F}^{p_i+m_i}; \; P_i \circ y_i = Q_i \circ u_i \right\}, \end{aligned} \tag{7.19}$$

where the matrices P_i and Q_i are computed via Theorem 3.2.22. We assume without loss of generality that the P_i are square.

The series connection of these two systems is given by the network

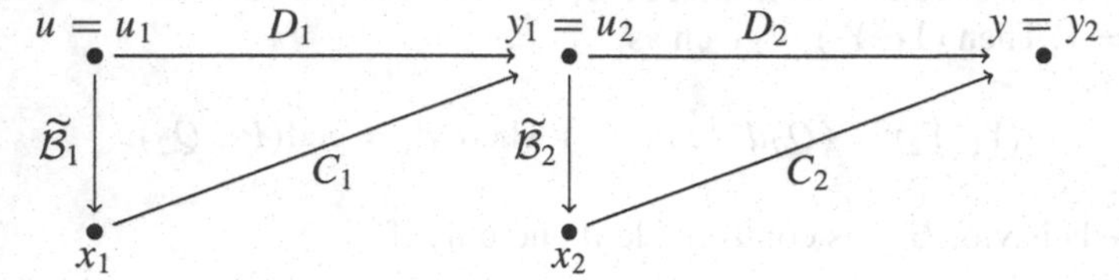

The equations of the network are

$$\begin{aligned} A_1 \circ x_1 &= B_1 \circ u, \\ y_1 &= C_1 \circ x_1 + D_1 \circ u, \\ A_2 \circ x_2 &= B_2 \circ y_1 = B_2 C_1 \circ x_1 + B_2 D_1 \circ u, \\ y &= C_2 \circ x_2 + D_2 \circ y_1 = C_2 \circ x_2 + D_2 C_1 \circ x_1 + D_2 D_1 \circ u, \end{aligned}$$

or, in Rosenbrock form,

$$\begin{aligned} \begin{pmatrix} A_1 & 0 \\ -B_2C_1 & A_2 \end{pmatrix} \circ \begin{pmatrix} x_1 \\ x_2 \end{pmatrix} &= \begin{pmatrix} B_1 \\ B_2 D_1 \end{pmatrix} \circ u, \\ y &= (D_2 C_1, C_2) \circ \begin{pmatrix} x_1 \\ x_2 \end{pmatrix} + D_2 D_1 \circ u. \end{aligned} \tag{7.20}$$

These equations give rise to the behaviors

$$\begin{aligned} \mathcal{B}_{x_1,x_2,u} &:= \left\{ \begin{pmatrix} x_1 \\ x_2 \\ u \end{pmatrix} \in \mathcal{F}^{n_1+n_2+m},\ \begin{pmatrix} A_1 & 0 \\ -B_2C_1 & A_2 \end{pmatrix} \circ \begin{pmatrix} x_1 \\ x_2 \end{pmatrix} = \begin{pmatrix} B_1 \\ B_2D_1 \end{pmatrix} \circ u \right\}, \\ \mathcal{B}_{y_1,y,u} &:= \begin{pmatrix} C_1 & 0 & D_1 \\ D_2C_1 & C_2 & D_2D_1 \\ 0 & 0 & \mathrm{id}_m \end{pmatrix} \circ \mathcal{B}_{x_1,x_2,u} \\ &= \left\{ \begin{pmatrix} y_1 \\ y \\ u \end{pmatrix} \in \mathcal{F}^{p_1+p+m};\ \begin{pmatrix} P_1 & 0 \\ -Q_2 & P_2 \end{pmatrix} \circ \begin{pmatrix} y_1 \\ y \end{pmatrix} = \begin{pmatrix} Q_1 \\ 0 \end{pmatrix} \circ u \right\}, \\ \mathcal{B}_{y,u} &:= \begin{pmatrix} 0 & \mathrm{id}_p & 0 \\ 0 & 0 & \mathrm{id}_m \end{pmatrix} \circ \mathcal{B}_{y_1,y,u} = \begin{pmatrix} 0 & \mathrm{id}_p & 0 \\ 0 & 0 & \mathrm{id}_m \end{pmatrix} \begin{pmatrix} C_1 & 0 & D_1 \\ D_2C_1 & C_2 & D_2D_1 \\ 0 & 0 & \mathrm{id}_m \end{pmatrix} \circ \mathcal{B}_{x_1,x_2,u} \\ &= \begin{pmatrix} D_2C_1 & C_2 & D_2D_1 \\ 0 & 0 & \mathrm{id}_m \end{pmatrix} \circ \mathcal{B}_{x_1,x_2,u}. \end{aligned}$$

$\mathcal{B}_{y_1,y,u}$ is the series connection of $\mathcal{B}_1$ and $\mathcal{B}_2$.

Corollary 7.3.4 (Observable series connection, cf. [2, Example 2.4–4 on pp. 151–152]) *If the Rosenbrock equations* (7.18) *are observable, the following statements are equivalent:*

1. *The Rosenbrock equations* (7.20) *are observable.*
2. $\mathcal{B}_{y_1,y,u}$ *is observable, i.e.,* $\begin{pmatrix} 0 & \mathrm{id}_p & 0 \\ 0 & 0 & \mathrm{id}_m \end{pmatrix} \cdot : \mathcal{B}_{y_1,y,u} \to \mathcal{B}_{y,u}$ *is bijective.*
3. *The matrix* $\begin{pmatrix} P_1 \\ Q_2 \end{pmatrix}$ *has a left inverse.*

Proof 1. $\Longleftrightarrow$ 2.: Consider the maps

$$\begin{pmatrix} D_2C_1 & C_2 & D_2D_1 \\ 0 & 0 & \mathrm{id}_m \end{pmatrix}\circ : \ \mathcal{B}_{x_1,x_2,u} \xrightarrow{\begin{pmatrix} C_1 & 0 & D_1 \\ D_2C_1 & C_2 & D_2D_1 \\ 0 & 0 & \mathrm{id}_m \end{pmatrix}\circ} \mathcal{B}_{y_1,y,u} \xrightarrow{\begin{pmatrix} 0 & \mathrm{id}_p & 0 \\ 0 & 0 & \mathrm{id}_m \end{pmatrix}\circ} \mathcal{B}_{y,u}.$$

All maps are clearly surjective. By Theorem and Definition 4.1.4, statement 1 is equivalent to the composite map being an isomorphism while statement 2 means that the second map is an isomorphism. We show that the first map is injective and thus an isomorphism. This implies 1. $\Longleftrightarrow$ 2.. Indeed, let

$$\begin{aligned}
&\begin{pmatrix} x_1 \\ x_2 \\ u \end{pmatrix} \in \ker\left(\begin{pmatrix} C_1 & 0 & D_1 \\ D_2C_1 & C_2 & D_2D_1 \\ 0 & 0 & \mathrm{id}_m \end{pmatrix}\circ : \ \mathcal{B}_{x_1,x_2,u} \to \mathcal{B}_{y_1,y,u}\right) \\
&\Longrightarrow u = 0, \ C_1 \circ x_1 = 0, \ D_2C_1 \circ x_1 + C_2 \circ x_2 = 0 \\
&\underset{\widetilde{\mathcal{B}}_1 \text{ observable}}{\Longrightarrow} \quad x_1 = 0, \ C_2 \circ x_2 = 0 \underset{\widetilde{\mathcal{B}}_2 \text{ observable}}{\Longrightarrow} \quad x_2 = 0.
\end{aligned}$$

2. $\Longleftrightarrow$ 3.: According to Theorem and Definition 4.1.4 item 2 holds if and only if the matrix $\begin{pmatrix} P_1 & 0 \\ -Q_2 & P_2 \\ 0 & \mathrm{id}_p \end{pmatrix}$ has a left inverse. This is equivalent to

$$\begin{pmatrix} P_1 & 0 \\ Q_2 & 0 \\ 0 & \mathrm{id}_p \end{pmatrix} = \begin{pmatrix} \mathrm{id}_{p_1} & 0 & 0 \\ 0 & -\mathrm{id}_p & P_2 \\ 0 & 0 & \mathrm{id}_p \end{pmatrix} \begin{pmatrix} P_1 & 0 \\ -Q_2 & P_2 \\ 0 & \mathrm{id}_p \end{pmatrix}$$

having a left inverse. This holds if and only if $\begin{pmatrix} P_1 \\ Q_2 \end{pmatrix}$ has a left inverse. □

Example 7.3.5 As a direct consequence, the series connection of two observable SISO systems is observable if and only if $\gcd(P_1, Q_2) = 1$.

◇

7.4 Parallel Connection

We consider the network

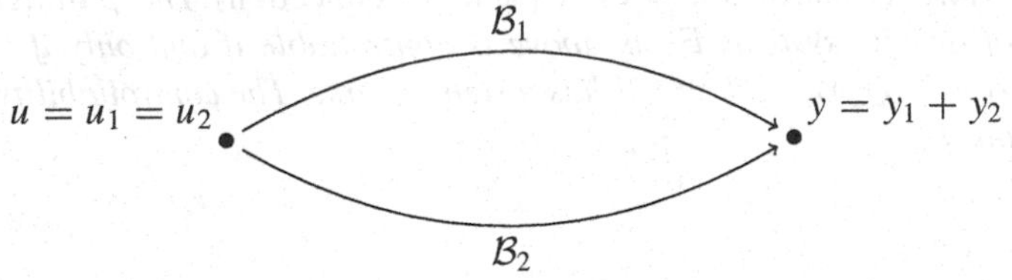

with the branch IO behaviors

$$\mathcal{B}_i = \left\{ \begin{pmatrix} y_i \\ u_i \end{pmatrix} \in \mathcal{F}^{p+m};\ P_i \circ y_i = Q_i \circ u_i \right\} \text{ with}$$
$$(P_i, -Q_i) \in \mathcal{D}^{p\times(p+m)},\ \operatorname{rank}(P_i) = p,\ H_i = P_i^{-1} Q_i.$$

The network equations are

$$\begin{aligned} &P_1 \circ y_1 = Q_1 \circ u,\ P_2 \circ y_2 = Q_2 \circ u,\ y = y_1 + y_2, \text{ or} \\ &A \circ \begin{pmatrix} y_1 \\ y_2 \end{pmatrix} = B \circ u,\ y = C \circ \begin{pmatrix} y_1 \\ y_2 \end{pmatrix} \text{ with} \\ &A := \begin{pmatrix} P_1 & 0 \\ 0 & P_2 \end{pmatrix},\ B := \begin{pmatrix} Q_1 \\ Q_2 \end{pmatrix},\ C := (\mathrm{id}_p, \mathrm{id}_p). \end{aligned} \tag{7.21}$$

These equations give rise to the IO systems

$$\begin{aligned} &\mathcal{B}_{y_1,y_2,u} := \left\{ \begin{pmatrix} y_1 \\ y_2 \\ u \end{pmatrix} \in \mathcal{F}^{2p+m};\ A \circ \begin{pmatrix} y_1 \\ y_2 \end{pmatrix} = B \circ u \right\} \text{ and} \\ &\mathcal{B}_{y,u} := \begin{pmatrix} C & 0 \\ 0 & \mathrm{id}_m \end{pmatrix} \circ \mathcal{B}_{y_1,y_2,u}. \end{aligned} \tag{7.22}$$

Both systems $\mathcal{B}_{y_1,y_2,u}$ and $\mathcal{B}_{y,u}$ are called the *parallel connection* or *parallel combination* of $\mathcal{B}_1$ and $\mathcal{B}_2$. The transfer matrices of $\mathcal{B}_{y_1,y_2,u}$ resp. of $\mathcal{B}_{y,u}$ are

$$\begin{aligned} &H := A^{-1}B = \begin{pmatrix} P_1^{-1} & 0 \\ 0 & P_2^{-1} \end{pmatrix} \begin{pmatrix} Q_1 \\ Q_2 \end{pmatrix} = \begin{pmatrix} H_1 \\ H_2 \end{pmatrix} \text{ resp.} \\ &CH = (\mathrm{id}_p, \mathrm{id}_p) \begin{pmatrix} H_1 \\ H_2 \end{pmatrix} = H_1 + H_2. \end{aligned} \tag{7.23}$$

Since $B_{y,u}$ is derived from the Rosenbrock equations (7.21), one constructs the equations of $\mathcal{B}_{y,u}$ according to Theorem 3.2.22 by means of a universal left annihilator

$$(Y_1, Y_2, -Y) \in \mathcal{D}^{p\times 3p} \text{ of } \begin{pmatrix} A \\ C \end{pmatrix} = \begin{pmatrix} P_1 & 0 \\ 0 & P_2 \\ \mathrm{id}_p & \mathrm{id}_p \end{pmatrix}.$$

Then, $Y_1 P_1 = Y_2 P_2 = Y$ and

$$\mathcal{B}_{y,u} = \left\{ \begin{pmatrix} y \\ u \end{pmatrix} \in \mathcal{F}^{p+m};\ Y \circ y = (Y_1, Y_2) \begin{pmatrix} Q_1 \\ Q_2 \end{pmatrix} \circ u = (Y_1 Q_1 + Y_2 Q_2) \circ u \right\}.$$

Alternatively, one constructs $(Y_1, -Y_2)$ as universal left annihilator of $\begin{pmatrix} P_1 \\ P_2 \end{pmatrix}$ and then defines $Y := Y_1 P_1 = Y_2 P_2$.

Corollary 7.4.1 (Controllability of a parallel connection) *The parallel connection $\mathcal{B}_{y,u}$ of two IO systems $\mathcal{B}_i$ as above is controllable if and only if the matrix $(Y, -(Y_1 Q_1 + Y_2 Q_2)) \in \mathcal{D}^{p\times(p+m)}$ has a right inverse. The controllability of the $\mathcal{B}_i$ is not assumed.*

Proof Theorem 4.3.5. □

Example 7.4.2 (*The SISO case*) Assume $m = p = 1$ in Corollary 7.4.1. Let $d := \gcd(P_1, P_2)$. Then the matrix $(Y_1, Y_2) := (P_2 d^{-1}, P_1 d^{-1})$ is a universal left annihilator of $\left(\begin{smallmatrix} P_1 \\ -P_2 \end{smallmatrix}\right)$. Hence, $\mathcal{B}_{y,u}$ is given by

$$\mathcal{B}_{y,u} = \left\{ \left(\begin{smallmatrix} y \\ u \end{smallmatrix}\right) \in \mathcal{F}^2;\ (P_1 P_2 d^{-1}) \circ y = (P_2 d^{-1} Q_1 + P_1 d^{-1} Q_2) \circ u \right\}.$$

The parallel connection is controllable if and only if

$$\gcd\left(\tfrac{P_1 P_2}{d},\ \tfrac{P_2 Q_1 + P_1 Q_2}{d}\right) = 1. \tag{7.24}$$

In [2, Problem 2.4–6.b on p. 156], it is stated that under the assumption that $\mathcal{B}_1$ and $\mathcal{B}_2$ are controllable, i.e., $\gcd(P_1, Q_1) = \gcd(P_2, Q_2) = 1$, then $\mathcal{B}_{y,u}$ is controllable $\Longleftrightarrow$ P_1 and P_2 are coprime, i.e., $d = \gcd(P_1, P_2) = 1$. While the implication "$\Longleftarrow$" holds, the implication "$\Longrightarrow$" does not, and we give a proof for the first one and a counterexample for the second one.
$\Longleftarrow$: With $d = \gcd(P_1, P_2) = 1$ we have to show $\gcd(P_1 P_2, P_2 Q_1 + P_1 Q_2) = 1$. Assume that a prime element $x \in \mathcal{D}$ divides the latter gcd. Then $x \mid P_1 P_2$, hence $x \mid P_1$ or $x \mid P_2$, for instance, $x \mid P_1$. Moreover

$$\begin{aligned} x \mid (P_2 Q_1 + P_1 Q_2) &\underset{x \mid P_1}{\Longrightarrow} x \mid P_2 Q_1 \underset{x \text{ prime}}{\Longrightarrow} x \mid P_2 \text{ or } x \mid Q_1 \\ &\Longrightarrow x \mid \gcd(P_1, P_2) = 1 \text{ or } x \mid \gcd(P_1, Q_1) = 1. \end{aligned}$$

This is a contradiction, hence $\gcd(P_1 P_2, P_2 Q_1 + P_1 Q_2) = 1$.
$\not\Longrightarrow$: Let a_1 and a_2 be nonzero nonunits and coprime. Let Q_1 and Q_2 be such that $Q_2 a_1 + Q_1 a_2 = 1$. Then also $\gcd(a_1, Q_1) = \gcd(a_2, Q_2) = 1$. Let $d \in \mathcal{D}$ be a nonzero nonunit with

$$\gcd(d, a_1) = \gcd(d, a_2) = \gcd(d, Q_1) = \gcd(d, Q_2) = 1$$

and define $P_1 := a_1 d$ and $P_2 := a_2 d$. Then $\mathcal{B}_1, \mathcal{B}_2$ are controllable since

$$\gcd(P_1, Q_1) = \gcd(a_1 d, Q_1) = \underbrace{\gcd(a_1, Q_1)}_{=1} \underbrace{\gcd(d, Q_1)}_{=1} = 1,\ \ \gcd(P_2, Q_2) = 1.$$

Moreover $\gcd(P_1, P_2) = \gcd(a_1 d, a_2 d) = \gcd(a_1, a_2) d = d$ and

$$\gcd\left(\tfrac{P_1 P_2}{d},\ \tfrac{P_2 Q_1 + P_1 Q_2}{d}\right) = \gcd\left(\tfrac{P_1 P_2}{d}, a_2 Q_1 + a_1 Q_2\right) = \gcd\left(\tfrac{P_1 P_2}{d}, 1\right) = 1.$$

Hence $\mathcal{B}_1, \mathcal{B}_2, \mathcal{B}_{y,u}$ are controllable, cf. (7.24), but $\gcd(P_1, P_2) \neq 1$.

For $\mathcal{D} = \mathbb{Z}$, an example of this kind is

$$a_1 = 2,\ a_2 = 5,\ Q_1 = 1,\ Q_2 = -2,\ d = 3,\ P_1 = 6,\ P_2 = 15.$$

To complement [2, Problem 2.4–6.b on p. 156], we present an example of a controllable parallel connection of two noncontrollable IO behaviors. Choose nonunits d_1, d_2 and $\widetilde{Q}_1$, $\widetilde{Q}_2$ with $\gcd(d_1\widetilde{Q}_1, d_2\widetilde{Q}_2) = 1$. Let $a_1, a_2 \in \mathcal{D}$ be such that $a_2 d_1 \widetilde{Q}_1 + a_1 d_2 \widetilde{Q}_2 = 1$. Then also $\gcd(a_1, a_2) = 1$. Define

$$d := d_1 d_2,\ P_1 := a_1 d,\ P_2 := a_2 d,\ Q_1 := \widetilde{Q}_1 d_1,\ \ Q_2 := \widetilde{Q}_2 d_2 \Longrightarrow$$
$$\gcd\left(\tfrac{P_1P_2}{d},\ \tfrac{P_2Q_1+P_1Q_2}{d}\right) = \gcd\left(\tfrac{P_1P_2}{d},\ a_2 d_1 \widetilde{Q}_1 + a_1 d_2 \widetilde{Q}_2\right) = \gcd\left(\tfrac{P_1P_2}{d},\ 1\right) = 1.$$

By (7.24) $\mathcal{B}_{y,u}$ is controllable, but $\mathcal{B}_1$ and $\mathcal{B}_2$ are not since

$$d_1 \mid \gcd(a_1 d_1 d_2,\ \widetilde{Q}_1 d_1) = \gcd(P_1, Q_1) \text{ and likewise}$$
$$d_2 \mid \gcd(a_2 d_1 d_2,\ \widetilde{Q}_2 d_2) = \gcd(P_2, Q_2).$$

◇

Corollary 7.4.3 (Observability of parallel connection, cf. [2, Problem 2.4-6.b]) *According to Theorem 4.1.4, the Rosenbrock equations* (7.21) *are observable if and only if the matrix*

$$\begin{pmatrix} A \\ C \end{pmatrix} = \begin{pmatrix} P_1 & 0 \\ 0 & P_2 \\ \mathrm{id}_p & \mathrm{id}_p \end{pmatrix}$$

is left invertible. This is equivalent to $\begin{pmatrix} P_1 \\ P_2 \end{pmatrix}$ *being left invertible.*
If $(X_1, X_2)\begin{pmatrix} P_1 \\ P_2 \end{pmatrix} = \mathrm{id}_p$, *the observability can be checked directly. Indeed, let*

$$y_i \in \mathcal{B}_i^0,\ i = 1, 2,\ \textit{with}\ y = y_1 + y_2 = 0 \Longrightarrow y_2 = -y_1,\ \begin{pmatrix} P_1 \\ P_2 \end{pmatrix} \circ y_1 = 0$$
$$\Longrightarrow y_1 = (X_1, X_2)\begin{pmatrix} P_1 \\ P_2 \end{pmatrix} \circ y_1 = 0 \Longrightarrow y_2 = -y_1 = 0.$$

Proof If $X_1 P_1 + X_2 P_2 = \mathrm{id}_p$, then

$$\begin{pmatrix} X_1 & -X_2 & X_2P_2 \\ -X_1 & X_2 & X_1P_1 \end{pmatrix} \begin{pmatrix} P_1 & 0 \\ 0 & P_2 \\ \mathrm{id}_p & \mathrm{id}_p \end{pmatrix} = \begin{pmatrix} \mathrm{id}_p & 0 \\ 0 & \mathrm{id}_p \end{pmatrix}.$$

Conversely, if

$$\begin{pmatrix} X_{11} & X_{12} & X_{13} \\ X_{21} & X_{22} & X_{23} \end{pmatrix} \begin{pmatrix} P_1 & 0 \\ 0 & P_2 \\ \mathrm{id}_p & \mathrm{id}_p \end{pmatrix} = \begin{pmatrix} \mathrm{id}_p & 0 \\ 0 & \mathrm{id}_p \end{pmatrix}$$
$$\Longrightarrow X_{11}P_1 + X_{13} = \mathrm{id}_p,\ X_{12}P_2 + X_{13} = 0 \Longrightarrow X_{11}P_1 - X_{12}P_2 = \mathrm{id}_p\,.$$

□

7.5 Feedback Interconnection

The feedback interconnection of a plant with a controller or compensator is the main tool for stabilization and control design in Chap. 10.

A feedback system is a network of the form

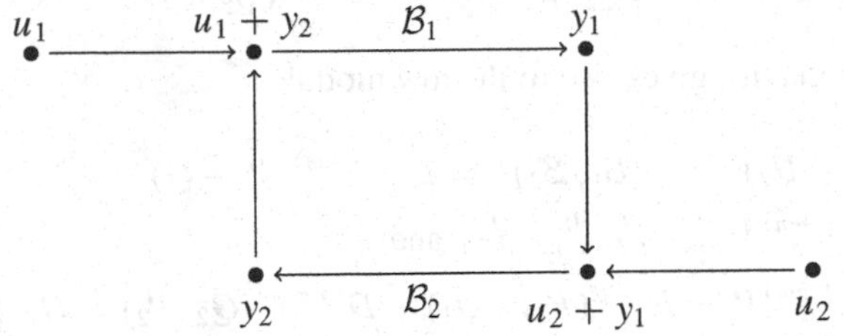

The common node dimension of u_1 and y_2 is denoted by m and that of u_2 and y_1 by p. The IO behavior $\mathcal{B}_1$ and its module of equations are given as

$$\begin{aligned}
&\mathcal{B}_1 := \left\{\left(\begin{smallmatrix} y_1 \\ u_1 \end{smallmatrix}\right) \in \mathcal{F}^{p+m};\ P_1 \circ y_1 = Q_1 \circ u_1\right\},\ (P_1, -Q_1) \in \mathcal{D}^{p\times(p+m)},\\
&\operatorname{rank}(P_1) = p,\ H_1 := P_1^{-1}Q_1,\ U_1 := \mathcal{B}_1^{\perp} = \mathcal{D}^{1\times p}(P_1, -Q_1) \subseteq \mathcal{D}^{1\times(p+m)},\\
&KU_1 = K^{1\times p}(P_1, -Q_1) = K^{1\times p}(\mathrm{id}_p, -H_1) \subseteq K^{1\times(p+m)}.
\end{aligned}$$

The corresponding data of $\mathcal{B}_2$ are

$$\begin{aligned}
&\mathcal{B}_2 := \left\{\left(\begin{smallmatrix} u_2 \\ y_2 \end{smallmatrix}\right) \in \mathcal{F}^{p+m};\ P_2 \circ y_2 = Q_2 \circ u_2\right\},\ (-Q_2, P_2) \in \mathcal{D}^{m\times(p+m)},\\
&\operatorname{rank}(P_2) = m,\ H_2 := P_2^{-1}Q_2,\ U_2 := \mathcal{B}_2^{\perp} = \mathcal{D}^{1\times m}(-Q_2, P_2) \subseteq \mathcal{D}^{1\times(p+m)},\\
&KU_2 = K^{1\times m}(-Q_2, P_2) = K^{1\times m}(-H_2, \mathrm{id}_m) \subseteq K^{1\times(p+m)}.
\end{aligned}$$

We write $\left(\begin{smallmatrix} u_2 \\ y_2 \end{smallmatrix}\right)$ instead of $\left(\begin{smallmatrix} y_2 \\ u_2 \end{smallmatrix}\right)$, etc. in order that $\left(\begin{smallmatrix} y_1 \\ u_1 \end{smallmatrix}\right), \left(\begin{smallmatrix} u_2 \\ y_2 \end{smallmatrix}\right) \in \mathcal{F}^{p+m}$. The equations of the interconnection behavior are

$$P_1 \circ y_1 = Q_1 \circ (u_1 + y_2) \quad\text{and}\quad P_2 \circ y_2 = Q_2 \circ (u_2 + y_1). \tag{7.25}$$

We define the signal vectors

$$y := \left(\begin{smallmatrix} y_1 \\ y_2 \end{smallmatrix}\right) \in \mathcal{F}^{p+m} \quad\text{and}\quad u := \left(\begin{smallmatrix} u_2 \\ u_1 \end{smallmatrix}\right) \in \mathcal{F}^{p+m}. \tag{7.26}$$

In the following equations we omit the nodes with the signals $u_1 + y_2$ and $u_2 + y_1$ because these signals are trivially determined by the signals at other nodes. We obtain the interconnected *feedback behavior* as

$$\begin{aligned}\mathcal{B} &:= \mathrm{fb}(\mathcal{B}_1, \mathcal{B}_2) := \left\{ \left(\begin{smallmatrix} y \\ u \end{smallmatrix}\right) \in \mathcal{F}^{p+m+p+m};\ \left(\begin{smallmatrix} y_1 \\ u_1+y_2 \end{smallmatrix}\right) \in \mathcal{B}_1,\ \left(\begin{smallmatrix} u_2+y_1 \\ y_2 \end{smallmatrix}\right) \in \mathcal{B}_2 \right\} \\ &= \left\{ \left(\begin{smallmatrix} y \\ u \end{smallmatrix}\right) \in \mathcal{F}^{p+m+p+m};\ \text{Eq. (7.25) is satisfied} \right\} \\ &= \left\{ \left(\begin{smallmatrix} y \\ u \end{smallmatrix}\right) \in \mathcal{F}^{(p+m)+(p+m)};\ \left(\begin{smallmatrix} P_1 & -Q_1 \\ -Q_2 & P_2 \end{smallmatrix}\right) \circ \left(\begin{smallmatrix} y_1 \\ y_2 \end{smallmatrix}\right) = \left(\begin{smallmatrix} 0 & Q_1 \\ Q_2 & 0 \end{smallmatrix}\right) \circ \left(\begin{smallmatrix} u_2 \\ u_1 \end{smallmatrix}\right) \right\} \\ &= \left\{ \left(\begin{smallmatrix} y \\ u \end{smallmatrix}\right) \in \mathcal{F}^{(p+m)+(p+m)};\ P \circ y = Q \circ u \right\} \\ &\qquad \text{with}\ \ P := \left(\begin{smallmatrix} P_1 & -Q_1 \\ -Q_2 & P_2 \end{smallmatrix}\right) \ \text{ and } \ Q := \left(\begin{smallmatrix} 0 & Q_1 \\ Q_2 & 0 \end{smallmatrix}\right). \end{aligned} \tag{7.27}$$

The feedback behavior gives rise to the row modules

$$\begin{aligned} U &:= \mathrm{fb}(U_1, U_2) := \mathrm{fb}(\mathcal{B}_1, \mathcal{B}_2)^{\perp} = \mathcal{D}^{1\times(p+m)}(P, -Q) \\ &= \mathcal{D}^{1\times(p+m)} \left(\begin{smallmatrix} P_1 & -Q_1 & 0 & -Q_1 \\ -Q_2 & P_2 & -Q_2 & 0 \end{smallmatrix}\right) \text{ and} \\ U^0 &:= \mathcal{D}^{1\times(p+m)} P = \mathcal{D}^{1\times p}(P_1, -Q_1) + \mathcal{D}^{1\times m}(-Q_2, P_2) = U_1 + U_2. \end{aligned} \tag{7.28}$$

Equation (7.27) suggests to consider $\mathcal{B}$ as an IO behavior with input $u = \left(\begin{smallmatrix} u_2 \\ u_1 \end{smallmatrix}\right)$. As Theorem and Definition 7.5.1 shows, this is often possible, but not always.

Theorem and Definition 7.5.1 (Well-posed feedback behaviors) *Under the preceding conditions the following properties are equivalent:*

1. *The feedback behavior* $\mathcal{B} = \mathrm{fb}(\mathcal{B}_1, \mathcal{B}_2)$ *is an IO behavior with input* $u = \left(\begin{smallmatrix} u_2 \\ u_1 \end{smallmatrix}\right)$.
2. $\mathrm{rank}\left(\begin{smallmatrix} P_1 & -Q_1 \\ -Q_2 & P_2 \end{smallmatrix}\right) = \dim_K(K^{1\times(p+m)}P) = p + m.$
3. $U^0 = U_1 + U_2 = U_1 \oplus U_2$, *i.e., the sum is direct. This is equivalent to*

$$\begin{aligned} &U_1 \cap U_2 = 0 \ \textit{or}\ KU_1 \cap KU_2 = 0 \ \textit{or} \\ &\dim_K(KU_1 + KU_2) = \dim_K(KU_1) + \dim_K(KU_2) = p + m. \end{aligned}$$

4. *The matrix* $\begin{pmatrix} \mathrm{id}_p & -H_1 \\ -H_2 & \mathrm{id}_m \end{pmatrix}$ *is invertible or, equivalently,* $\mathrm{id}_m - H_2H_1 \in \mathrm{Gl}_m(K)$ *or* $\mathrm{id}_p - H_1H_2 \in \mathrm{Gl}_p(K)$.
   The inverses of these three matrices are then connected as follows. Let

$$\begin{aligned} &X := (\mathrm{id}_m - H_2H_1)^{-1},\ Y := (\mathrm{id}_p - H_1H_2)^{-1} \\ &\Longrightarrow \begin{pmatrix} \mathrm{id}_p & -H_1 \\ -H_2 & \mathrm{id}_m \end{pmatrix}^{-1} = \begin{pmatrix} \mathrm{id}_p + H_1XH_2 & H_1X \\ XH_2 & X \end{pmatrix} = \begin{pmatrix} Y & YH_1 \\ H_2Y & \mathrm{id}_m + H_2YH_1 \end{pmatrix}. \end{aligned} \tag{7.29}$$

If these four conditions are satisfied the feedback behavior is called well-posed. *The transfer matrix of this IO behavior* $\mathrm{fb}(\mathcal{B}_1, \mathcal{B}_2)$ *is*

$$H := \begin{pmatrix} \mathrm{id}_p & -H_1 \\ -H_2 & \mathrm{id}_m \end{pmatrix}^{-1} \begin{pmatrix} 0 & H_1 \\ H_2 & 0 \end{pmatrix}. \tag{7.30}$$

The modules

$$\begin{aligned} U^0 &= \mathcal{D}^{1\times(p+m)} P \ \textit{resp.} \\ U &= U^0(\mathrm{id}_{p+m}, -H) = \mathcal{D}^{1\times(p+m)}(P, -Q) \subseteq \mathcal{D}^{1\times(p+m+p+m)} \end{aligned} \tag{7.31}$$

are the equation modules of $\mathcal{B}^0$ resp. of $\mathcal{B}$.

With $G := \left(\begin{smallmatrix} 0 & H_1 \\ H_2 & 0 \end{smallmatrix}\right)$, the identities

$$H = (\mathrm{id}_{p+m} - G)^{-1}G \quad and \quad (\mathrm{id}_{p+m} - G)^{-1} = \mathrm{id}_{p+m} + H \tag{7.32}$$

hold and imply that the coefficients of H belong to a subring of K if and only if those of $(\mathrm{id}_{p+m} - G)^{-1}$ do. This will be important in Sect. 10.1.5.

Proof 1. $\Longleftrightarrow$ 2.: By definition $\mathcal{B}$ is an IO behavior if and only if $\mathrm{rank}(P) = p + m$, i.e., the condition of 2 holds.
2. $\Longleftrightarrow$ 3.: From $U_1 + U_2 = \mathcal{D}^{1\times(p+m)}P$ follows $KU_1 + KU_2 = K^{1\times(p+m)}P$. The standard dimension formula implies

$$\begin{aligned}\mathrm{rank}(P) &= \dim_K(K^{1\times(p+m)}P) = \dim_K(KU_1 + KU_2)\\ &= \dim_K(KU_1) + \dim_K(KU_2) - \dim_K(KU_1 \cap KU_2)\\ &= p + m - \dim_K(KU_1 \cap KU_2).\end{aligned}$$

This shows the equivalence of item 2 and the conditions $KU_1 \cap KU_2 = 0$ from item 3. It remains to show the equivalence $U_1 \cap U_2 = 0 \Longleftrightarrow KU_1 \cap KU_2 = 0$ where $\Longleftarrow$ is obvious. Conversely, let

$$\begin{aligned}&\xi \in KU_1 \cap KU_2 \Longrightarrow \exists \alpha_i, \beta_i \in \mathcal{D},\ \beta_i \neq 0,\ \eta_i \in U_i,\ i = 1, 2,\ \text{with } \xi = \tfrac{\alpha_1}{\beta_1}\eta_1 = \tfrac{\alpha_2}{\beta_2}\eta_2\\ &\Longrightarrow \beta_1\beta_2\xi = \underbrace{\beta_2\alpha_1\eta_1}_{\in U_1} = \underbrace{\beta_1\alpha_2\eta_2}_{\in U_2} \in U_1 \cap U_2 = 0 \Longrightarrow \beta_1\beta_2\xi = 0 \Longrightarrow \xi = 0.\end{aligned}$$

2. $\Longleftrightarrow$ 4.: Since $\mathrm{rank}\left(\begin{smallmatrix} P_1 & 0 \\ 0 & P_2 \end{smallmatrix}\right) = \mathrm{rank}(P_1) + \mathrm{rank}(P_2) = p + m$ the equation

$$\begin{aligned}&P = \begin{pmatrix} P_1 & -Q_1 \\ -Q_2 & P_2 \end{pmatrix} = \begin{pmatrix} P_1 & 0 \\ 0 & P_2 \end{pmatrix}\begin{pmatrix} \mathrm{id}_p & -H_1 \\ -H_2 & \mathrm{id}_m \end{pmatrix} \text{ implies } \mathrm{rank}(P) = \mathrm{rank}\begin{pmatrix} \mathrm{id}_p & -H_1 \\ -H_2 & \mathrm{id}_m \end{pmatrix},\\ &\text{hence } \left(P \in \mathrm{Gl}_{p+m}(K) \Longleftrightarrow \begin{pmatrix} \mathrm{id}_p & -H_1 \\ -H_2 & \mathrm{id}_m \end{pmatrix} \in \mathrm{Gl}_{p+m}(K)\right).\end{aligned} \tag{7.33}$$

4.: Notice that

$$\begin{aligned}\begin{pmatrix} \mathrm{id}_p & -H_1 \\ -H_2 & \mathrm{id}_m \end{pmatrix} &= \begin{pmatrix} \mathrm{id}_p & 0 \\ -H_2 & \mathrm{id}_m \end{pmatrix}\begin{pmatrix} \mathrm{id}_p & -H_1 \\ 0 & \mathrm{id}_m - H_2H_1 \end{pmatrix}\\ &= \begin{pmatrix} \mathrm{id}_p & 0 \\ -H_2 & \mathrm{id}_m \end{pmatrix}\begin{pmatrix} \mathrm{id}_p & 0 \\ 0 & \mathrm{id}_m - H_2H_1 \end{pmatrix}\begin{pmatrix} \mathrm{id}_p & -H_1 \\ 0 & \mathrm{id}_m \end{pmatrix}\\ \begin{pmatrix} \mathrm{id}_p & -H_1 \\ -H_2 & \mathrm{id}_m \end{pmatrix} &= \begin{pmatrix} \mathrm{id}_p & -H_1 \\ 0 & \mathrm{id}_m \end{pmatrix}\begin{pmatrix} \mathrm{id}_p - H_1H_2 & 0 \\ 0 & \mathrm{id}_m \end{pmatrix}\begin{pmatrix} \mathrm{id}_p & 0 \\ -H_2 & \mathrm{id}_m \end{pmatrix}.\end{aligned} \tag{7.34}$$

The matrix $\begin{pmatrix} \mathrm{id}_p & 0 \\ -H_2 & \mathrm{id}_m \end{pmatrix}$ and the other matrices of this type encode elementary row or column operations and are invertible, indeed $\begin{pmatrix} \mathrm{id}_p & 0 \\ -H_2 & \mathrm{id}_m \end{pmatrix}^{-1} = \begin{pmatrix} \mathrm{id}_p & 0 \\ H_2 & \mathrm{id}_m \end{pmatrix}$.

Thus the three conditions of 4 are equivalent. From (7.34), we infer that

$$\begin{aligned}\begin{pmatrix}\mathrm{id}_p & -H_1\\ -H_2 & \mathrm{id}_m\end{pmatrix}^{-1} &= \begin{pmatrix}\mathrm{id}_p & H_1\\ 0 & \mathrm{id}_m\end{pmatrix}\begin{pmatrix}\mathrm{id}_p & 0\\ 0 & (\mathrm{id}_m - H_2H_1)^{-1}\end{pmatrix}\begin{pmatrix}\mathrm{id}_p & 0\\ H_2 & \mathrm{id}_m\end{pmatrix}\\ &= \begin{pmatrix}\mathrm{id}_p + H_1XH_2 & H_1X\\ XH_2 & X\end{pmatrix}.\end{aligned}$$

The equation involving Y follows likewise from (7.34).

It remains to prove the equations from (7.32). The first coincides with (7.30). The second is valid because

$$\begin{aligned}\mathrm{id}_{p+m} + H &= \begin{pmatrix}\mathrm{id}_p & -H_1\\ -H_2 & \mathrm{id}_m\end{pmatrix}^{-1}\begin{pmatrix}\mathrm{id}_p & -H_1\\ -H_2 & \mathrm{id}_m\end{pmatrix} + \begin{pmatrix}\mathrm{id}_p & -H_1\\ -H_2 & \mathrm{id}_m\end{pmatrix}^{-1}\begin{pmatrix}0 & H_1\\ H_2 & 0\end{pmatrix}\\ &= \begin{pmatrix}\mathrm{id}_p & -H_1\\ -H_2 & \mathrm{id}_m\end{pmatrix}^{-1}\left(\begin{pmatrix}\mathrm{id}_p & -H_1\\ -H_2 & \mathrm{id}_m\end{pmatrix} + \begin{pmatrix}0 & H_1\\ H_2 & 0\end{pmatrix}\right) = (\mathrm{id}_{p+m} - G)^{-1}\,\mathrm{id}_{p+m}.\end{aligned}$$

□

Example 7.5.2 For SISO behaviors, i.e., with $p = m = 1$, the well-posedness signifies that $1 \neq H_2H_1 = \frac{Q_1Q_2}{P_1P_2}$, i.e., $Q_1Q_2 \neq P_1P_2$. Then,

$$\begin{aligned}H &= \begin{pmatrix}1 & -H_1\\ -H_2 & 1\end{pmatrix}^{-1}\begin{pmatrix}0 & H_1\\ H_2 & 0\end{pmatrix}\\ &= \frac{1}{1-H_1H_2}\begin{pmatrix}1 & H_1\\ H_2 & 1\end{pmatrix}\begin{pmatrix}0 & H_1\\ H_2 & 0\end{pmatrix} = \frac{1}{1-H_1H_2}\begin{pmatrix}H_1H_2 & H_1\\ H_2 & H_1H_2\end{pmatrix}.\end{aligned}$$

◇

The following theorem establishes a canonical isomorphism of the feedback behavior $\mathcal{B} = \mathrm{fb}(\mathcal{B}_1, \mathcal{B}_2)$ and the product behavior $\mathcal{B}_1 \times \mathcal{B}_2 := \left\{\begin{pmatrix}w_1\\ w_2\end{pmatrix};\ w_i \in \mathcal{B}_i\right\}$. This will be used in Chap. 10. The modules of the product behavior are

$$\begin{aligned}U_1 \times U_2 &= (\mathcal{B}_1 \times \mathcal{B}_2)^{\perp} = \mathcal{D}^{1\times(p+m)}\begin{pmatrix}P_1 & -Q_1 & 0 & 0\\ 0 & 0 & -Q_2 & P_2\end{pmatrix} \subseteq \mathcal{D}^{1\times(p+m+p+m)},\\ \mathrm{M}(\mathcal{B}_1 \times \mathcal{B}_2) &= \mathcal{D}^{1\times(p+m+p+m)}/(U_1 \times U_2) \underset{\text{ident.}}{=} M(\mathcal{B}_1) \times M(\mathcal{B}_2).\end{aligned} \tag{7.35}$$

Theorem 7.5.3 *Define*

$$X := \begin{pmatrix}\mathrm{id}_p & 0 & 0 & 0\\ 0 & \mathrm{id}_m & 0 & \mathrm{id}_m\\ \mathrm{id}_p & 0 & \mathrm{id}_p & 0\\ 0 & \mathrm{id}_m & 0 & 0\end{pmatrix} \text{ and } Y := X^{-1} = \begin{pmatrix}\mathrm{id}_p & 0 & 0 & 0\\ 0 & 0 & 0 & \mathrm{id}_m\\ -\mathrm{id}_p & 0 & \mathrm{id}_p & 0\\ 0 & \mathrm{id}_m & 0 & -\mathrm{id}_m\end{pmatrix}.$$

Both matrices lie in $\mathrm{Gl}_{p+m+p+m}(F)$. *The identities*

$$(U_1 \times U_2)X = U \text{ and } UY = U_1 \times U_2 \tag{7.36}$$

hold. This implies that X *and* Y *induce inverse isomorphisms*

$$\mathrm{M}\left(\mathrm{fb}(\mathcal{B}_1,\mathcal{B}_2)\right) \underset{(\cdot X)_{\mathrm{ind}}}{\overset{(\cdot Y)_{\mathrm{ind}}}{\rightleftarrows}} \mathrm{M}(\mathcal{B}_1\times\mathcal{B}_2) = M(\mathcal{B}_1)\times M(\mathcal{B}_2), \tag{7.37}$$

and, by duality (Theorem 3.3.18), inverse behavior isomorphisms

$$\begin{aligned} &\mathrm{fb}(\mathcal{B}_1,\mathcal{B}_2) \underset{Y\circ}{\overset{X\circ}{\rightleftarrows}} \mathcal{B}_1\times\mathcal{B}_2, \quad \begin{pmatrix} y_1\\ y_2\\ u_2\\ u_1\end{pmatrix} \longmapsto \begin{pmatrix} y_1\\ u_1+y_2\\ u_2+y_1\\ y_2\end{pmatrix},\\ &\qquad\qquad \begin{pmatrix} y_1\\ y_2\\ v_2-y_1\\ v_1-y_2\end{pmatrix} \longleftarrow\!\shortmid \begin{pmatrix} y_1\\ v_1\\ v_2\\ y_2\end{pmatrix}. \end{aligned} \tag{7.38}$$

In particular, the feedback behavior $\mathrm{fb}(\mathcal{B}_1,\mathcal{B}_2)$ *is controllable if and only if both* $\mathcal{B}_1$ *and* $\mathcal{B}_2$ *have this property.*

Proof It is easy to check that $XY = YX = \mathrm{id}_{p+m+p+m}$,

$$\begin{pmatrix} P_1 & -Q_1 & 0 & 0\\ 0 & 0 & -Q_2 & P_2\end{pmatrix} X = (P,-Q) \text{ and } (P,-Q)Y = \begin{pmatrix} P_1 & -Q_1 & 0 & 0\\ 0 & 0 & -Q_2 & P_2\end{pmatrix}.$$

This obviously implies (7.36), (7.37), and (7.38).

Controllability of a behavior $\mathcal{B}$ is equivalent with the torsionfreeness of the module $\mathrm{M}(\mathcal{B})$, and this is preserved under isomorphism and under products. □

The assertions (7.37) and (7.38) of the preceding theorem are dual to each other. The advantage of (7.37) is that the signal module $\mathcal{F}$ does not enter in its formulation, and that will later be used in Theorem 10.1.20. For this reason, we formulate feedback in purely algebraic terms in the following corollary.

Corollary and Definition 7.5.4 (Feedback of modules) *Let* $\mathcal{D}$ *be a Noetherian integral domain with quotient field* K *and let*

$$U_1,\ U_2 \subseteq \mathcal{D}^{1\times(p+m)} = \mathcal{D}^{1\times p}\times\mathcal{D}^{1\times m}$$

be submodules with IO structures according to Theorem and Definition 6.2.2 such that there are unique transfer matrices H_1 *of* U_1 *and* H_2 *of* U_2 *with*

$$KU_1 = K^{1\times p}(\mathrm{id}_p,-H_1),\ \ KU_2 = K^{1\times m}(-H_2,\mathrm{id}_m) \subset K^{1\times(p+m)}.$$

Then the sum U_1+U_2 *is direct, i.e.,* $U_1+U_2 = U_1\oplus U_2$ *if and only if* $\begin{pmatrix}\mathrm{id}_p & -H_1\\ -H_2 & \mathrm{id}_m\end{pmatrix} \in \mathrm{Gl}_{p+m}(K)$.

If these equivalent conditions are satisfied, we say that the feedback module of U_1 *and* U_2 *is well-posed and we define*

$$\begin{aligned} &\mathrm{fb}(U_1,U_2) := (U_1\oplus U_2)(\mathrm{id}_{p+m},-H) \subset \mathcal{D}^{1\times(p+m+p+m)}, \text{ where}\\ &H := \begin{pmatrix}\mathrm{id}_p & -H_1\\ -H_2 & \mathrm{id}_m\end{pmatrix}^{-1}\begin{pmatrix} 0 & H_1\\ H_2 & 0\end{pmatrix}. \end{aligned}$$

Then the matrices X and Y of Theorem 7.5.3 induce the equalities

$$(U_1 \times U_2)X = \mathrm{fb}(U_1, U_2) \quad and \quad \mathrm{fb}(U_1, U_2)Y = U_1 \times U_2$$

and the mutually inverse isomorphisms

$$\mathcal{D}^{1\times(p+m+p+m)}/(U_1 \times U_2) \underset{(\cdot Y)_{\mathrm{ind}}}{\overset{(\cdot X)_{\mathrm{ind}}}{\rightleftarrows}} \mathcal{D}^{1\times(p+m+p+m)}/\mathrm{fb}(U_1, U_2).$$

Proof With representations $U_1 = \mathcal{D}^{1\times p}(P_1, -Q_1)$ and $U_2 = \mathcal{D}^{1\times m}(-Q_2, P_2)$ this corollary is a consequence of Theorems 7.5.1 and 7.5.3. □

Remark 7.5.5 1. In the section on control in a behavioral setting of [1, Sect. 10.8, pp. 363–368] the authors consider a (plant) behavior $\mathcal{B}_1 \subset \mathcal{F}^{m+p}$ with trajectories $w = \binom{u}{c}$ where the components u resp. c are called the exogeneous resp. the control component of the plant trajectory, and a (controller) behavior $\mathcal{B}_2 = \{c \in \mathcal{F}^p;\ R_c \circ c = 0\} \subset \mathcal{F}^p$. Their interconnection is defined as $\mathcal{B}_1 \wedge \mathcal{B}_2 := \{w = \binom{u}{c} \in \mathcal{B}_1;\ c \in \mathcal{B}_2\}$. In $\mathcal{B}_1 \wedge \mathcal{B}_2$ the controller restricts the control component of the plant trajectory and, hopefully, improves its properties. Consider the image behavior $\mathcal{B}_{1,c} := (0, \mathrm{id}_p)\mathcal{B}_1 \subseteq \mathcal{F}^p$ that is all of $\mathcal{F}^p$ if and only if the component c of w is free. The main difference to the interconnection of IO behaviors is that the component c of w is not assumed free. If c is not free there is a nonzero $P_c \in \mathcal{D}$ such that $P_c \circ c = 0$ for all $\binom{u}{c} \in \mathcal{B}_1$ and $\binom{R_c}{P_c \mathrm{id}_p} \circ c = 0$ for all $\binom{u}{c} \in \mathcal{B}_1 \wedge \mathcal{B}_2$. It can easily occur that this implies $c = 0$, and so the controller does not improve the control component, but annihilates it. We conclude that this more general concept of behavior interconnection has to be handled with care.
2. More general interconnections than those of IO behaviors in this book are the two-parameter compensators in [3, Sect. 5.6] and are also treated in [4] with references to previous papers in this area.

◊

References

1. J.W. Polderman, J.C. Willems, *Introduction to Mathematical Systems Theory. A Behavioral Approach*. Texts in Applied Mathematics, vol. 26 (Springer, New York, 1998)
2. T. Kailath, *Linear Systems* (Prentice-Hall, Englewood Cliffs, 1980)
3. M. Vidyasagar, *Control System Synthesis. A Factorization Approach*. MIT Press Series in Signal Processing, Optimization, and Control, vol. 7 (MIT Press, Cambridge, 1985)
4. I. Blumthaler, Stabilisation and control design by partial output feedback and by partial interconnection. Int. J. Control **85**(11), 1717–1736 (2012)

Chapter 8
The Transfer Matrix as Operator or Input/Output Map

In this chapter we show that in the standard discrete-time and continuous-time cases the transfer matrix H of an IO behavior acts as *transfer operator* or *input/output map* on interesting classes of signals and that this action is given by convolution in the most important cases. This operator generalizes the gain matrix from Sect. 6.3.6. We characterize external stability and, in particular, bounded input/bounded output (BIBO) stability of transfer operators. See the Introduction of Sect. 8.2 for more details in the continuous-time case.

8.1 Transfer Operators in the Discrete-Time Case

In Sect. 8.1 we show how a Laurent series $H \in F\langle\langle z\rangle\rangle$ in $z = s^{-1}$ and thus, in particular, a transfer function $H \in F(s)$ act on signals $u \in F^{\mathbb{N}}$. In this fashion a transfer matrix induces a transfer *operator*, which maps an input of an IO behavior to a corresponding output. In Sect. 8.1.2 we derive a closed-form representation of rational functions $H \in F(s) \subseteq F\langle\langle z\rangle\rangle$ as Laurent series. This enables to investigate (ℓ^p, ℓ^p)-stability or external stability of transfer matrices, $1 \le p \le \infty$, as well as BIBO stability of IO behaviors in Sect. 8.1.3.
In general, we assume an arbitrary field F, the polynomial algebra $\mathcal{D} := F[s]$, and the $F[s]$-injective cogenerator $\mathcal{F} := F^{\mathbb{N}}$ with the left shift action. For stability questions $F = \mathbb{C}, \mathbb{R}$ are required.

8.1.1 The Action of Laurent Series

We introduce an action of the algebra of Laurent series $F\langle\langle z\rangle\rangle$ on the discrete-time signals $\mathcal{F} = F^{\mathbb{N}}$, which is a natural extension of the left shift operation of $F[s]$ on $F^{\mathbb{N}}$, where $z = s^{-1}$. This is essential for interpreting transfer matrices as transfer

The original version of this chapter was revised: The errors in this chapter have been corrected. The correction to this chapter can be found at https://doi.org/10.1007/978-3-030-43936-1_14

U. Oberst et al., *Linear Time-Invariant Systems, Behaviors and Modules*, Differential-Algebraic Equations Forum,
https://doi.org/10.1007/978-3-030-43936-1_8

operators. We show that a rational function in $F(s)$ defines a causal operator if and only if it is a power series in $z = s^{-1}$ or, equivalently, proper. The field

$$K := F(s) = \left\{\tfrac{f}{g};\ f, g \in F[s],\ g \neq 0\right\} \tag{8.1}$$

of rational functions is the quotient field of $F[s]$. The element $z := s^{-1}$ is also transcendental or an indeterminate over F, and thus $F[z]$ is another polynomial subalgebra of $F(s)$ with $F(z) = F(s)$. The indeterminate z gives also rise to the *power series* algebra with its convolution multiplication, cf. (8.3),

$$F[\![z]\!] \underset{\text{ident.}}{=} F^{\mathbb{N}} \ni u = \sum_{t=0}^{\infty} u(t)z^t = (u(0), u(1), u(2), \ldots). \tag{8.2}$$

In certain situations it is preferable to write u_t instead of $u(t)$.

Remark 8.1.1 (*z-transform*) In *pure mathematics* a formal power series u is *defined* as the sequence $u = (u(t))_{t\in\mathbb{N}}$, i.e., as the function $u\colon\ \mathbb{N} \longrightarrow F, t \longmapsto u(t)$, whereas in *engineering mathematics* one usually distinguishes the sequence $u = (u(t))_{t\in\mathbb{N}}$ from its so-called z-transform $\sum_{t=0}^{\infty} u(t)z^t$. If $F = \mathbb{C}$ and if this power series is convergent for some nonzero $z_0 \in \mathbb{C}$, then u also denotes the holomorphic function $z \longmapsto \sum_{t=0}^{\infty} u(t)z^t$ in the interior of its disc of convergence. In this situation one identifies the indeterminate z with the complex variable z and one usually writes $u_t := u(t)$ and $u(z) := \sum_{t=0}^{\infty} u_t z^t$. Observe that $t \in \mathbb{N}$ is a *discrete* variable, whereas $z \in \mathbb{C}$ is a continuous one.
We will not use the z-transform terminology and will always identify the sequence and the formal power series. ◊

The *convolution multiplication* on $F[\![z]\!]$ is given by

$$\begin{aligned} *\colon\ & F[\![z]\!] \times F[\![z]\!] \longrightarrow F[\![z]\!],\\ & (u, v) \longmapsto uv = u * v \text{ with } (u * v)(t) := \sum_{i+j=t} u(i)\, v(j),\ t \in \mathbb{N}, \end{aligned} \tag{8.3}$$

and makes $F[\![z]\!]$ a principal ideal domain containing $F[z]$. The *order* $\mathrm{ord}_z(u)$ of a power series $u = \sum_{t=0}^{\infty} u(t)z^t$ is defined as

$$\mathrm{ord}_z(u) := \begin{cases} \min\{t \in \mathbb{N};\ u(t) \neq 0\} & \text{if } u \neq 0,\\ \infty & \text{if } u = 0. \end{cases} \tag{8.4}$$

The group of units or invertible elements of $F[\![z]\!]$ is

$$\mathrm{U}(F[\![z]\!]) = \{u \in F[\![z]\!];\ u(0) \neq 0,\ \text{i.e., } \mathrm{ord}_z(u) = 0\}. \tag{8.5}$$

The element z is the unique prime element of $F[\![z]\!]$, up to association. Indeed every nonzero $u \in F[\![z]\!]$ has the unique prime factor decomposition

$$u = \underbrace{\left(u(m) + u(m+1)z + u(m+2)z^2 + \cdots\right)}_{\in \mathrm{U}(F[\![z]\!])} z^m, \quad \text{where } m = \mathrm{ord}_z(u). \tag{8.6}$$

Every nonzero element v of the quotient field

$$F(\!(z)\!) := \mathrm{quot}(F[\![z]\!]) = \left\{\tfrac{u_1}{u_2};\ u_1, u_2 \in F[\![z]\!],\ u_2 \neq 0\right\} \tag{8.7}$$

has the unique representation

$$v = z^m u \text{ with } m \in \mathbb{Z} \text{ and } u \in \mathrm{U}(F[\![z]\!]). \tag{8.8}$$

We extend the order function by $\mathrm{ord}_z(v) := m \in \mathbb{Z}$. We identify v with the sequence $(v(t))_{t\in\mathbb{Z}} = (\ldots, 0, \overset{m}{u(0)}, u(1), u(2), \ldots) \in F^{\mathbb{Z}}$ of left bounded support $\mathrm{supp}(v) := \{t \in \mathbb{Z};\ v(t) \neq 0\}$ and obtain

$$\begin{aligned} F(\!(z)\!) &= \Big\{v = \sum_{t\in\mathbb{Z}} v(t)z^t := (v(t))_{t\in\mathbb{Z}} \in F^{\mathbb{Z}};\ \exists m \in \mathbb{Z}\ \forall t < m:\ v(t) = 0\Big\} \\ &= \left\{z^m a;\ m \in \mathbb{Z},\ a \in F[\![z]\!]\right\} \underset{s=z^{-1}}{=} \{s^m a;\ m \in \mathbb{N},\ a \in F[\![z]\!]\}. \end{aligned} \tag{8.9}$$

The multiplication of the sequences $(v(t))_{t\in\mathbb{Z}}$ is again the convolution (8.3). The sequences $v = \sum_{t\in\mathbb{Z}} v(t)z^t = \sum_{t\geq m} v(t)z^t$ are called *formal Laurent series*. The algebras just introduced satisfy the inclusion relations

$$\begin{array}{ccc} F[s] \subset F(s) = & F(z) & \subset F(\!(z)\!) \\ & \cup & \cup \\ & F[z] & \subset F[\![z]\!]. \end{array}$$

It is important to distinguish the following two multiplications on $F^{\mathbb{N}} = F[\![z]\!]$:

- The *convolution multiplication*

$$F[\![z]\!] \times F[\![z]\!] \longrightarrow F[\![z]\!],\ (u, v) \longmapsto u * v = uv,$$

makes $F[\![z]\!]$ a commutative domain and gives rise to the *right shift*

$$z * (y(0), y(1), \ldots) = z * \sum_{t=0}^{\infty} y(t)z^t = \sum_{t=0}^{\infty} y(t)z^{t+1} = (0, y(0), y(1), \ldots). \tag{8.10}$$

In this fashion $F^{\mathbb{N}}$ is a torsionfree $F[\![z]\!]$- and $F[z]$-module.

- The *left shift*

$$F[s] \times F^{\mathbb{N}} \longrightarrow F^{\mathbb{N}},\ (f, y) \longmapsto f \circ y,\ \text{with } (s^m \circ y)(t) = y(t+m),$$

makes $F^{\mathbb{N}}$ a module over $F[s]$. This scalar multiplication was used to define the $F^{\mathbb{N}}$-behaviors. With respect to this product $F^{\mathbb{N}}$ is not torsionfree. Indeed, the torsion submodule $\mathrm{t}({}_{F[s]}F^{\mathbb{N}})$ of linearly recursive sequences played an important part in the preceding chapters. We will derive its primary decomposition in Theorem 8.1.15.

Since

$$z * (s \circ y) = (0, y(1), y(2), \cdots) \neq y = s \circ (z * y),$$

the two actions do not commute. The relation $s \circ (z * y) = y$ implies

$$s^m \circ (z^n * y) = \begin{cases} s^{m-n} \circ y & \text{if } m \geq n \\ z^{n-m} * y & \text{if } n \geq m \end{cases} \quad \text{for } y \in F^{\mathbb{N}}. \tag{8.11}$$

In contrast, for the injective cogenerator $F^{\mathbb{Z}}$ over the commutative domain $F[s, z] = \oplus_{t \in \mathbb{Z}} F s^t \subset F(\!(z)\!)$ of *Laurent polynomials*, the left shift action is the inverse of the right shift action, but the convolution multiplication is not defined on all of $F^{\mathbb{Z}}$.

Theorem and Definition 8.1.2 *We define the new action*

$$F(\!(z)\!) \times F^{\mathbb{N}} \longrightarrow F^{\mathbb{N}},\quad (H, u) \longmapsto H \circ_{\text{new}} u := s^m \circ (a * u),$$
$$\text{where } H = z^{-m} a = s^m a \text{ with } m \geq 0 \text{ and } a \in F^{\mathbb{N}} = F[\![z]\!]. \tag{8.12}$$

1. *This action is well defined, i.e., independent of the choice of m and a. Obviously, it is F-bilinear. It induces the injective F-linear map*

$$F(\!(z)\!) \longrightarrow \mathrm{Hom}_F(F^{\mathbb{N}}, F^{\mathbb{N}}),\quad H \longmapsto H \circ_{\text{new}} (-). \tag{8.13}$$

   In this fashion, every Laurent series can be considered as a linear map. If $H \in F(\!(z)\!)$ is interpreted as the transfer function of a discrete-time SISO (single input/single output) system, the corresponding map $u \longmapsto H \circ_{\text{new}} u$ is called a transfer operator *or* input/output map.
2. *If $H \in F[\![z]\!]$, then $H \circ_{\text{new}} u = H * u$. If $H \in F[s]$, then $H \circ_{\text{new}} u = H \circ u$. Therefore, we may and do again write $H \circ u := H \circ_{\text{new}} u$ for every $H \in F(\!(z)\!)$.*
3. *The following partial associativity relation holds:*

$$P \circ (H \circ u) = (PH) \circ u \text{ for } P \in F[s],\ H \in F(\!(z)\!),\ \text{and } u \in F^{\mathbb{N}}.$$

   If, in particular, $H = \frac{Q(s)}{P(s)}$ is a rational function, then

$$P \circ (H \circ u) = (PH) \circ u = Q \circ u,\ u \in F^{\mathbb{N}},$$

shows that $y := H \circ u$ *is a solution of* $P \circ y = Q \circ u$.
4. *The* F*-bilinear action* (8.12) *is neither* $F[s]$*-linear nor* $F[z]$*-linear. In particular,* $F^{\mathbb{N}}$ *is not a vector space over the field* $F((z))$.

Proof 1. Assume $H = s^m a = s^n b$ with $m, n \geq 0$ and $a, b \in F[[z]]$, and let $u \in F^{\mathbb{N}}$. The following implications prove the asserted independence:

$$z^{-m}a = s^m a = s^n b = z^{-n}b \Longrightarrow z^n * a = z^n a = z^m b = z^m * b$$
$$\Longrightarrow z^n * (a * u) = z^m * (b * u) \in F^{\mathbb{N}} \Longrightarrow s^m \circ (a * u) \underset{(8.11)}{=} s^{m+n} \circ \left(z^n * (a * u)\right)$$
$$= s^{m+n} \circ \left(z^m * (b * u)\right) = s^m \circ (a * u) = s^n \circ (b * u).$$

To show that (8.13) is injective, assume that for all $u \in F^{\mathbb{N}} = F[[z]]$

$$H \circ_{\text{new}} u = s^m \circ (a * u) = 0 \Longrightarrow a \underset{(8.11)}{=} s^m \circ (z^m * a) = 0 \Longrightarrow H = 0.$$

2. This follows directly from the definition of $\circ_{\text{new}}$ and item 1.
3. It suffices to prove this for $P = s$. For arbitrary $P \in F[s]$, the assertion then follows by induction and linear extension. Assume $H = s^m a$. Then

$$s \circ (H \circ_{\text{new}} u) = s \circ \left(s^m \circ (a * u)\right) = s^{m+1} \circ (a * u)$$
$$= (s^{m+1} a) \circ_{\text{new}} u = (sH) \circ_{\text{new}} u.$$

4. Since the actions of $s \in F((z))$ and of $z \in F((z))$ do not commute, the map $s \circ_{\text{new}} (-) = s \circ (-)$ is not $F[z]$-linear and the map $z \circ_{\text{new}} (-) = z * (-)$ is not $F[s]$-linear. □

Corollary and Definition 8.1.3 *The unit element*

$$1 = 1_{F[[z]]} = z^0 = \delta_0 = (1, 0, 0, \dots) \in F[[z]] = F^{\mathbb{N}}$$

is interpreted as the unit impulse at time zero. *If* $H \in F((z))$ *is interpreted as a transfer function of a SISO system, the output* $h := H \circ \delta_0$ *is called the* impulse response *of* H.
If H *is a power series, i.e., a Laurent series without negative powers of* z*, then* $h = H * 1_{F[[z]]} = H$*, i.e., the transfer function and its impulse response coincide. In the continuous-time case, that we discuss in Sect. 8.2, the transfer function and its impulse response have to be carefully distinguished.*
If $H = \sum_i H_i s^i \in F[s]$ *is a polynomial in* s*, then* $H \circ z^0 = H_0 z^0$*, and hence the transfer function* H *cannot be reconstructed from the impulse response* $H \circ z^0$. ◇

We consider the ideals and direct decompositions

$$
\begin{aligned}
F[s]_+ &:= F[s]s < F[s], & F[s] &= F[s]_+ \oplus F,\\
F[z]_+ &= F[z]z < F[z], & F[z] &= F \oplus F[z]_+,\\
F[\![z]\!]_+ &:= F[\![z]\!]z = F[\![z]\!] \setminus \mathrm{U}(F[\![z]\!]) < F[\![z]\!], & F[\![z]\!] &= F \oplus F[\![z]\!]_+.
\end{aligned}
\tag{8.14}
$$

If ${}_A M$ is an A-module over a commutative ring A, we write $U \leq_A M$ resp. $U <_A M$ instead of $U \subseteq M$ to emphasize that U is an A-submodule of M.

With these we obtain, for $H = \sum_{t\in\mathbb{Z}} H_t z^t \in F(\!(z)\!)$, the decompositions

$$
\begin{aligned}
H \in F(\!(z)\!) &= F[s] \oplus F[\![z]\!]_+ = F[s]_+ \oplus F[\![z]\!]\\
H \quad &= H_{\mathrm{pol}} + H_{\mathrm{ps},+} = H_{\mathrm{pol},+} + H_{\mathrm{ps}}\\
\text{where}\quad H_{\mathrm{pol}} &:= \sum_{t\leq 0} H_t z^t = \sum_{t\geq 0} H_{-t} s^t \in F[s], \quad H_{\mathrm{pol},+} := H_{\mathrm{pol}} - H_0\\
H_{\mathrm{ps}} &:= \sum_{t\geq 0} H_t z^t \in F[\![z]\!], \qquad H_{\mathrm{ps},+} := H_{\mathrm{ps}} - H_0,
\end{aligned}
\tag{8.15}
$$

where H_{pol} resp. H_{ps} are the polynomial (in s) part resp. the power series (in z) part of H. With $s \circ 1 = 0$ (where $1 = z^0 = \delta_0$), we conclude that $H_{\mathrm{pol},+} \circ 1 = 0$. Therefore,

$$
h := H \circ 1 = H_{\mathrm{ps}} \circ 1 = H_{\mathrm{ps}} * 1 = H_{\mathrm{ps}}
\tag{8.16}
$$

is the impulse response of H. For $P \in F[s]$ and $H \in F^{\mathbb{N}} = F[\![z]\!]$, we get

$$
P \circ H = P \circ (H * 1) \underset{\text{Theorem 8.1.2, 3.}}{=} (PH) \circ 1 = (PH)_{\mathrm{ps}}.
\tag{8.17}
$$

Next we extend the action $\circ_{\mathrm{new}}$ and the results of Theorem and Definition 8.1.2 to matrices and, by this, to systems with multiple inputs and multiple outputs.

Corollary and Definition 8.1.4 *1. With $\mathcal{F} = F^{\mathbb{N}}$ the bilinear form* (8.12) *is extended to the bilinear form*

$$
F(\!(z)\!)^{p\times m} \times \mathcal{F}^m \longrightarrow \mathcal{F}^p,
$$

$$
(H, u) = \left(H, \begin{pmatrix} u_1 \\ \vdots \\ u_m \end{pmatrix}\right) \longmapsto y = \begin{pmatrix} y_1 \\ \vdots \\ y_p \end{pmatrix} = H \circ u, \text{ where } y_i := \sum_{j=1}^{m} H_{ij} \circ u_j.
$$

The induced F-linear map

$$
F(\!(z)\!)^{p\times m} \longrightarrow \mathrm{Hom}_F(\mathcal{F}^m, \mathcal{F}^p), \; H \longmapsto H\circ = (u \mapsto H \circ u),
$$

is again injective.

2. *Like in the SISO case the action satisfies the partial associativity law*

$$P \circ (H \circ u) = (PH) \circ u \; for \; P \in F[s]^{k \times p}, \; H \in F(\!(z)\!)^{p \times m}, \; u \in \mathcal{F}^m.$$

In particular, if $Q := PH \in F[s]^{k \times m}$ *is a polynomial matrix, then* $y := H \circ u \in \mathcal{F}^p$ *is a solution of* $P \circ y = Q \circ u$.

3. *Let* $\mathcal{B} = \left\{ \binom{y}{u} \in \mathcal{F}^{p+m}; \; P \circ y = Q \circ u \right\}$ *be an IO behavior with* $(P, -Q) \in F[s]^{p \times (p+m)}$ *and transfer matrix* $H = P^{-1}Q \in F(s)^{p \times m} \subseteq F(\!(z)\!)^{p \times m}$. *Then*

$$\mathcal{B} = \mathcal{B}^0 \oplus \left\{ \binom{H \circ u}{u}; \; u \in \mathcal{F}^m \right\}, \; where \; \mathcal{B}^0 \underset{ident.}{=} \left\{ \binom{y}{0} \in \mathcal{F}^{p+m}; \; P \circ y = 0 \right\}. \tag{8.18}$$

The map $u \longmapsto H \circ u$ *is called the* transfer operator *or* input/output map *of the transfer matrix* H *or of the IO behavior* $\mathcal{B}$.

Proof The first two items are obvious consequences of Theorem and Definition 8.1.2. The direct sum decomposition in item 3 follows from

$$P \circ (y - H \circ u) = Q \circ u - Q \circ u = 0 \quad \text{and} \quad \binom{y}{u} = \binom{y - H \circ u}{0} + \binom{H \circ u}{u}. \qquad \square$$

A transfer operator $H\circ\colon \; u \longmapsto y$ is called *causal* if for all time instances $t \in \mathbb{N}$ the value of the output $y(t) = (H \circ u)(t)$ at time t is determined by the values of the input in the past, i.e., by $u(j), \; j \le t$. In other terms, the operator $H\circ$ is causal if and only if

$$\forall u \in F^{\mathbb{N}} \; \forall t \in \mathbb{N} : \Big(u(0) = \cdots = u(t) = 0 \Longrightarrow (H \circ u)(t) = 0 \Big). \tag{8.19}$$

Theorem 8.1.5 *1. A transfer function* $H \in F(\!(z)\!)$ *defines a causal transfer operator* $H\circ$ *if and only if it is a power series in* $F[\![z]\!]$.

2. *A transfer matrix* $H \in F(\!(z)\!)^{p \times m}$ *defines a causal transfer operator* $H\circ\colon F[\![z]\!]^m \longrightarrow F[\![z]\!]^p$ *if and only if all its entries are power series.*

Proof Item 2 is an obvious consequence of item 1.
For the first item, assume a power series $H = \sum_{t=0}^{\infty} H_t z^t$. Then the output

$$(H \circ u)(t) = (H * u)(t) = \sum_{j=0}^{t} H_{t-j} u(j), \; t \in \mathbb{N},$$

is obviously determined by the values $u(j), \; j \le t$.
Assume, conversely, that $H \in F(\!(z)\!)$ is not a power series. Let $-m < 0$ be its order. Then H has the form

$$H = z^{-m} a = s^m a \quad \text{with } a \in \mathrm{U}(F[\![z]\!]), \text{ i.e., } a(0) \ne 0 \Longrightarrow z^m = (0, \cdots, 0, \overset{m}{1}, \cdots)$$
$$\text{and } H \circ z^m = s^m \circ (a * z^m) = a, \; z^m(0) = 0, \; (H \circ z^m)(0) = a(0) \ne 0.$$

Hence $H\circ$ is not causal. $\square$

Remark 8.1.6 The field $F(\!(z)\!)$ acts on itself by convolution and can be considered as an $F(\!(z)\!)$-module, and in particular, as an $F[s]$-signal module. But notice that for $y = \sum_{t\in\mathbb{N}} y(t)z^t \in F[\![z]\!] = F^{\mathbb{N}}$ and $y(0) \neq 0$

$$s \circ y = \sum_{t\geq 0} y(t+1)z^t \neq s * y = \sum_{t\geq -1} y(t+1)z^t$$

holds. This says that $F[\![z]\!]$ is not an $F[s]$-submodule of $F(\!(z)\!)$. Therefore, we will use $F(\!(z)\!)$ as field of operators only, but not as a signal space. ◊

We next show that rational functions $H \in F(s) \subset F(\!(z)\!)$ are proper if and only if they are power series in $F[\![z]\!]$ and thus, by Theorem 8.1.5, define a causal operator $H\circ\colon\ F^{\mathbb{N}} \longrightarrow F^{\mathbb{N}}$. We write $\deg(P) = \deg_s(P)$ for a polynomial $P \in F[s]$ and $\deg_z(a(z))$ for the degree of a polynomial $a \in F[z]$. The set

$$F(s)_{\mathrm{pr}} := \left\{H = \tfrac{Q(s)}{P(s)} \in F(s);\ \deg_s(H) = \deg_s(Q) - \deg_s(P) \leq 0\right\}$$

is the subdomain of $F(s)$ of proper rational functions, see Sect. 5.5.2.

Theorem 8.1.7 *The ring $F(s)_{\mathrm{pr}}$ of proper rational functions is given by*

$$F(s)_{\mathrm{pr}} = \underbrace{\left\{\tfrac{a(z)}{b(z)};\ a, b \in F[z],\ b(0) \neq 0\right\}}_{=:A_1} = \underbrace{F(s) \cap F[\![z]\!]}_{=:A_2}.$$

Proof $F(s)_{\mathrm{pr}} \subseteq A_1$: Let $H = \frac{Q}{P} \in F(s)_{\mathrm{pr}}$ with $\deg_s(Q) \leq \deg_s(P) =: m$. We write

$$P = P_m s^m + \cdots + P_0,\ P_m \neq 0, \text{ and } Q = Q_m s^m + \cdots + Q_0$$

$$\Longrightarrow H = \tfrac{Q}{P} = \tfrac{z^m Q}{z^m P} = \frac{\overbrace{Q_0 z^m + \cdots + Q_m}^{=:a}}{\underbrace{P_0 z^m + \cdots + P_m}_{=:b}} = \tfrac{a}{b},\ b(0) = P_m \neq 0 \Longrightarrow H \in A_1.$$

$F(s)_{\mathrm{pr}} \supseteq A_1$: Let $H = \frac{a}{b}$ with $a, b \in F[z]$ and $b_0 = b(0) \neq 0$. Choose $m \in \mathbb{N}$ with $\deg_z(a) \leq m$ and $\deg_z(b) \leq m$. Then

$$H = \tfrac{a}{b} = \tfrac{a_m z^m + \cdots + a_0}{b_m z^m + \cdots + b_0} = \tfrac{s^m a}{s^m b} = \tfrac{a_0 s^m + \cdots + a_m}{b_0 s^m + \cdots + b_m}$$
$$\Longrightarrow \deg_s(s^m a) \leq m = \deg_s(s^m b) \Longrightarrow H = \tfrac{s^m a}{s^m b} \in F(s)_{\mathrm{pr}}.$$

$A_1 \subseteq A_2$: Let $H = \frac{a}{b} \in A_1$. Clearly, $A_1 \subseteq F(z) = F(s)$. Moreover the condition $b(0) \neq 0$ implies $b \in \mathrm{U}(F[\![z]\!])$ and $ab^{-1} \in F[\![z]\!]$.
$A_1 \supseteq A_2$: Let $H = \frac{a}{b} \in F(z) \cap F[\![z]\!]$ with $a, b \in F[z]$ and $\gcd(a, b) = 1$. If $b(0) \neq 0$, then $H \in A_1$ by definition.
If, on the contrary, $b(0) = 0$, then $z \mid b$ thus $b = z^m u$ with $m > 0$ and $u \in \mathrm{U}(F[\![z]\!])$.

From $\gcd(a, b) = 1$, we infer that $z \nmid a$, hence $a(0) \neq 0$ and thus $a \in \mathrm{U}(F[\![z]\!])$. This implies the prime factor decomposition $H = (au^{-1})z^{-m}$ in $F(\!(z)\!)$ with $m > 0$ and $au^{-1} \in \mathrm{U}(F[\![z]\!])$. But $m > 0$ contradicts $H \in A_2 \subset F[\![z]\!]$, hence $b(0) = 0$ cannot occur. □

Corollary 8.1.8 *A rational function $H \in F(s)$ gives rise to the causal transfer operator $H\circ\colon\ F^{\mathbb{N}} \longrightarrow F^{\mathbb{N}}$ if and only if it is proper. The same equivalence holds for rational matrices.*

Proof As can be seen from the characterization $F(s)_{\mathrm{pr}} = F(s) \cap F[\![z]\!]$ from Theorem 8.1.7, H is proper if and only if it is a power series in z. Theorem 8.1.5 states that this is the case if and only if $H\circ$ is proper. □

Corollary 8.1.9 *The ideal*

$$F(s)_{\mathrm{spr}} := \left\{H = \tfrac{Q}{P} \in F(s);\ \deg_s(H) = \deg_s(Q) - \deg_s(P) < 0\right\} < F(s)_{\mathrm{pr}}$$

of strictly proper rational functions from Sect. 5.5.2, Eq. (5.85) has the representation

$$F(s)_{\mathrm{spr}} = \left\{\tfrac{a}{b} \in F(z);\ a, b \in F[z],\ a(0) = 0,\ b(0) \neq 0\right\} = F(s) \cap F[\![z]\!]_{+}.$$

Proof The proof is analogous to that of Theorem 8.1.7. □

Corollary 8.1.10 *The direct decomposition*

$$F(\!(z)\!) = F[s] \oplus F[\![z]\!]_{+} = F[s]_{+} \oplus F[\![z]\!] \ni H = H_{\mathrm{pol}} + H_{\mathrm{ps},+} = H_{\mathrm{pol},+} + H_{\mathrm{ps}}$$

from (8.15) *induces the direct decomposition*

$$\begin{aligned} F(s) \underset{F[s]\subset F(s)}{=} & F[s] \oplus \left(F(s) \cap F[\![z]\!]_{+}\right) = F[s]_{+} \oplus \left(F(s) \cap F[\![z]\!]\right) \\ = & F[s] \oplus F(s)_{\mathrm{spr}} = F[s]_{+} \oplus F(s)_{\mathrm{pr}} \end{aligned}$$

that coincides with that from Sect. 5.5.2. Hence, $H_{\mathrm{pr}} = H_{\mathrm{ps}}$ and $H_{\mathrm{spr}} = H_{\mathrm{ps},+}$ are the proper resp. the strictly proper part of $H \in F(s)$, and can be computed according to Lemma 5.5.5.
Moreover (8.17) *implies, for $P \in F[s]$ and $H \in F(s)_{\mathrm{pr}} \subset F^{\mathbb{N}} = F[\![z]\!]$, that*

$$P \circ H = P \circ (H * 1) = PH \circ 1 = (PH)_{\mathrm{pr}}. \tag{8.20}$$

◇

Consider $\mathrm{t}\left({}_{F[s]}\mathcal{F}\right) \subset \mathcal{F} = F^{\mathbb{N}} = F[\![z]\!]$. The sequences in $\mathrm{t}\left({}_{F[s]}\mathcal{F}\right)$ are called *linearly recursive*.

Theorem 8.1.11 $F(s)_{\mathrm{pr}} = \mathrm{t}\left({}_{F[s]}F[\![z]\!]\right)$.

Proof $\subseteq$: Let $\frac{a}{f} \in F(s)_{\mathrm{pr}} \subset F[\![z]\!]$ with $a, f \in F[s]$, $f \neq 0$. From (8.20) we infer that $f \circ \frac{a}{f} = \left(f\frac{a}{f}\right) \circ 1 = a \circ 1 = a(0)$. Therefore, $(sf) \circ \frac{a}{f} = s \circ \left(f \circ \frac{a}{f}\right) = s \circ a(0) = 0$, and this means that $\frac{a}{f} \in \mathrm{ann}_{\mathcal{F}}(sf) \subseteq \mathrm{t}(\mathcal{F})$.
$\supseteq$: Let $y \in \mathrm{t}(F[\![z]\!])$ and $f \in F[s]\backslash\{0\}$ with $f \circ y = 0$. From (8.17) we infer that $(fy)_{\mathrm{ps}} = f \circ y = 0$, hence, by (8.15), $fy = (fy)_{\mathrm{pol},+} \in F[s]$. We conclude that $y = \frac{fy}{f} \in F(s) \cap F[\![z]\!] = F(s)_{\mathrm{pr}}$. □

8.1.2 The Primary Decomposition of $\mathrm{t}({}_{F[s]}F[\![z]\!]) = F(s)_{\mathrm{pr}}$

We first derive the primary decomposition of $F(s)_{\mathrm{pr}} = \mathrm{t}({}_{F[s]}F^{\mathbb{N}})$ and thereby generalize, to arbitrary fields F, our results from Sect. 5.3.3, where we dealt with $F = \mathbb{C}$ and $F = \mathbb{R}$ only. In Theorem 8.1.17 we compute, for $F = \mathbb{C}$, the power series representation of a proper rational function from its partial fraction decomposition. This theorem will be crucial for the characterization of externally stable transfer matrices and of BIBO stable IO behaviors in Sect. 8.1.3.

The primary decomposition of the torsion submodule $\mathrm{t}({}_{F[s]}\mathcal{F})$ of $\mathcal{F} = F^{\mathbb{N}} = F[\![z]\!]$ generalizes the decomposition $\mathrm{t}({}_{\mathbb{C}[s]}\mathbb{C}^{\mathbb{N}}) = \mathbb{C}^{(\mathbb{N})} \oplus \bigoplus_{0\neq\lambda\in\mathbb{C}} \mathbb{C}[t]\lambda^t$ from Corollary 5.3.19 to arbitrary fields F. As we have shown in Theorem 5.3.2, the primary decomposition of $\mathrm{t}(\mathcal{F})$ is

$$\mathrm{t}({}_{F[s]}\mathcal{F}) = \bigoplus_{q\in\mathcal{P}} \mathcal{F}_q, \ \mathcal{F}_q = \bigcup_{m=1}^{\infty} \mathrm{ann}_{\mathcal{F}}(q^m) = \{y \in \mathcal{F};\ \exists m \in \mathbb{N}:\ q^m \circ y = 0\},$$

where $\mathcal{P}$ is a representative system of the irreducible polynomials in $F[s]$. Obviously $\mathcal{F}_s = F^{(\mathbb{N})}$ is the space of sequences y with finite support $\mathrm{supp}(y) = \{t \in \mathbb{N};\ y(t) \neq 0\}$. We need some preparations for the description of the $\mathcal{F}_q$ with $q(0) \neq 0$. For a polynomial $0 \neq f = f_m s^m + f_{m-1}s^{m-1} + \cdots + f_0 \in F[s]$ of degree m we define the so-called *inverse polynomial*

$$\widehat{f}(s) := f_0 s^m + f_1 s^{m-1} + \cdots + f_m = s^m f(z) \in F[s], \text{ hence } \widehat{f}(z) = z^m f(s), \tag{8.21}$$

of degree $\deg_s(\widehat{f}) \leq m = \deg_s(f)$. Obviously, the equivalences

$$\deg_s(\widehat{f}) = \deg_s(f) \Longleftrightarrow f_0 = f(0) \neq 0 \Longleftrightarrow s \nmid f \tag{8.22}$$

hold and imply $\widehat{\widehat{f}} = f$. Hence the map $f \longmapsto \widehat{f}$ is an involution of the set

$$T := \{f \in F[s];\ f(0) \neq 0\},$$

i.e., the map is bijective and coincides with its inverse. The equation

$$\widehat{f_1 f_2} = s^{\deg_s(f_1 f_2)}(f_1 f_2)(z) = s^{\deg_s(f_1)+\deg_s(f_2)}(f_1 f_2)(z)$$
$$= s^{\deg_s(f_1)} f_1(z) s^{\deg_s(f_2)} f_2(z) = \widehat{f_1}\widehat{f_2}$$

shows that the map is multiplicative.
The multiplicativity also implies that $f \in T$ is irreducible if and only if $\widehat{f}$ is irreducible. In contrast, $s \notin T$ is irreducible, but $1 = \widehat{s}$ is not. Let $\mathcal{P}$ denote the set of all monic irreducible polynomials. Then $\mathcal{P}\backslash\{s\} \subset T$, and therefore,

$$\{s\} \uplus \{\widehat{q};\ q \in \mathcal{P},\ q \neq s\} \tag{8.23}$$

is also a system of representatives of irreducible polynomials up to association. Observe that $\widehat{q}$ is not monic in general.
In Theorem 8.1.13 we describe $\operatorname{ann}_{\mathcal{F}}(f) = \{y \in F^{\mathbb{N}};\ f(s) \circ y = 0\}$ by means of $\widehat{f}$ and in Theorem 8.1.15 the primary decomposition of $\mathrm{t}(\mathcal{T})$.

Lemma 8.1.12 *Let $f = s^m + \cdots + f_0 \in F[s]$ be a monic polynomial with $f_0 = f(0) \neq 0$. For every signal $y \in F^{\mathbb{N}} = F[\![z]\!]$ we have*

$$\widehat{f}(z) * y - z^m * (f(s) \circ y) \in F[z]_{\leq m-1}, \tag{8.24}$$

where $F[z]_{\leq m-1}$ denotes the space of polynomials of degree at most $m-1$.

Proof The polynomials $f = s^m + f_{m-1}s^{m-1} + \cdots + f_0,\ f_0 \neq 0$, and $\widehat{f}$ satisfy

$$\widehat{f}(s) = s^m f(z) = \sum_{j=0}^{m} f_{m-j}s^j,\ \ \widehat{f}(z) = z^m f(s),\ \ m = \deg_s(f) = \deg_z(\widehat{f}(z)).$$

We have to show

$(\widehat{f}(z) * y)(t) - (z^m * (f(s) \circ y))(t) = 0$ for $t \geq m$. But indeed

$$(\widehat{f}(z) * y)(t) = \sum_{j=0}^{m} f_{m-j}y(t-j) \quad \text{and}$$
$$\big(z^m * \big(f(s) \circ y\big)\big)(t) = \big(f(s) \circ y\big)(t-m) = \big(f \circ y\big)(t-m)$$
$$= \sum_{i=0}^{m} f_i y(t-m+i) \underset{j:=m-i}{=} \sum_{j=0}^{m} f_{m-j}y(t-j).$$

□

Lemma 8.1.12 implies the following new representation of $\operatorname{ann}_{\mathcal{F}}(f)$.

Theorem 8.1.13 *As in Lemma 8.1.12 let $f = s^m + \cdots + f_0 \in F[s]$ be a monic polynomial of degree m with nonzero constant term $f(0) = f_0 \neq 0$. The following assertions hold:*

1. $\mathrm{ann}_{\mathcal{F}}(f(s)) = F[z]_{\leq m-1} * \widehat{f}(z)^{-1}$.
2. *If $a(s) \in F[s]_{\leq m}$ and thus $\frac{a}{f} \in F(s)_{\mathrm{pr}} = F(s) \cap F[\![z]\!]$, then*

$$\frac{a(s)}{f(s)} = \frac{a(0)}{f(0)} + \left(\frac{a(s)}{f(s)} - \frac{a(0)}{f(0)}\right) \in \mathrm{ann}_{\mathcal{F}}(s) \oplus \mathrm{ann}_{\mathcal{F}}(f(s)). \tag{8.25}$$

Proof 1. Equation 8.5 implies $\widehat{f}(z) = f_0 z^m + \cdots + 1 \in \mathrm{U}(F[\![z]\!])$ and hence $\widehat{f}(z)^{-1} \in F[\![z]\!]$ and $F[z]_{\leq m-1} * \widehat{f}(z)^{-1} \subseteq F[\![z]\!]$. Assume $y \in \mathrm{ann}_{\mathcal{F}}(f(s))$, hence $f \circ y = 0$. Then, by Lemma 8.1.12,

$$F[z]_{\leq m-1} \ni \widehat{f}(z) * y - z^m * \underbrace{\left(f(s) \circ y\right)}_{=0} = \widehat{f}(z) * y$$
$$\Longrightarrow y \in F[z]_{\leq m-1} * \widehat{f}(z)^{-1} \Longrightarrow \mathrm{ann}_{\mathcal{F}}(f(s)) \subseteq F[z]_{\leq m-1} * \widehat{f}(z)^{-1}.$$

The last two spaces have F-dimension m and therefore coincide. Indeed, $\dim_F(\mathrm{ann}_{\mathcal{F}}(f(s))) = m$ by Theorem 3.1.5. Moreover

$$F[z]_{\leq m-1} \cong F[z]_{\leq m-1} * \widehat{f}(z)^{-1}$$
$$\Longrightarrow \dim_F\left(F[z]_{\leq m-1} * \widehat{f}(z)^{-1}\right) = \dim_F\left(F[z]_{\leq m-1}\right) = m.$$

2. As in the proof of Theorem 8.1.11 we get

$$f \circ \tfrac{a}{f} = a(0) \text{ and, of course, } f \circ \frac{a(0)}{f(0)} = \frac{f(0)a(0)}{f(0)} = a(0)$$
$$\Longrightarrow f \circ \left(\tfrac{a}{f} - \tfrac{a(0)}{f(0)}\right) = 0 \Longrightarrow \tfrac{a}{f} = \tfrac{a(0)}{f(0)} + \left(\tfrac{a}{f} - \tfrac{a(0)}{f(0)}\right) \in \mathrm{ann}_{\mathcal{F}}(s) + \mathrm{ann}_{\mathcal{F}}(f).$$

Finally, $f(0) \neq 0$ implies that s and $f(s)$ are coprime. Due to Theorem 5.1.4 the latter sum is direct. □

We apply the preceding theorem to a power $f = q^m$ of a monic irreducible polynomial $q \neq s$.

Corollary 8.1.14 *Let $q = s^d + \cdots + q_0 \neq s$ be a monic irreducible polynomial of degree d with nonzero $q_0 = q(0) \neq 0$, and let $m \in \mathbb{N}$, $m > 0$.*

1. *Then $\widehat{q}(z) = q_0 z^d + \cdots + 1$ and $\mathrm{ann}_{\mathcal{F}}(q^m) = \bigoplus_{j=1}^m F[z]_{\leq d-1}\widehat{q}(z)^{-j}$.*
2. *Let $a(s) \in F[s]_{\leq md}$. Then*

$$\frac{a(s)}{q(s)^m} = \frac{a(0)}{q(0)^m} + \left(\frac{a(s)}{q(s)^m} - \frac{a(0)}{q(0)^m}\right) \in \mathrm{ann}_{\mathcal{F}}(s) \oplus \mathrm{ann}_{\mathcal{F}}(q(s)^m).$$

Proof 1. Since $\deg(q^m) = dm$, $g \longmapsto \widehat{g}$ is multiplicative and hence $\widehat{q^m} = \widehat{q}^m$, Theorem 8.1.13 furnishes $\mathrm{ann}_{\mathcal{F}}(q^m) = F[z]_{\leq dm-1} * \widehat{q}(z)^{-m}$. The $\widehat{q}(z)$-adic representation of polynomials in $F[z]$, cf. Lemma 5.5.7, implies

$$F[z]_{\leq dm-1} = \bigoplus_{i=0}^{m-1} F[z]_{\leq d-1}\widehat{q}(z)^i \Longrightarrow \mathrm{ann}_{\mathcal{F}}(q^m) = F[z]_{\leq dm-1} * \widehat{q}(z)^{-m}$$
$$= \bigoplus_{i=0}^{m-1} F[z]_{\leq d-1}\widehat{q}(z)^{i-m} \underset{j:=m-i}{=} \bigoplus_{j=1}^{m} F[z]_{\leq d-1}\widehat{q}(z)^{-j}.$$

2. This follows directly from the second item of Theorem 8.1.13. □

Theorem 8.1.15 *The primary decomposition of the torsion submodule of* $\mathcal{F} = F^{\mathbb{N}} = F[\![z]\!]$ *is* $\mathrm{t}({}_{F[s]}F[\![z]\!]) = \bigoplus_{q \in \mathcal{P}} \mathcal{F}_q$, *where*

$$\mathcal{F}_q = \begin{cases} F^{(\mathbb{N})} = F[z] & \text{if } q = s, \\ \bigoplus_{j=1}^{\infty} F[z]_{\leq \deg_s(q)-1}\widehat{q}(z)^{-j} & \text{if } q \neq s. \end{cases}$$

This theorem extends the discrete-time case of Corollary 5.3.19 for $\mathbb{C}$ and $\mathbb{R}$ to arbitrary fields, in particular, to finite ones.

Proof This follows from $\mathcal{F}_s = F^{(\mathbb{N})}$ and from Corollary 8.1.14 via

$$\mathcal{F}_q = \bigcup_{m=1}^{\infty} \mathrm{ann}_{\mathcal{F}}(q^m) = \bigcup_{m=1}^{\infty} \bigoplus_{j=1}^{m} F[z]_{\leq d-1}\widehat{q}(z)^{-j} = \bigoplus_{j=1}^{\infty} F[z]_{\leq d-1}\widehat{q}(z)^{-j}.$$ □

Next we compute the power series representation of a proper rational function $H \in F(s)_{\mathrm{pr}} \subset F[\![z]\!]$. The partial fraction decomposition gives

$$\begin{aligned} F(s)_{\mathrm{pr}} &= F \oplus F(s)_{\mathrm{spr}} \underset{\text{Theorem 5.5.8}}{=} F \oplus \bigoplus_{q \in \mathcal{P}} \bigoplus_{j=1}^{\infty} F[s]_{\leq \deg_s(q)-1} q(s)^{-j} \\ &= \underbrace{F \oplus \bigoplus_{j=1}^{\infty} F s^{-j}}_{=\bigoplus_{j=0}^{\infty} F z^j = F[z]} \oplus \bigoplus_{q \in \mathcal{P}\setminus\{s\}} \bigoplus_{j=1}^{\infty} F[s]_{\leq \deg_s(q)-1} q(s)^{-j} \\ &= F[z] \oplus \bigoplus_{q \in \mathcal{P}\setminus\{s\}} \bigoplus_{j=1}^{\infty} F[s]_{\leq \deg_s(q)-1} q(s)^{-j}. \end{aligned} \tag{8.26}$$

It thus suffices to compute the power series representations of

$$\frac{a(s)}{q(s)^m} = \frac{a(0)}{q(0)^m} + \left(\frac{a}{q^m} - \frac{a(0)}{q(0)^m}\right) \underset{\text{Corollary 8.1.14, 2.}}{\in} Fz^0 \bigoplus \mathrm{ann}_{\mathcal{F}}(q^m) \subset Fz^0 \bigoplus \mathcal{F}_q, \tag{8.27}$$

where $q(0) \neq 0$ and $\deg_s(a) \leq \deg_s(q) - 1$. These can be computed in general for perfect fields via Galois theory, cf. [1, Sect. 5], but we will do this only for linear polynomials

$$\begin{aligned} &q(s) = s - \lambda \quad \text{with } \lambda \neq 0 \Longrightarrow \widehat{q}(z) = 1 - \lambda z \text{ and} \\ &(s-\lambda)^{-1} = z(1-\lambda z)^{-1} = \sum_{t=0}^{\infty} \lambda^t z^{t+1} = z * e_{\lambda,0}, \text{ where} \\ &e_{\lambda,k} = \left(\binom{t}{k}\lambda^{t-k}\right)_{t\in\mathbb{N}} = \sum_{t=0}^{\infty} \binom{t}{k}\lambda^{t-k}z^t \in F^{\mathbb{N}} = F[\![z]\!],\ \lambda \neq 0,\ k \in \mathbb{N}, \end{aligned} \tag{8.28}$$

from Lemma and Definition 5.2.7. This lemma and Corollary 8.1.14 imply

$$\begin{aligned} \mathrm{ann}_{\mathcal{F}}\left((s-\lambda)^m\right) &= \bigoplus_{k=0}^{m-1} Fe_{\lambda,k} = \bigoplus_{j=1}^{m} F(1-\lambda z)^{-j}, \quad \text{and} \\ \mathcal{F}_{s-\lambda} &= \bigoplus_{k=0}^{\infty} Fe_{\lambda,k} = \bigoplus_{j=1}^{\infty} F(1-\lambda z)^{-j}. \end{aligned} \tag{8.29}$$

In characteristic 0, e.g., for $F = \mathbb{C}$, Corollary 5.3.19 gives

$$\mathrm{ann}_{\mathcal{F}}\left((s-\lambda)^m\right) = F[t]_{\leq m-1}\lambda^t \subset \mathcal{F}_{s-\lambda} = F[t]\lambda^t. \tag{8.30}$$

In this formula and also in the sequel we write $y(t)$ for y, i.e.,

$$y = y(t) = (y(t))_{t\in\mathbb{N}} = \sum_{t=0}^{\infty} y(t)z^t \in F^{\mathbb{N}} = F[\![z]\!].$$

Lemma 8.1.16 *Let $0 \neq m \in \mathbb{N}$ and $0 \neq \lambda \in F$. Then the following holds.*

1. $(1-\lambda z)^{-m} = \sum_{t=0}^{\infty} \binom{t+m-1}{m-1}\lambda^t z^t$.
2. $(s-\lambda)^{-m} = z^m(1-\lambda z)^{-m} = z * e_{\lambda,m-1}$.
3. $(s-\lambda)^{-m} = (-\lambda)^{-m} + \left((s-\lambda)^{-m} - (-\lambda)^{-m}\right) \in Fz^0 \oplus \mathcal{F}_{s-\lambda}$ *and* $(s-\lambda)^{-m} - (-\lambda)^{-m} = \sum_{t=0}^{\infty} \binom{t-1}{m-1}\lambda^{-m}\lambda^t z^t \in \mathrm{ann}_{\mathcal{F}}\left((s-\lambda)^m\right)$.

For $F = \mathbb{C}$ the function $\binom{t-1}{m-1}\lambda^{-m}$ is a polynomial in t of degree $m-1$.

Proof Recall that for $m, t \in \mathbb{N}$ the identity

$$\binom{-m}{t} := \frac{(-m)(-m-1)\cdots(-m-t+1)}{t!} = (-1)^t\frac{(m+t-1)\cdots m}{t!} = (-1)^t\binom{t+m-1}{m-1} \in \mathbb{Z}$$

holds, hence $\binom{-m}{t}\lambda$ is well defined for all F and $\lambda \in F$.
1. The binomial series for $(1-\lambda z)^{-m}$ is

$$(1-\lambda z)^{-m} = \sum_{t=0}^{\infty} \binom{-m}{t}(-\lambda)^t z^t = \sum_{t\in\mathbb{N}} \binom{t+m-1}{m-1}\lambda^t z^t. \tag{8.31}$$

2. With $s-\lambda = s(1-\lambda z)$, we conclude that

$$\begin{aligned}(s-\lambda)^{-m} &= z^m(1-\lambda z)^{-m} = \sum_{t=0}^{\infty}\binom{t+m-1}{m-1}\lambda^t z^{t+m} \underset{t+m-1\mapsto t}{=} \\ z * \sum_{t\geq m-1} &\binom{t}{m-1}\lambda^{t-(m-1)}z^t = z*\sum_{t\geq 0}\binom{t}{m-1}\lambda^{t-(m-1)}z^t = z*e_{\lambda,m-1},\end{aligned} \tag{8.32}$$

since $\binom{t}{m-1} = 0$ if $0 \leq t < m-1$.
3. The first equation follows from (8.27), applied to $(s-\lambda)^{-m}$. Moreover

$$(z*e_{\lambda,m-1})(t) = \begin{cases}\binom{t-1}{m-1}\lambda^{t-1-(m-1)} & \text{for } t>0 \\ 0 & \text{for } t=0\end{cases}$$

$$\Longrightarrow \left((s-\lambda)^{-m} - (-\lambda)^{-m}\right)(t) = \left(z*e_{\lambda,m-1} - (-\lambda)^{-m}z^0\right)(t)$$

$$= \begin{cases}\binom{t-1}{m-1}\lambda^{t-m} & \text{for } t>0 \\ -(-\lambda)^{-m} & \text{for } t=0\end{cases} = \binom{t-1}{m-1}\lambda^{-m}\lambda^t \text{ since}$$

$$\binom{0-1}{m-1} = \binom{-1}{m-1} = -(-1)^m\binom{1+(m-1)-1}{1-1} = -(-1)^m,\ \binom{0-1}{m-1}\lambda^{0-m} = -(-\lambda)^{-m}. \ \square$$

Theorem 8.1.17 *1. Let $H \in \mathbb{C}(s)_{\mathrm{pr}} = \mathbb{C}(s) \cap \mathbb{C}[\![z]\!]$ be any complex proper rational function with its partial fraction decomposition*

$$H = H_0 + \sum_{\lambda\in\mathrm{pole}(H)}\sum_{k=1}^{m(\lambda)} a_{\lambda,k}(s-\lambda)^{-k},\ H_0,\ a_{\lambda,k}\in\mathbb{C},\ m(\lambda)\geq 1,\ a_{\lambda,m(\lambda)}\neq 0.$$

Let $m(0) := 0$ if $0 \notin \mathrm{pole}(H)$. Then H has the power series representation

$$\begin{aligned}H &= H_0 + \sum_{k=1}^{m(0)} a_{0,k}z^k + z\sum_{\lambda\in\mathrm{pole}(H),\ \lambda\neq 0}\ \sum_{k=1}^{m(\lambda)} a_{\lambda,k}e_{\lambda,k-1} \\ &= H_0 + \sum_{k=1}^{m(0)} a_{0,k}z^k + \sum_{\lambda\in\mathrm{pole}(H),\ \lambda\neq 0}\ \sum_{t\in\mathbb{N}} g_\lambda(t)\lambda^t z^t \text{ where} \\ g_\lambda(t) &= a_{\lambda,m(\lambda)}t^{m(\lambda)-1} + \cdots \in \mathbb{C}[t],\ \deg_t(g_\lambda) = m(\lambda)-1\geq 0,\ g_\lambda(0) = 0.\end{aligned} \tag{8.33}$$

Equation (8.70) *is the continuous-time analogue of* (8.33).

2. *The analogous representation is valid for* $H \in \mathbb{C}(s)_{\mathrm{pr}}^{p\times m} = \left(\mathbb{C}[s] \cap \mathbb{C}[\![z]\!]\right)^{p\times m}$.

Proof The proof follows directly from Lemma 8.1.16, 2. If $0 \in \mathrm{pole}(H)$ then $s^{-k} = z^k = \delta_k = (0, \dots, 0, \overset{k}{1}, 0, \dots)$. This explains the special role of the second summand. The form of g_λ results from that of $e_{\lambda,m(\lambda)-1}$. □

Corollary 8.1.18 *1. Let*

$$H = H_{\mathrm{pol},+} + H_{\mathrm{pr}} \in \mathbb{C}(s),\ u \in \mathbb{C}^{\mathbb{N}} = \mathbb{C}[\![z]\!].\ \textit{Then}\ H \circ u = H_{\mathrm{pol},+} \circ u + H_{\mathrm{pr}} * u,$$

where H_{pr} *can be computed according to* (8.33).

2. *If* u *itself lies in* $\mathbb{C}(s)_{\mathrm{pr}} = \mathbb{C}(s) \cap \mathbb{C}[\![z]\!]$ *then, by* (8.20),

$$H \circ u = H_{\mathrm{pol},+} \circ u + H_{\mathrm{pr}} * u = \left(H_{\mathrm{pol},+} u\right)_{\mathrm{pr}} + H_{\mathrm{pr}} u = (Hu)_{\mathrm{pr}}.$$

Thus the action of H *on* u *can be easily computed with* (5.86).

3. *The analogous result for rational matrices holds too.*

◊

8.1.3 External Stability of Discrete-Time Transfer Operators

In Theorem and Definition 8.1.25 and Theorem 8.1.26 we study (ℓ^p, ℓ^p)-stability, $1 \le p \le \infty$, of discrete-time transfer operators. For $p = \infty$ one obtains BIBO (bounded input/bounded output) stability. In Theorem 8.1.27 we characterize BIBO stability of IO behaviors and thus generalize Theorem 6.4.8 to inputs which are not polynomial-exponential.

We assume that $F := \mathbb{C}$ is the complex field. The case of real transfer functions and real signals is, of course, included in this case. The signal space $\mathcal{F} := \mathbb{C}^{\mathbb{N}} = \mathbb{C}[\![z]\!]$ is the space of sequences with complex entries or of power series with complex coefficients.

Reminder 8.1.19 (Convergent power series) Let $y = \sum_{t=0}^{\infty} y_t z^t \in \mathbb{C}[\![z]\!]$ be a formal power series with its *convergence radius* $R(y) := \left(\limsup_{t\in\mathbb{N}} \sqrt[t]{|y_t|}\right)^{-1}$. The series y converges for $|z| < R(y)$ and diverges for $|z| > R(y)$. The $\mathbb{C}$-subspace of $\mathbb{C}[\![z]\!]$ of *(locally) convergent power series* is

$$\mathbb{C}\langle z\rangle := \{y \in \mathbb{C}[\![z]\!];\ R(y) > 0\} = \left\{y \in \mathbb{C}[\![z]\!];\ \exists M > 0, R > 0\colon\ |y_t| \le MR^t\right\}. \tag{8.34}$$

If $R(y) > 0$, the function

$$y\colon \{z \in \mathbb{C};\ |z| < R(y)\} \longrightarrow \mathbb{C},\quad z \longmapsto y(z) := \sum_{t=0}^{\infty} y_t z^t,$$

is holomorphic. Here we identify the formal power series y in the indeterminate z with the holomorphic function of the complex variable z in the interior of the disc of convergence. This is possible because of the identity theorem of holomorphic functions. ◊

We will see in Theorem 8.1.23 that the set $\mathbb{C}\langle z\rangle$ is closed under convolution. Thus it is not only a vector subspace, but even a subalgebra of $\mathbb{C}[\![z]\!]$. For the proof of Theorem 8.1.23, we need standard arguments from analysis, cf., e.g., [2, Sect. 1.2]. Let $p \in \mathbb{R} \uplus \{\infty\}$ with $1 \le p$, and let $y = (y_t)_{t\in\mathbb{N}} \in \mathbb{C}^{\mathbb{N}}$ be a sequence. We define the quantity

$$\|y\|_p := \begin{cases} \left(\sum_{t\in\mathbb{N}} |y_t|^p\right)^{1/p} & \text{if } p < \infty \\ \sup_{t\in\mathbb{N}} |y_t| & \text{if } p = \infty \end{cases} \in [0,\infty] := [0,\infty) \uplus \{\infty\} \subseteq \mathbb{R} \uplus \{\infty\}. \tag{8.35}$$

We use the computation rules

$$a + \infty := \infty \text{ for } a \in \mathbb{R} \uplus \{\infty\},\ a\infty := \infty \text{ for } a \neq 0,\ \frac{1}{\infty} := 0.$$

Lemma 8.1.20 *Let* $x = (x_t)_{t\in\mathbb{N}}, y = (y_t)_{t\in\mathbb{N}} \in \mathbb{C}^{\mathbb{N}},\ a \in \mathbb{C},\ p \in [1,\infty]$.

1. *(i) If* $\|y\|_p = 0$, *then* $y = 0$. *(ii)* $\|ay\|_p = |a|\,\|y\|_p$.
2. *Let* $p, q \in [1,\infty]$ *with* $\frac{1}{p} + \frac{1}{q} = 1$. *Then the* Hölder inequality *holds:*

$$\|xy\|_1 = \sum_{t\in\mathbb{N}} |x_t|\,|y_t| \le \|x\|_p\,\|y\|_q,\ \textit{for } x, y \in \mathbb{C}^{\mathbb{N}},\ xy := (x_t y_t)_{t\in\mathbb{N}}. \tag{8.36}$$

With the embedding $\mathbb{C}^n \subset \mathbb{C}^{\mathbb{N}},\ y = (y_1, \ldots, y_n) := (y_1, \ldots, y_n, 0, \ldots),\ n \in \mathbb{N}$, *Hölder's inequality also holds in* $\mathbb{C}^n$, *where*

$$\|y\|_p = \begin{cases} \left(\sum_{t=1}^{n} |y_t|^p\right)^{1/p} & \textit{if } p < \infty \\ \max_{t\in\{1,\ldots,n\}} |y_t| & \textit{if } p = \infty \end{cases}.$$

3. *The* triangle or Minkowski inequality $\|x + y\|_p \le \|x\|_p + \|y\|_p$ *holds.*

Proof 1. This follows directly from the definition of $\|-\|_p$ in (8.35).
2. For $p = 1$ and $q = \infty$ (or vice versa), this is obvious. So let $p, q \in (1,\infty)$. If $\|x\|_p = 0$, then $x = 0$ and $\|xy\|_1 = \|x\|_p\|y\|_q = 0$. Thus we assume $\|x\|_p > 0$ and $\|y\|_q > 0$. If $\|x\|_p = \infty$ or $\|y\|_q = \infty$, the assertion holds obviously.
The case $p, q \in (1,\infty)$ and $\|x\|_p, \|y\|_q \in (0,\infty)$ remains to be shown. For this we use that the logarithm ln is a concave function, hence for $a, b \in (0,\infty)$ and $t \in [0,1]$, the inequality $(1-t)\ln(a) + t\ln(b) \le \ln\big((1-t)a + tb\big)$ holds. We infer

$$\ln(ab) = \ln(a) + \ln(b) = p^{-1}\ln(a^p) + q^{-1}\ln(b^q) \le \ln\left(p^{-1}a^p + q^{-1}b^q\right),\ t := q^{-1},$$

$$\underset{\exp(\cdot)}{\Longrightarrow} ab \le \tfrac{a^p}{p} + \tfrac{b^q}{q} \Longrightarrow \text{with } a := \tfrac{|x_t|}{\|x\|_p},\ b = \tfrac{|y_t|}{\|y\|_q}:$$

$$\begin{aligned}
&\frac{|x_t|}{\|x\|_p}\frac{|y_t|}{\|y\|_q} \le \frac{1}{p}\frac{|x_t|^p}{\|x\|_p^p} + \frac{1}{q}\frac{|y_t|^q}{\|y\|_q^q}\\
\underset{\sum_{t\in\mathbb{N}}}{\Longrightarrow}\ &\frac{\|xy\|_1}{\|x\|_p\,\|y\|_q} = \sum_{t\in\mathbb{N}}\frac{|x_t|\,|y_t|}{\|x\|_p\,\|y\|_q} \le \sum_{t\in\mathbb{N}}\frac{1}{p}\frac{|x_t|^p}{\|x\|_p^p} + \sum_{t\in\mathbb{N}}\frac{1}{q}\frac{|y_t|^q}{\|y\|_q^q}\\
&= \frac{1}{p}\frac{\|x\|_p^p}{\|x\|_p^p} + \frac{1}{q}\frac{\|y\|_q^q}{\|y\|_q^q} == \frac{1}{p} + \frac{1}{q} = 1.
\end{aligned} \tag{8.37}$$

3. The triangle inequalities for $p = 1$ and for $p = \infty$ are obvious and follow from the triangle inequality of the absolute value. Now assume $p \in (1, \infty)$, hence $q := \frac{p}{p-1} \in (1, \infty)$, and $x + y \neq 0$. Then

$$\begin{aligned}
\|x+y\|_p^p &= \sum_{t\in\mathbb{N}}|x_t+y_t|^p = \sum_{t\in\mathbb{N}}|x_t+y_t|\,|x_t+y_t|^{p-1}\\
&\le \sum_{t\in\mathbb{N}}\left(|x_t|+|y_t|\right)|x_t+y_t|^{p-1}\\
&= \left(\sum_{t\in\mathbb{N}}|x_t|\,|x_t+y_t|^{p-1}\right) + \left(\sum_{t\in\mathbb{N}}|y_t|\,|x_t+y_t|^{p-1}\right)\\
&\underset{2.}{\le} \|x\|_p\left\|\left(|x_t+y_t|^{p-1}\right)_{t\in\mathbb{N}}\right\|_q + \|y\|_p\left\|\left(|x_t+y_t|^{p-1}\right)_{t\in\mathbb{N}}\right\|_q\\
&= \left(\|x\|_p + \|y\|_p\right)\left(\sum_{t\in\mathbb{N}}|x_t+y_t|^{(p-1)q}\right)^{\frac{1}{q}}\\
&\underset{q(p-1)=p}{=} \left(\|x\|_p + \|y\|_p\right)\left(\sum_{t\in\mathbb{N}}|x_t+y_t|^{p}\right)^{1-\frac{1}{p}}\\
&= \left(\|x\|_p + \|y\|_p\right)\frac{\|x+y\|_p^p}{\|x+y\|_p} \Longrightarrow 1 \le \frac{\|x\|_p+\|y\|_p}{\|x+y\|_p}.
\end{aligned}$$

□

Lemma and Definition 8.1.21 *For $1 \le p \le \infty$, the sets*

$$\ell^p := \{y \in \mathbb{C}^{\mathbb{N}};\ \|y\|_p < \infty\} \tag{8.38}$$

are $\mathbb{C}$-subspaces of $\mathbb{C}^{\mathbb{N}} = \mathbb{C}[\![z]\!]$, and indeed Banach spaces with the norm $\|-\|_p$.

If the coefficient sequence of a formal power series is bounded, then it converges in a suitable neighborhood of zero. Therefore, the inclusion $\ell^\infty \subset \mathbb{C}\langle z\rangle \subset \mathbb{C}^{\mathbb{N}} = \mathbb{C}[\![z]\!]$ holds.

Proof We treat the case $p < \infty$; the case $p = \infty$ is analogous. The equality $\|ax\|_p = |a|\|x\|_p$ and the triangle inequality $\|x+y\|_p \le \|x\|_p + \|y\|_p$ imply that ℓ^p is a subspace of $\mathbb{C}^{\mathbb{N}}$ and that $\|-\|_p$ is a norm on ℓ^p.

To show that ℓ^p is a Banach space, i.e., complete, let $(y^{(m)})_{m\in\mathbb{N}}$ be a Cauchy sequence in ℓ^p, i.e.,

$$\forall \epsilon > 0\, \exists N \in \mathbb{N}\, \forall m, n \geq N\colon\ \|y^{(m)} - y^{(n)}\|_p \leq \epsilon. \tag{8.39}$$

For $t \in \mathbb{N}$, we infer that

$$\left|y_t^{(m)} - y_t^{(n)}\right| \leq \left(\sum_{s\in\mathbb{N}} |y_s^{(m)} - y_s^{(n)}|^p\right)^{1/p} = \left\|y^{(m)} - y^{(n)}\right\|_p \leq \epsilon.$$

This implies that the sequence $\left(y_t^{(m)}\right)_{m\in\mathbb{N}}$ is a Cauchy sequence in $\mathbb{C}$ and thus converges to a limit $y_t := \lim_{m\to\infty} y_t^{(m)}$. We define $y := (y_t)_{t\in\mathbb{N}}$ and show that $y = \lim_{m\to\infty} y^{(m)} \in \ell^p$.
For every $T \in \mathbb{N}$, the assumption (8.39) implies

$$\left(\sum_{t=0}^{T} |y_t^{(m)} - y_t^{(n)}|^p\right)^{1/p} \leq \left(\sum_{t\in\mathbb{N}} |y_t^{(m)} - y_t^{(n)}|^p\right)^{1/p} = \|y^{(m)} - y^{(n)}\|_p \leq \epsilon,$$

and consequently, by taking the limit over n, that

$$\begin{aligned}\epsilon &\geq \lim_{n\to\infty}\left(\sum_{t=0}^{T} |y_t^{(m)} - y_t^{(n)}|^p\right)^{1/p} = \left(\sum_{t=0}^{T} |y_t^{(m)} - \lim_{n\to\infty} y_t^{(n)}|^p\right)^{1/p}\\ &= \left(\sum_{t=0}^{T} |y_t^{(m)} - y_t|^p\right)^{1/p}.\end{aligned}$$

Since this inequality holds for all $T \in \mathbb{N}$, we infer that

$$\epsilon \geq \lim_{T\to\infty}\left(\sum_{t=0}^{T} |y_t^{(m)} - y_t|^p\right)^{1/p} = \left(\sum_{t\in\mathbb{N}} |y_t^{(m)} - y_t|^p\right)^{1/p} = \|y^{(m)} - y\|_p. \tag{8.40}$$

This implies $y^{(m)} - y \in \ell^p,\ y = y^{(m)} + (y - y^{(m)}) \in \ell^p,\ y = \lim_{m\to\infty} y^{(m)}$. □

Lemma 8.1.22 *If* $1 \leq p < q \leq \infty$ *and* $y \in \mathbb{C}^{\mathbb{N}}$, *then* $\|y\|_q \leq \|y\|_p$ *and* $\ell^p \subsetneq \ell^q$ *hold.*

Proof The inclusion $\ell^p \subseteq \ell^\infty$ is obvious. If $y \notin \ell^p \subseteq \ell^\infty$, i.e., $\|y\|_p = \infty$, then $\|y\|_q \leq \|y\|_p = \infty$. If $y \in \ell^p$, define

$$\begin{aligned}&Y := \|y\|_p^{-1} y \in \ell^p \subseteq \ell^\infty,\ Y_t = \|y\|_p^{-1} y_t,\ |Y_t| \leq 1\\ &\Longrightarrow \|Y\|_p = 1,\ \|Y\|_\infty = \tfrac{\|y\|_\infty}{\|y\|_p} \leq 1 \text{ and } \|y\|_\infty \leq \|y\|_p.\end{aligned}$$

If $q < \infty$, then $|Y_t| \leq 1$ and $p \leq q$ imply $|Y_t|^p \geq |Y_t|^q$ for all t and

$$1 = \|Y\|_p^p = \sum_{t\in\mathbb{N}} |Y_t|^p \geq \sum_{t\in\mathbb{N}} |Y_t|^q = \|Y\|_q^q = \frac{\|y\|_q^q}{\|y\|_p^q} \Longrightarrow \|y\|_q \leq \|y\|_p \Longrightarrow \ell^p \subseteq \ell^q.$$

Moreover $\ell^p \subsetneq \ell^q$ since $\left(\frac{1}{(t+1)^{1/p}}\right)_{t\in\mathbb{N}} \in \ell^q \setminus \ell^p$. □

Theorem 8.1.23 *For $p \in [1, \infty]$, $y \in \ell^p$, and $a \in \ell^1$* Young's inequality *holds:*

$$\|y * a\|_p \leq \|y\|_p \, \|a\|_1. \tag{8.41}$$

It implies $\ell^p * \ell^1 \subseteq \ell^p$, i.e., ℓ^1 is a Banach subalgebra of $\mathbb{C}[[z]]$ and the ℓ^p, $p \geq 1$, are ℓ^1-submodules. This also implies $\mathbb{C}\langle z\rangle * \mathbb{C}\langle z\rangle \subseteq \mathbb{C}\langle z\rangle$.

Proof We write $|y| := (|y_t|)_{t\in\mathbb{N}}$ and $|a| := (|a_t|)_{t\in\mathbb{N}}$. Then $\|y\|_p = \|\,|y|\,\|_p$ and $\|a\|_1 = \|\,|a|\,\|_1$, as well as

$$\|y * a\|_p = \left(\sum_{t\in\mathbb{N}} \left|\sum_{\tau\leq t} y_\tau a_{t-\tau}\right|^p\right)^{1/p} \leq \left(\sum_{t\in\mathbb{N}} \left(\sum_{\tau\leq t} |y_\tau|\,|a_{t-\tau}|\right)^p\right)^{1/p} = \|\,|y| * |a|\,\|_p.$$

Thus we assume without loss of generality that $y_t \geq 0$ and $a_t \geq 0$ for all $t \in \mathbb{N}$. Hence all the following sums converge absolutely, and the order of their summands is arbitrary. If $p = \infty$, then

$$\|y * a\|_\infty = \sup_{t\in\mathbb{N}} \sum_{\tau\leq t} y_{t-\tau}\, a_\tau \leq \sup_{t\in\mathbb{N}} \sum_{\tau\leq t} \|y\|_\infty\, a_\tau = \|y\|_\infty \sum_{\tau=0}^{\infty} a_\tau = \|y\|_\infty \, \|a\|_1.$$

For $p = 1$ we obtain

$$\|y * a\|_1 = \sum_{t\in\mathbb{N}} \sum_{\tau+\tau_1=t} y_\tau\, a_{\tau_1} = \sum_{\tau,\tau_1\in\mathbb{N}} y_\tau\, a_{\tau_1} = \left(\sum_{\tau\in\mathbb{N}} y_\tau\right)\left(\sum_{\tau_1\in\mathbb{N}} a_{\tau_1}\right) = \|y\|_1 \, \|a\|_1.$$

For $p \in (1, \infty)$ let $q \in (1, \infty)$ be such that $\frac{1}{p} + \frac{1}{q} = 1$. For fixed $t \in \mathbb{N}$, Hölder's inequality (8.36) for finite sequences furnishes

$$\begin{aligned}(y * a)_t &= \sum_{\tau=0}^{t} y_{t-\tau}\, a_\tau = \sum_{\tau=0}^{t} \left(y_{t-\tau}\, a_\tau^{1/p}\right) a_\tau^{1/q} \\ &\overset{(8.36)}{\leq} \left(\sum_{\tau=0}^{t} y_{t-\tau}^p\, a_\tau\right)^{1/p} \left(\sum_{\tau=0}^{t} a_\tau\right)^{1/q} \leq \left(\sum_{\tau=0}^{t} y_{t-\tau}^p\, a_\tau\right)^{1/p} \|a\|_1^{1/q}.\end{aligned}$$

With this we obtain

$$\|y * a\|_p^p = \sum_{t\in\mathbb{N}}\left(\sum_{\tau=0}^{t} y_{t-\tau}a_\tau\right)^p \le \sum_{t\in\mathbb{N}}\sum_{\tau=0}^{t} y_{t-\tau}^p a_\tau \|a\|_1^{p/q}$$
$$= \sum_{\tau\in\mathbb{N}} a_\tau \underbrace{\sum_{t\ge\tau} y_{t-\tau}^p}_{=\sum_{t\in\mathbb{N}} y_t^p = \|y\|_p^p} \|a\|_1^{p/q} \le \sum_{\tau\in\mathbb{N}} a_\tau \|y\|_p^p \|a\|_1^{p/q} = \|a\|_1 \|y\|_p^p \|a\|_1^{p/q}$$
$$\underset{(-)^{1/p}}{\Longrightarrow} \|y * a\|_p \le \|a\|_1^{1/p} \|y\|_p \|a\|_1^{1/q} = \|y\|_p \|a\|_1^{1/p+1/q} = \|y\|_p \|a_1\|. \qquad \square$$

Definition and Corollary 8.1.24 Let $H \in \mathbb{C}((z))$ be a Laurent series and let

$$H\circ:\ \mathbb{C}^{\mathbb{N}} = \mathbb{C}[[z]] \longrightarrow \mathbb{C}^{\mathbb{N}},\quad u \longmapsto H \circ u,$$

be its transfer operator or input/output map. For $p \in [1, \infty]$ the sequence H and the operator $H\circ$ are called *(ℓ^p, ℓ^p)-stable* if $H \circ \ell^p \subseteq \ell^p$. For $p = \infty$ this is also called *BIBO (bounded input / bounded output) stability*.
According to Theorem 8.1.23, a transfer function $H \in \ell^1 \subset \mathbb{C}[[z]]$ is (ℓ^p, ℓ^p)-stable for all $p \in [1, \infty]$. The same trivially holds for $H \in \mathbb{C}[s]$, and then also for $H \in \mathbb{C}[s] + \ell^1$. Notice that $\mathbb{C}^{(\mathbb{N})} = \mathbb{C}[z] \subset \ell^1$.
The analogous definition and result apply to matrices, namely, if $H \in \left(\mathbb{C}[s] + \ell^1\right)^{q\times m}$, then $H \circ (\ell^p)^m \subseteq (\ell^p)^q$, i.e., H is (ℓ^p, ℓ^p)-stable. ◊

The next theorem is the principal result on external stability of discrete-time transfer operators.

Theorem and Definition 8.1.25 (External stability of H) *Let $H \in \mathbb{C}(s)$ be a complex rational function. As in Theorem 8.1.17, we write $H = H_{\mathrm{pol},+} + H_{\mathrm{pr}}$ with $H_{\mathrm{pr}} \in \mathbb{C}(s)_{\mathrm{pr}} \subseteq \mathbb{C}[[z]] = \mathbb{C}^{\mathbb{N}}$. The following statements are equivalent:*

1. *Every pole of H belongs to the open unit disc $\{\lambda \in \mathbb{C};\ |\lambda| < 1\}$.*
2. *$H_{\mathrm{pr}} \in \ell^1$.*
3. *H is $(\ell^\infty, \ell^\infty)$-stable or BIBO stable.*
4. *H is (ℓ^1, ℓ^1)-stable.*
5. *H is (ℓ^p, ℓ^p)-stable for all $1 \le p \le \infty$.*

If these conditions are satisfied, the transfer function H and transfer operator $H\circ$ are called externally stable.

Proof We use the decomposition from (8.33):

$$H = H_{\mathrm{pol}} + \sum_{k=1}^{m(0)} a_{0,k} z^k + \sum_{\lambda\in\mathrm{pole}(H),\ \lambda\neq 0} g_\lambda(t)\lambda^t z^t \text{ where}$$
$$0 \neq g_\lambda(t) = a_{\lambda,m(\lambda)} t^{m(\lambda)-1} + \cdots \in \mathbb{C}[t],\ g_\lambda(0) = 0.$$

The left shift operator H_{pol} and the right shifts $s^{-k} = z^k$ obviously satisfy all conditions 1–5, and hence can be omitted in the proof. We thus assume that H is strictly

proper, that $0 \notin \operatorname{pole}(H)$ and hence $H = \sum_{\lambda \in \operatorname{pole}(H)} g_\lambda(t)\lambda^t z^t$.
1. $\Longrightarrow$ 2.: It suffices to show $(g_\lambda(t)\lambda^t)_{t\in\mathbb{N}} \in \ell^1$ for $|\lambda| < 1$. Let $|\lambda| < \rho < 1$, hence $|\lambda\rho^{-1}| < 1$. The geometric sequence $\left((\lambda\rho^{-1})^t\right)_{t\in\mathbb{N}}$ decreases faster than the polynomial sequence $g_\lambda(t)$ grows. Hence

$$\begin{aligned} M &:= \max_{t\in\mathbb{N}} |g_\lambda(t)|\,|\lambda|^t \rho^{-t} < \infty \Longrightarrow \forall t \in \mathbb{N}:\ |g_\lambda(t)|\,|\lambda|^t \le M\rho^t \\ &\Longrightarrow \sum_t |g_\lambda(t)\lambda^t| \le M \sum_t \rho^t = M(1-\rho)^{-1} < \infty \Longrightarrow (g_\lambda(t)\lambda^t)_t \in \ell^1. \end{aligned}$$

2. $\Longrightarrow$ 5.: This follows directly from Theorem 8.1.23.
5. $\Longrightarrow$ 3.: Obvious.
3. $\Longrightarrow$ 1.: From $1_{\mathbb{C}[\![z]\!]} = z^0 \in \ell^\infty$, we infer that

$$H = H * 1_{\mathbb{C}[\![z]\!]} = \sum_{\lambda\in\operatorname{pole}(H)} \sum_{t\in\mathbb{N}} g_\lambda(t)\lambda^t z^t \in \ell^\infty$$

is a bounded polynomial-exponential sequence. With Theorem 6.4.4, 2, we infer that for all $\lambda \in \operatorname{pole}(H)$ either $|\lambda| < 1$ or $\big(|\lambda| = 1$ and $g_\lambda = g_\lambda(0) \in \mathbb{C}\big)$ holds. In the second case $g_\lambda(0) = 0$ implies $g_\lambda = 0$, a contradiction, and hence this case cannot occur. Therefore $|\lambda| < 1$ for all $\lambda \in \operatorname{pole}(H)$, i.e., condition 1.
5. $\Longrightarrow$ 4.: Obvious.
4. $\Longrightarrow$ 2.: Since $1_{\mathbb{C}[\![z]\!]} = z^0 \in \ell^1$ and H is (ℓ^1, ℓ^1)-stable, we infer that $H_{\mathrm{pr}} = H \circ 1_{\mathbb{C}[\![z]\!]} \in \ell^1$. □

We obtain the extension of the preceding theorem to matrices $H \in \mathbb{C}(s)^{q\times m}$ by applying it to all entries of H. Recall that the pole set of a rational matrix is $\operatorname{pole}(H) = \bigcup_{i,j} \operatorname{pole}(H_{ij})$. The decomposition of every entry $H_{ij} \in \mathbb{C}(s)$ according to Theorem 8.1.17 implies the decomposition

$$\begin{aligned} H &= H_{\mathrm{pol}} + \sum_{k=1}^{m(0)} a_{0,k} z^k + \sum_{\lambda\in\operatorname{pole}(H),\ \lambda\neq 0} g_\lambda(t)\lambda^t z^t \text{ where } H_{\mathrm{pol}} \in \mathbb{C}[s]^{q\times m}, \\ a_{0,k} &\in \mathbb{C}^{q\times m},\ 0 \neq g_\lambda(t) = a_{\lambda,m(\lambda)} t^{m(\lambda)-1} + \cdots \in \mathbb{C}[t]^{q\times m},\ g_\lambda(0) = 0. \end{aligned}$$

Theorem 8.1.26 (Stability of rational transfer matrices) *For a rational transfer matrix $H \in \mathbb{C}(s)^{q\times m}$, the following statements are equivalent:*

1. *Every pole of H belongs to the open unit disc $\{\lambda \in \mathbb{C};\ |\lambda| < 1\}$.*
2. *$H_{\mathrm{pr}} \in (\ell^1)^{q\times m}$.*
3. *$H \circ (\ell^\infty)^m \subseteq (\ell^\infty)^q$.*
4. *$H \circ (\ell^1)^m \subseteq (\ell^1)^q$.*
5. *$H \circ (\ell^p)^m \subseteq (\ell^p)^q$ for all $1 \le p \le \infty$.*

If these conditions are satisfied, H is called externally stable. ◇

We finally derive the analogue of Theorem 6.4.8.

Theorem 8.1.27 (BIBO stable IO behaviors) *We assume the signal space* $\mathcal{F} := \mathbb{C}^{\mathbb{N}} = \mathbb{C}[\![z]\!]$ *and an IO behavior* $\mathcal{B} = \left\{ \binom{y}{u} \in \mathcal{F}^{q+m};\ P \circ y = Q \circ u \right\}$ *with* $(P, -Q) \in \mathbb{C}[s]^{q \times (q+m)}$, $\mathrm{rank}(P) = q$ *and* $H = P^{-1}Q$. *The following properties of* $\mathcal{B}$ *are equivalent:*

1. *If* $\binom{y}{u} \in \mathcal{B}$ *and* $u \in (\ell^\infty)^m$, *then also* $y \in (\ell^\infty)^q$.
2. (a) *The autonomous part* $\mathcal{B}^0 = \{y \in \mathcal{F}^q;\ P \circ y = 0\}$ *of* $\mathcal{B}$ *is stable, cf. Theorem and Definition 5.4.17, i.e., every pole* $\lambda \in \mathrm{char}(\mathcal{B}^0)$ *satisfies*

$$|\lambda| < 1 \ or\ \left(|\lambda| = 1 \ and\ \mathrm{index}(\mathcal{B}^0, s - \lambda) = 1\right).$$

   (b) *H is externally stable according to Theorem 8.1.26.*

Recall from Theorem 6.3.18 that $\mathrm{pole}(H) = \mathrm{char}(\mathcal{B}^0_{\mathrm{cont}}) \subseteq \mathrm{char}(\mathcal{B}^0)$ *and notice that while* $\mathcal{B}^0$ *is allowed to have certain poles on the unit circle, H must not have a pole there.*

Proof 1. $\Longrightarrow$ 2.: This follows as in Theorem 6.4.8.
2. $\Longrightarrow$ 1.: Let $\binom{y}{u} \in \mathcal{B}$ with $u \in (\ell^\infty)^m$. By Corollary and Definition 8.1.4, $H \circ u \in \mathcal{F}^q$ is well defined and $\binom{H \circ u}{u} \in \mathcal{B}$. Thus, $\binom{y - H \circ u}{0} = \binom{y}{u} - \binom{H \circ u}{u} \in \mathcal{B}$, i.e., $y - H \circ u \in \mathcal{B}^0$. Since $\mathcal{B}^0$ is stable, $y - H \circ u$ is bounded by Theorem and Definition 5.4.17. Since H is externally stable and u is bounded, also $H \circ u$ and hence $y = H \circ u + (y - H \circ u)$ are bounded. □

8.2 Transfer Operators in Continuous Time

In Sects. 8.2.1–8.2.3 we introduce and discuss the space $C^{-\infty} := C^{-\infty}(\mathbb{R}, \mathbb{C})$ of complex-valued distributions of finite order as the least signal space that is closed under differentiation and thus a $\mathbb{C}[s]$-module and that contains all piecewise continuous signals. The latter are needed, for instance, if an electrical network is switched on. The corresponding real data are, of course, also treated. In Sect. 8.2.4 we consider the subspace $C_+^{-\infty} := C_+^{-\infty}(\mathbb{R}, \mathbb{C})$ of distributions with left bounded support that is naturally a $\mathbb{C}(s)$-vector space. It contains Dirac's δ-distribution as *impulse signal* and gives rise to the *inverse Laplace transform* isomorphism $\mathcal{L}^{-1} : \mathbb{C}(s) \xrightarrow{\cong} \mathbb{C}(s) \circ \delta,\ H \mapsto H \circ \delta$, where $H \circ \delta$ is the *impulse response* of H. The inverse $\mathcal{L}$ of $\mathcal{L}^{-1}$ is the *Laplace transform*. In Sect. 8.2.7 we extend the Laplace transform to general *Laplace transformable* distributions. In Sect. 8.2.8 we also introduce and discuss the space $\mathcal{P}^{-\infty}$ of periodic distributions with period $T > 0$, $\omega := 2\pi/T$, with the natural action of the subring $\mathbb{C}(s)_{\mathrm{per}} := \{H \in \mathbb{C}(s);\ \mathrm{pole}(H) \cap \mathbb{Z}j\omega = \emptyset\}$ on it. Matrices $H \in \mathbb{C}(s)^{p \times m}$ resp. $H \in \mathbb{C}(s)_{\mathrm{per}}^{p \times m}$ induce the *transfer operators* or *input/output maps*

$$H\circ:\ \left(\mathrm{C}_+^{-\infty}\right)^m \to \left(\mathrm{C}_+^{-\infty}\right)^p \text{ resp. } H\circ:\ \left(\mathcal{P}^{-\infty}\right)^m \to \left(\mathcal{P}^{-\infty}\right)^p.$$

These transfer operators are basic tools for the study of IO behaviors, especially for electrical and mechanical networks. We describe these applications in Sects. 8.3 and 8.4. The convolution multiplication on $\mathrm{C}_+^{-\infty}$ enables to describe the transfer operator $H\circ:\ \left(\mathrm{C}^{-\infty}\right)^m \to \left(\mathrm{C}^{-\infty}\right)^p$ as convolution with the impulse response, cf. Sect. 8.2.5. In Sect. 8.2.6 we characterize bounded input/bounded output (BIBO) stability and, more generally, $(\mathrm{L}^p, \mathrm{L}^p)$-stability of the proper transfer operator $H\circ$ by the condition $\mathrm{pole}(H) \subset \mathbb{C}_- := \{\lambda \in \mathbb{C};\ \Re(\lambda) < 0\}$.

8.2.1 *The Space $\mathcal{D}^*(\mathbb{R}, \mathbb{C})$*

Motivation 8.2.1 Consider the series connection of a voltage source, a resistor, and a capacitor, i.e., the electrical network

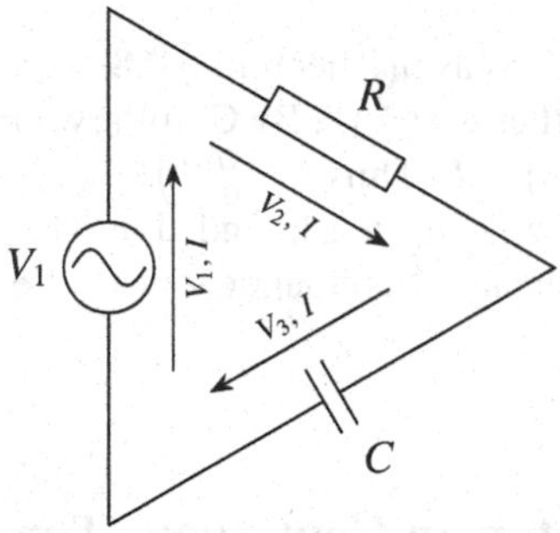

with the current I, the branch voltages V_1, V_2, and V_3, and the equations

$$V_1 + V_2 + V_3 = 0, \quad V_2 = RI, \quad \text{and} \quad I = C\tfrac{\mathrm{d}V_3}{\mathrm{d}t}.$$

We interpret the voltage $u := -V_1$ at the source as input and the voltage $y := V_3$ at the capacitor as output, and obtain the IO equation

$$\tfrac{\mathrm{d}y}{\mathrm{d}t} + \tfrac{1}{RC}y = \tfrac{1}{RC}u.$$

We assume that the system is at rest before time 0, i.e., that $V_1(t) = V_2(t) = V_3(t) = I(t) = 0$ for $t < 0$, and that a voltage $u_0 > 0$ is switched on at $t = 0$.
With the piecewise continuous *Heaviside function*

$$Y: \mathbb{R} \longrightarrow \mathbb{R}, \quad t \longmapsto Y(t) := \begin{cases} 1 & \text{if } t \geq 0, \\ 0 & \text{if } t < 0, \end{cases} \tag{8.42}$$

the equation for y becomes

$$\tfrac{\mathrm{d}y}{\mathrm{d}t} + \tfrac{1}{RC}y = \tfrac{1}{RC}u_0 Y(t).$$

Since the right side is not continuous, the solution y cannot be continuously differentiable.
In the standard first courses on analysis such equations are usually not treated. On the other hand, signals of the type Y are obviously significant, as the example shows. ◊

In the following we construct the least signal space that contains all continuous functions as well as the Heaviside function and its shifted versions, and is closed under differentiation, suitably defined. This is the space $\mathrm{C}^{-\infty}(\mathbb{R}, \mathbb{C})$ of *distributions of finite order*, see Lemma and Definition 8.2.25. We show that this space is an injective cogenerator over $\mathbb{C}[s]$. Therefore, most of the theorems of the preceding chapters are applicable to this new and larger signal space.
We start with the usual function spaces

$$\begin{array}{ccccc}
\mathrm{C}_0^{\infty}(\mathbb{R},\mathbb{R}) & & \subset & & \mathrm{C}_0^{0}(\mathbb{R},\mathbb{R}) \\
\cap & & & & \cap \\
\mathrm{C}^{\infty}(\mathbb{R},\mathbb{R}) & \subset & \mathrm{C}^{1}(\mathbb{R},\mathbb{R}) & \subset & \mathrm{C}^{0}(\mathbb{R},\mathbb{R}) \\
\cap & & \cap & & \cap \\
\mathrm{C}^{\infty}(\mathbb{R},\mathbb{C}) & \subset & \mathrm{C}^{1}(\mathbb{R},\mathbb{C}) & \subset & \mathrm{C}^{0}(\mathbb{R},\mathbb{C}) \\
\cup & & & & \cup \\
\mathrm{C}_0^{\infty}(\mathbb{R},\mathbb{C}) & & \subset & & \mathrm{C}_0^{0}(\mathbb{R},\mathbb{C})
\end{array}$$

where the lower index 0 signifies that the functions have compact support. Since every complex function $u\colon \mathbb{R} \longrightarrow \mathbb{C}$ is the sum of its real and its imaginary part, the direct sum decomposition

$$\begin{aligned}
&\mathrm{C}_0^{\infty}(\mathbb{R},\mathbb{C}) = \mathrm{C}_0^{\infty}(\mathbb{R},\mathbb{R}) \oplus j\,\mathrm{C}_0^{\infty}(\mathbb{R},\mathbb{R}) \ni u = \Re(u) + j\Im(u), \text{ with} \\
&j = \sqrt{-1},\ \Re(u)(t) := \Re(u(t)),\ \Im(u)(t) = \Im(u(t)),
\end{aligned} \tag{8.43}$$

holds, and likewise for the other spaces. We develop the following theory for the complex case that includes the case of real-valued functions or distributions. Whenever the real situation deviates from the complex one, we will point this out. We will often use the short notation $\mathrm{C}^0 := \mathrm{C}^0(\mathbb{R}, \mathbb{C})$ etc.

Remark 8.2.2 The space $\mathcal{D}'(\mathbb{R}, \mathbb{C})$ of complex distributions is usually defined as the space of all *continuous* $\mathbb{C}$-linear functions $T\colon\ \mathrm{C}_0^{\infty}(\mathbb{R}, \mathbb{C}) \longrightarrow \mathbb{C}$. Continuity requires a topology or at least a suitable notion of convergence on the space C_0^{∞}. For our construction of $\mathrm{C}^{-\infty}(\mathbb{R}, \mathbb{C}) \subset \mathcal{D}'(\mathbb{R}, \mathbb{C})$ this topology is not needed. This is a significant simplification. ◊

Definition and Corollary 8.2.3 We define the dual $\mathbb{C}$-vector space

$$\mathcal{D}^* := \mathcal{D}^*_{\mathbb{C}} := \mathrm{Hom}_{\mathbb{C}}(\mathrm{C}_0^\infty(\mathbb{R}, \mathbb{C}), \mathbb{C})$$

of all $\mathbb{C}$-linear maps (also called *functionals*) $T\colon\ \mathrm{C}_0^\infty(\mathbb{R}, \mathbb{C}) \longrightarrow \mathbb{C}$. The real counterpart of $\mathcal{D}^*_{\mathbb{C}}$ is $\mathcal{D}^*_{\mathbb{R}} := \mathrm{Hom}_{\mathbb{R}}(\mathrm{C}_0^\infty(\mathbb{R}, \mathbb{R}), \mathbb{R})$. With the usual identifications, the direct sum decomposition (8.43) implies

$$\mathcal{D}^*_{\mathbb{C}} = \mathcal{D}^*_{\mathbb{R}} \oplus j\mathcal{D}^*_{\mathbb{R}} \ni T = U + jV,$$
$$\text{where } T(u + jv) = (U + jV)(u + jv) = \big(U(u) - V(v)\big) + j\big(U(v) + V(u)\big).$$

Thus the complex-valued maps T have the form $T = U + jV$ with two real-valued maps U, V and, in particular, generalize these. ◊

Definition 8.2.4 For $a \in \mathbb{R}$, we define the δ-distribution $\delta_a \in \mathcal{D}^*_{\mathbb{C}}$ at a by

$$\delta_a\colon\ \mathrm{C}_0^\infty \longrightarrow \mathbb{C},\quad \varphi \longmapsto \delta_a(\varphi) := \varphi(a).$$

The functional δ_a is even defined on $\mathrm{C}_0^0(\mathbb{R}, \mathbb{C}) \supset \mathrm{C}_0^\infty(\mathbb{R}, \mathbb{C})$. For $a = 0$ we write $\delta := \delta_0$. Engineers often prefer the standard terminology of a *δ-function* instead of δ-distribution and write

$$\varphi(a) = \int_{-\infty}^{+\infty} \varphi(t)\, \delta(t - a)\, \mathrm{d}t.$$

Notice, however, that δ is not a function $\mathbb{R} \to \mathbb{R}$, and the meaning of this integral is exactly the definition we have just given. ◊

The prominent role of the Heaviside step function Y as signal requires the introduction of *piecewise continuous* functions. The signal space $\mathrm{C}^{-\infty}(\mathbb{R}, \mathbb{C})$ of distributions, introduced in Lemma and Definition 8.2.25, contains all reasonable function spaces and is an injective cogenerator.

Piecewise continuous functions in the sense of this book have only finitely many singularities of a simple kind. If $S = \{t_1 < t_2 < \cdots < t_n\}$, $t_i \in \mathbb{R}$, is a finite subset of the real line, we introduce $t_0 := -\infty$ and $t_{n+1} := +\infty$ and obtain the disjoint decomposition

$$\mathbb{R}\backslash S = \biguplus_{i=0}^{n} (t_i, t_{i+1}), \tag{8.44}$$

where (t_i, t_{i+1}) denotes the open interval from t_i to t_{i+1}.

Definition and Corollary 8.2.5 A function $y(t)$ is *piecewise continuous* if there is a finite subset $S = \{t_1 < t_2 < \cdots < t_n\}$, $t_i \in \mathbb{R}$, of $\mathbb{R}$ such that

1. $y \in \mathrm{C}^0(\mathbb{R}\backslash S, \mathbb{C})$, i.e., the function is defined on the disjoint union of open intervals (8.44) and is continuous there, and

2. for $i = 1, \ldots, n$, the left-sided and the right-sided limits

$$y(t_i-) := \lim_{t \to t_i,\ t<t_i} y(t) \quad \text{and} \quad y(t_i+) := \lim_{t \to t_i,\ t>t_i} y(t)$$

   exist.

With $y(t) := y(t+)$, we may consider y as a function on all of $\mathbb{R}$.
The equality of two piecewise continuous functions $y_i \in \mathrm{C}^0(\mathbb{R}\backslash S_i, \mathbb{C})$, $i = 1, 2$, is defined as

$$\begin{aligned} y_1 = y_2 :&\Longleftrightarrow y_1|_{\mathbb{R}\backslash(S_1\cup S_2)} = y_2|_{\mathbb{R}\backslash(S_1\cup S_2)} \\ &\Longleftrightarrow \forall t \in \mathbb{R}\colon\ y_1(t-) = y_2(t-) \text{ and } y_1(t+) = y_2(t+). \end{aligned}$$

A continuous function is obviously also piecewise continuous. The set of all piecewise continuous functions is a $\mathbb{C}$-vector space with the usual pointwise addition and scalar multiplication and is denoted by $\mathrm{C}^{0,\mathrm{pc}} := \mathrm{C}^{0,\mathrm{pc}}(\mathbb{R}, \mathbb{C})$. Piecewise continuous functions are Riemann integrable on every finite interval.
For $0 \le m \le \infty$, we introduce the space $\mathrm{C}^{m,\mathrm{pc}} := \mathrm{C}^{m,\mathrm{pc}}(\mathbb{R}, \mathbb{C})$ of piecewise m-times continuously differentiable functions in an analogous manner. They belong to some $\mathrm{C}^m(\mathbb{R}\backslash S, \mathbb{C})$ as before, i.e., they are m-times continuously differentiable on a disjoint union of open intervals, and satisfy the additional condition that for all $k = 0, \ldots, m$ and all $i = 1, \ldots, n$, the left- and the right-sided limits

$$y^{(k)}(t_i-) = \lim_{t \to t_i,\ t<t_i} y^{(k)}(t) \quad \text{and} \quad y^{(k)}(t_i+) = \lim_{t \to t_i,\ t>t_i} y^{(k)}(t)$$

exist, where $y^{(k)} = \frac{\mathrm{d}^k y}{\mathrm{d}t^k}$.
The definition of pc may be generalized to countable sets of singularities $S = \{t_1 < t_2 < \cdots < t_k < \cdots\}$ that are discrete or have no accumulation points. $\Diamond$

The standard example of a function in $\mathrm{C}^{\infty,\mathrm{pc}}$ is the Heaviside function Y with a jump singularity at 0, namely,

$$Y(t) := \begin{cases} 1 & \text{if } t \ge 0 \\ 0 & \text{if } t < 0 \end{cases}, \quad Y(0+) = 1, \ Y(0-) = 0. \tag{8.45}$$

The shifted functions

$$Y_a\colon \mathbb{R} \longrightarrow \mathbb{R}, \quad t \longmapsto Y_a(t) := Y(t-a), \quad a \in \mathbb{R},$$

have their jump at a. The next theorem shows that all piecewise continuous functions are uniquely composed of a continuous function and finitely many jumps.

Theorem 8.2.6 *The space* $C^{0,pc}(\mathbb{R}, \mathbb{C})$ *has the direct decomposition*

$$C^{0,pc}(\mathbb{R}, \mathbb{C}) = \qquad C^0(\mathbb{R}, \mathbb{C}) \qquad \bigoplus \oplus_{a\in\mathbb{R}} \mathbb{C} Y_a,$$

$$y = \left(y - \sum_{i=1}^{n} \big(y(t_i+) - y(t_i-)\big) Y_{t_i} \right) + \sum_{i=1}^{n} \big(y(t_i+) - y(t_i-)\big) Y_{t_i},$$

where y *is piecewise continuous with possible jumps at* $t_1 < t_2 < \cdots < t_n$.

Proof First we show that $C^{0,pc}(\mathbb{R}, \mathbb{C})$ is the sum of the spaces on the right-hand side. For this purpose, it suffices to show that the function

$$u := y - \sum_{i=1}^{n} \big(y(t_i+) - y(t_i-)\big) Y_{t_i}$$

is continuous. Clearly, y, Y_{t_i} for all i, and consequently also u are continuous on $\mathbb{R}\backslash S$, where $S = \{t_1, \ldots, t_n\}$. Let $i \in \{1, \ldots, n\}$. In order to show that u is continuous at t_i, it is sufficient to show that $u(t_i+) = u(t_i-)$. But since $t_i \neq t_j$ for $i \neq j$, we have that

$$\begin{aligned} x :=& \sum_{j\in\{1,\ldots,n\}\backslash\{i\}} \big(y(t_j+) - y(t_j-)\big)\, Y_{t_j}(t_i) = \\ =& \sum_{j\in\{1,\ldots,n\}\backslash\{i\}} \big(y(t_j+) - y(t_j-)\big)\, Y_{t_j}(t_i+) \\ =& \sum_{j\in\{1,\ldots,n\}\backslash\{i\}} \big(y(t_j+) - y(t_j-)\big)\, Y_{t_j}(t_i-), \end{aligned}$$

and hence,

$$u(t_i+) = y(t_i+) - \big(y(t_i+) - y(t_i-)\big) \underbrace{Y_{t_i}(t_i+)}_{=Y(0+)=1} - x = y(t_i-) - x \quad \text{and}$$

$$u(t_i-) = y(t_i-) - \big(y(t_i+) - y(t_i-)\big) \underbrace{Y_{t_i}(t_i-)}_{=Y(0-)=0} - x = y(t_i-) - x = u(t_i+).$$

To prove that the sum is direct, let

$$0 = u + \sum_{i=1}^{n} a_i Y_{t_i} \text{ with } t_1 < \cdots < t_n,$$

where u and thus $y := -u = \sum_{i=1}^{n} a_i Y_{t_i}$ are continuous. But

$$Y_{t_n}(t_n+) = Y(0+) = 1, \quad Y_{t_n}(t_n-) = Y(0-) = 0 \text{ and}$$
$$Y_{t_i}(t_n+) = Y_{t_i}(t_n-) = Y(t_n - t_i) = 1, \ i < n,$$
$$\Longrightarrow \sum_{i=1}^{n} a_i = \sum_{i=1}^{n} a_i Y_{t_i}(t_n+) = y(t_n+) = y(t_n-) = \sum_{i=1}^{n} a_i Y_{t_i}(t_n-) = \sum_{i=1}^{n-1} a_i$$
$$\Longrightarrow a_n = 0 \underset{\text{recursion}}{\Longrightarrow} a_{n-1} = \cdots = a_1 = 0 \Longrightarrow u = 0 \Longrightarrow \text{sum direct.}$$
□

Lemma 8.2.7 *If y is piecewise continuous, the Riemann integral $\int_{t_0}^{t} y(x)\,\mathrm{d}x$ is a continuous function of t for all $t_0 \in \mathbb{R}$.* ◇

The next theorem establishes the important embedding of functions into distributions. We first define the $\mathbb{C}$-bilinear map

$$\begin{aligned} &\mathrm{C}^{0,\mathrm{pc}} \times \mathrm{C}_0^0(\mathbb{R}, \mathbb{C}) \longrightarrow \mathbb{C}, \\ &(u, \varphi) \longmapsto \langle u, \varphi \rangle := \int_{-\infty}^{\infty} u(t)\,\varphi(t)\,\mathrm{d}t = \int_a^b u(t)\,\varphi(t)\,\mathrm{d}t, \end{aligned} \tag{8.46}$$

where the support of φ is contained in the closed finite interval $[a, b]$. This bilinear map induces the $\mathbb{C}$-linear map

$$\mathrm{C}^{0,\mathrm{pc}}(\mathbb{R}, \mathbb{C}) \longrightarrow \mathrm{Hom}_{\mathbb{C}}(\mathrm{C}_0^\infty(\mathbb{R}, \mathbb{C}), \mathbb{C}) = \mathcal{D}^*, \quad u \longmapsto \langle u, - \rangle. \tag{8.47}$$

Theorem and Definition 8.2.8 *The map* (8.47) *is a monomorphism. Thus we will mostly, but not always,* identify

$$\mathrm{C}^{0,\mathrm{pc}}(\mathbb{R}, \mathbb{C}) \subset \mathcal{D}^* \ni u = \langle u, - \rangle, \quad \text{via } u(\varphi) = \langle u, \varphi \rangle = \int_{-\infty}^{\infty} u(t)\,\varphi(t)\,\mathrm{d}t.$$

Notice that, in this fashion, u has two different meanings, namely, as a (in general nonlinear) complex-valued function $u(t)$ on $\mathbb{R}$ and as a $\mathbb{C}$-linear function $u(\varphi)$ on $\mathrm{C}_0^\infty(\mathbb{R}, \mathbb{C})$. The arguments t and φ show which interpretation is meant.

Proof Since the map (8.47) is linear, it suffices to show that its kernel is zero. So let u be piecewise continuous with possible jumps at $t_1 < \cdots < t_n$ and assume that $\langle u, - \rangle = 0$. We are going to show that $u = 0$ or, equivalently, that $u(t) = 0$ for all $t \neq t_i$. Let $t_0 := -\infty$ and $t_{n+1} := +\infty$. Let $u = v + jw$ be the decomposition of u into its real and imaginary part and assume that, for instance, $v \neq 0$, that $t_i < \tau < t_{i+1}$, and that $v(\tau) > 0$ without loss of generality. We choose ϵ such that

$$[\tau - \epsilon, \tau + \epsilon] \subset (t_i, t_{i+1}), \quad \text{and} \quad v(t) > 0 \text{ for } t \in [\tau - \epsilon, \tau + \epsilon].$$

According to (4.59), let $\varphi := \varphi_{\tau,\epsilon} \in \mathrm{C}_0^\infty(\mathbb{R}, \mathbb{R})$ with

$$\varphi(t) > 0 \text{ for } |t - \tau| < \epsilon \quad \text{and} \quad \varphi(t) = 0 \text{ for } |t - \tau| \geq \epsilon.$$

Since $\langle u, - \rangle = 0$, we have that

$$0 = \langle u, \varphi \rangle = \langle v, \varphi \rangle + j \langle w, \varphi \rangle \in \mathbb{C} = \mathbb{R} \oplus j\mathbb{R},$$

hence, in particular, $\langle v, \varphi \rangle = 0$. On the other hand, $v(t) > 0$ and $\varphi(t) > 0$ for $t \in (\tau - \epsilon, \tau + \epsilon)$, and $\varphi(t) = 0$ for $t \notin (\tau - \epsilon, \tau + \epsilon)$ implies that

$$\langle v, \varphi \rangle = \int_{-\infty}^{\infty} v(t)\, \varphi(t)\, \mathrm{d}t = \int_{\tau-\epsilon}^{\tau+\epsilon} v(t)\, \varphi(t)\, \mathrm{d}t > 0.$$

This contradiction refutes the assumption and thus proves the theorem. □

In engineering mathematics, the δ-distribution is interpreted as an idealized impulse of length zero. In order to formulate this properly in Theorem 8.2.10, we need a notion of convergence on the space of distributions.

Lemma 8.2.9 *The space $\mathcal{D}^*$ is closed with respect to weak convergence in the following sense. Assume that $u_n \in \mathcal{D}^*$, $n \geq 1$, is a sequence such that for all $\varphi \in C_0^\infty(\mathbb{R}, \mathbb{C})$ the pointwise limit $u(\varphi) := \lim_{n\to\infty} u_n(\varphi) \in \mathbb{C}$ exists. Then also $u \in \mathcal{D}^*$, and $\lim_{n\to\infty} u_n := u$ is called the limit of the sequence u_n.*

Proof The map u is $\mathbb{C}$-linear since the limit has this property. □

Theorem 8.2.10 (The δ-distribution as impulse) *For $\epsilon > 0$ we consider $\varphi_{0,\epsilon}$ from* (4.59) *with*

$$\varphi_{0,\epsilon}(t) \begin{cases} > 0 & \text{for } |t| < \epsilon \\ 0 & \text{for } |t| \geq \epsilon \end{cases} \quad \text{and define } u_\epsilon := \frac{\varphi_{0,\epsilon}}{\int_{-\infty}^{\infty} \varphi_\epsilon(t)\, \mathrm{d}t} \in C^\infty(\mathbb{R}, \mathbb{R}) \subset \mathcal{D}^*$$
$$\Longrightarrow \int_{-\infty}^{\infty} u_\epsilon(t)\, \mathrm{d}t = 1.$$

The function u_ϵ is interpreted as an impulse of length 2ϵ and mass 1.
If $(\epsilon_n)_{n\in\mathbb{N}}$ is a zero sequence of positive real numbers, then $(u_{\epsilon_n})_{n\in\mathbb{N}}$ is a sequence in $C_0^\infty(\mathbb{R}, \mathbb{R}) \subset \mathcal{D}^*$ that converges weakly in $\mathcal{D}^*$ to δ. This result suggests to interpret δ as an idealized impulse of length zero.

Proof For $\varphi \in C_0^\infty$, we have to show $\lim_{n\to\infty} u_{\epsilon_n}(\varphi) = \delta(\varphi) = \varphi(0)$.
(i) First assume $\varphi(0) = 0$. Let $\epsilon' > 0$. Then

$$\exists n_0\ \forall n \geq n_0\ \forall t \in [-\epsilon_n, \epsilon_n]:\ |\varphi(t)| \leq \epsilon'$$

$$\Longrightarrow |u_{\epsilon_n}(\varphi)| = \left|\int_{-\infty}^{\infty} u_{\epsilon_n}(t)\,\varphi(t)\,\mathrm{d}t\right| = \left|\int_{-\epsilon_n}^{\epsilon_n} u_{\epsilon_n}(t)\,\varphi(t)\,\mathrm{d}t\right|$$

$$\leq \int_{-\epsilon_n}^{\epsilon_n} u_{\epsilon_n}(t)\,\underbrace{|\varphi(t)|}_{\leq \epsilon'}\,\mathrm{d}t \leq \epsilon' \int_{-\epsilon_n}^{\epsilon_n} u_{\epsilon_n}(t)\,\mathrm{d}t = \epsilon'$$

$$\underset{\epsilon' \text{ arbitrary}}{\Longrightarrow} \lim_{n\to\infty} u_{\epsilon_n}(\varphi) = 0 = \varphi(0).$$

(ii) For arbitrary $\varphi \in \mathrm{C}_0^\infty$ choose $0 < \epsilon_n < 1$, $\epsilon' > 0$ and $\psi_{\epsilon'}$ from (4.61). Then

$$\psi(t) := \psi_{\epsilon'}(t-1)\psi_{\epsilon'}(-t-1) \in \mathrm{C}_0^\infty,\ \psi|_{[-1,1]} = 1,\ u_{\epsilon_n}(\psi) = 1 \text{ and}$$
$$\varphi = \varphi(0)\psi + (\varphi - \varphi(0)\psi) \text{ with } (\varphi - \varphi(0)\psi)(0) = 0$$
$$\underset{(i)}{\Longrightarrow} u_{\epsilon_n}(\varphi) = \varphi(0)u_{\epsilon_n}(\psi) + u_{\epsilon_n}(\varphi - \varphi(0)\psi) \underset{n\to\infty}{\longrightarrow} \varphi(0) = \delta(\varphi). \quad \square$$

8.2.2 Actions on Functionals

We generalize differentiation, shifting, and multiplication by smooth functions to functionals in $\mathcal{D}^*$ and derive properties of these actions. In particular, $\mathcal{D}^*$ becomes a $\mathbb{C}[s]$-module whose torsion submodule coincides with that of C^∞, i.e., consists of the polynomial-exponential functions.

Reminder 8.2.11 Assume a field F, a commutative F-algebra A, an A-module $(V, \circ)$, and an F-vector space W. Then $\mathrm{Hom}_F(V, W)$ is an F-vector space and even an A-module with the scalar multiplication

$$(a \circ y)(v) := y(a \circ v) \quad \text{for } y \in \mathrm{Hom}_F(V, W),\ a \in A, \text{ and } v \in V.$$

This applies in particular to the dual vector space $V^* = \mathrm{Hom}_F(V, F)$. ◇

We apply this to the $\mathbb{C}[s]$-submodule $\mathrm{C}_0^\infty(\mathbb{R}, \mathbb{C})$ of $\mathrm{C}^\infty(\mathbb{R}, \mathbb{C})$ with the scalar multiplication from the preceding chapters, i.e., given by $s \circ y = y' = \frac{\mathrm{d}y}{\mathrm{d}t}$. This gives rise to the following scalar multiplication $\circ_1$:

$$(s \circ_1 y)(\varphi) := y(s \circ \varphi),\ y \in \mathcal{D}^*,\ \varphi \in \mathrm{C}_0^\infty(\mathbb{R}, \mathbb{C}).$$

This $\mathbb{C}[s]$-structure on $\mathcal{D}^*$ has the disadvantage that the embedding (8.47) is not $\mathbb{C}[s]$-linear on C^1-functions. Therefore we define the final $\mathbb{C}[s]$-structure on $\mathcal{D}^*$ via

$$(s \circ y)(\varphi) := y(-s \circ \varphi) = -y(\tfrac{d\varphi}{dt}) \text{ for } y \in \mathcal{D}^* \text{ and } \varphi \in C_0^\infty(\mathbb{R}, \mathbb{C}), \; y' := s \circ y. \tag{8.48}$$

The functional y' is called the *distributional derivative* of y. For the higher derivatives we obtain

$$\begin{aligned} &y^{(m)} := s^m \circ y, \quad \text{i.e., } y^{(m)}(\varphi) = (-1)^m y(\varphi^{(m)}), \text{ and for} \\ &f = \sum_{m=0}^{\deg_s(f)} a_m s^m \in \mathbb{C}[s]: \; (f \circ y)(\varphi) = y\left(\sum_{m=0}^{\deg_s(f)} a_m (-1)^m \varphi^{(m)} \right). \end{aligned} \tag{8.49}$$

Corollary 8.2.12 *Let $a \in \mathbb{R}$. The derivative of the shifted Heaviside function $Y_a\colon t \mapsto Y(t-a)$ is δ_a.*

Proof Let $\varphi \in C_0^\infty(\mathbb{R}, \mathbb{C})$. Let $[A, B] \subset \mathbb{R}$ be a finite interval which contains a and such that $\operatorname{supp}(\varphi) \subseteq (A, B)$. Then

$$\begin{aligned} Y_a'(\varphi) &= -Y_a(\varphi') = -\int_{-\infty}^{\infty} Y(t-a)\,\varphi'(t)\,\mathrm{d}t = -\int_a^B \varphi'(t)\,\mathrm{d}t \\ &= -\big(\varphi(B) - \varphi(a)\big) = \varphi(a) = \delta_a(\varphi). \end{aligned}$$

□

In the sequel we will shorten proofs like the preceding one by not mentioning the interval $[A, B]$ containing the support of φ explicitly. The preceding proof, for instance, becomes

$$\begin{aligned} Y_a'(\varphi) &= -Y_a(\varphi') = -\int_{-\infty}^{\infty} Y(t-a)\,\varphi'(t)\,\mathrm{d}t = -\int_a^\infty \varphi'(t)\,\mathrm{d}t \\ &= -\big(\varphi(\infty) - \varphi(a)\big) = \varphi(a) = \delta_a(\varphi), \end{aligned}$$

where we use $y(\infty) := \lim_{t\to\infty} y(t)$ and $y(-\infty) := \lim_{t\to-\infty} y(t)$ if these limits exist. However, the introduction of $[A, B]$ is always possible in this book, and therefore the existence of infinite integrals $\int_{-\infty}^{\infty} u(t)\,\mathrm{d}t$ is never a problem.

Lemma 8.2.13 *Let $u \in C^1(\mathbb{R}, \mathbb{C})$ be continuously differentiable. Then*

$$s \circ \langle u, - \rangle = \langle s \circ u, - \rangle.$$

With the identification $u = \langle u, - \rangle \in C^1(\mathbb{R}, \mathbb{C}) \subset C^{0,\mathrm{pc}} \subset \mathcal{D}^*$ from Theorem and Definition 8.2.8, this signifies that the customary derivative and the distributional derivative of a C^1-function coincide.

Proof For $u \in \mathrm{C}^1(\mathbb{R}, \mathbb{C})$ and $\varphi \in \mathrm{C}_0^\infty(\mathbb{R}, \mathbb{C})$, partial integration yields

$$\begin{aligned}
\big(s \circ \langle u, -\rangle\big)(\varphi) &= -\langle u, \varphi'\rangle = -\int_{-\infty}^{\infty} u(t)\,\varphi'(t)\,\mathrm{d}t \\
&= -(u\varphi)\Big|_{-\infty}^{\infty} + \int_{-\infty}^{\infty} u'(t)\,\varphi(t)\,\mathrm{d}t \\
&= -u(\infty)\,\varphi(\infty) + u(-\infty)\,\varphi(-\infty) + \int_{-\infty}^{\infty} u'(t)\,\varphi(t)\,\mathrm{d}t \\
&\underset{\varphi(-\infty)=\varphi(\infty)=0}{=} \langle u', \varphi\rangle = \langle u', -\rangle(\varphi).
\end{aligned}$$

□

In the next theorem we differentiate arbitrary functions in $\mathrm{C}^{1,\mathrm{pc}}(\mathbb{R}, \mathbb{C})$.

Theorem 8.2.14 *Let $u \in \mathrm{C}^{1,\mathrm{pc}}(\mathbb{R}, \mathbb{C})$ be a function with possible singularities at $t_1 < \cdots < t_n$. Let $w \in \mathrm{C}^{0,\mathrm{pc}}(\mathbb{R}, \mathbb{C})$ be defined by $w|_{(t_i,t_{i+1})} := u'|_{(t_i,t_{i+1})}$, i.e., w is the ordinary first derivative of u where this is defined. Then the distributional derivative of u is*

$$u' = w + \sum_{i=1}^{n} \big(u(t_i+) - u(t_i-)\big)\delta_{t_i}.$$

Proof (i) Assume first that u is continuous and $\varphi \in \mathrm{C}_0^\infty$. We have to show that $u'(\varphi) = w(\varphi)$. Let $t_0 < t_1$ and $t_n < t_{n+1}$ be such that $\mathrm{supp}(\varphi) \subseteq [t_0, t_{n+1}]$. This means, in particular, that $\varphi(t_0) = \varphi(t_{n+1}) = 0$. We use partial integration for the functions $u|_{[t_i,t_{i+1}]} \in \mathrm{C}^1([t_i, t_{i+1}], \mathbb{C})$

$$\begin{aligned}
u'(\varphi) &= -u(\varphi') = -\int_{-\infty}^{\infty} u(t)\,\varphi'(t)\,\mathrm{d}t = -\int_{t_0}^{t_{n+1}} u(t)\,\varphi'(t)\,\mathrm{d}t \\
&= -\sum_{i=0}^{n} \int_{t_i}^{t_{i+1}} u(t)\,\varphi'(t)\,\mathrm{d}t = -\sum_{i=0}^{n} \left((u\varphi)\Big|_{t_i}^{t_{i+1}} - \int_{t_i}^{t_{i+1}} \underbrace{u'(t)}_{=w(t)}\varphi(t)\,\mathrm{d}t \right) \\
&= \sum_{i=0}^{n} \big(-u(t_{i+1})\,\varphi(t_{i+1}) + u(t_i)\,\varphi(t_i)\big) + \int_{t_0}^{t_{n+1}} w(t)\,\varphi(t)\,\mathrm{d}t \\
&\underset{\varphi(t_0)=\varphi(t_{n+1})=0}{=} \int_{-\infty}^{\infty} w(t)\,\varphi(t)\,\mathrm{d}t = w(\varphi).
\end{aligned}$$

(ii) According to Theorem 8.2.6 an arbitrary $u \in \mathrm{C}^{1,\mathrm{pc}}(\mathbb{R}, \mathbb{C})$ can be written as

$$u = v + \sum_{i=1}^{n} \big(u(t_i+) - u(t_i-)\big)\,Y_{t_i},\quad v \text{ continuous},\quad S := \{t_1 < \cdots < t_n\}. \tag{8.50}$$

The second summand is contained in $\mathrm{C}^{1,\mathrm{pc}}(\mathbb{R}\backslash S, \mathbb{C})$ and has derivative zero there. Therefore the derivatives of u and v on $\mathbb{R}\backslash S$ coincide and are equal to w. From (i) we infer $v' = w$ and from Corollary 8.2.12 $Y_{t_i}' = \delta_{t_i}$. Hence

$$u' = v' + \sum_{i=1}^{n}(u(t_i+) - u(t_i-))Y'_{t_i} = v' + \sum_{i=1}^{n}(u(t_i+) - u(t_i-))\delta_{t_i}. \qquad \square$$

Corollary 8.2.15 *Let $y \in \mathrm{C}^{0,\mathrm{pc}}(\mathbb{R}, \mathbb{C})$ be piecewise continuous and define*

$$u\colon\ \mathbb{R} \longrightarrow \mathbb{C},\quad t \longmapsto \int_{t_0}^{t} y(x)\,\mathrm{d}x.$$

Then $u \in \mathrm{C}^{1,\mathrm{pc}}(\mathbb{R}, \mathbb{C})$ and $u' = y$. This implies in particular that $Y = u'$ with

$$u(t) := \int_0^t Y(x)\,\mathrm{d}x = \begin{cases} t & if\, t \geq 0 \\ 0 & if\, t \leq 0 \end{cases} = tY(t).$$

Proof This follows from the preceding theorem since u is continuous and its derivative coincides with y in the nonsingular points of y. $\square$

Definition and Corollary 8.2.16 (*The shift action*) The action

$$\mathbb{R} \times \mathbb{R} \longrightarrow \mathbb{R},\quad t \longmapsto t + a,$$

of the abelian group $\mathbb{R}$ on itself induces the $\mathbb{C}$-linear action

$$\circ_{\mathrm{s}}\colon\ \mathbb{R} \times \mathrm{C}^{0,\mathrm{pc}}(\mathbb{R}, \mathbb{C}) \longrightarrow \mathrm{C}^{0,\mathrm{pc}}(\mathbb{R}, \mathbb{C}),\quad (a, y) \longmapsto a \circ_{\mathrm{s}} y,\quad (a \circ_{\mathrm{s}} y)(t) := y(t - a),$$

of the abelian group $\mathbb{R}$ on the vector space $\mathrm{C}^{0,\mathrm{pc}}(\mathbb{R}, \mathbb{C})$, and it induces similar actions on all other function spaces introduced above. The action is called $\mathbb{C}$-*linear* since for fixed $a \in \mathbb{R}$, the map

$$a \circ_{\mathrm{s}} (-)\colon\ \mathrm{C}^{0,\mathrm{pc}}(\mathbb{R}, \mathbb{C}) \longrightarrow \mathrm{C}^{0,\mathrm{pc}}(\mathbb{R}, \mathbb{C})$$

has this property. If $a > 0$, the function $(a \circ_{\mathrm{s}} y)(t) = y(t - a)$ is a copy of y, but shifted a units to the right.
By Reminder 8.2.11 and in analogy to the $\mathbb{C}[s]$-action on $\mathcal{D}$, this action on $\mathrm{C}_0^{\infty}(\mathbb{R}, \mathbb{C})$ induces the $\mathbb{C}$-linear action

$$\circ_1\colon\ \mathbb{R} \times \mathcal{D}^* \longrightarrow \mathcal{D}^*,\quad (a, y) \longmapsto a \circ_1 y,\quad (a \circ_1 y)(\varphi) := y(a \circ_{\mathrm{s}} \varphi),$$

on $\mathcal{D}^* = \mathrm{Hom}_{\mathbb{C}}(\mathrm{C}_0^{\infty}(\mathbb{R}, \mathbb{C}), \mathbb{C})$. Again, this action is not quite the right one because the embedding $\mathrm{C}^{0,\mathrm{pc}} \longrightarrow \mathcal{D}^*$ is not equivariant, i.e., it does not preserve the action. Therefore, we define the final action on $\mathcal{D}^*$ as

$$\circ_{\mathrm{s}}\colon\ \mathbb{R} \times \mathcal{D}^* \longrightarrow \mathcal{D}^*,\quad (a, y) \longmapsto a \circ_{\mathrm{s}} y,\quad (a \circ_{\mathrm{s}} y)(\varphi) := y((-a) \circ_{\mathrm{s}} \varphi), \tag{8.51}$$

and call it the distributional shift action. $\diamondsuit$

Lemma 8.2.17 *The embedding*

$$C^{0,pc}(\mathbb{R}, \mathbb{C}) \longrightarrow \mathcal{D}^*, \quad u \longmapsto \langle u, -\rangle,$$

is equivariant, i.e., $a \circ_s \langle u, -\rangle = \langle a \circ_s u, -\rangle$ *for all* $u \in C^{0,pc}(\mathbb{R}, \mathbb{C})$ *and* $a \in \mathbb{R}$. *With the identification* $u = \langle u, -\rangle$, *this signifies that the distributional shift and the functional shift of a piecewise continuous function coincide.*

Proof Let $u \in C^{0,pc}$ and $a \in \mathbb{R}$, and let $\varphi \in C_0^\infty$. Then

$$\begin{aligned}\big(\langle a \circ_s u, -\rangle\big)(\varphi) &= \langle a \circ_s u, \varphi\rangle = \int_{-\infty}^{\infty} u(t-a)\,\varphi(t)\,\mathrm{d}t \\ &\overset{\tau := t-a}{=\!=\!=} \int_{-\infty}^{\infty} u(\tau)\,\varphi(\tau + a)\,\mathrm{d}\tau = \int_{-\infty}^{\infty} u(\tau)\big((-a)\circ_s \varphi\big)(\tau)\,\mathrm{d}\tau \\ &= \langle u, (-a)\circ_s \varphi\rangle = \big(a \circ_s \langle u, -\rangle\big)(\varphi).\end{aligned}$$

□

Lemma 8.2.18 *The shift action on* $\mathcal{D}^*$ *commutes with differentiation. This means that the shift action is* $\mathbb{C}[s]$*-linear, i.e.,* $a \circ_s (f \circ y) = f \circ (a \circ_s y)$ *for all* $y \in \mathcal{D}^*$, $a \in \mathbb{R}$, *and* $f \in \mathbb{C}[s]$.

Proof It suffices to prove this for $f = s$. Recall $y' := s \circ y,\ y \in \mathcal{D}^*$. Clearly, $a \circ_s \varphi' = (a \circ_s \varphi)'$ for $\varphi \in C_0^\infty(\mathbb{R}, \mathbb{C})$. But this implies

$$\begin{aligned}\big(a \circ_s (s \circ y)\big)(\varphi) &= (a \circ_s y')(\varphi) = y'\big((-a)\circ_s \varphi\big) = -y\big(((-a)\circ_s \varphi)'\big) \\ &= -y\big((-a)\circ_s \varphi'\big) = -(a \circ_s y)(\varphi') = (a \circ_s y)'(\varphi) \\ &= \big(s \circ (a \circ_s y)\big)(\varphi).\end{aligned}$$

□

The space $C^\infty(\mathbb{R}, \mathbb{C})$ is a $\mathbb{C}$-algebra with the argumentwise multiplication $(\alpha\varphi)(t) := \alpha(t)\,\varphi(t)$ for $\alpha, y \in C^\infty(\mathbb{R}, \mathbb{C})$, and $C_0^\infty(\mathbb{R}, \mathbb{C})$ is a $C^\infty(\mathbb{R}, \mathbb{C})$-submodule or ideal of this ring. According to Reminder 8.2.11 the action

$$C^\infty(\mathbb{R}, \mathbb{C}) \times \mathcal{D}^* \longrightarrow \mathcal{D}^*, \quad (\alpha, y) \longmapsto \alpha y, \quad (\alpha y)(\varphi) := y(\alpha\varphi), \tag{8.52}$$

makes $\mathcal{D}^*$ a $C^\infty(\mathbb{R}, \mathbb{C})$-module.

Lemma 8.2.19 *The embedding* (8.47) *of* $C^{0,pc}(\mathbb{R}, \mathbb{C})$ *into* $\mathcal{D}^*$ *is* C^∞*-linear, i.e.,* $\langle \alpha u, -\rangle = \alpha\langle u, -\rangle$ *for* $\alpha \in C^\infty(\mathbb{R}, \mathbb{C})$, $u \in C^{0,pc}(\mathbb{R}, \mathbb{C})$. *Moreover the product rule holds, i.e.,*

$$(\alpha y)' = \alpha' y + \alpha y' \text{ for } \alpha \in C^\infty \text{ and } y \in \mathcal{D}^*.$$

◇

The proof is simple. It can be shown that there is no multiplication on $\mathcal{D}^*$ which extends the multiplication from (8.52). Later we will, however, define the convolution product $u_1 * u_2$ of distributions with left bounded support.

Example 8.2.20 Let $\alpha \in \mathrm{C}^\infty(\mathbb{R}, \mathbb{C})$ and $a \in \mathbb{R}$.

1. $\alpha\delta_a = \alpha(a)\delta_a$ since for $\varphi \in \mathrm{C}_0^\infty(\mathbb{R}, \mathbb{C})$,

$$(\alpha\delta_a)(\varphi) = \delta_a(\alpha\varphi) = \alpha(a)\,\varphi(a) = \alpha(a)\,\delta_a(\varphi) = \big(\alpha(a)\,\delta_a\big)(\varphi).$$

2. $(\alpha Y_a)' = \alpha' Y_a + \alpha(a)\,\delta_a$. ◊

Although $\mathcal{D}^*$ is a much bigger $\mathbb{C}[s]$-module than C^∞, we are going to show next that their torsion submodules coincide, i.e., also the torsion submodule of $\mathcal{D}^*$ is the space of polynomial-exponential functions.

Lemma 8.2.21 *Consider the constant function* $1 \in \mathrm{C}^\infty(\mathbb{R}, \mathbb{C}) \subset \mathrm{C}^{0,\mathrm{pc}}$ *and the* $\mathbb{C}$-*linear functional*

$$T := \langle 1, -\rangle \colon\ \mathrm{C}_0^\infty(\mathbb{R}, \mathbb{C}) \longrightarrow \mathbb{C},\ \varphi \longmapsto T(\varphi) = \langle 1, \varphi\rangle = \int_{-\infty}^{\infty} \varphi(t)\,\mathrm{d}t.$$

Every functional $u \in \mathcal{D}^*$ *with zero derivative is constant, i.e., it is a multiple of* T, *in other words* $\mathrm{ann}_{\mathcal{D}^*}(s) = \{u \in \mathcal{D}^*;\ u' = 0\} = \mathbb{C}T$. *With the identification* $\mathrm{C}^{0,\mathrm{pc}} \subset \mathcal{D}^*$, *this signifies that*

$$\{u \in \mathcal{D}^*;\ u' = 0\} = \{u \in \mathrm{C}^\infty(\mathbb{R}, \mathbb{C});\ u' = 0\} = \mathbb{C}.$$

Proof In this proof we do not identify the constant function 1 with its associated linear functional T.

(i) We derive a direct sum decomposition of $\mathrm{C}_0^\infty(\mathbb{R}, \mathbb{C})$. Let $\varphi_1 \geq 0$ be an arbitrary nonzero function in $\mathrm{C}_0^\infty(\mathbb{R}, \mathbb{R})$. Then $T(\varphi_1) > 0$, hence

$$\varphi_0 := \tfrac{\varphi_1}{T(\varphi_1)} \in \mathrm{C}_0^\infty(\mathbb{R}, \mathbb{C}) \ \text{ with } \ T(\varphi_0) = 1.$$

Therefore, the $\mathbb{C}$-linear function $T\colon\ \mathrm{C}_0^\infty(\mathbb{R}, \mathbb{C}) \longrightarrow \mathbb{C}$ has the right inverse

$$S\colon\ \mathbb{C} \longrightarrow \mathrm{C}_0^\infty(\mathbb{R}, \mathbb{C}),\quad \alpha \longmapsto \alpha\varphi_0.$$

Theorem and Definition 4.2.18 implies the direct sum decomposition

$$\mathrm{C}_0^\infty(\mathbb{R}, \mathbb{C}) = \mathbb{C}\varphi_0 \oplus \ker(T) \ni \varphi = T(\varphi)\,\varphi_0 + \psi \text{ where } \psi := \varphi - T(\varphi)\,\varphi_0.$$

(ii) The function ψ satisfies $T(\psi) = \int_{-\infty}^{\infty} \psi(t)\,\mathrm{d}t = 0$. The integral $\psi_1(t) := \int_{-\infty}^{t} \psi(x)\,\mathrm{d}x$ has the derivative $\psi_1' = \psi$ and belongs to C^∞. If $\psi(x) = 0$ for $|x| \geq a$ then

$$\psi_1(t) = \int_{-\infty}^{t} \psi(x)\,\mathrm{d}x = \begin{cases} 0 & \text{if } t \leq -a \\ \int_{-\infty}^{\infty} \psi(x)\,\mathrm{d}x = T(\psi) = 0 & \text{if } t \geq a \end{cases} \Longrightarrow \psi_1 \in \mathrm{C}_0^\infty.$$

Let now $u \in \mathcal{D}^*$ with $u' = 0$ and $\varphi \in C_0^\infty(\mathbb{R}, \mathbb{C})$ with the ensuing

$$\psi := \varphi - T(\varphi)\varphi_0 \text{ and } \psi_1(t) := \int_{-\infty}^{t} \psi(x)\,dx \Longrightarrow u(\psi) = u(\psi_1') = -u'(\psi_1) = 0$$
$$\Longrightarrow u(\varphi) = u\big(T(\varphi)\varphi_0 + \psi\big) = T(\varphi)u(\varphi_0) + u(\psi) = \big(u(\varphi_0)\,T\big)(\varphi)$$
$$\Longrightarrow u = u(\varphi_0)T \Longrightarrow \big\{u \in \mathcal{D}^*;\ u' = 0\big\} = \mathbb{C}T. \qquad \square$$

In the sequel we will always identify a piecewise continuous function $u \in C^{0,\mathrm{pc}}(\mathbb{R}, \mathbb{C})$ with its associated functional $\langle u, -\rangle \in \mathcal{D}^*_{\mathbb{C}}$.

Theorem 8.2.22 (The torsion submodule of $\mathcal{D}^*$) *The modules* C^∞ *and* $\mathcal{D}^*$ *have the same torsion submodule*

$$\mathrm{t}(\mathcal{D}^*) := \{y \in \mathcal{D}^*;\ \exists 0 \neq f \in \mathbb{C}[s] \text{ with } f \circ y = 0\} = \mathrm{t}\left(C^\infty(\mathbb{R}, \mathbb{C})\right) = \bigoplus_{\lambda \in \mathbb{C}} \mathbb{C}[t]e^{\lambda t}.$$

Proof The last equality was shown in Corollary 5.3.19.
Because of the primary decomposition of Theorem 5.3.2, it suffices to show that for all $\lambda \in \mathbb{C}$, the $(s - \lambda)$-primary components and, more precisely, the annihilators of $(s - \lambda)^m$ for every m coincide, i.e.,

$$\mathrm{ann}_{\mathcal{D}^*}\left((s - \lambda)^m\right) = \mathrm{ann}_{C^\infty(\mathbb{R},\mathbb{C})}\left((s - \lambda)^m\right) = \mathbb{C}[t]_{\le m-1}e^{\lambda t}.$$

First we prove, for $\lambda = 0$,

$$\mathrm{ann}_{\mathcal{D}^*}(s^m) = \{u \in \mathcal{D}^*;\ s^m \circ u = 0\} = \mathbb{C}[t]_{\le m-1}.$$

The right side is obviously contained in the left side. For $m = 0$, both sides are zero. Lemma 8.2.21 contains the assertion for $m = 1$. We prove the assertion for higher m by induction. Let $m > 1$ and $u \in \mathcal{D}^*$ with $s^m \circ u = u^{(m)} = 0$. From $s^m \circ u = s^{m-1} \circ (s \circ u) = 0$, i.e., $s \circ u \in \mathrm{ann}_{\mathcal{D}^*}(s^{m-1})$, we infer by the induction hypothesis that $f := s \circ u \in \mathbb{C}[t]_{\le m-2}$. Let $g \in \mathbb{C}[t]_{\le m-1}$ be any indefinite integral of f. Then $s \circ u = f = s \circ g$, which implies that $s \circ (u - g) = 0$, i.e., $u - g \in \mathrm{ann}_{\mathcal{D}^*}(s) = \mathbb{C}$ by Lemma 8.2.21. Hence,

$$u = (u - g) + g \in \mathbb{C} + \mathbb{C}[t]_{\le m-1} = \mathbb{C}[t]_{\le m-1}.$$

For arbitrary $\lambda \in \mathbb{C}$, the function $e^{\lambda t}$ is a unit in $C^\infty(\mathbb{R}, \mathbb{C})$ and induces, as in (5.13), the $C^\infty(\mathbb{R}, \mathbb{C})$-linear isomorphism

$$e^{\lambda t}\cdot:\ \mathcal{D}^* \overset{\cong}{\longrightarrow} \mathcal{D}^*, \quad y \longmapsto e^{\lambda t}y.$$

By the product rule, we have $s \circ (e^{\lambda t}y) = \lambda e^{\lambda t}y + e^{\lambda t}y'$, i.e., $(s - \lambda) \circ (e^{\lambda t}y) = e^{\lambda t}(s \circ y)$ and, for higher powers, $(s - \lambda)^m \circ (e^{\lambda t}y) = e^{\lambda t}(s^m \circ y)$. Therefore, the isomorphism $e^{\lambda t}\cdot$ induces the isomorphism

$$e^{\lambda t}\cdot:\ \mathrm{ann}_{\mathcal{D}^*}(s^m) \xrightarrow{\cong} \mathrm{ann}_{\mathcal{D}^*}\left((s-\lambda)^m\right),\quad y \longmapsto e^{\lambda t}y,$$

and with $\mathrm{ann}_{\mathcal{D}^*}(s^m) = \mathbb{C}[t]_{\leq m-1}$ from the first part of the proof follows that $\mathrm{ann}_{\mathcal{D}^*}\left((s-\lambda)^m\right) = \mathbb{C}[t]_{\leq m-1}e^{\lambda t}$.
The primary decomposition (Theorem 5.3.2) finally yields

$$\begin{aligned}\mathrm{t}(\mathcal{D}^*) &= \bigoplus_{\lambda\in\mathbb{C}} \mathrm{t}_{s-\lambda}(\mathcal{D}^*) = \bigoplus_{\lambda\in\mathbb{C}} \bigcup_{m=0}^{\infty} \mathrm{ann}_{\mathcal{D}^*}\left((s-\lambda)^m\right)\\ &= \bigoplus_{\lambda\in\mathbb{C}} \bigcup_{m=0}^{\infty} \mathbb{C}[t]_{\leq m-1}e^{\lambda t} = \bigoplus_{\lambda\in\mathbb{C}} \mathbb{C}[t]e^{\lambda t}.\end{aligned}$$

□

We finally treat the parameter transformation $t \mapsto -t$. For a function $f(t)$ we define $\check{f}(t) := f(-t)$. If $f : \mathbb{R} \to \mathbb{C}$ is a piecewise continuous function with compact support the equation $\int_{-\infty}^{\infty} \check{f}(t)dt = \int_{-\infty}^{\infty} f(t)dt$ holds, hence

$$\begin{aligned}&\forall f \in \mathrm{C}^{0,\mathrm{pc}}(\mathbb{R},\mathbb{C})\forall \varphi \in \mathrm{C}_0^{\infty}:\\ &\qquad \int_{-\infty}^{\infty} \check{f}(t)\varphi(t)dt = \int_{-\infty}^{\infty} f(t)\check{\varphi}(t)dt \text{ or } \langle \check{f},\varphi\rangle = \langle f,\check{\varphi}\rangle.\end{aligned} \tag{8.53}$$

In the last equation f, $\check{f}$ are considered as distributions in $\mathcal{D}^*$. This suggests to define

$$\begin{aligned}&\forall T \in \mathcal{D}^*\forall \varphi \in \mathrm{C}_0^{\infty}:\ \check{T}(\varphi) := T\check{\ }(\varphi) := \langle \check{T},\varphi\rangle := \langle T,\check{\varphi}\rangle = T(\check{\varphi})\\ &\Longrightarrow \forall f \in \mathrm{C}^{0,\mathrm{pc}}:\ \langle \check{f},-\rangle = \langle f,-\rangle\check{\ }\\ &\Longrightarrow \forall T \in \mathcal{D}^*:\ (T\check{\ })\check{\ } = T,\ s\circ\check{T} = ((-s)\circ T)\check{\ }\\ &\Longrightarrow \forall P \in \mathbb{C}[s]\forall T \in \mathcal{D}^*:\ P(s)\circ\check{T} = (P(-s)\circ T)\check{\ }.\end{aligned} \tag{8.54}$$

For example, we obtain

$$\check{\delta} = \delta,\ \check{Y}(t) = Y(-t) = \begin{cases} 1 & \text{if } t \leq 0\\ 0 & \text{if } t > 0\end{cases},\ s\circ\check{Y} = -(s\circ Y)\check{\ } = -\check{\delta} = -\delta. \tag{8.55}$$

8.2.3 The Space $\mathrm{C}^{-\infty}(\mathbb{R},\mathbb{C})$ of Distributions of Finite Order

In this section we introduce the signal space $\mathrm{C}^{-\infty}(\mathbb{R},\mathbb{C})$ of distributions of finite order, which consists of arbitrary derivatives in $\mathcal{D}^*$ of continuous functions. In particular, it is contained in the space $\mathcal{D}'$ of distributions, i.e., of continuous linear functionals $T\colon\ C_0^\infty(\mathbb{R},\mathbb{C}) \longrightarrow \mathbb{C}$ according to Schwartz [3], which is a subspace of $\mathcal{D}^*$. We will neither introduce the space $\mathcal{D}'$ nor prove that $C^{-\infty}(\mathbb{R},\mathbb{C})$ is the space of distributions of finite order, and refer the reader to [4, Theorem 4.4.7 on p. 114] for this. We show that the space $C^{-\infty}(\mathbb{R},\mathbb{C})$ is an injective cogenerator. It contains all signals that are needed in engineering and is therefore especially well suited for systems theory.

Lemma and Definition 8.2.23 *The differentiation operator*

$$s\circ\colon\ C^1(\mathbb{R},\mathbb{C}) \longrightarrow C^0(\mathbb{R},\mathbb{C})$$

has the right inverse

$$I_a\colon C^0(\mathbb{R},\mathbb{C}) \longrightarrow C^1(\mathbb{R},\mathbb{C}),\quad u \longmapsto I_a(u) := \int_a^t u(x)\,\mathrm{d}x,$$

where $a \in \mathbb{R}$ *is arbitrary, i.e.,* $I_a(u)' = u$ *is satisfied. Additionally, for* $0 \le m \le \infty$ *and* $k < \infty$, *we have that*

$$(I_a)^k\big(C^m(\mathbb{R},\mathbb{C})\big) \subset C^{m+k}(\mathbb{R},\mathbb{C}) \quad\textit{and}\quad s^k \circ C^{m+k}(\mathbb{R},\mathbb{C}) = C^m(\mathbb{R},\mathbb{C}).$$

Proof These are standard results from analysis. □

Corollary 8.2.24 $C^m(\mathbb{R},\mathbb{C}) = \{u \in \mathcal{D}^*;\ s^m \circ u = u^{(m)} \in C^0(\mathbb{R},\mathbb{C})\},\ m \in \mathbb{N}.$

Proof By definition of C^m, the left side is contained in the right side.
Conversely, let $u \in \mathcal{D}^*$ with $v := s^m \circ u \in C^0(\mathbb{R},\mathbb{C})$. The preceding lemma furnishes $u_1 := (I_0)^m(v) \in C^m(\mathbb{R},\mathbb{C})$ with $s^m \circ u_1 = v$, hence $s^m \circ (u - u_1) = 0$. We infer from Theorem 8.2.22 that $u - u_1 \in \mathbb{C}[t]_{\le m-1}$ and conclude that

$$u = (u - u_1) + u_1 \in \mathbb{C}[t]_{\le m-1} + C^m(\mathbb{R},\mathbb{C}) = C^m(\mathbb{R},\mathbb{C}).$$ □

For $m \in \mathbb{N}$ we define the subspace

$$C^{-m} := C^{-m}(\mathbb{R},\mathbb{C}) := s^m \circ C^0(\mathbb{R},\mathbb{C}) \subset \mathcal{D}^*. \tag{8.56}$$

The elements of C^{-m} are obtained as m-th derivatives of continuous functions and are called *distributions of order* m. The distributions $v \in C^{-m}(\mathbb{R},\mathbb{C})$ satisfy

$$v(\varphi) = (s^m \circ u)(\varphi) = (-1)^m\, u(\varphi^{(m)}) = (-1)^m \int_{-\infty}^{\infty} u(t)\,\varphi^{(m)}(t)\,\mathrm{d}t, \tag{8.57}$$

where $\varphi \in C_0^\infty(\mathbb{R},\mathbb{C})$ and $u \in C^0(\mathbb{R},\mathbb{C})$ with $v = s^m \circ u$.

Lemma and Definition 8.2.25 (Distributions of finite order) *The inclusion* $\mathrm{C}^1(\mathbb{R}, \mathbb{C}) \subset \mathrm{C}^0(\mathbb{R}, \mathbb{C})$ *implies* $\mathrm{C}^0(\mathbb{R}, \mathbb{C}) = s \circ \mathrm{C}^1(\mathbb{R}, \mathbb{C}) \subset s \circ \mathrm{C}^0(\mathbb{R}, \mathbb{C}) = \mathrm{C}^{-1}(\mathbb{R}, \mathbb{C})$ *and then, by induction, the ascending sequence of* $\mathbb{C}$*-subspaces*

$$\mathrm{C}^0 \subset \mathrm{C}^{-1} \subset \cdots \subset \mathrm{C}^{-m} = s^m \circ \mathrm{C}^0 \subset \mathrm{C}^{-(m+1)} = \mathrm{C}^{-m-1} \subset \cdots \subset \mathrm{C}^{-\infty},$$
$$\text{where } \mathrm{C}^{-\infty} := \mathrm{C}^{-\infty}(\mathbb{R}, \mathbb{C}) := \bigcup_{m=0}^{\infty} s^m \circ \mathrm{C}^0(\mathbb{R}, \mathbb{C}) \subset \mathcal{D}^*.$$

As an increasing union of subspaces $\mathrm{C}^{-\infty}(\mathbb{R}, \mathbb{C})$ *is also a subspace of* $\mathcal{D}^*$. *Moreover, it is obviously closed under differentiation, i.e.,* $s \circ \mathrm{C}^{-\infty}(\mathbb{R}, \mathbb{C}) \subset \mathrm{C}^{-\infty}(\mathbb{R}, \mathbb{C})$, *and therefore even a* $\mathbb{C}[s]$*-submodule of* $\mathcal{D}^*$. *The elements of* $\mathrm{C}^{-\infty}(\mathbb{R}, \mathbb{C})$ *are called (complex-valued)* distributions of finite order.
It follows directly from (8.57) *that a distribution* $v = s^m \circ u$, $u \in \mathrm{C}^0$, *of order* m *is continuous in the following sense: Assume that functions* $\varphi_1, \varphi_2, \ldots$ *and* φ *in* $\mathrm{C}_0^\infty(\mathbb{R}, \mathbb{C})$ *have their support in a fixed finite interval* $[A, B]$ *and that* $\lim_{k\to\infty} \varphi_k^{(m)} = \varphi^{(m)}$, *where the limit is uniform on the interval* $[A, B]$. *Then* $v(\varphi) = \lim_{k\to\infty} v(\varphi_k)$.
Real-valued distributions are obtained as derivatives of real-valued functions:

$$\mathrm{C}^{-\infty}(\mathbb{R}, \mathbb{R}) := \bigcup_{m=0}^{\infty} s^m \circ \mathrm{C}^0(\mathbb{R}, \mathbb{R}) \subset \mathcal{D}_{\mathbb{R}}^* \cap \mathrm{C}^{-\infty}(\mathbb{R}, \mathbb{C}) \subset \mathrm{C}^{-\infty}(\mathbb{R}, \mathbb{C}).$$

The space $\mathrm{C}^{-\infty}(\mathbb{R}, \mathbb{R})$ *is also closed under differentiation and therefore an* $\mathbb{R}[s]$*-submodule of* $\mathcal{D}_{\mathbb{R}}^*$.

Proof The proof is already contained in the formulation of the lemma. □

Corollary 8.2.26 *All piecewise continuous functions, all delta distributions* δ_a, $a \in \mathbb{R}$, *and their derivatives are distributions of finite order.*

Proof Let u be piecewise continuous. Then u_1 defined by $u_1(t) := \int_a^t u(x)\,\mathrm{d}x$ for an arbitrary $a \in \mathbb{R}$ is continuous and, according to Lemma 8.2.15, its distributional derivative is u. Hence $u \in s \circ \mathrm{C}^0(\mathbb{R}, \mathbb{C}) = \mathrm{C}^{-1}(\mathbb{R}, \mathbb{C})$.
The Heaviside function Y_a with jump at $a \in \mathbb{R}$ is piecewise continuous and hence $Y_a \in s \circ \mathrm{C}^0(\mathbb{R}, \mathbb{C})$. Since its derivative is δ_a, we conclude that

$$\delta_a = s \circ Y_a \in s \circ (s \circ \mathrm{C}^0(\mathbb{R}, \mathbb{C})) = \mathrm{C}^{-2}(\mathbb{R}, \mathbb{C}).$$

More explicitly, we define the continuous function

$$\begin{aligned} u(t) = \int_a^t Y(x-a)\,\mathrm{d}x &= \begin{cases} 0 & \text{if } t \le a \\ \int_a^t 1\,\mathrm{d}x = t - a & \text{if } t \ge a \end{cases} \\ &= (t-a)\,Y(t-a) = (t-a)\,Y_a(t) \end{aligned}$$

and conclude that $s \circ \big((t-a)\,Y_a(t)\big) = Y_a(t)$ and

$$s^2 \circ \left((t-a)\,Y_a\right) = s \circ Y_a(t) = \delta_a. \qquad \square$$

The space $C^{-\infty}(\mathbb{R}, \mathbb{C})$ is the largest signal space in continuous time that we are going to use. The signal space L^1_{loc}, used in [5], is not contained in $C^{-\infty}$, cf. [3, Chap. 3, Theorem III], and unsuitable in our opinion, cf. (2.53).

Corollary 8.2.27 *The equality $s^m \circ C^n(\mathbb{R}, \mathbb{C}) = C^{n-m}(\mathbb{R}, \mathbb{C})$ holds for $m \in \mathbb{N}$ and $n \in \mathbb{Z}$.*

Proof For $n \le 0$, this follows from the definition of $C^n(\mathbb{R}, \mathbb{C})$, for $0 \le m \le n$ from Lemma 8.2.23, and for $0 \le n < m$ from

$$s^m \circ C^n(\mathbb{R}, \mathbb{C}) = s^{m-n} \circ \left(s^n \circ C^n(\mathbb{R}, \mathbb{C})\right) = s^{m-n} \circ C^0(\mathbb{R}, \mathbb{C}) = C^{n-m}(\mathbb{R}, \mathbb{C}). \; \square$$

Theorem 8.2.28 *The $\mathbb{C}[s]$-module $C^{-\infty}(\mathbb{R}, \mathbb{C})$ is an injective cogenerator. Likewise, $C^{-\infty}(\mathbb{R}, \mathbb{R})$ is an injective $\mathbb{R}[s]$-cogenerator.*
Many theorems of the preceding chapters were proven for arbitrary injective cogenerators over principal ideal domains and therefore hold for these new signal spaces. The inclusion $C^\infty(\mathbb{R}, \mathbb{C}) \subset C^{-\infty}(\mathbb{R}, \mathbb{C}) \subset \mathcal{D}^*$ and Theorem 8.2.22 imply

$$\mathrm{t}\left(C^{-\infty}(\mathbb{R}, \mathbb{C})\right) = \mathrm{t}\left(C^\infty(\mathbb{R}, \mathbb{C})\right) = \bigoplus_{\lambda \in \mathbb{C}} \mathbb{C}[t]e^{\lambda t},$$

and likewise $\mathrm{t}\left({}_{\mathbb{R}[s]}C^{-\infty}(\mathbb{R}, \mathbb{R})\right) = \mathrm{t}\left(C^\infty(\mathbb{R}, \mathbb{R})\right)$.

Proof We treat the complex case. The proof in the real case is analogous.
Since $C^\infty(\mathbb{R}, \mathbb{C})$ is a cogenerator and contained in $C^{-\infty}(\mathbb{R}, \mathbb{C})$ also the latter module is a cogenerator according to Lemma and Definition 3.3.2.
Since $\mathbb{C}[s]$ is a principal ideal domain, $C^{-\infty}(\mathbb{R}, \mathbb{C})$ is injective if it is divisible, cf. Corollary 3.2.11. Thus let $0 \ne f \in \mathbb{C}[s]$ with $d := \deg(f) \ge 0$ and $u = s^m \circ u_0 \in C^{-m}$, $m \ge 0$, $u_0 \in C^0$. Consider the differential equation $f \circ y = u$. From the standard theory we know that the differential equation $f \circ y_0 = u_0$ with continuous u_0 has a solution $y_0 \in C^d(\mathbb{R}, \mathbb{C})$. Then

$$\begin{aligned} &y := s^m \circ y_0 \in s^m \circ C^d(\mathbb{R}, \mathbb{C}) = C^{d-m}(\mathbb{R}, \mathbb{C}) = C^{-m+d} \subseteq C^{-\infty}(\mathbb{R}, \mathbb{C}) \text{ and} \\ &f \circ y = f \circ (s^m \circ y_0) = s^m \circ (f \circ y_0) = s^m \circ u_0 = u \end{aligned}$$

imply that $C^{-\infty}$ is divisible and thus injective. $\square$

Corollary 8.2.29 *Let $f \in \mathbb{C}[s] \setminus \{0\}$ with $\deg(f) = d \ge 0$ and $u \in C^m(\mathbb{R}, \mathbb{C})$ with $-\infty \le m \le \infty$. Then every solution $y \in \mathcal{D}^*$ of $f \circ y = u$ is contained in $C^{m+d}(\mathbb{R}, \mathbb{C})$.*

Proof According to the proof of Theorem 8.2.28 there is a solution $y \in \mathrm{C}^{m+d}(\mathbb{R}, \mathbb{C})$. Let $z \in \mathcal{D}^*$ be an arbitrary solution of $f \circ y = u$. Then $f \circ (z - y) = u - u = 0$, hence $z - y \in \mathrm{t}(\mathcal{D}^*) = \mathrm{t}\big(\mathrm{C}^\infty(\mathbb{R}, \mathbb{C})\big) \subset \mathrm{C}^\infty(\mathbb{R}, \mathbb{C})$ and we conclude that

$$z = y + (z - y) \in \mathrm{C}^{m+d}(\mathbb{R}, \mathbb{C}) + \mathrm{C}^\infty(\mathbb{R}, \mathbb{C}) = \mathrm{C}^{m+d}(\mathbb{R}, \mathbb{C}). \qquad \square$$

Lemma 8.2.30 *The space* $\mathrm{C}^{-\infty}(\mathbb{R}, \mathbb{C})$ *is closed under the shift action and is a* $\mathrm{C}^\infty(\mathbb{R}, \mathbb{C})$*-submodule of* $\mathcal{D}^*$. *In other words, both* $\mathbb{R}$ *and* $\mathrm{C}^\infty(\mathbb{R}, \mathbb{C})$ *act on* $\mathrm{C}^{-\infty}(\mathbb{R}, \mathbb{C})$ *like on* $\mathcal{D}^*$. *The corresponding result also holds for real signals.*

Proof Let $m \in \mathbb{N}$, $v \in \mathrm{C}^{-m}(\mathbb{R}, \mathbb{C})$, and $a \in \mathbb{R}$. Let $u \in \mathrm{C}^0(\mathbb{R}, \mathbb{C})$ be such that $s^m \circ u = v$. Since differentiation and the shift action commute, we obtain

$$a \circ_{\mathrm{s}} v = a \circ_{\mathrm{s}} (s^m \circ u) = s^m \circ (a \circ_{\mathrm{s}} u) \in \mathrm{C}^{-m}(\mathbb{R}, \mathbb{C}).$$

Let, additionally, $\alpha \in \mathrm{C}^\infty(\mathbb{R}, \mathbb{C})$ and $\varphi \in \mathrm{C}_0^\infty(\mathbb{R}, \mathbb{C})$. Then

$$\begin{aligned}
(\alpha v)(\varphi) &= \big(\alpha(s^m \circ u)\big)(\varphi) = (s^m \circ u)(\alpha\varphi) = (-1)^m u\big(s^m \circ (\alpha\varphi)\big) \\
&= (-1)^m u\left(\sum_{i=0}^{m} \tbinom{m}{i} \alpha^{(m-i)} (s^i \circ \varphi)\right) \\
&= \sum_{i=0}^{m} (-1)^{m-i} \tbinom{m}{i} \big(\alpha^{(m-i)} u\big)\big((-1)^i (s^i \circ \varphi)\big) \\
&= \left(\sum_{i=0}^{m} (-1)^{m-i} \tbinom{m}{i} s^i \circ \big(\alpha^{(m-i)} u\big)\right)(\varphi),
\end{aligned}$$

hence $\alpha v = \sum_{i=0}^{m} (-1)^{m-i} \binom{m}{i} s^i \circ \big(\alpha^{(m-i)} u\big) \in \mathrm{C}^{-m}(\mathbb{R}, \mathbb{C})$. $\square$

For a matrix $R \in \mathbb{C}[s]^{k \times l}$, we have to distinguish the behaviors

$$\begin{aligned}
\mathcal{B}_{\mathrm{C}^\infty(\mathbb{R},\mathbb{C})} &= \{w \in \mathrm{C}^\infty(\mathbb{R}, \mathbb{C})^l;\ R \circ w = 0\} \quad \text{and} \\
\mathcal{B}_{\mathrm{C}^{-\infty}(\mathbb{R},\mathbb{C})} &= \{w \in \mathrm{C}^{-\infty}(\mathbb{R}, \mathbb{C})^l;\ R \circ w = 0\}.
\end{aligned} \tag{8.58}$$

In general, a larger signal space also leads to more trajectories in the behavior, i.e., $\mathcal{B}_{\mathrm{C}^\infty(\mathbb{R},\mathbb{C})} \subset \mathcal{B}_{\mathrm{C}^{-\infty}(\mathbb{R},\mathbb{C})}$. However, for autonomous systems this is not so, as the following theorem shows.

Theorem 8.2.31 *Consider the behaviors from* (8.58) *and the factor module* $M = \mathbb{C}[s]^{1 \times l} / \mathbb{C}[s]^{1 \times k} R$. *The following properties are equivalent:*

1. *The module M is a torsion module.*
2. $\mathcal{B}_{\mathrm{C}^\infty(\mathbb{R},\mathbb{C})}$ *is autonomous.*
3. $\mathcal{B}_{\mathrm{C}^{-\infty}(\mathbb{R},\mathbb{C})}$ *is autonomous.*

4. *$\mathcal{B}_{\mathrm{C}^\infty(\mathbb{R},\mathbb{C})} = \mathcal{B}_{\mathrm{C}^{-\infty}(\mathbb{R},\mathbb{C})}$.*
5. *Every trajectory in $\mathcal{B}_{\mathrm{C}^{-\infty}(\mathbb{R},\mathbb{C})}$ is analytic. Indeed, it is contained in the space $\bigoplus_{\lambda\in\mathbb{C}} \mathbb{C}[t]^l e^{\lambda t}$ of polynomial-exponential functions and thus a power series which converges everywhere.*

The equivalences also hold for the $\mathbb{R}[s]$-signal modules $\mathrm{C}^\infty(\mathbb{R},\mathbb{R}) \subset \mathrm{C}^{-\infty}(\mathbb{R},\mathbb{R})$ with the appropriate change of item 5 according to Corollary 5.3.19.

Proof 1. $\Longleftrightarrow$ 2. and 1. $\Longleftrightarrow$ 3.: Since both signal spaces are injective cogenerators, these equivalences follow directly from Definition 4.2.4.
1. $\Longrightarrow$ 4.: Lemma 4.2.6 and Theorem 8.2.29 imply that

$$\mathcal{B}_{\mathrm{C}^{-\infty}} \subseteq \mathrm{t}(\mathrm{C}^{-\infty})^l = \mathrm{t}(\mathrm{C}^\infty)^l \subseteq (\mathrm{C}^\infty)^l.$$

Therefore, $\mathcal{B}_{\mathrm{C}^\infty} = (\mathrm{C}^\infty)^l \cap \mathcal{B}_{\mathrm{C}^{-\infty}} = \mathcal{B}_{\mathrm{C}^{-\infty}}$.
4. $\Longrightarrow$ 3.: If $\mathcal{B}_{\mathrm{C}^{-\infty}}$ is not autonomous, then there is a surjective observable $\xi\circ\colon \mathcal{B}_{\mathrm{C}^{-\infty}} \longrightarrow \mathrm{C}^{-\infty}$ with $\xi \in \mathbb{C}[s]^{1\times l}$. Thus, $\xi \circ \mathcal{B}_{\mathrm{C}^{-\infty}} = \mathrm{C}^{-\infty} \supsetneq \mathrm{C}^\infty = \xi \circ \mathcal{B}_{\mathrm{C}^\infty}$, which implies that $\mathcal{B}_{\mathrm{C}^{-\infty}} \supsetneq \mathcal{B}_{\mathrm{C}^\infty}$.
2. $\Longrightarrow$ 5.: $\mathcal{B}_{\mathrm{C}^{-\infty}} \subseteq \mathrm{t}(\mathrm{C}^\infty)^l = \oplus_{\lambda\in\mathbb{C}}\mathbb{C}[t]^l e^{\lambda t}$.
5. $\Longrightarrow$ 4.: Obvious. □

8.2.4 Distributions with Left Bounded Support, Impulse Response, and Laplace Transform

In this section we show how a transfer matrix H acts as (transfer) operator on the space of input signal vectors with left bounded support (Theorem 8.2.37 and Corollary and Definition 8.2.40). Such signals describe a system which is initially relaxed or initially at rest. We will see in Theorem and Definition 8.2.41 that the initially relaxed part of an IO behavior is completely determined by its transfer matrix H. In Theorems 8.2.47, 8.2.52, and 8.2.53 we compute the impulse response or inverse Laplace transform of any transfer matrix H and explain the principal application of these notions in systems theory and electrical engineering. Finally, in Theorems 8.2.59 and 8.2.60, we characterize the properness of H by the piecewise continuity of its step response and by the action of H on the spaces of (piecewise) continuous signals with left bounded support. Again, we develop the theory for complex and for real signals and use $\mathcal{F} := \mathrm{C}^{-\infty} := \mathrm{C}^{-\infty}(\mathbb{R},\mathbb{C})$.
The support of a piecewise continuous function u is called *left bounded* if there is some time instant $t_0 \in \mathbb{R}$ such that

$$\mathrm{supp}(u) := \overline{\{t \in \mathbb{R};\ u(t) \neq 0\}} \subseteq t_0 + \mathbb{R}_{\geq 0} = [t_0, \infty), \quad \text{i.e., } u(t) = 0 \text{ for } t < t_0.$$

For obvious reasons a signal with left bounded support is also called *initially relaxed* or *at rest*. Such signals are often used in electrical engineering, for instance, if an

electrical network is switched on at some time by connecting it with a voltage source. We will not define the support of a distribution $T \in \mathrm{C}^{-\infty} = \mathrm{C}^{-\infty}(\mathbb{R}, \mathbb{C})$ and refer the reader to [3, Chap. I, Sect. 3, pp. 26–28] for this. We will, however, define when the support of a distribution is left bounded.
By definition the support of a function $\varphi \in \mathrm{C}_0^\infty$ is compact and hence $t_1 := \max \operatorname{supp}(\varphi) \in \operatorname{supp}(\varphi)$ is well defined and $\operatorname{supp}(\varphi) \subset (-\infty, t_1]$. For $t_0 \in \mathbb{R}$ this implies the equivalence

$$\begin{aligned} \operatorname{supp}(\varphi) \subset (-\infty, t_0) = \{t \in \mathbb{R};\ t < t_0\} &\Longleftrightarrow \exists t_1 < t_0 :\ \operatorname{supp}(\varphi) \subset (-\infty, t_1] \\ &\Longleftrightarrow \exists t_1 < t_0\ \forall t \geq t_1 :\ \varphi(t) = 0. \end{aligned} \tag{8.59}$$

As an example, recall the function $\varphi_{0,1} \in \mathrm{C}_0^\infty$ from (4.59) with $\varphi_{0,1}(t) > 0$ if $|t| < 1$ and $\varphi_{0,1} = 0$ if $|t| \geq 1$. The support of the function $\varphi(t) := \varphi_{0,1}(t - t_0 + 1)$ is contained in $(-\infty, t_0]$, but not in $(-\infty, t_0)$. We define

$$\begin{aligned} &\mathrm{C}^{-\infty}\big([t_0, \infty)\big) := \big\{T \in \mathrm{C}^{-\infty};\ \forall \varphi \in \mathrm{C}_0^\infty \text{ with } \operatorname{supp}(\varphi) \subset (-\infty, t_0) :\ \langle T, \varphi\rangle = 0\big\} \\ &\subset \mathrm{C}_+^{-\infty} := \bigcup_{t_0 \in \mathbb{R}} \mathrm{C}^{-\infty}\big([t_0, \infty)\big) \text{ with } \mathrm{C}^{-\infty}\big([t_1, \infty)\big) \subseteq \mathrm{C}^{-\infty}\big([t_0, \infty)\big) \text{ for } t_0 \leq t_1. \end{aligned} \tag{8.60}$$

We also introduce $\mathrm{C}_+^{0,\mathrm{pc}} := \mathrm{C}_+^{0,\mathrm{pc}}(\mathbb{R}, F)$ as

$$\mathrm{C}_+^{0,\mathrm{pc}} := \mathrm{C}^{0,\mathrm{pc}} \cap \mathrm{C}_+^{-\infty} = \big\{u \in \mathrm{C}^{0,\mathrm{pc}};\ \exists t_0 \forall t \leq t_0 :\ u(t) = 0\big\}. \tag{8.61}$$

The distributions in $\mathrm{C}_+^{-\infty}$ are said to have *left bounded support* due to the following.

Lemma 8.2.32 *For a piecewise continuous function u and its associated distribution* $\langle u, -\rangle$, *the following equivalence holds:*

$$\operatorname{supp}(u) \subseteq [t_0, \infty) \Longleftrightarrow \langle u, -\rangle \in \mathrm{C}^{-\infty}\big([t_0, \infty)\big). \tag{8.62}$$

Proof $\Longrightarrow$. This is obvious since $u\varphi = 0$ implies $\langle u, \varphi\rangle = \int_{-\infty}^{\infty} u(\tau)\,\varphi(\tau)\,\mathrm{d}\tau = 0$.
$\Longleftarrow$. Assume that $t_1 < t_0$ and $u(t_1) \neq 0$ where u is, without loss of generality, continuous in t_1. As in the proof of Theorem and Definition 8.2.8 we construct a $\varphi \in \mathrm{C}_0^\infty$ with $t_1 \in \operatorname{supp}(\varphi) \subset (-\infty, t_0)$ and $\langle u, \varphi\rangle \neq 0$. This is a contradiction. □

For all $t_0 \in \mathbb{R}$ and $m \in \mathbb{Z}$ we also introduce the spaces

$$\begin{aligned} &\mathrm{C}^m\big([t_0, \infty)\big) := \mathrm{C}^m \cap \mathrm{C}^{-\infty}\big([t_0, \infty)\big) \subset \mathrm{C}_+^m := \mathrm{C}^m \cap \mathrm{C}_+^{-\infty} = \bigcup_{t_0 \in \mathbb{R}} \mathrm{C}^m\big([t_0, \infty)\big) \\ &\underset{\mathrm{C}^m \supset \mathrm{C}^{m+1}}{\Longrightarrow} \mathrm{C}^m\big([t_0, \infty)\big) \supset \mathrm{C}^{m+1}\big([t_0, \infty)\big) \ \text{ and } \ \mathrm{C}_+^m \supset \mathrm{C}_+^{m+1}. \end{aligned} \tag{8.63}$$

Lemma 8.2.33 *For all $t_0 \in \mathbb{R}$ and $m \in \mathbb{Z}$ the identity*

$$s \circ \mathrm{C}^m\big([t_0, \infty)\big) = \mathrm{C}^{m-1}\big([t_0, \infty)\big) \tag{8.64}$$

holds. By (8.63) *this implies directly*

$$s \circ \mathrm{C}_+^m = \mathrm{C}_+^{m-1},\ s \circ \mathrm{C}^{-\infty}\big([t_0, \infty)\big) = \mathrm{C}^{-\infty}\big([t_0, \infty)\big),\ s \circ \mathrm{C}_+^{-\infty} = \mathrm{C}_+^{-\infty}.$$

Induction furnishes for all $d \in \mathbb{N}$ that

$$s^d \circ \mathrm{C}^m\big([t_0, \infty)\big) = \mathrm{C}^{m-d}\big([t_0, \infty)\big) \text{ and } s^d \circ \mathrm{C}^0\big([t_0, \infty)\big) = \mathrm{C}^{-d}\big([t_0, \infty)\big).$$

In particular, $\mathrm{C}^{-\infty}\big([t_0, \infty)\big)$ and $\mathrm{C}_+^{-\infty}$ are $\mathbb{C}[s]$-submodules of $\mathrm{C}^{-\infty}$.

Proof We have to show (8.64) only.
$\subseteq$: Let $u \in \mathrm{C}^m\big([t_0, \infty)\big)$ and let $\varphi \in \mathrm{C}_0^\infty$ with $\operatorname{supp}(\varphi) \subseteq (-\infty, t_0)$, hence

$$\begin{aligned}&\operatorname{supp}(\varphi') \subseteq (-\infty, t_0) \Longrightarrow \langle s \circ u, \varphi\rangle = -\langle u, \varphi'\rangle = 0 \Longrightarrow s \circ u \in \mathrm{C}^{-\infty}\big([t_0, \infty)\big)\\ &\underset{s\circ \mathrm{C}^m = \mathrm{C}^{m-1}}{\Longrightarrow} s \circ u \in \mathrm{C}^{m-1}([t_0, \infty)) \Longrightarrow s \circ \mathrm{C}^m([t_0, \infty)) \subseteq \mathrm{C}^{m-1}([t_0, \infty)).\end{aligned}$$

$\supseteq$: Let $u \in \mathrm{C}^{m-1}\big([t_0, \infty)\big)$. If $m > 0$, then u is a continuous function. We define $v(t) := \int_{t_0}^t u(\tau)\,\mathrm{d}\tau \in \mathrm{C}^m\big([t_0, \infty)\big)$ and obtain $u = s \circ v \in s \circ \mathrm{C}^m\big([t_0, \infty)\big)$.
If $m \le 0$, there is a $v \in \mathrm{C}^m$ with $u = s \circ v$ since $\mathrm{C}^{m-1} = s \circ \mathrm{C}^m$. The example $v = 1$ and $u = s \circ 1 = 0$ shows that v does not necessarily have left bounded support. Let $\varphi \in \mathrm{C}_0^\infty$ with $\operatorname{supp}(\varphi) \subset (-\infty, t_0)$. Choose another test function $\varphi_0 \in \mathrm{C}_0^\infty$ with $\operatorname{supp}(\varphi_0) \subset (-\infty, t_0)$ and $\langle 1, \varphi_0\rangle = \int_{-\infty}^\infty \varphi_0(\tau)\,\mathrm{d}\tau = 1$. As in the proof of Lemma 8.2.21, we decompose

$$\begin{aligned}&\varphi = \langle 1, \varphi\rangle\varphi_0 + \varphi_1 \text{ with } \varphi_1 := \varphi - \langle 1, \varphi\rangle\varphi_0 \in \mathrm{C}_0^\infty \Longrightarrow \operatorname{supp}(\varphi_1) \subseteq (-\infty, t_0)\\ &\Longrightarrow \exists t_1 < t_0 \text{ with } \operatorname{supp}(\varphi_1) \subseteq (-\infty, t_1],\ \langle 1, \varphi_1\rangle = \langle 1, \varphi\rangle - \langle 1, \varphi\rangle\langle 1, \varphi_0\rangle = 0\\ &\Longrightarrow \psi_1(t) := \int_{-\infty}^t \varphi_1(\tau)\,\mathrm{d}\tau \in \mathrm{C}_0^\infty,\ \operatorname{supp}(\psi_1) \subset (-\infty, t_0),\ \psi_1' = \varphi_1\\ &\Longrightarrow \langle v, \varphi_1\rangle = \langle v, \psi_1'\rangle = -\langle s \circ v, \psi_1\rangle = -\langle u, \psi_1\rangle \underset{u \in \mathrm{C}^{m-1}([t_0, \infty])}{=} 0\\ &\Longrightarrow \langle v, \varphi\rangle \underset{\varphi = \langle 1, \varphi\rangle\varphi_0 + \varphi_1}{=} \langle 1, \varphi\rangle\langle v, \varphi_0\rangle \Longrightarrow \big\langle v - \langle v, \varphi_0\rangle 1, \varphi\big\rangle = 0.\end{aligned}$$

Since φ with $\operatorname{supp}(\varphi) \subset (-\infty, t_0)$ was arbitrary, we conclude that

$$\begin{aligned}&v_1 := v - \langle v, \varphi_0\rangle \in \mathrm{C}^m \cap \mathrm{C}^{-\infty}\big([t_0, \infty)\big) = \mathrm{C}^m\big([t_0, \infty)\big)\\ &\underset{s\circ 1 = 0}{\Longrightarrow} u = s \circ v = s \circ v_1 \in s \circ \mathrm{C}^m\big([t_0, \infty)\big) \Longrightarrow s \circ \mathrm{C}^m\big([t_0, \infty)\big) \supseteq \mathrm{C}^{m-1}([t_0, \infty)).\end{aligned}$$

□

Example 8.2.34 For $a \in \mathbb{R}$, the function $(t-a)\,Y_a = (t-a)\,Y(t-a)$ belongs to $\mathrm{C}^0\big([a,\infty)\big)$ and hence

$$Y_a = s \circ (t-a)\,Y_a \in s \circ \mathrm{C}^0\big([a,\infty)\big) = \mathrm{C}^{-1}\big([a,\infty)\big) \quad \text{and}$$
$$\delta_a = s \circ Y_a \in s^2 \circ \mathrm{C}^0\big([a,\infty)\big) = \mathrm{C}^{-2}\big([a,\infty)\big).$$

Hence all derivatives of Y_a and δ_a belong to $\mathrm{C}^{-\infty}\big([a,\infty)\big)$.
If, more generally, u is piecewise continuous and $\mathrm{supp}(u) \subseteq [t_0,\infty)$, then $v(t) = \int_{t_0}^{t} u(\tau)\,\mathrm{d}\tau \in \mathrm{C}^0\big([t_0,\infty)\big)$ and $u = s \circ v \in \mathrm{C}^{-1}\big([t_0,\infty)\big)$. ◊

We are going to show in Theorem 8.2.37 that the $\mathbb{C}[s]$-module $\mathrm{C}_+^{-\infty}$ is even a vector space over the field $\mathbb{C}(s)$ of rational functions. The following lemma is the main ingredient of the proof.

Lemma 8.2.35 *1. If $\mathcal{B} \subset (\mathrm{C}^{-\infty})^q = \mathrm{C}^{-\infty}(\mathbb{R},\mathbb{C})^q$ is an autonomous behavior, then $\mathcal{B} \cap (\mathrm{C}_+^{-\infty})^q = 0$.*
2. For $u \in \mathrm{C}^0\big([t_0,\infty)\big)^q$ and $A \in \mathbb{C}^{q\times q}$, the linear differential system $x' = Ax + u$ has a unique solution in $(\mathrm{C}_+^{-\infty})^q$, namely,

$$x(t) = \int_{t_0}^{t} e^{(t-\tau)A} u(\tau)\,\mathrm{d}\tau \in \mathrm{C}^1\big([t_0,\infty)\big)^q.$$

3. For $u \in \mathrm{C}^0\big([t_0,\infty)\big)$ and a polynomial $f = s^n + f_{n-1}s^{n-1} + \cdots + f_0 \in \mathbb{C}[s]$ of degree n, the differential equation $f \circ y = u$ has a unique solution in $\mathrm{C}_+^{-\infty}$, and indeed $y \in \mathrm{C}^n\big([t_0,\infty)\big)$. In particular, there is a unique function $y \in \mathrm{C}_+^{-\infty}$ such that $s \circ y = y' = u$, namely, $y(t) = \int_{t_0}^{t} u(x)\,\mathrm{d}x \in \mathrm{C}^1\big([t_0,\infty)\big)$.

Proof 1. Let $w \in \mathcal{B} \cap (\mathrm{C}_+^{-\infty})^q$. Since $\mathcal{B}$ is autonomous, w is analytic by Theorem 8.2.31. On the other hand, w has left bounded support by assumption and thus is zero since an everywhere convergent power series vanishes if it vanishes on a nonempty open interval.
2. From the standard theory of differential equations we know that the given $x \in \mathrm{C}^1\big([t_0,\infty)\big)^q$ is such a solution. If $x_1 \in (\mathrm{C}_+^{-\infty})^q$ is another solution, then $x_1 - x$ is a trajectory in the autonomous behavior $\mathcal{B} := \{z \in (\mathrm{C}^{-\infty})^q;\ z' = Az\}$ and has left bounded support. Item 1 implies $x_1 - x = 0$.
3. We transform the equation $f \circ y = u$ into the system $x' = Ax + \begin{pmatrix} 0 \\ \vdots \\ 0 \\ u \end{pmatrix}$, where $A \in \mathbb{C}^{n\times n}$ is the companion matrix of the polynomial f. Then the assertion follows from item 2. □

Lemma 8.2.36 *The $\mathbb{C}[s]$-module $\mathrm{C}_+^{-\infty}$ is stable under translation and a C^∞-submodule of $\mathrm{C}^{-\infty}$. Indeed, $u \in \mathrm{C}^{-\infty}\big([t_0,\infty)\big)$, $a \in \mathbb{R}$, and $\alpha \in \mathrm{C}^\infty$ imply*

$$a \circ_{\mathrm{s}} u \in \mathrm{C}^{-\infty}\big([a+t_0,\infty)\big) \text{ and } \alpha u \in \mathrm{C}^{-\infty}\big([t_0,\infty)\big).$$

Proof For $\psi \in C_0^\infty$ we have

$$(a \circ_s \psi)(t) = \psi(t-a) \Longrightarrow \operatorname{supp}(a \circ_s \psi) = a + \operatorname{supp}(\psi)$$
$$\Longrightarrow \operatorname{supp}((-a) \circ_s \psi) = -a + \operatorname{supp}(\psi).$$

For $\varphi \in C_0^\infty$ with $\operatorname{supp}(\varphi) \subset (-\infty, t_0 + a)$ this implies

$$\operatorname{supp}((-a) \circ_s \varphi) \subset -a + (-\infty, t_0 + a) = (-\infty, t_0)$$
$$\Longrightarrow \langle a \circ_s u, \varphi \rangle = \langle u, (-a) \circ_s \varphi \rangle = 0 \Longrightarrow a \circ_s u \in C^{-\infty}\big([t_0 + a, \infty)\big).$$

The second assertion follows easily from $\operatorname{supp}(\alpha\varphi) \subseteq \operatorname{supp}(\varphi)$. □

Theorem 8.2.37 *The $\mathbb{C}[s]$-signal module $C_+^{-\infty}(\mathbb{R}, \mathbb{C})$ is a $\mathbb{C}(s)$-vector space, i.e., an injective and torsionfree $\mathbb{C}[s]$-module, with a unique scalar multiplication $\circ$ that extends the given action of $\mathbb{C}[s]$. In other terms, for all nonzero $P \in \mathbb{C}[s]$ the map $P\circ\colon C_+^{-\infty} \longrightarrow C_+^{-\infty}$ is an isomorphism.*
A rational function $H = \frac{Q}{P} \in \mathbb{C}(s)$ gives rise to the $\mathbb{C}(s)$-linear map or operator

$$H\circ\colon C_+^{-\infty}(\mathbb{R}, \mathbb{C}) \longrightarrow C_+^{-\infty}(\mathbb{R}, \mathbb{C}),\ u \longmapsto y := H \circ u,$$

where y is the unique solution in $C_+^{-\infty}$ of the differential equation $P \circ y = Q \circ u$. For nonzero H, the operator $H\circ$ is, of course, an isomorphism. The operators $H\circ$ commute with the shift action.
Likewise, the $\mathbb{C}[s]$-module $C^{-\infty}\big([t_0, \infty]\big)$ is a $\mathbb{C}(s)$-vector space, and the $\mathbb{R}[s]$-module $C_+^{-\infty}(\mathbb{R}, \mathbb{R})$ is an $\mathbb{R}(s)$-vector space.

Proof Let $P \in \mathbb{C}[s]$ with $\deg(P) = n \geq 0$. The behavior $\mathcal{B} := \{y \in C^{-\infty};\ P \circ y = 0\}$ is autonomous and has only the zero trajectory in $C_+^{-\infty}$ according to item 1 of Lemma 8.2.35. Thus $P\circ$ is injective on $C_+^{-\infty}$. Let now $m \geq 0$ and

$$u = s^m \circ u_0 \in C_+^{-m}\big([t_0, \infty)\big) \underset{\text{Lemma 8.2.33}}{=} s^m \circ C_+^0\big([t_0, \infty)\big) \subset C_+^{-\infty}\big([t_0, \infty)\big)$$

with $u_0 \in C_+^0\big([t_0, \infty)\big)$. According to item 3 of Lemma 8.2.35, there is a solution $y_0 \in C_+^n\big([t_0, \infty)\big)$ of $P \circ y_0 = u_0$, hence

$$y := s^m \circ y_0 \in s^m \circ C_+^n\big([t_0, \infty)\big) = C_+^{n-m}\big(t_0, \infty)\big) \subset C_+^{-\infty}\big([t_0, \infty)\big) \text{ and}$$
$$P \circ y = P \circ s^m \circ y_0 = s^m \circ P \circ y_0 = s^m \circ u_0 = u.$$

Therefore, $P \circ C_+^{-\infty}\big([t_0, \infty)\big) = C_+^{-\infty}\big([t_0, \infty)\big)$, i.e., $P\circ$ is also surjective and thus indeed bijective on $C_+^{-\infty}\big([t_0, \infty)\big)$ and on $C_+^{-\infty}$.
We have shown in Lemma 8.2.18 that differentiation and the shift action commute. For the preceding data and $a \in \mathbb{R}$, this implies

$$P \circ (a \circ_s y) = a \circ_s (P \circ y) = a \circ_s (Q \circ u) = Q \circ (a \circ_s u),$$

hence $H \circ (a \circ_s u) = a \circ_s y = a \circ_s (H \circ u)$, i.e., the shift action commutes with the scalar multiplication of the $\mathbb{C}(s)$-vector space $\mathrm{C}_+^{-\infty}(\mathbb{R}, \mathbb{C})$. □

Corollary 8.2.38 *If $P \in \mathbb{C}[s]$ is a polynomial of degree n, then*

$$P^{-1} \circ \mathrm{C}^0\big([t_0, \infty)\big) = \mathrm{C}^n\big([t_0, \infty)\big) \quad and \quad P^{-1} \circ \mathrm{C}_+^0 = \mathrm{C}_+^n .$$

Proof This follows directly from item 3 of Lemma 8.2.35. □

Corollary 8.2.39 *1. Every nonzero element in $\mathrm{C}_+^{-\infty}(\mathbb{R}, \mathbb{C})$ is $\mathbb{C}(s)$-linearly independent, i.e., $H \circ w = 0$ implies $H = 0$ or $w = 0$. In particular,*

$$\mathbb{C}[s] \cong \mathbb{C}[s] \circ w, \ f \mapsto f \circ w, \ for\ 0 \neq w \in \mathrm{C}_+^{-\infty} \ and$$
$$\mathbb{C}[s] \cong \mathbb{C}[s] \circ \delta_a, \ s^n \mapsto s^n \circ \delta_a = \delta_a^{(n)}, \ for\ a \in \mathbb{R},$$

i.e., $\mathbb{C}[s] \circ \delta_a$ has the $\mathbb{C}$-basis $\delta_a^{(n)}$, $n \in \mathbb{N}$.
2. The distributions δ_a, $a \in \mathbb{R}$, are $\mathbb{C}[s]$- and hence $\mathbb{C}(s)$-linearly independent, i.e., $\sum_{a\in\mathbb{R}} \mathbb{C}(s) \circ \delta_a = \bigoplus_{a\in\mathbb{R}} \mathbb{C}(s) \circ \delta_a$.

Proof 1. A $\mathbb{C}(s)$-vector space is $\mathbb{C}[s]$−torsionfree.
2. Assume $\sum_{i=1}^n f_i \circ \delta_{a_i} = 0$ with $f_i \in \mathbb{C}[s]$ and pairwise distinct $a_i \in \mathbb{R}$. We have to show $f_i = 0$ for all i. By means of (4.61) we choose functions $\varphi_i \in \mathrm{C}_0^\infty$ with $\varphi_i(t) = \begin{cases} 1 & \text{if } t \text{ near } a_i, \\ 0 & \text{if } t \text{ near } a_j \text{ and } j \neq i. \end{cases}$ For arbitrary $\varphi \in \mathrm{C}_0^\infty$ this implies

$$(f_i \circ \delta_{a_i})(\varphi) = (f_i \circ \delta_{a_i})(\varphi_i\varphi) \text{ and } (f_j \circ \delta_{a_j})(\varphi_i\varphi) = 0 \text{ for } j \neq i$$
$$\Longrightarrow \forall i: \ (f_i \circ \delta_{a_i})(\varphi) = \sum_{j=1}^n (f_j \circ \delta_{a_j})(\varphi_i\varphi) = \left(\sum_{j=1}^n f_j \circ \delta_{a_j}\right)(\varphi_i\varphi) = 0$$
$$\Longrightarrow \forall i: \ f_i \circ \delta_{a_i} = 0 \underset{1.}{\Longrightarrow} \forall i: \ f_i = 0.$$
□

Corollary and Definition 8.2.40 (Continuous-time transfer operator and impulse response) *Since the distributions $\mathrm{C}_+^{-\infty}(\mathbb{R}, \mathbb{C})$ with left bounded support form a $\mathbb{C}(s)$-vector space, a rational matrix $H \in \mathbb{C}(s)^{p\times m}$ induces the $\mathbb{C}(s)$-linear map*

$$H\circ: \ \mathrm{C}_+^{-\infty}(\mathbb{R}, \mathbb{C})^m \longrightarrow \mathrm{C}_+^{-\infty}(\mathbb{R}, \mathbb{C})^p,$$
$$u \longmapsto y := H \circ u, \quad where\ y_i := \sum_{j=1}^m H_{ij} \circ u_j. \tag{8.65}$$

This is the transfer operator *or* input/output map *defined by H. The signals u and $H \circ u$ are called the* input *and the* output, *respectively. The matrix*

$$h = (h_{ij})_{i,j} := H \circ \delta := (H_{ij} \circ \delta)_{i,j} \in \mathrm{C}_+^{-\infty}(\mathbb{R}, \mathbb{C})^{p\times m}$$

is called the impulse response *of H for obvious reasons. For the input $u = (0, \dots, 0, \overset{j}{\delta}, 0, \dots, 0)^\top$ the output is $y := H \circ u = (H_{ij} \circ \delta)_{i=1,\dots,p} = (h_{ij})_{i=1,\dots,p}$. We will show in Sect. 8.2.5 that the action of the transfer operator is given by convolution with the impulse response.* ◊

Theorem and Definition 8.2.41 (The initially relaxed part of an IO behavior) *Let $\mathcal{B} := \left\{ \binom{y}{u} \in C^{-\infty}(\mathbb{R}, \mathbb{C})^{p+m};\ P \circ y = Q \circ u \right\}$ with $P \in \mathbb{C}[s]^{p \times p}$ and* rank$(P) =$ *p be an IO behavior with transfer matrix $H = P^{-1}Q \in \mathbb{C}(s)^{p \times m}$. Then the graph of $H\circ$ induces the $\mathbb{C}(s)$-isomorphism*

$$\binom{H}{\mathrm{id}_m}\circ\colon\ C_+^{-\infty}(\mathbb{R}, \mathbb{C})^m \overset{\cong}{\to} \mathcal{B} \cap C_+^{-\infty}(\mathbb{R}, \mathbb{C})^{p+m},\quad u \longmapsto \binom{H \circ u}{u},$$

where $\mathcal{B} \cap C_+^{-\infty}(\mathbb{R}, \mathbb{C})^{p+m} = \left\{ \binom{y}{u} \in C_+^{-\infty}(\mathbb{R}, \mathbb{C})^{p+m};\ y = H \circ u \right\}$.
Hence the part of the behavior which is initially relaxed *or* initially at rest, *namely, $\mathcal{B} \cap C_+^{-\infty}(\mathbb{R}, \mathbb{C})^{p+m}$, is the graph of the transfer operator $H\circ$ and, in particular, uniquely determined by the transfer matrix H.*

Proof If $u \in C_+^{-\infty}(\mathbb{R}, \mathbb{C})^m$, then $P \circ (H \circ u) = (PH) \circ u = Q \circ u$ and $\binom{H \circ u}{u} \in \mathcal{B} \cap C_+^{-\infty}(\mathbb{R}, \mathbb{C})^{p+m}$. This shows that the map is well defined and a $\mathbb{C}(s)$-monomorphism. If, conversely, $\binom{y}{u} \in \mathcal{B} \cap C_+^{-\infty}(\mathbb{R}, \mathbb{C})^{p+m}$ then

$$P \circ y = Q \circ u \Longrightarrow y = P^{-1}P \circ y = P^{-1} \circ P \circ y = P^{-1} \circ Q \circ u = P^{-1}Q \circ u = H \circ u.$$

□

In Theorem 8.2.47 we compute the impulse response $h = H \circ \delta$ for arbitrary rational functions $H \in \mathbb{C}(s)$. For this we need the subspace $\mathcal{F}_2 \subseteq C_+^{-\infty}(\mathbb{R}, \mathbb{C})$ which we introduce in Theorem 8.2.43. In contrast to $C_+^{-\infty}(\mathbb{R}, \mathbb{C})$ the subspaces $\mathcal{F}_3$ and $\mathcal{F}_4$ of $C^{-\infty}(\mathbb{R}, \mathbb{C})$ from Theorem 8.2.44 are injective cogenerators, and the signals in $\mathcal{F}_4$ are given by finitely many data and can thus be handled by computer.

Lemma 8.2.42 *A $\mathbb{C}[s]$- and $\mathrm{t}(C^\infty(\mathbb{R}, \mathbb{C}))$-submodule $\mathcal{F}$ of $\mathcal{D}^*$ is a $\mathbb{C}(s)$-vector space, i.e., divisible and torsionfree, if and only if the differentiation operator $s\circ\colon\ \mathcal{F} \longrightarrow \mathcal{F}$ is bijective.*

Proof The condition of the theorem is obviously necessary.
Since $s\circ\colon\ \mathcal{F} \longrightarrow \mathcal{F}$ is bijective, so are all maps $s^m\circ\colon\ \mathcal{F} \longrightarrow \mathcal{F}$ for $m \in \mathbb{N}$. Recall that $\mathrm{t}(C^\infty)$ consists of the polynomial-exponential functions in particular that $e^{\lambda t}.\colon\ \mathcal{F} \longrightarrow \mathcal{F}$ is defined and bijective. The equation $(s - \lambda) \circ (e^{\lambda t} v) = e^{\lambda t}(s \circ v)$ implies the factorization

$$(s - \lambda)^m\circ\colon\ \mathcal{F} \overset{e^{-\lambda t}\cdot}{\longrightarrow} \mathcal{F} \overset{s^m\circ}{\longrightarrow} \mathcal{F} \overset{e^{\lambda t}\cdot}{\longrightarrow} \mathcal{F} \tag{8.66}$$

into bijective maps, and hence $(s - \lambda)^m\circ\colon\ \mathcal{F} \longrightarrow \mathcal{F}$ is bijective too. Since every nonzero $f \in \mathbb{C}[s]$ is a product of such powers $(s - \lambda)^m$ up to a constant, we conclude that $f\circ\colon \mathcal{F}\longrightarrow\mathcal{F}$ is also an isomorphism. This signifies that $\mathcal{F}$ is divisible and torsionfree, and therefore it is a $\mathbb{C}(s)$-space by Theorem 5.3.9. □

Theorem 8.2.43 *1. The space $\mathcal{F}_1 := \mathcal{F}_{1,\mathbb{C}} := \mathrm{C}^\infty(\mathbb{R},\mathbb{C})Y + \mathbb{C}[s]\circ\delta,\ \delta=\delta_0$, is a $\mathbb{C}(s)$-subspace of $\mathrm{C}_+^{-\infty}(\mathbb{R},\mathbb{C})$. Moreover the sum is direct, i.e.,*

$$\mathcal{F}_1 = \mathcal{F}_{1,\mathbb{C}} = \mathrm{C}^\infty(\mathbb{R},\mathbb{C})Y \oplus \mathbb{C}[s]\circ\delta \subset \mathrm{C}_+^{-\infty}(\mathbb{R},\mathbb{C}).$$

The so-called impulsive trajectories *of a behavior have components in this signal space and will be discussed in Sect. 8.2.10.*

2. *Likewise,*

$$\mathcal{F}_2 := \mathcal{F}_{2,\mathbb{C}} := \mathrm{t}(\mathrm{C}^\infty(\mathbb{R},\mathbb{C}))Y \oplus \mathbb{C}[s]\circ\delta$$

with $\mathrm{t}(\mathrm{C}^\infty(\mathbb{R},\mathbb{C})) = \bigoplus_{\lambda\in\mathbb{C}}\mathbb{C}[t]e^{\lambda t}$ is a $\mathbb{C}(s)$-subspace of $\mathrm{C}_+^{-\infty}(\mathbb{R},\mathbb{C})$. Its elements can be described by finitely many complex numbers. Via the Laplace transform it is isomorphic to $\mathbb{C}(s)$, i.e., $\mathbb{C}(s)$-one-dimensional, cf. Theorem 8.2.47, and hence the least $\mathbb{C}(s)$-subspace of $\mathcal{F}_\mathbb{C}$ containing δ.

3. *There are the corresponding real $\mathbb{R}(s)$-vector spaces*

$$\mathcal{F}_{1,\mathbb{R}} = \mathrm{C}^\infty(\mathbb{R},\mathbb{R})Y \oplus \mathbb{R}[s]\circ\delta \supset \mathcal{F}_{2,\mathbb{R}} = \mathrm{t}(\mathrm{C}^\infty(\mathbb{R},\mathbb{R}))Y \oplus \mathbb{R}[s]\circ\delta.$$

Proof We only show 1; the items 2 and 3 are proven analogously.
To show that $\mathcal{F}_1$ is a $\mathbb{C}[s]$-submodule, it suffices to prove $s\circ(\mathrm{C}^\infty\, Y)\subseteq\mathcal{F}_1$. For $\alpha Y\in\mathrm{C}^\infty\, Y$, we get

$$s\circ(\alpha Y) = \alpha' Y + \alpha Y' = \alpha' Y + \alpha\delta = \alpha' Y + \alpha(0)\delta \in \mathcal{F}_1.$$

As in the proof of Lemma 8.2.30 one sees that $\mathbb{C}[s]\circ\delta$ is a C^∞-submodule of $\mathrm{C}^{-\infty}$. The same holds trivially for $\mathrm{C}^\infty\, Y$ and then also for their sum.
According to the preceding lemma it remains to prove that the differentiation operator $s\circ\ \mathcal{F}_1\longrightarrow\mathcal{F}_1$ is bijective. It is injective since $\mathrm{C}_+^{-\infty}$ is a $\mathbb{C}(s)$-space. The $\mathbb{C}$-basis vectors $\delta = s\circ Y$ and $\delta^{(i)} = s\circ\delta^{(i-1)},\ i\geq 1$, of $\mathbb{C}[s]\circ\delta$ are contained in $s\circ\mathcal{F}_1$, therefore $\mathbb{C}[s]\circ\delta\subseteq s\circ\mathcal{F}_1$. Let $u=\alpha Y\in\mathrm{C}^\infty\, Y$ and $\beta\in\mathrm{C}^\infty$ an indefinite integral of α, hence $\beta'=\alpha$. Then

$$\begin{aligned}\alpha Y &= \beta' Y = \beta' Y + \big(\beta-\beta(0)\big)\,\delta = \beta' Y + \big(\beta-\beta(0)\big)\,Y' = \big((\beta-\beta(0))\,Y\big)' \\ &= s\circ\big((\beta-\beta(0))\,Y\big)\in s\circ\mathcal{F}_1 \Longrightarrow s\circ\mathcal{F}_1 = \mathcal{F}_1.\end{aligned}$$

To show that the sum is direct, assume that $\alpha Y = f\circ\delta$ with $\alpha\in\mathrm{C}^\infty$ and $f\in\mathbb{C}[s]$. We have to show $\alpha Y=0$, i.e., $\alpha(t)=0$ for $t>0$. Assume $\alpha(t_1)\neq 0$ for some $t_1>0$. We construct, as in the proof of Theorem and Definition 8.2.8, a test function $\varphi\in\mathrm{C}_0^\infty$ with $t_1\in\operatorname{supp}(\varphi)\subset(0,\infty)$ and $\langle\alpha,\varphi\rangle\neq 0$. Thus, $\alpha\varphi=\alpha Y\varphi$. We conclude $\langle\alpha,\varphi\rangle = \langle\alpha Y,\varphi\rangle = \langle f\circ\delta,\varphi\rangle = 0$, since $\operatorname{supp}(\varphi)\subset(0,\infty)$ and thus $\varphi^{(m)}(0)=0,\ m\in\mathbb{N}$. This is a contradiction and thus $\alpha Y=0$. □

A $\mathbb{C}(s)$-vector space V, i.e., a divisible and torsionfree $\mathbb{C}[s]$-module, is never a cogenerator since $\mathrm{Hom}_{\mathbb{C}[s]}(M, V) = 0$ for all torsion modules M. In particular, $\mathrm{C}_+^{-\infty}$ is not a cogenerator.

Theorem 8.2.44 *1. The $\mathbb{C}[s]$-submodule*

$$\mathcal{F}_3 := \mathcal{F}_{3,\mathbb{C}} := \mathrm{C}^{\infty}(\mathbb{R}, \mathbb{C}) + \mathrm{C}_+^{-\infty}(\mathbb{R}, \mathbb{C}) \subset \mathcal{F} := \mathcal{F}_{\mathbb{C}} = \mathrm{C}^{-\infty}(\mathbb{R}, \mathbb{C})$$

of sums of smooth functions and distributions with left bounded support is an injective cogenerator. It contains many relevant signals, but not all piecewise continuous periodic ones, cf. Sect. 8.2.8.

2. *The injective cogenerator $\mathcal{F}_3$ contains the injective cogenerator*

$$\mathcal{F}_4 := \mathcal{F}_{4,\mathbb{C}} := \mathrm{t}(\mathcal{F}_{\mathbb{C}}) + \mathcal{F}_2 = \mathrm{t}(\mathcal{F}_{\mathbb{C}}) \oplus \mathrm{t}(\mathcal{F}_{\mathbb{C}})Y \oplus \mathbb{C}[s] \circ \delta, \ \mathrm{t}(\mathcal{F}_{\mathbb{C}}) = \bigoplus_{\lambda \in \mathbb{C}} \mathbb{C}[t]e^{\lambda t}.$$

All signals in $\mathcal{F}_4$ are determined by finitely many complex numbers and can therefore be easily handled by computer. Many signals of electrical or mechanical engineering belong to $\mathcal{F}_4$. The advantage of $\mathcal{F}_4$ compared to $\mathcal{F}_2$ is that it is also a cogenerator.

3. *There are the corresponding $\mathbb{R}[s]$-injective cogenerators*

$$\begin{aligned}
\mathcal{F}_{\mathbb{R}} &:= \mathrm{C}^{-\infty}(\mathbb{R}, \mathbb{R}) \supset \mathcal{F}_{3,\mathbb{R}} := \mathrm{C}^{\infty}(\mathbb{R}, \mathbb{R}) + \mathrm{C}_+^{-\infty}(\mathbb{R}, \mathbb{R}) \\
&\supset \mathcal{F}_{4,\mathbb{R}} := \mathrm{t}(\mathcal{F}_{\mathbb{R}}) + \mathcal{F}_{2,\mathbb{R}} = \mathrm{t}(\mathcal{F}_{\mathbb{R}}) \oplus \mathrm{t}(\mathcal{F}_{\mathbb{R}})Y \oplus \mathbb{R}[s] \circ \delta \\
&\supset \mathrm{t}(\mathcal{F}_{\mathbb{R}}) = \oplus_{\rho \in \mathbb{R}} \mathbb{R}[t]e^{\rho t} \oplus \oplus_{\rho, \omega \in \mathbb{R}, \omega > 0} \mathbb{R}[t]e^{\rho t} \left(\mathbb{R}\cos(\omega t) \oplus \mathbb{R}\sin(\omega t)\right).
\end{aligned}$$

Proof 1. Since C^{∞} is a cogenerator, so is the larger module $\mathcal{F}_3$. This follows directly from Definition and Lemma 3.3.2. Both summands are divisible and hence so is their sum $\mathcal{F}_3$. Thus $\mathcal{F}_3$ is injective by Corollary 3.2.11.
2. Since $\mathrm{t}(\mathcal{F})$ is an injective cogenerator the same property of $\mathcal{F}_4$ follows as in item 1. Moreover

$$\mathbb{C}[s] \circ \delta \cap \mathrm{C}^{0,\mathrm{pc}} = 0 \Longrightarrow \mathcal{F}_4 = (\mathrm{t}(\mathcal{F}) + \mathrm{t}(\mathcal{F})Y) \oplus \mathbb{C}[s] \circ \delta.$$

Analytic functions, in particular the polynomial-exponential functions in $\mathrm{t}(\mathcal{F})$, are zero if they are zero on a small interval. This implies

$$\mathrm{t}(\mathcal{F}) \cap \mathrm{t}(\mathcal{F})Y = 0 \Longrightarrow \mathcal{F}_4 = \mathrm{t}(\mathcal{F}) \oplus \mathrm{t}(\mathcal{F})Y \oplus \mathbb{C}[s] \circ \delta.$$ □

Remark 8.2.45 1. For the *open* interval $(0, \infty)$ analogues of the preceding considerations hold. One obtains the injective $\mathbb{C}[s]$-cogenerators

$$\mathrm{t}(\mathcal{F}) \underset{\text{ident.}}{=} \mathrm{t}(\mathcal{F})|_{(0,\infty)} \subset \mathrm{C}^{\infty}(0, \infty) \subset \mathrm{C}^{-\infty}(0, \infty) := \mathrm{C}^{-\infty}((0, \infty), \mathbb{C}). \quad (8.67)$$

For IO equations $P \circ y = Q \circ u,\ P \in \mathbb{C}[s]^{p\times p},\ \mathrm{rank}(P) = p,\ Q \in \mathbb{C}[s]^{p\times m}$, and any input $u \in \mathrm{t}(\mathcal{F})^m \underset{\text{ident.}}{=} \left(\mathrm{t}(\mathcal{F})|_{(0,\infty)}\right)^m$ there is an output $y \in \mathrm{t}(\mathcal{F})^p$. Every solution $\widetilde{y} \in \mathrm{C}^{-\infty}(0,\infty)^p$ of $P \circ \widetilde{y} = Q \circ u$ has the form $\widetilde{y} = y + z$ with $P \circ z = 0$ and hence $z, \widetilde{y} \in \mathrm{t}(\mathcal{F})^p$. In particular, $\widetilde{y}$ can be uniquely analytically extended to $\mathbb{R}$, the values $\widetilde{y}_k^{(\mu)}(0) = \widetilde{y}_k^{(\mu)}(0+),\ k \le p,\ \mu \ge 0$, exist and can be prescribed as initial condition at $t = 0$, cf. Theorem 8.2.53, 2–3.
These signal spaces are useful, for instance, for the modeling of electrical networks that are switched on at time $t = 0$ with an initial condition.
2. The equation $s \circ \ln(t) = t^{-1}$ on $(0,\infty)$ with $\ln(0+) = -\infty$ shows that for general $u \in \mathrm{C}^\infty(0,\infty)$ the initial values $u(0+),\ y(0+)$ do not necessarily exist.
3. For the often used closed time interval $[0,\infty)$ and a piecewise continuous function $u : [0,\infty) \to \mathbb{C}$, the meaning of $P(d/dt)y(t)$ and $Q(d/dt)u(t)$ for $t = 0$, cf. [6, p. 92, Chap. 3], [7, p. 55, Chap. 4] is not obvious without the distribution space $\mathrm{C}^{-\infty}(\mathbb{R},\mathbb{C})$ on *all* of $\mathbb{R}$. The expressions $u(0-) := \lim_{t\to 0, t<0} u(t)$ or even $u^{(i)}(0-),\ i > 0$, are not defined, unless the function u is at least defined on an open interval containing 0 and not on $[0,\infty)$ only. Nor are $u(0)$ or $u(0-)$ defined for a distribution u that is not a function. ◊

The following lemma serves as a preparation for Theorem 8.2.47, where we compute the impulse response of a rational function.

Lemma 8.2.46 *We use the functions $e_{\lambda,m}(t) = \frac{t^m}{m!}e^{\lambda t}$ from (5.15). For $\lambda \in \mathbb{C}$ and $m \in \mathbb{N}$, the identity*

$$(s-\lambda)^{-m} \circ \delta = \begin{cases} \delta & \text{if } m = 0, \\ e_{\lambda,m-1}Y & \text{if } m \ge 1 \end{cases}$$

holds, with the special case $(s-\lambda)^{-1} \circ \delta = e^{\lambda t}Y(t)$. Observe that

$$(s-\lambda)^{-m} \circ \delta \ \ \text{is} \ \begin{cases} \text{a distribution} & \text{if } m = 0, \\ \text{piecewise continuous with a jump at } t = 0 & \text{if } m = 1, \\ \text{continuous} & \text{if } m \ge 2. \end{cases}$$

Proof The proof proceeds by induction, the case $m = 0$ being obvious.
Since $\mathrm{C}_+^{-\infty}(\mathbb{R},\mathbb{C})$ is a $\mathbb{C}(s)$-vector space, the equations $f \circ y = u$ and $y = f^{-1} \circ u$ are equivalent for $0 \ne f \in \mathbb{C}[s]$. For $m = 1$, the product rule yields

$$(s-\lambda) \circ (e_{\lambda,0}Y) = (e^{\lambda t}Y)' - \lambda e^{\lambda t}Y = \lambda e^{\lambda t}Y + e^{\lambda t}\delta - \lambda e^{\lambda t}Y = e^{\lambda 0}\delta = \delta,$$

hence $(s-\lambda)^{-1} \circ \delta = e_{\lambda,0}Y$. Now let $m > 1$ and hence $e_{\lambda,m-1}(0) = 0$. Recall from Corollary 5.2.1 that $(s-\lambda)^k \circ e_{\lambda,m} = e_{\lambda,m-k}$ for $k \le m$. The product rule and the induction hypothesis furnish

$$(s-\lambda)\circ(e_{\lambda,m-1}Y)=e'_{\lambda,m-1}Y+e_{\lambda,m-1}(0)\delta-\lambda e_{\lambda,m-1}Y$$
$$=\big((s-\lambda)\circ e_{\lambda,m-1}\big)Y\underset{\text{Corollary 5.2.1}}{=}e_{\lambda,m-2}Y$$
$$\underset{\text{induction}}{=}(s-\lambda)^{-(m-1)}\circ\delta\Longrightarrow e_{\lambda,m-1}Y=(s-\lambda)^{-m}\circ\delta.$$ □

In the following theorems we are going to use the signal spaces from Theorems 8.2.43 and 8.2.44.

Theorem and Definition 8.2.47 (Impulse response and Laplace transform, cf. [5, Theorem 3.5.2], [8, Sect. 5.3])

1. *Consider the complex signal spaces*

$$\mathcal{F}:=\mathcal{F}_{\mathbb{C}}:=\mathrm{C}^{-\infty}(\mathbb{R},\mathbb{C})\supset\mathcal{F}_2:=\mathcal{F}_{2,\mathbb{C}}:=\mathbb{C}[s]\circ\delta\oplus\mathrm{t}(\mathcal{F}_{\mathbb{C}})Y,\ \mathrm{t}(\mathcal{F}_{\mathbb{C}})=\bigoplus_{\lambda\in\mathbb{C}}\mathbb{C}[t]e^{\lambda t} \tag{8.68}$$

and $H\in\mathbb{C}(s)$ with the constructive partial fraction decomposition

$$\mathbb{C}(s)=\mathbb{C}[s]\oplus\mathbb{C}(s)_{\mathrm{spr}}=\mathbb{C}[s]\oplus\bigoplus_{\lambda\in\mathbb{C}}\bigoplus_{k=1}^{\infty}\mathbb{C}(s-\lambda)^{-k}$$
$$H=H_{\mathrm{pol}}+H_{\mathrm{spr}}=\sum_{i=0}^{m}a_is^i+\sum_{\lambda\in\mathrm{pole}(H)}\sum_{k=1}^{m_\lambda}a_{\lambda,k}(s-\lambda)^{-k} \tag{8.69}$$
$$\text{with } m_\lambda\geq 1,\ a_{\lambda,m_\lambda}\neq 0 \text{ for all } \lambda\in\mathrm{pole}(H)$$

from Corollary 5.5.9. Then the impulse response $h:=H\circ\delta$ of H is

$$h=H_{\mathrm{pol}}\circ\delta+H_{\mathrm{spr}}\circ\delta=\sum_{i=0}^{m}a_i\delta^{(i)}+\alpha Y\in\mathcal{F}_2 \text{ with}$$
$$\alpha:=\sum_{\lambda\in\mathrm{pole}(H)}\sum_{k=1}^{m_\lambda}a_{\lambda,k}e_{\lambda,k-1},\ e_{\lambda,k-1}(t)=\frac{t^{k-1}}{(k-1)!}e^{\lambda t}. \tag{8.70}$$

Since the partial fraction decomposition is implemented in all computer algebra systems, the computation of the impulse response is easy.

2. *(a) The impulse response $H\circ\delta$ is piecewise continuous, with a possible jump at $t=0$ only, if and only if H is strictly proper, i.e.,* $\deg_s(H)\leq -1$.
(b) It is even continuous or, equivalently, continuous at $t=0$, if and only if in addition $\alpha(0)=0$ or, equivalently, sH is strictly proper or $\deg_s(H)\leq -2$.
3. *The map*

$$\mathbb{C}(s)\to\mathcal{F}_2=\mathbb{C}[s]\circ\delta\oplus\oplus_{\lambda\in\mathbb{C}}\mathbb{C}[t]e^{\lambda t},\ H\mapsto H\circ\delta, \tag{8.71}$$

is a $\mathbb{C}(s)$-isomorphism. The inverse $\mathbb{C}(s)$-isomorphisms

$$\mathcal{L}:\mathcal{F}_2\xrightarrow{\cong}\mathbb{C}(s),\ u=\sum_{i=0}^{\infty}a_i\delta^{(i)}+\left(\sum_{\lambda\in\mathbb{C}}\sum_{k=1}^{\infty}a_{\lambda,k}e_{\lambda,k-1}\right)Y$$
$$\mapsto\mathcal{L}(u):=\sum_{i=0}^{\infty}a_is^i+\sum_{\lambda\in\mathbb{C}}\sum_{k=1}^{\infty}a_{\lambda,k}(s-\lambda)^{-k},$$
$$\mathcal{L}^{-1}:\mathbb{C}(s)\xrightarrow{\cong}\mathcal{F}_2,\ H\mapsto\mathcal{L}^{-1}(H):=H\circ\delta,\tag{8.72}$$

are called the Laplace *resp. the* inverse Laplace transform *as are the images* $\mathcal{L}(u)$ *resp.* $\mathcal{L}^{-1}(H)$. *Equations* (8.69)–(8.72) *contain algorithms for the computation of* $\mathcal{L}$ *and* $\mathcal{L}^{-1}$. *So essentially, the transform pairs* $u\leftrightarrow\mathcal{L}(u)$ *are computed by means of the partial fraction decomposition of rational functions. In electrical or mechanical engineering the study of the signal* $u=H\circ\delta=\mathcal{L}^{-1}(H)$ *resp. of its isomorphic image* $H=\mathcal{L}(u)$ *is said to belong to the* time domain *resp. the* frequency domain.

In particular, a polynomial-exponential signal that starts at $t=0$, *i.e., a signal in* $\mathrm{t}(\mathcal{F})Y=\left(\bigoplus_{\lambda\in\mathbb{C}}\mathbb{C}[t]e^{\lambda t}\right)Y$, *can be generated as the step response* $\widetilde{H}\circ Y=\widetilde{H}s^{-1}\circ\delta$ *of a proper rational function* $\widetilde{H}$, *for instance by a suitable electrical one-port with transfer function* $\widetilde{H}$ *and a DC current input of one Ampère starting at* $t=0$, *cf. Sect. 8.3.*

4. *The isomorphisms* $\mathcal{L}$ *and* $\mathcal{L}^{-1}$ *are extended to matrices componentwise. If*

$$H_1\in\mathbb{C}(s)^{p\times m},\ H_2\in\mathbb{C}(s)^m,\ u_2:=H_2\circ\delta\in\mathcal{F}_2^m,\ H_2=\mathcal{L}(u_2),\ \text{then}$$
$$H_1\circ u_2=H_1\circ(H_2\circ\delta)=(H_1H_2)\circ\delta=\mathcal{L}^{-1}(H_1H_2)\tag{8.73}$$

where $(H_1H_2)\circ\delta$ *is computed according to (i).*

In words: *If* H_1 *is the transfer matrix of an IO behavior and* $u_2\in\mathcal{F}_2^m$ *is an input trajectory with components in* $\mathcal{F}_2$, *then the output* $H_1\circ u_2$ *can be easily computed by partial fraction decomposition of* $H_1H_2=\mathcal{L}(H_1\circ u_2)$. *Since* $\mathcal{F}_2$ *is large, this computation is often applicable.*

5. *If* $H\in\mathbb{C}(s)_{\mathrm{spr}}$ *and* $u\in\mathrm{C}_+^{0,\mathrm{pc}}$ *then* $H\circ u\in\mathrm{C}_+^0$.

Proof 1. Since $\mathcal{F}_2\subset\mathrm{C}_+^{-\infty}(\mathbb{R},\mathbb{C})$ are $\mathbb{C}(s)$-vector spaces and the distributive law holds, the formula for the impulse response follows directly from (8.69) and Lemma 8.2.46.

2. (a) Due to the direct sum decomposition of $\mathcal{F}_2$ the impulse response $H\circ\delta$ is piecewise continuous if and only if $H_{\mathrm{pol}}\circ\delta=0$. Since $\mathcal{F}_2$ is a $\mathbb{C}(s)$-vector space, this is the case if and only if $H_{\mathrm{pol}}=0$, i.e., if and only if H is strictly proper.

(b) By (a), sH is strictly proper if and only if $sH\circ\delta=s\circ(\alpha Y)=\alpha'Y+\alpha(0)\delta$ is piecewise continuous. This holds if and only if $\alpha(0)=0$ or αY is continuous.

3. It is injective since δ is nonzero and $\mathbb{C}(s)$ is a field. It is surjective since *all* polynomial-exponential functions have the form $\alpha=\sum_{\lambda\in\mathbb{C},k\geq 1}a_{\lambda,k}e_{\lambda,k-1}$.

4. $\mathcal{F}_2$ is a $\mathbb{C}(s)$-vector space.

5. It suffices to show this for $H=(s-\lambda)^{-k}$, $k\geq 1$. If $\lambda=0$ then $s^{-1}\circ u=\int_{-\infty}^{t}u(\tau)\mathrm{d}\tau$ is continuous and hence $s^{-k}\circ u\in\mathrm{C}_+^{k-1}\subseteq\mathrm{C}_+^0$.

For $\lambda \in \mathbb{C}$ there are the mutually inverse isomorphisms

$$(s-\lambda)^k\circ:\ \mathrm{C}_+^{-\infty}\xrightarrow{e^{-\lambda t}\cdot}\mathrm{C}_+^{-\infty}\xrightarrow{s^k\circ}\mathrm{C}_+^{-\infty}\xrightarrow{e^{\lambda t}\cdot}\mathrm{C}_+^{-\infty}\ \text{and}$$
$$(s-\lambda)^{-k}\circ:\ \mathrm{C}_+^{-\infty}\xrightarrow{e^{-\lambda t}\cdot}\mathrm{C}_+^{-\infty}\xrightarrow{s^{-k}\circ}\mathrm{C}_+^{-\infty}\xrightarrow{e^{\lambda t}\cdot}\mathrm{C}_+^{-\infty}$$
$$\Longrightarrow \forall u\in \mathrm{C}_+^{0,\mathrm{pc}}:\ e^{-\lambda t}u\in \mathrm{C}_+^{0,\mathrm{pc}}\ \text{and}\ (s-\lambda)^{-k}\circ u=e^{\lambda t}s^{-k}\circ(e^{-\lambda t}u)\in \mathrm{C}_+^{0}.\ \square$$

Remark 8.2.48 Nonrational Laplace transforms $\mathcal{L}(u)$ are usually defined for more general distributions or functions u, see Theorem 8.2.89. Nonrational transfer functions occur for other types of linear systems, cf. Theorem 8.2.73 and [9], but are not studied in this book. ◊

In the next theorem we consider the impulse response on $\mathbb{R}(s)$ and the Laplace transform on the *real* space $\mathcal{F}_{2,\mathbb{R}}$ via complexification as in Sect. 6.3.7. We use the following signal spaces:

$$\begin{aligned}
&\mathcal{F}:=\mathcal{F}_{\mathbb{C}}\supset \mathrm{t}(\mathcal{F}_{\mathbb{C}})=\oplus_{\lambda\in\mathbb{C}}\mathbb{C}[t]e^{\lambda t}\\
&\qquad\qquad\supset \mathrm{t}_+(\mathcal{F}_{\mathbb{C}}):=\oplus_{\rho\in\mathbb{R}}\mathbb{R}[t]e^{\rho t}\oplus\oplus_{\rho,\omega\in\mathbb{R},\ \omega>0}\mathbb{C}[t]e^{(\rho+j\omega)t},\\
&\mathcal{F}_{5,\mathbb{C}}:=\mathbb{R}[s]\circ\delta\oplus \mathrm{t}_+(\mathcal{F}_{\mathbb{C}})Y\oplus \mathrm{t}_+(\mathcal{F}_{\mathbb{C}})\subset\mathcal{F}_{4,\mathbb{C}}=\mathbb{C}[s]\circ\delta\oplus \mathrm{t}(\mathcal{F}_{\mathbb{C}})Y\oplus \mathrm{t}(\mathcal{F}_{\mathbb{C}}),\\
&\mathcal{F}_{\mathbb{R}}:=\mathrm{C}^{-\infty}(\mathbb{R},\mathbb{R})=\{u\in\mathcal{F}_{\mathbb{C}};\ u=\overline{u}\}\supset\mathcal{F}_{4,\mathbb{R}}=\mathbb{R}[s]\circ\delta\oplus \mathrm{t}(\mathcal{F}_{\mathbb{R}})Y\oplus \mathrm{t}(\mathcal{F}_{\mathbb{R}})\\
&\qquad\supset \mathrm{t}(\mathcal{F}_{\mathbb{R}})=\oplus_{\rho\in\mathbb{R}}\mathbb{R}[t]e^{\rho t}\oplus\oplus_{\rho,\omega\in\mathbb{R},\ \omega>0}\mathbb{R}[t]e^{\rho t}\left(\mathbb{R}\cos(\omega t)\oplus\mathbb{R}\sin(\omega t)\right).
\end{aligned}\tag{8.74}$$

Theorem 8.2.49 *1. The Laplace transform $\mathcal{L}$ on $\mathcal{F}_{2,\mathbb{C}}$ and its inverse $\mathcal{L}^{-1}$ induce the inverse $\mathbb{R}(s)$-isomorphisms*

$$\begin{aligned}
&\mathcal{L}^{-1}:\mathbb{R}(s)=\left\{H\in\mathbb{C}(s);\ H=\overline{H}\right\}\overset{\cong}{\longleftrightarrow}\mathcal{F}_{2,\mathbb{R}}=\left\{u\in\mathcal{F}_{2,\mathbb{C}};\ u=\overline{u}\right\}:\mathcal{L}\ \textit{where}\\
&H=fg^{-1},\ f=\sum_i f_is^i,\ g\in\mathbb{C}[s],\ u=\sum_{i=0}^{\infty}a_i\delta^{(i)}+\left(\sum_{\lambda\in\mathbb{C}}\sum_{k=1}^{\infty}a_{\lambda,k}e_{\lambda,k-1}\right)Y,\\
&\overline{H}:=\overline{f}\overline{g}^{-1},\ \overline{f}:=\sum_i\overline{f_i}s^i,\ \overline{u}:=\sum_{i=0}^{\infty}\overline{a_i}\delta^{(i)}+\left(\sum_{\lambda\in\mathbb{C}}\sum_{k=1}^{\infty}\overline{a_{\lambda,k}}e_{\overline{\lambda},k-1}\right)Y.\ \textit{Moreover}\\
&\overline{\mathcal{L}^{-1}(H)}=\mathcal{L}^{-1}\left(\overline{H}\right),\ \overline{\mathcal{L}(U)}=\mathcal{L}(\overline{U}),\\
&\underset{\Re(z)=2^{-1}(z+\overline{z})}{\Longrightarrow}\ \Re\left(\mathcal{L}^{-1}(H)\right)=\mathcal{L}^{-1}\left(\Re(H)\right),\ \Re\left(\mathcal{L}(U)\right)=\mathcal{L}(\Re(U)).
\end{aligned}\tag{8.75}$$

2. *(a) Let $j:=\sqrt{-1}$. The* complexification map

$$
\begin{aligned}
\mathfrak{C}:\mathrm{t}(\mathcal{F}_{\mathbb{R}}) \to \mathrm{t}(\mathcal{F}_{\mathbb{C}}):\ u = & \sum_{\rho\in\mathbb{R},\ k\in\mathbb{N}} a_{\rho,k}t^k e^{\rho t}\\
& + \sum_{\rho,\omega\in\mathbb{R},\ \omega>0,\ k\in\mathbb{N}} t^k e^{\rho t}\left(a_{\rho,\omega,k}\cos(\omega t) + b_{\rho,\omega,k}\sin(\omega t)\right)\\
\mapsto \mathfrak{C}(u) := & \sum_{\rho\in\mathbb{R},\ k\in\mathbb{N}} a_{\rho,k}t^k e^{\rho t}\\
& + \sum_{\rho,\omega\in\mathbb{R},\ \omega>0,\ k\in\mathbb{N}} t^k\left(a_{\rho,\omega,k} - jb_{\rho,\omega,k}\right)e^{(\rho+j\omega)t}
\end{aligned}
\tag{8.76}
$$

with $a_{\rho,k}, a_{\rho,\omega,k}, b_{\rho,\omega,k} \in \mathbb{R}$ is an $\mathbb{R}[s]$-monomorphism with the image $\mathrm{t}_+(\mathcal{F}_{\mathbb{C}})$ and the $\mathbb{R}[s]$-linear retraction

$$\Re : \mathrm{t}(\mathcal{F}_{\mathbb{C}}) \to \mathrm{t}(\mathcal{F}_{\mathbb{R}}),\ U \mapsto \Re(U).$$

(b) The map $\mathfrak{C}$ is extended to the $\mathbb{R}$-monomorphism

$$\mathfrak{C} : \mathcal{F}_{4,\mathbb{R}} \to \mathcal{F}_{4,\mathbb{C}},\ T + uY + v \mapsto T + \mathfrak{C}(u)Y + \mathfrak{C}(v), \tag{8.77}$$

with $T \in \mathbb{R}[s] \circ \delta$, $u, v \in \mathrm{t}(\mathcal{F}_{\mathbb{R}})$, the image $\mathcal{F}_{5,\mathbb{C}}$ and the $\mathbb{R}[s]$-linear retraction $\Re : U \mapsto \Re(U)$. The extended map $\mathfrak{C}$ is not $\mathbb{R}[s]$-linear since

$$
\begin{aligned}
&\mathfrak{C}(\sin(\omega t)Y) = -je^{j\omega t}Y,\ s\circ(\sin(\omega t)Y) = \omega\cos(\omega t)Y\\
&\mathfrak{C}\left(s\circ(\sin(\omega t)Y)\right) = \omega e^{j\omega t}Y \neq \omega e^{j\omega t}Y - j\delta\\
&\qquad = s\circ\left(-je^{j\omega t}Y\right) = s\circ\mathfrak{C}(\sin(\omega t)Y).
\end{aligned}
$$

3. *If $\mathcal{B}_{\mathbb{R}} = \left\{w \in \mathcal{F}_{\mathbb{R}}^l;\ R\circ w = 0\right\}$, $R \in \mathbb{R}[s]^{k\times l}$, is a real behavior and $\mathcal{B}_{\mathbb{C}} := \left\{w \in \mathcal{F}_{\mathbb{C}}^l;\ R\circ w = 0\right\}$ its complexification, the map $\Re$ induces $\mathbb{R}[s]$-epimorphisms*

$$\mathcal{B}_{\mathbb{C}} \to \mathcal{B}_{\mathbb{R}},\ \mathcal{F}_{4,\mathbb{C}}^l \cap \mathcal{B}_{\mathbb{C}} \to \mathcal{F}_{4,\mathbb{R}}^l \cap \mathcal{B}_{\mathbb{R}},\ \mathrm{t}(\mathcal{F}_{\mathbb{C}})^l \cap \mathcal{B}_{\mathbb{C}} \to \mathrm{t}(\mathcal{F}_{\mathbb{R}})^l \cap \mathcal{B}_{\mathbb{R}}.$$

The map $\mathfrak{C}$ maps $\mathrm{t}(\mathcal{F}_{\mathbb{R}})^l \cap \mathcal{B}_{\mathbb{R}}$ into $\mathrm{t}(\mathcal{F}_{\mathbb{C}})^l \cap \mathcal{B}_{\mathbb{C}}$, but not $\mathcal{F}_{4,\mathbb{R}}^l \cap \mathcal{B}_{\mathbb{R}}$ into $\mathcal{F}_{4,\mathbb{C}}^l \cap \mathcal{B}_{\mathbb{C}}$.

4. *Polynomial-exponential functions are obviously closed under multiplication. The multiplication of complex ones is simple due to $\left(t^k e^{\lambda t}\right)\left(t^l e^{\mu t}\right) = t^{k+l}e^{(\lambda+\mu)t}$. If $u_1, u_2 \in \mathrm{t}(\mathcal{F}_{\mathbb{R}})Y$ and $U_k := \mathfrak{C}(u_k) \in \mathrm{t}(\mathcal{F}_{\mathbb{C}})Y$, then $u_k = \Re(U_k)$ and*

$$\mathcal{L}(u_1u_2) = 2^{-1}\Re\left(\mathcal{L}\left(U_1\overline{U_2} + U_1U_2\right)\right) \in \mathbb{R}(s)_{\mathrm{spr}}. \tag{8.78}$$

This furnishes a simple algorithm for the computation of the Laplace transform on the real space $\mathrm{t}(\mathcal{F}_{\mathbb{R}})Y$.

5. *In Theorem 8.2.47, 4, consider*

$$H_1 \in \mathbb{R}(s)^{p\times m},\ u_2 \in \mathcal{F}^m_{2,\mathbb{R}},\ U_2 := \mathfrak{C}(u_2),\ \Re(U_2) = u_2,\ H_2 := \mathcal{L}(U_2).\ \text{Then}$$
$$H_1 \circ u_2 = H_1 \circ \Re(H_2 \circ \delta) = H_1 \circ \Re(H_2) \circ \delta = H_1\Re(H_2) \circ \delta = \Re(H_1 H_2) \circ \delta$$
$$= \Re(H_1 H_2 \circ \delta) \Longrightarrow H_1 \circ u_2 = \Re\left(\mathcal{L}^{-1}(H_1 H_2)\right) \tag{8.79}$$

where $\mathcal{L}^{-1}(H_1 H_2) = H_1 H_2 \circ \delta$ *is computed via Theorem 8.2.47.*

In electrical engineering *parts of this theorem are regularly applied to simplify the computation of real systems via complexification.*

Proof 1. This follows from

$$\overline{a(s-\lambda)^{-k} \circ \delta} = \overline{ae_{\lambda,k-1}Y} = \overline{a}e_{\overline{\lambda},k-1}Y = \overline{a}(s-\overline{\lambda})^{-k} \circ \delta = \overline{a(s-\lambda)^{-k}} \circ \delta$$
$$\Longrightarrow \forall H \in \mathbb{C}(s),\ U := H \circ \delta:\ \overline{H \circ \delta} = \overline{H} \circ \delta,\ \overline{\mathcal{L}^{-1}(H)} = \mathcal{L}^{-1}\left(\overline{H}\right),\ \mathcal{L}(\overline{U}) = \overline{\mathcal{L}(U)}$$
$$\Longrightarrow \Re(H \circ \delta) = 2^{-1}(H \circ \delta + \overline{H \circ \delta}) = \Re(H) \circ \delta,$$
$$\mathcal{L}(\Re(U)) = \Re\left(\mathcal{L}(U)\right),\ \mathcal{L}^{-1}\left(\Re(H)\right) = \Re\left(\mathcal{L}^{-1}(H)\right). \tag{8.80}$$

2. (a) Obviously $\mathfrak{C}$ is an $\mathbb{R}$-monomorphism with the image $\mathrm{t}_+(\mathcal{F}_{\mathbb{C}})$ and the retraction $\Re$. The equation $s \circ (\Re(U) + j\Im(U)) = s \circ \Re(U) + js \circ \Im(U)$ implies $s \circ \Re(U) = \Re(s \circ U)$ and hence that $\Re$ is $\mathbb{R}[s]$-linear. For $f := t^k e^{\rho t}$, hence $f' = kt^{k-1}e^{\rho t} + t^k \rho e^{\rho t}$, and $\omega > 0$ define

$$u = f\cos(\omega t) \Longrightarrow s \circ u = f'\cos(\omega t) - f\omega\sin(\omega t)$$
$$\Longrightarrow \mathfrak{C}(s \circ u) = f'e^{j\omega t} + j\omega f e^{j\omega t}$$
$$s \circ \mathfrak{C}(u) = s \circ (fe^{j\omega t}) = (f' + j\omega f)e^{j\omega t} = \mathfrak{C}(s \circ u).$$

Likewise one proves $\mathfrak{C}(s \circ u) = s \circ \mathfrak{C}(u)$ for $u := f\sin(\omega t)$. This shows that $\mathfrak{C}$ is $\mathbb{R}[s]$-linear.

(b) This follows directly from (a).

3. $\Re(R \circ U) = R \circ \Re(U)$ since $R \in \mathbb{R}[s]^{k\times l}$, $\Re|_{\mathcal{B}_{\mathbb{R}}} = \mathrm{id}_{\mathcal{B}_{\mathbb{R}}}$.

4. $\Re(z)\Re(w) = \Re(z\Re(w)) = 2^{-1}\Re(z\overline{w} + zw),\ z, w \in \mathbb{C}$,

$\Longrightarrow u_1 u_2 = 2^{-1}\Re\left(U_1\overline{U_2} + U_1 U_2\right)$.

5. This follows from (8.80). □

Remark 8.2.50 Let

$$U \in \mathrm{C}^{0,\mathrm{pc}}[0,\infty),\ \overline{U}(t) := \overline{U(t)} \text{ and } \mathcal{L}(U)(s) := \int_0^\infty U(t)e^{-st}\mathrm{d}t$$
$$\Longrightarrow \overline{\mathcal{L}(U)(s)} = \mathcal{L}(\overline{U})(\overline{s}),\ s \in \mathbb{C},\ \Re(s) \text{ large enough.}$$

This seems to contradict $\overline{\mathcal{L}(U)(s)} = \mathcal{L}(\overline{U})(s)$ from (8.75), but does not really: s in (8.80) is an indeterminate, and $H \mapsto \overline{H}$, $H \in \mathbb{C}(s)$, keeps s invariant, i.e., $\overline{s} = s$ and $\overline{(s-\lambda)^{-1}} = (s-\overline{\lambda})^{-1}$. This definition cannot be extended to general, nonrational H. ◇

Example 8.2.51 *Laplace transform pairs* $u \leftrightarrow H := \mathcal{L}(u),\ u = H \circ \delta$:
(i) The simplest pairs are

$$u := (s-\lambda)^{-k} \circ \delta = e_{\lambda,k-1}Y = \tfrac{t^{k-1}}{(k-1)!}e^{\lambda t}Y,\ \mathcal{L}(u) = (s-\lambda)^{-k},\ \lambda \in \mathbb{C},\ k \geq 1.$$

(ii) Let $\lambda_1 \neq \lambda_2$ and

$$\begin{aligned} H &:= (s-\lambda_1)^{-1}(s-\lambda_2)^{-1} = (\lambda_1-\lambda_2)^{-1}\left((s-\lambda_1)^{-1} - (s-\lambda_2)^{-1}\right) \\ \Longrightarrow u &:= H \circ \delta = (\lambda_1-\lambda_2)^{-1}\left(e^{\lambda_1 t} - e^{\lambda_2 t}\right)Y,\ \mathcal{L}(u) = H. \end{aligned}$$

(iii) Let $H = \sum_{\lambda\in\mathbb{C},k\geq 1} a_{\lambda,k}(s-\lambda)^{-k}$ be strictly proper, $\mu \in \mathbb{C}$ and

$$\begin{aligned} u &:= H \circ \delta = \sum_{\lambda,k} a_{\lambda,k}e_{\lambda,k-1}Y,\ H = \mathcal{L}(u) \Longrightarrow e^{\mu t}u = \sum_{\lambda,k} a_{\lambda,k}e_{\lambda+\mu,k-1}Y \\ &= \sum_{\lambda\in\mathbb{C},k\geq 1} a_{\lambda,k}(s-(\lambda+\mu))^{-k} \circ \delta = H(s-\mu) \circ \delta \Longrightarrow \mathcal{L}\left(e^{\mu t}u\right) = \mathcal{L}(u)(s-\mu). \end{aligned}$$

A simple proof shows this equation for $u \in \mathbb{C}[s] \circ \delta$ too.
(iv) For H and u as in (iii) and $n \geq 0$ we consider

$$\begin{aligned} t^n u &= \sum_{\lambda,k} a_{\lambda,k}\tfrac{(n+k-1)!}{(k-1)!}\tfrac{t^{n+k-1}}{(n+k-1)!}e^{\lambda t}Y \\ &= \sum_{\lambda,k} a_{\lambda,k}\tfrac{(n+k-1)!}{(k-1)!}(s-\lambda)^{-(n+k)} \circ \delta \\ \Longrightarrow \mathcal{L}(t^n u) &= \sum_{\lambda,k} a_{\lambda,k}\tfrac{(n+k-1)!}{(k-1)!}(s-\lambda)^{-(n+k)}. \end{aligned}$$

(v) Let $\omega > 0$ and

$$\begin{aligned} u &:= \cos(\omega t)Y = \Re\left(e^{j\omega t}Y\right) = \Re\left((s-j\omega)^{-1} \circ \delta\right) = \Re\left((s-j\omega)^{-1}\right) \circ \delta \\ &= \Re\left((s+j\omega)(s^2+\omega^2)^{-1}\right) \circ \delta = s(s^2+\omega^2)^{-1} \circ \delta \Longrightarrow \mathcal{L}(\cos(\omega t)) = s(s^2+\omega^2)^{-1}. \end{aligned}$$

(vi) Let $\omega > 0$ and

$$\begin{aligned}
u &:= \cos^2(\omega t)Y = (\cos(\omega t)Y)(\cos(\omega t)Y) = \Re(e^{j\omega t}Y)\Re(e^{j\omega t}Y)\\
&= 2^{-1}\Re\left(e^{j\omega t}e^{-j\omega t}Y + e^{j\omega t}e^{j\omega t}Y\right) = 2^{-1}\Re\left(Y + e^{j2\omega t}Y\right)\\
&= 2^{-1}\Re\left(s^{-1}\circ\delta + (s-j2\omega)^{-1}\circ\delta\right) = 2^{-1}\Re\left(s^{-1} + (s-j2\omega)^{-1}\right)\circ\delta\\
&\Longrightarrow \mathcal{L}(\cos^2(\omega t)) = 2^{-1}\Re\left(s^{-1} + (s-j2\omega)^{-1}\right) = (s^2+2\omega^2)s^{-1}(s^2+4\omega^2)^{-1}.
\end{aligned}$$

◇

In *electrical engineering* the signal space $\mathcal{F}_{4,F}$ is used if a network is switched on at the time $t = 0$. Initial conditions at $t = 0$ play a part in this theory. The next theorem deals with this situation. We use the following lemma whose *nonconstructive resp. constructive* proofs are given in Theorems 6.3.11 resp. 12.3.12 in context with the *canonical observability state space realization*.

Lemma 8.2.52 (Canonical observability realization) *Consider the matrix* $P \in F[s]^{p\times p}$ *of rank* p *and the autonomous behavior* $\mathcal{B}^0 := \left\{z \in \mathcal{F}_F^p;\ P\circ z = 0\right\} \subseteq \mathrm{t}(\mathcal{F}_F)^p$. *Then there are* observability indices *and matrices*

$$\begin{aligned}
&d^{\mathrm{ob}}(k) \geq 0,\ k = 1,\ldots,p,\ \Gamma^{\mathrm{ob}} := \left\{(k,\mu);\ 1\leq k\leq p,\ 0\leq\mu\leq d^{\mathrm{ob}}(k)-1\right\},\\
&A^{\mathrm{ob}} \in F^{\Gamma^{\mathrm{ob}}\times\Gamma^{\mathrm{ob}}},\ C^{\mathrm{ob}} \in F^{p\times\Gamma^{\mathrm{ob}}},\ x = \left(x_{k,\mu}\right)_{(k,\mu)\in\Gamma^{\mathrm{ob}}} \in \mathcal{F}_F^{\Gamma^{\mathrm{ob}}},
\end{aligned} \tag{8.81}$$

and isomorphisms

$$\begin{aligned}
&\mathcal{B}_s^0 := \left\{x \in \mathcal{F}_F^{\Gamma^{\mathrm{ob}}};\ s\circ x = A^{\mathrm{ob}}x\right\} \cong \mathcal{B}^0,\ x \leftrightarrow z = C^{\mathrm{ob}}x,\ x_{k,\mu} = z_k^{(\mu)} = s^\mu\circ z_k,\\
&\mathcal{B}^0 \cong \mathcal{B}_s^0 \cong F^{\Gamma^{\mathrm{ob}}},\ z \leftrightarrow x \leftrightarrow x(0),\ \textit{where}\\
&x = e^{A^{\mathrm{ob}}t}x(0),\ x(0) = \left(z_k^{(\mu)}(0)\right)_{(k,\mu)\in\Gamma^{\mathrm{ob}}},\ z = C^{\mathrm{ob}}x.
\end{aligned} \tag{8.82}$$

Then $x(0)$ *is the vector of* canonical initial conditions *of* $\mathcal{B}^0$. *For* $p = 1$ *and* $d^{\mathrm{ob}}(1) := \deg_s(P)$ *this is the standard Theorem 3.1.5.*
Notice that for $d(k) = 0$ *the index* $(k,0)$ *does not belong to* Γ^{ob}, $x_{k,0}$ *is not defined, and* $z_k = x_{k,0}$ *does not hold. Instead the equation* $z = C^{\mathrm{ob}}x$ *implies* $z_k = \sum_{(l,\mu)\in\Gamma^{\mathrm{ob}}} C^{\mathrm{ob}}_{k,(l,\mu)}x_{l,\mu}$. *The equation* $d(k) = 0$ *may hold but is often ignored in the engineering literature.* ◇

In the following theorem we assume, for $F = \mathbb{C}, \mathbb{R}$, an IO behavior

$$\mathcal{B} := \left\{\binom{y}{u} \in \mathcal{F}_F^{p+m};\ P_1 \circ y = Q_1 \circ u\right\},\ \mathcal{B}^0 := \left\{y \in \mathcal{F}^p;\ P_1 \circ y = 0\right\} \subseteq \mathrm{t}(\mathcal{F}_F)^p,$$
$$\text{with } P_1 \in F[s]^{p\times p},\ Q_1 \in F[s]^{p\times m},\ \operatorname{rank}(P_1) = p,\ H_1 = P_1^{-1}Q_1. \tag{8.83}$$

Assume $H_2 \in F(s)^m$ with strictly proper H_1H_2 and the input

$$\begin{aligned} &u = H_2 \circ \delta \in \mathcal{F}_2^m, \text{ hence}\\ &y := H_1 \circ u = H_1H_2 \circ \delta = \beta Y \in \mathrm{t}(\mathcal{F}_F)^p Y,\ \binom{y}{u} \in \mathcal{F}_{2,F}^{p+m} \cap \mathcal{B}. \end{aligned} \tag{8.84}$$

The function β exists uniquely according to Theorem 8.2.47. Any other output $\widetilde{y}$ to the input u, with $\binom{\widetilde{y}}{u} \in \mathcal{B}$, then has the form

$$\begin{aligned} &\widetilde{y} = y + z = \beta Y + z \text{ with } z \in \mathcal{B}^0 \subseteq \mathrm{t}(\mathcal{F}_F)^p\\ &\Longrightarrow \widetilde{y},\ y = \beta Y,\ z \in \mathrm{C}^\infty((0,\infty), F)^p\\ &\Longrightarrow \forall (k,\mu) \in \Gamma^{\mathrm{ob}}:\ \widetilde{y}_k^{(\mu)}(0+) = y_k^{(\mu)}(0+) + z_k^{(\mu)}(0+) = \beta_k^{(\mu)}(0) + z_k^{(\mu)}(0) \text{ where}\\ &\widetilde{y}_k^{(\mu)}(0+) := \lim_{t\to 0,\, t>0} \widetilde{y}_k^{(\mu)}(t),\ y_k^{(\mu)}(0+) := \lim_{t\to 0,\, t>0} y_k^{(\mu)}(t) = \beta_k^{(\mu)}(0). \end{aligned} \tag{8.85}$$

If H_2 is strictly proper too and hence $H_2 \circ \delta = \alpha Y$, $\alpha \in \mathrm{t}(\mathcal{F})^m$, then the injective cogenerator property of $\mathrm{t}(\mathcal{F}_F)$ implies the existence of a trajectory

$$\widehat{y} \in \mathrm{t}(\mathcal{F})^p \text{ with } P_1 \circ \widehat{y} = Q_1 \circ \alpha. \tag{8.86}$$

The simplest but important example for this in electrical engineering is

$$\alpha = \alpha(0)e^{j\omega t},\ \alpha(0) \in \mathbb{C}^m,\ \omega > 0,\ j\omega \notin \operatorname{pole}(H_1),\ \widehat{y} := H_1(j\omega)\alpha.$$

Theorem 8.2.53 (Initial value problem via Laplace transform) *Consider the data from* (8.83)–(8.86) *with strictly proper H_1H_2 and $H_1H_2 \circ \delta = \beta Y$ where β is easily computed by partial fraction decomposition of H_1H_2. For*

$$\xi = \left(\xi_k^{(\mu)}\right)_{(k,\mu)\in\Gamma^{\mathrm{ob}}} \in F^{\Gamma^{\mathrm{ob}}} \textit{ define } z := C^{\mathrm{ob}} e^{A^{\mathrm{ob}}t}\left(\xi_k^{(\mu)} - \beta_k^{(\mu)}(0)\right)_{(k,\mu)\in\Gamma^{\mathrm{ob}}} \in \mathcal{B}^0.$$

The matrices A^{ob} and C^{ob} are computed by the algorithm in Corollary 12.3.5. Then

1. *$P_1 \circ (\beta Y) = (P_1 \circ \beta)Y + T_y$, $T_y \in F[s]^p \circ \delta$. If H_2 is strictly proper too and thus $H_2 \circ \delta = \alpha Y$, $\alpha \in \mathrm{t}(\mathcal{F}_F)^m$, then $P_1 \circ \beta = Q_1 \circ \alpha$ and thus $\widehat{y} - \beta \in \mathcal{B}^0$.*

2. *Every solution $\widetilde{y} \in \left(\mathrm{C}^{-\infty}(\mathbb{R}, F)\right)^p$ of $P_1 \circ \widetilde{y} = Q_1 \circ (H_2 \circ \delta)$ satisfies* (8.85). *The unique solution of the initial value problem*

$$P_1 \circ \widetilde{y} = Q_1 \circ (H_2 \circ \delta) \text{ with } \widetilde{y}_k^{(\mu)}(0+) = \xi_k^{(\mu)} \text{ for } (k,\mu) \in \Gamma^{\text{ob}} \text{ is}$$
$$\widetilde{y} := \beta Y + z = \beta Y + C^{\text{ob}} e^{A^{\text{ob}} t} \left(\xi_k^{(\mu)} - \beta_k^{(\mu)}(0) \right)_{(k,\mu)\in\Gamma^{\text{ob}}}. \tag{8.87}$$

3. *Every solution* $\widehat{y} \in \left(\mathrm{C}^{-\infty}(\mathbb{R}, F)\right)^p$ *of* $P_1 \circ \widehat{y} = Q_1 \circ \alpha$ *belongs to* $\mathrm{t}(\mathcal{F}_F)^p \subset (\mathrm{C}^{\infty}(\mathbb{R}, F))^p$. *The unique solution of the initial value problem*

$$P_1 \circ \widehat{y} = Q_1 \circ \alpha \text{ with } \widehat{y}_k^{(\mu)}(0) = \xi_k^{(\mu)} \text{ for } (k,\mu) \in \Gamma^{\text{ob}} \text{ is}$$
$$\widehat{y} := \beta + z = \beta + C^{\text{ob}} e^{A^{\text{ob}} t} \left(\xi_k^{(\mu)} - \beta_k^{(\mu)}(0) \right)_{(k,\mu)\in\Gamma^{\text{ob}}}. \tag{8.88}$$

If $\mathcal{B}^0$ *is asymptotically stable, the inclusions* $\widetilde{y} - (\beta Y),\ \widehat{y} - \beta \in \mathcal{B}^0$ *imply*

$$\lim_{t\to\infty} (\widetilde{y} - \beta)(t) = \lim_{t\to\infty} (\widehat{y} - \beta)(t) = 0,$$

such that $\widetilde{y}$, $\widehat{y}$ *and* β *can often be identified for* $t \to \infty$ *in practical situations.*

4. *The assertions from 1, 2, 3 also hold with the more general data*

$$u = \alpha Y \in \left(\mathrm{C}^{\infty}\right)^m Y \subset \mathcal{F}_{1,F}^m = F[s]^m \circ \delta \oplus \left(\mathrm{C}^{\infty}\right)^m Y,\ H_1 \text{ proper}$$
$$y = H_1 \circ (\alpha Y) = \beta Y \in \left(\mathrm{C}^{\infty}\right)^p Y,\ P_1 \circ \beta|_{(0,\infty)} = Q_1 \circ \alpha|_{(0,\infty)},$$
$$\binom{\widetilde{y}}{u} \in \mathcal{B},\ \widetilde{y} = \beta Y + z,\ z \in \mathcal{B}^0 \subseteq \mathrm{t}(\mathcal{F}_F)^p,\ \widetilde{y}_k^{(\mu)}(0+) = \beta_k^{(\mu)}(0) + z_k^{(\mu)}(0).$$

In particular, these u are sufficiently differentiable *in the sense of [6, Sect. 3.2.1]. Variants of this method for solving an initial value problem with the inverse Laplace transform* $\beta Y = H_1 H_2 \circ \delta = \mathcal{L}^{-1}(H_1 \mathcal{L}(H_2 \circ \delta))$ *play an important part in the engineering literature, cf. [6, Sect. 3.2.1], [10, Sect. 6.6], [11, p. 155].*

Proof 1. Recall

$$s \circ (\beta Y) = (s \circ \beta)Y + \beta(s \circ Y) = (s \circ \beta)Y + \beta\delta = (s \circ \beta)Y + \beta(0)\delta$$
$$\underset{\text{induction}}{\Longrightarrow} P_1 \circ (\beta Y) = (P_1 \circ \beta)Y + T_y,\ T_y \in F[s]^p \circ \delta \text{ and likewise}$$
$$Q_1 \circ (\alpha Y) = (Q_1 \circ \alpha)Y + T_u,\ T_u \in F[s]^p \circ \delta$$
$$\Longrightarrow 0 = P_1 \circ (\beta Y) - Q_1 \circ (\alpha Y) = (P_1 \circ \beta - Q_1 \circ \alpha)Y + (T_y - T_u)$$
$$\in \mathcal{F}_{2,F}^p = \mathrm{t}(\mathcal{F}_F)^p Y \oplus F[s]^p \circ \delta$$
$$\Longrightarrow (P_1 \circ \beta - Q_1 \circ \alpha)Y = T_y - T_u = 0 \underset{P_1\circ\beta - Q_1\circ\alpha \text{ analytic}}{\Longrightarrow} P_1 \circ \beta = Q_1 \circ \alpha$$
$$\Longrightarrow \binom{\beta}{\alpha},\ \binom{\widetilde{y}}{\alpha} \in \mathcal{B} \Longrightarrow \widetilde{y} - \beta \in \mathcal{B}^0.$$

2. 3. The first assertions follow from $\alpha \in \mathrm{t}(\mathcal{F}_F)^m$ resp. $\alpha Y \in \mathcal{F}_{4,F}^m$ and the injective cogenerator property of these function modules. The rest follows directly from Lemma 8.2.52 and Eq. (8.85).

4. The arguments of 1, 2, and 3 hold with slight changes for the injective cogenerators

$\mathbb{C}^\infty$ and $F[s] \circ \delta \oplus (\mathbb{C}^\infty Y + \mathbb{C}^\infty)$ instead of $\mathrm{t}(\mathcal{F}_F)$ and $\mathcal{F}_{4,F} = \mathbb{C}[s] \circ \delta \oplus \mathrm{t}(\mathcal{F})Y + \oplus\, \mathrm{t}(\mathcal{F})$. □

Example 8.2.54 Consider

$$\begin{aligned}
&\widetilde{y}'' + 2\widetilde{y}' = s(s+2) \circ y = u,\ H_1 = \tfrac{1}{s(s+2)},\ \widetilde{y}(0+) := 4,\ \widetilde{y}'(0+) := -1\\
&H_2 := \tfrac{s^2}{s-1} = 1 + s + \tfrac{1}{s-1},\ u := H_2 \circ \delta = \delta + \delta' + e^t Y,\\
&H_1 H_2 = \tfrac{s}{(s-1)(s+2)} = \tfrac{1}{3}\left(\tfrac{1}{s-1} + 2\tfrac{1}{s+2}\right),\ H_1 H_2 \circ \delta = \beta Y,\ \beta = \tfrac{1}{3}\left(e^t + 2e^{-2t}\right)\\
&\widetilde{y} = \tfrac{1}{3}\left(e^t + 2e^{-2t}\right) Y + 2 + e^{-2t}.
\end{aligned}$$
◇

Remark 8.2.55 1. If in Theorem 8.2.53, 2–3, $\mathcal{B}^0$ is asymptotically stable, i.e., $\mathrm{spec}(A^{\mathrm{ob}}) = \mathrm{char}(\mathcal{B}^0) \subset \mathbb{C}_-$, then $\lim_{t\to\infty}(\widetilde{y} - \beta Y)(t) = \lim_{t\to\infty} z(t) = 0$, and z is called the *transient*. For its *exact* computation from the initial conditions one needs the initial data

$$\forall (k,\mu) \in \Gamma^{\mathrm{ob}} :\ \xi_k^{(\mu)} = z_k^{(\mu)}(0) = \widetilde{y}_k^{(\mu)}(0+) - \beta_k^{(\mu)}(0),$$

where the $\widetilde{y}_k^{(\mu)}(0+)$ have to be measured or prepared, whereas the $\beta_k^{(\mu)}(0)$ can be computed with chosen precision. Notice that, in general, high derivatives of signals can neither be measured nor prepared, so the exact trajectory for given initial conditions at $t = 0$ or, equivalently, its transient cannot be determined. This is, however, possible if in Lemma 8.2.52 the $d^{\mathrm{ob}}(k)$ and thus the occurring μ are small, say $d^{\mathrm{ob}}(k) \le 2$.
2. If $\mathcal{B}^0$ is not asymptotically stable, then there are essentially different trajectories $z_1, z_2 \in \mathcal{B}^0$ with nearby initial vectors. The same holds for the corresponding outputs $\widetilde{y}_i = \beta Y + z_i,\ i = 1, 2$. The initial vectors arise from measurements. Small errors in these may give rise to essentially different trajectories $\widetilde{y}_i,\ i = 1, 2$. Hence, the computed $\widetilde{y}$ may not represent the real trajectory. ◇

Corollary 8.2.56 *In Theorem 8.2.53 assume that $\mathcal{B}^0$ is asymptotically stable. Consider $\alpha := \alpha(0)e^{j\omega t}$, $\alpha(0) \in \mathbb{C}^m$, $\omega \ge 0$, and hence $\begin{pmatrix} H_1(j\omega)\alpha(0) \\ \alpha(0) \end{pmatrix} e^{j\omega t} \in \mathcal{B}$. If y is any output to α or to $\alpha Y = \alpha(0)(s - j\omega)^{-1} \circ \delta$, then*

$$y - H_1(j\omega)\alpha(0)e^{j\omega t} \in \mathcal{B}^0 \text{ and } \lim_{t\to\infty}\left(y(t) - H_1(j\omega)\alpha(0)e^{j\omega t}\right) = 0.$$

Hence $H_1(j\omega)\alpha(0)e^{j\omega t}$ is a steady-state solution and y can be identified with it. ◇

Example 8.2.57 In Theorem 8.2.53, 1, use the following data:

$$\begin{aligned}
&P_1 := s-\lambda,\ Q_1 := 1,\ H_1 = (s-\lambda)^{-1},\ H_2 := (s-j\omega)^{-1},\ \lambda \neq j\omega,\ \omega > 0,\\
&\Longrightarrow H_2 \circ \delta = \alpha Y,\ \alpha = e^{j\omega t},\ y_1 := H_1(j\omega)\alpha = -(\lambda - j\omega)^{-1} e^{j\omega t}\\
&H_1 H_2 = (s-\lambda)^{-1}(s-j\omega)^{-1} = (\lambda - j\omega)^{-1}\left((s-\lambda)^{-1} - (s-j\omega)^{-1}\right),\\
&y := H_1 H_2 \circ \delta = \mathcal{L}^{-1}(H_1 H_2) = \beta Y,\ \beta = (\lambda - j\omega)^{-1}\left(e^{\lambda t} - e^{j\omega t}\right)\\
&\Longrightarrow \beta - y_1 = (\lambda - j\omega)^{-1} e^{\lambda t},\ P_1 \circ (\beta - y_1) = 0.
\end{aligned}$$

If $\Re(\lambda) < 0$, i.e., if $\mathcal{B}^0$ is asymptotically stable, then y and y_1 are steady-state outputs for their respective inputs and

$$\lim_{t\to\infty}(y-y_1)(t) = \lim_{t\to\infty}(\beta - y_1)(t) = \lim_{t\to\infty}(\lambda - j\omega)^{-1}e^{\lambda t} = 0.$$

If $\Re(\lambda) > 0$ then $\lim_{t\to\infty}(y-y_1)(t) = \infty$. If, finally,

$$\Re(\lambda) = 0 \text{ or } \lambda = j\omega_1,\ \omega_1 \neq \omega,\ \text{then } (y-y_1)(t) = (j(\omega_1 - \omega))^{-1} e^{j\omega_1 t},$$

oscillates.
Interpretation: In electrical engineering the output y_1 is used in connection with the impedance for sinusoidal inputs. For more complicated inputs the Laplace transform is used for the computation of y. Unless the system is asymptotically stable, these two methods give essentially different results. Asymptotic stability is an essential assumption for most applications and measurements in engineering. ◊

In the next theorem we discuss possible impulsive trajectories, cf. Sect. 8.2.10 for a broader treatment.

Theorem 8.2.58 *The notations are those from Theorem 8.2.53. We assume that H_2 is strictly proper, but we make no assumption on $H_1 H_2$. Consider $\binom{y}{u} \in \mathcal{B}$ where*

$$u = \alpha Y = H_2 \circ \delta,\ H_2 := \mathcal{L}(\alpha Y) \in F(s)^m_{\mathrm{spr}},\ y := H_1 \circ u = H_1 H_2 \circ \delta = \mathcal{L}^{-1}(H_1 H_2).$$

1. *Since $\mathcal{F}_{4,F} = \mathrm{t}(\mathcal{F}_F) \oplus \mathrm{t}(\mathcal{F}_F)Y \oplus F[s] \circ \delta$ is an injective cogenerator and $\mathcal{B}^0 \subset \mathrm{t}(\mathcal{F}_F)^p$, every trajectory $\binom{w}{\alpha Y}$ or $\binom{w}{\alpha(1-Y)}$ of $\mathcal{B} \subseteq \mathcal{F}_F^{p+m}$ belongs to $\mathcal{F}_{4,F}^{p+m}$.*
2. *The following properties are equivalent:*
   (a) *$H_1 H_2$ is not strictly proper, i.e., $(H_1 H_2)_{\mathrm{pol}} \neq 0$ or $\deg_s(H_1 H_2) \geq 0$.*
   (b) *The output $y = \mathcal{L}^{-1}(H_1 H_2)$ or* every *output w to the input αY has a nonzero component in $F[s]^p \circ \delta$ or, in other words, is* impulsive.
   (c) Every *output w to the input $\alpha(1-Y)$ is impulsive.*
3. *If $H_1 H_2$ is strictly proper or $\deg_s(H_1 H_2) \leq -1$, and thus the outputs from 2 are piecewise continuous with a possible jump at $t = 0$, the following properties are equivalent:*

(a) sH_1H_2 *is not strictly proper, i.e.,* $\deg(H_1H_2) = -1$ *or, by Theorem 8.2.47, 2,* $y = \mathcal{L}^{-1}(H_1H_2)$ *has a nonzero jump at* $t = 0$.
(b) All *outputs from 2, (a), (b), have a* nonzero *jump at* $t = 0$.

4. *If* H_1 *is proper resp. strictly proper, then all outputs in 2 resp. 3 for* $H_2 \in F(s)^m_{\mathrm{spr}}$ *are piecewise continuous resp. even continuous.*

In words: 2. All *outputs* w *to these inputs are impulsive if* H_1H_2 *is not strictly proper. Except in special cases, such situations have to be avoided in technical situations by the choice of* H_2 *with strictly proper* H_1H_2*. In electrical engineering the inputs* αY, $\alpha \in \mathrm{t}(\mathcal{F}_F)^m$, *resp.* $\alpha(1-Y)$ *mean that at time zero the input is switched on resp. off.*
4. Continuous outputs are sometimes required for physical reasons.

Proof 1. Obvious.
2. Consider $y := H_1 \circ (\alpha Y) = H_1H_2 \circ \delta$ with $P_1 \circ y = Q_1 \circ (\alpha Y)$ and

$$
\begin{aligned}
&H_1H_2 = (H_1H_2)_{\mathrm{pol}} + (H_1H_2)_{\mathrm{spr}} \in F[s]^p \oplus F(s)^p_{\mathrm{spr}}\\
&\Longrightarrow y = H_1H_2 \circ \delta = (H_1H_2)_{\mathrm{pol}} \circ \delta + (H_1H_2)_{\mathrm{spr}} \circ \delta \in F[s]^p \circ \delta \oplus \mathrm{t}(\mathcal{F}_F)^p Y.
\end{aligned}
$$

(a) $\Longleftrightarrow$ (b): Let $\binom{w_1}{\alpha Y} \in \mathcal{B}$, i.e., $P_1 \circ w_1 = Q_1 \circ (\alpha Y)$. From $\binom{y}{\alpha Y} \in \mathcal{B}$ we infer

$$
\begin{aligned}
&w_2 := w_1 - y \in \mathcal{B}^0 \subseteq \mathrm{t}(\mathcal{F}_F)^p \Longrightarrow w_1 = y + w_2\\
&= (H_1H_2)_{\mathrm{pol}} \circ \delta + (H_1H_2)_{\mathrm{spr}} \circ \delta + w_2 \in \mathcal{F}^p_{4,F} = F[s]^p \circ \delta \oplus \mathrm{t}(\mathcal{F}_F)^p Y \oplus \mathrm{t}(\mathcal{F}_F)^p.
\end{aligned}
$$

Hence $(H_1H_2)_{\mathrm{pol}} \circ \delta$ is the component of w_1 in $F[s]^p \circ \delta$. This is nonzero if and only if $(H_1H_2)_{\mathrm{pol}} \neq 0$.
(a) $\Longleftrightarrow$ (c): Let $\binom{w_3}{\alpha(1-Y)} \in \mathcal{B}$ and $y_1 \in \mathrm{t}(\mathcal{F}_F)^p$ with $P_1 \circ y_1 = Q_1 \circ \alpha$. Then

$$
\begin{aligned}
&P_1 \circ w_3 = Q_1 \circ (\alpha(1-Y)) = Q_1 \circ \alpha - Q_1 \circ (\alpha Y) = P_1 \circ (y_1 - y)\\
&\Longrightarrow w_4 := w_3 - y_1 + y \in \mathcal{B}^0 \subseteq \mathrm{t}(\mathcal{F}_F)^p\\
&\Longrightarrow w_3 = -y + (w_4 + y_1) = -(H_1H_2)_{\mathrm{pol}} \circ \delta - (H_1H_2)_{\mathrm{spr}} \circ \delta + (w_4 + y_1)\\
&\qquad\qquad \in \mathcal{F}^p_{4,F} = F[s]^p \circ \delta \oplus \mathrm{t}(\mathcal{F}_F)^p Y \oplus \mathrm{t}(\mathcal{F}_F)^p.
\end{aligned}
$$

Hence $-(H_1H_2)_{\mathrm{pol}} \circ \delta$ is the component of w_3 in $F[s]^p \circ \delta$. This is nonzero if and only if $(H_1H_2)_{\mathrm{pol}} \neq 0$.
3. The proof is the same as that of 2.
4. The assertion follows from 2 and 3 since $\deg_s(H_2) \leq -1$ implies

$$
\begin{aligned}
&\big(\deg_s(H_1) \leq 0 \Longrightarrow \deg_s(H_1H_2) \leq -1\big) \text{ and}\\
&\big(\deg_s(H_1) \leq -1 \Longrightarrow \deg_s(H_1H_2) \leq -2 \Longrightarrow \deg_s(sH_1H_2) \leq -1\big).
\end{aligned}
$$

□

We have seen that the impulse response $h := H \circ \delta = \mathcal{L}^{-1}(H)$ can be computed easily. Due to the engineering difficulties of generating the ideal impulse δ as input and then h as output, the measurement of h is a demanding task. Since technical generation of the step function, i.e., the Heaviside function $Y = s^{-1} \circ \delta$, is easy, engineers often prefer the *step response*

$$\begin{aligned} H \circ Y &= Hs^{-1} \circ \delta = s^{-1} \circ (H \circ \delta) = s^{-1} \circ h \text{ with } h := H \circ \delta \\ \Longrightarrow (H \circ Y)(t) &= \int_0^t h(\tau)\mathrm{d}\tau \text{ if } H \in \mathbb{C}(s)_{\mathrm{spr}},\ h = H \circ \delta = \alpha Y \in \mathrm{C}_+^{0,\mathrm{pc}}. \end{aligned} \tag{8.89}$$

The step response of H is the impulse response of Hs^{-1} and can thus also be easily computed by the latter's partial fraction decomposition. The step response characterizes the properness of H in the following fashion.

Theorem 8.2.59 (Characterization of proper transfer functions) *Let* $H = H_{\mathrm{pol}} + H_{\mathrm{spr}} \in \mathbb{C}(s)$. *The following properties are equivalent:*

1. *H is proper, i.e., $H_{\mathrm{pol}} \in \mathbb{C}$ or $\deg_s(H) \le 0$ or $Hs^{-1} \in F(s)_{\mathrm{spr}}$.*
2. *The step response $H \circ Y = Hs^{-1} \circ \delta$ is a piecewise continuous function.*
3. *$H \circ \mathrm{C}_+^{0,\mathrm{pc}} \subseteq \mathrm{C}_+^{0,\mathrm{pc}}$.*
4. *$H \circ \mathrm{C}_+^{0} \subseteq \mathrm{C}_+^{0}$.*

Proof Consider the decomposition $H = H_{\mathrm{pol}} + H_{\mathrm{spr}}$.
1. $\Longleftrightarrow$ 2.: The following equivalences hold:

$$H \text{ is proper} \Longleftrightarrow Hs^{-1} \text{ is strictly proper} \underset{\text{Theorem 8.2.47,2.}}{\Longleftrightarrow} H \circ Y = Hs^{-1}\delta \in \mathrm{C}_+^{0,\mathrm{pc}}.$$

1. $\Longrightarrow$ 3. and 1. $\Longrightarrow$ 4.: $H_{\mathrm{pol}} \in \mathbb{C}$ and Theorem 8.2.47,5, imply 3. and 4.
3. $\Longrightarrow$ 2.: The statement in item 2. is a special case of that in item 3.
4. $\Longrightarrow$ 1.: Assume that $H_{\mathrm{pol}} = a_m s^m + \cdots + a_0$ is not constant, i.e., $m > 0$ and $a_m \neq 0$, and define $u := s^{-(m+1)} \circ \delta = s^{-m} \circ Y = \frac{t^m}{m!} Y \in \mathrm{C}_+^0$. Then

$$H_{\mathrm{pol}} \circ u = a_m Y + \left(\sum_{i=0}^{m-1} a_i s^{-(m+1-i)} \circ \delta \right) \in \mathbb{C}Y + \mathrm{C}_+^0$$

and since $a_m Y \neq 0$, the function $H_{\mathrm{pol}} \circ u$ is not continuous.
On the other hand the continuity of u and item 4. imply that $H \circ u$ is continuous. By Theorem 8.2.47, 5, also $H_{\mathrm{spr}} \circ u$ and thus $H_{\mathrm{pol}} \circ u = H \circ u - H_{\mathrm{spr}} \circ u$ are continuous. This is a contradiction to $a_m \neq 0$. Hence $H_{\mathrm{pol}} \in \mathbb{C}$. □

Theorem 8.2.60 (Characterization of proper transfer matrices) *Let $H \in \mathbb{C}(s)^{p\times m}$ and write it componentwise as*

$$H = H_{\text{pol}} + H_{\text{spr}} \in \mathbb{C}[s]^{p\times m} \oplus \mathbb{C}(s)^{p\times m}_{\text{spr}}.$$

Then the following properties are equivalent:

1. *H is proper, i.e., $H_{\text{pol}} \in \mathbb{C}^{p\times m}$.*
2. *$H \circ Y = (H_{ij} \circ Y)_{i,j} \in \left(\mathrm{C}^{0,\text{pc}}\right)^{p\times m}$.*
3. *$H \circ (\mathrm{C}_+^{0,\text{pc}})^m \subseteq (\mathrm{C}_+^{0,\text{pc}})^p$.*
4. *$H \circ (\mathrm{C}_+^0)^m \subseteq (\mathrm{C}_+^0)^p$.*

Proof This follows directly from Theorem 8.2.59. □

Corollary 8.2.61 *According to Theorem 8.2.37, the action of $H \in \mathbb{C}(s)$ and the shift action of $a \in \mathbb{R}$ commute. In particular,*

$$H \circ \delta_a = H \circ (a \circ_{\text{s}} \delta) = a \circ_{\text{s}} (H \circ \delta) \quad \text{and} \quad H \circ Y_a = H \circ (a \circ_{\text{s}} Y) = a \circ_{\text{s}} (H \circ Y).$$

Hence the actions of H on the shifted distribution δ_a resp. on the shifted Heaviside function Y_a are obtained by shifting the impulse resp. step response. ◇

8.2.5 The Convolution Multiplication

In Theorem 8.2.37 we have shown that any transfer matrix $H \in \mathbb{C}(s)^{p\times m}$ acts as transfer operator $H\circ : \left(\mathrm{C}_+^{-\infty}\right)^m \to \left(\mathrm{C}_+^{-\infty}\right)^p$ on signals with left bounded support. In this section we show $H \circ u = h * u,\ h = H \circ \delta$, i.e., that the output is the *convolution of the impulse response with the input*. If H is strictly proper and thus has a piecewise continuous impulse response h with left bounded support and if also u is piecewise continuous with left bounded support, then $H \circ u = h * u$ is continuous and given by the integral $(h * u)(t) = \int_{-\infty}^{\infty} h(t - x)u(x)\,\mathrm{d}x$. This is the generalization of the standard solution integral (3.63) for state space systems. The simplest and often applied computation of $h * u = H \circ u$ is described in Theorem 8.2.47, 4. The actual computation of the integral for complicated u is rarely needed. The convolution integral is decisive in Sect. 8.2.6 on *external stability*.

We reduce the convolution of distributions in $\mathrm{C}_+^{-\infty}$ to that of functions in C_+^0 with the advantage that only the Riemann integral of continuous functions on finite intervals, but not the Lebesgue theory is needed. The simplicity is paid for by simple, but lengthier arguments. We refer to [3] and [4] for more advanced and elegant derivations.

In the following lemmas, we use two continuous functions $u_i \in \mathrm{C}_+^0$, $i = 1, 2$, with left bounded support in $[t_i, \infty)$, i.e., $\text{supp}(u_i) \subseteq [t_i, \infty)$ for $i = 1, 2$. Let $M := \max(|t_1|, |t_2|)$ and $T > 0$.

Lemma 8.2.62 *For $t \in [-T, T]$ we have that*

$$\operatorname{supp}\big(x \mapsto u_1(t-x)\,u_2(x)\big) \subseteq [-(T+M), T+M].$$

Proof If $x \le -(T+M)$, then $x \le -M \le t_2$ and hence $u_2(x) = 0$ since $\operatorname{supp}(u_2) \subseteq [t_2, \infty)$.
If $x \ge M + T$, i.e., $-x \le -M - T$, we use that $t \le T$ and obtain that

$$t - x \le T + (-M - T) = -M \le t_1.$$

Therefore, $u_1(t-x) = 0$. □

Lemma and Definition 8.2.63 (Convolution) *With the preceding data, an arbitrary finite interval $[A, B] \supseteq [-(M+T), M+T]$—which depends on T—and $t \in [-T, +T]$, we define the convolution*

$$\begin{aligned}(u_1 * u_2)(t) &:= \int_{-\infty}^{\infty} u_1(t-x)\,u_2(x)\,\mathrm{d}x = \int_{-(T+M)}^{T+M} u_1(t-x)\,u_2(x)\,\mathrm{d}x \\ &= \int_A^B u_1(t-x)\,u_2(x)\,\mathrm{d}x.\end{aligned}$$

Because of the preceding lemma the three integrals coincide for $t \in [-T, +T]$, i.e., they are finite integrals of continuous functions.
By the theorem on continuous dependence of (finite) Riemann integrals on parameters, the convolution $u_1 * u_2$ is a continuous function on $[-T, T]$. Since T is arbitrary, the convolution is actually defined and continuous for the whole time axis $\mathbb{R}$, where $[A, B]$ has to be enlarged with increasing T. The convolution has also left bounded support, indeed

$$\operatorname{supp}(u_1 * u_2) \subseteq [t_1 + t_2, \infty).$$

Proof Only the last statement has still to be proven. Let $t \le t_1 + t_2$. Since $t = (t - x) + x$, this implies that $t - x \le t_1$ or $x \le t_2$. Because of $\operatorname{supp}(u_i) \subseteq [t_i, \infty)$ we infer that $u_1(t-x) = 0$ or $u_2(x) = 0$. Consequently, $0 = u_1(t-x)u_2(x) = (u_1 * u_2)(t)$. □

Lemma 8.2.64 *The convolution is commutative, i.e., $u_1 * u_2 = u_2 * u_1$.*

Proof We give two proofs, a shorter one, which uses the infinite integrals, and a longer one, which uses finite integrals only. In the sequel we always use the shorter method, which, however, can always be reduced to the manipulation of *finite* Riemann integrals.

1.

$$
\begin{aligned}
(u_2 * u_1)(t) &= \int_{-\infty}^{+\infty} u_2(t-x)\, u_1(x)\, \mathrm{d}x \underset{y:=t-x}{=} -\int_{+\infty}^{-\infty} u_1(t-y)\, u_2(y)\, \mathrm{d}y \\
&= \int_{-\infty}^{+\infty} u_1(t-y)\, u_2(y)\, \mathrm{d}y = (u_1 * u_2)(t).
\end{aligned}
$$

2. For $t \in \mathbb{R}$, we choose $T > 0$ with $t \in [-T, T]$. For t_i with $\mathrm{supp}(u_i) \subseteq [t_i, \infty)$, $i = 1, 2$, let $M := \max(|t_1|, |t_2|)$. We define $A_2 := -(2T + M)$ and $B_2 := 2T + M$. In particular, this implies $[-(T + M), T + M] \subseteq [A_2, B_2]$. This interval can be used to compute $(u_2 * u_1)(t)$ for $t \in [-T, T]$. We infer that

$$
\begin{aligned}
(u_2 * u_1)(t) &= \int_{A_2}^{B_2} u_2(t-x)\, u_1(x)\, \mathrm{d}x \underset{y:=t-x}{=} -\int_{t-A_2}^{t-B_2} u_1(t-y)\, u_2(y)\, \mathrm{d}y \\
&= \int_{t-B_2}^{t-A_2} u_1(t-y)\, u_2(y)\, \mathrm{d}y.
\end{aligned}
$$

From $t \le T$, we infer that $t - B_2 \le T - (2T + M) = -(T + M)$, and $t \ge -T$ implies that $t - A_2 \ge -T - (-(2T + M)) = T + M$. Thus, $\mathrm{supp}(u_1 * u_2) \subseteq [-(T + M), T + M] \subseteq [t - B_2, t - A_2]$, and we conclude that

$$
\begin{aligned}
(u_2 * u_1)(t) &= \int_{t-B_2}^{t-A_2} u_1(t-y)\, u_2(y)\, \mathrm{d}y = \int_{-(T+M)}^{T+M} u_1(t-y)\, u_2(y)\, \mathrm{d}y \\
&= (u_1 * u_2)(t).
\end{aligned}
$$

□

Theorem 8.2.65 *With its standard $\mathbb{C}$-vector space structure and the convolution product, the set $\mathrm{C}^0_+(\mathbb{R}, \mathbb{C})$ is a commutative $\mathbb{C}$-algebra. We will see in Theorem 8.2.67 that the neutral element of the convolution product is the δ-distribution. Thus, $\mathrm{C}^0_+(\mathbb{R}, \mathbb{C})$ is an algebra without identity element.*
This ring has no zero divisors, i.e., it is an integral domain [12, p. 327], but we do not need this result and will not prove it. It is the starting point of Mikusinski's operational calculus, which we explain in Sect. 8.2.9.

Proof The distributivity of the multiplication is obvious since the argumentwise multiplication of functions has this property. It remains to prove the convolution product is associative. But

$$
\begin{aligned}
\big((u_1 * u_2) * u_3\big)(t) &= \int_{-\infty}^{\infty} (u_1 * u_2)(t-y)\, u_3(y)\, \mathrm{d}y \\
&= \int_{-\infty}^{\infty}\int_{-\infty}^{\infty} u_1(t-y-x)\, u_2(x)\, u_3(y)\, \mathrm{d}x\, \mathrm{d}y \\
&\underset{z:=x+y}{=} \int_{-\infty}^{\infty}\int_{-\infty}^{\infty} u_1(t-z)\, u_2(z-y)\, u_3(y)\, \mathrm{d}y\, \mathrm{d}z
\end{aligned}
$$

$$= \int_{-\infty}^{\infty} u_1(t-z)\,(u_2 * u_3)(z)\,\mathrm{d}z = \big(u_1 * (u_2 * u_3)\big)(t).$$

□

Lemma 8.2.66 *If $u_1 \in \mathrm{C}^1_+$ and $u_2 \in \mathrm{C}^0_+$, then*

$$u_1 * u_2 \in \mathrm{C}^1_+ \quad \text{and} \quad (u_1 * u_2)' = u_1' * u_2.$$

Induction then implies for $u_1 \in \mathrm{C}^n_+$, $u_2 \in \mathrm{C}^0_+$, and $0 \le m \le n$ that

$$\begin{aligned} &u_1 * u_2 \in \mathrm{C}^n_+ \quad \text{and} \\ &s^m \circ (u_1 * u_2) = (u_1 * u_2)^{(m)} = u_1^{(m)} * u_2 = (s^m \circ u_1) * u_2 \in \mathrm{C}^{n-m}_+ . \end{aligned}$$

Proof This is the standard result in Riemann integration theory concerning differentiation with respect to a parameter, here t, under the integral sign. Notice that $u_1(t-x)$ is continuously differentiable with respect to t and that $\frac{\mathrm{d}u_1(t-x)}{dt} = u_1'(t-x)$ is continuous. With A, B, and T from above, we obtain for $t \in (-T, T)$ that

$$\begin{aligned} \frac{\mathrm{d}(u_1 * u_2)}{\mathrm{d}t}(t) &= \frac{\mathrm{d}}{\mathrm{d}t}\left(\int_A^B u_1(t-x)\,u_2(x)\,\mathrm{d}x\right) \\ &= \int_A^B u_1'(t-x)\,u_2(x)\,\mathrm{d}x = (u_1' * u_2)(t). \end{aligned}$$

□

Now we are going to extend the convolution to the space $\mathrm{C}^{-\infty}_+(\mathbb{R}, \mathbb{C}) = \bigcup_{m=0}^{\infty} \mathrm{C}^{-m}_+$ of distributions of finite order with left bounded support. For $i = 1, 2$, let $u_i \in \mathrm{C}^{-m_i}_+ \subset \mathrm{C}^{-\infty}_+$ with $m_i \in \mathbb{N}$, and let $v_i := s^{-m_i} \circ u_i \in \mathrm{C}^0_+$, hence $u_i = s^{m_i} \circ v_i$. We define

$$u_1 * u_2 := s^{m_1+m_2} \circ (v_1 * v_2), \tag{8.90}$$

where $v_1 * v_2$ is the convolution product of continuous functions with left bounded support from Lemma and Definition 8.2.63, and the resulting function is identified with the distribution $\langle v_1 * v_2, -\rangle$ as in Theorem and Definition 8.2.8.

Theorem 8.2.67 (Convolution in $\mathrm{C}^{-\infty}_+(\mathbb{R}, \mathbb{C})$) *The convolution from* (8.90) *is well defined and makes $\mathrm{C}^{-\infty}_+(\mathbb{R}, \mathbb{C})$ a commutative ring with identity element δ. It is even an integral domain like C^0_+ in Theorem 8.2.65, but we do not need this and will not prove it.*
From the defining equation (8.90) *follows directly that the convolution of distributions extends the convolution of continuous functions from Lemma and Definition 8.2.63, i.e., the equality $\langle u_1 * u_2, -\rangle = \langle u_1, -\rangle * \langle u_2, -\rangle$ holds for $u_1, u_2 \in \mathrm{C}^0_+$.*

Proof *Well defined*: For $i = 1, 2$, let $u_i = s^{m_i} \circ v_i = s^{n_i} \circ w_i$ for $m_i, n_i \in \mathbb{N}$ and $v_i, w_i \in \mathrm{C}^0_+$. Without loss of generality, we assume that $m_1 \leq n_1$. Then $s^{m_1} \circ v_1 = s^{n_1} \circ w_1$ implies that $w_1 = s^{-(n_1-m_1)} \circ v_1 \in \mathrm{C}_+^{n_1-m_1}$, and with Lemma 8.2.66, we obtain that also $w_1 * v_2 \in \mathrm{C}_+^{n_1-m_1}$. Hence,

$$\begin{aligned} s^{m_1+m_2} \circ (v_1 * v_2) &= s^{m_1+m_2} \circ \big((s^{n_1-m_1} \circ w_1) * v_2\big) \\ &\underset{\text{Lemma 8.2.66}}{=} s^{m_1+m_2} \circ s^{n_1-m_1} \circ (w_1 * v_2) \\ &= s^{n_1+m_2} \circ (w_1 * v_2). \end{aligned}$$

The same argument, applied to $s^{m_2} \circ v_2 = s^{n_2} \circ w_2$, furnishes $s^{n_1+m_2} \circ (w_1 * v_2) = s^{n_1+n_2} \circ (w_1 * w_2)$; hence, we conclude that $s^{m_1+m_2} \circ (v_1 * v_2) = s^{n_1+n_2} \circ (w_1 * w_2)$.
$\mathrm{C}_+^{-\infty}(\mathbb{R}, \mathbb{C})$ *is a commutative ring*: This follows from (8.90) and Theorem 8.2.65.
δ *is the identity element*: We have to show that $\delta * u_2 = u_2$ for all $u_2 \in \mathrm{C}_+^{-\infty}$. Let $m_2 \in \mathbb{N}$ and $v_2 \in \mathrm{C}^0_+$ such that $u_2 = s^{m_2} * v_2$. We use that $\delta = s \circ Y = s^2 \circ (tY)$. Thus,

$$\begin{aligned} \delta * u_2 = s^{2+m_2} \circ \big((tY) * v_2\big) &= s^{2+m_2} \circ \left(\int_{-\infty}^{\infty} (t-x)\, Y(t-x)\, v_2(x)\, \mathrm{d}x\right) \\ &= s^{m_2} \circ \left(s^2 \circ \left(\int_{-\infty}^{t} (t-x)\, v_2(x)\, \mathrm{d}x\right)\right) \\ &= s^{m_2} \circ \left(s \circ \left((t-t)\, v_2(t) + \int_{-\infty}^{t} v_2(x)\, \mathrm{d}x\right)\right) \\ &= s^{m_2} \circ \left(s \circ \left(\int_{-\infty}^{t} v_2(x)\, \mathrm{d}x\right)\right) = s^{m_2} \circ v_2 = u_2. \end{aligned}$$

□

Theorem 8.2.68 (Convolution, impulse response, and Laplace transform)

1. *For all* $u_1, u_2 \in \mathrm{C}_+^{-\infty}(\mathbb{R}, \mathbb{C})$ *and* $H \in \mathbb{C}(s)$ *we have the identity*

$$H \circ (u_1 * u_2) = (H \circ u_1) * u_2 \Longrightarrow H \circ u_2 = h * u_2 \text{ with } h := H \circ \delta. \quad (8.91)$$

Hence the action of $H \in \mathbb{C}(s)$ *on* $u_2 \in \mathrm{C}_+^{-\infty}$ *is given by convolution with the impulse response* $h = H \circ \delta$ *of* H.
2. *If* $H = H_0 + H_{\mathrm{spr}}$ *is proper,* $h_{\mathrm{spr}} := H_{\mathrm{spr}} \circ \delta \in \mathrm{t}(\mathcal{F})Y$, $u \in \mathrm{C}^{0,\mathrm{pc}}[0, \infty)$ *and* $y := H \circ u = H_0 u + H_{\mathrm{spr}} \circ u$, *then* $H_{\mathrm{spr}} \circ u$ *is continuous and*

$$(H_{\mathrm{spr}} \circ u)(t) = (h_{\mathrm{spr}} * u)(t) = \int_0^t h_{\mathrm{spr}}(t-\tau) u(\tau) \mathrm{d}\tau \in \mathrm{C}^0[0, \infty).$$

3. *The subspace*

$$\mathcal{F}_2 = \mathbb{C}[s] \circ \delta \oplus \bigoplus_{\lambda \in \mathbb{C}} \mathbb{C}[t] e^{\lambda t} Y \subset \mathrm{C}_+^{-\infty}(\mathbb{R}, \mathbb{C})$$

is a subfield, and the Laplace transform isomorphism and its inverse are field isomorphisms

$$\mathcal{L}^{-1} : \mathbb{C}(s) \xrightarrow{\cong} \mathcal{F}_2,\ H \mapsto H \circ \delta, \quad \mathcal{L} : \mathcal{F}_2 \xrightarrow{\cong} \mathbb{C}(s).$$

4. C_+^{∞} *is an ideal of* $\mathrm{C}_+^{-\infty}$, *i.e.,* $\mathrm{C}_+^{-\infty} * \mathrm{C}_+^{\infty} \subseteq \mathrm{C}_+^{\infty}$.

Proof 1. We first prove the identity $H \circ (u_1 * u_2) = (H \circ u_1) * u_2$ for $H = s$. Let $u_i = s^{m_i} \circ v_i$ with continuous v_i. Then

$$s \circ (u_1 * u_2) = s^{1+m_1+m_2} \circ (v_1 * v_2) = (s^{1+m_1} \circ v_1) * (s^{m_2} \circ v_2) = (s \circ u_1) * u_2.$$

By induction and linear extension we conclude the formula for a polynomial H. Since $\mathrm{C}_+^{-\infty}$ is a $\mathbb{C}(s)$-vector space (Theorem 8.2.37), the result for arbitrary rational functions follows immediately. As special case we get

$$H \circ u_2 = H \circ (\delta * u) = (H \circ \delta) * u = h * u_2.$$

2. See Lemma 8.2.70.
3. It suffices to show that the isomorphism $\mathcal{L}^{-1} :\ H \mapsto H \circ \delta$ is multiplicative. But

$$1 \circ \delta = \delta \text{ and } (H_1 \circ \delta) * (H_2 \circ \delta) \underset{(i)}{=} (H_1 H_2) \circ (\delta * \delta) = (H_1 H_2) \circ \delta.$$

4. To show $\mathrm{C}_+^{-\infty} * \mathrm{C}_+^{\infty} \subseteq \mathrm{C}_+^{\infty}$, let $u_1 \in \mathrm{C}_+^{-\infty}$ and $u_2 \in \mathrm{C}_+^{\infty}$. Let $m_1 \in \mathbb{N}$ and $v_1 \in \mathrm{C}_+^0$ such that $u_1 = s^{m_1} \circ v_1$. Then we have $v_1 * u_2 \in \mathrm{C}_+^{\infty}$ by Lemma 8.2.66, and we conclude that

$$u_1 * u_2 = (s^{m_1} \circ v_1) * u_2 \underset{(8.90)}{=} s^{m_1} \circ (v_1 * u_2) \in s^{m_1} \circ \mathrm{C}_+^{\infty} \subseteq \mathrm{C}_+^{\infty}.$$

□

Lemma 8.2.69 *For* $a \in \mathbb{R}$ *and* $u \in \mathrm{C}_+^{-\infty}(\mathbb{R}, \mathbb{C})$, *the identity*

$$a \circ_s u = \delta_a * u \tag{8.92}$$

holds. For $b \in \mathbb{R}$ and $u_1, u_2 \in \mathrm{C}_+^{-\infty}(\mathbb{R}, \mathbb{C})$, this implies the identities

$$\delta_{a+b} = \delta_a * \delta_b \quad \text{and} \quad a \circ_s (u_1 * u_2) = (a \circ_s u_1) * u_2. \tag{8.93}$$

Proof Let $u = s^m \circ v$ with $m \in \mathbb{N}$ and $v \in \mathrm{C}_+^0$. With $Y_a = a \circ_s Y$ we have

$$\begin{aligned}\big(((t-a)Y_a) * v\big)(t) &= \int_{-\infty}^{\infty} (t-a-x)\, Y(t-a-x)\, v(x)\, \mathrm{d}x \\ &\underset{y:=x+a}{=} \int_{-\infty}^{\infty} (t-y)\, Y(t-y)\, v(y-a)\, \mathrm{d}y \\ &= \int_{-\infty}^{\infty} (t-y)\, Y(t-y)\, (a \circ_s v)(y)\, \mathrm{d}y = \big((tY) * (a \circ_s v)\big)(t).\end{aligned}$$

This and $\delta_a = s \circ Y_a = s^2 \circ \big((t-a)Y_a\big)$ imply

$$\begin{aligned}\delta_a * u &= s^{m+2} \circ \big(((t-a)Y_a) * v\big) = s^{m+2} \circ \big((tY) * (a \circ_s v)\big) \\ &= s^m \circ \big(\delta * (a \circ_s v)\big) = s^m \circ (a \circ_s v) \underset{\text{Lemma 8.2.18}}{=} a \circ_s (s^m \circ v) = a \circ_s u.\end{aligned}$$

With this, we obtain $\delta_a * \delta_b = a \circ_s \delta_b = \delta_{a+b}$ as well as

$$a \circ_s (u_1 * u_2) = \delta_a * (u_1 * u_2) = (\delta_a * u_1) * u_2 = (a \circ_s u_1) * u_2. \qquad \square$$

Lemma 8.2.70 *The defining integral representation of the convolution of continuous functions from Lemma and Definition 8.2.63 also holds for piecewise continuous functions. In particular, the convolution of two piecewise continuous functions is continuous.*

Proof Since by Theorem 8.2.6, every piecewise continuous function is a $\mathbb{C}$-linear combination of a continuous function and shifted step functions $Y_a = a \circ_s Y$, where $Y_a(t) = Y(t-a)$, we have to show only that

$$(u_1 * u_2)(t) = \int_{-\infty}^{\infty} u_1(t-x)\, u_2(x)\, \mathrm{d}x \quad \text{for } u_1, u_2 \in \mathrm{C}_+^0 \cup \{Y_a;\ a \in \mathbb{R}\}.$$

For continuous functions $u_1, u_2 \in \mathrm{C}_+^0$, this follows directly from the defining equation (8.90), compare Theorem 8.2.67.

For $u_1 = Y_a$, $a \in \mathbb{R}$, and $u_2 \in \mathrm{C}^0_+$, we get

$$\begin{aligned} Y_a * u_2 &= s \circ \big(((t-a)\,Y_a) * u_2\big) = \left(\int_{-\infty}^{\infty} (t-a-x)\,Y(t-a-x)\,u_2(x)\,\mathrm{d}x\right)' \\ &= \left(\int_{-\infty}^{t-a} (t-a-x)\,u_2(x)\,\mathrm{d}x\right)' = \int_{-\infty}^{t-a} u_2(x)\,\mathrm{d}x \\ &= \int_{-\infty}^{\infty} Y(t-a-x)\,u_2(x)\,\mathrm{d}x = \int_{-\infty}^{\infty} Y_a(t-x)\,u_2(x)\,\mathrm{d}x. \end{aligned}$$

We see from the last expression in the second line of this computation that $Y_a * u_2$ is a continuous function.
For $u_1 = Y_a$ and $u_2 = Y_b$, $a, b \in \mathbb{R}$, and the distributional convolution we use that $Y * Y = (s^{-1} \circ \delta) * (s^{-1} \circ \delta) = s^{-2} \circ \delta = tY$, and obtain

$$Y_a * Y_b = (a \circ_{\mathrm{s}} Y) * (b \circ_{\mathrm{s}} Y) \overset{(8.93)}{=\!=\!=} (a+b) \circ_{\mathrm{s}} (Y * Y) = (a+b) \circ_{\mathrm{s}} (tY).$$

For the convolution as functions we

$$\begin{aligned} \int_{-\infty}^{\infty} Y(t-a-x)\,Y(x-b)\,\mathrm{d}x &\underset{y:=x-b}{=} \int_{-\infty}^{\infty} Y(t-(a+b)-y)\,Y(y)\,\mathrm{d}y \\ = \int_{-\infty}^{t-(a+b)} Y(y)\,\mathrm{d}y &= \begin{cases} 0 & \text{if } t-(a+b) < 0 \\ t-(a+b) & \text{if } t-(a+b) \geq 0 \end{cases} \\ = \big(t-(a+b)\big)\,Y\big(t-(a+b)\big) &= (a+b) \circ_{\mathrm{s}} (tY). \end{aligned}$$

These computations also imply that $\mathrm{C}^{0,\mathrm{pc}}_+ * \mathrm{C}^{0,\mathrm{pc}}_+ \subseteq \mathrm{C}^0_+$. □

Example 8.2.71 (*State space equations*) We consider the state space equations

$$s \circ x = Ax + Bu, \;\; y = Cx + D \circ u$$
$$\text{with } A \in \mathbb{C}^{n\times n}, \;\; B \in \mathbb{C}^{n\times m}, \;\; C \in \mathbb{C}^{p\times n}, \;\; D \in \mathbb{C}[s]^{p+m}$$

with the transfer matrix $H = H_{\mathrm{pol}} + H_{\mathrm{spr}} = D + C(s\,\mathrm{id}_n - A)^{-1}B$. Since

$$\begin{aligned} &(s\,\mathrm{id}_n - A) \circ (e^{tA}Y) = Ae^{tA}Y + e^{tA}\delta - Ae^{tA}Y = e^{0A}\delta = \mathrm{id}_n\,\delta, \\ &\Longrightarrow \mathcal{L}^{-1}\left((s\,\mathrm{id}_n - A)^{-1}\right) = (s\,\mathrm{id}_n - A)^{-1} \circ \delta = e^{tA}Y \\ &\quad\;\; \mathcal{L}^{-1}(H_{\mathrm{spr}}) = H_{\mathrm{spr}} \circ \delta = Ce^{tA}BY \\ &\Longrightarrow \forall u \in (\mathrm{C}^{0,\mathrm{pc}}_+)^m : \; H \circ u = D \circ u + \int_{-\infty}^{t} Ce^{(t-x)A}Bu(x)\mathrm{d}x. \end{aligned}$$

This is a variant of the standard solution (3.63). However, here it defines a map $u \mapsto H \circ u$. ◇

The Laplace transform can be easily extended to a larger space of distributions. Recall $\delta_a = a \circ \delta \in \mathrm{C}_+^{-\infty}$ where $\delta_a(\varphi) = \varphi(a)$ for $\varphi \in \mathrm{C}_0^\infty$.

Lemma 8.2.72 *For $F = \mathbb{C}, \mathbb{R}$ the direct sum subspace*

$$\mathfrak{L}_F := \bigoplus_{a \in \mathbb{R}} F(s) \circ \delta_a \subset \mathrm{C}_+^{-\infty}(\mathbb{R}, F) \tag{8.94}$$

from Corollary 8.2.39 is a unital subalgebra of $\mathrm{C}_+^{-\infty}(\mathbb{R}, F)$. The elements of $\mathfrak{L}_F$ are finite sums $\sum_{a \in \mathbb{R}} H_a \circ \delta_a,\ H_a \in F(s)$.

Proof Obviously $1_{\mathrm{C}_+^{-\infty}} = \delta \in \mathfrak{L}_F$. Moreover, by Theorem 8.2.68 and Lemma 8.2.69,

$$(H_a \circ \delta_a) * (H_b \circ \delta_b) = (H_a H_b) \circ (\delta_a * \delta_b) = (H_a H_b) \circ \delta_{a+b} \in \mathfrak{L}_F. \qquad \square$$

Let $\epsilon_a,\ a \in \mathbb{R}$, be the standard basis of the $F(s)$-space $F(s)[\mathbb{R}] := F(s)^{(\mathbb{R})} = \bigoplus_{a \in \mathbb{R}} F(s)\epsilon_a$. Since $\mathbb{R}$ is an abelian group, the space $F(s)[\mathbb{R}]$ is a commutative $F(s)$-algebra with the product $\epsilon_a \epsilon_b := \epsilon_{a+b}$, the one-element ϵ_0 and the subfield $F(s)$. It is called the *group algebra* of $\mathbb{R}$ with coefficients in $F(s)$.

Theorem 8.2.73 *There are the inverse algebra isomorphisms*

$$\mathcal{L}^{-1} : F(s)[\mathbb{R}] \cong \mathfrak{L}_F : \mathcal{L},\ \mathcal{L}(u) = H = \sum_{a \in \mathbb{R}} H_a \epsilon_a \leftrightarrow \mathcal{L}^{-1}(H) = u = \sum_{a \in \mathbb{R}} H_a \circ \delta_a, \tag{8.95}$$

that extend the Laplace transform on $\mathcal{F}_{2,F} = F(s) \circ \delta = F[s] \circ \delta \oplus \mathrm{t}(\mathcal{F}_F)Y$ resp. its inverse, and keep the same name. The isomorphism $\mathcal{L}$ induces the module action

$$\circ : F(s)[\mathbb{R}] \times \mathrm{C}_+^{-\infty} \to \mathrm{C}_+^{-\infty},\ H \circ u_2 := \mathcal{L}^{-1}(H) * u_2.$$

For $H = \sum_{a \in \mathbb{R}} H_a \epsilon_a,\ H_a \in F(s),\ H^{(1)},\ H^{(2)} \in F(s)[\mathbb{R}],\ u_2 \in \mathrm{C}_+^{-\infty}$ this implies

$$\begin{aligned} &\mathcal{L}^{-1}(H) = H \circ \delta,\ H \circ u_2 = \left(\sum_{a \in \mathbb{R}} H_a \circ \delta_a \right) * u_2 = \sum_{a \in \mathbb{R}} a \circ_{\mathrm{s}} (H_a \circ u_2), \\ &H^{(1)} \circ \left(H^{(2)} \circ \delta\right) = \left(H^{(1)} \circ \delta\right) * \left(H^{(2)} \circ \delta\right) = \left(H^{(1)} H^{(2)}\right) \circ \delta. \end{aligned} \tag{8.96}$$

In particular, for $\lambda \in \mathbb{C},\ k > 0,\ a \in \mathbb{R}$:

$$\begin{aligned} &\mathcal{L}^{-1}\left((s-\lambda)^{-k}\epsilon_a\right) = a \circ_{\mathrm{s}} (s-\lambda)^{-k} \circ \delta = a \circ_{\mathrm{s}} (e_{\lambda,k-1}Y) \\ &\Longrightarrow \mathcal{L}^{-1}\left((s-\lambda)^{-k}\epsilon_a\right)(t) = \tfrac{(t-a)^{k-1}}{(k-1)!} e^{\lambda(t-a)} Y(t-a). \end{aligned} \tag{8.97}$$

With the obvious changes, the assertions of Theorem 8.2.68 hold.

Proof The proof follows from Lemma 8.2.72 and the explicit formulas. □

Remark 8.2.74 (i) The elements $H \in F(s)[\mathbb{R}]$ acting on $\mathrm{C}_+^{-\infty}$ are called *differential-difference* operators and include *delay-differential* operators like $((s\epsilon_0 + \epsilon_a) \circ u_2)(t) = u_2'(t) + u_2(t-a),\ u_2 \in \mathrm{C}^{1,\mathrm{pc}}$. The solution of differential-difference systems

$$R \circ w = u,\ R \in F(s)[\mathbb{R}]^{k\times l},\ w \in \left(\mathrm{C}_+^{-\infty}\right)^l,\ u \in \left(\mathrm{C}_+^{-\infty}\right)^k,\ \text{or}$$
$$R \circ w = u,\ R \in F[s][\mathbb{R}]^{k\times l},\ w \in \left(\mathrm{C}^{-\infty}\right)^l,\ u \in \left(\mathrm{C}^{-\infty}\right)^k$$

is more difficult than that of pure differential systems and not subject of the present book, cf. [9]. So nonrational transfer functions $H^{(1)}$ play no part in this book.
(ii) However, signals of the form

$$u_2 := \mathcal{L}^{-1}(H^{(2)}) = H^{(2)} \circ \delta = \sum_{a\in\mathbb{R}} H_a \circ \delta_a,\ H^{(2)} = \sum_{a\in\mathbb{R}} H_a \epsilon_a \in F(s)[\mathbb{R}],\ H_a \in F(s), \tag{8.98}$$

are important, cf. [13, p. 483]. The standard decomposition $H_a = H_{a,\mathrm{pol}} + H_{a,\mathrm{spr}}$ implies

$$\begin{aligned} &H_a \circ \delta = H_{a,\mathrm{pol}} \circ \delta + \alpha_a Y \in F[s] \circ \delta \oplus \mathrm{t}(\mathcal{F}_F) Y \\ &\Longrightarrow u_2 = \sum_{a\in\mathbb{R}} H_a \circ \delta_a = \sum_{a\in\mathbb{R}} a \circ_s (H_a \circ \delta) = \sum_{a\in\mathbb{R}} a \circ_s (H_{a,\mathrm{pol}} \circ \delta + \alpha_a Y) \\ &= \sum_{a\in\mathbb{R}} \left(H_{a,\mathrm{pol}} \circ \delta_a + \alpha_a(t-a) Y(t-a)\right). \end{aligned} \tag{8.99}$$

For a transfer function $H^{(1)} \in F(s)$ the output with left bounded support to this input u_2 is

$$H^{(1)} \circ u_2 = (H^{(1)} H^2) \circ \delta = \sum_{a\in\mathbb{R}} a \circ_s \left((H^{(1)} H_a) \circ \delta\right) \tag{8.100}$$

and can thus be computed as in (8.99) with the partial fraction decomposition of the $H^{(1)} H_a,\ a \in \mathbb{R},\ H_a \neq 0$. For matrices the analogous result holds.
(iii) In the usual treatment of the Laplace transform one identifies

$$\begin{aligned} &\epsilon_a = e^{-as},\ \epsilon_a \epsilon_b = e^{-as} e^{-bs} = e^{-(a+b)s} = \epsilon_{a+b} \\ &\Longrightarrow F(s)[\mathbb{R}] = \bigoplus_{a\in\mathbb{R}} F(s) e^{-as} \ni \mathcal{L}\left(\sum_a H_a \circ \delta_a\right) = \sum_a H_a(s) e^{-as}. \end{aligned} \tag{8.101}$$

◊

Example 8.2.75 Choose $0 \neq \lambda \in \mathbb{C}$ and $0 < a \in \mathbb{R}$ and define

$$H^{(1)} := (s-\lambda)^{-1},\ u_2 := Y - Y_a,\ Y_a := a \circ Y,\ Y_a(t) = \begin{cases} 0 & \text{if } t < a \\ 1 & \text{if } t \geq a \end{cases}$$

$$\Longrightarrow u_2(t) = \begin{cases} 0 & \text{if } t < 0 \text{ or } t \geq a \\ 1 & \text{if } 0 \leq t < a \end{cases},$$

$$u_2 = s^{-1} \circ \delta - a \circ (s^{-1} \circ \delta) = s^{-1} \circ \delta - s^{-1} \circ \delta_a \Longrightarrow \mathcal{L}(u_2) = s^{-1}(\epsilon_0 - \epsilon_a),$$

$$H^{(1)} \circ u_2 = H^{(1)} s^{-1} \circ \delta - a \circ \left(H^{(1)} s^{-1} \circ \delta\right).$$

With $\epsilon_0 = 1$ and $\epsilon_a = e^{-as}$ the usual form is $\mathcal{L}(u_2) = s^{-1}(1 - e^{-as})$. Further

$$\begin{aligned} H^{(1)} s^{-1} &= \lambda^{-1}\left((s-\lambda)^{-1} - s^{-1}\right) \\ &\Longrightarrow \mathcal{L}^{-1}(H^{(1)} s^{-1}) = H^{(1)} s^{-1} \circ \delta = \lambda^{-1}(e^{\lambda t} - 1) Y, \\ a \circ \left(H^{(1)} s^{-1} \circ \delta\right) &= \lambda^{-1}(e^{\lambda(t-a)} - 1) Y_a \\ &\Longrightarrow \left(H^{(1)} \circ u_2\right)(t) = \begin{cases} 0 & \text{if } t < 0 \\ \lambda^{-1}(e^{\lambda t} - 1) & \text{if } 0 \leq t < a \\ \lambda^{-1} e^{\lambda t}\left(1 - e^{-\lambda a}\right) & \text{if } t \geq a. \end{cases} \end{aligned}$$

Since $H^{(1)}$ is strictly proper and u_2 is piecewise continuous, the function $H^{(1)} \circ u_2$ is continuous according to Theorem 8.2.47, 5, and indeed $H^{(1)} \circ u_2$ is continuous both in 0 and in a. ◊

In the next section and in context with the Laplace transform, we need the following.

Lemma 8.2.76 $(e^{\lambda t} u_1) * (e^{\lambda t} u_2) = e^{\lambda t}(u_1 * u_2),\ \lambda \in \mathbb{C},\ u_1, u_2 \in \mathrm{C}_+^{-\infty}$.

Proof 1. For continuous u_1 and u_2 we get

$$\begin{aligned} \left((e^{\lambda t} u_1) * (e^{\lambda t} u_2)\right)(t) &= \int_{-\infty}^{\infty} e^{\lambda(t-x)} u_1(t-x) e^{\lambda x} u_2(x) \mathrm{d}x \\ &= e^{\lambda t} \int_{-\infty}^{\infty} u_1(t-x) u_2(x) \mathrm{d}x = e^{\lambda t}(u_1 * u_2)(t). \end{aligned}$$

2. Consider arbitrary distributions with left bounded support

$$\begin{aligned} & s^{n_i} \circ u_i,\ u_i \in \mathrm{C}_+^0,\ n_i \geq 0,\ i = 1, 2 \Longrightarrow (s-\lambda) \circ (e^{\lambda t} u_i) = e^{\lambda t} s \circ u_i \\ & \Longrightarrow (s-\lambda)^{n_i} \circ (e^{\lambda t} u_i) = e^{\lambda t} s^{n_i} \circ u_i \\ & \Longrightarrow (e^{\lambda t} s^{n_1} \circ u_1) * (e^{\lambda t} s^{n_2} \circ u_2) \\ & \quad = \left((s-\lambda)^{n_1} \circ (e^{\lambda t} u_1)\right) * \left((s-\lambda)^{n_2} \circ (e^{\lambda t} u_2)\right) \end{aligned}$$

$$= (s-\lambda)^{n_1+n_2} \circ \left((e^{\lambda t}u_1) * (e^{\lambda t}u_2)\right) \underset{1.}{=} (s-\lambda)^{n_1+n_2} \circ \left(e^{\lambda t}(u_1 * u_2)\right)$$
$$= e^{\lambda t}s^{n_1+n_2} \circ (u_1 * u_2) = e^{\lambda t}(s^{n_1} \circ u_1) * (s^{n_2} \circ u_2).$$

□

8.2.6 External Stability for Continuous-Time Transfer Operators

In this section, we study the external stability of continuous-time transfer operators $H\circ$ as analogue to Sect. 8.1.3 and characterize it by properties of pole(H).

Definition 8.2.77 In continuous time, a rational function $H \in \mathbb{C}(s)$ is called bounded input/bounded output (BIBO) stable if a bounded, piecewise continuous input $u \in \mathrm{C}_+^{0,\mathrm{pc}}$ induces a bounded, piecewise continuous output

$$y := H \circ u = H_{\mathrm{pol}} \circ u + H_{\mathrm{spr}} \circ u = H_{\mathrm{pol}} \circ u + (\alpha Y) * u,\ \ H_{\mathrm{spr}} \circ \delta = \alpha Y \in \mathrm{t}(\mathcal{F}_{\mathbb{C}})Y,$$

with continuous $H_{\mathrm{spr}} \circ u$, cf. 8.2.47 and 8.2.68. ◊

Theorem 8.2.59 implies the following.

Corollary 8.2.78 *Every BIBO stable rational function is proper.* ◊

Example 8.2.79 Consider the nonproper transfer function $H = s$, which maps an input $u \in \mathrm{C}_+^{-\infty}(\mathbb{R}, \mathbb{C})$ to its derivative $y = s \circ u = u'$. From (4.61) use

$$\psi = \psi_\epsilon \in \mathrm{C}^\infty \text{ with } 0 \le \psi(t) \le 1,\ \ \psi|_{(-\infty,0)} = 0,\ \ \psi|_{(1,\infty)} = 1.$$

Then $u := e^{it^2}\psi Y \in \mathrm{C}_+^\infty$ has its support in $[0, \infty)$ and is bounded, indeed $|u(t)| = 1$ for $t \ge 1$. Its derivative is

$$s \circ u = u' = (2it\psi + \psi')e^{it^2}Y \in \mathrm{C}_+^{0,\mathrm{pc}} \text{ with } |u'(t)| = 2t \text{ for } t \ge 1.$$

So $s \circ u$ is not bounded and differentiation is not BIBO stable. ◊

Similarly as in Sect. 8.1.3 we define and analyze various function spaces before we present the theorems on external stability.

Definition 8.2.80 For $p \in \mathbb{R}$, $1 \le p \le \infty$, we define

$$\|-\|_p\colon\ \mathrm{C}^{0,\mathrm{pc}} \longrightarrow [0,\infty],\quad u \longmapsto \begin{cases} \left(\int_{-\infty}^{\infty}|u(t)|^p\,\mathrm{d}t\right)^{1/p} & \text{if } p < \infty, \\ \sup_{t\in\mathbb{R}}|u(t)| & \text{if } p = \infty. \end{cases}$$

Notice that we allow $\|u\|_p = \infty$. Moreover we define the function spaces

$$\begin{aligned} \mathrm{L}^p &:= \{u \in \mathrm{C}^{0,\mathrm{pc}};\ \|u\|_p < \infty\} \quad \text{for } p < \infty \text{ and} \\ \mathrm{L}^\infty &:= \{u \in \mathrm{C}^{0,\mathrm{pc}};\ \|u\|_\infty < \infty\}. \end{aligned} \tag{8.102}$$

The space L^∞ consists of the bounded piecewise continuous functions. ◊

In Sect. 10.2.4 we will introduce the completion $\mathfrak{L}^q$ of L^q that is a *Banach space* and can be described as a space of equivalence classes of Lebesgue integrable functions on $\mathbb{R}$ modulo almost zero functions. In this book we do not use Lebesgue's integration theory and the latter description. We need, however, the norms $\|-\|_p$ on L^p, see below.

Lemma 8.2.81 *Let $u, v \in \mathrm{C}^{0,\mathrm{pc}}$ and let $p, q \in [1, \infty]$ with $\frac{1}{p} + \frac{1}{q} = 1$. Then the Hölder inequality*

$$\|uv\|_1 = \int_{-\infty}^{\infty} |u(t)|\,|v(t)|\,\mathrm{d}t \le \|u\|_p\,\|v\|_q \tag{8.103}$$

and the triangle inequality

$$\|u + v\|_p \le \|u\|_p + \|v\|_p \tag{8.104}$$

hold. Hence the L^p are normed spaces with the Minkowski norm $\|-\|_p$.

Proof We start with the Hölder inequality. For $p = \infty$ and $q = 1$ the proof is obvious. Assume that $p, q \in (1, \infty)$ with $\frac{1}{p} + \frac{1}{q} = 1$. On the finitely many points t where u is discontinuous, we set $u(t) := u(t+)$, and similarly for v. For $T > 0$ we define

$$a(T) := \left(\int_{-T}^{T} |u(t)|^p\,\mathrm{d}t\right)^{1/p} \quad \text{and} \quad b(T) := \left(\int_{-T}^{T} |v(t)|^q\,\mathrm{d}t\right)^{1/q}.$$

As in (8.37) the concavity of the logarithm implies

$$\begin{aligned} &\forall t \in [-T, T]:\ \frac{|u(t)|}{a(T)}\frac{|v(t)|}{b(T)} \le \frac{1}{p}\frac{|u(t)|^p}{a(T)^p} + \frac{1}{q}\frac{|v(t)|^q}{b(T)^q} \\ &\Longrightarrow \frac{1}{a(T)b(T)}\int_{-T}^{T} |u(t)|\,|v(t)|\,\mathrm{d}t \le \frac{1}{p} + \frac{1}{q} = 1 \\ &\Longrightarrow \int_{-T}^{T} |u(t)|\,|v(t)|\,\mathrm{d}t \le a(T)b(T) = \left(\int_{-T}^{T} |u(t)|^p\,\mathrm{d}t\right)^{1/p}\left(\int_{-T}^{T} |v(t)|^q\,\mathrm{d}t\right)^{1/q} \\ &\underset{T\to\infty}{\Longrightarrow} \|uv\|_1 \le \|u\|_p\|v\|_q. \end{aligned} \tag{8.105}$$

The proof of the triangle inequality is analogous to that of Lemma 8.1.20,3, by replacing $\sum_{t\in\mathbb{N}}$ by $\int_{-T}^{T}$ and then letting $T \to \infty$. □

Let $\mathrm{L}^q_+ := \mathrm{L}^q \cap \mathrm{C}^{0,\mathrm{pc}}_+$ be the space of functions in L^q with left bounded support.

Theorem 8.2.82 *For $p \in [1, \infty]$, $u \in \mathrm{L}_+^p$, and $v \in \mathrm{L}_+^1$,* Young's inequality

$$\|u * v\|_p \leq \|u\|_p \, \|v\|_1 \tag{8.106}$$

holds. This inequality implies that $\mathrm{L}_+^p * \mathrm{L}_+^1 \subseteq \mathrm{L}_+^p$. Hence L_+^1 is a (nonunital) normed subalgebra of $\mathrm{C}_+^{0,\mathrm{pc}}$ and the L_+^p, $p \in [1, \infty]$, are L_+^1-submodules of $\mathrm{C}_+^{0,\mathrm{pc}}$. In particular, the convolution $\mathrm{L}_+^p \longrightarrow \mathrm{L}_+^p$, $u \longmapsto u * v$, is a continuous map.

Proof Let $M > 0$ be such that $\mathrm{supp}(u) \cup \mathrm{supp}(v) \subseteq [-M, \infty)$, and let $T > 0$. Recall from Lemma and Definition 8.2.63 and Lemma 8.2.70 that for $t \in [-T, T]$ the integral representation

$$(u * v)(t) = \int_{-\infty}^{\infty} u(t-x)\, v(x)\, \mathrm{d}x = \int_{-(T+M)}^{T+M} u(t-x)\, v(x)\, \mathrm{d}x$$

holds also for piecewise continuous functions. Notice that with the function $|u|$ defined by $|u|(t) := |u(t)|$, the relation

$$|u * v|(t) = \left| \int_{-\infty}^{\infty} u(t-x)\, v(x)\, \mathrm{d}x \right| \leq \int_{-\infty}^{\infty} |u(t-x)|\, |v(x)|\, \mathrm{d}x = (|u| * |v|)(t)$$

holds. Thus we assume without loss of generality that $u(t) \geq 0$ and $v(t) \geq 0$ for all $t \in \mathbb{R}$. If $p = \infty$, then

$$\begin{aligned}(u * v)(t) &= \int_{-\infty}^{\infty} u(t-x)\, v(x)\, \mathrm{d}x \leq \int_{-\infty}^{\infty} \sup_{\tau \in \mathbb{R}} u(\tau)\, v(x)\, \mathrm{d}x \\ &= \sup_{\tau \in \mathbb{R}} u(\tau) \int_{-\infty}^{\infty} v(x)\, \mathrm{d}x = \|u\|_\infty \, \|v\|_1\end{aligned}$$

and thus $\|u * v\|_\infty \leq \|u\|_\infty \, \|v\|_1$. For $p = 1$, we obtain

$$\begin{aligned}\int_{-T}^{T} (u * v)(t)\, \mathrm{d}t &= \int_{-T}^{T} \left(\int_{-(T+M)}^{T+M} u(t-x)\, v(x)\, \mathrm{d}x \right) \mathrm{d}t \\ &= \int_{-(T+M)}^{T+M} v(x) \left(\int_{-T}^{T} u(t-x)\, \mathrm{d}t \right) \mathrm{d}x \\ &\leq \int_{-(T+M)}^{T+M} v(x) \underbrace{\left(\int_{-\infty}^{\infty} u(t-x)\, \mathrm{d}t \right)}_{= \|u\|_1} \mathrm{d}x \leq \|u\|_1 \int_{-\infty}^{\infty} v(x)\, \mathrm{d}x = \|u\|_1 \, \|v\|_1 \\ \underset{T \to \infty}{\Longrightarrow} \; \|u * v\|_1 &\leq \|u\|_1 \, \|v\|_1.\end{aligned}$$

For $p \in (1, \infty)$, let q be such that $\frac{1}{p} + \frac{1}{q} = 1$. For $T > 0$ and $t \in [-T, T]$, the equations

$$
\begin{aligned}
(u * v)(t) &= \int_{-(T+M)}^{T+M} u(t-x)\, v(x)\, \mathrm{d}x = \int_{-(T+M)}^{T+M} u(t-x)\, v(x)^{1/p}\, v(x)^{1/q}\, \mathrm{d}x \\
&\underset{(8.105)}{\leq} \left(\int_{-(T+M)}^{T+M} u(t-x)^p\, v(x)\, \mathrm{d}x \right)^{1/p} \left(\int_{-(T+M)}^{T+M} v(x)\, \mathrm{d}x \right)^{1/q} \\
&\leq \left(\int_{-(T+M)}^{T+M} u(t-x)^p\, v(x)\, \mathrm{d}x \right)^{1/p} \|v\|_1^{1/q}
\end{aligned}
$$

hold. Taking the p-th power and integrating furnishes

$$
\begin{aligned}
\int_{-T}^{T} (u * v)(t)^p\, \mathrm{d}t &\leq \|v\|_1^{p/q} \int_{-T}^{T} \left(\int_{-(T+M)}^{T+M} u(t-x)^p\, v(x)\, \mathrm{d}x \right) \mathrm{d}t \\
&= \|v\|_1^{p/q} \int_{-(T+M)}^{T+M} v(x) \left(\int_{-T}^{T} u(t-x)^p\, \mathrm{d}t \right) \mathrm{d}x \\
&\leq \|v\|_1^{p/q} \int_{-(T+M)}^{T+M} v(x) \underbrace{\left(\int_{-\infty}^{\infty} u(t-x)^p\, \mathrm{d}t \right)}_{=\|u\|_p^p} \mathrm{d}x \\
&\leq \|v\|_1^{p/q}\, \|u\|_p^p \int_{-\infty}^{\infty} v(x)\, \mathrm{d}x = \|v\|_1^{p/q}\, \|u\|_p^p\, \|v\|_1.
\end{aligned}
$$

By taking the limit $T \to \infty$ and then the p-th root we get

$$
\|u * v\|_p = \left(\int_{-\infty}^{\infty} (u * v)(t)^p\, \mathrm{d}t \right)^{1/p} \leq \|u\|_p\, \|v\|_1^{1/q}\, \|v\|_1^{1/p} = \|u\|_p\, \|v\|_1. \quad \square
$$

The next theorem is the continuous-time analogue of Theorem and Definition 8.1.25. Due to Corollary 8.2.78 we assume a *proper* transfer function $H \in \mathbb{C}(s)_{\mathrm{pr}}$, hence $H \circ C_+^{0,\mathrm{pc}} \subseteq C_+^{0,\mathrm{pc}}$ according to Theorem 8.2.59.
By Theorem 8.2.47 the partial fraction decomposition of H, namely,

$$
\begin{aligned}
H = H_0 + H_{\mathrm{spr}} = H_0 + \sum_{\lambda \in \mathrm{pole}(H)} \sum_{k=1}^{m_\lambda} a_{\lambda,k} (s-\lambda)^{-k} \text{ with} \\
H_0 \in \mathbb{C} \text{ and } \forall \lambda \in \mathrm{pole}(H) : \; m_\lambda \geq 1, \; a_{\lambda,m_\lambda} \neq 0
\end{aligned}
\tag{8.107}
$$

gives rise to the concrete representation of the impulse response

$$h = H \circ \delta = H_0\delta + H_{\rm spr} \circ \delta = H_0\delta + \alpha Y \in \mathbb{C}\delta \oplus \left(\bigoplus_{\lambda\in\mathbb{C}} \mathbb{C}[t]e^{\lambda t}\right)Y,$$

$$\text{where } \alpha := \sum_{\lambda\in\mathrm{pole}(H)} \sum_{k=1}^{m_\lambda} a_{\lambda,k}\tfrac{t^{k-1}}{(k-1)!}e^{\lambda t} = \sum_{\lambda\in\mathrm{pole}(H)} g_\lambda(t)e^{\lambda t} \tag{8.108}$$

$$\text{and } g_\lambda(t) := \sum_{k=1}^{m_\lambda} a_{\lambda,k}\tfrac{t^{k-1}}{(k-1)!} \in \mathbb{C}[t] \quad \text{with} \quad \deg(g_\lambda) = m_\lambda - 1.$$

Theorem and Definition 8.2.83 (External stability of proper rational functions) *For the proper rational function H from* (8.107) *and its impulse response h from* (8.108) *the following statements are equivalent:*

1. $\mathrm{pole}(H) \subset \mathbb{C}_- = \{\lambda \in \mathbb{C};\ \Re(\lambda) < 0\}$.
2. $H_{\rm spr} \circ \delta = \alpha Y \in \mathrm{L}^1_+$.
3. *H is BIBO stable, i.e., $H \circ \mathrm{L}^\infty_+ = h * \mathrm{L}^\infty_+ \subseteq \mathrm{L}^\infty_+$.*
4. *H is $(\mathrm{L}^1_+, \mathrm{L}^1_+)$- stable, i.e., $H \circ \mathrm{L}^1_+ = h * \mathrm{L}^1_+ \subseteq \mathrm{L}^1_+$.*
5. *H is $(\mathrm{L}^p_+, \mathrm{L}^p_+)$- stable for all $1 \le p \le \infty$, i.e., $H \circ \mathrm{L}^p_+ = h * \mathrm{L}^p_+ \subseteq \mathrm{L}^p_+$.*

If these equivalent properties are satisfied, then the rational function H and the associated operator $H \circ (-) = h * (-)$ are called externally stable.

Proof Since the constant H_0 satisfies all five statements, we assume in the proof without loss of generality that H is strictly proper, i.e., that $H_0 = 0$ and thus

$$h = H \circ \delta = \alpha Y, \quad \text{with} \quad \alpha = \sum_{\lambda\in\mathrm{pole}(H)} g_\lambda(t)e^{\lambda t}.$$

1. $\Longrightarrow$ 2.: It suffices to prove that $\left(t \mapsto g_\lambda(t)e^{\lambda t}Y(t)\right) \in \mathrm{L}^1_+$ for $\Re(\lambda) < 0$. Choose ρ with $\Re(\lambda) < \rho < 0$ and hence $\Re(\lambda - \rho) < 0$. For $t \to \infty$ the exponential function $e^{(\lambda-\rho)t}$ decreases faster than the polynomial $g_\lambda(t)$ grows. Therefore

$$M := \sup_{t\ge 0}\left|g_\lambda(t)e^{(\lambda-\rho)t}\right| < \infty \Longrightarrow \forall t \ge 0:\ \left|g_\lambda(t)e^{\lambda t}\right| \le M e^{\rho t}$$

$$\Longrightarrow \left\|g_\lambda(t)e^{\lambda t}Y(t)\right\|_1 = \int_0^\infty \left|g_\lambda(t)e^{\lambda t}\right| \mathrm{d}t \le M\int_0^\infty e^{\rho t}\,\mathrm{d}t = \frac{M}{\rho}e^{\rho t}\Big|_0^\infty = \tfrac{M}{-\rho} < \infty$$

$$\Longrightarrow g_\lambda(t)e^{\lambda t}Y(t) \in \mathrm{L}^1_+ .$$

2. $\Longrightarrow$ 5.: This is a direct consequence of Young's inequality (Theorem 8.2.82).
5. $\Longrightarrow$ 3, 4.: Items 3. and 4. are special cases of item 5.
3. $\Longrightarrow$ 1.: We assume that H has poles λ with $\Re(\lambda) \ge 0$ and show that this contradicts 3. We write the strictly proper rational function H as

$$H = H^{<} + H^{\geq} \quad \text{with } H^{<} := \sum_{\lambda \in \text{pole}(H),\ \Re(\lambda)<0} \sum_{k=1}^{m_\lambda} a_{\lambda,k}(s-\lambda)^{-k}$$
$$\text{and } H^{\geq} := \sum_{\lambda \in \text{pole}(H),\ \Re(\lambda)\geq 0} \sum_{k=1}^{m_\lambda} a_{\lambda,k}(s-\lambda)^{-k}.$$

From the already proven implication 1. $\Longrightarrow$ 3. we infer that $H^{<} \circ \mathrm{L}_+^\infty \subseteq \mathrm{L}_+^\infty$. With item 3 this yields $H^{\geq} \circ \mathrm{L}_+^\infty = (H - H^{<}) \circ \mathrm{L}_+^\infty \subseteq \mathrm{L}_+^\infty$, i.e., that $H^{\geq}$ satisfies condition 3 too. Therefore we may and do assume $H = H^{\geq} \neq 0$ in the rest of this proof. Condition 3 and $Y \in \mathrm{L}_+^\infty$ imply $H \circ Y = Hs^{-1} \circ \delta \in \mathrm{L}_+^\infty$. Since $\text{pole}(H) \subseteq \text{pole}(Hs^{-1}) \subseteq \{0\} \cup \text{pole}(H)$ we get the partial fraction decomposition

$$Hs^{-1} = \sum_{\lambda \in \{0\}\cup\text{pole}(H)} \sum_{k=1}^{n_\lambda} b_{\lambda,k}(s-\lambda)^{-k},\ b_{\lambda,n_\lambda} \neq 0,$$

where $n_\lambda \geq 1$ for all λ, and $n_0 \geq 2$ if $0 \in \text{pole}(H)$. We deduce that

$$H \circ Y = Hs^{-1} \circ \delta = \beta Y,\ \beta(t) := \sum_{\lambda \in \{0\}\cup\text{pole}(H)} h_\lambda(t)e^{\lambda t} \text{ where}$$
$$h_\lambda(t) = \sum_{k=1}^{n_\lambda} b_{\lambda,k}\frac{t^{k-1}}{(k-1)!} = b_{\lambda,n_\lambda}t^{n_\lambda-1} + \cdots \in \mathbb{C}[t],\ \deg_t(h_\lambda) = n_\lambda - 1.$$

Since $H \circ Y = \beta Y$ is bounded so is $|\beta(t)|$ for $t \geq 0$. Theorem 6.4.4 implies

$$\beta \in \oplus_{\mu\in\mathbb{C},\Re(\mu)<0}\mathbb{C}[t]e^{\mu t} \bigoplus \oplus_{\mu\in\mathbb{C},\Re(\mu)=0}\mathbb{C}e^{\mu t} \underset{\Re(\lambda)\geq 0}{\Longrightarrow} \forall \lambda \in \{0\}\cup\text{pole}(H):$$
$$n_\lambda \leq 1,\ \Re(\lambda) = 0 \Longrightarrow 0 \notin \text{pole}(H) \text{ and } Hs^{-1} = \sum_{\lambda \in \{0\}\uplus\text{pole}(H)} b_{\lambda,1}(s-\lambda)^{-1}$$
$$\underset{\frac{s}{s-\lambda}=1+\frac{\lambda}{s-\lambda}}{\Longrightarrow} H = (Hs^{-1})s = \sum_{\lambda \in \{0\}\uplus\text{pole}(H)} b_{\lambda,1} + \sum_{\lambda\in\text{pole}(H)} b_{\lambda,1}\lambda(s-\lambda)^{-1}$$
$$\underset{H \text{ strictly proper}}{\Longrightarrow} H = \sum_{\lambda\in\text{pole}(H)} b_{\lambda,1}\lambda(s-\lambda)^{-1},\ \forall\lambda \in \text{pole}(H):\ \lambda \neq 0,\ \Re(\lambda) = 0$$
$$\Longrightarrow \forall \lambda \in \text{pole}(H):\ \lambda \neq 0,\ \Re(\lambda) = 0, a_\lambda := b_{\lambda,1}\lambda \neq 0,\ H = \sum_{\lambda\in\text{pole}(H)} a_\lambda(s-\lambda)^{-1}. \tag{8.109}$$

By assumption the pole set of H is not empty. Let $\mu \in \text{pole}(H)$. Hence $\Re(\mu) = 0$ and therefore the input $u := (s-\mu)^{-1} \circ \delta = e^{\mu t}Y \in \mathrm{L}_+^\infty$ is bounded. For $\lambda \neq \mu$ the partial fraction decomposition $\frac{1}{(s-\lambda)(s-\mu)} = \frac{1}{\lambda-\mu}\left(\frac{1}{s-\lambda} - \frac{1}{s-\mu}\right)$ implies

$$(s-\lambda)^{-1} \circ u = \left((s-\lambda)(s-\mu)\right)^{-1} \circ \delta$$
$$= \tfrac{1}{\lambda-\mu}\left((s-\lambda)^{-1} \circ \delta - (s-\mu)^{-1} \circ \delta\right) = \tfrac{1}{\lambda-\mu}\left(e^{\lambda t} - e^{\mu t}\right)Y.$$

If $\lambda \in \mathrm{pole}(H)$ and thus $\Re(\lambda) = 0$, the last function is bounded and contained in L_+^∞. In contrast, $(s-\mu)^{-1} \circ u = (s-\mu)^{-2} \circ \delta = te^{\mu t}Y$ is unbounded. We conclude that also

$$H \circ u = a_{\mu,1}(s-\mu)^{-1} \circ u + \sum_{\lambda \in \mathrm{pole}(H)\setminus\{\mu\}} a_{\lambda,1}(s-\lambda)^{-1} \circ u$$

is unbounded, and this contradicts condition 3.
4. $\Longrightarrow$ 1. : Let $\mu \in \mathbb{C}\setminus \mathrm{pole}(H)$ with $\Re(\mu) < 0$. Then the function $(s-\mu)^{-1} \circ \delta = e^{\mu t}Y$ belongs to L_+^1. The rational function $H(s-\mu)^{-1}$ is strictly proper since H is proper. Condition 4 implies that also $H(s-\mu)^{-1} \circ \delta = H \circ (e^{\mu t}Y)$ belongs to L_+^1. This means that $H(s-\mu)^{-1}$ satisfies condition 2, which is equivalent to condition 1. Since the poles of H are also poles of $H(s-\mu)^{-1}$, condition 1 holds for H too. □

We extend the preceding theorem to rational matrices. Since, above, we used the letter p as index for L^p, we consider $H \in \mathbb{C}(s)^{q\times m}$ here.

Corollary and Definition 8.2.84 (Externally stable rational matrices) *For a proper rational matrix*

$$H = H_0 + H_{\mathrm{spr}} \in \mathbb{C}(s)_{\mathrm{pr}}^{q\times m} = \mathbb{C}^{q\times m} \oplus \mathbb{C}(s)_{\mathrm{spr}}^{q\times m},$$

the following properties are equivalent:

1. $\mathrm{pole}(H) \subset \mathbb{C}_- = \{z \in \mathbb{C};\ \Re(z) < 0\}$.
2. *The (piecewise continuous) impulse response $H_{\mathrm{spr}} \circ \delta$ belongs to $(\mathrm{L}_+^1)^{q\times m}$.*
3. *The operator $H\circ$ is BIBO stable, i.e., $H \circ (\mathrm{L}_+^\infty)^m \subseteq (\mathrm{L}_+^\infty)^q$.*
4. *The operator $H\circ$ is $(\mathrm{L}_+^1, \mathrm{L}_+^1)$-stable, i.e., $H \circ (\mathrm{L}_+^1)^m \subseteq (\mathrm{L}_+^1)^q$.*
5. *The operator $H\circ$ is $(\mathrm{L}_+^p, \mathrm{L}_+^p)$-stable for all $1 \le p \le \infty$, i.e., $H \circ (\mathrm{L}_+^p)^m \subseteq (\mathrm{L}_+^p)^q$.*

If these properties are satisfied, then the matrix H and the operator $H\circ$ are called externally stable.

Proof The asserted equivalences follow directly from Theorem and Definition 8.2.83. □

Corollary 8.2.85 *Let $\mathcal{B} := \left\{\binom{y}{u} \in \mathcal{F}^{q+m};\ P \circ y = Q \circ u\right\}$ be an asymptotically stable IO behavior with proper transfer matrix $H = P^{-1}Q$, i.e.,* $\mathrm{pole}(H) \subseteq \mathrm{char}(\mathcal{B}^0) \subset \mathbb{C}_-$ *and $H \in \mathbb{C}(s)_{\mathrm{pr}}^{q\times m}$. Let $\alpha \in \mathcal{F}^m$ be polynomial-exponential and $u := \alpha Y$. Then $H_2 := \mathcal{L}(\alpha Y)$ and HH_2 are strictly proper and $H \circ u = HH_2 \circ \delta = \beta Y$ with polynomial-exponential $\beta \in \mathcal{F}^q$. Any other solution y of $P \circ y = Q \circ u$ has the form $y = H \circ u + z$ with $z \in \mathcal{B}^0$.*
If $u = \alpha Y$ belongs to $\left(\mathrm{L}_+^p\right)^m$, $1 \le p \le \infty$, then $H \circ u$ and $yY = H \circ u + zY$ belong to $\left(\mathrm{L}_+^p\right)^q$.

Proof Due to $\mathrm{char}(\mathcal{B}^0) \subset \mathbb{C}_-$, the trajectories zY belong to all $\left(\mathrm{L}_+^p\right)^q$. □

In the following we extend Theorem 6.4.5 on BIBO stability of IO behaviors from polynomial-exponential inputs to general piecewise continuous ones.

Theorem 8.2.86 ([5, Theorem 7.6.2 on p. 265]) *For an IO behavior*

$$\mathcal{B} := \left\{ \left(\begin{smallmatrix} y \\ u \end{smallmatrix}\right) \in \mathrm{C}^{-\infty}(\mathbb{R}, \mathbb{C})^{p+m};\ P \circ y = Q \circ u \right\} \tag{8.110}$$

with transfer matrix $H \in \mathbb{C}(s)^{p \times m}$ the following properties are equivalent:

1. *For every trajectory $\left(\begin{smallmatrix} y \\ u \end{smallmatrix}\right) \in \mathcal{B}$ where u is piecewise continuous and bounded for $t \to \infty$, i.e., $\sup_{t \geq 0} \max \left\{ |u_i(t)|,\ i = 1, \ldots, m \right\} < \infty$, also the output y is piecewise continuous and bounded for $t \to \infty$.*
2. *(a) The autonomous part $\mathcal{B}^0 = \left\{ y \in \mathrm{C}^{-\infty}(\mathbb{R}, \mathbb{C})^p;\ P \circ y = 0 \right\} \subseteq \mathrm{C}^{\infty}(\mathbb{R}, \mathbb{C})^p$ is stable. By Theorem and Definition 5.4.17 this means that*

$$\forall \lambda \in \mathrm{char}(\mathcal{B}^0) :\ \Re(\lambda) < 0 \text{ or } \left(\Re(\lambda) = 0 \text{ and } \mathrm{index}(\mathcal{B}^0, s - \lambda) = 1\right).$$

   (b) The transfer matrix H is proper and externally stable, i.e., it satisfies the equivalent conditions of Corollary and Definition 8.2.84.

Proof 1. $\Longrightarrow$ 2. Let $y \in \mathcal{B}^0$. Then $\left(\begin{smallmatrix} y \\ 0 \end{smallmatrix}\right) \in \mathcal{B}$, and condition 1 implies that y is bounded for $t \to \infty$. By Theorem and Definition 5.4.17 this signifies that $\mathcal{B}^0$ is stable. For every $i \in \{1, \ldots, m\}$, the piecewise continuous vector

$$u := (0, \ldots, 0, \overset{i}{Y}, 0, \ldots, 0)^\top \in (\mathrm{C}_+^{0,\mathrm{pc}})^m$$

is bounded for $t \to \infty$, and the trajectory $\left(\begin{smallmatrix} H \circ u \\ u \end{smallmatrix}\right)$ belongs to $\mathcal{B}$. By condition 1 the output $H \circ u$ is also piecewise continuous, hence so are the entries of the matrix $H \circ Y$. By Theorem 8.2.60 this is equivalent to H being proper.
For arbitrary $u \in (\mathrm{L}_+^\infty)^m$, the trajectory $\left(\begin{smallmatrix} H \circ u \\ u \end{smallmatrix}\right)$ belongs to $\mathcal{B}$. Condition 1 implies that $H \circ u \in (\mathrm{L}_+^\infty)^p$. Hence, $H \circ (\mathrm{L}_+^\infty)^m \subseteq (\mathrm{L}_+^\infty)^p$, i.e., H is BIBO stable and externally stable by Corollary and Definition 8.2.84.
2. $\Longrightarrow$ 1. (i) Let $\left(\begin{smallmatrix} y \\ u \end{smallmatrix}\right) \in \mathcal{B}$ with piecewise continuous $u \in (\mathrm{C}^{0,\mathrm{pc}})^m$ that is bounded for $t \to \infty$. We decompose

$$\begin{aligned} &u = u_+ + u_- \text{ with } u_+ := uY \text{ and } u_- := u\,(1 - Y) \\ &\Longrightarrow u_+ \in \left(\mathrm{L}_+^\infty\right)^m \underset{2.(b)}{\Longrightarrow} y_+ := H \circ u_+ \in \left(\mathrm{L}_+^\infty\right)^p. \end{aligned}$$

(ii) Consider Eq. (8.54). The transfer matrix $H(-s) = P(-s)^{-1} Q(-s)$ is proper like $H = H(s) = P(s)^{-1} Q(s)$. Define

$$\begin{aligned} &u_1 := (u_-)^\vee \in \left(\mathrm{C}_+^{0,\mathrm{pc}}\right)^m,\ u_1(t) := u_-(-t) \Longrightarrow y_1 := H(-s) \circ u_1 \in \left(\mathrm{C}_+^{0,\mathrm{pc}}\right)^p \\ &\Longrightarrow P(-s) \circ y_1 = Q(-s) \circ u_1 \Longrightarrow (P(-s) \circ y_1)^\vee = (Q(-s) \circ u_1)^\vee \\ &\underset{(8.54)}{\Longrightarrow} P(s) \circ (y_1)^\vee = Q(s) \circ (u_1)^\vee. \end{aligned}$$

With $y_- := (y_1)\check{} \in \left(\mathrm{C}_-^{0,\mathrm{pc}}\right)^p$, $y := y_+ + y_-$, $u_1 = (u_-)\check{}$, $u_- = (u_1)\check{}$ we infer

$$P \circ y = P \circ y_+ + P \circ y_- = Q \circ u_+ + Q \circ u_- = Q \circ u, \ \binom{y}{u} \in \mathcal{B}.$$

Here the support of y_1 resp. of $y_- = (y_1)\check{}$ is bounded to the left resp. to the right and hence there is t_0 with $y_-(t) = 0$ and $y(t) = y_+(t)$ for $t \geq t_0$. But $y_+ \in \left(\mathrm{L}_+^\infty\right)^p$ and hence y is bounded for $t \to \infty$. Let

$$\binom{\widetilde{y}}{u} \in \mathcal{B} \Longrightarrow \widetilde{y} - y \in \mathcal{B}^0 \underset{2.(a)}{\Longrightarrow} \widetilde{y} - y \text{ bounded for } t \to \infty$$
$$\Longrightarrow \widetilde{y} = y + (\widetilde{y} - y) \text{ bounded for } t \to \infty \Longrightarrow \text{ condition 1.}$$

□

8.2.7 The General Laplace Transform

Theorem 8.2.82 enables the extension of the Laplace transform to more general distributions and to prove its properties as already indicated in Sect. 2.3. We define the $\mathbb{C}$-spaces

$$\begin{aligned} \mathfrak{A}_+^{0,\,\mathrm{pc}} &:= \left\{u \in \mathrm{C}_+^{0,\mathrm{pc}}(\mathbb{R}, \mathbb{C});\ \exists \sigma > 0 \text{ such that } ue^{-\sigma t} \in \mathrm{L}_+^\infty\right\}, \\ \mathfrak{A}_+^0 &:= \mathfrak{A}_+^{0,\,\mathrm{pc}} \cap \mathrm{C}^0 \text{ where } \mathrm{L}_+^\infty = \left\{u \in \mathrm{C}_+^{0,\mathrm{pc}};\ \sup_{t \in \mathbb{R}} |u(t)| < \infty\right\}. \end{aligned} \tag{8.111}$$

For $u \in \mathfrak{A}_+^{0,\mathrm{pc}}$ and $s \in \mathbb{C}$ with $\Re(s) > \sigma$ this implies

$$\begin{aligned} &|u(t)e^{-st}| = |u(t)e^{-\sigma t}|\, e^{-(\Re(s)-\sigma)t} \in L_+^1 \text{ and } u(\infty)e^{-s\infty} := \lim_{t\to\infty} u(t)e^{-st} = 0 \\ &\quad \text{where } L_+^1 := \left\{u \in \mathrm{C}_+^{0,\mathrm{pc}};\ \int_{-\infty}^{\infty} |u(t)|dt < \infty\right\} \\ &\Longrightarrow \widetilde{\mathcal{L}}(u)(s) := \int_{-\infty}^{\infty} u(t)e^{-st}dt \text{ holomorphic for } s \in \mathbb{C},\ \Re(s) > \sigma. \end{aligned} \tag{8.112}$$

The holomorphic function $\widetilde{\mathcal{L}}(u)(s)$ is the *Laplace transform* of u as usually defined, but we assume left bounded support of u only instead of $\mathrm{supp}(u) \subseteq [0, \infty)$. Recall

$$(s - \lambda)^{-n} \circ \delta = \tfrac{t^{n-1}}{(n-1)!}e^{\lambda t}Y \in \mathfrak{A}_+^{0,\mathrm{pc}},\ \mathcal{L}((s - \lambda)^{-n} \circ \delta) = (s - \lambda)^{-n}.$$

For $s \in \mathbb{C}$ with $\Re(s) > \max(0, \Re(\lambda))$ we get

$$\widetilde{\mathcal{L}}((s - \lambda)^{-1} \circ \delta) = \int_0^\infty e^{-(s-\lambda)t}dt = -(s - \lambda)^{-1}[e^{-t(s-\lambda)}]_0^\infty = (s - \lambda)^{-1}.$$

Partial integration and induction give $\widetilde{\mathcal{L}}((s-\lambda)^{-n}\circ\delta)=(s-\lambda)^{-n}$ for $n>0$. Since any strictly proper H is a finite $\mathbb{C}$-linear combination of functions $(s-\lambda)^{-n},\ n>0,\ \lambda\in\mathbb{C}$, we conclude for strictly proper H

$$H\circ\delta=\alpha Y\in\mathfrak{A}_+^{0,\mathrm{pc}},\ \ \alpha\in\oplus_{\lambda\in\mathbb{C}}\mathbb{C}[t]e^{\lambda t},\ \ \widetilde{\mathcal{L}}(H\circ\delta)=H\quad (H \text{ strictly proper})\,. \tag{8.113}$$

Assume $u_1,u_2\in\mathfrak{A}_+^{0,\mathrm{pc}}$ with $u_k e^{-\sigma_k t}\in\mathrm{L}_+^\infty,\ k=1,2$. Theorem 8.2.82 implies

$$\begin{aligned}&\forall\sigma>\max(\sigma_1,\sigma_2):\ u_1e^{-\sigma t}*u_2e^{-\sigma t}\in\mathrm{L}_+^1\cap\mathrm{L}_+^\infty \text{ and } \left((u_1e^{-\sigma t})*(u_2e^{-\sigma t})\right)(t)\\&\quad=\int_{-\infty}^{\infty}u_1(x)e^{-\sigma x}u_2(t-x)e^{-\sigma(t-x)}dx=(u_1*u_2)(t)e^{-\sigma t}\in\mathrm{L}_+^1\cap\mathrm{L}_+^\infty\\&\Longrightarrow u_1*u_2\in\mathfrak{A}_+^0.\end{aligned} \tag{8.114}$$

Moreover, for $\Re(s)>\sigma$,

$$\begin{aligned}\widetilde{\mathcal{L}}(u_1*u_2)(s)&=\int_{\mathbb{R}(t)}e^{-st}\int_{\mathbb{R}(x)}u_1(x)u_2(t-x)\mathrm{d}x\mathrm{d}t\\&=\int_{\mathbb{R}(t)}\left(\int_{\mathbb{R}(x)}u_1(x)e^{-sx}u_2(t-x)e^{-s(t-x)}\mathrm{d}x\right)\mathrm{d}t\\&=\int_{\mathbb{R}(x)}u_1(x)e^{-sx}\left(\int_{\mathbb{R}(t)}u_2(t-x)e^{-s(t-x)}\mathrm{d}t\right)\mathrm{d}x\\&=\int_{\mathbb{R}(x)}u_1(x)e^{-sx}\left(\int_{\mathbb{R}(t)}u_2(t)e^{-st}\mathrm{d}t\right)\mathrm{d}x\\&=\left(\int_{\mathbb{R}(x)}u_1(x)e^{-sx}dx\right)\left(\int_{\mathbb{R}(t)}u_2(t)e^{-st}\mathrm{d}t\right).\end{aligned} \tag{8.115}$$

Notice that finite Riemann integrals are needed for the proof of (8.115) only. Summing up, we obtain the *exchange theorem*

$$\begin{aligned}&\forall u_1,u_2\in\mathfrak{A}_+^{0,\mathrm{pc}}\exists\sigma>0 \text{ such that for } s\in\mathbb{C},\ \Re(s)>\sigma:\\&\widetilde{\mathcal{L}}(u_1*u_2)(s)=\widetilde{\mathcal{L}}(u_1)(s)\widetilde{\mathcal{L}}(u_2)(s).\end{aligned} \tag{8.116}$$

For strictly proper $H\in\mathbb{C}(s)_{\mathrm{spr}}$ and $u\in\mathfrak{A}_+^{0,\mathrm{pc}}$ this implies

$$H\circ u=(H\circ\delta)*u\in\mathfrak{A}_+^{0,\mathrm{pc}} \text{ and } \widetilde{\mathcal{L}}(H\circ u)=\widetilde{\mathcal{L}}(H\circ\delta)\widetilde{\mathcal{L}}(u)=H\widetilde{\mathcal{L}}(u). \tag{8.117}$$

If $u\in\mathrm{C}_+^1$ is continuously differentiable and

$$\sigma > 0,\ ue^{-\sigma t}, u'e^{-\sigma t} \in \mathrm{L}^{\infty}_{+},\ u, u' \in \mathfrak{A}^{0}_{+},\ \text{then for } \Re(s) > \sigma$$

$$\begin{aligned}\widetilde{\mathcal{L}}(u')(s) &= \int_{-\infty}^{\infty} u'(t)e^{-st}dt = [u(t)e^{-st}]_{-\infty}^{\infty} + s\int_{-\infty}^{\infty} u(t)e^{-st}dt \\ &\underset{u(-\infty)=0,\ u(\infty)e^{-s\infty}=0}{=} s\int_{-\infty}^{\infty} u(t)e^{-st}dt \Longrightarrow \widetilde{\mathcal{L}}(s\circ u)(s) = s\widetilde{\mathcal{L}}(u)(s).\end{aligned} \tag{8.118}$$

Let, more generally,

$$\begin{aligned}& d > 0,\ u \in \mathrm{C}^{d}_{+},\ u, u^{(d)} = s^{d}\circ u \in \mathfrak{A}^{0}_{+} \\ & \Longrightarrow \text{for } 0 \le i \le d:\ u^{(i)} = s^{i}\circ u = s^{-(d-i)}\circ s^{d}\circ u \\ & \underset{(8.117)}{\Longrightarrow} s^{i}\circ u \in \mathfrak{A}^{0}_{+}\cap \mathrm{C}^{d-i},\ s^{i+1}\circ u = s\circ(s^{i}\circ u) \in \mathfrak{A}^{0}_{+}\cap \mathrm{C}^{d-i-1} \\ & \underset{(8.118)}{\Longrightarrow} \widetilde{\mathcal{L}}(s^{i+1}\circ u) = s\widetilde{\mathcal{L}}(s^{i}\circ u) \underset{\text{induction}}{\Longrightarrow} \widetilde{\mathcal{L}}(s^{d}\circ u) = s^{d}\widetilde{\mathcal{L}}(u).\end{aligned} \tag{8.119}$$

After these preparations we can prove the main result on the general Laplace transform.

Theorem 8.2.87 *Define*

$$\begin{aligned}\mathfrak{A}_{+} :&= \bigcup_{n=0}^{\infty} s^{n}\circ \mathfrak{A}^{0}_{+} \\ &= \left\{u \in \mathrm{C}^{-\infty}_{+};\ \exists n \ge 0 \exists \sigma > 0 \text{ with } s^{-n}\circ u \in \mathrm{C}^{0}_{+},\ e^{-\sigma t}s^{-n}\circ u \in \mathrm{L}^{\infty}_{+}\right\}.\end{aligned} \tag{8.120}$$

For $u \in \mathfrak{A}_{+}$ and $s^{-n}\circ u \in \mathfrak{A}^{0}_{+}$ define the holomorphic function $\widetilde{\mathcal{L}}(u)(s) := s^{n}\widetilde{\mathcal{L}}(s^{-n}\circ u)(s)$ in the open half-plane $\{s \in \mathbb{C};\ \Re(s) > \sigma\}$. The subset $\mathfrak{A}_{+} \subset \mathrm{C}^{-\infty}_{+}$ is called the space of Laplace transformable distributions *with left bounded support. Then*

1. *$\widetilde{\mathcal{L}}(u)$ is well defined, i.e., independent of the choice of $n \ge 0$, and coincides, for $u \in \mathfrak{A}^{0,\mathrm{pc}}_{+}$, with $\widetilde{\mathcal{L}}(u)$ from (8.112).*
2. *$\widetilde{\mathcal{L}}(H\circ\delta) = H = \mathcal{L}(H\circ\delta),\ H \in \mathbb{C}(s) \Longrightarrow \widetilde{\mathcal{L}}|_{\mathcal{F}_2} = \mathcal{L}$. Thus $\widetilde{\mathcal{L}}$ is an extension of $\mathcal{L}$ from Theorem 8.2.47, and we can thus write $\mathcal{L} := \widetilde{\mathcal{L}}$ in the sequel.*
3. *$\mathfrak{A}_{+}$ is a unital subalgebra of $\left(\mathrm{C}^{-\infty}_{+}, *\right)$, and the exchange theorem holds, i.e.,*

$$\begin{aligned}&\forall u_1, u_2 \in \mathfrak{A}_{+}\exists\sigma > 0 \text{ such that} \\ &\widetilde{\mathcal{L}}(u_1 * u_2)(s) = \widetilde{\mathcal{L}}(u_1)(s)\widetilde{\mathcal{L}}(u_2)(s) \text{ for } s \in \{z \in \mathbb{C};\ \Re(z) > \sigma\}.\end{aligned}$$

4. *$\mathfrak{A}_{+}$ is a $\mathbb{C}(s)$-subspace of $\mathrm{C}^{-\infty}_{+}$ and $\widetilde{\mathcal{L}}$ is $\mathbb{C}(s)$-linear, i.e.,*

$$\begin{aligned}&\forall H \in \mathbb{C}(s)\forall u \in \mathfrak{A}_{+}\exists\sigma > 0 \text{ such that} \\ &\widetilde{\mathcal{L}}(H\circ u)(s) = H(s)\widetilde{\mathcal{L}}(u)(s) \text{ for } s \in \{z \in \mathbb{C};\ \Re(z) > \sigma\}.\end{aligned}$$

5. *As usual the Laplace transform $\mathcal{L}$ is extended to matrices componentwise. For any IO behavior as in (8.110) and input $u \in \mathfrak{A}^{m}_{+}$ the unique solution $y = H\circ u \in$*

$(\mathbb{C}_+^{-\infty})^p$ of $P \circ y = Q \circ u$ belongs to $\mathfrak{A}_+^p$ and for some $\sigma > 0$

$$\mathcal{L}(y)(s) = H(s)\mathcal{L}(u)(s),\ P(s)\mathcal{L}(y)(s) = Q(s)\mathcal{L}(u)(s),\ s \in \{z \in \mathbb{C};\ \Re(z) > \sigma\}.$$

6. *The Laplace transform is injective, i.e., $\mathcal{L}(u) = 0$ implies $u = 0$. If $ue^{-\sigma t} \in \mathrm{L}_+^\infty$, if $\rho > \sigma$, and if $\mathcal{L}(u)(\rho + j\omega)$ is absolutely integrable as function of ω, then the following inversion formula holds:*

$$u(t) = (2\pi j)^{-1} \int_{\rho - j\infty}^{\rho + j\infty} \mathcal{L}(u)(s)e^{st}\,ds. \tag{8.121}$$

The proof of 6 by means of the Fourier inversion formula for functions (not for distributions) will be given in Theorem 10.2.51.

Proof 1. Assume

$$\begin{aligned}
&s^{-n_k} \circ u \in \mathfrak{A}_+^0,\ k = 1, 2,\ n_1 < n_2\\
&\Longrightarrow s^{-n_2} \circ u = s^{n_1 - n_2} \circ s^{-n_1} \circ u \in \mathbb{C}^{n_2 - n_1},\ s^{n_2 - n_1} \circ s^{-n_2} \circ u = s^{-n_1} \circ u\\
&\underset{(8.119)}{\Longrightarrow} s^{n_2 - n_1}\widetilde{\mathcal{L}}(s^{-n_2} \circ u) = \widetilde{\mathcal{L}}(s^{-n_1} \circ u)\\
&\Longrightarrow s^{n_1}\widetilde{\mathcal{L}}(s^{-n_1} \circ u) = s^{n_1}s^{n_2 - n_1}\widetilde{\mathcal{L}}(s^{-n_2} \circ u) = s^{n_2}\widetilde{\mathcal{L}}(s^{-n_2} \circ u).
\end{aligned}$$

This establishes the asserted independence. Moreover $s^{-n} \circ u \in \mathfrak{A}_+^0$ implies $s^{-(n+k)} \circ u \in \mathfrak{A}_+^0$ and $\widetilde{\mathcal{L}}(u) = s^{n+k}\widetilde{\mathcal{L}}(s^{-(n+k)}u)$. Also see the end of 2.
2. From (8.113) we know $\widetilde{\mathcal{L}}(H \circ \delta) = H$ for strictly proper H. Since $H = H_{\mathrm{pol}} + H_{\mathrm{spr}}$ it remains to show $\widetilde{\mathcal{L}}(s^k \circ \delta) = s^k$ for $k \geq 0$, see 3 (i). But

$$\begin{aligned}
s^{-2} \circ \delta = tY \in \mathfrak{A}_+^0 \Longrightarrow \widetilde{\mathcal{L}}(s^k \circ \delta) &= \widetilde{\mathcal{L}}(s^{k+2} \circ (s^{-2} \circ \delta))\\
&= s^{k+2}\widetilde{\mathcal{L}}(s^{-2} \circ \delta) = s^{k+2}s^{-2} = s^k.
\end{aligned}$$

If $u \in \mathfrak{A}_+^{0,\mathrm{pc}}$ then $s^{-2} \circ u = (s^{-2} \circ \delta) * u \in \mathfrak{A}_+^0$ and hence

$$\widetilde{\mathcal{L}}(u) = s^2\widetilde{\mathcal{L}}\left((s^{-2} \circ \delta) * u\right) \underset{(8.117)}{=} s^2 s^{-2}\widetilde{\mathcal{L}}(u) = \widetilde{\mathcal{L}}(u).$$

This proves the last assertion of 1.
3. (i) For $u_k \in \mathfrak{A}_+$, $k = 1, 2$, choose, by 1, the same $n \geq 0$ with $s^{-n} \circ u_k \in \mathfrak{A}_+^0$. Then

$$\begin{aligned}
&s^{-n} \circ (u_1 + u_2) = s^{-n} \circ u_1 + s^{-n} \circ u_2 \in \mathfrak{A}_+^0 \Longrightarrow u_1 + u_2 \in \mathfrak{A}_+ \text{ and}\\
&\widetilde{\mathcal{L}}(u_1 + u_2) = s^n\left(\widetilde{\mathcal{L}}(s^{-n} \circ u_1) + \widetilde{\mathcal{L}}(s^{-n} \circ u_2)\right) = \widetilde{\mathcal{L}}(u_1) + \widetilde{\mathcal{L}}(u_2).
\end{aligned}$$

Hence $\mathfrak{A}_+$ is a $\mathbb{C}$-subspace of $\mathbb{C}_+^{-\infty}$ and $\widetilde{\mathcal{L}}$ id $\mathbb{C}$-linear on $\mathfrak{A}_+$. A special case was used in 2 already.

(ii) Let

$$
\begin{aligned}
& u_k \in \mathfrak{A}_+,\ k=1,2, \text{ and } s^{-n_k}\circ u_k \in \mathfrak{A}_+^0 \\
& \underset{(8.116)}{\Longrightarrow} s^{-(n_1+n_2)}\circ(u_1 * u_2) = (s^{-n_1}\circ u_1) * (s^{-n_2}\circ u_2) \in \mathfrak{A}_+^0 \text{ and} \\
& \widetilde{\mathcal{L}}\left(s^{-(n_1+n_2)}\circ(u_1 * u_2)\right) = \widetilde{\mathcal{L}}(s^{-n_1}\circ u_1)\widetilde{\mathcal{L}}(s^{-n_2}\circ u_2) \\
& \Longrightarrow \widetilde{\mathcal{L}}(u_1 * u_2) = s^{n_1+n_2}\widetilde{\mathcal{L}}\left(s^{-(n_1+n_2)}\circ(u_1 * u_2)\right) \\
& = \left(s^{n_1}\widetilde{\mathcal{L}}(s^{-n_1}\circ u_1)\right)\left(s^{n_2}\widetilde{\mathcal{L}}(s^{-n_2}\circ u_2)\right) = \widetilde{\mathcal{L}}(u_1)\widetilde{\mathcal{L}}(u_2).
\end{aligned}
$$

This says that $\mathfrak{A}_+$ is a subalgebra of $\left(\mathrm{C}_+^{-\infty}, *\right)$ and that $\widetilde{\mathcal{L}}$ is multiplicative on $\mathfrak{A}_+$.
4. For $H \in \mathbb{C}(s)$ and $u \in \mathfrak{A}_+$ items 2 and 3 imply

$$
\begin{aligned}
& H\circ\delta \in \mathfrak{A}_+,\ H\circ u = (H\circ\delta) * u \in \mathfrak{A}_+ \text{ and} \\
& \widetilde{\mathcal{L}}(H\circ u) = \widetilde{\mathcal{L}}((H\circ\delta) * u) \underset{3.}{=} \widetilde{\mathcal{L}}(H\circ\delta)\widetilde{\mathcal{L}}(u) \underset{2.}{=} H\widetilde{\mathcal{L}}(u).
\end{aligned}
$$

This says that $\mathfrak{A}_+$ is a $\mathbb{C}(s)$-subspace of $\mathrm{C}_+^{-\infty}$ and $\mathcal{L} := \widetilde{\mathcal{L}}$ is $\mathbb{C}(s)$-linear.
5. Obvious. □

For the actual computation of $y = H\circ u$ the expressions $y = HH_2\circ\delta$ for $u = H_2\circ\delta$ or, more generally, $y = H\circ u = (H\circ\delta) * u$ are simpler than (8.121) applied to y, i.e., for suitable $\rho > 0$ and under suitable assumptions,

$$
y(t) = (2\pi j)^{-1}\int_{\rho-j\infty}^{\rho+j\infty} H(s)\mathcal{L}(u)(s)e^{st}ds.
$$

8.2.8 *Periodic Distributions and Fourier Series*

In this section we generalize Corollaries 6.3.24 and 6.3.25 to infinite Fourier series $u = \sum_{\mu\in\mathbb{Z}} u_\mu e^{j\mu\omega t}$ and present the theory of the title. The following derivations are a simplified form of [3, pp. 224–231]. Theorems 8.2.102, 8.2.107, and 8.2.110 are the main results for applications in electrical and mechanical engineering.
We consider the base field $\mathbb{C}$, a frequency $\omega > 0$, and the period $T := 2\pi\omega^{-1}$. Let $\mathcal{P}^0 := \mathcal{P}(T)$ denote the space of T-periodic continuous functions $u : \mathbb{R} \to \mathbb{C}$, i.e., with $T\circ_s u = u$, where $(T\circ_s u)(t) = u(t-T)$, and, for all $d \geq 0$, $\mathcal{P}^d := \mathcal{P}^0 \cap \mathrm{C}^d(\mathbb{R},\mathbb{C})$. Let $S^1 := \{z \in \mathbb{C};\ |z| = 1\}$ be the unit circle. Notice $\overline{z} = z^{-1}$ for $z \in S^1$. The unit circle is a compact topological space and gives rise to the commutative normed algebra $\mathrm{C}^0(S^1,\mathbb{C})$ of continuous functions $g : S^1 \to \mathbb{C}$ with the norm $\|g\|_\infty := \max_{z\in S^1}|g(z)|$. The corresponding topology is the uniform one, and therefore $\mathrm{C}^0(S^1,\mathbb{C})$ is complete or a *Banach algebra*. As usual, the letter z is used both for the elements $z \in S^1$ and the injection $z := \mathrm{inj} : S^1 \overset{\subseteq}{\to} \mathbb{C}$.
The continuous surjection $\epsilon :\ \mathbb{R} \to S^1,\ t \mapsto e^{j\omega t}$, is a group epimorphism with ker-

nel $\mathbb{Z}T$ and a local homeomorphism, indeed a homeomorphism on each open interval (t_0, t_1) with $t_0 < t_1 < t_0 + T$.

Lemma 8.2.88 *The map* $C^0(S^1, \mathbb{C}) \to \mathcal{P}^0$, $g \mapsto g\epsilon$, *is a norm-preserving algebra isomorphism, hence* $\mathcal{P}^0$ *can be studied via* $C^0(S^1, \mathbb{C})$.

Proof (i) The map is obviously well defined and an algebra monomorphism.
(ii) *The map is surjective*: Let $f \in \mathcal{P}^0, z \in S^1$ and $\epsilon(t) = \epsilon(t') = z$. Then $t' = t + \mu T$, $\mu \in \mathbb{Z}$, and hence $f(t') = f(t + \mu T) = f(t)$. Thus $g(z) := f(t)$ is well defined, and obviously $g = f\epsilon$. Since ϵ is a local homeomorphism, the continuity of f implies that of g.
(iii) $\|f\|_\infty := \max_{t\in\mathbb{R}} |f(t)| = \max_{t\in[0,T]} |f(t)| = \max_{z\in S^1} |g(z)| = \|g\|_\infty$. □

Result 8.2.89 (Cf. [Wikipedia: https://en.wikipedia.org/wiki/Stone-Weierstrass_theorem, September 8, 2019]) *Consider a compact topological space S and a (unital) subalgebra* $A \subseteq C^0(S, \mathbb{R})$. *If A separates the points of S, i.e.,*

$$\forall z, z' \in S \text{ with } z \neq z' \exists g \in A \text{ with } g(z) \neq g(z'),$$

then A is dense in $C^0(S, \mathbb{R})$, *i.e., for all* $g \in C^0(S, \mathbb{R})$ *and* $\epsilon > 0$ *there is* $g_1 \in A$ *with* $\|g - g_1\|_\infty \leq \epsilon$.
The proof of this important theorem is simple and standard. ◊

Corollary 8.2.90 (Stone-Weierstraß, complex case) *In the situation of Result 8.2.89 assume that B is a* $\mathbb{C}$*-subalgebra of* $C^0(S, \mathbb{C})$ *that separates the points. In addition assume that B is closed under conjugation, i.e.,* $(g \in B \Longrightarrow \overline{g} \in B)$ *or* $B = \overline{B}$ *where* $\overline{g}(z) := \overline{g(z)}$. *Then B is dense in* $C^0(S, \mathbb{C})$.

Proof Obviously $C^0(S, \mathbb{R})$ is an $\mathbb{R}$-subalgebra of $C^0(S, \mathbb{C})$ and the real part induces an $\mathbb{R}$-epimorphism $\Re : C^0(S, \mathbb{C}) \to C^0(S, \mathbb{R})$. The equations $\Re(g) = 2^{-1}(g + \overline{g})$ and $\Re(g_1)\Re(g_2) = 2^{-1}\Re(\overline{g_1}g_2 + g_1g_2)$ imply that $A := \Re(B)$ is an $\mathbb{R}$-subalgebra of B, hence $A = \{a \in B;\ \Re(a) = a\}$. The equation $\Im(g) = \Re(-jg)$ implies $\Im(B) \subseteq A$. If

$$\begin{aligned} &z, z' \in S,\ z \neq z', g = \Re(g) + j\Im(g) \in B,\ \Re(g), \Im(g) \in A, \text{ and } g(z) \neq g(z') \\ &\Longrightarrow \Re(g)(z) \neq \Re(g)(z') \text{ or } \Im(g)(z) \neq \Im(g)(z'). \end{aligned}$$

Hence also A separates the points of S, and thus is dense in $C^0(S, \mathbb{R})$ by Result 8.2.89. Let $\epsilon > 0$ and

$$\begin{aligned} &g = g_1 + jg_2 \in C^0(S, \mathbb{C}),\ g_i \in C^0(S, \mathbb{R}), \text{ and } f_i \in A,\ i = 1, 2, \text{ with} \\ &\|g_i - f_i\|_\infty \leq \epsilon/2 \Longrightarrow f := f_1 + jf_2 \in B \text{ and} \\ &\|g - f\|_\infty \leq \|g_1 - f_1\|_\infty + \|g_2 - f_2\|_\infty \leq \epsilon/2 + \epsilon/2 = \epsilon. \end{aligned}$$

This shows that B is dense in $C^0(S, \mathbb{C})$. □

In the obvious fashion we consider the complex *Laurent polynomial algebra* $B := \mathbb{C}[z, z^{-1}] = \oplus_{\mu\in\mathbb{Z}}\mathbb{C}z^{\mu}$ as subalgebra of $C^0(S^1, \mathbb{C})$. The function $z : S^1 \to \mathbb{C},\ z \mapsto z$, obviously separates points. Recall $\overline{z} = z^{-1}$ for $z \in S^1$. Then

$$\begin{aligned}
&\overline{\sum_{\mu\in\mathbb{Z}} u_\mu z^\mu} = \sum_{\mu\in\mathbb{Z}} \overline{u_\mu}\overline{z}^\mu = \sum_{\mu\in\mathbb{Z}} \overline{u_\mu} z^{-\mu} = \sum_{\mu\in\mathbb{Z}} \overline{u_{-\mu}} z^\mu \in \mathbb{C}[z, z^{-1}] \text{ and} \\
&\Re(\mathbb{C}[z, z^{-1}]) = \left\{\sum_{\mu\in\mathbb{Z}} u_\mu z^\mu;\ \forall \mu \in \mathbb{Z} :\ u_{-\mu} = \overline{u_\mu}\right\}.
\end{aligned} \tag{8.122}$$

According to Corollary 8.2.90 the algebra $B = \mathbb{C}[z, z^{-1}]$ is dense in $C^0(S^1, \mathbb{C})$. The $\mathbb{C}$-subspace $B' := \oplus_{\mu\in\mathbb{Z}}\mathbb{C}e^{j\mu\omega t} \subset \mathcal{P}^0$ is obviously a $\mathbb{C}$-subalgebra, and the isomorphism $\mathcal{P}^0 \cong C^0(S^1, \mathbb{C})$ from Lemma 8.2.88 induces the isomorphism $B' \cong B,\ e^{j\mu\omega t} \mapsto z^\mu$. Corollary 8.2.90 thus implies the following.

Corollary 8.2.91 *The subalgebra* $\oplus_{\mu\in\mathbb{Z}}\mathbb{C}e^{j\mu\omega t}$ *of* $\mathcal{P}^0$ *of* finite Fourier or trigonometric series *is closed under conjugation and dense in* $\mathcal{P}^0$*. Its real part of* finite trigonometric polynomials *is a dense subalgebra of* $\mathcal{P}^0_{\mathbb{R}} := \mathcal{P}^0 \cap C^0(\mathbb{R}, \mathbb{R})$*, and has the form*

$$\begin{aligned}
\Re\left(\oplus_{\mu\in\mathbb{Z}}\mathbb{C}e^{j\mu\omega t}\right) &= \left\{u = \sum_{\mu\in\mathbb{Z}} u_\mu e^{j\mu\omega t};\ \forall \mu \in \mathbb{Z};\ \overline{u_\mu} = u_{-\mu}\right\} \\
&= \mathbb{R} \oplus \oplus_{\mu=1}^{\infty} (\mathbb{R}\cos(\mu\omega t) \oplus \mathbb{R}\sin(\mu\omega t)) \text{ where} \\
a_0 := u_0,\ a_\mu - jb_\mu &:= 2u_\mu,\ a_0, a_\mu, b_\mu \in \mathbb{R}, \text{ and} \\
u = \sum_{\mu\in\mathbb{Z}} u_\mu e^{j\mu\omega t} &= a_0 + \sum_{\mu=1}^{\infty}(a_\mu\cos(\mu\omega t) + b_\mu\sin(\mu\omega t)),
\end{aligned} \tag{8.123}$$

Proof Only the last equation has yet to be shown. But $u_0 = u_{-0} = \overline{u_0}$ implies that $a_0 = u_0$ is real and $u_0 e^{j0\omega t} = a_0 \in \mathbb{R}$. For $\mu \geq 1$ we get

$$\begin{aligned}
&u_\mu e^{j\mu\omega t} + u_{-\mu}e^{-j\mu\omega t} = u_\mu e^{j\mu\omega t} + \overline{u_\mu e^{j\mu\omega t}} \\
&= 2\Re(u_\mu e^{j\mu\omega t}) = 2\Re(2^{-1}(a_\mu - jb_\mu)e^{j\mu\omega t}) = a_\mu\cos(\mu\omega t) + b_\mu\sin(\mu\omega t) \\
&\Longrightarrow u = \sum_{\mu\in\mathbb{Z}} u_\mu e^{j\mu\omega t} = u_0 + \sum_{\mu=1}^{\infty}\left(u_\mu e^{j\mu\omega t} + u_{-\mu}e^{-j\mu\omega t}\right) \\
&= a_0 + \sum_{\mu=1}^{\infty}(a_\mu\cos(\mu\omega t) + b_\mu\sin(\mu\omega t)).
\end{aligned}$$

□

Let $\mathcal{P}^{0,\mathrm{pc}} \subset C^{0,\mathrm{pc}}(\mathbb{R}, \mathbb{C})$ denote the space of piecewise continuous T-periodic functions. On $\mathcal{P}^{0,\mathrm{pc}}$ we use the scalar product $\langle f, g\rangle = T^{-1}\int_0^T \overline{f(t)}g(t)dt$ from (6.116) and define the $\mathbb{C}$-linear Fourier coefficients and transform

$$\begin{aligned}
&\mathbb{F}: \mathcal{P}^{0,\mathrm{pc}} \to \mathbb{C}^{\mathbb{Z}},\ u \mapsto \mathbb{F}(u) = (\mathbb{F}(u)(\mu))_{\mu\in\mathbb{Z}},\\
&\mathbb{F}(u)(\mu) := \langle e^{j\mu\omega t}, u\rangle = T^{-1}\int_0^T e^{-j\mu\omega t}u(t)dt.
\end{aligned} \tag{8.124}$$

We also have to apply the analogues of the Banach spaces in Lemma 8.1.21, i.e.,

$$\begin{aligned}
\mathbb{C}^{(\mathbb{Z})} \subset \ell^1(\mathbb{Z}) &:= \left\{x \in \mathbb{C}^{\mathbb{Z}};\ \|x\|_1 := \sum_{\mu\in\mathbb{Z}} |x(\mu)| < \infty\right\}\\
\subset \ell^2(\mathbb{Z}) &:= \left\{x \in \mathbb{C}^{\mathbb{Z}};\ \|x\|_2 = \left(\sum_{\mu\in\mathbb{Z}} |x(\mu)|^2\right)^{1/2} < \infty\right\}\\
\subset \ell^\infty(\mathbb{Z}) &:= \left\{x \in \mathbb{C}^{\mathbb{Z}};\ \|x\|_\infty = \sup_{\mu\in\mathbb{Z}} |x(\mu)| < \infty\right\}.
\end{aligned} \tag{8.125}$$

For $x \in \mathbb{C}^{\mathbb{Z}}$ let $\overline{x}(\mu) := \overline{x(\mu)}$. Recall that $\ell^2(\mathbb{Z})$ is a Hilbert space with the scalar product $\langle x, y\rangle := \sum_{\mu\in\mathbb{Z}} \overline{x}(\mu)y(\mu)$ and the norm $\|x\|_2 = \langle x, x\rangle^{1/2}$. Schwarz' inequality $|\langle x, y\rangle| \le \|x\|_2\|y\|_2$ holds and implies

$$xy \in \ell^1,\ \sum_{\mu\in\mathbb{Z}} |x(\mu)||y(\mu)| \le \|x\|_2\|y\|_2. \tag{8.126}$$

Theorem 8.2.92 *For* $u, v \in \mathcal{P}^{0,\mathrm{pc}}$ Parzeval's equation *holds, i.e.,*

$$\begin{aligned}
&\mathbb{F}(u) \in \ell^2(\mathbb{Z}),\ \|\mathbb{F}(u)\|_2 = \|u\|_2,\ \langle u, v\rangle = \langle \mathbb{F}(u), \mathbb{F}(v)\rangle.\ \textit{Moreover}\\
&\lim_{n\to\infty} \|u - \sum_{\mu=-n}^{n} \mathbb{F}(u)(\mu)e^{j\mu\omega t}\|_2 = 0\ \textit{or}\ u = \sum_{\mu\in\mathbb{Z}} \mathbb{F}(u)(\mu)e^{j\mu\omega t}
\end{aligned} \tag{8.127}$$

where the last sum converges with respect to $\| - \|_2$. *Hence* $\mathbb{F}$ *induces an, automatically injective, isometry* $\mathbb{F}: \mathcal{P}^{0,\mathrm{pc}} \to \ell^2(\mathbb{Z})$.

Proof For $n \ge 0$ let $f_n := \sum_{\mu=-n}^{n} \mathbb{F}(u)(\mu)e^{j\mu\omega t}$. According to Corollary 6.3.25 we obtain

$$\|f_n\|_2^2 = \sum_{\mu=-n}^{n} |\mathbb{F}(u)(\mu)|^2, \; u - f_n \perp f_n \underset{\text{Pythagoras}}{\Longrightarrow} \|f_n\|_2^2 + \|u - f_n\|_2^2 = \|u\|_2^2$$
$$\Longrightarrow \sum_{\mu=-n}^{n} |\mathbb{F}(u)(\mu)|^2 \leq \|u\|_2^2$$
$$\Longrightarrow \|\mathbb{F}(u)\|_2^2 = \sum_{\mu\in\mathbb{Z}} |\mathbb{F}(u)(\mu)|^2 \leq \|u\|_2^2, \; \mathbb{F}(u) \in \ell^2(\mathbb{Z}), \; \|\mathbb{F}(u)\|_2 \leq \|u\|_2.$$

Assume that $\epsilon := \|u\|_2 - \|\mathbb{F}(u)\|_2 > 0$. Obviously $\mathcal{P}^0$ is $\| - \|_2$-dense in $\mathcal{P}^{0,\mathrm{pc}}$. Since $\oplus_\mu \mathbb{C}e^{j\mu\omega t}$ is $\| - \|_\infty$-dense in $\mathcal{P}^0$, it is also $\| - \|_2$-dense in $\mathcal{P}^0$ and hence in $\mathcal{P}^{0,\mathrm{pc}}$. Thus there is an f_n as above such that $\|u - f_n\|_2 \leq \epsilon/2$. This implies the contradiction

$$\|\mathbb{F}(u)\|_2 \geq \|f_n\|_2 \geq \|u\|_2 - \|u - f_n\|_2 \geq \|u\|_2 - \epsilon/2 > \|\mathbb{F}(u)\|_2.$$

Hence $\|\mathbb{F}(u)\|_2 = \|u_2\|$. The remaining assertions follow easily. □

Remark 8.2.93 (Finite) trigonometric series $\Re\left(\sum_{\mu\in\mathbb{Z}} \widehat{u}(\mu)e^{j\mu\omega t}\right)$, $\widehat{u} \in \mathbb{C}^{(\mathbb{Z})}$, can be easily generated in electrical engineering. This is not the case for signals in form of *infinite* Fourier series. Therefore the preceding theorem is used in practical situations to approximate an arbitrary $u \in \mathcal{P}_{\mathbb{R}}^{0,c}$ by realizable signals $\Re\left(\sum_{\mu=-n}^{n} \mathbb{F}(u)(\mu)e^{j\mu\omega t}\right)$. ◊

Lemma 8.2.94 (Cf. [Wikipedia: https://de.wikipedia.org/wiki/Fourierreihe, September 8, 2019]) *Let $u \in \mathcal{P}^{0,\mathrm{pc}}$, hence $\mathbb{F}(u) \in \ell^2(\mathbb{Z})$.*

1. If $\mathbb{F}(u) \in \ell^1(\mathbb{Z})$, in particular if $\mathbb{F}(u)(\mu) = O(\mu^{-2})$, then

$$\lim_{n\to\infty} \|u - \sum_{\mu=-n}^{n} \mathbb{F}(u)(\mu)e^{j\mu\omega t}\|_\infty = 0, \text{ hence } u = \sum_{\mu\in\mathbb{Z}} \mathbb{F}(u)(\mu)e^{j\mu\omega t} \in \mathcal{P}^0. \tag{8.128}$$

So in this case the Fourier series converges uniformly.

2. If $u \in \mathcal{P}^1$, then

$$\mathbb{F}(s \circ u)(\mu) = j\mu\omega\mathbb{F}(u)(\mu), \; \mathbb{F}(u) \in \ell^1(\mathbb{Z}). \tag{8.129}$$

3. If $u \in \mathcal{P}^2$, then

$$\mathbb{F}\left((1 - s^2) \circ u\right)(\mu) \underset{(ii)}{=} (1 + \mu^2\omega^2)\mathbb{F}(u)(\mu). \tag{8.130}$$

Proof The proofs are simple and standard.
1. Let $\widehat{u} := \mathbb{F}(u)$. Due to $\sum_\mu |\widehat{u}(\mu)| < \infty$ the sum $v := \sum_\mu \widehat{u}(\mu)e^{j\mu\omega t}$ is uniformly convergent in $\mathcal{P}^0$ and thus continuous. Moreover

$$\mathbb{F}(v)(\nu) = \langle e^{j\nu\omega t}, v\rangle = \sum_\mu \widehat{u}(\mu)\langle e^{j\nu\omega t}, e^{j\mu\omega t}\rangle = \sum_\mu \widehat{u}(\mu)\delta_{\nu,\mu} = \widehat{u}(\nu)$$

$$\Longrightarrow \mathbb{F}(u) = \mathbb{F}(v) \underset{\mathbb{F}|_{\mathcal{P}^{0,\mathrm{pc}}}\text{ injective}}{\Longrightarrow} u = v = \sum_\mu \widehat{u}(\mu)e^{j\mu\omega t}.$$

If $\mathbb{F}(u)(\mu) = O(\mu^{-2})$, i.e., $|\mathbb{F}(u)(\mu)| \leq M\mu^{-2},\ M > 0, \mu \neq 0$, then

$$\sum_{\mu\in\mathbb{Z}} |\mathbb{F}(u)(\mu)| \leq |\mathbb{F}(u)(0)| + M\sum_{\mu\neq 0}\mu^{-2} < \infty \Longrightarrow \mathbb{F}(u) \in \ell^1(\mathbb{Z}).$$

2. For $\mu = 0$ we have $\mathbb{F}(s\circ u)(0) = T^{-1}\int_0^T u'(t)dt = T^{-1}(u(T) - u(0)) = 0$. For $\mu \neq 0$ partial integration of $\mathbb{F}(u')(\mu) = T^{-1}\int_0^T e^{-j\mu\omega t}u'(t)dt$ implies the assertion, hence $|\mathbb{F}(u)(\mu)| = |(j\mu\omega)|^{-1}|\mathbb{F}(s\circ u)(\mu)|$ for $\mu \neq 0$. Define

$$x, y \in \ell^2(\mathbb{Z}) :\ x(\mu) := \begin{cases} |\mathbb{F}(u)(0)| & \text{if } \mu = 0 \\ |(j\mu\omega)|^{-1} & \text{if } \mu \neq 0 \end{cases},\ y(\mu) := \begin{cases} 1 & \text{if } \mu = 0 \\ |\mathbb{F}(s\circ u)(\mu)| & \text{if } \mu \neq 0 \end{cases}$$

$$\Longrightarrow |\mathbb{F}(u)(\mu)| = x(\mu)y(\mu) \underset{(8.126)}{\Longrightarrow} \mathbb{F}(u) \in \ell^1(\mathbb{Z}).$$

3. Obvious. □

Remark 8.2.95 (Cf. [Wikipedia: https://de.wikipedia.org/wiki/Fourierreihe, September 8, 2019]) Lemma 8.2.94 covers the convergence of most Fourier series in applications, cf. Theorem 8.2.110. In that theorem u is given as a piecewise continuous, periodic input of an IO behavior with strictly proper transfer matrix. The Fourier series of the periodic output y is computed from $\mathbb{F}(u)$ and is indeed uniformly convergent and continuous according to Lemma 8.2.94, 1, whereas, for general $u \in \mathcal{P}^{0,\mathrm{pc}}$, we proved $\|-\|_2$-convergence of $u = \sum_\mu \mathbb{F}(u)(\mu)e^{j\mu\omega t}$ only. There is a famous and difficult result, Carleson's theorem [loc.cit.], that states that for a Lebesgue square-integrable, in particular for a piecewise continuous, periodic function u the equation $u(t) = \lim_{n\to\infty}\sum_{\mu=-n}^{n}\mathbb{F}(u)(\mu)e^{j\mu\omega t}$ holds for almost all t. There are various earlier results of this type. For practical signals this deep result is without significance. ◇

Lemma 8.2.94, 2, suggests to make the $\mathbb{C}$-algebra $\mathbb{C}^{\mathbb{Z}}$ (with the componentwise multiplication) a $\mathbb{C}[s]$-module via

$$s\circ : \mathbb{C}^{\mathbb{Z}} \to \mathbb{C}^{\mathbb{Z}},\ x \mapsto (s\circ x)(\mu) := j\mu\omega x(\mu). \tag{8.131}$$

Recall that for $x \in \mathbb{C}^{\mathbb{Z}}$ and $k \in \mathbb{Z}$ the condition $x(\mu) = O((1+\mu^2\omega^2)^k)$ means that there is $M > 0$ such that $|x(\mu)| \leq M((1+\mu^2\omega^2)^k$ for all $\mu \in \mathbb{Z}$. Item (iii) of the lemma suggests to define the increasing sequence of subspaces

$$\cdots \subset \mathfrak{s}^{-2d} := \left\{x \in \mathbb{C}^{\mathbb{Z}};\ x(\mu) = O((1+\mu^2\omega^2)^d)\right\} \subset \mathfrak{s}^{-2(d+1)} \subset \cdots \subset \mathbb{C}^{\mathbb{Z}}, d \in \mathbb{Z}, \tag{8.132}$$

and their union

$$\mathfrak{s}^{-\infty} := \bigcup_{d\in\mathbb{Z}} \mathfrak{s}^{-2d} = \bigcup_{d=0}^{\infty} \mathfrak{s}^{-2d} = \left\{x \in \mathbb{C}^{\mathbb{Z}};\ \exists k \geq 0 \text{ with } |x(\mu)| = O(|\mu|^k)\right\}. \tag{8.133}$$

The sequences in $\mathfrak{s}^{-\infty}$ grow at most polynomially and are called *slowly increasing*. The space $\mathfrak{s}^{-\infty}$ is obviously a subalgebra of $\mathbb{C}^{\mathbb{Z}}$. The space $\mathfrak{s}^{-\infty}$ contains

$$\begin{aligned} &\ell^1(\mathbb{Z}) \subset \ell^2(\mathbb{Z}) \subset \ell^\infty(\mathbb{Z}) \subset \mathfrak{s}^{-\infty} \text{ with} \\ &\ell^\infty \ell^i \subseteq \ell^i,\ i = 1, 2, \infty,\quad \ell^2(\mathbb{Z})\ell^2(\mathbb{Z}) \subseteq \ell^1(\mathbb{Z}), \end{aligned} \tag{8.134}$$

the last inclusion following from (8.126). We define the subalgebra $\mathbb{C}(s)_{\text{per}}$ of $\mathbb{C}(s)$ of all rational functions without poles in $\mathbb{Z}j\omega$, i.e.,

$$\begin{aligned} \mathbb{C}(s)_{\text{per}} &:= \{H \in \mathbb{C}(s);\ \text{pole}(H) \cap \mathbb{Z}j\omega = \emptyset\} \\ &= \left\{fg^{-1};\ f, g \in \mathbb{C}[s],\ V_{\mathbb{C}}(g) \subset \mathbb{C}\setminus\mathbb{Z}j\omega\right\} \\ &= \mathbb{C}[s] \bigoplus \oplus_{\lambda\in\mathbb{C}\setminus\mathbb{Z}j\omega} \oplus_{k=1}^{\infty} \mathbb{C}(s-\lambda)^{-k}, \end{aligned} \tag{8.135}$$

the last equation following from the partial fraction decomposition. In Theorem 9.2.2, it will be shown that $\mathbb{C}(s)_{\text{per}}$ is a principal ideal domain with the primes $s - j\mu\omega,\ \mu \in \mathbb{Z}$. By definition $H(j\mu\omega) \in \mathbb{C}$ is defined for all $H \in \mathbb{C}(s)_{\text{per}}$ and $\mu \in \mathbb{Z}$. It is obvious that

$$\forall H \in \mathbb{C}[s]:\ h := (H(j\mu\omega))_{\mu\in\mathbb{Z}} \in \mathfrak{s}^{-\infty} \text{ and } \forall \lambda \notin \mathbb{Z}j\omega,\ \forall k \geq 0,\ H := (s-\lambda)^{-k}:$$

$$h = (H(j\mu\omega))_{\mu\in\mathbb{Z}} = \left((j\mu\omega - \lambda)^{-k}\right)_{\mu\in\mathbb{Z}} \in \begin{cases} \ell^\infty & \text{if } k - 0 \\ \ell^2(\mathbb{Z}) & \text{if } k = 1 \text{ since } \sum_{0\neq\mu\in\mathbb{Z}} \mu^{-2} < \infty. \\ \ell^1(\mathbb{Z}) & \text{if } k \geq 2 \end{cases}$$

This implies that

$$\begin{aligned} &\forall H \in \mathbb{C}(s)_{\text{per}} \text{ and } h := (H(j\mu\omega))_{\mu\in\mathbb{Z}}:\ h \in \mathfrak{s}^{-\infty} \text{ and} \\ &h \in \begin{cases} \ell^\infty(\mathbb{Z}) & \text{if } H \text{ is proper} \\ \ell^2(\mathbb{Z}) & \text{if } H \text{ is strictly proper}. \\ \ell^1(\mathbb{Z}) & \text{if } \deg_s(H) \leq -2 \end{cases} \end{aligned} \tag{8.136}$$

The map $\mathbb{C}(s)_{\text{per}} \to \mathfrak{s}^{-\infty} \subset \mathbb{C}^{\mathbb{Z}},\ H \mapsto h$, is obviously an algebra monomorphism, and therefore $\mathfrak{s}^{-\infty}$ and $\mathbb{C}^{\mathbb{Z}}$ are $\mathbb{C}(s)_{\text{per}}$-modules with the scalar multiplication

$$H \circ \widehat{u} = h \cdot \widehat{u},\ H \in \mathbb{C}(s)_{\text{per}},\ h := (H(j\mu\omega))_{\mu\in\mathbb{Z}},\ \widehat{u} \in \mathbb{C}^{\mathbb{Z}},\ (H \circ \widehat{u})(\mu) = H(j\mu\omega)\widehat{u}(\mu). \tag{8.137}$$

Generalization to matrices yields

$$\forall H \in \mathbb{C}(s)_{\mathrm{per}}^{p\times m}:\ h := (H(j\mu\omega))_\mu \in \left(\mathfrak{s}^{-\infty}\right)^{p\times m} \text{ and}$$
$$h \in \begin{cases} \ell^\infty(\mathbb{Z})^{p\times m} \text{ if } H \text{ is proper} \\ \ell^2(\mathbb{Z})^{p\times m} \text{ if } H \text{ is strictly proper} \\ \ell^1(\mathbb{Z})^{p\times m} \text{ if } \deg_s(H) \le -2 \end{cases} \tag{8.138}$$
$$\Longrightarrow H\circ := h\cdot : \left(\mathfrak{s}^{-\infty}\right)^m \to \left(\mathfrak{s}^{-\infty}\right)^p,\ \widehat{u} \mapsto \widehat{y} := H\circ\widehat{u} = h\cdot\widehat{u}.$$

If

$$\begin{aligned} &H := P^{-1}Q \in \mathbb{C}(s)_{\mathrm{per}}^{p\times m},\ \widehat{u}\in\mathfrak{s}^{-\infty},\ \widehat{y} := H\circ\widehat{u}, \text{ then} \\ &P\circ\widehat{y} = PH\circ\widehat{u} = Q\circ\widehat{u},\ \forall\mu:\ P(j\mu\omega)\widehat{y}(\mu) = Q(j\mu\omega)\widehat{u}(\mu). \end{aligned} \tag{8.139}$$

The map $H\circ$ induces

$$\begin{aligned} &H\circ = h\cdot : \begin{cases} \ell^i(\mathbb{Z})^m \to \ell^i(\mathbb{Z})^p,\ i=\infty,1,2, & \text{if } H \text{ is proper} \\ \ell^2(\mathbb{Z})^m \to \ell^1(\mathbb{Z})^p, & \text{if } H \text{ is strictly proper} \end{cases} \text{ since} \\ &h\ell^i(\mathbb{Z})^m \subseteq \ell^\infty(\mathbb{Z})^{p\times m}\ell^i(\mathbb{Z})^m \subseteq \ell^i(\mathbb{Z})^p \text{ resp.} \\ &h\ell^2(\mathbb{Z})^m \subseteq \ell^2(\mathbb{Z})^{p\times m}\ell^2(\mathbb{Z})^m \subseteq \ell^1(\mathbb{Z})^p. \end{aligned} \tag{8.140}$$

Assume, in the last line of (8.139), the sharper condition

$$\begin{aligned} &P \in \mathrm{Gl}_p\left(\mathbb{C}(s))_{\mathrm{per}}\right) \text{ or } \det(P)^{-1}\in\mathbb{C}(s)_{\mathrm{per}} \text{ or } \mathrm{V}_{\mathbb{C}}(\det(P)) \subset \mathbb{C}\setminus\mathbb{Z}j\omega \\ &\Longrightarrow \forall\mu\in\mathbb{Z}:\ P(j\mu\omega)\in\mathrm{Gl}_p(\mathbb{C}), \\ &\quad P\circ: \left(\mathbb{C}^{\mathbb{Z}}\right)^p \overset{\cong}{\to} \left(\mathbb{C}^{Z}\right)^p,\ P\circ: \left(\mathfrak{s}^{-\infty}\right)^p \overset{\cong}{\to} \left(\mathfrak{s}^{-\infty}\right)^p, \\ &\quad \forall\widehat{u}\in\left(\mathfrak{s}^{-\infty}\right)^m \exists_1\widehat{y}\in\left(\mathfrak{s}^{-\infty}\right)^p \text{ with } P\circ\widehat{y} = Q\circ\widehat{u},\ \text{viz. } \widehat{y} = H\circ\widehat{u}. \end{aligned} \tag{8.141}$$

Since $\mathrm{V}_{\mathbb{C}}(1-s^2) = \{\pm 1\} \subset \mathbb{C}\setminus\mathbb{Z}j\omega$ we conclude the isomorphisms

$$\begin{aligned} &(1-s^2)\circ:\ \mathbb{C}^{\mathbb{Z}} \cong \mathbb{C}^{\mathbb{Z}},\ \widehat{x}\mapsto \left((1+\mu^2\omega^2)\widehat{x}(\mu)\right)_{\mu\in\mathbb{Z}},\ (1-s^2)\circ:\ \mathfrak{s}^{-\infty}\cong\mathfrak{s}^{-\infty}, \\ &(1-s^2)\circ\mathfrak{s}^{-2d} = \mathfrak{s}^{-2(d+1)},\ (1-s^2)^{-1}\circ\mathfrak{s}^{-2(d+1)} = \mathfrak{s}^{-2d},\ d\in\mathbb{Z}. \end{aligned} \tag{8.142}$$

The next lemma is decisive for the definition and properties of *periodic distributions* in $\mathcal{F} := \mathrm{C}^{-\infty}(\mathbb{R},\mathbb{C})$. For $u\in\mathcal{F}$ we define $\Delta(u) := T\circ_s u - u$ where $\circ_s$ is the shift action from Definition 8.2.16. For continuous u this means $\Delta(u)(t) = u(t-T) - u(t)$. For $d\ge 0$ we consider the autonomous behavior

$$\begin{aligned} &\mathcal{B}_d := \mathrm{ann}_{\mathcal{F}}((1-s^2)^d) = \left\{w\in\mathcal{F};\ (1-s^2)^d\circ w = 0\right\} \subset \mathrm{t}(\mathcal{F}), \text{ hence} \\ &\mathcal{B}_d \cong \mathbb{C}^{2d},\ w\mapsto \left(w^{(i)}(0)\right)_{i=0,\cdots,2d-1},\ w^{(i)} = s^i\circ w. \end{aligned} \tag{8.143}$$

As usual we define the real analogues

$$\mathcal{P}^d_{\mathbb{R}} := \mathrm{C}^d(\mathbb{R},\mathbb{R})\bigcap\mathcal{P}^0,\ d\ge 0. \tag{8.144}$$

Lemma 8.2.96 *1. For $d \geq 0$ the following map is a $\mathbb{C}$-isomorphism:*

$$\phi_d : \mathcal{B}_d \to \mathbb{C}^{2d}, \; w \mapsto \left(\Delta(w)^{(i)}(0)\right)_{i=0,\ldots,2d-1}.$$

2. The map $(1-s^2)^d \circ : \mathcal{P}^{2d} := \mathcal{C}^{2d} \cap \mathcal{P}^0 \to \mathcal{P}^0$ is an isomorphism. For periodic functions, this is the analogue of the isomorphism

$$P\circ : \mathcal{C}^d_+ \cong \mathcal{C}^0_+, \; P \in \mathbb{C}[s], \; \deg_s(P) = d \geq 0$$

from Corollary 8.2.38.

3. For the real base field there are the analogous isomorphisms

$$\phi_d : \left\{w \in \mathcal{F}_{\mathbb{R}};\; (1-s^2)^d \circ w = 0\right\} \cong \mathbb{R}^{2d} \text{ and } (1-s^2)^d \circ : \; \mathcal{P}^{2d}_{\mathbb{R}} \cong \mathcal{P}^0_{\mathbb{R}}.$$

Proof 1. The proof proceeds by induction on $d \geq 0$. For $d = 0$ and $(1-s^2)^0 = 1$ the assertion is trivial.
$d = 1$: There is the isomorphism

$$\mathcal{B}_1 = \mathbb{C}e^t \oplus \mathbb{C}e^{-t} \cong \mathbb{C}^2, \; w = ae^t + be^{-t} \mapsto \begin{pmatrix} a \\ b \end{pmatrix}.$$

Moreover

$$\begin{aligned}
&\begin{pmatrix} \Delta(w)(t) \\ \Delta(w)'(t) \end{pmatrix} = \begin{pmatrix} ae^{t-T}+be^{-t+T}-ae^t-be^{-t} \\ ae^{t-T}-be^{-t+T}-ae^t+be^{-t} \end{pmatrix} \\
&\Longrightarrow \begin{pmatrix} \Delta(w)(0) \\ \Delta(w)'(0) \end{pmatrix} = M\begin{pmatrix} a \\ b \end{pmatrix}, \; M := \begin{pmatrix} e^{-T}-1 & e^T-1 \\ e^{-T}-1 & -(e^T-1) \end{pmatrix} \\
&\Longrightarrow \det(M) = -2(e^{-T}-1)(e^T-1) \underset{T\neq 0}{\neq} 0, \; M \in \mathrm{Gl}_2(\mathbb{R}) \\
&\Longrightarrow \phi_1 : \mathcal{B}_1 \cong \mathbb{C}^2 \cong \mathbb{C}^2, \; ae^t + be^{-t} \mapsto \begin{pmatrix} a \\ b \end{pmatrix} \mapsto M\begin{pmatrix} a \\ b \end{pmatrix} = \begin{pmatrix} \Delta(w)(0) \\ \Delta(w)'(0) \end{pmatrix}.
\end{aligned}$$

$d > 1$: We assume that the assertion is true for $d-1$, and thus get the isomorphism

$$\phi_{d-1} : \; \mathcal{B}_{d-1} \cong \mathbb{C}^{2(d-1)}, \; w \mapsto \left(\Delta(w)^{(i)}(0)\right)_{0\leq i\leq 2(d-1)-1}.$$

The map ϕ_d is injective : Let $w_d \in \mathcal{B}_d$ and $\xi := \phi_d(w_d)$. Then

$$\begin{aligned}
w_{d-1} &:= (1-s^2)\circ w_d \in \mathcal{B}_{d-1} \text{ and } (1-s^2)\circ\Delta(w_d) = (1-s^2)\circ(T\circ_s w_d - w_d) \\
&= T\circ_s(1-s^2)\circ w_d - (1-s^2)\circ w_d = \Delta(w_{d-1}) \\
&\Longrightarrow \text{with } \eta := \phi_{d-1}(w_{d-1}) : \; \forall i = 0,\ldots,2(d-1)-1 : \; \xi_i - \xi_{i+2} = \eta_i.
\end{aligned} \tag{8.145}$$

Now assume $\xi = 0$ and hence $\phi_{d-1}(w_{d-1}) = \eta = 0$. Since ϕ_{d-1} is an isomorphism, this implies $w_{d-1} = 0$ and hence $(1-s^2)\circ w_d = 0$ and $w_d \in \mathcal{B}_1$ with $(\Delta(w_d)(0), \Delta(w_d)'(0))^\top = (\xi_0, \xi_1)^\top = 0$. From the proven case $d = 1$ we infer

$w_d = 0$ and the injectivity of ϕ_d.
The map ϕ_d is surjective: Assume $\xi \in \mathbb{C}^{2d}$. Define

$$\eta \in \mathbb{C}^{2(d-1)} \text{ by } \eta_i := \xi_i - \xi_{i+2},\ 0 \le i \le 2(d-1)-1,\ w_{d-1} := \phi_{d-1}^{-1}(\eta) \in \mathcal{B}_{d-1}.$$

Let y be any solution of $(1-s^2)\circ y = w_{d-1}$, hence $y \in \mathcal{B}_d$. Let $w_d := y + w$ where $w \in \mathcal{B}_1$ has to be suitably chosen. Then

$$\Delta(w_d) = \Delta(y) + \Delta(w) \Longrightarrow \begin{pmatrix} \Delta(w_d)(0) \\ \Delta(w_d)'(0) \end{pmatrix} = \begin{pmatrix} \Delta(y)(0) \\ \Delta(y)'(0) \end{pmatrix} + \begin{pmatrix} \Delta(w)(0) \\ \Delta(w)'(0) \end{pmatrix}.$$

According to the case $d = 1$ we choose the unique $w \in \mathcal{B}_1$ such that

$$\begin{pmatrix} \Delta(w)(0) \\ \Delta(w)'(0) \end{pmatrix} = \begin{pmatrix} \xi_0 \\ \xi_1 \end{pmatrix} - \begin{pmatrix} \Delta(y)(0) \\ \Delta(y)'(0) \end{pmatrix} \Longrightarrow \begin{pmatrix} \Delta(w_d)(0) \\ \Delta(w_d)'(0) \end{pmatrix} = \begin{pmatrix} \xi_0 \\ \xi_1 \end{pmatrix} \text{ for } w_d := y + w \in \mathcal{B}_d.$$

Then $(1-s^2)\circ w_d = w_{d-1}$. Define

$$\begin{aligned}
&\zeta := \left(\Delta(w_d)^{(i)}(0)\right)_{0\le i\le 2d-1} \in \mathbb{C}^{2d} \Longrightarrow \begin{pmatrix} \zeta_0 \\ \zeta_1 \end{pmatrix} = \begin{pmatrix} \Delta(w_d)(0) \\ \Delta(w_d)'(0) \end{pmatrix} = \begin{pmatrix} \xi_0 \\ \xi_1 \end{pmatrix} \text{ and} \\
&\forall i = 0, \dots, 2(d-1)-1:\ \zeta_i - \zeta_{i+2} \underset{(8.145)}{=} \phi_{d-1}(w_{d-1})_i = \eta_i = \xi_i - \xi_{i+2} \\
&\underset{\text{induction}}{\Longrightarrow} \xi = \zeta = \left(\Delta(w_d)^{(i)}(0)\right)_{0\le i\le 2d-1} = \phi_d(w_d) \Longrightarrow \phi_d \text{ surjective.}
\end{aligned}$$

2. *Injective*: Assume that $w \in \mathcal{P}^{2d}$ and $(1-s^2)^d \circ w = 0$. Then $w \in \mathcal{B}_d$, $\Delta(w) = T\circ w - w = 0$ and hence $\phi_d(w) = 0$. From (1) we infer $w = 0$.
Surjective: Let $u \in \mathcal{P}^0$ and $y_1 \in \mathrm{C}^{2d}$ any solution of $(1-s^2)^d \circ y_1 = u$. Define $y := y_1 + w \in \mathrm{C}^{2d}$ with a suitable $w \in \mathcal{B}_d$ and thus $(1-s^2)^d \circ y = u$. It suffices to choose w such that y is periodic. The equations

$$\begin{aligned}
&\Delta(y) = \Delta(y_1) + \Delta(w),\ (1-s^2)^d \circ \Delta(y) = (1-s^2)^d \circ \Delta(y_1) = \Delta(u) = 0 \text{ and} \\
&\left(\Delta(y)^{(i)}(0)\right)_{0\le i\le 2d-1} = \left(\Delta(y_1)^{(i)}(0)\right)_{0\le i\le 2d-1} + \phi_d(w)
\end{aligned}$$

are obvious. According to 1 choose

$$w \in \mathcal{B}_d \text{ with } \phi_d(w) = -\left(\Delta(y_1)^{(i)}(0)\right)_{0\le i\le 2d-1} \Longrightarrow \left(\Delta(y)^{(i)}(0)\right)_{0\le i\le 2d-1} = 0.$$

Since $\Delta(y)$ is determined by the initial condition $\left(\Delta(y)^{(i)}(0)\right)_{0\le i\le 2d-1} = 0$ we conclude $\Delta(y) = 0$ and thus $y \in \mathcal{P}^{2d}$ with $(1-s^2)^d \circ y = u$.
3. $1 - s^2 \in \mathbb{R}[s]$. □

We are now going to define periodic distributions. For $d \ge 0$ we define

$$\mathcal{P}^{-2d} := (1-s^2)^d \circ \mathcal{P}^0 \subset (1-s^2)^d \circ \mathrm{C}^0 \subset \mathrm{C}^{-2d} \subset \mathrm{C}^{-\infty},\ d \ge 0. \tag{8.146}$$

Theorem 8.2.97 *1. There is the increasing sequence of subspaces of* $\mathcal{F} = \mathrm{C}^{-\infty}(\mathbb{R}, \mathbb{C})$

$$\cdots \subseteq \mathcal{P}^2 \subseteq \mathcal{P}^0 \subseteq \cdots \subset \mathcal{P}^{-2d} \subseteq \mathcal{P}^{-2(d+1)} \subseteq \cdots \subseteq \mathcal{P}^{-\infty} := \bigcup_{d=0}^{\infty} \mathcal{P}^{-2d} \subseteq \mathcal{F} = \mathrm{C}^{-\infty}. \tag{8.147}$$

The space $\mathcal{P}^{-\infty}$ *is a* $\mathbb{C}[s]$*-submodule of* $\mathrm{C}^{-\infty}$*, and* $(1-s^2)\circ : \mathcal{P}^{-\infty} \to \mathcal{P}^{-\infty}$ *is an isomorphism with* $(1-s^2)\circ \mathcal{P}^{-2d} = \mathcal{P}^{-2(d+1)}$ *for all* $d \in \mathbb{Z}$.
2. The analogous result holds for the real case, i.e.,

$$\begin{aligned} \mathcal{P}_{\mathbb{R}}^{-2d} &:= (1-s^2)^d \circ \mathcal{P}_{\mathbb{R}}^0 = \mathcal{P}^{-2d} \bigcap \mathcal{F}_{\mathbb{R}} = \mathcal{P}^{-2d} \bigcap \mathrm{C}^{-\infty}(\mathbb{R}, \mathbb{R}) \\ \mathcal{P}_{\mathbb{R}}^{-\infty} &:= \bigcup_{d\geq 0} \mathcal{P}_{\mathbb{R}}^{-2d} = \mathcal{P}^{-\infty} \bigcap \mathcal{F}_{\mathbb{R}},\ (1-s^2)\circ :\ \mathcal{P}_{\mathbb{R}}^{-\infty} \cong \mathcal{P}_{\mathbb{R}}^{-\infty}. \end{aligned} \tag{8.148}$$

Proof 1. (i) From Lemma 8.2.96 we get

$$\begin{aligned} &(1-s^2)\circ \mathcal{P}^2 = \mathcal{P}^0 \Longrightarrow \forall d \geq 0: \\ &\mathcal{P}^{-2d} = (1-s^2)^d \circ \mathcal{P}^0 = (1-s^2)^{d+1} \circ \mathcal{P}^2 \subseteq (1-s^2)^{d+1} \circ \mathcal{P}^0 = \mathcal{P}^{-2(d+1)}. \end{aligned}$$

As union of the increasing sequence $\mathcal{P}^{-2d}$ also $\mathcal{P}^{-\infty}$ is a $\mathbb{C}$-subspace.
(ii) $\mathcal{P}^{-\infty}$ *is a* $\mathbb{C}[s]$*-submodule of* $\mathrm{C}^{-\infty}$*, i.e.,* $s \circ \mathcal{P}^{-\infty} \subseteq \mathcal{P}^{-\infty}$, since for $d \geq 0$

$$\begin{aligned} s \circ \mathcal{P}^{-2d} &= s(1-s^2)^d \circ \mathcal{P}^0 = s(1-s^2)^{d+1} \circ \mathcal{P}^2 \\ &= (1-s^2)^{d+1} \circ (s \circ \mathcal{P}^2) \subseteq (1-s^2)^{d+1} \circ \mathcal{P}^0 = \mathcal{P}^{-2(d+1)}. \end{aligned}$$

(iii) $(1-s^2) : \mathcal{P}^{-\infty} \to \mathcal{P}^{-\infty}$ *is injective*: Let

$$\begin{aligned} &y \in \mathcal{P}^0,\ d \geq 0,\ u := (1-s^2)^d \circ y \in \mathcal{P}^{-2d} \text{ and } 0 = (1-s^2)\circ u = \\ &= (1-s^2)^{d+1} \circ y = 0 \underset{\text{Lemma 8.2.96,2}}{\Longrightarrow} y \in \mathcal{P}^{2(d+1)},\ y = 0 \Longrightarrow u = 0. \end{aligned}$$

(iv) $(1-s^2) : \mathcal{P}^{-\infty} \to \mathcal{P}^{-\infty}$ *is surjective* since $(1-s^2)\circ \mathcal{P}^{-2d} = \mathcal{P}^{-2(d+1)}$ and

$$\begin{aligned} &(1-s^2)\circ \mathcal{P}^{-\infty} = (1-s^2)\circ \bigcup_{d=0}^{\infty} \mathcal{P}^{-2d} \\ &= \bigcup_{d=0}^{\infty} (1-s^2)\circ \mathcal{P}^{-2d} = \bigcup_{d=0}^{\infty} \mathcal{P}^{-2(d+1)} = \mathcal{P}^{-\infty}. \end{aligned}$$

2. $1 - s^2 \in \mathbb{R}[s]$. ☐

Theorem 8.2.98 $\mathcal{P}^{-\infty} = \left\{u \in \mathcal{F} = \mathrm{C}^{-\infty}(\mathbb{R}, \mathbb{C});\ T \circ_s u = u\right\}$.
Therefore the distributions in $\mathcal{P}^{-\infty}$ *are called* T*-periodic.*
In particular, $\mathcal{P}^{0,\mathrm{pc}} \subset \mathcal{P}^{-2}$. *The real analogue is* $\mathcal{P}_{\mathbb{R}}^{-\infty} = \{u \in \mathcal{F}_{\mathbb{R}};\ T \circ_s u = u\}$.

Proof (i) All $u = (1-s^2)^d \circ y,\ y \in \mathcal{P}^0$, are T-periodic since

$$T \circ_s u = T \circ_s ((1-s^2)^d \circ y) = (1-s^2)^d \circ (T \circ_s y) = (1-s^2)^d \circ y = u.$$

(ii) Let $u \in \mathrm{C}^{-2d}$, $d > 0$, be T-periodic, i.e., $T \circ_s u = u$, and let $y_1 \in \mathrm{C}^0$ satisfy $(1-s^2)^d \circ y_1 = u$, cf. Corollary 8.2.29, hence

$$(1-s^2)^d \circ \Delta(y_1) = \Delta((1-s^2)^d \circ y_1) = \Delta(u) \underset{T\text{-periodic}}{=} 0$$
$$\underset{\text{Lemma 8.2.96}}{\Longrightarrow} \Delta(y_1) \in \mathcal{B}_d.$$

Define $y := y_1 + w$ with a suitable $w \in \mathcal{B}_d = \left\{w;\ (1-s^2)^d \circ w = 0\right\}$. Then

$$\begin{aligned}&(1-s^2)^d \circ y = (1-s^2)^d \circ y_1 + (1-s^2)^d \circ w = (1-s^2)^d \circ y_1 = u \text{ and}\\ &\Delta(y) = \Delta(y_1) + \Delta(w) \in \mathcal{B}_d,\\ &\left(\Delta(y)^{(i)}(0)\right)_{0\le i\le 2d-1} = \left(\Delta(y_1)^{(i)}(0)\right)_{0\le i\le 2d-1} + \phi_d(w).\end{aligned}$$

According to Lemma 8.2.96 we choose the unique $w \in \mathcal{B}_d$ with

$$\begin{aligned}&\phi_d(w) = -\left(\Delta(y_1)^{(i)}(0)\right)_{0\le i\le 2d-1} \Longrightarrow \left(\Delta(y)^{(i)}(0)\right)_{0\le i\le 2d-1} = 0\\ &\underset{\Delta(y)\in\mathcal{B}_d}{\Longrightarrow} \Delta(y) = 0 \Longrightarrow y \in \mathcal{P}^0,\ u = (1-s^2)^d \circ y \in \mathcal{P}^{-2d} \subset \mathcal{P}^{-\infty}.\end{aligned}$$

□

We finally construct the Fourier transform isomorphisms.

Theorem 8.2.99 (Fourier transform and series)
For $u = (1-s^2)^d \circ y \in \mathcal{P}^{-\infty},\ y = (1-s^2)^{-d} \circ u \in \mathcal{P}^0$, *define*

$$\begin{aligned}\mathbb{F}(u) := (1-s^2)^d \circ \mathbb{F}(y) = (1-s^2)^d \circ \mathbb{F}\left((1-s^2)^{-d} \circ u\right) \in \mathfrak{s}^{-\infty}. \textit{ Then}\\ \mathbb{F} : \mathcal{P}^{-\infty} \to \mathfrak{s}^{-\infty}\end{aligned} \tag{8.149}$$

is a well-defined $\mathbb{C}[s]$*-isomorphism, called the* Fourier transform.

Proof (i) *Well defined, i.e., independent of* d: Assume

$$\begin{aligned}&d < d_1,\ u \in \mathcal{P}^{-2d_1},\ y_1 := (1-s^2)^{-d_1} \circ u \in \mathcal{P}^0\\ &\Longrightarrow y_1 = (1-s^2)^{-d_1} \circ u = (1-s^2)^{-(d_1-d)} \circ (1-s^2)^{-d} \circ u\\ &\quad = (1-s^2)^{-(d_1-d)} \circ y \in \mathcal{P}^{2(d_1-d)} \subseteq \mathcal{P}^2\end{aligned}$$

$$\Longrightarrow y_1 \in \mathcal{P}^{2(d_1-d)},\ y = (1-s^2)^{d_1-d} \circ y_1$$
$$\underset{\text{Lemma 8.2.94,3}}{\Longrightarrow} \mathbb{F}(y) = (1-s^2)^{d_1-d} \circ \mathbb{F}(y_1)$$
$$\Longrightarrow \mathbb{F}(u) = (1-s^2)^d \circ \mathbb{F}(y) = (1-s^2)^{d_1} \circ \mathbb{F}(y_1).$$

(ii) By construction

$$\mathbb{F}|_{\mathcal{P}^{-2d}} : \mathcal{P}^{-2d} \xrightarrow{(1-s^2)^{-d}\circ} \mathcal{P}^0 \xrightarrow{\mathbb{F}|_{\mathcal{P}^0}} \mathfrak{s}^{-\infty} \xrightarrow{(1-s^2)^d\circ} \mathfrak{s}^{-\infty}$$

is a $\mathbb{C}$-monomorphism, and the same is obviously inherited by $\mathbb{F}$ on $\mathcal{P}^{-\infty}$.
(iii) $\mathbb{F}$ *is surjective*: Let

$$d \geq 0,\ \widehat{u} \in \mathfrak{s}^{-2d},\ \text{hence } \widehat{y} := (1-s^2)^{-(d+1)} \circ \widehat{u} \in \mathfrak{s}^{-2d+2(d+1)} = \mathfrak{s}^2$$
$$\Longrightarrow \exists M > 0 \forall \mu \in \mathbb{Z}:\ |\widehat{y}(\mu)| \leq M(1+\mu^2\omega^2)^{-1} \Longrightarrow \widehat{y} \in \ell^1(\mathbb{Z})$$
$$\underset{\text{Lemma 8.2.94,1}}{\Longrightarrow} y := \sum_{\mu\in\mathbb{Z}} \widehat{y}(\mu)e^{j\mu\omega t} \in \mathcal{P}^0 \text{ and } \mathbb{F}(y) = \widehat{y}$$
$$\Longrightarrow \mathbb{F}((1-s^2)^{d+1} \circ y) = (1-s^2)^{d+1} \circ \mathbb{F}(y) = (1-s^2)^{d+1} \circ \widehat{y} = \widehat{u} \in \operatorname{im}(\mathbb{F}).$$

(iv) $\mathbb{F}$ is $\mathbb{C}[s]$-*linear, i.e.*, $\mathbb{F}(s \circ u) = s \circ \mathbb{F}(u)$: Let $u \in \mathcal{P}^{-2d}$, $d \geq 0$, and

$$y := (1-s^2)^{-(d+1)} \circ u \in \mathcal{P}^2 = \mathrm{C}^2 \bigcap \mathcal{P}^0 \underset{\text{Lemma 8.2.94,2.}}{\Longrightarrow} \mathbb{F}(s \circ y) = s \circ \mathbb{F}(y),$$
$$s \circ u = s(1-s^2)^{d+1} \circ y = (1-s^2)^{d+1} \circ (s \circ y)$$
$$\Longrightarrow \mathbb{F}(s \circ u) = (1-s^2)^{d+1} \circ \mathbb{F}(s \circ y) = (1-s^2)^{d+1} s \circ \mathbb{F}(y)$$
$$= s \circ ((1-s^2)^{d+1} \circ \mathbb{F}(y)) = s \circ \mathbb{F}(u).$$

□

Corollary 8.2.100 *Since* $\mathfrak{s}^{-\infty}$ *is a* $\mathbb{C}(s)_{\text{per}}$*-module and* $\mathbb{F} : \mathcal{P}^{-\infty} \cong \mathfrak{s}^{-\infty}$ *a* $\mathbb{C}[s]$*-isomorphism, the scalar multiplication*

$$H \circ u := \mathbb{F}^{-1}(H \circ \mathbb{F}(u)),\ H \in \mathbb{C}(s)_{\text{per}},\ u \in \mathcal{P}^{-\infty}, \tag{8.150}$$

extends that for $H \in \mathbb{C}[s]$ *and makes* $\mathcal{P}^{-\infty}$ *a* $\mathbb{C}(s)_{\text{per}}$*-module and* $\mathbb{F}$ *a* $\mathbb{C}(s)_{\text{per}}$*-isomorphism.*

◇

Example 8.2.101 Let $\widehat{u} \in \ell^\infty(\mathbb{Z})$, $\widehat{u}(\mu) := 1$, denote the one-element of $\mathfrak{s}^{-\infty} \subset \mathbb{C}^{\mathbb{Z}}$ and $u := \mathbb{F}^{-1}(\widehat{u})$. The standard notation for u is $u = \sum_{\mu\in\mathbb{Z}} e^{j\mu\omega t}$, but we have not introduced this. Instead, let $\lambda \notin \mathbb{Z}j\omega$. Then we can write

$$u = (s-\lambda)^2 \circ (s-\lambda)^{-2} \circ \mathbb{F}^{-1}(\widehat{u}) = (s-\lambda)^2 \circ \mathbb{F}^{-1}\left((s-\lambda)^{-2} \circ \widehat{u}\right)$$
$$= (s-\lambda)^2 \circ \mathbb{F}^{-1}\left(\left((j\mu\omega-\lambda)^{-2}\right)_{\mu\in\mathbb{Z}}\right) = (s-\lambda)^2 \circ \sum_{\mu\in\mathbb{Z}}(j\mu\omega-\lambda)^{-2}e^{j\mu\omega t} \text{ with}$$
$$\left((j\mu\omega-\lambda)^{-2}\right)_{\mu\in\mathbb{Z}} \in \ell^1(\mathbb{Z}),\ \sum_{\mu\in\mathbb{Z}}(j\mu\omega-\lambda)^{-2}e^{j\mu\omega t} \in \mathcal{P}^0,\ u \in \mathcal{P}^{-2}.$$

Likewise one obtains $u = 1 + s^2 \circ \sum_{0\neq\mu\in\mathbb{Z}}(j\mu\omega)^{-2}e^{j\mu\omega t}$. ◇

The following theorem is a direct consequence of the $\mathbb{C}(s)_{\text{per}}$-isomorphism $\mathbb{F}$ and the main application of Fourier series in *electrical and mechanical engineering.*

Theorem 8.2.102 *1. If $\mathcal{B} = \left\{w \in \mathcal{F}^l;\ R \circ w = 0\right\}$, $R \in \mathbb{C}[s]^{[k\times l]}$, is any behavior, then the Fourier isomorphism $\mathbb{F}$ induces the $\mathbb{C}(s)_{\text{per}}$-isomorphism*

$$\mathbb{F} : \mathcal{B}\bigcap\left(\mathcal{P}^{-\infty}\right)^l \cong \left\{\widehat{x} \in \left(\mathfrak{s}^{-\infty}\right)^l;\ R \circ \widehat{x} = 0\right\},\ w \mapsto \widehat{w} := \mathbb{F}(w). \quad (8.151)$$

2. Let $\mathcal{B} = \left\{\binom{y}{u} \in \mathcal{F}^{p+m};\ P \circ y = Q \circ u\right\}$ be an IO behavior with transfer matrix $H = P^{-1}Q \in \mathbb{C}(s)_{\text{per}}^{p\times m}$. Then there is the $\mathbb{C}(s)_{\text{per}}$-monomorphism

$$\left(\mathcal{P}^{-\infty}\right)^m \to \mathcal{B}\bigcap\left(\mathcal{P}^{-\infty}\right)^{p+m},\ u \mapsto \binom{H\circ u}{u}. \quad (8.152)$$

Hence, for $H \in \mathbb{C}(s)_{\text{per}}^{p\times m}$, $H\circ : \left(\mathcal{P}^{-\infty}\right)^m \to \left(\mathcal{P}^{-\infty}\right)^p$, $u \mapsto H \circ u$, is another transfer operator besides $H : \left(\mathrm{C}_+^{-\infty}\right)^m \to \left(\mathrm{C}_+^{-\infty}\right)^p$ from (8.65).
If, in addition, $P \in \mathrm{Gl}_p(\mathbb{C}(s)_{\text{per}})$ or $\mathrm{V}_{\mathbb{C}}(\det(P)) \subset \mathbb{C}\backslash\mathbb{Z}j\omega$, then $H \circ u$ is the unique T-periodic *solution of $P \circ y = Q \circ u$.*
3. Assume in 2 that $u \in \left(\mathcal{P}^{0,\text{pc}}\right)^m$ and hence $\widehat{u} := \mathbb{F}(u) \in \ell^2(\mathbb{Z})^m$. Also assume that $H = H_0 + H_{\text{spr}},\ H_0 \in \mathbb{C}^{p\times m}$, is proper.
(a) If $H = H_{\text{spr}}$ is strictly proper, then $\widehat{y} = H \circ \widehat{u} \in \ell^1(\mathbb{Z})^p$ and thus

$$y = H \circ u = \sum_{\mu\in\mathbb{Z}} H(j\mu\omega)\widehat{u}(\mu)e^{j\mu\omega t} \in \left(\mathcal{P}^0\right)^p \quad (8.153)$$

is uniformly convergent and continuous according to Lemma 8.2.94.
(b) For general proper H this implies

$$y = H \circ u = H_0 u + H_{\text{spr}} \circ u \in \left(\mathcal{P}^{0,\text{pc}}\right)^p,\ H_{\text{spr}} \circ u \in \left(\mathcal{P}^0\right)^p. \quad (8.154)$$

Proof 1., 2. $\mathbb{F}$ is a $\mathbb{C}(s)_{\text{per}}$-isomorphism, and $P \circ y = PH \circ u = Q \circ u$.
3. (a) $\ell^2(\mathbb{Z})\ell^2(\mathbb{Z}) \underset{(8.134)}{\subseteq} \ell^1(\mathbb{Z})$ and (8.128). (b) Obvious. □

Example 8.2.103 Consider

$$\begin{aligned}
&P := \operatorname{diag}(s-1, s^2+\omega^2) = \begin{pmatrix} s-1 & 0 \\ 0 & s^2+\omega^2 \end{pmatrix}, \ Q := \operatorname{diag}(1, s^2+\omega^2),\\
&\det(P)(j\omega) = 0, \ j\omega \in \mathrm{V}_{\mathbb{C}}(\det(P)) \cap \mathbb{Z}j\omega, \ P(j\omega) = \begin{pmatrix} j\omega-1 & 0 \\ 0 & 0 \end{pmatrix} \notin \mathrm{Gl}_2(\mathbb{C}),\\
&H := P^{-1}Q = \operatorname{diag}((s-1)^{-1}, 1) \in \mathbb{C}(s)^{2\times 2}_{\mathrm{per}}, \ u := \begin{pmatrix} 1 \\ 0 \end{pmatrix},\\
&\mathrm{V}_{\mathbb{C}}(s-1) \cap \mathbb{Z}j\omega = \emptyset, \ (s-1)\circ 1 = -1 \Longrightarrow (s-1)^{-1} \circ 1 = -1\\
&\Longrightarrow H \circ u = \begin{pmatrix} -1 \\ 0 \end{pmatrix} = -u \Longrightarrow P \circ (-u) = Q \circ u.
\end{aligned}$$

Moreover

$$(s^2+\omega^2) \circ e^{j\omega t} = 0 \Longrightarrow P \circ \left(-u + \begin{pmatrix} 0 \\ e^{j\omega t} \end{pmatrix}\right) = P \circ \begin{pmatrix} -1 \\ e^{j\omega t} \end{pmatrix} = Q \circ u.$$

Hence the equation $P \circ y = Q \circ u$ has more than one periodic solution y. ◊

Remark 8.2.104 (i) In electrical engineering the equation

$$P \circ \left(\sum_{\mu \in \mathbb{Z}} H(j\mu\omega)\widehat{u}(\mu)e^{j\mu\omega t} \right) = Q \circ \left(\sum_{\mu \in \mathbb{Z}} \widehat{u}(\mu)e^{j\mu\omega t} \right)$$

is often inferred without proof from the so-called *superposition principle* and the obvious equations $P \circ H(j\mu\omega)\widehat{u}(\mu)e^{j\mu\omega t} = Q \circ \widehat{u}(\mu)e^{j\mu\omega t}$. Theorem 8.2.102 proves this principle. In general, the meaning of the differential expressions $Q \circ \left(\sum_\mu \widehat{u}(\mu)e^{j\mu\omega t}\right)$ etc. and the type of convergence of $\sum_{\mu \in \mathbb{Z}} H(j\mu\omega)\widehat{u}(\mu)e^{j\mu\omega t}$ are not discussed, if $u = \sum_\mu \widehat{u}(\mu)e^{j\mu\omega t}$ is piecewise continuous, but not sufficiently differentiable. In contrast to $\mathcal{P}^{0,\mathrm{pc}}$ the $\mathbb{C}(s)_{\mathrm{per}}$-module $\mathcal{P}^{-\infty}$ is closed under differentiation and thus makes all these expressions well defined.
(ii) Notice that

$$\mathcal{P}^{0,\mathrm{pc}} \subseteq \mathrm{L}^\infty(\mathbb{R}, \mathbb{C}), \ \text{but } \mathcal{P}^{0,\mathrm{pc}} \bigcap \mathrm{L}^p(\mathbb{R}, \mathbb{C}) = 0, \ p = 1, 2.$$

Therefore periodic signals need a special treatment. For instance, Theorem and Definition 8.2.83 on L^p, $p = 1, 2$, cannot be applied to periodic signals. This also holds for the results in Sect. 10.2.5 on the Fourier transform (integral) on $\mathfrak{L}^p \supset \mathrm{L}^p$, $p = 1, 2$. ◊

Recall

$$\begin{aligned}
&\mathcal{P}^{-\infty} \subset \mathrm{C}^{-\infty} \subset \operatorname{Hom}_{\mathbb{C}}(\mathrm{C}_0^\infty, \mathbb{C}) \text{ and } \forall u \in \mathrm{C}^{0,\mathrm{pc}}, \ \varphi \in \mathrm{C}_0^\infty, \ P \in \mathbb{C}[s]:\\
&u(\varphi) = \int_{-\infty}^{\infty} u(t)\varphi(t)dt, \ (P(s) \circ u)(\varphi) = u(P(-s) \circ \varphi).
\end{aligned} \tag{8.155}$$

The analogue of C_0^∞ for T-periodic functions is

$$\mathcal{P}^{\infty} := \mathcal{P}^{\infty}(\mathbb{R}, \mathbb{C}) = \mathcal{P}^0 \bigcap \mathrm{C}^{\infty}. \tag{8.156}$$

Equation (8.130) implies

$$\mathbb{F}(\mathcal{P}^{\infty}) \subset \mathfrak{s}^{\infty} := \left\{\widehat{u} \in \mathbb{C}^{\mathbb{Z}};\ \forall d \geq 0:\ \mathbb{F}(u)(\mu) = O\left((1+\mu^2\omega^2)^{-d}\right)\right\} \subset \mathfrak{s}^{-\infty}. \tag{8.157}$$

Then $\mathfrak{s}^{\infty}$ is obviously an ideal of $\mathfrak{s}^{-\infty}$. Moreover

$$\mathbb{F}(\mathcal{P}^{\infty}) = \mathfrak{s}^{\infty} \text{ since } \forall \widehat{u} \in \mathfrak{s}^{\infty}:\ u := \sum_{\mu \in \mathbb{Z}} \widehat{u}(\mu) e^{j\mu\omega t} \in \mathcal{P}^{\infty},\ \mathbb{F}(u) = \widehat{u}. \tag{8.158}$$

In analogy to (8.155) we define

$$\begin{aligned} &\forall y \in \mathcal{P}^0,\ d \geq 0,\ u := (1-s^2)^d \circ y \in \mathcal{P}^{-2d},\ \varphi \in \mathcal{P}^{\infty}: \\ &y(\varphi) = T^{-1} \int_0^T y(t)\varphi(t)dt,\ u(\varphi) := y((1-s^2)^d \circ \varphi). \end{aligned} \tag{8.159}$$

Some simple arguments show that the map

$$\mathcal{P}^{-\infty} \to \mathrm{Hom}_{\mathbb{C}}(\mathcal{P}^{\infty}, \mathbb{C}),\ u \mapsto (\varphi \mapsto u(\varphi)),\ \text{with } u(e^{-j\mu\omega t}) = \mathbb{F}(u)(\mu) \tag{8.160}$$

is a $\mathbb{C}[s]$-monomorphism. In [3, pp. 224–231] it is shown that the image of the preceding map is the space $(\mathcal{D}')_{S^1}$ of all $\mathbb{C}$-linear maps $U : \mathcal{P}^{\infty} \to \mathbb{C}$ that are continuous with respect to a suitable topology on $\mathcal{P}^{\infty}$. This result is not used in electrical engineering or in this book.

Corollary 8.2.105 *If $u \in \mathcal{P}^{-\infty}$ and $\alpha \in \mathcal{P}^{\infty}$, then $\alpha u \in \mathcal{P}^{-\infty}$, i.e., $\mathcal{P}^{-\infty}$ is a $\mathcal{P}^{\infty}$-module.*

Proof With Theorem 8.2.98 we have to show $T \circ_s (\alpha u) = \alpha u$. If $\varphi \in \mathrm{C}_0^{\infty}$, then

$$\begin{aligned} &(T \circ_s (\alpha u))(\varphi) = (\alpha u)((-T) \circ_s \varphi) = u(\alpha((-T) \circ_s \varphi)) = u((-T) \circ_s ((T \circ_s \alpha)\varphi))) \\ &\underset{\alpha \text{ per.}}{=} (T \circ_s u)(\alpha\varphi) \underset{u \text{ per.}}{=} u(\alpha\varphi) = (\alpha u)(\varphi) \Longrightarrow T \circ_s (\alpha u) = \alpha u \in \mathcal{P}^{-\infty}. \end{aligned}$$

□

We finally discuss the *computation* of $\mathbb{F}(u)$ and $\mathbb{F}^{-1}(\widehat{u})$ for most signals in electrical and mechanical engineering. The following lemma is often used.

Lemma 8.2.106 *1.* $\mathbb{F}\left(e^{j\mu_0\omega t}u\right)(\mu) = \mathbb{F}(u)(\mu - \mu_0),\ u \in \mathcal{P}^{-\infty},\ \mu_0, \mu \in \mathbb{Z}.$
2. $\mathbb{F}(a \circ_s u)(\mu) = e^{-j\mu\omega a}\mathbb{F}(u)(\mu),\ a \in \mathbb{R},\ \mu \in \mathbb{Z}.$

Proof 1. Since any $u = (1-s^2)^d \circ v,\ v \in \mathcal{P}^0$, is a linear combination of derivatives $s^i \circ v$, it suffices to show the assertion for $s^i \circ v$ by induction on $i \in \mathbb{N}$.
$i = 0$: $\mathbb{F}(e^{j\mu_0\omega t}v)(\mu) = T^{-1}\int_0^T v(t)e^{-j(\mu-\mu_0)\omega t}dt = \mathbb{F}(v)(\mu - \mu_0)$.

$i \Longrightarrow i+1$: Assume $u = s^{i+1} \circ v = s \circ w,\ w := s^i \circ v$. By induction the assertion holds for w. Then

$$\begin{aligned}
& s \circ (e^{j\mu_0\omega t} w) = j\mu_0\omega e^{j\mu_0\omega t} w + e^{j\mu_0\omega t} u \\
& \Longrightarrow \mathbb{F}\left(s \circ (e^{j\mu_0\omega t} w)\right)(\mu) = j\mu_0\omega \mathbb{F}\left(e^{j\mu_0\omega t} w\right)(\mu) + \mathbb{F}\left(e^{j\mu_0\omega t} u\right)(\mu) \\
& \underset{\text{induction}}{\Longrightarrow} j\mu\omega\mathbb{F}(w)(\mu-\mu_0) = j\mu_0\omega\mathbb{F}(w)(\mu-\mu_0) + \mathbb{F}\left(e^{j\mu_0\omega t} u\right)(\mu) \\
& \Longrightarrow \mathbb{F}\left(e^{j\mu_0\omega t} u\right)(\mu) = j(\mu-\mu_0)\omega\mathbb{F}(w)(\mu-\mu_0) \\
& \qquad = \mathbb{F}(s \circ w)(\mu-\mu_0) = \mathbb{F}(u)(\mu-\mu_0).
\end{aligned}$$

2. Assume

$$v \in \mathcal{P}^0,\ u = (1-s^2)^d \circ v \in \mathcal{P}^{-2d},\ d \geq 0, \Longrightarrow a \circ_s u = (1-s^2)^d \circ (a \circ_s v).$$

$$\begin{aligned}
\mathbb{F}(a \circ_s v)(\mu) &= T^{-1} \int_0^T v(t-a) e^{-j\mu\omega t} dt \\
&= T^{-1} \int_{-a}^{T-a} v(t) e^{-j\mu\omega(t+a)} dt = e^{-j\mu\omega a}\mathbb{F}(v)(\mu)
\end{aligned}$$

$$\begin{aligned}
\Longrightarrow \mathbb{F}(a \circ_s u)(\mu) &= (1+\mu^2\omega^2)^d \mathbb{F}(a \circ_s v)(\mu) \\
&= (1+\mu^2\omega^2)^d e^{-j\mu\omega a}\mathbb{F}(v)(\mu) = e^{-j\mu\omega a}\mathbb{F}(u)(\mu).
\end{aligned}$$

□

We extend any piecewise continuous function u on an interval $[T_0, T_0+T]$ of length T to a periodic function $\widetilde{u} \in \mathcal{P}^{0,\mathrm{pc}}$ by

$$\begin{aligned}
& \forall \mu \in \mathbb{Z} \forall T_0 \leq t < T_0 + T : \ \widetilde{u}(t+\mu T) := u(t). \text{ Define} \\
& \widehat{u} := \mathbb{F}(u) := \mathbb{F}(\widetilde{u}),\ \widehat{u}(\mu) := T^{-1} \int_0^T \widetilde{u}(t) e^{-j\mu\omega t} dt \\
& \qquad = T^{-1} \int_{T_0}^{T_0+T} u(t) e^{-j\mu\omega t} dt \\
& \Longrightarrow \mathbb{F} : \ \mathrm{C}^{0,\mathrm{pc}}[T_0, T_0+T] \to \ell^2(\mathbb{Z}) \ \mathbb{C}\text{-linear, injective.}
\end{aligned} \tag{8.161}$$

Notice $\widetilde{u}((T_0+T)-) = u((T_0+T)-),\ \widetilde{u}((T_0+T)+) = u(T_0+)$. For

$$\begin{aligned}
& \alpha \in \mathrm{t}(\mathcal{F}) = \oplus_{\lambda \in \mathbb{C}} \mathbb{C}[t] e^{\lambda t} \text{ and } T_0 \leq T_1 < T_2 \leq T_0 + T \text{ define} \\
& u(t) := \begin{cases} \alpha(t) & \text{if } T_1 \leq t < T_2 \\ 0 & \text{if } T_0 \leq t < T_1 \text{ or } T_2 \leq t < T_0 + T \end{cases} \Longrightarrow u \in \mathrm{C}^\infty[T_1, T_2),\ u \in \mathrm{C}^{0,\mathrm{pc}} \\
& \Longrightarrow u|_{[T_1,T_2)} = \alpha|_{[T_1,T_2)},\ u(T_1+) = \alpha(T_1),\ u(T_2-) = \alpha(T_2), \\
& \quad \mathbb{F}(u)(\mu) := \mathbb{F}(\widetilde{u})(\mu) = T^{-1} \int_{T_1}^{T_2} \alpha(t) e^{-j\mu\omega t} dt.
\end{aligned} \tag{8.162}$$

Note that u depends on α and on the choice of T_0, T_1, T_2. Almost all periodic signals in electrical engineering are finite sums of such u. Since all $\alpha \in \mathrm{t}(\mathcal{F})$ are finite $\mathbb{C}$-linear combinations of the $e_{\lambda,k}(t) = t^k\,(k!)^{-1}\,e^{\lambda t}$, $\lambda \in \mathbb{C}$, $k \geq 0$, it suffices to compute $\mathbb{F}(u)$ for $\alpha = e_{\lambda,k}$.

Theorem 8.2.107 *Assume*

$$\lambda \notin \mathbb{Z}j\omega,\ k \geq 0,\ \alpha := e_{\lambda,k},\ u_k := u_{\lambda,k} \text{ from (8.162) with } u_{\lambda,k}|_{[T_1,T_2)} = e_{\lambda,k}|_{[T_1,T_2)}. \tag{8.163}$$

The Fourier sequence $\widehat{u}_k := \mathbb{F}(u_k)$ of u_k is given by

$$\begin{aligned}
&\widehat{u}_k(\mu) = \sum_{i=0}^{k}(j\mu\omega-\lambda)^{-(k+1-i)}\widehat{x}_\lambda^{(i)}(\mu),\ \widehat{u}_k = \sum_{i=0}^{k}(s-\lambda)^{-(k+1-i)}\circ\widehat{x}_\lambda^{(i)} \text{ where}\\
&\widehat{x}_\lambda^{(i)}(\mu) := T^{-1}(i!)^{-1}\left(T_1^i e^{(\lambda-j\mu\omega)T_1} - T_2^i e^{(\lambda-j\mu\omega)T_2}\right)\\
&\qquad = T^{-1}(i!)^{-1}\left(T_1^i e^{\lambda T_1}e^{-j\mu\omega T_1} - T_2^i e^{\lambda T_2}e^{-j\mu\omega T_2}\right) \text{ and}\\
&\widehat{x}_\lambda^{(i)} \in \ell^\infty(\mathbb{Z}),\ \left((j\mu\omega-\lambda)^{-(k+1-i)}\widehat{x}^{(i)}(\mu)\right)_{\mu\in\mathbb{Z}} \in \begin{cases}\ell^2(\mathbb{Z}) & \text{if } i=k\\ \ell^1(\mathbb{Z}) & \text{if } 0\leq i<k.\end{cases}
\end{aligned} \tag{8.164}$$

For u_k this implies $u_k = \sum_{i=0}^{k} u_{k,i}$ where

$$\text{for } 0 \leq i < k:\ u_{k,i} := \sum_{\mu\in\mathbb{Z}}(j\mu\omega-\lambda)^{-(k+1-i)}\widehat{x}_\lambda^{(i)}(\mu)e^{j\mu\omega t} \in \mathcal{P}^0$$

$$\text{for } i = k:\ u_{k,k} := u_k - \sum_{i=0}^{k-1} u_{k,i} = \sum_{\mu\in\mathbb{Z}}(j\mu\omega-\lambda)^{-1}\widehat{x}_\lambda^{(k)}(\mu) \in \mathcal{P}^{0,\mathrm{pc}}.$$

The Fourier series $u_{k,i}$, $i < k$, are uniformly convergent and thus continuous in $\mathcal{P}^0$. In contrast, $u_{k,k}$ is contained in $\mathcal{P}^{0,\mathrm{pc}}$, and its Fourier series converges in the $\|-\|_2$-norm only.

Proof The assumption $\lambda \notin \mathbb{Z}j\omega$ implies the isomorphism

$$(s-\lambda)\circ = (j\mu\omega-\lambda)_\mu\cdot :\ \mathfrak{s}^{-\infty} \cong \mathfrak{s}^{-\infty},\ (s-\lambda)^{-1}\circ :\ \mathfrak{s}^{-\infty} \cong \mathfrak{s}^{-\infty}.$$

The proof proceeds by induction on k.

For $k = 0$ we obtain

$$\begin{aligned}\widehat{u}_0(\mu) &= T^{-1}\int_{T_1}^{T_2} e^{(\lambda - j\mu\omega)t}\,dt \\ &= T^{-1}(j\mu\omega - \lambda)^{-1}\left(e^{(\lambda - j\mu\omega)T_1} - e^{(\lambda - j\mu\omega)T_2}\right) = (j\mu\omega - \lambda)^{-1}\widehat{x}_\lambda^{(0)}(\mu).\end{aligned}$$

For $k > 0$ partial integration furnishes

$$\begin{aligned}\widehat{u}_k(\mu) &= T^{-1}\int_{T_1}^{T_2} t^k (k!)^{-1} e^{(\lambda - j\mu\omega)t}\,dt \\ &= T^{-1}(j\mu\omega - \lambda)^{-1}(k!)^{-1}\left(T_1^k e^{(\lambda - j\mu\omega)T_1} - T_2^k e^{(\lambda - j\mu\omega)T_2}\right) \\ &\quad + T^{-1}(j\mu\omega - \lambda)^{-1}\int_{T_1}^{T_2} t^{k-1}((k-1)!)^{-1} e^{(\lambda - j\mu\omega)t}\,dt \\ &= (j\mu\omega - \lambda)^{-1}\left(\widehat{x}_\lambda^{(k)}(\mu) + \widehat{u}_{k-1}(\mu)\right) \\ &\underset{\text{induction}}{=} (j\mu\omega - \lambda)^{-1}\left(\widehat{x}_\lambda^{(k)}(\mu) + \sum_{i=0}^{k-1}(j\mu\omega - \lambda)^{-(k-i)}\widehat{x}_\lambda^{(i)}(\mu)\right) \\ &= \sum_{i=0}^{k}(j\mu\omega - \lambda)^{-(k+1-i)}\widehat{x}_\lambda^{(i)}(\mu).\end{aligned}$$

The remaining assertions follow from

$$\begin{aligned}u_k &= \mathbb{F}^{-1}\mathbb{F}(u_k) = \mathbb{F}^{-1}(\widehat{u}_k) = \mathbb{F}^{-1}\left(\sum_{i=0}^{k}\left((j\mu\omega - \lambda)^{-(k+1-i)}\widehat{x}_\lambda^{(i)}(\mu)\right)_\mu\right) \\ &= \sum_{i=0}^{k}\mathbb{F}^{-1}\left(\left((j\mu\omega - \lambda)^{-(k+1-i)}\widehat{x}_\lambda^{(i)}(\mu)\right)_\mu\right) \\ &= \sum_{i=0}^{k}\sum_{\mu\in\mathbb{Z}}(j\mu\omega - \lambda)^{-(k+1-i)}\widehat{x}_\lambda^{(i)}(\mu)e^{j\mu\omega t}.\end{aligned}$$

□

For $\lambda = 0$ and $e_{0,k} = t^k/k!$ we have $j0\omega - \lambda = 0$ and hence $(j0\omega - \lambda)^{-1}$ is not defined. For $\mu \neq 0$ the derivations of the preceding theorem remain valid.

Corollary 8.2.108 *1. For $\alpha := e_{0,k} = t^k/k!$ the assertions of the preceding theorem remain valid for the indices $\mu \neq 0$. One gets the modified equations*

$$\widehat{u}_k(0) = T^{-1}((k+1)!)^{-1}\left(T_2^{k+1} - T_1^{k+1}\right),$$

$$\forall \mu \neq 0: \ \widehat{u}_k(\mu) = \sum_{i=0}^{k}(j\mu\omega)^{-(k+1-i)}\widehat{x_0^{(i)}}(\mu) \text{ where}$$

$$\widehat{x_0^{(i)}}(\mu) = T^{-1}(i!)^{-1}\left(T_1^i e^{-j\mu\omega T_1} - T_2^i e^{-j\mu\omega T_2}\right)$$

$$u_k = \widehat{u}_k(0) + \sum_{i=0}^{k}\sum_{\mu\in\mathbb{Z},\ \mu\neq 0}(j\mu\omega)^{-(k+1-i)}\widehat{x_0^{(i)}}(\mu)e^{j\mu\omega t},$$

$$\sum_{\mu\neq\mu_0}(j(\mu-\mu_0)\omega)^{-(k+1-i)}\widehat{x_0^{(i)}}(\mu-\mu_0)e^{j\mu\omega t} \in \begin{cases}\mathcal{P}^0 & \text{if } 0\le i\le k-1\\ \mathcal{P}^{0,\mathrm{pc}} & \text{if } i=k.\end{cases} \tag{8.165}$$

The Fourier series in the last line are uniformly resp. $\|-\|_2$*-convergent for* $i<k$ *resp.* $i=k$.

2. *For* $\lambda = j\mu_0\omega \in \mathbb{Z}j\omega$ *and*

$$v_k(t) := \begin{cases} t^k(k!)^{-1}e^{j\mu_0\omega t} & \text{if } T_1\le t<T_2\\ 0 & \text{if } T_0\le t<T_1 \text{ or } T_2\le t<T_0+T\end{cases} \tag{8.166}$$

we obtain $v_k = u_k e^{j\mu_0\omega t}$ *with* u_k *from 1 and* $\mathbb{F}(v_k)(\mu) = \mathbb{F}(u_k)(\mu-\mu_0)$ *by Lemma 8.2.106.*

Proof 1. $\widehat{u}_k(0) = T^{-1}\int_{T_1}^{T_2} t^k(k!)^{-1}dt = T^{-1}((k+1)!)^{-1}\left(T_2^{k+1}-T_1^{k+1}\right)$.
2. Obvious. □

Corollary 8.2.109 *1. We have thus computed the Fourier series of all signals*

$$u\in\mathcal{P}^{0,\mathrm{pc}},\ u(t) := \begin{cases} t^k(k!)^{-1}e^{\lambda t} & \text{if } T_1\le t<T_2\\ 0 & \text{if } T_0\le t<T_1 \text{ or } T_2\le t<T_0+T\end{cases},$$
$$\text{where } \lambda\in\mathbb{C}, k\in\mathbb{N},\ T_0\le T_1<T_2\le T_0+T, \tag{8.167}$$

and hence also of their $\mathbb{C}$*-linear combinations. Practically all periodic signals from electrical engineering have this form, cf. [8, pp. 250–253]. Their Fourier series follow constructively from Theorem 8.2.107 and Corollary 8.2.108.*

2. *The results of 1 are extended to periodic vector functions componentwise.*

◊

For the inputs from Corollary 8.2.109 also the steady-state outputs $H\circ u$ can be easily computed. We consider an IO behavior $\mathcal{B}$ with proper transfer matrix $H = H_0 + H_{\mathrm{spr}} = P^{-1}Q \in \mathbb{C}(s)_{\mathrm{per}}^{p\times m}$ as in Theorem 8.2.102, but for notational simplicity we first assume the SISO case $p=m=1,\ H\in\mathbb{C}(s)_{\mathrm{per}}$, only.

Theorem 8.2.110 (The Fourier series of the steady-state output)

1. *Consider a SISO behavior $\mathcal{B}$ with proper transfer function $H = H_0 + H_{spr} = P^{-1}Q \in \mathbb{C}(s)_{per}$, the periodic input u from* (8.163) *and* (8.164) *and its steady-state output $y := H \circ u = H_0 u + H_{spr} \circ u$. Then*

$$\widehat{y} := \mathbb{F}(y),\ \widehat{y}(\mu) = H_0\widehat{u}(\mu) + H_{spr}(j\mu\omega)\widehat{u}(\mu) = H_0\widehat{u}(\mu) +$$

$$+\sum_{i=0}^{k} T^{-1}(i!)^{-1}(j\mu\omega - \lambda)^{-(k+1-i)} H_{spr}(j\mu\omega)\left(T_1^i e^{(\lambda-j\mu\omega)T_1} - T_2^i e^{(\lambda-j\mu\omega)T_2}\right),$$

$$y = H_0 u + \sum_{i=0}^{k}\sum_{\mu\in\mathbb{Z}} T^{-1}(i!)^{-1}(j\mu\omega - \lambda)^{-(k+1-i)} H_{spr}(j\mu\omega)\cdot$$

$$\cdot\left(T_1^i e^{(\lambda-j\mu\omega)T_1} - T_2^i e^{(\lambda-j\mu\omega)T_2}\right) e^{j\mu\omega t} \in \mathcal{P}^{0,\mathrm{pc}}. \tag{8.168}$$

For all $i = 0, \ldots, k$ the Fourier series $\sum_\mu$ are uniformly convergent and thus continuous. A possible discontinuity of y can only arise from u and the summand $H_0 u$ of y.

2. *For $\lambda = 0$ Eq.* (8.168) *is valid except for the summands for $\mu = 0$. The steady-state output is given by*

$$y = H_0 u + H_{spr}(0)T^{-1}((k+1)!)^{-1}(T_2^{k+1} - T_1^{k+1}) + \sum_{i=0}^{k}\ \sum_{\mu\in\mathbb{Z},\ \mu\neq 0} X, \tag{8.169}$$

where X is the expression from (8.168) *for $\lambda = 0$.*

3. *For $\lambda = j\mu_0\omega$ the signal u from* (8.167) *has the form $u = \widetilde{u}e^{j\mu_0\omega t}$ where $\widetilde{u}$ is the signal from 2. With Lemma 8.2.106 we conclude*

$$\mathbb{F}(u)(\mu) = \mathbb{F}(\widetilde{u})(\mu - \mu_0),\ y = H_0 u + \sum_{\mu\in\mathbb{Z}} H_{spr}(j\mu\omega)\mathbb{F}(u)(\mu)e^{j\mu\omega t} =$$

$$= H_0 u + \sum_{\mu\in\mathbb{Z}} H_{spr}(j\mu\omega)\mathbb{F}(\widetilde{u})(\mu - \mu_0)e^{j\mu\omega t}. \tag{8.170}$$

4. *The generalization to MIMO systems with proper $H = H_0 + H_{spr} = P^{-1}Q \in \mathbb{C}(s)_{per}^{p\times m}$ is obvious. In 1, 2, 3 one replaces u by $u\xi$, $\xi \in \mathbb{C}^m$, and H_0 resp. $H_{spr}(j\mu\omega)$ by $H_0\xi$ resp. $H_{spr}(j\mu\omega)\xi$. So for arbitrary finite sums of such inputs, hence for almost all periodic input signals of MIMO behaviors that are used in electrical engineering, the corresponding steady-state output can be computed algorithmically.*

Proof 1. From Theorem 8.2.107 and Corollary 8.2.108 we know the expression for $\widehat{u}(\mu)$. From (8.153) we infer

$$\widehat{y}(\mu) = H(j\mu\omega)\widehat{u}(\mu) = H_0\widehat{u}(\mu) + H_{spr}(j\mu\omega)\widehat{u}(\mu).$$

Together we obtain (8.168). Moreover, for $i = 0, \dots, k$,

$$\left((j\mu\omega - \lambda)^{-(k+1-i)} H_{\rm spr}(j\mu\omega)\left(T_1^i e^{(\lambda - j\mu\omega)T_1} - T_2^i e^{(\lambda - j\mu\omega)T_2}\right)\right)_{\mu\in\mathbb{Z}} \in \ell^2(\mathbb{Z})\cdot\ell^2(\mathbb{Z})\cdot\ell^\infty(\mathbb{Z}) \subset \ell^1(\mathbb{Z}).$$

This implies the uniform convergence and the continuity of the Fourier series.
2., 3. follow likewise from Corollary 8.2.108.
4. Obvious. □

Many signals in applications are real and have symmetry properties. The following theorem holds for these. According to (8.53)–(8.55) we consider

$$\sigma(u) := \check{u},\ \sigma(u)(t) := u(-t) \text{ if } u \in \mathrm{C}^0,\ \sigma(u)(\varphi) := u(\sigma(\varphi)) \text{ if } u \in \mathrm{C}^{-\infty},\ \varphi \in \mathrm{C}_0^\infty. \tag{8.171}$$

A distribution u with $\sigma(u) = u$ resp. $\sigma(u) = -u$ is called *even resp. odd.*

Theorem 8.2.111 (Fourier transform for real and symmetric periodic signals)

1. *Consider*

$$u = (1 - s^2)^d \circ v \in \mathcal{P}^{-2d},\ d \geq 0,\ v \in \mathcal{P}^0,$$
$$\widehat{u} := \mathbb{F}(u),\ \widehat{v} := \mathbb{F}(v),\ \widehat{u}(\mu) = (1 + \mu^2\omega^2)^d \widehat{v}(\mu).$$

(i) Then

$$u \in \mathcal{P}_{\mathbb{R}}^{-2d},\ \text{i.e., } u \text{ is real or } v \in \mathcal{P}_{\mathbb{R}}^0 \Longleftrightarrow \forall \mu \in \mathbb{Z}:\ \mathbb{F}(u)(-\mu) = \overline{\mathbb{F}(u)(\mu)}. \tag{8.172}$$

(ii) For $u \in \mathcal{P}_{\mathbb{R}}^{0,\mathrm{pc}}$ this implies, as in Corollary 8.2.91,

$$u = a_0 + \sum_{\mu=1}^{\infty}\left(a_\mu \cos(\mu\omega t) + b_\mu \sin(\mu\omega t)\right),\ a_0 := \widehat{u}(0),\ a_\mu - jb_\mu := 2\widehat{u}(\mu). \tag{8.173}$$

2. *Assume $u \in \mathcal{P}^{-\infty}$. Then*

$$\begin{aligned}
&\sigma(u) \in \mathcal{P}^{-\infty},\ \mathbb{F}(\sigma(u))(\mu) = \mathbb{F}(u)(-\mu) \text{ and}\\
&(\sigma(u) = u \Longleftrightarrow \forall \mu \in \mathbb{Z}:\ \mathbb{F}(u)(-\mu) = \mathbb{F}(u)(\mu))\\
&(\sigma(u) = -u \Longleftrightarrow \forall \mu \in \mathbb{Z}:\ \mathbb{F}(u)(-\mu) = -\mathbb{F}(u)(\mu))\,.
\end{aligned} \tag{8.174}$$

For real $u \in \mathcal{P}_{\mathbb{R}}^{-\infty}$ this implies

$$\begin{aligned}
&\sigma(u) = u \Longleftrightarrow \forall \mu \in \mathbb{Z}:\ \overline{\mathbb{F}(u)(\mu)} = \mathbb{F}(u)(\mu),\ \text{i.e., } \mathbb{F}(u)(\mu) \in \mathbb{R}\\
&\sigma(u) = -u \Longleftrightarrow \forall \mu \in \mathbb{Z}:\ \overline{\mathbb{F}(u)(\mu)} = -\mathbb{F}(u)(\mu),\ \text{i.e., } \mathbb{F}(u)(\mu) \in \mathbb{R}j.
\end{aligned} \tag{8.175}$$

For real $u \in \mathcal{P}^{0,\mathrm{pc}}$ *this means*

$$u \text{ even} \Longleftrightarrow u = a_0 + \sum_{\mu=1}^{\infty} a_\mu \cos(\mu\omega t),\ a_\mu \in \mathbb{R},\ \text{where}$$
$$a_0 := \widehat{u}(0),\ \forall \mu \geq 1:\ a_\mu := 2\widehat{u}(\mu), \tag{8.176}$$
$$u \text{ odd} \Longleftrightarrow u = \sum_{\mu=1}^{\infty} b_\mu \sin(\mu\omega t),\ \forall \mu \geq 1:\ b_\mu := 2j\widehat{u}(\mu) \in \mathbb{R}.$$

3. *If* $u \in \left(\mathcal{P}_{\mathbb{R}}^{-\infty}\right)^m$ *and* $H \in \left(\mathbb{R}(s) \bigcap \mathbb{C}(s)_{\mathrm{per}}\right)^{p\times m}$ *are real, then also the periodic output* $y := H \circ u$ *with* $\mathbb{F}(y)(\mu) = H(j\mu\omega)\mathbb{F}(u)(\mu)$ *is real.*

Proof 1.(i) $\Longrightarrow$:

$$\widehat{v}(-\mu) = T^{-1}\int_0^T v(t)e^{j\mu\omega t}dt = \overline{T^{-1}\int_0^T v(t)e^{-j\mu\omega t}dt} = \overline{\widehat{v}(\mu)}$$
$$\Longrightarrow \widehat{u}(-\mu) = (1+\mu^2\omega^2)^d\widehat{v}(-\mu) = \overline{(1+\mu^2\omega^2)^d\widehat{v}(\mu)} = \overline{\widehat{u}(\mu)}$$

$\Longleftarrow$: Let

$$w := (1-s^2)^{-1} \circ v = (1-s^2)^{-(d+1)} \circ u \in \mathcal{P}^2,\ \widehat{w} := \mathbb{F}(w),$$
$$\Longrightarrow \widehat{w}(-\mu) = (1+\mu^2\omega^2)^{-(d+1)}\widehat{u}(-\mu) = (1+\mu^2\omega^2)^{-(d+1)}\overline{\widehat{u}(\mu)}$$
$$= \overline{(1+\mu^2\omega^2)^{-(d+1)}\widehat{u}(\mu)} = \overline{\widehat{w}(\mu)}$$
$$\Longrightarrow w = \sum_{\mu\in\mathbb{Z}} \widehat{w}(\mu)e^{j\mu\omega t} = \overline{w} \in \mathcal{P}_{\mathbb{R}}^2 \Longrightarrow u = (1-s^2)^{d+1} \circ w \in \mathcal{P}_{\mathbb{R}}^{-2d}.$$

2. For continuous u the equation follows from

$$\mathbb{F}(\sigma(u))(\mu) = T^{-1}\int_0^T u(-t)e^{-j\mu\omega t}dt = T^{-1}\int_0^{-T} u(t)e^{j\mu\omega t}(-dt)$$
$$= T^{-1}\int_{-T}^{0} u(t)e^{j\mu\omega t}dt = T^{-1}\int_0^T u(t)e^{j\mu\omega t}dt = \mathbb{F}(u)(-\mu).$$

For $u \in \mathcal{P}^{-\infty}$ the same argument as in 1 applies.
3. $\overline{\mathbb{F}(y)(-\mu)} = \overline{H(-j\mu\omega)\mathbb{F}(u)(-\mu)} \underset{H\text{ real}}{=} H(j\mu\omega)\mathbb{F}(u)(\mu) = \mathbb{F}(y)(\mu).$ □

Example 8.2.112 (Cf. [8, p.119]) Consider the piecewise continuous function $u \in \mathcal{P}^{0,\mathrm{pc}}$, defined by

$$u(t) := \begin{cases} 1 & \text{if } 0 \leq t < T/2 \\ -1 & \text{if } T/2 \leq t < T \end{cases} \Longrightarrow u = u_1 - u_2 \text{ where} \tag{8.177}$$
$$u_1|_{[0,T/2)} = e_{0,0}|_{[0,T/2)},\ u_2|_{[T/2,T)} = e_{0,0}|_{[T/2,T)},\ u(-t) = -u(t).$$

With $\widehat{u} := \mathbb{F}(u)$, $\widehat{u}_1 := \mathbb{F}(u_1)$, $\widehat{u}_2 := \mathbb{F}(u_2)$ we get $\widehat{u}(\mu) = \widehat{u}_1(\mu) - \widehat{u}_2(\mu)$. From (8.165) we infer

$$
\begin{aligned}
&\widehat{u}_1(0) = T^{-1}(T/2 - 0) = 1/2,\ \widehat{u}_2(0) = T^{-1}(T - T/2) = 1/2 \text{ and } \forall \mu \neq 0 : \\
&\widehat{u}_1(\mu) = (j\mu\omega)^{-1}T^{-1}\left(e^{-j\mu\omega 0} - e^{-j\mu\omega T/2}\right) \\
&\quad \underset{\omega T = 2\pi,\ e^{-j\omega T/2} = -1}{=} \begin{cases} 2\pi^{-1}\mu^{-1}(2j)^{-1} & \text{if } \mu \in 1 + 2\mathbb{Z} \\ 1/2 & \text{if } \mu = 0 \\ 0 & \text{if } \mu \in 2\mathbb{Z}\setminus\{0\} \end{cases} \\
&\widehat{u}_2(\mu) \underset{\text{likewise}}{=} \begin{cases} -2\pi^{-1}\mu^{-1}(2j)^{-1} & \text{if } \mu \in 1 + 2\mathbb{Z} \\ 1/2 & \text{if } \mu = 0 \\ 0 & \text{if } \mu \in 2\mathbb{Z}\setminus\{0\} \end{cases}
\end{aligned} \tag{8.178}
$$

We infer

$$
\begin{aligned}
u_1 &= 1/2 + 2\pi^{-1} \sum_{\mu \in 1+2\mathbb{N}} \mu^{-1}(2j)^{-1}\left(e^{j\mu\omega t} - e^{-j\mu\omega t}\right) \\
&= 1/2 + 2\pi^{-1} \sum_{\mu \in 1+2\mathbb{N}} \mu^{-1}\sin(\mu\omega t) \\
u_2 &= 1/2 - 2\pi^{-1} \sum_{\mu \in 1+2\mathbb{N}} \mu^{-1}\sin(\mu\omega t),\ u = 4\pi^{-1} \sum_{\mu \in 1+2\mathbb{N}} \mu^{-1}\sin(\mu\omega t).
\end{aligned} \tag{8.179}
$$

The Fourier series converges in the $\| - \|_2$-norm. ◊

Example 8.2.113 (Cf. [8, p. 252, (13)] Consider the periodic function $u(t) = |\sin(\omega t)|$ and $\widehat{u} := \mathbb{F}(u)$. Since

$$
|\sin(\omega t)|_{[T/2,T)} = -\sin(\omega t)_{[T/2,T)} \text{ and } \sin(\omega t) = (2j)^{-1}\left(e^{j\omega t} - e^{-j\omega t}\right)
$$

we get, with u_1, $\widehat{u}_1$, u_2, $\widehat{u}_2$ from Example 8.2.112,

$$
\begin{aligned}
u &= (2j)^{-1}\left(e^{j\omega t}u_1 - e^{-j\omega t}u_1 - e^{j\omega t}u_2 + e^{j\omega t}u_2\right) \\
&\underset{\text{Lemma 8.2.106}}{\Longrightarrow} \widehat{u}(\mu) = (2j)^{-1}\left(\widehat{u}_1(\mu - 1) - \widehat{u}_1(\mu + 1) - \widehat{u}_2(\mu - 1) + \widehat{u}_2(\mu + 1)\right).
\end{aligned}
$$

Since u is real, it suffices to compute $\widehat{u}(\mu)$, $\mu \geq 0$. But

$$\begin{aligned}
\widehat{u}(0) &= (2j)^{-1}2\pi^{-1}(2j)^{-1}(-1-1-1-1) = 2\pi^{-1}\\
\widehat{u}(1) &= (2j)^{-1}\left(\widehat{u}_1(0)-\widehat{u}_1(2)-\widehat{u}_2(0)+\widehat{u}_2(2)\right)\\
&= (2j)^{-1}\left(2^{-1}-0-2^{-1}+0\right) = 0\\
\widehat{u}(\mu) &= 0 \text{ for } \mu \in 3+2\mathbb{N} \text{ since } \widehat{u}_i(\mu-1) = \widehat{u}_i(\mu+1) = 0,\ i = 1,2\\
\widehat{u}(\mu) &= (2j)^{-1}2\pi^{-1}(2j)^{-1}\left((\mu-1)^{-1}-(\mu+1)^{-1}+(\mu-1)^{-1}-(\mu+1)^{-1}\right)\\
&= -2\pi^{-1}((\mu-1)(\mu+1))^{-1} \in \mathbb{R} \text{ for } \mu \in 2+2\mathbb{N}.
\end{aligned}$$

Since u is real and even we infer

$$\begin{aligned}
u &= \widehat{u}(0) + \sum_{\mu\in 2+2\mathbb{N}} \widehat{u}(\mu)\left(e^{j\mu\omega t}+e^{-j\mu\omega t}\right)\\
&= 2\pi^{-1} - 4\pi^{-1}\sum_{\mu\in 2+2\mathbb{N}} ((\mu-1)(\mu+1))^{-1}\cos(\mu\omega t) \in \mathcal{P}^0.
\end{aligned} \tag{8.180}$$

This Fourier series converges uniformly in $\mathcal{P}^0$. ◊

Example 8.2.114 (Cf. [8, p.117]) Consider the real and even continuous function $u \in \mathcal{P}^0$, defined by

$$\begin{aligned}
&u(t) := \begin{cases} t & \text{if } 0 \le t \le T/2 \\ T-t & \text{if } T/2 \le t \le T \end{cases},\ \widehat{u} := \mathbb{F}(u), \Longrightarrow u = u_3 - u_4 + Tu_2 \text{ with}\\
&u_3|_{[0,T/2)} = e_{0,1}|_{[0,T/2)},\ u_4|_{[T/2,T)} = e_{0,1}|_{[T/2,T)},\ u_2 \text{ from (8.177)},\ e_{0,1}(t) = t\\
&\Longrightarrow \widehat{u} = \widehat{u}_3 - \widehat{u}_4 + T\widehat{u}_2,\ \widehat{u}_i := \mathbb{F}(u_i).
\end{aligned} \tag{8.181}$$

Since u is real and even, Lemma 8.2.111 implies

$$\widehat{u}(-\mu) = \overline{\widehat{u}(\mu)} = \widehat{u}(\mu) = \Re(\widehat{u}_3(\mu)) - \Re(\widehat{u}_4(\mu)) + T\Re(\widehat{u}_2(\mu)) \in \mathbb{R}.$$

We apply (8.165) and (8.178). For $\mu = 0$ we get

$$\begin{aligned}
\widehat{u}(0) &= \widehat{u}_3(0) - \widehat{u}_4(0) + T\widehat{u}_2(0)\\
&= (2T)^{-1}((T/2)^2 - 0^2) - (2T)^{-1}(T^2 - (T/2)^2) + TT^{-1}(T - T/2) = T/4.
\end{aligned}$$

For $\mu > 0$ Eq. (8.165) gives

$$\begin{aligned}
\widehat{u}_3(\mu) &= (j\mu\omega)^{-2}T^{-1}\left(e^{-j\mu\omega 0} - e^{-j\mu\omega T/2}\right)\\
&\quad + (j\mu\omega)^{-1}T^{-1}\left(0e^{-j\mu\omega 0} - (T/2)e^{-j\mu\omega T/2}\right)\\
&\underset{e^{j\omega T/2}=-1}{\Longrightarrow} \Re(\widehat{u}_3(\mu)) = -(\mu\omega)^{-2}T^{-1}(1-(-1)^{\mu}),\\
\widehat{u}_4(\mu) &= (j\mu\omega)^{-2}T^{-1}\left(e^{-j\mu\omega T/2} - e^{-j\mu\omega T}\right)\\
&\quad + (j\mu\omega)^{-1}T^{-1}\left((T/2)e^{-j\mu\omega T/2} - Te^{-j\mu\omega T}\right)
\end{aligned}$$

$$\Longrightarrow \Re(\widehat{u}_4(\mu)) = -(\mu\omega)^{-2}T^{-1}((-1)^{\mu} - 1).$$

Finally (8.178) implies $T\widehat{u}_2(\mu) \in \mathbb{R}j$ and hence $T\Re(\widehat{u}_2(\mu)) = 0,\ \mu \neq 0$. Summing up we obtain

$$\begin{aligned}
&\forall \mu \geq 1 : \widehat{u}(\mu) = -(\mu^2\omega^2)^{-1}T^{-1}\left(2(1-(-1)^{\mu})\right) = -T(2\pi)^{-2}\mu^{-2}\left(2(1-(-1)^{\mu})\right)\\
&\Longrightarrow u = T/2\left(1/2 - 4\pi^{-2}\sum_{\mu\in 1+2\mathbb{N}}\mu^{-2}\cos(\mu\omega t)\right).
\end{aligned} \tag{8.182}$$

Again this series converges uniformly in $\mathcal{P}^0$. ◊

Example 8.2.115 Consider

$$\begin{aligned}
&T_0 := T_1 := -T/2 < T_2 = T_0 + T = T/2,\ u|_{[T_1,T_2)}(t) := t^3\\
&\Longrightarrow u|_{[T_1,T_2)} = 6e_{0,3}|_{[T_1,T_2)},\ e_{0,3} = t^3(3!)^{-1},\ \widehat{u} := \mathbb{F}(u) = 6\mathbb{F}(e_{0,3}).
\end{aligned}$$

The periodic function u is real and odd, and hence

$$\begin{aligned}
&\forall \mu \in \mathbb{Z} : \widehat{u}(-\mu) = \overline{\widehat{u}(\mu)} = -\widehat{u}(\mu)\\
&\Longrightarrow \widehat{u}(0) = 0,\ \forall \mu \geq 1 : \widehat{u}(\mu) \in \mathbb{R}j,\ \widehat{u}(-\mu) = -\widehat{u}(\mu),\\
&\underset{(8.176)}{\Longrightarrow} u = \sum_{\mu=1}^{\infty} b_{\mu}\sin(\mu\omega t),\ b_{\mu} := 2j\widehat{u}(\mu).
\end{aligned}$$

For $\mu \geq 1$ Eq. (8.165) gives

$$\begin{aligned}
\widehat{u}(\mu) &= \sum_{i=0}^{3}(j\mu\omega)^{-(i+1)}x_0^{(3-i)}(\mu) = (j\mu\omega)^{-1}x_0^{(3)}(\mu) + (j\mu\omega)^{-3}x_0^{(1)}(\mu) \text{ where}\\
x_0^{(3-i)}(\mu) &= 6T^{-1}((3-i)!)^{-1}\left((-T/2)^{3-i}e^{-j\mu\omega(-T/2)} - (T/2)^{3-i}e^{-j\mu\omega(T/2)}\right)\\
&\underset{e^{j\mu\omega T/2}=-1}{=} \begin{cases} 6T^{-1}((3-i)!)^{-1}(-2)(T/2)^{3-i}(-1)^{\mu} & \text{if } i = 0, 2\\ 0 & \text{if } i = 1, 3 \end{cases}\\
\Longrightarrow \widehat{u}(\mu) &= -j\mu^{-1}\omega^{-1}6T^{-1}6^{-1}(-2)(T/2)^3(-1)^{\mu}\\
&\quad + j\mu^{-3}\omega^{-3}6T^{-1}(-2)(T/2)(-1)^{\mu}\\
&\underset{\omega^{-1}=T(2\pi)^{-1}}{=} j(-1)^{\mu}(T/2\pi)^3\left(\mu^{-3}(\pi^2\mu^2 - 6)\right)\\
\Longrightarrow b_{\mu} &= 2j\widehat{u}(\mu) = -2(-1)^{\mu}(T/(2\pi))^3\mu^{-3}\left(\pi^2\mu^2 - 6\right),\\
u &= -2(T/(2\pi))^3\sum_{\mu=1}^{\infty}(-1)^{\mu}\mu^{-3}\left(\pi^2\mu^2 - 6\right)\sin(\mu\omega T).
\end{aligned}$$

The partial sums converge to u in the $\| - \|_2$-norm. ◊

8.2.9 *Mikusinski's Calculus*

Whereas we have developed a simplified one-dimensional distribution theory, some systems theorists, for instance M. Fliess, use *Mikusinski's calculus* [14] of generalized functions. In this section we explain the connection of these two theories, see also [3, p. 173].
We start with the fact, unproven in this book (compare Theorem 8.2.65), that the commutative ring $\mathrm{C}_+^0 = \mathrm{C}_+^0(\mathbb{R}, \mathbb{C})$ with the convolution multiplication $*$ has no zero divisors. Therefore, its has a quotient field

$$\mathcal{K} := \mathrm{quot}(\mathrm{C}_+^0) := \left\{\tfrac{u}{v};\ u, v \in \mathrm{C}_+^0,\ v \neq 0\right\}$$

which is constructed as usual, and where C_+^0 is embedded via $\mathrm{C}_+^0 \ni u = \frac{u*v}{v} \in \mathcal{K}$ for an arbitrary nonzero $v \in \mathrm{C}_+^0$. The multiplication in the quotient field is also denoted by $*$. Mikusinski introduced $\mathcal{K}$ as his space of *generalized functions* that is adapted to time functions on the time interval $[0, \infty)$ or, more generally, with left bounded support. In systems theory $\mathcal{K}$ is used both as signal space and as ring of operators.

Theorem 8.2.116 *Let φ be any nonzero function in C_+^∞. Then the map*

$$\mathrm{C}_+^{-\infty}(\mathbb{R}, \mathbb{C}) \longrightarrow \mathcal{K},\quad u \longmapsto \tfrac{u*\varphi}{\varphi}, \tag{8.183}$$

is well defined and a ring monomorphism. Moreover, the map is independent of the choice of φ. We identify $\mathrm{C}_+^{-\infty}(\mathbb{R}, \mathbb{C}) \subseteq \mathcal{K}$ by $u = \frac{u*\varphi}{\varphi}$. In Theorem 8.2.68 we have seen that the map $\mathbb{C}(s) \longrightarrow \mathrm{C}_+^{-\infty}(\mathbb{R}, \mathbb{C})$, $H \longmapsto H \circ \delta$, where H is mapped to its impulse response, is also a ring monomorphism. Hence we identify

$$\mathbb{C}(s) \subseteq \mathrm{C}_+^{-\infty} \subseteq \mathcal{K} \ni H = H \circ \delta = \tfrac{H \circ \varphi}{\varphi} \text{ where } 0 \neq \varphi \in \mathrm{C}_+^\infty .$$

More generally, one gets

$$H = \tfrac{H \circ u}{u} = \tfrac{u}{H^{-1} \circ u} \text{ for } 0 \neq H \in \mathbb{C}(s) \text{ and } 0 \neq u \in \mathrm{C}_+^\infty$$

and, with $tY, t^2Y \in \mathrm{C}_+^0$, the special identifications

$$1 = \delta = \tfrac{tY}{tY},\quad s = s \circ \delta = \delta' = \tfrac{2s^{-2} \circ \delta}{2s^{-3} \circ \delta} = \tfrac{2tY}{t^2Y},\ \text{and } s^{-1} = \tfrac{2s^{-3} \circ \delta}{2s^{-2} \circ \delta} = \tfrac{t^2Y}{2tY}.$$

The space $\mathrm{C}_+^{-\infty}$ is a proper *subset of $\mathcal{K}$, indeed no $\varphi^{-1} = \frac{\varphi}{\varphi^2}$ with $0 \neq \varphi \in \mathrm{C}_+^\infty$ is a distribution.*

Proof According to Theorem 8.2.68, the function $u * \varphi$ is contained in C_+^∞, hence $\frac{u*\varphi}{\varphi} \in \mathrm{quot}(\mathrm{C}_+^0) = \mathcal{K}$. The map (8.183) is $\mathbb{C}$-linear since $u \mapsto u * \varphi$ has this property. To show that the map is injective, assume that $\frac{u*\varphi}{\varphi} = 0$ for some nonzero $\varphi \in \mathrm{C}_+^\infty$. This implies that $u * \varphi = 0$ and thus $u = 0$ since $\varphi \neq 0$ and C_+^0 and hence $\mathrm{C}_+^{-\infty}$ are integral domains as stated, but not proven above. From

$$((u_1 * u_2) * \varphi) * \varphi * \varphi = (u_1 * \varphi) * (u_2 * \varphi) * \varphi$$

follows $\frac{(u_1*u_2)*\varphi}{\varphi} = \frac{u_1*\varphi}{\varphi} * \frac{u_2*\varphi}{\varphi}$, and hence (8.183) is multiplicative. It maps $\delta = 1_{C_+^{-\infty}}$ onto $\frac{\delta*\varphi}{\varphi} = \frac{\varphi}{\varphi} = 1_{\mathcal{K}}$.
The map (8.183) is independent of the choice of φ. Indeed, for $0 \neq \varphi_2 \in C_+^\infty$, the equation $(u * \varphi) * \varphi_2 = (u * \varphi_2) * \varphi$ implies $\frac{u*\varphi}{\varphi} = \frac{u*\varphi_2}{\varphi_2}$. The other equalities follow in the same manner.
Let $\varphi \in C_+^\infty$ and assume that there is a distribution $u \in C_+^{-\infty}$ with $u = \varphi^{-1}$. Then

$$\frac{\varphi}{\varphi^2} = \varphi^{-1} = u = \frac{u*\varphi}{\varphi},$$

which signifies that $(u * \varphi) * \varphi^2 = \varphi^2 = \delta * \varphi^2$ holds. We cancel φ^2 and obtain $u * \varphi = \delta \in C_+^{-\infty}$. But since $\varphi \in C_+^\infty$, also $u * \varphi \in C_+^\infty$ by Theorem 8.2.68, which is a contradiction. □

With the identification

$$H = H \circ \delta = \frac{(H\circ\delta)*\varphi}{\varphi} = \frac{H\circ\varphi}{\varphi} \in \mathbb{C}(s) \subset C_+^{-\infty} \subset \mathcal{K}, \text{ where } 0 \neq \varphi \in C_+^\infty,$$

$H * u \in \mathcal{K}$ is defined for all $u \in \mathcal{K}$. Therefore, the differential equation $P * y = Q * u$ with $P, Q \in \mathbb{C}[s]$, $P \neq 0$, has the unique solution $y = P^{-1}Q * u$ for given $u \in \mathcal{K}$.

Remark 8.2.117 Compared to the space $C^{-\infty}$ of distributions of finite order, which is the least space containing all continuous functions and their derivatives, Mikusinski's space $\mathcal{K}$ of generalized functions has two disadvantages that make it less suitable for systems theory.

1. It contains too many elements without a technical interpretation, e.g., the elements φ^{-1} for $0 \neq \varphi \in C_+^\infty$, which are not even distributions.
2. Since $\mathcal{K}$ is a field and therefore torsionfree, the differential equations $P * y = 0$ with $0 \neq P \in \mathbb{C}[s]$ have only the zero solution in $\mathcal{K}$. Therefore, autonomous systems are trivial for this signal space. Matters like stability of autonomous systems cannot be treated using $\mathcal{K}$. ◊

8.2.10 Impulsive Behaviors

This section was inspired by Bourlès' paper [15] where also the history of this subject is presented. Vardulakis [7, Sect. 4.2 on pp. 176–216] exposes this material too. The following presentation is different from and independent of these quoted sources. We expose the theory for the complex ground field $\mathbb{C}$, but it is equally valid for the real field $\mathbb{R}$. Impulsive trajectories were discussed in Theorem 8.2.58. Here we construct the associated impulsive behaviors, cf. [16].
As signal space we use the $\mathbb{C}(s)$-subspace

$$\mathcal{F}_1 = \mathrm{C}^\infty(\mathbb{R}, \mathbb{C})Y \oplus \mathbb{C}[s] \circ \delta \subset \mathrm{C}_+^{-\infty}(\mathbb{R}, \mathbb{C})$$

from Theorem 8.2.43, 1, or, alternatively from Theorem 8.2.43, 2,

$$\mathcal{F}_2 = \mathrm{t}(\mathrm{C}^\infty)Y \bigoplus \mathbb{C}[s] \circ \delta = \left(\bigoplus_{\lambda \in \mathbb{C}} \mathbb{C}[t]e^{\lambda t}\right)Y \oplus \mathbb{C}[s] \circ \delta \subset \mathrm{C}_+^{-\infty}(\mathbb{R}, \mathbb{C}).$$

Let $\mathcal{B}$ be an IO behavior with transfer matrix H, assume that $\binom{y}{u} \in \mathcal{B} \cap \mathcal{F}_1^{p+m}$ is in the part of $\mathcal{B}$ which is initially at rest, and so, by Theorem and Definition 8.2.41, $y = H \circ u$. In this section we choose inputs of the form $u = vY \in (\mathrm{C}^\infty)^m Y \subset \mathcal{F}_1^m$, which typically occur when a machine is switched on at time $t = 0$. The *impulsive part* of the output $y = H \circ u \in \mathcal{F}_1^p = (\mathrm{C}^\infty)^p Y \oplus \mathbb{C}[s]^p \circ \delta$ is its component in $\mathbb{C}[s]^p \circ \delta$. This part is potentially damaging for a technical device. However, it is zero if the transfer matrix is proper, cf. Theorem 8.2.59. In Theorem 8.2.120 we show that it is determined by the values of v and of finitely many of its derivatives v at zero. In Theorem 8.2.124 we characterize the impulsive parts of all $H \circ u$ with $u \in (\mathrm{C}^\infty)^m Y$ by a discrete behavior $\mathcal{B}_{\mathrm{imp}}$, the *impulsive behavior* of $\mathcal{B}$. Let

$$\mathrm{proj}\colon \mathcal{F}_1 \longrightarrow \mathbb{C}[s] \circ \delta \quad \text{with} \quad \ker(\mathrm{proj}) = \mathrm{C}^\infty Y \tag{8.184}$$

be the projection onto the second direct summand of $\mathcal{F}_1$. We use the same notation proj for the componentwise extension $\mathrm{proj}\colon \mathcal{F}_1^p \longrightarrow \mathbb{C}[s]^p \circ \delta$.

Lemma and Definition 8.2.118 *If $H \in \mathbb{C}(s)$ is proper, then $H \circ (\mathrm{C}^\infty Y) \subseteq \mathrm{C}^\infty Y$. In contrast, if H is not proper, then according to Theorem 8.2.59, $H \circ Y = H \circ (1Y)$ has a nonzero component in $\mathbb{C}[s] \circ \delta$. Therefore, if $v \in \mathrm{C}^\infty$ and*

$$H = H_{\mathrm{pol},+} + H_{\mathrm{pr}} \in \mathbb{C}(s) = \mathbb{C}[s]_+ \oplus \mathbb{C}(s)_{\mathrm{pr}}, \ \ \mathbb{C}[s]_+ := \mathbb{C}[s]s, \ \ \textit{then}$$
$$\mathrm{proj}\left(H \circ (vY)\right) = \mathrm{proj}\left(H_{\mathrm{pol},+} \circ (vY)\right) \in \mathbb{C}[s] \circ \delta$$

is a linear combination of derivatives of δ and called the impulsive component *of $H \circ (vY)$.*

Proof According to Theorem 8.2.59, the properness of H implies $H \circ \mathrm{C}_+^{0,\mathrm{pc}} \subseteq \mathrm{C}_+^{0,\mathrm{pc}}$ and therefore,

$$H \circ \mathrm{C}^\infty Y \subseteq \mathrm{C}_+^{0,\mathrm{pc}} \cap \left(\mathrm{C}^\infty Y \oplus \mathbb{C}[s] \circ \delta\right)$$
$$\underset{\mathrm{C}^\infty Y \subset \mathrm{C}_+^{0,\mathrm{pc}}}{=} \mathrm{C}^\infty Y \oplus \left(\mathrm{C}_+^{0,\mathrm{pc}} \cap \mathbb{C}[s] \circ \delta\right) = \mathrm{C}^\infty Y.$$

□

The following technical lemma is a preparation for Theorem 8.2.120.

Lemma 8.2.119 *For $w \in \mathrm{C}^\infty$, $i \in \mathbb{N}$, and $j \in \{0, \dots, i\}$: $s^j \circ (w\frac{t^i}{i!}Y) \in \mathrm{C}^\infty Y$.*

Proof From $\frac{t^i}{i!}Y = s^{-(i+1)} \circ \delta$, we infer, for $j \leq i$, that

$$s^j \circ \left(w\frac{t^i}{i!}Y\right) = s^j \circ \left(w(s^{-(i+1)} \circ \delta)\right) = \sum_{k=0}^{j} \binom{j}{k} w^{(j-k)}(s^{-(i+1-k)} \circ \delta)$$
$$= \left(\sum_{k=0}^{j} \binom{j}{k} w^{(j-k)} \frac{t^{i-k}}{(i-k)!}\right) Y \in \mathrm{C}^\infty Y.$$

□

The impulsive component $\mathrm{proj}(H \circ (vY))$ of $H \circ (vY)$ can be explicitly computed as follows. It is zero if H is proper. Therefore assume

$$H = H_{\mathrm{pol},+} + H_{\mathrm{pr}} = H_d s^d + \dots + H_1 s + H_{\mathrm{pr}} \in \mathbb{C}[s]_+^{p\times m} \oplus \mathbb{C}(s)_{\mathrm{pr}}^{p\times m}, \text{ with}$$
$$H_i \in \mathbb{C}^{p\times m},\ d := \deg_s(H_{\mathrm{pol},+}) > 0,\ H_d \neq 0. \tag{8.185}$$

Theorem 8.2.120 *Consider H from* (8.185) *and $v \in (\mathrm{C}^\infty)^m$ with the Taylor series $v = \sum_{i=0}^{d-1} v^{(i)}(0)\frac{t^i}{i!} + \frac{t^d}{d!}w$, $w \in (\mathrm{C}^\infty)^m$, up to order d. The impulsive component of the signal vector vY is*

$$\mathrm{proj}\left(H \circ (vY)\right) = \sum_{k=0}^{d-1} \delta^{(k)} \left(\sum_{i=0}^{d-1-k} H_{k+1+i} v^{(i)}(0)\right) \in \bigoplus_{k=0}^{d-1} \mathbb{C}^p \delta^{(k)} \subset \mathbb{C}[s]^p \circ \delta. \tag{8.186}$$

Proof Recall $\frac{t^i}{i!}Y = s^{-(i+1)} \circ \delta$. Lemma 8.2.119 implies

$$\begin{aligned}
&\mathrm{proj}\left(s^j \circ \left(\tfrac{t^i}{i!}Y\right)\right) = 0 \text{ for } j \leq i \text{ and } \mathrm{proj}\left(s^j \circ \left(\tfrac{t^d}{d!}wY\right)\right) = 0 \text{ for } j \leq d \\
&\Longrightarrow \mathrm{proj}\left(H \circ (vY)\right) \underset{\text{Lemma 8.2.118}}{=} \mathrm{proj}\left(H_{\mathrm{pol},+} \circ (vY)\right) \\
&\quad = \mathrm{proj}\left(\sum_{j=1}^{d} H_j s^j \circ \left(\sum_{i=0}^{d-1} v^{(i)}(0)\frac{t^i}{i!}Y + \frac{t^d}{d!}wY\right)\right) \\
&\quad = \sum_{i=0}^{d-1} \sum_{j=i+1}^{d} H_j v^{(i)}(0)\, \delta^{(j-(i+1))} \\
&\quad \underset{k:=j-i-1}{=} \sum_{k=0}^{d-1} \delta^{(k)} \left(\sum_{i=0}^{d-1-k} H_{k+i+1} v^{(i)}(0)\right).
\end{aligned} \tag{8.187}$$

□

We reformulate (8.186) in matrix form. We define

$$V := \begin{pmatrix} v \\ v^{(1)} \\ \vdots \\ v^{(d-1)} \end{pmatrix} \in (C^\infty)^{md} \quad \text{and} \quad \mathcal{H} := \begin{pmatrix} H_1 & H_2 & \cdots & H_{d-1} & H_d \\ H_2 & H_3 & \cdots & H_d & 0 \\ \vdots & \vdots & & \vdots & \vdots \\ H_d & 0 & \cdots & 0 & 0 \end{pmatrix} \in \mathbb{C}^{pd \times md}. \tag{8.188}$$

We also identify

$$\bigoplus_{k=0}^{d-1} \mathbb{C}^p \delta^{(k)} = \mathbb{C}^{pd} \ni \sum_{k=0}^{d-1} a_k \delta^{(k)} = \begin{pmatrix} a_0 \\ \vdots \\ a_{d-1} \end{pmatrix}.$$

With this, (8.186) obtains the form

$$\begin{aligned} \operatorname{proj}\big(H \circ (vY)\big) &= \sum_{k=0}^{d-1} \delta^{(k)} \big(H_{k+1} v(0) + H_{k+2} v^{(1)}(0) + \cdots + H_d v^{(d-1-k)}(0)\big) \\ &= \sum_{k=0}^{d-1} \delta^{(k)} \big(\mathcal{H} V(0)\big)_k = \mathcal{H} V(0) \in \bigoplus_{k=0}^{d-1} \mathbb{C}^p \delta^{(k)} = \mathbb{C}^{pd}. \end{aligned} \tag{8.189}$$

Corollary 8.2.121 *With the identification* $\mathbb{C}^{pd} = \bigoplus_{k=0}^{d-1} \mathbb{C}^p \delta^{(k)} \subset \mathbb{C}[s]^p \circ \delta$ *and the matrices* V *and* $\mathcal{H}$ *from* (8.188) *the* $\mathbb{C}$*-linear map*

$$\begin{array}{ccccccc} (C^\infty)^m & \xrightarrow{\cdot Y} & (C^\infty)^m Y & \xrightarrow{H \circ} & \mathcal{F}_1^p & \xrightarrow{\operatorname{proj}} & \mathbb{C}[s]^p \circ \delta \\ v & \longmapsto & vY & \longmapsto & H \circ (vY) & \longmapsto & \operatorname{proj}\big(H \circ (vY)\big) \end{array}$$

factorizes as

$$\begin{array}{ccccccc} (C^\infty)^m & \longrightarrow & (C^\infty)^{md} & \xrightarrow{\text{initial value}} & \mathbb{C}^{md} & \xrightarrow{\mathcal{H}} & \mathbb{C}^{pd} = \oplus_{k=0}^{d-1} \mathbb{C}^p \circ \delta^{(k)} \\ v & \longmapsto & V & \longmapsto & V(0) & \longmapsto & \mathcal{H} V(0) \end{array}$$

i.e., the impulsive component can be computed as

$$\operatorname{proj}\big(H \circ (vY)\big) = \mathcal{H} V(0) \underset{\text{identification}}{=} \sum_{k=0}^{d-1} \delta^{(k)} \big(\mathcal{H} V(0)\big)_k.$$

Since the map $(C^\infty)^m \longrightarrow \mathbb{C}^{md}$, $v \longmapsto V(0)$, *is surjective, the space of all impulsive components is*

$$\operatorname{proj}\left(H\circ(\mathrm{C}^\infty)^m Y\right)=\mathcal{H}\mathbb{C}^{md}\subset\mathbb{C}^{pd}=\bigoplus_{k=0}^{d-1}\mathbb{C}^p\delta^{(k)},$$

i.e., the column space of $\mathcal{H}$. A basis of $\operatorname{proj}\left(H\circ(\mathrm{C}^\infty)^m Y\right)$ *can be obtained by means of the Gauß algorithm, applied to $\mathcal{H}$. Notice that the dimension of the used spaces and the size of V and $\mathcal{H}$ depend on the degree d of $H_{\mathrm{pol},+}$.*
The trajectory $H\circ(vY)$ is piecewise continuous if and only if $\mathcal{H}V(0)=0$. This is a linear condition on the initial values $v(0), v'(0), \ldots, v^{(d-1)}(0)$.

Proof The assertion holds since for any vector $x := \begin{pmatrix} x_0\\ \vdots\\ x_{d-1}\end{pmatrix}\in\mathbb{C}^{md}$, the function $v := \sum_{i=0}^{d-1} x_i\frac{t^i}{i!}\in(\mathrm{C}^\infty)^m$ has the initial vector $V(0)=x$. □

The subspace $\operatorname{proj}\left(H\circ(\mathrm{C}^\infty)^m Y\right)$ of the $\mathbb{C}$-infinite dimensional $\mathbb{C}(s)$-space $\mathcal{F}_1^p$ is not a subbehavior of $\mathcal{F}_1^p$ because it is finite-dimensional over $\mathbb{C}$. Therefore, we do *not* call $\operatorname{proj}\left(H\circ(\mathrm{C}^\infty)^m Y\right)$ the impulsive behavior of H as, for instance, Bourlès does in [15]. Instead we are going to show that the subspace $\operatorname{proj}\left(H\circ(\mathrm{C}^\infty)^m Y\right)\subset\mathbb{C}[s]^p\circ\delta$ is isomorphic to a *discrete-time* behavior $\mathcal{B}_{\mathrm{imp}}$ that we call the *impulsive behavior* of H.
We consider the sequence space $\mathbb{C}^{\mathbb{N}}$ with the left shift action of the polynomial algebra $\mathbb{C}[s^{-1}]$ on it, i.e., $(s^{-k}\circ\widetilde{y})(l)=\widetilde{y}(l+k)$, which makes it an injective cogenerator. We are primarily interested in the submodule

$$\mathrm{t}_{s^{-1}}(\mathbb{C}^{\mathbb{N}})=\mathbb{C}^{(\mathbb{N})}=\bigoplus_{k=0}^{\infty}\mathbb{C}\delta_k,\quad\text{where } \delta_k=(0,\ldots,0,\overset{k}{1},0,\ldots),$$

of finite sequences, since the impulsive behavior consists of such sequences. By Corollary 5.3.12, $\mathbb{C}^{(\mathbb{N})}=\mathrm{t}_{s^{-1}}(\mathbb{C}^{\mathbb{N}})$ is an injective $\mathbb{C}[s^{-1}]$-module. Let

$$\phi_{\mathrm{ind}}:\ \mathbb{C}[s]\circ\delta\overset{\cong}{\longrightarrow}\mathbb{C}^{(\mathbb{N})},\quad \sum_{k=0}^{\infty}a_k\delta^{(k)}\longmapsto\sum_{k=0}^{\infty}a_k\delta_k=(a_0,a_1,\ldots),\tag{8.190}$$

denote the canonical isomorphism. Define $\phi:=\phi_{\mathrm{ind}}\operatorname{proj}$, hence

$$\phi:\ \mathcal{F}_1=\mathrm{C}^\infty Y\oplus\mathbb{C}[s]\circ\delta\to\mathbb{C}^{(\mathbb{N})},\ vY+\sum_{k=0}^{\infty}a_k\delta^{(k)}\mapsto(a_0,a_1,\ldots),\ \text{with}$$
$$\ker(\phi)=\ker(\operatorname{proj})=\mathbb{C}^\infty Y,\ \operatorname{im}(\phi)=\mathbb{C}^{(\mathbb{N})}.\tag{8.191}$$

The componentwise extension $\phi:\ \mathcal{F}_1^p\to(\mathbb{C}^{(\mathbb{N})})^p$ is denoted by the same letter, and likewise for ϕ_{ind}.
In Lemma 8.2.118 we have shown that $\mathrm{C}^\infty Y$ is a $\mathbb{C}[s^{-1}]$-submodule of $\mathcal{F}_1$, and $\mathbb{C}^{(\mathbb{N})}$ is also a $\mathbb{C}[s^{-1}]$-module. Since $s^{-1}\circ\delta=Y\notin\mathbb{C}[s]\circ\delta$, clearly, $\mathbb{C}[s]\circ\delta$ is not a $\mathbb{C}[s^{-1}]$-submodule of $\mathcal{F}_1$ and hence ϕ_{ind} is not $\mathbb{C}[s^{-1}]$-linear. But ϕ is $\mathbb{C}[s^{-1}]$-linear, as we will show in the next lemma.

Lemma 8.2.122 *The map ϕ is $\mathbb{C}[s^{-1}]$-linear.*

Proof It suffices to show that $\phi(s^{-1} \circ y) = s^{-1} \circ \phi(y)$ for $y \in \mathrm{C}^\infty Y$ and for $y = \delta^{(k)}$, $k \in \mathbb{N}$. If $y = vY$ with $v \in \mathrm{C}^\infty$, then $\phi(y) = 0$ and hence also $s^{-1} \circ \phi(y) = 0$. Moreover also $s^{-1} \circ y \in \mathrm{C}^\infty Y$ and thus $\phi(s^{-1} \circ y) = 0$.
For the derivatives $y = \delta^{(k)}$, $k \in \mathbb{N}$, of δ holds

$$s^{-1} \circ \delta^{(k)} = \begin{cases} \delta^{(k-1)} & \text{if } k > 0 \\ Y & \text{if } k = 0 \end{cases} \text{ and hence}$$

$$\phi(s^{-1} \circ \delta^{(k)}) = \begin{cases} \delta_{k-1} & \text{if } k > 0 \\ 0 & \text{if } k = 0 \end{cases} = s^{-1} \circ \delta_k = s^{-1} \circ \phi(\delta^{(k)}).$$ □

We have seen in the proof of Lemma 8.2.122 that the left shift of $(a_0, a_1, \ldots) \in \mathbb{C}^{(\mathbb{N})}$ corresponds to the application of s^{-1} to $\sum_{k\in\mathbb{N}} a_k \delta^{(k)}$. Therefore we defined $s^{-1}\circ$ instead of $s\circ$ as the left shift on $\mathbb{C}^{\mathbb{N}}$.
We consider an IO behavior

$$\mathcal{B} = \left\{ \binom{y}{u} \in (\mathrm{C}^{-\infty})^{p+m};\ P \circ y = Q \circ u \right\}$$

with $\mathrm{rank}(P) = p$ and transfer matrix $H \in \mathbb{C}(s)^{p\times m}$ with $PH = Q$. Since $\mathcal{F}_1$ is a $\mathbb{C}(s)$-space by Theorem and Definition 8.2.43, any $u \in \mathcal{F}_1^m$ gives rise to $H \circ u \in \mathcal{F}_1^p$. Theorem 8.2.41 furnishes the isomorphism

$$\mathcal{F}_1^m = \left(\mathrm{C}^\infty Y \oplus \mathbb{C}[s] \circ \delta\right)^m \xrightarrow[\mathbb{C}(s)]{\cong} \mathcal{B} \cap \mathcal{F}_1^{p+m} = \left\{ \binom{y}{u} \in \mathcal{F}_1^{p+m};\ y = H \circ u \right\}. \tag{8.192}$$

Due to $\mathbb{C}(s) = \mathbb{C}(s^{-1})$ we consider H as a rational matrix in s^{-1}. As in Theorem 6.2.4 we construct $(A, -B) \in \mathbb{C}[s^{-1}]^{p\times(p+m)}$ with

$$\begin{aligned} &\mathbb{C}[s^{-1}]^{1\times p} A = \{\xi \in \mathbb{C}[s^{-1}]^{1\times p};\ \xi H \in \mathbb{C}[s^{-1}]^{1\times m}\},\ \mathrm{rank}(A) = p,\ B = AH, \\ &\Longrightarrow \exists \binom{X_A}{X_B} \in \mathbb{C}[s^{-1}]^{(p+m)\times p} \text{ with } \mathrm{id}_p = (A, -B)\binom{X_A}{X_B} = AX_A - BX_B. \end{aligned} \tag{8.193}$$

Since $\mathcal{F}_1$ is a $\mathbb{C}(s)$-space and P, A belong to $\mathrm{Gl}_p(\mathbb{C}(s))$, the equivalences

$$\begin{aligned} \binom{y}{u} \in \mathcal{B} &\Longleftrightarrow P \circ y = Q \circ u = P \circ H \circ u \underset{\mathrm{rank}(P)=p}{\Longleftrightarrow} y = H \circ u \\ &\underset{\mathrm{rank}(A)=p}{\Longleftrightarrow} A \circ y = A \circ (H \circ u) = (AH) \circ u = B \circ u \end{aligned} \tag{8.194}$$

hold for $\binom{y}{u} \in \mathcal{F}_1^{p+m}$ and imply

$$\mathcal{B} \cap \mathcal{F}_1^{p+m} = \left\{ \binom{y}{u} \in \mathcal{F}_1^{p+m};\ y = H \circ u \right\} = \left\{ \binom{y}{u} \in \mathcal{F}_1^{p+m};\ A \circ y = B \circ u \right\}.$$

With $A \in \mathbb{C}[s^{-1}]^{p\times p}$ of rank p the autonomous discrete-time behavior

$$\mathcal{B}_{\mathrm{imp}} := \{\widetilde{y} \in (\mathbb{C}^{(\mathbb{N})})^p;\ A \circ \widetilde{y} = A(s^{-1}) \circ \widetilde{y} = 0\} \tag{8.195}$$

is called the *impulsive behavior* of $\mathcal{B}$ or H. We will show in Theorem 8.2.124 that the space $\mathrm{proj}\left(H \circ (\mathrm{C}^{\infty})^m Y\right)$ of all impulsive components and the discrete behavior $\mathcal{B}_{\mathrm{imp}}$ are isomorphic $\mathbb{C}$-vector spaces.
The elements of $\mathcal{B}_{\mathrm{imp}}$ are finite sequences, and next we give an upper bound for their length.

Lemma 8.2.123 *As before we write $H = H_{\mathrm{pol},+} + H_{\mathrm{pr}} \in \mathbb{C}[s]_+^{p\times m} \oplus \mathbb{C}(s)_{\mathrm{pr}}^{p\times m}$. If $\deg_s(H_{\mathrm{pol},+}) = d$ then $s^{-d} \circ \mathcal{B}_{\mathrm{imp}} = 0$, i.e.,*

$$\mathcal{B}_{\mathrm{imp}} \subseteq \{\widetilde{y} \in (\mathbb{C}^{(\mathbb{N})})^p;\ \forall t \geq d:\ \widetilde{y}(t) = 0\}.$$

Proof The degree $d = \deg_s(H_{\mathrm{pol},+})$ implies $s^{-d}H \in \mathbb{C}(s)_{\mathrm{pr}}^{p\times m}$. From Theorem 8.1.7 we infer the existence of $B_1 \in \mathbb{C}[s^{-1}]^{p\times m}$ and $b \in \mathbb{C}[s^{-1}]$ with nonzero constant term $b(0) \neq 0$ such that $s^{-d}H = B_1 b^{-1}$, hence $s^{-d}bH = B_1 \in \mathbb{C}[s^{-1}]^{p\times p}$. By (8.193) there is $X \in \mathbb{C}[s^{-1}]^{p\times p}$ with $s^{-d}b\,\mathrm{id}_p = XA$, hence

$$b \circ (s^{-d} \circ \mathcal{B}_{\mathrm{imp}}) = s^{-d}b \circ \mathcal{B}_{\mathrm{imp}} = XA \circ \mathcal{B}_{\mathrm{imp}} = 0. \tag{8.196}$$

Because of $b(0) \neq 0$, the polynomials $b = B(s^{-1})$ and s^{-1} are coprime in $\mathbb{C}[s^{-1}]$. Lemma 5.1.5 and $\mathbb{C}^{(\mathbb{N})} = \mathrm{t}_{s^{-1}}(\mathbb{C}^{\mathbb{N}})$ imply the $\mathbb{C}[s^{-1}]$-isomorphism

$$b\circ\colon\ (\mathbb{C}^{(\mathbb{N})})^p \xrightarrow{\cong} (\mathbb{C}^{(\mathbb{N})})^p,\quad \widetilde{y} \longmapsto b \circ \widetilde{y} \underset{(8.196)}{\Longrightarrow} s^{-d} \circ \mathcal{B}_{\mathrm{imp}} = 0. \qquad \square$$

Theorem 8.2.124 (Impulsive behavior) *Let $d := \deg_s(H_{\mathrm{pol},+})$. The identity*

$$\phi\left(H \circ \left(\mathrm{C}^{\infty}\right)^m Y\right) = \mathcal{B}_{\mathrm{imp}}$$

holds and induces the $\mathbb{C}$-isomorphism

$$\phi_{\mathrm{ind}}\colon\ \mathrm{proj}\left(H \circ (\mathrm{C}^{\infty})^m Y\right) \xrightarrow[\mathbb{C}]{\cong} \mathcal{B}_{\mathrm{imp}},\ \sum_{k\in\mathbb{N}} a_k\delta^{(k)} \leftrightarrow \widetilde{y} = (a_0, a_1, \ldots).$$

If $\widetilde{y} = (\widetilde{y}(0), \ldots, \widetilde{y}(d-1), 0, 0, \ldots) \in \mathcal{B}_{\mathrm{imp}}$ and $\widehat{y} := \sum_{k=0}^{d-1} \widetilde{y}(k)\delta^{(k)}$ then

$$v := -X_B A \circ \widehat{y} \in \left(\mathrm{C}^{\infty}\right)^m Y \text{ and } \widetilde{y} = \phi(H \circ v). \tag{8.197}$$

A $\mathbb{C}$-basis of the autonomous behavior $\mathcal{B}_{\mathrm{imp}}$ and then of $\mathrm{proj}\left(H \circ (\mathrm{C}^{\infty})^m Y\right)$ *can be computed by means of Theorem 5.4.11, cf. also Corollary 8.2.121.*

Proof $\subseteq$: Recall that ϕ is $\mathbb{C}[s^{-1}]$-linear. Let

$$\begin{aligned} &u \in \left(\mathrm{C}^\infty\right)^m Y = \ker(\phi) \subset \mathcal{F}_1^m, \ y := H \circ u \in \mathcal{F}_1^p \\ &\Longrightarrow A \circ y = B \circ u \underset{B(s^{-1})\text{ proper}}{\in} \left(\mathrm{C}^\infty\right)^p Y \\ &\Longrightarrow A \circ \phi(y) = \phi(A \circ y) = 0 \Longrightarrow \phi(y) \in \mathcal{B}_{\mathrm{imp}}. \end{aligned}$$

$\supseteq$: Let $\widetilde{y} = (\widetilde{y}(0), \widetilde{y}(1), \ldots, \widetilde{y}(d-1), 0, \ldots) \in \mathcal{B}_{\mathrm{imp}} \underset{\text{Lemma 8.2.123}}{\subseteq} \mathrm{ann}_{(\mathbb{C}^{\mathbb{N}})^p}(s^{-d})$, and define $\widehat{y} := \sum_{k=0}^{d-1} \widetilde{y}(k)\, \delta^{(k)} \in \mathbb{C}[s]^p \circ \delta$. This implies

$$\phi(\widehat{y}) = \widetilde{y} \Longrightarrow \phi(A \circ \widehat{y}) = A \circ \phi(\widehat{y}) = A \circ \widetilde{y} = 0 \Longrightarrow A \circ \widehat{y} \in \ker(\phi) = (\mathrm{C}^\infty)^p Y.$$

With (8.193) we infer

$$A \circ \widehat{y} = AX_A \circ (A \circ \widehat{y}) - BX_B \circ (A \circ \widehat{y}) \Longrightarrow A \circ (\underbrace{-\widehat{y} + X_A A \circ \widehat{y}}_{=:y}) = B \circ (\underbrace{X_B A \circ \widehat{y}}_{=:u}).$$

With (8.194) we get $\binom{y}{u} \in \mathcal{B}$ and $y = H \circ u$. Moreover $A \circ \widehat{y} \in (\mathrm{C}^\infty)^p Y$ implies $u = X_B \circ (A \circ \widehat{y}),\ v := -u \underset{X_B(s^{-1})\text{ proper}}{\in} (\mathrm{C}^\infty)^m Y$. We conclude

$$\begin{aligned} &\phi(H \circ u) = \phi(y) = \phi(-\widehat{y} + X_A A \circ \widehat{y}) = -\phi(\widehat{y}) + X_A A \circ \phi(\widehat{y}) \\ &= -\widetilde{y} + X_A \circ \underbrace{(A \circ \widetilde{y})}_{=0} = -\widetilde{y} \Longrightarrow \widetilde{y} = \phi(H \circ (-u)) = \phi(H \circ v) \in \phi\left(H \circ \left(\mathrm{C}^\infty\right)^m Y\right) \\ &\Longrightarrow \phi\left(H \circ \left(\mathrm{C}^\infty\right)^m Y\right) = \mathcal{B}_{\mathrm{imp}}. \end{aligned}$$

Since $\phi = \phi_{\mathrm{ind}}\,\mathrm{proj}$ and ϕ_{ind} is an isomorphism, we infer the isomorphism

$$\begin{aligned} \phi_{\mathrm{ind}}: \ \mathrm{proj}\left(H \circ ((\mathrm{C}^\infty)^m Y)\right) &\overset{\cong}{\longrightarrow} \phi_{\mathrm{ind}}\left(\mathrm{proj}\left(H \circ ((\mathrm{C}^\infty)^m Y)\right)\right) \\ &= \phi\left(H \circ (\mathrm{C}^\infty)^m Y\right) = \mathcal{B}_{\mathrm{imp}}. \end{aligned}$$

□

8.3 Electrical Networks

8.3.1 The Input/Output Behavior of an Electrical Network

In this section we apply the theory of Sect. 8.2 to the electrical networks from Sect. 3.1.3. Electrical and mechanical engineering are the principal applications of the Sects. 8.2.4 and 8.2.8.

The base field is $F = \mathbb{R}$ or $F = \mathbb{C}$, and the signal space is $\mathcal{F} := \mathcal{F}_F := C^{-\infty}(\mathbb{R}, F)$. The underlying graph (V, K) with vertex (node) set V and edge (branch) set K is connected. Let $n := \sharp(K)$ resp. $m := \sharp(V)$ be the number of edges resp. nodes.
A real behavior $\mathcal{B}_{\mathbb{R}} = \{w \in \mathcal{F}_{\mathbb{R}}^l;\ R \circ w = 0\}$, $R \in \mathbb{R}[s]^{k\times l}$, is often studied by means of its complexification

$$\mathcal{B}_{\mathbb{C}} = \{w \in \mathcal{F}_{\mathbb{C}}^l;\ R \circ w = 0\} \text{ with } \mathcal{B}_{\mathbb{R}} = \mathcal{B}_{\mathbb{C}} \bigcap \mathcal{F}_{\mathbb{R}}^l = \{w \in \mathcal{B}_{\mathbb{C}}^l;\ w = \overline{w}\}. \tag{8.198}$$

According to Lemmas 3.1.18 and 3.1.19 there is the exact sequence

$$0 \to \ker(d) = \sum_{\omega \text{ circuit}} \mathbb{Z}c(\omega) \overset{\subseteq}{\to} C_1(\mathbb{Z}) := \mathbb{Z}^K \xrightarrow{d=A\cdot} C_0(\mathbb{Z}) := \mathbb{Z}^V \overset{\epsilon}{\to} \mathbb{Z} \to 0$$

with $d(k) = \mathrm{dom}(k) - \mathrm{cod}(k)$, $\epsilon(v) = 1$,

$$A_{vk} = A(v,k) = \begin{cases} 1 & \text{if } v = \mathrm{dom}(k) \neq \mathrm{cod}(k) \\ -1 & \text{if } v = \mathrm{cod}(k) \neq \mathrm{dom}(k) \\ 0 & \text{otherwise} \end{cases}$$

$$\Longrightarrow r := \mathrm{rank}(A) = \dim_{\mathbb{Z}}(\mathrm{im}(d)) = \dim_{\mathbb{Z}}(\mathbb{Z}^V) - 1 = m - 1,$$
$$\dim_{\mathbb{Z}}(\ker(d)) = n - r = n - m + 1. \tag{8.199}$$

We also need the exact sequence (8.199) with other coefficients, namely, F and $\mathcal{F}$. We derive this in the following fashion.

Remark 8.3.1 1. If S is a commutative ring, we write S^K as $S^K = \oplus_{k\in K} Sk$ where $k \underset{\text{ident.}}{=} \delta_k = (0, \ldots, \overset{k}{1}, \ldots, 0)$ and $(k)_{k\in K} \in \left(S^K\right)^{1\times K}$ is the standard basis. If N is any S-module there is the isomorphism and subsequent identification

$$\begin{aligned} &\mathrm{Hom}_S(S^K, N) \cong N^K,\ \varphi \mapsto U = (U_k)_{k\in K},\ U_k := \varphi(k), \\ &\Longrightarrow \mathrm{Hom}_S(S^K, N) \underset{\text{ident.}}{=} N^K,\ \varphi = U,\ \varphi(k) = U(k) = U_k. \end{aligned} \tag{8.200}$$

Recall that $\mathrm{Hom}_S(-, N)$ is exact if and only if ${}_SN$ is injective.
2. (Cf. Remark 3.1.16) In the following we use matrix modules

$$N^{V\times K} \ni X = (X(v,k))_{v,k},\ X_{vk} := X_{v,k} := X(v,k),\ N^K = N^{K\times 1},\ N^{1\times K}, \tag{8.201}$$

where V and K are arbitrary finite index sets. Addition and multiplication of these more general matrices are the natural ones. If computer calculations with these matrices are needed, they are transformed into usual matrices in the following fashion: The sets V and K are enumerated by chosen bijections

$$\begin{aligned} &\alpha : \{1, \ldots, m\} \overset{\cong}{\longrightarrow} V,\ \beta : \{1, \ldots, n\} \overset{\cong}{\longrightarrow} K \\ &\Longrightarrow N^{V\times K} \underset{\text{ident.}}{=} N^{m\times n},\ X = (X_{vk})_{v,k} = (X_{\mu\nu})_{\mu,\nu},\ X_{\mu\nu} := X_{\alpha(\mu)\beta(\nu)}. \end{aligned} \tag{8.202}$$

◇

Let $B \in \mathbb{Z}^{K\times(n-r)}$ be chosen such that the columns $B(-,\rho)$, $\rho = 1,\dots,n-r$, are a $\mathbb{Z}$-basis of $\ker(d)$, hence $\ker(d) = \oplus_{\rho=1}^{n-r}\mathbb{Z}B(-,\rho)$. Also let $E := (1)_{v\in V} \in \mathbb{Z}^{1\times V}$ denote the matrix of ϵ, i.e., $\epsilon(\eta) = E\eta$ for $\eta \in \mathbb{Z}^V$. The exact sequence (8.199) implies the exact sequence

$$0 \to \mathbb{Z}^{n-r} \xrightarrow{B\cdot} \mathbb{Z}^K \xrightarrow{d=A\cdot} \mathbb{Z}^V \xrightarrow{\epsilon=E\cdot} \mathbb{Z} \to 0. \tag{8.203}$$

Now let ${}_{\mathbb{Z}}N$ be injective, i.e., divisible. In particular, this is the case for each F-space N, for instance, F and $\mathcal{F}$. Application of the exact functor $\mathrm{Hom}_{\mathbb{Z}}(-,N)$ to (8.203) implies the dual exact sequence

$$0 \leftarrow N^{n-r} \xleftarrow{B^\top\cdot} N^K \xleftarrow{d^*=A^\top\cdot} N^V \xleftarrow{\epsilon^*=E^\top\cdot} N \leftarrow 0 \tag{8.204}$$

with $d^* := \mathrm{Hom}(d,N) \underset{\text{ident.}}{=} A^\top\cdot$. Since ${}_F F$ is injective, application of the exact functor $\mathrm{Hom}_F(-,F)$ to (8.204) with $N := F$ furnishes the exact sequence

$$\begin{aligned}
&0 \to F^{n-r} \xrightarrow{B\cdot} F^K \xrightarrow{d=A\cdot} F^V \xrightarrow{\epsilon=E\cdot} F \to 0 \Longrightarrow \ker(d : F^K \to F^V) = \\
&= \left\{\xi \in F^K;\ A\xi = 0\right\} = BF^{n-r} = F(B\mathbb{Z}^{n-r}) \underset{(8.199)}{=} \sum_{\omega \text{ circuit}} Fc(\omega) \\
&\Longrightarrow F^K/\ker(d) \cong \mathrm{im}(d) = AF^K = \sum_{k\in K} FA(-,k),\ k+\ker(d) \mapsto A(-,k), \\
&\ker(d) = \sum_{\omega \text{ circuit}} Fc(\omega).
\end{aligned} \tag{8.205}$$

This is the sequence (8.199) with F instead of $\mathbb{Z}$.
Since $\mathrm{rank}(A) = r$, there is a disjoint decomposition $K = K_1 \uplus K_2$ with $\sharp(K_1) = r$ such that the columns $A(-,k_1)$, $k_1 \in K_1$, are a basis of $\mathrm{im}(d) = AF^K$. Indeed, the Gauß algorithm, i.e., a series of elementary row operations and column permutations, furnishes the *echelon* form of A in block form as

$$\begin{aligned}
&XA = \begin{array}{c} r \\ m-r \end{array}\!\!\overset{\begin{array}{cc} K_1 & K_2 \end{array}}{\begin{pmatrix} \mathrm{id}_r & M \\ 0 & 0 \end{pmatrix}} \in F^{(r+(m-r))\times(K_1\uplus K_2)},\ m = \sharp(V),\ X \in \mathrm{Gl}_m(F),\ \text{with} \\
&K = K_1 \uplus K_2,\ \sharp(K_1) = \mathrm{rank}(A) = r = m-1, \\
&M := (M(k_1,k_2))_{k_1,k_2} = \left(M_{k_1,k_2}\right)_{k_1,k_2} \in F^{K_1\times K_2} = F^{r\times(n-r)}, \\
&\Longrightarrow \forall k_2 \in K_2 :\ XA(-,k_2) = (XA)(-,k_2) = \sum_{k_1\in K_1} XA(-,k_1)M(k_1,k_2), \\
&\underset{X\in\mathrm{Gl}_m(F)}{\Longrightarrow} A|_{K_2} = A|_{K_1}M,\ A|_{K_i} := (A(v,k_i))_{v,k_i} \in F^{V\times K_i}.
\end{aligned} \tag{8.206}$$

The columns $A(-,k_1)$, $k_1 \in K_1$, are an F-basis of AF^K, hence $\mathrm{rank}(A|_{K_1}) = \sharp(K_1) = r$. The isomorphism $F^K/\ker(d) \cong AF^K$, $k + \ker(d) \mapsto A(-,k)$ then implies the following.

Corollary 8.3.2 *With the data from* (8.206), *i.e.,*

$$K = K_1 \uplus K_2,\ r = \mathrm{rank}(A) = \mathrm{rank}\left(A|_{K_1}\right) = \sharp(K_1) = m-1,\ M \in F^{K_1 \times K_2}, \tag{8.207}$$

the space

$$F^K/\ker(d) = \sum_{k\in K} F\overline{k},\ \overline{k} := k + \ker(d),\ d = A\cdot,$$

has the F-basis $\overline{k_1}$, $k_1 \in K_1$, and the basis representations

$$\forall k_2 \in K_2 : \overline{k_2} = \sum_{k_1\in K_1} \overline{k_1} M(k_1,k_2) \text{ or } \left(\overline{k_2}\right)_{k_2\in K_2} = \left(\overline{k_1}\right)_{k_1\in K_1} M. \tag{8.208}$$

Notice that the decomposition $K = K_1 \uplus K_2$ and the matrix $M \in F^{K_1\times K_2}$ are not unique, but depend on the applied Gauß algorithm. Indeed, each subset $K_1 \subseteq K$ of r branches with F-linearly independent columns $A(-,k_1)$, $k_1 \in K_1$, can be obtained. Often M has all its entries in $\mathbb{Z}$. ◊

According to (3.35) we obtain

$$\begin{aligned}
&\mathrm{Hom}_F(F^K,\mathcal{F}) \underset{\text{identification}}{=} \mathcal{F}^K,\ U = (U(k))_{k\in K} = (U_k)_{k\in K},\ \text{and}\\
&\mathfrak{U} := \mathrm{Hom}_F(F^K/\ker(d),\mathcal{F})\\
&= \left\{U \in \mathcal{F}^K; \forall c = \sum_k c_k k \in \ker(d) : U(c) = \sum_k c_k U(k) = 0\right\}\\
&\underset{\ker(d)=\sum F c(\omega)}{=} \left\{U \in \mathcal{F}^K; \forall \text{ circuit } \omega : U(c(\omega)) = 0\right\}.
\end{aligned} \tag{8.209}$$

As usual we identify the vector $(U(k))_{k\in K} \in \mathfrak{U}$ with the linear map

$$U : F^K/\ker(d) \to \mathcal{F},\ \overline{k} \mapsto U(k) = U_k,\ \text{hence } U(k) = U(\overline{k}).$$

According to Kirchhoff's voltage law (KVL) from Definition 3.1.20 a voltage vector $U = (U_k)_{k\in K}$ of the network satisfies $U(c(\omega)) = 0$ for every circuit ω and therefore $U(c) = 0$ for every closed 1-chain $c \in \ker(d) \subseteq F^K$. Hence U belongs to $\mathfrak{U}$. We decompose any $U \in \mathcal{F}^K$ by

$$U = \begin{pmatrix} U_{K_1} \\ U_{K_2} \end{pmatrix} \in \mathcal{F}^K = \mathcal{F}^{K_1 \uplus K_2},\ U_{K_i} = (U_{k_i})_{k_i \in K_i}. \tag{8.210}$$

Corollary 8.3.2 immediately implies

Corollary 8.3.3 *The maps*

$$\begin{aligned}
&\mathfrak{U} = \mathrm{Hom}_F(F^K/\ker(d), \mathcal{F}) \cong \mathcal{F}^{K_1},\ U = (U_k)_{k\in K} = \begin{pmatrix} U_{K_1} \\ U_{K_2} \end{pmatrix} \leftrightarrow U_{K_1},\ \textit{with}\\
&U_k = U(k) = U(\overline{k}),\ U_{K_2} = M^T U_{K_1},\ \forall k_2 \in K_2:\ U_{k_2} = \sum_{k_1\in K_1} M(k_1,k_2)U_{k_1}
\end{aligned} \tag{8.211}$$

are well-defined inverse isomorphisms. In particular, $\mathfrak{U} \subseteq \mathcal{F}^K$ *is a controllable subbehavior of* $\mathcal{F}^K$*. The components* $U_{k_1},\ k_1 \in K_1$*, of* $U \in \mathfrak{U}$ *can be chosen arbitrarily and then determine the components* $U_{k_2},\ k_2 \in K_2$*, uniquely.*

Proof The equations $\overline{k_2} = \sum_{k_1\in K_1} \overline{k_1} M(k_1,k_2),\ k_2 \in K_2$, imply

$$U_{k_2} = U(\overline{k_2}) = \sum_{k_1\in K_1} U(\overline{k_1})M(k_1,k_2) = \sum_{k_1} M^\top(k_2,k_1)U_{k_1} \Longrightarrow U_{K_2} = M^\top U_{K_1}.$$

□

Next we establish the consequences of Kirchhoff's current law (KCL). We apply the exact functor $\mathrm{Hom}_F(-,\mathcal{F})$ to the exact sequence (8.204) with $N := F$ and obtain the exact sequence

$$\begin{aligned}
&0 \to \mathcal{F}^{n-r} \xrightarrow{B\cdot} \mathcal{F}^K \xrightarrow{A\cdot} \mathcal{F}^V \xrightarrow{E\cdot} \mathcal{F} \to 0 \text{ with}\\
&(AI)_v = \sum_k A_{vk} I_k = \sum_{k:v\to} I_k - \sum_{k:\to v} I_k.
\end{aligned} \tag{8.212}$$

Define

$$\begin{aligned}
\mathfrak{I} &:= \ker(A\cdot : \mathcal{F}^K \to \mathcal{F}^V) = \{I \in \mathcal{F}^K;\ AI = 0\}\\
&= \left\{ I \in \mathcal{F}^K;\ \forall v \in V:\ \sum_{k:v\to} I_k - \sum_{k:\to v} I_k = 0 \right\}.
\end{aligned} \tag{8.213}$$

According to KCL any current vector $I = (I_k)_{k\in K}$ of the network belongs to $\mathfrak{I}$. We again decompose $I = \begin{pmatrix} I_{K_1} \\ I_{K_2} \end{pmatrix},\ I_{K_i} := (I_{k_i})_{k_i\in K_i}$.

Corollary 8.3.4 *The maps*

$$\begin{aligned}
&\mathfrak{I} \cong \mathcal{F}^{K_2},\ I = (I_k)_{k\in K} \leftrightarrow I_{K_2} := (I_{k_2})_{k_2\in K_2},\ \textit{with}\\
&I_{K_1} = -MI_{K_2}\ \textit{or}\ \forall k_1 \in K_1:\ I_{k_1} = -\sum_{k_2\in K_2} M(k_1,k_2)I_{k_2}
\end{aligned} \tag{8.214}$$

are well-defined inverse isomorphisms. Hence the space $\mathfrak{I}$ *is a controllable subbehavior of* $\mathcal{F}^K$ *of* $\mathrm{rank}(\mathfrak{I}) = \sharp(K_2) = n - m + 1$ *where* $n = \sharp(K)$ *and* $m = \sharp(V)$*. The components* $I_{k_2},\ k_2 \in K_2$*, of* $I \in \mathfrak{I}$ *can be chosen arbitrarily and then determine uniquely the components* $I_{k_1},\ k_1 \in K_1$*.*

Proof Recall

$$A = (A|_{K_1}, A|_{K_2}) = A|_{K_1}(\mathrm{id}_{K_1}, M) \in F^{V\times(K_1 \uplus K_2)},\ \mathrm{rank}(A|_{K_1}) = \sharp(K_1) = r.$$

Hence $A|_{K_1} \in F^{V\times K_1}$ can be cancelled as a left factor. This implies

$$\begin{aligned} I \in \mathfrak{I} &\Longleftrightarrow 0 = AI = A|_{K_1}(\mathrm{id}_{K_1}, M)\begin{pmatrix} I_{K_1} \\ I_{K_2} \end{pmatrix} \\ &\Longleftrightarrow 0 = (\mathrm{id}_{K_1}, M)\begin{pmatrix} I_{K_1} \\ I_{K_2} \end{pmatrix} = I_{K_1} + MI_{K_2} \Longleftrightarrow I_{K_1} = -MI_{K_2}. \end{aligned}$$

□

Corollaries 8.3.3 and 8.3.4 imply the following.

Theorem 8.3.5 *The voltage vector $U = (U_k)_{k\in K}$ and the current vector $I = (I_k)_{k\in K}$ of the network satisfy the equations*

$$U_{K_2} - M^{\top}U_{K_1} = 0,\ I_{K_1} + MI_{K_2} = 0 \text{ or } \begin{pmatrix} \mathrm{id}_{K_1} & 0 & M & 0 \\ 0 & \mathrm{id}_{K_2} & 0 & -M^{\top} \end{pmatrix}\begin{pmatrix} I_{K_1} \\ U_{K_2} \\ I_{K_2} \\ U_{K_1} \end{pmatrix} = 0. \tag{8.215}$$

These are equivalent to Kirchhoff's voltage and current law. Altogether these are $n = \sharp(K) = \sharp(K_1) + \sharp(K_2)$ equations for the $2n$ components of $\begin{pmatrix} U \\ I \end{pmatrix}$ that are uniquely determined by $\begin{pmatrix} U_{K_1} \\ I_{K_2} \end{pmatrix}$.
Similar equations are usually derived by means of graph theory, cf. [10, Sects. 3.1, 3.2], cf. [17, Chap. 3]. Corollary 8.3.2 also belongs to this field. In this approach the branch sets K_1 resp. K_2 correspond to a tree *of the graph resp. its* tree complement *or* cotree, *and, essentially, I_{K_2} resp. U_{K_1} are the so-called* mesh or loop currents *resp.* node potentials. *The derivation of* (8.215) *is easier than the standard arguments from the literature.* ◊

Remark 8.3.6 In the literature it is often assumed that all branches $k \in K$ occur in a circuit. This is justified by the following considerations.
Let $K' \subseteq K$ denote the set of all branches that are contained in some circuit and $V' := \bigcup_{k'\in K'} \{\mathrm{dom}(k'), \mathrm{cod}(k')\}$. Then (V', K') is a connected graph with all branches in circuits. The definition of K' implies

$$\ker(d) = \sum_{\omega \text{ circuit}} Fc(\omega) \subseteq F^{K'} = \oplus_{k'\in K'} Fk' \subset F^{K}.$$

The equations $\overline{k_2} = \sum_{k_1\in K_1} \overline{k_1} M(k_1, k_2),\ k_2 \in K_2$, imply

$$k_2 - \sum_{k_1 \in K_1} k_1 M(k_1, k_2) \in \ker(d) \subseteq F^{K'}$$
$$\underset{\text{comparison of coefficients}}{\Longrightarrow} K_2 \subseteq K',\ \forall k_1 \in K_1 \backslash K' :\ M(k_1, k_2) = 0$$
$$\underset{(8.215)}{\Longrightarrow} \forall k_2 \in K_2 :\ U_{k_2} = \sum_{k_1 \in K_1 \cap K'} M(k_1, k_2) U_{k_1} \tag{8.216}$$
$$\forall k_1 \in K_1 :\ I_{k_1} = \begin{cases} -\sum_{k_2 \in K_2} M(k_1, k_2) I_{k_2} & \text{if } k_1 \in K_1 \cap K' \\ 0 & \text{if } k_1 \in K_1 \backslash K'. \end{cases}$$

Due to $K_2 \subseteq K'$, hence $K' = (K_1 \cap K') \uplus K_2$, and (8.216) all branch voltages and currents $U_k, I_k, k \in K$, are fully determined by $U_k, I_k, k \in K'$. Therefore it is sufficient to investigate the network with graph (V', K') where all branches belong to circuits. ◇

We generalize the electrical networks from Definition 3.1.20.

Definition and Corollary 8.3.7 A *passive one-port* is modeled as a SISO IO behavior $\mathcal{B} = \left\{\binom{U}{I} \in \mathcal{F}^2;\ P \circ U = Q \circ I\right\}$ where $\binom{U}{I}$ is a voltage-current pair and $P, Q \in F[s]$ are given nonzero polynomials. Recall that U and I are, possibly piecewise continuous or even generalized, functions of t. The one-port gives rise to the two nonzero transfer functions $H = P^{-1}Q$ and $H^{-1} = Q^{-1}P$, both U and I can be considered as inputs. The corresponding impulse responses are $h := H \circ \delta$ resp. $h^- := H^{-1} \circ \delta$ that satisfy the equations $h * h^- = \delta$. If $\binom{U}{I} \in C_+^{-\infty}(\mathbb{R}, \mathbb{C})$, i.e., if the voltage and the current vanish for $t \le t_0 \in \mathbb{R}$, then U and I determine each other via the convolution equations $U = h * I$ and $I = h^- * U$.
Technically the one-port is a box with one *port, i.e., one pair of poles or terminals.* A voltage U between the terminals and a current I through them are related by the equation $P \circ U = Q \circ I$.
Note that real voltages and currents are at least piecewise continuous, jumps resulting from switchings, mostly at time $t = 0$. However, differentiation of such signals as in $P \circ U = Q \circ I$ may produce distributions, and therefore the latter are indispensible. If components of U or I turn out to be distributions, this indicates that the solutions destroy the network, and that this has to be modified. Also note that resistors R, inductors L, and capacitors C are the simplest passive one-ports. We will show below how one-ports arise from RLC networks. ◇

Definition 8.3.8 We define a (generalized) electrical network. We keep the data from Definition 3.1.20, but replace the RLC branches from that definition by general passive one-ports $P_k \circ U_k = Q_k \circ I_k$ along the branch k with the transfer functions $H_k = P_k^{-1} Q_k$ and $H_k^{-1} = Q_k^{-1} P_k$. In addition to one-port branches, the network has source branches where voltages or currents can be supplied to the network, hence the decomposition $K = K_s \uplus K_p$ where K_s resp. K_p contain the source resp. the one-port branches. *Kirchhoff's laws hold for these more general networks too.* ◇

Remark 8.3.9 Assume a network with two nodes v, w and two branches $k_1, k_2 : v \to w$ where k_1 is a source branch with $\binom{U_1}{I_1}$ and k_2 is the branch of a passive generalized one-port with the equations $P_2 \circ U_2 = Q_2 \circ I_2$. The direction of the voltage U_i and of the current I_i, $i = 1, 2$, are that of the arrow $k_i : v \to w$. The Kirchhoff laws furnish $U_1 = U_2$ and $I_1 + I_2 = 0$ (node equation at v). The connection of the source voltage U_1 and current I_1 follows as $P_2 \circ U_1 = -Q_2 \circ I_1$. If the direction of k_1 was reversed, i.e., $k_1 : w \to v$, then the Kirchhoff laws would give $U_1 + U_2 = 0$ and $I_1 = I_2$ and again $P_2 \circ U_1 = -Q_2 \circ I_1$. In order to avoid this minus sign in I_1, it is customary in electrical engineering to reverse the direction of the current in source branches, i.e., to denote the current $-I_1$ from w to v as I_1 and to obtain $I_1 = I_2$ and $U_1 = U_2$. Since this is a deviation from the standard convention that the voltage and current along a branch point in the same direction, we do *not* use this deviation in general. ◇

We now assume such a generalized network and investigate its equations. We are going to show that, generically, the network is an IO behavior with the free inputs $\binom{I_{K_{2,s}}}{U_{K_{1,s}}}$, cf. (8.218). Define

$$V_s := \{\mathrm{dom}(k), \mathrm{cod}(k);\ k \in K_s\},\ n_s := \sharp(K_s),\ n_p = \sharp(K_p),\ m_s := \sharp(V_s), \tag{8.217}$$

hence $\sharp(K) = n = n_s + n_p$. The set V_s is called the set of *terminals or poles* of the network where voltages and currents can be supplied to the network from outside or where the network can be connected with other networks. The network is also called an m_s*-pole* or n_s-port. In the literature the latter name is used only if $m_s = 2n_s$, i.e., if the terminals $\mathrm{dom}(k), \mathrm{cod}(k),\ k \in K_s$, are pairwise different. Then $k \in K_s$ is called a *port. Most of the following considerations hold without any further assumption on the source branches*, the IO condition, cf. Theorem 8.3.11, implying all necessary properties. In particular, instead of two-ports with four terminals arbitrary networks with two source branches can be considered. In the literature, the network is called *passive* if its source inputs as external excitations are omitted. We always include these into the computations.
With the additional decomposition $K = K_1 \uplus K_2$ from (8.206) we obtain

$$\begin{aligned} K_1 &= K_{1,s} \uplus K_{1,p} := (K_1 \cap K_s) \uplus (K_1 \cap K_p), \\ K_2 &= K_{2,s} \uplus K_{2,p} := (K_2 \cap K_s) \uplus (K_2 \cap K_p). \end{aligned} \tag{8.218}$$

Recall the matrices $M \in F^{K_1 \times K_2}$, $M^\top \in F^{K_2 \times K_1}$ from (8.206). If $L_i \subseteq K_i$, $i = 1, 2$, we define

$$\begin{aligned} M_{L_1,L_2} &:= (M(k_1, k_2))_{(k_1,k_2)\in L_1\times L_2} \in F^{L_1\times L_2} \\ (M^\top)_{L_2,L_1} &:= \left(M^T(k_2, k_1)\right)_{(k_2,k_1)\in L_2\times L_1} = \left(M_{L_1,L_2}\right)^\top . \end{aligned} \tag{8.219}$$

In addition to the Kirchhoff laws the passive port branches give rise to the equations $P_k \circ U_k = Q_k \circ I_k$. For every subset $L \subset K_p$ we get the equations

$$P_L \circ U_L = Q_L \circ I_L,\ P_L = \operatorname{diag}(P_k;\ k \in L) \in F[s]^{L\times L},\ Q_L = \operatorname{diag}(Q_k;\ k \in L),$$
$$H_L = P_L^{-1} Q_L = \operatorname{diag}(H_k;\ k \in L) \in F(s)^{L\times L},\ H_L^{-1} = Q_L^{-1} P_L. \tag{8.220}$$

The equations of the network, i.e., the Kirchhoff and the passive one-port equations, are now written in the matrix form

$$\begin{array}{c} \\ K_{1,s}\\ K_{1,p}\\ K_{2,s}\\ K_{2,p}\\ K_{1,p}\\ K_{2,p}\end{array}
\begin{array}{c}
\begin{array}{cccccc} K_{1,s} & K_{1,p} & K_{2,s} & K_{2,p} & K_{1,p} & K_{2,p}\end{array}\\
\begin{pmatrix}
\mathrm{id}_{K_{1,s}} & 0 & 0 & 0 & 0 & M_{K_{1,s},K_{2,p}}\\
0 & \mathrm{id}_{K_{1,p}} & 0 & 0 & 0 & M_{K_{1,p},K_{2,p}}\\
0 & 0 & \mathrm{id}_{K_{2,s}} & 0 & -\left(M_{K_{1,p},K_{2,s}}\right)^{\top} & 0\\
0 & 0 & 0 & \mathrm{id}_{K_{2,p}} & -\left(M_{K_{1,p},K_{2,p}}\right)^{\top} & 0\\
0 & -Q_{K_{1,p}} & 0 & 0 & P_{K_{1,p}} & 0\\
0 & 0 & 0 & P_{K_{2,p}} & 0 & -Q_{K_{2,p}}
\end{pmatrix}\end{array}
\begin{pmatrix} I_{K_{1,s}}\\ I_{K_{1,p}}\\ U_{K_{2,s}}\\ U_{K_{2,p}}\\ U_{K_{1,p}}\\ I_{K_{2,p}}\end{pmatrix}$$

$$= \begin{array}{c} \\ K_{1,s}\\ K_{1,p}\\ K_{2,s}\\ K_{2,p}\\ K_{1,p}\\ K_{2,p}\end{array}
\begin{array}{c}
\begin{array}{cc} K_{2,s} & K_{1,s}\end{array}\\
\begin{pmatrix}
-M_{K_{1,s},K_{2,s}} & 0\\
-M_{K_{1,p},K_{2,s}} & 0\\
0 & \left(M_{K_{1,s},K_{2,s}}\right)^{\top}\\
0 & \left(M_{K_{1,s},K_{2,p}}\right)^{\top}\\
0 & 0\\
0 & 0
\end{pmatrix}\end{array}
\begin{pmatrix} I_{K_{2,s}}\\ U_{K_{1,s}}\end{pmatrix},\quad y := \begin{pmatrix} I_{K_{1,s}}\\ I_{K_{1,p}}\\ U_{K_{2,s}}\\ U_{K_{2,p}}\\ U_{K_{1,p}}\\ I_{K_{2,p}}\end{pmatrix},\quad u := \begin{pmatrix} I_{K_{2,s}}\\ U_{K_{1,s}}\end{pmatrix} \tag{8.221}$$

$$\Longrightarrow P \circ y = Q \circ u,$$

where $P \in F[s]^{(n+n_p)\times(n+n_p)}$ and $Q \in F[s]^{(n+n_p)\times n_s}$ denote the 6×6- resp. 6×2-block matrices in the preceding equation. Note that U resp. I are the voltage resp. current vectors of the network, whereas $u := \begin{pmatrix} I_{K_{2,s}}\\ U_{K_{1,s}}\end{pmatrix}$ will be considered as its input vector and y as the output. The network behavior is

$$\mathcal{B} := \left\{\left(\begin{smallmatrix} y\\ u\end{smallmatrix}\right) \in \mathcal{F}^{(n+n_p)+n_s};\ P \circ y = Q \circ u\right\}. \tag{8.222}$$

By the $F[s]$-linear elementary row operations on $(P, -Q)$

$$\text{5th row} + Q_{K_{1,p}} \times \text{2nd row},\ \text{6th row} - P_{K_{2,p}} \times \text{4th row}$$

we obtain the equivalent equations

$$P' \circ y = Q' \circ u \text{ where}$$
$$P' := \left(\begin{smallmatrix} \mathrm{id}_n & P_I\\ 0 & P_{II}\end{smallmatrix}\right) \in F[s]^{(n+n_p)\times(n+n_p)},\ Q' := \left(\begin{smallmatrix} Q_I\\ Q_{II}\end{smallmatrix}\right) \in F[s]^{(n+n_p)\times n_s} \tag{8.223}$$

and

$$
\begin{aligned}
P_I &:= \begin{pmatrix} 0 & M_{K_{1,s},K_{2,p}} \\ 0 & M_{K_{1,p},K_{2,p}} \\ -\left(M_{K_{1,p},K_{2,s}}\right)^\top & 0 \\ -\left(M_{K_{1,p},K_{2,p}}\right)^\top & 0 \end{pmatrix} \in F^{n\times n_p}, \\
Q_I &:= \begin{pmatrix} -M_{K_{1,s},K_{2,s}} & 0 \\ -M_{K_{1,p},K_{2,s}} & 0 \\ 0 & \left(M_{K_{1,s},K_{2,s}}\right)^\top \\ 0 & \left(M_{K_{1,s},K_{2,p}}\right)^\top \end{pmatrix} \in F^{n\times n_s} \\
P_{II} &:= \begin{pmatrix} p_{K_{1,p}} & q_{K_{1,p}} M_{K_{1,p},K_{2,p}} \\ p_{K_{2,p}}\left(M_{K_{1,p},K_{2,p}}\right)^\top & -q_{K_{2,p}} \end{pmatrix} \in F[s]^{n_p\times n_p} \\
Q_{II} &= \begin{pmatrix} -q_{K_{1,p}} M_{K_{1,p},K_{2,s}} & 0 \\ 0 & -p_{K_{2,p}}\left(M_{K_{1,s},K_{2,p}}\right)^\top \end{pmatrix} \in F[s]^{n_p\times n_s}.
\end{aligned}
\tag{8.224}
$$

Correspondingly we decompose y by

$$
\begin{aligned}
&y := \begin{pmatrix} y_I \\ y_{II} \end{pmatrix} \in \mathcal{F}^{n+n_p},\ y_I = \left(I_{K_{1,s}}, I_{K_{1,p}}, U_{K_{2,s}}, U_{K_{2,p}}\right)^\top \in \mathcal{F}^n, \\
&y_{II} = \left(U_{K_{1,p}}, I_{K_{2,p}}\right)^\top \in \mathcal{F}^{n_p},\ u = \left(I_{K_{2,s}}, U_{K_{1,s}}\right)^\top \in \mathcal{F}^{n_s}.
\end{aligned}
\tag{8.225}
$$

We then get the following equivalent equations:

$$
\begin{aligned}
&P \circ y = Q \circ u \Longleftrightarrow P' \circ y = Q' \circ u \Longleftrightarrow \begin{pmatrix} \mathrm{id}_n & P_I \\ 0 & P_{II} \end{pmatrix} \circ \begin{pmatrix} y_I \\ y_{II} \end{pmatrix} = \begin{pmatrix} Q_I \\ Q_{II} \end{pmatrix} \circ u \\
&\Longleftrightarrow y_I = -P_I y_{II} + Q_I u \quad \text{and} \quad P_{II} \circ y_{II} = Q_{II} \circ u.
\end{aligned}
\tag{8.226}
$$

Notice that P_I, Q_I have coefficients in F. Hence y_I is completely determined by y_{II} and u. This suggests to define the behavior $\mathcal{B}_{II}$ and the projection C_{II} by

$$
\begin{aligned}
&\mathcal{B}_{II} := \left\{ \begin{pmatrix} y_{II} \\ u \end{pmatrix} \in \mathcal{F}^{n_p+n_s};\ P_{II} \circ y_{II} = Q_{II} \circ u \right\}, \\
&C_{II} = (0, \mathrm{id}_{n_p}) \in F^{n_p\times(n+n_p)},\ C_{II}\cdot : \mathcal{F}^{n+n_p} \to \mathcal{F}^{n_p},\ \begin{pmatrix} y_I \\ y_{II} \end{pmatrix} \mapsto y_{II}.
\end{aligned}
\tag{8.227}
$$

The equations (8.226) obviously imply the following.

Corollary 8.3.10 *The projection C_{II} induces the isomorphism*

$$
\begin{pmatrix} C_{II} & 0 \\ 0 & \mathrm{id}_{n_p} \end{pmatrix} \cdot : \mathcal{B} \stackrel{\cong}{\to} \mathcal{B}_{II},\ \begin{pmatrix} y \\ u \end{pmatrix} = \begin{pmatrix} y_I \\ y_{II} \\ u \end{pmatrix} \mapsto \begin{pmatrix} y_{II} \\ u \end{pmatrix},\ y_{II} = C_{II} y. \tag{8.228}
$$

Its inverse is given as

$$
\mathcal{B}_{II} \stackrel{\cong}{\to} \mathcal{B},\ \begin{pmatrix} y_{II} \\ u \end{pmatrix} \mapsto \begin{pmatrix} y \\ u \end{pmatrix},\ y = \begin{pmatrix} y_I \\ y_{II} \end{pmatrix} = \begin{pmatrix} -P_I \\ \mathrm{id}_{n_p} \end{pmatrix} y_{II} + \begin{pmatrix} Q_I \\ 0 \end{pmatrix} u, \tag{8.229}
$$

where P_I and Q_I have entries in F. ◇

Theorem 8.3.11 *Consider an electrical network with a connected graph* (V, K) *and the decomposition* $K = K_s \uplus K_p$ *into source branches and passive one-port branches. The Kirchhoff laws furnish the decomposition* $K = K_1 \uplus K_2$ *from* (8.206). *The network behavior from* (8.222) *is*

$$\begin{aligned}
&\mathcal{B} := \left\{\begin{pmatrix} y \\ u \end{pmatrix} \in \mathcal{F}^{2n} = \mathcal{F}^{(n+n_p)+n_s};\ P \circ y = Q \circ u\right\},\\
&y := \left(I_{K_{1,s}}, I_{K_{1,p}}, U_{K_{2,s}}, U_{K_{2,p}}, U_{K_{1,p}}, I_{K_{2,p}}\right)^\top \in \mathcal{F}^{n+n_p},\ u = \begin{pmatrix} I_{K_{2,s}} \\ U_{K_{1,s}} \end{pmatrix} \in \mathcal{F}^{n_s},\\
&\mathcal{B}^0 := \left\{y \in \mathcal{F}^{n+n_p};\ P \circ y = 0\right\},\quad n := \sharp(K), n_s := \sharp(K_s),\ n_p := \sharp(K_p).
\end{aligned} \tag{8.230}$$

Via Corollary 8.3.10 it gives rise to the isomorphic behavior

$$\begin{aligned}
&\mathcal{B}_{II} := \left\{\begin{pmatrix} y_{II} \\ u \end{pmatrix} \in \mathcal{F}^{n} = \mathcal{F}^{n_p+n_s};\ P_{II} \circ y_{II} = Q_{II} \circ u\right\},\\
&y_{II} := \begin{pmatrix} U_{K_{1,p}} \\ I_{K_{2,p}} \end{pmatrix} \in \mathcal{F}^{n_p},\ u = \begin{pmatrix} I_{K_{2,s}} \\ U_{K_{1,s}} \end{pmatrix} \in \mathcal{F}^{n_s},\\
&\mathcal{B}^0_{II} := \left\{y_{II} \in \mathcal{F}^{n_p};\ P_{II} \circ y_{II} = 0\right\}.
\end{aligned} \tag{8.231}$$

1. *The behaviors* $\mathcal{B}_{II}$ *and thus* $\mathcal{B}$ *are IO behaviors if and only if*

$$\operatorname{rank}(P_{II}) = n_p \Longleftrightarrow \det(P_{II}) \neq 0 \Longleftrightarrow \operatorname{rank}(P) = n + n_p \Longleftrightarrow \det(P) \neq 0.$$

This is also equivalent to the unique solvability over $F(s)$ *of* $P_{II}H_{II} = Q_{II}$ *and thus of* $PH = Q$. *The state space representation in Theorem 8.3.32 below implies the IO property for a suitable choice of* $K = K_1 \uplus K_2$. *In the sequel we assume* $\det(P_{II}) \neq 0$.

2. *The transfer matrices* $H := P^{-1}Q \in F(s)^{(n+n_p)\times n_s}$ *and* $H_{II} = P_{II}^{-1}Q_{II} \in F(s)^{n_p \times n_s}$ *are related by*

$$H_{II} = C_{II}H = \begin{pmatrix} H_{U_{K_{1,p}},u} \\ H_{I_{K_{2,p}},u} \end{pmatrix},\quad H = \begin{pmatrix} Q_I - P_I H_{II} \\ H_{II} \end{pmatrix}. \tag{8.232}$$

Let $\mathcal{G}$ *denote the signal space* $\mathrm{C}_+^{-\infty}$ *or* $\mathcal{P}^{-\infty}$, *the latter under the condition* $\operatorname{char}(\mathcal{B}^0) \cap \mathbb{Z}j\omega = \emptyset$. *Theorems 8.2.41 and 8.2.102 imply the isomorphisms*

$$\begin{array}{ccccc}
\mathcal{B} \cap \mathcal{G}^{(n+n_p)+n_s} & \cong & \mathcal{B}_{II} \cap \mathcal{G}^{n_p+n_s} & \cong & \mathcal{G}^{n_s}\\
\begin{pmatrix} H \circ u \\ u \end{pmatrix} & \leftrightarrow & \begin{pmatrix} H_{II} \circ u \\ u \end{pmatrix} & \leftrightarrow & u.
\end{array} \tag{8.233}$$

3. *Let* e_{II} *resp.* e_{II}^{uc} (uc *for uncontrollable) denote the* n_p*-th elementary divisors of* P_{II} *resp. of* $(P_{II}, -Q_{II})$. *Then the characteristic varieties are*

$$\begin{aligned}\operatorname{char}(\mathcal{B}^0) &= \operatorname{char}(\mathcal{B}^0_{II}) = \mathrm{V}_{\mathbb{C}}(e_{II})\\ &= \mathrm{V}_{\mathbb{C}}(\det(P_{II})) = \left\{\lambda \in \mathbb{C};\ \operatorname{rank}(P_{II}(\lambda)) < n_p\right\}\\ \supset \operatorname{char}(\mathcal{B}) &= \operatorname{char}(\mathcal{B}_{II}) = \mathrm{V}_{\mathbb{C}}(e^{\mathrm{uc}}_{II})\\ &= \left\{\lambda \in \mathbb{C};\ \operatorname{rank}(P_{II}(\lambda), -Q_{II}(\lambda)) < n_p\right\}.\end{aligned} \tag{8.234}$$

The behaviors $\mathcal{B}^0$ and $\mathcal{B}^0_{II}$ are asymptotically stable if and only if $\operatorname{char}(\mathcal{B}^0_{II}) \subset \mathbb{C}_- = \{\lambda \in \mathbb{C};\ \Re(\lambda) < 0\}$. *These behaviors are stable if and only if for all* $\lambda \in \operatorname{char}(\mathcal{B}^0)$

$$\Re(\lambda) < 0 \text{ or } \lambda = j\omega \in \mathbb{R}j \text{ and } m(j\omega) = \operatorname{index}(\mathcal{B}^0, s - j\omega) = 1. \tag{8.235}$$

Here $m(j\omega)$ is the multiplicity of the root $j\omega$ of e_{II}, cf. Theorem 5.4.8.

4. *$\mathcal{B} \cong \mathcal{B}_{II}$ are controllable if and only if $e^{\mathrm{uc}}_{II} = 1$.*
5. *The detailed study of $\mathcal{B}$ proceeds by the methods from Sects. 8.2.4 and 8.2.8, in particular, by Theorems 8.2.53, 8.2.58, 8.2.102, and 8.2.107. Assume that $\mathcal{B}^0_{II}$ and thus $\mathcal{B}^0$ are asymptotically stable and that H_{II} and thus H are proper. The asymptotic stability implies that every output y_{ss} to a fixed input u is a steady-state output, i.e., for every other output y to u there is a polynomial-exponential z with $y = y_{ss} + z$ and* $\lim_{t\to\infty} z(t) = 0$.
   (i) If $u = \begin{pmatrix} I_{K_{2,s}} \\ U_{K_{1,s}} \end{pmatrix} = \alpha Y \in \mathcal{F}_F^{K_s}$ is an input with polynomial-exponential vector α and $H_2 := \mathcal{L}(u) \in F(s)^{K_s}_{\mathrm{spr}}$, then $y_{ss} := H \circ u = H H_2 \circ \delta = \beta Y$ with polynomial-exponential β is an output for u, and β is an output for the input α. The vector $\begin{pmatrix} y_{ss} \\ u \end{pmatrix} \in \mathcal{F}^{K \uplus K}$ coincides with $\begin{pmatrix} U \\ I \end{pmatrix}$ up to a permutation of the components and contains all *branch voltages and currents in the form γY, where γ is a computed polynomial-exponential function.*
   (ii) If $u \in \left(\mathcal{P}^{0,\mathrm{pc}}\right)^{K_s}$ is a piecewise continuous, T-periodic input, $T > 0$, $\omega := 2\pi T^{-1}$, then $y_{ss} := H \circ u$ with $\mathbb{F}(y_{ss})(\mu) = H(j\mu\omega)\mathbb{F}(u)(\omega)$, $\mu \in \mathbb{Z}$, is the unique T-periodic output to u and also piecewise continuous. It is even continuous with uniformly convergent Fourier series if H_{II} and thus H are strictly proper. Then y_{ss} is called the (unique) *stationary or steady-state output.*

Proof 1. This follows directly from the isomorphism $\mathcal{B} \cong \mathcal{B}_{II}$.
2. The isomorphic behaviors $\mathcal{B} \cong \mathcal{B}_{II}$ are related by the Rosenbrock equations

$$\begin{aligned}&P \circ y = Q \circ u,\ y_{II} = C_{II} y,\ \mathcal{B}_{II} = \begin{pmatrix} C_{II} & 0 \\ 0 & \mathrm{id}_{n_p} \end{pmatrix} \mathcal{B} \underset{\text{Theorem 6.3.1}}{\Longrightarrow} H_{II} = C_{II} H,\\ &P_{II} \circ y_{II} = Q_{II} \circ u,\ y = \begin{pmatrix} -P_I \\ \mathrm{id}_{n_p} \end{pmatrix} y_{II} + \begin{pmatrix} Q_I \\ 0 \end{pmatrix} u\\ &\qquad \underset{\text{Theorem 6.3.1}}{\Longrightarrow} H = \begin{pmatrix} -P_I \\ \mathrm{id}_{n_p} \end{pmatrix} H_{II} + \begin{pmatrix} Q_I \\ 0 \end{pmatrix} = \begin{pmatrix} Q_I - P_I H_{II} \\ H_{II} \end{pmatrix}.\end{aligned}$$

3. This follows from the definition of the characteristic varieties and the isomorphism $\mathcal{B} \cong \mathcal{B}_{II}$.
4. Theorem 4.2.12.

5. Cf. the quoted sections. Since $\mathcal{B}^0 \cong \mathcal{B}^0_{II}$ is assumed asymptotically stable, $\mathcal{B}^0$ and H have no poles in $\mathbb{Z}j\omega$ and H has coefficients in $\mathbb{C}(s)_{\mathrm{per}}$. Therefore, in (ii), the output $y_{ss} := H \circ u$ is well defined and indeed the unique T-periodic output to the input u. □

Corollary 8.3.12 (The port condition) *Assume that $\mathcal{B}$ is an IO behavior and that $k : v \to w$ is a source branch such that no other source branch has terminals v or w. Such a k is sometimes called a* port. *Define*

$$\begin{aligned}
&K' := \left\{k' \in K \uplus (-K);\ \mathrm{cod}(k') = v,\ k' \neq k\right\} \subseteq K_p \uplus (-K_p),\\
&K'' := \left\{k'' \in K \uplus (-K);\ \mathrm{dom}(k'') = w,\ k'' \neq k\right\} \subseteq K_p \uplus (-K_p)\\
&\underset{\text{current law at } v \text{ and } w}{\Longrightarrow} \quad I_k = \sum_{k' \in K'} I_{k'} = \sum_{k'' \in K''} I_{k''}.
\end{aligned} \tag{8.236}$$

Thus I_k is the current that flows from v to w through the source branch k or out of the network into terminal v or from terminal w into the network. The equation in the third row of (8.236) *is called the* port condition *[17, p. 105], [Wikipedia; https://en.wikipedia.org/wiki/Port_(circuit_theory), May 24, 2018]. The IO property implies the* existence *of all branch voltages and currents U_l, I_l, $l \in K$, for given input $u = (I_{K_{2,s}}, U_{K_{1,s}})^\top$ and, in particular, that of I_k. The current law then implies the port condition* (8.236). *It is implied by the IO condition* $\det(P_{II}) \neq 0$ *that can be trivially checked. It is needed if several networks are connected and Kirchhoff's laws are applied to the connected behavior, cf. Theorem 8.3.29 and Examples 8.3.30 and 8.3.31.* ◊

Remark 8.3.13 1. The IO condition says that the input components $u = \begin{pmatrix} I_{K_{2,s}} \\ U_{K_{1,s}} \end{pmatrix}$ are maximally free. In electrical engineering the word *free* is often replaced by *independent*, however, in general, without the precise mathematical condition $\mathrm{rank}(P) = n + n_p = 2n - n_s$ of this terminology.

2. An ideal *resistor, capacitor resp. inductor one-port* is characterized by the equations

$$\begin{aligned}
&1 \circ U_R = RI_R,\ H_R = R,\ H_R^{-1} = R^{-1},\ \mathrm{V}_{\mathbb{C}}(1) = \emptyset,\\
&Cs \circ U_C = I_C,\ H_C = \tfrac{1}{Cs},\ H_C^{-1} = Cs,\ \mathrm{V}_{\mathbb{C}}(Cs) = \{0\}\,,\ \text{resp.}\\
&1 \circ U_L = Ls \circ I_L,\ H_L = Ls,\ H_L^{-1} = \tfrac{1}{Ls},\ \mathrm{V}_{\mathbb{C}}(1) = \emptyset.
\end{aligned} \tag{8.237}$$

The ideal capacitor is not asymptotically stable, but a real capacitor always is, since it has a resistive component. Therefore one may always assume $\mathrm{V}_{\mathbb{C}}(P_k) \subset \mathbb{C}_-$, $k \in K_p$.
The equations are controllable. Notice that H_C^{-1} and H_L are nonconstant polynomials and hence not proper and have the distributional impulse responses $H_C^{-1} \circ \delta = C\delta'$, $H_L \circ \delta = L\delta'$. Thus electrical networks with nonproper transfer matrix H can easily occur. See Theorem 8.2.58 for the consequences.

3. More complicated networks can be treated by replacing the diagonal matrices P_{K_p} and Q_{K_p} by more complicated matrices, cf. Corollary 8.3.22.
(i) *Ideal transformer and ideal gyrator*: These are given by two branches $k_1, k_2 \in K_p$ with the two equations

$$\begin{aligned}&U_{k_2} = aU_{k_1} \text{ and } I_{k_2} = a^{-1}I_{k_1} \text{ resp. } U_{k_2} = aI_{k_1} \text{ and } I_{k_2} = -a^{-1}U_{k_1},\ a > 0,\\ &\text{instead of } P_{k_1} \circ U_{k_1} = Q_{k_1} \circ I_{k_1} \text{ and } P_{k_2} \circ U_{k_2} = Q_{k_2} \circ I_{k_2}.\end{aligned} \tag{8.238}$$

(ii) *Voltage- and current-controlled voltage and current sources*: This is given by

$$k_1 \in K,\ k_2 \in K_p,\ k_1 \neq k_2,\ u_1 \in \{U_{k_1}, I_{k_1}\},\ y_2 \in \{U_{k_2}, I_{k_2}\},\ f_2 \circ y_2 = f_1 \circ u_1, \tag{8.239}$$

where f_1, f_2 are nonzero polynomials. The equation $f_2 \circ y_2 = f_1 \circ u_1$ or $y_2 = (f_2^{-1} f_1) \circ u_1$ in steady state replaces the equation $P_{k_2} \circ U_{k_2} = Q_{k_2} \circ I_{k_2}$. If y_2 is a voltage and u_1 a voltage (current), one talks about a voltage- (current)-controlled voltage source and likewise for a current source. Note, however, that k_2 belongs to K_p and is not a source branch and y_2 *cannot be freely chosen*, so the term voltage (current) *source* is misleading in our convention.
4. (Cf. [10, Sect. 6.3, pp. 283–296]) In Theorem 8.3.11, item 5, transients can, in general, not be determined, cf. Remark 8.2.55, because they require the generally unknown initial data of the autonomous equations that follow from the value at $t = 0$ of higher derivatives of u. However, for the state space representations in Theorem 8.3.32 this is possible.
5. *All* components of u are *voltage* sources if and only if $K_s = K_{1,s}$, i.e., $K_s \subseteq K_1$. Such a decomposition $K = K_1 \uplus K_2$ exists if and only if $\operatorname{rank}(A_{K_s}) = \sharp(K_s) = n_s$ or the columns $A_{-,k}$, $k \in K_s$, of the incidence matrix are $\mathbb{R}$-linearly independent. Then K_1 can be obtained by extending these columns to an $\mathbb{R}$-basis of $A\mathbb{R}^K$ by further columns of A .
All components of u are *current* sources if and only if $K_s = K_{2,s}$, i.e., $K_s \subseteq K_2$ or $K_1 \subseteq K \setminus K_s$. Such a decomposition $K = K_1 \uplus K_2$ exists if and only if $\operatorname{rank}(A_{K\setminus K_s}) = \operatorname{rank}(A)$ or $A_{K\setminus K_s}\mathbb{R}^{K\setminus K_s} = A\mathbb{R}^K$. Then $K_1 \subseteq K\setminus K_s$ is chosen such that $A_{K_1}\mathbb{R}^{K_1} = A_{K\setminus K_s}\mathbb{R}^{K\setminus K_s} = A\mathbb{R}^K$.

◇

Corollary 8.3.14 *If transformers or controlled sources are used in addition to one-port passive elements, the network behavior $\mathcal{B}$ is best described as*

$$\begin{aligned}&\mathcal{B} := \{w \in \mathcal{F}^{2n};\ R \circ w = 0\} \text{ with}\\ &w := (U_{K_1}, U_{K_2}, I_{K_1}, I_{K_2})^\top \in \mathcal{F}^{K_1 \uplus K_2 \uplus K_1 \uplus K_2} = \mathcal{F}^{2n},\\ &R := \begin{pmatrix} -M^\top & \mathrm{id}_{K_2} & 0 & 0\\ 0 & 0 & \mathrm{id}_{K_1} & M\\ R_{p,1} & R_{p,2} & R_{p,3} & R_{p,4}\end{pmatrix} \in F[s]^{(K_2 \uplus K_1 \uplus K_p)\times(K_1 \uplus K_2 \uplus K_1 \uplus K_2)} = F[s]^{(n+n_p)\times 2n},\end{aligned} \tag{8.240}$$

where $R_p = (R_{p,1}, R_{p,2}, R_{p,3}, R_{p,4}) \in F[s]^{K_p \times (K_1 \uplus K_2 \uplus K_1 \uplus K_2)}$ *and* $R_p \circ w = 0$ *consists of the row equations* $P_k \circ U_k = Q_k \circ I_k$ *(one-port),* (8.238) *(transformer or gyrator) and* (8.239) *(controlled source). One chooses the IO decomposition*

$$w = \begin{pmatrix} y \\ u \end{pmatrix} \in \mathcal{F}^{(n+n_p)+n_s}, \ u := \begin{pmatrix} I_{K_{2,s}} \\ U_{K_{1,s}} \end{pmatrix}, \ R = (P, -Q), \ P \in F[s]^{(n+n_p)\times(n+n_p)}. \tag{8.241}$$

The IO condition $\operatorname{rank}(P) = n + n_p$ *and the asymptotic stability condition* $V_{\mathbb{C}}(\det(P)) \subset \mathbb{C}_-$ *can be easily checked. With the obvious changes Theorem 8.3.11 holds.*
Notice that the smaller matrix $(P_{II}, -Q_{II}) \in F[s]^{n_p \times (n_p+n_s)}$ *is not used here.* ◊

Example 8.3.15 (*Real voltage and current sources with load*) A real voltage source with a load is modeled by a network behavior $\mathcal{B}$ with, cf. [17, Sect. 1.2],

$$\begin{aligned} &V = \{1, 2, 3\}, \ K = \{k_s : 1 \to 3, \ k_i : 1 \to 2, k_e : 2 \to 3\}, \\ &\text{branch voltages } U_s, U_i, U_e, \ \text{ branch currents } I_s, I_i, I_e \\ &P_i \circ U_i = Q_i \circ I_i, \ P_e \circ U_e = Q_e \circ I_e, \ 0 \neq P_i, P_e, Q_i, Q_e \in F[s], \\ &H_i := P_i^{-1} Q_i, \ H_e := P_e^{-1} Q_e. \end{aligned} \tag{8.242}$$

The voltage source U_s is an ideal one. The one-port along k_i is the internal (index i) one-port of the source, i.e., always connected with it. The one-port along k_e is an external load (index e). The Kirchhoff laws are trivial here and furnish the network equations

$$\begin{aligned} &U_s = U_i + U_e, \ -I_s = I_i = I_e =: I, \ P_i \circ U_i = Q_i \circ I, \ P_e \circ U_e = Q_e \circ I \\ &\Longleftrightarrow P \circ \begin{pmatrix} U_i \\ U_e \\ I \end{pmatrix} := \begin{pmatrix} 1 & 1 & 0 \\ P_i & 0 & Q_i \\ 0 & P_e & -Q_e \end{pmatrix} \circ \begin{pmatrix} U_i \\ U_e \\ I \end{pmatrix} = \begin{pmatrix} 1 \\ 0 \\ 0 \end{pmatrix} U_s \\ &\underset{\text{2nd row}-P_i\cdot\text{1st row}}{\Longleftrightarrow} \begin{pmatrix} 1 & 1 & 0 \\ 0 & -P_i & -Q_i \\ 0 & P_e & -Q_e \end{pmatrix} \circ \begin{pmatrix} U_i \\ U_e \\ I \end{pmatrix} = \begin{pmatrix} 1 \\ -P_i \\ 0 \end{pmatrix} \circ U_s \\ &\Longrightarrow \det(P) = P_i Q_e + P_e Q_i. \end{aligned} \tag{8.243}$$

In order that U_s is a voltage source, i.e., an admissible input of $\mathcal{B}$, we require

$$\begin{aligned} &\det(P) = P_i Q_e + P_e Q_i = P_i P_e (H_i + H_e) \neq 0 \Longrightarrow H_i + H_e \neq 0, \ H := P^{-1} Q, \\ &\mathcal{B}^0 \cong \left\{ \begin{pmatrix} U_i \\ I \end{pmatrix} \in \mathcal{F}^2; \ \begin{pmatrix} P_i & -Q_i \\ -P_e & -Q_e \end{pmatrix} \circ \begin{pmatrix} U_i \\ I \end{pmatrix} = 0 \right\}, \ \begin{pmatrix} U_i \\ -U_i \\ I \end{pmatrix} \leftrightarrow \begin{pmatrix} U_i \\ I \end{pmatrix}, \\ &\operatorname{char}(\mathcal{B}^0) = V_{\mathbb{C}}(\det(P)) = V_{\mathbb{C}}(P_i Q_e + P_e Q_i) = V_{\mathbb{C}}(P_i P_e (H_i + H_e)). \end{aligned} \tag{8.244}$$

So the computation of $\operatorname{char}(\mathcal{B}^0)$ requires P_i, P_e besides H_i, H_e.
In order that steady-state computations are valid, we assume asymptotic stability of $\mathcal{B}$, i.e., $\operatorname{char}(\mathcal{B}^0) \subset \mathbb{C}_-$. Then an input voltage $U_s \in \mathcal{F}_+ := C_+^{-\infty}(\mathbb{R}, F)$ induces its steady-state output

$$
\begin{aligned}
&(U_i, U_e, I)^\top = H \circ U_s \in \mathcal{F}_+^3 \Longrightarrow w := (U_i, U_e, I, -U_s)^\top \in \mathcal{B} \cap \mathcal{F}_+^4\\
&\Longrightarrow P_i \circ U_i = Q_i \circ I,\ U_i = H_i \circ I,\ U_e = H_e \circ I,\ U_s = U_i + U_e = (H_i + H_e) \circ I,\\
&I = (H_i + H_e)^{-1} \circ U_s,\ U_i = H_i(H_i + H_e)^{-1} \circ U_s,\ U_e = H_e(H_i + H_e)^{-1} \circ U_s\\
&\Longrightarrow H = (H_i + H_e)^{-1}(H_i, H_e, 1)^\top,\ I_s = -(H_i + H_e)^{-1} \circ U_s.
\end{aligned}
\tag{8.245}
$$

If $\widetilde{w} := (\widetilde{U}_i, \widetilde{U}_e, \widetilde{I}, -U_s)^\top \in \mathcal{B}$ is any other (real) trajectory with input U_s then $\lim_{t\to\infty}(\widetilde{w} - w)(t) = 0$ and $\widetilde{w}$ and w can be identified in practice. Hence the restriction to steady-state components is justified if the stability condition $\mathrm{char}(\mathcal{B}^0) \subset \mathbb{C}_-$ is satisfied. If there is no load or U_e is shorted, then

$$
P_e = 1,\ Q_e = 0,\ H_e = 0,\ U_e = 0,\ I_s = -I = -H_i^{-1} \circ U_s. \tag{8.246}
$$

Secondly assume a period $T > 0$ with frequency $\omega = 2\pi T^{-1}$ and well-defined internal and external *impedances* $Z_i := H_i(j\omega),\ Z_e := H_e(j\omega)$ for this frequency. With $j\omega \notin \mathrm{char}(\mathcal{B}^0) = \mathrm{V}_{\mathbb{C}}(P_i P_e(H_i + H_e)) \subset \mathbb{C}_-$ we get

$$
0 \neq \det(P(j\omega)) = P_i(j\omega)P_e(j\omega)(Z_i + Z_e) \Longrightarrow P_i(j\omega),\ P_e(j\omega),\ Z_i + Z_e \neq 0. \tag{8.247}
$$

For a sinusoidal input U_s and its steady-state outputs, cf. [17, (1.8)], we get

$$
\begin{aligned}
&U_s := U_s(0)e^{j\omega t},\ (U_i, U_e, I)^\top := H(j\omega)U_s \Longrightarrow (U_i, U_e, I, -U_s)^\top \in \mathcal{B} \cap \mathbb{C}^4 e^{j\omega t}\\
&\Longrightarrow P_i(j\omega)U_i = Q_i(j\omega)I_i,\ U_i = Z_i I,\ U_e = Z_e I,\ U_s = U_i + U_e = (Z_i + Z_e)I,\\
&I = (Z_i + Z_e)^{-1}U_s,\ U_i = H_i(j\omega)(H_i(j\omega) + H_e(j\omega))^{-1}U_s = \tfrac{Z_i}{Z_i + Z_e}U_s,\\
&U_e = \tfrac{Z_e}{Z_i + Z_e}U_s.
\end{aligned}
\tag{8.248}
$$

In analogy to (8.242) a real *current* source with a load is modeled by a network behavior with the data and equations

$$
\begin{aligned}
&V = \{1, 2\},\ K = \{k_s, k_i, k : 1 \to 2\},\ -I_s = I_i + I_e,\ U := U_i = U_e,\\
&\begin{pmatrix} 1 & 1 & 0\\ Q_i & 0 & -P_i\\ 0 & Q_e & -P_e \end{pmatrix} \circ \begin{pmatrix} I_i\\ I_e\\ U \end{pmatrix} = \begin{pmatrix} -1\\ 0\\ 0 \end{pmatrix} I_s \Longleftrightarrow \begin{pmatrix} 1 & 1 & 0\\ 0 & -Q_i & -P_i\\ 0 & Q_e & -P_e \end{pmatrix} \circ \begin{pmatrix} I_i\\ I_e\\ U \end{pmatrix} = \begin{pmatrix} -1\\ Q_i\\ 0 \end{pmatrix} \circ I_s.
\end{aligned}
\tag{8.249}
$$

This is an asymptotically stable IO behavior with input I_s if and only if

$$
Q_i P_e + Q_e P_i \neq 0,\ \mathrm{V}_{\mathbb{C}}(P_i Q_e + P_e Q_i) \subset \mathbb{C}_-. \tag{8.250}
$$

This condition coincides with that for the real voltage source, cf. (8.244), and is assumed in the sequel. Choose the current $I_s \in \mathcal{F}_+$ and the induced steady-state output $(I_i, I_e, U)^\top \in \mathcal{F}_+^3$. As in (8.245) we obtain

$$\begin{aligned}I_i &= H_i^{-1}\circ U,\ I_e = H_e^{-1}\circ U,\ -I_s = (H_i^{-1}+H_e^{-1})\circ U,\\ U &= -(H_i^{-1}+H_e^{-1})^{-1}\circ I_s,\\ I_e &= -H_e^{-1}(H_i^{-1}+H_e^{-1})^{-1}\circ I_s = -H_i(H_i+H_e)^{-1}\circ I_s,\\ U_e &= -(H_i^{-1}+H_e^{-1})^{-1}\circ I_s.\end{aligned} \tag{8.251}$$

Now we use the source current $I_s := -H_i^{-1}\circ U_s$ from (8.246) and conclude

$$I_e = (H_i+H_e)^{-1}\circ U_s,\ U_e = H_e(H_i+H_e)^{-1}\circ U_s. \tag{8.252}$$

The pairs (U_e, I_e) from (8.245) and (8.252) coincide. Therefore the real voltage source (8.242) resp. current source (8.249) with current $I_s = -H_i^{-1}\circ U_s$ are considered equivalent (sometimes called dual). Notice that in the model of the real voltage source the internal one-port $P_i\circ U_i = Q_i\circ I_i$ is connected in series with the ideal voltage source and the load, whereas in the equivalent current source it is connected in parallel. Thus any real voltage source of U_s can be transformed into a real current source of $I_s = -H_i^{-1}\circ U_s$.
If $U_s = U_s(0)e^{j\omega t}$ is sinusoidal and the impedance $Z_i := H_i(j\omega)$ is well defined and nonzero, then $I_s = -Z_i^{-1}U_s$. The internal impedance Z_i of the real voltage source can be easily measured and therefore the equivalent current source is known. The transfer matrix H_i and hence the general current $I_s = -H_i^{-1}\circ U_s$ cannot be determined so easily. ◇

Example 8.3.16 (*Helmholtz/Thévenin and Mayer/Norton equivalent*, cf. [10, Sects. 4.2.1, 4.2.2], [17, Sect. 1.4]) Let the behavior from Theorem 8.3.11 be an IO behavior and asymptotically stable, i.e., $\mathrm{pole}(H)\subseteq \mathrm{char}(\mathcal{B}^0)\subset \mathbb{C}_-$, and $u = \begin{pmatrix} I_{K_{2,s}} \\ U_{K_{1,s}}\end{pmatrix}\in \mathcal{F}_+^{K_s}$, $\mathcal{F}_+ := \mathrm{C}_+^{-\infty}(\mathbb{R}, F)$, an input with the steady-state output $y = H\circ u$. Let $\ell\in K_{2,s}$ be a source branch so that I_ℓ is a component of u and U_ℓ one of y. The entries of H are denoted by H_{U_ℓ,U_k}, etc. The steady-state equation $y = H\circ u$ implies the linear row relation, cf. [10, (4.19)], [17, (1.17)],

$$\begin{aligned}U_\ell &= \sum_{k\in K_{1,s}} H_{U_\ell,U_k}\circ U_k + \sum_{k\in K_{2,s},\ k\neq\ell} H_{U_\ell,I_k}\circ I_k + H_{U_\ell,I_\ell}\circ I_\ell\\ &= U_{\mathrm{Th}} + H_{\mathrm{Th}}\circ I_\ell \text{ where } H_{\mathrm{Th}} := H_{U_\ell,I_\ell} \text{ and}\\ U_{\mathrm{Th}} &:= \sum_{k\in K_{1,s}} H_{U_\ell,U_k}\circ U_k + \sum_{k\in K_{2,s},\ k\neq\ell} H_{U_\ell,I_k}\circ I_k.\end{aligned} \tag{8.253}$$

Of course, the U_k and I_k, $k\in K_s$, have left bounded support or are periodic. In the literature the network behavior $\mathcal{B}$ with these specifications is often called a *one-port with source branch* (input port) ℓ and *independent internal* voltage and current sources $U_k, k\in K_{1,s}$, and I_k, $\ell\neq k\in K_{2,s}$. The equation $U_\ell = U_{\mathrm{Th}} + H_{\mathrm{Th}}\circ I_\ell$ is called the *Helmholtz/Thévenin equivalent* of $\mathcal{B}$ for the chosen source branch ℓ. We call U_{Th}, $H_{\mathrm{Th}} := H_{U_\ell,I_\ell}$ resp. $H_{\mathrm{Th}}(j\omega)$, $\omega > 0$, the *Helmholtz/Thévenin voltage, transfer function resp. impedance at the frequency* ω. Obviously

$$\begin{cases} U_\ell = H_{\mathrm{Th}} \circ I_\ell & \text{if all internal sources are zero, hence } U_{\mathrm{Th}} = 0, \\ U_\ell = U_{\mathrm{Th}} & \text{if } I_\ell = 0, \text{ i.e., the one-port } \ell \text{ is idle (open circuit)}, \\ U_{\mathrm{Th}} = -H_{\mathrm{Th}} \circ I_\ell & \text{if } U_\ell = 0, \text{ i.e., the one-port } \ell \text{ is shorted (short circuit).} \end{cases} \tag{8.254}$$

Both U_{Th} and H_{Th} follow directly from (8.253) if the network $\mathcal{B}$ and its internal sources are known. For sinusoidal voltages and currents of frequency ω the $U_{\mathrm{Th}}, U_\ell, I_\ell$ can be measured and $H_{\mathrm{Th}}(j\omega)$ can be computed without knowledge of the complete transfer matrix H.
We now replace the source (U_l, I_l) by a one-port load $P_\ell \circ U_\ell = Q_\ell \circ I_\ell,\ H_\ell := P_\ell^{-1} Q_\ell$, i.e., connect the load with the two terminals of ℓ and add the equation $U_\ell = H_\ell \circ I_\ell$ in steady state. Then (8.253) becomes

$$U_{\mathrm{Th}} = (-H_{\mathrm{Th}} + H_\ell) \circ I_\ell. \tag{8.255}$$

This is the steady-state equation of a series connection of the voltage U_{Th} with two one-ports, given by their transfer functions $-H_{\mathrm{Th}}$ resp. H_ℓ.
If $\omega > 0$ is a frequency, if also H_ℓ has no pole in $j\omega$ and if the input u is sinusoidal with frequency ω and factor $e^{j\omega t}$, then Eq. (8.254) becomes $U_{\mathrm{Th}} = (-H_{\mathrm{Th}}(j\omega) + H_\ell(j\omega))I_\ell$.
If all voltages and currents are direct (and real), i.e., $\omega = 0$, then $U_{\mathrm{Th}} = (-H_{\mathrm{Th}}(0) + H_\ell(0))I_\ell$ represents Ohm's law with the real resistance $-H_{\mathrm{Th}}(0) + H_\ell(0)$.
If $\ell \in K_2 \backslash K_s$ is not a source branch, we proceed as follows: We modify $\mathcal{B}$ to $\widetilde{\mathcal{B}}$ by enlarging K_s to $\widetilde{K_s} := K_s \uplus \{\ell\}$, hence $\ell \in \widetilde{K}_{2,s}$, and by omitting the one-port equation $P_\ell \circ U_\ell = Q_\ell \circ I_\ell,\ H_\ell = P_\ell^{-1} Q_\ell$, cf. [10, Bild 4.8]. We assume that $\widetilde{\mathcal{B}}$ is an IO behavior with input $\widetilde{u} = (u, I_\ell)^\top$ and transfer matrix $\widetilde{H}$ and is asymptotically stable. Notice that the graph decompositions $K = K_1 \uplus K_2$ for $\mathcal{B}$ and $\widetilde{\mathcal{B}}$ coincide. We connect the two terminals of ℓ with the previously omitted one-port $P_l \circ U_\ell = Q_\ell \circ I_\ell$ and reobtain $\mathcal{B}$ from $\widetilde{\mathcal{B}}$ and

$$\begin{aligned} &\widetilde{U}_{\mathrm{Th}} = (-\widetilde{H}_{\mathrm{Th}} + H_\ell) \circ I_\ell \text{ with } \widetilde{H}_{\mathrm{Th}} := \widetilde{H}_{U_\ell, I_\ell} \text{ and} \\ &\widetilde{U}_{\mathrm{Th}} := \sum_{k \in K_{1,s}} \widetilde{H}_{U_\ell, U_k} \circ U_k + \sum_{k \in K_{2,s}} \widetilde{H}_{U_\ell, I_k} \circ I_k. \end{aligned} \tag{8.256}$$

We finally discuss the case $\ell \in K_{1,s}$. The network equations and the behavior $\mathcal{B}$ remain unchanged; however, its IO structure differs and gives rise to the input (output, transfer matrix) $\widehat{u}$ ($\widehat{y}$, $\widehat{H}$) such that U_ℓ resp. I_ℓ are components of $\widehat{u}$ resp. $\widehat{y}$. The analogue of the Helmholtz/Thévenin equation (8.253) is the *Mayer/Norton equation or equivalent*

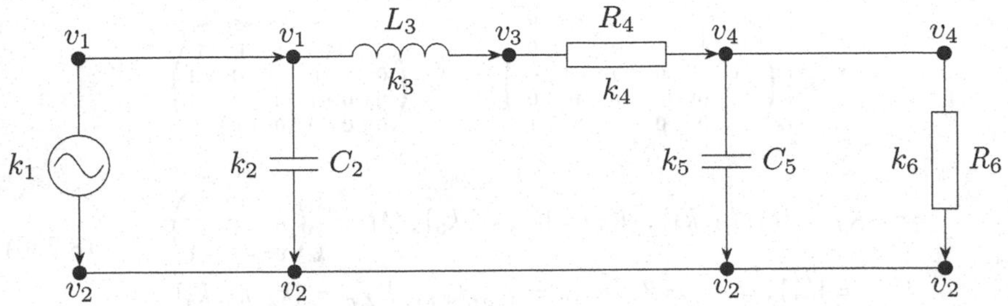

Fig. 8.1 The one-port of Example 8.3.17

$$I_\ell = I_N + H_N \circ U_\ell \text{ where } H_N := \widehat{H}_{I_\ell,U_\ell} \text{ and}$$
$$I_N := \sum_{k\in K_{1,s},\, k\neq \ell} \widehat{H}_{I_\ell,U_k} \circ U_k + \sum_{k\in K_{2,s}} \widehat{H}_{I_\ell,I_k} \circ I_k. \tag{8.257}$$

We replace the source (U_ℓ, I_ℓ) by a one-port load with equation

$$P_\ell \circ U_\ell = Q_\ell \circ I_\ell,\ H_\ell := P_\ell^{-1} Q_\ell \neq 0,\ I_\ell = H_\ell^{-1} \circ U_\ell \text{ (steady state), and get}$$
$$I_N = (-H_N + H_\ell^{-1}) \circ U_\ell. \tag{8.258}$$

This is the steady-state equation of a parallel connection of a current source I_N with two one-ports, given by the transfer functions $H_N^{-1} = H_{I_\ell,U_\ell}^{-1}$ and $-H_\ell$. The signs in these equations may vary if different sign conventions are used as often in the engineering literature. ◇

Example 8.3.17 Consider the one-port of Fig. 8.1, cf. [17, p. 72, (b)] that represents an equivalent circuit for a diode for small voltages and currents. The data are

$$V = \{v_1, \ldots, v_4\},\ K = \{k_1, \ldots, k_6\},\ K_s = \{k_1\},\ K_p = \{k_2, \ldots, k_6\},$$
$$U_i := U_{k_i},\ I_i := I_{k_i},\ i = 1, \ldots, 6,$$
$$C_2 s \circ U_2 = I_2,\ U_3 = L_3 s \circ I_3,\ U_4 = R_4 I_4,\ C_5 s \circ U_5 = I_5,\ U_6 = R_6 I_6. \tag{8.259}$$

The incidence matrix A with one echelon form $\widetilde{A}$ is

$$
A=\begin{array}{c}\\ v_1\\ v_2\\ v_3\\ v_4\end{array}\!\!\begin{array}{c}\begin{array}{cccccc}k_1&k_2&k_3&k_4&k_5&k_6\end{array}\\ \begin{pmatrix}1&1&1&0&0&0\\-1&-1&0&0&-1&-1\\0&0&-1&1&0&0\\0&0&0&-1&1&1\end{pmatrix}\end{array},\quad
\widetilde{A}=\begin{array}{c}\begin{array}{cccccc}k_1&k_3&k_4&k_2&k_5&k_6\end{array}\\ \begin{pmatrix}1&0&0&1&1&1\\0&1&0&0&-1&-1\\0&0&1&0&-1&-1\\0&0&0&0&0&0\end{pmatrix}\end{array}
$$

$$
\Longrightarrow K_1=\{k_1,k_3,k_4\},\ K_2=\{k_2,k_5,k_6\},\ M=\begin{array}{c}\\ k_1\\ k_3\\ k_4\end{array}\!\!\begin{array}{c}\begin{array}{ccc}k_2&k_5&k_6\end{array}\\ \begin{pmatrix}1&1&1\\0&-1&-1\\0&-1&-1\end{pmatrix}\end{array}
$$
$$
K_{1,s}=\{k_1\},\ K_{2,s}=\emptyset,\ K_{1,p}=\{k_3,k_4\},\ K_{2,p}=\{k_2,k_5,k_6\} \tag{8.260}
$$
$$
A|_{K_2}=\begin{array}{c}\begin{array}{ccc}k_2&k_5&k_6\end{array}\\ \begin{pmatrix}1&0&0\\-1&-1&-1\\0&0&0\\0&1&1\end{pmatrix}\end{array},\quad
A|_{K_1}=\begin{array}{c}\begin{array}{ccc}k_1&k_3&k_4\end{array}\\ \begin{pmatrix}1&1&0\\-1&0&0\\0&-1&1\\0&0&-1\end{pmatrix}\end{array},\quad
A|_{K_2}=A|_{K_1}M.
$$

The last equation confirms the correctness of the computations.
Next we compute the matrices that appear as blocks in (8.221):

$$
\begin{aligned}
&M_{K_{1,s},K_{2,p}}=\begin{array}{c}\\ k_1\end{array}\!\!\begin{array}{c}\begin{array}{ccc}k_2&k_5&k_6\end{array}\\ \begin{pmatrix}1&1&1\end{pmatrix}\end{array}, &&M_{K_{1,p},K_{2,p}}=\begin{array}{c}\\ k_3\\ k_4\end{array}\!\!\begin{array}{c}\begin{array}{ccc}k_2&k_5&k_6\end{array}\\ \begin{pmatrix}0&-1&-1\\0&-1&-1\end{pmatrix}\end{array}\\
&M_{K_{1,s},K_{2,s}}\underset{K_{2,s}=\emptyset}{=}\emptyset, &&M_{K_{1,p},K_{2,s}}=\emptyset\\
&P_{K_{1,p}}=\begin{array}{c}\\ k_3\\ k_4\end{array}\!\!\begin{array}{c}\begin{array}{cc}k_3&k_4\end{array}\\ \begin{pmatrix}1&0\\0&1\end{pmatrix}\end{array}, &&Q_{K_{1,p}}=\begin{array}{c}\\ k_3\\ k_4\end{array}\!\!\begin{array}{c}\begin{array}{cc}k_3&k_4\end{array}\\ \begin{pmatrix}L_3s&0\\0&R_4\end{pmatrix}\end{array}\\
&P_{K_{2,p}}=\begin{array}{c}\\ k_2\\ k_5\\ k_6\end{array}\!\!\begin{array}{c}\begin{array}{ccc}k_2&k_5&k_6\end{array}\\ \begin{pmatrix}C_2s&0&0\\0&C_5s&0\\0&0&1\end{pmatrix}\end{array}, &&Q_{K_{2,p}}=\begin{array}{c}\\ k_2\\ k_5\\ k_6\end{array}\!\!\begin{array}{c}\begin{array}{ccc}k_2&k_5&k_6\end{array}\\ \begin{pmatrix}1&0&0\\0&1&0\\0&0&R_6\end{pmatrix}\end{array}
\end{aligned} \tag{8.261}
$$
$$
y=\begin{pmatrix}y_I\\y_{II}\end{pmatrix}=(I_1,I_3,I_4,U_2,U_5,U_6,U_3,U_4,I_2,I_5,I_6)^\top,\quad u=U_1.
$$

With these blocks one forms the block matrices from (8.221), viz.

$$
P=\begin{array}{c}\\ K_{1,s}=k_1\\ K_{1,p}\\ K_{2,p}\\ K_{1,p}\\ K_{2,p}\end{array}\!\!\begin{array}{c}\begin{array}{ccccc}K_{1,s}=k_1&K_{1,p}&K_{2,p}&K_{1,p}&K_{2,p}\end{array}\\ \begin{pmatrix}1&0&0&0&M_{K_{1,s},K_{2,p}}\\0&\mathrm{id}_2&0&0&M_{K_{1,p},K_{2,p}}\\0&0&\mathrm{id}_3&-M^\top_{K_{1,p},K_{2,p}}&0\\0&-Q_{K_{1,p}}&0&P_{K_{1,p}}&0\\0&0&P_{K_{2,p}}&0&-Q_{K_{2,p}}\end{pmatrix}\end{array},
$$
$$
Q=\begin{array}{c}\\ K_{1,s}=k_1\\ K_{1,p}\\ K_{2,p}\\ K_{1,p}\\ K_{2,p}\end{array}\!\!\begin{array}{c}K_{1,s}=k_1\\ \begin{pmatrix}0\\0\\M^\top_{K_{1,s},K_{2,p}}\\0\\0\end{pmatrix}\end{array},\qquad R:=(P,-Q), \tag{8.262}
$$
$$
y=\begin{pmatrix}y_I\\y_{II}\end{pmatrix}=(I_1,I_3,I_4,U_2,U_5,U_6,U_3,U_4,I_2,I_5,I_6)^\top,\qquad u=U_1,
$$
$$
\mathcal{B}=\left\{\begin{pmatrix}y\\u\end{pmatrix}\in\mathcal{F}^{12};\ P\circ y=Q\circ u\right\}.
$$

The matrices from (8.224) and (8.225) are

$$
P_{II} = \begin{array}{c} \\ k_3 \\ k_4 \\ k_2 \\ k_5 \\ k_6 \end{array}\!\!\begin{array}{c} \begin{array}{ccccc} k_3 & k_4 & k_2 & k_5 & k_6 \end{array} \\ \left(\begin{array}{cc|ccc} 1 & 0 & 0 & -L_3 s & -L_3 s \\ 0 & 1 & 0 & -R_4 & -R_4 \\ \hline 0 & 0 & -1 & 0 & 0 \\ -C_5 s & -C_5 s & 0 & -1 & 0 \\ -1 & -1 & 0 & 0 & -R_6 \end{array}\right) \end{array}, \qquad P_I = \begin{array}{c} \\ k_1 \\ K_{1,p} \\ K_{2,p} \end{array}\!\!\begin{array}{c} \begin{array}{cc} K_{1,p} & K_{2,p} \end{array} \\ \begin{pmatrix} 0 & M_{K_{1,s},K_{2,p}} \\ 0 & M_{1_{1,p},K_{2,p}} \\ -M^\top_{K_{1,p},K_{2,p}} & 0 \end{pmatrix} \end{array},
$$

$$
Q_{II} = \begin{array}{c} \\ k_3 \\ k_4 \\ k_2 \\ k_5 \\ k_6 \end{array}\!\!\begin{array}{c} k_1 \\ \left(\begin{array}{c} 0 \\ 0 \\ \hline -C_2 s \\ -C_5 s \\ -1 \end{array}\right) \end{array}, \qquad Q_I = \begin{array}{c} \\ k_1 \\ K_{1,p} \\ K_{2,p} \end{array}\!\!\begin{array}{c} k_1 \\ \begin{pmatrix} 0 \\ 0 \\ M^\top_{K_{1,s},K_{2,p}} \end{pmatrix} \end{array},
$$

$$
y_I = (I_1, I_3, I_4, U_2, U_5, U_6)^\top, \qquad y_{II} = (U_3, U_4, I_2, I_5, I_6)^\top
$$

$$
\mathcal{B}_{II} = \left\{ \binom{y_{II}}{u} \in \mathcal{F}^6;\ P_{II} \circ y = Q_{II} \circ u \right\} \cong \mathcal{B},\ \binom{y_{II}}{u} \mapsto \binom{-P_I y_{II} + Q_I u}{y_{II}}. \tag{8.263}
$$

The behaviors $\mathcal{B}_{II} \cong \mathcal{B}$ are IO behaviors since

$$
\det(P_{II}) = -\left(C_5 L_3 R_6 s^2 + (C_5 R_4 R_6 + L_3)s + (R_6 + R_4)\right) \neq 0. \tag{8.264}
$$

The 5th elementary divisor e^{uc}_{II} of $(P_{II}, -Q_{II}) \in F[s]^{5\times 6}$ is 1 and shows that $\mathcal{B}_{II}$ and $\mathcal{B}$ are controllable.

We compute the transfer matrices

$$
H_{II} = P_{II}^{-1} Q_{II}
$$

$$
= \begin{pmatrix} (H_{II})_{U_3,U_1} \\ (H_{II})_{U_4,U_1} \\ (H_{II})_{I_2,U_1} \\ (H_{II})_{I_5,U_1} \\ (H_{II})_{I_6,U_1} \end{pmatrix} = \frac{1}{\det(P_{II})} \begin{array}{c} \\ k_3 \\ k_4 \\ k_2 \\ k_5 \\ k_6 \end{array}\!\!\begin{array}{c} k_1 \\ \left(\begin{array}{c} -C_5 L_3 R_6 s^2 - L_3 s \\ -C_5 R_4 R_6 s - R_4 \\ \hline -C_2 C_5 L_3 R_6 s^3 - (C_5 R_4 R_6 + L_3) C_2 s^2 - (R_6 + R_4) C_2 s \\ -C_5 R_6 s \\ -1 \end{array}\right) \end{array} \tag{8.265}
$$

and

$$H = P^{-1}Q = \left(H_{I_1,U_1}, H_{I_3,U_1}, \ldots, H_{I_6,U_1}\right)^\top$$

$$= \frac{1}{\det(P_{II})} \begin{array}{c} \\ k_1 \\ k_3 \\ k_4 \\ k_2 \\ k_5 \\ k_6 \\ k_3 \\ k_4 \\ k_2 \\ k_5 \\ k_6 \end{array} \begin{pmatrix} k_1 \\ C_2C_5L_3R_6s^3+(C_2L_3+C_2C_5R_4R_6)s^2+(C_2R_6+C_5R_6+C_2R_4)s+1 \\ \hline -C_5R_6s-1 \\ -C_5R_6s-1 \\ \hline -C_5L_3R_6s^2-(C_5R_4R_6+L_3)s-R_4-R_6 \\ -R_6 \\ -R_6 \\ \hline -C_5L_3R_6s^2-L_3s \\ -C_5R_4R_6s-R_4 \\ \hline -C_2C_5L_3R_6s^3-(C_5R_4R_6+L_3)C_2s^2-(R_4+R_6)C_2s \\ -C_5R_6s \\ -1 \end{pmatrix}. \tag{8.266}$$

We check for the correctness of the computations resp. read out

$$H - \begin{pmatrix} Q_I - P_I H_{II} \\ H_{II} \end{pmatrix} = 0 \text{ resp. } H_{I_1,U_1} = \det(P_{II})^{-1} N \text{ with}$$
$$N := C_2C_5L_3R_6s^3 + (C_2L_3 + C_2C_5R_4R_6)s^2 + (C_2R_6 + C_5R_6 + C_2R_4)s + 1. \tag{8.267}$$

The components H_{I_1,U_1} and H_{I_2,U_1} have a polynomial part of degree $3-2=1$ and hence are not proper. Choose the DC voltage input $U_1 := \alpha Y = \alpha s^{-1} \circ \delta$ with $\alpha > 0$ and a jump at $t = 0$. Then

$$\begin{aligned} &H_{I_1,U_1}s^{-1} \in -C_2 + F(s)_{\text{spr}},\ H_{I_2,U_1}s^{-1} \in C_2 + F(s)_{\text{spr}} \\ &\Longrightarrow I_1 = H_{I_1,U_1} \circ U_1 = H_{I_1,U_1}s^{-1} \circ \alpha\delta \in -\alpha C_2\delta + \mathrm{t}(\mathcal{F})Y, \\ &\quad I_2 = H_{I_2,U_1} \circ U_1 = H_{2_1,U_1}s^{-1} \circ \alpha\delta \in \alpha C_2\delta + \mathrm{t}(\mathcal{F})Y. \end{aligned} \tag{8.268}$$

The nonzero distributional components $-/+\alpha C_2\delta$ of I_1 and I_2 may destroy the network and suggest to modify it or to consider I_1 instead of U_1 as input as in Example 8.3.25. Since the network is asymptotically stable, see below, y and especially I_1 and I_2 are steady-state outputs.

For the consideration of stability we compute the 5th elementary divisor e_{II} of P_{II}, namely

$$e_{II} = s^2 + \frac{C_5R_4R_6+L_3}{C_5L_3R_6}s + \frac{R_4+R_6}{C_5L_3R_6}$$

and its two roots

$$\mathrm{V}_{\mathbb{C}}(e_{II}) = \mathrm{char}(\mathcal{B}^0) = \left\{ -\frac{C_5R_6R_6+L_3}{2C_5L_3R_6} \pm \frac{\sqrt{(C_5R_4R_6-L_3)^2-4C_5L_3R_6^2}}{2C_5L_3R_6} \right\}. \tag{8.269}$$

If the radicand is smaller than zero, then the two roots are conjugate complex numbers with negative real part. If the radicand is greater than zero, then

$$(C_5R_4R_6 + L_3)^2 > (C_5R_4R_6 - L_3)^2 - 4C_5L_3R_6^2 > 0 \text{ and}$$
$$C_5R_4R_6 + L_3 > \sqrt{(C_5R_4R_6 - L_3)^2 - 4C_5L_3R_6^2}$$

and again both roots are negative. Hence $\mathcal{B}^0$ is asymptotically stable.
A real sinusoidal input $U_1 = U_1(0)\cos(\omega t) = \Re(e^{j\omega t})U_1(0),\ \omega > 0$, gives rise to the real steady-state output

$$\begin{aligned} &y = \Re\left(H(j\omega)e^{j\omega t}\right)U_1(0) = \left(\Re\left(H(j\omega)\right)\cos(\omega t) - \Im\left(H(j\omega)\right)\sin(\omega t)\right)U_1(0),\\ &\text{in particular } I_1 = \Re\left(H_{I_1,U_1}(j\omega)e^{j\omega t}\right)U_1(0). \end{aligned} \tag{8.270}$$

The numbers

$$H_{I_1,U_1}(j\omega),\ \Re\left(H_{I_1,U_1}(j\omega)\right) \text{ resp. } \Im\left(H_{I_1,U_1}(j\omega)\right) \tag{8.271}$$

are called the *admittance, conductance* resp. *susceptance* of the one-port at the frequency ω, cf. Definition 8.3.19. ◊

8.3.2 Elimination of the Internal One-Ports of the Network and Other IO Structures

The data of Sect. 8.3.1 remain in force. We assume that the network behavior $\mathcal{B}$ is an IO behavior, i.e., $\det(P_{II}) \neq 0$. *Elimination of the one-port branches* furnishes a new behavior $\mathcal{B}_s$ in the following fashion: Define the output component y_s of y and the matrix C by

$$\begin{aligned} &y_s := \begin{pmatrix} I_{K_{1,s}} \\ U_{K_{2,s}} \end{pmatrix} \in \mathcal{F}^{n_s} = \mathcal{F}^{K_{1,s}\uplus K_{2,s}} \text{ and } C := \begin{pmatrix} \mathrm{id}_{K_{1,s}} & 0 & 0 & 0\,0\,0 \\ 0 & 0 & \mathrm{id}_{K_{2,s}} & 0\,0\,0 \end{pmatrix}\\ &\quad \in F^{n_s\times(n+n_p)} = F^{(K_{1,s}\uplus K_{2,s})\times(K_{1,s}\uplus K_{1,p}\uplus K_{2,s}\uplus K_{2,p}\uplus K_{1,p}\uplus K_{2,p})} \Longrightarrow y_s = Cy. \end{aligned} \tag{8.272}$$

Due to Theorem 6.3.1 we obtain the image IO behavior

$$\begin{aligned} &\mathcal{B}_s := \begin{pmatrix} C & 0 \\ 0 & \mathrm{id}_{n_s} \end{pmatrix}\mathcal{B} = \left\{\begin{pmatrix} y_s \\ u \end{pmatrix} \in \mathcal{F}^{n_s\times n_s};\ P_s \circ y_s = Q_s \circ u\right\} \text{ with}\\ &\begin{pmatrix} y_s \\ u \end{pmatrix} = \begin{pmatrix} (U_k)_{k\in K_s} \\ (I_k)_{k\in K_s} \end{pmatrix} \text{ (up to the order of the components),} \end{aligned} \tag{8.273}$$

and suitable P_s and Q_s. Obviously $\mathcal{B}_s$ describes the voltages and currents at the source branches only, but not those of the interior branches, and is therefore often called a *black box*.
The computation of $\mathcal{B}_s$ is simplified by means of the isomorphism $\mathcal{B}_{II} \cong \mathcal{B}$ and the resulting epimorphism

$$\begin{aligned}
&\mathcal{B}_{II} \to \mathcal{B}_s,\ \binom{y_{II}}{u} \mapsto \binom{y_s}{u},\ \text{with}\\
&y_s = C\left(\begin{pmatrix} -P_I \\ \mathrm{id}_{n_p}\end{pmatrix} y_{II} + \begin{pmatrix} Q_I \\ 0\end{pmatrix} u\right) = C_s y_{II} + D_s u \text{ where}\\
&C_s = \begin{pmatrix} 0 & -M_{K_{1,s},K_{2,p}} \\ \left(M_{K_{1,p},K_{2,s}}\right)^{\top} & 0 \end{pmatrix} \in F^{n_s\times n_p},\\
&D_s = \begin{pmatrix} -M_{K_{1,s},K_{2,s}} & 0 \\ 0 & \left(M_{K_{1,s},K_{2,s}}\right)^{\top} \end{pmatrix} \in F^{n_s\times n_s}.
\end{aligned} \tag{8.274}$$

In other words, $\mathcal{B}_s$ is described by the Rosenbrock equations

$$P_{II}\circ y_{II} = Q_{II}\circ u,\ \ y_s = C_s y_{II} + D_s u.$$

We apply Theorem 6.3.1 to compute the equations of $\mathcal{B}_s$. Since

$$\begin{pmatrix} P_{II} \\ C_s \end{pmatrix} \in F[s]^{(n_p+n_s)\times n_p} \text{ and } \operatorname{rank}\begin{pmatrix} P_{II} \\ C_s \end{pmatrix} = n_p$$

the universal left annihilator of $\begin{pmatrix} P_{II} \\ C_s \end{pmatrix}$ has the form

$$\begin{aligned}
&(-X_s, P_s) \in F[s]^{n_s\times(n_p+n_s)},\ \operatorname{rank}(P_s) = n_s\\
&\Longrightarrow X_s P_{II} = P_s C_s,\ P_s^{-1}X_s = C_s P_{II}^{-1},\ Q_s := X_s Q_{II} + P_s D_s.
\end{aligned} \tag{8.275}$$

In particular, $P_s^{-1}X_s$ is the left coprime factorization of $C_s P_{II}^{-1}$. With these matrices Theorem 8.3.11 implies the following.

Theorem and Definition 8.3.18 *1. Assume that the network behavior $\mathcal{B}$ is an IO behavior and the data from (8.274) and (8.275). Then the image behavior $\mathcal{B}_s := \begin{pmatrix} C & 0 \\ 0 & \mathrm{id}_{n_s}\end{pmatrix}\mathcal{B}$, obtained from $\mathcal{B}$ by elimination of the $U_k, I_k,\ k \in K_p$, is also an IO behavior, and indeed*

$$\begin{aligned}
&\mathcal{B}_s = \left\{\binom{y_s}{u} \in \mathcal{F}^{n_s+n_s};\ P_s\circ y_s = Q_s\circ u\right\},\ y_s = \begin{pmatrix} I_{K_{1,s}} \\ U_{K_{2,s}}\end{pmatrix},\ u = \begin{pmatrix} I_{K_{2,s}} \\ U_{K_{1,s}}\end{pmatrix},\\
&H_s := P_s^{-1}Q_s = CH = \begin{pmatrix} H_{I_{K_{1,s}},u} \\ H_{U_{K_{2,s}},u}\end{pmatrix} = D_s + C_s H_{II}.
\end{aligned} \tag{8.276}$$

If currents $I_k,\ k \in K_{2,s}$, and voltages $U_k,\ k \in K_{1,s}$, are supplied at the source branches $k \in K_s$, then the currents $I_k,\ k \in K_{1,s}$, and voltages $U_k,\ k \in K_{2,s}$, at the source branches satisfy the equations $P_s \circ \begin{pmatrix} I_{K_{1,s}} \\ U_{K_{2,,s}}\end{pmatrix} = Q_s \circ \begin{pmatrix} I_{K_{2,s}} \\ U_{K_{1,s}}\end{pmatrix}$.

2. With $\mathcal{G}$ from Theorem 8.3.11,2, there are the isomorphisms

$$\begin{aligned}
&\begin{matrix} \mathcal{B}\cap\mathcal{G}^{(n+n_p)+n_s} & \cong & \mathcal{B}_{II}\cap\mathcal{G}^{n_p+n_s} & \cong & \mathcal{B}_s\cap\mathcal{G}^{n_s+n_s} & \cong & \mathcal{G}^{n_s} \\ \binom{H\circ u}{u} & \leftrightarrow & \binom{H_{II}\circ u}{u} & \leftrightarrow & \binom{H_s\circ u}{u} & \leftrightarrow & u \end{matrix} \quad \text{where}\\
&H_{II} = C_{II}H = \begin{pmatrix} H_{U_{K_{1,p}},u} \\ H_{I_{K_{2,p}},u}\end{pmatrix},\ H = \begin{pmatrix} Q_I - P_I H_{II} \\ H_{II}\end{pmatrix},\ H_s \text{ as in (8.276).}
\end{aligned} \tag{8.277}$$

3. *Theorem 8.3.11, 5, is applicable to* $\mathcal{B}_s$.

Proof Theorem 6.3.1. □

Definition 8.3.19 (*Impedance, admittance*) In the situation of Theorem 8.3.18 with $F = \mathbb{R}$ assume a one-port, i.e.,

$$n_s = 1,\ K_s = \{k_s\},\ k_s \in K_2 \text{ (for instance)},\ K_{2,s} = \{k_s\},\ K_{1,s} = \emptyset$$
$$\mathcal{B}_s = \left\{\begin{pmatrix} U_{k_s} \\ I_{k_s}\end{pmatrix} \in \mathcal{F}^2;\ P_s \circ U_{k_s} = Q_s \circ I_{k_s}\right\},\ H_s = P_s^{-1}Q_s \in \mathbb{R}(s),$$

with input I_{k_s} and output U_{k_s}. Assume that $\mathcal{B}_s^0$ is asymptotically stable, i.e., $\mathrm{char}(\mathcal{B}_s^0) = V_{\mathbb{C}}(P_s) \subset \mathbb{C}_-$. Then $H_s(j\omega) \in \mathbb{C}$ is defined for $\omega > 0$. In the representation

$$Z(\omega) := H_s(j\omega) = \Re(Z(\omega)) + j\Im(Z(\omega)) \tag{8.278}$$

the numbers $Z(\omega)$, $\Re(Z(\omega))$ resp. $\Im(Z(\omega))$ are rational functions of ω, and are called the (complex) *impedance*, the (real) *resistance* resp. the (real) *reactance* of the one-port at the frequency ω.
If Q_s or H_s are nonzero, then $\mathcal{B}_s$ is also an IO behavior with input U_{k_s}, output I_{k_s}, and transfer function $Q_s^{-1}P_s = H_s^{-1}$. In the representation

$$Y(\omega) := H_s(j\omega)^{-1} = \Re(Y(\omega)) + j\Im(Y(\omega)) = |Z(\omega)|^{-2}\left(\Re(Z(\omega)) - j\Im(Z(\omega))\right)$$

the numbers

$$Y(\omega),\ \Re(Y(\omega)) = |Z(\omega)|^{-2}\Re(Z(\omega)) \text{ resp. } \Im(Y(\omega)) = -|Z(\omega)|^{-2}\Im(Z(\omega)) \tag{8.279}$$

are called the (complex) *admittance*, (real) *conductance* resp. (real) *susceptance* of the one-port at ω. A real sinusoidal input

$$I_{k_s} = |I_{k_s}(0)|\cos(\omega t + \varphi) = \Re\left(I_{k_s}(0)e^{j\omega t}\right),\ \omega > 0,\ I_{k_s}(0) = |I_{k_s}(0)|e^{j\varphi} \in \mathbb{C},$$

gives rise to the real steady-state output

$$\begin{aligned} U_{k_s} &= \Re\left(Z(\omega)e^{j(\omega t+\varphi)}\right)|I_{k_s}(0)| \\ &= \left(\Re(Z(\omega))\cos(\omega t + \varphi) - \Im(Z(\omega))\sin(\omega t + \varphi)\right)|I_{k_s}(0)|. \end{aligned}$$ ◇

Example 8.3.20 We illustrate Theorem 8.3.18 with the series resonant RLC circuit from Fig. 8.2 with vertex (node) set V, branch set K, incidence matrix A, and the standard port equations

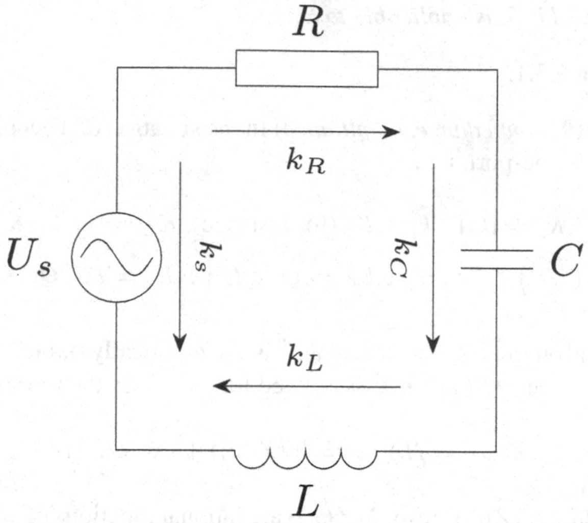

Fig. 8.2 The series connection of a source, a resistor, a capacitor, and an inductor

$$\begin{aligned}
&K = \{k_s,\ k_R,\ k_C,\ k_L\},\ V = \{v_1,\ v_2,\ v_3,\ v_4\},\\
&k_s : v_1 \to v_2,\ k_R : v_1 \to v_3,\ k_C : v_3 \to v_4,\ k_L : v_4 \to v_2,\\
&K_s = \{k_s\},\ K_p = \{k_R, k_C, k_L\},\\
&n := \sharp(K) = 4,\ n_s = \sharp(K_s) = 1,\ n_p := \sharp(K_p) = 3,\ n + n_p = 7,\\
&A = \begin{array}{c} \\ v_1 \\ v_2 \\ v_3 \\ v_4 \end{array}\!\!\begin{array}{c} \begin{array}{cccc} k_s & k_R & k_C & k_L \end{array} \\ \begin{pmatrix} 1 & 1 & 0 & 0 \\ -1 & 0 & 0 & -1 \\ 0 & -1 & 1 & 0 \\ 0 & 0 & -1 & 1 \end{pmatrix} \end{array} \in F^{V\times K} = F^{4\times 4}.
\end{aligned} \tag{8.280}$$

By elementary row operations the incidence matrix A is transformed to

$$\begin{aligned}
&\begin{pmatrix} 1\,0\,0 & 1 \\ 0\,1\,0 & -1 \\ 0\,0\,1 & -1 \\ 0\,0\,0 & 0 \end{pmatrix} = \begin{pmatrix} \mathrm{id}_3 & M \\ 0 & 0 \end{pmatrix} \in F^{4\times(3+1)},\ M := \begin{array}{c} \\ k_s \\ k_R \\ k_C \end{array}\!\!\begin{array}{c} k_L \\ \begin{pmatrix} 1 \\ -1 \\ -1 \end{pmatrix} \end{array}\\
&\Longrightarrow K_1 = \{k_s, k_R, k_C\},\ K_2 = \{k_L\},\\
&K_{1,s} = \{k_s\},\ K_{2,s} = \emptyset,\ K_{1,p} = \{k_R, k_C\},\ K_{2,p} = \{k_L\},\\
&M_{K_{1,p},K_{2,p}} = \begin{pmatrix} -1 \\ -1 \end{pmatrix},\ M_{k_s,K_{2,p}} = 1,\\
&y_I = (I_s, I_R, I_C, U_L)^\top,\ y_{II} = (U_R, U_C, I_L)^\top,\ y_s = I_s,\ u = U_s.
\end{aligned} \tag{8.281}$$

The port-branch equations are

$$
\begin{aligned}
&U_R = RI_R,\ Cs\circ U_C = I_C,\ U_L = Ls\circ I_L,\\
&\Longrightarrow P_{K_{1,p}} := \operatorname{diag}(1, Cs),\ Q_{K_{1,p}} := \operatorname{diag}(R, 1),\ P_{K_{2,p}} = 1,\ Q_{K_{2,p}} = Ls,\\
&P_{II} = \begin{pmatrix} 1 & 0 & -R\\ 0 & Cs & -1\\ -1 & -1 & -Ls \end{pmatrix},\ Q_{II} = \begin{pmatrix} 0\\ 0\\ -1 \end{pmatrix},\ P_I = \begin{pmatrix} 0 & 0 & 1\\ 0 & 0 & -1\\ 0 & 0 & -1\\ 1 & 1 & 0 \end{pmatrix},\ Q_I = \begin{pmatrix} 0\\ 0\\ 0\\ 1 \end{pmatrix},\\
&C_s = (0, 0, -1).
\end{aligned}
\tag{8.282}
$$

The network equations are

$$
\begin{aligned}
&P_{II}\circ y_{II} = Q_{II}\circ u,\ y_I = -P_I y_{II} + Q_I u\\
&\Longleftrightarrow \begin{pmatrix} 1 & 0 & -R\\ 0 & Cs & -1\\ -1 & -1 & -Ls \end{pmatrix}\begin{pmatrix} U_R\\ U_C\\ I_L \end{pmatrix} = \begin{pmatrix} 0\\ 0\\ -1 \end{pmatrix} U_s,\ \begin{pmatrix} I_s\\ I_R\\ I_C\\ U_L \end{pmatrix} = -\begin{pmatrix} 0 & 0 & 1\\ 0 & 0 & -1\\ 0 & 0 & -1\\ 1 & 1 & 0 \end{pmatrix}\begin{pmatrix} U_R\\ U_C\\ I_L \end{pmatrix} + \begin{pmatrix} 0\\ 0\\ 0\\ 1 \end{pmatrix} U_s\\
&\Longleftrightarrow U_R - RI_L = 0,\ Cs\circ U_C - I_L = 0,\ -U_R - U_C - Ls\circ I_L = -U_s,\\
&\qquad I_s = -I_L,\ I_R = I_L,\ I_C = I_L,\ U_L = -U_R - U_C + U_s\\
&\Longrightarrow I := I_R = I_C = I_L = -I_s,\ U_s = U_R + U_C + U_L\\
&\qquad U_R = RI,\ C(s\circ U_C) = I,\ L(s\circ I) = U_L.
\end{aligned}
\tag{8.283}
$$

These equations can, of course, be obtained directly from Kirchhoff's laws and the one-port equations, but have been derived here as the simplest case of the algorithm from Theorem 8.3.11 and Definition 8.3.19.
The determinant of P_{II} is

$$
\begin{aligned}
&\det(P_{II}) = -N = -LC\left(s^2 + RL^{-1}s + (LC)^{-1}\right) \neq 0,\ N := LCs^2 + RCs + 1,\\
&\Longrightarrow \operatorname{char}(\mathcal{B}_{II}^0) = \mathrm{V}_{\mathbb{C}}(\det(P_{II})) = \left\{\lambda_{-/+} = -\tfrac{R}{2L} - / + \sqrt{\left(\tfrac{R}{2L}\right)^2 - \tfrac{1}{LC}}\right\}.
\end{aligned}
\tag{8.284}
$$

This shows that $\mathcal{B} \cong \mathcal{B}_{II}$ are IO behaviors and that

$$
\operatorname{char}(\mathcal{B}^0) = \operatorname{char}(\mathcal{B}_{II}^0) = \left\{\lambda_{-/+}\right\} \subset \mathbb{C}_- \text{ since } R, C, L > 0. \tag{8.285}
$$

Thus $\mathcal{B}^0 \cong \mathcal{B}_{II}^0$ is asymptotically stable. The transfer matrix of $\mathcal{B}_{II}$ is the unique solution H_{II} of $P_{II}H_{II} = Q_{II}$ and given by

$$
\begin{aligned}
&H_{II} = N^{-1}(RCs, 1, Cs)^\top,\ N := CLs^2 + CRs + 1,\\
&\Longrightarrow H = \begin{pmatrix} Q_I - P_I H_{II}\\ H_{II} \end{pmatrix} = N^{-1}(-Cs, Cs, Cs, CLs^2, RCs, 1, Cs)^\top \in \mathbb{C}(s)^7_{\mathrm{pr}}\\
&y = \begin{pmatrix} y_I\\ y_{II} \end{pmatrix} = (I_s, I_R, I_C, U_L, U_R, U_C, I_L)^\top = H\circ u = H\circ U_s.
\end{aligned}
\tag{8.286}
$$

Since H is proper and $\mathcal{B}^0$ is asymptotically stable, Theorem 8.3.11, 5, is applicable. Consider an input $U_s = H_2\circ\delta = \alpha Y,\ H_2 \in \mathbb{C}(s)_{\mathrm{spr}}$, with polynomial-exponential α and the output $y := H\circ U_s = HH_2\circ\delta$. All components of H except $H_{U_L,U_s} = 1 - N^{-1}(CRs + 1)$ are strictly proper, and H_{U_L,U_s} is proper. Therefore all components of y except U_L are continuous, but $U_L = U_s - N^{-1}(CR + 1)H_2\circ\delta$ is piecewise continuous with the same jump at $t = 0$ as U_s. Similarly, if U_s is a piecewise continuous T-periodic input, $T > 0$, $\omega := 2\pi T^{-1}$, and $y := H \circ U_s$ is the unique T-periodic output then all components of y except U_L have a uniformly convergent and continuous Fourier series, whereas U_L is given as

$$U_L = U_s - \sum_{\mu \in \mathbb{Z}} N(j\mu\omega)^{-1}(CRj\mu\omega + 1)e^{j\mu\omega t}$$

where again $\sum_{\mu\in\mathbb{Z}}$ is uniformly convergent and continuous. The Smith form of

$$\begin{aligned}&(P_{II}, -Q_{II}) \text{ is } \begin{pmatrix} 1\,0 & 0 & 0\\ 0\,1 & 0 & 0\\ 0\,0 & s^2+RL^{-1}s+(CL)^{-1} & 0\end{pmatrix} \Longrightarrow e^{uc}_{II} = s^2 + RL^{-1}s + (CL)^{-1}\\ &\Longrightarrow \operatorname{char}(\mathcal{B}_{II}) = \mathrm{V}_{\mathbb{C}}(e^{\mathrm{uc}}_{II}) = \operatorname{char}(\mathcal{B}^0_{II}) = \left\{\lambda_{-/+}\right\}.\end{aligned} \tag{8.287}$$

Hence $\mathcal{B}_{II}$ is not controllable with the uncontrollable poles $\lambda_{-/+}$.
We proceed to $\mathcal{B}_s$. The Smith form $X\begin{pmatrix} P_{II}\\ C_s\end{pmatrix}$, $X \in \mathrm{Gl}_4(F[s])$, of $\begin{pmatrix} P_{II}\\ C_s\end{pmatrix}$ is obtained as

$$\begin{pmatrix} 1 & 0 & 0 & -R\\ -1 & 0 & -1 & R+Ls\\ 0 & 0 & 0 & -1\\ Cs & 1 & Cs & -N\end{pmatrix}\begin{pmatrix} 1 & 0 & -R\\ 0 & Cs & -1\\ -1 & -1 & -Ls\\ 0 & 0 & -1\end{pmatrix} = \begin{pmatrix} 1&0&0\\ 0&1&0\\ 0&0&1\\ 0&0&0\end{pmatrix},\quad N = CLs^2 + CRs + 1.$$

This shows that $\begin{pmatrix} P_{II}\\ C_s\end{pmatrix}$ is left invertible, i.e., that $C_s : \mathcal{B}^0_{II} \to \mathcal{B}^0_s$ is injective. Therefore $\mathcal{B}_{II}$ and $\mathcal{B}$ are observable from $\mathcal{B}_s$. This follows from (8.283) directly.
According to Theorem 3.2.18 the universal left annihilator of $\begin{pmatrix} P_{II}\\ C_s\end{pmatrix}$ is

$$\begin{aligned}&(Cs, 1, Cs, -N) \Longrightarrow (Cs, 1, Cs)P_{II} = NC_s\\ &\Longrightarrow N \circ I_s = NC_s \circ y_{II} = (Cs, 1, Cs)Q_{II} \circ u = -Cs \circ U_s\\ &\underset{\text{Theorem 6.3.1}}{\Longrightarrow} \mathcal{B}_s = \left\{\begin{pmatrix} I_s\\ U_s\end{pmatrix} \in \mathcal{F}^2;\ N \circ I_s = -Cs \circ U_s\right\},\\ &\qquad H_s = H_{I_s,U_s} = -CsN^{-1} = -L^{-1}s\left(s^2 + RL^{-1}s + (CL)^{-1}\right)^{-1},\\ &\qquad H_s^{-1} = -C^{-1}s^{-1}N.\end{aligned} \tag{8.288}$$

The component $H_s \circ U_s = H_{I_s,U_s} \circ U_s$ has been treated above.
Now consider $\mathcal{B}_s$ as controllable IO behavior with input I_s, i.e.,

$$\begin{aligned}&\mathcal{B}_s = \left\{\begin{pmatrix} U_s\\ I_s\end{pmatrix} \in \mathcal{F}^2;\ -Cs \circ U_s = N \circ I_s\right\},\ H_s^{-1} = -(Cs)^{-1}N,\\ &\mathcal{B}^{0,U_s}_s := \{U_s \in \mathcal{F};\ s \circ U_s = 0\} = \mathbb{C},\ \operatorname{char}(\mathcal{B}^{0,U_s}) = \{0\}.\end{aligned} \tag{8.289}$$

Thus the behavior $\mathcal{B}^{0,U_s}_s$ is not asymptotically stable and any $\begin{pmatrix} U_s\\ I_s\end{pmatrix} \in \mathcal{B}_s$ gives rise to $\begin{pmatrix} U_s+U_0\\ I_s\end{pmatrix} \in \mathcal{B}_s$, $U_0 \in \mathbb{C}$. Mostly $U_0 = 0$ is assumed.
Since $\deg_s(H_s^{-1}) = 1$, H_s^{-1} is not proper and therefore admits impulsive solutions. Indeed, assume $I_s = H_2 \circ \delta = \alpha Y$, $H_2 \in \mathbb{C}(s)_{\mathrm{spr}}$, with polynomial-exponential α. Then $U_s := H_s^{-1} \circ I_s = H_s^{-1}H_2 \circ \delta$ with $\deg_s(H_s^{-1}H_2) \le 0$ has an impulsive com-

ponent if and only if $H_s^{-1}H_2 \notin \mathbb{C}(s)_{\rm spr}$, i.e., $\deg_s(H_2) = -1$, cf. Theorem 8.2.58. For instance, for

$$\begin{aligned}&H_2 := s^{-1},\ H_s^{-1}s^{-1} = -C^{-1}s^{-2}N = -L - Rs^{-1} - C^{-1}s^{-2},\ I_s = s^{-1}\delta = Y,\\ &U_s = H_s^{-1}Y = -L\delta - RY - C^{-1}tY.\end{aligned}$$

All branch voltages of this input current result from

$$\begin{aligned}y = \begin{pmatrix} y_I \\ y_{II}\end{pmatrix} &= (I_s, I_R, I_C, U_L, U_R, U_C, I_L)^\top\\ &= N^{-1}(-Cs, Cs, Cs, CLs^2, RCs, 1, Cs)^\top(-C^{-1}s^{-2}N)\circ\delta\\ &= (s^{-1}, -s^{-1}, -s^{-1}, -L, -Rs^{-1}, -C^{-1}s^{-2}, -s^{-1})^\top \circ \delta\\ &= (Y, -Y, -Y, -L\delta, -RY, -C^{-1}tY, -Y)^\top,\end{aligned}$$

cf. Theorem 8.3.21. The components $U_L = -L\delta$ and U_s create a problem. Its solution via a modification of the network is discussed in [8, p. 160]. ◇

Next we study other IO decompositions of $\mathcal{B}_s$ and ensuing new IO structures of $\mathcal{B}$. In Theorem 8.3.18 define the vector w_s and matrix R_s by

$$\begin{aligned}w_s = (w_{s,1}, \ldots, w_{s,2n_s})^\top &:= \begin{pmatrix} y_s \\ u\end{pmatrix} = \left((I_{K_{1,s}}, U_{K_{2,s}}), (I_{K_{2,s}}, U_{K_{1,s}})\right)^\top\\ &\in \mathcal{F}^{K_s \uplus K_s} = \mathcal{F}^{2n_s},\\ R_s := (P_s, -Q_s) &\in F[s]^{n_s\times(n_s+n_s)} = F^{K_s\times(K_s\uplus K_s)},\\ \Longrightarrow \operatorname{rank}(R_s) &= \operatorname{rank}(P_s) = n_s,\ \mathcal{B}_s = \left\{w \in \mathcal{F}^{2n_s};\ R_s \circ w = 0\right\}.\end{aligned}\tag{8.290}$$

The decomposition $R_s = (P_s, -Q_s)$ is *one* IO structure of $\mathcal{B}_s$, but *all* linearly independent n_s columns of R_s give rise to different IO structures. To systematize this, consider any permutation $\sigma = \begin{pmatrix} 1 & 2 & \cdots & 2n_s \\ \sigma(1) & \sigma(2) & \cdots & \sigma(2n_s)\end{pmatrix} \in \mathfrak{S}_{2n_s}$, i.e., a reordering of the numbers 1 to $2n_s$. The corresponding permutation matrix is

$$\begin{aligned}&\Pi(\sigma) \in \mathrm{Gl}_{2n_s}(\mathbb{Z}),\ \Pi(\sigma)_{\mu\nu} := \delta_{\mu,\sigma(\nu)} \Longrightarrow \Pi(\sigma)\Pi(\tau) = \Pi(\sigma\tau),\ \Pi(\sigma^{-1}) = \Pi(\sigma)^{-1}\\ &\Longrightarrow \forall 2n_s \times \bullet \text{ matrices } X:\ (\Pi(\sigma)X)_{\mu,-} = X_{\sigma^{-1}(\mu),-},\\ &\quad \forall \bullet \times 2n_s \text{ matrices } X:\ (X\Pi(\sigma))_{-,\nu} = X_{-,\sigma(\nu)}\\ &\quad R_s \circ w = R_s\Pi(\sigma)\Pi(\sigma^{-1})\circ w = R_s^\sigma \circ w_s^\sigma \text{ where}\\ &\quad R_s^\sigma := R_s\Pi(\sigma),\ (R_s^\sigma)_{-,\nu} = (R_s)_{-,\sigma(\nu)},\\ &\quad w_s^\sigma := \Pi(\sigma^{-1})w_s = \Pi(\sigma^{-1})\begin{pmatrix} y_s \\ u\end{pmatrix},\ w_{s,\mu}^\sigma = w_{s,\sigma(\mu)}.\end{aligned}\tag{8.291}$$

Now decompose the matrix R_s^σ and signal vector w_s^σ as

$$\begin{aligned}
&R_s^\sigma =: (P_s^\sigma, -Q_s^\sigma) \in F[s]^{n_s\times(n_s+n_s)},\ w_s^\sigma := \begin{pmatrix} y_s^\sigma \\ u^\sigma \end{pmatrix} \in \mathcal{F}^{n_s+n_s}\\
&\Longrightarrow \mathcal{B}_s^\sigma := \left\{\begin{pmatrix} y_s^\sigma \\ u^\sigma \end{pmatrix} \in \mathcal{F}^{n_s+n_s};\ P_s^\sigma \circ y_s^\sigma = Q_s^\sigma \circ u^\sigma\right\} \text{ where}\\
&P_s^\sigma = ((R_s)_{-,\sigma(1)}, \dots, (R_s)_{-,\sigma(n_s)}),\ Q^\sigma = -((R_s)_{-,\sigma(n_s+1)}, \dots, (R_s)_{-,\sigma(2n_s)})\\
&\Longrightarrow \mathcal{B}_s \cong \mathcal{B}_s^\sigma,\ w_s = \begin{pmatrix} y_s \\ u \end{pmatrix} \mapsto w_s^\sigma = \Pi(\sigma^{-1})\begin{pmatrix} y_s \\ u \end{pmatrix} = \begin{pmatrix} y_s^\sigma \\ u^\sigma \end{pmatrix}.
\end{aligned} \tag{8.292}$$

Notice that $\begin{pmatrix} U_{K_s} \\ I_{K_s} \end{pmatrix}$, $\begin{pmatrix} y_s \\ u \end{pmatrix}$, and $\begin{pmatrix} y_s^\sigma \\ u^\sigma \end{pmatrix}$ and also $\mathcal{B}_s^\sigma$ and $\mathcal{B}_s$ coincide up to the order of the components.

Theorem 8.3.21 *With the data from Theorem and Definition 8.3.18 and from (8.290)–(8.292) assume* $\operatorname{rank}(P_s^\sigma) = n_s$, *i.e., that the* n_s *columns* $(R_s^\sigma)_{-\nu} = (R_s)_{-\sigma(\nu)}$ *for* $\nu = 1, \dots, n_s$ *are* $F(s)$*-linearly independent.*

1. *The behavior* $\mathcal{B}_s^\sigma$ *is an IO behavior with the IO decomposition* $R_s^\sigma = (P_s^\sigma, -Q_s^\sigma)$, *the input* u^σ, *and the transfer matrix* $H_s^\sigma = (P_s^\sigma)^{-1}Q_s^\sigma \in F(s)^{n_s\times n_s}$.
2. *The total network behavior* $\mathcal{B}$ *is also an IO behavior with the input vector* u^σ. *As such it can be written as*

$$\mathcal{B}^\sigma := \left\{\begin{pmatrix} y^\sigma \\ u^\sigma \end{pmatrix} \in \mathcal{F}^{(n+n_p)+n_s};\ P^\sigma \circ y^\sigma = Q^\sigma \circ u^\sigma\right\},\ H^\sigma := (P^\sigma)^{-1}Q^\sigma,$$

where $\begin{pmatrix} y^\sigma \\ u^\sigma \end{pmatrix} = \begin{pmatrix} y \\ u \end{pmatrix} = \begin{pmatrix} (U_k)_{k\in K} \\ (I_k)_{k\in K} \end{pmatrix}$ *and* $(P, -Q) = (P^\sigma, -Q^\sigma)$, up to the appropriate reordering of the components resp. columns. *Notice that* P^σ, Q^σ *can be easily read off* $(P, -Q)$. *In general,* $\operatorname{char}(\mathcal{B}^{\sigma,0}) = \mathrm{V}_{\mathbb{C}}(\det(P^\sigma))$ *differs from* $\operatorname{char}(\mathcal{B}^0)$ *and is not contained in* $\mathbb{C}_-$, *cf.* (8.289).
3. *With* $\mathcal{G} = \mathrm{C}_+^{-\infty}$ *or* ($\mathcal{G} = \mathcal{P}^{-\infty}$ *and* $\det(P^\sigma) \in \mathrm{U}(\mathbb{C}(s)_{\mathrm{per}})$) *there are the isomorphisms*

$$\begin{array}{ccccccc}
\mathcal{B}\cap\mathcal{G}^{(n+n_p)+n_s} & \cong & \mathcal{B}_s\cap\mathcal{G}^{n_s+n_s} & & \cong \mathcal{B}_s^\sigma\cap\mathcal{G}^{n_s+n_s} & \cong & \mathcal{G}^{n_s}\\
\begin{pmatrix} H\circ u \\ u \end{pmatrix} & \leftrightarrow & \begin{pmatrix} H_s\circ u \\ u \end{pmatrix} = \Pi(\sigma)\begin{pmatrix} H_s^\sigma\circ u^\sigma \\ u^\sigma \end{pmatrix} & \leftrightarrow & \begin{pmatrix} H_s^\sigma\circ u^\sigma \\ u^\sigma \end{pmatrix} & \leftrightarrow & u^\sigma,
\end{array}$$

i.e., for given $u^\sigma \in \mathcal{G}^{n_s}$ *there is a unique trajectory* $\begin{pmatrix} H\circ u \\ u \end{pmatrix}$ *in* $\mathcal{B}\cap\mathcal{G}^{(n+n_p)+n_s}$ *with* u^σ *as subcomponent, where* $\begin{pmatrix} y_s \\ u \end{pmatrix} := \Pi(\sigma)\begin{pmatrix} H_s^\sigma\circ u^\sigma \\ u^\sigma \end{pmatrix}$. *Then* $\begin{pmatrix} H^\sigma\circ u^\sigma \\ u^\sigma \end{pmatrix} = \begin{pmatrix} H\circ u \\ u \end{pmatrix}$, up to the order of the components.
4. *Theorem 8.3.11, 5, is applicable to* $\mathcal{B}^\sigma$ *if* H^σ *is proper and* $\mathrm{V}_{\mathbb{C}}(\det(P^\sigma)) \subset \mathbb{C}_-$. *If the latter conditions do not hold, as, for instance, in* (8.289), *the network has to be redesigned or a different input* u^σ *has to be chosen.*

Proof 1. Follows from $\operatorname{rank}(P_s^\sigma) = n_s$.
2. Consider the canonical epimorphisms

$$\begin{array}{ccccccc}
\mathcal{B} & \xrightarrow{\text{proj}} & \mathcal{B}_s & \cong & \mathcal{B}_s^\sigma & \xrightarrow{\text{proj}} & \mathcal{F}^{n_s}\\
\begin{pmatrix} y \\ u \end{pmatrix} & \mapsto & \begin{pmatrix} y_s \\ u \end{pmatrix} & \mapsto & \Pi(\sigma^{-1})\begin{pmatrix} y_s \\ u \end{pmatrix} = \begin{pmatrix} y_s^\sigma \\ u^\sigma \end{pmatrix} & \mapsto & u^\sigma
\end{array}.$$

The exact transfer space functor from Theorem 6.1.4 implies the surjections

$$\mathcal{B}_K \xrightarrow{\text{proj}} (\mathcal{B}_s)_K \cong (\mathcal{B}_s^\sigma)_K \to K^{n_s}, \ K = F(s). \text{ Since}$$
$$\operatorname{rank}(\mathcal{B}) = (n+n) - (n+n_p) = n_s = \dim_K(K^{n_s})$$

the composed map is an isomorphism. Theorem 6.2.2, 4, implies that $\mathcal{B}$ is an IO behavior with input u^σ.
3. See Theorem 8.3.18, 2. □

Corollary 8.3.22 *In Theorem 8.3.21 the matrix $R^\sigma = (P^\sigma, -Q^\sigma)$ of $\mathcal{B}^\sigma$ can be directly derived from (8.240), i.e., from $R \circ w = 0$ with $w = (U_K, I_K)^\top = (U_{K_1}, U_{K_2}, I_{K_1}, I_{K_2})^\top$. In particular, transformers and controlled sources are permitted in the network. One chooses the subvector $u^\sigma \in \mathcal{F}^{n_s}$ of $(U_{K_s}, I_{K_s})^\top \in \mathcal{F}^{2n_s}$ and the IO decomposition*

$$w = \begin{pmatrix} y^\sigma \\ u^\sigma \end{pmatrix} \in \mathcal{F}^{(n+n_p)+n_s}, \ R = (P^\sigma, -Q^\sigma) \in \mathbb{F}[s]^{(n+n_p)\times((n+n_p)+n_s)},$$
$$\mathcal{B}^\sigma = \left\{ \begin{pmatrix} y^\sigma \\ u^\sigma \end{pmatrix} \in \mathcal{F}^{(n+n_p)+n_s}; \ P^\sigma \circ y^\sigma = Q^\sigma \circ u^\sigma \right\}.$$

The IO condition $\operatorname{rank}(P^\sigma) = n + n_p$ *or* $\det(P^\sigma) \neq 0$, *hence* $H^\sigma = (P^\sigma)^{-1} Q^\sigma$, *and the asymptotic stability condition* $\operatorname{char}(\mathcal{B}^{\sigma,0}) = \operatorname{V}_{\mathbb{C}}(\det(P^\sigma)) \subset \mathbb{C}_-$ *can be easily checked. These necessary and sufficient conditions for steady state and superposition principle considerations are often ignored in the electrical engineering literature, since P^σ is not readily available. With the obvious changes Theorem 8.3.11 holds. The matrices $(P_{II}, -Q_{II})$ are not used.*
Unless $\mathcal{B}^\sigma$ is controllable or, equivalently, R^σ is right invertible or $\operatorname{char}(\mathcal{B}^\sigma) = \emptyset$, *the asymptotic stability cannot be concluded from that of H^σ.* ◇

In addition to $\sigma \in \mathfrak{S}_{2n_s} = \mathfrak{S}(K_s \uplus K_s)$ choose a permutation $\Sigma \in \mathfrak{S}_{2n} = \mathfrak{S}(K \uplus K)$ that extends σ, preferably

$$\Sigma(U_k) = \begin{cases} \sigma(U_k) & \text{if } k \in K_s \\ U_k & \text{if } k \notin K_s \end{cases}, \ \Sigma(I_k) = \begin{cases} \sigma(I_k) & \text{if } k \in K_s, \\ I_k & \text{if } k \notin K_s \end{cases}, \tag{8.293}$$

where the $k \in K \uplus K$ and $U_k, I_k \in \{U_l, I_l;\ l \in K\}$ are considered as component indices of the vector $\begin{pmatrix} U \\ I \end{pmatrix} = \begin{pmatrix} (U_k)_{k\in K} \\ (I_k)_{k\in K} \end{pmatrix}$. Then

$$\begin{aligned}
&\begin{pmatrix} y_s^\sigma \\ u^\sigma \end{pmatrix} := \Pi(\sigma^{-1}) \begin{pmatrix} y_s \\ u \end{pmatrix}, \ R_s^\sigma = R_s \Pi(\sigma) = (P_s, -Q_s)\Pi(\sigma), \\
&\begin{pmatrix} y^\sigma \\ u^\sigma \end{pmatrix} := \Pi(\Sigma^{-1}) \begin{pmatrix} y \\ u \end{pmatrix}, \ R^\sigma = (P^\sigma, -Q^\sigma) := R\Pi(\Sigma) = (P, -Q)\Pi(\Sigma) \\
&\mathcal{B}_s = \left\{ \begin{pmatrix} y_s \\ u \end{pmatrix} \in \mathcal{F}^{n_s+n_s}; \ P_s \circ y_s = Q_s \circ u \right\} \\
&\quad \cong \mathcal{B}_s^\sigma = \left\{ \begin{pmatrix} y_s^\sigma \\ u^\sigma \end{pmatrix} \in \mathcal{F}^{n_s+n_s}; \ P^\sigma \circ y_s^\sigma = Q^\sigma \circ u^\sigma \right\}, \ \begin{pmatrix} y_s \\ u \end{pmatrix} \mapsto \begin{pmatrix} y_s^\sigma \\ u^\sigma \end{pmatrix}, \\
&\mathcal{B} = \left\{ \begin{pmatrix} y \\ u \end{pmatrix} \in \mathcal{F}^{2n}; \ P \circ y = Q \circ u \right\} \\
&\quad \cong \mathcal{B}^\sigma = \left\{ \begin{pmatrix} y^\sigma \\ u^\sigma \end{pmatrix} \in \mathcal{F}^{2n}; \ P^\sigma \circ y^\sigma = Q^\sigma \circ u^\sigma \right\}, \ \begin{pmatrix} y \\ u \end{pmatrix} \mapsto \begin{pmatrix} y^\sigma \\ u^\sigma \end{pmatrix}.
\end{aligned} \tag{8.294}$$

Notice that σ and Σ are not uniquely determined because the order of the components of $y, u, y^\sigma, y_s^\sigma, u^\sigma$ etc. can be chosen arbitrarily.

Theorem 8.3.23 *In Theorem 8.3.21 consider the data from* (8.294). *Assume* $\operatorname{rank}(P_s^\sigma) = n_s$ *and the transfer matrix* $H_s^\sigma = (P_s^\sigma)^{-1} Q_s^\sigma$. *Then the transfer matrix* $H^\sigma = (P^\sigma)^{-1} Q^\sigma$ *of* $\mathcal{B}^\sigma$ *with input* u^σ *is given by*

$$H^\sigma = \underbrace{(\mathrm{id}_{n+n_p}, 0)}_{(n+n_p)\times 2n} \underbrace{\Pi(\Sigma^{-1})}_{2n\times 2n} \underbrace{\begin{pmatrix} H \\ \mathrm{id}_{n_s} \end{pmatrix}}_{2n\times n_s} \underbrace{(0, \mathrm{id}_{n_s})}_{n_s\times 2n_s} \underbrace{\Pi(\sigma)}_{2n_s\times 2n_s} \underbrace{\begin{pmatrix} H_s^\sigma \\ \mathrm{id}_{n_s} \end{pmatrix}}_{2n_s\times n_s}. \tag{8.295}$$

Proof We compute in the transfer spaces, i.e., with $\binom{y}{u} = \begin{pmatrix} H \\ \mathrm{id}_{n_s} \end{pmatrix} \circ u \in \mathcal{B}_{F(s)} \subseteq F(s)^{2n}$ and $H \circ u = Hu$, and likewise for the other behaviors. Equation (8.294) furnishes

$$\begin{aligned}
H^\sigma u^\sigma &= (\mathrm{id}_{n+n_p}, 0) \begin{pmatrix} H^\sigma u^\sigma \\ u^\sigma \end{pmatrix} = (\mathrm{id}_{n+n_p}, 0) \begin{pmatrix} H^\sigma \\ \mathrm{id}_{n_s} \end{pmatrix} u^\sigma \\
&= (\mathrm{id}_{n+n_p}, 0)\Pi(\Sigma^{-1}) \begin{pmatrix} H \\ \mathrm{id}_{n_s} \end{pmatrix} u, \\
\begin{pmatrix} H_s u \\ u \end{pmatrix} &= \Pi(\sigma) \begin{pmatrix} H_s^\sigma u^\sigma \\ u^\sigma \end{pmatrix} \Longrightarrow u = (0, \mathrm{id}_{n_s})\Pi(\sigma) \begin{pmatrix} H_s^\sigma \\ \mathrm{id}_{n_s} \end{pmatrix} u^\sigma \\
\Longrightarrow H^\sigma u^\sigma &= (\mathrm{id}_{n+n_p}, 0)\Pi(\Sigma^{-1}) \begin{pmatrix} H \\ \mathrm{id}_{n_s} \end{pmatrix} (0, \mathrm{id}_{n_s})\Pi(\sigma) \begin{pmatrix} H_s^\sigma \\ \mathrm{id}_{n_s} \end{pmatrix} u^\sigma \\
\Longrightarrow H^\sigma &= (\mathrm{id}_{n+n_p}, 0)\Pi(\Sigma^{-1}) \begin{pmatrix} H \\ \mathrm{id}_{n_s} \end{pmatrix} (0, \mathrm{id}_{n_s})\Pi(\sigma) \begin{pmatrix} H_s^\sigma \\ \mathrm{id}_{n_s} \end{pmatrix}.
\end{aligned}$$

□

Remark 8.3.24 1. Theorem 8.3.23 only requires $\operatorname{rank}(P) = n + n_p = \sharp(K) + \sharp(K_p)$ or, equivalently, $\operatorname{rank}(P_{II}) = n_p$ and $\operatorname{rank}(P_s^\sigma) = n_s = \sharp(K_s)$. The first condition means that the network behavior $\mathcal{B}$ is an IO behavior with input $u = \begin{pmatrix} I_{K_{2,s}} \\ U_{K_{1,s}} \end{pmatrix}$ and transfer matrix $H = P^{-1}Q$. Then also the black box behavior $\mathcal{B}_s$ with the trajectories $\begin{pmatrix} U_{K_s} \\ I_{K_s} \end{pmatrix}$ is an IO behavior with input u and output $y_s = \begin{pmatrix} I_{K_{1,s}} \\ U_{K_{2,s}} \end{pmatrix}$. The second condition ensures that $\mathcal{B}_s^\sigma$ and also $\mathcal{B}^\sigma$ are IO behaviors with input u^σ and transfer matrices $H_s^\sigma = (P_s^\sigma)^{-1} Q_s^\sigma$ resp. $H^\sigma = (P^\sigma)^{-1} Q^\sigma$. The u^σ with $\operatorname{rank}(P_s^\sigma) = n_s$ are all permissible inputs of $\mathcal{B}$. Assume that $\mathcal{B}^\sigma$ is asymptotically stable, i.e., $\mathrm{V}_{\mathbb{C}}(\det(P^\sigma)) \subset \mathbb{C}_-$. If the input u^σ has left bounded support or is periodic of period $T > 0$, then the components of $y^\sigma := H^\sigma \circ u^\sigma$ are the steady-state voltages and currents of *all* branches except the given excitations in u^σ. According to (8.295) the computation of $H^\sigma \in F(s)^{(n+n_p)\times n_s}$ requires $H = P^{-1}Q$ and $H_s^\sigma = (P_s^\sigma)^{-1} Q_s^\sigma$. We have not found Theorems 8.3.21 and 8.3.23 in the engineering literature. In particular, these theorems are applicable to two-ports, i.e., $n_s = 2$, that are intensively studied in the literature.

2. The order of the components in y_s^σ and u^σ is irrelevant; hence, there are only $\binom{2n_s}{n_s} = ((2n_s)!)(n_s!)^{-2}$ essentially different decompositions $\binom{y_s^\sigma}{u^\sigma}$. This is the number of subsets of size n_s of the set $\{U_k, I_k;\ k \in K_s\}$ of size $2n_s$. They are IO decompositions if and only if $\operatorname{rank}(P_s^\sigma) = n_s$ or $\det(P_s^\sigma) \neq 0$.

3. The components of $w_s^\sigma = \binom{y_s^\sigma}{u^\sigma}$ appear pairwise as $\binom{U_k}{I_k}$, $k \in K_s$. In particular,

there are $\sigma \in \mathfrak{S}_{2n_s}$ such that for all $k \in K_s$ one component of $\binom{U_k}{I_k}$ is an input component, i.e., a component of u^σ, and the other one an output component of y_s^σ. This holds for $(y_s, u)^\top = \left(I_{K_{1,s}}, U_{K_{2,s}}, I_{K_{2,s}}, U_{K_{1,s}}\right)^\top$ and for a two-port with $w_s^\sigma = (U_1, U_2, I_1, I_2)^\top$, but not for $w_s^\tau = (y_s^\tau, u^\tau)^\top = (U_2, I_2, U_1, I_1)^\top$. ◇

Example 8.3.25 Consider the data and notations from Example 8.3.17 and from Theorem 8.3.21. The behavior $\mathcal{B}_s$ has the trajectories $\binom{y_s}{u} = \binom{I_1}{U_1}$ with U_1 as input, I_1 as output and transfer function H_{I_1,U_1}. The latter is nonzero and hence $\mathcal{B}_s$ and $\mathcal{B}$ are also IO behaviors with I_1 as input. With the transposition $\sigma = \binom{1\ 2}{2\ 1} = \sigma^{-1}$ we obtain

$$\mathcal{B}_s \cong \mathcal{B}_s^\sigma,\ \binom{I_1}{U_1} \mapsto \binom{U_1}{I_1} = \Pi(\sigma^{-1})\binom{I_1}{U_1},\ H_s = H_{I_1,U_1},\ H_s^\sigma = H_s^{-1} = H_{I_1,U_1}^{-1}. \tag{8.296}$$

We use the transposition $\Sigma = \Sigma^{-1} := \begin{pmatrix} 1 & 2 & \cdots & 11 & 12 \\ 12 & 2 & \cdots & 11 & 1 \end{pmatrix}$, the inputs $u = U_1$ and $u^\sigma = I_1$ and the trajectories

$$\begin{aligned}
&\binom{y}{u} := \begin{pmatrix} I_1 \\ z \\ U_1 \end{pmatrix} := (I_1, I_3, I_4, U_2, U_5, U_6, U_3, U_4, I_2, I_5, I_6, U_1)^\top \in \mathcal{F}^{12},\\
&z := (I_3, I_4, U_2, U_5, U_6, U_3, U_4, I_2, I_5, I_6)^\top,\ \binom{y^\sigma}{u^\sigma} := \Pi(\Sigma^{-1})\binom{y}{u} = \begin{pmatrix} U_1 \\ z \\ I_1 \end{pmatrix}.
\end{aligned} \tag{8.297}$$

We write

$$\begin{aligned}
&R = (P, -Q) = (R_{-,1}, R_{-,2}, \dots, R_{-,11}, R_{-,12}) \in \mathbb{R}[s]^{11\times 12}\\
&\Longrightarrow R^\sigma = (P^\sigma, -Q^\sigma) = R\Pi(\Sigma) = \left(R_{-,12}, R_{-,2}\dots, R_{-,11} R_{-,1}\right) \in \mathbb{R}[s]^{11\times 12},\\
&R^\sigma_{-,1} = R_{-,12} = -Q,\ Q^\sigma = -R^\sigma_{-,12} = -R_{-,1},\ R^\sigma_{-,\mu} = R_{-,\mu},\ \mu = 2, \cdots, 10,
\end{aligned} \tag{8.298}$$

and obtain

$$\begin{aligned}
&\mathcal{B}^\sigma = \left\{\binom{y^\sigma}{I_1} \in \mathcal{F}^{12};\ P^\sigma \circ y^\sigma = Q^\sigma \circ I_1\right\},\\
&(\mathcal{B}^\sigma)^0 = \left\{y^\sigma \in \mathcal{F}^{11};\ P^\sigma \circ y^\sigma = 0\right\},\ H^\sigma = (P^\sigma)^{-1} Q^\sigma.
\end{aligned} \tag{8.299}$$

From (8.295) we get

$$\begin{aligned}
H^\sigma &= (\mathrm{id}_{11}, 0)\Pi(\Sigma^{-1})\begin{pmatrix} H_{I_1,U_1} \\ H_{z,U_1} \\ 1 \end{pmatrix}(0, 1)\Pi(\sigma)\binom{H_s^\sigma}{1}\\
&= (\mathrm{id}_{11}, 0)\begin{pmatrix} 1 \\ H_{z,U_1} \\ H_{I_1,U_1} \end{pmatrix}(0, 1)\binom{1}{H_s^\sigma} = \binom{1}{H_{z,U_1}} H_s^\sigma \text{ where,}\\
H_s^\sigma &= H_{I_1,U_1}^{-1},\ H_{z,U_1} = (H_{I_3,U_1}, \dots, H_{I_6,U_1})^\top.
\end{aligned} \tag{8.300}$$

The test of $H^\sigma - (P^\sigma)^{-1} Q^\sigma = 0$ confirms the correctness of the computations. The properness of H^σ resp. the stability of $\mathcal{B}^{\sigma,0}$ can be read of (8.300) resp.

$$\mathrm{char}(\mathcal{B}^{\sigma,0}) = \mathrm{V}_{\mathbb{C}}(\det(P^\sigma)) = \mathrm{V}_{\mathbb{C}}(e^\sigma) \tag{8.301}$$

where e^σ is the 11th elementary divisor of P^σ. The data from Example 8.3.17 furnish

$$\begin{aligned}
\det(P^\sigma) &= -C_2C_5L_3R_6s^3 - (L_3 + C_5R_4R_6)C_2s^2 - (C_2R_4 + C_2R_6 + C_5R_6)s - 1,\\
H_s^\sigma &= H_{I_1,U_1}^{-1} = \tfrac{C_5L_3R_6s^2+(C_5R_4R_6+L_3)s+R_4+R_6}{\det(P^\sigma)},\\
H^\sigma &= \begin{pmatrix} 1 \\ H_{z,U_1}\end{pmatrix} H_{I_1,U_1}^{-1} = \tfrac{1}{\det(P^\sigma)}\begin{pmatrix} C_5L_3R_6s^2+(C_5R_4R_6+L_3)s+R_4+R_6\\ C_5R_6s+1\\ C_5R_6s+1\\ C_5L_3R_6s^2+(C_5R_4R_6+L_3)s+R_4+R_6\\ R_6\\ R_6\\ C_5L_3R_6s^2+L_3s\\ C_5R_4R_6+R_4\\ C_2C_5L_3R_6s^3+(C_5R_4R_6+L_3)C_2s^2+(R_4+R_6)C_2s\\ C_5R_6s\\ 1\end{pmatrix},\\
e^\sigma &= -\tfrac{\det(P^\sigma)}{C_2C_5L_3R_6}.
\end{aligned} \tag{8.302}$$

Since $\mathcal{B} \cong \mathcal{B}^\sigma$ and $\mathcal{B}$ is controllable, so is $\widehat{\mathcal{B}}^\sigma$. The component H_{I_2,I_1}^σ is of degree $3-3=0$ and thus proper with $\left(H_{I_2,I_1}^\sigma\right)_{\text{pol}} = -1$, all other components are strictly proper. Hence, if the current I_1 is of the form $I_1 = \alpha Y$ with polynomial-exponential α and a jump at $t=0$, then I_2 has a jump too, but all other branch voltages and currents are continuous.
If all constants $C_2 = L_3 = \cdots = R_6 = 1$ are 1, we get

$$\begin{aligned}
e^\sigma &= -s^3 - 2s^2 - 3s - 1,\\
\operatorname{char}\left((\mathcal{B}^\sigma)^0\right) &= \mathrm{V}_{\mathbb{C}}(e^\sigma) \approx \{-0.4302,\ -0.7849 \pm 1.3071j\}.
\end{aligned} \tag{8.303}$$

Thus the behavior $\mathcal{B}^\sigma$, i.e., $\mathcal{B}$ with input I_1, is asymptotically stable for this choice of constants. The steady-state language is applicable, i.e., all outputs to an input I_1 coincide for $t \to \infty$. ◇

Example 8.3.26 Choose $\sigma \in \mathfrak{S}_{2n_s}$ such that

$$\begin{aligned}
&\begin{pmatrix} y_s^\sigma \\ u^\sigma \end{pmatrix} = \begin{pmatrix} U_{K_s} \\ I_{K_s}\end{pmatrix} \in \mathcal{F}^{n_s+n_s},\ R_s^\sigma = (P_s^\sigma, -Q_s^\sigma) \in \mathbb{R}[s]^{n_s\times(n_s+n_s)},\\
&\mathcal{B}_s^\sigma = \left\{\begin{pmatrix} U_{K_s} \\ I_{K_s}\end{pmatrix} \in \mathcal{F}^{n_s+n_s};\ P_s^\sigma \circ U_{K_s} = Q_s^\sigma \circ I_{K_s}\right\},\ H_s^\sigma = \left(P_s^\sigma\right)^{-1} Q_s^\sigma,\\
&\mathcal{B}^\sigma = \left\{\begin{pmatrix} y^\sigma \\ I_{K_s}\end{pmatrix} \in \mathcal{F}^{(n+n_p)+n_s};\ P^\sigma \circ y^\sigma = Q^\sigma \circ I_{K_s}\right\},\ H^\sigma = (P^\sigma)^{-1} Q^\sigma.
\end{aligned} \tag{8.304}$$

Assume $\operatorname{rank}(P_s^\sigma) = n_s$ and $\mathrm{V}_{\mathbb{C}}(\det(P^\sigma)) \subset \mathbb{C}_-$, i.e., that $\mathcal{B}^\sigma$ is an asymptotically stable IO behavior with input I_{K_s}. We call $H_s^\sigma = (P_s^\sigma)^{-1}Q_s^\sigma$ the *impedance transfer matrix*. If I_{K_s} has left bounded support, the steady-state voltage vector is $U_{K_s} = H_s^\sigma \circ I_{K_s}$. If I_{K_s} is sinusoidal with factor $e^{j\omega t}$, then the steady-state voltage vector is $U_{K_s} = H_s^\sigma \circ I_{K_s} = H_s^\sigma(j\omega)I_{K_s}$, and $H_s^\sigma(j\omega)$ is called the *impedance matrix* of the network at the frequency ω.
Assume in addition

$$H_s^\sigma \in \mathrm{Gl}_{n_s}(F(s)),\ \text{i.e., } \operatorname{rank}(Q_s^\sigma) = \operatorname{rank}(H_s^\sigma) = n_s,$$

and choose $\tau \in \mathfrak{S}_{2n_s}$ such that

$$\begin{aligned}
&\left(\begin{smallmatrix} y_s^\tau \\ u^\tau \end{smallmatrix}\right) = \left(\begin{smallmatrix} I_{K_s} \\ U_{K_s} \end{smallmatrix}\right), \; R_s^\tau = (P_s^\tau, -Q_s^\tau) = (-Q_s^\sigma, P_s^\sigma), \\
&\mathcal{B}_s^\tau = \left\{ \left(\begin{smallmatrix} I_{K_s} \\ U_{K_s} \end{smallmatrix}\right) \in \mathcal{F}^{n_s+n_s}; \; Q_s^\sigma \circ I_{K_s} = P_s^\sigma \circ U_{K_s} \right\}, \\
&H_s^\tau = \left(-Q_s^\sigma\right)^{-1} (-P_s^\sigma) = \left(H_s^\sigma\right)^{-1}.
\end{aligned} \tag{8.305}$$

Then $H_s^\tau = (H_s^\sigma)^{-1}$ is called the *admittance transfer matrix* of the network. If $\mathcal{B}^\tau$ is asymptotically stable then $H_s^\tau(j\omega) = H_s^\sigma(j\omega)^{-1}$ is called the *admittance matrix*. ◊

Example 8.3.27 (i) In Theorem 8.3.21 consider a standard two-port, i.e., $n_s = 2$, $K_s = \{k_1, k_2\}$, $R_s = (P_s, -Q_s)$ where most often k_1 resp. k_2 are called the *input resp. the output* port. Since we deal with the black box behaviors $\mathcal{B}_s^\sigma$ only, we omit, for notational simplicity, the index s in

$$R_s^\upsilon = (P_s^\upsilon, -Q_s^\upsilon) \in F[s]^{2\times(2+2)} \text{ and } H_s^\upsilon = \left(P_s^\upsilon\right)^{-1} Q_s^\upsilon \in F(s)^{2\times 2},$$

in particular, in $R_s = (P_s, -Q_s) = (P_s^{\mathrm{id}}, -Q_s^{\mathrm{id}})$ where id is the identity permutation. In this new notation we obtain $R^\sigma = R\Pi(\sigma)$, $R^\sigma_{\mu,\nu} = R_{\mu,\sigma(\nu)}$, and

$$\begin{aligned}
&\left(\begin{smallmatrix} y_s^\sigma \\ u^\sigma \end{smallmatrix}\right) = \Pi(\sigma^{-1}) \left(\begin{smallmatrix} y_s \\ u \end{smallmatrix}\right) = \Pi(\sigma^{-1}) \left(I_{K_{1,s}}, U_{K_{2,s}}, I_{K_{2,s}}, U_{K_{1,s}}\right)^\top \\
&\mathcal{B}_s^\sigma = \left\{ \left(\begin{smallmatrix} y_s^\sigma \\ u^\sigma \end{smallmatrix}\right) \in \mathcal{F}^2; \; P^\sigma \circ y_s^\sigma = Q^\sigma \circ u^\sigma \right\}.
\end{aligned}$$

If $D^\sigma := \det(P^\sigma) \neq 0$ we get

$$\begin{aligned}
&P^\sigma = \begin{pmatrix} R_{1,\sigma(1)} & R_{1,\sigma(2)} \\ R_{2,\sigma(1)} & R_{2,\sigma(2)} \end{pmatrix}, \; Q^\sigma = -\begin{pmatrix} R_{1,\sigma(3)} & R_{1,\sigma(4)} \\ R_{2,\sigma(3)} & R_{2,\sigma(4)} \end{pmatrix}, \\
&(P^\sigma)^{-1} = (D^\sigma)^{-1} \begin{pmatrix} R_{2,\sigma(2)} & -R_{1,\sigma(2)} \\ -R_{2,\sigma(1)} & R_{1,\sigma(1)} \end{pmatrix}, \; D^\sigma := R_{1,\sigma(1)} R_{2,\sigma(2)} - R_{1,\sigma(2)} R_{2,\sigma(1)}, \\
&H^\sigma = \begin{pmatrix} H_{11}^\sigma & H_{12}^\sigma \\ H_{21}^\sigma & H_{22}^\sigma \end{pmatrix}, \begin{cases} H_{11}^\sigma := -(D^\sigma)^{-1} \left(R_{2,\sigma(2)} R_{1,\sigma(3)} - R_{1,\sigma(2)} R_{2,\sigma(3)}\right) \\ H_{12}^\sigma := -(D^\sigma)^{-1} \left(R_{2,\sigma(2)} R_{1,\sigma(4)} - R_{1,\sigma(2)} R_{2,\sigma(4)}\right) \\ H_{21}^\sigma := -(D^\sigma)^{-1} \left(-R_{2,\sigma(1)} R_{1,\sigma(3)} + R_{1,\sigma(1)} R_{2,\sigma(3)}\right) \\ H_{22}^\sigma := -(D^\sigma)^{-1} \left(-R_{2,\sigma(1)} R_{1,\sigma(4)} + R_{1,\sigma(1)} R_{2,\sigma(4)}\right) \end{cases}
\end{aligned} \tag{8.306}$$

Many entries $H^\sigma_{\mu\nu}(s)$ or $H^\sigma_{\mu\nu}(j\omega)$, $\sigma \in \mathfrak{S}_4$, $\mu, \nu = 1, 2$, are provided with their special names.

(ii) For $\left(\begin{smallmatrix} y_s^\sigma \\ u^\sigma \end{smallmatrix}\right) = \left(U_{k_2}, I_{k_2}, U_{k_1}, I_{k_1}\right)^\top$ we get the steady-state equation

$$\left(\begin{smallmatrix} U_2 \\ I_2 \end{smallmatrix}\right) = H^\sigma \circ \left(\begin{smallmatrix} U_1 \\ I_1 \end{smallmatrix}\right), \; \left(\begin{smallmatrix} U_2 \\ I_2 \end{smallmatrix}\right) := \left(\begin{smallmatrix} U_{k_2} \\ I_{k_2} \end{smallmatrix}\right), \; \left(\begin{smallmatrix} U_1 \\ I_1 \end{smallmatrix}\right) := \left(\begin{smallmatrix} U_{k_1} \\ I_{k_1} \end{smallmatrix}\right). \tag{8.307}$$

The components H_{11}^σ resp. $(H^\sigma)_{11}^{-1}$ with $U_2 \underset{I_1=0}{=} H_{11}^\sigma \circ U_1$ and $U_1 = \left(H_{11}^\sigma\right)^{-1} \circ U_2$ are called the *forward resp. reverse transfer function* of the two-port.

(iii) Assume, with suitable σ, the steady-state equation $\binom{I_1}{U_2} = H^\sigma \circ \binom{U_1}{I_2}$, cf. [17, p. 126]. We consider the network as a one-port with current source I_2 and internal voltage source U_1. Its Helmholtz/Thévenin equivalent for the source branch k_2 according to (8.253) is

$$U_2 = H^\sigma_{21} \circ U_1 + H^\sigma_{22} \circ I_2 = U_{\mathrm{Th}} + H_{\mathrm{Th}} \circ I_2, \ U_{\mathrm{Th}} := H^\sigma_{21} \circ U_1, \ H_{\mathrm{Th}} := H^\sigma_{22}.$$

If k_2 is connected with a load $U_2 = H_2 \circ I_2$, one gets the equation

$$U_{\mathrm{Th}} = (-H_{\mathrm{Th}} + H_2) \circ I_2, \ U_{\mathrm{Th}} := H^\sigma_{21} \circ U_1, \ H_{\mathrm{Th}} = H^\sigma_{22}.$$

◇

We finally discuss various interconnections of two multiports.

Example 8.3.28 (*Mixed series and parallel connection*). Consider two n_s-ports $\mathcal{B}^{(1)}$ and $\mathcal{B}^{(2)}$ as black boxes, i.e., the components of their trajectories are the source voltages and currents only. The data of $\mathcal{B}^{(1)}$ and $\mathcal{B}^{(2)}$ are distinguished by upper indices 1 and 2. We have to assume $m_s = 2n_s$, i.e., that the terminals of the source branches of the $\mathcal{B}^{(i)}$ are pairwise different, since we need the port condition from Corollary 8.3.12. We choose a one-one correspondence $K_s^{(1)} \cong K_s^{(2)}$, $k^{(1)} \mapsto k^{(2)}$, and then identify $K_s := K_s^{(1)} = K_s^{(2)}$. For $i = 1, 2$ and $k \in K_s$ let $v_k^{(i)}$ resp. $w_k^{(i)}$ denote the input resp. output terminal of $\mathcal{B}^{(i)}$ of the source $\binom{U_k}{I_k}$. Let $\binom{y^{(i)}}{u^{(i)}}$ denote the trajectories of $\mathcal{B}^{(i)}$. For each $k \in K_s$ we assume that one component of $\binom{U_k}{I_k}$ belongs to $u^{(i)}$ for both $i = 1, 2$ and the other one to $y^{(i)}$. If it is I_k, then this is used as current input to both $\mathcal{B}^{(1)}$ and $\mathcal{B}^{(2)}$, and likewise for U_k. The equations of the $\mathcal{B}^{(i)}$ are

$$\begin{aligned} &P^{(i)} \circ y^{(i)} = Q^{(i)} \circ u^{(i)}, \ i = 1, 2, \ y^{(i)}, u^{(i)} \in \mathcal{F}^{n_s}, \\ &\operatorname{rank}\left(P^{(i)}\right) = n_s, \ H^{(i)} = \left(P^{(i)}\right)^{-1} Q^{(i)}. \end{aligned} \tag{8.308}$$

According to the standard convention, but in contrast to most electrical engineering books, U_k and the current I_k have the same direction. We connect the $\mathcal{B}^{(i)}$ with the sources as follows:

1. *Voltage input, parallel connection*: Assume that the input components $u_k^{(1)}$ and hence $u_k^{(2)}$ are voltages. The two terminal pairs associated with k and the source are parallel connected according to Fig. 8.3a. The current $-I_k^{(i)}$ flows from the terminal $v_k^{(i)}$ into $\mathcal{B}^{(i)}$. The node equation at $v_k^{(i)}$ is $I_k - I_k^{(1)} - I_k^{(2)} = 0$. We infer

$$U_k = u_k^{(1)} = u_k^{(2)} = U_k^{(1)} = U_k^{(2)}, \ I_k = I_k^{(1)} + I_k^{(2)} = y_k^{(1)} + y_k^{(2)}. \tag{8.309}$$

2. *Current input, series connection*: Assume that the input components $u_k^{(1)}$ and hence $u_k^{(2)}$ are currents. The two terminal pairs associated with k and the source are series connected according to Fig. 8.3b. We conclude

Fig. 8.3 The parallel and the series connection

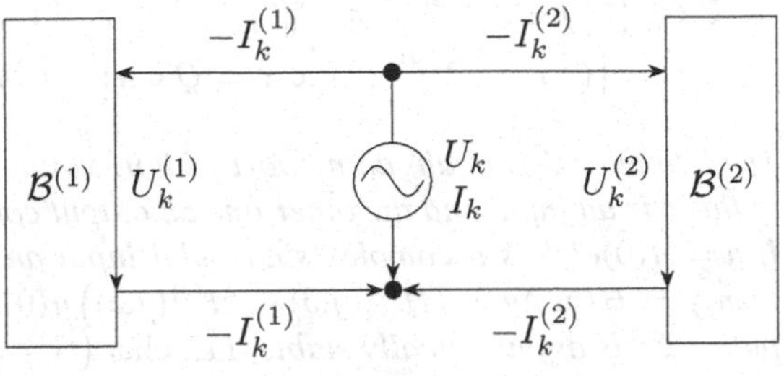

(a) The parallel connection.

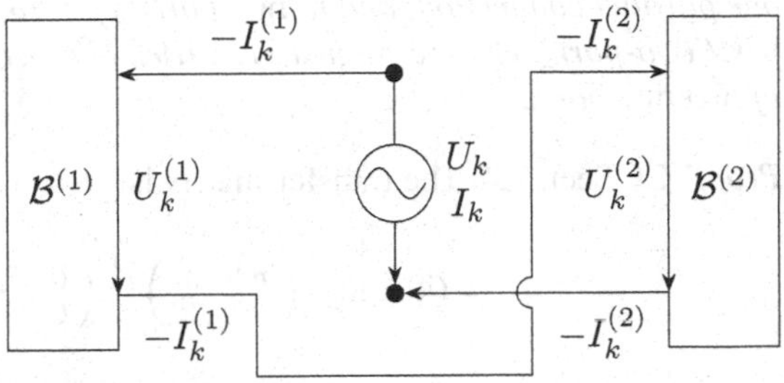

(b) The series connection.

$$I_k = u_k^{(1)} = u_k^{(2)} = I_k^{(1)} = I_k^{(2)}, \; U_k = U_k^{(1)} + U_k^{(2)} = y_k^{(1)} + y_k^{(2)}. \tag{8.310}$$

In (8.309) and (8.310) we have used the port condition for $k \in K_s^{(i)}, \; i = 1, 2$. These connections are realized for all $k \in K_s$. Thus the equations of the connected behavior are the Rosenbrock equations

$$\begin{aligned} &\begin{pmatrix} P^{(1)} & 0 \\ 0 & P^{(2)} \end{pmatrix} \circ \begin{pmatrix} y^{(1)} \\ y^{(2)} \end{pmatrix} = \begin{pmatrix} Q^{(1)} \\ Q^{(2)} \end{pmatrix} u, \; u := u^{(1)} = u^{(2)}, \\ &y := y^{(1)} + y^{(2)} = (\mathrm{id}_n, \mathrm{id}_n) \begin{pmatrix} y^{(1)} \\ y^{(2)} \end{pmatrix}. \end{aligned} \tag{8.311}$$

If the two n_s-ports are asymptotically stable, then so is the interconnected n_s-port. Notice that the components of u are voltage or current sources that are used for all terminal pairs of both $\mathcal{B}^{(1)}$ an $\mathcal{B}^{(2)}$.

Theorem 8.3.29 *Assume the Rosenbrock equations* (8.311) *with their mixed series and parallel connection. Compute a universal left annihilator*

$$\begin{aligned} &\left(-X^{(1)}, -X^{(2)}, P\right) \in F[s]^{n \times 3n} \; \textit{of} \begin{pmatrix} P^{(1)} & 0 \\ 0 & P^{(2)} \\ \mathrm{id}_n & \mathrm{id}_n \end{pmatrix}, \; \textit{i.e.,} \\ &P = X^{(1)} P^{(1)} = X^{(2)} P^{(2)}. \; \textit{Define } Q := X^{(1)} Q^{(1)} + X^{(2)} Q^{(2)}. \end{aligned} \tag{8.312}$$

Then the connected behavior is an IO behavior

$$\begin{aligned}\mathcal{C} &:= \left\{\binom{y}{u} \in \mathcal{F}^{K_s \uplus K_s};\ \exists\ \binom{y^{(i)}}{u} \in \mathcal{B}^{(i)},\ i=1,2,\ with\ y = y^{(1)} + y^{(2)}\right\}\\ &= \left\{\binom{y}{u} \in \mathcal{F}^{K_s \uplus K_s};\ P \circ y = Q \circ u\right\},\ with\ H = P^{-1}Q = H^{(1)} + H^{(2)}.\end{aligned} \tag{8.313}$$

The behavior $\mathcal{C}$ is again an n_s-port with n_s voltage-current pairs U_k, I_k where one of these is an input and the other one an output component.
If $u = u(0)e^{j\omega t}$ is a complex sinusoidal input and $j\omega \notin \operatorname{pole}(H^{(1)}) \cup \operatorname{pole}(H^{(2)})$, then $y = H(j\omega)u = \left(H^{(1)}(j\omega) + H^{(2)}(j\omega)\right) u(0)e^{j\omega t}$ is a complex sinusoidal output. If $\mathcal{B}^0$ is asymptotically stable, i.e., $\operatorname{char}\left(\mathcal{B}^0\right) = \mathrm{V}_{\mathbb{C}}(\det(P)) \subset \mathbb{C}_-$, *then y is a steady-state output.*
The theorem especially applies to two one-ports where there are only one series and one parallel connection, and to two-ports, cf. [Wikipedia, https://en.wikipedia.org/wiki/Two-port_network, August 31, 2019], [17, Sect. 6.6.5] for various special cases of two two-ports.

Proof Cf. Sect. 7.4. The transfer matrix is

$$H = (\mathrm{id}_n, \mathrm{id}_n)\begin{pmatrix} P^{(1)} & 0 \\ 0 & P^{(2)} \end{pmatrix}^{-1}\begin{pmatrix} Q^{(1)} \\ Q^{(2)} \end{pmatrix} = H^{(1)} + H^{(2)}. \qquad \square$$

Note that the series connection in (8.310) differs from that in Sect. 7.3, where the output U_1 of the first behavior is taken as the input I_2 of the second one and the transfer function is H_2H_1. The identification $I_2 = U_1$ of a current with a voltage is an error in electrical engineering. ◇
There are many other interconnections of two multiports. We mention two examples.

Example 8.3.30 (*Chain or cascade connection*, cf. [17, Sect. 6.6.5, p.164]) Consider two two-ports with the IO equations

$$\begin{aligned}&P^{(1)} \circ \begin{pmatrix} U_2^{(1)} \\ I_2^{(1)} \end{pmatrix} = Q^{(1)} \circ \begin{pmatrix} U_1^{(1)} \\ I_1^{(1)} \end{pmatrix}, \quad P^{(2)} \circ \begin{pmatrix} U_2^{(2)} \\ I_2^{(2)} \end{pmatrix} = Q^{(2)} \circ \begin{pmatrix} U_1^{(2)} \\ I_1^{(2)} \end{pmatrix},\\ &H^{(i)} := \left(P^{(i)}\right)^{-1} Q^{(i)}.\end{aligned} \tag{8.314}$$

In the literature, the entries of $\operatorname{diag}(1,-1)H^{(i)}(j\omega)$ are called the *chain or cascade* parameters of the two-ports. The -1 in the diagonal matrix is due to different choices of signs in certain source currents. The two-ports are connected according to Fig. 8.4.

Recall that $-I_2^{(1)}$ resp. $-I_1^{(2)}$ are the currents from their common node into the first resp. the second two-port, and hence $I_1^{(2)} = -I_2^{(1)}$ and $U_2^{(1)} = U_1^{(2)}$. Hence the equations of the interconnection are

$$\begin{aligned}&P^{(1)} \circ \begin{pmatrix} U_2^{(1)} \\ I_2^{(1)} \end{pmatrix} = Q^{(1)} \circ \begin{pmatrix} U_1^{(1)} \\ I_1^{(1)} \end{pmatrix},\ P^{(2)} \circ \begin{pmatrix} U_2^{(2)} \\ I_2^{(2)} \end{pmatrix} = \widetilde{Q}^{(2)} \circ \begin{pmatrix} U_2^{(1)} \\ I_2^{(1)} \end{pmatrix},\ \text{with}\\ &\widetilde{Q}^{(2)} = Q^{(2)} \operatorname{diag}(1,-1) = \begin{pmatrix} Q_{11}^{(2)} & -Q_{12}^{(2)} \\ Q_{21}^{(2)} & -Q_{22}^{(2)} \end{pmatrix},\\ &\widetilde{H}^{(2)} = \left(P^{(2)}\right)^{-1}\widetilde{Q}^{(2)} = H^{(2)} \operatorname{diag}(1,-1) = \begin{pmatrix} H_{11}^{(2)} & -H_{12}^{(2)} \\ H_{21}^{(2)} & -H_{22}^{(2)} \end{pmatrix}.\end{aligned} \tag{8.315}$$

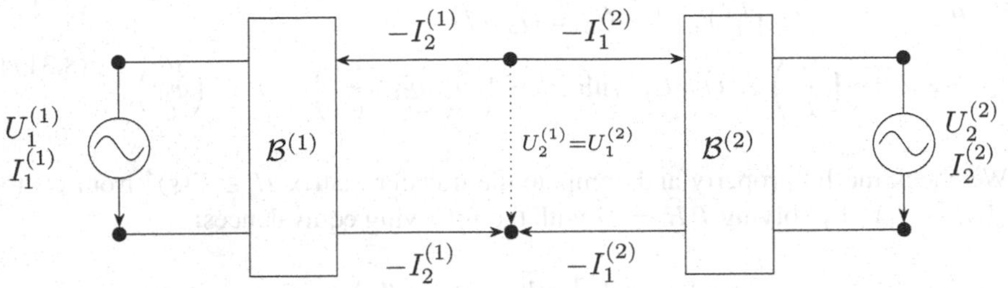

Fig. 8.4 The cascade connection from Example 8.3.30

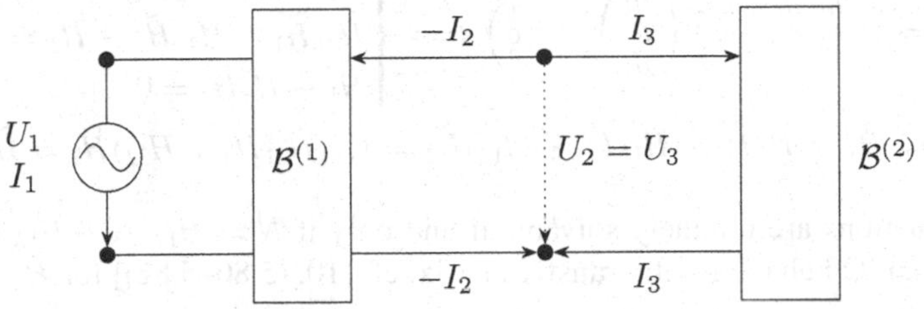

Fig. 8.5 The connection of a two-port and a one-port from Example 8.3.31

These are the equations of a series interconnection according to Sect. 7.3. The resulting behaviors are IO behaviors and asymptotically stable if the two two-ports have this property. The transfer matrix of the cascade connection is $H = \widetilde{H}^{(2)} H^{(1)} = H^{(2)} \operatorname{diag}(1, -1) H^{(1)}$. ◊

Example 8.3.31 (Cf. [17, Sect. 6.8], [10, Sect. 5.5.2]) Assume a two-port and a one-port with the IO equations

$$\begin{aligned} &P \circ \begin{pmatrix} U_1 \\ I_1 \end{pmatrix} = Q \circ \begin{pmatrix} U_2 \\ I_2 \end{pmatrix},\ H = P^{-1} Q, \text{ and } P_3 \circ U_3 = Q_3 \circ I_3,\ H_3 := P_3^{-1} Q_3, \\ &P = (P_{\mu\nu})_{\mu,\nu},\ Q = (Q_{\mu\nu})_{\mu,\nu} \in F[s]^{2\times 2},\ \operatorname{rank}(P) = 2, \\ &H = (H_{\mu\nu})_{\mu,\nu} \in F(s)^{2\times 2},\ 0 \neq P_3,\ Q_3 \in F[s]. \end{aligned} \tag{8.316}$$

Obviously it is assumed that $\mathcal{B}^{(1)}$ has the admissible input $\begin{pmatrix} U_2 \\ I_2 \end{pmatrix}$. The two networks are connected according to Fig. 8.5.

Then $I_2 = I_3$ and $U_2 = U_3$. Recall that in our convention, in contrast to the engineering usage, the current I_2 of the two-port flows in the direction of the voltage U_2 and therefore $-I_2$ from the input node into $\mathcal{B}^{(1)}$. In the language of Example 8.3.16 the behavior $\mathcal{B}^{(1)}$ is a one-port with the external source branch $\begin{pmatrix} U_2 \\ I_2 \end{pmatrix}$ and the internal source U_1 and that is loaded with $P_3 \circ U_2 = Q_3 \circ I_2$. The following computations give details to those of Example 8.3.16. The equations of the interconnected behavior $\widetilde{\mathcal{B}}$ are

$$P \circ \binom{U_1}{I_1} = Q \circ \binom{U_2}{I_2},\ P_3 \circ U_2 = Q_3 \circ I_2$$
$$\iff \widetilde{P} \circ \begin{pmatrix} U_2 \\ I_2 \\ I_1 \end{pmatrix} = \widetilde{Q} \circ U_1 \text{ with } \widetilde{P} := \begin{pmatrix} Q_{11} & Q_{12} & -P_{12} \\ Q_{21} & Q_{22} & -P_{22} \\ P_3 & -Q_3 & 0 \end{pmatrix},\ \widetilde{Q} := \begin{pmatrix} P_{11} \\ P_{21} \\ 0 \end{pmatrix}. \tag{8.317}$$

We check the IO property and compute the transfer matrix $\widetilde{H} \in F(s)^3$ from U_1 to $(U_2, I_2, I_1)^\top$ by solving $\widetilde{P}\widetilde{H} = \widetilde{Q}$ with the following equivalences:

$$\widetilde{P}\widetilde{H} = \widetilde{Q} \iff \begin{pmatrix} P^{-1} & 0 \\ 0 & P_3^{-1} \end{pmatrix} \begin{pmatrix} Q_{11} & Q_{12} & -P_{12} \\ Q_{21} & Q_{22} & -P_{22} \\ P_3 & -Q_3 & 0 \end{pmatrix} \begin{pmatrix} \widetilde{H}_1 \\ \widetilde{H}_2 \\ \widetilde{H}_3 \end{pmatrix} = \begin{pmatrix} P^{-1} & 0 \\ 0 & P_3^{-1} \end{pmatrix} \begin{pmatrix} P_{11} \\ P_{21} \\ 0 \end{pmatrix}$$
$$\iff \begin{pmatrix} H_{11} & H_{12} & 0 \\ H_{21} & H_{22} & -1 \\ 1 & -H_3 & 0 \end{pmatrix} \begin{pmatrix} \widetilde{H}_1 \\ \widetilde{H}_2 \\ \widetilde{H}_3 \end{pmatrix} = \begin{pmatrix} 1 \\ 0 \\ 0 \end{pmatrix} \iff \begin{cases} H_{11}\widetilde{H}_1 + H_{12}\widetilde{H}_2 = 1 \\ H_{21}\widetilde{H}_1 + H_{22}\widetilde{H}_2 - \widetilde{H}_3 = 0 \\ \widetilde{H}_1 - H_3\widetilde{H}_2 = 0 \end{cases}$$
$$\iff \widetilde{H}_1 = H_3\widetilde{H}_2,\ (H_{11}H_3 + H_{12})\widetilde{H}_2 = 1,\ (H_{21}H_3 + H_{22})\widetilde{H}_2 = \widetilde{H}_3.$$

These equations are uniquely solvable if and only if $N = H_{11}H_3 + H_{12} \neq 0$, and then $\widetilde{\mathcal{B}}$ is an IO behavior with transfer matrix, cf. [10, (5.80–5.82)] for $\widetilde{H}_2$,

$$\widetilde{H} = (\widetilde{H}_1, \widetilde{H}_2, \widetilde{H}_3)^\top = N^{-1}(H_3, 1, H_{21}H_3 + H_{22})^\top,\ N := H_{11}H_3 + H_{12}. \tag{8.318}$$

In order that steady-state computations are admissible we assume the asymptotic stability of $\widetilde{\mathcal{B}}$, i.e., $\mathrm{V}_{\mathbb{C}}(\det(\widetilde{P})) \subset \mathbb{C}_-$. For U_1 with left bounded support the corresponding steady-state output is

$$(U_2, I_2, I_1)^\top := \widetilde{H} \circ U_1 = N^{-1}(H_3, 1, H_{21}H_3 + H_{22})^\top \circ U_1,\ N := H_{11}H_3 + H_{12}.$$

Therefore $\widetilde{H}_1 = N^{-1}H_3 = \frac{H_3}{H_{11}H_3 + H_{12}}$ is called the *voltage transfer function*. If $\widetilde{H}_3 = N^{-1}(H_{21}H_3 + H_{22})$ is nonzero, i.e., $H_{21}H_3 + H_{22} \neq 0$, then

$$U_1 = \widetilde{H}_3^{-1} \circ I_1,\ I_2 = \widetilde{H}_2 \circ U_1 = \widetilde{H}_2\widetilde{H}_3^{-1} \circ I_1 \text{ with}$$
$$\widetilde{H}_3^{-1} = \frac{H_{11}H_3 + H_{12}}{H_{21}H_3 + H_{22}},\ \widetilde{H}_2\widetilde{H}_3^{-1} = \frac{1}{H_{21}H_3 + H_{22}}.$$

Then $\widetilde{H}_3^{\,-1}$ resp. $\widetilde{H}_2\widetilde{H}_3^{-1}$ are called the *input resp. current transfer functions*. Assume in addition that $T > 0$ is a period with $\omega := 2\pi T^{-1}$ and that $\widetilde{H}$ is proper. If U_1 is piecewise continuous and T-periodic, it has the $\|-\|_2$-convergent Fourier series $U_1 = \sum_{\mu \in \mathbb{Z}} \mathbb{F}(U_1)(\mu)e^{j\mu\omega t}$. The steady-state output has the same properties with the $\|-\|_2$-convergent Fourier series

$$(U_2, I_2, I_1)^\top = \sum_{\mu \in \mathbb{Z}} \left(\widetilde{H}_1(j\mu\omega), \widetilde{H}_2(j\mu\omega), \widetilde{H}_3(j\mu\omega)\right)^\top \mathbb{F}(U_1)(\mu)e^{j\mu\omega t}. \tag{8.319}$$

The latter is even uniformly convergent if $\widetilde{H}$ is strictly proper, cf. Theorem 8.2.102. ◊

8.3.3 State Space Representations for Electrical Networks

We construct the representations of the title by a variant of the method in [10, Sect. 3.4], cf. also [18]. The notations and assumptions are those from Sect. 8.3.1 with the additional assumption that the network is a pure RLC network.
First we construct a special decomposition $K = K_1 \uplus K_2$. Recall $K = K_s \uplus K_p$. We also need the finer decomposition $K_p = K_C \uplus K_R \uplus K_L$ where K_C consists of the capacitor branches, etc. We use the isomorphism

$$F^K / \ker(d) = \sum_{k\in K} F\overline{k} \cong AF^K,\ \overline{k} \mapsto A(-,k),\ \text{cf. (8.205), and construct}$$

$$\begin{aligned} &K_{1,s} \subseteq K_s,\ K_{1,C} \subseteq K_C,\ K_{1,R} \subseteq K_R,\ K_{1,L} \subseteq K_L,\\ &\qquad K_1 := K_{1,C} \uplus K_{1,R} \uplus K_{1,L} \uplus K_{1,s},\ \text{such that}\\ &\sum_{k\in K_s} \mathbb{R}\overline{k} = \bigoplus_{k\in K_{1,s}} \mathbb{R}\overline{k},\ \sum_{k\in K_s\uplus K_C} \mathbb{R}\overline{k} = \bigoplus_{k\in K_{1,s}\uplus K_{1,C}} \mathbb{R}\overline{k},\\ &\sum_{k\in K_s\uplus K_C\uplus K_R} \mathbb{R}\overline{k} = \bigoplus_{k\in K_{1,s}\uplus K_{1,C}\uplus K_{1,R}} \mathbb{R}\overline{k},\ \sum_{k\in K} \mathbb{R}\overline{k} = \bigoplus_{k\in K_1} \mathbb{R}\overline{k}. \end{aligned} \tag{8.320}$$

In other words, we use a special Gauß algorithm: We first compute a basis $A(-,k)$, $k \in K_{1,s}$, of the column space $A|_{K_s}\mathbb{R}^{K_s}$ and then extend it to an $\mathbb{R}$-basis $A(-,k)$, $k \in K_1$, of AF^K, first by a maximal number of columns $A(-,k)$ with $k \in K_C$ and then likewise with $k \in K_R$ and finally with $k \in K_L$. We infer the further decompositions

$$\begin{aligned} &K = K_1 \uplus K_2,\ K_2 = K_{2,C} \uplus K_{2,R} \uplus K_{2,L} \uplus K_{2,s},\\ &K_{2,X} := K_2 \cap K_X,\ X = C, R, L, s. \end{aligned} \tag{8.321}$$

From (8.206) recall $A|_{K_2} = A|_{K_1}M$. With the finer structure here this implies

$$A|_{K_2} = \left(A|_{K_{2,C}}, A|_{K_{2,R}}, A|_{K_{2,L}}, A|_{K_{2,s}}\right) = \left(A|_{K_{1,C}}, A|_{K_{1,R}}, A|_{K_{1,L}}, A|_{K_{1,s}}\right) M$$

$$\text{where } M = \begin{array}{c|cccc} & K_{2,C} & K_{2,R} & K_{2,L} & K_{2,s}\\ \hline K_{1,C} & M_{CC} & M_{CR} & M_{CL} & M_{Cs}\\ K_{1,R} & 0 & M_{RR} & M_{RL} & M_{Rs}\\ K_{1,L} & 0 & 0 & M_{LL} & M_{Ls}\\ K_{1,s} & M_{sC} & M_{sR} & M_{sL} & M_{ss} \end{array},\ M^\top = \begin{array}{c|cccc} & K_{1,C} & K_{1,R} & K_{1,L} & K_{1,s}\\ \hline K_{2,C} & M_{CC}^\top & 0 & 0 & M_{sC}^\top\\ K_{2,R} & M_{CR}^\top & M_{RR}^\top & 0 & M_{sR}^\top\\ K_{2,L} & M_{CL}^\top & M_{RL}^\top & M_{LL}^\top & M_{sL}^\top\\ K_{2,s} & M_{Cs}^\top & M_{Rs}^\top & M_{Ls}^\top & M_{ss}^\top \end{array}. \tag{8.322}$$

The zero blocks in M are a direct consequence of the choice of the $K_{1,C}, K_{1,R}, K_{1,L}$ in (8.320). Correspondingly we define

$$\begin{aligned} &U_{i,X} := U_{K_{i,X}},\ I_{i,X} := I_{K_{i,X}},\ i = 1, 2,\ X = C, R, L, s\\ &C := \operatorname{diag}(C_1, C_2),\ C_i = \operatorname{diag}(C_k; k \in K_{i,C}),\ i = 1, 2,\ \text{and likewise}\\ &R := \operatorname{diag}(R_1, R_2),\ L := \operatorname{diag}(L_1, L_2), \end{aligned} \tag{8.323}$$

where the C_k are the capacitances of the capacitor branches and likewise for R_k and L_k. Notice that we use C, R, L resp. s both for an index and for a matrix resp. an indeterminate. With these data the network equations are

$$\begin{aligned}
&\begin{pmatrix} U_{2,C} \\ U_{2,R} \\ U_{2,L} \\ U_{2,s} \end{pmatrix} = \begin{pmatrix} M_{CC}^\top & 0 & 0 & M_{sC}^\top \\ M_{CR}^\top & M_{RR}^\top & 0 & M_{sR}^\top \\ M_{CL}^\top & M_{RL}^\top & M_{LL}^\top & M_{sL}^\top \\ M_{Cs}^\top & M_{Rs}^\top & M_{Ls}^\top & M_{ss}^\top \end{pmatrix} \begin{pmatrix} U_{1,C} \\ U_{1,R} \\ U_{1,L} \\ U_{1,s} \end{pmatrix}, \\
&\begin{pmatrix} I_{1,C} \\ I_{1,R} \\ I_{1,L} \\ I_{1,s} \end{pmatrix} = -\begin{pmatrix} M_{CC} & M_{CR} & M_{CL} & M_{Cs} \\ 0 & M_{RR} & M_{RL} & M_{Rs} \\ 0 & 0 & M_{LL} & M_{Ls} \\ M_{sC} & M_{sR} & M_{sL} & M_{ss} \end{pmatrix} \begin{pmatrix} I_{2,C} \\ I_{2,R} \\ I_{2,L} \\ I_{2,s} \end{pmatrix} \\
&I_{i,C} = C_i s \circ U_{i,C},\ U_{i,R} = R_i I_{i,R},\ U_{i,L} = L_i s \circ I_{i,L},\ i = 1, 2.
\end{aligned} \tag{8.324}$$

The first resp. second equation in (8.324) are Kirchhoff's voltage resp. current law from Theorem 8.3.5. The last row contains the branch equations with positive coefficients. We arrange the components of $\left(\begin{smallmatrix} U \\ I \end{smallmatrix}\right) = \left(\left(\begin{smallmatrix} U_k \\ I_k \end{smallmatrix}\right)\right)_{k \in K} \in \mathcal{B}$ as

$$\begin{aligned}
&\widetilde{y} := (U_{1,R}, U_{1,L}, U_{2,C}, U_{2,R}, U_{2,L}, U_{2,s}, I_{1,C}, I_{1,R}, I_{1,L}, I_{1,s}, I_{2,C}, I_{2,R})^\top, \\
&\widetilde{x} := \begin{pmatrix} U_{1,C} \\ I_{2,L} \end{pmatrix},\ \widetilde{u} := \begin{pmatrix} U_{1,s} \\ I_{2,s} \end{pmatrix},\ \begin{pmatrix} U \\ I \end{pmatrix} = \begin{pmatrix} \widetilde{x} \\ \widetilde{y} \\ \widetilde{u} \end{pmatrix}.
\end{aligned} \tag{8.325}$$

Notice that the determination of $k \in K_s = K_{1,s} \uplus K_{2,s}$ as voltage or current source branch follows from the choice of $K_{1,s}$, and is not given a priori.

Theorem 8.3.32 *1. (Cf. [10, Sect. 3.4, (3.57), (3.61)]) Assume the real, pure RLC network $\mathcal{B}$ with the data from (8.320)–(8.325). There are real matrices $\widetilde{A}$, $\widetilde{B}_0$, $\widetilde{B}_1$, $\widetilde{C}$, $\widetilde{D}_0$, $\widetilde{D}_1$ such that the network equations (8.324) have the form*

$$\begin{aligned}
&s \circ \widetilde{x} = \widetilde{A}\widetilde{x} + \widetilde{B} \circ \widetilde{u},\ \widetilde{y} := \widetilde{C}\widetilde{x} + \widetilde{D} \circ \widetilde{u},\ \textit{with} \\
&\qquad\qquad \widetilde{B} := \widetilde{B}_1 s + \widetilde{B}_0,\ \widetilde{D} := \widetilde{D}_1 s + \widetilde{D}_0,\ \textit{or, equivalently,} \\
&s \circ \widetilde{x},\ \widetilde{y} \in \mathbb{R}^{\bullet\times\bullet}\widetilde{x} + \mathbb{R}^{\bullet\times\bullet}\widetilde{u}, +\mathbb{R}^{\bullet\times\bullet} s \circ \widetilde{u}\ \textit{or, equivalently,} \\
&\mathcal{B} = \left\{\begin{pmatrix} \widetilde{x} \\ \widetilde{y} \\ \widetilde{u} \end{pmatrix};\ \widetilde{P} \circ \begin{pmatrix} \widetilde{x} \\ \widetilde{y} \end{pmatrix} = \widetilde{Q} \circ \widetilde{u},\right\},\ \widetilde{P} := \begin{pmatrix} s\,\mathrm{id} - \widetilde{A} & 0 \\ -\widetilde{C} & \mathrm{id} \end{pmatrix},\ \widetilde{Q} := \begin{pmatrix} \widetilde{B} \\ \widetilde{D} \end{pmatrix}, \\
&\Longrightarrow \begin{pmatrix} \mathrm{id} & 0 \\ \widetilde{C} & \widetilde{D} \\ 0 & \mathrm{id} \end{pmatrix} \circ :\ \mathcal{B}_s := \left\{\begin{pmatrix} \widetilde{x} \\ \widetilde{u}, \end{pmatrix};\ s \circ \widetilde{x} = \widetilde{A}\widetilde{x} + \widetilde{B} \circ \widetilde{u},\right\} \cong \mathcal{B},\ \begin{pmatrix} \widetilde{x} \\ \widetilde{u}, \end{pmatrix} \mapsto \begin{pmatrix} \widetilde{x} \\ \widetilde{C}\widetilde{x} + \widetilde{D} \circ \widetilde{u}, \\ \widetilde{u} \end{pmatrix}, \\
&\qquad \begin{pmatrix} \mathrm{id} \\ \widetilde{C} \end{pmatrix} \cdot :\ \mathcal{B}_s^0 = \left\{\widetilde{x};\ s \circ \widetilde{x} = \widetilde{A}\widetilde{x}\right\} \cong \mathcal{B}^0 = \left\{\begin{pmatrix} \widetilde{x} \\ \widetilde{y} \end{pmatrix};\ \widetilde{P} \circ \begin{pmatrix} \widetilde{x} \\ \widetilde{y} \end{pmatrix} = 0\right\},\ \widetilde{x} \mapsto \begin{pmatrix} \widetilde{x} \\ \widetilde{C}\widetilde{x} \end{pmatrix}, \\
&\qquad \det\left(\widetilde{P}\right) = \det\left(s\,\mathrm{id} - \widetilde{A}\right) \neq 0,\ \mathrm{char}(\mathcal{B}^0) = \mathrm{V}_{\mathbb{C}}\left(\det(\widetilde{P})\right) = \mathrm{spec}(\widetilde{A}).
\end{aligned} \tag{8.326}$$

The isomorphism $\mathcal{B}_s^0 \cong \mathcal{B}^0$ implies that $\mathcal{B}_s^0$ is a minimal state representation of dimension

$$\sharp(K_{1,C}) + \sharp(K_{2,L}) = \dim_{\mathbb{R}}(\mathcal{B}_s^0) = \dim_{\mathbb{R}}(\mathcal{B}^0) = \deg_s\left(\det(\widetilde{P})\right). \tag{8.327}$$

The precise expressions of the matrices $\widetilde{A}$, etc. as rational functions of M, C, R, L follow from the equations in the proof. In particular, $\mathcal{B}$ is an IO behavior with input $\widetilde{u}$. The representation of $\mathcal{B}$ is the special form of (8.230).

2. *The transfer matrix of $\mathcal{B}$ is*

$$\widetilde{H} := \widetilde{P}^{-1}\widetilde{Q} = \begin{pmatrix} (s\,\mathrm{id} - \widetilde{A})^{-1}\widetilde{B} \\ \widetilde{D} + \widetilde{C}(s\,\mathrm{id} - \widetilde{A})^{-1}\widetilde{B} \end{pmatrix}. \tag{8.328}$$

Since $(s\,\mathrm{id} - \widetilde{A})^{-1}$ is strictly proper and $\deg_s(\widetilde{B}) \leq 1$, the matrix $(s\,\mathrm{id} - \widetilde{A})^{-1}\widetilde{B}$ is proper, and so is $\widetilde{H}$ if and only if $\widetilde{D}_1 = 0$.

3. *(Cf. [10, Sect. 6.6]) (a) (Cf. Theorem 8.2.53) Assume that the input $\widetilde{u} = \begin{pmatrix} U_{1,s} \\ I_{2,s} \end{pmatrix}$ is given as $\widetilde{u} = H_2 \circ \delta = \alpha Y$, $H_2 = \mathcal{L}(\widetilde{u}) \in \mathbb{R}(s)^{\bullet}_{\mathrm{spr}}$, with polynomial-exponential α. Then one output $\begin{pmatrix} \widetilde{x}_1 \\ \widetilde{y}_1 \end{pmatrix}$ to $\widetilde{u}$ has the form*

$$\begin{aligned} \widetilde{x}_1 &= (s\,\mathrm{id} - \widetilde{A})^{-1}\widetilde{B}H_2 \circ \delta = \mathcal{L}^{-1}\left((s\,\mathrm{id} - \widetilde{A})^{-1}(\widetilde{B}_1 s + \widetilde{B}_0)H_2\right) = \beta Y \\ \widetilde{y}_1 &= \widetilde{D}_1\alpha(0)\delta + \left(\widetilde{D}_1(s \circ \alpha) + \widetilde{D}_0\alpha + \widetilde{C}\beta\right)Y, \end{aligned} \tag{8.329}$$

again with polynomial-exponential β. All other outputs to $\widetilde{u}$ are

$$\begin{pmatrix} \widetilde{x} \\ \widetilde{y} \end{pmatrix} = \begin{pmatrix} \widetilde{x}_1 \\ \widetilde{y}_1 \end{pmatrix} + \begin{pmatrix} \mathrm{id} \\ \widetilde{C} \end{pmatrix} z,\ z := e^{t\widetilde{A}}z(0),\ z(0) := \widetilde{x}(0+) - \beta(0). \tag{8.330}$$

If $\mathcal{B}^0$ is asymptotically stable, i.e., $\mathrm{spec}(\widetilde{A}) \subset \mathbb{C}_-$, *then $\begin{pmatrix} \widetilde{x}_1 \\ \widetilde{y}_1 \end{pmatrix}$ is called a (the)* steady state *and z a (the)* transient. *If $\widetilde{D}_1\alpha(0) \neq 0$, then $\widetilde{D}_1\alpha(0)\delta$ is an impulsive component of $\widetilde{y}_1$.*

(b) (Cf. Theorem 8.2.102) Assume a period $T > 0$, $\omega := 2\pi T^{-1}$, a piecewise continuous, T-periodic input $\widetilde{u}$ with the Fourier series $\widetilde{u} = \sum_{\mu\in\mathbb{Z}} \mathbb{F}(\widetilde{u})(\mu)e^{j\mu\omega t}$, and $\mathbb{Z}j\omega \cap$ $\mathrm{spec}(\widetilde{A}) = \emptyset$. *Then the unique periodic output to $\widetilde{u}$ is*

$$\begin{aligned} \widetilde{x}_1 &:= \sum_{\mu\in\mathbb{Z}} (j\mu\omega\,\mathrm{id} - \widetilde{A})^{-1}\left(\widetilde{B}_1 j\mu\omega + \widetilde{B}_0\right)\mathbb{F}(\widetilde{u})(\mu)e^{j\mu\omega t}, \\ \widetilde{y}_1 &:= \widetilde{D}_1(s \circ \widetilde{u}) + \widetilde{D}_0\widetilde{u} + \widetilde{C}\widetilde{x}_1, \end{aligned} \tag{8.331}$$

where $\widetilde{x}_1$ and $\widetilde{D}_0\widetilde{u}$ are piecewise continuous, but $\widetilde{D}_1(s \circ \widetilde{u})$ may be a periodic distribution. If $\widetilde{B}_1 = 0$, then $\widetilde{x}_1$ is continuous with uniformly convergent Fourier series. All other outputs have the form (8.330).

4. *If $\widetilde{u}$ is any* C^1*-input, for instance, a sinusoid, then $\widetilde{x}$ and $\widetilde{y}$ are given by*

$$\begin{aligned} \widetilde{x} &= e^{t\widetilde{A}}\widetilde{x}(0) + \int_0^t e^{(t-\tau)\widetilde{A}}\left(\widetilde{B} \circ \widetilde{u}\right)(\tau)\mathrm{d}\tau,\ \widetilde{x}(0) = \begin{pmatrix} U_{1,C}(0) \\ I_{2,L}(0) \end{pmatrix}, \\ \widetilde{y} &= \widetilde{D}_1(s \circ \widetilde{u}) + \widetilde{D}_0\widetilde{u} + \widetilde{C}e^{t\widetilde{A}}\widetilde{x}(0) + \widetilde{C}\int_0^t e^{(t-\tau)\widetilde{A}}\left(\widetilde{B} \circ \widetilde{u}\right)(\tau)\mathrm{d}\tau. \end{aligned} \tag{8.332}$$

5. *(Cf. [10, (3.62)]) The equivalence*

$$M_{sC}=0,\ M_{Ls}=0 \Longleftrightarrow \widetilde{B}_1=0,\ \widetilde{D}_1=0 \tag{8.333}$$

holds and implies that H is proper. Then (8.326) *are standard state space equations for $\mathcal{B}$. They are observable since the state vector $\widetilde{x}$ is a component of the output $\binom{\widetilde{x}}{y}$ of $\mathcal{B}$.*

Proof 1. (i) The equations (8.324) directly furnish

$$\begin{aligned} &U_{2,C}=M_{CC}^\top U_{1,C}+M_{sC}^\top U_{1,s},\ I_{1,L}=-M_{LL}I_{2,L}-M_{Ls}I_{2,s}\\ &\qquad\in \mathbb{R}^{\bullet\times\bullet}\begin{pmatrix}U_{1,C}\\ I_{2,L}\end{pmatrix}+\mathbb{R}^{\bullet\times\bullet}\begin{pmatrix}U_{1,s}\\ I_{2,s}\end{pmatrix}=\mathbb{R}^{\bullet\times\bullet}\widetilde{x}+\mathbb{R}^{\bullet\times\bullet}\widetilde{u}.\end{aligned} \tag{8.334}$$

(ii) The equations (8.324) also imply the components U_R, I_R as

$$\begin{aligned} &U_{1,R}=R_1I_{1,R}\underset{(8.324)}{=}R_1\left(-M_{RR}I_{2,R}-M_{RL}I_{2,L}-M_{Rs}I_{2,s}\right)\\ &=R_1\left(-M_{RR}R_2^{-1}U_{2,R}-M_{RL}I_{2,L}-M_{Rs}I_{2,s}\right)\\ &\underset{(8.324)}{=}R_1\left(-M_{RR}R_2^{-1}\left(M_{CR}^\top U_{1,C}+M_{RR}^\top U_{1,R}+M_{s,R}^\top U_{1,s}\right)-M_{RL}I_{2,L}-M_{Rs}I_{2,s}\right)\\ &\Longrightarrow (R_1+R_1M_{RR}R_2^{-1}M_{RR}^\top R_1)I_{1,R}\\ &\qquad\in \mathbb{R}^{\bullet\times\bullet}U_{1,C}+\mathbb{R}^{\bullet\times\bullet}I_{2,L}+\mathbb{R}^{\bullet\times\bullet}U_{1,s}+\mathbb{R}^{\bullet\times\bullet}I_{2,s}\subseteq\mathbb{R}^{\bullet\times\bullet}\widetilde{x}+\mathbb{R}^{\bullet\times\bullet}\widetilde{u}.\end{aligned} \tag{8.335}$$

The precise coefficient matrices in front of $\widetilde{x}$ and $\widetilde{u}$ as rational expressions in M, C, R, L follow directly from the explicit expressions in (8.335) but are lengthy and omitted here.
The matrices R_1, R_2^{-1} are diagonal and positive-definite. Hence

$$R_1M_{RR}R_2^{-1}M_{RR}^\top R_1=(R_1M_{RR})R_2^{-1}(R_1M_{RR})^\top \tag{8.336}$$

is positive semi-definite. This implies the definiteness of

$$\begin{aligned} &R_1+R_1M_{RR}R_2^{-1}M_{RR}^\top R_1\in \mathrm{Gl}_\bullet(\mathbb{R})\text{ and}\\ &I_{1,R}\in\mathbb{R}^{\bullet\times\bullet}\widetilde{x}+\mathbb{R}^{\bullet\times\bullet}\widetilde{u},\ U_{1,R}=R_1I_{1,R}\in\mathbb{R}^{\bullet\times\bullet}\widetilde{x}+\mathbb{R}^{\bullet\times\bullet}\widetilde{u},\\ &U_{2,R}=M_{CR}^\top U_{1,C}+M_{RR}^\top U_{1,R}+M_{s,R}^\top U_{1,s},\ I_{2,R}=R_2^{-1}U_{2,R}\\ &\qquad\in\mathbb{R}^{\bullet\times\bullet}\widetilde{x}+\mathbb{R}^{\bullet\times\bullet}\widetilde{u}.\end{aligned} \tag{8.337}$$

Again the precise coefficients follow directly from the equations but are not explicitly stated here.
(iii) Next we derive equations for $U_{1,C}$, $I_{1,C}$, $I_{2,C}$ by

$$
\begin{aligned}
C_1(s \circ U_{1,C}) = I_{1,C} &\underset{(8.324)}{=} -M_{CC}I_{2,C} - M_{CR}I_{2,R} - M_{CL}I_{2,L} - M_{Cs}I_{2,s} \\
&= -M_{CC}C_2 s \circ U_{2,C} - M_{CR}I_{2,R} - M_{CL}I_{2,L} - M_{Cs}I_{2,s} \\
&\underset{(8.324)}{=} -M_{CC}C_2 s \circ \left(M_{CC}^\top U_{1,C} + M_{sC}^\top U_{1,s}\right) \\
&\quad - M_{CR}I_{2,R} - M_{CL}I_{2,L} - M_{Cs}I_{2,s} \\
&\Longrightarrow \left(C_1 + M_{CC}C_2M_{CC}^\top\right) s \circ U_{1,C} \in \mathbb{R}^{\bullet\times\bullet}\widetilde{x} + \mathbb{R}^{\bullet\times\bullet}\widetilde{u} + \mathbb{R}^{\bullet\times\bullet}s \circ \widetilde{u}.
\end{aligned}
\tag{8.338}
$$

But $\left(C_1 + M_{CC}C_2M_{CC}^\top\right) \in \mathrm{Gl}_\bullet(\mathbb{R})$ as in (8.337), and hence we obtain

$$
\begin{aligned}
&s \circ U_{1,C} \in \mathbb{R}^{\bullet\times\bullet}\widetilde{x} + \mathbb{R}^{\bullet\times\bullet}\widetilde{u} + \mathbb{R}^{\bullet\times\bullet}s \circ \widetilde{u}, \; I_{1,C} = C_1 s \circ U_{1,C} \\
&I_{2,C} = C_2 s \circ U_{2,C} = C_2 s \circ \left(M_{CC}^\top U_{1,C} + M_{sC}^\top U_{1,s}\right) \\
&\Longrightarrow s \circ U_{1,C}, \; I_{1,C}, \; I_{2,C} \in \mathbb{R}^{\bullet\times\bullet}\widetilde{x} + \mathbb{R}^{\bullet\times\bullet}\widetilde{u} + \mathbb{R}^{\bullet\times\bullet}s \circ \widetilde{u}.
\end{aligned}
\tag{8.339}
$$

Moreover

$$
s \circ U_{1,C}, \; I_{1,C}, \; I_{2,C} \in \mathbb{R}^{\bullet\times\bullet}\widetilde{x} + \mathbb{R}^{\bullet\times\bullet}\widetilde{u} \Longleftrightarrow M_{sC} = 0. \tag{8.340}
$$

(iv) The equation for $I_{2,L}$ is obtained as

$$
\begin{aligned}
L_2 s \circ I_{2,L} = U_{2,L} &\underset{(8.324)}{=} M_{CL}^\top U_{1,C} + M_{RL}^\top U_{1,R} + M_{LL}^\top U_{1,L} + M_{sL}^\top U_{1,s} \\
&\underset{(8.334))}{=} M_{CL}^\top U_{1,C} + M_{RL}^\top U_{1,R} + M_{LL}^\top L_1 s \circ \left(-M_{LL}I_{2,L} - M_{Ls}I_{2,s}\right) + M_{sL}^\top U_{1,s} \\
&\Longrightarrow \left(L_2 + M_{LL}^\top L_1 M_{LL}\right) s \circ I_{2,L} \underset{(8.335)}{\in} \mathbb{R}^{\bullet\times\bullet}\widetilde{x} + \mathbb{R}^{\bullet\times\bullet}\widetilde{u} + \mathbb{R}^{\bullet\times\bullet}s \circ \widetilde{u} \\
&\Longrightarrow s \circ I_{2,L} \in \mathbb{R}^{\bullet\times\bullet}\widetilde{x} + \mathbb{R}^{\bullet\times\bullet}\widetilde{u} + \mathbb{R}^{\bullet\times\bullet}s \circ \widetilde{u} \in \mathbb{R}^{\bullet\times\bullet}\widetilde{x} + \mathbb{R}^{\bullet\times\bullet}\widetilde{u} + \mathbb{R}^{\bullet\times\bullet}s \circ \widetilde{u}.
\end{aligned}
\tag{8.341}
$$

The equations for $U_{1,L}$, $U_{2,L}$, $I_{1,L}$ follow as

$$
\begin{aligned}
&I_{1,L} = -M_{LL}I_{2,L} - M_{Ls}I_{2,s}, \; U_{2,L} = L_2 s \circ I_{2,L}, \\
&U_{1,L} = L_1 s \circ I_{1,L} = -L_1 M_{LL} s \circ I_{2,L} - L_1 M_{Ls} s \circ I_{2,s}, \\
&\Longrightarrow U_{1,L}, \; U_{2,L}, \; I_{1,L} \in \mathbb{R}^{\bullet\times\bullet}\widetilde{x} + \mathbb{R}^{\bullet\times\bullet}\widetilde{u} + \mathbb{R}^{\bullet\times\bullet}s \circ \widetilde{u}.
\end{aligned}
\tag{8.342}
$$

Moreover

$$
s \circ I_{2,L}, \; U_{2,L}, \; U_{1,L}, \; I_{1,L} \in \mathbb{R}^{\bullet\times\bullet}\widetilde{x} + \mathbb{R}^{\bullet\times\bullet}\widetilde{u} \Longleftrightarrow M_{Ls} = 0. \tag{8.343}
$$

(v) The equations for $U_{2,s}$ and $I_{1,s}$ are

$$
\begin{aligned}
U_{2,s} &= M_{Cs}^\top U_{1,C} + M_{Rs}^\top U_{1,R} + M_{Ls}^\top U_{1,L} + M_{ss}^\top U_{1,s} \\
I_{1,s} &= -M_{sC}I_{2,C} - M_{sR}I_{2,R} - M_{sL}I_{2,L} - M_{ss}I_{2,s} \\
&\in \mathbb{R}^{\bullet\times\bullet}\widetilde{x} + \mathbb{R}^{\bullet\times\bullet}\widetilde{u} + \mathbb{R}^{\bullet\times\bullet}s \circ \widetilde{u}.
\end{aligned}
\tag{8.344}
$$

(vi) Summing up (i)–(v), we obtain

$$\begin{aligned}
&s\circ\widetilde{x},\widetilde{y}\in\mathbb{R}^{\bullet\times\bullet}\widetilde{x}+\mathbb{R}^{\bullet\times\bullet}\widetilde{u}+\mathbb{R}^{\bullet\times\bullet}s\circ\widetilde{u}\text{ and (8.326). Moreover}\\
&M_{sC}=0,\,M_{Ls}=0\Longleftrightarrow s\circ\widetilde{x},\widetilde{y}\in\mathbb{R}^{\bullet\times\bullet}\widetilde{x}+\mathbb{R}^{\bullet\times\bullet}\widetilde{u}\\
&\qquad\Longleftrightarrow \widetilde{B}_1=0,\ \widetilde{D}_1=0\Longleftrightarrow \text{(8.334)}.
\end{aligned}\tag{8.345}$$

2. Simple.
3. See the quoted theorems. Moreover

$$\begin{aligned}
\widetilde{y}_1&=\widetilde{D}\circ\widetilde{u}+\widetilde{C}\widetilde{x}_1=\widetilde{D}_1 s\circ(\alpha Y)+\widetilde{D}_0\alpha Y+\widetilde{C}\widetilde{x}_1\\
&=\widetilde{D}_1\alpha(0)\delta+\left(\widetilde{D}_1(s\circ\alpha)+\widetilde{D}_0\alpha\right)Y+\widetilde{C}\widetilde{x}_1.
\end{aligned}$$

□

Remark 8.3.33 The special feature of (8.326) is that certain components of the output vector of the IO behavior $\mathcal{B}$, here $\widetilde{x}=\begin{pmatrix}U_{1,C}\\ I_{2,L}\end{pmatrix}$, can be used as state vector, whereas, in general, also higher derivatives of these components have to be employed, cf. Lemma 8.2.52. The proof of Theorem 8.3.32 is surprising. ◊

Example 8.3.34 We apply the procedure from (8.320) to the electrical network from Example 8.3.17 with its incidence matrix A and obtain

$$\begin{aligned}
&K_1=\{k_5,k_4,k_1\},\ K_2=\{k_2,k_6,k_3\},\ K_{1,C}=\{k_5\},\ K_{1,R}=\{k_4\},\ K_{1,L}=\emptyset\\
&\qquad K_{1,s}=\{k_1\},\ K_{2,C}=\{k_2\},\ K_{2,R}=\{k_6\},\ K_{2,L}=\{k_3\},\ K_{2,s}=\emptyset
\end{aligned}$$

$$A=\begin{array}{c}\\ v_1\\ v_2\\ v_3\\ v_4\end{array}\!\!\begin{array}{c}\begin{matrix}k_1 & k_2 & k_3 & k_4 & k_5 & k_6\end{matrix}\\ \begin{pmatrix}1&1&1&0&0&0\\-1&-1&0&0&-1&-1\\0&0&-1&1&0&0\\0&0&0&-1&1&1\end{pmatrix}\end{array},\quad A|_{K_2}=A|_{K_1}M\text{ with}$$

$$A|_{K_1}=\begin{array}{c}\\ v_1\\ v_2\\ v_3\\ v_4\end{array}\!\!\begin{array}{c}\begin{matrix}k_5 & k_4 & k_1\end{matrix}\\ \begin{pmatrix}0&0&1\\-1&0&-1\\0&1&0\\1&-1&0\end{pmatrix}\end{array},\quad A|_{K_2}=\begin{array}{c}\\ v_1\\ v_2\\ v_3\\ v_4\end{array}\!\!\begin{array}{c}\begin{matrix}k_2 & k_6 & k_3\end{matrix}\\ \begin{pmatrix}1&0&1\\-1&-1&0\\0&0&-1\\0&1&0\end{pmatrix}\end{array}$$

$$M=\begin{array}{c}\\ k_5\\ k_4\\ k_1\end{array}\!\!\begin{array}{c}\begin{matrix}k_2 & k_6 & k_3\end{matrix}\\ \begin{pmatrix}0&1&-1\\0&0&-1\\1&0&1\end{pmatrix}\end{array},\quad M^\top=\begin{array}{c}\\ k_2\\ k_6\\ k_3\end{array}\!\!\begin{array}{c}\begin{matrix}k_5 & k_4 & k_1\end{matrix}\\ \begin{pmatrix}0&0&1\\1&0&0\\-1&-1&1\end{pmatrix}\end{array}.\tag{8.346}$$

The decisive property $M_{k_4,k_2}=0$ from (8.322) holds. Notice that $K_{1,L}=\emptyset$, that K_1 is *uniquely* determined by the algorithm from (8.320), and that $k_1\in K_1$ is determined as a *voltage* source branch, but not given as such. We write

$$U_i := U_{k_i},\ I_i := I_{k_i} \Longrightarrow U_{K_1} = (U_5, U_4, U_1)^\top,\ U_{K_2} = (U_2, U_6, U_3)^\top$$
$$I_{K_1} = (I_5, I_4, I_1)^\top,\ I_{K_2} = (I_2, I_6, I_3)^\top$$
$$\Longrightarrow \begin{pmatrix} U_2 \\ U_6 \\ U_3 \end{pmatrix} = \begin{matrix} & k_5 & k_4 & k_1 \\ k_2 & 0 & 0 & 1 \\ k_6 & 1 & 0 & 0 \\ k_3 & -1 & -1 & 1 \end{matrix} \begin{pmatrix} U_5 \\ U_4 \\ U_1 \end{pmatrix},\ \begin{pmatrix} I_5 \\ I_4 \\ I_1 \end{pmatrix} = - \begin{matrix} & k_2 & k_6 & k_3 \\ k_5 & 0 & 1 & -1 \\ k_4 & 0 & 0 & -1 \\ k_1 & 1 & 0 & 1 \end{matrix} \begin{pmatrix} I_2 \\ I_6 \\ I_3 \end{pmatrix}. \tag{8.347}$$

Of course, the k_i outside the matrices are the row resp. column indices. The two last equations are Kirchhoff's voltage resp. current law. Written out they have the form

$$U_2 = U_1,\ U_6 = U_5,\ U_3 = -U_5 - U_4 + U_1,$$
$$I_5 = -I_6 + I_3,\ I_4 = I_3,\ I_1 = -I_2 - I_3 \tag{8.348}$$

that can also be obtained directly. The equations of the one-ports are

$$C_2 s \circ U_2 = I_2,\ L_3 s \circ I_3 = U_3,\ U_4 = R_4 I_4,\ C_5 s \circ U_5 = I_5,\ U_6 = R_6 I_6. \tag{8.349}$$

According to (8.325) we obtain

$$\widetilde{x} = \begin{pmatrix} U_5 \\ I_3 \end{pmatrix},\ \widetilde{u} = U_1,\ \widetilde{y} = (I_1, U_2, I_2, U_3, U_4, I_4, I_5, U_6, I_6)^\top, \tag{8.350}$$

with a different order of the components of $\widetilde{y}$.

Now we proceed according to Eqs. (8.334)–(8.344). In contrast to the general case with its inversion of positive-definite matrices the resulting equations are very simple here. Equations (8.334), (8.335), and (8.335) furnish the already known equations

$$U_2 = U_1,\ I_4 = I_3,\ U_4 = R_4 I_4 = R_4 I_3,\ U_6 = U_5. \tag{8.351}$$

Notice that U_5 and I_3 are the components of the state vector. Equations (8.338) and (8.339) give

$$C_5 s \circ U_5 = I_5 \underset{(8.348)}{=} -I_6 + I_3 = -R_6^{-1} U_6 + I_3 = -R_6^{-1} U_5 + I_3 \text{ and}$$
$$L_3 s \circ I_3 = U_3 \underset{(8.348)}{=} -U_5 - U_4 + U_1 = -U_5 - R_4 I_3 + U_1$$
$$\Longrightarrow s \circ \begin{pmatrix} U_5 \\ I_3 \end{pmatrix} = \begin{pmatrix} -C_5^{-1} R_6^{-1} & C_5^{-1} \\ -L_3^{-1} & -L_3^{-1} R_4 \end{pmatrix} \begin{pmatrix} U_5 \\ I_3 \end{pmatrix} + \begin{pmatrix} 0 \\ L_3^{-1} \end{pmatrix} U_1 = \widetilde{A}\widetilde{x} + \widetilde{B}\widetilde{u} \text{ with}$$
$$\widetilde{A} := \begin{pmatrix} -C_5^{-1} R_6^{-1} & C_5^{-1} \\ -L_3^{-1} & -L_3^{-1} R_4 \end{pmatrix},\ \widetilde{B} := \begin{pmatrix} 0 \\ L_3^{-1} \end{pmatrix},\ \widetilde{x} = \begin{pmatrix} U_5 \\ I_3 \end{pmatrix},\ \widetilde{u} = U_1. \tag{8.352}$$

The other branch voltages and currents are

$$
\begin{aligned}
&U_2 = U_1,\ I_2 = C_2 s \circ U_1,\ I_1 \underset{(8.348)}{=} -I_2 - I_3 = -C_2 s \circ U_1 - I_3,\\
&I_4 = I_3,\ U_4 = R_4 I_3,\ U_3 \underset{(8.348)}{=} -U_5 - U_4 + U_1 = -U_5 - R_4 I_3 + U_1,\\
&U_6 = U_5,\ I_6 = R_6^{-1} U_5,\ I_5 \underset{(8.348)}{=} -I_6 + I_3 = -R_6^{-1} U_5 + I_3\\
&\Longrightarrow \widetilde{y} = \widetilde{C}\widetilde{x} + \widetilde{D} \circ U_1 \text{ with} \qquad (8.353)\\
&\widetilde{y} := \begin{pmatrix} I_1\\ U_2\\ I_2\\ U_3\\ U_4\\ I_4\\ I_5\\ U_6\\ I_6 \end{pmatrix},\ \widetilde{C} := \begin{pmatrix} 0 & -1\\ 0 & 0\\ 0 & 0\\ -1 & -R_4\\ 0 & R_4\\ 0 & 1\\ -R_6^{-1} & 1\\ 1 & 0\\ R_6^{-1} & 0 \end{pmatrix},\ \widetilde{D} := \begin{pmatrix} -C_2 s\\ 1\\ C_2 s\\ 1\\ 0\\ 0\\ 0\\ 0\\ 0 \end{pmatrix}.
\end{aligned}
$$

According to Theorem 8.3.32 we obtain the behavior isomorphism

$$
\begin{aligned}
&\begin{pmatrix} \mathrm{id}_2 & 0\\ \widetilde{C} & \widetilde{D}\\ 0 & 1 \end{pmatrix} \circ : \mathcal{B}_s = \left\{ \begin{pmatrix} \widetilde{x}\\ U_1 \end{pmatrix};\ s \circ \widetilde{x} = \widetilde{A}\widetilde{x} + \widetilde{B} U_1 \right\}\\
&\qquad \cong \mathcal{B} = \left\{ \begin{pmatrix} \widetilde{x}\\ \widetilde{y}\\ U_1 \end{pmatrix};\ \widetilde{P} \circ \begin{pmatrix} \widetilde{x}\\ \widetilde{y} \end{pmatrix} = \widetilde{Q} \circ U_1 \right\},\ \begin{pmatrix} \widetilde{x}\\ U_1 \end{pmatrix} \mapsto \begin{pmatrix} \widetilde{x}\\ \widetilde{C}\widetilde{x} + \widetilde{D} \circ U_1\\ U_1 \end{pmatrix},\ \text{with}\\
&\widetilde{P} := \begin{pmatrix} s\,\mathrm{id}_2 - \widetilde{A} & 0\\ -\widetilde{C} & 1 \end{pmatrix},\ \widetilde{Q} := \begin{pmatrix} \widetilde{B}\\ \widetilde{D} \end{pmatrix},\ \widetilde{H} := \widetilde{P}^{-1}\widetilde{Q} = \begin{pmatrix} (s\,\mathrm{id}_2 - \widetilde{A})^{-1}\widetilde{B}\\ \widetilde{D} + \widetilde{C}(s\,\mathrm{id}_2 - \widetilde{A})^{-1}\widetilde{B} \end{pmatrix}.
\end{aligned}
\tag{8.354}
$$

Here $\mathcal{B}$ is the IO behavior of the network from Example 8.3.17 with $\begin{pmatrix} (U_k)_{k\in K}\\ (I_k)_{k\in K} \end{pmatrix} = \begin{pmatrix} \widetilde{x}\\ \widetilde{y}\\ U_1 \end{pmatrix}$, $y = \begin{pmatrix} \widetilde{x}\\ \widetilde{y} \end{pmatrix}$ and $H = \widetilde{H}$, *up to the order of the components*. Notice that $\widetilde{C}(s\,\mathrm{id}_2 - \widetilde{A})^{-1}\widetilde{B}$ is proper, but $\widetilde{D}$ and $H = \widetilde{H}$ are not proper, cf. Example 8.3.17. We check that, up to the order of the components, H from Example 8.3.17 indeed coincides with $\widetilde{H} = \begin{pmatrix} (s\,\mathrm{id}_2 - \widetilde{A})^{-1}\widetilde{B}\\ \widetilde{D} + \widetilde{C}(s\,\mathrm{id}_2 - \widetilde{A})^{-1}\widetilde{B} \end{pmatrix}$. A simple calculation shows

$$
\begin{aligned}
&(s\,\mathrm{id}_2 - \widetilde{A})^{-1}\widetilde{B} = N^{-1}\begin{pmatrix} R_6\\ C_5 R_6 s + 1 \end{pmatrix} \text{ with}\\
&\qquad N = L_3 C_5 R_6 s^2 + (L_3 + R_4 C_5 R_6)s + (R_4 + R_6).
\end{aligned}
\tag{8.355}
$$

This implies

$$
\widetilde{H} = \begin{pmatrix} (s\,\mathrm{id}_2 - \widetilde{A})^{-1}\widetilde{B}\\ \widetilde{D} + \widetilde{C}(s\,\mathrm{id}_2 - \widetilde{A})^{-1}\widetilde{B} \end{pmatrix} = \begin{pmatrix} 0\\ 0\\ \widetilde{D} \end{pmatrix} + N^{-1}\begin{pmatrix} \mathrm{id}_2\\ \widetilde{C} \end{pmatrix}\begin{pmatrix} R_6\\ C_5 R_6 s + 1 \end{pmatrix}
\tag{8.356}
$$

and then

$$\begin{aligned}
\widetilde{H} &= \begin{pmatrix} H_{U_5,U_1} \\ H_{I_3,U_1}, \\ H_{I_1,U_1} \\ H_{U_2,U_1} \\ H_{I_2,U_1} \\ H_{U_3,U_1} \\ H_{U_4,U_1} \\ H_{I_4,U_1} \\ H_{I_5,U_1} \\ H_{U_6,U_1} \\ H_{I_6,U_1} \end{pmatrix} = \begin{pmatrix} 0 \\ 0 \\ -C_2 s \\ 1 \\ C_2 s \\ 1 \\ 0 \\ 0 \\ 0 \\ 0 \\ 0 \end{pmatrix} + N^{-1} \begin{pmatrix} 1 & 0 \\ 0 & 1 \\ 0 & -1 \\ 0 & 0 \\ 0 & 0 \\ -1 & -R_4 \\ 0 & R_4 \\ 0 & 1 \\ -R_6^{-1} & 1 \\ 1 & 0 \\ R_6^{-1} & 0 \end{pmatrix} \begin{pmatrix} R_6 \\ C_5 R_6 s+1 \end{pmatrix} \\
&= \begin{pmatrix} N^{-1} R_6 \\ N^{-1}(C_5 R_6 s+1) \\ -C_2 s - N^{-1}(C_5 R_6 s+1) \\ 1 \\ C_2 s \\ 1+N^{-1}(-R_6 - R_4(C_5 R_6 s+1)) \\ N^{-1} R_4 (C_5 R_6 s+1) \\ N^{-1}(C_5 R_6 s+1) \\ N^{-1} C_5 R_6 s \\ N^{-1} R_6 \\ N^{-1} \end{pmatrix} \\
&= N^{-1} \begin{pmatrix} R_6 \\ C_5 R_6 s+1 \\ -C_2 L_3 C_5 R_6 s^3 - (C_2 L_3 + C_2 R_4 C_5 R_6) s^2 - (C_2 R_4 + C_2 R_6 + C_5 R_6) s - 1 \\ L_3 C_5 R_6 s^2 + (L_3 + R_4 C_5 R_6) s + (R_4 + R_6) \\ C_2 L_3 C_5 R_6 s^3 + (C_2 L_3 + C_2 R_4 C_5 R_6) s^2 + (C_2 R_4 + C_2 R_6) s \\ L_3 C_5 R_6 s^2 + L_3 s \\ R_4 C_5 R_6 s + R_4 \\ C_5 R_6 s+1 \\ C_5 R_6 s \\ R_6 \\ 1 \end{pmatrix}.
\end{aligned} \tag{8.357}$$

Comparison of (8.357) with (8.266) confirms the equality $H = \widetilde{H}$, up to the order of the components. This is also a strong hint at the validity of the algorithms contained in Theorems 8.3.11 and 8.3.32.

If the DC voltage input $U_1 = Y$ (Y Heaviside) is chosen, (8.353) implies

$$\begin{aligned}
& s \circ \begin{pmatrix} U_5 \\ I_3 \end{pmatrix} = \widetilde{A} \begin{pmatrix} U_5 \\ I_3 \end{pmatrix} + \widetilde{B} U_1, \; I_1 = -I_3 - C_2 s \circ U_1 \\
\Longrightarrow & \begin{pmatrix} U_5(t) \\ I_3(t) \end{pmatrix} = e^{t\widetilde{A}} \begin{pmatrix} U_5(0) \\ I_3(0) \end{pmatrix} + \int_0^t e^{(t-\tau)\widetilde{A}} \widetilde{B} \mathrm{d}\tau, \; I_1 = -I_3 - C_2 \delta.
\end{aligned}$$

This shows that there is no initial value expression for I_1, and that the network is destroyed by this input. ◊

Remark 8.3.35 1. Recall that K_1 resp. K_2 correspond to a tree resp. cotree (tree complement) of the graph in the literature and that the columns A_{-,k_1}, $k_1 \in K_1$, are a basis of the column space $A\mathbb{R}^K$. In [10, p. 131, below] it is assumed that K_1 contains all voltage, but no current source branches, i.e., $K_s = K_{1,s}$, $K_{2,s} = \emptyset$. This means that the columns $A(-,k)$, $k \in K_s$, are linearly independent. This is not true in general. In [19, p. 53] the algorithm (C1)–(C4) assumes that $K_R \subset K_1$ and hence the linear independence of the columns $A(-,k)$, $k \in K_R$, which again does not

generally hold. Therefore the algorithms from [10, Sect. 3.4] and [19, pp. 53–55] do not apply to arbitrary electrical networks.
2. In general, the resulting transfer matrix $H = \widetilde{H}$ of the algorithm is not proper, see Example 8.3.34. This is in contrast to [19, p. 55] and to [10, (3.62a), (3.62b)] which therefore do not hold generally. Since H depends only on the network and its IO structure, i.e., the choice $\begin{pmatrix} U_{1,s} \\ I_{2,s} \end{pmatrix}$ of its input, the nonproperness cannot be avoided by choosing a different state space representation. Therefore the (unproven) algorithm (C1)–(C4) from [19, pp. 53–55] does not apply to all networks. Also recall from Theorem 8.3.32 that, in general, the state equations are $s \circ \widetilde{x} = \widetilde{A}\widetilde{x} + \widetilde{B} \circ \widetilde{u}$ with $\widetilde{B} = \widetilde{B}_1 s + \widetilde{B}_0$, i.e., they do not have Kalman's standard form. The algorithm from [19, pp. 53–55] differs from the proven one in [10, Sect. 3.4].
3. If the behavior $\mathcal{B}^0$ is asymptotically stable and if only the steady state is of interest, the state space representation is not needed. ◇

8.3.4 Electric Power and Reciprocity

In this section we prove four important theorems on electric power with the methods of Sects. 8.3.1 and 8.3.2. The assumptions of these sections remain in force with $F := \mathbb{R}$. If $\binom{U}{I} \in \mathcal{B}_{\mathbb{R}}$, $U = (U_k)_{k\in K}$, $I = (I_k)_{k\in K}$, are real and piecewise continuous trajectories of the behavior $\mathcal{B}$, then the product $U_k(t)I_k(t)$, $k \in K$, is the instantaneous power along k. Tellegen's theorem and the reciprocity theorem are more generally proven for $F = \mathbb{C}$.

Theorem 8.3.36 (Theorem of Tellegen, cf. [17, p. 51], [10, Sect. 4.4]) *Assume the behavior from Theorem 8.3.11 and trajectories* $\binom{U}{I}$, $\binom{\widetilde{U}}{\widetilde{I}} \in \mathcal{B}$ *with piecewise continuous and possibly complex* $U = (U_k)_{k\in K}$, $I = (I_k)_{k\in K}$, $\widetilde{U}, \widetilde{I} \in \left(\mathrm{C}^{0,\mathrm{pc}}\right)^K$. *Then*

$$U^\top I = \sum_{k\in K} U_k I_k = \widetilde{U}^\top I = U^\top \widetilde{I} = 0. \tag{8.358}$$

Notice that $U, I, \widetilde{U}, \widetilde{I}$ *are not assumed sinusoidal.*

Proof It obviously suffices to show $\widetilde{U}^\top I = 0$. The theorem 8.3.5 implies

$$\begin{aligned} &\widetilde{U}_{K_2} = M^\top \widetilde{U}_{K_1},\ \widetilde{U}_{K_2}^\top = \widetilde{U}_{K_1}^\top M,\ I_{K_1} = -MI_{K_2} \\ &\Longrightarrow \widetilde{U}^\top I = \widetilde{U}_{K_2}^\top I_{K_2} + \widetilde{U}_{K_1}^\top I_{K_1} = \widetilde{U}_{K_1}^\top M I_{K_2} + \widetilde{U}_{K_1}^\top (-M) I_{K_2} = 0. \end{aligned}$$

□

Corollary 8.3.37 *The previous Theorem also holds if* $\binom{U}{I}$ *resp.* $\binom{\widetilde{U}}{\widetilde{I}}$ *belong to different network behaviors* $\mathcal{B}$ *resp.* $\widetilde{\mathcal{B}}$ with the same graph (V, K). ◇

The preceding theorem implies the following one for sinusoidal inputs only. We assume that the notations and conditions of Theorem 8.3.11, 1–5, hold. In particular, $K = K_s \uplus K_p$ is decomposed into source and passive one-port branches. The equations of the latter are

$$\forall k \in K_p : \ P_k \circ U_k = Q_k \circ I_k, \ 0 \neq P_k, Q_k \in \mathbb{C}[s], \ H_k := P_k^{-1} Q_k. \tag{8.359}$$

Let $\mathcal{F}_- := \bigoplus_{\lambda \in \mathbb{C}, \, \Re(\lambda) < 0} \mathbb{C}[t]e^{\lambda t}$ denote the space of polynomial-exponential signals u with $\lim_{t \to \infty} u(t) = 0$. An autonomous behavior $\mathcal{C} \subset \mathcal{F}^m$ is asymptotically stable if and only if $\mathcal{C} \subset \mathcal{F}_-^m$. We choose two sinusoidal inputs of the network behavior $\mathcal{B}$, i.e.,

$$u := \begin{pmatrix} I_{K_{2,s}} \\ U_{K_{1,s}} \end{pmatrix} := u(0)e^{j\omega t}, \ \widetilde{u} = \begin{pmatrix} \widetilde{I}_{K_{2,s}} \\ \widetilde{U}_{K_{1,s}} \end{pmatrix} = \widetilde{u}(0)e^{j\omega t}, \ u(0), \ \widetilde{u}(0) \in \mathbb{C}^{K_s}. \tag{8.360}$$

Theorem 8.3.38 (Reciprocity theorem, cf. [17, Sect. 2.8], [10, Sect. 4.4.2]) *Assume the IO behavior $\mathcal{B}$ from Theorem 8.3.11 and that $\mathcal{B}^0$ is asymptotically stable, i.e.,* $\mathrm{char}(\mathcal{B}^0) \subset \mathbb{C}_-$. *Consider the sinusoidal inputs $u, \widetilde{u}$ from* (8.360), *and let $\binom{U}{I} \in \mathcal{B}$ and $\binom{\widetilde{U}}{\widetilde{I}} \in \mathcal{B}$ be, necessarily polynomial-exponential, trajectories of $\mathcal{B}$ for these inputs, i.e.,* $\begin{pmatrix} I_{K_{2,s}} \\ U_{K_{1,s}} \end{pmatrix} = u, \ \begin{pmatrix} \widetilde{I}_{K_{2,s}} \\ \widetilde{U}_{K_{1,s}} \end{pmatrix} = \widetilde{u}$. *Then*

$$\begin{aligned} &\widetilde{U}_{K_s}^\top I_{K_s} - U_{K_s}^\top \widetilde{I}_{K_s} = \sum_{k \in K_s} \widetilde{U}_k I_k - \sum_{k \in K_s} U_k \widetilde{I}_k \in \mathcal{F}_-, \ \textit{hence} \\ &\lim_{t \to \infty} \left(\widetilde{U}_{K_s}^\top I_{K_s} - U_{K_s}^\top \widetilde{I}_{K_s}\right)(t) = 0, \end{aligned} \tag{8.361}$$

i.e., practically, $\widetilde{U}_{K_s}^\top I_{K_s}$ and $U_{K_s}^\top \widetilde{I}_{K_s}$ coincide.
Notice that $\mathcal{B}$ need not be an RLC network as is usually required in the literature. But it is essential that only one-port equations $P_k \circ U_k = Q_k \circ I_k, \ k \in K_p$, occur.

Proof (i) Theorem 8.3.36 furnishes $U^\top \widetilde{I} = \widetilde{U}^\top I = 0$, hence

$$\sum_{k \in K_s} \widetilde{U}_k I_k - \sum_{k \in K_s} U_k \widetilde{I}_k = -\sum_{k \in K_p} \left(\widetilde{U}_k I_k - U_k \widetilde{I}_k\right).$$

Thus it suffices to show $\widetilde{U}_k I_k - U_k \widetilde{I}_k \in \mathcal{F}_-$ for $k \in K_p$.

(ii) We define the outputs

$$
\begin{aligned}
&y := H(j\omega)u,\ \widetilde{y} := H(j\omega)\widetilde{u},\\
&\begin{pmatrix} U^1 \\ I^1 \end{pmatrix} := \begin{pmatrix} y \\ u \end{pmatrix} = \begin{pmatrix} U^1(0) \\ I^1(0) \end{pmatrix} e^{j\omega t} \in \mathcal{B},\ \begin{pmatrix} \widetilde{U}^1 \\ \widetilde{I}^1 \end{pmatrix} := \begin{pmatrix} \widetilde{y}^1 \\ \widetilde{u} \end{pmatrix} = \begin{pmatrix} \widetilde{U}^1(0) \\ \widetilde{I}^1(0) \end{pmatrix} e^{j\omega t} \in \mathcal{B},\\
&\Longrightarrow \forall k \in K_p:\ P_k \circ U_k^1 = Q_k \circ I_k^1,\ P_k \circ \widetilde{U}_k^1 = Q_k \circ \widetilde{I}_k^1\\
&\Longrightarrow P_k(j\omega)U_k^1(0) = Q_k(j\omega)I_k^1(0),\ P_k(j\omega)\widetilde{U}_k^1(0) = Q_k(j\omega)\widetilde{I}_k^1(0).
\end{aligned}
\tag{8.362}
$$

Since $\begin{pmatrix} U \\ I \end{pmatrix}$ and $\begin{pmatrix} U^1 \\ I^1 \end{pmatrix}$ have the same input u, their difference lies in $\mathcal{B}^0$ and hence $\begin{pmatrix} U \\ I \end{pmatrix} - \begin{pmatrix} U^1 \\ I^1 \end{pmatrix} \in \mathcal{F}_-^{K \uplus K}$ and likewise $\begin{pmatrix} \widetilde{U} \\ \widetilde{I} \end{pmatrix} - \begin{pmatrix} \widetilde{U}^1 \\ \widetilde{I}^1 \end{pmatrix} \in \mathcal{F}_-^{K \uplus K}$. Moreover $u,\ \widetilde{u}$ are bounded (for $t \to \infty$), and so are $\begin{pmatrix} U \\ I \end{pmatrix}, \begin{pmatrix} U^1 \\ I^1 \end{pmatrix}, \begin{pmatrix} \widetilde{U} \\ \widetilde{I} \end{pmatrix}, \begin{pmatrix} \widetilde{U}^1 \\ \widetilde{I}^1 \end{pmatrix}$ by Theorem 6.4.5.
The only nonzero entries of the row P_{k-}, $k \in K_p$, of P are (P_k, Q_k). Since $\mathcal{B}^0$ is asymptotically stable and thus $P(j\omega) \in \mathrm{Gl}_\bullet(\mathbb{C})$, the vector $(P_k(j\omega), Q_k(j\omega))$ is nonzero. If $k \in K_p$ and

$$
\begin{cases} P_k(j\omega) \neq 0 \Longrightarrow U_k^1(0) = H_k(j\omega)I_k^1(0),\ \widetilde{U}_k^1(0) = H_k(j\omega)\widetilde{I}_k^1(0) \\ P_k(j\omega) = 0 \Longrightarrow Q_k(j\omega) \neq 0 \underset{(8.362)}{\Longrightarrow} I_k^1(0) = \widetilde{I}_k^1(0) = 0 \Longrightarrow I_k^1 = \widetilde{I}_k^1 = 0. \end{cases}
$$

In particular, if $P_k(j\omega) = 0$, then $\widetilde{U}_k I_k - U_k \widetilde{I}_k = 0 \in \mathcal{F}_-$, cf. (i).
For $k \in K_p$ with $P_k(j\omega) \neq 0$ we obtain

$$
U_k\widetilde{I}_k - H_k(j\omega)I_k^1\widetilde{I}_k^1 = U_k\widetilde{I}_k - U_k^1\widetilde{I}_k^1 = (U_k - U_k^1)\widetilde{I}_k + U_k^1\left(\widetilde{I}_k - \widetilde{I}_k^1\right) \in \mathcal{F}_- \tag{8.363}
$$

since all signals in this equation are polynomial-exponential and bounded and the differences $U_k - U_k^1$ and $\widetilde{I}_k - \widetilde{I}_k^1$ belong to $\mathcal{F}_-$. In the same fashion we get

$$
\forall k \in K_p \text{ with } P_k(j\omega) \neq 0:\ \widetilde{U}_k I_k - H_k(j\omega)I_k^1\widetilde{I}_k^1 \in \mathcal{F}_- \underset{(8.363)}{\Longrightarrow} U_k\widetilde{I}_k - \widetilde{U}_k I_k \in \mathcal{F}_-.
$$

According to (i) this was to be shown. □

Remark 8.3.39 In the literature Theorem 8.3.38 and its proof differ in the following aspects:
(i) The condition $\mathrm{char}(\mathcal{B}^0) \subset \mathbb{C}_-$ is not explicitly stated, but implicitly assumed by the existence of steady (stationary) state trajectories.
(ii) The equations $U_k = H_k(j\omega)I_k,\ \widetilde{U}_k = H_k(j\omega)\widetilde{I}_k,\ k \in K_p$, cf. [17, p. 52], [10, (4.45a,b)], are often assumed, but do not hold in general since steady or stationary state solutions are unique only up to summands in $\mathcal{F}_-^\bullet$. ◇

Theorem 8.3.40 (Reciprocal networks) *In Theorem 8.3.38 assume in addition that $\mathcal{B}$ admits the source current vector I_{K_s} as input, cf. Theorem 8.3.21. Then the transfer matrix H_s^σ from I_{K_s} to U_{K_s} is symmetric, i.e., $\left(H_s^\sigma\right)^\top = H_s^\sigma$, and the network is called* reciprocal. *Recall that for* $j\omega \notin \mathrm{pole}(H_s^\sigma)$ *the matrix* $Z(\omega) := H_s^\sigma(j\omega)$ *is called the* impedance matrix.

Proof We use the data from Theorem 8.3.21, in particular, the transfer matrix H^σ and its submatrix H_s^σ. For $\xi, \widetilde{\xi} \in \mathbb{C}^{K_s}$ and $\omega \in \mathbb{R}$, $j\omega \notin \mathrm{pole}(H^\sigma)$, define the sinusoidal inputs $u^\sigma = I_{K_s} := \xi e^{j\omega t}$ and $\widetilde{u}^\sigma = \widetilde{I}_{K_s} := \widetilde{\xi} e^{j\omega t}$ and the trajectories

$$
\begin{aligned}
&\begin{pmatrix} U \\ I \end{pmatrix} := \begin{pmatrix} H^\sigma(j\omega)\xi \\ \xi \end{pmatrix} e^{j\omega t},\ \begin{pmatrix} \widetilde{U} \\ \widetilde{I} \end{pmatrix} := \begin{pmatrix} H^\sigma(j\omega)\widetilde{\xi} \\ \widetilde{\xi} \end{pmatrix} e^{j\omega t} \in \mathcal{B},\\
&\Longrightarrow U_{K_s} = Z(\omega)\xi e^{j\omega t},\ U_{K_s}^\top = \xi^\top Z(\omega)^\top e^{j\omega t},\ Z(\omega) := H_s^\sigma(j\omega)\\
&\widetilde{U}_{K_s} = Z(\omega)\widetilde{\xi} e^{j\omega t},\ \widetilde{U}_{K_s}^\top = \widetilde{\xi}^\top Z(\omega)^\top \widetilde{\xi} e^{j\omega t}.
\end{aligned}
$$

From Theorem 8.3.38 we infer

$$
\begin{aligned}
&0 = \lim_{t\to\infty} \left(\widetilde{U}_{K_s}^\top I_{K_s} - U_{K_s}^\top \widetilde{I}_{K_s}\right)(t) = \lim_{t\to\infty} \left(\widetilde{\xi}^\top Z(\omega)^\top \xi - \xi^\top Z(\omega)^\top \widetilde{\xi}\right) e^{j\omega t}\\
&\Longrightarrow 0 = \widetilde{\xi}^\top Z(\omega)^\top \xi - \xi^\top Z(\omega)^\top \widetilde{\xi} = \widetilde{\xi}^\top \left(Z(\omega)^\top - Z(\omega)\right)\xi\\
&\quad \text{since } \xi^\top Z(\omega)^\top \widetilde{\xi} = \left(\xi^\top Z(\omega)^\top \widetilde{\xi}\right)^\top = \widetilde{\xi}^\top Z(\omega)\xi\\
&\underset{\xi,\widetilde{\xi}\ \text{arbitrary}}{\Longrightarrow} H_s^\sigma(j\omega)^\top = Z(\omega)^\top = Z(\omega) = H_s^\sigma(j\omega)\\
&\underset{\omega\ \text{arbitrary}}{\Longrightarrow} \left(H_s^\sigma\right)^\top = H_s^\sigma \quad (\omega \in \mathbb{R},\ j\omega \notin \mathrm{pole}(H^\sigma)).
\end{aligned}
$$

□

Remark 8.3.41 A network with two source branches k_1, k_2 is reciprocal if only if $h := H_{k_2 k_1}^\sigma = H_{k_1 k_2}^\sigma$ or $U_{k_2}|_{I_{k_2}=0} = h \circ I_{k_1},\ U_{k_1}|_{I_{k_1}=0} = h \circ I_{k_2}$. The network is called *symmetric* if it is reciprocal and if $H_{k_1 k_1}^v = H_{k_2 k_2}^v$ or, equivalently, if

$$
\left(\begin{pmatrix} U_{k_2} \\ U_{k_1} \end{pmatrix} = H_s^\sigma \circ \begin{pmatrix} I_{k_2} \\ I_{k_1} \end{pmatrix} \text{ and } \begin{pmatrix} U_{k_1} \\ U_{k_2} \end{pmatrix} = H_s^\sigma \circ \begin{pmatrix} I_{k_1} \\ I_{k_2} \end{pmatrix}\right) \text{ or } \begin{pmatrix} 0 & 1 \\ 1 & 0 \end{pmatrix} H^\sigma \begin{pmatrix} 0 & 1 \\ 1 & 0 \end{pmatrix} = H^\sigma.
$$

◊

We finally compute the *average power* of the asymptotically stable IO behavior $\mathcal{B}^\sigma$ from Theorem 8.3.40 for piecewise continuous, T-periodic *real* inputs, but with the following modified notation. Real current and voltage vectors resp. complex ones are denoted by $i = (i_k)_{k\in K_s}, u \in \mathcal{F}_{\mathbb{R}}^{K_s}$ resp. I, U. The behavior $\mathcal{B}_s^\sigma$ is given as

$$
\begin{aligned}
&\mathcal{B} := \mathcal{B}_{s,\mathbb{C}}^\sigma := \left\{\begin{pmatrix} U \\ I \end{pmatrix} \in \mathcal{F}_{\mathbb{C}}^{K_s \uplus K_s};\ P_s^\sigma \circ U = Q_s^\sigma \circ I\right\}\\
&\supset \mathcal{B}_{s,\mathbb{R}}^\sigma = \left\{\begin{pmatrix} u \\ i \end{pmatrix} \in \mathcal{F}_{\mathbb{R}}^{K_s \uplus K_s};\ P_s^\sigma \circ u = Q_s^\sigma \circ i\right\},\\
&P_s^\sigma, Q_s^\sigma \in \mathbb{R}[s]^{K_s\times K_s},\ H_s^\sigma = (P_s^\sigma)^{-1} Q_s^\sigma \in \mathbb{R}(s)^{K_s\times K_s},\\
&H_s^\sigma \text{ proper, } \operatorname{char}\left((\mathcal{B}_s^\sigma)^0\right) \subset \mathbb{C}_-.
\end{aligned}
\tag{8.364}
$$

Since H_s^σ is real, the equation $\overline{H_s^\sigma(jx)} = H_s^\sigma(-jx)$, $x \in \mathbb{R}$, holds. We do not assume that H_s^σ is symmetric as in Theorem 8.3.40. Nonsymmetric H_s^σ occur if the network behavior has more complicated equations as indicated in Remark 8.3.13, 4–5.
We choose a basic frequency $\omega > 0$, the corresponding period $T = 2\pi\omega^{-1}$ and a real piecewise continuous, T-periodic input i with the Fourier series

$$\begin{aligned} i &= a_0 + \sum_{\mu=1}^{\infty}(a_\mu \cos(\mu\omega t) + b_\mu \sin(\mu\omega t)), a_0 = (a_{0,k})_{k\in K_s}, a_\mu, b_\mu \in \mathbb{R}^{K_s} \\ \Longrightarrow i &= \sum_{\mu\in\mathbb{Z}} \mathbb{F}(i)(\mu)e^{j\mu\omega t} \text{ with } \mathbb{F}(i)(0) = a_0 \text{ and} \\ &\forall \mu > 0: \ 2\mathbb{F}(i)(\mu) = a_\mu - jb_\mu,\ 2\mathbb{F}(i)(-\mu) = 2\overline{\mathbb{F}(i)(\mu)} = a_\mu + jb_\mu, \end{aligned} \tag{8.365}$$

cf. Theorems 8.2.102 and 8.2.111. The Fourier series of i converges in the norm $\|-\|_2$. Note that i is real, whereas $\mathbb{F}(i)(\mu)$, $\mu \neq 0$, is complex in general. The assumed asymptotic stability of $\mathcal{B}_s^\sigma$ implies $\mathbb{Z}j\omega \cap \text{pole}(H_s^\sigma) = \emptyset$. According to Theorems 8.2.102 and 8.2.111 the real input i gives rise to the unique T-periodic real output

$$u = H_s^\sigma \circ i = \sum_{\mu\in\mathbb{Z}} \mathbb{F}(u)(\mu)e^{j\mu\omega t},\ \mathbb{F}(u)(\mu) := H_s^\sigma(j\mu\omega)\mathbb{F}(i)(\mu),\ \binom{u}{i} \in \mathcal{B}_{s,\mathbb{R}}^\sigma, \tag{8.366}$$

that is also piecewise continuous. This Fourier series converges in the $\|-\|_2$-norm. If H_s^σ is strictly proper, the series converges uniformly and is continuous. The matrix $Z(\mu\omega) := H_s^\sigma(j\mu\omega)$ is the impedance matrix of $\mathcal{B}_s^\sigma$ and depends on $\mu\omega$.
The assumed asymptotic stability implies that u is a steady-state output to the input i. Assume that $\binom{\widetilde{u}}{i} \in \mathcal{B}_{s,\mathbb{R}}^\sigma$ is any trajectory with input i. Then $\widetilde{u}^\top(t)i(t) = \sum_{k\in K_s} \widetilde{u}_k(t)i_k(t)$ is the total instantaneous power at the n_s source branches, and $T^{-1}\int_{t_0}^{t_0+T} \widetilde{u}^\top(t)i(t)dt$ is the mean value of the power in the time interval $[t_0, t_0 + T]$. Since $\widetilde{u}$ is not periodic in general, this mean value depends on t_0, whereas $T^{-1}\int_{t_0}^{t_0+T} u^\top(t)i(t)dt = T^{-1}\int_0^T u^\top(t)i(t)dt$ does not. However, i is bounded and $\widetilde{u} - u \in \mathcal{F}_-^{K_s} = \oplus_{\lambda\in\mathbb{C},\ \Re(\lambda)<0}\mathbb{C}[t]^{K_s}e^{\lambda t}$. This implies

$$\begin{aligned} &\exists \rho > 0 \exists c_1 > 0 \forall t \geq 0: \ \left|\left((\widetilde{u}^\top - u^\top)i\right)(t)\right| \leq c_1 e^{-\rho t} \\ \Longrightarrow\ &\exists c_2 > 0 \forall t_0 \geq 0: \ T^{-1}\left|\int_{t_0}^{t_0+T} (\widetilde{u}^\top(t) - u^\top(t))i(t)dt\right| \leq c_2 e^{-\rho t_0} \\ \Longrightarrow\ &\lim_{t_0\to\infty} T^{-1}\int_{t_0}^{t_0+T} \widetilde{u}^\top(t)i(t)dt = T^{-1}\int_0^T u^\top(t)i(t)dt. \end{aligned}$$

Therefore, it is customary to compute the average power with the periodic u instead of the actual $\widetilde{u}$. The number

$$\mathcal{P}_r := T^{-1}\int_0^T u^\top(t)i(t)\mathrm{d}t = \sum_{k\in K_s} T^{-1}\int_0^T u_k(t)i_k(t)\mathrm{d}t \tag{8.367}$$

is called the *real or active (average) power* of $\binom{H_s^\sigma \circ i}{i} \in \mathcal{B}_{s,\mathbb{R}}^\sigma$.

Remark 8.3.42 Assume the simple network with a source branch $k_s : v \to w$ and a resistive branch $k_R : v \to w$ with $u_{k_R} = Ri_{k_R},\ R > 0$. Then $u_{k_s} = u_{k_R}$ and $i_{k_s} = -i_{k_R}$. Hence the power

$$u_{k_s}i_{k_s} = -u_{k_R}i_{k_R} = -Ri_{k_R}^2 < 0$$

is negative in this case. This means that at time t energy flows from the source to the port. We extend this interpretation to all multiports. In the literature often $-i_{k_s} = i_{k_R}$ is called the source current. Then the sign of the source power changes, and positivity of the source power means a flow of energy from the source to the port. An analogous interpretation applies to $\mathcal{P}_r$. ◇

Recall the adjoint matrix $M^* := \overline{M}^\top$ of a complex matrix (or vector) M with the rule

$$(M_1M_2)^* = M_2^*M_1^*,\ (M^\top)^* = \overline{M},$$

and also the scalar product of two complex-valued, piecewise continuous, T-periodic functions f, g according to (6.116), i.e.,

$$\langle f, g\rangle = T^{-1}\int_0^T \overline{f}(t)g(t)\mathrm{d}t \underset{(8.127)}{=} \langle \mathbb{F}(f), \mathbb{F}(g)\rangle = \sum_{\mu\in\mathbb{Z}} \overline{\mathbb{F}(f)}(\mu)\mathbb{F}(g)(\mu), \tag{8.368}$$

where $\overline{f}(t) := \overline{f(t)}$, $\overline{\mathbb{F}(f)}(\mu) := \overline{\mathbb{F}(f)(\mu)}$. We conclude

$$\begin{aligned}\mathcal{P}_r &= T^{-1}\int_0^T u^\top(t)i(t)\mathrm{d}t = \sum_{k\in K_s}\langle u_k, i_k\rangle = \sum_{k\in K_s}\langle \mathbb{F}(u_k), \mathbb{F}(i_k)\rangle \\ &= \langle \mathbb{F}(u)^\top, \mathbb{F}(i)\rangle = \sum_{\mu\in\mathbb{Z}} \mathbb{F}(u)(\mu)^*\mathbb{F}(i)(\mu).\end{aligned} \tag{8.369}$$

Since H_s^σ, i and u are real, we obtain, cf. Theorem 8.2.111,

$$\overline{H_s^\sigma(j\mu\omega)} = H_s^\sigma(-j\mu\omega),\ \mathbb{F}(i)(-\mu) = \overline{\mathbb{F}(i)(\mu)},\ \mathbb{F}(u)(-\mu) = \overline{\mathbb{F}(u)(\mu)}.$$

Theorem 8.3.43 (Cf. [10, Sect. 5.5.4] for one-ports) *Consider the IO behavior $\mathcal{B}_s^\sigma$ from* (8.364), *the real, T-periodic, piecewise continuous input current vector i from* (8.365), *the real, periodic output voltage vector u from* (8.366), *and the active or real power $\mathcal{P}_r$ from* (8.367). *Recall*

$$i = a_0 + \sum_{\mu=1}^{\infty} \left(a_\mu \cos(\mu\omega t) + b_\mu \sin(\mu\omega t)\right) = \sum_{\mu\in\mathbb{Z}} \mathbb{F}(i)(\mu) e^{j\mu\omega t},$$

$$a_\mu, b_\mu \in \mathbb{R}^{K_s},\ a_0 = \mathbb{F}(i)(0),\ \forall \mu > 0:\ 2\mathbb{F}(i)(\mu) = a_\mu - jb_\mu,$$

$$\|i\|_2^2 = \langle i^\top, i\rangle = \langle \mathbb{F}(i)^\top, \mathbb{F}(i)\rangle = \sum_{\mu\in\mathbb{Z}} \|\mathbb{F}(i)(\mu)\|_2^2$$

$$= \|a_0\|_2^2 + 2^{-1} \sum_{\mu=1}^{\infty} \left(\|a_\mu\|_2^2 + \|b_\mu\|_2^2\right) \tag{8.370}$$

$$\|u\|_2^2 = \langle u^\top, u\rangle = \langle \mathbb{F}(u)^\top, \mathbb{F}(u)\rangle = \sum_{\mu\in\mathbb{Z}} \| H_s^\sigma(j\mu\omega)\mathbb{F}(i)(\mu)\|_2^2$$

$$= \| H_s^\sigma(0)a_0\|_2^2 + 2\sum_{\mu=1}^{\infty} \| H_s^\sigma(j\mu\omega)\mathbb{F}(i)(\mu)\|_2^2$$

$$= \| H_s^\sigma(0)a_0\|_2^2 + 2\sum_{\mu=1}^{\infty} \mathbb{F}(i)(\mu)^* H_s^\sigma(j\mu\omega)^* H_s^\sigma(j\mu\omega)\mathbb{F}(i)(\mu).$$

The real (average) power is

$$\mathcal{P}_r = a_0^\top H_s^\sigma(0)a_0 + \sum_{\mu=1}^{\infty} \mathbb{F}(i)(\mu)^* \left(H_s^\sigma(j\mu\omega) + H_s^\sigma(j\mu\omega)^*\right) \mathbb{F}(i)(\mu). \tag{8.371}$$

Note that the matrices $K(\omega) := H_s^\sigma(j\mu\omega) + H_s^\sigma(j\mu\omega)^*$ *are Hermitian, i.e.,* $K(\omega)^* = K(\omega)$*, and hence* $\xi^* K(\omega)\xi$ *is real for all* $\xi \in \mathbb{C}^{K_s}$*, especially* $\xi = \mathbb{F}(i)(\mu)$*.*

Proof Recall that $\mathbb{F}(i)(0) = a_0$ and $H(0)$ are real. We use

$$\mathbb{F}(u)(\mu) = H_s^\sigma(j\mu\omega)\mathbb{F}(i)(\mu) \Longrightarrow \mathbb{F}(u)(\mu)^* = \mathbb{F}(i)(\mu)^* H_s^\sigma(j\mu\omega)^*,$$
$$\Longrightarrow \mathbb{F}(u)(\mu)^*\mathbb{F}(i)(\mu) = \mathbb{F}(i)(\mu)^* H_s^\sigma(j\mu\omega)^*\mathbb{F}(i)(\mu)$$
$$\Longrightarrow \mathbb{F}(u)(0)^*\mathbb{F}(i)(0) = \mathbb{F}(i)(0)^* H_s^\sigma(0)^*\mathbb{F}(i)(0) = \mathbb{F}(i)(0)^\top H_s^\sigma(0)^\top\mathbb{F}(i)(0).$$

For $\mu > 0$ we compute

$$\mathbb{F}(i)(-\mu)^* H_s^\sigma(-j\mu\omega)^*\mathbb{F}(i)(-\mu) = \overline{\mathbb{F}(i)(\mu)^* H_s^\sigma(j\mu\omega)^* \mathbb{F}(i)(\mu)}$$
$$= \mathbb{F}(i)(\mu)^\top H_s^\sigma(j\mu\omega)^\top \overline{\mathbb{F}(i)(\mu)} \underset{(-)^\top}{=} \mathbb{F}(i)(\mu)^* H_s^\sigma(j\mu\omega)\mathbb{F}(i)(\mu)$$
$$\Longrightarrow \mathbb{F}(i)(\mu)^* H_s^\sigma(j\mu\omega)^*\mathbb{F}(i)(\mu) + \mathbb{F}(i)(-\mu)^* H_s^\sigma(-j\mu\omega)^*\mathbb{F}(i)(-\mu)$$
$$= \mathbb{F}(i)(\mu)^* \left(H_s^\sigma(j\mu\omega)^* + H_s^\sigma(j\mu\omega)\right)\mathbb{F}(i)(\mu)$$
$$\Longrightarrow \mathcal{P}_r \underset{(8.369)}{=} \sum_{\mu\in\mathbb{Z}} \mathbb{F}(i)(\mu)^* H_s^\sigma(j\mu\omega)^*\mathbb{F}(i)(\mu)$$

$$= a_0^\top H_s^\sigma(0)^\top a_0 + \sum_{\mu=1}^{\infty} \mathbb{F}(i)(\mu)^* \left(H_s^\sigma(j\mu\omega) + H_s^\sigma(j\mu\omega)^*\right)\mathbb{F}(i)(\mu).$$

□

Remark 8.3.44 (Cf. [Wikipedia; https://en.wikipedia.org/wiki/AC_power, July 18, 2019]), [20, Sect. 3.3.4]) Theorem 8.3.43 admits a geometric interpretation in the space $\mathcal{H}_\mathbb{R} := \left(\mathcal{P}_\mathbb{R}^{0,\mathrm{pc}}\right)^{K_s}$ with the scalar product

$$\langle u^\top, i\rangle = \sum_{k\in K_s} T^{-1}\int_0^T u_k(t)i_k(t)\mathrm{d}t,\ \ \|u\|_2^2 = \langle u^\top, u\rangle,$$

which gives rise to various further notions from electrical engineering. We assume $i \neq 0$ and $u \neq 0$. The norm $\|u\|_2$ is called the rms (*root mean square*) or effective voltage of $\mathcal{B}^\sigma$ and likewise for $\|i\|_2$, mainly for the case $n_s = \sharp(K_s) = 1$. If $u \in \mathbb{R}i$, i.e., u and i are $\mathbb{R}$-linearly dependent, then $\mathcal{P}_r = \langle u^\top, i\rangle = +/- \|u\|_2\|i\|_2$. If, in particular, $u = -Ri,\ R > 0$, as for a pure resistor, where u, i are the source data and $-i$ the current through the resistor, one says that u and i are *in phase*, and one obtains $\mathcal{P}_r = -\|u\|_2\|i\|_2 = -R\|i\|_2^2$, the $-$-sign meaning that, on average, energy flows from the source to the resistor. In general, $\mathcal{P}_{app} := \|u\|_2\|i\|_2$ is called the *apparent power* of the pair (u, i).
Now assume that u and i are $\mathbb{R}$-linearly independent, such that $E := \mathbb{R}i + \mathbb{R}u$ is a Euclidean subplane in $(\mathcal{P}_\mathbb{R}^{0,\mathrm{pc}})^{K_s}$. The vectors $u, i \in E$ are an $\mathbb{R}$-basis of E and give rise to the unit vectors $e_u := \|u\|_2^{-1}u$ and $e_i := \|i\|_2^{-1}i$. There are exactly two unit vectors $e_u^\perp$ in the Euclidean plane $E = \mathbb{R}u + \mathbb{R}i$ which, together with e_u, form an orthonormal basis of E. After the choice of $e_u^\perp$ one obtains

$$\begin{aligned}&(e_u, e_i) = (e_u, e_u^\perp)A,\ A := \begin{pmatrix}1 & \cos(\Theta)\\ 0 & \sin(\Theta)\end{pmatrix},\ e_i = e_u\cos(\Theta) + e_u^\perp\sin(\Theta),\\ &0 < \Theta < 2\pi,\ \Theta \neq \pi,\ -1 < \cos(\Theta) < 1,\ \det(A) = \sin(\Theta) \neq 0.\end{aligned}$$

We choose the unique $e_u^\perp$ with $\det(A) = \sin(\Theta) > 0$ or $0 < \Theta < \pi$. This means that the $\mathbb{R}$-bases (u, i), (e_u, e_i) and $(e_u, e_u^\perp)$ of E have the same orientation. We define the *reactive power* $\mathcal{P}_{react} := \mathcal{P}_{app}\sin(\Theta) > 0$ and conclude

$$\begin{aligned}&\cos(\Theta) = \langle e_u^\top, e_i\rangle = \langle u^\top, i\rangle\|u\|_2^{-1}\|i\|_2^{-1} = \mathcal{P}_r\mathcal{P}_{app}^{-1},\\ &\Longrightarrow e_i = e_u\mathcal{P}_r\mathcal{P}_{app}^{-1} + e_u^\perp\mathcal{P}_{react}\mathcal{P}_{app}^{-1},\ \mathcal{P}_{app}^2 = \mathcal{P}_r^2 + \mathcal{P}_{react}^2.\end{aligned} \tag{8.372}$$

The number $\cos(\Theta) = \mathcal{P}_r\mathcal{P}_{app}^{-1}$ is called the *power factor*.
Define $|u|_k(t) := |u_k(t)|$, hence $|u| \in \mathcal{F}_\mathbb{R}^{K_s}$ and likewise $|i| \in \mathcal{F}_\mathbb{R}^{K_s}$. Then

$$\||u|\|_2 = \|u\|_2,\ \||i|\|_2 = \|i\|_2$$
$$\Longrightarrow 0 \le \langle |u|^\top, |i|\rangle = \sum_{k\in K_s} T^{-1}\int_0^T |u_k(t)||i_k(t)|dt \le \|u\|_2\|i\|_2 = \mathcal{P}_{app}. \tag{8.373}$$

So $\mathcal{P}_{app}$ is an upper bound for the average of the instantaneous powers

$$\sum_{k\in K_s} |p_k(t)| = \sum_{k\in K_s} |u_k(t)||i_k(t)|.$$

Even if $\cos(\Theta) = 0$ and hence $\mathcal{P}_r = 0$, there may be large instantaneous powers. Hence $\mathcal{P}_{app}$ and $\mathcal{P}_{react}^2 = \mathcal{P}_{app}^2 - \mathcal{P}_r^2$ are important numbers for the construction of large networks. ◊

Corollary 8.3.45 (Cf. [20, (3.68)]) *If $n_s = \sharp(K_s) = 1$, i.e., if $\mathcal{B}^\sigma$ is a one-port with $0 \ne P^\sigma, Q^\sigma \in \mathbb{R}[s]$ and $H^\sigma = (P^\sigma)^{-1} Q^\sigma$, then Eqs.* (8.370) *and* (8.371) *are simpler, viz.,*

$$\begin{aligned}
\|i\|_2^2 &= \sum_{\mu\in\mathbb{Z}} |\mathbb{F}(i)(\mu)|^2 = a_0^2 + 2^{-1}\sum_{\mu=1}^{\infty}(a_\mu^2 + b_\mu^2)\\
\|u\|_2^2 &= \sum_{\mu\in\mathbb{Z}} |H^\sigma(j\mu\omega)|^2|\mathbb{F}(i)(\mu)|^2 = H^\sigma(0)^2 a_0^2 + 2^{-1}\sum_{\mu=1}^{\infty}|H^\sigma(j\mu\omega)|^2(a_\mu^2 + b_\mu^2)\\
\mathcal{P}_r &= H^\sigma(0)a_0^2 + \sum_{\mu=1}^{\infty}\left(H^\sigma(j\mu\omega) + \overline{H^\sigma(j\mu\omega)}\right)|\mathbb{F}(i)(\mu)|^2\\
&= H^\sigma(0)a_0^2 + 2^{-1}\sum_{\mu=1}^{\infty}\Re\left(H^\sigma(j\mu\omega)\right)(a_\mu^2 + b_\mu^2).
\end{aligned} \tag{8.374}$$

The impedance *of the one-port at the frequency $\mu\omega$ is*

$$Z(\mu\omega) := H^\sigma(j\mu\omega) = \Re(H^\sigma(j\mu\omega)) + j\Im(H^\sigma(j\mu\omega)) =: R(\mu\omega) + jX(\mu\omega), \tag{8.375}$$

where its real resp. imaginary part $R(\mu\omega)$ resp. $X(\mu\omega)$ are called the resistance *resp. the* reactance *of the port. The equation for $\|u\|_2^2$ resp. for $\mathcal{P}_r$ contains the factors $|Z(\mu\omega)|^2 = R(\mu\omega)^2 + X(\mu\omega)^2$ resp. $R(\mu\omega)$. The expression from* (8.374) *for $\mathcal{P}_r$ is a far-reaching generalization of the power formula $\mathcal{P}_r = -R\|i\|_2^2 = -R^{-1}\|u\|_2^2$, $R > 0$, for a pure resistance.* ◊

If in Corollary 8.3.45 $-i$ is considered as the source current, as is frequently done, then i, H^σ, and $\mathcal{P}_r$ change their sign, but the derived equations still hold.

Example 8.3.46 (Cf. [20, Sect. 2.7]) In the situation of Corollary 8.3.45 we consider the simplest, but very important standard case of a one-port $\mathcal{B}^\sigma$ with the sinusoidal current input

$$
\begin{aligned}
i &= a_1\cos(\omega t) + b_1\sin(\omega t) = \mathbb{F}(i)(1)e^{j\omega t} + \mathbb{F}(i)(-1)e^{-j\omega t} \\
&= \Re\left(2\mathbb{F}(i)(1)e^{j\omega t}\right) = \Re\left((a_1 - jb_1)e^{j\omega t}\right) \text{ with} \\
2\mathbb{F}(i)(1) &= a_1 - jb_1,\ \|i\|_2^2 = 2^{-1}(a_1^2 + b_1^2).
\end{aligned}
$$

The impedance at ω is

$$
\begin{aligned}
&Z := H(j\omega) =: R + jX = |Z|e^{j\Theta},\ |Z| = \left(R^2 + X^2\right)^{1/2}, \\
&\cos(\Theta) = R/|Z|,\ \sin(\Theta) = X/|Z|.
\end{aligned}
$$

We infer

$$
\begin{aligned}
u &= H(j\omega)\mathbb{F}(i)(1)e^{j\omega t} + H(-j\omega)\mathbb{F}(i)(-1)e^{-j\omega t} = \Re\left(2Z\mathbb{F}(i)(1)e^{j\omega t}\right) \\
&= |Z|\Re(2\mathbb{F}(i)(1)e^{j(\omega t+\Theta)}),\ \|u\|_2^2 = 2^{-1}|Z|^2(a_1^2 + b_1^2), \\
\mathcal{P}_r &= 2^{-1}R(a_1^2 + b_1^2),\ \mathcal{P}_{app} = \|i\|_2\|u\|_2 = |Z|2^{-1}(a_1^2 + b_1^2), \\
\mathcal{P}_r\mathcal{P}_{app}^{-1} &= R/|Z| = \cos(\Theta),\ \mathcal{P}_{react}^2 = \mathcal{P}_{app}^2 - \mathcal{P}_r^2 = \mathcal{P}_{app}^2\sin^2(\Theta).
\end{aligned}
$$

Notice that $(a_1 - jb_1)e^{j\omega t}$ resp. $Z(a_1 - jb_1)e^{j\omega t}$ are the complexifications of the real current resp. voltage i resp. u according to Lemma 6.3.28.
Up to possibly the sign, $\sin(\Theta)$ coincides with the positive one from Remark 8.3.44. Only in the special case of the present example with only one frequency ω and one impedance $Z = R + jX = |Z|e^{j\Theta}$ one could *define* $\mathcal{P}_{react} := \mathcal{P}_{app}X|Z|^{-1}$ that coincides with the positive one from Remark 8.3.44, up to possibly the sign. ◊

8.4 Mechanical Networks

In this section we show that certain mechanical networks satisfy the same equations as electrical networks, and, above all, Kirchhoff's circuit and node laws. We give the electrical quantities a new mechanical meaning. Cf. [21], [20, Sect. 3.1, pp. 65–103, Chap. 4, pp. 135–142], [22, Sect. 2.3.4], [Wikipedia; https://en.wikipedia.org/wiki/Mechanical-electrical_analogies, August 12, 2019]. We discuss *translational mechanical networks* in detail only.

The notations and mathematical assumptions are those of the real electrical networks in Sect. 8.3:

$$
(V, K'),\ \binom{u}{i} \in \mathcal{F}_F^{K' \uplus K'},\ u = (u_k)_{k\in K'},\ i = (i_k)_{k\in K'},\ F := \mathbb{R}. \tag{8.376}
$$

Here (V, K') is a connected directed graph with branches (edges) $k : v \to w$ in K' between nodes (vertices) v and w from V. The space $F^{K'}$ has the standard basis $(k)_{k\in K'} \in F^{K'} \subset \mathcal{F}_F^{K'}$, hence $i = (i_k)_{k\in K'} = \sum_{k\in K'} i_k k$. The u_k resp. i_k will be reinterpreted as velocities resp. forces. This furnishes the so-called *mobility or*

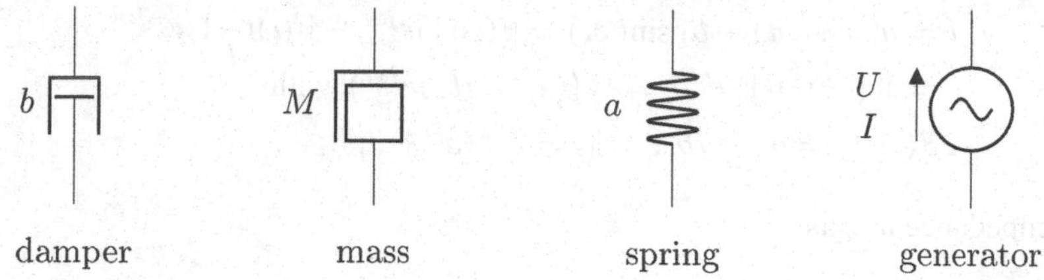

Fig. 8.6 The symbols of the standard components of mechanical networks

Firestone analogy of electrical networks, cf. [21]. Recall that real networks are mostly studied via complexification.

As with electrical networks, also the mechanical ones are visualized as a valued, connected, directed graph. The standard symbols for the branches are shown in Fig. 8.6.

1. **The graph**: We assume that (V, K') is connected and that for every node v there are at least two arrows (directed edges or branches) with domain or codomain v, but without arrows $k : v \to v$. In particular, this implies that every branch belongs to a *circuit or loop*. Moreover we assume a distinguished node v_0 that is called *ground* or *earth*. It is either fixed in an inertial reference frame or moves with constant velocity in it. Every arrow $k : v \to w$ gives rise to its inverse $-k : w \to v$ in $F^{K'}$.

2. **Points on the real line where forces are applied**: We assume a map

$$V \to \mathrm{C}^1(\mathbb{R}, \mathbb{R}),\ v \mapsto x_v,\ x_v(t) \in \mathbb{R}. \tag{8.377}$$

Then $x_v(t)$ is interpreted as the *position* of the node v on the real axis $\mathbb{R}$ at time t to which a force is applied, see below. The points $x_v(t) \in \mathbb{R}$ are those of the *inertial reference frame* $\mathbb{R}$. By physical conditions we ensure that $x_v(t) \neq x_w(t)$ for $v \neq w$. Mathematically, this is ensured by appropriate initial conditions. Solutions (see below) with $x_w(t) = x_v(t)$ are collisions and destroy the mechanical network. This corresponds to undesired impulsive solutions in electrical networks, cf. Sect. 8.2.10. That all $x_v(t)$ lie on one real line $\mathbb{R}$ and not in the three-dimensional space $\mathbb{R}^3$ explains the attribute *translational* of the treated mechanical networks.

By assumption, the coordinate x_{v_0} satisfies

$$s^2 \circ x_{v_0} = 0 \text{ or even } s \circ x_{v_0} = 0. \tag{8.378}$$

We also assume

$$x_{v_0} < x_v,\ v \neq v_0. \tag{8.379}$$

3. **Masses**: At each point $x_v(t)$, $v \neq v_0$, there is a mass $M_v \geq 0$. The condition $M_v = 0$ is interpreted as absence of the mass. For each $M_v > 0$ we add a branch $k_v : v_0 \to v$ and obtain the final branch set

$$K := K' \uplus \{k_v;\ v_0 \neq v \in V,\ M_v > 0\},\ n := \sharp\{v \in V;\ v_0 \neq v,\ M_v > 0\}. \tag{8.380}$$

So we deal with an n-body problem, where the bodies move along a line.

4. **Distances, velocities, and Kirchhoff's circuit law**: If $k : v \to w$ is an arrow we define the distances resp. velocities

$$\begin{aligned} &x_k(t) := x_w(t) - x_v(t),\ u_k := s \circ x_k = x_k',\ x_{-k} := -x_k,\ u_{-k} := -u_k,\\ &x := (x_k)_{k\in K},\ u := (u_k)_{k\in K} \in \mathcal{F}_F^K \underset{\text{ident.}}{=} \mathrm{Hom}_F(F^K, \mathcal{F}_F). \end{aligned} \tag{8.381}$$

The velocity difference u_k is the analogue of the voltage (=potential difference) between the points v and w. In standard electrical networks the *position* of the terminals in space is not considered.

Lemma 8.4.1 (Kirchhoff's circuit law) *If*

$$v_1 \xrightarrow{k_1} v_2 \longrightarrow \cdots \xrightarrow{k_{n-1}} v_n = v_1,\ k_\mu \in K \cup (-K),\ c := \sum_{\mu=1}^{n-1} k_\mu \in \mathbb{Z}^K \subset F^K,$$

is a closed path, then

$$x(c) := \sum_{\mu=1}^{n-1} x_{k_\mu} = 0,\ \textit{hence}\ u(c) := \sum_{\mu=1}^{n-1} u_{k_\mu} = 0.$$

Thus both x and u satisfy the analogues of Kirchhoff's voltage law.

Proof $\sum_{\mu=1}^{n-1} x_{k_\mu} = \sum_{\mu=1}^{n-1}\left(x_{v_{\mu+1}} - x_{v_\mu}\right) = x_{v_n} - x_{v_1} \underset{v_n=v_1}{=} 0.$ □

Often there is another node v_1 like v_0 with the constant velocity $s \circ x_{v_1} = s \circ x_{v_0}$ in the inertial reference frame. In this situation we always assume

$$x_{v_0} < x_v < x_{v_1},\ v \neq v_0, v_1. \tag{8.382}$$

Thus the mechanical network is situated between two points x_{v_0} and x_{v_1} that are fixed or move with the same constant velocity in an inertial reference frame. If $\widetilde{k} : v \to v_1$ is any arrow, then

$$u_{\widetilde{k}} = s \circ x_{\widetilde{k}} = s \circ (x_{v_1} - x_v) = s \circ (x_{v_0} - x_v).$$

In this situation we *replace* $\widetilde{k}$ by a new branch

$$k : v \to v_0 \Longrightarrow u_k := s \circ (x_{v_0} - x_v) = s \circ (x_{v_1} - x_v) = u_{\widetilde{k}}. \tag{8.383}$$

In K' we omit $\widetilde{k}$ and add k. This permits to replace v_1 by v_0 in the equations. This is important for the assumption that each node v has two arrows with domain or

codomain v and for Kirchhoff's node law, see below.
5. **Forces**: In the pairs (x_k, i_k), (u_k, i_k), $k \in K$, the function $i_k(t)$ is interpreted as a force on the point $x_v(t)$. We write

$$i_k = f_k \epsilon_k,\ \epsilon_k := (x_w(t) - x_v(t))|x_w(t) - x_v(t)|^{-1},\ \epsilon_{-k} := -\epsilon_k,\ \epsilon_k \in \{+/-1\}. \tag{8.384}$$

Note that 1 and -1 are the unique unit vectors in $\mathbb{R}$. Then i_k is interpreted as a force on the point x_v of size f_k in direction ϵ_k. If $f_k < 0$ then $i_k = (-f_k)\epsilon_{-k}$ is a positive force of size $-f_k$ in direction ϵ_{-k}. Recall

$$-k : w \to v,\ \epsilon_{-k} = -\epsilon_k \Longrightarrow -i_k = i_{-k} = -f_k\epsilon_k = f_k\epsilon_{-k}.$$

According to Newton's *action=reaction* law the force $-i_k = f_k\epsilon_{-k}$ is of size f_k on $x_w(t)$ in direction $x_v(t)$ or, more precisely, in direction $-\epsilon_k = \epsilon_{-k}$.
For $v \neq v_0$ the total force i_v on $x_v(t)$ (in direction 1) resp. *Newton's law* are

$$\begin{aligned} i_v :&= \sum \left\{i_k;\ k : v \to \bullet,\ k \in K'\right\} - \sum \left\{i_k;\ k : \bullet \to v,\ k \in K'\right\} \\ &= \sum \left\{i_k;\ k : v \to \bullet,\ k \in K' \uplus (-K')\right\} \text{ resp. } M_v s^2 \circ x_v = i_v. \end{aligned} \tag{8.385}$$

Consider, in particular, an arrow $\widetilde{k} : v \to v_1$ with $x_{v_0} < x_v < x_{v_1}$ as in (8.383) and its replacement $k : v \to v_0$ with $u_{\widetilde{k}} = u_k$. We define $i_k := i_{\widetilde{i_k}}$. Then (8.385) remains true with $k : v \to v_0$ instead of $\widetilde{k} : v \to v_1$. Summing up, we obtain

$$u_k = u_{\widetilde{k}},\ \epsilon_{\widetilde{k}} = 1,\ \epsilon_k = -1,\ i_k = i_{\widetilde{k}} = (-i_k)\epsilon_k \text{ if } \widetilde{k} : v \to v_1,\ k : v \to v_0. \tag{8.386}$$

In the sequel we consider graphs (V, K) with *one distinguished* node v_0, called *ground*, only.
6. **Kirchhoff's node law**: For every $v \neq v_0$ and $M_v > 0$ we use the additional arrow $k_v : v_0 \to v$ in K and obtain

$$\begin{aligned} &x_{k_v} := x_v - x_{v_0},\ u_{k_v} = s \circ x_{k_v} := s \circ x_v - s \circ x_{v_0}, \\ &s \circ u_{k_v} = s^2 \circ x_{k_v} \underset{s^2 \circ x_{v_0} = 0}{=} s^2 \circ x_v. \end{aligned} \tag{8.387}$$

We define the force i_{k_v} on x_{v_0} in direction x_v as

$$i_{k_v} := M_v s^2 \circ x_v = M_v s^2 \circ x_{k_v} = M_v s \circ u_{k_v}. \tag{8.388}$$

The reactive *inertial force* on x_v in direction v_0 is $-i_{k_v} = -M_v s^2 \circ x_v$. Equation (8.385) becomes

$$
\begin{aligned}
0 &= -M_v s^2 \circ x_v + i_v \underset{-i_{k_v}=i_{-k_v}}{=} i_{-k_v} + \sum \{i_k;\ k : v \to \bullet,\ k \in K' \uplus (-K')\} \\
&= \sum \{i_k;\ k : v \to \bullet,\ k \in K \uplus (-K)\},\ v \neq v_0,\ M_v > 0.
\end{aligned}
\tag{8.389}
$$

This implies Kirchhoff's node law for all $v \neq v_0$ and $M_v > 0$. For $v \neq v_0$ and $M_v = 0$ the node law for v follows directly from (8.385). For the exceptional node v_0 we use

$$
\begin{aligned}
&\sum_{k \in K \uplus (-K)} i_k \underset{i_{-k}=-i_k}{=} \sum_{k \in K} (i_k - i_k) = 0 \\
&\Longrightarrow 0 = \sum_{v \in V} \sum \{i_k;\ k : v \to \bullet,\ k \in K \uplus (-K)\} \\
&\qquad = \sum \{i_k;\ k : v_0 \to \bullet,\ k \in K \uplus (-K)\} \\
&\qquad + \sum_{v_0 \neq v \in V} \sum \{i_k;\ k : v \to \bullet,\ k \in K \uplus (-K)\} \\
&\qquad \underset{(8.389)}{=} \sum \{i_k;\ k : v_0 \to \bullet,\ k \in K \uplus (-K)\} \\
&\Longrightarrow \sum \{i_k;\ k : v_0 \to \bullet,\ k \in K \uplus (-K)\} = 0.
\end{aligned}
\tag{8.390}
$$

Corollary 8.4.2 *Kirchhoff's circuit law holds for the distance vector $x = (x_k)_{k \in K} \in \mathcal{F}_F^K$ and the velocity vector $u := (u_k)_{k \in K} \in \mathcal{F}_F^K$. Kirchhoff's node law holds for the force vector $i = (i_k)_{k \in K} \in \mathcal{F}_F^K$.* ◊

Remark 8.4.3 For mechanical networks Kirchhoff's circuit law results from its mathematical definition, and not from physics. The node law is a consequence of Newton's equation (8.385), and the purely mathematical introduction of the additional (artificial) arrow $k_v : v_0 \to v$ with $M_v > 0$ and the force $i_{k_v} := M_v s^2 \circ x_v$. ◊

Next we discuss the forces $i_k,\ k \in K',\ k \neq k_v$, that do not arise from a mass.
7. **Viscuous friction elements, damping**: A branch $k : v \to w$ in K' is called a *viscuous friction or damping branch* if a *friction element* is situated between the points $x_v(t)$ and $x_w(t)$. It is characterized by the equation

$$
b u_k = i_k,\ b > 0,\ k : v \to w. \tag{8.391}
$$

The numbers b resp. b^{-1} are called the *friction* resp. *lubricity* of the friction element. The equation $u_k = b^{-1} i_k$ corresponds to Ohm's law. If $u_k(t) = (s \circ x_k)(t)$ is positive and thus $x_k(t) = x_w(t) - x_v(t)$ increases, the force i_k on x_v is positive in direction to x_w. There are other types of friction that are not considered in this model.
Assume a friction branch $\widetilde{k} : v \to v_1$ and $i_{\widetilde{k}} = b u_{\widetilde{k}},\ b > 0$, and its replacement

$$
k : v \to v_0 \text{ with } u_k = u_{\widetilde{k}} \text{ and } i_k = i_{\widetilde{k}} = b u_{\widetilde{k}} = b u_k. \tag{8.392}
$$

Hence the equation for k is the same as for $\widetilde{k}$.
8. **Springs, spring branches**: A *spring branch* $k : v \to w$ is given by a *spring* that is situated between the points $x_v(t)$ and $x_w(t)$. The corresponding force is given by Hooke's law

$$i_k = a(x_k - x_{k,0}),\ a > 0,\ x_{k,0} \in \mathbb{R},\ \text{or}\ au_k = s \circ i_k. \tag{8.393}$$

The numbers a resp. a^{-1} are called the *spring constant* resp. the *compliance* of the spring. Obviously $u_k = a^{-1} s \circ i_k$ corresponds to an inductivity.
In particular, assume that $\widetilde{k} : v \to v_1$ has been replaced by $k : v \to v_0$. By definition $s \circ i_k = s \circ i_{\widetilde{k}} = au_{\widetilde{k}} = au_k$ holds. The equation for i_k is

$$\begin{aligned} i_k &:= i_{\widetilde{k}} = a\left(x_{\widetilde{k}} - x_{\widetilde{k},0}\right) = a\left(x_k - x_{k,0}\right) \\ \text{with } x_{\widetilde{k}} &= x_{v_1} - x_v,\ x_k = x_{v_0} - x_v,\ x_{k,0} := x_{\widetilde{k},0} + x_{v_0} - x_{v_1}. \end{aligned}$$

9. **Force or velocity generators**: In most practical situations, these are supported in v_0 and therefore, like the masses, belong to arrows $k : v_0 \to v$. Then $i_k > 0$ means a positive force on x_{v_0} in direction v and the reactive negative force $-i_k$ on x_v. The velocity resp. force generator furnishes u_k resp. i_k, whereas the corresponding i_k resp. u_k follow from the network equations. The generator branches are the analogues of the source branches in electrical networks.

Example 8.4.4 1. The replacement of arrows $\widetilde{k} : v \to v_1$ by $k : v \to v_0$ is necessary in order that the node law holds. Indeed, consider a mechanical network with three nodes and three branches, i.e.,

$$\begin{aligned} &V = \{v_0, v, v_1\},\ K = \left\{k_v : v_0 \to v,\ k_1 : v_0 \to v,\ \widetilde{k}_2 : v \to v_1\right\}, \\ &s \circ x_{v_0} = s \circ x_{v_1} = 0,\ x_{v_0} < x_v < x_{v_1}. \end{aligned}$$

Assume a mass branch k_v and two spring branches $k_1, \widetilde{k}_2$ with the equations

$$i_{k_v} = Ms \circ u_{k_v},\ s \circ i_{k_1} = a_1 u_{k_1},\ s \circ i_{\widetilde{k}_2} = a_2 u_{\widetilde{k}_2}.$$

The node law does not hold for this graph, since v_1 with only one branch to it would imply $i_{\widetilde{k}_2} = 0$. Instead, we replace

$$\begin{aligned} &\widetilde{k}_2 \text{ by } k_2 : v \to v_0 \text{ with } u_{k_2} = u_{\widetilde{k}_2},\ s \circ i_{k_2} = s \circ i_{\widetilde{k}_2} = a_2 u_{k_2} \\ &\Longrightarrow V = \{v_0, v\},\ K = \{k_v : v_0 \to v,\ k_1 : v_0 \to v,\ k_2 : v \to v_0\}. \end{aligned}$$

Kirchhoff's node law at v_0 and circuit law imply

$$\begin{aligned}
&0 = i_{k_v} + i_{k_1} - i_{k_2},\ u := u_{k_v} = u_{k_1} = -u_{k_2}\\
&\Longrightarrow 0 = s \circ i_{k_v} + s \circ i_{k_1} - s \circ i_{k_2} = Ms^2 \circ u + a_1 u - a_2(-u) \Longrightarrow\\
&\left(s^2 + \omega^2\right) \circ u = 0,\ \omega^2 := M^{-1}(a_1 + a_2),\ u = u_+ e^{j\omega t} + u_- e^{-j\omega t},\ \overline{u_+} = u_-,\\
&\Longrightarrow x_v(t) = x_v(0) + (j\omega)^{-1}\left(u_+ e^{j\omega t} - u_- e^{-j\omega t}\right) - (j\omega)^{-1}(u_+ - u_-).
\end{aligned}$$

Of course, the node equation at v arises by differentiation from Newton's equation $Mx_v'' = -a_1\left(x_v - x_{v_0}\right) + a_2\left(x_{v_1} - x_v\right)$, but the derivation above applies and confirms the general algorithm.
2. Consider the simple graph with two branches $k_1, k_2 : v_0 \to v$ where k_1 is a mass branch with the equation $ms \circ u_1 = i_1$ and k_2 a generator branch with input i_2. Let $x_0 = x_{v_0},\ x := x_v$, hence $u_1 = s \circ (x - x_0)$. Then $u := u_1 = u_2,\ i := i_2 = -i_1,\ ms \circ u = -i$. For $i = -mg$ with the earth gravity constant g these data furnish

$$s \circ u = g,\ u(t) = u(0) + gt,\ x(t) - x_0(t) = x(0) - x_0(0) + u(0)t + 2^{-1}gt^2.$$

Notice that $ms \circ u = mg$ only holds if $x_0(t)$ is a point with constant velocity in an *inertial reference frame*. ◊

Remark 8.4.5 1. All results of Sect. 8.3 can be applied to the mechanical networks described above. The variables u, i.e., voltage or velocity, resp. i, i.e., current or force, are also called the *difference (across, effort)* variable resp. the *through (flow)* variable [22, p. 95]. Newton's *action=reaction* law for a branch $k : v \to w$, i.e., , that $i_k = f\epsilon_k$ is the force on x_v in direction x_w and $-i_k = f\epsilon_{-k}$ is the reactive force on x_w in direction x_v, enables the interpretation of the force as a through variable like the current. Since in mechanics the force i rather than the velocity has the physical, but not the mathematical, aspect of an effort, certain derived notions get different names. If, for instance, $\binom{u}{i}$ are the variables of the source branch of a one-port and H is the transfer function from i to u and H^{-1} that from u to i, then $H(j\omega)^{-1}$ is called the *electrical admittance*, but the *mechanical impedance*. In the correspondence $voltage \leftrightarrow force$ one therefore talks about the *impedance analogy*, cf. [Wikipedia; https://en.wikipedia.org/wiki/Impedance_analogy, April 12, 2019]. Notice, however, that the mathematics of Sects. 8.3 and 8.4 does not change at all by this different terminology. Due to the special role of the ground node v_0 in mechanical networks, the latter are more special than general electrical networks.
2. With the basic SI units, cf. [Wikipedia; https://en.wikipedia.org/wiki/International_System_of_Units, September 7, 2019], second s, meter m, kilogram kg, and ampère A, the units of velocity, force, energy, power, and voltage are

$$\begin{aligned}
&\text{velocity: m} \cdot \text{s}^{-1},\\
&\text{force: newton } 1 \cdot \text{N} = 1 \cdot \text{kg} \cdot \text{m} \cdot \text{s}^{-2},\\
&\text{energy, work, heat: joule } 1 \cdot \text{J} = 1 \cdot \text{N} \cdot \text{m} = 1 \cdot \text{kg} \cdot \text{m}^2 \cdot \text{s}^{-2},\\
&\text{power: watt } 1 \cdot \text{W} = 1 \cdot \text{J} \cdot \text{s}^{-1} = 1 \cdot \text{kg} \cdot \text{m}^2 \cdot \text{s}^{-3},\\
&\text{voltage: volt } 1 \cdot \text{V} = 1 \cdot \text{W} \cdot \text{A}^{-1}.
\end{aligned} \tag{8.394}$$

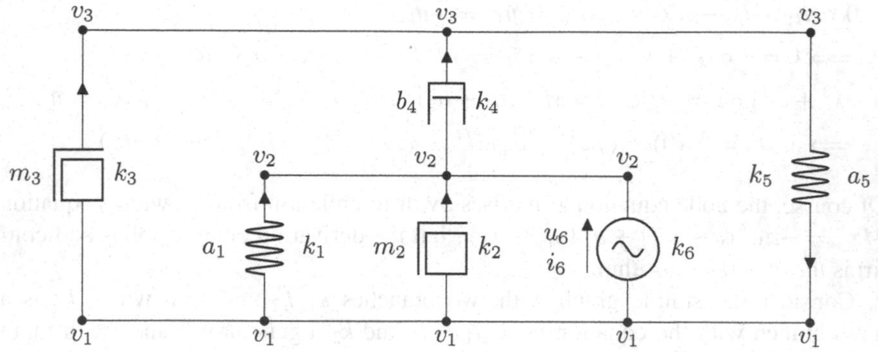

Fig. 8.7 The mechanical network from Example 8.4.6

3. Kirchoff's circuit law also holds for the position vector $x = (x_k)_{k \in K}$. However, if $k : v \to w$ is a spring branch and $i_k - a(x_k - x_{k,0}) = 0$ and $s \circ i_k - a u_k = 0$, then the second equation is of the behavioral type $P \circ \binom{i_k}{x_k} = 0,\ P \in \mathbb{F}[s]^{1\times 2}$, but the first is not. Therefore the trajectories $\binom{x}{i}$ in mechanics are not analogous to the trajectories $\binom{u}{i}$ of an electrical network, and Sect. 8.3 is not generally applicable to them.
4. Since $s \circ x_k = u_k$ and hence

$$x_k(t) = x_k(0) + \int_0^t u_k(\tau)\mathrm{d}\tau,\ x_v(t) = x_{v_0}(t) + \sum_{i=1}^{n} x_{k_i}(t) \text{ where}$$
$$v \in V,\ v_0 \xrightarrow{k_1} v_1 \xrightarrow{k_2} \cdots \xrightarrow{k_n} v_n = v,\ k_i \in K \uplus (-K), \tag{8.395}$$

the absolute positions $x_v(t)$ can be computed from $\binom{u}{i}$.
5. Smith (Cambridge, 2002) invented mechanical branch elements with the equations $i_k = M s \circ u_k$ of the mass branches, where, however, $k : v \to w$ is any arrow with arbitrary moving points x_v, x_w and $u_v = s \circ x_v$ is *not constant*. ◊

Example 8.4.6 We consider the mechanical network from Fig. 8.7 and its graph. The data are

$$V = \{v_1, \ldots, v_4\},\ s \circ x_{v_1} = s \circ x_{v_4} = 0,\ K = \{k_1, \ldots, k_6\}. \tag{8.396}$$

Here $x_{v_1} < x_{v_4}$ are the fixed left and right boundary of the network. The arrows k_μ follow from Fig. 8.7. The arrow $k_5 : v_3 \to v_1$ replaces the arrow $\widetilde{k_5} : v_3 \to v_4$ according to (8.383). The branches are specified as follows:

$$
\begin{aligned}
&\{k_2 : v_1 \to v_2, k_3 : v_1 \to v_3\} = \{\text{mass branches}\}\,,\\
&\{k_1 : v_1 \to v_2, k_5 : v_3 \to v_1\} = \{\text{spring branches}\}\,,\\
&(k_4 : v_2 \to v_3) = \text{friction branch},\ \ (k_6 : v_1 \to v_2) = \text{generator (source) branch.}
\end{aligned}
\tag{8.397}
$$

The velocity resp. the force of the branch k_μ is denoted by $u_\mu := u_{k_\mu}$ resp. $i_\mu := i_{k_\mu}$. The branch equations of the nongenerator branches $k_1, \ldots, k_5$ are

$$
m_2 s \circ u_2 = i_2,\ m_3 s \circ u_3 = i_3,\ a_1 u_1 = s \circ i_1,\ b_4 u_4 = i_4,\ a_5 u_5 = s \circ i_5,
\tag{8.398}
$$

where the m_2, m_3 are positive masses, a_1, a_5 are positive spring constants, and b_4 is a positive friction. The behavior of this network is

$$
\mathcal{B} := \left\{ \binom{u}{i} \in \mathcal{F}^{6+6};\ (8.398) \text{ and Kirchhoff's laws hold} \right\}.
\tag{8.399}
$$

The incidence matrix $A \in \mathbb{Z}^{V \times K}$ according to (8.199) is

$$
A := \begin{array}{c} \\ v_1 \\ v_2 \\ v_3 \end{array}
\begin{array}{c} \begin{array}{cccccc} k_1 & k_2 & k_3 & k_4 & k_5 & k_6 \end{array} \\
\begin{pmatrix} 1 & 1 & 1 & 0 & -1 & 1 \\ -1 & -1 & 0 & 1 & 0 & -1 \\ 0 & 0 & -1 & -1 & 1 & 0 \end{pmatrix} \end{array}.
\tag{8.400}
$$

In the sequel we use the results and notations from (8.206) to (8.235). A sequence of simple elementary row operations and column permutations on A furnishes the matrix

$$
\begin{aligned}
&\widetilde{A} := \begin{array}{c} \begin{array}{cccccc} k_1 & k_3 & k_2 & k_4 & k_5 & k_6 \end{array} \\
\begin{pmatrix} 1 & 0 & 1 & -1 & 0 & 1 \\ 0 & 1 & 0 & 1 & -1 & 0 \\ 0 & 0 & 0 & 0 & 0 & 0 \end{pmatrix} \end{array} \Longrightarrow \operatorname{rank}(A) = 2,\\
&K = K_1 \uplus K_2,\ K_1 = \{k_1, k_3\}\,,\ K_2 = \{k_2, k_4, k_5, k_6\}\,,\\
&M := \begin{array}{c} \\ k_1 \\ k_3 \end{array}
\begin{array}{c} \begin{array}{cccc} k_2 & k_4 & k_5 & k_6 \end{array} \\
\begin{pmatrix} 1 & -1 & 0 & 1 \\ 0 & 1 & -1 & 0 \end{pmatrix} \end{array} \in \mathbb{R}^{K_1 \times K_2}.
\end{aligned}
\tag{8.401}
$$

We verify $A|_{K_2} = A|_{K_1} M$ with

$$
\begin{aligned}
&A|_{K_2} := \begin{array}{c} \\ v_1 \\ v_2 \\ v_3 \end{array}
\begin{array}{c} \begin{array}{cccc} k_2 & k_4 & k_5 & k_6 \end{array} \\
\begin{pmatrix} 1 & 0 & -1 & 1 \\ -1 & 1 & 0 & -1 \\ 0 & -1 & 1 & 0 \end{pmatrix} \end{array},\ \
A|_{K_1} := \begin{array}{c} \\ v_1 \\ v_2 \\ v_3 \end{array}
\begin{array}{c} \begin{array}{cc} k_1 & k_3 \end{array} \\
\begin{pmatrix} 1 & 1 \\ -1 & 0 \\ 0 & -1 \end{pmatrix} \end{array},\\
&M = \begin{array}{c} \\ k_1| \\ k_3| \end{array}
\begin{array}{c} \begin{array}{cccc} k_2 & k_4 & k_5 & k_6 \end{array} \\
\begin{pmatrix} 1 & -1 & 0 & 1 \\ 0 & 1 & -1 & 0 \end{pmatrix} \end{array},\ \
M^\top = \begin{array}{c} \\ k_2 \\ k_4 \\ k_5 \\ k_6| \end{array}
\begin{array}{c} \begin{array}{cc} k_1 & k_3 \end{array} \\
\begin{pmatrix} 1 & 0 \\ -1 & 1 \\ 0 & -1 \\ 1 & 0 \end{pmatrix} \end{array}
\end{aligned}
\tag{8.402}
$$

The velocity and force vectors $u = (u_\mu)_\mu,\ i = (i_\mu)_\mu \in \mathcal{F}^6$ are decomposed as

$$u_{K_1} := \begin{pmatrix} u_1 \\ u_3 \end{pmatrix},\ u_{K_2} := (u_2, u_4, u_5, u_6)^\top,\ i_{K_1} = \begin{pmatrix} i_1 \\ i_3 \end{pmatrix},\ i_{K_2} = (i_2, i_4, i_5, i_6)^\top. \tag{8.403}$$

The Kirchhoff laws from Corollary 8.3.5 furnish

$$\begin{aligned} (u_2, u_4, u_5, u_6)^\top &= u_{K_2} = M^\top u_{K_1} = (u_1, -u_1 + u_3, -u_3, u_1)^\top, \\ (i_1, i_3)^\top &= i_{K_1} = -M i_{K_2} = (-i_2 + i_4 - i_6, -i_4 + i_5)^\top. \end{aligned} \tag{8.404}$$

The unique generator branch is k_6, and hence (8.218) has the form

$$\begin{aligned} &K_s := \{k_6\},\ K_p := \{k_1, \ldots, k_5\},\ K_1 = \{k_1, k_3\},\ K_2 = \{k_2, k_4, k_5, k_6\}, \\ &K_{1,s} = \emptyset,\ K_{1,p} = K_1 = \{k_1, k_3\},\ K_{2,s} = K_s = \{k_6\},\ K_{2,p} = \{k_2, k_4, k_5\}. \end{aligned} \tag{8.405}$$

The network behavior $\mathcal{B}$ is isomorphic to the behavior $\mathcal{B}_{II}$ from (8.227). We therefore compute the matrices P_{II} and Q_{II} of $\mathcal{B}_{II}$ from (8.224), given by

$$\begin{aligned} P_{II} &:= \begin{pmatrix} P_{K_{1,p}} & Q_{K_{1,p}} M_{K_{1,p},K_{2,p}} \\ P_{K_{2,p}} \left(M_{K_{1,p},K_{2,p}}\right)^\top & -Q_{K_{2,p}} \end{pmatrix} \in F[s]^{(K_{1,p} \uplus K_{2,p}) \times (K_{1,p} \uplus K_{2,p})} \\ Q_{II} &= \begin{pmatrix} -Q_{K_{1,p}} M_{K_{1,p},K_{2,s}} & 0 \\ 0 & -P_{K_{2,p}} \left(M_{K_{1,s},K_{2,p}}\right)^\top \end{pmatrix} \in F[s]^{(K_{1,p} \uplus K_{2,p}) \times (K_{1,s} \uplus K_{2,s})}. \end{aligned} \tag{8.406}$$

The needed matrices are

$$\begin{aligned} P_{K_{1,p}} &= \begin{array}{c|cc} & k_1 & k_3 \\ \hline k_1 & a_1 & 0 \\ k_3 & 0 & m_3 s \end{array},\quad Q_{K_{1,p}} = \begin{array}{c|cc} & k_1 & k_3 \\ \hline k_1 & s & 0 \\ k_3 & 0 & 1 \end{array}, \\ P_{K_{2,p}} &= \begin{array}{c|ccc} & k_2 & k_4 & k_5 \\ \hline k_2 & m_2 s & 0 & 0 \\ k_4 & 0 & b_4 & 0 \\ k_5 & 0 & 0 & a_5 \end{array},\quad Q_{K_{2,p}} = \begin{array}{c|ccc} & k_2 & k_4 & k_5 \\ \hline k_2 & 1 & 0 & 0 \\ k_4 & 0 & 1 & 0 \\ k_5 & 0 & 0 & s \end{array}, \\ M_{K_{1,p},K_{2,p}} &= \begin{array}{c|ccc} & k_2 & k_4 & k_5 \\ \hline k_1 & 1 & -1 & 0 \\ k_3 & 0 & 1 & -1 \end{array},\quad M^\top_{K_{1,p},K_{2,p}} = \begin{array}{c|cc} & k_1 & k_3 \\ \hline k_2 & 1 & 0 \\ k_4 & -1 & 1 \\ k_5 & 0 & -1 \end{array}, \\ Q_{K_{1,p}} M_{K_{1,p},K_{2,p}} &= \begin{array}{c|ccc} & k_2 & k_4 & k_5 \\ \hline k_1 & s & -s & 0 \\ k_3 & 0 & 1 & -1 \end{array},\quad P_{K_{2,p}} \left(M_{K_{1,p},K_{2,p}}\right)^\top = \begin{array}{c|cc} & k_1 & k_3 \\ \hline k_2 & m_2 s & 0 \\ k_4 & -b_4 & b_4 \\ k_5 & 0 & -a_5 \end{array}. \end{aligned} \tag{8.407}$$

Inserting these matrices into P_{II} and more simply into Q_{II} gives

$$P_{II} = \begin{array}{c} \\ k_1 \\ k_3 \\ k_2 \\ k_4 \\ k_5 \end{array} \begin{array}{c} \begin{array}{ccccc} k_1 & k_3 & k_2 & k_4 & k_5 \end{array} \\ \begin{pmatrix} a_1 & 0 & s & -s & 0 \\ 0 & m_3 s & 0 & 1 & -1 \\ m_2 s & 0 & -1 & 0 & 0 \\ -b_4 & b_4 & 0 & -1 & 0 \\ 0 & -a_5 & 0 & 0 & -s \end{pmatrix} \end{array}, \quad Q_{II} = -s \begin{array}{c} k_6 \\ \begin{pmatrix} 1 \\ 0 \\ 0 \\ 0 \\ 0 \end{pmatrix} \end{array}, \tag{8.408}$$

$$y_{II} = \begin{pmatrix} u_{K_{1,p}} \\ i_{K_{2,p}} \end{pmatrix} = (u_1, u_3, i_2, i_4, i_5)^\top, \quad \begin{pmatrix} i_{K_{2,s}} \\ u_{K_{1,s}} \end{pmatrix} = i_6.$$

The equation $\begin{pmatrix} i_{K_{2,s}} \\ u_{K_{1,s}} \end{pmatrix} = i_6$ means that a force generator produces the input i_6. By different elementary operations $A \mapsto \widetilde{A}$ one could easily obtain $k_6 \in K_1$, so that a velocity generator would be used. Theorem 8.3.11 implies the isomorphism

$$\mathcal{B}_{II} = \left\{ \begin{pmatrix} y_{II} \\ i_6 \end{pmatrix} \in \mathcal{F}^6;\ P_{II} \circ y_{II} = (-s \circ i_6, 0, \ldots, 0)^\top \right\} \cong \mathcal{B} \subset \mathcal{F}^{6+6}, \ \begin{pmatrix} y_{II} \\ i_6 \end{pmatrix} \mapsto \begin{pmatrix} y_I \\ y_{II} \\ i_6 \end{pmatrix},$$

where $\begin{pmatrix} y_I \\ y_{II} \end{pmatrix} := (i_1, i_3, u_6, u_2, u_4, u_5, u_1, u_3, i_2, i_4, i_5)^\top$, $y_I = -P_I y_{II} + Q_I i_6$ and

$$P_I = \begin{pmatrix} 0 & 0 & 1 & -1 & 0 \\ 0 & 0 & 0 & 1 & -1 \\ -1 & 0 & 0 & 0 & 0 \\ -1 & 0 & 0 & 0 & 0 \\ 1 & -1 & 0 & 0 & 0 \\ 0 & 1 & 0 & 0 & 0 \end{pmatrix}, \quad Q_I = \begin{pmatrix} -1 \\ 0 \\ 0 \\ 0 \\ 0 \\ 0 \end{pmatrix}. \tag{8.409}$$

By computer one obtains

$$\begin{aligned} D &:= \det(P_{II}) = -m_2 m_3 s^4 - (m_2 + m_3) b_4 s^3 \\ &\qquad - (a_1 m_3 + a_5 m_2) s^2 - (a_1 + a_5) b_4 s - a_1 a_5 \text{ and} \\ H_{II} &= D^{-1} \begin{pmatrix} m_3 s^3 + b_4 s^2 + a_5 s \\ b_4 s^2 \\ m_2 m_3 s^4 + b_4 m_2 s^3 + a_5 m_2 s^2 \\ -b_4 m_3 s^3 - a_5 b_4 s \\ -a_5 b_4 s \end{pmatrix}, \ \deg_s(H_{II}) = 0. \end{aligned} \tag{8.410}$$

Hence $\mathcal{B}$ is an IO behavior with input i_6 and proper transfer matrix.
Substitution of all constants by 1 furnishes

$$D = \det(P_{II}) = -s^4 - 2s^3 - 2s^2 - 2s - 1, \ V_{\mathbb{C}}(\det(P_{II})) = \{-1, \ -j, \ j\} \text{ and}$$

$$H_{II} = -(s^4 + 2s^3 + 2s^2 + 2s + 1)^{-1} \begin{pmatrix} s^3 + s^2 + s \\ s^2 \\ s^4 + s^3 + s^2 \\ -s^3 - s \\ -s \end{pmatrix}.$$

Hence, in this case, $\mathcal{B}$ is stable, but not asymptotically stable. The Smith form of $(P_{II}, -Q_{II})$ is $(\mathrm{id}_5, 0)$; hence, $\mathcal{B}$ is controllable and $\mathrm{pole}(H) = \mathrm{char}(\mathcal{B}^0) = \{-1, j, -j\}$ for the case of constants 1.
According to Theorem 8.3.11 the transfer matrix $\mathcal{B}$ is $H = \begin{pmatrix} Q_I - P_I H_{II} \\ H_{II} \end{pmatrix}$ and implies

$$H=\begin{pmatrix} H_{i_1,i_6}\\ H_{i_3,i_6}\\ H_{u_6,i_6}\\ H_{u_2,i_6}\\ H_{u_4,i_6}\\ H_{u_5,i_6}\\ H_{u_1,i_6}\\ H_{u_3,i_6}\\ H_{i_2,i_6}\\ H_{i_4,i_6}\\ H_{i_5,i_6}\end{pmatrix} = D^{-1}\begin{pmatrix} a_1m_3s^2+a_1b_4s+a_1a_5\\ b_4m_3s^3\\ m_3s^3+b_4s^2+a_5s\\ m_3s^3+b_4s^2+a_5s\\ -m_3s^3-a_5s\\ -b_4s^2\\ m_3s^3+b_4s^2+a_5s\\ b_4s^2\\ m_2m_3s^4+b_4m_2s^3+a_5m_2s^2\\ -b_4m_3s^3-a_5b_4s\\ -a_5b_4s\end{pmatrix}.$$

The component $H_{u_6,i_6} = D^{-1}(m_3s^3 + b_4s^2 + a_5s)$ is also the transfer function from i_6 to u_6 of the derived one-port. Its *mechanical admittance* at the frequency $\omega > 0$ is the number $H_{u_6,i_6}(j\omega)$, whereas, in an electrical network, this is the impedance. If all constants are 1, one gets

$$H=\begin{pmatrix} H_{i_1,i_6}\\ H_{i_3,i_6}\\ H_{u_6,i_6}\\ H_{u_2,i_6}\\ H_{u_4,i_6}\\ H_{u_5,i_6}\\ H_{u_1,i_6}\\ H_{u_3,i_6}\\ H_{i_2,i_6}\\ H_{i_4,i_6}\\ H_{i_5,i_6}\end{pmatrix} = D^{-1}\begin{pmatrix} s^2+s+1\\ s^3\\ s^3+s^2+s\\ s^3+s^2+s\\ -s^3-s\\ -s^2\\ s^3+s^2+s\\ s^2\\ s^4+s^3+s^2\\ -s^3-s\\ -s\end{pmatrix}.$$

The partial fraction decomposition of H is

$$H=\begin{pmatrix} H_{i_1,i_6}\\ H_{i_3,i_6}\\ H_{u_6,i_6}\\ H_{u_2,i_6}\\ H_{u_4,i_6}\\ H_{u_5,i_6}\\ H_{u_1,i_6}\\ H_{u_3,i_6}\\ H_{i_2,i_6}\\ H_{i_4,i_6}\\ H_{i_5,i_6}\end{pmatrix} = \begin{pmatrix}0\\0\\0\\0\\0\\0\\0\\0\\-1\\0\\0\end{pmatrix} + \begin{pmatrix} j/4\\ -j/4\\ -1/4\\ -1/4\\ 0\\ 1/4\\ -1/4\\ -1/4\\ -j/4\\ 0\\ -j/4\end{pmatrix}(s-j)^{-1} + \begin{pmatrix} -j/4\\ j/4\\ -1/4\\ -1/4\\ 0\\ 1/4\\ -1/4\\ -1/4\\ j/4\\ 0\\ j/4\end{pmatrix}(s+j)^{-1}$$

$$+\begin{pmatrix}0\\-1\\-1/2\\-1/2\\1\\-1/2\\-1/2\\1/2\\1\\1\\0\end{pmatrix}(s+1)^{-1} + \begin{pmatrix}-1/2\\1/2\\1/2\\1/2\\-1\\1/2\\1/2\\-1/2\\-1/2\\-1\\-1/2\end{pmatrix}(s+1)^{-2}.$$

The impulse response $H \circ \delta$ follows from this directly with

$$(s-j)^{-1}\circ\delta = e^{jt}Y,\ (s+j)^{-1}\circ\delta = e^{-jt}Y,\ (s+1)^{-1}\circ\delta = e^{-t}Y,\ (s+1)^{2}\circ\delta = te^{-t}Y.$$

There are the sinusoidal, not asymptotically stable autonomous trajectories

$$y_{II}^0 := \begin{pmatrix} -j \\ -j \\ 1 \\ 0 \\ 1 \end{pmatrix} e^{jt} \in \mathcal{B}_{II}^0 \text{ since } P_{II}(s) \circ y_{II}^0 = \left(P_{II}(j) \begin{pmatrix} -j \\ -j \\ 1 \\ 0 \\ 1 \end{pmatrix} \right) e^{jt} = 0 e^{jt} = 0 \text{ and}$$

$$y^0 := \begin{pmatrix} y_I^0 \\ y_{II}^0 \end{pmatrix} = \begin{pmatrix} -P_I y_{II}^0 \\ y_{II}^0 \end{pmatrix} = (i_1^0, i_3^0, u_6^0, u_2^0, u_4^0, u_5^0, u_1^0, u_3^0, i_2^0, i_4^0, i_5^0)^\top$$
$$= y^0(0) e^{jt} \in \mathcal{B}^0 \text{ with } y^0(0) = (-1, 1, -j, -j, 0, 0, -j, -j, 1, 0, 1)^\top,$$

and likewise their complex conjugates. Notice $u_4^0 = 0$ and $i_4^0 = 0$, i.e., in the trajectory y^0 the friction element does not move, does not dissipate energy, and thus admits an oscillating movement of the network. Of course, this can only happen with the indicated initial condition.
If i_6 is a piecewise continuous periodic input with period

$$T = 2\pi\omega^{-1} \notin 2\pi\mathbb{Z} \text{ and hence } 1 \notin \mathbb{Z}\omega,\ j, -j \notin \mathbb{Z}j\omega,\ \text{pole}(H) \cap \mathbb{Z}j\omega = \emptyset$$

then, according to Eq. (8.141), $H \circ i_6$ is the unique T-periodic output to the input i_6, i.e., with $\binom{H \circ i_6}{i_6} \in \mathcal{B}$. Notice that y^0 is 2π-periodic, but not T-periodic. But $H \circ i_6$ is not a steady-state output in the sense that all other trajectories $\binom{y_1}{i_6} \in \mathcal{B}$ satisfy $\lim_{t\to\infty}(y_1(t) - (H \circ i_6)(t)) = 0$.

◇

Example 8.4.7 We again consider the mechanical network from Example 8.4.6. If $\frac{a_1}{m_2} = \frac{a_5}{m_3}$, the behavior is not asymptotically stable since

$$\mathrm{V}_{\mathbb{C}}\big(\det(P_{II})\big) = \left\{ \pm\sqrt{\frac{a_5}{m_3}}\,j,\ \frac{-(m_2+m_3)b_4 \pm \sqrt{(m_2+m_3)^2 b_4^2 - 4a_5 m_2^2 m_3}}{2m_2 m_3} \right\}.$$

Avoiding this case we choose $m_3 := 2$ and all other constants 1 and obtain

$$\det(P_{II}) = -2s^4 - 3s^3 - 3s^2 - 2s - 1,\ \mathrm{V}_{\mathbb{C}}\big(\det(P_{II})\big) \approx \left\{\lambda_1, \overline{\lambda_1}, \lambda_2, \overline{\lambda_2}\right\} \text{ with}$$
$$\lambda_1 := -0.721 - 0.483j,\ \lambda_2 := -0.029 - 0.814j$$

$$H = \begin{pmatrix} H_{i_1,i_6} \\ H_{i_3,i_6} \\ H_{u_6,i_6} \\ H_{u_2,i_6} \\ H_{u_4,i_6} \\ H_{u_5,i_6} \\ H_{u_1,i_6} \\ H_{u_3,i_6} \\ H_{i_2,i_6} \\ H_{i_4,i_6} \\ H_{i_5,i_6} \end{pmatrix} = (-2s^4 - 3s^3 - 3s^2 - 2s - 1)^{-1} \begin{pmatrix} 2s^2+s+1 \\ 2s^3 \\ 2s^3+s^2+s \\ 2s^3+s^2+s \\ -2s^3-s \\ -s^2 \\ 2s^3+s^2+s \\ s^2 \\ 2s^4+s^3+s^2 \\ -2s^3-s \\ -s \end{pmatrix}.$$

Hence $\mathcal{B}$ is asymptotically stable. The partial fraction decomposition of H is

$$H \approx \begin{pmatrix} 0 \\ 0 \\ 0 \\ 0 \\ 0 \\ 0 \\ 0 \\ 0 \\ -1 \\ 0 \\ 0 \end{pmatrix} + (s-\lambda_1)^{-1} \begin{pmatrix} 0.105-0.562j \\ -0.544+0.253j \\ -0.347+0.355j \\ -0.347+0.355j \\ 0.526-0.650j \\ -0.179+0.295j \\ -0.347+0.355j \\ 0.179-0.295j \\ 0.422-0.088j \\ 0.526-0.650j \\ -0.018-0.398j \end{pmatrix} + \left(s-\overline{\lambda_1}\right)^{-1} \begin{pmatrix} 0.105+0.562j \\ -0.544-0.253j \\ -0.347-0.355j \\ -0.347-0.355j \\ 0.526+0.650j \\ -0.179-0.295j \\ -0.347-0.355j \\ 0.179+0.295j \\ 0.422+0.088j \\ 0.526+0.650j \\ -0.018+0.398j \end{pmatrix}$$

$$+ (s-\lambda_2)^{-1} \begin{pmatrix} -0.105-0.192j \\ 0.044+0.291j \\ -0.153+0.091j \\ -0.153+0.091j \\ -0.026-0.070j \\ 0.179-00.021j \\ -0.153+0.091j \\ -0.179+0.021j \\ 0.078+0.122j \\ -0.026-0.070j \\ 0.018+0.221j \end{pmatrix} + \left(s-\overline{\lambda_2}\right)^{-1} \begin{pmatrix} -0.105+0.192j \\ 0.044-0.291j \\ -0.153-0.091j \\ -0.153-0.091j \\ -0.026+0.070j \\ 0.179+0.021j \\ -0.153-0.091j \\ -0.179-0.021j \\ 0.078-0.122j \\ -0.026+0.070j \\ 0.018-0.221j \end{pmatrix}.$$

The sinusoidal inputs $\alpha = e^{j\omega t},\ \omega > 0$, or $e^{j\omega t}Y$ give rise to the steady-state output $H(j\omega)e^{j\omega t}$, i.e., every output y to α or αY satisfies, cf. Corollary 8.2.56,

$$\begin{aligned} &\lim_{t\to\infty}\left(y(t) - H(j\omega)e^{j\omega t}\right) = 0,\ \lim_{t\to\infty}\left(u_6(t) - H_{u_6,i_6}(j\omega)e^{j\omega t}\right) = 0 \text{ where} \\ &H_{u_6,i_6}(j\omega) = -(2(j\omega)^4 + 3(j\omega)^3 + 3(j\omega)^2 + 2(j\omega) + 1)^{-1} \\ &\qquad\qquad\qquad \cdot (2(j\omega)^3 + (j\omega)^2 + (j\omega)). \end{aligned}$$

One also says that the output $u_6(t)$ *tracks* the reference signal $H_{u_6,i_6}(j\omega)e^{j\omega t}$. General tracking will be treated in Sect. 10.2.1. ◊

Of course, the mathematics of Sect. 8.3.4 is applicable to the products $p_k(t) = u_k(t)i_k(t)$ that are powers in the case of electrical networks. It turns out that this also holds in the mechanical case. Hence $p_k = u_k i_k$ has the dimension $[m \cdot s^{-1} \cdot N] = [J \cdot s^{-1}] = [W]$ of a power. Both in the electrical and in the mechanical situations the u_k (voltages, velocities) and i_k (currents, forces) are called *conjugate power variables*.
The integral of power with respect to time is an energy. The energy supply of the branch element of branch k in the time interval $[t_0, t]$ is $E_k := \int_{t_0}^{t} p_k(\tau)\mathrm{d}\tau$. It may be positive or negative. In the mechanical situation the energy of the various branches is
(i) for a mass branch:

$$\begin{aligned} &i_k = m_k s \circ u_k : \ p_k = u_k m_k s \circ u_k = s \circ \left(2^{-1} m_k u_k^2\right) \\ &\Longrightarrow E_k(t) = \int_{t_0}^{t} p_k(\tau)\mathrm{d}\tau = 2^{-1} m_k \left(u_k(t)^2 - u_k(t_0)^2\right), \end{aligned} \tag{8.411}$$

(ii) for a spring branch:

$$\begin{aligned} & i_k = a(x_k - x_{k,0}), \ s \circ i_k = au_k : \\ & \quad p_k = u_k i_k = s \circ \left(2^{-1} a^{-1} i_k^2\right) = s \circ \left(2^{-1} a (x_k - x_{k,0})^2\right) \\ & \Longrightarrow E_k(t) = 2^{-1} a \left((x_k(t) - x_{k,0})^2 - (x_k(t_0) - x_{k,0})^2\right), \end{aligned} \tag{8.412}$$

(iii) for a friction branch:

$$i_k = bu_k, \ p_k = bu_k^2, \ E_k = b \int_{t_0}^{t} u_k(\tau)^2 \mathrm{d}\tau. \tag{8.413}$$

The energies in cases (i), (ii) resp. (iii) are called *kinetic, potential resp. dissipative*. The mass and the spring can store energy, whereas the friction element consumes or dissipates energy. Note that $x_{k,0}$ is the resting position of the spring with zero force, whereas $x_k(t_0)$ is the position at time t_0.

Remark 8.4.8 1. There are *rotational mechanical networks* where the objects rotate around one given axis and where force, velocity, and mass are replaced by *torque, angular velocity, and angular momentum*. The mathematics of the electrical and translational mechanical networks is applicable without change.
The same mathematics can be applied to *acoustical networks* where u is volume flow and i is pressure, cf. [Wikipedia; https://en.wikipedia.org/wiki/Mechanical-electrical_analogies, August 12, 2019], [20].
2. *Transducers* transport energy from one domain to another, and enable the considerations of systems that have both mechanical and electrical parts, like an electric motor. Mathematically, the transducer is a linear map from a pair of power conjugate variables in one energy domain to a pair in another domain, for instance, from (voltage, current) to (velocity, force). A transducer is a generalization of a two-port in electrical engineering, with its two-ports in different energy domains. The transducers play an important part in *mechatronics*, cf. [20], [22]. ◊

References

1. U. Oberst, Variations on the fundamental principle for linear systems of partial differential and difference equations with constant coefficients. AAECC **6**, 211–243 (1995)
2. H.-J. Schmeißer, *Höhere Analysis. Lecture Notes* (Universität Jena, 2005)
3. L. Schwartz. *Théorie des distributions*. Publications de l'Institut de Mathématique de l'Université de Strasbourg, No. IX–X. (Hermann, Paris, nouvelle édition, entiérement corrigée, refondue et augmentée edition, 1966)
4. L.Hörmander, *The Analysis of Linear Partial Differential Operators. I. Distribution Theory and Fourier Analysis*, vol. 256 of *Grundlehren der Mathematischen Wissenschaften* (Springer, Berlin, 1983)
5. J.W. Polderman, J.C. Willems, *Introduction to Mathematical Systems Theory. A Behavioral Approach*. Texts in Applied Mathematics, vol. 26 (Springer, New York, 1998)
6. F.M. Callier, C.A. Desoer, *Multivariable Feedback Systems* (Springer Texts in Electrical Engineering. Springer, New York, 1982)

7. A.I.G. Vardulakis, *Linear Multivariable Control. Algebraic Analysis and Synthesis Methods* (Wiley, Chichester, 1991)
8. M. Albach, *Grundlagen der Elektrotechnik 2. Periodische und nichtperiodische Signalformen* (Pearson Studium, München, 2005)
9. H. Bourlès, U. Oberst, Generalized convolution behaviors and topological algebra. Acta Appl. Math. **141**, 107–148 (2016)
10. R. Unbehauen, *Elektrische Netzwerke. Eine Einführung in die Analyse* (Springer, Berlin, 1987)
11. P.J. Antsaklis, A.N. Michel, *Linear Systems*, 2nd edn. (Birkhäuser, Boston, 2006)
12. E.C. Titchmarsh, *Introduction to the Theory of Fourier Integrals* (Oxford University Press, Oxford, 1967)
13. F.M. Callier, C.A. Desoer, *Linear System Theory* (Springer, New York, 1991)
14. J. Mikusiński, *Operational Calculus*, vol. 8, *International Series of Monographs on Pure and Applied Mathematics* (Pergamon Press, New York, 1959)
15. H. Bourlès, Impulsive systems and behaviors in the theory of linear dynamical systems. Forum Math. **17**(5), 781–807 (2005)
16. C. Bargetz, Impulsive Solutions of Differential Behaviors. Master's thesis, University of Innsbruck, 2008
17. L.P. Schmidt, G. Schaller, S. Martius, *Grundlagen der Elektrotechnik 3. Netzwerke* (Pearson Studium, München, 2006)
18. R.W. Newcomb, *Network Theory. The State-Space Approach* (Librairie Universitaire, Louvain, 1969)
19. D. Hinrichsen, A.J. Pritchard, *Mathematical Systems Theory I. Modelling, State Space Analysis, Stability and Robustness*. Texts in Applied Mathematics, vol. 48 (Springer, Berlin, 2005)
20. R.G. Ballas, G. Pfeifer, R. Werthschützky, *Elektromechanische Systeme in der Mikrotechnik und Mechatronik* (Springer, Berlin, 2009)
21. F.A. Firestone, A new analogy between mechanical and electrical systems. J. Acoust. Soc. Am. **4**(3), 249–267 (1933)
22. K. Jantschek, *Mechatronic Systems Design. Methods, Models, Concepts* (Springer, Berlin, 2012)

Chapter 9
Stability via Quotient Modules

In this chapter we present the details concerning quotient behaviors and stability that were introduced in Sect. 2.8. In pure mathematics the method of this chapter is also called *localization*.

9.1 Quotient Rings and Modules

The quotient field of a commutative integral domain, e.g., the field of rational functions as the quotient field of the polynomial ring, is usually constructed in the first algebra course. Quotient *modules* require the introduction of zero divisors of the module in the denominators and the modification of the usual construction. In this section we introduce quotient modules for arbitrary commutative rings $\mathcal{D}$ and multiplicative submonoids

$$T \subset \mathcal{D} \text{ with } 1 \in T, \ \forall t_1, t_2 \in T : \ t_1 t_2 \in T, \tag{9.1}$$

but soon specialize the results to principal ideal domains and saturated submonoids (Sect. 9.2). The monoid T is called *saturated* if $ab \in T$ implies $a, b \in T$. Without loss of generality saturated monoids can always be assumed, see below. We establish the standard properties of quotient rings and modules.

Example 9.1.1 1. Every $t \in \mathcal{D}$ generates the submonoid $\langle t \rangle := \{1, t, t^2, \dots\}$ of all powers of t. Except for trivial cases, this submonoid is not saturated.

2. An ideal $\mathfrak{p}$ of $\mathcal{D}$ is called a *prime ideal* if $T(\mathfrak{p}) := \mathcal{D} \setminus \mathfrak{p}$ is a submonoid of $\mathcal{D}$, i.e., if $1 \in T(\mathfrak{p})$ or $\mathfrak{p} \subsetneq \mathcal{D}$, and $t_1, t_2 \notin \mathfrak{p}$ implies that $t_1 t_2 \notin \mathfrak{p}$. Since $\mathfrak{p}$ is an ideal, the set $T(\mathfrak{p})$ is also saturated.
If $\mathfrak{p} = \mathcal{D}q$, $q \in \mathcal{D}$, is a principal ideal, then $\mathfrak{p}$ is a prime ideal if and only if q is a prime element. Then

U. Oberst et al., *Linear Time-Invariant Systems, Behaviors and Modules*, Differential-Algebraic Equations Forum,
https://doi.org/10.1007/978-3-030-43936-1_9

$$T(q) := T(\mathcal{D}q) = \mathcal{D} \setminus \mathcal{D}q = \{t \in \mathcal{D};\ q \text{ does not divide } t\}. \tag{9.2}$$

◊

In the following lemma we characterize all pairs $(\mathcal{D}, T)$ which are relevant for the one-dimensional systems theory of this book. Let $\mathcal{D}$ be a principal ideal domain and let $\mathcal{P} \subset \mathcal{D}$ be a representative system of the *prime or irreducible* elements of $\mathcal{D}$, up to association. The standard case is the polynomial algebra $\mathcal{D} := F[s]$ over a field F with the set $\mathcal{P}$ of monic irreducible polynomials. Every nonzero element t of $\mathcal{D}$ has a unique prime factor decomposition

$$t = u \prod_{q \in \mathcal{P}} q^{\mathrm{ord}_q(t)} \quad \text{with } u \in \mathrm{U}(\mathcal{D}) \text{ and } \mathrm{ord}_q(t) \in \mathbb{N}, \tag{9.3}$$

where almost all, i.e., all but finitely many, orders $\mathrm{ord}_q(t)$ are zero. Here $\mathrm{U}(\mathcal{D})$ is the group of *multiplicatively invertible elements or units* of $\mathcal{D}$.

Lemma and Definition 9.1.2 *Let $\mathcal{D}$ be a principal ideal domain with its representative set $\mathcal{P}$ of prime elements. A multiplicative submonoid $T \subseteq \mathcal{D}$ induces the disjoint decomposition $\mathcal{P} = \mathcal{P}_1(T) \uplus \mathcal{P}_2(T)$ of the prime elements of $\mathcal{D}$ where*

$$\mathcal{P}_1(T) := \{q \in \mathcal{P};\ \exists t \in T:\ q \mid t\} \quad \text{and} \quad \mathcal{P}_2(T) := \mathcal{P} \setminus \mathcal{P}_1(T).$$

Conversely, a disjoint decomposition $\mathcal{P} = \mathcal{P}_1 \uplus \mathcal{P}_2$ induces the saturated multiplicative submonoid

$$\begin{aligned} T(\mathcal{P}_1) &:= \left\{t \in \mathcal{D} \setminus \{0\};\ \forall q \in \mathcal{P}_2:\ \mathrm{ord}_q(t) = 0,\ \text{i.e., } q \nmid t\right\} \\ &= \bigcap_{q \in \mathcal{P}_2} (\mathcal{D} \setminus \mathcal{D}q) = \left\{t = u \prod_{q \in \mathcal{P}_1} q^{\mathrm{ord}_q(t)} \in \mathcal{D};\ u \in \mathrm{U}(\mathcal{D}),\ \mathrm{ord}_q(t) \in \mathbb{N}\right\} \end{aligned} \tag{9.4}$$

where almost all $\mathrm{ord}_q(t)$ are zero. Then the identity $\mathcal{P}_1(T(\mathcal{P}_1)) = \mathcal{P}_1$ holds for all $\mathcal{P}_1 \subseteq \mathcal{P}$. Conversely, if a submonoid $T \subseteq \mathcal{D}$ is given, then

$$T(\mathcal{P}_1(T)) = T' := \left\{t' \in \mathcal{D};\ \exists t \in T:\ t' \mid t\right\},$$

and this is obviously the saturation of T, i.e., the smallest saturated monoid containing T. We infer that there is a one-to-one correspondence between the saturated submonoids of $\mathcal{D}$ and the disjoint decompositions $\mathcal{P} = \mathcal{P}_1 \uplus \mathcal{P}_2$ of the representative system of prime elements.

Proof (i) The set $T(\mathcal{P}_1)$ from (9.4) is clearly closed under multiplication, and it contains 1. It is saturated because the elements of $\mathcal{P}_1$ are irreducible, and hence all divisors of elements in $T(\mathcal{P}_1)$ lie in $T(\mathcal{P}_1)$ too.

(ii) $\mathcal{P}_1 = \mathcal{P}_1(T(\mathcal{P}_1))$: Simple.
(iii) $T(\mathcal{P}_1(T)) \subseteq T'$: Let $t \in T(\mathcal{P}_1(T))$. Then t has a prime factor decomposition $t = u \prod_{q\in\mathcal{P}_1(T)} q^{\mathrm{ord}_q(t)}$. For every $q \in \mathcal{P}_1(T)$, we choose a $t_q \in T$ with $q \mid t_q$ and obtain $t | t_1 := \prod_{q\in\mathcal{P}_1(T)} t_q^{\mathrm{ord}_q(t)} \in T \Longrightarrow t \in T'$.
(iv) $T(\mathcal{P}_1(T)) \supseteq T'$: If $t' | t = u \prod_{q\in\mathcal{P}_1(T)} q^{\mathrm{ord}_q(t)} \in T$ then

$$t' = u' \prod_{q\in\mathcal{P}_1(T)} q^{\mathrm{ord}_q(t')},\ \mathrm{ord}_q(t') \le \mathrm{ord}_q(t) \Longrightarrow t' \in T(\mathcal{P}_1(T)).$$

□

In (9.2) and (9.4) note

$$\forall q_2 \in \mathcal{P} : T(q_2) = T(\mathcal{P}\setminus\{q_2\}) = \mathcal{D}\setminus\mathcal{D}q_2. \tag{9.5}$$

In context with stability the elements of a saturated multiplicative submonoid T of a principal ideal domain $\mathcal{D}$ are called *T-stable*. If M is a $\mathcal{D}$-module the submodule

$$\mathrm{t}_T(M) := \{x \in M;\ \exists t \in T \text{ with } tx = 0\} \subseteq \mathrm{t}(M) \tag{9.6}$$

is called the T-torsion submodule of M. If ${}_{\mathcal{D}}\mathcal{F}$ is an injective cogenerator, the elements $\mathrm{t}_T(\mathcal{F})$ are also called T-stable.

Example 9.1.3 1. The usual complex stable polynomials arise according to Lemma 9.1.2. The canonical system of irreducible polynomials is

$$\mathcal{P}_{\mathbb{C}} = \{s-\lambda;\ \lambda\in\mathbb{C}\} \overset{\cong}{\longleftrightarrow} \mathbb{C},\quad s-\lambda \longleftrightarrow \lambda. \tag{9.7}$$

Hence a decomposition of $\mathcal{P}_{\mathbb{C}} = \mathcal{P}_{\mathbb{C},1} \uplus \mathcal{P}_{\mathbb{C},2}$ corresponds to a decomposition

$$\mathbb{C} = \Lambda_1 \uplus \Lambda_2 \text{ with } \mathcal{P}_{\mathbb{C},i} := \{s-\lambda;\ \lambda\in\Lambda_i\},\ i = 1, 2. \tag{9.8}$$

The sets Λ_1 resp. Λ_2 are called *the stable resp. the unstable region* of $\mathbb{C}$ with respect to (9.8). The saturated monoid $T_{\mathbb{C}} := T(\mathcal{P}_{\mathbb{C},1})$ according to (9.4) is

$$\begin{aligned} T := T_{\mathbb{C}} &= \{t\in\mathbb{C}[s]\setminus\{0\};\ \forall\lambda\in\Lambda_2\colon\ s-\lambda \text{ does not divide } t\} \\ &= \{t\in\mathbb{C}[s]\setminus\{0\};\ \forall\lambda\in\Lambda_2\colon\ t(\lambda)\neq 0\} \\ &= \{t\in\mathbb{C}[s]\setminus\{0\};\ \mathrm{V}_{\mathbb{C}}(t)\subseteq\Lambda_1\}, \end{aligned} \tag{9.9}$$

where $\mathrm{V}_{\mathbb{C}}(t)$ is the set of zeros of t. According to Theorem 5.2.2 and Corollary 5.3.19 the T-torsion submodule of $\mathrm{C}^{-\infty}(\mathbb{R}, \mathbb{C})$ is

$$\mathrm{t}_T(\mathrm{C}^{-\infty}(\mathbb{R},\mathbb{C})) = \oplus_{\lambda\in\Lambda_1}\mathbb{C}[t]e^{\lambda t}. \tag{9.10}$$

The standard case is the open left half-plane $\Lambda_1 := \mathbb{C}_-$ as stable region. The T-stable signals w satisfy $\lim_{t\to\infty} w(t) = 0$ if and only if $\Lambda_1 \subseteq \mathbb{C}_-$. These signals can be neglected in many engineering applications.
In the discrete-time case with $\mathcal{F} = \mathbb{C}^{\mathbb{N}}$ the T-torsion submodule is

$$t_T(\mathbb{C}^{\mathbb{N}}) = \oplus_{\lambda\in\Lambda_1} t_{s-\lambda}(\mathbb{C}^{\mathbb{N}}),\ t_{s-\lambda}(\mathbb{C}^{\mathbb{N}}) = \begin{cases} \mathbb{C}[t](\lambda^t)_{t\in\mathbb{N}} & \text{if } \lambda \neq 0 \\ \mathbb{C}^{(\mathbb{N})} & \text{if } \lambda = 0 \end{cases}. \tag{9.11}$$

The T-stable signals w satisfy $\lim_{t\to\infty} w(t) = 0$ if and only if $\Lambda_1 \subseteq \mathbb{D}_1$ where $\mathbb{D}_1 := \{\lambda \in \mathbb{C};\ |\lambda| < 1\}$ is the open unit disc. The signals in $\mathbb{C}^{(\mathbb{N})}$ are the $T(\{s\})$-stable signals and called *deadbeat signals*.
2. In the real case the data are the same as in the complex case with the additional assumption that the Λ_i are invariant under complex conjugation, i.e., $\overline{\Lambda_i} = \Lambda_i$. However, we use the ring $\mathcal{D} = \mathbb{R}[s]$ now. The canonical set of irreducible real polynomials is

$$\mathcal{P}_{\mathbb{R}} = \{s-\lambda;\ \lambda \in \mathbb{R}\} \uplus \{(s-\lambda)(s-\overline{\lambda});\ \lambda \in \mathbb{C},\ \Im(\lambda) > 0\}.$$

Like $\mathcal{P}_{\mathbb{C}}$ also $\mathcal{P}_{\mathbb{R}}$ decomposes as

$$\begin{aligned} &\mathcal{P}_{\mathbb{R}} = \mathcal{P}_{\mathbb{R},1} \uplus \mathcal{P}_{\mathbb{R},2} \text{ with } \mathcal{P}_{\mathbb{R},1} = \{q \in \mathcal{P}_{\mathbb{R}};\ V_{\mathbb{C}}(q) \subseteq \Lambda_1\} \\ &\Longrightarrow T_{\mathbb{R}} := T(\mathcal{P}_{\mathbb{R},1}) = \{t \in \mathbb{R}[s] \setminus \{0\};\ V_{\mathbb{C}}(t) \subseteq \Lambda_1\}. \end{aligned} \tag{9.12}$$

Again the T-stable signals satisfy $\lim_{t\to\infty} w(t) = 0$ and can thus be neglected in most engineering applications if and only if $\Lambda_1 \subseteq \mathbb{C}_-$ in the continuous-time case and $\Lambda_1 \subseteq \mathbb{D}_1 = \{\lambda \in \mathbb{C};\ |\lambda| < 1\}$ in the discrete-time case.
3. If $T = \mathcal{D} \setminus \{0\}$, then $\mathcal{D}_T = \operatorname{quot}(\mathcal{D}) = K$, the T-stable behaviors are just the autonomous ones. ◊

In the sequel we construct quotient rings and modules with respect to a submonoid T. The prototype for this construction is the quotient field K of an integral domain $\mathcal{D}$, namely,

$$K := \big(\mathcal{D} \times (\mathcal{D} \setminus \{0\})\big)/\sim\ \ni \tfrac{a}{t} := \overline{(a,t)},$$

where $\sim$ is the equivalence relation

$$(a_1, t_1) \sim (a_2, t_2) :\Longleftrightarrow t_2 a_1 = t_1 a_2 \tag{9.13}$$

and $\overline{(a,t)}$ denotes the equivalence class containing (a, t).

Lemma and Definition 9.1.4 *Let $\mathcal{D}$ be a commutative ring, let $T \subseteq \mathcal{D}$ be a multiplicative submonoid, and let ${}_{\mathcal{D}}M$ be a $\mathcal{D}$-module. On the set $M \times T$ we define the relation*

$$(x_1, t_1) \sim (x_2, t_2) :\Longleftrightarrow \exists t \in T \text{ with } t t_2 x_1 = t t_1 x_2. \tag{9.14}$$

Compared to (9.13), the definition (9.14) contains an additional factor $t \in T$. This is necessary because M may contain torsion elements and thus $tx = 0$ for nonzero $t \in T$ and $x \in M$ is possible, see item 7. For $M = \mathcal{D}$ this means that $\mathcal{D}$ has zero divisors, see item 5.

The following statements hold:

1. *The relation from (9.14) is an equivalence relation. The equivalence classes and the set of these classes are denoted by*

$$\tfrac{x}{t} := \overline{(x,t)} \in M_T := (M \times T)/\!\sim \text{ where } (x,t) \in M \times T.$$

From this definition follows directly that elements in T can be cancelled, i.e.,

$$\tfrac{x}{t} = \tfrac{sx}{st} \text{ for } x \in M \text{ and } s, t \in T.$$

The presence of $t \in T$ in (9.14) is crucial for the transitivity of $\sim$.

2. *The addition*

$$+\colon\ M_T \times M_T \longrightarrow M_T,\quad \left(\tfrac{x_1}{t_1}, \tfrac{x_2}{t_2}\right) \longmapsto \tfrac{x_1}{t_1} + \tfrac{x_2}{t_2} := \tfrac{t_2x_1+t_1x_2}{t_1t_2},$$

is well defined and makes M_T an abelian group with zero element $0_{M_T} = \frac{0}{1} = \frac{0}{t}$ for arbitrary $t \in T$. The additive inverse of $\frac{x}{t}$ is $-\frac{x}{t} = \frac{-x}{t}$.

3. *On $\mathcal{D}_T$ the multiplication*

$$\cdot\colon\ \mathcal{D}_T \times \mathcal{D}_T \longrightarrow \mathcal{D}_T,\quad \left(\tfrac{d_1}{t_1}, \tfrac{d_2}{t_2}\right) \longmapsto \tfrac{d_1}{t_1}\tfrac{d_2}{t_2} := \tfrac{d_1d_2}{t_1t_2},$$

*is well defined and makes $\mathcal{D}_T$ a commutative ring with the addition from item 2. Its identity element is $1_{\mathcal{D}_T} = \frac{1}{1} = \frac{t}{t}$ for arbitrary $t \in T$. In context with stability the elements of $\mathcal{D}_T$ are called T-*stable fractions

4. *The elements $\frac{t_1}{t_2}$, $t_1, t_2 \in T$, are invertible in $\mathcal{D}_T$ and $\left(\frac{t_1}{t_2}\right)^{-1} = \frac{t_2}{t_1}$. If T is saturated, then all invertible elements of $\mathcal{D}_T$ have this form.*

5. *The canonical map*

$$\mathrm{can} := \mathrm{can}_{\mathcal{D}}\colon\ \mathcal{D} \longrightarrow \mathcal{D}_T,\quad d \longmapsto \tfrac{d}{1},$$

is a ring homomorphism with $\mathrm{can}(T) \subseteq \mathrm{U}(\mathcal{D}_T)$. Its kernel is the ideal

$$\mathrm{t}_T(\mathcal{D}) := \ker(\mathrm{can}_{\mathcal{D}}) = \{d \in \mathcal{D};\ \exists t \in T\colon\ td = 0\},$$

which is a subset of the set of zero divisors of $\mathcal{D}$.

6. *The scalar multiplication*

$$\cdot\colon\ \mathcal{D}_T \times M_T \longrightarrow M_T,\quad \left(\tfrac{d}{t_1}, \tfrac{x}{t_2}\right) \longmapsto \tfrac{d}{t_1}\tfrac{x}{t_2} := \tfrac{dx}{t_1t_2},$$

is well defined and makes M_T a $\mathcal{D}_T$-module.

Since $\mathrm{can}_{\mathcal{D}}\colon\ \mathcal{D} \longrightarrow \mathcal{D}_T$ *is a ring homomorphism, every $\mathcal{D}_T$-module N can be considered as $\mathcal{D}$-module with the scalar multiplication*

$$dn := \mathrm{can}_{\mathcal{D}}(d)\,n = \tfrac{d}{1}n \text{ for } d \in \mathcal{D} \text{ and } n \in N,$$

and this will always be done. In particular, it makes sense to talk about the $\mathcal{D}$-linearity of the map can_M *in the following item:*

7. *The canonical map*

$$\mathrm{can} := \mathrm{can}_M\colon\ M \longrightarrow M_T,\quad x \longmapsto \tfrac{x}{1},$$

*is $\mathcal{D}$-linear. Its kernel is the T-*torsion submodule

$$\mathrm{t}_T(M) := \ker(\mathrm{can}_M) = \mathrm{can}_M^{-1}(0) = \{x \in M;\ \exists t \in T\colon\ tx = 0\}. \tag{9.15}$$

*The canonical map is thus injective if and only if its T-torsion submodule is zero or, in other words, if and only if $t\cdot\colon\ M \longrightarrow M,\quad x \longmapsto tx$, is a monomorphism for all $t \in T$. In general, the canonical map is neither injective nor surjective. If $\mathcal{D}$ is an integral domain and $0 \notin T$, then the T-torsion submodule is contained in the torsion submodule. Hence, if $M = \mathrm{t}_T(M)$, then the module M is a torsion module and it is called a T-*torsion module.

Proof 1. The relation $\sim$ is clearly reflexive and symmetric. To show that it is also transitive, let $(x_i, t_i) \in M \times T$ for $i = 1, 2, 3$ with

$$\begin{array}{lll}(x_1, t_1) \sim (x_2, t_2) & \text{i.e., } st_2x_1 = st_1x_2 & \text{for some } s \in T \quad \text{and} \\ (x_2, t_2) \sim (x_3, t_3) & \text{i.e., } tt_3x_2 = tt_2x_3 & \text{for some } t \in T.\end{array}$$

Then $stt_2 \in T$ and

$$(stt_2)t_3x_1 = tt_3(st_2x_1) = tt_3(st_1x_2) = st_1(tt_3x_2) = st_1(tt_2x_3) = (stt_2)t_1x_3,$$

hence $(x_1, t_1) \sim (x_3, t_3)$.

2. We have to show that the addition is well defined, i.e., that it is independent of the choice of the representatives. Let

$$\tfrac{x_1}{t_1} = \tfrac{x_1'}{t_1'} \text{ and } \tfrac{x_2}{t_2} = \tfrac{x_2'}{t_2'},\ \text{i.e.,}$$
$$s_1t_1'x_1 = s_1t_1x_1' \text{ and } s_2t_2'x_2 = s_2t_2x_2' \text{ for some } s_1, s_2 \in T.$$

Then

$$\begin{aligned}\frac{t_2x_1+t_1x_2}{t_1t_2} &= \frac{s_1s_2t_1't_2'(t_2x_1+t_1x_2)}{s_1s_2t_1't_2't_1t_2} = \frac{s_2t_2't_2(s_1t_1'x_1)+s_1t_1't_1(s_2t_2'x_2)}{s_1s_2t_1't_2't_1t_2}\\ &= \frac{s_2t_2't_2(s_1t_1x_1')+s_1t_1't_1(s_2t_2x_2')}{s_1s_2t_1't_2't_1t_2} = \frac{s_1s_2t_1t_2(t_2'x_1'+t_1'x_2')}{s_1s_2t_1't_2't_1t_2}\\ &= \frac{t_2'x_1'+t_1'x_2'}{t_1't_2'}.\end{aligned}$$

The proof of the remaining statements is simple.
3. The proof is analogous to that of 2, but easier.
4. From $\frac{t_1}{t_2}\frac{t_2}{t_1} = \frac{t_1t_2}{t_2t_1} = \frac{1}{1} = 1_{\mathcal{D}_T}$ follows that $\left(\frac{t_1}{t_2}\right)^{-1} = \frac{t_2}{t_1}$. Assume that T is saturated, and let $\frac{d_1}{t_1}, \frac{d_2}{t_2} \in \mathcal{D}_T$ with $\frac{d_2}{t_2} = \left(\frac{d_1}{t_1}\right)^{-1}$. Then $\frac{d_1d_2}{t_1t_2} = 1_{\mathcal{D}_T} = \frac{1}{1}$, i.e.,

$$\exists s \in T:\ s\cdot 1\cdot d_1d_2 = st_1t_2\cdot 1 \in T.$$

Since the product lies in the saturated set T, so do its factors. In particular, $d_1, d_2 \in T$ and we conclude that $\mathrm{U}(\mathcal{D}) = \left\{\frac{t_1}{t_2};\ t_1, t_2 \in T\right\}$.
5. That $\mathrm{can}_{\mathcal{D}}$ is a ring homomorphism and that $\mathrm{can}(T) \subseteq \mathrm{U}(\mathcal{D}_T)$ follow directly from the definition. If $d \in \mathcal{D}$ with $\mathrm{can}_{\mathcal{D}}(d) = \frac{d}{1} = 0_{\mathcal{D}_T} = \frac{0_{\mathcal{D}}}{1}$, then there is a $t \in T$ with $td = 0_{\mathcal{D}}$. Conversely,

$$d \in \mathcal{D},\ t \in T,\ td = 0 \text{ imply } \mathrm{can}_{\mathcal{D}}(d) = \frac{d}{1} = \frac{td}{t} = \frac{0_{\mathcal{D}}}{t} = 0_{\mathcal{D}_T}.$$

6. This follows like 2 and 3.
7. That can_M is $\mathcal{D}$-linear is obvious. The expression for $\ker(\mathrm{can}_M)$ follows like that for $\ker(\mathrm{can}_{\mathcal{D}})$ in 5. □

Remark 9.1.5 (*Localization*) If $\mathfrak{p}$ is a prime ideal of $\mathcal{D}$ and $T := T(\mathfrak{p}) = \mathcal{D}\setminus\mathfrak{p}$ then $\mathcal{D}_{\mathfrak{p}} := \mathcal{D}_{T(\mathfrak{p})}$ is called the *local ring of $\mathcal{D}$ at $\mathfrak{p}$*. One also talks about the *localization* of $\mathcal{D}$ at $\mathfrak{p}$. This terminology comes from higher dimensional algebraic geometry where the set of prime ideals is interpreted as the set of points of a scheme (variety). In the case of $\mathcal{D} = \mathbb{C}[s]$ these are the prime ideals $\mathbb{C}[s](s-\lambda)$, $\lambda \in \mathbb{C}$, and 0. We do not use the notation $\mathcal{D}_{\mathfrak{p}}$ for $\mathcal{D}_{T(\mathfrak{p})}$, since $\mathfrak{p}$ is also multiplicatively closed and as such induces $\mathcal{D}_{\mathfrak{p}} = 0$. ◊

Corollary and Definition 9.1.6 *In the situation of Example 9.1.3, the elements of the quotient rings*

$$\mathbb{C}[s]_{T_{\mathbb{C}}} = \left\{\frac{d(s)}{t(s)} \in \mathbb{C}(s);\ d, t \in \mathbb{C}[s],\ \mathrm{V}_{\mathbb{C}}(t) \subseteq \Lambda_1\right\} \text{ and}$$
$$\mathbb{R}[s]_{T_{\mathbb{R}}} = \left\{\frac{d(s)}{t(s)} \in \mathbb{R}(s);\ d, t \in \mathbb{R}[s],\ \mathrm{V}_{\mathbb{C}}(t) \subseteq \Lambda_1\right\}$$

are those rational functions whose complex poles belong to Λ_1. *They are called* $T_{\mathbb{C}}$-stable *or* $T_{\mathbb{R}}$-stable *or simply* Λ_1-stable. *In the standard cases these are the rational functions whose poles lie in the left complex half-plane in the continuous-time case and in the interior of the unit disc in the discrete-time case.* ◊

Theorem 9.1.7 (The universal property of quotient objects)

1. *The quotient ring $\mathcal{D}_T$, more precisely the pair $(\mathcal{D}_T, \mathrm{can}_{\mathcal{D}})$, has the following* universal property*:*
   (a) $\mathrm{can}_{\mathcal{D}}(T) \subseteq \mathrm{U}(\mathcal{D}_T)$.
   (b) *If $\varphi\colon\ \mathcal{D} \longrightarrow B$ is a homomorphism of commutative rings with $\varphi(T) \subseteq \mathrm{U}(B)$, then there is a unique ring homomorphism $\psi\colon\ \mathcal{D}_T \longrightarrow B$ with*
   $$\psi\,\mathrm{can}_{\mathcal{D}} = \varphi, \quad \textit{namely,}\quad \psi\left(\tfrac{d}{t}\right) = \varphi(t)^{-1}\,\varphi(d)\ \textit{for}\ \tfrac{d}{t} \in \mathcal{D}_T.$$
2. *The module M_T, more precisely the pair (M_T, can_M), has the following universal property:*
   (a) *The module M_T is a $\mathcal{D}_T$-module and $\mathrm{can}_M\colon\ M \to M_T$ is $\mathcal{D}$-linear.*
   (b) *If N is any $\mathcal{D}_T$-module and $\varphi\colon\ M \longrightarrow N$ is $\mathcal{D}$-linear, there is a unique $\mathcal{D}_T$-linear map $\psi\colon\ M_T \longrightarrow N$ such that*
   $$\psi\,\mathrm{can}_M = \varphi, \quad \textit{namely,}\quad \psi\left(\tfrac{x}{t}\right) = \tfrac{\varphi(x)}{t}\ \textit{for}\ \tfrac{x}{t} \in M_T.$$

   In other terms, the map
   $$\mathrm{Hom}_{\mathcal{D}_T}(M_T, N) \overset{\cong}{\longrightarrow} \mathrm{Hom}_{\mathcal{D}}(M, N), \quad \psi \longmapsto \psi\,\mathrm{can}_M,$$
   is bijective, and indeed a $\mathcal{D}$- and even a $\mathcal{D}_T$-isomorphism.

Proof 1. (a) Lemma and Definition 9.1.4, item 5.
(b) *Uniqueness.* Let $\widetilde{\psi}\colon\ \mathcal{D}_T \longrightarrow B$ be a ring homomorphism with $\widetilde{\psi}\,\mathrm{can}_{\mathcal{D}} = \varphi$. Since ring homomorphisms preserve inverse elements, we get

$$\begin{aligned}\widetilde{\psi}\left(\tfrac{d}{t}\right) &= \widetilde{\psi}\left(\left(\tfrac{t}{1}\right)^{-1}\tfrac{d}{1}\right) = \widetilde{\psi}\left(\tfrac{t}{1}\right)^{-1}\,\widetilde{\psi}\left(\tfrac{d}{1}\right)\\ &= \widetilde{\psi}\big(\mathrm{can}_{\mathcal{D}}(t)\big)^{-1}\,\widetilde{\psi}(\mathrm{can}_{\mathcal{D}}(d)) = \varphi(t)^{-1}\,\varphi(d) \quad \text{for } \tfrac{d}{t} \in \mathcal{D}_T.\end{aligned}$$

This shows that the given map $\psi\colon\ \mathcal{D}_T \longrightarrow B$ is the *unique* ring homomorphism with $\psi\,\mathrm{can}_{\mathcal{D}} = \varphi$.
Existence. One has to verify that the map $\psi\left(\frac{d}{t}\right) := \varphi(t)^{-1}\varphi(d)$ is well defined, i.e., $\varphi(t)^{-1}\varphi(d) \in B$ and that $\psi\left(\frac{d}{t}\right)$ is independent of the representation of $\frac{d}{t} \in \mathcal{D}_T$, and that it is indeed a ring homomorphism which satisfies $\psi\,\mathrm{can}_{\mathcal{D}} = \varphi$, i.e., $\psi\left(\frac{d}{1}\right) = \varphi(d)$ for $d \in \mathcal{D}$. We omit the details.
2. The ring homomorphism $\mathrm{can}_{\mathcal{D}}\colon\ \mathcal{D} \longrightarrow \mathcal{D}_T$ makes every $\mathcal{D}_T$-module N a $\mathcal{D}$-module via $dn := \frac{d}{1}n$. Therefore, $\mathrm{Hom}_{\mathcal{D}}(M, N)$ is a $\mathcal{D}$-module with the scalar multiplication $(d\varphi)(x) := d\varphi(x) = \varphi(dx)$ for $d \in \mathcal{D}$, $\varphi \in \mathrm{Hom}_{\mathcal{D}}(M, N)$, and $x \in M$. Since N is a $\mathcal{D}_T$-module, $\mathrm{Hom}_{\mathcal{D}}(M, N)$ is even a $\mathcal{D}_T$-module with the scalar multiplication $\left(\frac{d}{t}\varphi\right)(x) := \frac{d}{t}\varphi(x)$. The rest of the proof for modules is analogous to that for rings in item 1 and is omitted. □

Corollary 9.1.8 *1. Let $T_1 \subseteq T_2 \subseteq \mathcal{D}$ be two multiplicative submonoids. Then $\mathrm{can}_{T_2} \colon \mathcal{D} \longrightarrow \mathcal{D}_{T_2}$ induces the ring homomorphism*

$$\mathrm{can}\colon \mathcal{D}_{T_1} \longrightarrow \mathcal{D}_{T_2}, \quad \tfrac{d}{t_1} \longmapsto \tfrac{d}{t_1}.$$

2. If T_2 and thus T_1 contain no zero divisors, the map can *is injective, and we identify $\mathcal{D} \subseteq \mathcal{D}_{T_1} \subseteq \mathcal{D}_{T_2}$.*
This applies in particular if $\mathcal{D}$ is an integral domain and $T_1 \subseteq T_2 := \mathcal{D} \setminus \{0\}$. Then $K := \mathcal{D}_{T_2}$ is the quotient field of $\mathcal{D}$ and $\mathcal{D} \subseteq \mathcal{D}_T \subseteq K = \mathrm{quot}(\mathcal{D})$.

3. Let T_1 be an arbitrary multiplicative submonoid and let

$$T_2 := \{\text{all divisors of elements in } T_1\} = \{t_2 \in \mathcal{D};\ \exists d \in \mathcal{D}\colon\ dt_2 \in T_1\}$$

be the saturation of T_1. Then can: $\mathcal{D}_{T_1} \xrightarrow{\cong} \mathcal{D}_{T_2}$ is an isomorphism. Therefore, we may assume without loss of generality that the submonoid T is saturated, and we will always do this except when stated otherwise.

Proof 1. The map $\varphi := \mathrm{can}_{T_2}\colon\ \mathcal{D} \longrightarrow \mathcal{D}_{T_2}$ satisfies $\varphi(T_1) \subseteq \varphi(T_2) \subseteq \mathrm{U}(\mathcal{D}_{T_2})$. According to the universal property of $\mathrm{can}_{T_1}\colon\ \mathcal{D} \longrightarrow \mathcal{D}_{T_1}$, it induces the homomorphism $\mathrm{can} := \psi\colon \mathcal{D}_{T_1} \longrightarrow \mathcal{D}_{T_2}, \frac{d}{t_1} \longmapsto \frac{d}{t_1}$.
2. Let $\frac{d}{t_1} \in \mathcal{D}_{T_1}$ with $0_{\mathcal{D}_{T_2}} = \mathrm{can}\left(\frac{d}{t_1}\right) = \frac{d}{t_1}$. Then there is a $t_2 \in T_2$ with $t_2 d = 0$. By assumption T_2 contains no zero divisors, hence $d = 0$ and $\frac{d}{t_1} = 0_{\mathcal{D}_{T_1}}$.
3. Obviously T_2 is a saturated submonoid and contains T_1.
To show that can is surjective, let $\frac{d_2}{t_2} \in \mathcal{D}_{T_2}$. Then

$$\exists d \in \mathcal{D} \text{ with } t_1 := dt_2 \in T_1 \Longrightarrow d_2 t_1 = d_2 d t_2 \Longrightarrow \mathrm{can}\left(\frac{dd_2}{t_1}\right) = \frac{d_2}{t_2} \in \mathcal{D}_{T_2}.$$

To show that can is injective, let

$$\frac{d_1}{t_1} \in \mathcal{D}_{T_1} \text{ with } \mathrm{can}\left(\frac{d_1}{t_1}\right) = \frac{d_1}{t_1} = 0_{\mathcal{D}_{T_2}} \Longrightarrow \exists t_2 \in T_2 \text{ with } t_2 d_1 = 0 \text{ and}$$
$$\exists d \in \mathcal{D} \text{ with } t_1' = dt_2 \in T_1 \Longrightarrow t_1' d_1 = dt_2 d_1 = 0 \Longrightarrow \frac{d_1}{t_1} = \frac{d_1 t_1'}{t_1 t_1'} = \frac{0}{t_1 t_1'} = 0_{\mathcal{D}_{T_1}}.$$

□

Lemma 9.1.9 (The functor $(-)_T$)

1. For $f \in \mathrm{Hom}_{\mathcal{D}}(M, N)$ the map $f_T\colon\ M_T \longrightarrow N_T,\ \frac{x}{t} \longmapsto \frac{f(x)}{t}$, is well defined and $\mathcal{D}_T$-linear, and the diagram

$$\begin{array}{ccc} M & \xrightarrow{\ \mathrm{can}_M\ } & M_T \\ \big\downarrow{\scriptstyle f} & & \big\downarrow{\scriptstyle f_T} \\ N & \xrightarrow{\ \mathrm{can}_N\ } & N_T \end{array}$$

is commutative, i.e., $\mathrm{can}_N\, f = f_T\, \mathrm{can}_M$.

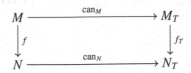

2. *The assignment*

$$(-)_T\colon\ \operatorname{mod}_{\mathcal{D}} \longrightarrow \operatorname{mod}_{\mathcal{D}_T}, \qquad \begin{array}{c} M \\ \Big\downarrow f \\ N \end{array} \longmapsto \begin{array}{c} M_T \\ \Big\downarrow f_T \\ N_T \end{array} \qquad \begin{array}{c} \frac{x}{t} \\ \Big\downarrow \\ f_T\left(\frac{x}{t}\right) = \frac{f(x)}{t} \end{array}$$

is an additive covariant functor, i.e.,

(a) $(\mathrm{id}_M)_T = \mathrm{id}_{M_T}$,
(b) if $M_1 \xrightarrow{f} M_2 \xrightarrow{g} M_3$, *then* $(gf)_T = g_T f_T$, *and*
(c) if $f, g\colon M \longrightarrow N$, *then* $(f+g)_T = f_T + g_T$.

Proof To show that f_T is well defined, let $\frac{x}{t} = \frac{x'}{t'} \in M_T$. Then there is an $s \in T$ with $st'x = stx'$. Since f is $\mathcal{D}$-linear, this implies $st'f(x) = stf(x')$, i.e., $\frac{f(x)}{t} = \frac{f(x')}{t'}$. All the other assertions of the lemma can be easily checked. □

The following theorem is the most important result about quotient modules.

Theorem 9.1.10 (The exactness of $(-)_T$) *The functor* $(-)_T$ *is exact, i.e., if*

$$M_1 \xrightarrow{f_1} M_2 \xrightarrow{f_2} M_3 \tag{9.16}$$

is an exact sequence, then so is

$$M_{1,T} \xrightarrow{f_{1,T}} M_{2,T} \xrightarrow{f_{2,T}} M_{3,T}. \tag{9.17}$$

In particular, if f *is a monomorphism then so is* f_T. *Therefore, if* $M_1 \subseteq M_2$ *is a submodule and thus the map* $\mathrm{inj}\colon\ M_1 \xrightarrow{\subseteq} M_2$ *is injective, then also* $\mathrm{inj}_T\colon\ M_{1,T} \longrightarrow M_{2,T}$ *is injective, and we identify*

$$M_{1,T} \subseteq M_{2,T} \text{ via } \tfrac{x_1}{t} = \mathrm{inj}_T\left(\tfrac{x_1}{t}\right) = \tfrac{\mathrm{inj}(x_1)}{t} = \tfrac{x_1}{t}.$$

Proof The sequence (9.17) is a complex, i.e., $\mathrm{im}\left(f_{1,T}\right) \subseteq \ker\left(f_{2,T}\right)$, since

$$f_{2,T} f_{1,T} = (f_2 f_1)_T \underset{f_2 f_1 = 0}{=} \left(M_1 \xrightarrow{0} M_3\right)_T = \left(M_{1,T} \xrightarrow{0} M_{3,T}\right).$$

To show that $\ker(f_{2,T}) \subseteq \mathrm{im}(f_{1,T})$, let $\frac{x_2}{t_2} \in \ker(f_{2,T})$, hence

$$0_{M_{3,T}} = f_{2,T}\left(\tfrac{x_2}{t_2}\right) = \tfrac{f_2(x_2)}{t_2} \Longrightarrow \exists t_3 \in T \text{ with } 0_{M_3} = t_3 f_2(x_2) = f_2(t_3 x_2)$$
$$\Longrightarrow t_3 x_2 \in \ker(f_2) = \operatorname{im}(f_1) \Longrightarrow \exists x_1 \text{ with } t_3 x_2 = f(x_1)$$
$$\Longrightarrow \tfrac{x_2}{t_2} = \tfrac{t_3 x_2}{t_3 t_2} = \tfrac{f_1(x_1)}{t_3 t_2} = f_{1,T}\left(\tfrac{x_1}{t_3 t_2}\right) \in \operatorname{im}(f_{1,T}).$$
□

Corollary 9.1.11 *If $V \subseteq M_T$ is a $\mathcal{D}_T$-submodule, then $\operatorname{can}_M^{-1}(V)_T = V$.*

Proof $\subseteq$: Let $x \in \operatorname{can}_M^{-1}(V)$, $t \in T \Longrightarrow \frac{x}{t} = \frac{1}{t}\frac{x}{1} = \frac{1}{t}\operatorname{can}_M(x) \in \mathcal{D}_T V = V$.
$\supseteq$: Let $\frac{x}{t} \in V$. Then $\operatorname{can}_M(x) = \frac{x}{1} = \frac{t}{1}\frac{x}{t} \in \mathcal{D}_T V = V$, hence $x \in \operatorname{can}_M^{-1}(V)$ and $\frac{x}{t} \in \operatorname{can}_M^{-1}(V)_T$. □

Corollary 9.1.12 *If M is a finitely generated resp. Noetherian $\mathcal{D}$-module, then M_T is a finitely generated resp. Noetherian $\mathcal{D}_T$-module.*

Proof (i) If $M = \mathcal{D}x_1 + \cdots + \mathcal{D}x_m$, then $M_T = \mathcal{D}_T\frac{x_1}{1} + \cdots + \mathcal{D}_T\frac{x_m}{1}$.
(ii) A module is Noetherian if and only if all its submodules are finitely generated, see Definition 3.2.13. Let ${}_{\mathcal{D}}M$ be Noetherian and $V \subseteq M_T$ a $\mathcal{D}_T$-submodule. Then $\operatorname{can}_M^{-1}(V)$ is $\mathcal{D}$-finitely generated since ${}_{\mathcal{D}}M$ is Noetherian. By item 1, the $\mathcal{D}_T$-module $V \underset{\text{Corollary 9.1.11}}{=} \operatorname{can}_M^{-1}(V)_T$ is also finitely generated. □

Corollary 9.1.13 *If M' is a $\mathcal{D}$-submodule of M, the map*

$$(M/M')_T \xrightarrow{\cong} M_T/M'_T, \quad \frac{x+M'}{t} \longmapsto \frac{x}{t} + M'_T, \tag{9.18}$$

is well defined and a $\mathcal{D}_T$-isomorphism. Thus we identify $(M/M')_T = M_T/M'_T$.

Proof The sequence $0 \longrightarrow M' \xrightarrow{\text{inj}} M \xrightarrow{\text{can}} M/M' \longrightarrow 0$ is exact and hence, by Theorem 9.1.10, so is the sequence $0 \to M'_T \xrightarrow{\text{inj}_T} M_T \xrightarrow{\text{can}_T} (M/M')_T \to 0$. The homomorphism theorem (Theorem 3.1.31) furnishes the isomorphism (9.18). □

Corollary 9.1.14 *1. The equivalence $\mathcal{D}_T = 0 \Longleftrightarrow 0 \in T$ holds. In general, we will therefore assume that $0 \notin T$.*
2. The equivalence $M_T = 0 \Longleftrightarrow \mathrm{t}_T(M) = M$ holds. If $\mathcal{D}$ is an integral domain and $0 \notin T$, this signifies that M is a T-torsion module.
3. Assume that $\mathcal{D}$ is an integral domain and that ${}_{\mathcal{D}}M$ is finitely generated. Then $M_T = 0$ if and only if $T \cap \operatorname{ann}_{\mathcal{D}}(M) \neq \emptyset$.

Proof 1. $\mathcal{D}_T = 0 \Longleftrightarrow \frac{1}{1} = 1_{\mathcal{D}_T} = 0_{\mathcal{D}_T} = \frac{0}{1} \Longleftrightarrow \exists t \in T\colon\ t = t\cdot 1 = t \cdot 0 = 0$.
2. If $M_T = 0$, then $\mathrm{t}_T(M) = \operatorname{can}_M^{-1}(0) = \ker(\operatorname{can}_M\colon\ M \to M_T) \underset{M_T=0}{=} M$.

If, conversely, $M = \mathrm{t}_T(M) = \operatorname{can}_M^{-1}(0)$, then $M_T = \operatorname{can}_M^{-1}(0)_T \underset{\text{Corollary 9.1.11}}{=} 0$.
3. Let $M = \mathcal{D}x_1 + \cdots + \mathcal{D}x_n$. If $M_T = 0$, then $\mathrm{t}_T(M) = M$ by item 2. Therefore, there are $t_i \in T$ with $t_i x_i = 0$ for all i. Then the finite product $t := t_1 \cdots t_n \in T$ annihilates all x_i, and thus $tM = 0$, i.e., $t \in T \cap \operatorname{ann}_{\mathcal{D}}(M)$. If, conversely $t \in T \cap \operatorname{ann}_{\mathcal{D}}(M)$ and $\frac{x}{t_1} \in M_T$, then $\frac{x}{t_1} = \frac{tx}{tt_1} = \frac{0}{tt_1} = 0_{M_T}$. □

Theorem 9.1.15 *For a $\mathcal{D}$-module M the following properties are equivalent:*

1. *M is also a $\mathcal{D}_T$-module such that $\frac{d}{1}x = dx$ for all $d \in \mathcal{D}$ and $x \in M$.*
2. *For every $t \in T$ the map $t\cdot\colon\ M \longrightarrow M,\ x \longmapsto tx$, is an isomorphism.*
3. *The canonical map $\mathrm{can}_M\colon\ M \longrightarrow M_T$ is an isomorphism.*

Proof 1. $\Longrightarrow$ 2.: For $t \in T$ the map $t\cdot\colon\ M \longrightarrow M,\ x \mapsto tx = \frac{t}{1}x$, is an isomorphism with inverse $x \mapsto \frac{1}{t}x$.
2. $\Longrightarrow$ 3.: Let $x \in M$ with $0_{M_T} = \mathrm{can}_M(x) = \frac{x}{1}$. Then there is a $t \in T$ with $tx = 0$. Since $t\cdot$ is injective, this implies $x = 0$. Hence can_M is injective.
To prove its surjectivity let $\frac{x}{t} \in M_T$. By 2 there is an $x' \in M$ with $x = tx'$. Therefore, $\frac{x}{t} = \frac{tx'}{t} = \frac{x'}{1} = \mathrm{can}_M(x')$.
3. $\Longrightarrow$ 1.: Since $\mathrm{can}_M\colon\ M \longrightarrow M_T$ is an isomorphism and M_T is a $\mathcal{D}_T$-module, so is M by transport of structure. In detail, the $\mathcal{D}_T$-scalar multiplication on M with the asserted property is

$$\frac{d}{t}x := \mathrm{can}_M^{-1}\left(\frac{d}{t}\,\mathrm{can}_M(x)\right) = \mathrm{can}_M^{-1}\left(\frac{dx}{t}\right) \text{ for } x \in M,\ d \in \mathcal{D},\ t \in T. \qquad \square$$

9.2 Quotient Modules for Principal Ideal Domains

In this section we specialize the theory of Sect. 9.1 *to the case of a principal ideal domain $\mathcal{D}$ with its quotient field $K := \mathcal{D}_{\mathcal{D}\setminus\{0\}}$. In Theorem* 9.2.10 *and Corollary* 9.2.14 *we characterize left and right invertibility of matrices $R \in \mathcal{D}^{k\times l}$ over quotient rings $\mathcal{D}_T$ and establish the structure of the quotient module M_T for $M = \mathcal{D}^{1\times l}/\mathcal{D}^{1\times k}R$.*

Let $\mathcal{P} \subset \mathcal{D}$ be a representative system of the prime elements of $\mathcal{D}$, up to association. Such a representative system $\mathcal{P}$ is characterized by the property that the map

$$\mathrm{U}(\mathcal{D}) \times \mathbb{Z}^{(\mathcal{P})} \overset{\cong}{\longleftrightarrow} K \setminus \{0\},\quad \left(u, (\mu_q)_{q\in\mathcal{P}}\right) \longleftrightarrow a = u\prod_{q\in\mathcal{P}} q^{\mu_q}, \tag{9.19}$$

is a group isomorphism. The number $\mu_q \in \mathbb{Z}$ is the *order* of a with respect to $q \in \mathcal{P}$, and it is denoted by $\mathrm{ord}_q(a) := \mu_q$. Of course, almost all orders, i.e., all but finitely many, are zero. This representation is the unique prime factor decomposition of elements of the quotient field. The homomorphism property implies $\mathrm{ord}_q(ab) = \mathrm{ord}_q(a) + \mathrm{ord}_q(b)$ for $a, b \neq 0$. Moreover, we have that

$$\begin{aligned} a \in \mathcal{D}\setminus\{0\} &\Longleftrightarrow \forall q \in \mathcal{P}\colon\ \mathrm{ord}_q(a) \geq 0 \quad \text{and} \\ a \in \mathrm{U}(\mathcal{D}) &\Longleftrightarrow \forall q \in \mathcal{P}\colon\ \mathrm{ord}_q(a) = 0. \end{aligned} \tag{9.20}$$

We have shown in Lemma and Definition 9.1.2 that a saturated submonoid $T \subseteq \mathcal{D}$ gives rise to the disjoint decomposition

$$\mathcal{P} = \mathcal{P}_1 \uplus \mathcal{P}_2, \quad \text{where } \mathcal{P}_1 := \mathcal{P}_1(T) = \{q \in \mathcal{P};\ \exists t \in T:\ q \mid t\} = T \cap \mathcal{P} \ \text{ and}$$
$$\mathcal{P}_2 := \mathcal{P}_2(T) = \mathcal{P} \setminus \mathcal{P}_1 = \mathcal{P} \setminus T.$$

The submonoid T can be reobtained from $\mathcal{P}_1$ and $\mathcal{P}_2$ via

$$\begin{aligned} T = T(\mathcal{P}_1) &= \left\{t \in K \setminus \{0\};\ \forall q \in \mathcal{P}_1:\ \operatorname{ord}_q(t) \geq 0,\ \forall q \in \mathcal{P}_2:\ \operatorname{ord}_q(t) = 0\right\} \\ &= \left\{t = u \prod_{q \in \mathcal{P}_1} q^{\operatorname{ord}_q(t)};\ u \in \mathrm{U}(\mathcal{D}),\ \operatorname{ord}_q(t) \geq 0\right\} \subseteq \mathcal{D} \setminus \{0\}. \end{aligned} \tag{9.21}$$

Lemma 9.2.1 *1. The inclusions $\mathcal{D} \subset \mathcal{D}_T \subset K$ hold via the identification $d = \frac{d}{1}$ for $d \in \mathcal{D}$.*
2. $\mathcal{D}_T \setminus \{0\} = \{a \in K \setminus \{0\};\ \forall q \in \mathcal{P}_2:\ \operatorname{ord}_q(a) \geq 0\}$.
3. By restricting the isomorphism (9.19) to the units of $\mathcal{D}_T$ we obtain

$$\begin{aligned} \mathrm{U}(\mathcal{D}) \times \mathbb{Z}^{(\mathcal{P}_1)} &\overset{\cong}{\longleftrightarrow} \mathrm{U}(\mathcal{D}_T) = \left\{\tfrac{t_1}{t_2};\ t_1, t_2 \in T\right\} \\ &\qquad\qquad = \{a \in K \setminus \{0\};\ \forall q \in \mathcal{P}_2:\ \operatorname{ord}_q(a) = 0\}, \\ \left(u, (\mu_q)_{q \in \mathcal{P}_1}\right) &\longleftrightarrow u \prod_{q \in \mathcal{P}_1} q^{\mu_q}. \end{aligned}$$

4. $T = \mathcal{D} \cap \mathrm{U}(\mathcal{D}_T)$.

Proof These properties follow directly from the constructions of T and $\mathcal{D}_T$, and from the unique prime factor decomposition (9.19). □

In the sequel we will always consider $\mathcal{D}$ as a subring of $\mathcal{D}_T$ and of K and identify $d = \frac{d}{1}$ for $d \in \mathcal{D}$ like $\mathbb{Z} \subset \mathbb{Q}$, $5 = \frac{5}{1}$, or $F[s] \subset F(s)$.

Theorem 9.2.2 *Assume a multiplicative submonoid T and let $\mathcal{P} = \mathcal{P}_1 \uplus \mathcal{P}_2$ be the induced decomposition from Lemma and Definition 9.1.2. Then the integral domain $\mathcal{D}_T$ is a principal ideal domain with quotient field $K = \operatorname{quot}(\mathcal{D}) = \operatorname{quot}(\mathcal{D}_T)$. In particular, $\mathcal{D}_T$ is a factorial ring. The set $\mathcal{P}_2$ is a representative system of prime elements of $\mathcal{D}_T$, up to association. Let $0 \neq a \in K$ and let $a = u \prod_{q \in \mathcal{P}} q^{\operatorname{ord}_q(a)}$ be the prime factor decomposition of $a \in K$ with respect to $\mathcal{D}$ as in (9.19). Then*

$$a = \left(u \prod_{q \in \mathcal{P}_1} q^{\operatorname{ord}_q(a)}\right) \prod_{q \in \mathcal{P}_2} q^{\operatorname{ord}_q(a)} \quad \text{with} \quad u \prod_{q \in \mathcal{P}_1} q^{\operatorname{ord}_q(a)} \in \mathrm{U}(\mathcal{D}_T)$$

is the prime factor decomposition of $a \in K$ with respect to $\mathcal{D}_T$.

Proof Let $\mathfrak{b} \subseteq \mathcal{D}_T$ be an ideal and $\mathfrak{a} := \operatorname{can}_{\mathcal{D}}^{-1}(\mathfrak{b})$ its inverse image. Since $\mathcal{D}$ is a principal ideal domain, there is some $a \in \mathfrak{a}$ with $\mathfrak{a} = \mathcal{D}a$. This implies

$$\mathfrak{b} \underset{\text{Corollary 9.1.11}}{=} \operatorname{can}_{\mathcal{D}}^{-1}(\mathfrak{b})_T = \mathfrak{a}_T = \mathcal{D}_T a.$$

Hence $\mathfrak{b}$ is a principal ideal and $\mathcal{D}_T$ a principal ideal domain.
The disjoint decomposition $\mathcal{P} = \mathcal{P}_1 \uplus \mathcal{P}_2$ induces the group isomorphism

$$\mathbb{Z}^{(\mathcal{P})} \overset{\cong}{\longleftrightarrow} \mathbb{Z}^{(\mathcal{P}_1)} \times \mathbb{Z}^{(\mathcal{P}_2)}, \quad \mu = (\mu_q)_{q\in\mathcal{P}} \longleftrightarrow (\mu^1, \mu^2), \quad \text{where } \mu^i := (\mu_q)_{q\in\mathcal{P}_i}.$$

With item 3 of Lemma 9.2.1 we infer the group isomorphism

$$\begin{array}{c} \mathrm{U}(\mathcal{D}_T) \times \mathbb{Z}^{(\mathcal{P}_2)} \overset{\cong}{\longleftrightarrow} \left(\mathrm{U}(\mathcal{D}) \times \mathbb{Z}^{(\mathcal{P}_1)}\right) \times \mathbb{Z}^{(\mathcal{P}_2)} \overset{\cong}{\longleftrightarrow} \mathrm{U}(\mathcal{D}) \times \mathbb{Z}^{(\mathcal{P})} \overset{\cong}{\longleftrightarrow} K \setminus \{0\}, \\ \left(u \prod_{q\in\mathcal{P}_1} q^{\mu_q}, (\mu_q)_{q\in\mathcal{P}_2}\right) \longleftrightarrow u \prod_{q\in\mathcal{P}} q^{\mu_q}. \end{array}$$

This implies that $\mathcal{P}_2$ is a representative system of the prime elements of $\mathcal{D}_T$. □

Corollary and Definition 9.2.3 (Discrete valuation ring) *Let $q_2 \in \mathcal{P}$. We consider the decomposition $\mathcal{P} = \left(\mathcal{P} \setminus \{q_2\}\right) \uplus \{q_2\}$ that gives rise to*

$$\begin{aligned} T(q_2) &:= T(\mathcal{P} \setminus \{q_2\}) = \{t \in \mathcal{D};\ q_2 \nmid t\} = \mathcal{D} \setminus \mathcal{D}q_2 \text{ and} \\ \mathcal{D}_{T(q_2)} &= \left\{\tfrac{d}{t} \in K;\ d, t \in \mathcal{D},\ q_2 \nmid t\right\}. \end{aligned}$$

In the literature on commutative algebra this quotient ring is usually denoted by $\mathcal{D}_{\mathcal{D}q_2}$. This notation is confusing and we will not use it.
The ring $\mathcal{D}_{T(q_2)}$ is a principal ideal domain with the unique prime element q_2, up to association, and is also called a discrete valuation ring. *Let $a = u \prod_{q\in\mathcal{P}} q^{\mathrm{ord}_q(a)} \in K \setminus \{0\}$ be the prime factor decomposition with respect to $\mathcal{D}$. Then that with respect to $\mathcal{D}_{T(q_2)}$ is*

$$a = \left(u \prod_{q\in\mathcal{P}\setminus\{q_2\}} q^{\mathrm{ord}_q(a)}\right) q_2^{\mathrm{ord}_{q_2}(a)}, \quad \text{where } u \prod_{q\in\mathcal{P}\setminus\{q_2\}} q^{\mathrm{ord}_q(a)} \in \mathrm{U}(\mathcal{D}_{T(q_2)}). \tag{9.22}$$

The map $\mathrm{ord}_{q_2}:\ K \setminus \{0\} \longrightarrow \mathbb{Z}$ is called the discrete valuation *associated with the prime element q_2.* ◊

Example 9.2.4 (*Proper rational functions*) Let F be a field, s an indeterminate, and $z := s^{-1} \in F(s) = F(z)$ its inverse. In Theorem 8.1.7 we have shown that the ring of proper rational functions can be represented as

$$\begin{aligned} F(s)_{\mathrm{pr}} &= \left\{\tfrac{a(z)}{b(z)};\ a, b \in F[z],\ b(0) \neq 0\right\} = F[z]_{T(z)} \text{ where } z \in F[z] \text{ prime}, \\ T(z) &:= F[z] \setminus F[z]z = \{b \in F[z];\ z \nmid b\} = \{b \in F[z];\ b(0) \neq 0\}. \end{aligned} \tag{9.23}$$

In particular, the ring of proper rational function is a discrete valuation ring with the unique prime element z (up to association) and the discrete valuation

$$\begin{aligned}\operatorname{ord}_z:\ & F(z)\setminus\{0\}\longrightarrow \mathbb{Z},\\ r=\tfrac{a_m s^m+\cdots}{b_n s^n+\cdots}=\tfrac{a_m+a_{m-1}z+\cdots}{b_n+b_{n-1}z+\cdots}z^{n-m}&\longmapsto \operatorname{ord}_z(r)=n-m=-\deg_s(r),\end{aligned}\tag{9.24}$$

where $a_m\neq 0$ and $b_n\neq 0$, and $\frac{a_m+a_{m-1}z+\cdots}{b_n+b_{n-1}z+\cdots}$ is a unit in $F(s)_{\mathrm{pr}}$. By Corollary 8.1.9 the ideal $F(s)_{\mathrm{pr}}z$ is the ideal $F(s)_{\mathrm{spr}}$ of strictly proper rational functions. ◇

Next we discuss the Smith/McMillan form with respect to a quotient ring $\mathcal{D}_T$, where T is the multiplicative submonoid from (9.21), and we trace it back to the Smith form with respect to $\mathcal{D}$.

Corollary 9.2.5 *Consider the inclusions $\mathcal{D}\subseteq\mathcal{D}_T\subseteq K=\operatorname{quot}(\mathcal{D})$. Let $R\in K^{k\times l}$ be any matrix with its Smith/McMillan form with respect to $\mathcal{D}$ according to Corollary 3.2.19, i.e.,*

$$\begin{aligned}&XRY=\begin{pmatrix}\operatorname{diag}(e_1,\ldots,e_p) & 0\\ 0 & 0\end{pmatrix}=S\in K^{k\times l}\\ &\text{with } p=\operatorname{rank}(R),\ X\in \operatorname{Gl}_k(\mathcal{D})\subseteq \operatorname{Gl}_k(\mathcal{D}_T),\ Y\in\operatorname{Gl}_l(\mathcal{D})\subseteq\operatorname{Gl}_l(\mathcal{D}_T)\\ &\text{and } e_1\underset{\mathcal{D}}{\mid} e_2\underset{\mathcal{D}}{\mid}\cdots\underset{\mathcal{D}}{\mid} e_p,\ 0\neq e_i\in K,\ \text{hence } e_1\underset{\mathcal{D}_T}{\mid} e_2\underset{\mathcal{D}_T}{\mid}\cdots\underset{\mathcal{D}_T}{\mid} e_p.\end{aligned}\tag{9.25}$$

These equations and the uniqueness of the Smith form imply that S is the Smith form of R both with respect to $\mathcal{D}$ and to $\mathcal{D}_T$.

This is applicable to the polynomial algebra $\mathcal{D}=F[s]\subseteq K=F(s)$ over a field F and especially to the ring $F(s)_{\mathrm{pr}}=F[z]_{T(z)}$ of proper rational functions from Example 9.2.4 and, later, to the ring of proper and stable rational functions. ◇

In the following lemma we use the Smith form for the solution of linear matrix equations where the given data have entries in K, but the solutions have entries in $\mathcal{D}$, and apply this to $\mathcal{D}_T$.

Lemma 9.2.6 *We consider the matrix equation*

$$Q=ZR\quad\text{with given } Q\in K^{m\times l},\ R\in K^{k\times l}\ \text{and unknown } Z\in\mathcal{D}^{m\times k},\tag{9.26}$$

and the Smith/McMillan form of R according to (9.25). The equation $Q=ZR$ is equivalent to

$$QY=ZX^{-1}XRY=Z'\begin{pmatrix}\operatorname{diag}(e_1,\ldots,e_p) & 0\\ 0 & 0\end{pmatrix}\quad\text{with } Z':=ZX^{-1}\in\mathcal{D}^{m\times k},$$

or, written differently, to

$$QY_{-j}=\begin{cases}Z'_{-j}e_j & \text{if } j\in\{1,\ldots,p\}\\ 0 & \text{if } j\in\{p+1,\ldots,l\}\end{cases}.\tag{9.27}$$

1. *With the data from (9.26) the equation $Q=ZR$ has a solution $Z\in\mathcal{D}^{m\times k}$ if and only if (9.27) has a solution $Z'\in\mathcal{D}^{m\times k}$. This is the case if and only if*

$$QY_{-j} \in \mathcal{D}^m e_j \quad \textit{for } j \in \{1, \dots, p\} \textit{ and} \\ QY_{-j} = 0 \quad\quad \textit{for } j \in \{p+1, \dots, l\}. \tag{9.28}$$

If (9.28) holds, one solution $Z'^0 \in \mathcal{D}^{m\times k}$ of (9.27) is given by

$$Z'^0_{-j} := \begin{cases} QY_{-j}e_j^{-1} \in \mathcal{D}^m & \text{if } j \in \{1,\dots,p\}, \\ 0 \in \mathcal{D}^m & \text{if } j \in \{p+1,\dots,l\}. \end{cases} \tag{9.29}$$

Consequently, one solution of $Q = ZR$ is $Z^0 := Z'^0 X \in \mathcal{D}^{m\times k}$.

2. *With $X = \begin{pmatrix} X_I \\ X_{II} \end{pmatrix} \in \mathcal{D}^{(p+(k-p))\times l}$ the matrix X_{II} is a universal left annihilator of R by Theorem 3.2.18, hence $\left\{\xi \in \mathcal{D}^{1\times k};\ \xi R = 0\right\} = \mathcal{D}^{1\times(k-p)}X_{II}$ and*

$$\{Z \in \mathcal{D}^{m\times k};\ Q = ZR\} = Z^0 + \mathcal{D}^{m\times(k-p)}X_{II} \\ = \{Z^0 + MX_{II};\ M \in \mathcal{D}^{m\times(k-p)}\}. \tag{9.30}$$

Proof The proof follows directly from the indicated equations. □

Corollary 9.2.7 *In particular, Lemma 9.2.6 implies that matrix equations $Q = ZR$ where Q and R have entries in K and Z has entries in $\mathcal{D}_T$ can be computed by means of the Smith form of R with respect to $\mathcal{D}$. The matrix X_{II} is a universal left annihilator over $\mathcal{D}$ and over $\mathcal{D}_T$.* ◇

Lemma 9.2.8 *Assume a matrix $R \in K^{k\times l}$ with its Smith/McMillan form $S = XRY = \begin{pmatrix} E & 0 \\ 0 & 0 \end{pmatrix}$, $E := \operatorname{diag}(e_1, \dots, e_p)$, from (9.25).*

1. *The following statements are equivalent:*

   (a) $\operatorname{rank}(R) = l$ *and* $e_l^{-1} \in \mathcal{D}$.
   (b) $\operatorname{rank}(R) = l$ *and* $E^{-1} \in \mathcal{D}^{l\times l}$.
   (c) The matrix R has a left inverse in $\mathcal{D}^{l\times k}$. One of these left inverses is $Z^0 := Y(E^{-1}, 0)X$.

2. *If the conditions of item 1 are satisfied with Z^0 from (c), then all left inverses in $\mathcal{D}^{l\times k}$ form the affine submodule*

$$Z^0 + \mathcal{D}^{l\times(k-l)}X_{II} \subseteq \mathcal{D}^{l\times k} \text{ where } X = \begin{pmatrix} X_I \\ X_{II} \end{pmatrix} \in \mathcal{D}^{(l+(k-l))\times k}.$$

If $Q \in \mathcal{D}^{m\times l}$, then the solution set of the matrix equation $Q = YR$, $Y \in \mathcal{D}^{m\times k}$, is $QZ^0 + \mathcal{D}^{m\times(k-l)}X_{II}$.

For $R \in \mathcal{D}^{k\times \ell}$ we reobtain the result from Theorem 3.3.23.

Proof 1. All three conditions imply that $p = l$; hence, we assume this and conclude that $S = \begin{pmatrix} E \\ 0 \end{pmatrix}$ and $(E^{-1}, 0)S = \operatorname{id}_l$, where $E^{-1} = \operatorname{diag}(e_1^{-1}, \dots, e_\ell^{-1})$. Then the lemma is the special case of Lemma 9.2.6 with $m = l$ and $Q = \operatorname{id}_l$.

$(b) \Longrightarrow (a)$: This is obvious.
$(a) \Longrightarrow (b)$: The conditions $e_{i-1} \underset{\mathcal{D}}{|} e_i$ imply that $e_i^{-1} \underset{\mathcal{D}}{|} e_{i-1}^{-1}$ for all i. Hence

$$e_l^{-1} \in \mathcal{D} \Longleftrightarrow \forall i = 1, \dots, l : e_i^{-1} \in \mathcal{D} \Longleftrightarrow E^{-1} \in \mathcal{D}^{l \times l}.$$

$(b) \Longleftrightarrow (c)$: According to (9.27) there is Z with $\mathrm{id}_l = Q = ZR$ if and only if

$$YE^{-1} \in \mathcal{D}^{l \times l} \underset{Y \in \mathrm{Gl}_l(\mathcal{D})}{\Longleftrightarrow} E^{-1} \in \mathcal{D}^{l \times l} \text{ and then}$$

$$Z'^0 = (YE^{-1}, 0) = Y(E^{-1}, 0),\ Z = Z'^0 X = Y(E^{-1}, 0)X,\ \mathrm{id}_l = ZR.$$

2. This is a direct consequence of Lemma 9.2.6, item 2. □

Lemma 9.2.9 *Assume a matrix $R \in K^{k \times l}$ with its Smith form $S = XRY$ from (9.25). Then R is contained in $\mathcal{D}_T^{k \times l}$ if and only if its elementary divisors e_i belong to $\mathcal{D}_T$, i.e., by Theorem 9.2.2, if and only if*

$$\mathrm{ord}_q(e_i) \geq 0 \ \textit{ for all } i = 1, \dots, p \textit{ and } q \in \mathcal{P}_2.$$

Proof Since $X \in \mathrm{Gl}_k(\mathcal{D})$, also $X^{-1} \in \mathrm{Gl}_k(\mathcal{D}) \subseteq \mathrm{Gl}_k(\mathcal{D}_T)$, and likewise, $Y^{-1} \in \mathrm{Gl}_l(\mathcal{D}_T)$. If the e_i belong to $\mathcal{D}_T$, this implies

$$R = X^{-1} \begin{pmatrix} \mathrm{diag}(e_1, \dots, e_p) & 0 \\ 0 & 0 \end{pmatrix} Y^{-1} \in \mathcal{D}_T^{k \times l}.$$

If, conversely,

$$R \in \mathcal{D}_T^{k \times l} \Longrightarrow \begin{pmatrix} \mathrm{diag}(e_1, \dots, e_p) & 0 \\ 0 & 0 \end{pmatrix} = XRY \in \mathcal{D}_T^{k \times l} \Longrightarrow \forall i = 1, \cdots, p : \ e_i \in \mathcal{D}_T.$$

□

In the sequel we assume $R \in \mathcal{D}^{k \times l}$ with its Smith form from (9.25). The matrix R gives rise to its behavior $\mathcal{B}$ and the row resp. factor modules

$$\begin{aligned} &\mathcal{B} := \left\{ w \in \mathcal{F}^l;\ R \circ w = 0 \right\}, \\ &U := \mathcal{D}^{1 \times k} R \subseteq \mathcal{D}^{1 \times l} \text{ and } U_T = \mathcal{D}_T^{1 \times k} R \subseteq \mathcal{D}_T^{1 \times l} \text{ resp.} \\ &M := \mathcal{D}^{1 \times l} / \mathcal{D}^{1 \times k} R \text{ and } M_T = \left(\mathcal{D}^{1 \times l} / U \right)_T \underset{\text{Corollary 9.1.13}}{=} \mathcal{D}_T^{1 \times l} / U_T. \end{aligned} \tag{9.31}$$

In Theorem 3.3.23 and Corollary 3.3.24 we characterized when R is left resp. right invertible, and in item 2 of Theorem 4.2.12 when M is torsionfree. Applying these results to $R \in \mathcal{D}_T^{k \times l}$ and to M_T furnishes the following theorem, which turns out to be significant for the construction of compensators and observers in Chaps. 10 and 11.

Theorem 9.2.10 *Let $R \in \mathcal{D}^{k \times l}$ be a matrix, and let $e_1 \underset{\mathcal{D}}{|} \cdots \underset{\mathcal{D}}{|} e_p$ be its nonzero elementary divisors in $\mathcal{D}$, where $p :=$ rank(R). Let $T \subseteq \mathcal{D}$ be a saturated multiplicative submonoid. The following characterizations hold:*

1. *R has a left inverse in $\mathcal{D}_T^{l\times k}$ if and only if $p = l$ and $e_p \in T$.*
2. *R has a right inverse in $\mathcal{D}_T^{l\times k}$ if and only if $p = k$ and $e_p \in T$.*
3. *The following statements are equivalent:*

   (a) *The $\mathcal{D}_T$-module M_T is torsionfree.*
   (b) *The $\mathcal{D}_T$-module M_T is free.*
   (c) *The row module $U_T = \mathcal{D}_T^{1\times k} R$ is a direct summand of $\mathcal{D}_T^{1\times l}$.*
   (d) *$e_p \in T$.*

The condition $e_p \in T$ signifies that all prime factors of e_p belong to $\mathcal{P}_1$.

Proof 1. For $e_p \in \mathcal{D}$ the equivalence $e_p^{-1} \in \mathcal{D}_T \iff e_p \in \mathcal{D} \cap \mathrm{U}(\mathcal{D}_T) = T$ holds, cf. Lemma 9.2.1, 4. The assertion then follows by applying Lemma 9.2.8 to the principal ideal domain $\mathcal{D}_T$.
2. This follows from item 1 by transposition.
3. The equivalences $(a) \iff (b) \iff (d)$ follow from item 2 of Theorem 4.2.12, applied to $\mathcal{D}_T$.
For the equivalence $(a) \iff (c)$, we assume without loss of generality that $\mathrm{rank}(R) = k$, cf. Theorem and Definition 4.2.8, item 3. This signifies that the map $\cdot R\colon\ \mathcal{D}_T^{1\times k} \longrightarrow \mathcal{D}_T^{1\times l}$ is a monomorphism. By item 2 of Theorem 4.2.12 the module M is torsionfree if and only if R has a right inverse $Y \in \mathcal{D}_T^{l\times k}$ with $RY = \mathrm{id}_k$, in other words, the map

$$\cdot RY\colon\ \mathcal{D}_T^{1\times k} \xrightarrow{\cdot R} \mathcal{D}_T^{1\times l} \xrightarrow{\cdot Y} \mathcal{D}_T^{1\times k}$$

is the identity on $\mathcal{D}_T^{1\times k}$, i.e., the monomorphism $\cdot R$ has the linear retraction $\cdot Y$. By item 1 of Theorem and Definition 4.2.18 this is equivalent to $\mathrm{im}(\cdot R) = \mathcal{D}_T^{1\times k} R = U_T$ being a direct summand of $\mathcal{D}_T^{1\times l}$.

□

Remark 9.2.11 Assume $\mathcal{D} = F[s]$ with $F = \mathbb{R}$ or $F = \mathbb{C}$, a decomposition $\mathbb{C} = \Lambda_1 \uplus \Lambda_2$ and the associated monoid $T := \{t \in F[s] \setminus \{0\};\ \mathrm{V}_{\mathbb{C}}(t) \subseteq \Lambda_1\}$ of stable polynomials according to item 1 of Example 9.1.3 be given. The characteristic variety of the behavior $\mathcal{B}$ according to Definition and Corollary 5.4.1 is $\mathrm{char}(\mathcal{B}) := \mathrm{V}_{\mathbb{C}}(e_p)$. Hence $e_p \in T$ in Theorem 9.2.10 means $\mathrm{char}(\mathcal{B}) \subseteq \Lambda_1$. ◇

Lemma 9.2.12 *Let N_i, $i \in I$, be a family of $\mathcal{D}$-modules. Then $(\bigoplus_{i\in I} N_i)_T = \bigoplus_{i\in I} N_{i,T}$. If, in particular, $V = \bigoplus_{i\in I} V_i$ is a direct sum of submodules of a module N, then $V_T = \bigoplus_{i\in I} V_{i,T} \subseteq N_T$.* ◇

Lemma 9.2.13 *For a $\mathcal{D}$-module N, the identity $\mathrm{t}({}_{\mathcal{D}}N)_T = \mathrm{t}({}_{\mathcal{D}_T}N_T)$ holds.*

Proof $\subseteq$. Let $\frac{x}{t} \in \mathrm{t}({}_{\mathcal{D}}N)_T$, and let $0 \neq d \in \mathcal{D}$ with $dx = 0$. Then $\frac{d}{1}\frac{x}{t} = \frac{dx}{t} = \frac{0}{t} = 0$, i.e., $\frac{x}{t} \in \mathrm{t}({}_{\mathcal{D}_T}N_T)$.
$\supseteq$. Conversely, let $\frac{x}{t} \in \mathrm{t}({}_{\mathcal{D}_T}N_T)$, and let $0 \neq \frac{d_1}{t_1} \in \mathcal{D}_T$ with $0 = \frac{d_1}{t_1}\frac{x}{t} = \frac{d_1 x}{t_1 t}$. Then there is a $t_2 \in T$ with $0 = t_2(d_1 x) = (t_2 d_1)x$, and hence $x \in \mathrm{t}({}_{\mathcal{D}}N)$ and $\frac{x}{t} \in \mathrm{t}({}_{\mathcal{D}}N)_T$.

□

Corollary 9.2.14 *The $\mathcal{D}_T$-module $M_T = (\mathcal{D}^{1\times l}/\mathcal{D}^{1\times k}R)_T$ decomposes as*

$$M_T = \bigoplus_{i=1}^{l} \mathcal{D}_T \overline{Y_{i-}^{-1}} \overset{[}{\longleftrightarrow} \mathcal{D}_T] \cong \prod_{i=1}^{p} \mathcal{D}_T/\mathcal{D}_T e_i \times \mathcal{D}_T^{1\times(l-p)},$$
$$\overline{\eta Y^{-1}} \longleftrightarrow (\overline{\eta_1}, \ldots, \overline{\eta_p}, \eta_{p+1}, \ldots, \eta_l). \tag{9.32}$$

Via this map, we obtain the representations

$$\mathrm{t}(M_T) \underset{\text{Lemma 9.2.13}}{=} \mathrm{t}(M)_T = \bigoplus_{i=1}^{p} \mathcal{D}_T \overline{Y_{i-}^{-1}} \overset{\cong}{\longleftrightarrow} \prod_{i=1}^{p} \mathcal{D}_T/\mathcal{D}_T e_i,$$

of the torsion module $\mathrm{t}(M_T)$ *and*

$$M_T/\mathrm{t}(M_T) \underset{\text{Lemma 9.2.13}}{=} M_T/\mathrm{t}(M)_T \underset{\text{Corollary 9.1.13}}{=} (M/\mathrm{t}(M))_T \overset{\cong}{\longleftrightarrow} \mathcal{D}_T^{1\times(l-p)}$$

of the torsionfree factor module $M_T/\mathrm{t}(M_T)$.
In particular, this implies the following statements:

1. *M_T is a torsion module if and only if $p = \mathrm{rank}(R) = l$. Then M itself is a torsion module.*
2. *M_T is torsionfree, hence free, if and only if $e_p \in T$.*
3. *M is a T-torsion module (see (9.15)) if and only if $p = l$ and $e_p \in T$. By the last two items of Corollary 9.1.14, M is a T-torsion module if and only if the equivalent conditions $M_T = 0$, $M = \mathrm{t}_T(M)$, and $T \cap \mathrm{ann}_{\mathcal{D}}(M) \neq \emptyset$ hold.*

Proof Since $M_T = \left(\mathcal{D}^{1\times l}/\mathcal{D}^{1\times k}R\right)_T \underset{\text{Corollary 9.1.13}}{=} \mathcal{D}_T^{1\times l}/\mathcal{D}_T^{1\times k}R$ and since by (9.25) the matrix $XRY = \begin{pmatrix} \mathrm{diag}(e_1,\ldots,e_p) & 0 \\ 0 & 0 \end{pmatrix}$ is the Smith form of R with respect to both $\mathcal{D}$ and $\mathcal{D}_T$, application of Theorem and Definition 4.2.8 to the $\mathcal{D}_T$-module M_T furnishes (9.32). The decompositions of $\mathrm{t}(M_T)$ and $M_T/\mathrm{t}(M_T)$ are direct consequences.
1. This follows from the decomposition (9.32) and Theorem 4.2.12, item 1.
2. Theorem 9.2.10, item 3 (d).
3. From the isomorphism (9.32) follows directly that $M_T = 0$ if and only if $l = p$ and $\mathcal{D}_T/\mathcal{D}_T e_i = 0$ for all i. The latter condition is equivalent to the e_i being units in $\mathcal{D}_T$, i.e., $e_i \in \mathcal{D} \cap \mathrm{U}(\mathcal{D}_T) = T$, as we have already seen in the proof of Theorem 9.2.10, item 1 (a). Since T is saturated and $e_1 \mid \cdots \mid e_p$, this is equivalent to $e_p \in T$. □

9.3 Quotients of Signal Modules and Behaviors

The assumptions and notations of Sect. 9.2 remain in force. Additionally, we assume an injective cogenerator signal $\mathcal{D}$-module $\mathcal{F}$.

We study the T-torsion submodule $\mathrm{t}_T(\mathcal{F})$ and the quotient module $\mathcal{F}_T$. In the standard cases, $\mathrm{t}_T(\mathcal{F})$ is the space of polynomial-exponential functions or sequences that converge to zero for $t \to \infty$ and can therefore be neglected in most engineering applications. We derive the decomposition $\mathcal{F} = \mathrm{t}_T(\mathcal{F}) \oplus \mathcal{F}_T$, up to isomorphism, and interpret $\mathcal{F}_T$ as the module of steady states.

Unless $T = \mathcal{D} \setminus \{0\}$ and $\mathcal{F}$ is a torsion module, the $\mathcal{D}_T$-module $\mathcal{F}_T$ is an injective cogenerator in the category of $\mathcal{D}_T$-modules, and therefore the behavioral systems theory can be applied to this situation without change. Furthermore, the exact functor $(-)_T$ maps every behavior $\mathcal{B}$ to its steady-state part $\mathcal{B}_T$, the module of equations $U = \mathcal{D}^{1\times k} R$ of $\mathcal{B}$ to the module of equations $U_T = \mathcal{D}_T^{1\times k} R$ of $\mathcal{B}_T$, and the factor module $M = \mathcal{D}^{1\times l}/U$ of $\mathcal{B}$ to the factor module $M_T = \mathcal{D}_T^{1\times l}/U_T$ of $\mathcal{B}_T$.

This enables to investigate $\mathcal{F}$-behaviors $\mathcal{B}$ up to their negligible parts. In the standard cases this means that behaviors are considered equal if their characteristics are the same for large times t. This point of view is essential for our treatment of compensators and observers.

We first study $\mathrm{t}_T(M)$ for a not necessarily finitely generated $\mathcal{D}$-module M and then apply this to the signal module $\mathcal{F}$. From Theorem 5.3.2 and with Lemmas 9.2.12 and 9.2.13 we get the primary decomposition

$$\mathrm{t}(M) = \bigoplus_{q\in\mathcal{P}} \mathrm{t}_q(M) \quad \text{and} \quad \mathrm{t}(M_T) = \mathrm{t}(M)_T = \bigoplus_{q\in\mathcal{P}} \mathrm{t}_q(M)_T \text{ where}$$
$$\mathrm{t}_q(M) = \bigcup_{k=1}^{\infty} \mathrm{ann}_M(q^k) \quad \text{and} \quad \mathrm{ann}_M(q^k) = \{x \in M;\ q^k x = 0\}. \tag{9.33}$$

Theorem 9.3.1 (Quotients of torsion submodules) *Let M be a $\mathcal{D}$-module with its torsion submodule $\mathrm{t}(M)$. Then the following statements hold:*

1. *For $q \in \mathcal{P}_1 = \mathcal{P} \cap T$ the identity $\mathrm{t}_q(M)_T = 0$ holds. For $q \in \mathcal{P}_2$ one gets*

$$\mathrm{t}_q(M)_T \overset{\cong}{\longleftrightarrow} \mathrm{t}_q(M), \quad \frac{x_q}{t_q} \longmapsto x'_q, \quad \frac{x'_q}{1} \longleftarrow x'_q,$$

where $x'_q \in \mathrm{t}_q(M)$ is unique with $t_q x'_q = x_q$. We identify $\mathrm{t}_q(M)_T = \mathrm{t}_q(M)$ for $q \in \mathcal{P}_2$ and consequently also

$$\mathrm{t}(M)_T \underset{(9.33)}{=} \bigoplus_{q\in\mathcal{P}} \mathrm{t}_q(M)_T \underset{\text{ident.}}{=} \bigoplus_{q\in\mathcal{P}_2} \mathrm{t}_q(M) \subset \mathrm{t}(M).$$

2. *The torsion module $\mathrm{t}(M)$ has the direct sum decomposition*

$$\mathrm{t}(M) = \left(\bigoplus_{q\in\mathcal{P}_1} \mathrm{t}_q(M)\right) \oplus \left(\bigoplus_{q\in\mathcal{P}_2} \mathrm{t}_q(M)\right) = \mathrm{t}_T(M) \oplus \mathrm{t}(M)_T.$$

Proof 1. If $q \in \mathcal{P}_1 \subset T$ and $\frac{x}{t} \in \mathrm{t}_q(M)_T$ there is a number $k \in \mathbb{N} \setminus \{0\}$ with $q^k x = 0$ and thus $\frac{x}{t} = \frac{q^k x}{q^k t} = \frac{0}{q^k t} = 0$.

If q is contained in $\mathcal{P}_2$, it is coprime to all elements $t \in T$. From Lemma 5.1.5 we conclude that for all $t \in T$ and all $k \geq 1$ the maps

$$t\cdot:\ \mathrm{ann}_M(q^k) \xrightarrow{\cong} \mathrm{ann}_M(q^k),\ x \longmapsto tx, \text{ and hence } t\cdot:\ \mathrm{t}_q(M) \xrightarrow{\cong} \mathrm{t}_q(M)$$

are isomorphisms. By Theorem 9.1.15 this is equivalent to the isomorphy of

$$\mathrm{can}_{\mathrm{t}_q(M)}:\ \mathrm{t}_q(M) \xrightarrow{\cong} \mathrm{t}_q(M)_T,\quad x \longmapsto \tfrac{x}{1}.$$

From this we deduce the identification

$$\mathrm{t}_q(M) = \mathrm{t}_q(M)_T \ni x' = \tfrac{x'}{1} = \tfrac{x}{t} \quad \text{if } tx' = x.$$

2. Since $\mathrm{t}(M)_T$ is a $\mathcal{D}_T$-module, the map $t\cdot:\ \mathrm{t}(M)_T \longrightarrow \mathrm{t}(M)_T$ is an isomorphism with inverse $t^{-1}\cdot$. In contrast, every element of $\mathrm{t}_T(M)$ is annihilated by some $t \in T$, hence, $\mathrm{t}_T(M) \cap \mathrm{t}(M)_T = 0$.

For $q \in \mathcal{P}_1 = T \cap \mathcal{P}$, we have that $q^k \in T$ for all k. Therefore, the inclusions $\mathrm{ann}_M(q^k) \subseteq \mathrm{t}_T(M)$ and thus also $\mathrm{t}_q(M) \subseteq \mathrm{t}_T(M)$ hold. We infer that

$$\mathrm{t}(M) = \underbrace{\left(\bigoplus_{q\in\mathcal{P}_1} \mathrm{t}_q(M)\right)}_{\subseteq \mathrm{t}_T(M)} \oplus \underbrace{\left(\bigoplus_{q\in\mathcal{P}_2} \mathrm{t}_q(M)\right)}_{\underset{1.}{=}\mathrm{t}(M)_T} \subseteq \mathrm{t}_T(M) \oplus \mathrm{t}(M)_T \subseteq \mathrm{t}(M).$$

Therefore, all inclusions are indeed equalities. □

We are now going to apply the preceding considerations to the injective cogenerator ${}_{\mathcal{D}}\mathcal{F}$ and in particular to all signal modules over $F[s]$ which were discussed in the previous sections.

Recall from Theorem 5.3.10 that $\mathrm{t}(\mathcal{F}) = \bigoplus_{q\in\mathcal{P}} \mathrm{t}_q(\mathcal{F})$ is itself an injective cogenerator and a direct summand of $\mathcal{F}$, and the factor module $\mathcal{F}/\,\mathrm{t}(\mathcal{F})$ is injective. Thus there is a direct decomposition

$$\mathcal{F} = \mathrm{t}(\mathcal{F}) \oplus \mathcal{F}_K = \bigoplus_{q\in\mathcal{P}} \mathrm{t}_q(\mathcal{F}) \oplus \mathcal{F}_K \quad \text{with} \quad \mathcal{F}_K \underset{K}{\overset{\cong}{\longleftrightarrow}} \mathcal{F}/\,\mathrm{t}(\mathcal{F}), \tag{9.34}$$

where $\mathcal{F}_K$ is a torsionfree injective module and therefore a vector space over $K = \mathrm{quot}(\mathcal{D})$ by Theorem 5.3.9, hence, the notation $\mathcal{F}_K$. The elements of $\mathcal{F}_K$ form a representative system of the equivalence classes in $\mathcal{F}$ modulo $\mathrm{t}(\mathcal{F})$.

While the direct summands $\mathrm{t}_q(\mathcal{F})$ are well-known modules of polynomial-exponential functions or sequences in the standard cases, the direct complement $\mathcal{F}_K$ of $\mathrm{t}(\mathcal{F})$ is not unique and no such $\mathcal{F}_K$ can be constructed if $\mathcal{F}$ is one of the standard function modules. This is due to the use of Zorn's lemma in the proof of the decomposition (9.34). In other words, for $y \in \mathcal{F}$ there is a representation

$$y = \sum_{q\in\mathcal{P}} y_q + y_K \in \mathcal{F} = \bigoplus_{q\in\mathcal{P}} \mathrm{t}_q(\mathcal{F}) \oplus \mathcal{F}_K. \tag{9.35}$$

But neither the component $y_K \in \mathcal{F}_K$ nor the components y_q can be constructively determined. We will see, however, that the preceding decomposition is nevertheless valuable. The torsion submodule $\mathrm{t}(\mathcal{F})$ of $\mathcal{F}$ contains the T-torsion submodule $\mathrm{t}_T(\mathcal{F}) = \bigoplus_{q\in\mathcal{P}_1} \mathrm{t}_q(\mathcal{F})$ and, by the identification from Theorem 9.3.1, the quotient module $\mathrm{t}(\mathcal{F})_T = \mathrm{t}(\mathcal{F}_T) \underset{\text{ident.}}{=} \bigoplus_{q\in\mathcal{P}_2} \mathrm{t}_q(\mathcal{F})$.

Theorem 9.3.2 (Direct decompositions of injective cogenerators) *The following direct sum decompositions of the injective $\mathcal{D}$-cogenerator $\mathcal{F}$ hold:*

1. $\mathrm{t}(\mathcal{F}) = \left(\bigoplus_{q\in\mathcal{P}_1} \mathrm{t}_q(\mathcal{F})\right) \oplus \left(\bigoplus_{q\in\mathcal{P}_2} \mathrm{t}_q(\mathcal{F})\right) \underset{\text{ident.}}{=} \mathrm{t}_T(\mathcal{F}) \oplus \mathrm{t}(\mathcal{F})_T$.

2. $\mathcal{F} = \mathrm{t}(\mathcal{F}) \oplus \mathcal{F}_K \underset{\text{ident.}}{=} \mathrm{t}_T(\mathcal{F}) \oplus \mathrm{t}(\mathcal{F})_T \oplus \mathcal{F}_K$.

3. $\mathcal{F}_T = \mathrm{t}(\mathcal{F})_T \oplus \mathcal{F}_K \underset{\text{ident.}}{=} \bigoplus_{q\in\mathcal{P}_2} \mathrm{t}_q(\mathcal{F}) \oplus \mathcal{F}_K \cong \bigoplus_{q\in\mathcal{P}_2} \mathrm{t}_q(\mathcal{F}) \oplus \mathcal{F}/\mathrm{t}(\mathcal{F})$.

4. $\mathcal{F}_T \overset{\cong}{\longleftrightarrow} \mathcal{F}/\mathrm{t}_T(\mathcal{F})$, $\frac{y}{1} \longleftarrow\!\shortmid \overline{y}$. *We identify these two modules and obtain the nonconstructive direct sum decomposition*

$$\mathcal{F} = \mathrm{t}_T(\mathcal{F}) \oplus \mathcal{F}_T \ni y = (y - y_T) + y_T, \ y_T = \tfrac{y}{1} = \overline{y} \in \mathcal{F}_T = \mathcal{F}/\mathrm{t}_T(\mathcal{F}). \tag{9.36}$$

All involved $\mathcal{D}$-modules are injective and thus satisfy the fundamental principle.

As factor module the module $\mathcal{F}_T = \mathcal{F}/\mathrm{t}_T(\mathcal{F})$ is, of course, uniquely determined. However, as direct summand of $\mathcal{F}$ it is unique only after the nonconstructive choice of the direct summand $\mathcal{F}_K$. The component $y_T \in \mathcal{F}_T \subseteq \mathcal{F}$ of $y \in \mathcal{F}$ cannot be computed in general, but for $x, y \in \mathcal{F}$ the equality $x_T = y_T$ can be decided constructively since it is equivalent to $x - y \in \mathrm{t}_T(\mathcal{F})$ or, in other words, to the existence of a $t \in T$ with $t \circ (x - y) = 0$.

Proof 1. Theorem 9.3.1, item 2.
2. The second decomposition is a consequence of item 1 and (9.34).
3.
$$\begin{aligned}\mathcal{F}_T &= (\mathrm{t}_T(\mathcal{F}) \oplus \mathrm{t}(\mathcal{F})_T \oplus \mathcal{F}_K)_T \\ &= (\mathrm{t}_T(\mathcal{F}))_T \oplus (\mathrm{t}(\mathcal{F})_T)_T \oplus (\mathcal{F}_K)_T = 0 \oplus \mathrm{t}(\mathcal{F})_T \oplus \mathcal{F}_K.\end{aligned}$$
4. This follows directly from the decompositions in items 2 and 3.
Corollary 5.3.6 states that if a direct sum is injective, then all the summands are injective too. Therefore, the injectivity of $\mathcal{F}$ and the direct sum representations

$$\mathcal{F} = \mathrm{t}(\mathcal{F}) \oplus \mathcal{F}_K = \mathrm{t}_T(\mathcal{F}) \oplus \mathrm{t}(\mathcal{F})_T \oplus \mathcal{F}_K = \mathrm{t}_T(\mathcal{F}) \oplus \mathcal{F}_T$$

imply that all appearing modules are injective. □

Corollary 9.3.3 *For $T := \mathcal{D}\setminus\{0\}$, the preceding theorem implies the identification of K-vector spaces*

$$\mathcal{F}_K = \mathcal{F}_{\mathcal{D}\setminus\{0\}} = \mathcal{F}/\operatorname{t}(\mathcal{F}) \ni y = \frac{y}{1} = \overline{y}.$$

$\Diamond$

Definition and Corollary 9.3.4 *In the situation of Theorem 9.3.2, the signals in $\operatorname{t}_T(\mathcal{F})$ are called* T-stable, T-small, *or* T-negligible. *Two signals $x, y \in \mathcal{F}$ are called* T-equivalent *if they satisfy the equivalent conditions*

$$\begin{aligned} x - y \in \operatorname{t}_T(\mathcal{F}) &\Longleftrightarrow \overline{x} = \overline{y} \in \mathcal{F}/\operatorname{t}_T(\mathcal{F}) \\ &\Longleftrightarrow \tfrac{x}{1} = \tfrac{y}{1} \in \mathcal{F}_T = \left\{\tfrac{z}{t};\ z \in \mathcal{F},\ t \in T\right\} \\ &\Longleftrightarrow x_T = y_T \in \mathcal{F}_T \subseteq \operatorname{t}_T(\mathcal{F}) \oplus \mathcal{F}_T = \mathcal{F}. \end{aligned}$$

The same terminology applies to vector signals $x, y \in \mathcal{F}^p = \mathcal{F}_T^p \oplus \operatorname{t}_T(\mathcal{F})^p$.

In engineering terminology T-equivalent signals are said to have the same T-steady state *or are called* T-estimates *of each other.*

In general, there is no unique steady or stationary state of a signal in $\mathcal{F}$, as is often suggested by engineering language. However, after the nonconstructive choice of $\mathcal{F}_K$ with $\mathcal{F} = \mathcal{F}_K \oplus \operatorname{t}(\mathcal{F})$ we call x_T in the decomposition $x = x_T + (x - x_T) \in \mathcal{F} = \mathcal{F}_T \oplus \operatorname{t}_T(\mathcal{F})$ the steady state *of x. It cannot be computed in general, but it differs from x only by the T-small signal $x - x_T$. In the standard cases $\operatorname{t}_T(\mathcal{F})$ consists of certain polynomial-exponential functions, and T-equivalence can be checked constructively. If $y = \sum_{q\in\mathcal{P}} y_q \in \operatorname{t}(\mathcal{F}) = \bigoplus_{q\in\mathcal{P}} \operatorname{t}_q(\mathcal{F})$, then $y_T = \sum_{q\in\mathcal{P}_2} y_q \in \operatorname{t}(\mathcal{F})_T = \bigoplus_{q\in\mathcal{P}_2} \operatorname{t}_q(\mathcal{F})$ is its steady state.* $\Diamond$

Example 9.3.5 We check that the just introduced *steady-state* terminology coincides with the usual one in the standard engineering cases. We focus on the complex continuous case. Let $\mathcal{D} = \mathbb{C}[s]$ with the quotient field $K = \mathbb{C}(s)$, and assume the disjoint decomposition $\mathbb{C} = \Lambda_1 \uplus \Lambda_2$ with $\Lambda_1 = \mathbb{C}_-$ and the signal space $\mathcal{F} = \mathrm{C}^\infty(\mathbb{R}, \mathbb{C})$ or $\mathcal{F} = \mathrm{C}^{-\infty}(\mathbb{R}, \mathbb{C})$. In both cases the torsion module of $\mathcal{F}$ is $\operatorname{t}(\mathcal{F}) = \bigoplus_{\lambda\in\mathbb{C}} \mathbb{C}[t]e^{\lambda t}$. The decomposition $\mathbb{C} = \Lambda_1 \uplus \Lambda_2$ induces the saturated submonoid $T = \{t \in \mathbb{C}[s];\ \mathrm{V}_\mathbb{C}(t) \subseteq \Lambda_1\}$ of stable polynomials and the quotient ring $\mathbb{C}[s]_T$ of stable rational functions.

Even for $\mathcal{F} = \mathrm{C}^\infty(\mathbb{R}, \mathbb{C})$ the $\mathbb{C}(s)$-space $\mathcal{F}_{\mathbb{C}(s)}$ cannot be constructed. The decomposition of $\operatorname{t}(\mathcal{F})$, however, is explicit, namely,

$$\operatorname{t}(\mathcal{F}) = \left(\bigoplus_{\lambda\in\mathbb{C}_-} \mathbb{C}[t]e^{\lambda t}\right) \oplus \left(\bigoplus_{\lambda\in\Lambda_2} \mathbb{C}[t]e^{\lambda t}\right) = \operatorname{t}_T(\mathcal{F}) \oplus \operatorname{t}(\mathcal{F})_T.$$

Theorem and Definition 5.4.16 provides us with the characterization

$$\mathrm{t}_T(\mathcal{F}) = \bigoplus_{\lambda \in \mathbb{C}_-} \mathbb{C}[t]e^{\lambda t} = \{y \in \mathrm{t}(\mathcal{F});\ \lim_{t\to\infty} y(t) = 0\}.$$

Therefore a signal is T-stable if and only if it is polynomial-exponential and converges to zero for $t \to \infty$. Two trajectories x and y in $\mathcal{F}$ have the same steady state if their difference $x - y$ is a polynomial-exponential function with $\lim_{t\to\infty}(x(t) - y(t)) = 0$. Notice that this does not imply the existence of $\lim_{t\to\infty} x(t)$ or of $\lim_{t\to\infty} y(t)$. If $x = \sum_{\lambda\in\mathbb{C}} f_\lambda(t)e^{\lambda t}$ is a polynomial-exponential signal, then its steady state is $x_T = \sum_{\lambda\in\Lambda_2} f_\lambda(t)e^{\lambda t}$.

The real and discrete analogues of this complex continuous case follow likewise from Corollary 5.3.19 and Theorem and Definition 5.4.16. ◇

Based on the following theorem, we will formulate the theory of steady-state behaviors in Theorem and Definition 9.3.7.

Theorem 9.3.6 *Assume that $T \subsetneq \mathcal{D} \setminus \{0\}$, i.e., that $\mathcal{D}_T \neq K$, or that $\mathcal{F}$ is not a torsion module. Then the module $\mathcal{F}_T$ is an injective cogenerator in the category* $\mathrm{mod}_{\mathcal{D}_T}$ *of $\mathcal{D}_T$-modules. Therefore, it induces the standard duality theory between finitely generated $\mathcal{D}_T$-modules with given generators and ${}_{\mathcal{D}_T}\mathcal{F}_T$-behaviors, see Theorems 3.3.10 and 3.3.18. By Theorem 5.3.10 also $\mathrm{t}(\mathcal{F}_T) = \mathrm{t}(\mathcal{F})_T$ is an injective $\mathcal{D}_T$-cogenerator. If $\mathcal{F}$ is a torsion module and $T = \mathcal{D} \setminus \{0\}$ then $\mathcal{F}_T = 0$ is not a cogenerator.*

Proof *Injectivity*: In Theorem 9.3.2, we have shown that $\mathcal{F}_T$ is an injective $\mathcal{D}$-module and thus divisible by Corollary 3.2.11, i.e., all maps $d\circ\colon\ \mathcal{F}_T \to \mathcal{F}_T,\ 0 \neq d \in \mathcal{D}$, are surjective. Since $T \subseteq \mathrm{U}(\mathcal{D}_T)$, the map $t^{-1}\cdot :\ \mathcal{F}_T \to \mathcal{F}_T,\ t \in T$, is an isomorphism. Hence $\frac{d}{t}\cdot = d \cdot t^{-1}\cdot : \mathcal{F}_T \to \mathcal{F}_T,\ 0 \neq \frac{d}{t} \in \mathcal{D}_T$, is surjective too. This means that $\mathcal{F}_T$ is $\mathcal{D}_T$-divisible and thus injective.
Cogenerator: By Theorem 3.3.5, we have to show that $\mathcal{F}_T$ contains all simple $\mathcal{D}_T$-modules, up to isomorphism.
(i) If $\mathcal{D}_T = K$ and $\mathcal{F}$ is not a torsion module then $\mathcal{F}_T \neq 0$, and this contains the unique simple K-module K.
(ii) If $T \subsetneq \mathcal{D} \setminus \{0\}$ then $\mathcal{D}_T$ is not a field. Since $\mathcal{P}_2$ is a nonempty representative system of the prime elements of $\mathcal{D}_T$ (Theorem 9.2.2), every simple $\mathcal{D}_T$-module is isomorphic to $\mathcal{D}_T/\mathcal{D}_T q_2$ with $q_2 \in \mathcal{P}_2$, see Lemma 3.3.6. Since $\mathcal{F}$ is a $\mathcal{D}$-cogenerator, it contains $\mathcal{D}/\mathcal{D}q_2$ up to isomorphism, i.e., there is a monomorphism $\varphi\colon\ \mathcal{D}/\mathcal{D}q_2 \longrightarrow \mathcal{F}$. Since the functor $(-)_T$ is exact (Theorem 9.1.10), this $\mathcal{D}$-monomorphism induces the $\mathcal{D}_T$-monomorphism $\mathcal{D}_T/\mathcal{D}_T q_2 = (\mathcal{D}/\mathcal{D}q_2)_T \xrightarrow{\varphi_T} \mathcal{F}_T$. □

In the rest of this section we assume $T \subsetneq \mathcal{D} \setminus \{0\}$.

Theorem and Definition 9.3.7 (Quotients of behaviors)

1. *The direct sum decompositions of Theorem 9.3.2 induce analogous decompositions of the behavior $\mathcal{B} := \{w \in \mathcal{F}^l;\ R \circ w = 0\}$, where $R \in \mathcal{D}^{k\times l}$. Indeed, since $\mathcal{B} \subseteq \mathcal{F}^l = \mathrm{t}_T(\mathcal{F})^l \oplus \mathcal{F}_T^l = \mathrm{t}(\mathcal{F})^l \oplus \mathcal{F}_K^l = \mathrm{t}_T(\mathcal{F})^l \oplus \mathrm{t}(\mathcal{F})_T^l \oplus \mathcal{F}_K^l$, the identities*

$$\mathrm{t}(\mathcal{B}) = \mathcal{B} \cap \mathrm{t}(\mathcal{F})^l, \qquad \mathrm{t}_q(\mathcal{B}) = \mathcal{B} \cap \mathrm{t}_q(\mathcal{F})^l,$$
$$\mathrm{t}_T(\mathcal{B}) = \mathcal{B} \cap \mathrm{t}_T(\mathcal{F})^l, \ \textit{and}\quad \mathrm{t}(\mathcal{B})_T = \mathcal{B} \cap \mathrm{t}(\mathcal{F})_T^l$$

hold. The decompositions from Theorem 9.3.2 imply the new ones

$$\begin{aligned}
&\mathrm{t}_T(\mathcal{B}) = \bigoplus_{q\in\mathcal{P}_1}\mathrm{t}_q(\mathcal{B}),\ \ \mathrm{t}(\mathcal{B})_T = \bigoplus_{q\in\mathcal{P}_2}\mathrm{t}_q(\mathcal{B}),\ \ \mathrm{t}(\mathcal{B}) = \mathrm{t}_T(\mathcal{B})\oplus\mathrm{t}(\mathcal{B})_T \text{ and}\\
&\mathcal{B} = \mathrm{t}(\mathcal{B})\oplus\mathcal{B}_K = \mathrm{t}_T(\mathcal{B})\oplus\mathrm{t}(\mathcal{B})_T\oplus\mathcal{B}_K \text{ with } \mathcal{B}_K := \mathcal{B}\cap\mathcal{F}_K^l,\\
&\Longrightarrow \mathcal{B}_T = \{w\in\mathcal{F}_T^l;\ R\circ w = 0\} = \mathrm{t}(\mathcal{B})_T\oplus\mathcal{B}_K,\\
&\mathcal{B} = \mathrm{t}_T(\mathcal{B})\oplus\mathcal{B}_T \text{ and } \mathcal{B}_T \underset{\text{ident.}}{=} \mathcal{B}/\mathrm{t}_T(\mathcal{B})
\end{aligned}$$

with the identifications introduced for $\mathcal{F}$ in Theorem 9.3.2. In particular, we infer that $\mathcal{B}_T$ is an $\mathcal{F}_T$-behavior.

We call $\mathrm{t}_T(\mathcal{B})$ *the* T-small part *or* T-autonomous part *of $\mathcal{B}$, and $\mathcal{B}_T$ is the* T-steady state *of $\mathcal{B}$.*

2. *The $\mathcal{D}_T$-system module of $\mathcal{B}_T$ is the quotient module M_T of the system module $M = \mathcal{D}^{1\times l}/U$ of $\mathcal{B}$, indeed*

$$\begin{aligned}
&\mathrm{Hom}_{\mathcal{D}_T}(M_T,\mathcal{F}_T) \overset{\cong}{\longleftrightarrow} \mathrm{Hom}_{\mathcal{D}}(M,\mathcal{F}_T) \overset{\cong}{\longleftrightarrow} \{w\in\mathcal{F}_T^l;\ R\circ w = 0\} = \mathcal{B}_T,\\
&\Longrightarrow (U_T)^\perp = (U^\perp)_T = \mathcal{B}_T \text{ and } U_T = U_T^{\perp\perp} = \mathcal{B}_T^\perp = \mathcal{D}_T^{1\times k}R.
\end{aligned}$$

3. *The following properties are equivalent:*
   (a) There is a $t\in T$ such that $t\circ\mathcal{B} = 0$.
   (b) $\mathcal{B} = \mathrm{t}_T(\mathcal{B})$, i.e., $\mathcal{B}\subseteq\mathrm{t}_T(\mathcal{F})^l$, or, by item 2 of Corollary 9.1.14, $\mathcal{B}_T = 0$.
   (c) The $\mathcal{D}_T$-module M_T is zero, i.e., like (b), $M = \mathrm{t}_T(M)$.
   (d) The matrix R has a left inverse in $\mathcal{D}_T^{l\times k}$, i.e., by item 1 (a) of Theorem 9.2.10, $\mathrm{rank}(R) = l$ *and $e_l\in T$.*

   Under these equivalent conditions, $\mathcal{B}$ is called T-small *or* T-autonomous*.*
4. *An exact sequence of behaviors and behavior morphisms (see Definition and Corollary 3.3.16)*

$$\mathcal{B}_1 \overset{P\circ}{\longrightarrow} \mathcal{B}_2 \overset{Q\circ}{\longrightarrow} \mathcal{B}_3,$$

i.e., $P\circ\mathcal{B}_1 = \{w\in\mathcal{B}_2;\ Q\circ w = 0\}$, induces the exact sequence of T-steady-state behaviors

$$\mathcal{B}_{1,T} \overset{P\circ}{\longrightarrow} \mathcal{B}_{2,T} \overset{Q\circ}{\longrightarrow} \mathcal{B}_{3,T},$$

i.e., $(P\circ\mathcal{B}_1)_T = P\circ\mathcal{B}_{1,T} = \{w\in\mathcal{B}_{2,T};\ Q\circ w = 0\}$.

Proof 1. The decomposition $\mathcal{F}^l = \mathrm{t}(\mathcal{F})^l\oplus\mathcal{F}_K^l$ does, in general, not imply an analogous decomposition $\mathcal{B} = (\mathcal{B}\cap\mathrm{t}(\mathcal{F})^l)\oplus(\mathcal{B}\cap\mathcal{F}_K^l)$ for an arbitrary $\mathcal{D}$-submodule $\mathcal{B}\subseteq\mathcal{F}^l$ (the sum is always direct, but in general only a submodule of $\mathcal{B}$). However, if $w = w_1 + w_2\in\mathcal{F}^l = \mathrm{t}(\mathcal{F})^l\oplus\mathcal{F}_K^l$, then

$$R\circ w = R\circ w_1 + R\circ w_2 \in \mathrm{t}(\mathcal{F})^k\oplus\mathcal{F}_K^k$$

and hence

$$\begin{aligned} w \in \mathcal{B} &\iff R \circ w = 0 \iff R \circ w_1 = R \circ w_2 = 0 \\ &\iff w_1 \in \mathcal{B} \cap \mathrm{t}(\mathcal{F})^l = \mathrm{t}(\mathcal{B}) \text{ and } w_2 \in \mathcal{B} \cap \mathcal{F}_K^l = \mathcal{B}_K. \end{aligned}$$

Therefore, $\mathcal{B} = \mathrm{t}(\mathcal{B}) \oplus \mathcal{B}_K$. The same argument applies to all other decompositions.
2. The second isomorphism in item 2 is the Malgrange isomorphism from Theorem 3.1.36 for the $\mathcal{D}$-module $\mathcal{F}_T$, and the first one comes from the universal property of M_T (Theorem 9.1.7), namely,

$$\mathrm{Hom}_{\mathcal{D}_T}(M_T, \mathcal{F}_T) \overset{\cong}{\longleftrightarrow} \mathrm{Hom}_{\mathcal{D}}(M, \mathcal{F}_T),\ \psi \longleftrightarrow \varphi,\ \psi\left(\tfrac{\overline{\xi}}{t}\right) = \tfrac{\varphi(\overline{\xi})}{t},\ \xi \in \mathcal{D}^{1\times l}.$$

3. $(a) \Longrightarrow (b)$: This follows directly from the definition of $\mathrm{t}_T(\mathcal{B})$ in (9.15).
$(b) \iff (c)$:

$$\mathcal{B} = \mathrm{t}_T(\mathcal{B}) \underset{\text{Corollary 9.1.14}}{\iff} 0 = \mathcal{B}_T \underset{\text{item 2}}{\cong} \mathrm{Hom}_{\mathcal{D}_T}(M_T, \mathcal{F}_T) \underset{\mathcal{F}_T\text{cogenerator}}{\iff} M_T = 0.$$

$(c) \iff (a)$: By Corollary 9.1.14, item 3, $M_T = 0 \iff T \cap \mathrm{ann}_{\mathcal{D}}(M) \neq \emptyset$. Lemma 4.2.6, item 1, implies $\mathrm{ann}_{\mathcal{D}}(M) = \mathrm{ann}_{\mathcal{D}}(\mathcal{B})$. Therefore, $M_T = 0$ if and only if there is a $t \in T$ with $t \circ \mathcal{B} = 0$.
$(c) \iff (d)$: The module $M_T = \mathcal{D}_T^{1\times l}/\mathcal{D}_T^{1\times k} R$ is zero if and only if $\mathcal{D}_T^{1\times k} R = \mathcal{D}_T^{1\times l}$. This holds if and only if R has a left inverse in $\mathcal{D}_T^{l\times k}$.
4. This follows from the exactness of $(-)_T$, cf. Theorem 9.1.10. □

Remark 9.3.8 The T-autonomous part $\mathrm{t}_T(\mathcal{B})$ of a behavior consists of T-negligible trajectories that are often also negligible from an engineering point of view, whereas the T-steady-state part $\mathcal{B}_T = \mathcal{B}/\mathrm{t}_T(\mathcal{B})$ contains the important technical information. Theorems 9.3.6 and 9.3.7 imply that the behavior theory of the preceding chapters can be applied to the $\mathcal{F}_T$-behavior $\mathcal{B}_T$. ◊

Remark 9.3.9 Consider the case $T = \mathcal{D} \setminus \{0\}$ and $K = \mathcal{D}_T$. The space $\mathcal{F}_K$ is a K-vector space, and hence there is an index set I such that $\mathcal{F}_K \underset{K}{\cong} K^{(I)}$. For a behavior $\mathcal{B} = \{w \in \mathcal{F}^l;\ R \circ w = 0\}$ we infer the K-isomorphism

$$\mathcal{B}_K = \{w \in \mathcal{F}_K^l;\ R \circ w = 0\} \cong \{\widetilde{w} \in K^l;\ R\widetilde{w} = 0\}^{(I)}$$

where $\{\widetilde{w} \in K^l;\ R\widetilde{w} = 0\}$ is the transfer space of $\mathcal{B}$ according to Lemma and Definition 6.1.1. In particular, Theorem 9.3.7, 4 implies Theorem 6.1.4, i.e., the exactness of the transfer space. ◊

9.4 Trivial and T-stable IO Behaviors

This section is devoted to T-stable IO behaviors that generalize the standard IO systems with asymptotically stable autonomous part. We assume a principal ideal domain $\mathcal{D}$ with quotient field K, a saturated multiplicative submonoid $T \subsetneq \mathcal{D} \setminus \{0\}$ and an injective $\mathcal{D}$-cogenerator ${}_{\mathcal{D}}\mathcal{F}$.

We consider an IO behavior $\mathcal{B}$, its autonomous part $\mathcal{B}^0$, and its modules

$$\begin{aligned}
&\mathcal{B} = \left\{ \binom{y}{u} \in \mathcal{F}^{p+m};\ P \circ y = Q \circ u \right\},\ \mathcal{B}^0 = \left\{ y \in \mathcal{F}^p;\ P \circ y = 0 \right\},\ \text{with}\\
&R := (P, -Q) \in \mathcal{D}^{p\times(p+m)},\ \text{rank}(P) = p,\ H := P^{-1}Q \in K^{p\times m},\\
&U = \mathcal{D}^{1\times p}R,\ M = \mathcal{D}^{1\times(p+m)}/U,\ U^0 = \mathcal{D}^{1\times p}P,\ M^0 = \mathcal{D}^{1\times p}/U^0.
\end{aligned} \tag{9.37}$$

We use the Smith forms of R and P, viz.,

$$X^R R Y^R = \begin{pmatrix} \text{diag}(e_1^R, \dots, e_p^R) & 0 \\ 0 & 0 \end{pmatrix} \in \mathcal{D}^{p\times(p+m)},\ X^P P Y^P = \begin{pmatrix} \text{diag}(e_1^P, \dots, e_p^P) \\ 0 \end{pmatrix} \in \mathcal{D}^{p\times p}. \tag{9.38}$$

The corresponding quotient objects are

$$\begin{aligned}
&\mathcal{B}_T = \left\{ \binom{y}{u} \in \mathcal{F}_T^{p+m};\ P \circ y = Q \circ u \right\},\ \mathcal{B}_T^0 = \left\{ y \in \mathcal{F}_T^p;\ P \circ y = 0 \right\},\\
&U_T = \mathcal{D}_T^{1\times p}R,\ M_T = \mathcal{D}_T^{1\times(p+m)}/U_T,\ U_T^0 = \mathcal{D}_T^{1\times p}P,\ M_T^0 = \mathcal{D}_T^{1\times p}/U_T^0.
\end{aligned} \tag{9.39}$$

Clearly, $\mathcal{B}_T \subseteq \mathcal{F}_T^l$ is an IO behavior with the same IO structure as $\mathcal{B} \subseteq \mathcal{F}^l$.

The T-stable behaviors are reduced to *trivial behaviors*, which we introduce first.

Lemma and Definition 9.4.1 (Trivial IO behaviors) *For the IO behavior $\mathcal{B}$ from (9.37) the following properties are equivalent:*

1. *$\mathcal{B}^0 = 0$. By duality, this is equivalent to*

$$M^0 = 0 \iff \mathcal{D}^{1\times p}P = \mathcal{D}^{1\times p} \iff P \in \text{Gl}_p(\mathcal{D}) \iff e_p^P \in \text{U}(\mathcal{D}).$$

2. *(a) $\mathcal{B}$ is controllable. By Theorem 4.2.12, 2, this is equivalent to*

$$M \text{ is torsionfree, i.e., free} \iff R \text{ has a right inverse} \iff e_p^R \in \text{U}(\mathcal{D}).$$

   (b) $H \in \mathcal{D}^{p\times m}$.

If these equivalent conditions are satisfied, then we call the IO behavior trivial.

Proof 1. $\Longrightarrow$ 2.: The exact sequence (6.11) together with the equivalence 1. $\iff$ 6. from Theorem and Definition 6.2.2 furnishes the exact sequence

$$0 \longrightarrow \mathcal{D}^{1\times m} \xrightarrow{(\cdot(0,\text{id}_m))_{\text{ind}}} M \xrightarrow{\left(\cdot\binom{\text{id}_p}{0}\right)_{\text{ind}}} M^0 \longrightarrow 0.$$

Since $M^0 = 0$, this implies $\mathcal{D}^{1\times m} \cong M$ and thus the freeness of M. Moreover $P^{-1} \in \mathcal{D}^{p\times p}$ and $H = P^{-1}Q \in \mathcal{D}^{p\times m}$.
2. $\Longrightarrow$ 1.: Since $\mathcal{B}$ is controllable, it is the unique controllable realization of H according to Theorem 6.2.4 and $U^0 = U^0_{\text{cont}} = \{\xi \in \mathcal{D}^{1\times p};\ \xi H \in \mathcal{D}^{1\times m}\}$ by (6.17). Since the entries of H lie in $\mathcal{D}$, the module U^0 coincides with $\mathcal{D}^{1\times p}$. □

The following theorem is an essential tool in Sect. 10.1.

Theorem and Definition 9.4.2 (T-stable IO behaviors) *For the IO behavior $\mathcal{B}$ from (9.37) and the associated data from (9.39) the following properties are equivalent:*

1. *The behavior $\mathcal{B}^0$ is T-autonomous, i.e., $\mathcal{B}^0_T = 0$, or, equivalently,*

$$M^0_T = 0 \Longleftrightarrow P \in \mathrm{Gl}_p(\mathcal{D}_T) \Longleftrightarrow \det(P) \in T = \mathcal{D} \cap \mathrm{U}(\mathcal{D}_T) \Longleftrightarrow e^P_p \in T.$$

2. (a) *The $\mathcal{F}_T$-behavior $\mathcal{B}_T$ is controllable, i.e., the module M_T is torsionfree and thus free or, equivalently, $e^R_p \in T = \mathcal{D} \cap \mathrm{U}(\mathcal{D}_T)$.*
   (b) *$H \in \mathcal{D}_T^{p\times m}$, i.e., the entries of H are T-stable elements of K.*

IO behaviors $\mathcal{B}$ with these properties are called T-stable.

Proof The equivalences of this theorem follow directly from those of Lemma and Definition 9.4.1, applied to the principal ideal domain $\mathcal{D}_T$, the injective $\mathcal{D}_T$-cogenerator $\mathcal{F}_T$, and the IO $\mathcal{F}_T$-behavior $\mathcal{B}_T$. □

Corollary 9.4.3 (Stable behaviors in the standard cases) *Assume the situation of items 1 or 2 of Example 9.1.3, i.e., $F = \mathbb{C}$ or $F = \mathbb{R}$, the operator ring $\mathcal{D} = F[s]$, the multiplicative submonoid $T = \{t \in F[s] \setminus \{0\};\ \mathrm{V}_{\mathbb{C}}(t) \subseteq \Lambda_1\}$ which results from a disjoint decomposition $\mathbb{C} = \Lambda_1 \uplus \Lambda_2$ with $\Lambda_1 = \overline{\Lambda_1}$ for $F = \mathbb{R}$, and the quotient ring $F[s]_T \subseteq K$ of T-stable rational functions. An IO behavior $\mathcal{B}$ is T-stable if and only if it satisfies the following equivalent conditions:*

1. *The characteristic variety* $\mathrm{char}(\mathcal{B}^0) = \mathrm{V}_{\mathbb{C}}(e^P_p)$ *of $\mathcal{B}^0$ is contained in Λ_1.*
2. (a) *The characteristic variety* $\mathrm{char}(\mathcal{B}) = \mathrm{V}_{\mathbb{C}}(e^R_p)$ *of $\mathcal{B}$ is contained in Λ_1.*
   (b) *The entries of H are T-stable rational functions.*

For the standard function spaces and stable regions

$$\begin{aligned} \mathcal{F} &= F^{\mathbb{N}}, & \Lambda_{1,\text{std}} &= \{\lambda \in \mathbb{C};\ |\lambda| < 1\} \text{ for discrete time and}\\ \mathcal{F} &= \mathrm{C}^{-\infty}(\mathbb{R}, F), & \Lambda_{1,\text{std}} &= \{\lambda \in \mathbb{C};\ \Re(\lambda) < 0\} \text{ for continuous time,} \end{aligned}$$

a T_{std}-stable behavior is exactly one with asymptotically stable autonomous part. If its transfer matrix H is proper, it is also externally stable according to Theorems 8.1.26 and 8.2.83. Of course, these properties hold too if $\Lambda_1 \subseteq \Lambda_{1,\text{std}}$.

Proof In Definition and Corollary 5.4.1, we defined the characteristic variety as the vanishing set of the largest elementary divisor. Therefore, the statements of this theorem are just rewritings of those from Theorem and Definition 9.4.2. □

Remark 9.4.4 Engineers prefer T-stable IO behaviors for suitably chosen $\Lambda_1 \subseteq \Lambda_{1,\mathrm{std}}$ with also proper transfer matrix for the following reasons:

1. The properness of H implies that the IO system admits a state space realization with only constant matrices A, B, C, and D; and those can be built with adders, multipliers, and integrators (in the continuous-time case) or delay elements (in the discrete-time case), see Sects. 6.3.4 and 7.2. In the continuous-time case no differentiators are needed, and in the discrete-time case the system is causal, cf. Corollary 8.1.8.
   In the continuous-time case a nonproper transfer matrix causes impulsive solutions that may destroy the system, see Sect. 8.2.10.
2. According to Theorem 8.2.86 every trajectory $\binom{y}{u}$ with bounded input for $t \to \infty$ also has a bounded output y for $t \to \infty$. Again "explosive" outputs, which may destroy the system, are prevented.

In the continuous-time case the stable systems according to Theorem 8.2.86 are *not* characterized by T-stability for some T. ◊

9.5 Proper Stable Rational Functions

In this section we describe the principal ideal domain $\mathcal{S}$ of proper and T-stable rational functions, which is of greatest significance for stabilization theory. It is used in the books [1] *resp.* [2] *for the feedback stabilization of transfer matrices resp. of Rosenbrock systems. We are going to use it for the construction of proper stabilizing feedback compensators of IO behaviors and of proper observers. The main result (Theorem and Definition 9.5.4) states that the ring of proper and stable rational functions is a quotient ring of a suitable polynomial ring. Therefore the constructive methods from above, which use the Smith form over a polynomial algebra only, can be applied to it.*

Properness is defined for rational functions, and therefore we assume that $\mathcal{D} := F[s]$ is the polynomial algebra over a field F with its set $\mathcal{P}$ of monic irreducible polynomials. In addition we assume a disjoint decomposition $\mathcal{P} = \mathcal{P}_1 \uplus \mathcal{P}_2$ and the associated saturated multiplicative submonoid

$$T := \left\{ t = u \prod_{q \in \mathcal{P}_1} q^{\mathrm{ord}_q(t)};\ u \in F \setminus \{0\},\ \mathrm{ord}_q(t) \geq 0 \right\}. \tag{9.40}$$

In (5.85), we have defined the ring of proper rational functions as

$$F(s)_{\mathrm{pr}} = \left\{ a = \tfrac{f(s)}{g(s)} \in F(s);\ \deg_s(a) = \deg_s(f) - \deg_s(g) \leq 0 \right\}.$$

Definition 9.5.1 The intersection

$$\mathcal{S} := F[s]_T \cap F(s)_{\rm pr} = \left\{a = \tfrac{f}{t};\ t \in T,\ \deg_s(a) \le 0\right\}$$

is called the ring of *proper (and) T-stable rational functions.* ◊

In general, the intersection of two rings does not inherit their (good) properties. The ring $\mathcal{S}$, however, is an exception.

The standard method in systems theory as, for instance, in [1] and [2], is to show directly that this is a Euclidean ring. Here we proceed differently. We assume additionally that there is a linear polynomial $s - \alpha \in \mathcal{P}_1 \subset T,\ \alpha \in F$. This induces the substitution isomorphism

$$F[s] \overset{\cong}{\longleftrightarrow} F[s],\ f(s) \longmapsto f_\alpha := f(s+\alpha),\ g(s-\alpha) \longleftarrow\!\!\dashv g. \tag{9.41}$$

The rational function $\widehat{s} := \frac{1}{s-\alpha}$ is transcendental over F, i.e., it is an indeterminate. In other words, the map

$$F[s] \overset{\cong}{\longrightarrow} \widehat{\mathcal{D}} := F[\widehat{s}] = F\left[\tfrac{1}{s-\alpha}\right],\ \ g \longmapsto g(\widehat{s}) = g\left(\tfrac{1}{s-\alpha}\right), \tag{9.42}$$

is an algebra isomorphism. Therefore, $\widehat{\mathcal{D}} = F\left[\frac{1}{s-\alpha}\right]$ is a polynomial algebra, and a principal ideal domain to which the considerations of the preceding sections can be applied.

Example 9.5.2 In the standard continuous-time case with $F = \mathbb{R}$ or $F = \mathbb{C}$ and the stability decomposition $\mathbb{C} = \Lambda_1 \uplus \Lambda_2$ with $\Lambda_1 = \{\lambda \in \mathbb{C};\ \Re(\lambda) < 0\}$, the submonoid T is

$$T = \{t \in F[s];\ \mathrm{V}_{\mathbb{C}}(t) \subseteq \Lambda_1\}.$$

In this situation $s + 1 = s - (-1) \in T$, i.e., we can choose $\alpha = -1$.
The standard decomposition in the discrete-time case is $\Lambda = \Lambda_1 \uplus \Lambda_2$ with $\Lambda_1 = \{\lambda \in \mathbb{C};\ |\lambda| < 1\}$, and thus we can choose $\alpha = 0$. ◊

With the submonoid $\langle s - \alpha\rangle := \{(s-\alpha)^k,\ k \ge 0\}$ of $F[s] = F[s-\alpha]$, we construct the quotient ring

$$\mathcal{L}_\alpha := F[s-\alpha, (s-\alpha)^{-1}] := F[s]_{s-\alpha} := F[s]_{\langle s-\alpha\rangle} = \bigoplus_{k\in\mathbb{Z}} F(s-\alpha)^k \subset F(s) \tag{9.43}$$

of *Laurent polynomials* in the indeterminate $s - \alpha$. The monoid $\langle s - \alpha\rangle$ is not saturated, and the saturated set of all its divisors is $\{\beta(s-\alpha)^k;\ 0 \ne \beta \in F,\ k \ge 0\}$. Every nonzero Laurent polynomial g in $s - \alpha$ has a unique representation

$$g = \sum_{i=m}^{n} a_i (s-\alpha)^i = (s-\alpha)^m \sum_{i=0}^{n-m} a_{m+i}(s-\alpha)^i$$
$$\text{with} \quad m \le n \in \mathbb{Z}, \quad a_m, a_n \neq 0 \quad \text{and} \quad \deg_s(g) = m + (n-m) = n.$$

Hence we can write $\widehat{\mathcal{D}}$ as

$$\begin{aligned}\widehat{\mathcal{D}} &= F[\widehat{s}] = F[(s-\alpha)^{-1}] = \{g \in \mathcal{L}_\alpha;\ \deg_s(g) \le 0\} = \mathcal{L}_\alpha \cap F(s)_{\mathrm{pr}} \\ &\subseteq \mathcal{S} = F[s]_T \cap F(s)_{\mathrm{pr}},\end{aligned} \tag{9.44}$$

where the last inclusion follows from $s - \alpha \in T$. The group of units of $\mathcal{L}_\alpha$ is

$$\mathrm{U}(\mathcal{L}_\alpha) = \left\{\beta(s-\alpha)^k;\ 0 \neq \beta \in F,\ k \in \mathbb{Z}\right\}. \tag{9.45}$$

According to Example 9.2.4 the prime element $\widehat{s}$ of $F[\widehat{s}]$ gives rise to the saturated submonoid $T(\widehat{s}) = F[\widehat{s}] \setminus F[\widehat{s}]\widehat{s}$ of $F[\widehat{s}]$ and the discrete valuation ring $F[\widehat{s}]_{T(\widehat{s})}$ with the unique prime element $\widehat{s}$.

Lemma 9.5.3 *1. The identity $\mathcal{S}_{\widehat{s}} = F[s]_T$ holds, where $\mathcal{S}_{\widehat{s}}$ denotes the quotient ring of $\mathcal{S}$ with respect to the multiplicative submonoid*

$$\{\widehat{s}^k = (s-\alpha)^{-k};\ k \ge 0\} \subset \mathcal{S},$$

see item 1 of Example 9.1.1. Hence the ring $F[s]_T$ of T-stable rational functions is a quotient ring of the ring $\mathcal{S}$ of proper T-stable rational functions.
2. The ring of proper rational functions satisfies

$$F(s)_{\mathrm{pr}} = F[\widehat{s}]_{T(\widehat{s})}$$

where

$$T(\widehat{s}) := \left\{g \in F[\widehat{s}];\ \widehat{s} \nmid g\right\} = \{g \in F[\widehat{s}];\ g(0) \neq 0\}$$

is the saturated submonoid $T(\widehat{s}) = F[\widehat{s}] \setminus F[\widehat{s}]\widehat{s}$ of $F[\widehat{s}]$, cf. Example 9.2.4, and $F[\widehat{s}]_{T(\widehat{s})}$ is a principal ideal domain with the unique prime element $\widehat{s}$.

Proof 1. $\subseteq$: This inclusion holds because $\mathcal{S}$ is contained in $F[s]_T$ and $\widehat{s}$ is a unit in $F[s]_T$.
$\supseteq$: Let $\frac{f}{t} \in F[s]_T$, and let $m \in \mathbb{N}$ be such that $\deg_s(f) \le \deg_s(t) + m$. Then $\frac{f}{t(s-\alpha)^m} \in \mathcal{S}$ and $\frac{f}{t} = \frac{\frac{f}{t(s-\alpha)^m}}{\frac{1}{(s-\alpha)^m}} \in \mathcal{S}_{\widehat{s}}$.
2. According to Corollary and Definition 9.2.3 every nonzero rational function $a \in F(s) = F(\widehat{s})$ has a unique prime factor decomposition with respect to $F[\widehat{s}]_{T(\widehat{s})}$, namely,

$$\begin{aligned}&a = \widehat{s}^k \tfrac{f}{g},\ f, g \in T(\widehat{s}),\ \tfrac{f}{g} \in \mathrm{U}(F[\widehat{s}]_{T(\widehat{s})}),\ k \in \mathbb{Z},\\ &\deg_s(f) = \deg_s(g) = 0,\ \deg_s(\widehat{s}) = -1,\ \deg_s(a) = -k.\end{aligned}$$

Hence

$$a \in F(s)_{\mathrm{pr}} \Longleftrightarrow \deg_s(a) \le 0 \Longleftrightarrow k \ge 0 \Longleftrightarrow a \in F[\widehat{s}]_{T(\widehat{s})}. \qquad \square$$

Our next goal is to describe the ring $\mathcal{S}$ of proper and stable rational functions as a quotient ring of the polynomial algebra $F[\widehat{s}]$. For this we need the relation between the prime elements of $F[s]$ and those of $F[\widehat{s}]$; in particular, we have to find a suitable disjoint decomposition of the prime elements of $F[\widehat{s}]$.

Let $0 \neq f \in \mathcal{D} = F[s]$. We write the Taylor polynomial of f at α as

$$f = f_\alpha(s-\alpha) = \sum_{i=m}^{n} a_i (s-\alpha)^i, \ 0 \le m \le n = \deg_s(f), a_m \neq 0, \ a_n \neq 0, \tag{9.46}$$

and define

$$\widehat{f_\alpha} := \frac{f}{(s-\alpha)^{\deg_s(f)}} = a_n + \cdots + a_m \widehat{s}^{n-m} \in \widehat{\mathcal{D}} := F[\widehat{s}].$$

Hence

$$\deg_{\widehat{s}}(\widehat{f_\alpha}) = \deg_s(f) - m \quad \text{and} \quad \left(\deg_{\widehat{s}}(\widehat{f_\alpha}) = \deg_s(f) \Longleftrightarrow f(\alpha) \neq 0\right).$$

The sets

$$\mathcal{D} \setminus \mathcal{D}(s-\alpha) = \{f \in F[s];\ f(\alpha) \neq 0\} \text{ resp. } \widehat{\mathcal{D}} \setminus \widehat{\mathcal{D}}\widehat{s} = \{g \in F[\widehat{s}];\ g(0) \neq 0\} \tag{9.47}$$

are saturated submonoids of $\mathcal{D}$ resp. $\widehat{\mathcal{D}}$. Moreover the maps

$$\mathcal{D} \setminus \mathcal{D}(s-\alpha) \overset{\cong}{\longleftrightarrow} \widehat{\mathcal{D}} \setminus \widehat{\mathcal{D}}\widehat{s}, \ f \longmapsto \widehat{f_\alpha}(\widehat{s}) = \widehat{s}^{\deg_s(f)} f, \quad (s-\alpha)^{\deg_{\widehat{s}}(g)} g \longleftarrow g, \tag{9.48}$$

are inverse monoid isomorphisms by definition. The sets

$$\begin{aligned} \mathcal{P} &:= \{f \in F[s];\ f \text{ monic and irreducible}\} \text{ resp.} \\ \widehat{\mathcal{P}} &:= \{\widehat{s}\} \uplus \{g \in F[\widehat{s}];\ g \text{ irreducible},\ g(0) = 1\} \end{aligned} \tag{9.49}$$

are representative systems of the primes of $\mathcal{D}$ resp. $\widehat{\mathcal{D}}$. The isomorphism (9.48) induces the bijection

$$\mathcal{P} \setminus \{s-\alpha\} \overset{\cong}{\longrightarrow} \widehat{\mathcal{P}} \setminus \{\widehat{s}\}, \quad q(s) = (s-\alpha)^{\deg_s(q)} \widehat{q_\alpha} \leftrightarrow \widehat{q_\alpha}(\widehat{s}) = \frac{q}{(s-\alpha)^{\deg_s(q)}}, \tag{9.50}$$

with $\deg_s(q) = \deg_{\widehat{s}}(\widehat{q_\alpha})$. In contrast, the prime $q = s - \alpha$ is mapped onto $\widehat{q_\alpha} = 1$ and $\widehat{s} \neq \widehat{q_\alpha}$ for all α. Hence $s - \alpha$ and $\widehat{s} = (s-\alpha)^{-1}$ require a special treatment.

For nonzero $a \in F(s)$ consider the prime factor decomposition

$$
\begin{aligned}
&a = u\prod_{q\in\mathcal{P}} q(s)^{\mathrm{ord}_q(a)} \in F(s) \text{ with } u \in F\setminus\{0\} \text{ and } \left(\mathrm{ord}_q(a)\right)_q \in \mathbb{Z}^{(\mathcal{P})} \\
&\Longrightarrow \deg_s(a) = \sum_{q\in\mathcal{P}} \mathrm{ord}_q(a)\deg_s(q) \in \mathbb{Z}.
\end{aligned}
\tag{9.51}
$$

The equations $\widehat{q}_\alpha(\widehat{s}) = \frac{q(s)}{(s-\alpha)^{\deg_s(q)}} = q(s)\widehat{s}^{\deg_s(q)}$ then imply

$$
\begin{aligned}
a &= u\prod_{q\in\mathcal{P}} q^{\mathrm{ord}_q(a)} = u\prod_{q\in\mathcal{P}} \widehat{q}_\alpha(\widehat{s})^{\mathrm{ord}_q(a)}\widehat{s}^{-\mathrm{ord}_q(a)\deg_s(q)} \\
&\underset{(s-\alpha)_\alpha=1}{=} u\widehat{s}^{-\deg_s(a)} \prod_{q\in\mathcal{P}\setminus\{s-\alpha\}} \widehat{q}_\alpha(\widehat{s})^{\mathrm{ord}_q(a)} \in F(\widehat{s}) = F(s).
\end{aligned}
\tag{9.52}
$$

This is the prime factor decomposition of a with respect to $\widehat{\mathcal{P}}$ and implies

$$
\mathrm{ord}_{\widehat{q}_\alpha}(a) = \mathrm{ord}_q(a) \quad \text{for} \quad q \in \mathcal{P}\setminus\{s-\alpha\} \quad \text{and} \quad \mathrm{ord}_{\widehat{s}}(a) = -\deg_s(a).
$$

Via the bijection (9.50) the decomposition $\mathcal{P} = \mathcal{P}_1 \uplus \mathcal{P}_2$ with $s-\alpha \in \mathcal{P}_1$ induces the decomposition

$$
\begin{aligned}
\widehat{\mathcal{P}} = \widehat{\mathcal{P}}_1 \uplus \widehat{\mathcal{P}}_2, \quad &\text{where } \widehat{\mathcal{P}}_1 := \left\{\widehat{q}_\alpha;\ q \in \mathcal{P}_1\setminus\{s-\alpha\}\right\} \cong \mathcal{P}_1\setminus\{s-\alpha\} \text{ and} \\
&\widehat{\mathcal{P}}_2 := \widehat{\mathcal{P}}\setminus\widehat{\mathcal{P}}_1 \text{ with } \widehat{\mathcal{P}}_2\setminus\{\widehat{s}\} = \{\widehat{q}_\alpha;\ q\in\mathcal{P}_2\} \cong \mathcal{P}_2.
\end{aligned}
\tag{9.53}
$$

According to Lemma and Definition 9.1.2, this decomposition, in turn, generates the saturated submonoid

$$
\begin{aligned}
\widehat{T} &:= \left\{u\prod_{q\in\mathcal{P}_1\setminus\{s-\alpha\}} \widehat{q}_\alpha(\widehat{s})^{\mu(q)};\ u\in F\setminus\{0\},\ \mu(q)\geq 0\right\} \\
&= \left\{\widehat{t}_\alpha(\widehat{s}) = \frac{t}{(s-\alpha)^{\deg_s(t)}};\ t\in T\right\},
\end{aligned}
\tag{9.54}
$$

where we have used (9.40) and the correspondence (9.52) and (9.53) of the prime factor decompositions with respect to $\mathcal{P}$ and $\widehat{\mathcal{P}}$.

Theorem and Definition 9.5.4 ($\mathcal{S}$ as polynomial quotient ring)

1. *The identity $\mathcal{S} = F[\widehat{s}]_{\widehat{T}}$ holds, i.e., the ring $\mathcal{S}$ of proper and T-stable rational functions is a quotient ring of the polynomial algebra $F[\widehat{s}] = F[\frac{1}{s-\alpha}]$. In particular, it is a principal ideal domain. Therefore, Theorem 9.2.2 and Remark 9.2.5 are applicable to $\mathcal{S}$ and, in particular, the Smith form with respect to $\mathcal{S}\subset F(s)$ can be computed by means of that with respect to $F[\widehat{s}]$.*
2. *The elements $\widehat{q}_\alpha(\widehat{s}) = \frac{q}{(s-\alpha)^{\deg_s(q)}}$, $q\in\mathcal{P}_2$, and $\widehat{s} = \frac{1}{s-\alpha}$ form a system of representatives of primes of $\mathcal{S}$, up to association.*
3. *The group of units of $\mathcal{S} = F(s)_{\mathrm{pr}} \cap F[s]_T$ is*

$$\begin{aligned}\mathrm{U}(\mathcal{S}) &= \left\{\tfrac{t_1}{t_2};\ t_1, t_2 \in T,\ \deg_s(t_1) = \deg_s(t_2)\right\} = \left\{\tfrac{\widehat{t}_{1,\alpha}}{\widehat{t}_{2,\alpha}};\ \widehat{t}_{1,\alpha}, \widehat{t}_{2,\alpha} \in \widehat{T}\right\}\\ &= \left\{a = u \prod_{q\in\mathcal{P}_1} q^{\operatorname{ord}_q(a)};\ \deg_s(a) = \sum_{q\in\mathcal{P}_1} \operatorname{ord}_q(a)\deg_s(q) = 0\right\}\\ &= \left\{a = u \prod_{\widehat{q}_\alpha\in\widehat{\mathcal{P}}_1} \widehat{q}_\alpha^{\mu_q};\ \mu_q \in \mathbb{Z}\right\},\end{aligned}$$

and the identity $\widehat{T} = \mathrm{U}(\mathcal{S}) \cap F[\widehat{s}]$ holds.

Proof 1. For $a = \frac{f}{t} \in \mathcal{S}$ with $t \in T$ and $\deg_s(f) \le \deg_s(t)$ one gets

$$\frac{f}{t} = \widehat{s}^{\deg_s(t)-\deg_s(f)} \frac{\frac{f}{(s-\alpha)^{\deg_s(f)}}}{\frac{t}{(s-\alpha)^{\deg_s(t)}}} = \frac{\widehat{s}^{\deg_s(t)-\deg_s(f)}\widehat{f}_\alpha(\widehat{s})}{\widehat{t}_\alpha(\widehat{s})} \underset{(9.54)}{\in} F[\widehat{s}]_{\widehat{T}} \Longrightarrow \mathcal{S} \subseteq F[\widehat{s}]_{\widehat{T}}.$$

Conversely, $F[\widehat{s}] \subseteq \mathcal{S}$ and $\widehat{T} \subseteq \mathrm{U}(\mathcal{S})$ imply $F[\widehat{s}]_{\widehat{T}} \subseteq \mathcal{S}$.
2. This is a consequence of Theorem 9.2.2 and (9.54).
3. This follows from Lemma 9.2.1. □

Example 9.5.5 For $F = \mathbb{C}$ or $F = \mathbb{R}$, we consider $\mathcal{D} = F[s]$ with its quotient field $K = F(s)$. Let $\mathbb{C} = \Lambda_1 \uplus \Lambda_2$ be a disjoint decomposition with its induced multiplicative submonoid $T = \{t \in F[s];\ \mathrm{V}_{\mathbb{C}}(t) \subseteq \Lambda_1\}$.

We first consider $F = \mathbb{C}$. We assume that $\Lambda_1 \neq \emptyset$ and $\Lambda_2 \neq \emptyset$, and we choose $\alpha \in \Lambda_1$. The induced decomposition $\mathcal{P} = \mathcal{P}_1 \uplus \mathcal{P}_2$ is given by $\mathcal{P}_i = \{s - \lambda;\ \lambda \in \Lambda_i\}$. If $F = \mathbb{C}$ and $0 \neq a = u\prod_{\lambda\in\mathbb{C}}(s-\lambda)^{\operatorname{ord}_{s-\lambda}(a)} \in F(s)$, then

$$\begin{cases}\operatorname{ord}_{s-\lambda}(a)\\ -\operatorname{ord}_{s-\lambda}(a)\end{cases} \text{is the multiplicity of the } \begin{cases}\text{zero } \lambda \text{ of } a & \text{if } \operatorname{ord}_{s-\lambda}(a) > 0,\\ \text{pole } \lambda \text{ of } a & \text{if } \operatorname{ord}_{s-\lambda}(a) < 0.\end{cases}$$

For $q = s - \lambda$, $\lambda \neq \alpha$, the polynomial $\widehat{q}_\alpha(\widehat{s}) \in F[\widehat{s}]$ from (9.48) is

$$\widehat{q}_\alpha(\widehat{s}) = \widehat{s}^{\deg_s(q)} q = \widehat{s}(s-\lambda) = \widehat{s}\left(\tfrac{1}{\widehat{s}} + \alpha - \lambda\right) = 1 + (\alpha-\lambda)\widehat{s}.$$

Hence, by (9.53) and (9.54), the induced data are

$$\begin{aligned}\widehat{\mathcal{P}}_1 &= \left\{\tfrac{s-\lambda}{s-\alpha} = 1 + (\alpha-\lambda)\widehat{s};\ \lambda \in \Lambda_1 \setminus \{\alpha\}\right\},\\ \widehat{\mathcal{P}}_2 &= \left\{1 + (\alpha-\lambda)\widehat{s};\ \lambda \in \Lambda_2\right\} \uplus \{\widehat{s}\}, \quad \text{and}\\ \widehat{T} &= \left\{u \prod_{\lambda\in\Lambda_1\setminus\{\alpha\}} \left(1 + (\alpha-\lambda)\widehat{s}\right)^{m_\lambda};\ m_\lambda \ge 0\right\}.\end{aligned}$$

The set $\widehat{\mathcal{P}}_2$ is a representative system of the prime elements of $\mathcal{S}$.

For $F = \mathbb{R}$, we assume in addition that $\mathbb{R} \cap \Lambda_1 \neq 0$ and choose α from this intersection. Then $\mathcal{P} = \mathcal{P}_1 \uplus \mathcal{P}_2$ is given by

$$\mathcal{P}_i = \{s - \lambda;\ \lambda \in \mathbb{R} \cap \Lambda_i\} \uplus \{(s - \lambda)(s - \overline{\lambda});\ \lambda \in \Lambda_i,\ \Im(\lambda) > 0\},\ i = 1, 2.$$

For $q = (s - \lambda)(s - \overline{\lambda}),\ \Im(\lambda) > 0$, we have

$$\begin{aligned}
\widehat{q}_\alpha &= \widehat{(s - \lambda)}_\alpha \widehat{(s - \overline{\lambda})}_\alpha = \left(1 + (\alpha - \lambda)\widehat{s}\right)\left(1 + (\alpha - \overline{\lambda})\widehat{s}\right) \\
&= 1 + 2(\alpha - \Re(\lambda))\widehat{s} + \left(\alpha^2 - 2\alpha\Re(\lambda) + |\lambda|^2\right)\widehat{s}^2 \Longrightarrow \widehat{\mathcal{P}} = \widehat{\mathcal{P}}_1 \uplus \widehat{\mathcal{P}}_2 \text{ with} \\
\widehat{\mathcal{P}}_1 &= \left\{1 + (\alpha - \lambda)\widehat{s};\ \lambda \in \mathbb{R} \cap (\Lambda_1 \setminus \{\alpha\})\right\} \\
&\quad \uplus \left\{\widehat{q}_\alpha;\ q = (s - \lambda)(s - \overline{\lambda}),\ \lambda \in \Lambda_1,\ \Im(\lambda) > 0\right\} \quad \text{and} \\
\widehat{\mathcal{P}}_2 &= \{\widehat{s}\} \uplus \left\{1 + (\alpha - \lambda)\widehat{s};\ \lambda \in \mathbb{R} \cap (\Lambda_2 \setminus \{\alpha\})\right\} \\
&\quad \uplus \left\{1 + 2(\alpha - \Re(\lambda))\widehat{s} + \left(\alpha^2 - 2\alpha\Re(\lambda) + |\lambda|^2\right)\widehat{s}^2;\ \lambda \in \Lambda_2,\ \Im(\lambda) > 0\right\}.
\end{aligned}$$

◇

The following theorem will later be used for the construction of *proper* observers and controllers. We need the representations

$$\begin{aligned}
\mathcal{S} &= \widehat{\mathcal{D}}_{\widehat{T}} = \left\{g\widehat{t}^{-1},\ g \in \widehat{\mathcal{D}},\ \widehat{t} \in \widehat{T}\right\} \quad \text{and} \\
\mathcal{D}_T &= \mathcal{S}_{\widehat{s}} = \left\{g\widehat{t}^{-1}\widehat{s}^{-n},\ g \in \widehat{\mathcal{D}},\ \widehat{t} \in \widehat{T},\ n \in \mathbb{N}\right\} \text{ with} \\
ft^{-1} &= \widehat{f}_\alpha \widehat{t}_\alpha^{-1} \widehat{s}^{\deg_s(t) - \deg_s(f)},\ \widehat{f}_\alpha = f\widehat{s}^{\deg_s(f)} \in \widehat{\mathcal{D}},\ \widehat{t}_\alpha = t\widehat{s}^{\deg_s(t)} \in \widehat{T}.
\end{aligned}$$

Theorem 9.5.6 *For arbitrary $m \in \mathbb{N}$ and nonzero $e \in \widehat{\mathcal{D}}$ the equation $\mathcal{D}_T = \mathcal{D}_T e + \mathcal{S}\widehat{s}^m$ holds. More precisely, let $l := \mathrm{ord}_{\widehat{s}}(e)$ and hence $e = e_1\widehat{s}^l$ with $e_1 \in \widehat{\mathcal{D}}$ and $\widehat{s} \nmid e_1$, and let $0 \neq r \in \mathcal{D}_T$ be arbitrary with the specifications*

$$r = g\widehat{t}^{-1}\widehat{s}^{-n} \in \mathcal{D}_T = \mathcal{S}_{\widehat{s}} \text{ with } g \in \widehat{\mathcal{D}},\ \widehat{t} \in \widehat{T},\ n \in \mathbb{N}.$$

1. *Let $a, b \in \widehat{\mathcal{D}}$ be such that $1 = ae_1 + b\widehat{s}^{m+n}$. Such a representation can be computed with the Euclidean algorithm for $\widehat{\mathcal{D}} = F[\widehat{s}]$. Then*

$$\begin{aligned}
r &= r(ae_1 + b\widehat{s}^{m+n}) = (ar\widehat{s}^{-l})e + (br\widehat{s}^n)\widehat{s}^m \\
&= (ar\widehat{s}^{-l})e + (bg\widehat{t}^{-1})\widehat{s}^m \in \mathcal{D}_T e + \mathcal{S}\widehat{s}^m,
\end{aligned} \tag{9.55}$$

i.e., for arbitrary $m \geq 0$, $0 \neq e \in \widehat{\mathcal{D}}$ and $r = g\widehat{t}^{-1}\widehat{s}^{-n} \in \mathcal{D}_T$, the pair

$$(c^0, d^0) := (ar\widehat{s}^{-l}, br\widehat{s}^n) \in \mathcal{D}_T \times \mathcal{S} \tag{9.56}$$

is a solution of

$$r = c^0 e + d^0\widehat{s}^m \in \mathcal{D}_T = \mathcal{D}_T e + \mathcal{S}\widehat{s}^m.$$

Notice that $\mathcal{S}\widehat{s}$ consists of the strictly proper and T-stable rational functions.

2. *The set of all other solutions* $(c, d) \in \mathcal{D}_T \times \mathcal{S}$ *of* $r = ce + d\widehat{s}^m$ *is*

$$(c^0, d^0) + \mathcal{S}\widehat{s}^{m-l}(1, -\widehat{s}^{-m}e).$$

Proof 1. Equation (9.55) is checked easily.
2. The equivalences

$$\begin{aligned} & ce + d\widehat{s}^m = r = c^0e + d^0\widehat{s}^m \quad \text{with } (c, d) \in \mathcal{D}_T \times \mathcal{S} \\ \iff & \underbrace{(c - c^0)}_{\in \mathcal{D}_T} e\widehat{s}^{-m} = d^0 - d \in \mathcal{S} \\ \iff & (c, d) = (c^0, d^0) + h(1, -e\widehat{s}^{-m}) \text{ with } h := c - c^0 \in \mathcal{D}_T \text{ and } he\widehat{s}^{-m} \in \mathcal{S} \end{aligned}$$

hold. Therefore, it suffices to prove $\{h \in \mathcal{D}_T;\ he\widehat{s}^{-m} \in \mathcal{S}\} = \mathcal{S}\widehat{s}^{m-l}$.
$\supseteq$: $\mathcal{S}\widehat{s}^{m-l} \subseteq \mathcal{S}_{\widehat{s}} = \mathcal{D}_T$ and $\mathcal{S}\widehat{s}^{m-l}e\widehat{s}^{-m} = \mathcal{S}\widehat{s}^{-l}e = \mathcal{S}e_1 \underset{e_1 \in \widehat{\mathcal{D}} \subseteq \mathcal{S}}{\subseteq} \mathcal{S}$.
$\subseteq$: Assume $h \in \mathcal{D}_T$ and $he\widehat{s}^{-m} \in \mathcal{S}$. By Theorem and Definition 9.5.4 the element $\widehat{s} \in \widehat{\mathcal{P}}_2 \subset \widehat{\mathcal{D}} \subset \mathcal{S}$ is a prime of $\mathcal{S}$. Therefore h has a unique representation

$$h = g\widehat{s}^q \text{ with } g \in \mathcal{S},\ q := \mathrm{ord}_{\widehat{s}}(h) \in \mathbb{Z},\ \text{and } \widehat{s} \nmid g.$$

Hence

$$\mathcal{S} = \{h \in \mathcal{S}_{\widehat{s}};\ \mathrm{ord}_{\widehat{s}}(h) \geq 0\} \ni he\widehat{s}^{-m} = g\widehat{s}^q e_1 \widehat{s}^l \widehat{s}^{-m} = ge_1\widehat{s}^{q+l-m}$$

with $ge_1 \in \mathcal{S},\ \widehat{s} \nmid ge_1$. This is equivalent to $q + l - m \geq 0$ and to

$$h = g\widehat{s}^q = g\widehat{s}^{q+l-m}\widehat{s}^{m-l} \in \mathcal{S}\widehat{s}^{m-l}.$$

□

References

1. M. Vidyasagar, *Control System Synthesis. A Factorization Approach*. MIT Press Series in Signal Processing, Optimization, and Control, vol. 7 (MIT Press, Cambridge, 1985)
2. A.I.G. Vardulakis, *Linear Multivariable Control. Algebraic Analysis and Synthesis Methods* (Wiley, Chichester, 1991)

Chapter 10
Compensators

1. *Stabilization* of an unstable IO behavior $\mathcal{B}_1$ (the *plant*) by output feedback means to construct a suitable IO behavior $\mathcal{B}_2$ such that the feedback or closed-loop behavior $\mathcal{B} := \mathrm{fb}(\mathcal{B}_1, \mathcal{B}_2)$ is well-posed and T-stable for a suitable monoid T of stable polynomials. The behavior $\mathcal{B}_2$ is then called a *T-stabilizing compensator* or *T-stabilizing controller* for $\mathcal{B}_1$. In Sect. 10.1, we characterize T-stabilizability of $\mathcal{B}_1$, i.e., the existence of a T-stabilizing compensator, and then parametrize all stabilizing compensators. Our main emphasis is on IO behaviors $\mathcal{B}_2$ where both $\mathcal{B}_2$ and $\mathcal{B}$ are proper, whereas $\mathcal{B}_1$ is not assumed proper, cf. Theorems 10.1.20 and 10.1.34.
2. *Compensator design* means to construct stabilizing compensators $\mathcal{B}_2$ that realize desired tasks like *tracking*, *disturbance rejection* or *model matching*. We discuss the existence and parametrization of such restricted compensators in detail in Sect. 10.2, especially in Theorems 10.2.8 and 10.2.11. Compensator design belongs to the most important subjects of systems theory and is treated in many textbooks, but always in advanced chapters, see, e.g., [1–6]. Among the early outstanding researchers in this important field were Kučera [7], Youla [8], Desoer, Vidyasagar and coworkers, and many further colleagues. Whereas they studied the stabilization of and compensator design for Rosenbrock equations or transfer matrices, we construct compensators for IO behaviors with the new technique of quotient modules from Chap. 9 that was developed in Blumthaler's (Ingrid Scheicher's) thesis. The use of modules and split exact sequences for the proofs is due to Quadrat [9]. The books [4] and [10] and further suggestions by Bourlès were essential for the presentation in this book.
3. In the standard situations we also show that the constructed compensators are *robust*. This signifies that $\mathcal{B}_2$ is also a valid compensator for all nearby plants $\mathcal{B}_1$. This obviously requires to topologize the set of all plants $\mathcal{B}_1$. In the continuous-time case we first introduce *two* norms $\|h\|$ and $\|h\|_1$ on the algebra $\mathcal{S}$ of

The original version of this chapter was revised: The errors in this chapter have been corrected. The correction to this chapter can be found at https://doi.org/10.1007/978-3-030-43936-1_14

U. Oberst et al., *Linear Time-Invariant Systems, Behaviors and Modules*, Differential-Algebraic Equations Forum,
https://doi.org/10.1007/978-3-030-43936-1_10

proper and stable rational functions and the induced norms on $\mathcal{S}^{1\times\bullet}$; the norm $\|h\|_1$ is finer. For a proper and stable transfer matrix $H \in \mathcal{S}^{p\times m}$ this enables to compute the operator norms of $\|H\circ : (L_+^\infty)^m \to (L_+^\infty)^p\|$ (Theorem 10.2.36) and $\|H\circ : (L_+^2)^m \to (L_+^2)^p\|$ (Theorem 10.2.46) where L_+^∞ resp. L_+^2 are the normed spaces of piecewise continuous functions with left bounded support that are bounded resp. square-integrable. For the L^2-case this requires the Fourier transform that we present to the necessary extent with complete proofs in Sect. 10.2.5.

4. As indicated above the two norms on $\mathcal{S}$ are used to topologize the set of controllable IO $\mathcal{F}_T$-behaviors $\mathcal{B}_1$. If $\mathcal{B}_2$ is a valid compensator of a plant $\mathcal{B}_1$ then it has the same property for all plants $\widetilde{\mathcal{B}}_1$ in a suitable neighborhood of $\mathcal{B}_1$, cf. Theorems 10.2.33, 10.2.47 and 10.2.50. If moreover H resp. $\widetilde{H}$ denote the proper and stable closed-loop transfer matrices of $\mathcal{B} := \mathrm{fb}(\mathcal{B}_1, \mathcal{B}_2)$ resp. of $\widetilde{\mathcal{B}} := \mathrm{fb}(\widetilde{\mathcal{B}}_1, \mathcal{B}_2)$ then the assignments
$$\widetilde{\mathcal{B}}_1 \mapsto \left(\widetilde{H}\circ : (\mathrm{L}_+^\infty)^{p+m} \to (\mathrm{L}_+^\infty)^{p+m}\right),\ \left(\widetilde{H}\circ : (\mathrm{L}_+^2)^{p+m} \to (\mathrm{L}_+^2)^{p+m}\right)$$
are continuous, i.e., $\widetilde{H}\circ$ is near $H\circ$ if $\widetilde{\mathcal{B}}_1$ is near $\mathcal{B}_1$. The proofs require some preparations on continuous linear operators between Banach spaces that are explained in Sect. 10.2.4.
5. In Sect. 12.5 we treat the famous stabilization of state behaviors by Luenberger observers and state feedback that was the first stabilization method for multivariable systems. This method furnishes far fewer compensators than that of Sect. 10.1.5 of the present chapter and does not imply a parametrization and a tracking result. It has, however, the advantage that the state dimensions of plant and compensator are equal. As far as we see, the results of Sect. 12.5 are not simple corollaries of those in Sect. 10.1.5. In Sect. 10.2.6 we treat the problem of model matching.

10.1 Stabilizing Compensators

In Sect. 10.1, *we characterize T-stabilizability by the property that the highest nonzero elementary divisor of $\mathcal{B}_1$ is T-stable, i.e., belongs to T. We then construct and parametrize all such compensators. The results of Sect.* 10.1.1 *hold for general principal ideal domains $\mathcal{D}$.*

In the subsequent sections we specialize the results to polynomial rings $\mathcal{D} = F[s]$ over a field F. In Sect. 10.1.2 *we parametrize the compensators with proper feedback behavior. In Sect.* 10.1.4 *we add the design goal that also the compensator should be proper. Section* 10.1.3 *is devoted to pole placement, which means that the feedback behavior's poles must be contained in a given finite set. Finally, in Sect.* 10.1.5, *we reformulate the results for state space systems, i.e., we consider state space realizations of the plant, the compensator and the feedback behavior.*

10.1.1 *T-stabilizable IO Behaviors*

We use the assumptions and the data from Sect. 9.1, in particular, a principal ideal domain $\mathcal{D}$ with its quotient field $K = \mathrm{quot}(\mathcal{D})$, as well as a saturated submonoid

$$T = \left\{ u \prod_{q \in \mathcal{P}_1} q^{\mu(q)};\ 0 \neq u \in \mathrm{U}(\mathcal{D}),\ \mu(q) \in \mathbb{N} \right\} \subsetneq \mathcal{D} \setminus \{0\} \tag{10.1}$$

which is induced by the proper subset $\mathcal{P}_1 \subsetneq \mathcal{P}$ of a representative system $\mathcal{P}$ of prime elements of $\mathcal{D}$. If ${}_{\mathcal{D}}\mathcal{F}$ is an injective cogenerator then so is ${}_{\mathcal{D}_T}\mathcal{F}_T$.

We also use the feedback behavior $\mathrm{fb}(\mathcal{B}_1, \mathcal{B}_2)$ from Sect. 7.5 and briefly recall its properties. The interconnection diagram of the feedback behavior is

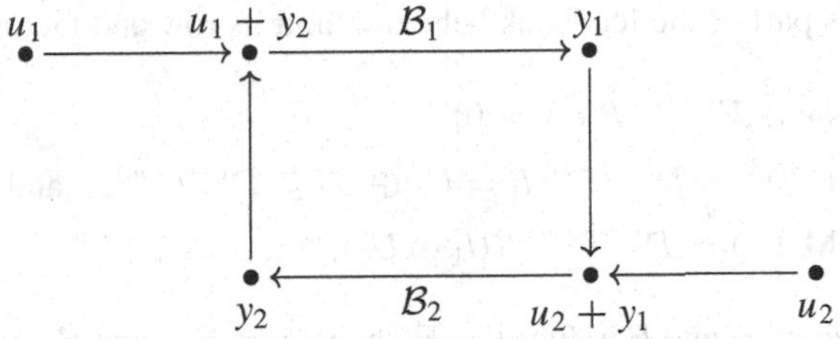

The plant $\mathcal{B}_1$ and its modules and matrices are given as

$$\begin{aligned} &\mathcal{B}_1 = \left\{ \left(\begin{smallmatrix} y_1 \\ u_1 \end{smallmatrix}\right) \in \mathcal{F}^{p+m};\ P_1 \circ y_1 = Q_1 \circ u_1 \right\} = U_1^{\perp},\ \ (P_1, -Q_1) \in \mathcal{D}^{p \times (p+m)}, \\ &U_1 := \mathcal{D}^{1 \times p}(P_1, -Q_1) \subseteq \mathcal{D}^{1 \times (p+m)},\ M_1 := \mathrm{M}(\mathcal{B}_1) = \mathcal{D}^{1 \times (p+m)} / U_1, \\ &\text{with } \mathrm{rank}(P_1) = p \text{ or } P_1 \in \mathrm{Gl}_p(K),\ H_1 = P_1^{-1} Q_1,\ K U_1 = K^{1 \times p}(\mathrm{id}_p, -H_1). \end{aligned} \tag{10.2}$$

The compensator $\mathcal{B}_2$ is an IO system with the complementary IO structure. It and its modules and matrices are given as

$$\begin{aligned} &\mathcal{B}_2 = \left\{ \left(\begin{smallmatrix} u_2 \\ y_2 \end{smallmatrix}\right) \in \mathcal{F}^{p+m};\ P_2 \circ y_2 = Q_2 \circ u_2 \right\} = U_2^{\perp},\ (-Q_2, P_2) \in \mathcal{D}^{m \times (p+m)}, \\ &U_2 := \mathcal{D}^{1 \times m}(-Q_2, P_2) \subseteq \mathcal{D}^{1 \times (p+m)},\ M_2 = \mathrm{M}(\mathcal{B}_2) = \mathcal{D}^{1 \times (p+m)} / U_2, \\ &\text{with } \mathrm{rank}(P_2) = m \text{ or } P_2 \in \mathrm{Gl}_m(K),\ H_2 = P_2^{-1} Q_2,\ K U_2 = K^{1 \times m}(-H_2, \mathrm{id}_m). \end{aligned} \tag{10.3}$$

With the signal vectors $y = \left(\begin{smallmatrix} y_1 \\ y_2 \end{smallmatrix}\right) \in \mathcal{F}^{p+m}$ and $u := \left(\begin{smallmatrix} u_2 \\ u_1 \end{smallmatrix}\right) \in \mathcal{F}^{p+m}$, the resulting *feedback behavior* or *closed-loop behavior* and its modules are

$$\begin{aligned} &\mathcal{B} := \mathrm{fb}(\mathcal{B}_1, \mathcal{B}_2) = \left\{ \left(\begin{smallmatrix} y \\ u \end{smallmatrix}\right) \in \mathcal{F}^{(p+m)+(p+m)};\ P \circ y = Q \circ u \right\} \\ &U := \mathcal{B}^{\perp} = \mathcal{D}^{1 \times (p+m)}(P, -Q) \subseteq \mathcal{D}^{1 \times (p+m+p+m)}, \quad \text{and} \\ &M := \mathrm{M}(\mathcal{B}) = \mathcal{D}^{1 \times (p+m+p+m)} / U \\ &\text{with } P := \left(\begin{smallmatrix} P_1 & -Q_1 \\ -Q_2 & P_2 \end{smallmatrix}\right),\ Q := \left(\begin{smallmatrix} 0 & Q_1 \\ Q_2 & 0 \end{smallmatrix}\right) \in \mathcal{D}^{(p+m) \times (p+m)}. \end{aligned} \tag{10.4}$$

The feedback behavior is well-posed (see Theorem and Definition 7.5.1) if and only if

$$\mathcal{D}^{1\times(p+m)}P = U_1 + U_2 = U_1 \oplus U_2 \tag{10.5}$$

holds, i.e., if and only if $U_1 \cap U_2 = 0$. If it is well-posed, then it is an IO behavior with input u and output y, its transfer matrix is

$$H = \begin{pmatrix} \mathrm{id}_p & -H_1 \\ -H_2 & \mathrm{id}_m \end{pmatrix}^{-1} \begin{pmatrix} 0 & H_1 \\ H_2 & 0 \end{pmatrix},$$

and its equation module U can be written as

$$U = (U_1 \oplus U_2)(\mathrm{id}_{p+m}, -H).$$

The autonomous part of the feedback behavior, and its row and factor modules are

$$\begin{aligned} \mathcal{B}^0 &= \{y \in \mathcal{F}^{p+m};\ P \circ y = 0\}, \\ U^0 &= (\mathcal{B}^0)^\perp = \mathcal{D}^{1\times(p+m)}P = U_1 \oplus U_2 \subseteq \mathcal{D}^{1\times(p+m)}, \quad \text{and} \\ M^0 &= \mathrm{M}(\mathcal{B}^0) = \mathcal{D}^{1\times(p+m)}/(U_1 \oplus U_2). \end{aligned} \tag{10.6}$$

The $\mathcal{F}$-behaviors $\mathcal{B}_i$ and $\mathcal{B}$ induce the $\mathcal{F}_T$-behaviors $\mathcal{B}_{i,T}$ and $\mathcal{B}_T$ and corresponding quotient modules $U_{i,T}$, $M_{i,T}$, U_T, and M_T.

Lemma 10.1.1 *Assume that the feedback behavior from* (10.4) *is well-posed. Then the following statements are equivalent:*

1. *The feedback behavior $\mathcal{B}$ is T-stable, i.e., its autonomous part $\mathcal{B}^0 = \{y \in \mathcal{F}^{p+m};\ P \circ y = 0\}$ is T-small or, in other words, $\mathcal{B}^0_T = 0$.*
2. *The identity*

$$U_{1,T} \oplus U_{2,T} = \mathcal{D}_T^{1\times(p+m)}P = \mathcal{D}_T^{1\times(p+m)} \tag{10.7}$$

holds or, equivalently, $P \in \mathrm{Gl}_{p+m}(\mathcal{D}_T)$.

3. *(a) The $\mathcal{F}_T$-behavior $\mathcal{B}_T$ is controllable, i.e., M_T is $\mathcal{D}_T$-torsionfree thus $\mathcal{D}_T$-free, and*
   (b) $H \in \mathcal{D}_T^{(p+m)\times(p+m)}$.

These conditions imply that the quotient behaviors $\mathcal{B}_{1,T}$ and $\mathcal{B}_{2,T}$ are controllable or, in other words, that $M_{1,T} \cong U_{2,T}$ and $M_{2,T} \cong U_{1,T}$ are free.

Proof The equivalences follow directly from Theorem and Definition 9.4.2, applied to $\mathrm{fb}(\mathcal{B}_1, \mathcal{B}_2)$. The direct decomposition in item 2 implies the isomorphisms

$$M_{1,T} = \mathcal{D}_T^{1\times p}/U_{1,T} \cong U_{2,T} \quad \text{and} \quad M_{2,T} = \mathcal{D}_T^{1\times m}/U_{2,T} \cong U_{1,T}.$$

Hence, $M_{1,T}$ and $M_{2,T}$ are torsionfree and thus free. □

Definition 10.1.2 A given plant $\mathcal{B}_1$ is called *T-stabilizable* if there exists a compensator $\mathcal{B}_2$ such that the feedback behavior is well-posed and T-stable. Such a $\mathcal{B}_2$ is then called a *T-stabilizing compensator* or *T-stabilizing controller* of $\mathcal{B}_1$. If $\mathcal{B}_2$ is also stable, then it is called a *strong* T-stabilizing compensator.

For $T = \{1\}$ and $\mathcal{D}_T = \mathcal{D}$, we talk of a *trivializing* compensator. Hence $\mathcal{B}_2$ is a T-stabilizing compensator of $\mathcal{B}_1$ if and only if $\mathcal{B}_{2,T}$ is a trivializing compensator of $\mathcal{B}_{1,T}$. ◊

Assumption 10.1.3 In the sequel, we assume the necessary condition for T-stabilizability of $\mathcal{B}_1$ that $\mathcal{B}_{1,T}$ is controllable or, equivalently, that $M_{1,T} = \mathcal{D}_T^{1\times(p+m)}/U_{1,T}$ is $\mathcal{D}_T$-free. ◊

Before we give equivalent conditions for the controllability of $\mathcal{B}_{1,T}$ in terms of the highest elementary divisor and the characteristic variety in Corollaries 10.1.4 and 10.1.5, respectively, we amend the properties of the quotient functor $(-)_T$ by the following technical lemma.

Corollary 10.1.4 *Let $e_p^1 \in \mathcal{D}$ be the highest nonzero elementary divisor of $(P_1, -Q_1)$. Then Assumption 10.1.3 is satisfied (i.e., $\mathcal{B}_{1,T}$ is controllable) if and only if $e_p^1 \in T$.*

Proof According to Corollary 9.2.5, the Smith form of $(P_1, -Q_1)$ with respect to $\mathcal{D}$ is also the Smith form with respect to $\mathcal{D}_T$. Therefore, e_p^1 is also the highest nonzero elementary divisor of $(P_1, -Q_1)$ with respect to $\mathcal{D}_T$. By Theorem 4.2.12, the behavior $\mathcal{B}_{1,T}$ is controllable if and only if e_p^1 is a unit in $\mathcal{D}_T$, i.e.,

$$e_p^1 \in \mathcal{D} \cap \mathrm{U}(\mathcal{D}_T) \underset{\text{Lemma 9.2.1, 4}}{=} T.$$

□

Corollary 10.1.5 *Let $\mathcal{D} = \mathbb{C}[s]$, and let $T \subset \mathbb{C}[s]$ be defined by a disjoint decomposition $\mathbb{C} = \Lambda_1 \uplus \Lambda_2$, $\emptyset \neq \Lambda_i$, $i = 1, 2$, as*

$$T = \{t \in F[s];\ \mathrm{V}_{\mathbb{C}}(t) \subseteq \Lambda_1\},$$

compare Example 9.1.3, item 1. Then the behavior $\mathcal{B}_{1,T}$ is controllable if and only if $\mathrm{char}(\mathcal{B}_1) \subseteq \Lambda_1$.

By Definition 6.2.8, the elements of $\mathrm{char}(\mathcal{B}_1)$ *are the uncontrollable poles of $\mathcal{B}_1$. Therefore, the quotient behavior $\mathcal{B}_{1,T}$ is controllable if and only if all uncontrollable poles of $\mathcal{B}_1$ are T-stable.*

The assertion also holds in the real case $\mathcal{D} = \mathbb{R}[s]$ if we assume in addition that the Λ_i are invariant under complex conjugation, i.e., $\overline{\Lambda_i} = \Lambda_i$, $i = 1, 2$, see Example 9.1.3, item 2.

Proof According to Corollary 10.1.4, the behavior $\mathcal{B}_{1,T}$ is controllable if and only if the highest nonzero elementary divisor e_p^1 lies in T, i.e., $\mathrm{V}_{\mathbb{C}}(e_p^1) \subseteq \Lambda_1$ holds. By

Definition and Corollary 5.4.1, the characteristic variety of $\mathcal{B}_1$ is $\operatorname{char}(\mathcal{B}_1) = \mathrm{V}_{\mathbb{C}}(e_p^1)$, which proves the claim. □

We have seen in Lemma 10.1.1 that a T-stabilizing compensator gives rise to a direct complement of the quotient module $U_{1,T}$ in $\mathcal{D}_T^{1\times(p+m)}$. In Theorem 10.1.13, we will show that every T-stabilizing compensator can be obtained via such a direct complement. For this purpose we adapt the characterization of direct summands from Theorem and Definition 4.2.18 and Theorem 6.2.5 on coprime factorizations to the current setting.

We conduct the following derivations for the principal ideal domain $\mathcal{D}$. In the sequel, we apply these to $\mathcal{D}_T$, $\widehat{\mathcal{D}}$, $\mathcal{S} = \widehat{\mathcal{D}}_{\widehat{T}}$, and $\mathcal{D}_T = \mathcal{S}_{\widehat{\mathcal{S}}}$, cf. Sect. 9.5.

Let $H_1 = \widehat{P}_1^{-1}\widehat{Q}_1$ be a left coprime factorization of H_1 over $\mathcal{D}$, i.e., with

$$(\widehat{P}_1, -\widehat{Q}_1) \in \mathcal{D}^{p\times(p+m)},\ \operatorname{rank}(\widehat{P}_1) = p,\ (\widehat{P}_1, -\widehat{Q}_1) \text{ right invertible over } \mathcal{D}, \tag{10.8}$$

cf. Theorem 6.2.5. The left coprime factorization is essentially unique, i.e., for any two left coprime factorizations $H_1 = \widehat{P}_1^{-1}\widehat{Q}_1 = (\widehat{P}_1')^{-1}\widehat{Q}_1'$, there is an invertible matrix

$$S_1 \in \mathrm{Gl}_p(\mathcal{D}) \ \text{ such that } \ (\widehat{P}_1', -\widehat{Q}_1') = S_1(\widehat{P}_1, -\widehat{Q}_1). \tag{10.9}$$

The left coprime factorizations give rise to the controllable realizations of H_1.

Furthermore, let $H_1 = \widehat{N}_1\widehat{D}_1^{-1}$ be a right coprime factorization with

$$\begin{pmatrix} \widehat{N}_1 \\ \widehat{D}_1 \end{pmatrix} \in \mathcal{D}^{(p+m)\times m},\ \operatorname{rank}(\widehat{D}_1) = m,\ \begin{pmatrix} \widehat{N}_1 \\ \widehat{D}_1 \end{pmatrix} \text{ left invertible over } \mathcal{D}. \tag{10.10}$$

By Theorem 6.2.5, 3, the sequence

$$0 \longrightarrow \mathcal{D}^{1\times p} \xrightarrow{\cdot(\widehat{P}_1, -\widehat{Q}_1)} \mathcal{D}^{1\times(p+m)} \xrightarrow{\cdot\begin{pmatrix} \widehat{N}_1 \\ \widehat{D}_1 \end{pmatrix}} \mathcal{D}^{1\times m} \longrightarrow 0 \tag{10.11}$$

is exact. By Theorem 4.2.18, the sequence is even split exact because $\begin{pmatrix} \widehat{N}_1 \\ \widehat{D}_1 \end{pmatrix}$ has a left inverse inverse. Furthermore, by item 2 of Theorem and Definition 4.2.18, a right inverse $\begin{pmatrix} \widehat{D}_2^0 \\ \widehat{N}_2^0 \end{pmatrix}$ of $(\widehat{P}_1, -\widehat{Q}_1)$ and a left inverse $(-\widehat{Q}_2^0, \widehat{P}_2^0)$ of $\begin{pmatrix} \widehat{N}_1 \\ \widehat{D}_1 \end{pmatrix}$ can be chosen such that also the sequence

$$0 \longleftarrow \mathcal{D}^{1\times p} \xleftarrow{\cdot\begin{pmatrix} \widehat{D}_2^0 \\ \widehat{N}_2^0 \end{pmatrix}} \mathcal{D}^{1\times(p+m)} \xleftarrow{\cdot(-\widehat{Q}_2^0, \widehat{P}_2^0)} \mathcal{D}^{1\times m} \longleftarrow 0 \tag{10.12}$$

is exact, and thus $\ker\left(\cdot\begin{pmatrix} \widehat{D}_2^0 \\ \widehat{N}_2^0 \end{pmatrix}\right) = \operatorname{im}\left(\cdot(-\widehat{Q}_2^0, \widehat{P}_2^0)\right)$ is a direct complement of $\operatorname{im}\left(\cdot(\widehat{P}_1, -\widehat{Q}_1)\right) = \ker\left(\cdot\begin{pmatrix} \widehat{N}_1 \\ \widehat{D}_1 \end{pmatrix}\right)$.

Lemma 10.1.6 *Assume an IO behavior*

$$\mathcal{B}_1 := \left\{\begin{pmatrix} y_1 \\ u_1 \end{pmatrix} \in \mathcal{F}^{p+m};\ P_1 \circ y_1 = Q_1 \circ u_1\right\} \text{ with}$$
$$R_1 := (P_1, -Q_1) \in \mathcal{D}^{p\times(p+m)},\ \operatorname{rank}(P_1) = p,\ H_1 := P_1^{-1}Q_1.$$

Let

$$XR_1Y = (E, 0),\ X \in \mathrm{Gl}_p(\mathcal{D}),\ Y \in \mathrm{Gl}_{p+m}(\mathcal{D}),$$
$$E = \operatorname{diag}(e_1, \dots, e_p),\ e_1 | \cdots | e_p \neq 0$$

be the Smith form of R_1. *Write* Y *and* Y^{-1} *in block form as*

$$\begin{aligned} Y &= (Y_I, Y_{II}),\ Y^{-1} = \begin{pmatrix} Y_I^{-1} \\ Y_{II}^{-1} \end{pmatrix} \in \mathcal{D}^{(p+m)\times(p+m)} \\ &\Longrightarrow Y^{-1}Y = \begin{pmatrix} Y_I^{-1}Y_I & Y_I^{-1}Y_{II} \\ Y_{II}^{-1}Y_I & Y_{II}^{-1}Y_{II} \end{pmatrix} = \mathrm{id}_{p+m} = \begin{pmatrix} \mathrm{id}_p & 0 \\ 0 & \mathrm{id}_m \end{pmatrix}, \\ YY^{-1} &= Y_IY_I^{-1} + Y_{II}Y_{II}^{-1} = \begin{pmatrix} \mathrm{id}_p & 0 \\ 0 & \mathrm{id}_m \end{pmatrix}, \end{aligned} \tag{10.13}$$

and define the block matrices

$$\begin{pmatrix} \widehat{P}_1 & -\widehat{Q}_1 \\ -\widehat{Q}_2^0 & \widehat{P}_2^0 \end{pmatrix} := Y^{-1} \in \mathrm{Gl}_{p+m}(\mathcal{D}),\ \begin{pmatrix} \widehat{D}_2^0 & \widehat{N}_1 \\ \widehat{N}_2^0 & \widehat{D}_1 \end{pmatrix} := \begin{pmatrix} \widehat{P}_1 & -\widehat{Q}_1 \\ -\widehat{Q}_2^0 & \widehat{P}_2^0 \end{pmatrix}^{-1} := Y. \tag{10.14}$$

The just defined matrices satisfy the conditions (10.8)–(10.12), *in particular* $H_1 = \widehat{P}_1^{-1}\widehat{Q}_1 = \widehat{N}_1\widehat{D}_1^{-1}$ *are the essentially unique coprime factorizations over* $\mathcal{D}$. *This lemma holds for any principal ideal domain* $\mathcal{D}$.

Proof The equations (10.13) and (10.14) directly imply the split exactness of the sequences (10.11) and (10.12). The equation $X(P_1, -Q_1)Y = (E, 0)$ implies

$$(P_1, -Q_1)\begin{pmatrix} \widehat{N}_1 \\ \widehat{D}_1 \end{pmatrix} = 0 \Longrightarrow \widehat{N}_1 = H_1\widehat{D}_1,\ \begin{pmatrix} \widehat{N}_1 \\ \widehat{D}_1 \end{pmatrix} = \begin{pmatrix} H_1 \\ \mathrm{id}_m \end{pmatrix}\widehat{D}_1 \Longrightarrow \operatorname{rank}(\widehat{D}_1) = m.$$

According to Theorem 6.2.5, 4, $H_1 = \widehat{P}_1^{-1}\widehat{Q}_1 = \widehat{N}_1\widehat{D}_1^{-1}$ are coprime factorizations. □

Now we switch over to the quotient ring $\mathcal{D}_T$. It is clear from (10.8) and (10.10) that left or right coprime factorizations over $\mathcal{D}$ have the same property over $\mathcal{D}_T$. Together with the fact that the quotient functor $(-)_T$ is exact (Theorem 9.1.10), this signifies that the sequences

$$\begin{aligned} &0 \longrightarrow \mathcal{D}_T^{1\times p} \xrightarrow{\cdot(\widehat{P}_1, -\widehat{Q}_1)} \mathcal{D}_T^{1\times(p+m)} \xrightarrow{\cdot\begin{pmatrix} \widehat{N}_1 \\ \widehat{D}_1 \end{pmatrix}} \mathcal{D}_T^{1\times m} \longrightarrow 0 \quad \text{and} \\ &0 \longleftarrow \mathcal{D}_T^{1\times p} \xleftarrow{\cdot\begin{pmatrix} \widehat{D}_2^0 \\ \widehat{N}_2^0 \end{pmatrix}} \mathcal{D}_T^{1\times(p+m)} \xleftarrow{\cdot(-\widehat{Q}_2^0, \widehat{P}_2^0)} \mathcal{D}_T^{1\times m} \longleftarrow 0 \end{aligned} \tag{10.15}$$

are split exact.

The link between these results and the T-stabilizing compensators of $\mathcal{B}_1 = U_1^{\perp}$ is formed by Assumption 10.1.3, namely, the controllability of the quotient behavior $\mathcal{B}_{1,T} = (U_{1,T})^{\perp}$. Since the controllable realization $\left(\mathcal{D}_T^{1\times p}(\widehat{P}_1, -\widehat{Q}_1)\right)^{\perp}$ of H_1 in the category of ${}_{\mathcal{D}_T}\mathcal{F}_T$-behaviors is unique (Theorem 6.2.4, item 2.) and thus equal to $\mathcal{B}_{1,T}$, we have that

$$U_{1,T} = \mathcal{D}_T^{1\times p}(P_1, -Q_1) = \mathcal{D}_T^{1\times p}(\widehat{P}_1, -\widehat{Q}_1). \tag{10.16}$$

In the following lemma, we use Theorem and Definition 4.2.18 for parametrizing all direct complements of $U_{1,T}$.

Lemma 10.1.7 *We use the data we have just introduced. Then the following maps are bijective:*

$$\begin{array}{lcc}
\left\{\widehat{U}_2 \le_{\mathcal{D}_T} \mathcal{D}_T^{1\times(p+m)};\ \mathcal{D}_T^{1\times(p+m)} = \mathcal{D}_T^{1\times p}(\widehat{P}_1, -\widehat{Q}_1) \oplus \widehat{U}_2\right\} & \ni & \widehat{U}_2 \\
\updownarrow & & \updownarrow \\
\left\{\begin{pmatrix}\widehat{D}_2\\ \widehat{N}_2\end{pmatrix} \in \mathcal{D}_T^{(p+m)\times p};\ (\widehat{P}_1, -\widehat{Q}_1)\begin{pmatrix}\widehat{D}_2\\ \widehat{N}_2\end{pmatrix} = \mathrm{id}_p\right\} & \ni & \begin{pmatrix}\widehat{D}_2\\ \widehat{N}_2\end{pmatrix} \\
\updownarrow & & \updownarrow \\
\left\{(-\widehat{Q}_2, \widehat{P}_2) \in \mathcal{D}_T^{m\times(p+m)};\ (-\widehat{Q}_2, \widehat{P}_2)\begin{pmatrix}\widehat{N}_1\\ \widehat{D}_1\end{pmatrix} = \mathrm{id}_m\right\} & \ni & (-\widehat{Q}_2, \widehat{P}_2) \\
\updownarrow & & \updownarrow \\
\mathcal{D}_T^{m\times p} & \ni & X,
\end{array} \tag{10.17}$$

where the maps are given by the equations

$$\begin{aligned}
\widehat{U}_2 &= \mathcal{D}_T^{1\times m}(-\widehat{Q}_2, \widehat{P}_2) = \ker\left(\cdot \begin{pmatrix}\widehat{D}_2\\ \widehat{N}_2\end{pmatrix}\right), \\
(-\widehat{Q}_2, \widehat{P}_2) &= (-\widehat{Q}_2^0, \widehat{P}_2^0) + X(\widehat{P}_1, -\widehat{Q}_1), \\
&\quad \text{i.e., } \widehat{Q}_2 = \widehat{Q}_2^0 - X\widehat{P}_1 \text{ and } \widehat{P}_2 = \widehat{P}_2^0 - X\widehat{Q}_1, \text{ and} \\
\begin{pmatrix}\widehat{D}_2\\ \widehat{N}_2\end{pmatrix} &= \begin{pmatrix}\widehat{D}_2^0\\ \widehat{N}_2^0\end{pmatrix} - \begin{pmatrix}\widehat{N}_1\\ \widehat{D}_1\end{pmatrix} X, \\
&\quad \text{i.e., } \widehat{D}_2 = \widehat{D}_2^0 - \widehat{N}_1 X \text{ and } \widehat{N}_2 = \widehat{N}_2^0 - \widehat{D}_1 X.
\end{aligned} \tag{10.18}$$

Moreover, we get

$$\mathrm{id}_{p+m} = \begin{pmatrix}\widehat{P}_1 & -\widehat{Q}_1\\ -\widehat{Q}_2 & \widehat{P}_2\end{pmatrix}\begin{pmatrix}\widehat{D}_2 & \widehat{N}_1\\ \widehat{N}_2 & \widehat{D}_1\end{pmatrix} = \begin{pmatrix}\widehat{D}_2 & \widehat{N}_1\\ \widehat{N}_2 & \widehat{D}_1\end{pmatrix}\begin{pmatrix}\widehat{P}_1 & -\widehat{Q}_1\\ -\widehat{Q}_2 & \widehat{P}_2\end{pmatrix}, \tag{10.19}$$

i.e., $\begin{pmatrix}\widehat{P}_1 & -\widehat{Q}_1\\ -\widehat{Q}_2 & \widehat{P}_2\end{pmatrix}^{-1} = \begin{pmatrix}\widehat{D}_2 & \widehat{N}_1\\ \widehat{N}_2 & \widehat{D}_1\end{pmatrix}$.
This lemma holds for arbitrary principal ideal domains $\mathcal{D}'$ instead of $\mathcal{D}_T$ for matrices

with entries in $\mathcal{D}'$, *especially for* $\widehat{D}$, $\mathcal{S} = \widehat{D}_{\widehat{T}}$ *and* $\mathcal{D}_T = \mathcal{S}_{\widehat{s}}$ *where* $\mathcal{D}'^{1\times p}(\widehat{P}_1, -\widehat{Q}_1)$ *is a given direct summand of* $\mathcal{D}'^{1\times(p+m)}$.

Proof This follows directly from Theorem and Definition 4.2.18 with

$$M' = \mathcal{D}^{1\times p}, \quad M = \mathcal{D}^{1\times(p+m)}, \quad M'' = \mathcal{D}^{1\times m},$$
$$f_1 = \cdot(\widehat{P}_1, -\widehat{Q}_1), \quad g_1 = \cdot\begin{pmatrix}\widehat{N}_1\\ \widehat{D}_1\end{pmatrix}, \quad f_2 = \cdot(-\widehat{Q}_2, \widehat{P}_2),$$
$$g_2 = \cdot\begin{pmatrix}\widehat{D}_2\\ \widehat{N}_2\end{pmatrix}, \quad h = \cdot X.$$

Note that, for instance, $\mathrm{id}_p = g_2 f_1 = \cdot\left((\widehat{P}_1, -\widehat{Q}_1)\begin{pmatrix}\widehat{D}_2\\ \widehat{N}_2\end{pmatrix}\right)$, i.e., the order of the corresponding factors in the product changes because $f_1\colon \xi \longmapsto \xi(\widehat{P}_1, -\widehat{Q}_1)$ is the multiplication of a *row* vector with a matrix from the *right*. The first equality of (10.19) follows from

$$\begin{aligned}
\mathrm{id}_p &= (\widehat{P}_1, -\widehat{Q}_1)\begin{pmatrix}\widehat{D}_2\\ \widehat{N}_2\end{pmatrix} = \widehat{P}_1\widehat{D}_2 - \widehat{Q}_1\widehat{N}_2,\\
0 &= (\widehat{P}_1, -\widehat{Q}_1)\begin{pmatrix}\widehat{N}_1\\ \widehat{D}_1\end{pmatrix} = \widehat{P}_1\widehat{N}_1 - \widehat{Q}_1\widehat{D}_1,\\
\mathrm{id}_m &= (-\widehat{Q}_2, \widehat{P}_2)\begin{pmatrix}\widehat{N}_1\\ \widehat{D}_1\end{pmatrix} = -\widehat{Q}_2\widehat{N}_1 + \widehat{P}_2\widehat{D}_1, \text{ and}\\
0 &= (-\widehat{Q}_2, \widehat{P}_2)\begin{pmatrix}\widehat{D}_2\\ \widehat{N}_2\end{pmatrix} = -\widehat{Q}_2\widehat{D}_2 + \widehat{P}_2\widehat{D}_1.
\end{aligned}$$

The second equality in (10.19) follows from the fact that for square matrices over a commutative ring, the identity $AB = \mathrm{id}_p$ implies that also $BA = \mathrm{id}_p$. □

Lemma 10.1.7 furnishes direct sum decompositions

$$\mathcal{D}_T^{1\times(p+m)} = U_{1,T} \oplus \widehat{U}_2 \quad \text{with} \quad \widehat{U}_2 = \mathcal{D}_T^{1\times m}(-\widehat{Q}_2, \widehat{P}_2). \tag{10.20}$$

In order that $\widehat{U}_2$ can coincide with the module $U_{2,T}$ of an IO compensator $\mathcal{B}_2 = U_2^{\perp}$, where $U_2 \subseteq \mathcal{D}^{1\times(p+m)}$, we need that the matrix $\widehat{P}_2 \in \mathcal{D}_T^{m\times m}$ has $\mathrm{rank}(\widehat{P}_2) = m$ or, in other words, that $\det(\widehat{P}_2) \neq 0$. This is *generically* the case, in the sense we define in Definition 10.1.8.

Definition 10.1.8 Let $\mathcal{D}$ be an infinite integral domain, let I be a finite set and let $\mathcal{D}[s_i;\ i \in I]$ be the polynomial algebra in the indeterminates s_i, $i \in I$. We say that a family of propositions $(\mathfrak{A}(v))_{v\in\mathcal{D}^I}$ holds for *almost all* $v \in \mathcal{D}^I$ or that it holds *generically* if there is a nonzero polynomial $f \in \mathcal{D}[s_i;\ i \in I]$ such that $\mathfrak{A}(v)$ holds for all v in the complement of the zero set of f, i.e., for all

$$v \in D_{\mathcal{D}^I}(f) := \mathcal{D}^I \setminus \mathrm{V}_{\mathcal{D}^I}(f) \quad \text{where } \mathrm{V}_{\mathcal{D}^I}(f) := \{v \in \mathcal{D}^I;\ f(v) = 0\}. \qquad \Diamond$$

If I consists of only one element, we identify $\mathcal{D}^I = \mathcal{D}$. In this case, the vanishing set $V_{\mathcal{D}}(f)$ is finite and $D_{\mathcal{D}}(f)$ is its infinite complement. If $\mathcal{D}$ is the real or complex field, the vanishing set $V_{\mathbb{C}^I}(f)$ has measure zero and is nowhere dense.

Lemma 10.1.9 *For an infinite principal ideal domain $\mathcal{D}$, almost all $\widehat{P}_2 \in \mathcal{D}_T^{m\times m}$ which arise from the constructions in Lemma 10.1.7 have nonzero determinant* $\det(\widehat{P}_2)$.

Proof By (10.18), the matrix $\widehat{P}_2$ is constructed as $\widehat{P}_2 = \widehat{P}_2^0 - X\widehat{Q}_1$, where $\widehat{P}_2^0$ and $\widehat{Q}_1$ are matrices over $\mathcal{D}$, while $X \in \mathcal{D}_T^{m\times p}$. We regard X as an $m \times p$-matrix of indeterminates and consider the polynomial

$$f(X) := \det(\widehat{P}_2^0 - X\widehat{Q}_1) \in \mathcal{D}[X] := \mathcal{D}\big[X_{ij};\ i \in \{1, \dots, m\},\ j \in \{1, \dots, p\}\big].$$

According to Definition 10.1.8, we have to show that the polynomial $f \in \mathcal{D}[X]$ is nonzero, and we do this by showing that $f(X^0) \neq 0$ for a specific $X^0 \in K^{m\times p}$.

From (10.10) or Lemma 10.1.6 we have $\det(\widehat{D}_1) \neq 0$. Therefore, the inverse $\widehat{D}_1^{-1} \in K^{m\times m}$ exists over the quotient field K and consequently, $(0, \widehat{D}_1^{-1})$ is a left inverse of $\begin{pmatrix} \widehat{N}_1 \\ \widehat{D}_1 \end{pmatrix}$ over K. By Lemma 10.1.7 applied to the field $\mathcal{D}_{\mathcal{D}\setminus\{0\}} = K$ there exists a matrix $X^0 \in K^{m\times p}$ with $\widehat{D}_1^{-1} = \widehat{P}_2^0 - X^0\widehat{Q}_1$. Hence

$$f(X^0) = \det(\widehat{P}_2^0 - X^0\widehat{Q}_1) = \det(\widehat{D}_1^{-1}) \neq 0.$$

□

In the following theorem, we describe how to obtain a T-stabilizing compensator from $\widehat{U}_2$. Then we parametrize all stabilizing compensators in Theorem 10.1.13 and after that, we give some remarks on the constructiveness of these results.

Theorem 10.1.10 *Assume the IO plant $\mathcal{B}_1$ with transfer matrix $H_1 = P_1^{-1}Q_1$ from* (10.2) *and that the quotient behavior $\mathcal{B}_{1,T}$ is controllable. By Lemma 10.1.6 compute the matrices*

$$Y = \begin{pmatrix} \widehat{D}_2^0 & \widehat{N}_1 \\ \widehat{N}_2^0 & \widehat{D}_1 \end{pmatrix}, Y^{-1} = \begin{pmatrix} \widehat{P}_1 & -\widehat{Q}_1 \\ -\widehat{Q}_2^0 & \widehat{P}_2^0 \end{pmatrix} \in \mathrm{Gl}_{p+m}(\mathcal{D}) \text{ such that}$$
$$H_1 = \widehat{P}_1^{-1}\widehat{Q}_1 = \widehat{N}_1\widehat{D}_1^{-1}$$

are the essentially unique coprime factorizations. It suffices to choose $(-\widehat{Q}_2^0, \widehat{P}_2^0)$ and $\begin{pmatrix} \widehat{D}_2^0 \\ \widehat{N}_2^0 \end{pmatrix}$ with entries in $\mathcal{D}_T$ instead of $\mathcal{D}$. Let $X \in \mathcal{D}_T^{m\times p}$ and define

$$(-\widehat{Q}_2, \widehat{P}_2) := (-\widehat{Q}_2^0, \widehat{P}_2^0) + X(\widehat{P}_1, -\widehat{Q}_1) \in \mathcal{D}_T^{m\times(p+m)}. \tag{10.21}$$

Assume $\det(\widehat{P}_2) \neq 0$ *which is generically the case according to Lemma 10.1.9.*

Then the unique controllable realization of $H_2 := \widehat{P}_2^{-1}\widehat{Q}_2$ in $\mathcal{F}^{p+m}$, i.e., the IO behavior $\mathcal{B}_2 = U_2^{\perp} \subseteq \mathcal{F}^{p+m}$ with

$$U_2 = \mathcal{D}^{1\times m}(-Q_2, P_2) := \ker\left(\cdot\begin{pmatrix}\mathrm{id}_p\\ H_2\end{pmatrix} : \mathcal{D}^{1\times(p+m)} \longrightarrow K^{1\times p}\right) \quad (10.22)$$

(see Theorem 6.2.4) is a T-stabilizing compensator of $\mathcal{B}_1$. Hence $H_2 = P_2^{-1}Q_2$ is the essentially unique left coprime factorization of H_2 over $\mathcal{D}$. The quotient behavior of the feedback behavior

$$\begin{aligned}&\mathcal{B} = \mathrm{fb}(\mathcal{B}_1, \mathcal{B}_2) = \left\{\begin{pmatrix}y\\u\end{pmatrix} \in \mathcal{F}^{(p+m)+(p+m)};\ P \circ y = Q \circ u\right\}\\ &\text{with } P = \begin{pmatrix}P_1 & -Q_1\\ -Q_2 & P_2\end{pmatrix},\ Q = \begin{pmatrix}0 & Q_1\\ Q_2 & 0\end{pmatrix} \in \mathcal{D}^{(p+m)\times(p+m)}\end{aligned} \quad (10.23)$$

from (10.4) *with respect to T is*

$$\begin{aligned}\mathcal{B}_T &= \left\{\begin{pmatrix}y\\u\end{pmatrix} \in \mathcal{F}_T^{(p+m)+(p+m)};\ P \circ y = Q \circ u\right\}\\ &= \left\{\begin{pmatrix}y\\u\end{pmatrix} \in \mathcal{F}_T^{(p+m)+(p+m)};\ \widehat{P} \circ y = \widehat{Q} \circ u\right\}\\ \text{with } \widehat{P} &:= \begin{pmatrix}\widehat{P}_1 & -\widehat{Q}_1\\ -\widehat{Q}_2 & \widehat{P}_2\end{pmatrix},\ \widehat{Q} := \begin{pmatrix}0 & \widehat{Q}_1\\ \widehat{Q}_2 & 0\end{pmatrix} \in \mathcal{D}_T^{(p+m)\times(p+m)}.\end{aligned} \quad (10.24)$$

The transfer matrix of $\mathcal{B}$ is

$$H = \widehat{P}^{-1}\widehat{Q} = \begin{pmatrix}\widehat{D}_2 & \widehat{N}_1\\ \widehat{N}_2 & \widehat{D}_1\end{pmatrix}\begin{pmatrix}0 & \widehat{Q}_1\\ \widehat{Q}_2 & 0\end{pmatrix} = \begin{pmatrix}\widehat{N}_1\widehat{Q}_2 & \widehat{D}_2\widehat{Q}_1\\ \widehat{D}_1\widehat{Q}_2 & \widehat{N}_2\widehat{Q}_1\end{pmatrix}. \quad (10.25)$$

Proof (i) By Lemma 10.1.7, the matrices $\widehat{P}_2$ and $\widehat{Q}_2$ from (10.21) give rise to the direct sum decomposition

$$\mathcal{D}_T^{1\times(p+m)} = U_{1,T} \oplus \widehat{U}_2, \quad \text{where } \widehat{U}_2 = \mathcal{D}_T^{1\times m}(-\widehat{Q}_2, \widehat{P}_2). \quad (10.26)$$

We define

$$\begin{aligned}&U_2' := \mathcal{D}^{1\times(p+m)} \bigcap \widehat{U}_2,\ \mathcal{B}_2' := U_2'^{\perp} \subseteq \mathcal{F}^{p+m}\\ \underset{\text{Corollary 9.1.11}}{\Longrightarrow}\ &U_{2,T}' = \widehat{U}_2 \Longrightarrow \mathcal{D}_T^{1\times(p+m)} = U_{1,T} \oplus U_{2,T}',\\ &KU_2' = KU_{2,T}' = K\widehat{U}_2 = K^{1\times m}(-\widehat{Q}_2, \widehat{P}_2).\end{aligned} \quad (10.27)$$

Since $\det(\widehat{P}_2) \neq 0$ by assumption, the $\mathcal{F}_T$-behavior

$$\widehat{\mathcal{B}}_2 := \widehat{U}_2^{\perp} = \left\{\begin{pmatrix}u_2\\y_2\end{pmatrix} \in \mathcal{F}_T^{p+m};\ \widehat{P}_2 \circ y_2 = \widehat{Q}_2 \circ u_2\right\}$$

is an IO behavior with input u_2, output y_2, and transfer matrix $H_2 := \widehat{P}^{-1}\widehat{Q}_2$. By item 1 of Theorem and Definition 6.2.2, the IO structure and the transfer matrix depend only on the vector space $K\widehat{U}_2$. The last line of (10.27) thus implies that $\mathcal{B}_2'$ has the same IO structure and transfer matrix H_2 as $\widehat{\mathcal{B}}_2$. The second line of (10.27) then implies that $\mathcal{B}_2'$ is a T-stabilizing compensator of $\mathcal{B}_1$.

(ii) The canonical map

$$\mathcal{D}^{1\times(p+m)}/U_2' = \mathcal{D}^{1\times(p+m)}/\left(\mathcal{D}^{1\times(p+m)}\bigcap \widehat{U}_2\right)$$
$$\longrightarrow \mathcal{D}_T^{1\times(p+m)}/\widehat{U}_2 \cong U_{1,T},\ \xi + U_2' \mapsto \xi + \widehat{U}_2 \mapsto \operatorname{proj}(\xi)$$

is injective and implies that $\mathcal{D}^{1\times(p+m)}/U_2'$ like $U_{1,T}$ is torsionfree. Hence $\mathcal{B}_2'$ is controllable and thus the unique controllable realization of H_2. We conclude that $\mathcal{B}_2' = \mathcal{B}_2$ and that $\mathcal{B}_2$ is a controllable T-stabilizing compensator of $\mathcal{B}_1$ as asserted.
(iii) The equations (10.23)–(10.25) follow from the identities

$$U_{1,T} = \mathcal{D}_T^{1\times p}(P_1, -Q_1) = \mathcal{D}_T^{1\times p}(\widehat{P}_1, -\widehat{Q}_1),$$
$$U_{2,T} = \mathcal{D}_T^{1\times m}(-Q_2, P_2) = \mathcal{D}_T^{1\times m}(-\widehat{Q}_2, \widehat{P}_2).$$

These imply the existence of $S_1 \in \mathrm{Gl}_p(\mathcal{D}_T)$ and $S_2 \in \mathrm{Gl}_m(\mathcal{D}_T)$ such that

$$(\widehat{P}_1, -\widehat{Q}_1) = S_1(P_1, -Q_1),\ (-\widehat{Q}_2, \widehat{P}_2) = S_2(-Q_2, P_2)$$
$$\Longrightarrow S := \operatorname{diag}(S_1, S_2) \in \mathrm{Gl}_{p+m}(\mathcal{D}_T),\ (\widehat{P}, -\widehat{Q}) = S(P, -Q)$$
$$\Longrightarrow H = P^{-1}Q = \widehat{P}^{-1}\widehat{Q}.$$

From (10.19) we have

$$\widehat{P}^{-1} = \begin{pmatrix} \widehat{D}_2 & \widehat{N}_1 \\ \widehat{N}_2 & \widehat{D}_1 \end{pmatrix} \Longrightarrow H = \widehat{P}^{-1}\widehat{Q} = \begin{pmatrix} \widehat{D}_2 & \widehat{N}_1 \\ \widehat{N}_2 & \widehat{D}_1 \end{pmatrix}\begin{pmatrix} 0 & \widehat{Q}_1 \\ \widehat{Q}_2 & 0 \end{pmatrix} = \begin{pmatrix} \widehat{N}_1\widehat{Q}_2 & \widehat{D}_2\widehat{Q}_1 \\ \widehat{D}_1\widehat{Q}_2 & \widehat{N}_2\widehat{Q}_1 \end{pmatrix}.$$

□

Corollary 10.1.11 *The behavior $\mathcal{B}_1$ is T-stabilizable, i.e., has a T-stabilizing compensator, if and only if the highest nonzero elementary divisor e_p^1 of $(P_1, -Q_1)$ lies in T.*

Proof By Corollary 10.1.4 $\mathcal{B}_{1,T}$ is controllable if and only if $e_p^1 \in T$. If $\mathcal{B}_1$ is T-stabilizable, then $\mathcal{B}_{1,T}$ is controllable by Lemma 10.1.1. Assume, conversely, that $\mathcal{B}_{1,T}$ is controllable. We choose a matrix $X \in \mathcal{D}_T^{m\times p}$ such that $\det(\widehat{P}_2) \neq 0$ in (10.21). Then Theorem 10.1.10 furnishes a T-stabilizing compensator of $\mathcal{B}_1$. □

Corollary 10.1.12 *For the computation of the coprime factorization $H_2 = \widehat{P}_2^{-1}\widehat{Q}_2$ from $(-\widehat{Q}_2, \widehat{P}_2)$ in Theorem 10.1.10 one can apply Lemma 10.1.6 to $(\widehat{P}_2, -\widehat{Q}_2)$ in $\mathcal{D}_T^{m\times(m+p)}$ and $\operatorname{rank}(\widehat{P}_2) = m$ instead of $(P_1, -Q_1)$ in the following steps:*

1. *Choose $0 \neq f \in \mathcal{D}$ such that $f(-\widehat{Q}_2, \widehat{P}_2) \in \mathcal{D}^{m\times(p+m)}$, for instance, the least common denominator $f \in T$ of the entries of $(-\widehat{Q}_2, \widehat{P}_2) \in \mathcal{D}_T^{m\times(p+m)}$.*
2. *Compute the Smith form over $\mathcal{D}$ of $f(\widehat{P}_2, -\widehat{Q}_2) \in \mathcal{D}^{m\times(m+p)}$, i.e.,*

$$Xf(\widehat{P}_2, -\widehat{Q}_2)Y = (E, 0),\ E \in \mathcal{D}^{m\times m},\ \operatorname{rank}(E) = m,$$
$$X \in \mathrm{Gl}_m(\mathcal{D}),\ Y \in \mathrm{Gl}_{m+p}(\mathcal{D}).$$

3. *Compute* $Y^{-1} = \begin{pmatrix} P_2 & -Q_2 \\ * & * \end{pmatrix}$. *Then* $H_2 = P_2^{-1} Q_2$ *is the required coprime factorization over* $\mathcal{D}$. ◇

Next we parametrize all controllable T-stabilizing compensators and then all T-stabilizing compensators of $\mathcal{B}_1$.

Theorem 10.1.13 (Parametrization of stabilizing compensators) *We use the same data as in Theorem 10.1.10.*

1. *Every controllable T-stabilizing compensator is obtained by the method of Theorem 10.1.10. Indeed, the map*

$$\begin{aligned} &\{X \in \mathcal{D}_T^{m\times p};\ \det(\widehat{P}_2^0 - X\widehat{Q}_1) \neq 0\} \\ &\to \{\text{controllable stabilizing compensators of } \mathcal{B}_1\},\ X \mapsto \mathcal{B}_2, \end{aligned} \tag{10.28}$$

is a bijection. In other words, the matrices X with $\det(\widehat{P}_2^0 - X\widehat{Q}_1) \neq 0$ parametrize *the set of controllable compensators of* $\mathcal{B}_1$.
2. *All other T-stabilizing compensators $\widetilde{\mathcal{B}}_2 = \widetilde{U}_2^{\perp}$ of $\mathcal{B}_1$ are obtained as $\widetilde{U}_2 = \mathcal{D}^{1\times m} A(-Q_2, P_2)$, where $U_2 = \mathcal{D}^{1\times m}(-Q_2, P_2)$ defines a controllable compensator $\mathcal{B}_2 = U_2^{\perp}$ as in item 1 and $A \in \mathcal{D}^{m\times m}$ satisfies* $\det(A) \in T$.
Then the two quotient behaviors $\mathrm{fb}(\mathcal{B}_1, \widetilde{\mathcal{B}}_2)_T$ *and* $\mathrm{fb}(\mathcal{B}_1, \mathcal{B}_2)_T$, *the transfer matrices of $\widetilde{\mathcal{B}}_2$ and of $\mathcal{B}_2$ and of* $\mathrm{fb}(\mathcal{B}_1, \widetilde{\mathcal{B}}_2)$ *and* $\mathrm{fb}(\mathcal{B}_1, \mathcal{B}_2)$ *coincide.*

Proof 1. (i) *The map from* (10.28) *is surjective*: Let $\mathcal{B}_2 = U_2^{\perp}$ with

$$U_2 = \mathcal{D}^{1\times m}(-Q_2, P_2),\ (-Q_2, P_2) \in \mathcal{D}^{m\times(p+m)},\ \mathrm{rank}(P_2) = m,$$

be a controllable T-stabilizing compensator of $\mathcal{B}_1$. Since $\mathcal{B}_2$ is T-stabilizing, Lemma 10.1.1 implies the direct sum decomposition $U_{1,T} \oplus U_{2,T} = \mathcal{D}_T^{1\times(p+m)}$. By the parametrization of the direct complements from Lemma 10.1.7, there is a unique $X \in \mathcal{D}_T^{m\times p}$ such that

$$\begin{aligned} U_{2,T} &= \mathcal{D}_T^{1\times m}(-\widehat{Q}_2, \widehat{P}_2) \text{ where} \\ (-\widehat{Q}_2, \widehat{P}_2) &= (-\widehat{Q}_2^0, \widehat{P}_2^0) + X(\widehat{P}_1, -\widehat{Q}_1). \end{aligned}$$

Moreover $U_2 = \mathcal{D}^{1\times m}(-Q_2, P_2)$ implies $U_{2,T} = \mathcal{D}_T^{1\times m}(-Q_2, P_2)$. From these two representations, we infer that there is $S_2 \in \mathrm{Gl}_m(\mathcal{D}_T)$ with $S_2(-Q_2, P_2) = (-\widehat{Q}_2, \widehat{P}_2)$. In particular, this implies

$$\det(S_2) \neq 0 \text{ and } S_2 P_2 = \widehat{P}_2 \Longrightarrow \det(\widehat{P}_2) = \det(S_2)\det(P_2) \neq 0.$$

According to Theorem 10.1.10 it remains to show

$$U_2 = U_2' := U_{2,T} \cap \mathcal{D}^{1\times(p+m)}.$$

The controllability of $\mathcal{B}_2$ means that $\mathcal{D}^{1\times(p+m)}/U_2$ is torsionfree and hence T-torsionfree. Item 7 of Lemma and Definition 9.1.4 implies that the map

$$\begin{aligned}\mathcal{D}^{1\times(p+m)}/U_2 &\longrightarrow \left(\mathcal{D}^{1\times(p+m)}/U_2\right)_T \underset{\text{Corollary 9.1.13}}{=} \mathcal{D}_T^{1\times(p+m)}/U_{2,T},\\ \xi+U_2 &\longmapsto \tfrac{\xi+U_2}{1},\ \tfrac{\xi+U_2}{t} \longmapsto \tfrac{\xi}{t}+U_{2,T},\end{aligned}$$

is injective and thus that $U_2 = U_{2,T}\bigcap \mathcal{D}^{1\times(p+m)}$ as asserted.
(ii) The map (10.28) is injective since X is uniquely determined by $U_{2,T}$ according to Lemma 10.1.7.
2. Let $\widetilde{\mathcal{B}}_2 = \widetilde{U}_2^{\perp}$ be any T-stabilizing compensator, and hence

$$\mathcal{D}_T^{1\times(p+m)} = U_{1,T}\oplus \widetilde{U}_{2,T}.$$

Define

$$U_2 := \widetilde{U}_{2,T}\cap \mathcal{D}^{1\times(p+m)} = \mathcal{D}^{1\times m}(-Q_2, P_2) \supseteq \widetilde{U}_2 \text{ and } \mathcal{B}_2 := U_2^{\perp}.$$

By part 1. the behavior $\mathcal{B}_2$ is a controllable T-stabilizing compensator of $\mathcal{B}_1$ with $U_{2,T} = \widetilde{U}_{2,T}$. Moreover U_2, $U_{2,T} = \widetilde{U}_{2,T}$, and $\widetilde{U}_2$ have the same IO structure and thus the same dimension m. This and the inclusion $\widetilde{U}_2 \subseteq U_2$ imply that there is $A\in\mathcal{D}^{m\times m}$ with $\det(A)\neq 0$ and $\widetilde{U}_2 = \mathcal{D}^{1\times m}A(-Q_2, P_2)$. Then

$$\mathcal{D}_T^{1\times m}(-Q_2, P_2) = U_{2,T} = \widetilde{U}_{2,T} = \mathcal{D}_T^{1\times m}A(-Q_2, P_2)$$

implies $A\in \mathrm{Gl}_m(\mathcal{D}_T)$ and thus $\det(A)\in\mathcal{D}\cap \mathrm{U}(\mathcal{D}_T) = T$ by Lemma 9.2.1. □

Remark 10.1.14 Consider the standard operator rings $\mathcal{D} = F[s]$ with $F=\mathbb{R}$ or $F=\mathbb{C}$ and the standard signal spaces $\mathrm{C}^{\infty}(\mathbb{R}, F)$, $\mathrm{C}^{-\infty}(\mathbb{R}, F)$, or $F^{\mathbb{N}}$. With the notations of item 2 of Theorem 10.1.13, the canonical surjective map

$$\mathrm{can}\colon \mathcal{D}^{1\times m}/\mathcal{D}^{1\times m}AP_2 \longrightarrow \mathcal{D}^{1\times m}/\mathcal{D}^{1\times m}P_2,\quad \xi+\mathcal{D}^{1\times m}AP_2 \longmapsto \xi+\mathcal{D}^{1\times m}P_2,$$

implies the relation

$$\begin{aligned}\dim_F(\widetilde{\mathcal{B}}_2^0) &= \dim_F\left(\mathcal{D}^{1\times m}/\mathcal{D}^{1\times m}(AP_2)\right) = \deg(\det(A)) + \deg(\det(P_2))\\ &\geq \deg(\det(P_2)) = \dim_F\left(\mathcal{D}^{1\times m}/\mathcal{D}^{1\times m}P_2\right) = \dim_F(\mathcal{B}_2^0)\end{aligned}$$

of the F-dimensions of the autonomous parts of a noncontrollable and the corresponding controllable compensator. The dimension $\dim_F(\mathcal{B}_2^0)$ is also the state space dimension of the essentially unique observable and controllable or minimal state space realization of $\mathcal{B}_2$, cf. Theorem and Definition 6.3.10 and Theorem 6.3.11. For complexity and cost reasons this dimension should be small, and therefore, it is generally advantageous to choose $\mathcal{B}_2$ controllable as in Theorem 10.1.10 and in item 1 of Theorem 10.1.13. ◊

Example 10.1.15 For $\mathcal{D} = F[s]$ with $F = \mathbb{R}$ or $F = \mathbb{C}$, a decomposition $\mathbb{C} = \Lambda_1 \uplus \Lambda_2$ and the induced multiplicative submonoid $T = \{t \in F[s];\ \mathrm{V}_{\mathbb{C}}(t) \subseteq \Lambda_1\}$ consider an observable state space system

$$s \circ x = Ax + Bu, \quad y = Cx + D \circ u$$

with the associated behaviors

$$\begin{aligned} \mathcal{B}^{\mathrm{s}} &:= \left\{\left(\begin{smallmatrix} x \\ u \end{smallmatrix}\right) \in \mathcal{F}^{n+m};\ s \circ x = Ax + Bu\right\} \quad \text{and} \\ \mathcal{B} &:= \left(\begin{smallmatrix} C & D \\ 0 & \mathrm{id}_m \end{smallmatrix}\right) \circ \mathcal{B}_s = \left\{\left(\begin{smallmatrix} y \\ u \end{smallmatrix}\right) \in \mathcal{F}^{p+m};\ P \circ y = Q \circ u\right\}, \end{aligned}$$

the isomorphism $\left(\begin{smallmatrix} C & D \\ 0 & \mathrm{id}_m \end{smallmatrix}\right) \circ \colon \mathcal{B}^{\mathrm{s}} \xrightarrow{\cong} \mathcal{B}$, and the characteristic varieties

$$\mathrm{char}(\mathcal{B}) = \mathrm{char}(\mathcal{B}^{\mathrm{s}}) = \left\{\lambda \in \mathbb{C};\ \mathrm{rank}(s\,\mathrm{id}_n - A, -B) < n\right\}.$$

By Corollaries 10.1.5 and 10.1.11, the behaviors $\mathcal{B}^{\mathrm{s}}$ and $\mathcal{B}$ are T-stabilizable if and only if this set of uncontrollable poles is contained in Λ_1, compare [6, Definition 2.2 on p. 330]. ◊

Remark 10.1.16 (*Constructiveness*) The constructions of Lemma 10.1.6 and of Theorems 10.1.10 and 10.1.13 need the following tools only:

1. Computations in the principal ideal domain $\mathcal{D}$, in particular, the Smith form algorithm. If $\mathcal{D}$ is the polynomial ring over a finite field, over the rational field $\mathbb{Q} \subset \mathbb{R}$, or over the Gaussian field $\mathbb{Q}[j] \subset \mathbb{C}$, $j := \sqrt{-1}$, then all these computations can be carried without numerical errors in all computer algebra systems.
2. A method for deciding whether an element $t \in \mathcal{D}$ lies in T.

Consider, for instance, the standard continuous-time case

$$\begin{aligned} &F = \mathbb{R} \text{ or } F = \mathbb{C}, \quad \mathcal{D} = F[s], \quad \mathcal{F} = \mathrm{C}^{\infty}(\mathbb{R}, F) \text{ or } \mathcal{F} = \mathrm{C}^{-\infty}(\mathbb{R}, F), \\ &\Lambda_{1,\mathrm{cont}} = \mathbb{C}_- = \{\lambda \in \mathbb{C};\ \Re(\lambda) < 0\}, \quad T_{\mathrm{cont}} = \{t \in F[s];\ \mathrm{V}_{\mathbb{C}}(t) \subseteq \Lambda_{1,\mathrm{cont}}\} \end{aligned} \tag{10.29}$$

and the standard discrete-time case

$$\begin{aligned} &F = \mathbb{R} \text{ or } F = \mathbb{C}, \quad \mathcal{D} = F[s], \quad \mathcal{F} = F^{\mathbb{N}}, \\ &\Lambda_{1,\mathrm{dis}} = \mathbb{D} = \{\lambda \in \mathbb{C};\ |\lambda| < 1\}, \quad T_{\mathrm{dis}} = \{t \in F[s];\ \mathrm{V}_{\mathbb{C}}(t) \subseteq \Lambda_{1,\mathrm{dis}}\}. \end{aligned} \tag{10.30}$$

The polynomials in T_{cont} and T_{dis} are called *Hurwitz* and *Schur* polynomials, respectively. In [11, Sect. 3.4, pp. 296–368] Hinrichsen and Pritchard give a comprehensive account on results and constructive tests for $t \in T_{\mathrm{cont}}$ and for $t \in T_{\mathrm{dis}}$. These tests are implemented in computer algebra systems, and we do not treat them in this book, cf. MAPLE, Mathematica, Wolfram MathWorld. ◊

Next we illustrate Theorems 10.1.10 and 10.1.13 by constructing T-stabilizing compensators of a given plant. In order that we can write down all expressions, we treat SISO plant, i.e., with $p = m = 1$, but we do not use the computational shortcuts that come with this special situation. Thus it can be seen how our algorithms work also in the more complicated MIMO case.

Example 10.1.17 We use the standard setting for the continuous-time case of (10.29) with $\Lambda_1 = \mathbb{C}_-$ and the corresponding monoid T. We apply Theorem 10.1.10 to the plant

$$\mathcal{B}_1 = \left(\mathcal{D}(P_1, -Q_1)\right)^{\perp}$$

with $P_1 := (s+1)(s+2)(s-3)(s-4) = s^4 - 4s^3 - 7s^2 + 22s + 24$ and
$Q_1 := (s+1)(s+2)(s-5) = s^3 - 2s^2 - 13s - 10.$

The Smith form of $(P_1, -Q_1) = (s+1)(s+2)((s-3)(s-4), -(s-5))$ is

$$U(P_1, -Q_1)V = (e^1, 0) \quad \text{with} \quad U = 1, \quad e^1 = (s+1)(s+2), \text{ and}$$

$$V = \begin{pmatrix} \widehat{D}_2^0 & \widehat{N}_1 \\ \widehat{N}_2^0 & \widehat{D}_1 \end{pmatrix} = \begin{pmatrix} \frac{1}{2} & -s+5 \\ \frac{1}{2}s-1 & -(s-3)(s-4) \end{pmatrix},$$

$$V^{-1} = \begin{pmatrix} \widehat{P}_1 & -\widehat{Q}_1 \\ -\widehat{Q}_2^0 & \widehat{P}_2^0 \end{pmatrix} = \begin{pmatrix} -(s-3)(s-4) & s-5 \\ -\frac{1}{2}s+1 & \frac{1}{2} \end{pmatrix}$$

where already $\det(\widehat{P}_2^0) = \frac{1}{2} \neq 0$. The highest elementary divisor of $(P_1, -Q_1)$ is $(s+1)(s+2) \in T$ and therefore $\mathcal{B}_1$ is T-stabilizable. Hence $(-\widehat{Q}_2^0, \widehat{P}_2^0)$ defines a stabilizing compensator.

We choose the parameter $X = \frac{1}{s+3} \in \mathcal{D}_T^{1\times 1} = \mathcal{D}_T$ and obtain the compensator

$$(-\widehat{Q}_2, \widehat{P}_2) = (-\widehat{Q}_2^0, \widehat{P}_2^0) + X(\widehat{P}_1, -\widehat{Q}_1) = -\left(\frac{3s^2-13s+18}{2(s+3)}, \frac{-3s+7}{2(s+3)}\right) \text{ with}$$

$$H_2 := \frac{\widehat{Q}_2}{\widehat{P}_2} = \frac{3s^2-13s+18}{3s-7} \text{ and } \gcd(3s^2 - 13s + 18, -3s + 7) = 1.$$

The unique controllable realization of H_2 is

$$\mathcal{B}_2 = \left(\mathcal{D}(-Q_2, P_2)\right)^{\perp} \quad \text{with} \quad P_2 := 3s - 7 \quad \text{and} \quad Q_2 := 3s^2 - 13s + 18.$$

The closed-loop or feedback behavior is then given by the matrices

$$P = \begin{pmatrix} P_1 & -Q_1 \\ -Q_2 & P_2 \end{pmatrix} = \begin{pmatrix} e^1(s-3)(s-4) & -e^1(s-5) \\ -3s^2+13s-18 & 3s-7 \end{pmatrix} \text{ and}$$

$$Q = \begin{pmatrix} 0 & Q_1 \\ Q_2 & 0 \end{pmatrix} = \begin{pmatrix} 0 & e^1(s-5) \\ 3s^2-13s+18 & 0 \end{pmatrix}.$$

It is T-stable since $\det(P) = 2(s+1)(s+2)(s+3) \in T$. Its transfer matrix is

$$H = P^{-1}Q = \frac{1}{2(s+3)} \begin{pmatrix} (s-5)(3s^2-13s+18) & (s-5)(3s-7) \\ (s-3)(s-4)(3s^2-13s+18) & (s-5)(3s^2-13s+18) \end{pmatrix} \in \mathcal{D}_T^{2\times 2}.$$

Neither H nor H_2 are proper. In Examples 10.1.22 and 10.1.36, we will construct compensators such that the transfer matrices of the feedback system and of the compensator are proper. ◊

10.1.2 Proper Feedback Behavior

For the reasons we have explained in Remark 9.4.4 *engineers prefer proper and* T*-stable IO behaviors. In particular, the compensator* $\mathcal{B}_2$ *should give rise to a* T*-stable feedback behavior* $\mathcal{B} = \mathrm{fb}(\mathcal{B}_1, \mathcal{B}_2)$ *with proper transfer matrix* H*. We characterize those feedback behaviors in Theorem* 10.1.20*. We have to apply the technique from Sect.* 9.5*.*

Let $\mathcal{D} = F[s]$ for a field F and let $T \subseteq \mathcal{D} \setminus \{0\}$ be a saturated multiplicative submonoid. As in Sect. 9.5, we assume that T contains a linear polynomial $s - \alpha$, $\alpha \in F$. We define $\widehat{s} := \frac{1}{s-\alpha}$ and use the polynomial ring $\widehat{\mathcal{D}} := F[\widehat{s}]$ with $K := F(s) = F(\widehat{s})$. For nonzero $f \in F[s]$ we define the polynomial $\widehat{f_\alpha}(\widehat{s}) := \widehat{s}^{\deg_s(f)} f \in F[\widehat{s}]$ with $\widehat{f_\alpha}(0) \neq 0$. According to Theorem and Definition 9.5.4 we have

$$\mathcal{S} := \mathcal{D}_T \cap F(s)_{\mathrm{pr}} = F[\widehat{s}]_{\widehat{T}} \ni \frac{f}{t} = \frac{\widehat{s}^{\deg_s(t)-\deg_s(f)}\widehat{f_\alpha}(\widehat{s})}{\widehat{t_\alpha}(\widehat{s})},\ \widehat{T} = \left\{\widehat{t_\alpha}(\widehat{s});\ t \in T\right\}. \tag{10.31}$$

In Lemma 9.5.3 we also proved

$$\mathcal{D}_T = \mathcal{S}_{\widehat{s}} \ni \frac{f}{t} = \frac{\frac{f}{t(s-\alpha)^m}}{\widehat{s}^m} = \frac{\frac{\widehat{s}^{\deg_s(t)+m-\deg_s(f)}\widehat{f_\alpha}(\widehat{s})}{\widehat{t_\alpha}(\widehat{s})}}{\widehat{s}^m},\ \deg_s(f) \leq \deg_s(t) + m. \tag{10.32}$$

Assumption 10.1.18 As in Assumption 10.1.3 we start with the IO behavior

$$\begin{aligned} &\mathcal{B}_1 = U_1^\perp \subseteq \mathcal{F}^{p+m} \text{ where } U_1 = \mathcal{D}^{1\times p}R_1,\ R_1 = (P_1, -Q_1) \in \mathcal{D}^{p\times(p+m)}, \\ &\mathrm{rank}(P_1) = p,\ H_1 := P_1^{-1}Q_1,\ M_1 = \mathcal{D}^{1\times(p+m)}/U_1,\ M_{1,T} = \mathcal{D}_T^{1\times(p+m)}/U_{1,T} \end{aligned} \tag{10.33}$$

and assume that $\mathcal{B}_1$ is T-stabilizable. This means that $M_{1,T}$ is free or that $(P_1, -Q_1)$ has a right inverse in $\mathcal{D}_T^{(p+m)\times p}$.

Consider the Smith/McMillan form of $R_1 \in \mathcal{D}^{p\times(p+m)} \subseteq F(\widehat{s})^{p\times(p+m)} = F(s)^{p\times(p+m)}$ over $\widehat{\mathcal{D}} = F[\widehat{s}]$, i.e.,

$$\begin{aligned} &\widehat{X}R_1\widehat{Y} = (\widehat{E}, 0),\ \widehat{X} \in \mathrm{Gl}_p(\widehat{\mathcal{D}}),\ \widehat{Y} \in \mathrm{Gl}_{p+m}(\widehat{\mathcal{D}}), \\ &\widehat{E} = \mathrm{diag}(\widehat{e_1}, \dots, \widehat{e_p}),\ 0 \neq \widehat{e_i} \underset{\mathcal{D},\widehat{\mathcal{D}}\subset\mathcal{D}_T}{\in} \mathcal{D}_T,\ \widehat{e_1} \underset{\widehat{\mathcal{D}}}{|} \cdots \underset{\widehat{\mathcal{D}}}{|} \widehat{e_p}, \\ &\begin{pmatrix} \widehat{P_1} & -\widehat{Q}_1 \\ -\widehat{Q}_2^0 & \widehat{P}_2^0 \end{pmatrix} := \widehat{Y}^{-1},\ \begin{pmatrix} \widehat{D}_2^0 & \widehat{N}_1 \\ \widehat{N}_2^0 & \widehat{D}_1 \end{pmatrix} := \widehat{Y} \in \widehat{\mathcal{D}}^{(p+m)\times(p+m)}. \end{aligned} \tag{10.34}$$

This is also the Smith form of R_1 over $\mathcal{D}_T \supseteq \widehat{\mathcal{D}}$. Therefore the right invertibility of R_1 over $\mathcal{D}_T$ is equivalent to $\widehat{e}_p \in \mathrm{U}(\mathcal{D}_T)$. ◇

Corollary 10.1.19 *The simplest method for the computation of $\widehat{Y}$ is to choose $d \geq \deg_s(R_1)$ and to compute the Smith form of $\widehat{s}^d R_1 \in \widehat{\mathcal{D}}^{p\times(p+m)}$ over $\widehat{\mathcal{D}}$, i.e.,*

$$\begin{aligned} &\widehat{X}(\widehat{s}^d R_1)\widehat{Y} = (E', 0) \in \widehat{\mathcal{D}}^{p\times(p+m)} \text{ where}\\ &E' = \operatorname{diag}(e'_1, \ldots, e'_p) = \widehat{s}^d \widehat{E},\ 0 \neq e'_i = \widehat{s}^d \widehat{e}_i \in \widehat{\mathcal{D}},\ e'_1 \underset{\widehat{\mathcal{D}}}{|} \cdots \underset{\widehat{\mathcal{D}}}{|} e'_p. \end{aligned} \tag{10.35}$$

The element $\widehat{s}^{-m_p} e'_p = (s-\alpha)^{m_p} e'_p$, $m_p := \deg_{\widehat{s}}(e'_p)$, belongs to $\mathcal{D}$ and $\widehat{s} = (s-\alpha)^{-1}$ is a unit of $\mathcal{D}_T$. The controllability of $\mathcal{B}_{1,T}$ or right invertibility of R_1 over $\mathcal{D}_T$ is thus equivalent to

$$\widehat{e}_p \in \mathrm{U}(\mathcal{D}_T) \Longleftrightarrow e'_p \in \mathrm{U}(\mathcal{D}_T) \Longleftrightarrow (s-\alpha)^{m_p} e'_p \in T = \mathcal{D} \cap \mathrm{U}(\mathcal{D}_T). \tag{10.36}$$

This furnishes a new algorithm to decide the T-stabilizability of $\mathcal{B}_1$. ◇

As in Theorem 6.2.5 or in Lemma 10.1.6 the condition $\operatorname{rank}(P_1) = p$ implies that $H_1 = \widehat{P}_1^{-1}\widehat{Q}_1 = \widehat{N}_1\widehat{D}_1^{-1}$ are the coprime factorizations of $H_1 = P_1^{-1}Q_1$ over $\widehat{\mathcal{D}}$ and therefore also over the quotient ring $\mathcal{D}_T$ of $\widehat{\mathcal{D}}$. Since both

$$\mathcal{B}_{1,T} = \left\{ \begin{pmatrix} y_1 \\ u_1 \end{pmatrix} \in \mathcal{F}_T^{p+m};\ P_1 \circ y_1 = Q_1 \circ u_1 \right\} \text{ and}$$
$$\left\{ \begin{pmatrix} y_1 \\ u_1 \end{pmatrix} \in \mathcal{F}_T^{p+m};\ \widehat{P}_1 \circ y_1 = \widehat{Q}_1 \circ u_1 \right\}$$

are the unique controllable realization of $H_1 = P_1^{-1}Q_1 = \widehat{P}_1^{-1}\widehat{Q}_1$ in $\mathcal{F}_T^{p+m}$ they coincide, and hence

$$U_{1,T} = \mathcal{D}_T^{1\times p}(P_1, -Q_1) = \mathcal{D}_T^{1\times p}(\widehat{P}_1, -\widehat{Q}_1). \tag{10.37}$$

With the data from (10.33)–(10.37) we now apply Lemma 10.1.7 and Theorems 10.1.10 and 10.1.13 to the ring $\mathcal{D}_T$ instead of $\mathcal{D}$. The controllable realization of H_2 in (10.22) still has to be computed over the ring $\mathcal{D} = F[s]$ in order to obtain the T-stabilizing compensator.

Theorem 10.1.20 (Proper feedback behavior, compare [4, Sect. 5.2], [5, Sect. 6.5], [6, Sect. 7.4]) *With the data from* (10.33)–(10.37) *let $\mathcal{B}_2 = U_2^{\perp}$ be a T-stabilizing compensator of $\mathcal{B}_1$ with feedback behavior $\mathcal{B} := \mathrm{fb}(\mathcal{B}_1, \mathcal{B}_2)$, especially $\mathcal{D}_T^{1\times(p+m)} = U_{1,T} \oplus U_{2,T}$. Let $X \in \mathcal{D}_T^{m\times p}$ be the unique matrix from Lemma 10.1.7 with*

$$\begin{aligned} &(-\widehat{Q}_2, \widehat{P}_2) = (-\widehat{Q}_2^0, \widehat{P}_2^0) + X(\widehat{P}_1, -\widehat{Q}_1),\ \det(\widehat{P}_2) \neq 0 \text{ and}\\ &U_{2,T} = \mathcal{D}_T^{1\times m}(-\widehat{Q}_2, \widehat{P}_2),\ \mathcal{B}_{2,T} = \left\{ \begin{pmatrix} u_2 \\ y_2 \end{pmatrix} \in \mathcal{F}_T^{p+m};\ \widehat{P}_2 \circ y_2 = \widehat{Q}_2 \circ u_2 \right\}. \end{aligned} \tag{10.38}$$

Also consider the transfer matrix H of $\mathcal{B}$ and $\mathcal{B}_T$, i.e.,

$$H = \widehat{P}^{-1}\widehat{Q} \quad with \quad \widehat{P} = \begin{pmatrix} \widehat{P}_1 & -\widehat{Q}_1 \\ -\widehat{Q}_2 & \widehat{P}_2 \end{pmatrix}, \; \widehat{Q} = \begin{pmatrix} 0 & \widehat{Q}_1 \\ \widehat{Q}_2 & 0 \end{pmatrix}.$$

Then the following properties are equivalent:

1. $H \in \mathcal{S}^{(p+m)\times(p+m)}$, *i.e., the transfer matrix H is not only stable, but also proper.*
2. $X \in \mathcal{S}^{m\times p}$ *and* $\widehat{P} \in \mathrm{Gl}_{p+m}(\mathcal{S})$.

Notice that Corollary 8.2.85 is applicable to this $\mathcal{B}$ *and* H *in the continuous-time standard case.*

Proof 2. $\Longrightarrow$ 1.: The assumption $X \in \mathcal{S}^{m\times p}$ and Eqs. (10.34) and (10.38) imply $\widehat{Q} \in \mathcal{S}^{(p+m)\times(p+m)}$. From $\widehat{P} \in \mathrm{Gl}_{p+m}(\mathcal{S})$ we infer $\widehat{P}^{-1} \in \mathrm{Gl}_{p+m}(\mathcal{S})$ and $H = \widehat{P}^{-1}\widehat{Q} \in \mathcal{S}^{(p+m)\times(p+m)}$.

1. $\Longrightarrow$ 2.: According to Theorem and Definition 9.4.2, the T-stability of $\mathcal{B}^0$ means that $\widehat{P} \in \mathrm{Gl}_{p+m}(\mathcal{D}_T)$ and that $H \in \mathcal{D}_T^{(p+m)\times(p+m)}$. Condition 1 gives $H \in \mathcal{S}^{(p+m)\times(p+m)}$.

(i) First we describe the quotient behaviors $\mathcal{B}_{1,T}$ and $\mathcal{B}_{2,T}$ by matrices over $\mathcal{S}$. Since $\widehat{\mathcal{D}} \subset \mathcal{S}$, the matrix $(\widehat{P}_1, -\widehat{Q}_1)$ of $\mathcal{B}_{1,T}$ already has all entries in $\widehat{\mathcal{D}}$ and hence in $\mathcal{S}$. By construction $(-\widehat{Q}_2, \widehat{P}_2)$ belongs to $\mathcal{D}_T^{m\times(p+m)}$, but not necessarily to $\mathcal{S}^{m\times(p+m)}$. Consider a left coprime factorization of the compensator's transfer matrix $H_2 = \widehat{P}_2^{-1}\widehat{Q}_2$ over $\widehat{\mathcal{D}}$, i.e., a representation

$$H_2 = \widetilde{P}_2^{-1}\widetilde{Q}_2 \text{ with } (-\widetilde{Q}_2, \widetilde{P}_2) \in \widehat{\mathcal{D}}^{m\times(p+m)}, \; \mathrm{rank}(\widetilde{P}_2) = m \text{ and over } \widehat{\mathcal{D}} \text{ right invertible } (-\widetilde{Q}_2, \widetilde{P}_2).$$

Since $\mathcal{D}_T \supseteq \widehat{\mathcal{D}}$, this is also a left coprime factorization over $\mathcal{D}_T$. The quotient behavior $\mathcal{B}_{2,T}$ is controllable by construction, and hence the unique controllable realization of H_2 in $\mathcal{F}_T^{p+m}$. The $\mathcal{F}_T$-IO behavior defined by $(-\widetilde{Q}_2, \widetilde{P}_2)$ has the same property. This implies

$$\begin{aligned} \mathcal{B}_{2,T} &= \left\{\begin{pmatrix} u_2 \\ y_2 \end{pmatrix} \in \mathcal{F}_T^{p+m};\; \widetilde{P}_2 \circ y_2 = \widetilde{Q}_2 \circ u_2\right\} \\ &= \left\{\begin{pmatrix} u_2 \\ y_2 \end{pmatrix} \in \mathcal{F}_T^{p+m};\; \widehat{P}_2 \circ y_2 = \widehat{Q}_2 \circ u_2\right\} \\ U_{2,T} &= \mathcal{D}_T^{1\times m}(-\widetilde{Q}_2, \widetilde{P}_2) = \mathcal{D}_T^{1\times m}(-\widehat{Q}_2, \widehat{P}_2). \end{aligned}$$

We define the two matrices

$$\widetilde{P} := \begin{pmatrix} \widehat{P}_1 & -\widehat{Q}_1 \\ -\widetilde{Q}_2 & \widetilde{P}_2 \end{pmatrix}, \; \widetilde{Q} := \begin{pmatrix} 0 & \widehat{Q}_1 \\ \widetilde{Q}_2 & 0 \end{pmatrix} \in \widehat{\mathcal{D}}^{(p+m)\times(p+m)} \subset \mathcal{S}^{(p+m)\times(p+m)}.$$

Then the quotient behavior of the feedback behavior is

$$\mathcal{B}_T = \mathrm{fb}(\mathcal{B}_{1,T}, \mathcal{B}_{2,T}) = \left\{\begin{pmatrix} y \\ u \end{pmatrix} \in \mathcal{F}_T^{(p+m)\times(p+m)};\; \widetilde{P} \circ y = \widetilde{Q} \circ u\right\}$$

with transfer matrix $H = \widetilde{P}^{-1}\widetilde{Q} \in \mathcal{S}^{(p+m)\times(p+m)}$.

(ii) The entries of $(\widetilde{P}, -\widetilde{Q})$ belong to $\mathcal{S}$, hence, we can apply Corollary and Definition 7.5.4 over $\mathcal{S}$ and obtain the isomorphism

$$\begin{aligned}&\mathcal{S}^{1\times 2(p+m)}/\mathcal{S}^{1\times(p+m)}(\widetilde{P},-\widetilde{Q})\\&\quad\cong\left(\mathcal{S}^{1\times(p+m)}/\mathcal{S}^{1\times p}(\widehat{P}_1,-\widehat{Q}_1)\right)\times\left(\mathcal{S}^{1\times(p+m)}/\mathcal{S}^{1\times m}(-\widetilde{Q}_2,\widetilde{P}_2)\right).\end{aligned}\tag{10.39}$$

Since $H_1=\widehat{P}_1^{-1}\widehat{Q}_1$ and $H_2=\widetilde{P}_2^{-1}\widetilde{Q}_2$ are left coprime factorizations over $\widehat{\mathcal{D}}$ and thus also over the quotient ring $\mathcal{S}=\widehat{\mathcal{D}}_{\widehat{T}}$, the matrices $(\widehat{P}_1,-\widehat{Q}_1)$ and $(-\widetilde{Q}_2,\widetilde{P}_2)$ are right invertible over $\mathcal{S}$ and the two factors in the second line of (10.39) are free $\mathcal{S}$-modules. Therefore, also the module in the first line of (10.39) is free, and hence the matrix $(\widetilde{P},-\widetilde{Q})$ has a right inverse Z over $\mathcal{S}$. Since $H\in\mathcal{S}^{(p+m)\times(p+m)}$ by assumption 1, we also get

$$\begin{aligned}&(\mathrm{id}_{p+m},-H)Z\in\mathcal{S}^{(p+m)\times(p+m)}\\&\Longrightarrow\widetilde{P}(\mathrm{id}_{p+m},-H)Z=(\widetilde{P},-\widetilde{Q})Z=\mathrm{id}_{p+m}\in\mathcal{S}^{(p+m)\times(p+m)}\\&\Longrightarrow\widetilde{P}\in\mathrm{Gl}_{p+m}(\mathcal{S}),\ \widetilde{P}^{-1}=(\mathrm{id}_{p+m},-H)Z\text{ and}\\&\qquad\mathcal{S}^{1\times(p+m)}=\mathcal{S}^{1\times(p+m)}\widetilde{P}=\mathcal{S}^{1\times p}(\widehat{P}_1,-\widehat{Q}_1)\oplus\mathcal{S}^{1\times m}(-\widetilde{Q}_2,\widetilde{P}_2).\end{aligned}\tag{10.40}$$

But notice that $(-\widetilde{Q}_2,\widetilde{P}_2)$ is *not* necessarily a left inverse of $\begin{pmatrix}\widehat{N}_1\\\widehat{D}_1\end{pmatrix}$.
(iii) By applying Lemma 10.1.7 to the direct sum from (10.40) over the ring $\mathcal{S}$, we obtain matrices

$$\begin{aligned}&X'\in\mathcal{S}^{m\times p}\text{ and}\\&(-\widehat{Q}_2',\widehat{P}_2'):=(-\widehat{Q}_2^0,\widehat{P}_2^0)+X'(\widehat{P}_1,-\widehat{Q}_1)\in\mathcal{S}^{m\times(p+m)}\text{ such that}\\&\mathcal{S}^{1\times m}(-\widetilde{Q}_2,\widetilde{P}_2)=\mathcal{S}^{1\times m}(-\widehat{Q}_2',\widehat{P}_2').\end{aligned}$$

Since $\mathcal{D}_T=\mathcal{S}_{\widetilde{S}}$, we also get

$$\begin{aligned}&U_{2,T}=\mathcal{D}_T^{1\times m}(-\widehat{Q}_2,\widehat{P}_2)=\mathcal{D}_T^{1\times m}(-\widetilde{Q}_2,\widetilde{P}_2)=\mathcal{D}_T^{1\times m}(-\widehat{Q}_2',\widehat{P}_2')\text{ where}\\&(-\widehat{Q}_2',\widehat{P}_2')=(-\widehat{Q}_2^0,\widehat{P}_2^0)+X'(\widehat{P}_1,-\widehat{Q}_1)\in\mathcal{D}_T^{m\times(p+m)}\\&(-\widehat{Q}_2,\widehat{P}_2)=(-\widehat{Q}_2^0,\widehat{P}_2^0)+X(\widehat{P}_1,-\widehat{Q}_1)\in\mathcal{D}_T^{m\times(p+m)}.\end{aligned}$$

By application of Lemma 10.1.7 to the ring $\mathcal{D}_T$ these two representations of $U_{2,T}$ coincide, and hence

$$\begin{aligned}&X=X'\in\mathcal{S}^{m\times p},\ (-\widehat{Q}_2,\widehat{P}_2)=(-\widehat{Q}_2',\widehat{P}_2')\in\mathcal{S}^{m\times(p+m)}\text{ and}\\&\widehat{P}\in\mathcal{S}^{(p+m)\times(p+m)}.\end{aligned}$$

The direct sum from (10.40) implies

$$\begin{aligned}&\mathcal{S}^{1\times(p+m)}=\mathcal{S}^{1\times p}(\widehat{P}_1,-\widehat{Q}_1)\oplus\mathcal{S}^{1\times m}(-\widehat{Q}_2,\widehat{P}_2)=\mathcal{S}^{1\times(p+m)}\widehat{P}\text{ and}\\&\widehat{P}\in\mathrm{Gl}_{p+m}(\mathcal{S}).\end{aligned}\tag{10.41}$$

Together with $X=X'\in\mathcal{S}^{m\times p}$ this proves item 2. □

We specialize Theorems 10.1.10, 10.1.13, and 10.1.20 to the continuous-time and discrete-time standard cases.

Remark 10.1.21 Let $\mathcal{D} = F[s]$ with $F = \mathbb{R}$ or $F = \mathbb{C}$, let

$$\mathcal{F} := \begin{cases} \mathrm{C}^{-\infty}(\mathbb{R}, F) & \text{(continuous time)}, \\ F^{\mathbb{N}} & \text{(discrete time)}, \end{cases}$$

and let $T = \{t \in F[s];\ \mathrm{V}_{\mathbb{C}}(t) \subseteq \Lambda_{1,\mathrm{std}}\}$ be given by the disjoint decomposition

$$\mathbb{C} = \Lambda_{1,\mathrm{std}} \uplus \Lambda_{2,\mathrm{std}} \quad \text{where}$$
$$\Lambda_{1,\mathrm{std}} = \begin{cases} \mathbb{C}_- := \{\lambda \in \mathbb{C};\ \Re(\lambda) < 0\} & \text{(continuous time)}, \\ \mathbb{D} := \{\lambda \in \mathbb{C};\ |\lambda| < 1\} & \text{(discrete time)}, \end{cases}$$

compare Example 9.1.3, items 1 and 2.

If $\mathcal{B}_1$ is T-stabilizable, then Theorems 10.1.10 and 10.1.13 construct all T-stabilizing compensators $\mathcal{B}_2$ with feedback (closed-loop) behavior $\mathcal{B} := \mathrm{fb}(\mathcal{B}_1, \mathcal{B}_2)$ and $\mathrm{char}(\mathcal{B}^0) \subseteq \Lambda_{1,\mathrm{std}}$, and Theorem 10.1.20 constructs those $\mathcal{B}$ with additionally proper (and T-stable) transfer matrix. In particular, all trajectories of $\mathcal{B}^0$ are asymptotically stable (Corollary 9.4.3). Moreover, the closed-loop transfer matrix H as operator $H\circ$ is BIBO-stable or even $(\mathrm{L}^p_+, \mathrm{L}^p_+)$-stable for continuous time and (ℓ^p, ℓ^p)-stable for discrete time for all p with $1 \le p \le \infty$, cf. Theorems 8.2.83 and 8.1.26. ◊

Example 10.1.22 We apply Assumption 10.1.18 and Theorem 10.1.20 to the plant from Example 10.1.17, i.e.,

$$\mathcal{B}_1 = \left(\mathcal{D}(P_1, -Q_1)\right)^{\perp} \text{ with } P_1 := e^1(s-3)(s-4),\ Q_1 := e^1(s-5),$$
$$e^1 = (s+1)(s+2),\ R_1 = (P_1, -Q_1),\ H_1 = P_1^{-1}Q_1 = \tfrac{s-5}{(s-3)(s-4)}.$$

But now we are going to stabilize it with a proper closed-loop transfer matrix.

We choose $\alpha := -1$, i.e., $\widehat{s} = (s+1)^{-1}$, and use (10.35). Since $\deg_s(R_1) = 4$, we compute the Smith form over $\widehat{\mathcal{D}} = F[\widehat{s}]$ of

$$\widehat{R}_1 := \widehat{s}^4 R_1 = ((\widehat{s}+1)(4\widehat{s}-1)(5\widehat{s}-1), \widehat{s}(\widehat{s}+1)(6\widehat{s}-1)) \in \widehat{\mathcal{D}}^{1\times 2} \text{ as}$$
$$\widehat{X}\widehat{R}_1\widehat{Y} = (\widehat{s}+1, 0) \text{ with } \widehat{X} = 1,\ \widehat{Y} \in \mathrm{Gl}_2(\widehat{\mathcal{D}}),\ \widehat{s}+1 = \tfrac{s+2}{s+1} \in \mathrm{U}(\mathcal{D}_T),$$
$$\widehat{Y} = \begin{pmatrix} 102\widehat{s}+1 & -6\widehat{s}^2+\widehat{s} \\ -340\widehat{s}+93 & 20\widehat{s}^2-9\widehat{s}+1 \end{pmatrix},\ \widehat{Y}^{-1} = \begin{pmatrix} \widehat{P}_1 & -\widehat{Q}_1 \\ -\widehat{Q}_2^0 & \widehat{P}_2^0 \end{pmatrix} = \begin{pmatrix} 20\widehat{s}^2-9\widehat{s}+1 & 6\widehat{s}^2-\widehat{s} \\ 340\widehat{s}-93 & 102\widehat{s}+1 \end{pmatrix}.$$

The condition $\widehat{s}+1 \in \mathrm{U}(\mathcal{D}_T)$ means that $\mathcal{B}_1$ is T-stabilizable. The left coprime factorization of H_1 over $F[\widehat{s}]$ is

$$H_1 = P_1^{-1}Q_1 = \widehat{P}_1^{-1}\widehat{Q}_1 = -\tfrac{(6\widehat{s}-1)\widehat{s}}{(4\widehat{s}-1)(5\widehat{s}-1)},\ (\widehat{P}_1, -\widehat{Q}_1) = (20\widehat{s}^2 - 9\widehat{s} + 1, 6\widehat{s}^2 - \widehat{s}).$$

By choosing the parameter $X := 93$, which is a proper and T-stable rational function, we obtain

$$(-\widehat{Q}_2, \widehat{P}_2) = (-\widehat{Q}_2^0, \widehat{P}_2^0) + X(\widehat{P}_1, -\widehat{Q}_1) = ((1860\widehat{s} - 497)\widehat{s}, 558\widehat{s}^2 + 9\widehat{s} + 1) \text{ and}$$
$$H_2 := \widehat{P}_2^{-1}\widehat{Q}_2 = -\frac{(1860\widehat{s}-497)\widehat{s}}{558\widehat{s}^2+9\widehat{s}+1} = \frac{497s-1363}{s^2+11s+568} \in F(\widehat{s}) = F(s).$$

The unique controllable realization of H_2 furnishes the compensator

$$\mathcal{B}_2 = \left(\mathcal{D}(-Q_2, P_2)\right)^{\perp} \quad \text{with } P_2 = s^2 + 11s + 568 \text{ and } Q_2 = 497s - 1363.$$

The closed-loop or feedback behavior is then described by the matrices

$$P = \begin{pmatrix} P_1 & -Q_1 \\ -Q_2 & P_2 \end{pmatrix} = \begin{pmatrix} e^1(s-3)(s-4) & -e^1(s-5) \\ -497s+1363 & s^2+11s+568 \end{pmatrix} \quad \text{and}$$
$$Q = \begin{pmatrix} 0 & Q_1 \\ Q_2 & 0 \end{pmatrix} = \begin{pmatrix} 0 & e^1(s-5) \\ 497s-1363 & 0 \end{pmatrix}.$$

It is T-stable since $\det(P) = (s+1)^5(s+2) \in T$ and its transfer matrix

$$H = P^{-1}Q = \frac{1}{(s+1)^4}\begin{pmatrix} (s-5)(497s-1363) & (s-5)(s^2+11s+568) \\ (s-3)(s-4)(497s-1363) & (s-5)(497s-1363) \end{pmatrix}$$

is (even strictly) proper.

Notice that also the compensator $\mathcal{B}_2$ is proper. This is not an accident; in Example 10.1.36 we will see that the given plant $\mathcal{B}_1$ has only proper compensators. ◊

Example 10.1.23 We consider the linearized pendulum from Example 3.1.12 in the unstable upright position with the Eq. (3.22), i.e.,

$$0 = y_1'' - \frac{(M+m)g}{ML}y_1 - \frac{1}{ML}u_1 = (P_1, -Q_1) \circ \begin{pmatrix} y_1 \\ u_1 \end{pmatrix}, \; y_1 := \psi, \; u_1 := u, \text{ where}$$
$$P_1 = s^2 - \lambda^2, \; \lambda := +\sqrt{\frac{(M+m)g}{ML}} > 0, \; Q_1 := \frac{1}{ML} > 0,$$
$$\mathcal{B}_1 = \left\{\begin{pmatrix} y_1 \\ u_1 \end{pmatrix} \in \mathcal{F}^2;\; P_1 \circ y_1 = Q_1 \circ u_1\right\}, \; H_1 := P_1^{-1}Q_1, \; \deg_s(H_1) = -2.$$

The polynomial P_1 has the root $\lambda > 0$, and hence $\mathcal{B}_1^0$ is $\mathbb{C}_-$-unstable as the physical model suggests. Since $\gcd(P_1, Q_1) = 1$, the behavior $\mathcal{B}_1$ is controllable and thus $\mathbb{C}_-$-stabilizable. We apply Theorem 10.1.20. Since $\deg_s(P_1, -Q_1) = 2$, we compute the Smith form of $\widehat{s}^2(P_1, -Q_1) = \widehat{s}^2(s^2 - \lambda^2, -(ML)^{-1})$ as

$$\widehat{X}\widehat{s}^2(P_1, -Q_1)\widehat{Y} = (1, 0) \text{ with } \widehat{X} = 1, \; \widehat{Y} \in \mathrm{Gl}_2(\widehat{\mathcal{D}}),$$
$$\widehat{Y} = \begin{pmatrix} \widehat{D}_2^0 & \widehat{N}_1 \\ \widehat{N}_2^0 & \widehat{D}_1 \end{pmatrix} = \begin{pmatrix} 2\widehat{s}+1 & \frac{\widehat{s}^2}{ML} \\ 2\big(ML-g(M+m)\big)\widehat{s}-3ML-g(M+m) & \frac{\big(ML-g(M+m)\big)}{ML}\widehat{s}^2-2\widehat{s}+1 \end{pmatrix},$$

$$\widehat{Y}^{-1} = \begin{pmatrix} \widehat{P}_1 & -\widehat{Q}_1 \\ -\widehat{Q}_2^0 & \widehat{P}_2^0 \end{pmatrix} = \begin{pmatrix} \frac{\left(ML-g(M+m)\right)}{ML}\widehat{s}^2-2\widehat{s}+1 & -\frac{\widehat{s}^2}{ML} \\ 2\left(g(M+m)-ML\right)\widehat{s}+3ML+g(M+m) & 2\widehat{s}+1 \end{pmatrix}.$$

Since $\widehat{P}_2^0 \neq 0$, $(-\widehat{Q}_2^0, \widehat{P}_2^0)$ itself defines a controllable stabilizing compensator. All others are obtained by

$$\begin{aligned} &(-\widehat{Q}_2, \widehat{P}_2) = (-\widehat{Q}_2^0, \widehat{P}_2^0) + X(\widehat{P}_1, -\widehat{Q}_1) \text{ where } X \in \mathcal{S},\ \widehat{P}_2 = \widehat{P}_2^0 - X\widehat{Q}_1 \neq 0, \\ &H_2 = \widehat{P}_2^{-1}\widehat{Q}_2 = P_2^{-1}Q_2,\ P_2, Q_2 \in \mathcal{D},\ \gcd_{\mathcal{D}}(P_2, Q_2) = 1, \\ &\mathcal{B}_2 = \left\{\binom{u_2}{y_2} \in \mathcal{F}^2;\ P_2 \circ y_2 = Q_2 \circ u_2\right\}. \end{aligned}$$

With various choices for X we get

$$\begin{aligned} &\text{(i)}\quad X := 0:\ P_2 = -s-3,\ Q_2 = \big(3ML + g(M+m)\big)s + ML + 3g(M+m) \\ &\text{(ii)}\quad X := ML:\ P_2 = -s^2 - 4s - 2, \\ &\qquad\qquad Q_2 = \big(4ML + g(M+m)\big)s^2 + 4\big(ML + g(M+m)\big)s \\ &\qquad\qquad\quad + ML + 2g(M+m) \\ &\text{(iii)}\quad X := \tfrac{ML}{s+1} = ML\widehat{s}:\ P_2 = -s^3 - 5s^2 - 7s - 2, \\ &\qquad Q_2 = \big(3ML + g(M+m)\big)s^3 + \big(8ML + 5g(M+m)\big)s^2 \\ &\qquad\quad + \big(5ML + 7g(M+m)\big)s + ML + 2g(M+m). \end{aligned}$$

The transfer function H_2 is proper since H_1 is strictly proper, cf. Theorem 10.1.37 below. The stable autonomous part of the closed-loop behavior is

$$\mathcal{B}^0 = \mathrm{fb}(\mathcal{B}_1, \mathcal{B}_2)^0 = \left\{\binom{y_1}{y_2} \in \mathcal{F}^2;\ (s^2 - \lambda^2) \circ y_1 = Q_1 y_2,\ P_2 \circ y_2 = Q_2 \circ y_1\right\}.$$

The horizontal force $u := y_2$ on the base of the pendulum stabilizes it in the upright position, characterized by the angle $\psi = y_1 = 0$. The state space dimension n_2 of an observable state space representation

$$\begin{aligned} &s \circ x_2 = A_2 x_2 + B_2 u_2,\ y_2 = C_2 x_2 + D_2,\ x_2 \in \mathcal{F}^{n_2}, \text{ with} \\ &H_2 = P_2^{-1} Q_2 = D_2 + C_2 \left(s\, \mathrm{id}_{n_2} - A_2\right)^{-1} B_2 \end{aligned}$$

is $n_2 = \deg_s(P_2)$, hence $n_2 = 1, 2$ resp. 3 in cases (i), (ii) resp. (iii).

The state space dimension of $\mathcal{B}_1$ is $n_1 = \dim_F(\mathcal{B}_1^0) = \deg_s(P_1) = 2$. Notice that a stabilizing compensator that is composed of a Luenberger observer and state feedback, always has state dimension $n_2 = n_1 = 2$, cf. Theorem 12.5.2, [12, Theorem 10.5.1]. The compensator from case (i) is simpler, cf. [12, (10.16)]. ◊

10.1.3 Pole Placement

For a given plant $\mathcal{B}_1$, pole placement means to construct a compensator such that the poles of the feedback behavior $\mathcal{B}$ belong to a pre-selected finite subset Λ_1 of $\Lambda_{1,\text{std}}$, and thus to influence the speed of convergence to 0 of the trajectories of $\mathcal{B}^0$. The uncontrollable poles of $\mathcal{B}_1$ have to belong to Λ_1.

We continue with the feedback stabilization setting of Sect. 10.1.1 where the operator ring $\mathcal{D}$ is an arbitrary principal ideal domain. In Corollary 10.1.11, we have shown that the plant $\mathcal{B}_1$ is T-stabilizable if and only if the highest nonzero elementary divisor e_p^1 of $(P_1, -Q_1)$ lies in T. Let $\mathcal{P}$ be a set of representatives of prime elements of $\mathcal{D}$. We define

$$\begin{aligned}\mathcal{P}_{\min} &:= \{q \in \mathcal{P};\ q \mid e_p^1\} \quad \text{and} \\ T_{\min} &:= \left\{u \prod_{q \in \mathcal{P}_{\min}} q^{\mu(q)};\ u \in \mathrm{U}(\mathcal{D}),\ \mu(q) \geq 0\right\} \\ &= \left\{t \in \mathcal{D};\ \exists k \in \mathbb{N}:\ t \mid (e_p^1)^k\right\}.\end{aligned} \tag{10.42}$$

The set $T_{\min}$ is the least saturated monoid containing e_p^1, see Sect. 9.2, and $\mathcal{D}_{T_{\min}} = \mathcal{D}_{e_p^1}$ by item 3 of Corollary 9.1.8, where $\mathcal{D}_{e_p^1} = \left\{\frac{f}{(e_p^1)^k};\ f \in \mathcal{D},\ k \in \mathbb{N}\right\}$.

Corollary 10.1.24 *1. The monoid $T_{\min}$ is the least saturated monoid T' for which $\mathcal{B}_1$ is T'-stabilizable. Hence the statements of Theorems 10.1.10 and 10.1.13 are true for this minimal monoid $T_{\min}$ instead of T.*

2. For an arbitrary saturated T the behavior $\mathcal{B}_1$ is T-stabilizable if and only if $T_{\min} \subseteq T$.

Proof According to Corollary 10.1.11, the behavior $\mathcal{B}_1$ if T-stabilizable if and only if $e_p^1 \in T$. By definition this holds for $T_{\min}$. □

We specialize Corollary 10.1.24 to the situation where $\mathcal{D} = F[s]$ with $F = \mathbb{R}$ or $F = \mathbb{C}$, and where the submonoid T is given via a disjoint decomposition $\mathbb{C} = \Lambda_1 \uplus \Lambda_2$ as

$$T = \{t \in F[s];\ \mathrm{V}_{\mathbb{C}}(t) \subseteq \Lambda_1.$$

In the real case $F = \mathbb{R}$, we assume in addition that $\overline{\Lambda_i} = \Lambda_i$. Hence, by definition of T, the elementary divisor e_p^1 belongs to T if and only if the characteristic variety $\mathrm{char}(\mathcal{B}_1) = \mathrm{V}_{\mathbb{C}}(e_p^1)$ of uncontrollable poles of $\mathcal{B}_1$ (see Definition and Corollary 5.4.1) is contained in Λ_1. Also recall that the feedback behavior $\mathcal{B} := \mathrm{fb}(\mathcal{B}_1, \mathcal{B}_2)$ is T-stable if and only if $\mathrm{char}(\mathcal{B}^0) \subseteq \Lambda_1$, see Corollary 9.4.3. The poles of the feedback behavior $\mathcal{B}$, i.e., the elements of $\mathrm{char}(\mathcal{B}^0)$, are called the *closed-loop poles* in standard engineering language. We also assume $\alpha \in \Lambda_1 \cap F$ and define, in analogy to (10.42),

$$\Lambda_{\min} := \{\alpha\} \cup \mathrm{V}_{\mathbb{C}}(e_p^1) = \{\alpha\} \cup \mathrm{char}(\mathcal{B}_1),$$
$$\mathcal{P}_{\min} := \{s-\alpha\} \cup \{q \in \mathcal{P};\ q \mid e_p^1\}, \quad \text{and}$$
$$T_{\min} := \left\{t \in \mathcal{D};\ \exists k \in \mathbb{N}:\ t \mid ((s-\alpha)e_p^1)^k\right\} = \{t \in F[s];\ \mathrm{V}_{\mathbb{C}}(t) \subseteq \Lambda_{\min}\}. \tag{10.43}$$

Recall that $\mathrm{char}(\mathcal{B}_1)$ contains the uncontrollable poles of $\mathcal{B}_1$.

Corollary 10.1.25 *For the saturated submonoid $T := \{t \in F[s];\ \mathrm{V}_{\mathbb{C}}(t) \subseteq \Lambda_1\}$ with $\alpha \in \Lambda_1$ and $s - \alpha \in T$ the IO behavior $\mathcal{B}_1$ is T-stabilizable if and only if*

$$\mathrm{char}(\mathcal{B}_1) \subseteq \Lambda_1 \ \textit{i.e.,}\ \Lambda_{\min} \subseteq \Lambda_1 \ \textit{or}\ T_{\min} \subseteq T.$$

In particular, $\mathcal{B}_1$ is always $T_{\min}$-stabilizable and $T_{\min}$ is the least saturated monoid T' which contains $s-\alpha$ and for which $\mathcal{B}_1$ is T'-stabilizable. ◊

Theorem 10.1.26 (Pole placement or pole assignment) *Let $t_0 \in F[s]$ be a nonzero polynomial and define*

$$f := (s-\alpha)e_p^1 t_0 \in F[s],$$
$$\Lambda_1 := \mathrm{V}_{\mathbb{C}}(f) = \{\alpha\} \cup \mathrm{V}_{\mathbb{C}}(e_p^1) \cup \mathrm{V}_{\mathbb{C}}(t_0), \quad \textit{and} \tag{10.44}$$
$$T := \{t \in F[s];\ \mathrm{V}_{\mathbb{C}}(t) \subseteq \mathrm{V}_{\mathbb{C}}(f)\} = \{t \in F[s];\ \exists k \in \mathbb{N}:\ t \mid f^k\}.$$

Then $\mathcal{B}_1$ is T-stabilizable. Let $\mathcal{B}_2$ be the controllable compensator according to Theorem 10.1.20 given by the parameters

$$X := \widehat{t}^{-1}\widehat{X} \in \mathcal{S}^{m\times p} = \widehat{\mathcal{D}}_{\widehat{T}}^{m\times p} \ \textit{with}\ \widehat{\mathcal{D}} = F[\widehat{s}], \widehat{s} = (s-\alpha)^{-1},$$
$$t \in T,\ \widehat{t} = t\widehat{s}^{\deg(t)} \in \widehat{T} \ \textit{and}\ \widehat{X} \in \widehat{\mathcal{D}}^{m\times p}, \tag{10.45}$$

as the controllable realization of $H_2 = \widehat{P}_2^{-1}\widehat{Q}_2$, where

$$(-\widehat{Q}_2, \widehat{P}_2) := (-\widehat{Q}_2^0, \widehat{P}_2^0) + X(\widehat{P}_1, -\widehat{Q}_1), \quad \det(\widehat{P}_2) \neq 0. \tag{10.46}$$

Then the feedback system's poles satisfy $\mathrm{char}(\mathrm{fb}(\mathcal{B}_1, \mathcal{B}_2)^0) \subseteq \Lambda_1$ *and its transfer matrix $H \in (\mathcal{D}_T \cap F(s)_{\mathrm{pr}})^{m\times p}$ is proper.*

In the engineering language, this means that the closed-loop poles, i.e., the elements of $\mathrm{char}(\mathrm{fb}(\mathcal{B}_1, \mathcal{B}_2)^0)$, *can be* placed into *or* assigned to *the finite set* $\mathrm{V}_{\mathbb{C}}(f)$. *This is done by choosing the parameters $X = \widehat{t}^{-1}\widehat{X}$ as in* (10.45), *in particular, by choosing $t \in T$, where T is the set from* (10.44).

Proof The sets Λ_1 and T from (10.44) satisfy all conditions from Corollary 10.1.25, hence $\mathcal{B}_1$ is T-stabilizable. The remaining assertions are direct consequences of Theorem 10.1.20 and the definition of T. □

Remark 10.1.27 With the continuous resp. discrete standard sets

$$\Lambda_{1,\mathrm{std}} = \mathbb{C}_- = \{z \in \mathbb{C};\ \Re(z) < 0\} \ \text{resp.}\ \Lambda_{1,\mathrm{std}} = \mathbb{D} = \{z \in \mathbb{C};\ |z| < 1\}$$

and $T_{\mathrm{std}} = \left\{t \in F[s];\ \mathrm{V}_{\mathbb{C}}(t) \subset \Lambda_{1,\mathrm{std}}\right\}$ one usually requires $\Lambda_1 \subseteq \Lambda_{1,\mathrm{std}}$, i.e., that

- $\mathrm{V}_{\mathbb{C}}(e_1^p) \subseteq \Lambda_{1,\mathrm{std}}$, i.e., that $e_1^p \in T_{\mathrm{std}}$, and thus the plant $\mathcal{B}_1$ is asymptotically stabilizable (see Remark 10.1.21),
- $\alpha \in \Lambda_{1,\mathrm{std}}$ and
- $\mathrm{V}_{\mathbb{C}}(t_0) \subseteq \Lambda_{1,\mathrm{std}}$, i.e., that t_0 is a Hurwitz or Schur polynomial.

The resulting feedback system is then asymptotically stable. ◇

Remark 10.1.28 (*Constructiveness*) In the situation of Theorem 10.1.26, the inclusion $t \in T$ can be easily decided by Euclidean division because of the equivalences

$$t \in T \iff \exists k \in \mathbb{N}:\ t \mid f^k \iff t \mid f^{\deg(t)}. \tag{10.47}$$

Assume, as in all practical cases, that $e_p^1 \in \mathbb{Q}[j](s),\ j := \sqrt{-1}$, i.e., that the coefficients of e_p^1 are rational or Gaussian numbers. Also assume that

$$\alpha \in \mathbb{Q}[j], \quad \text{and} \quad t_0 = \prod_{i=1}^{n}(s - \alpha_i) \in \mathbb{Q}[j][s] \text{ with } \alpha_i \in \mathbb{Q}[j]. \tag{10.48}$$

Then $f = e_p^1(s - \alpha)t_0 \in \mathbb{Q}[j][s]$ and $\Lambda_1 = \mathrm{V}_{\mathbb{C}}(e_p^1) \cup \{\alpha, \alpha_1, \ldots, \alpha_n\} \subseteq \mathbb{Q}[j]$. In general one requires $\Lambda_1 \subseteq \mathbb{Q}[j] \cap \Lambda_{1,\mathrm{std}}$. For $t \in \mathbb{Q}[j][s]$, the test (10.47) can then be carried out without numerical errors and it is much easier than that for general $\Lambda_1 \subseteq \mathbb{C}$, compare Remark 10.1.16. ◇

Example 10.1.29 We illustrate Theorem 10.1.26 on the data from Example 10.1.22. The uncontrollable poles of the plant $\mathcal{B}_1$ are the zero set of $e^1 = (s+1)(s+2)$, i.e., $\{-2, -1\}$, and we chose $\alpha = -1$ in order to treat properness. This means that $\Lambda_{\min} = \{-2, -1\}$ for the given plant. The poles of the feedback system with the compensator from Example 10.1.22 lie in $\Lambda_{\min}$, as can be seen from $\det(P) = (s+1)^5(s+2)$.

In order to compute a different compensator whose poles lie in $\Lambda_1 := \Lambda_{\min} \cup \{5\}$, we use the polynomial $t_0 := s + 5$ and define

$$f := (s - \alpha)e^1 t_0 = (s+1)^2(s+2)(s+5).$$

Then $\Lambda_1 = \mathrm{V}_{\mathbb{C}}(f)$ and the set T from (10.44) is

$$T = \left\{a(s+1)^{\mu_{-1}}(s+2)^{\mu_{-2}}(s+5)^{\mu_{-5}};\ a \in F \setminus \{0\},\ \mu_{-1}, \mu_{-2}, \mu_{-5} \in \mathbb{N}\right\}.$$

By setting $t := s + 5 \in T$ and $\widehat{X} := 1 \in F[\widehat{s}]$ in (10.45), we obtain $\widehat{t} = 4\widehat{s} + 1$ and the parameter $X = \frac{\widehat{X}}{\widehat{t}} = \frac{1}{4\widehat{s}+1}$. We insert this into (10.46) and then compute

$$H_2 = \frac{\widehat{Q}_2}{\widehat{P}_2} = \frac{1340\widehat{s}^2 - 23\widehat{s} - 94}{402\widehat{s}^2 + 107\widehat{s} + 1} = \frac{94s^2 + 211s - 1223}{s^2 + 109s + 510}.$$

The controllable realization of H_2, namely,

$$\mathcal{B}_2 = \left(\mathcal{D}(-Q_2, P_2)\right)^\perp \text{ with}$$
$$(-Q_2, P_2) = (-(94s^2 + 211s - 1223), s^2 + 109s + 510)$$

is a T-stabilizing compensator which leads to a proper feedback system, described by matrices P and Q as in Example 10.1.22. The set of closed-loop poles is $\mathrm{V}_{\mathbb{C}}(P) = \mathrm{V}_{\mathbb{C}}((s+5)(s+2)(s+1)^4) = \{-1, -2, -5\}$ as intended. ◊

10.1.4 Proper Compensators

In this section we characterize those compensators $\mathcal{B}_2$ that T-stabilize the plant and have a proper transfer matrix H_2 and a proper closed-loop transfer matrix H. We do this via a condition on the parameter matrix X from Theorem 10.1.20 *and show that almost all choices of the parameters lead to proper compensators (Theorem* 10.1.34*). In Theorem* 10.1.37*, we also characterize the strictly proper compensators. We start with some technical preparations.*

As before we use a field F, $\mathcal{D} = F[s]$, $\alpha \in F$ with $s - \alpha \in T$ and

$$\begin{aligned}
&\widehat{\mathcal{D}} = F[\widehat{s}] \subset \mathcal{S} = \widehat{\mathcal{D}}_{\widehat{T}} = \left\{\tfrac{\widehat{a}}{\widehat{t}};\ \widehat{a} \in \widehat{\mathcal{D}},\ \widehat{t} \in \widehat{T}\right\} \\
&\qquad \subset F(s)_{\mathrm{pr}} = \widehat{\mathcal{D}}_{T(\widehat{s})} = \left\{\tfrac{\widehat{a}}{\widehat{b}};\ \widehat{a}, \widehat{b} \in \widehat{\mathcal{D}},\ \widehat{b}(0) \neq 0\right\}, \\
&\text{where } \widehat{T} := \{\widehat{t}_\alpha(\widehat{s}) = t(s-\alpha)^{-\deg(t)} = t\widehat{s}^{\deg(t)};\ t \in T\} \\
&\qquad \subset T(\widehat{s}) := \{\widehat{b} \in \widehat{\mathcal{D}};\ \widehat{b}(0) \neq 0\} = \widehat{\mathcal{D}} \setminus \widehat{\mathcal{D}}\widehat{s}
\end{aligned} \tag{10.49}$$

from Lemma 9.5.3 and Theorem and Definition 9.5.4. The rings from (10.49) are principal ideal domains and have the prime $\widehat{s}$. In Example 9.2.4 we have seen that $z = \frac{1}{s}$ is the unique prime of $F(s)_{\mathrm{pr}}$ up to association and that $F(s)_{\mathrm{spr}} = F(s)_{\mathrm{pr}} z$ is the ideal of strictly proper rational functions. Then

$$\begin{aligned}
&z = \tfrac{1}{s} = \tfrac{s-\alpha}{s}\tfrac{1}{s-\alpha} = \tfrac{s-\alpha}{s}\widehat{s} \text{ and } \left(\tfrac{s-\alpha}{s}\right)^{-1} = \tfrac{s}{s-\alpha} \in F(s)_{\mathrm{pr}} \\
&\Longrightarrow \tfrac{s-\alpha}{s} \in \mathrm{U}\left(F(s)_{\mathrm{pr}}\right) \Longrightarrow F(s)_{\mathrm{spr}} = F(s)_{\mathrm{pr}} z = F(s)_{\mathrm{pr}} \widehat{s}.
\end{aligned}$$

We extend $\widehat{\mathcal{D}} \longrightarrow F, \widehat{a}(\widehat{s}) \longmapsto \widehat{a}(0)$ to the F-algebra epimorphism

$$\widehat{\nu}\colon \widehat{\mathcal{D}}_{T(\widehat{s})} = F(s)_{\mathrm{pr}} \longrightarrow F, \quad \widehat{f} = \tfrac{\widehat{a}(\widehat{s})}{\widehat{b}(\widehat{s})} \longmapsto \widehat{\nu}(\widehat{f}) = \widehat{f}(0) := \tfrac{\widehat{a}(0)}{\widehat{b}(0)}. \tag{10.50}$$

Lemma 10.1.30 *1. Lemma 5.5.5 furnishes the direct decomposition $F(s)_{\mathrm{pr}} = F \oplus F(s)_{\mathrm{spr}}$ with $F(s)_{\mathrm{spr}} = F(s)_{\mathrm{pr}}\widehat{s} = \ker(\widehat{\nu})$. The decomposition of $\widehat{f} = \frac{\widehat{a}}{\widehat{b}}$ with $\widehat{a}, \widehat{b} \in \widehat{\mathcal{D}}$ and $\widehat{b}(0) \neq 0$ is given by*

$$\widehat{f}=\widehat{f}(0)+(\widehat{f}-\widehat{f}(0)) \text{ with } \widehat{f}-\widehat{f}(0)=\tfrac{\widehat{a}-\widehat{a}(0)\,\widehat{b}(0)^{-1}\widehat{b}}{\widehat{b}}\in F(s)_{\rm spr}=F(s)_{\rm pr}\widehat{s}$$
$$\Longrightarrow \widehat{f}=\widehat{f}(0)+\widehat{s}\,\widehat{g},\ \widehat{g}:=\widehat{c}\,\widehat{b}^{-1}\in F(s)_{\rm pr},\ \widehat{a}-\widehat{a}(0)\widehat{b}(0)^{-1}\widehat{b}=\widehat{s}\,\widehat{c},\ \widehat{c}\in\widehat{\mathcal{D}}. \tag{10.51}$$

2. *The group of units of $F(s)_{\rm pr}$ is*

$$\begin{aligned}\mathrm{U}(F(s)_{\rm pr})&=\underbrace{\{\widehat{f}(\widehat{s})\in F(s)_{\rm pr};\ \widehat{f}(0)\neq 0\}}_{=:A}\\ &=\underbrace{\{\widehat{f_0}+\widehat{s}\widehat{g};\ 0\neq\widehat{f_0}\in F,\ \widehat{g}\in F(s)_{\rm pr}\}}_{=:B}.\end{aligned} \tag{10.52}$$

3. *For matrices in $F(s)_{\rm pr}^{m\times m}$, items 1 and 2 imply*

$$\begin{aligned}&F(s)_{\rm pr}^{m\times m}=F^{m\times m}\oplus F(s)_{\rm spr}^{m\times m},\ F(s)_{\rm spr}^{m\times m}=F(s)_{\rm pr}^{m\times m}\widehat{s} \text{ and}\\ &\mathrm{Gl}_m(F(s)_{\rm pr})=\left\{\widehat{P_2}\in F(s)_{\rm pr}^{m\times m};\ \widehat{P_2}(0)\in\mathrm{Gl}_m(F),\ \text{i.e., } \det(\widehat{P_2}(0))\neq 0\right\}\\ &\qquad=\left\{A+\widehat{s}B;\ A\in\mathrm{Gl}_m(F),\ B\in F(s)_{\rm pr}^{m\times m}\right\}.\end{aligned} \tag{10.53}$$

In particular, if F is infinite, almost all $A\in F^{m\times m}$ and $\widehat{P_2}\in F(s)_{\rm pr}^{m\times m}$ are invertible, compare Definition 10.1.8 and Lemma 10.1.9.

Proof 1. It suffices to notice $\widehat{a}-\widehat{a}(0)\,\widehat{b}(0)^{-1}\widehat{b}\in\widehat{\mathcal{D}},\ \left(\widehat{a}-\widehat{a}(0)\,\widehat{b}(0)^{-1}\widehat{b}\right)(0)=0$.
2. $\mathrm{U}(F(s)_{\rm pr})\subseteq A$: If $\widehat{f}\in\mathrm{U}(F(s)_{\rm pr})$, then $\widehat{f}\widehat{f^{-1}}=1$. The algebra homomorphism (10.52) furnishes $\widehat{f}(0)\widehat{f^{-1}}(0)=1$, hence $\widehat{f}(0)\neq 0$.
$A\subseteq\mathrm{U}(F(s)_{\rm pr})$: The condition $\widehat{f}=\frac{\widehat{a}}{\widehat{b}}\in A,\ \widehat{f}(0)=\widehat{a}(0)\widehat{b}(0)^{-1}\neq 0$ implies $\widehat{a}(0)\neq 0$ and $\widehat{f^{-1}}=\frac{\widehat{b}}{\widehat{a}}\in F(s)_{\rm pr}$, hence $\widehat{f}\in\mathrm{U}\left(F(s)_{\rm pr}\right)$.
$A\subseteq B$: $\widehat{f}\in A\Longrightarrow\widehat{f}(0)\neq 0\underset{(10.51)}{\Longrightarrow}\widehat{f}=\widehat{f}(0)+\widehat{s}\widehat{g}\in B$.
$B\subseteq A$: obvious.
3. This follows directly from 1 and 2, the fact that a square matrix is invertible if and only if its determinant is a unit, and from $\det(\widehat{P_2})(0)=\det(\widehat{P_2}(0))$. This equality holds since $\widehat{f}\mapsto\widehat{f}(0)$ is an algebra homomorphism. □

As a direct consequence of the preceding lemma for the ring of proper and T-stable rational functions $\mathcal{S}$, we obtain the following corollary.

Corollary 10.1.31 *There is the direct decomposition*

$$\mathcal{S}=F\oplus\mathcal{S}\widehat{s} \text{ with } \mathcal{S}\widehat{s}=\ker(\widehat{\nu})\cap\mathcal{S}. \tag{10.54}$$

The decomposition of $\widehat{f}=\frac{\widehat{a}}{\widehat{t}}\in\mathcal{S}$ with $\widehat{a}\in\widehat{\mathcal{D}}$ and $\widehat{t}\in\widehat{T}\subset\widehat{\mathcal{D}}$ is given by

$$\begin{aligned}&\widehat{f}=\widehat{f}(0)+(\widehat{f}-\widehat{f}(0)) \text{ with } \widehat{f}-\widehat{f}(0)=\tfrac{\widehat{a}-\widehat{a}(0)\,\widehat{t}(0)^{-1}\widehat{t}}{\widehat{t}}\in\mathcal{S}\widehat{s} \text{ and}\\ &\widehat{a}-\widehat{a}(0)\,\widehat{t}(0)^{-1}\widehat{t}=\widehat{s}\widehat{c},\ \widehat{c}\in\widehat{\mathcal{D}}\Longrightarrow\widehat{f}=\widehat{f}(0)+\widehat{s}\widehat{g},\ \widehat{g}:=\tfrac{\widehat{c}}{\widehat{t}}\in\mathcal{S}.\end{aligned} \tag{10.55}$$

The decomposition (10.54) *induces the decomposition for $m \times p$-matrices*

$$\mathcal{S}^{m\times p} = F^{m\times p} \oplus \widehat{s}\mathcal{S}^{m\times p} \ni X = X_0 + \widehat{s}Y \text{ with } X_0 \in F^{m\times p},\ Y \in \mathcal{S}^{m\times p}. \quad (10.56)$$

◊

The proper rational functions are often characterized as those rational functions which do not have "poles at infinity". In the following remark we explain this terminology.

Remark 10.1.32 We consider F as an affine line and extend it to a projective line $\overline{F} := F \uplus \{\infty\}$ by a point at infinity. The addition and the multiplication are extended by

$$\alpha + \infty := \infty \text{ for all } \alpha \in F \text{ and } \alpha \cdot \infty := \tfrac{\alpha}{0} := \infty \text{ for all } \alpha \in F \setminus \{0\}.$$

Let

$$\widehat{f} = \tfrac{\widehat{a}}{\widehat{b}} \in F(s) = F(\widehat{s}) \quad \text{with} \quad \widehat{a}, \widehat{b} \in F[\widehat{s}] \text{ and } \gcd(\widehat{a}, \widehat{b}) = 1$$

be any nonzero rational function in coprime representation. Then 0 is not a common zero of $\widehat{a}$ and $\widehat{b}$, and therefore the map

$$\widehat{\nu}\colon\ F(\widehat{s}) \longrightarrow \overline{F},\ \widehat{f} = \tfrac{\widehat{a}}{\widehat{b}} \longmapsto \widehat{f}(0) := \tfrac{\widehat{a}(0)}{\widehat{b}(0)}$$

is well defined and obviously an extension of $\widehat{\nu}$ of (10.50).

If a nonzero $f \in F(s)$ is written as quotient of polynomials in s with nonzero leading coefficients a_m and b_n, one obtains

$$f = \tfrac{a}{b} = \tfrac{a_m s^m + \cdots}{b_n s^n + \cdots} = \tfrac{a_m (s-\alpha)^m + \cdots}{b_n (s-\alpha)^n + \cdots} = \tfrac{a_m}{b_n} \widehat{s}^{\,n-m} \tfrac{1+\cdots}{1+\cdots}$$

hence

$$f(\infty) := \widehat{\nu}(f) = \begin{cases} \tfrac{a_m}{b_m} & \text{if } m = \deg_s(a) = \deg_s(b) = n, \text{ i.e., } f \in \mathrm{U}(F(s)_{\mathrm{pr}}), \\ 0 & \text{if } \deg_s(a) < \deg_s(b), \text{ i.e., } f \in F(s)_{\mathrm{spr}}, \\ \infty & \text{if } \deg_s(a) > \deg_s(b), \text{ i.e., } f \notin F(s)_{\mathrm{pr}}. \end{cases}$$

The notation $f(\infty)$ is motivated by $\widehat{s} = \frac{1}{s-\alpha}$, i.e., $s = \alpha + \frac{1}{\widehat{s}}$ and $\infty = \alpha + \frac{1}{0}$. For $F = \mathbb{R}$ or $F = \mathbb{C}$ we get

$$f(\infty) = \widehat{\nu}(f) = \lim_{s\to\infty} f(s). \quad (10.57)$$

In engineering books properness of a rational function $f \in F(s)$ for $F = \mathbb{R}$ or $F = \mathbb{C}$ is often *defined* by the condition $\lim_{s\to\infty} f(s) \in F$, cf. [13, p. 382]. ◊

We will need the following technical preparation.

Lemma 10.1.33 *Let* $\binom{B}{C} \in F^{(p+m)\times m}$ *be a matrix of rank m. Then there is a matrix* $A \in F^{m\times p}$ *such that* $\det(C + AB) \neq 0$. *In other words, the polynomial* $f(X) := \det(C - XB) \in F[X]$ *with an* $m \times p$*-matrix X of indeterminates is nonzero since* $f(-A) \neq 0$. *If F is an infinite field, this implies that* almost all *matrices* $X_0 \in F^{m\times p}$ *satisfy* $\det(C - X_0B) \neq 0$, *see Definition 10.1.8.*

Proof Let $r := \operatorname{rank}(C)$ and $C_{II} \in F^{r\times m}$ an $r \times m$-submatrix of C of rank r, hence $F^{1\times r}C_{II} = F^{1\times m}C$. Let $C_I \in F^{(m-r)\times m}$ be the submatrix of C of the complementary rows. There is a permutation matrix $P \in \mathrm{Gl}_m(F)$ such that $P\binom{C_I}{C_{II}} = C$. From $\operatorname{rank}\binom{B}{C} = m$ we infer

$$F^{1\times m} = F^{1\times(p+m)}\binom{B}{C} = F^{1\times p}B + F^{1\times m}C = F^{1\times p}B + F^{1\times r}C_{II}.$$

Hence there is a submatrix $B_{II} \in F^{(m-r)\times m}$ of B such that the rows of $\binom{B_{II}}{C_{II}} \in F^{m\times m}$ are a basis of $F^{1\times m}$ or $\binom{B_{II}}{C_{II}} \in \mathrm{Gl}_m(F)$. Again we consider $B_I \in F^{(p-(m-r))\times m}$ and a permutation matrix $Q \in \mathrm{Gl}_p(F)$ such that $Q\binom{B_I}{B_{II}} = B$. Since $\operatorname{rank}(C_{II}) = \operatorname{rank}(C) = r$ there is a matrix $E \in F^{(m-r)\times r}$ such that $C_I = EC_{II}$. This implies $R := \begin{pmatrix} \mathrm{id}_{m-r} & E \\ 0 & \mathrm{id}_r \end{pmatrix} \in \mathrm{Gl}_m(F) = \mathrm{Gl}_{(m-r)+r}(F)$ and

$$\begin{aligned}
R\binom{B_{II}}{C_{II}} &= \begin{pmatrix} \mathrm{id}_{m-r} & E \\ 0 & \mathrm{id}_r \end{pmatrix}\binom{B_{II}}{C_{II}} = \binom{B_{II}+EC_{II}}{C_{II}} \\
&= \binom{C_I}{C_{II}} + \binom{B_{II}}{0} = \binom{C_I}{C_{II}} + \begin{pmatrix} 0 & \mathrm{id}_{m-r} \\ 0 & 0 \end{pmatrix}\binom{B_I}{B_{II}} \\
\Longrightarrow PR\binom{B_{II}}{C_{II}} &= P\binom{C_I}{C_{II}} + P\begin{pmatrix} 0 & \mathrm{id}_{m-r} \\ 0 & 0 \end{pmatrix}Q^{-1}Q\binom{B_I}{B_{II}} \\
&= C + AB,\ A := P\begin{pmatrix} 0 & \mathrm{id}_{m-r} \\ 0 & 0 \end{pmatrix}Q^{-1} \\
\Longrightarrow C + AB &= PR\binom{B_{II}}{C_{II}} \in \mathrm{Gl}_m(F),\ \det(C + AB) \neq 0.
\end{aligned}$$

□

Theorem 10.1.34 (Proper compensators) *We assume the situation of Theorem 10.1.20, in particular, a T-stabilizing compensator* $\mathcal{B}_2 = U_2^{\perp}$ *with proper closed-loop transfer matrix H. We also need the representation*

$$\begin{aligned}
U_{2,T} &= \mathcal{D}_T^{1\times m}(-\widehat{Q}_2, \widehat{P}_2) \quad \text{with} \\
(-\widehat{Q}_2, \widehat{P}_2) &= (-\widehat{Q}_2^0, \widehat{P}_2^0) + X(\widehat{P}_1, -\widehat{Q}_1) \in \mathcal{D}_T^{m\times(p+m)} \quad \text{and } \det(\widehat{P}_2) \neq 0.
\end{aligned}$$

By Assumption 10.1.18 and Theorem 10.1.20 the properness of H implies $X \in \mathcal{S}^{m\times p}$, $(-\widehat{Q}_2, \widehat{P}_2) \in \mathcal{S}^{m\times(p+m)}$ *and hence, by (10.56),*

$$X = X_0 + \widehat{s}Y \in \mathcal{S}^{m\times p} \subseteq F(s)_{\mathrm{pr}}^{m\times p},\ X_0 = X(0) \in F^{m\times p} \text{ and } Y \in \mathcal{S}^{m\times p}.$$

The entries of the matrices $(-\widehat{Q}_2^0, \widehat{P}_2^0)$ *and* $(\widehat{P}_1, -\widehat{Q}_1)$ *lie in* $\widehat{\mathcal{D}} = F[\widehat{s}] \subseteq F(s)_{\mathrm{pr}}$. *Hence all these matrices can be evaluated at* $\widehat{s} = 0$ *via* $\widehat{\nu}$ *from (10.50).*

Then $\mathcal{B}_2$ *is proper, i.e.,* $H_2 = \widehat{P}_2^{-1}\widehat{Q}_2$ *is proper, if and only if*

$$\det\left(\widehat{P}_2^0(0) - X_0\,\widehat{Q}_1(0)\right) \neq 0. \tag{10.58}$$

If F is infinite, for instance, in the standard cases $F = \mathbb{R}$ or $F = \mathbb{C}$, this condition holds for almost all X_0.

Proof (i) By (10.53), we have that

$$\widehat{P}_2 \in \mathrm{Gl}_m(F(s)_{\mathrm{pr}}) \Longleftrightarrow 0 \neq \det(\widehat{P}_2(0)) = \det\left(\widehat{P}_2^0(0) - X(0)\,\widehat{Q}_1(0)\right),$$

i.e., we just have to show that $H_2 \in F(s)_{\mathrm{pr}}^{m\times p} \Longleftrightarrow \widehat{P}_2 \in \mathrm{Gl}_m(F(s)_{\mathrm{pr}})$ holds.
$\Longleftarrow$: $\widehat{Q}_2 \in \mathcal{S}^{m\times p} \subseteq F(s)_{\mathrm{pr}}^{m\times p}$, $\widehat{P}_2 \in \mathrm{Gl}_m(F(s)_{\mathrm{pr}}) \Longrightarrow H_2 = \widehat{P}_2^{-1} Q_2 \in F(s)_{\mathrm{pr}}^{m\times p}$.
$\Longrightarrow$: By (10.1.7) the matrix $(-\widehat{Q}_2, \widehat{P}_2)$ has the right inverse $\begin{pmatrix} \widehat{N}_1 \\ \widehat{D}_1 \end{pmatrix} \in \widehat{\mathcal{D}}^{(p+m)\times m} \subseteq F(s)_{\mathrm{pr}}^{(p+m)\times m}$ and hence

$$\mathrm{id}_m = (-\widehat{Q}_2, \widehat{P}_2)\begin{pmatrix} \widehat{N}_1 \\ \widehat{D}_1 \end{pmatrix} = \widehat{P}_2 Z,\ Z := (-H_2, \mathrm{id}_m)\begin{pmatrix} \widehat{N}_1 \\ \widehat{D}_1 \end{pmatrix}.$$

By assumption H_2 is proper and so is Z. Thus $\widehat{P}_2 \in \mathcal{S}^{m\times m} \subseteq F(s)_{\mathrm{pr}}^{m\times m}$ has the proper right inverse Z. Since $\widehat{P}_2$ is square this implies

$$\widehat{P}_2 \in \mathrm{Gl}_m(F(s)_{\mathrm{pr}}) \text{ or } \det(\widehat{P}_2(0)) = \det\left(\widehat{Q}_2^0(0) - X(0)\widehat{P}_2(0)\right) \neq 0.$$

(ii) It remains to show that (10.58) holds for almost all X_0. By construction

$$\begin{pmatrix} \widehat{P}_1 & -\widehat{Q}_1 \\ -\widehat{Q}_2^0 & \widehat{P}_2^0 \end{pmatrix} \in \mathrm{Gl}_{p+m}(\widehat{\mathcal{D}}) \subset \mathrm{Gl}_m(F(s)_{\mathrm{pr}}).$$

This implies $\begin{pmatrix} \widehat{P}_1(0) & -\widehat{Q}_1(0) \\ -\widehat{Q}_2^0(0) & \widehat{P}_2^0(0) \end{pmatrix} \in \mathrm{Gl}_{p+m}(F)$ and thus $\operatorname{rank}\begin{pmatrix} \widehat{Q}_1(0) \\ \widehat{P}_2^0(0) \end{pmatrix} = m$. By Lemma 10.1.33 the polynomial $\det\left(\widehat{P}_2^0(0) - X'\,\widehat{Q}_1(0)\right)$ with an $m \times p$-matrix X' of indeterminates is nonzero. If F is infinite, this implies that

$$\det\left(\widehat{P}_2(0)\right) = \det\left(\widehat{P}_2^0(0) - X_0\,\widehat{Q}_1(0)\right) \neq 0 \text{ and } \widehat{P}_2 \in \mathrm{Gl}_m(F(s)_{\mathrm{pr}})$$

for almost all $X_0 \in F^{m\times p}$ and the ensuing $\widehat{P}_2$. □

Remark 10.1.35 Theorem 10.1.34 contains an algorithm for constructing all compensators $\mathcal{B}_2$ of a T-stabilizable IO behavior $\mathcal{B}_1$ for which both the feedback (closed-loop) behavior and the compensator have proper transfer matrices. The properness of $\mathcal{B}_1$ is not assumed. ◇

Example 10.1.36 We use again the setting of Example 10.1.22. The given plant is

$$\mathcal{B}_1 = \left(\mathcal{D}(P_1, -Q_1)\right)^{\perp} \quad \text{with } P_1 := e^1(s-3)(s-4) \text{ and } Q_1 := e^1(s-5),$$

where $e^1 = (s+1)(s+2)$. Recall that

$$\widehat{Q}_1 = (6\widehat{s} - 1)\widehat{s}, \quad \text{therefore} \quad \widehat{Q}_1(0) = 0, \quad \text{and}$$
$$\widehat{P}_2^0 = -102\widehat{s} - 1, \quad \text{therefore} \quad \widehat{P}_2^0(0) = -1.$$

Hence the condition $\widehat{P}_2^0(0) - X_0\widehat{Q}_1(0) \neq 0$ holds for all $X_0 \in F$ and hence for all $X \in \mathcal{S}$, and this means that all T-stabilizing compensators which induce a proper feedback system are automatically proper themselves. We characterize this phenomenon in the following theorem. ◊

Theorem 10.1.37 (Cf. [4, Lemma 25, Proposition 27, on pp. 114–115]) *Assume the data from Theorem 10.1.34. By construction the compensator $\mathcal{B}_2$ is proper, i.e., $\widehat{P}_2(0) \in \mathrm{Gl}_m(F)$. Likewise H_1 is proper if and only if $\widehat{P}_1(0) \in \mathrm{Gl}_p(F)$, see part (i) of the proof of Theorem 10.1.34. The following assertions hold:*

1. *There is a strictly proper compensator $\mathcal{B}_2$ if and only if $\mathcal{B}_1$ is proper. If this holds strictly proper compensators arise with the parameters*

$$X = X_0 + \widehat{s}Y, \; X_0 := \widehat{Q}_2^0(0)\widehat{P}_1(0)^{-1}, \; Y \in \mathcal{S}^{m\times p}. \tag{10.59}$$

2. *If $\mathcal{B}_1$ is strictly proper, then* all *matrices $X \in \mathcal{S}^{m\times p}$*

   and the ensuing matrices $(-\widehat{Q}_2, \widehat{P}_2) = (-\widehat{Q}_2^0, \widehat{P}_2^0) + X(\widehat{P}_1, -\widehat{Q}_1)$ according to Theorem 10.1.20 give rise to proper compensators.

Proof 1. The proper compensator $\mathcal{B}_2$ is strictly proper if and only if

$$H_2(0) = \widehat{P}_2(0)^{-1}\widehat{Q}_2(0) = 0 \Longleftrightarrow \widehat{Q}_2^0(0) - X_0\widehat{P}_1(0) = \widehat{Q}_2(0) = 0. \tag{10.60}$$

(i) If $\mathcal{B}_1$ is proper, and hence $\widehat{P}_1(0) \in \mathrm{Gl}_p(F)$ the equivalent equations (10.59) and (10.60) give rise to strictly proper compensators.
(ii) If $\mathcal{B}_2$ is a strictly proper compensator and thus $\widehat{Q}_2(0) = 0$ the invertible matrix

$$\widehat{P}(0) = \begin{pmatrix} \widehat{P}_1(0) & -\widehat{Q}_1(0) \\ -\widehat{Q}_2(0) & \widehat{P}_2(0) \end{pmatrix} = \begin{pmatrix} \widehat{P}_1(0) & -\widehat{Q}_1(0) \\ 0 & \widehat{P}_2(0) \end{pmatrix}$$

implies $\widehat{P}_1(0) \in \mathrm{Gl}_p(F)$, and hence that $\mathcal{B}_1$ is proper.

2. If $H_1 = \widehat{P}_1^{-1}\widehat{Q}_1$ is strictly proper then $\widehat{Q}_1(0) = \widehat{P}_1(0)H_1(0) = \widehat{P}_1(0)0 = 0$. By construction

$$\widehat{P}^0 := \begin{pmatrix} \widehat{P}_1 & -\widehat{Q}_1 \\ -\widehat{Q}_2^0 & \widehat{P}_2^0 \end{pmatrix} \in \mathrm{Gl}_{p+m}(\widehat{\mathcal{D}}) \subset \mathrm{Gl}_{p+m}(F(s)_{\mathrm{pr}})$$
$$\Longrightarrow \widehat{P}^0(0) = \begin{pmatrix} \widehat{P}_1(0) & -\widehat{Q}_1(0) \\ -\widehat{Q}_2^0(0) & \widehat{P}_2^0(0) \end{pmatrix} = \begin{pmatrix} \widehat{P}_1(0) & 0 \\ -\widehat{Q}_2^0(0) & \widehat{P}_2^0(0) \end{pmatrix} \in \mathrm{Gl}_{p+m}(F)$$
$$\Longrightarrow \widehat{P}_2^0(0) \in \mathrm{Gl}_m(F)$$
$$\Longrightarrow \forall X \in \mathcal{S}^{m\times p}: \; \widehat{P}_2(0) = \widehat{P}_2^0(0) - X(0)\,\widehat{Q}_1(0) = \widehat{P}_2^0(0) \in \mathrm{Gl}_m(F)$$
$$\Longrightarrow \forall X \in \mathcal{S}^{m\times p}: \; H_2 \in F(s)_{\mathrm{pr}}^{m\times p}.$$

□

Example 10.1.38 In the situation of Examples 10.1.22 and 10.1.36, the plant's transfer matrix $H_1 = \frac{s-5}{(s-3)(s-4)}$ is strictly proper. Hence all compensators which produce a proper feedback system are proper themselves. The only parameters $X \in \mathcal{S}$ which induce a strictly proper compensator in this situation are those with $X_0 = X(0) = \frac{\widehat{Q}_2^0(0)}{\widehat{P}_1(0)} = \frac{0}{-1} = 0$.

For a nonproper compensator which T-stabilizes the plant via a proper feedback system, we have to start with a plant which is not strictly proper. For example, let us exchange the roles of P_1 and Q_1 in the plant of Example 10.1.22, i.e., we consider the plant

$$\mathcal{B}_1 = (\mathcal{D}(P_1, -Q_1))^{\perp} \quad \text{with } P_1 := e^1(s-5) \text{ and } Q_1 := e^1(s-3)(s-4)$$

and the nonproper transfer matrix $H_1 = \frac{(s-3)(s-4)}{s-5}$. All the computations of Example 10.1.22 are being conducted with the entries of the matrices swapped and the result is the parametrization

$$\begin{aligned}(-\widehat{Q}_2, \widehat{P}_2) &= (-\widehat{Q}_2^0, \widehat{P}_2^0) + X(\widehat{P}_1, -\widehat{Q}_1) \\ &= (102\widehat{s} + 1, 340\widehat{s} - 93) + X((6\widehat{s} - 1)\widehat{s}, (5\widehat{s} - 1)(4\widehat{s} - 1))\end{aligned}$$

where $X \in \mathcal{S}$. All X with $X_0 = X(0) = \frac{\widehat{P}_2^0(0)}{\widehat{Q}_1(0)} = \frac{-93}{-1} = -93$ lead to nonproper compensators. In fact, for $X = 93$, we obtain the nonproper transfer matrix

$$H_2 = -\frac{558\widehat{s}^2 + 9\widehat{s} + 1}{1860\widehat{s}^2 - 497\widehat{s}} = \frac{s^2 + 11s + 568}{497s - 1363}.$$

Its controllable realization

$$\mathcal{B}_2 = (\mathcal{D}(-Q_2, P_2))^{\perp} \quad \text{with } P_2 = 497s - 1363 \text{ and } Q_2 = s^2 + 11s + 568$$

leads to a feedback system with $\det(P) = (s+1)^4 \in T$ and T-stable and proper transfer matrix

$$H = P^{-1}Q = \frac{-1}{(s+1)^4}\begin{pmatrix} (s-3)(s-4)(s^2+11s+568) & (s-3)(s-4)(497s-1363) \\ (s-5)(s^2+11s+568) & (s-3)(s-4)(s^2+11s+568) \end{pmatrix}.$$

◇

10.1.5 *Stabilization of State Space Behaviors*

In this section we reformulate Theorem 10.1.34 *for state space equations. In Theorem* 10.1.39 *we construct a state space realization of the feedback behavior from the given plant in state space form and a state space realization of the compensator from Theorem* 10.1.34.

We consider state space equations

$$\begin{aligned} &s \circ x_1 = A_1x_1 + B_1u_1,\ y_1 = C_1x_1 + D_1u_1,\\ &\text{where } x_1 \in \mathcal{F}^{n_1},\ u_1 \in \mathcal{F}^m,\ y_1 \in \mathcal{F}^p,\ \text{and } A_1, B_1, C_1, D_1 \in F^{\bullet\times\bullet}, \end{aligned} \tag{10.61}$$

together with the IO behaviors

$$\begin{aligned} \mathcal{B}_1^{\mathrm{s}} &:= \left\{\left(\begin{smallmatrix} x_1\\ u_1\end{smallmatrix}\right) \in \mathcal{F}^{n_1+m};\ (s\,\mathrm{id}_{n_1} - A_1) \circ x_1 = B_1u_1\right\} \quad \text{and}\\ \mathcal{B}_1 &:= \left(\begin{smallmatrix} C_1 & D_1\\ 0 & \mathrm{id}_m\end{smallmatrix}\right) \circ \mathcal{B}_1^{\mathrm{s}} = \left\{\left(\begin{smallmatrix} y_1\\ u_1\end{smallmatrix}\right) \in \mathcal{F}^{p+m};\ P_1 \circ y_1 = Q_1 \circ u_1\right\}, \end{aligned}$$

and their transfer matrices

$$\begin{aligned} H_1^{\mathrm{s}} &= (s\,\mathrm{id}_{n_1} - A_1)^{-1}B_1 \in F(s)_{\mathrm{spr}}^{n_1\times m} \quad \text{and}\\ H_1 &= P_1^{-1}Q_1 = D_1 + C_1(s\,\mathrm{id}_{n_1} - A_1)^{-1}B_1 \in F(s)_{\mathrm{pr}}^{p\times m}. \end{aligned}$$

The transfer matrix H_1 is proper since $D_1 \in F^{p\times m}$. By evaluating H_1 at infinity in the sense of Remark 10.1.32, we obtain

$$\begin{aligned} &H_1(s = \infty) = H_1(\widehat{s} = 0) := \widehat{\nu}(H_1) = D_1, \quad \text{where}\\ &\widehat{\nu}\colon F(s)_{\mathrm{pr}} \longrightarrow F,\quad f(s) = \widehat{f}(\widehat{s}) \longmapsto \widehat{f}(0) =: f(\widehat{s} = 0) =: f(\infty) =: f(s = \infty). \end{aligned}$$

We assume that the Kalman system (10.61) is *T-observable (-detectable)*, i.e., that

$$\left(\begin{smallmatrix} C_1 & D_1\\ 0 & \mathrm{id}_m\end{smallmatrix}\right) \circ\colon\ \mathcal{B}_{1,T}^{\mathrm{s}} \xrightarrow{\cong} \mathcal{B}_{1,T} \text{ and } C_1\circ\colon\ \mathcal{B}_{1,T}^{\mathrm{s},0} \xrightarrow{\cong} \mathcal{B}_{1,T}^{0} \tag{10.62}$$

are isomorphisms or, equivalently, if the n_1-th elementary divisor of $\left(\begin{smallmatrix} s\,\mathrm{id}_{n_1} - A\\ C\end{smallmatrix}\right)$ belongs to T. In particular, this holds if the Eq. (10.61) are observable, for instance, if they are the essentially unique observable state space realization of the IO behavior $\mathcal{B}_1$. See Chap. 11 for a full discussion of T-observability and T-observers.

We further assume that $\mathcal{B}_1$ is T-stabilizable, i.e., that $\mathcal{B}_{1,T}^{s} \cong \mathcal{B}_{1,T}$ is controllable or that $(s\,\mathrm{id}_{n_1} - A_1, -B_1)$ and $(P_1, -Q_1)$ have right inverses over $\mathcal{D}_T$. This is the case if and only if the highest elementary divisors of these matrices belong to T. For the standard cases with $F = \mathbb{R}$ or $F = \mathbb{C}$, $\Lambda_1 \subset \mathbb{C}$ and $T = \{t \in F[s];\ \mathrm{V}_{\mathbb{C}}(t) \subseteq \Lambda_1\}$, this is also equivalent to

$$\mathrm{char}(\mathcal{B}_1^{s}) = \left\{\lambda \in \mathrm{spec}(A_1);\ \mathrm{rank}(\lambda\,\mathrm{id}_{n_1} - A_1, -B_1) < n_1\right\} \subseteq \Lambda_1. \tag{10.63}$$

Theorems 10.1.34 and 10.1.37 are applicable to $\mathcal{B}_1$ and describe how to obtain proper or even strictly proper T-stabilizing compensators such that also the feedback system is proper. Let

$$\mathcal{B}_2 = \left\{\left(\begin{smallmatrix} u_2\\ y_2\end{smallmatrix}\right) \in \mathcal{F}^{p+m};\ P_2 \circ y_2 = Q_2 \circ u_2\right\},\ (-Q_2, P_2) \in F[s]^{m\times(p+m)}, \tag{10.64}$$

with proper transfer matrix $H_2 = P_2^{-1}Q_2 \in F(s)_{\rm pr}^{m\times p}$ be a controllable T-stabilizing compensator of $\mathcal{B}_1$ according to Theorem 10.1.34. Since $\mathcal{B}_2$ is controllable, there are observable and controllable, in other words, minimal state space equations for $\mathcal{B}_2$, see Theorem 6.3.11 and the algorithms in Theorems 12.3.5 and 12.4.5 below. In Theorem and Definition 6.3.10 we have shown that these realizations are unique up to similarity. The state space equations have the form

$$\begin{aligned} &s \circ x_2 = A_2x_2 + B_2u_2,\ \ y_2 = C_2x_2 + D_2u_2, \\ &\text{where } x_2 \in \mathcal{F}^{n_2},\ u_2 \in \mathcal{F}^p,\ y_2 \in \mathcal{F}^m, \text{ and } A_2, B_2, C_2, D_2 \in F^{\bullet\times\bullet} \end{aligned} \tag{10.65}$$

with the ensuing state behavior and isomorphism

$$\begin{aligned} &\mathcal{B}_2^{\rm s} := \left\{ \begin{pmatrix} u_2 \\ x_2 \end{pmatrix} \in \mathcal{F}^{p+n_2};\ (s\,{\rm id}_{n_2} - A_2)\circ x_2 = B_2u_2 \right\}, \\ &\begin{pmatrix} 0 & {\rm id}_m \\ D_2 & C_2 \end{pmatrix}\circ \colon \mathcal{B}_2^{\rm s} \xrightarrow{\cong} \mathcal{B}_2,\ \begin{pmatrix} u_2 \\ x_2 \end{pmatrix} \mapsto \begin{pmatrix} u_2 \\ C_2x_2 + D_2u_2 \end{pmatrix}. \end{aligned} \tag{10.66}$$

The state space dimension n_2 of this realization is

$$n_2 = \dim_F \left(F[s]^{1\times m} / F[s]^{1\times m} P_2 \right) = \deg(\det(P_2)). \tag{10.67}$$

The transfer matrices of $\mathcal{B}_2^{\rm s}$ resp. $\mathcal{B}_2$ are

$$\begin{aligned} &H_2^{\rm s} = (s\,{\rm id}_{n_2} - A_2)^{-1}B_2 \in F(s)_{\rm spr}^{n_2\times p} \text{ resp.} \\ &H_2 = P_2^{-1}Q_2 = D_2 + C_2(s\,{\rm id}_{n_2} - A_2)^{-1}B_2 \in F(s)_{\rm pr}^{m\times p},\ D_2 = H_2(\infty). \end{aligned}$$

Define $y := \begin{pmatrix} y_1 \\ y_2 \end{pmatrix}$, $u := \begin{pmatrix} u_2 \\ u_1 \end{pmatrix} \in \mathcal{F}^{p+m}$. The feedback behavior and its proper transfer matrix H are

$$\begin{aligned} \mathcal{B} &= \left\{ \begin{pmatrix} y \\ u \end{pmatrix} \in \mathcal{F}^{2(p+m)};\ P_1\circ y_1 = Q_1\circ(u_1+y_2),\ P_2\circ y_2 = Q_2\circ(u_2+y_1) \right\} \\ &= \left\{ \begin{pmatrix} y \\ u \end{pmatrix} \in \mathcal{F}^{2(p+m)};\ P\circ y = Q\circ u \right\} \text{ and} \\ H &= P^{-1}Q = ({\rm id}_{p+m} - G)^{-1}G, \text{ where} \\ P &:= \begin{pmatrix} P_1 & -Q_1 \\ -Q_2 & P_2 \end{pmatrix},\ Q := \begin{pmatrix} 0 & Q_1 \\ Q_2 & 0 \end{pmatrix},\ G := \begin{pmatrix} 0 & H_1 \\ H_2 & 0 \end{pmatrix},\ {\rm id}_{p+m} - G = \begin{pmatrix} {\rm id}_p & -H_1 \\ -H_2 & {\rm id}_m \end{pmatrix}, \end{aligned} \tag{10.68}$$

see Sect. 7.5. From Theorem and Definition 7.5.1 we know that

$$({\rm id}_{p+m} - G)^{-1} = {\rm id}_{p+m} + H = \begin{pmatrix} {\rm id}_p + H_1ZH_2 & H_1Z \\ ZH_2 & Z \end{pmatrix} \text{ with } Z := ({\rm id}_m - H_2H_1)^{-1}. \tag{10.69}$$

By assumption and construction, H_1, H_2, and H are proper and so are ${\rm id}_{p+m} - G$ and $({\rm id}_{p+m} - G)^{-1}$. Hence ${\rm id}_{p+m} - G \in {\rm Gl}_{p+m}(F(s)_{\rm pr})$ and

$$
\begin{aligned}
\mathrm{Gl}_{p+m}(F) \ni \mathrm{id}_{p+m} - G(\infty) &= \begin{pmatrix} \mathrm{id}_p & -H_1(\infty) \\ -H_2(\infty) & \mathrm{id}_m \end{pmatrix} = \begin{pmatrix} \mathrm{id}_p & -D_1 \\ -D_2 & \mathrm{id}_m \end{pmatrix} \\
&= \begin{pmatrix} \mathrm{id}_p & 0 \\ -D_2 & \mathrm{id}_m \end{pmatrix} \begin{pmatrix} \mathrm{id}_p & -D_1 \\ 0 & \mathrm{id}_m - D_2 D_1 \end{pmatrix} \\
\Longrightarrow D_0 &:= \mathrm{id}_m - D_2 D_1 \in \mathrm{Gl}_m(F) \\
\Longrightarrow \mathrm{id}_m - H_2(\infty)H_1(\infty) &= D_0 \in \mathrm{Gl}_m(F) \Longrightarrow \mathrm{id}_m - H_2 H_1 \in \mathrm{Gl}_m(F(s)_{\mathrm{pr}}) \\
\Longrightarrow Z &= (\mathrm{id}_m - H_2 H_1)^{-1} \in \mathrm{Gl}_m(F(s)_{\mathrm{pr}}),\ Z(0) = D_0^{-1} \\
\underset{(10.69)}{\Longrightarrow} \begin{pmatrix} \mathrm{id}_p & -D_1 \\ -D_2 & \mathrm{id}_m \end{pmatrix}^{-1} &= \mathrm{id}_{p+m} + H(\infty) = \begin{pmatrix} \mathrm{id}_p + D_1 D_0^{-1} D_2 & D_1 D_0^{-1} \\ D_0^{-1} D_2 & D_0^{-1} \end{pmatrix}.
\end{aligned}
\tag{10.70}
$$

Next we construct state space equations for the feedback behavior $\mathcal{B} = \mathrm{fb}(\mathcal{B}_1, \mathcal{B}_2)$. Let $x := \begin{pmatrix} x_1 \\ x_2 \end{pmatrix} \in \mathcal{F}^n$, where $n := n_1 + n_2$. We define the matrices

$$
\begin{aligned}
A &:= \begin{pmatrix} A_1 & 0 \\ 0 & A_2 \end{pmatrix} + \begin{pmatrix} 0 & B_1 \\ B_2 & 0 \end{pmatrix} \left(\mathrm{id}_{p+m} + H(\infty)\right) \begin{pmatrix} C_1 & 0 \\ 0 & C_2 \end{pmatrix} \in F^{(n_1+n_2)\times(n_1+n_2)}, \\
B &:= \begin{pmatrix} 0 & B_1 \\ B_2 & 0 \end{pmatrix} \left(\mathrm{id}_{p+m} + H(\infty)\right) \in F^{(n_1+n_2)\times(p+m)}, \\
C &:= \left(\mathrm{id}_{p+m} + H(\infty)\right) \begin{pmatrix} C_1 & 0 \\ 0 & C_2 \end{pmatrix} \in F^{(p+m)\times(n_1+n_2)}, \\
D &:= H(\infty) \in F^{(p+m)\times(p+m)},\ \text{where } \mathrm{id}_{p+m} + H(\infty) = \begin{pmatrix} \mathrm{id}_p & -D_1 \\ -D_2 & \mathrm{id}_m \end{pmatrix}^{-1}.
\end{aligned}
\tag{10.71}
$$

We will show in the following theorem that the associated state space equations

$$
s \circ x = Ax + Bu,\ y = Cx + Du,\ x = \begin{pmatrix} x_1 \\ x_2 \end{pmatrix},\ y = \begin{pmatrix} y_1 \\ y_2 \end{pmatrix},\ u = \begin{pmatrix} u_2 \\ u_1 \end{pmatrix} \tag{10.72}
$$

with the state space behavior $\mathcal{B}^s$ and output map

$$
\begin{pmatrix} C & D \\ 0 & \mathrm{id}_{p+m} \end{pmatrix} \circ:\ \mathcal{B}^s = \left\{ \begin{pmatrix} x \\ u \end{pmatrix} \in \mathcal{F}^{(n_1+n_2)+(p+m)};\ s \circ x = Ax + Bu \right\} \longrightarrow \mathcal{F}^{2(p+m)} \tag{10.73}
$$

are a state space realization of the feedback behavior $\mathcal{B}$.

Theorem 10.1.39 (Cf. [6, (3.78) on p. 575]) *For the data from* (10.72) *and* (10.73) *the equality* $\begin{pmatrix} C & D \\ 0 & \mathrm{id}_{p+m} \end{pmatrix} \circ \mathcal{B}^s = \mathcal{B} = \mathrm{fb}(\mathcal{B}_1, \mathcal{B}_2)$ *holds, i.e., the system* (10.72) *realizes the feedback IO behavior* $\mathcal{B}$ *and* $H = D + C(s\,\mathrm{id}_n - A)^{-1}B$. *The state space system* (10.72) *is* T*-observable, i.e.,* $\begin{pmatrix} C & D \\ 0 & \mathrm{id}_{p+m} \end{pmatrix} \circ:\ \mathcal{B}^s_T \xrightarrow{\cong} \mathcal{B}_T$. *Since* $\mathcal{B}$ *is proper and* T*-stable so is* $\mathcal{B}^s$, *hence* $\det(s\,\mathrm{id}_n - A) \in T$. *For the standard situation* $F = \mathbb{R}$ *or* $F = \mathbb{C}$ *and* $T = \{t \in F[s];\ \mathrm{V}_{\mathbb{C}}(t) \subseteq \Lambda_1\}$, *this means that* $\mathrm{spec}(A) \subseteq \Lambda_1$.

The systems (10.65) *represent all possible proper state space equations that* T*-stabilize the given system* (10.61) *by output feedback. Pole placement according to Sect. 10.1.3 is applicable. The isomorphism* $\begin{pmatrix} 0 & \mathrm{id}_m \\ D_2 & C_2 \end{pmatrix} \circ:\ \mathcal{B}_2^s \xrightarrow{\cong} \mathcal{B}_2$ *can be realized by standard elementary building blocks as we have shown in Example 7.2.2.*

Proof (i) We only prove the inclusion $\begin{pmatrix} C & D \\ 0 & \mathrm{id}_{p+m} \end{pmatrix} \circ \mathcal{B}^s \supseteq \mathcal{B}$, the proof of the converse inclusion is similar, but simpler and omitted.

Let $\begin{pmatrix} y \\ u \end{pmatrix} \in \mathcal{B}$ with $y = \begin{pmatrix} y_1 \\ y_2 \end{pmatrix} \in \mathcal{F}^{p+m}$ and $u = \begin{pmatrix} u_2 \\ u_1 \end{pmatrix} \in \mathcal{F}^{p+m}$. The first two lines of (10.68) imply $\begin{pmatrix} y_1 \\ v_1 \end{pmatrix} := \begin{pmatrix} y_1 \\ u_1 + y_2 \end{pmatrix} \in \mathcal{B}_1$ and $\begin{pmatrix} v_2 \\ y_2 \end{pmatrix} := \begin{pmatrix} u_2 + y_1 \\ y_2 \end{pmatrix} \in \mathcal{B}_2$. Since $\mathcal{B}_1^s$ and $\mathcal{B}_2^s$ are state space realizations of $\mathcal{B}_1$ and $\mathcal{B}_2$, respectively, there are $x_1 \in \mathcal{F}^{n_1}$ and $x_2 \in \mathcal{F}^{n_2}$ with

$$\begin{aligned}
&\begin{pmatrix} x_1 \\ v_1 \end{pmatrix} \in \mathcal{B}_1^{\mathrm{s}},\ \begin{pmatrix} y_1 \\ v_1 \end{pmatrix} = \begin{pmatrix} C_1 & D_1 \\ 0 & \mathrm{id}_m \end{pmatrix} \circ \begin{pmatrix} x_1 \\ v_1 \end{pmatrix},\ \begin{pmatrix} v_2 \\ x_2 \end{pmatrix} \in \mathcal{B}_2^{\mathrm{s}},\ \begin{pmatrix} v_2 \\ y_2 \end{pmatrix} = \begin{pmatrix} 0 & \mathrm{id}_p \\ D_2 & C_2 \end{pmatrix} \circ \begin{pmatrix} v_2 \\ x_2 \end{pmatrix} \\
&\Longrightarrow s \circ x_1 = A_1 x_1 + B_1 v_1,\ y_1 = C_1 x_1 + D_1 v_1, \\
&\quad s \circ x_2 = A_2 x_2 + B_2 v_2,\ y_2 = C_2 x_2 + D_2 v_2, \\
&\quad s \circ \begin{pmatrix} x_1 \\ x_2 \end{pmatrix} = \begin{pmatrix} A_1 & 0 \\ 0 & A_2 \end{pmatrix} \begin{pmatrix} x_1 \\ x_2 \end{pmatrix} + \begin{pmatrix} 0 & B_1 \\ B_2 & 0 \end{pmatrix} \begin{pmatrix} v_2 \\ v_1 \end{pmatrix}.
\end{aligned} \tag{10.74}$$

With $y_1 = v_2 - u_2$ and $y_2 = v_1 - u_1$, the map (10.73) implies

$$\begin{aligned}
&v_2 - u_2 = C_1 x_1 + D_1 v_1,\ v_1 - u_1 = C_2 x_2 + D_2 v_2 \\
&\Longrightarrow \begin{pmatrix} \mathrm{id}_p & -D_1 \\ -D_2 & \mathrm{id}_m \end{pmatrix} \begin{pmatrix} v_2 \\ v_1 \end{pmatrix} = \begin{pmatrix} u_2 \\ u_1 \end{pmatrix} + \begin{pmatrix} C_1 & 0 \\ 0 & C_2 \end{pmatrix} \begin{pmatrix} x_1 \\ x_2 \end{pmatrix} \\
&\Longrightarrow v := \begin{pmatrix} v_2 \\ v_1 \end{pmatrix} = \underbrace{\begin{pmatrix} \mathrm{id}_p & -D_1 \\ -D_2 & \mathrm{id}_m \end{pmatrix}^{-1}}_{=\mathrm{id}_{p+m} + H(\infty)} u + \underbrace{\begin{pmatrix} \mathrm{id}_p & -D_1 \\ -D_2 & \mathrm{id}_m \end{pmatrix}^{-1}}_{=\mathrm{id}_{p+m} + H(\infty)} \begin{pmatrix} C_1 & 0 \\ 0 & C_2 \end{pmatrix} x \\
&\Longrightarrow s \circ x = \left(\begin{pmatrix} A_1 & 0 \\ 0 & A_2 \end{pmatrix} + \begin{pmatrix} 0 & B_1 \\ B_2 & 0 \end{pmatrix} \left(\mathrm{id}_{p+m} + H(\infty)\right) \begin{pmatrix} C_1 & 0 \\ 0 & C_2 \end{pmatrix} \right) x \\
&\quad + \begin{pmatrix} 0 & B_1 \\ B_2 & 0 \end{pmatrix} \left(\mathrm{id}_{p+m} + H(\infty)\right) u = Ax + Bu \Longrightarrow \begin{pmatrix} x \\ u \end{pmatrix} \in \mathcal{B}^{\mathrm{s}}.
\end{aligned} \tag{10.75}$$

The output of (10.72) for $\begin{pmatrix} x \\ u \end{pmatrix} \in \mathcal{B}^s$ is

$$\begin{aligned}
Cx + Du &= \left(\mathrm{id}_{p+m} + H(\infty)\right) \begin{pmatrix} C_1 & 0 \\ 0 & C_2 \end{pmatrix} x + H(\infty) u \\
&= \left(\mathrm{id}_{p+m} + H(\infty)\right) \left(\begin{pmatrix} C_1 & 0 \\ 0 & C_2 \end{pmatrix} x + u \right) - u \underset{(10.75)}{=} v - u = \begin{pmatrix} y_1 \\ y_2 \end{pmatrix}.
\end{aligned}$$

This completes the proof of $\begin{pmatrix} C & D \\ 0 & \mathrm{id}_{p+m} \end{pmatrix} \circ \mathcal{B}^s \supseteq \mathcal{B}$.

(ii) In order to show that the system (10.72) is T-observable (-detectable), we have to prove that the induced epimorphisms of the quotient behaviors

$$\begin{pmatrix} C & D \\ 0 & \mathrm{id}_{p+m} \end{pmatrix} \circ :\ \mathcal{B}_T^s \longrightarrow \mathcal{B}_T \text{ and } C\circ :\ \mathcal{B}_T^{s,0} \longrightarrow \mathcal{B}_T^0 \tag{10.76}$$

are also injective and thus isomorphisms. It suffices to show this for $C\circ$. So let $x = \begin{pmatrix} x_1 \\ x_2 \end{pmatrix} \in \mathcal{F}_T^{n_1+n_2}$ with

$$\begin{aligned}
&s \circ x = Ax = \begin{pmatrix} A_1 & 0 \\ 0 & A_2 \end{pmatrix} x + \begin{pmatrix} 0 & B_1 \\ B_2 & 0 \end{pmatrix} \begin{pmatrix} \mathrm{id}_p & -D_1 \\ -D_2 & \mathrm{id}_m \end{pmatrix}^{-1} \begin{pmatrix} C_1 & 0 \\ 0 & C_2 \end{pmatrix} x \text{ and} \\
&0 = Cx = \begin{pmatrix} \mathrm{id}_p & -D_1 \\ -D_2 & \mathrm{id}_m \end{pmatrix}^{-1} \begin{pmatrix} C_1 & 0 \\ 0 & C_2 \end{pmatrix} x \\
&\Longrightarrow s \circ x_1 = A_1 x_1,\ s \circ x_2 = A_2 x_2,\ C_1 x_1 = 0,\ C_2 x_2 = 0.
\end{aligned}$$

Since $\mathcal{B}_1$ is T-observable by assumption, cf. (10.62), and $\mathcal{B}_2$ is even observable by construction, we infer $x_1 = 0$ and $x_2 = 0$. Hence (10.76) is injective.

(iii) By construction $\mathcal{B}_T$ is T-stable or, equivalently, $\mathcal{B}_T^0 = 0$. The isomorphism (10.76) implies $\mathcal{B}_T^{s,0} \cong \mathcal{B}_T^0 = 0$ and therefore $\mathcal{B}^s$ and the system (10.72) are also T-stable, i.e., $\det\left(s\,\mathrm{id}_{n_1+n_2} - A\right) \in T$. □

Corollary 10.1.40 (Cf. [13, Sect. 7.2, p. 522], [12, Sect. 10.5], [6, p. 575]).

1. *Assume that $\mathcal{B}_2$ is strictly proper, and hence $D_2 = 0$. We have constructed such compensators in Theorem 10.1.37, item 1. Then the matrices A, B, C, and $D = H(\infty)$ from* (10.71) *have the simpler form*

$$
\begin{aligned}
&\mathrm{id}_{p+m} + H(\infty) = \begin{pmatrix} \mathrm{id}_p & -D_1 \\ 0 & \mathrm{id}_m \end{pmatrix}^{-1} = \begin{pmatrix} \mathrm{id}_p & D_1 \\ 0 & \mathrm{id}_m \end{pmatrix}, \ H(\infty) = \begin{pmatrix} 0 & D_1 \\ 0 & 0 \end{pmatrix},\\
&A = \begin{pmatrix} A_1 & 0 \\ 0 & A_2 \end{pmatrix} + \begin{pmatrix} 0 & B_1 \\ B_2 & 0 \end{pmatrix}\begin{pmatrix} \mathrm{id}_p & D_1 \\ 0 & \mathrm{id}_m \end{pmatrix}\begin{pmatrix} C_1 & 0 \\ 0 & C_2 \end{pmatrix} = \begin{pmatrix} A_1 & B_1C_2 \\ B_2C_1 & A_2+B_2D_1C_2 \end{pmatrix},\\
&B = \begin{pmatrix} 0 & B_1 \\ B_2 & 0 \end{pmatrix}\begin{pmatrix} \mathrm{id}_p & D_1 \\ 0 & \mathrm{id}_m \end{pmatrix} = \begin{pmatrix} 0 & B_1 \\ B_2 & B_2D_1 \end{pmatrix},\\
&C = \begin{pmatrix} \mathrm{id}_p & D_1 \\ 0 & \mathrm{id}_m \end{pmatrix}\begin{pmatrix} C_1 & 0 \\ 0 & C_2 \end{pmatrix} = \begin{pmatrix} C_1 & D_1C_2 \\ 0 & C_2 \end{pmatrix}.
\end{aligned}
\tag{10.77}
$$

2. *Assume that $\mathcal{B}_1$ is strictly proper, i.e., $D_1 = 0$. Then the matrices A, B, C, D simplify to*

$$
\begin{aligned}
&\mathrm{id}_{p+m} + H(\infty) = \begin{pmatrix} \mathrm{id}_p & 0 \\ D_2 & \mathrm{id}_m \end{pmatrix}, \ D = H(\infty) = \begin{pmatrix} 0 & 0 \\ D_2 & 0 \end{pmatrix},\\
&A = \begin{pmatrix} A_1+B_1D_2C_1 & B_1C_2 \\ B_2C_1 & A_2 \end{pmatrix}, \ B = \begin{pmatrix} B_1D_2 & B_1 \\ B_2 & 0 \end{pmatrix}, \ C = \begin{pmatrix} C_1 & 0 \\ D_2C_1 & C_2 \end{pmatrix}.
\end{aligned}
\tag{10.78}
$$

   The autonomous part of this closed-loop behavior is treated in [12, Sect. 10.5]. In particular, Theorem 10.1.39 and Eq. (10.78) *solve a variant of the open problem from [12, (10.16)] since the matrices A describe* all *closed-loop state space behaviors with* $\mathrm{spec}(A) \subseteq \Lambda_1$.
3. *If both $\mathcal{B}_1$ and $\mathcal{B}_2$ are strictly proper and thus $D_1 = 0$ and $D_2 = 0$ then* (10.77) *simplifies to*

$$
A = \begin{pmatrix} A_1 & B_1C_2 \\ B_2C_1 & A_2 \end{pmatrix}, \ B = \begin{pmatrix} 0 & B_1 \\ B_2 & 0 \end{pmatrix}, \ C = \begin{pmatrix} C_1 & 0 \\ 0 & C_2 \end{pmatrix}, \ D = \begin{pmatrix} 0 & 0 \\ 0 & 0 \end{pmatrix}. \tag{10.79}
$$

◇

Remark 10.1.41 In the literature, e.g., in [13, (3a), (3b) on p. 523], [12, (10.19), Theorem 10.5.1 on p. 353] and [6, (4.5)–(4.9) on pp. 364–365], state space compensators of state space plants are often constructed by means of *static state feedback* and of Luenberger *state estimators* of the state x_1 where the plant and compensator states have the same dimension. This is only a small share of the compensators from Corollary 10.1.40. We are going to discuss this special case in Sect. 12.5.

In this book we do not address the interesting problem of finding those compensators from Corollary 10.1.40 *with minimal state dimension* $n_2 = \deg_s(\det(P_2))$. ◇

The plant and the compensator in Theorem 10.1.39 are proper, but not strictly proper, whereas the behaviors in our running Example 10.1.22 are strictly proper. Hence we use different behaviors for the following example.

Example 10.1.42 Again, we use the standard setting of the continuous-time case as in Example 10.1.14. We also apply the algorithmic canonical observability state space realization from Sect. 12.3. We consider the SISO plant

$$\mathcal{B}_1 = \left(\mathcal{D}(P_1, -Q_1)\right)^{\perp} \text{ with } \begin{aligned} P_1 &:= (s-3)(s+1)(s+2)(s+3) \text{ and} \\ Q_1 &:= (s-2)(s+1)(s+2)^2, \end{aligned}$$

with proper transfer matrix $H_1 = \frac{s^2-4}{s^2-9}$. The plant is T-stabilizable because $\gcd(P_1, Q_1) = (s+1)(s+2) \in T$. With the method from Theorem 10.1.20 and $\alpha := -1$, we obtain

$$(-\widehat{Q}_2, \widehat{P}_2) = \left(\tfrac{16}{5}\widehat{s} + \tfrac{12}{5}, \tfrac{6}{5}\widehat{s} - \tfrac{7}{5}\right) + X\left(-8\widehat{s}^2 - 2\widehat{s} + 1, 3\widehat{s}^2 + 2\widehat{s} - 1\right),$$

where the parameter X is a proper and T-stable rational function. By choosing $X = \frac{1}{5}$, we get

$$(-\widehat{Q}_2, \widehat{P}_2) = \tfrac{1}{5}\left(-8\widehat{s}^2 + 14\widehat{s} + 13, 3\widehat{s}^2 - 4\widehat{s} - 8\right),$$

hence the (proper) transfer matrix of the compensator is

$$H_2 = \frac{\frac{8}{5}\widehat{s}^2 - \frac{14}{5}\widehat{s} - \frac{13}{5}}{\frac{3}{5}\widehat{s}^2 - \frac{4}{5}\widehat{s} - \frac{8}{5}} = \frac{13s^2+40s+19}{8s^2+20s+9}$$

and the corresponding controllable compensator is $\mathcal{B}_2 = \left(\mathcal{D}(-Q_2, P_2)\right)^{\perp}$ with

$$(-Q_2, P_2) = (-13s^2 - 40s - 19, 8s^2 + 20s + 9).$$

Next we compute state space realizations of the plant, the compensator, and the feedback system. With the method from Theorem and Definition 12.3.5 and Example 12.3.10, we compute the observability realization

$$A_1^{\mathrm{ob}} = \begin{pmatrix} 0 & 1 & 0 & 0 \\ 0 & 0 & 1 & 0 \\ 0 & 0 & 0 & 1 \\ 18 & 27 & 7 & -3 \end{pmatrix}, \quad B_1^{\mathrm{ob}} = \begin{pmatrix} 0 \\ 5 \\ 0 \\ 45 \end{pmatrix}, \quad C_1^{\mathrm{ob}} = (1\,0\,0\,0), \quad D_1^{\mathrm{ob}} = (1)$$

of the plant. The observability realization is always observable, and therefore in particular T-observable, which is a prerequisite for Theorem 10.1.39. The observability realization of the compensator is

$$A_2^{\mathrm{ob}} = \begin{pmatrix} 0 & 1 \\ -\frac{9}{8} & -\frac{5}{2} \end{pmatrix}, \quad B_2^{\mathrm{ob}} = \begin{pmatrix} \frac{15}{16} \\ -\frac{115}{64} \end{pmatrix}, \quad C_2^{\mathrm{ob}} = (1, 0), \quad D_2^{\mathrm{ob}} = \left(\tfrac{13}{8}\right).$$

Since the compensator is controllable, so is its observability realization (Theorem and Definition 12.3.5, item 4). The compensator's observability realization is observable by construction; hence, it is the unique (up to similarity) state space realization of $\mathcal{B}_2$.

The matrices from (10.70) are

$$D_0 = 1 - D_2^{\mathrm{ob}} D_1^{\mathrm{ob}} = 1 - \tfrac{13}{8} \cdot 1 = -\tfrac{5}{8} \quad \text{and}$$
$$\mathrm{id}_{p+m} + H(\infty) = \begin{pmatrix} \mathrm{id}_p + D_1^{\mathrm{ob}} D_0^{-1} D_2^{\mathrm{ob}} & D_1^{\mathrm{ob}} D_0^{-1} \\ D_0^{-1} D_2^{\mathrm{ob}} & D_0^{-1} \end{pmatrix} = \begin{pmatrix} -\frac{8}{5} & -\frac{8}{5} \\ -\frac{13}{5} & -\frac{8}{5} \end{pmatrix}.$$

With this, we can compute the matrices

$$A = \begin{pmatrix} 0 & 1 & 0 & 0 & 0 & 0 \\ -13 & 0 & 1 & 0 & -8 & 0 \\ 0 & 0 & 0 & 1 & 0 & 0 \\ -99 & 27 & 7 & -3 & -72 & 0 \\ -\frac{3}{2} & 0 & 0 & 0 & -\frac{3}{2} & 1 \\ \frac{23}{8} & 0 & 0 & 0 & \frac{7}{4} & -\frac{5}{2} \end{pmatrix}, \quad B = \begin{pmatrix} 0 & 0 \\ -13 & -8 \\ 0 & 0 \\ -117 & -72 \\ \frac{-3}{2} & -\frac{3}{2} \\ \frac{23}{8} & \frac{23}{8} \end{pmatrix},$$
$$C = \begin{pmatrix} -\frac{8}{5} & 0 & 0 & 0 & -\frac{8}{5} & 0 \\ -\frac{13}{5} & 0 & 0 & 0 & -\frac{8}{5} & 0 \end{pmatrix}, \quad \text{and} \quad D = \begin{pmatrix} -\frac{13}{5} & -\frac{8}{5} \\ -\frac{13}{5} & -\frac{13}{5} \end{pmatrix}$$

from (10.71), which constitute a state space realization of the feedback behavior by Theorem 10.1.39. This can be verified by computing the image

$$\begin{pmatrix} C & D \\ 0 & \mathrm{id}_2 \end{pmatrix} \circ \left\{ \begin{pmatrix} x \\ u \end{pmatrix} \in \mathcal{F}^{6+2};\ (s\,\mathrm{id}_6 - A, -B) \circ \begin{pmatrix} x \\ u \end{pmatrix} = 0 \right\}$$

via Theorem 3.2.20 and by checking item 1 of Theorem 3.3.10 that this image behavior is indeed equal to $\mathcal{B} = \left\{ \begin{pmatrix} y \\ u \end{pmatrix} \in \mathcal{F}^{2+2};\quad (P, -Q) \circ \begin{pmatrix} y \\ u \end{pmatrix} = 0 \right\}$, where $P = \begin{pmatrix} P_1 & -Q_1 \\ -Q_2 & P_2 \end{pmatrix}$ and $Q = \begin{pmatrix} 0 & Q_1 \\ Q_2 & 0 \end{pmatrix}$. ◊

10.2 Compensator Design

10.2.1 Tracking and Disturbance Rejection

The following theory is a variant of that in [4, Sect. 7.5] *and* [10, Chap. 6]. *Whereas Vidyasagar describes systems by their transfer matrix and uses the Laplace transform implicitly, we treat behaviors and essentially use injective cogenerator quotient signal modules for the proofs, cf. Sect. 9.3. We also present an algorithm for the construction of the desired compensators. Also the topological and robustness considerations are inspired by, but differ from those of* [4].

We consider the interconnection behavior from Fig. 10.1. We assume the data from Theorem 10.1.20 and the notations used in Sect. 10.1. Again $T \subsetneq F[s] \setminus \{0\}$ is a saturated monoid that gives rise to T-stability notions. Recall the standard situation:

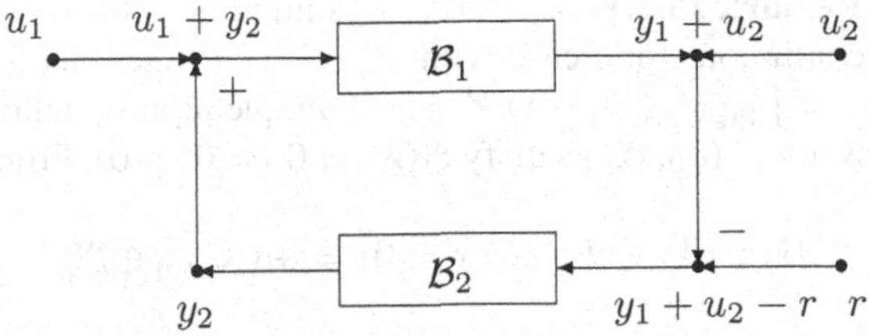

Fig. 10.1 The tracking behavior $\mathcal{T}$

$$
\begin{aligned}
&F := \mathbb{R}, \mathbb{C},\ \mathcal{F} := \begin{cases} C^{+/-\infty}(\mathbb{R}, F) & \text{(continuous time)} \\ F^{\mathbb{N}} & \text{(discrete time)} \end{cases} \\
&\mathbb{C} = \Lambda_1 \uplus \Lambda_2,\ \Lambda_1 \subseteq \begin{cases} \mathbb{C}_- & \text{(continuous time)} \\ \mathbb{D} & \text{(discrete time)} \end{cases},\ \Lambda_1 = \overline{\Lambda_1} \text{ if } F = \mathbb{R} \\
&T := \{t \in T;\ \mathrm{V}_{\mathbb{C}}(t) \subseteq \Lambda_1\}.
\end{aligned}
\tag{10.80}
$$

The continuous or discrete time is denoted by τ since t is the generic name of a polynomial in the monoid T. In this standard situation an *autonomous* behavior $\mathcal{B}'$ is T-stable if and only $\text{char}(\mathcal{B}') \subseteq \Lambda_1$ or, in other words, if the poles (=characteristic values) of $\mathcal{B}'$ belong to Λ_1. Then all trajectories $w \in \mathcal{B}'$ are asymptotically stable, i.e., $\lim_{\tau\to\infty} w(\tau) = 0$. Moreover an IO behavior $\mathcal{B}$ is T-stable if and only its autonomous part $\mathcal{B}^0$ is T-stable. Then the transfer matrix H of $\mathcal{B}$ is (L^p, L^p)-stable in continuous time and (ℓ^p, ℓ^p)-stable in discrete time for $1 \le p \le \infty$. The behavior is T-stabilizable if $\text{char}(\mathcal{B}_1) \subseteq \Lambda_1$, i.e., if its uncontrollable poles lie in Λ_1.

The plant $\mathcal{B}_1$ is a given IO behavior with possibly nonproper transfer matrix H_1. The behavior $\mathcal{B}_2$ is a controllable T-stabilizing compensator of $\mathcal{B}_1$ according to Theorem 10.1.20, i.e., the feedback behavior $\mathcal{B} := \text{fb}(\mathcal{B}_1, \mathcal{B}_2)$ is well-posed and T-stable and its transfer matrix H is proper.

The signals u_1 resp. u_2 are interpreted as *disturbances* of the input resp. of the output of the plant $\mathcal{B}_1$; hence, $y_1 + u_2$ is the disturbed output signal. The signal r is a *reference signal* that the plant should follow. Therefore $e := y_1 + u_2 - r$ is the error signal, i.e., the difference between the actual disturbed and the desired output of the plant, that should be T-stable, i.e., $t \circ e = 0$ for some $t \in T$. In the standard situation this implies the desired approximation $\lim_{\tau\to\infty} e(\tau) = 0$. The disturbance and reference signals are not free, but a nonzero polynomial $\phi \in \mathcal{D} = F[s]$ is assumed with

$$\phi \circ u_1 = 0,\ \phi \circ u_2 = \phi \circ r = 0,\ u_1 \in \mathcal{F}^m,\ u_2, r \in \mathcal{F}^p. \tag{10.81}$$

Thus the signals belong to special autonomous systems $\{v \in \mathcal{F}^\bullet;\ \phi \circ v = 0\}$ that are also called *signal generators*. The equations $s \circ r = 0$ (continuous time) resp. $\Delta \circ r = 0,\ \Delta = s - 1$, (discrete case) describe a constant reference signal, the equations $s^2 \circ r = 0$ resp. $\Delta^2 \circ r = 0$ describe linear signals $a\tau + b$, so-called ramps for $a > 0$.

Remark 10.2.1 *(Cf.* [10, Assumption 116, p. 163]*)* Consider the standard continuous-time case with $F = \mathbb{C}$ (for notational simplicity) and $\Lambda_1 = \mathbb{C}_-$. Then $\phi = \prod_{\lambda \in \mathrm{V}_{\mathbb{C}}(\phi)} (s-\lambda)^{n(\lambda)}$ has a unique representation $\phi = \phi_- \phi_0 \phi_+$ where the roots λ of ϕ_- (ϕ_0, ϕ_+) satisfy $\Re(\lambda) < 0$ $(= 0, > 0)$. From Sect. 5.2 we conclude

$$\mathcal{B}_\phi := \{y \in \mathcal{F};\ \phi \circ y = 0\} = \oplus_{\lambda \in \mathrm{V}_{\mathbb{C}}(\phi)} \oplus_{k=0}^{n(\lambda)-1} \mathbb{C}t^k e^{\lambda t} = \mathcal{B}_{\phi_-} \oplus \mathcal{B}_{\phi_0} \oplus \mathcal{B}_{\phi_+}.$$

Exponentially increasing resp. decreasing reference or disturbance signals are unrealistic resp. give no contribution for $t \to \infty$. Hence it is reasonable to assume that $\phi = \phi_0$ or $\mathrm{V}_{\mathbb{C}}(\phi) \subset \mathbb{R}j$ and $\mathcal{B}_\phi = \oplus_{j\omega \in \mathrm{V}_{\mathbb{C}}(\phi)} \oplus_{k=0}^{n(j\omega)-1} \mathbb{C}t^k e^{j\omega t}$.
The analogous remark applies to the discrete-time case. ◇

Under the condition in (10.81) the equations of the interconnection in Figure 10.1 are

$$P_1 \circ y_1 = Q_1 \circ (u_1 + y_2),\ P_2 \circ y_2 = Q_2 \circ (u_2 + y_1 - r),\ \phi \circ (u_2, u_1, r) = 0. \tag{10.82}$$

The associated *interconnection behavior* $\mathcal{C}$ and *error behavior* $\mathcal{E}$ are

$$\begin{aligned} \mathcal{C} &:= \left\{ \begin{pmatrix} y \\ u \\ r \end{pmatrix} \in \mathcal{F}^{2(p+m)+p};\ (10.82) \text{ holds} \right\}, \\ y &= \begin{pmatrix} y_1 \\ y_2 \end{pmatrix}, u = \begin{pmatrix} u_2 \\ u_1 \end{pmatrix} \in \mathcal{F}^{p+m},\ r \in \mathcal{F}^p \\ \mathcal{E} &:= \operatorname{im}\left(\mathcal{C} \longrightarrow \mathcal{F}^p,\ \begin{pmatrix} y \\ u \\ r \end{pmatrix} \mapsto e = y_1 + u_2 - r \right) \\ &= \left\{ e = y_1 + u_2 - r \in \mathcal{F}^p;\ \begin{pmatrix} y \\ u \\ r \end{pmatrix} \in \mathcal{C} \right\}. \end{aligned} \tag{10.83}$$

The behavior $\mathcal{C}$ is given by the equation

$$\mathcal{C} = \left\{ w = \begin{pmatrix} y_1 \\ y_2 \\ u_2 \\ u_1 \\ r \end{pmatrix} \in \mathcal{F}^\bullet;\ R \circ w = 0 \right\},\ R := \begin{pmatrix} P_1 & -Q_1 & 0 & -Q_1 & 0 \\ -Q_2 & P_2 & -Q_2 & 0 & Q_2 \\ 0 & 0 & \phi\mathrm{id} & 0 & 0 \\ 0 & 0 & 0 & \phi\mathrm{id} & 0 \\ 0 & 0 & 0 & 0 & \phi\mathrm{id} \end{pmatrix}. \tag{10.84}$$

Due to $e := y_1 + u_2 - r = Zw,\ Z := (\mathrm{id}, 0, \mathrm{id}, 0, -\mathrm{id})$, the equation of the error behavior $\mathcal{E}$ is obtained as

$$\mathcal{E} = \{e \in \mathcal{F}^p;\ E \circ e = 0\} \tag{10.85}$$

where $(-X, E)$ is a universal left annihilator of $\begin{pmatrix} R \\ Z \end{pmatrix}$, cf. Theorem 3.2.20.

Since the quotient functor $(-)_T$ is exact the epimorphism $\mathcal{C} \to \mathcal{E}$ induces the epimorphism $\mathcal{C}_T \to \mathcal{E}_T$. This implies

$$\begin{aligned} \mathcal{E}_T &:= \left\{ e := y_1 + u_2 - r \in \mathcal{F}_T^p;\ \begin{pmatrix} y \\ u \\ r \end{pmatrix} \in \mathcal{C}_T \right\} \text{ where} \\ \mathcal{C}_T &:= \left\{ \begin{pmatrix} y \\ u \\ r \end{pmatrix} \in \mathcal{F}_T^{2(p+m)+p};\ (10.82) \text{ holds} \right\}. \end{aligned} \tag{10.86}$$

Definition 10.2.2 The tracking and rejection condition is that the trajectories of $\mathcal{E}$ and therefore $\mathcal{E}$ itself are T-stable, i.e., $\mathcal{E}_T = 0$. ◊

Assumption 10.2.3 If $\phi \in T$, then $\{y \in \mathcal{F};\ \phi \circ y = 0\}$ and hence $\mathcal{C}$ and $\mathcal{E}$ are T-stable without any additional condition. This is uninteresting in Sect. 10.2, and hence we assume $\phi \notin T$ in the sequel. ◊

In the sequel we need the following matrices from Theorem 10.1.20:

$$\begin{aligned}
&P = \begin{pmatrix} P_1 & -Q_1 \\ -Q_2 & P_2 \end{pmatrix},\ Q = \begin{pmatrix} 0 & Q_1 \\ Q_2 & 0 \end{pmatrix},\ X \in \mathcal{S}^{m\times p},\\
&(-\widehat{Q}_2, \widehat{P}_2) = (-\widehat{Q}_2^0, \widehat{P}_2^0) + X(\widehat{P}_1, -\widehat{Q}_1),\ \begin{pmatrix} \widehat{D}_2 \\ \widehat{N}_2 \end{pmatrix} = \begin{pmatrix} \widehat{D}_2^0 \\ \widehat{N}_2^0 \end{pmatrix} - \begin{pmatrix} \widehat{N}_1 \\ \widehat{D}_1 \end{pmatrix} X \in \mathcal{S}^{\bullet\times\bullet},\\
&\widehat{P} = \begin{pmatrix} \widehat{P}_1 & -\widehat{Q}_1 \\ -\widehat{Q}_2 & \widehat{P}_2 \end{pmatrix} \in \mathrm{Gl}_{p+m}(\mathcal{S}),\ \widehat{P}^{-1} = \begin{pmatrix} \widehat{D}_2 & \widehat{N}_1 \\ \widehat{N}_2 & \widehat{D}_1 \end{pmatrix},\ \widehat{Q} = \begin{pmatrix} 0 & \widehat{Q}_1 \\ \widehat{Q}_2 & 0 \end{pmatrix} \in \mathcal{S}^{\bullet\times\bullet},\\
&S = \begin{pmatrix} S_1 & 0 \\ 0 & S_2 \end{pmatrix} \in \mathrm{Gl}_{p+m}(\mathcal{D}_T),\ S(P, -Q) = (\widehat{P}, -\widehat{Q}),\\
&\Longrightarrow \begin{pmatrix} \widehat{P}_1 & -\widehat{Q}_1 \\ -\widehat{Q}_2 & \widehat{P}_2 \end{pmatrix}\begin{pmatrix} \widehat{D}_2 & \widehat{N}_1 \\ \widehat{N}_2 & \widehat{D}_1 \end{pmatrix} = \begin{pmatrix} \widehat{D}_2 & \widehat{N}_1 \\ \widehat{N}_2 & \widehat{D}_1 \end{pmatrix}\begin{pmatrix} \widehat{P}_1 & -\widehat{Q}_1 \\ -\widehat{Q}_2 & \widehat{P}_2 \end{pmatrix} = \mathrm{id}_{p+m},\\
&\qquad H = P^{-1}Q = \widehat{P}^{-1}\widehat{Q} = \begin{pmatrix} \widehat{N}_1\widehat{Q}_2 & \widehat{D}_2\widehat{Q}_1 \\ \widehat{D}_1\widehat{Q}_2 & \widehat{N}_2\widehat{Q}_1 \end{pmatrix}.
\end{aligned}\tag{10.87}$$

Since $S\circ$, $P\circ$ and $\widehat{P}\circ$ act as isomorphisms on $\mathcal{F}_T^{p+m}$ we infer the following equivalences and consequences for $\begin{pmatrix} y \\ u \\ r \end{pmatrix} \in \mathcal{F}_T^{2(p+m)-p}$:

$$\begin{aligned}
&P_1 \circ y_1 = Q_1 \circ (u_1 + y_2),\ P_2 \circ y_2 = Q_2 \circ (u_2 + y_1 - r)\\
&\Longleftrightarrow P \circ y = Q \circ \begin{pmatrix} u_2 - r \\ u_1 \end{pmatrix} \Longleftrightarrow \widehat{P} \circ y = \widehat{Q} \circ \begin{pmatrix} u_2 - r \\ u_1 \end{pmatrix}\\
&\Longleftrightarrow y = \widehat{P}^{-1}\widehat{Q} \circ \begin{pmatrix} u_2 - r \\ u_1 \end{pmatrix} = H \circ \begin{pmatrix} u_2 - r \\ u_1 \end{pmatrix} \Longleftrightarrow y = \begin{pmatrix} \widehat{N}_1\widehat{Q}_2 & \widehat{D}_2\widehat{Q}_1 \\ \widehat{D}_1\widehat{Q}_2 & \widehat{N}_2\widehat{Q}_1 \end{pmatrix} \circ \begin{pmatrix} u_2 - r \\ u_1 \end{pmatrix}\\
&\Longrightarrow e = y_1 + u_2 - r = \widehat{N}_1\widehat{Q}_2 \circ (u_2 - r) + \widehat{D}_2\widehat{Q}_1 \circ u_1 + u_2 - r\\
&\underset{\widehat{D}_2\widehat{P}_1 = \mathrm{id}_p + \widehat{N}_1\widehat{Q}_2}{=} \widehat{D}_2\widehat{P}_1(u_2 - r) + \widehat{D}_2\widehat{Q}_1 \circ u_1 = (\widehat{D}_2\widehat{P}_1, \widehat{D}_2\widehat{Q}_1, -\widehat{D}_2\widehat{P}_1) \circ \begin{pmatrix} u \\ r \end{pmatrix}.
\end{aligned}\tag{10.88}$$

Equation 10.88 implies the isomorphism resp. epimorphism

$$\begin{aligned}
&\left\{\begin{pmatrix} u \\ r \end{pmatrix} \in \mathcal{F}_T^{p+m+p};\ \phi \circ \begin{pmatrix} u \\ r \end{pmatrix} = 0\right\} \cong \mathcal{C}_T,\ \begin{pmatrix} u \\ r \end{pmatrix} \mapsto \begin{pmatrix} H \circ \begin{pmatrix} u_2 - r \\ u_1 \end{pmatrix} \\ u \\ r \end{pmatrix},\\
&\left\{\begin{pmatrix} u \\ r \end{pmatrix} \in \mathcal{F}_T^{p+m+p};\ \phi \circ \begin{pmatrix} u \\ r \end{pmatrix} = 0\right\}\\
&\qquad \to \mathcal{E}_T = \left\{e = y_1 + u_2 - r;\ \begin{pmatrix} u \\ r \end{pmatrix} \in \mathcal{F}_T^{p+m+p};\ \phi \circ \begin{pmatrix} u \\ r \end{pmatrix} = 0\right\}.
\end{aligned}\tag{10.89}$$

In particular, $\mathcal{C}_T$ and hence $\mathcal{C}$ and $\mathcal{E}$ are autonomous.

Theorem 10.2.4 (Cf. [4, Lemma 7.5.8]) *Data and notations as in Theorems 10.1.20 and 10.1.34 and their consequences in* (10.82)–(10.89). *Assume a T-stabilizable behavior $\mathcal{B}_1$ with not necessarily proper transfer matrix and a T-stabilizing controllable compensator $\mathcal{B}_2$ with proper closed-loop transfer matrix H according to*

Theorem 10.1.20, defined by means of

$$(-\widehat{Q}_2, \widehat{P}_2) = (-\widehat{Q}_2^0, \widehat{P}_2^0) + X(\widehat{P}_1, -\widehat{Q}_1), \ X \in \mathcal{S}^{m\times p}.$$

The associated closed-loop behavior $\mathcal{B} = \mathrm{fb}(\mathcal{B}_1, \mathcal{B}_2)$ tracks the reference signal r and rejects disturbances u_1, u_2 with $\phi \circ (u_2, u_1, r) = 0$ if and only if

$$\phi^{-1}\widehat{D}_2 \in \mathcal{D}_T^{p\times p} \text{ or, due to } \widehat{s} \in \mathrm{U}(\mathcal{D}_T), \widehat{\phi}^{-1}\widehat{D}_2 \in \mathcal{D}_T^{p\times p}, \ \widehat{\phi} := \phi\widehat{s}^{\deg(\phi)} \in \widehat{\mathcal{D}}. \tag{10.90}$$

Proof The compensator $\mathcal{B}_2$ has the desired properties if and only if $\mathcal{E}_T = 0$. According to (10.89) and (10.88) this is equivalent to the implication

$$\forall \binom{u}{r} \in \mathcal{F}_T^{p+m+p} : \ \phi \circ \binom{u}{r} = 0 \Longrightarrow (\widehat{D}_2\widehat{P}_1, \widehat{D}_2\widehat{Q}_1, -\widehat{D}_2\widehat{P}_1) \circ \binom{u}{r} = 0.$$

Since $\mathcal{F}_T$ is an injective cogenerator over $\mathcal{D}_T$ this is equivalent to

$$\begin{aligned} &\exists Z \in \mathcal{D}_T^{p\times(p+m+p)} : \ (\widehat{D}_2\widehat{P}_1, \widehat{D}_2\widehat{Q}_1, -\widehat{D}_2\widehat{P}_1) = Z\phi \\ &\Longleftrightarrow \phi^{-1}(\widehat{D}_2\widehat{P}_1, \widehat{D}_2\widehat{Q}_1, -\widehat{D}_2\widehat{P}_1) \in \mathcal{D}_T^{p\times(p+m+p)} \\ &\Longleftrightarrow \phi^{-1}\widehat{D}_2(\widehat{P}_1, -\widehat{Q}_1) \in \mathcal{D}_T^{p\times(p+m)} \Longleftrightarrow \phi^{-1}\widehat{D}_2 \in \mathcal{D}_T^{p\times p}. \end{aligned}$$

The last equivalence holds since $\widehat{\mathcal{D}} \subset \mathcal{D}_T$ and since

$$(\widehat{P}_1, -\widehat{Q}_1) \in \widehat{\mathcal{D}}^{p\times(p+m)}, \ \begin{pmatrix} \widehat{D}_2^0 \\ \widehat{N}_2^0 \end{pmatrix} \in \widehat{\mathcal{D}}^{(p+m)\times p}, \ (\widehat{P}_1, -\widehat{Q}_1)\begin{pmatrix} \widehat{D}_2^0 \\ \widehat{N}_2^0 \end{pmatrix} = \mathrm{id}_p .$$

□

In the sequel we need the data introduced in Sect. 9.5: T-stability was introduced by a disjoint decomposition $\mathcal{P} = \mathcal{P}_1 \uplus \mathcal{P}_2$, $\mathcal{P}_1 \neq \emptyset \neq \mathcal{P}_2$, of the set $\mathcal{P}$ of monic irreducible polynomials of $\mathcal{D} = F[s]$ and an analogous decomposition $\widehat{\mathcal{P}} = \widehat{\mathcal{P}}_1 \uplus \widehat{\mathcal{P}}_2$ for the primes of $\widehat{\mathcal{D}}$. In the standard cases with $F = \mathbb{R}, \mathbb{C}$ and a disjoint decomposition $\mathbb{C} = \Lambda_1 \uplus \Lambda_2$, where $\emptyset \neq \Lambda_1 \subseteq \mathbb{C}_-$ (continuous time) or $\emptyset \neq \Lambda_1 \subseteq \mathbb{D}$ in (discrete time), the set $\mathcal{P}_1$ is defined as $\mathcal{P}_1 := \{q \in \mathcal{P}; \ \mathrm{V}_{\mathbb{C}}(q) \subseteq \Lambda_1\}$. According to (9.53) the sets $\mathcal{P}$ and $\widehat{\mathcal{P}}$ are connected by the bijections

$$\begin{array}{ccccccc} \mathcal{P} & = & \{s-\alpha\} \ \uplus & \mathcal{P}_1 \setminus \{s-\alpha\} & \uplus & \mathcal{P}_2 & \\ & & & \downarrow\cong & & \downarrow\cong & \\ \widehat{\mathcal{P}} & = & & \widehat{\mathcal{P}}_1 & \uplus & \widehat{\mathcal{P}}_2 \setminus \{\widehat{s}\} & \uplus \ \{\widehat{s}\} \end{array}$$

$$\text{with } \mathcal{P}_1 \setminus \{s-\alpha\} \cong \widehat{\mathcal{P}}_1, \ q \longmapsto \widehat{q}_\alpha = q\widehat{s}^{\deg_s(q)}, \ \ \mathcal{P}_2 \cong \widehat{\mathcal{P}}_2 \setminus \{\widehat{s}\}, \ q \longmapsto \widehat{q}_\alpha. \tag{10.91}$$

Recall that a polynomial $f = a_m(s-\alpha)^m + \cdots \in \mathcal{D}$, $a_m \neq 0$, gives rise to

$$\widehat{f} := \widehat{f_\alpha} = f\widehat{s}^{\deg_s(f)} = f\widehat{s}^m = a_m + a_{m-1}\widehat{s}^1 + \cdots \in \widehat{\mathcal{D}} = F[\widehat{s}], \ \widehat{f}(0) = a_m \neq 0. \tag{10.92}$$

The decomposition of primes gives rise to the saturated monoids

$$T = \left\{ u \prod_{q \in \mathcal{P}_1} q^{\mu(q)}; \ 0 \neq u \in F, \ \mu(q) \geq 0 \right\} \subset \mathcal{D},$$
$$\widehat{T} := \left\{ \widehat{t} := t\widehat{s}^{\deg(t)}; \ t \in T \right\} = \left\{ u \prod_{\widehat{q} \in \widehat{\mathcal{P}}_1} \widehat{q}^{\mu(\widehat{q})}; \ 0 \neq u \in F, \ \mu(\widehat{q}) \geq 0 \right\} \subset \widehat{\mathcal{D}}. \tag{10.93}$$

The monoids induce the quotient rings

$$F[\widehat{s}] = \widehat{\mathcal{D}} \subset \mathcal{S} = \widehat{\mathcal{D}}_{\widehat{T}} \subset \mathcal{D}_T = \mathcal{S}_{\widehat{s}} \supset \mathcal{D} = F[s]. \tag{10.94}$$

The set $\widehat{\mathcal{P}}_2 \subset \widehat{\mathcal{D}} \subset \mathcal{S} = \widehat{\mathcal{D}}_{\widehat{T}}$ is a representative system of the primes of $\mathcal{S}$, in particular $\mathcal{S}\widehat{q}, \ \widehat{q} \in \widehat{\mathcal{P}}_2$, is a maximal ideal of $\mathcal{S}$. For $\widehat{q} \in \widehat{\mathcal{P}}_2$ we also need the canonical algebra epimorphism

$$\nu_{\widehat{q}} : \widehat{\mathcal{D}} \to F(\widehat{q}) := \widehat{\mathcal{D}}/\widehat{\mathcal{D}}\widehat{q}, \ \widehat{f} \mapsto \widehat{f} + \widehat{\mathcal{D}}\widehat{q}. \tag{10.95}$$

This map is extended to matrices componentwise. An element $\widehat{e} \in \widehat{\mathcal{D}}$ is a unit of $\mathcal{S}$ if and only if $\nu_{\widehat{q}}(\widehat{e}) \neq 0$ for all $\widehat{q} \in \widehat{\mathcal{P}}_2$.

Corollary 10.2.5 *If in Theorem 10.2.4 the condition* (10.90) *is satisfied, i.e.,*

$$\text{if } \widehat{\phi}^{-1}\widehat{D}_2 \in \mathcal{D}_T^{p\times p} \text{ then indeed } \widehat{\phi}^{-1}\widehat{D}_2 \in \mathcal{S}^{p\times p}. \tag{10.96}$$

Proof Due to $\mathcal{D}_T = \mathcal{S}_{\widehat{s}}$ every nonzero element of $\mathcal{D}_T$ has a unique representation

$$h = g\widehat{s}^k, \ g \in \mathcal{S}, \ k \in \mathbb{Z}, \ \widehat{s} \nmid g \text{ or } g(0) \neq 0.$$

(i) If $0 \neq f \in \mathcal{S}$ and $h := \widehat{\phi}^{-1} f \in \mathcal{D}_T$ then indeed $\widehat{\phi}^{-1} f \in \mathcal{S}$: Consider

$$h := \widehat{\phi}^{-1} f = g\widehat{s}^k, \ k \in \mathbb{Z}, \ g \in \mathcal{S}, \ g(0) \neq 0.$$

If $k \geq 0$ then obviously $h \in \mathcal{S}$. If

$$k < 0 \Longrightarrow 0 = \widehat{s}^{-k}(0) f(0) = \widehat{\phi}(0) g(0) \underset{g(0)\neq 0}{\Longrightarrow} \widehat{\phi}(0) = 0.$$

This contradicts $\widehat{\phi}(0) \neq 0$ according to (10.92).
(ii) The equation $\begin{pmatrix} \widehat{D}_2 \\ \widehat{N}_2 \end{pmatrix} = \begin{pmatrix} \widehat{D}_2^0 \\ \widehat{N}_2^0 \end{pmatrix} - \begin{pmatrix} \widehat{N}_1 \\ \widehat{D}_1 \end{pmatrix} X, \ X \in \mathcal{S}^{m\times p}$ implies $\widehat{D}_2 \in \mathcal{S}^{p\times p}$. Application of (i) to all entries of $\widehat{D}_2$ then yields the asserted implication

$$\widehat{\phi}^{-1}\widehat{D}_2 \in \mathcal{D}_T^{p\times p} \Longrightarrow \widehat{\phi}^{-1}\widehat{D}_2 \in \mathcal{S}^{p\times p}.$$

□

Remark 10.2.6 (*Connection with the literature, cf.* [2, Chap. 7], [4, Sect. 7.5]) The quoted books work explicitly or implicitly with the Laplace transform and [4] only with transfer matrices. Therefore the disturbance and reference signals are not described by the condition $\phi\circ(u_2,u_1,r)=0$, but, cf. [2, p. 198, (17)], as

$$\begin{aligned}&(u_2,u_1,r)^\top = H_3\circ\delta = \alpha_3 Y,\ \alpha_3\in \mathrm{t}(\mathcal{F})^{p+m+p},\ \text{where}\\ &H_3=\mathcal{L}\left((u_2,u_1,r)^\top\right) := \phi^{-1}Q_3,\ Q_3\in\mathcal{D}^{p+m+p},\ d:=\deg_s(\phi)>\deg_s(Q_3).\end{aligned}\tag{10.97}$$

Then obviously, cf. Theorem 8.2.53,

$$\begin{aligned}&\mathrm{pole}(H_3)\subseteq \mathrm{V}_{\mathbb{C}}(\phi),\ \phi\circ(u_2,u_1,r)^\top = Q_3\circ\delta,\ \phi\circ\alpha_3=0,\\ &(u_2,u_1,r)^\top|_{[0,\infty)} = \alpha_3|_{[0,\infty)}.\end{aligned}\tag{10.98}$$

Hence the reference and disturbance signals of the literature and of our book coincide for $t\geq 0$ and are polynomial-exponential there. Assume (10.97) and (10.98) and a compensator according to Theorem 10.2.4. Then

$$H_3=\phi^{-1}Q_3=\left(\phi\widehat{s}^d\right)^{-1}\left(\widehat{s}^d Q_3\right)=\widehat{\phi}^{-1}\widehat{Q}_3,\ \widehat{Q}_3\in\widehat{\mathcal{D}}^{p+m+p},\ \widehat{Q}_3(0)=0.$$

Recall the closed-loop transfer matrix $H=\begin{pmatrix}\widehat{N}_1\widehat{Q}_2 & \widehat{D}_2\widehat{Q}_1\\ \widehat{D}_1\widehat{Q}_2 & \widehat{N}_2\widehat{Q}_1\end{pmatrix}\in\mathcal{S}^{(p+m)\times(p+m)}$ of $\mathcal{B}:=\mathrm{fb}(\mathcal{B}_1,\mathcal{B}_2)$ from (10.87) and $\widehat{\phi}^{-1}\widehat{D}_2\in\mathcal{S}^{p\times p}$ from (10.96). The signal $(u_2,u_1,r)^\top = H_3\circ\delta=\widehat{\phi}^{-1}\widehat{Q}_3\circ\delta$ gives rise to the output $\binom{y_1}{y_2}:=H\circ\binom{u_2-r}{u_1}$ of $\mathcal{B}$ and to the error, cf. (10.88),

$$\begin{aligned}e&=y_1+u_2-r=\left(\widehat{D}_2\widehat{P}_1,\widehat{D}_2\widehat{Q}_1,-\widehat{D}_2\widehat{P}_1\right)\circ(u_2,u_1,r)^\top\\ &=\left(\widehat{D}_2\widehat{P}_1,\widehat{D}_2\widehat{Q}_1,-\widehat{D}_2\widehat{P}_1\right)\circ\widehat{\phi}^{-1}\widehat{Q}_3\circ\delta=\left(\widehat{\phi}^{-1}\widehat{D}_2\right)(\widehat{P}_1,\widehat{Q}_1,-\widehat{P}_1)\widehat{Q}_3\circ\delta\\ &\Longrightarrow e=H_e\circ\delta,\ H_e:=\left(\widehat{\phi}^{-1}\widehat{D}_2\right)(\widehat{P}_1,\widehat{Q}_1,-\widehat{P}_1)\widehat{Q}_3\in\mathcal{S}^{p\times p},\ H_e(0)=0.\end{aligned}\tag{10.99}$$

Hence H_e is strictly proper and T-stable. The condition $H_e\in\mathcal{S}^{p\times p},\ H_e(0)=0$, is that from [4, p. 296, (R2)], [2, p. 206, (70)]. If $H_e=P_e^{-1}Q_e$ is the left coprime factorization over $\mathcal{D}$ this implies

$$\begin{aligned}&\det(P_e)\in T,\ e=H_e\circ\delta=\beta_e Y\in\mathrm{t}(\mathcal{F})^p Y,\\ &P_e\circ(\beta_e Y)=Q_e\circ\delta,\ P_e\circ\beta_e=0,\ \beta_e\in\mathrm{t}_T(\mathcal{F})^p.\end{aligned}\tag{10.100}$$

◇

In the next Theorem 10.2.8 we characterize the *existence* of compensators according to Theorem 10.2.4. We need the following preparatory

Lemma 10.2.7 *Let $R \in \widehat{\mathcal{D}}^{p\times l}$ be a matrix of rank p with its highest (p-th) elementary divisor $\widehat{e}_p \in \widehat{\mathcal{D}}$ and $\widehat{q} \in \widehat{\mathcal{P}}$. Then $\nu_{\widehat{q}}(R) \in F(\widehat{q})^{p\times l}$ has rank p too or, equivalently, is right invertible if and only if $\nu_{\widehat{q}}(\widehat{e}_p) \neq 0$.*

Proof R is equivalent to its Smith form for which the assertion is obvious. □

Theorem 10.2.8 (Cf. [3, Theorem 9–22]), [2, Chap. 7]), [4, Sect. 7.5], [10, Chap. 6]) *Under the conditions of Theorem 10.2.4 the following statements are equivalent:*

1. *$\mathcal{B}_1$ admits a controllable, proper, T-stabilizing compensator $\mathcal{B}_2$ such that the feedback behavior $\mathcal{B} = \mathrm{fb}(\mathcal{B}_1, \mathcal{B}_2)$ is proper, tracks r and rejects u_2, u_1 with $\phi \circ (u_2, u_1, r) = 0$.*
2. *The matrix $(\widehat{\phi}\widehat{P}_1, -\widehat{Q}_1)$ has a right inverse in $\mathcal{S}^{(p+m)\times p}$ or its highest (p-th) elementary divisor $\widehat{e} \in \widehat{\mathcal{D}}$ belongs to $\widehat{T} = \widehat{\mathcal{D}} \cap \mathrm{U}(\widehat{\mathcal{D}}_{\widehat{T}}) = \widehat{\mathcal{D}} \cap \mathrm{U}(\mathcal{S})$.*
3. *The matrix $(\widehat{\phi}\,\mathrm{id}_p, -\widehat{Q}_1)$ has a right inverse in $\mathcal{S}^{(p+m)\times p}$ or its highest (p-th) elementary divisor $\widehat{f} \in \widehat{\mathcal{D}}$ belongs to $\widehat{T} = \widehat{\mathcal{D}} \cap \mathrm{U}(\mathcal{S})$.*
4. $\mathrm{rank}(\widehat{N}_1)\left(= \mathrm{rank}(\widehat{Q}_1) = \mathrm{rank}(Q_1) = \mathrm{rank}(H_1)\right) = p$ *and the highest (p-th) elementary divisor $\widehat{g}$ of $\widehat{N}_1$ satisfies $\widehat{h} := \gcd_{\widehat{\mathcal{D}}}(\widehat{\phi}, \widehat{g}) \in \widehat{T}$.*
5. *Assume* $\mathrm{rank}(\widehat{N}_1) = p$. *With $\widehat{g}$ from 4. define*

$$g := (s-\alpha)^{\deg_{\widehat{s}}(\widehat{g})}\widehat{g} \in \mathcal{D},\ g(\alpha) \neq 0,\ h := \gcd_{\mathcal{D}}(\phi, g). \tag{10.101}$$

Then items 1 to 4 are equivalent to $h \in T$.

Notice that $\widehat{g} \in \widehat{\mathcal{D}}$ and $h \in \mathcal{D}$ resp. $h \in T$ are easy to compute resp. checked. See Theorem 10.2.17 for another equivalent description with the zeros of H_1.

Proof 1. $\Longrightarrow$ 2.: According to Theorem 10.2.4 and Corollary 10.2.5 the compensator satisfies $\widehat{\phi}^{-1}\widehat{D}_2 \in \mathcal{S}^{p\times p}$. The equation

$$(\widehat{\phi}\widehat{P}_1, -\widehat{Q}_1)\begin{pmatrix}\widehat{\phi}^{-1}\widehat{D}_2\\ \widehat{N}_2\end{pmatrix} = (\widehat{P}_1, -\widehat{Q}_1)\begin{pmatrix}\widehat{D}_2\\ \widehat{N}_2\end{pmatrix} = \mathrm{id}_p,$$

then implies that $\begin{pmatrix}\widehat{\phi}^{-1}\widehat{D}_2\\ \widehat{N}_2\end{pmatrix}$ is a right inverse of $(\widehat{\phi}\widehat{P}_1, -\widehat{Q}_1)$ in $\mathcal{S}^{(p+m)\times p}$.

2. $\Longleftrightarrow$ 3.: Since $H_1 = \widehat{P}_1^{-1}\widehat{Q}_1$ is a left coprime factorization over $\widehat{\mathcal{D}}$ the matrix $(\widehat{P}_1, -\widehat{Q}_1)$ has a right inverse in $\widehat{\mathcal{D}}^{(p+m)\times p}$, and hence, by Lemma 10.2.7,

$$\forall \widehat{q} \in \widehat{\mathcal{P}}:\ \mathrm{rank}\left(\nu_{\widehat{q}}(\widehat{P}_1), -\nu_{\widehat{q}}(\widehat{Q}_1)\right) = p. \tag{10.102}$$

Recall that $\widehat{\mathcal{P}}_2$ is a set of representatives of the primes of $\mathcal{S}$. By Theorem 4.2.12, 2. (c), (d), the following equivalences are valid:

$$\begin{aligned} &2.\text{ holds} \Longleftrightarrow \widehat{e} \in \mathrm{U}(\mathcal{S}) \Longleftrightarrow \widehat{e} \in \widehat{T} = \widehat{\mathcal{D}} \cap \mathrm{U}(\mathcal{S}) \Longleftrightarrow \forall \widehat{q} \in \widehat{\mathcal{P}}_2:\ \nu_{\widehat{q}}(\widehat{e}) \neq 0.\\ &3.\text{ holds} \Longleftrightarrow \widehat{f} \in \mathrm{U}(\mathcal{S}) \Longleftrightarrow \widehat{f} \in \widehat{T} \Longleftrightarrow \forall \widehat{q} \in \widehat{\mathcal{P}}_2:\ \nu_{\widehat{q}}(\widehat{f}) \neq 0.\end{aligned} \tag{10.103}$$

With (10.103) and Lemma 10.2.7 we infer the equivalences

$$
\begin{aligned}
&\text{2. holds} \iff \forall \widehat{q} \in \widehat{\mathcal{P}}_2 : \ \operatorname{rank}\left(\nu_{\widehat{q}}(\widehat{\phi})\nu_{\widehat{q}}(\widehat{P}_1), -\nu_{\widehat{q}}(\widehat{Q}_1)\right) = p \\
&\underset{(10.102)}{\iff} \forall \widehat{q} \in \widehat{\mathcal{P}}_2 : \ \text{If } \nu_{\widehat{q}}(\widehat{\phi}) = 0 \text{ or } \widehat{q} | \widehat{\phi} \text{ then } \operatorname{rank}\left(\nu_{\widehat{q}}(\widehat{Q}_1)\right) = p \\
&\iff \forall \widehat{q} \in \widehat{\mathcal{P}}_2 : \ \operatorname{rank}\left(\nu_{\widehat{q}}(\widehat{\phi})\,\mathrm{id}_p, -\nu_{\widehat{q}}(\widehat{Q}_1)\right) = p \underset{\text{analogous}}{\iff} \text{3. holds.}
\end{aligned}
\tag{10.104}
$$

3. $\iff$ 4.: Since $H_1 = \widehat{P}_1^{-1}\widehat{Q}_1 = \widehat{N}_1\widehat{D}_1^{-1}$ are coprime factorizations over $\widehat{\mathcal{D}}$, the matrices $\widehat{Q}_1$ and $\widehat{N}_1$ are equivalent and have the same elementary divisors, cf. Corollary and Definition 6.2.6, in particular $\operatorname{rank}\left(\nu_{\widehat{q}}(\widehat{N}_1)\right) = \operatorname{rank}\left(\nu_{\widehat{q}}(\widehat{Q}_1)\right)$ for all $\widehat{q} \in \widehat{\mathcal{P}}$. With (10.104) this implies that

$$
\text{3. holds} \iff \forall \widehat{q} \in \widehat{\mathcal{P}}_2 : \ \left(\nu_{\widehat{q}}(\widehat{\phi}) = 0 \Longrightarrow \operatorname{rank}\left(\nu_{\widehat{q}}(\widehat{N}_1)\right) = p\right). \tag{10.105}
$$

Since $\phi \notin T$ by Assumption 10.2.3 and thus $\widehat{\phi} \notin \widehat{T}$ there is $\widehat{q} \in \widehat{\mathcal{P}}_2$ with $\widehat{q} \mid \widehat{\phi}$ or $\nu_{\widehat{q}}(\widehat{\phi}) = 0$. By (10.105) this implies

$$
\begin{aligned}
&\operatorname{rank}\left(\nu_{\widehat{q}}(\widehat{N}_1)\right) = \operatorname{rank}\left(\nu_{\widehat{q}}(\widehat{Q}_1)\right) = p \Longrightarrow \operatorname{rank}(\widehat{N}_1) = p \le m \\
&\Longrightarrow \text{Smith form } \widehat{U}\widehat{N}_1\widehat{V} = (\widehat{E}, 0), \ \widehat{E} = \operatorname{diag}(\widehat{e}_1, \ldots, \widehat{e}_p = \widehat{g}) \in \widehat{\mathcal{D}}^{p \times p}.
\end{aligned}
\tag{10.106}
$$

Since $\operatorname{rank}\left(\nu_{\widehat{q}}(\widehat{N}_1)\right) = \operatorname{rank}\left(\nu_{\widehat{q}}(\widehat{E})\right)$ we infer

$$
\operatorname{rank}\left(\nu_{\widehat{q}}(\widehat{N}_1)\right) = p \iff \nu_{\widehat{q}}(\widehat{g}) \ne 0 \iff \widehat{q} \nmid \widehat{g}. \tag{10.107}
$$

Thus (10.105) and (10.107) imply

$$
\begin{aligned}
&\text{3. holds} \iff \left(\forall \widehat{q} \in \widehat{\mathcal{P}}_2 : \ \nu_{\widehat{q}}(\widehat{\phi}) = 0 \Longrightarrow \nu_{\widehat{q}}(\widehat{g}) \ne 0 \text{ or } \widehat{q} \nmid \widehat{g}\right) \\
&\iff \underset{\widehat{\mathcal{D}}}{\gcd}(\widehat{\phi}, \widehat{g}) \text{ has primes only in } \widehat{\mathcal{P}}_1 \iff \underset{\widehat{\mathcal{D}}}{\gcd}(\widehat{\phi}, \widehat{g}) \in \widehat{T} \iff 4.
\end{aligned}
\tag{10.108}
$$

4. $\iff$ 5.: (a) We use Sect. 9.5. Recall $\widehat{f}_\alpha := \widehat{s}^{\deg_s(f)} f, \ 0 \ne f \in \mathcal{D}, \ \widehat{f}_\alpha(0) \ne 0$ and $\widehat{(s-\alpha)}_\alpha = 1$, and likewise $g := (s-\alpha)^{\deg_{\widehat{s}}(\widehat{g})}\widehat{g}, \ g(\alpha) \ne 0$. Write

$$
\begin{aligned}
&\phi = (s-\alpha)^\mu \phi_1, \ \phi_1(\alpha) \ne 0, \ \widehat{g} = \widehat{s}^\nu \widehat{g}_1, \ \widehat{g}_1(0) \ne 0 \\
&\Longrightarrow \widehat{\phi} := \widehat{\phi}_\alpha = \widehat{s}^{\deg_s(\phi)} \phi = \widehat{s}^{\deg_s(\phi_1)} \phi_1 = \widehat{\phi}_{1,\alpha}, \ \widehat{g}_1 = \widehat{s}^{\deg_s(g)} g \\
&\Longrightarrow h = \underset{\mathcal{D}}{\gcd}(\phi, g) \underset{g(\alpha)\ne 0}{=} \underset{\mathcal{D}}{\gcd}(\phi_1, g), \ h(\alpha) \ne 0, \\
&\widehat{h} = \underset{\widehat{\mathcal{D}}}{\gcd}(\widehat{\phi}, \widehat{g}) \underset{\widehat{\phi}(0)\ne 0}{=} \underset{\widehat{\mathcal{D}}}{\gcd}(\widehat{\phi}, \widehat{g}_1), \ \widehat{h}(0) \ne 0.
\end{aligned}
$$

(b) Recall the inverse bijections

$$
\begin{aligned}
&\mathcal{P}\setminus\{s-\alpha\}\cong\widehat{\mathcal{P}}\setminus\{\widehat{s}\},\ q=(s-\alpha)^{\deg_s(q)}\widehat{q}\leftrightarrow\widehat{q}=\widehat{s}^{\deg_s(q)}q,\ \text{especially}\\
&\{q\in\mathcal{P}\cap T;\, q(\alpha)\neq 0\}=\{q\in\mathcal{P}_1;\ q(\alpha)\neq 0\}\cong\widehat{\mathcal{P}}_1=\widehat{\mathcal{P}}\cap\widehat{T}\Longrightarrow\ \text{for } q(\alpha)\neq 0:\\
&\left(q\mid h\Longleftrightarrow q\mid\phi_1 \text{ and } q\mid g\Longleftrightarrow\widehat{q}\mid\widehat{\phi}\text{ and }\widehat{q}\mid\widehat{g}_1\Longleftrightarrow\widehat{q}\mid\widehat{h}\right),\ \text{hence}\\
&h\in T\underset{s-\alpha\in T}{\Longleftrightarrow}\{q\in\mathcal{P}\setminus\{s-\alpha\};\ q\mid h\}\subseteq\mathcal{P}_1\\
&\qquad\Longleftrightarrow\{\widehat{q}\in\widehat{\mathcal{P}}\setminus\{\widehat{s}\};\ \widehat{q}\mid\widehat{h}\}\subseteq\widehat{\mathcal{P}}_1\underset{\widehat{s}\nmid\widehat{h}}{\Longleftrightarrow}\widehat{h}\in\widehat{T}.
\end{aligned}
$$

2. $\Longrightarrow$ 1.: (a) We use Lemma 10.1.6, consider the Smith form of $(\widehat{\phi P_1},-\widehat{Q}_1)\in\widehat{\mathcal{D}}^{(p+m)\times p}$ over $\widehat{\mathcal{D}}$ and hence over $\mathcal{S}$, and obtain

$$
\begin{aligned}
&\widetilde{U}(\widehat{\phi P_1},-\widehat{Q}_1)\widetilde{V}=(\widetilde{E},0),\ \widetilde{E}=\operatorname{diag}(\ldots,\widehat{e})\in\mathrm{Gl}_p(\mathcal{S})\ \text{where}\\
&\widetilde{U}\in\mathrm{Gl}_p(\widehat{\mathcal{D}}),\ \widetilde{V}=:\begin{pmatrix}\widetilde{D}_2&\widetilde{N}_1\\ \widetilde{N}_2&\widetilde{D}_1\end{pmatrix}\in\mathrm{Gl}_{p+m}(\widehat{\mathcal{D}})\Longrightarrow\widetilde{U}(\widehat{\phi P_1},-\widehat{Q}_1)\begin{pmatrix}\widetilde{D}_2\\ \widetilde{N}_2\end{pmatrix}=\widetilde{E}\\
&\underset{\widehat{e}\in\mathrm{U}(\mathcal{S})}{\Longrightarrow}(\widehat{\phi P_1},-\widehat{Q}_1)\begin{pmatrix}\widetilde{D}_2^0\\ \widetilde{N}_2^0\end{pmatrix}=\mathrm{id}_p\ \text{ with }\begin{pmatrix}\widetilde{D}_2^0\\ \widetilde{N}_2^0\end{pmatrix}:=\begin{pmatrix}\widetilde{D}_2\\ \widetilde{N}_2\end{pmatrix}\widetilde{E}^{-1}\widetilde{U}\in\mathcal{S}^{(p+m)\times p},\\
&\qquad\begin{pmatrix}\widetilde{D}_2^0&\widetilde{N}_1\\ \widetilde{N}_2^0&\widetilde{D}_1\end{pmatrix}=\begin{pmatrix}\widetilde{D}_2&\widetilde{N}_1\\ \widetilde{N}_2&\widetilde{D}_1\end{pmatrix}\begin{pmatrix}\widetilde{E}^{-1}\widetilde{U}&0\\ 0&\mathrm{id}_m\end{pmatrix}\in\mathrm{Gl}_{p+m}(\mathcal{S}),\\
&\Longrightarrow(\widehat{\phi P_1},-\widehat{Q}_1)\begin{pmatrix}\widetilde{D}_2^0&\widetilde{N}_1\\ \widetilde{N}_2^0&\widetilde{D}^1\end{pmatrix}=(\mathrm{id}_p,0),\ \operatorname{rank}\left(\widetilde{D}_2^0(0),\widetilde{N}_1(0)\right)=p.
\end{aligned}
\tag{10.109}
$$

By the transposed form of Lemma 10.1.33 there is a matrix $\widetilde{X}_0\in F^{m\times p}$ such that the inequality $\det\left(\widetilde{D}_2^0(0)-\widetilde{N}_1(0)\widetilde{X}_0\right)\neq 0$ holds. Then this holds for almost all $\widetilde{X}_0\in F^{m\times p}$. We choose such an $\widetilde{X}_0$ and define

$$
\begin{aligned}
&\begin{pmatrix}\widetilde{D}_2^1\\ \widetilde{N}_2^1\end{pmatrix}:=\begin{pmatrix}\widetilde{D}_2^0\\ \widetilde{N}_2^0\end{pmatrix}-\begin{pmatrix}\widetilde{N}_1\\ \widetilde{D}_1\end{pmatrix}\widetilde{X}_0\in\mathcal{S}^{(p+m)\times p}\\
&\Longrightarrow\det(\widetilde{D}_2^1(0))=\det\left(\widetilde{D}_2^0(0)-\widetilde{N}_1(0)\widetilde{X}_0\right)\neq 0,\ \widetilde{D}_2^1(0)\in\mathrm{Gl}_p(F),\\
&\qquad\widetilde{D}_2^1\in\mathrm{Gl}_p(F(s)_{\mathrm{pr}}),\ \begin{pmatrix}\widetilde{D}_2^1&\widetilde{N}_1\\ \widetilde{N}_2^1&\widetilde{D}_1\end{pmatrix}=\begin{pmatrix}\widetilde{D}_2^0&\widetilde{N}_1\\ \widetilde{N}_2^0&\widetilde{D}_1\end{pmatrix}\begin{pmatrix}\mathrm{id}_p&0\\ -\widetilde{X}_0&\mathrm{id}_m\end{pmatrix}\in\mathrm{Gl}_{p+m}(\mathcal{S}),\\
&\qquad(\widehat{\phi P_1},-\widehat{Q}_1)\begin{pmatrix}\widetilde{D}_2^1&\widetilde{N}_1\\ \widetilde{N}_2^1&\widetilde{D}_1\end{pmatrix}\underset{(10.109)}{=}(\mathrm{id}_p,0)\begin{pmatrix}\mathrm{id}_p&0\\ -\widetilde{X}_0&\mathrm{id}_m\end{pmatrix}=(\mathrm{id}_p,0)\\
&\Longrightarrow\mathrm{id}_p=(\widehat{\phi P_1},-\widehat{Q}_1)\begin{pmatrix}\widetilde{D}_2^1\\ \widetilde{N}_2^1\end{pmatrix}=(\widehat{P}_1,-\widehat{Q}_1)\begin{pmatrix}\widehat{D}_2^1\\ \widehat{N}_2^1\end{pmatrix},\ \begin{pmatrix}\widehat{D}_2^1\\ \widehat{N}_2^1\end{pmatrix}:=\begin{pmatrix}\widehat{\phi}\widetilde{D}_2^1\\ \widetilde{N}_2^1\end{pmatrix}\in\mathcal{S}^{(p+m)\times p}.
\end{aligned}
\tag{10.110}
$$

(b) The sequences

$$
\begin{aligned}
&0\to\widehat{\mathcal{D}}^{1\times p}\xrightarrow{\cdot(\widehat{P}_1,-\widehat{Q}_1)}\widehat{\mathcal{D}}^{1\times(p+m)}\xrightarrow{\cdot\begin{pmatrix}\widehat{N}_1\\ \widehat{D}_1\end{pmatrix}}\widehat{\mathcal{D}}^{1\times m}\to 0\ \text{and hence}\\
&0\to\mathcal{S}^{1\times p}\xrightarrow{\cdot(\widehat{P}_1,-\widehat{Q}_1)}\mathcal{S}^{1\times(p+m)}\xrightarrow{\cdot\begin{pmatrix}\widehat{N}_1\\ \widehat{D}_1\end{pmatrix}}\mathcal{S}^{1\times m}\to 0
\end{aligned}
\tag{10.111}
$$

are exact and $\begin{pmatrix}\widehat{D}_2^1\\ \widehat{N}_2^1\end{pmatrix} \in \mathcal{S}^{(p+m)\times p}$ is a right inverse of $(\widehat{P}_1, -\widehat{Q}_1)$. Lemma 10.1.7 implies a left inverse $(-\widehat{Q}_2^1, \widehat{P}_2^1)$ of $\begin{pmatrix}\widehat{N}_1\\ \widehat{D}_1\end{pmatrix}$ such that

$$\begin{aligned}
&\begin{pmatrix}\widehat{D}_2^1 & \widehat{N}_1\\ \widehat{N}_2^1 & \widehat{D}_1\end{pmatrix} \in \mathrm{Gl}_{p+m}(\mathcal{S}),\ \begin{pmatrix}\widehat{P}_1 & -\widehat{Q}_1\\ -\widehat{Q}_2^1 & \widehat{P}_2^1\end{pmatrix} = \begin{pmatrix}\widehat{D}_2^1 & \widehat{N}_1\\ \widehat{N}_2^1 & \widehat{D}_1\end{pmatrix}^{-1},\\
&\widehat{D}_2^1 = \widehat{\phi}\widetilde{D}_2^1,\ \widetilde{D}_2^1 \in \mathcal{S}^{p\times p},\ \widehat{D}_2^1(0) = \widehat{\phi}(0)\widetilde{D}_2^1(0) \in \mathrm{Gl}_p(F),\ \widehat{D}_2^1 \in \mathrm{Gl}_p(F(s)_{\mathrm{pr}})\\
&H_2^1 := \widehat{N}_2^1(\widehat{D}_2^1)^{-1} \in F(s)_{\mathrm{pr}}^{m\times p}.
\end{aligned} \tag{10.112}$$

As usual we get

$$\begin{aligned}
&\widehat{Q}_2^1\widehat{D}_2^1 = \widehat{P}_2^1\widehat{N}_2^1 = \widehat{P}_2^1 H_2^1\widehat{D}_2^1 \underset{\widehat{D}_2^1(0)\in\mathrm{Gl}_p(F)}{\Longrightarrow} \widehat{Q}_2^1 = \widehat{P}_2^1 H_2^1,\ \widehat{Q}_2^1(0) = \widehat{P}_2^1(0)H_2^1(0)\\
&\Longrightarrow (-\widehat{Q}_2^1, \widehat{P}_2^1) = \widehat{P}_2^1(-H_2^1, \mathrm{id}_m),\ (-\widehat{Q}_2^1(0), \widehat{P}_2^1(0)) = \widehat{P}_2^1(0)(-H_2^1(0), \mathrm{id}_m)\\
&\Longrightarrow \mathrm{rank}(\widehat{P}_2^1(0)) = \mathrm{rank}(-\widehat{Q}_2^1(0), \widehat{P}_2^1(0)) = m\\
&\Longrightarrow \widehat{P}_2^1(0) \in \mathrm{Gl}_m(F),\ \widehat{P}_2^1 \in \mathrm{Gl}_m(F(s)_{\mathrm{pr}}),\ H_2^1 = (\widehat{P}_2^1)^{-1}\widehat{Q}_2^1.
\end{aligned} \tag{10.113}$$

(c) Let $\mathcal{B}_2^1$ denote the controllable realization of H_2^1 in $\mathcal{F}^{p+m}$. Theorem 10.1.20 and Eq. (10.112) imply that $\mathcal{B}_2^1$ is a T-stabilizing compensator of $\mathcal{B}_1$ with proper feedback behavior. Equation (10.113) shows that H_2^1 is proper. Finally Theorem 10.2.4 and $\widehat{\phi}^{-1}\widehat{D}_2^1 = \widetilde{D}_2^1 \in \mathcal{S}^{p\times p}$ from (10.112) imply that $\mathcal{B}_2^1$ tracks signals r and rejects disturbances u_1, u_2 with $\phi \circ (u_2, u_1, r) = 0$. □

Corollary 10.2.9 *Assume that the conditions of Theorem 10.2.8 hold. Then $X \in \mathcal{S}^{m\times p}$ satisfies $\widehat{\phi}^{-1}\widehat{N}_1 X \in \mathcal{S}^{p\times p}$ if and only if there is $Z = \begin{pmatrix}Z_I\\ Z_{II}\end{pmatrix} \in \mathcal{S}^{(p+(m-p))\times p}$ with $X = \widehat{V}\begin{pmatrix}\widehat{\phi}Z_I\\ Z_{II}\end{pmatrix}$, $\widehat{V}$ from (10.106).*

Proof $\Longrightarrow$:

$$X = \widehat{V}\begin{pmatrix}\widehat{\phi}Z_I\\ Z_{II}\end{pmatrix} \underset{(10.106)}{\Longrightarrow} \widehat{\phi}^{-1}\widehat{N}_1 X = \widehat{\phi}^{-1}\widehat{U}^{-1}(\widehat{E}, 0)\widehat{V}^{-1}\widehat{V}\begin{pmatrix}\widehat{\phi}Z_I\\ Z_{II}\end{pmatrix} = \widehat{U}^{-1}\widehat{E}Z_I \in \mathcal{S}^{p\times p}.$$

$\Longrightarrow$: (i) By condition 4 of Theorem 10.2.8 the elements $\widehat{g} = \widehat{e}_p$ and $\widehat{\phi}$ are coprime in $\mathcal{S}$, and hence there are $\widehat{a}, \widehat{b} \in \mathcal{S}$ with $1 = \widehat{a}\widehat{e}_p + \widehat{b}\widehat{\phi}$. This implies

$$\begin{aligned}
&\forall i = 1, \dots, p:\ 1 = (\widehat{a}\widehat{e}_p\widehat{e}_i^{-1})\widehat{e}_i + \widehat{b}\widehat{\phi},\ \widehat{a}, \widehat{b}, \widehat{a}\widehat{e}_p\widehat{e}_i^{-1} \in \mathcal{S},\ \text{cf. (10.106)}\\
&\Longrightarrow \mathrm{id}_p = \widehat{A}\widehat{E} + \widehat{b}\widehat{\phi}\,\mathrm{id}_p,\ \widehat{A} := \widehat{a}\,\mathrm{diag}(\widehat{e}_p\widehat{e}_i^{-1};\ i = 1, \dots, p) \in \mathcal{S}^{p\times p}.
\end{aligned} \tag{10.114}$$

(ii) Assume $X \in \mathcal{S}^{m\times p}$ and $\widehat{\phi}^{-1}\widehat{N}_1 X \in \mathcal{S}^{p\times p}$. Define

$$\begin{aligned}
&Y := \begin{pmatrix}Y_I\\ Y_{II}\end{pmatrix} := \widehat{V}^{-1}X \in \mathcal{S}^{(p+(m-p))\times p} \Longrightarrow X = \widehat{V}\begin{pmatrix}Y_I\\ Y_{II}\end{pmatrix}\\
&\Longrightarrow \widehat{\phi}^{-1}\widehat{N}_1 X = \widehat{\phi}^{-1}\widehat{U}^{-1}(\widehat{E}, 0)\widehat{V}^{-1}\widehat{V}\begin{pmatrix}Y_I\\ Y_{II}\end{pmatrix} = \widehat{\phi}^{-1}\widehat{U}^{-1}\widehat{E}Y_I \Longrightarrow \widehat{\phi}^{-1}\widehat{E}Y_I \in \mathcal{S}^{p\times p}.
\end{aligned}$$

With (10.114) this implies

$$Z_I := \widehat{\phi}^{-1} Y_I = \mathrm{id}_p \widehat{\phi}^{-1} Y_I = (\widehat{A}\widehat{E} + \widehat{b}\widehat{\phi}\,\mathrm{id}_p)\widehat{\phi}^{-1} Y_I = \widehat{A}\left(\widehat{\phi}^{-1}\widehat{E} Y_I\right) + \widehat{b} Y_I \in \mathcal{S}^{p\times p}$$

$$\Longrightarrow X = \widehat{V}\left(\begin{smallmatrix} Y_I \\ Y_{II} \end{smallmatrix}\right) = \widehat{V}\left(\begin{smallmatrix} \widehat{\phi} Z_I \\ Z_{II} \end{smallmatrix}\right) \text{ with } Z_{II} := Y_{II}.$$

□

Corollary 10.2.10 (Algorithm) *We summarize the necessary constructive steps for the computation of $\mathcal{B}_2^1$ in Theorem 10.2.8. The plant*

$$\mathcal{B}_1 = \left\{\left(\begin{smallmatrix} y_1 \\ u_1 \end{smallmatrix}\right) \in \mathcal{F}^{p+m};\ P_1 \circ y_1 = Q_1 \circ u_1\right\} \textit{ with } H_1 = P_1^{-1} Q_1$$

and the nonzero polynomial $\phi \in \mathcal{D} = F[s]$ are given.
1. Let $d_1 := \deg_s(P_1, -Q_1)$, hence $\widehat{s}^{d_1}(P_1, -Q_1) \in \widehat{\mathcal{D}}^{p\times(p+m)}$. Compute the Smith form of this matrix over $\widehat{\mathcal{D}}$, i.e.,

$$U\left(\widehat{s}^{d_1}(P_1, -Q_1)\right) V = (E, 0),\ E = \mathrm{diag}(e_1, \dots, e_p) \in \widehat{\mathcal{D}}^{p\times p},\ e_1 \underset{\widehat{\mathcal{D}}}{\mid} \cdots \underset{\widehat{\mathcal{D}}}{\mid} e_p \neq 0,$$

$$U \in \mathrm{Gl}_p(\widehat{\mathcal{D}}),\ V \in \mathrm{Gl}_{p+m}(\widehat{\mathcal{D}}) \Longrightarrow U(P_1, -Q_1)V = (\widehat{s}^{-d_1} E, 0), \tag{10.115}$$

the latter equation being the Smith/McMillan form over $\widehat{\mathcal{D}}$ and over $\mathcal{D}_T = \mathcal{S}_{\widehat{s}}$ where all matrices have entries in $\mathcal{D}_T$. Check T-stabilizability of $\mathcal{B}_1$ by the condition $e_p \in$ U$(\mathcal{D}_T)$ or $(s-\alpha)^{\deg_{\widehat{s}}(e_p)} e_p \in T = \mathcal{D} \cap \mathrm{U}(\mathcal{D}_T)$, cf. (10.36). Assume this in the sequel. According to (10.34) we obtain

$$V = \begin{pmatrix} \widehat{D}_2^0 & \widehat{N}_1 \\ \widehat{N}_2^0 & \widehat{D}_1 \end{pmatrix},\ V^{-1} = \begin{pmatrix} \widehat{P}_1 & -\widehat{Q}_1 \\ -\widehat{Q}_2^0 & \widehat{P}_2^0 \end{pmatrix},\ H_1 = P_1^{-1} Q_1 = \widehat{P}_1^{-1}\widehat{Q}_1 = \widehat{N}_1 \widehat{D}_1^{-1}, \tag{10.116}$$

the latter representations being the coprime ones over $\widehat{\mathcal{D}}$. The matrices $\widehat{N}_1$ and $\widehat{Q}_1$ are equivalent over $\widehat{\mathcal{D}}$ and especially rank$(H_1) =$ rank$(\widehat{Q}_1) =$ rank$(\widehat{N}_1)$.
2. Compute the Smith form of $\widehat{N}_1$ (or $\widehat{Q}_1$) over $\widehat{\mathcal{D}}$ and especially rank$(\widehat{N}_1)$. *Assume the necessary condition $p =$ rank$(\widehat{N}_1)$ and let $\widehat{g} \in \widehat{\mathcal{D}}$ be its p-th elementary divisor. Compute $g := (s-\alpha)^{\deg_{\widehat{s}}(\widehat{g})}\widehat{g}$ and $h = \gcd_{\mathcal{D}}(\phi, g)$ according to (10.101). Check the condition $h \in T$, and assume this in the sequel. Then condition 5 of Theorem 10.2.8 and all equivalent conditions 1.–4. are satisfied. We now apply the part* 2. $\Longrightarrow$ 1. *of the theorem's proof to construct the desired compensators.*
3. Compute the Smith form of $(\widehat{\phi}\widehat{P}_1, -\widehat{Q}_1)$ over $\widehat{\mathcal{D}}$ as in (10.109), i.e.,

$$\widetilde{U}(\widehat{\phi}\widehat{P}_1, -\widehat{Q}_1)\widetilde{V} = (\widetilde{E}, 0),\ \widetilde{E} = \mathrm{diag}(\widetilde{e}_1, \dots, \widetilde{e}_p),\ \widetilde{e}_1 \underset{\widehat{\mathcal{D}}}{\mid} \cdots \underset{\widehat{\mathcal{D}}}{\mid} \widetilde{e}_p,\ 0 \neq \widetilde{e}_i \in \widehat{\mathcal{D}}$$

$$\widehat{\phi} := \widehat{s}^{\deg_s(\phi)}\phi \in \widehat{\mathcal{D}},\ \widetilde{U} \in \mathrm{Gl}_p(\widehat{\mathcal{D}}),\ \begin{pmatrix} \widetilde{D}_2 & \widetilde{N}_1 \\ \widetilde{N}_2 & \widetilde{D}_1 \end{pmatrix} := \widetilde{V} \in \mathrm{Gl}_{p+m}(\widehat{\mathcal{D}}). \tag{10.117}$$

4. Compute the matrix

$$\begin{pmatrix} \widetilde{D}_2^0 \\ \widetilde{N}_2^0 \end{pmatrix} := \begin{pmatrix} \widetilde{D}_2 \\ \widetilde{N}_2 \end{pmatrix} \widetilde{E}^{-1}\widetilde{U} \in \mathcal{S}^{(p+m)\times p} \Longrightarrow \begin{pmatrix} \widetilde{D}_2^0 & \widetilde{N}_1 \\ \widetilde{N}_2^0 & \widetilde{D}_1 \end{pmatrix} \in \mathrm{Gl}_{p+m}(\mathcal{S}). \tag{10.118}$$

5. Choose a matrix $\widetilde{X}_0 \in F^{m\times p}$ *such that* $\det\left(\widetilde{D}_2^0(0) - \widetilde{N}_1(0)\widetilde{X}_0\right) \neq 0$. *For almost all* $\widetilde{X}_0 \in F^{m\times p}$ *this is the case. Compute the matrices*

$$\begin{aligned} &\begin{pmatrix} \widetilde{D}_2^1 \\ \widetilde{N}_2^1 \end{pmatrix} := \begin{pmatrix} \widetilde{D}_2^0 \\ \widetilde{N}_2^0 \end{pmatrix} - \begin{pmatrix} \widetilde{N}_1 \\ \widetilde{D}_1 \end{pmatrix} \widetilde{X}_0 \in \mathcal{S}^{(p+m)\times p}, \begin{pmatrix} \widehat{D}_2^1 \\ \widehat{N}_2^1 \end{pmatrix} := \begin{pmatrix} \phi\widetilde{D}_2^1 \\ \widetilde{N}_2^1 \end{pmatrix}, \widehat{D}_2^1(0) \in \mathrm{Gl}_p(F) \\ &\Longrightarrow \begin{pmatrix} \widehat{D}_2^1 & \widehat{N}_1 \\ \widehat{N}_2^1 & \widehat{D}_1 \end{pmatrix}, \begin{pmatrix} \widehat{P}_1 & -\widehat{Q}_1 \\ -\widehat{Q}_2^1 & \widehat{P}_2^1 \end{pmatrix} := \begin{pmatrix} \widehat{D}_2^1 & \widehat{N}_1 \\ \widehat{N}_2^1 & \widehat{D}_1 \end{pmatrix}^{-1} \in \mathrm{Gl}_{p+m}(\mathcal{S}), \ \widehat{P}_2^1(0) \in \mathrm{Gl}_p(F), \\ &\Longrightarrow H_2^1 := \widehat{N}_2^1\left(\widehat{D}_2^1\right)^{-1} = \left(\widehat{P}_2^1\right)^{-1}\widehat{Q}_2^1 \in F(s)_{\mathrm{pr}}^{m\times p}. \end{aligned} \tag{10.119}$$

6. The controllable realization $\mathcal{B}_2^1$ *of* H_2^1 *is computed via Theorem 6.2.4, i.e.,*

$$\mathcal{B}_2^1 := \left\{ \begin{pmatrix} u_2 \\ y_2 \end{pmatrix} \in \mathcal{F}^{p+m};\ P_2^1 \circ y_2 = Q_2^1 \circ u_2 \right\} \tag{10.120}$$

where $(-Q_2^1, P_2^1) \in \mathcal{D}^{m\times(p+m)}$ *is the universal left annihilator of* $\begin{pmatrix} \mathrm{id}_p \\ H_2^1 \end{pmatrix} \in F(s)^{(p+m)\times m}$ *or of* $\begin{pmatrix} d\,\mathrm{id}_p \\ dH_2^1 \end{pmatrix} \in \mathcal{D}^{(p+m)\times m}$ *where* $d \in \mathcal{D}$ *is a common denominator of the entries of* H_2^1. *Then* $\mathcal{B}_2^1$ *is a controllable, proper,* T*-stabilizing compensator of* $\mathcal{B}_1$ *with proper feedback behavior* $\mathcal{B}$, *and the latter tracks signals* r *and rejects disturbances* u_2, u_1 *with* $\phi \circ (u_2, u_1, r) = 0$.

7. Constructivity*: Notice that all calculations except the tests of* $t \in T$ *only require the Smith form of polynomial matrices in* $F[s]$ *or* $F[\widehat{s}]$ *and the subsequent computation of their universal left annihilators; cf. the Euclidean ring computations in [4]. Assume, as in all practical cases, that* ϕ *and the coefficients of* $(P_1, -Q_1)$ *lie in the polynomial algebra* $\mathbb{Q}[j][s]$ *where* $\mathbb{Q}[j] \subset \mathbb{C}$, $j = \sqrt{-1}$, *is the field of Gaussian numbers. Then all matrices of the algorithm can be computed without any numerical error. However, the test of* $t \in T$, $t \in \mathbb{Q}[j][s]$, *for instance,* $\mathrm{V}_{\mathbb{C}}(h) \subseteq \Lambda_1$, *may use numerical approximations since the roots of* h *in* $\mathbb{C}$ *can be computed only by their approximations in* $\mathbb{Q}[j]$. *For special* Λ_1 *and* T *there are methods (Routh-Hurwitz criterion) to check* $t \in T$ *without computing the roots of* t. *We do not treat these methods and refer to [6] or [12] for a discussion. With today's computer algebra systems these criteria are less important than in the past.* ◇

In the next theorem we construct (parametrize) *all* compensators according to Theorem 10.2.4.

Theorem 10.2.11 (Parametrization, cf. [4, Theorems 7.5.2 and 7.5.32]) *Assume the equivalent conditions from Theorem 10.2.8, the compensator* $\mathcal{B}_2^1$ *from Corollary 10.2.10,* $\phi \notin T$, *hence* $\mathrm{rank}(\widehat{N}_1)$ $(= \mathrm{rank}(Q_1) = \mathrm{rank}(H_1)) = p \leq m$, *and the Smith form*

$$\widehat{U}\widehat{N}_1\widehat{V} = (\widehat{E}, 0),\ \widehat{U} \in \mathrm{Gl}_p(\widehat{\mathcal{D}}),\ \widehat{V} \in \mathrm{Gl}_m(\widehat{\mathcal{D}}),\ \widehat{E} \in \widehat{\mathcal{D}}^{p\times p},\ \mathrm{rank}(\widehat{E}) = p. \tag{10.121}$$

Choose matrices

$$Z = \begin{pmatrix} Z_I \\ Z_{II} \end{pmatrix} \in \mathcal{S}^{(p+(m-p))\times p} \text{ and define } X := \widehat{V}\begin{pmatrix} \widehat{\phi} Z_I \\ Z_{II} \end{pmatrix} \in \mathcal{S}^{(p+(m-p))\times p},$$
$$\begin{pmatrix} \widehat{D}_2 \\ \widehat{N}_2 \end{pmatrix} := \begin{pmatrix} \widehat{D}_2^1 \\ \widehat{N}_2^1 \end{pmatrix} - \begin{pmatrix} \widehat{N}_1 \\ \widehat{D}_1 \end{pmatrix} X, \ (-\widehat{Q}_2, \widehat{P}_2) := (-\widehat{Q}_2^1, \widehat{P}_2^1) + X(\widehat{P}_1, -\widehat{Q}_1). \tag{10.122}$$

Then $\det(\widehat{P}_2(0)) \neq 0$ *or* $\widehat{P}_2 \in \mathrm{Gl}_m(F(s)_{\mathrm{pr}})$ *for almost all* $Z(0) \in F^{m\times p}$. *We assume this and define* $H_2 := \widehat{P}_2^{-1}\widehat{Q}_2 = \widehat{N}_2\widehat{D}_2^{-1} \in F(s)_{\mathrm{pr}}^{m\times p}$. *Notice that* $\widehat{V}$, *but not* $\widehat{U}$ *and* $\widehat{E}$ *are used for the definition of* X.

1. *The unique controllable realization* $\mathcal{B}_2 \subseteq \mathcal{F}^{p+m}$ *of* H_2 *is a controllable, proper, T-stabilizing compensator of* $\mathcal{B}_1$ *with proper and T-stable feedback behavior* $\mathcal{B} := \mathrm{fb}(\mathcal{B}_1, \mathcal{B}_2)$ *such that* $\mathcal{B}$ *tracks all reference signals* r *and rejects all disturbances* u_2, u_1 *that satisfy* $\phi \circ (u_2, u_1, r) = 0$.
2. *Each compensator as in 1. is obtained by this construction, i.e., the matrices* $Z \in \mathcal{S}^{m\times p}$ *parametrize the set of all these compensators.*

Proof 1. We apply Theorem 10.1.34 to the behavior $\mathcal{B}_1$, but use the matrices

$$\begin{pmatrix} \widehat{P}_1 & -\widehat{Q}_1 \\ -\widehat{Q}_2^1 & \widehat{P}_2^1 \end{pmatrix} \in \mathrm{Gl}_{p+m}(\mathcal{S}), \ \begin{pmatrix} \widehat{D}_2^1 & \widehat{N}_1 \\ \widehat{N}_2^1 & \widehat{D}_1 \end{pmatrix} = \begin{pmatrix} \widehat{P}_1 & -\widehat{Q}_1 \\ -\widehat{Q}_2^1 & \widehat{P}_2^1 \end{pmatrix}^{-1} \text{ with}$$
$$\widehat{P}_2^1(0) \in \mathrm{Gl}_m(F), \ \widehat{P}_2^1 \in \mathrm{Gl}_m(F(s)_{\mathrm{pr}}) \text{ instead of}$$
$$\begin{pmatrix} \widehat{P}_1 & -\widehat{Q}_1 \\ -\widehat{Q}_2^0 & \widehat{P}_2^0 \end{pmatrix} \in \mathrm{Gl}_{p+m}(\widehat{\mathcal{D}}), \ \begin{pmatrix} \widehat{D}_2^0 & \widehat{N}_1 \\ \widehat{N}_2^0 & \widehat{D}_1 \end{pmatrix} = \begin{pmatrix} \widehat{P}_1 & -\widehat{Q}_1 \\ -\widehat{Q}_2^0 & \widehat{P}_2^0 \end{pmatrix}^{-1}.$$

The advantage is that the compensator $\mathcal{B}_2^1$ is proper by construction. Then

$$\widehat{D}_2 = \widehat{D}_2^1 - \widehat{N}_1 X, \ \widehat{\phi}^{-1}\widehat{D}_2 = \widehat{\phi}^{-1}\widehat{D}_2^1 - \widehat{\phi}^{-1}\widehat{N}_1 X \Longrightarrow \widehat{\phi}^{-1}\widehat{D}_2 \in \mathcal{S}^{p\times p} \text{ since}$$
$$\widehat{\phi}^{-1}\widehat{D}_1^2 \underset{(10.119)}{\in} \mathcal{S}^{p\times p}, \ \widehat{\phi}^{-1}\widehat{N}_1 X \underset{\text{Corollary 10.2.9}}{\in} \mathcal{S}^{p\times p}$$

Moreover $\det\left(\widehat{D}_2^1(0)\right) \neq 0$, hence for almost all

$$Z(0) \in F^{m\times p} \text{ and } X(0) = \widehat{V}(0)\begin{pmatrix} \widehat{\phi}(0)Z_I(0) \\ Z_{II}(0) \end{pmatrix} \in F^{m\times p} \text{ also}$$
$$\det\left(\widehat{D}_2(0)\right) = \det\left(\widehat{D}_2^1(0) - \widehat{N}_1(0)X(0)\right) \neq 0.$$

2. *One obtains all such compensators*: By construction $\mathcal{B}_2^1$ is one compensator with the asserted properties. Let $\mathcal{B}_2$ be any compensator of this type. According to Theorem 10.1.34 it is the controllable realization of the proper transfer matrix H_2 that is given by

$$
\begin{aligned}
&H_2 = \widehat{P}_2^{-1}\widehat{Q}_2 = \widehat{N}_2\widehat{D}_2^{-1} \text{ where}\\
&(-\widehat{Q}_2, \widehat{P}_2) = (-\widehat{Q}_2^1, \widehat{P}_2^1) + X(\widehat{P}_1, -\widehat{Q}_1),\ \begin{pmatrix}\widehat{D}_2\\ \widehat{N}_2\end{pmatrix} = \begin{pmatrix}\widehat{D}_2^1\\ \widehat{N}_2^1\end{pmatrix} - \begin{pmatrix}\widehat{N}_1\\ \widehat{D}_1\end{pmatrix} X,\ X \in \mathcal{S}^{m\times p}\\
&\widehat{\phi}^{-1}\widehat{D}_2 \underset{\text{Corollary 10.2.5}}{\in} \mathcal{S}^{p\times p},\ \widehat{P}_2(0) \in \mathrm{Gl}_m(F),\ \widehat{P}_2 \in \mathrm{Gl}_m(F(s)_{\mathrm{pr}})\\
&\Longrightarrow \widehat{\phi}^{-1}\widehat{N}_1 X = \widehat{\phi}^{-1}\widehat{D}_2 - \widehat{\phi}^{-1}\widehat{D}_2^1 \in \mathcal{S}^{p\times p}\\
&\underset{\text{Corollary 10.2.9}}{\Longrightarrow} X = \widehat{V}\begin{pmatrix}\widehat{\phi}z_I\\ z_{II}\end{pmatrix},\ \begin{pmatrix}z_I\\ z_{II}\end{pmatrix} \in \mathcal{S}^{m\times p}.
\end{aligned}
\tag{10.123}
$$

□

Example 10.2.12 0. *Construction of* $\mathcal{B}_1$: We choose the standard continuous-time case with

$$
\begin{aligned}
&F := \mathbb{R},\ \alpha := -1,\ \widehat{s} = (s+1)^{-1},\ T := \{t \in \mathbb{R}[s];\ \mathrm{V}_{\mathbb{C}}(t) \subset \mathbb{C}_-\} \text{ and}\\
&p := 2 < m := 3,\ \phi := s,\ \{r \in \mathcal{F};\ \phi \circ r = 0\} = \mathbb{R}.
\end{aligned}
$$

Hence the reference and disturbance signals have to be constant, a special, but often used case. We choose this simple ϕ in order to get printable results, and the matrices

$$
\begin{aligned}
&S_1 := \begin{pmatrix} 0 & 0 & s+2 & 0 & 0\\ 0 & 0 & 0 & 0 & s+3\end{pmatrix},\ V_1 := \begin{pmatrix} 1 & 0 & 0 & 0 & 0\\ s & 1 & 0 & 0 & 0\\ 1 & 0 & 1 & 0 & 0\\ 0 & 0 & 0 & 1 & 0\\ 0 & -1 & 0 & 0 & 1\end{pmatrix},\ V_2 := \begin{pmatrix} 1 & s-1 & 0 & 0 & 0\\ 0 & 1 & 0 & s & 0\\ 0 & 0 & 1 & 0 & -1\\ 0 & 0 & 0 & 1 & 0\\ 0 & 0 & 0 & 0 & 1\end{pmatrix} \in \mathrm{Gl}_5(\mathbb{R}[s]),\\
&V_3 := V_1V_2 = \begin{pmatrix} 1 & s-1 & 0 & 0 & 0\\ s & s^2-s+1 & 0 & s & 0\\ 1 & s-1 & 1 & 0 & -1\\ 0 & 0 & 0 & 1 & 0\\ 0 & -1 & 0 & -s & 1\end{pmatrix} \in \mathrm{Gl}_5(\mathbb{R}[s]),\\
&(P_1, -Q_1) := S_1V_3 \text{ where } P_1 = \begin{pmatrix} s+2 & s^2+s-2\\ 0 & -s-3\end{pmatrix},\ Q_1 = \begin{pmatrix} -s-2 & 0 & s+2\\ 0 & s^2+3s & -s-3\end{pmatrix},\\
&\mathcal{B}_1 = \left\{\begin{pmatrix} y_1\\ u_1\end{pmatrix} \in \mathcal{F}^{2+3} :\ P_1 \circ y_1 = Q_1 \circ u_1\right\}.
\end{aligned}
$$

By construction the Smith form of $(P_1, -Q_1)$ is $\begin{pmatrix} 1 & 0 & 0 & 0 & 0\\ 0 & (s+2)(s+3) & 0 & 0 & 0\end{pmatrix}$. Since $\det(P_1) = -(s+2)(s+3) \in T$, the behavior $\mathcal{B}_1$ is T-stable and thus T-stabilizable. The transfer matrix $H_1 = P_1^{-1}Q_1$ is polynomial of degree > 0 and not proper.
We now apply the algorithm of Corollary 10.2.10 to $(P_1, -Q_1)$ step by step.
1. Since $\deg_s(P_1, -Q_1) = 2$, we compute the Smith form of $\widehat{s}^2(P_1, -Q_1)$ over $\widehat{\mathcal{D}} = \mathbb{R}[\widehat{s}]$ according to (10.115) and (10.116) and obtain

$$
\begin{aligned}
&U\widehat{s}^2(P_1, -Q_1)V = (E, 0) \in \widehat{\mathcal{D}}^{2\times(2+3)},\ U \in \mathrm{Gl}_2(\widehat{\mathcal{D}}), V \in \mathrm{Gl}_5(\widehat{\mathcal{D}}),\\
&V = \begin{pmatrix}\widehat{D}_2^0 & \widehat{N}_1\\ \widehat{N}_2^0 & \widehat{D}_1\end{pmatrix} = \begin{pmatrix} 2 & 2\widehat{s}+\frac{3}{2} & -1 & 4\widehat{s}^2-6\widehat{s}+2 & 2\widehat{s}\\ 1 & \widehat{s}+\frac{1}{2} & 0 & 2\widehat{s}^2-2\widehat{s} & \widehat{s}\\ 0 & 0 & 1 & 0 & 0\\ 0 & -\frac{1}{2} & 0 & 2\widehat{s}^2 & \widehat{s}\\ 0 & \frac{1}{2} & 0 & 0 & 1\end{pmatrix} \in \mathrm{Gl}_5(\widehat{\mathcal{D}}) \subset \widehat{\mathcal{D}}^{(2+3)\times(2+3)},\\
&V^{-1} = \begin{pmatrix}\widehat{P}_1 & -\widehat{Q}_1\\ -\widehat{Q}_2^0 & \widehat{P}_2^0\end{pmatrix} = \begin{pmatrix} -2\widehat{s}^3-\widehat{s}^2+\widehat{s} & 4\widehat{s}^3+2\widehat{s}^2-2\widehat{s}+1 & -2\widehat{s}^3-\widehat{s}^2+\widehat{s} & -2\widehat{s}^2+\widehat{s}+1 & 2\widehat{s}^3-\widehat{s}^2-2\widehat{s}\\ 2\widehat{s}^2 & -4\widehat{s}^2 & 2\widehat{s}^2 & 2\widehat{s}-2 & -2\widehat{s}^2+2\widehat{s}\\ 0 & 0 & 1 & 0 & 0\\ \frac{1}{2}\widehat{s}+\frac{1}{2} & -\widehat{s}-1 & \frac{1}{2}\widehat{s}+\frac{1}{2} & \frac{1}{2} & -\frac{1}{2}\widehat{s}\\ -\widehat{s}^2 & 2\widehat{s}^2 & -\widehat{s}^2 & -\widehat{s}+1 & \widehat{s}^2-\widehat{s}+1\end{pmatrix}.
\end{aligned}
$$

2. We compute the Smith form of $\widehat{N}_1 \in \widehat{\mathcal{D}}^{2\times 3}$ over $\widehat{\mathcal{D}}$, see also item 8 below, or, sufficiently, its second elementary divisor $\widehat{g} = \widehat{s}$. Since it is nonzero, the rank condition $\mathrm{rank}(\widehat{N}_1) = 2$ follows. We compute $g = (s+1)^{\deg_{\widehat{s}}\widehat{s}} = 1 \in \mathcal{D}$ and conclude $h := \gcd_{\mathcal{D}}(\phi, g) = \gcd_{\mathcal{D}}(s, g) = 1 \in T$. Hence all equivalent conditions of Theorem 10.2.8 are satisfied.

3., 4. According to (10.117) and (10.118) the Smith form over $\widehat{\mathcal{D}}$ of $(\widehat{\phi P_1}, -\widehat{Q}_1) = ((1-\widehat{s})\widehat{P}_1, -\widehat{Q}_1)$ furnishes

$$\widetilde{U} = \begin{pmatrix} 0 & \frac{1}{2} \\ -2 & -\widehat{s}-1 \end{pmatrix}$$

$$\widetilde{V} = \begin{pmatrix} 0 & 0 & 0 & 1 & 0 \\ 0 & -\frac{1}{2} & 0 & 0 & 2\widehat{s} \\ 1 & 0 & \widehat{s}-1 & 0 & 0 \\ -\widehat{s}-1 & \frac{3}{2}\widehat{s} & -\widehat{s} & \widehat{s} & -6\widehat{s}^2+2\widehat{s} \\ 0 & -\widehat{s}+\frac{3}{2} & \widehat{s}-1 & -\widehat{s}+1 & 4\widehat{s}^2-6\widehat{s}+2 \end{pmatrix}$$

$$\begin{pmatrix} \widetilde{D}_2^0 \\ \widetilde{N}_2^0 \end{pmatrix} = \begin{pmatrix} 0 & 0 \\ 1 & \frac{1}{2}\widehat{s}+\frac{1}{2} \\ 0 & \frac{1}{2} \\ -3\widehat{s} & -\frac{3}{2}\widehat{s}^2-2\widehat{s}-\frac{1}{2} \\ 2\widehat{s}-3 & \widehat{s}^2-\frac{1}{2}\widehat{s}-\frac{3}{2} \end{pmatrix}$$

$$\begin{pmatrix} \widetilde{D}_2^0 & \widetilde{N}_1 \\ \widetilde{N}_2^0 & \widetilde{D}_1 \end{pmatrix} = \begin{pmatrix} 0 & 0 & 0 & 1 & 0 \\ 1 & \frac{1}{2}\widehat{s}+\frac{1}{2} & 0 & 0 & 2\widehat{s} \\ 0 & \frac{1}{2} & \widehat{s}-1 & 0 & 0 \\ -3\widehat{s} & -\frac{3}{2}\widehat{s}^2-2\widehat{s}-\frac{1}{2} & -\widehat{s} & \widehat{s} & -6\widehat{s}^2+2\widehat{s} \\ 2\widehat{s}-3 & \widehat{s}^2-\frac{1}{2}\widehat{s}-\frac{3}{2} & \widehat{s}-1 & -\widehat{s}+1 & 4\widehat{s}^2-6\widehat{s}+2 \end{pmatrix}.$$

5. We choose $\widetilde{X}_0 \in F^{3\times 2}$ with $\det\left(\widetilde{D}_2^0(0) - \widetilde{N}_1(0)\widetilde{X}_0\right) \neq 0$. Since $\det\left(\widetilde{D}_2^0(0)\right) = 0$ we choose

$$\widetilde{X}_0 := \begin{pmatrix} 0 & 0 \\ 1 & 0 \\ 0 & 0 \end{pmatrix}, \ \begin{pmatrix} \widetilde{D}_2^1 \\ \widetilde{N}_2^1 \end{pmatrix} := \begin{pmatrix} \widetilde{D}_2^0 \\ \widetilde{N}_2^0 \end{pmatrix} - \begin{pmatrix} \widetilde{N}_1 \\ \widetilde{D}_1 \end{pmatrix}\widetilde{X}_0 \in \mathcal{S}^{5\times 2} \text{ with}$$
$$\det\left(\widetilde{D}_2^0(0) - \widetilde{N}_1(0)\widetilde{X}_0\right) = -\frac{1}{2}$$

and obtain

$$\begin{pmatrix} \widetilde{D}_2^1 \\ \widetilde{N}_2^1 \end{pmatrix} = \begin{pmatrix} -1 & 0 \\ 1 & \frac{1}{2}\widehat{s}+\frac{1}{2} \\ 0 & \frac{1}{2} \\ -4\widehat{s} & -\frac{3}{2}\widehat{s}^2-2\widehat{s}-\frac{1}{2} \\ 3\widehat{s}-4 & \widehat{s}^2-\frac{1}{2}\widehat{s}-\frac{3}{2} \end{pmatrix},$$

$$\begin{pmatrix} \widehat{D}_2^1 \\ \widehat{N}_2^1 \end{pmatrix} = \begin{pmatrix} \phi\widetilde{D}_2^1 \\ \widetilde{N}_2^1 \end{pmatrix} = \begin{pmatrix} \widehat{s}-1 & 0 \\ -\widehat{s}+1 & -\frac{1}{2}\widehat{s}^2+\frac{1}{2} \\ 0 & \frac{1}{2} \\ -4\widehat{s} & -\frac{3}{2}\widehat{s}^2-2\widehat{s}-\frac{1}{2} \\ 3\widehat{s}-4 & \widehat{s}^2-\frac{1}{2}\widehat{s}-\frac{3}{2} \end{pmatrix}, \quad H_2^1 = \widehat{N}_2^1\left(\widehat{D}_2^1\right)^{-1}.$$

6. The left coprime factorization $\left(P_2^1\right)^{-1} Q_2^1 = H_2^1$ over $\mathbb{R}[s]$ furnishes the controllable realization of H_2^1 and is given by the matrices

$$P_2^1 = \begin{pmatrix} -s^3-2s^2 & -s^3-2s^2 & s \\ -4s^4-15s^3-15s^2-2s & -4s^4-15s^3-15s^2-2s & 4s^2+7s \\ 4s^4+16s^3+21s^2+10s & 4s^4+16s^3+21s^2+10s+1 & -4s^2-8s-4 \end{pmatrix},$$

$$Q_2^1 = \begin{pmatrix} 0 & 4s^2+4s-1 \\ -1 & 16s^3+44s^2+28s \\ 0 & -16s^3-48s^2-48s-16 \end{pmatrix}.$$

Here $(-Q_2^1, P_2^1) \in \mathcal{D}^{3\times(2+3)}$ is a universal left annihilator of $f_1\begin{pmatrix} \widehat{D}_2^1 \\ \widehat{N}_2^1 \end{pmatrix}$ where $f_1 \in \mathcal{D} = \mathbb{R}[s]$ is a common denominator of the entries of $\begin{pmatrix} \widehat{D}_2^1 \\ \widehat{N}_2^1 \end{pmatrix} \in F(s)^{(2+3)\times 2}$.

7. *Test for correctness*: According to (10.85) the error behavior is

$$\mathcal{E}^1 := \left\{e \in \mathcal{F}^2;\ E^1 \circ e = 0\right\} \text{ with}$$
$$E^1 = \begin{pmatrix} \frac{s^2+2s+1}{\frac{1}{2}s^5+\frac{15}{4}s^4+\frac{35}{4}s^3+\frac{25}{4}s^2+\frac{3}{4}s} & \frac{s^3+5s^2+7s+3}{\frac{1}{2}s^6+\frac{21}{4}s^5+21s^4+\frac{77}{2}s^3+\frac{63}{2}s^2\,\frac{49}{4}s+3} \end{pmatrix}.$$

The rank of E^1 is 2 and its largest elementary divisor is $(s+3)(s+1)^4$; hence, it is an autonomous T-stable behavior as required. The feedback behavior is a T-stable IO behavior and its transfer matrix is proper.

8. *Other parameters*: According to 10.121 we compute the Smith form of $\widehat{N}_1$ from 1, i.e.,

$$\widehat{U}\widehat{N}_1\widehat{V} = (\widehat{E}, 0) \in \widehat{\mathcal{D}}^{2\times(2+1)}, \ \widehat{V} = \begin{pmatrix} 1 & 0 & 0 \\ 0 & 1 & \widehat{s} \\ 0 & -2\widehat{s}+3 & -2\widehat{s}^2+3\widehat{s}-1 \end{pmatrix} \in \mathrm{Gl}_3(\widehat{\mathcal{D}}).$$

We choose the parameters

$$Z := \begin{pmatrix} \frac{1}{s+2} & 0 \\ 0 & 0 \\ 1 & 0 \end{pmatrix} \in \mathcal{S}^{(p+(m-p)\times p)} = \mathcal{S}^{(2+1)\times 2} \text{ and}$$

$$X := \widehat{V}\begin{pmatrix} \phi Z_I \\ Z_{II} \end{pmatrix} = \begin{pmatrix} \frac{-\widehat{s}^2+\widehat{s}}{\widehat{s}+1} & 0 \\ \widehat{s} & 0 \\ -2\widehat{s}^2+3\widehat{s}-1 & 0 \end{pmatrix}$$

to obtain

$$\begin{pmatrix} \widehat{D}_2^2 \\ \widehat{N}_2^2 \end{pmatrix} := \begin{pmatrix} \widehat{D}_2^1 \\ \widehat{N}_2^1 \end{pmatrix} - \begin{pmatrix} \widehat{N}_1 \\ \widehat{D}_1 \end{pmatrix} X \in \widehat{\mathcal{D}}^{(2+3)\times 2}, \ H_2^2 := \widehat{N}_2^2 \left(\widehat{D}_2^2\right)^{-1} = \left(P_2^2\right)^{-1} Q_2^2$$

where $(-Q_2^2, P_2^2) \in \mathcal{D}^{3\times(2+3)}$ is a universal left annihilator of $f_2 \begin{pmatrix} \widehat{D}_2^2 \\ \widehat{N}_2^2 \end{pmatrix} \in \mathcal{D}^{(2+3)\times 2}$ with a common denominator $f_2 \in \mathcal{D}$ of $\begin{pmatrix} \widehat{D}_2^2 \\ \widehat{N}_2^2 \end{pmatrix} \in \widehat{\mathcal{D}}^{(2+3)\times 2} \subset F(s)^{(2+3)\times 2}$. The computations give

$$\begin{pmatrix} \widehat{D}_2^2 \\ \widehat{N}_2^2 \end{pmatrix} = \begin{pmatrix} \frac{\widehat{s}-1}{\widehat{s}+1} & 0 \\ -\widehat{s}^2+1 & -\frac{1}{2}\widehat{s}^2+\frac{1}{2} \\ \frac{\widehat{s}^2-\widehat{s}}{\widehat{s}+1} & \frac{1}{2} \\ -3\widehat{s}^2-3\widehat{s} & -\frac{3}{2}\widehat{s}^2-2\widehat{s}-\frac{1}{2} \\ 2\widehat{s}^2-3 & \widehat{s}^2-\frac{1}{2}\widehat{s}-\frac{3}{2} \end{pmatrix}$$

$$P_2^2 = \begin{pmatrix} s^2+2s & 0 & s^3+4s^2+2s \\ -\frac{3}{2}s^6-\frac{21}{2}s^5-28s^4-34s^3-17s^2-2s & 0 & -\frac{3}{2}s^7-\frac{27}{2}s^6-46s^5-75s^4-59s^3-19s^2-s \\ -\frac{9}{2}s^8-\frac{81}{2}s^7-144s^6-249s^5-199s^4-43s^3+10s^2-6s+4 & -1 & -\frac{9}{2}s^9-\frac{99}{2}s^8-216s^7-474s^6-535s^5-264s^4-17s^3+4s^2-s+6 \end{pmatrix}$$

$$Q_2^2 = \begin{pmatrix} 0 & -3s^3-12s^2-8s-1 \\ -s-2 & \frac{9}{2}s^7+\frac{81}{2}s^6+141s^5+\frac{483}{2}s^4+\frac{421}{2}s^3+86s^2+12s \\ -3s^3-12s^2-10s+4 & \frac{27}{2}s^9+\frac{297}{2}s^8+657s^7+\frac{2979}{2}s^6+\frac{3597}{2}s^5+1047s^4+185s^3-10s^2-16 \end{pmatrix}.$$

As correctness test we use (10.85) to compute the error behavior $\mathcal{E}^2 := \{e \in \mathcal{F}^2;\ E^2 \circ e = 0\}$, and obtain $E^2 \in \mathcal{D}^{2\times 2}$ with

$$\begin{aligned}
E_{11}^2 &= -2s^2-8s-8 \\
E_{12}^2 &= -s^3-5s^2-7s-3 \\
E_{21}^2 &= -9s^{12}-135s^{11}-873s^{10}-3171s^9-7040s^8-9680s^7-7862s^6-3226s^5-359s^4+43s^3+31s^2+55s-6 \\
E_{22}^2 &= -\tfrac{9}{2}s^{13}-72s^{12}-\tfrac{999}{2}s^{11}-1968s^{10}-\tfrac{9671}{2}s^9-7634s^8-\tfrac{15299}{2}s^7-4574s^6-1388s^5-117s^4+27s^3+32s^2+14s-3.
\end{aligned}$$

The rank of E^2 is 2 and its largest elementary divisor is $(s+3)(s+2)(s+1)^5$; hence, it is an autonomous T-stable behavior. The feedback behavior is a T-stable IO behavior and its transfer matrix is proper.

It is unlikely that such results can be obtained by the Euclidean algorithm for S in [4, Sect. 2.3]. ◊

10.2.2 Connection with the Zeros of H_1

We describe the connection of conditions 4 and 5 of Theorem 10.2.8 *with the zeros of the transfer matrix of* H_1. *Cf.* [13, p. 446], [3, p. 627], [10, p. 42].

With the data of the preceding section we assume the standard case

$$F = \mathbb{C}, \mathbb{R},\ \mathcal{D} = F[s],\ \mathbb{C} = \Lambda_1 \uplus \Lambda_2,\ \Lambda_1 \subseteq \Lambda_{1,\mathrm{std}},\ T = \{t \in \mathcal{D};\ \mathrm{V}_{\mathbb{C}}(t) \subseteq \Lambda_1\},$$
$$\alpha \in F \cap \Lambda_1,\ \widehat{s} = (s-\alpha)^{-1},\ \widehat{\mathcal{D}} = F[\widehat{s}],\ 0 \neq f \in \mathcal{D},\ \widehat{f}_\alpha = (s-\alpha)^{-\deg_s(f)} f \in \widehat{\mathcal{D}},$$
$$p \leq m,\ H_1 = P_1^{-1}Q_1 = \widehat{P}_1^{-1}\widehat{Q}_1 = \widehat{N}_1\widehat{D}_1^{-1},\ \mathrm{rank}(H_1) = p. \tag{10.124}$$

For simplicity we assume that also $H_1 = P_1^{-1}Q_1$ is the coprime representation, i.e., that $\mathcal{B}_1 = \left\{\left(\begin{smallmatrix} y_1 \\ u_1 \end{smallmatrix}\right) \in \mathcal{F}^{p+m};\ P_1 \circ y_1 = Q_1 \circ u_1\right\}$ is the controllable realization of H_1. Recall that $p = \mathrm{rank}(H_1) = \mathrm{rank}(Q_1) = \mathrm{rank}(\widehat{Q}_1) = \mathrm{rank}(\widehat{N}_1)$.

Definition 10.2.13 The behavior

$$\mathrm{zd}(H_1) := \left\{u_1 \in \mathcal{F}^m;\ Q_1 \circ \circ u_1 = 0\right\} \cong \mathcal{B}_1 \cap \left(\{0\} \times \mathcal{F}^m\right),\ u_1 \leftrightarrow \left(\begin{smallmatrix} 0 \\ u_1 \end{smallmatrix}\right),$$

is called the *zero-dynamics* of H_1. Its characteristic variety

$$\mathrm{zero}(H_1) := \mathrm{char}(\mathrm{zd}(H_1)) = \{\lambda \in \mathbb{C};\ \mathrm{rank}(Q_1(\lambda)) < p = \mathrm{rank}(Q_1)\}$$

is called the set of *McMillan or transmission zeros* of $\mathcal{B}_1$ or H_1, cf. [13, p. 446], [10, p. 42]. ◊

Lemma 10.2.14 *The condition* $\mathrm{rank}(Q_1) = p$ *is equivalent to the surjectivity of* $\mathrm{proj} = (\mathrm{id}_p, 0)\circ :\ \mathcal{B}_1 \to \mathcal{F}^p,\ \left(\begin{smallmatrix} y_1 \\ u_1 \end{smallmatrix}\right) \mapsto y_1$.

Proof By duality we obtain the equivalences

$$\begin{aligned}
&\mathcal{B}_1 \xrightarrow{(\mathrm{id}_p,0)} \mathcal{F}^p \text{ surjective}\\
&\Longleftrightarrow \mathcal{D}^{1\times p} \xrightarrow{(\cdot(\mathrm{id}_p,0))_{\mathrm{ind}}} \mathcal{D}^{1\times(p+m)}/\mathcal{D}^{1\times p}(P_1, -Q_1),\ \xi \mapsto \overline{(\xi, 0)},\ \text{injective}\\
&\Longleftrightarrow \forall \xi, \eta \in \mathcal{D}^{1\times p}:\ ((\xi, 0) = \eta(P_1, -Q_1) = (\eta P_1, -\eta Q_1) \Longrightarrow \xi = 0)\\
&\Longleftrightarrow \forall \xi \in \mathcal{D}^{1\times p}:\ \left(\xi H_1 = \xi P_1^{-1}Q_1 = 0 \Longrightarrow \xi = 0\right) \Longleftrightarrow \mathrm{rank}(H_1) = p.
\end{aligned}$$

□

Since $\mathcal{B}_1$ is controllable, Theorem 6.2.9 implies $\{\lambda \in \mathbb{C};\ \mathrm{rank}(P_1(\lambda)) < p\} = \mathrm{pole}(H_1)$.

Lemma 10.2.15 *If* $F = \mathbb{C}$ *and* $\lambda \notin \mathrm{pole}(H_1)$ *and thus* $H_1(\lambda) \in \mathbb{C}^{p\times m}$, *then*

$$\lambda \in \mathrm{zero}(H_1) \Longleftrightarrow \exists 0 \neq \xi \in \mathbb{C}^{1\times p} \forall u = u(0)e_{\lambda,0}(t) \in \mathbb{C}^m e_{\lambda,0}(t):\ \xi H_1(\lambda)u = 0.$$

Here $u = u(0)e_{\lambda,0}$ *resp.* $H_1(\lambda)u$ *are a sinusoidal input resp. output, cf. (5.15) and (5.22). An analogous result holds for* $F = \mathbb{R}$.

Proof The matrix $P_1(\lambda)$ is invertible and $H_1(\lambda) = P_1(\lambda)^{-1}Q_1(\lambda)$ is defined with $\mathrm{rank}(Q_1(\lambda)) = \mathrm{rank}(H_1(\lambda))$ since λ is not a pole of H_1. Then

$$\lambda \in \operatorname{zero}(H_1) \Longleftrightarrow \operatorname{rank}(Q_1(\lambda)) = \operatorname{rank}(H_1(\lambda)) < p$$
$$\Longleftrightarrow \exists 0 \neq \xi \in \mathbb{C}^{1\times p} \text{ with } \xi H_1(\lambda) = 0$$
$$\Longleftrightarrow \exists 0 \neq \xi \in \mathbb{C}^{1\times p} \forall u(0) \in \mathbb{C}^m : \ \xi H_1(\lambda)u(0) = 0 \left(\Longleftrightarrow \xi H_1(\lambda)u(0)e_{\lambda,0} = 0\right).$$

□

Next we compare properties of the two coprime factorizations $P_1^{-1}Q_1 = \widehat{P}_1^{-1}\widehat{Q}_1 = H_1$ over $\mathcal{D}$ resp. $\widehat{\mathcal{D}}$. Recall the fractional-linear transformation

$$\mathbb{C} \uplus \{\infty\} \cong \mathbb{C} \uplus \{\infty\}, \ \lambda = \alpha + \widehat{\lambda}^{-1} \leftrightarrow \widehat{\lambda} = (\lambda - \alpha)^{-1}, \ \alpha \leftrightarrow \infty, \ \infty \leftrightarrow 0. \tag{10.125}$$

Lemma 10.2.16 *For* $\alpha \neq \lambda \in \mathbb{C}$ *and* $\widehat{\lambda} := (\lambda - \alpha)^{-1}$ *the identity* $\operatorname{rank}(Q_1(\lambda)) = \operatorname{rank}\left(\widehat{Q}_1(\widehat{\lambda})\right)$ *holds. This implies the bijection*

$$\begin{aligned} \operatorname{zero}(H_1) \setminus \{\alpha\} &= \{\lambda \in \mathbb{C};\ \lambda \neq \alpha,\ \operatorname{rank}(Q_1(\lambda)) < p\} \\ &\cong \left\{\widehat{\lambda} \in \mathbb{C};\ \widehat{\lambda} \neq 0,\ \operatorname{rank}(\widehat{Q}_1(\widehat{\lambda})) < p\right\}, \ \lambda \leftrightarrow \widehat{\lambda}. \end{aligned} \tag{10.126}$$

Proof Let

$$\begin{aligned} & d := \deg_s(P_1, -Q_1),\ \widehat{d} := \deg_{\widehat{s}}(\widehat{P}_1, -\widehat{Q}_1). \text{ Then} \\ & \widehat{s}^d(P_1, -Q_1) \in \widehat{\mathcal{D}}^{p\times(p+m)},\ H_1 = \left(\widehat{s}^d P_1\right)^{-1}\left(\widehat{s}^d Q_1\right) = \widehat{P}_1^{-1}\widehat{Q}_1 \\ & \underset{\widehat{P}_1^{-1}\widehat{Q}_1 \text{ coprime}}{\Longrightarrow} \quad \exists \widehat{A} \in \widehat{\mathcal{D}}^{p\times p} \text{ with } \widehat{s}^d(P_1, -Q_1) = \widehat{A}(\widehat{P}_1, -\widehat{Q}_1) \\ & \Longrightarrow (s-\alpha)^{-d}Q_1(s) = \widehat{A}\left((s-\alpha)^{-1}\right)\widehat{Q}_1\left((s-\alpha)^{-1}\right) \\ & \underset{s=\lambda}{\Longrightarrow} \widehat{\lambda}^d Q_1(\lambda) = \widehat{A}(\widehat{\lambda})\widehat{Q}_1(\widehat{\lambda}) \underset{\widehat{\lambda}\neq 0}{\Longrightarrow} \operatorname{rank}(Q_1(\lambda)) \leq \operatorname{rank}(\widehat{Q}_1(\widehat{\lambda})). \end{aligned}$$

Likewise $\left((s-\alpha)^{\widehat{d}}\widehat{P}_1\right)^{-1}\left((s-\alpha)^{\widehat{d}}\widehat{Q}_1\right) = P_1^{-1}Q_1$ implies $\operatorname{rank}(\widehat{Q}_1(\widehat{\lambda})) \leq \operatorname{rank}(Q_1(\lambda))$. □

By Corollary and Definition 6.2.6 $\widehat{Q}_1$ and $\widehat{N}_1$ are equivalent over $\widehat{\mathcal{D}}$; hence, $\widehat{g}$ is the p-th elementary divisor of $\widehat{N}_1$ and $\widehat{Q}_1$ and therefore

$$\left\{\widehat{\lambda} \in \mathbb{C};\ \operatorname{rank}\left(\widehat{Q}_1(\widehat{\lambda})\right) < p\right\} = \mathrm{V}_{\mathbb{C}}(\widehat{g}) := \left\{\widehat{\lambda} \in \mathbb{C};\ \widehat{g}(\widehat{\lambda}) = 0\right\}. \tag{10.127}$$

Also recall

$$\widehat{\Lambda}_1 := \left\{\widehat{\lambda} = (\lambda - \alpha)^{-1};\ \alpha \neq \lambda \in \Lambda_1\right\},\ \widehat{T} := \left\{\widehat{t} \in \mathbb{C}[\widehat{s}];\ \mathrm{V}_{\mathbb{C}}(\widehat{t}) \subseteq \widehat{\Lambda}_1\right\}. \tag{10.128}$$

Theorem 10.2.17 *The equivalent conditions of Theorem 10.2.8 are satisfied if and only if* $\mathrm{V}_{\mathbb{C}}(\phi) \cap \operatorname{zero}(H_1) \subseteq \Lambda_1$.

Proof Since $\alpha \in \Lambda_1$ the condition of the theorem is equivalent to

$$\mathrm{V}_{\mathbb{C}}(\phi) \cap (\mathrm{zero}(H_1) \setminus \{\alpha\}) \subseteq \Lambda_1. \text{ Assume } \lambda \in \mathrm{V}_{\mathbb{C}}(\phi) \cap (\mathrm{zero}(H_1) \setminus \{\alpha\}).$$

Recall

$$\begin{aligned}
&\widehat{\phi} = \widehat{s}^{\deg_s(\phi)}\phi \Longrightarrow \widehat{\phi}(\widehat{\lambda}) = \widehat{\lambda}^{\deg_s(\phi)}\phi(\lambda) \Longrightarrow \left(\phi(\lambda) = 0 \Longleftrightarrow \widehat{\phi}(\widehat{\lambda}) = 0\right)\\
&\Longrightarrow \mathrm{V}_{\mathbb{C}}(\phi) \cap (\mathrm{zero}(H_1) \setminus \{\alpha\}) \underset{(10.126),(10.127)}{\cong} \mathrm{V}_{\mathbb{C}}(\widehat{\phi}) \cap (\mathrm{V}_{\mathbb{C}}(\widehat{g}) \setminus \{0\})\\
&\underset{\widehat{\phi}(0)\neq 0}{=} \mathrm{V}_{\mathbb{C}}(\widehat{\phi}) \cap \mathrm{V}_{\mathbb{C}}(\widehat{g}) = \mathrm{V}_{\mathbb{C}}\left(\underset{\widehat{\mathcal{D}}}{\gcd}(\widehat{\phi}, \widehat{g})\right), \ \lambda \leftrightarrow \widehat{\lambda}.
\end{aligned}$$

With $\Lambda_1 \setminus \{\alpha\} \cong \widehat{\Lambda}_1, \ \lambda \leftrightarrow \widehat{\lambda}$, we infer

$$\begin{aligned}
&\mathrm{V}_{\mathbb{C}}(\phi_1) \cap (\mathrm{zero}(H_1) \setminus \{\alpha\}) \subseteq \Lambda_1 \Longleftrightarrow \mathrm{V}_{\mathbb{C}}\left(\underset{\widehat{\mathcal{D}}}{\gcd}(\widehat{\phi}, \widehat{g})\right) \subseteq \widehat{\Lambda}_1\\
&\Longleftrightarrow \underset{\widehat{\mathcal{D}}}{\gcd}(\widehat{\phi}, \widehat{g}) \in \widehat{T} \Longleftrightarrow \text{condition 4 of Theorem 10.2.8.}
\end{aligned}$$

□

Example 10.2.18 Consider the continuous-time case and $F = \mathbb{C}$. With the data from (10.87) and from Theorem 10.2.17 with the closed-loop (feedback) behavior $\mathcal{B}$ assume that the condition of the theorem is not satisfied. Therefore there is $\lambda \in (\mathrm{V}_{\mathbb{C}}(\phi) \cap \mathrm{zero}(H_1)) \setminus \Lambda_1$, hence $\lambda \neq \alpha$, $\widehat{\lambda} := (\lambda - \alpha)^{-1} \in \mathbb{C}$ and $\mathrm{rank}\left(\widehat{N}_1(\widehat{\lambda})\right) < p$. Choose $0 \neq \xi \in \mathbb{C}^{1\times p}$ with $\xi\widehat{N}_1(\widehat{\lambda}) = 0$, the disturbance signals $u_1 := 0, \ u_2 = 0$, the reference signal

$$r := \xi^* e^{\lambda t}, \ \xi^* := \overline{\xi}^{\top} \left(\Longrightarrow \phi \circ r = \phi(\lambda) r = 0, \ \xi r = \|\xi\|_2^2 e^{\lambda t} \neq 0\right)$$

and the input signal $\left(\begin{smallmatrix} u_2 - r \\ u_1 \end{smallmatrix}\right) = \left(\begin{smallmatrix} -r \\ 0 \end{smallmatrix}\right)$ of $\mathcal{B}$, cf. (10.88).

The closed-loop transfer matrix is $H = P^{-1}Q = \left(\begin{smallmatrix} \widehat{N}_1\widehat{Q}_2 & \widehat{D}_2\widehat{Q}_1 \\ \widehat{D}_1\widehat{Q}_2 & \widehat{N}_2\widehat{Q}_1 \end{smallmatrix}\right)$. The behavior $\mathcal{B}^0$ is T-stable by construction and $\lambda \notin \Lambda_1$; hence, $P(\lambda) \in \mathrm{Gl}_{p+m}(\mathbb{C})$ and $H(\lambda) \in \mathbb{C}^{(p+m)\times(p+m)}$ are defined. Therefore

$$\begin{aligned}
&\left(\begin{smallmatrix} y_1 \\ y_2 \end{smallmatrix}\right) := H(\lambda)\left(\begin{smallmatrix} -r \\ 0 \end{smallmatrix}\right) \in \mathcal{B}, \ y_1 = H_{y_1,u_2}(\lambda)(-r) = \widehat{N}_1(\widehat{\lambda})\widehat{Q}_2(\widehat{\lambda})(-r)\\
&\underset{\xi\widehat{N}_1(\widehat{\lambda})=0}{\Longrightarrow} \xi(y_1 - r) = -\xi r = -\|\xi\|_2^2 e^{\lambda t} \neq 0 \Longrightarrow y_1 - r \notin \oplus_{\mu\in\Lambda_1}\mathbb{C}[t]^p e^{\mu t},
\end{aligned}$$

i.e., the plant output y_1 does not track r although $\phi \circ r = 0$.

If $F = \mathbb{R}$ and $\phi \in \mathbb{R}[s]$ write $\lambda = \rho + j\omega$. Let $r = \xi^* e^{\lambda t}$ and y_1 be defined as before. Then the real (imaginary) part $\Re(y_1)$ ($\Im(y_1)$) is an output of the real behavior $\mathcal{B}$ to the real reference signal $\Re(r)$ ($\Im(r)$). As before we obtain

$$\phi \circ r = 0,\ \xi(y_1 - r) = -\|\xi\|_2^2 e^{\lambda t} \Longrightarrow \phi \circ \Re(r) = \phi \circ \Im(r) = 0 \text{ and}$$
$$\Re\left(\xi(y_1 - r)\right) = \Re(\xi)(\Re(y_1) - \Re(r)) - \Im(\xi)(\Im(y_1) - \Im(r)) = -\|\xi\|_2^2 \Re(e^{\lambda t}).$$

If both $\Re(y_1) - \Re(r)$ and $\Im(y_1) - \Im(r)$ were T-stable, then so would be $\|\xi\|_2^2 \Re(e^{\lambda t}) = \|\xi\|_2^2 e^{\rho t} \cos(\omega t)$ which contradicts $\lambda \notin \Lambda_1$. If, for instance, $\Re(y_1) - \Re(r)$ is not T-stable, then $\Re(y_1)$ does not track $\Re(r)$ although $\phi \circ \Re(r) = 0$. Hence the conditions of Theorem 10.2.8 are not satisfied. ◊

10.2.3 Graph Topology and Robustness

A stabilizing compensator $\mathcal{B}_2$ of a stabilizable behavior $\mathcal{B}_1$ is called robust if it is also a stabilizing compensator of all plants $\widetilde{\mathcal{B}_1}$ in the neighborhood of $\mathcal{B}_1$. Theorem 10.2.33 *and Example* 10.2.34 *show that the compensators from Theorem* 10.2.11 *are robust and that moreover the proper and stable closed-loop transfer matrix $\widetilde{H}$ of the feedback behavior* fb$(\widetilde{\mathcal{B}_1}, \mathcal{B}_2)$ *depends continuously on $\widetilde{\mathcal{B}_1}$. To discuss this one obviously needs notions of topology that we are going to introduce first. We repeat that our presentation is a variant of* [4, Sect. 7.2].

We consider systems with the standard continuous or discrete signal spaces $\mathcal{F}$ over the ground-field $F = \mathbb{C}, \mathbb{R}$ and define stability by means of a disjoint decomposition of

$$\mathbb{C} = \Lambda_1 \uplus \Lambda_2,\ \text{open } \Lambda_1 \subset \Lambda_{1,\mathrm{std}} = \begin{cases} \mathbb{C}_- = \{z \in \mathbb{C};\ \Re(z) < 0\} & \text{continuous time} \\ \mathbb{D} := \{z \in \mathbb{C};\ |z| < 1\} & \text{discrete time} \end{cases},$$
$$\Lambda_1 = \overline{\Lambda_1} \text{ if } F = \mathbb{R},\ T := \{t \in F[s];\ \mathrm{V}_{\mathbb{C}}(t) \subseteq \Lambda_1\}. \tag{10.129}$$

We extend the affine line $\mathbb{C}$ to the projective line $\mathbb{P} := \mathbb{C} \uplus \{\infty\}$. In the standard fashion $\mathbb{P}$ becomes a compact topological space. Bases of neighborhoods of the points $z_0 \in \mathbb{P}$ are

$$B(z_0, r) := \begin{cases} \{z \in \mathbb{C};\ |z - z_0| < r\} & \text{if } z_0 \in \mathbb{C} \\ \{z \in \mathbb{C};\ |z| > r\} \uplus \{\infty\} & \text{if } z_0 = \infty \end{cases},\ r > 0,\ z_0 \in \mathbb{P}. \tag{10.130}$$

Then $\mathbb{C}$ is open in $\mathbb{P}$ and every open subset of $\mathbb{C}$ is also open in $\mathbb{P}$. The decomposition $\mathbb{C} = \Lambda_1 \uplus \Lambda_2$ induces the decomposition

$$\mathbb{P} = \Lambda_1 \uplus \mathcal{K},\ \mathcal{K} := \Lambda_2 \uplus \{\infty\}. \tag{10.131}$$

Since Λ_1 is assumed open the set $\mathcal{K}$ is closed and therefore compact in the compact space $\mathbb{P}$.

The algebra $B := \mathrm{C}^0(\mathcal{K}) = \mathrm{C}^0(\mathcal{K}, \mathbb{C})$ of continuous functions from $\mathcal{K}$ to $\mathbb{C}$ is a commutative *normed* $\mathbb{C}$-*algebra* with the maximum norm

$$\|f\| := \sup\{|f(z)|;\ z \in \mathcal{K}\} = \max\{|f(z)|;\ z \in \mathcal{K}\} \text{ with } \|f+g\| \le \|f\| + \|g\|,$$
$$\|\alpha\| = |\alpha|,\ \|\alpha f\| = |\alpha|\|f\|,\ \|fg\| \le \|f\|\|g\|,\ f, g \in B = \mathrm{C}^0(\mathcal{K}, \mathbb{C}),\ \alpha \in F. \tag{10.132}$$

Convergence with respect to this norm is uniform convergence on $\mathcal{K}$. The algebra is complete, i.e., every Cauchy sequence converges, and therefore a *Banach* $\mathbb{C}$-*algebra*. Notice that even for $F = \mathbb{R}$ the functions in $\mathrm{C}^0(\mathcal{K})$ are complex-valued. A function $f \in B$ is a unit, i.e.,

$$f \in \mathrm{U}(B) \Longleftrightarrow \forall z \in \mathcal{K}:\ f(z) \neq 0 \Longleftrightarrow \min\{|f(z)|;\ z \in \mathcal{K}\} > 0. \tag{10.133}$$

If $0 \neq H = ft^{-1} \in \mathcal{D}_T = F[s]_T$ and thus $\mathrm{V}_{\mathbb{C}}(t) \subseteq \Lambda_1$ or $t(z) \neq 0$ for all $z \in \Lambda_2$ the *complex-valued* function $(ft^{-1})(z) := f(z)t(z)^{-1}$ is continuous on Λ_2. If in addition $H = ft^{-1}$ is proper then

$$\begin{aligned}
& ft^{-1} = \tfrac{f}{t} = \tfrac{f_m s^m + \cdots}{t_n s^n + \cdots},\ m \le n,\ t_n \neq 0 \\
\Longrightarrow\ & \lim_{z\to\infty} (ft^{-1})(z) = \begin{cases} \frac{f_m}{t_m} & \text{if } m = n \\ 0 & \text{if } m < n \end{cases} =: (ft^{-1})(\infty) \\
\Longrightarrow\ & H(z) = (ft^{-1})(z) \in B = \mathrm{C}^0(\mathcal{K}, \mathbb{C}),\ \|H\| = \sup_{z\in\mathcal{K}} |H(z)|.
\end{aligned} \tag{10.134}$$

We identify $H = ft^{-1}$ with the function $(ft^{-1})(z),\ z \in \mathcal{K}$, and obtain the inclusion

$$\mathcal{S} := F[s]_T \cap F(s)_{\mathrm{pr}} \subset B := \mathrm{C}^0(\mathcal{K}) = \mathrm{C}^0(\mathcal{K}, \mathbb{C}). \tag{10.135}$$

Under additional assumptions on Λ_2 the computation of $\|f\|$ can be sharpened by means of the maximum principle for holomorphic functions. If X' is a subset of a topological space then $\mathrm{int}_X(X')$ is its interior, $\mathrm{cl}_X(X')$ its closure and $\partial_X(X') := \mathrm{cl}_X(X') \setminus \mathrm{int}_X(X')$ its boundary. *We additionally assume that* Λ_2 *is unbounded and* $\mathrm{int}_{\mathbb{C}}(\Lambda_2)$ *is dense in* Λ_2 *and connected.* This is satisfied in all standard cases.

Lemma 10.2.19 *The decompositions*

$$\Lambda_2 = \mathrm{int}_{\mathbb{C}}(\Lambda_2) \uplus \partial_{\mathbb{C}}(\Lambda_2) \subset \mathcal{K} = \mathrm{int}_{\mathbb{P}}(\mathcal{K}) \uplus \partial_{\mathbb{P}}(\mathcal{K})$$

are related by

$$\begin{aligned}
& \mathrm{int}_{\mathbb{P}}(\mathcal{K}) = \begin{cases} \mathrm{int}_{\mathbb{C}}(\Lambda_2) & \textit{if (i): } \Lambda_1 \textit{ is unbounded} \\ \mathrm{int}_{\mathbb{C}}(\Lambda_2) \uplus \{\infty\} & \textit{if (ii): } \Lambda_1 \textit{ is bounded} \end{cases} \\
\Longrightarrow\ & \partial_{\mathbb{P}}(\mathcal{K}) = \begin{cases} \partial_{\mathbb{C}}(\Lambda_2) \uplus \{\infty\} & \textit{if (i): } \Lambda_1 \textit{ is unbounded} \\ \partial_{\mathbb{C}}(\Lambda_2) & \textit{if (ii): } \Lambda_1 \textit{ is bounded} \end{cases}
\end{aligned} \tag{10.136}$$

Proof (ii)

$$\begin{aligned}\infty \in \mathrm{int}_{\mathbb{P}}(\mathcal{K}) &\Longleftrightarrow \exists r > 0 \text{ with } B(\infty, r) \subset \mathcal{K} \\ &\Longleftrightarrow \exists r > 0 \text{ with } B(\infty, r) \cap \Lambda_1 = \emptyset \Longleftrightarrow \Lambda_1 \text{ bounded} \qquad (10.137)\\ \Longrightarrow \mathrm{int}_{\mathbb{P}}(\mathcal{K}) &= \mathrm{int}_{\mathbb{C}}(\Lambda_2) \uplus \{\infty\}, \ \partial_{\mathbb{P}}(\mathcal{K}) = \partial_{\mathbb{C}}(\Lambda_2)\end{aligned}$$

(i) If Λ_1 is unbounded or, equivalently, $\infty \notin \mathrm{int}_{\mathbb{P}}(\mathcal{K})$ then

$$\mathrm{int}_{\mathbb{P}}(\mathcal{K}) = \mathrm{int}_{\mathbb{C}}(\Lambda_2), \ \partial_{\mathbb{P}}(\mathcal{K}) = \partial_{\mathbb{C}}(\Lambda_2) \uplus \{\infty\}. \qquad (10.138)$$

□

The standard examples in continuous resp. discrete time are the unbounded open half-space $\Lambda_1 = \Lambda_{1,\mathrm{std}} := \mathbb{C}_- = \{z \in \mathbb{C};\ \Re(z) < 0\}$ resp. the bounded open disc $\Lambda_1 = \Lambda_{1,\mathrm{std}} = \mathbb{D} := \{z \in \mathbb{C};\ |z| < 1\}$. In the following we assume a continuous function $f \in B = \mathrm{C}^0(\mathcal{K})$ that is holomorphic on the open set $\mathrm{int}_{\mathbb{C}}(\Lambda_2)$ and compute its norm $\|f\| := \max\{|f(z)|;\ z \in \mathcal{K}\}$. The functions in $\mathcal{S} \subset \mathrm{C}^0(\mathcal{K})$ are of this type.

Theorem 10.2.20 *Consider the decomposition* $\mathbb{C} = \Lambda_1 \uplus \Lambda_2$ *from* (10.129) *with nonempty and open* Λ_1, *unbounded* Λ_2 *and such that* $\mathrm{int}_{\mathbb{C}}(\Lambda_2)$ *is dense in* Λ_2 *and connected. These assumptions hold in all interesting cases. Let* $f \in \mathrm{C}^0(\mathcal{K})$ *be holomorphic on* $\mathrm{int}_{\mathbb{C}}(\Lambda_2)$. *Then*

$$\|f\| = \max\{|f(z)|;\ z \in \partial_{\mathbb{P}}(\mathcal{K})\} = \sup\{|f(z)|;\ z \in \partial_{\mathbb{C}}(\Lambda_2)\}. \qquad (10.139)$$

This is applicable to all functions in $\mathcal{S} \subset \mathrm{C}^0(\mathcal{K})$.

Proof Let $U := \mathrm{int}_{\mathbb{C}}(\Lambda_2)$. This is a nonempty connected open set. If f is holomorphic in U and f assumes its absolute maximum in z_0, i.e., if $|f(z)| \le |f(z_0)|$ for all $z \in U$ then f is constant on U. This is the *maximum principle* for holomorphic functions. It is obvious that the connectedness of U is a necessary requirement.
(i) *Assume that* Λ_1 *is unbounded*: This implies $\mathrm{int}_{\mathbb{P}}(\mathcal{K}) = \mathrm{int}_{\mathbb{C}}(\Lambda_2) = U$. Let $z_0 \in \mathcal{K}$ with $|f(z_0)| = \max\{|f(z)|;\ z \in \mathcal{K}\}$. If $z_0 \in \partial_{\mathbb{P}}(\mathcal{K})$ then $\|f\| = |f(z_0)| = \max\{|f(z)|;\ z \in \partial_{\mathbb{P}}(\mathcal{K})\}$. Assume, in contrast, that $z_0 \in U$. From the maximum principle we conclude that f is constant on U. Since U is dense in $\Lambda_2 = \mathrm{cl}_{\mathbb{C}}(U)$ and f is continuous on Λ_2 the function f is constant on Λ_2, i.e., $f(z) = f(z_0),\ z \in \Lambda_2$. Since Λ_2 is unbounded there is a sequence of points $z_k \in \Lambda_2$ with $\lim_{k\to\infty} z_k = \infty$. Since f is continuous in ∞ we infer

$$f(\infty) = f(z_0) \text{ and } \|f\| = |f(\infty)| \underset{\infty \in \partial_{\mathbb{P}}(\mathcal{K})}{=} \max\{|f(z)|;\ z \in \partial_{\mathbb{P}}(\mathcal{K})\}.$$

The limit $\lim_k f(z_k) = f(\infty)$ and $\partial_{\mathbb{P}}(\mathcal{K}) = \partial_{\mathbb{C}}(\Lambda_2) \uplus \{\infty\}$ imply the remaining assertion

$$\|f\| = \max\{|f(z)|;\ z \in \partial_{\mathbb{P}}(\mathcal{K})\} = \sup\{|f(z)|;\ z \in \partial_{\mathbb{C}}(\Lambda_2)\}.$$

(ii) *Assume that* Λ_1 *is bounded*: Hence there is $r > 0$ such that

$$\Lambda_2 \supseteq U' := \{z \in \mathbb{C};\ |z| > r\} \Longrightarrow U' \subseteq U = \mathrm{int}_{\mathbb{C}}(\Lambda_2) \subset \mathrm{int}_{\mathbb{P}}(\mathcal{K}) = \mathrm{int}_{\mathbb{C}}(\Lambda_2) \uplus \{\infty\}.$$

Moreover $\partial_{\mathbb{C}}(\Lambda_2) = \partial_{\mathbb{P}}(\mathcal{K})$. Assume $z_0 \in \mathcal{K}$ with $\|f\| = |f(z_0)|$.
(a) If $z_0 \in \partial_{\mathbb{P}}(\mathcal{K})$ then again $\|f\| = \max\{|f(z)|;\ z \in \partial_{\mathbb{P}}(\mathcal{K})\}$.
(b) If $z_0 \in \mathrm{int}_{\mathbb{C}}(\Lambda_2)$ we conclude as in (i) that $f = f(z_0)$.
(c) Assume $z_0 = \infty$. Since f is holomorphic on $U = \mathrm{int}_{\mathbb{C}}(\Lambda_2)$ and therefore on U' the function $g(z) := f(z^{-1})$ is holomorphic in $\{z \in \mathbb{C};\ 0 < |z| < r^{-1}\}$. Since f is continuous in $\mathcal{K}$ the function g extended by $g(0) := f(\infty)$ is continuous in $V' := \{z \in \mathbb{C};\ |z| < r^{-1}\}$. Therefore 0 is a removable singularitiy of g and g is holomorphic in V'. Since $|f(\infty)| = |f(z_0)|$ is a maximum of $|f(z)|$ the point 0 is a maximum of $|g(z)|$ in V'. By the maximum principle applied to g on V' we infer that g is constant on V', and hence f is constant on U'. The identity theorem implies that f is constant on the connected open set $U = \mathrm{int}_{\mathbb{C}}(\Lambda_2) \supseteq U'$. As in (i) we conclude that f is constant, and hence $\|f\| = \max\{|f(z)|;\ z \in \partial_{\mathbb{P}}(\mathcal{K})\}$. Since $\partial_{\mathbb{P}}(\mathcal{K}) = \partial_{\mathbb{C}}(\Lambda_2)$ the additional equation in (10.139) is obvious. □

Example 10.2.21 (i) In *continuous time* with $\Lambda_1 := \mathbb{C}_-$ we get

$$\|f\| = \sup_{z \in \mathbb{C},\ \Re(z) \geq 0} |f(z)| = \max\{|f(z)|;\ z \in \mathbb{R}j \uplus \{\infty\}\} = \sup_{\omega \in \mathbb{R}} |f(j\omega)| \tag{10.140}$$

with $j := \sqrt{-1}$, if f is continuous on $\mathcal{K} = \{z \in \mathbb{C};\ \Re(z) \geq 0\} \uplus \{\infty\}$ and holomorphic on $\mathbb{C}_+$.
(ii) In *discrete time* with

$$\Lambda_1 := \mathbb{D} = \{z \in \mathbb{C};\ |z| < 1\},\ \mathcal{K} = \{z \in \mathbb{C};\ |z| \geq 1\} \uplus \{\infty\},$$
$$\partial_{\mathbb{P}}(\mathcal{K}) = \partial_{\mathbb{C}}(\Lambda_2) = S^1 := \{z \in \mathbb{C};\ |z| = 1\} \text{ we get } \|f\| = \max\{|f(z)|;\ z \in S^1\}$$

if f is continuous on $\mathcal{K}$ and holomorphic on $\{z \in \mathbb{C};\ |z| > 1\}$.

◊

The B-module $B^{1\times\ell}$, $\ell \in \mathbb{N}$, is a Banach space with the maximum norm

$$\|\xi\| := \|\xi\|_{\max} := \max\{\|\xi_i\|;\ i = 1, \dots, \ell\} \Longrightarrow \|f\xi\| \leq \|f\|\|\xi\|,\ f \in B. \tag{10.141}$$

In particular, $B^{k\times\ell}$ is a Banach space via

$$\|R\|_{\max} := \max_{i,j} \|R_{ij}\|,\ R = (R_{i,j})_{i,j} \in B^{k\times\ell}. \tag{10.142}$$

The B-module $B^{k\times\ell} \underset{\text{ident.}}{=} \mathrm{Hom}_B(B^{1\times k}, B^{1\times\ell})$ is also a Banach space via

$$\|R\| := \sup\left\{\frac{\|\xi R\|}{\|\xi\|};\ 0 \neq \xi \in B^{1\times k}\right\}$$
$$\Longrightarrow \|\xi R\| \leq \|\xi\|\|R\|,\ \|RS\| \leq \|R\|\|S\|,\ \xi \in B^{1\times k},\ R \in B^{k\times \ell},\ S \in B^{\ell\times m}. \tag{10.143}$$

In particular, the addition and multiplication of matrices is continuous.

Lemma 10.2.22 *The norms $\|R\|_{\max}$ and $\|R\|$ satisfy*

$$\|R\|_{\max} \leq \|R\| \leq k\|R\|_{\max}. \tag{10.144}$$

Hence they are equivalent and give rise to the same topology. In particular, $B^{k\times\ell}$ is a Banach space with $\|R\|$ as asserted.

Proof (i) For $\xi \in B^{1\times k}$ and $R \in B^{k\times\ell}$ we get $\xi R = (\sum_i \xi_i R_{ij})_j$, hence

$$\begin{aligned}\|\xi R\| &= \max_j \|\sum_i \xi_i R_{ij}\| \leq \max_j \sum_i \|\xi_i\|\|R_{ij}\| \\ &\leq k\|\xi\|\|R\|_{\max} \Longrightarrow \|R\| \leq k\|R\|_{\max}.\end{aligned}$$

(ii) For $\delta_i = (0, \ldots, 0, \overset{i}{1}, 0, \ldots, 0) \in B^{1\times k}$ we get

$$\|\delta_i R\| = \max_j \|R_{ij}\| \leq \|\delta_i\|\|R\| = \|R\| \Longrightarrow \|R\|_{\max} = \max_{i,j} \|R_{ij}\| \leq \|R\|.$$

□

Corollary 10.2.23 *The matrix algebra $A := B^{p\times p}$ with the norm $\|P\|$ is a non-commutative Banach algebra, i.e., a $\mathbb{C}$-algebra and a Banach space and the norm satisfies $\|\operatorname{id}_p\| = 1$ and $\|P_1P_2\| \leq \|P_1\|\|P_2\|$.* ◊

We also need that the group $\mathrm{Gl}_p(B)$ with the induced topology from $B^{p\times p}$ is a topological group, i.e., that the multiplication and the inversion of matrices are continuous. For the multiplication this is obvious. For the inversion we assume, more generally, any not necessarily commutative Banach algebra A with the elements $f, g, \ldots$ and its subgroup $U = \mathrm{U}(A)$ of invertible elements. We consider the open neighborhoods

$$B(f, r) := \{g \in A;\ \|g - f\| < r\},\ f \in A,\ r > 0. \tag{10.145}$$

Then for $f_1, f_2 \in A$ the multiplication with f_2 from the left or right is continuous and in particular

$$\begin{aligned} f_2\cdot:\ B(f_1, r) &\to B(f_2 f_1, \|f_2\|r),\ g \mapsto f_2 g, \\ \cdot f_2:\ B(f_1, r) &\to B(f_1 f_2, \|f_2\|r),\ g \mapsto g f_2. \end{aligned} \tag{10.146}$$

A main tool is the geometric series: Assume $0 < r < 1$ and $x \in A$ with $\|x\| \le r$. Then $\sum_{i=0}^{\infty} \|x\|^i = (1 - \|x\|)^{-1} \le (1-r)^{-1}$, and hence $\sum_{i=0}^{n} x^i$, $n \in \mathbb{N}$, is a Cauchy sequence and

$$\begin{aligned}
&\sum_{i=0}^{\infty} x^i \in A,\ \sum_{i=0}^{\infty} x^i = (1-x)^{-1},\ 1 - x \in U = \mathrm{U}(A),\ B(1,r) \subset U \\
&\|(1-x)^{-1}\| \le \sum_{i=0}^{\infty} \|x\|^i = (1 - \|x\|)^{-1} \le (1-r)^{-1}, \\
&(1-x)^{-1} - 1 = x(1-x)^{-1},\ \|(1-x)^{-1} - 1\| \le \|x\|(1 - \|x\|)^{-1} \le r(1-r)^{-1}.
\end{aligned} \tag{10.147}$$

Now let

$$\begin{aligned}
&r < \frac{1}{2} \Longrightarrow (1-r)^{-1} < 2,\ r(1-r)^{-1} \le 2r < 1 \\
&\underset{(10.147)}{\Longrightarrow} U \supset B(1,r) \xrightarrow{f = 1 - x \mapsto f^{-1} = (1-x)^{-1}} B(1, 2r) \subset U.
\end{aligned} \tag{10.148}$$

The last line of (10.148) shows that the inversion map $U \to U,\ f \mapsto f^{-1}$, is continuous in $1 \in U$. Let

$$\begin{aligned}
&r < 1,\ f \in U \text{ and } g \in B\left(f, r\|f^{-1}\|^{-1}\right) \Longrightarrow gf^{-1} - 1 = (g - f)f^{-1}, \\
&\|gf^{-1} - 1\| \le \|g - f\|\|f^{-1}\| < r\|f^{-1}\|^{-1}\|f^{-1}\| = r < 1 \\
&\underset{(10.147)}{\Longrightarrow} gf^{-1} \in U \Longrightarrow g = (gf^{-1})f \in U \Longrightarrow B\left(f, r\|f^{-1}\|^{-1}\right) \subset U.
\end{aligned} \tag{10.149}$$

This shows that U is an open subset of A.

Lemma 10.2.24 *The inversion map $U \to U,\ f \mapsto f^{-1}$, is continuous.*

Proof Let $f \in U$. Let $U \cap B(f^{-1}, r_4)$ be an open neighborhood of f^{-1} in U. We have to show that there is an open neighborhood $U \cap B(f, r_1)$ of f in U such that

$$U \cap B(f, r_1) \to U \cap B(f^{-1}, r_4),\ g \mapsto g^{-1}.$$

Choose $r_3, r_2, r_1 > 0$ in the following fashion: Let

$$0 < r_3 < \min(1, \|f^{-1}\|^{-1} r_4),\ 0 < r_2 < \frac{r_3}{2},\ 0 < r_1 < \|f^{-1}\|^{-1} r_2.$$

According to (10.146) and (10.149) we obtain, as required,

$$\begin{array}{ccccccc}
U \cap B(f, r_1) & \xrightarrow{\cdot f^{-1}} & B(1, r_2) & \xrightarrow{(-)^{-1}} & B(1, r_3) & \xrightarrow{f^{-1}\cdot} & U \cap B(f^{-1}, r_4) \\
g & \longmapsto & gf^{-1} & \longmapsto & (gf^{-1})^{-1} = fg^{-1} & \longmapsto & g^{-1} = f^{-1} fg^{-1}.
\end{array}$$

□

Corollary 10.2.25 *Consider the Banach algebras* $B = C^0(\mathcal{K}) = C^0(\mathcal{K}, \mathbb{C})$ *and* $A := B^{p\times p}$, $p > 0$. *Then the group* $\mathrm{Gl}_p(B) = \mathrm{U}(B^{p\times p})$ *of invertible matrices is open in* $B^{p\times p}$ *and a topological group, i.e., the multiplication and the inversion of invertible matrices are continuous.* ◊

Recall that we identify $\mathcal{S} = F[s]_T \cap F(s)_{\mathrm{pr}} \subset B = C^0(\mathcal{K})$. If

$$0 \neq h = ft^{-1} = (f_m s^m + \cdots)(t_n s^n + \cdots)^{-1},\ f, t \in F[s],\ m \leq n,\ f_m, t_n \neq 0 \text{ and}$$
$$\forall z \in \Lambda_2 = \mathcal{K} \setminus \{\infty\} : t(z) \neq 0 \text{ then}$$
$$h(z) := \begin{cases} f(z)/t(z) & \text{if } z \in \Lambda_2 = \mathcal{K} \setminus \{\infty\} \\ 0 & \text{if } z = \infty,\ m < n \\ f_m t_m^{-1} & \text{if } z = \infty,\ m = n \end{cases} \tag{10.150}$$

Lemma 10.2.26 *(i)* $\mathcal{S} \cap \mathrm{U}(B) = \mathrm{U}(\mathcal{S})$. *Hence* $\mathcal{S}$ *is a commutative normed algebra and its group* $\mathrm{U}(\mathcal{S})$ *of invertible elements is open in* $\mathcal{S}$ *and a topological group with the induced topology, i.e., the inversion is continuous.*
(ii) $\mathcal{S}^{p\times p} \bigcap \mathrm{Gl}_p(B) = \mathrm{Gl}_p(\mathcal{S})$, $p > 1$. *Hence* $\mathcal{S}^{p\times p}$ *is a noncommutative normed algebra and its group* $\mathrm{Gl}_p(\mathcal{S}) := \mathrm{U}(\mathcal{S}^{p\times p})$ *of invertible elements is open in* $\mathcal{S}^{p\times p}$ *and a topological group with the induced topology, i.e., the inversion is continuous.*

Proof (i) Let $h = ft^{-1} \in \mathcal{S}$ from (10.150) and assume $h \in \mathrm{U}(B)$, i.e., $h(z) \neq 0$ for all $z \in \mathcal{K}$. By (10.150) this implies $m = n$ and $f(z) \neq 0$ for all $z \in \Lambda_2 = \mathcal{K} \setminus \{\infty\}$ or $\mathrm{V}_{\mathbb{C}}(f) \subseteq \Lambda_1$. But this implies $f \in T$ and $t/f \in \mathcal{D}_T \cap F(s)_{\mathrm{pr}}$ and hence $h = f/t \in \mathrm{U}(\mathcal{S})$.
(ii) Let $P \in \mathcal{S}^{p\times p} \cap \mathrm{Gl}_p(B) \Longrightarrow \det(P) \in \mathcal{S} \cap \mathrm{U}(B) \underset{(i)}{=} \mathrm{U}(\mathcal{S}) \Longrightarrow P \in \mathrm{Gl}_p(\mathcal{S})$. □

Notice that Lemma 10.2.26 does not apply to $\widehat{\mathcal{D}} = F[\widehat{s}] \subset \mathcal{S}$, $\widehat{s} = (s - \alpha)^{-1}$, since $\widehat{\mathcal{D}} \cap \mathrm{U}(B) = \widehat{T} \supsetneq \mathrm{U}(\widehat{\mathcal{D}}) = F \setminus \{0\}$.

Robustness requires the notion of nearness of two behaviors $\mathcal{B}_1$ and $\mathcal{B}_1'$ that appear in the preceding theorems, i.e., we have to topologize the set of these behaviors. Let $K := F(s) = F(\widehat{s})$ be the field of rational functions. We apply the module-behavior duality to the injective cogenerator ${}_{\mathcal{D}_T}\mathcal{F}_T$.

For fixed $p, m \in \mathbb{N}$ the relevant set of behaviors is the set

$$\mathfrak{B} := \left\{\mathcal{B} \subseteq \mathcal{F}_T^{p+m};\ \mathcal{B} \text{ controllable IO behavior with input } u \in \mathcal{F}_T^m\right\}. \tag{10.151}$$

The required controllability of $\mathcal{B}$ gives rise to the bijection

$$\mathfrak{B} \cong K^{p\times m},\ \mathcal{B} := \text{controllable realization of } H \leftrightarrow H := \text{transfer matrix of } \mathcal{B}. \tag{10.152}$$

Let $H = \widehat{P}^{-1}\widehat{Q}$ be the left coprime factorization over $\widehat{\mathcal{D}}$ and hence over $\mathcal{D}_T = \mathcal{S}_{\widehat{s}} = (\widehat{\mathcal{D}}_{\widehat{T}})_{\widehat{s}}$. Then $\mathcal{B} = \left\{\binom{y}{u} \in \mathcal{F}_T^{p+m};\ \widehat{P} \circ y = \widehat{Q} \circ u\right\}$ and $(\widehat{P}, -\widehat{Q}) \in \widehat{\mathcal{D}}^{p\times(p+m)}$ has a right inverse in $\widehat{\mathcal{D}}^{(p+m)\times p} \subseteq \mathcal{S}^{(p+m)\times p}$. We define

$$\mathfrak{R} := \left\{\widehat{R} = (\widehat{P}, -\widehat{Q}) \in \mathcal{S}^{(p\times(p+m))};\ \det(\widehat{P}) \neq 0,\ \widehat{R} \text{ right invertible over } \mathcal{S}\right\} \text{ and}$$
$$\mathcal{B}(\widehat{R}) = \left\{\binom{y}{u} \in \mathcal{F}_T^{p+m};\ \widehat{P}\circ y = \widehat{Q}\circ u\right\} \Longrightarrow \mathfrak{R} \to \mathfrak{B},\ \widehat{R} \mapsto \mathcal{B}(\widehat{R}),\ \text{surjective.} \tag{10.153}$$

Due to Lemma 10.2.26 and the better topological properties of $\mathrm{Gl}_p(\mathcal{S})$ compared to those of $\mathrm{Gl}_p(\widehat{\mathcal{D}})$ we use $\mathcal{S}$ instead of $\widehat{\mathcal{D}}$ in the definition of $\mathfrak{R}$. Since ${}_{\mathcal{D}_T}\mathcal{F}_T$ is an injective cogenerator we know that $\mathcal{B}(\widehat{R}) = \mathcal{B}(\widehat{R}')$ if and only if there is $\widehat{U} \in \mathrm{Gl}_p(\mathcal{D}_T)$ with $\widehat{R}' = \widehat{U}\widehat{R}$.

Lemma 10.2.27 $\mathcal{B}(\widehat{R}') = \mathcal{B}(\widehat{R}) \Longleftrightarrow \exists \widehat{U} \in \mathrm{Gl}_p(\mathcal{S})$ *with* $\widehat{R}' = \widehat{U}\widehat{R}$.

Proof $\Longleftarrow$: obvious.
$\Longrightarrow$: There is $\widehat{U} \in \mathrm{Gl}_p(\mathcal{D}_T)$ such that $\widehat{R}' = \widehat{U}\widehat{R}$ and hence also $\widehat{R} = \widehat{U}^{-1}\widehat{R}'$. Let $\widehat{S} \in \mathcal{S}^{(p+m)\times p}$ be a right inverse of $\widehat{R} \in \mathfrak{R}$, hence

$$\widehat{U} = \widehat{U}\widehat{R}\widehat{S} = \widehat{R}'\widehat{S} \in \mathcal{S}^{p\times p} \text{ and, likewise, } \widehat{U}^{-1} \in \mathcal{S}^{p\times p} \Longrightarrow \widehat{U} \in \mathrm{Gl}_p(\mathcal{S}).$$

□

The preceding results can be reformulated by means of group actions: If G is a group, X a set and $\mu : G \times X \to X$, $(g, x) \mapsto gx := \mu(g, x)$, a group action (with $1_G x = x$ and $(g_1g_2)x = g_1(g_2x)$), there are the orbits $y := Gx$, $x \in X$, the orbit space $Y := G \setminus X := \{Gx;\ x \in X\}$ and the canonical surjection $\nu :=$ can $: X \to G \setminus X$, $x \mapsto Gx$. Then

$$X = \uplus_{y\in Y} y \text{ and } \left(\nu(x) = \nu(x') \Longleftrightarrow Gx = Gx' \Longleftrightarrow \exists g \in G \text{ with } x' = gx\right), \tag{10.154}$$

i.e., if x' lies in the orbit Gx of x. For subsets $X' \subseteq X$ resp. $Y' \subseteq Y$ we get

$$\nu^{-1}\nu(X') = GX' = \cup_{x\in X'} Gx \text{ and } \nu\nu^{-1}(Y') = Y' \tag{10.155}$$

since ν is surjective.

Assume in addition that X is a topological space, G is a topological group and $\mu : G \times X \to X$ is continuous. Then for $g \in G$ the map $g\cdot : X \to X, x \mapsto gx$, is a homeomorphism with inverse $x \mapsto g^{-1}x$ and and preserves open subsets in particular. The orbit space $G \setminus X$ becomes a topological space with the identification topology, i.e., the largest topology on $G \setminus X$ for which ν is continuous. By definition, $Y' \subseteq Y$ is open if and only if $\nu^{-1}(Y')$ is open. Then ν is an open map, i.e., the image $\nu(X')$ of an open subset $X' \subseteq X$ is open, since

$$X' \text{ open} \underset{gX' \text{ open}}{\Longrightarrow} \nu^{-1}\nu(X') = GX' = \cup_{g\in G} gX' \text{ open} \Longrightarrow \nu(X') \text{ open.} \tag{10.156}$$

There follows the canonical bijection

$$\begin{array}{ccc} \{X' \subseteq X;\ X' \text{ open},\ GX' = X'\} & \cong & \{Y' \subseteq Y;\ Y' \text{ open}\}, \\ X' = \nu^{-1}(Y') & \leftrightarrow & Y' = \nu(X') \end{array} \tag{10.157}$$

A basis of neighborhoods of $\nu(x)$ in Y consists of the open sets

$$\begin{aligned}&\nu(X') = \nu(GX') \ni \nu(x),\ X' \text{ open neighborhood of } x\\&\Longrightarrow \forall x'' \in X:\ \big(\nu(x'') = Gx'' \in \nu(X') \Longleftrightarrow \exists g \in G \text{ with } gx'' \in X'\big).\end{aligned} \tag{10.158}$$

Corollary 10.2.28 *For any topological space Y there is the bijection*

$$C^0(G \setminus X, Y) \cong \{\varphi \in C^0(X, Y);\ \forall g \in G \forall x \in X : \varphi(gx) = \varphi(x)\},\ \psi \leftrightarrow \varphi, \tag{10.159}$$

with $\varphi(x) = \psi(Gx)$, C^0 denoting the sets of continuous maps. Thus continuous maps ψ are defined by equivariant continuous maps φ. ◇

The spaces $X \times X$ and $Y \times Y$ are endowed with the product topology and the product map

$$\nu \times \nu : X \times X \to G \setminus X \times G \setminus X,\ (x, x') \mapsto (\nu(x), \nu(x')),\ \text{is continuous and open.}$$

A space Y is a Hausdorff space if and only if the diagonal $\Delta_Y := \{(y, y);\ y \in Y\}$ is closed in $Y \times Y$.

Lemma 10.2.29 *Assume that X is Hausdorff. Then $G \setminus X$ is Hausdorff too if and only if $R := \{(x, gx);\ x \in X, g \in G\}$ is closed in $X \times X$.*

Proof The following identities are obvious:

$$(\nu \times \nu)^{-1}(\Delta_Y) = R \text{ and hence } (\nu \times \nu)^{-1}((Y \times Y) \setminus \Delta_Y) = (X \times X) \setminus R.$$

Since $\nu \times \nu$ is continuous the first of these implies that R is closed if Δ_Y is closed, i.e., if Y is Hausdorff. The preceding identities further imply

$$(Y \times Y) \setminus \Delta_Y = (\nu \times \nu)(\nu \times \nu)^{-1}\ ((Y \times Y) \setminus \Delta_Y) = (\nu \times \nu)((X \times X) \setminus R).$$

If R is closed then $(X \times X) \setminus R$ is open. Since $\nu \times \nu$ is open so is the image $(Y \times Y) \setminus \Delta_Y$. Therefore Δ_Y is closed and Y is Hausdorff. □

Assume that X is a metric space with distance function $d(x, x')$, for instance $X = \mathfrak{R}$ from above. Then convergence in $G \setminus X$ can be reduced to convergence in X as the following lemma shows. It is used in the proof of Theorem 10.2.33 for the statements on limits. Every $x_0 \in X$ has the countable basis of neighborhoods $U(x_0, 1/k) := \{x \in X;\ d(x_0, x) < 1/k\},\ k \in \mathbb{N}$. Since $\nu = \mathrm{can} : X \to G \setminus X,\ x \mapsto Gx$, is continuous and open the images $\overline{U}(\overline{x}, 1/k) := \nu(U(x_0, 1/k)),\ k \in \mathbb{N}$, are a countable basis of neighborhoods of $\overline{x_0} = Gx_0$ in $G \setminus X$.

Lemma 10.2.30 *Let $\overline{x_n},\ n \geq 1$, be a convergent sequence in $G \setminus X$ with $\lim_{n\to\infty} \overline{x_n} = \overline{x_0}$. Then there is a convergent sequence $(y_n)_{n\geq 1}$ in X such that*

$$\forall n \geq 1: \ Gy_n = Gx_n \ or \ \overline{y_n} = \overline{x_n} \ and \ \lim_{n\to\infty} y_n = x_0. \tag{10.160}$$

Proof There is an increasing sequence

$$\begin{aligned} &n_1 < \ldots < n_k < n_{k+1} < \ldots \text{ such that} \\ &\forall n \geq n_k: \ \overline{x_n} \in \overline{U}(\overline{x_0}, 1/k) = GU(x_0, 1/k). \end{aligned} \tag{10.161}$$

In particular, for $n_k \leq n \leq n_{k+1} - 1$ there are $y_n \in U(x_0, 1/k)$ and $g_n \in G$ such that $x_n = g_n y_n$ and hence $\overline{y_n} = \overline{x_n}$. Obviously y_n converges to x_0 since $\forall n \geq n_k: \ y_n \in U(x_0, 1/k)$. □

We apply the preceding general considerations to the action

$$\begin{aligned} &X := \mathfrak{R} \subset \mathcal{S}^{p\times(p+m)}, \ G := \mathrm{Gl}_p(\mathcal{S}) \text{ and} \\ &\mu: \ \mathrm{Gl}_p(\mathcal{S}) \times \mathfrak{R} \to \mathfrak{R}, \ (\widehat{U}, \widehat{R}) \mapsto \widehat{U}\widehat{R}, \ \widehat{R} := (\widehat{P}, -\widehat{Q}). \end{aligned} \tag{10.162}$$

The action is continuous since the multiplication of matrices with coefficients in $\mathcal{S}$ has this property. The topology of $\mathcal{S} \subseteq B = \mathrm{C}^0(\mathcal{K}, \mathbb{C})$ is that induced from the Banach norm on B. The action is even free, i.e., the maps

$$\begin{aligned} &\mathrm{Gl}_p(\mathcal{S}) \to \mathfrak{R}, \ \widehat{U} \mapsto \widehat{U}\widehat{R}, \ \widehat{R} \in \mathfrak{R}, \text{ and} \\ &\mathrm{Gl}_p(\mathcal{S}) \times \mathfrak{R} \to \mathfrak{R} \times \mathfrak{R}, \ (\widehat{U}, \widehat{R}) \mapsto (\widehat{R}, \widehat{U}\widehat{R}), \end{aligned} \tag{10.163}$$

are injective.

Corollary and Definition 10.2.31 *There are the bijections*

$$\begin{array}{llll} \mathrm{Gl}_p(\mathcal{S}) \setminus \mathfrak{R} \cong & \mathfrak{B} & \cong & K^{p\times m} \\ \mathrm{Gl}_p(\mathcal{S})\widehat{R} & \mapsto \mathcal{B}(\widehat{R}) = \left\{ w \in \mathcal{F}_T^{p+m}; \ \widehat{R} \circ w = 0 \right\} & \mapsto & H = \widehat{P}^{-1}\widehat{Q} \end{array} \tag{10.164}$$

where $\widehat{R} = (\widehat{P}, -\widehat{Q}) \in \mathfrak{R}$. *Since* $\mathrm{Gl}_p(\mathcal{S}) \setminus \mathfrak{R}$ *is a topological space with the identification topology induced from that of* $\mathfrak{R}$ *the sets* $\mathfrak{B}$ *and* K^{p+m} *become topological spaces via* transport of structure *such that the bijections from* (10.164) *are homeomorphisms. This topology is called the* graph topology.

The neighborhoods of $\mathcal{B}_1 = \mathcal{B}(\widehat{R}_1) = \left\{ w \in \mathcal{F}_T^{p+m}; \ \widehat{R}_1 \circ w = 0 \right\} \in \mathfrak{B}, \ \widehat{R}_1 \in \mathfrak{R}$, *are constructed as follows: Choose an open neighborhood*

$$\begin{aligned} &\mathfrak{X} = \left\{ \widehat{R} \in \mathfrak{R}; \ \|\widehat{R} - \widehat{R}_1\| < \epsilon \right\}, \ \epsilon > 0, \ \textit{of } \widehat{R}_1 \textit{ and define} \\ &\mathfrak{Y} := \left\{ \mathcal{B}(\widehat{R}); \ \widehat{R} \in \mathfrak{X} \right\}, \ \mathcal{B}(\widehat{R}) := \left\{ w \in \mathcal{F}_T^{p+m}; \widehat{R} \circ w = 0 \right\}. \end{aligned} \tag{10.165}$$

By (10.158) *these* $\mathfrak{Y}$ *are a basis of open neighborhoods if* $\mathcal{B}_1$ *in* $\mathfrak{B}$.

Proof The bijections follow from (10.151) and Lemma 10.2.27. □

Lemma 10.2.32 *The graph topology is Hausdorff, i.e., according to Lemma 10.2.29, the subset*

$$\mathfrak{S} := \left\{(\widehat{R}, \widehat{U}\widehat{R});\ \widehat{R} \in \mathfrak{R},\ \widehat{U} \in \mathrm{Gl}_p(\mathcal{S})\right\} \subset \mathfrak{R} \times \mathfrak{R}$$

is closed.

Proof Assume

$$\begin{aligned}
&\lim_{k\to\infty} (\widehat{R}_k, \widehat{U}_k\widehat{R}_k) = (\widehat{R}, \widehat{R}'),\ \widehat{R}_k, \widehat{R}, \widehat{R}' \in \mathfrak{R},\ \widehat{U}_k \in \mathrm{Gl}_p(\mathcal{S})\\
&\Longrightarrow \lim_k \widehat{R}_k = \widehat{R},\ \lim_k \widehat{U}_k\widehat{R}_k = \widehat{R}'.
\end{aligned}$$

We have to show $(\widehat{R}, \widehat{R}') \in \mathfrak{S}$, i.e., that there is $\widehat{U} \in \mathrm{Gl}_p(\mathcal{S})$ with $\widehat{R}' = \widehat{U}\widehat{R}$. Choose

$$\widehat{S}, \widehat{S}' \in \mathcal{S}^{(p+m)\times p} \text{ with } \widehat{R}\widehat{S} = \widehat{R}'\widehat{S}' = \mathrm{id}_p \Longrightarrow \lim_k \widehat{R}_k\widehat{S} = \widehat{R}\widehat{S} = \mathrm{id}_p\,.$$

By Lemma 10.2.26 the group $\mathrm{Gl}_p(\mathcal{S})$ is open in $\mathcal{S}^{p\times p}$. Hence there is $k_0 \in \mathbb{N}$ with $\widehat{R}_k\widehat{S} \in \mathrm{Gl}_p(\mathcal{S}),\ k \geq k_0$. For notational convenience we assume $k_0 = 0$ without loss of generality. Define $\widehat{S}_k := \widehat{S}(\widehat{R}_k\widehat{S})^{-1}$ and hence $\widehat{R}_k\widehat{S}_k = \mathrm{id}_p$. Since $\mathrm{Gl}_p(\mathcal{S})$ is a topological group the limit $\lim_k \widehat{R}_k\widehat{S} = \mathrm{id}_p$ implies

$$\begin{aligned}
&\lim_k (\widehat{R}_k\widehat{S})^{-1} = \mathrm{id}_p \Longrightarrow \lim_k \widehat{S}_k = \lim_k \widehat{S}(\widehat{R}_k\widehat{S})^{-1} = \widehat{S}\,\mathrm{id}_p = \widehat{S}\\
&\Longrightarrow \lim_k \widehat{U}_k = \lim_k (\widehat{U}_k\widehat{R}_k)\widehat{S}_k = \lim_k (\widehat{U}_k\widehat{R}_k) \lim_k \widehat{S}_k = \widehat{R}'\widehat{S}\\
&\Longrightarrow \widehat{R}' = \lim_k \widehat{U}_k\widehat{R}_k = \lim_k \widehat{U}_k \lim_k \widehat{R}_k = \widehat{R}'\widehat{S}\widehat{R}\\
&\Longrightarrow \mathrm{id}_p = \widehat{R}'\widehat{S}' = (\widehat{R}'\widehat{S})(\widehat{R}\widehat{S}') \Longrightarrow \widehat{U} := \widehat{R}'\widehat{S} = \lim_k \widehat{U}_k \in \mathrm{Gl}_p(\mathcal{S})\\
&\Longrightarrow \widehat{R}' = \lim_k \widehat{U}_k\widehat{R}_k = \lim_k \widehat{U}_k \lim_k \widehat{R}_k = \widehat{U}\widehat{R} \Longrightarrow (\widehat{R}, \widehat{R}') \in \mathfrak{S}.
\end{aligned}$$

□

We finally prove the robustness of the compensator $\mathcal{B}_2$ in Theorem 10.2.11. We start with the T-stabilizable behavior $\mathcal{B}_1 \subseteq \mathcal{F}^{p+m}$ with transfer matrix $H_1 \in K^{p\times m}$ and apply the preceding topological considerations to

$$\mathcal{B}_{1,T} = \mathcal{B}(\widehat{P}_1, -\widehat{Q}_1) = \left\{\left(\begin{smallmatrix} y_1 \\ u_1 \end{smallmatrix}\right) \in \mathcal{F}_T^{p\times m};\ \widehat{P}_1 \circ y_1 = \widehat{Q}_1 \circ u_1\right\} \tag{10.166}$$

where $H_1 = \widehat{P}_1^{-1}\widehat{Q}_1$ is the coprime factorization over $\widehat{\mathcal{D}}$ or, sufficiently, over $\mathcal{S}$ and hence $(\widehat{P}_1, -\widehat{Q}_1) \in \mathfrak{R}$. We assume that $\mathcal{B}_2$ is a stabilizing compensator of $\mathcal{B}_1$ with the properties established in Theorems 10.2.8 and 10.2.11. These behaviors are described by matrices

$$\begin{pmatrix} \widehat{P}_1 & -\widehat{Q}_1 \\ -\widehat{Q}_2 & \widehat{P}_2 \end{pmatrix} \in \mathrm{Gl}_{p+m}(\mathcal{S}),\ \begin{pmatrix} \widehat{D}_2 & \widehat{N}_1 \\ \widehat{N}_2 & \widehat{D}_1 \end{pmatrix} = \begin{pmatrix} \widehat{P}_1 & -\widehat{Q}_1 \\ -\widehat{Q}_2 & \widehat{P}_2 \end{pmatrix}^{-1},\ \widehat{\phi}^{-1}\widehat{D}_2 \in \mathcal{S}^{p\times p}$$
$$\mathcal{B}_{2,T} = \mathcal{B}(-\widehat{Q}_2, \widehat{P}_2) = \left\{\begin{pmatrix} u_2 \\ y_2 \end{pmatrix} \in \mathcal{F}_T^{p+m};\ \widehat{P}_2 \circ y_2 = \widehat{Q}_2 \circ u_2\right\},\ H_2 := \widehat{P}_2^{-1}\widehat{Q}_2. \tag{10.167}$$

The compensator $\mathcal{B}_2$ is the unique controllable realization in $\mathcal{F}^{p+m}$ of the transfer matrix $H_2 := \widehat{P}_2^{-1}\widehat{Q}_2 = \widehat{N}_2\widehat{D}_2^{-1}$. In particular, the following equation holds:

$$(\widehat{P}_1, -\widehat{Q}_1)\begin{pmatrix} \widehat{D}_2 \\ \widehat{N}_2 \end{pmatrix} = \widehat{P}_1\widehat{D}_2 - \widehat{Q}_1\widehat{N}_2 = \mathrm{id}_p\,.$$

Matrix multiplication is continuous and id_p belongs to the group $\mathrm{Gl}_p(\mathcal{S})$ that is open in $\mathcal{S}^{p\times p}$. Notice that det is continuous on $\mathcal{S}^{p\times p}$. Choose ϵ with

$$\begin{aligned} &0 < \epsilon \le \left\| \begin{pmatrix} \widehat{D}_2 \\ \widehat{N}_2 \end{pmatrix} \right\|^{-1} \text{ and the valid implication} \\ &\left(\|\widetilde{P}_1 - \widehat{P}_1\| \le \epsilon \Longrightarrow \|\det(\widetilde{P}_1) - \det(\widehat{P}_1)\| \le 2^{-1}\|\det(\widehat{P}_1)\| \ne 0\right) \\ &\Longrightarrow \det(\widetilde{P}_1) \ne 0. \end{aligned} \tag{10.168}$$

With this $\epsilon > 0$ consider

$$\begin{aligned} &(\widetilde{P}_1, -\widetilde{Q}_1) \in \mathcal{S}^{p\times(p+m)} \text{ with } \|(\widetilde{P}_1, -\widetilde{Q}_1) - (\widehat{P}_1, \widehat{Q}_1)\| < \epsilon \Longrightarrow \det(\widetilde{P}_1) \ne 0 \text{ and} \\ &\left\|(\widetilde{P}_1, -\widetilde{Q}_1)\begin{pmatrix} \widehat{D}_2 \\ \widehat{N}_2 \end{pmatrix} - id_p\right\| = \left\|\left((\widetilde{P}_1, -\widetilde{Q}_1) - (\widehat{P}_1, -\widehat{Q}_1)\right)\begin{pmatrix} \widehat{D}_2 \\ \widehat{N}_2 \end{pmatrix}\right\| < \epsilon\left\|\begin{pmatrix} \widehat{D}_2 \\ \widehat{N}_2 \end{pmatrix}\right\| \le 1 \\ &\Longrightarrow \widehat{U} := (\widetilde{P}_1, -\widetilde{Q}_1)\begin{pmatrix} \widehat{D}_2 \\ \widehat{N}_2 \end{pmatrix} \in \mathcal{S}^{p\times p} \cap Gl_p(B) \underset{\text{Lemma 10.2.26}}{=} Gl_p(\mathcal{S}) \\ &\Longrightarrow \left(\widehat{U}^{-1}(\widetilde{P}_1, -\widetilde{Q}_1)\right)\begin{pmatrix} \widehat{D}_2 \\ \widehat{N}_2 \end{pmatrix} = (\widetilde{P}_1, -\widetilde{Q}_1)\left(\begin{pmatrix} \widehat{D}_2 \\ \widehat{N}_2 \end{pmatrix}\widehat{U}^{-1}\right) = id_p \\ &\Longrightarrow (\widetilde{P}_1, -\widetilde{Q}_1) \in \mathfrak{R},\ \widetilde{H}_1 = \widetilde{P}_1^{-1}\widetilde{Q}_1. \end{aligned} \tag{10.169}$$

If Y is any topological space, Corollary 10.2.28 establishes the bijection

$$\begin{aligned} &\mathrm{C}^0\left(\mathrm{Gl}_p(\mathcal{S})) \setminus \mathfrak{R}, Y\right) \\ &\cong \left\{\varphi \in \mathrm{C}^0(\mathfrak{R}, Y);\ \forall \widehat{U} \in \mathrm{Gl}_p(\mathcal{S}):\ \varphi(\widehat{U}(\widehat{P}, -\widehat{Q})) = \varphi((\widehat{P}, -\widehat{Q}))\right\},\ \text{where} \\ &\psi\left(\mathrm{Gl}_p(\mathcal{S})(\widehat{P}, -\widehat{Q})\right) = \varphi\left((\widehat{P}, -\widehat{Q})\right). \end{aligned} \tag{10.170}$$

In most situations it thus suffices to define and consider φ instead of ψ.

Theorem 10.2.33 (Robustness of compensators) *Assume a T-stabilizable plant $\mathcal{B}_1$ and a compensator that satisfies all properties established in Theorems 10.2.8 and 10.2.11.*

1. Consider the detailed data from (10.166) to (10.169). For $(\widetilde{P}_1, -\widetilde{Q}_1) \in \mathcal{S}^{p\times(p+m)}$ in the open ϵ-neighborhood of $(\widehat{P}_1, -\widehat{Q}_1)$ as in (10.169), the following holds:
(i) $(\widetilde{P}_1, -\widetilde{Q}_1) \in \mathfrak{R},\ \widehat{U} := (\widetilde{P}_1, -\widetilde{Q}_1)\begin{pmatrix} \widehat{D}_2 \\ \widehat{N}_2 \end{pmatrix} \in \mathrm{Gl}_p(\mathcal{S}),\ \widetilde{H}_1 = \widetilde{P}_1^{-1}\widetilde{Q}_1.$
(ii) There are matrices

$$\begin{pmatrix} \widehat{U}^{-1}\widetilde{P}_1 & -\widehat{U}^{-1}\widetilde{Q}_1 \\ -\widehat{Q}_2 & \widehat{P}_2 \end{pmatrix} \in Gl_{p+m}(\mathcal{S}),\ \begin{pmatrix} \widehat{D}_2 & \widetilde{N}_1 \\ \widehat{N}_2 & \widetilde{D}_1 \end{pmatrix} = \begin{pmatrix} \widehat{U}^{-1}\widetilde{P}_1 & -\widehat{U}^{-1}\widetilde{Q}_1 \\ -\widehat{Q}_2 & \widehat{P}_2 \end{pmatrix}^{-1}, \tag{10.171}$$

with $\widehat{\phi}^{-1}\widehat{D}_2 \in \mathcal{S}^{p\times p}$. *Let* $\widetilde{\mathcal{B}}_1 \subseteq \mathcal{F}^{p+m}$ *be any IO realization of* $\widetilde{H}_1$, *i.e.,*

$$\widetilde{\mathcal{B}}_{1,T} := \mathcal{B}(\widetilde{P}_1, -\widetilde{Q}_1) = \mathcal{B}(\widehat{U}^{-1}\widetilde{P}_1, -\widehat{U}^{-1}\widetilde{Q}_1).$$

Equation (10.171) *says that* $\mathcal{B}_2$ *is a* T*-stabilizing compensator of all* $\widetilde{\mathcal{B}}_1$ *with all properties from Theorems 10.2.8 and 10.2.11 where* $\widetilde{\mathcal{B}}_{1,T}$ *runs over all indicated* $\mathcal{F}_T$*-behaviors in the neighborhood of* $\mathcal{B}_{1,T}$. *Note that* $\widehat{U}$ *and* $\widehat{U}^{-1}$ *are continuous functions of* $(\widetilde{P}_1, -\widetilde{Q}_1)$.
Recall that $\widetilde{\mathcal{B}}_{1,T}$ *is in the neighborhood of* $\mathcal{B}_{1,T} = \mathcal{B}(\widehat{P}_1, -\widehat{Q}_1)$ *if and only if* $\widetilde{\mathcal{B}}_{1,T} = \mathcal{B}(\widetilde{P}_1, -\widetilde{Q}_1)$ *for some* $(\widetilde{P}_1, -\widetilde{Q}_1) \in \mathfrak{R}$ *in the neighborhood of* $(\widehat{P}_1, -\widehat{Q}_1)$.
2. *Consider the proper and stable transfer matrices of the closed-loop behaviors:*

$$\begin{cases} H := \begin{pmatrix} \widehat{P}_1 & -\widehat{Q}_1 \\ -\widehat{Q}_2 & \widehat{P}_2 \end{pmatrix}^{-1} \begin{pmatrix} 0 & \widehat{Q}_1 \\ \widehat{Q}_2 & 0 \end{pmatrix} \\ \widetilde{H} := \begin{pmatrix} \widehat{U}^{-1}\widetilde{P}_1 & -\widehat{U}^{-1}\widetilde{Q}_1 \\ -\widehat{Q}_2 & \widehat{P}_2 \end{pmatrix}^{-1} \begin{pmatrix} 0 & \widehat{U}^{-1}\widetilde{Q}_1 \\ \widehat{Q}_2 & 0 \end{pmatrix} \end{cases} \in \mathcal{S}^{(p+m)\times(p+m)}. \text{ Then}$$

$$\lim_{(\widetilde{P}_1,-\widetilde{Q}_1)\to(\widehat{P}_1,-\widehat{Q}_1)} \widetilde{H} = H, \text{ i.e., } \lim_{(\widetilde{P}_1,-\widetilde{Q}_1)\to(\widehat{P}_1,-\widehat{Q}_1)} \|\widetilde{H} - H\| = 0.$$

Hence the maps $\varphi:\ (\widetilde{P}_1, -\widetilde{Q}_1) \mapsto \widetilde{H}$ *and* $\psi:\ \mathcal{B}(\widetilde{P}_1, -\widetilde{Q}_1) \mapsto \widetilde{H}$ *are continuous on* $\mathfrak{R}$ *resp. on* $\mathrm{Gl}_p(\mathcal{S}) \setminus \mathfrak{R} \cong \mathfrak{B}$.

Proof 1. (i) Equation (10.169).
(ii) By construction the sequence

$$0 \leftarrow \mathcal{S}^{1\times p} \xleftarrow{\begin{pmatrix} \widehat{D}_2 \\ \widehat{N}_2 \end{pmatrix}} \mathcal{S}^{1\times(p+m)} \xleftarrow{\cdot(-\widehat{Q}_2, \widehat{P}_2)} \mathcal{S}^{1\times m} \leftarrow 0$$

is exact. By (10.169) $(\widehat{U}^{-1}\widetilde{P}_1, -\widehat{U}^{-1}\widetilde{Q}_1)$ is a left inverse of $\begin{pmatrix} \widehat{D}_2 \\ \widehat{N}_2 \end{pmatrix}$ in $\mathcal{S}^{p\times(p+m)}$. The matrices from (10.171) then follow from Lemma 10.1.7.
2. We write lim for the limit $\lim_{(\widetilde{P}_1,-\widetilde{Q}_1)\to(\widehat{P}_1,-\widehat{Q}_1)}$. This implies $\lim \mathcal{B}(\widetilde{P}_1, -\widetilde{Q}_1) = \mathcal{B}(\widehat{P}_1, -\widehat{Q}_1)$. Recall that the multiplication of matrices in $\mathcal{S}^{\bullet\times\bullet}$ and the inversion on $\mathrm{Gl}_p(\mathcal{S})$ are continuous. We also need that for

$$\begin{pmatrix} A & B \\ C & D \end{pmatrix} \in \mathcal{S}^{(p+m)\times(p+m)}:\ \left\| \begin{pmatrix} A & B \\ C & D \end{pmatrix} \right\| \le \max(\|A\| + \|C\|, \|B\| + \|D\|),$$

and that therefore the matrix $\begin{pmatrix} A & B \\ C & D \end{pmatrix}$ is a continuous function of (A, B, C, D). With these preparations we obtain the following implications:

$$\begin{aligned}&\lim(\widetilde{P}_1,-\widetilde{Q}_1)=(\widehat{P}_1,-\widehat{Q}_1)\\&\Longrightarrow \lim\widehat{U}=\lim(\widetilde{P}_1,-\widetilde{Q}_1)\begin{pmatrix}\widehat{D}_2\\ \widehat{N}_2\end{pmatrix}=(\widehat{P}_1,-\widehat{Q}_1)\begin{pmatrix}\widehat{D}_2\\ \widehat{N}_2\end{pmatrix}=\mathrm{id}_p\\&\Longrightarrow \lim\widehat{U}^{-1}=\mathrm{id}_p\Longrightarrow \lim\widehat{U}^{-1}(\widetilde{P}_1,-\widetilde{Q}_1)=(\widehat{P}_1,-\widehat{Q}_1)\\&\Longrightarrow \lim\begin{pmatrix}\widehat{U}^{-1}\widetilde{P}_1 & -\widehat{U}^{-1}\widetilde{Q}_1\\ -\widehat{Q}_2 & \widehat{P}_2\end{pmatrix}=\begin{pmatrix}\widehat{P}_1 & -\widehat{Q}_1\\ -\widehat{Q}_2 & \widehat{P}_2\end{pmatrix},\\&\quad\ \lim\begin{pmatrix}0 & \widehat{U}^{-1}\widetilde{Q}_1\\ \widehat{Q}_2 & 0\end{pmatrix}=\begin{pmatrix}0 & \widehat{Q}_1\\ \widehat{Q}_2 & 0\end{pmatrix}\\&\Longrightarrow \lim\widetilde{H}=\lim\begin{pmatrix}\widehat{U}^{-1}\widetilde{P}_1 & -\widehat{U}^{-1}\widetilde{Q}_1\\ -\widehat{Q}_2 & \widehat{P}_2\end{pmatrix}^{-1}\begin{pmatrix}0 & \widehat{U}^{-1}\widetilde{Q}_1\\ -\widehat{Q}_2 & \end{pmatrix}\\&\qquad=\begin{pmatrix}\widehat{P}_1 & -\widehat{Q}_1\\ -\widehat{Q}_2 & \widehat{P}_2\end{pmatrix}^{-1}\begin{pmatrix}0 & \widehat{Q}_1\\ \widehat{Q}_2 & 0\end{pmatrix}=H.\end{aligned}$$

□

Example 10.2.34 In the situation of Theorem 10.2.33 with the assumed $\epsilon > 0$ consider the following special perturbations, cf. [4, pp. 251, 272].
1. *Additive perturbations* are defined by a small $A \in \mathcal{S}^{p\times m}$ and

$$\begin{aligned}&(\widetilde{P}_1,-\widetilde{Q}_1):=(\widehat{P}_1,-(\widehat{Q}_1+\widehat{P}_1A))\in\mathcal{S}^{p\times m}\\&\Longrightarrow(\widetilde{P}_1,-\widetilde{Q}_1)-(\widehat{P}_1,-\widehat{Q}_1)=(0,-\widehat{P}_1A)\\&\Longrightarrow\|(\widetilde{P}_1,-\widetilde{Q}_1)-(\widehat{P}_1,-\widehat{Q}_1)\|\le\|\widehat{P}_1\|\|A\|\\&\Longrightarrow\lim_{A\to0}\|(\widetilde{P}_1,-\widetilde{Q}_1)-(\widehat{P}_1,-\widehat{Q}_1)\|=0,\ \widetilde{H}_1=H_1+A,\ \lim_{A\to0}\widetilde{H}_1=H_1=\widehat{P}_1^{-1}\widehat{Q}_1.\end{aligned}\tag{10.172}$$

If $\|A\| \le \epsilon\|\widehat{P}_1\|^{-1}$ this $(\widetilde{P}_1, -\widetilde{Q}_1)$ satisfies all properties of Theorem 10.2.33. The expression $\widetilde{H}_1 = H_1 + A$ justifies the term *additive*. Of course, the limit $\lim_{A\to0}\widetilde{H}_1 = H_1$ refers to the graph topology of $F(s)^{p\times m}$ from Corollary and Definition 10.2.31.
2. *Multiplicative perturbations* are defined by a small $M \in \mathcal{S}^{p\times m}$ and

$$\begin{aligned}&(\widetilde{P}_1,-\widetilde{Q}_1):=(\widehat{P}_1(\mathrm{id}_p+M),-\widehat{Q}_1)\Longrightarrow(\widetilde{P}_1,-\widetilde{Q}_1)-(\widehat{P}_1,-\widehat{Q}_1)=(\widehat{P}_1M,0)\\&\Longrightarrow\|(\widetilde{P}_1,-\widetilde{Q}_1)-(\widehat{P}_1,-\widehat{Q}_1)\|\le\|\widehat{P}_1\|\|M\|\\&\Longrightarrow\lim_{M\to0}\|(\widetilde{P}_1,-\widetilde{Q}_1)-(\widehat{P}_1,-\widehat{Q}_1)\|=0,\ \widetilde{H}_1=(\mathrm{id}+M)^{-1}H_1,\ \lim_{M\to0}\widetilde{H}_1=H_1.\end{aligned}\tag{10.173}$$

3. Assume

$$\begin{aligned}&r\in C^0(\mathcal{K})=C^0(\overline{\mathbb{C}_+}\uplus\{\infty\}),\ \det(\widetilde{P}_1)\ne0\text{ and}\\&\forall s\in\mathcal{K}:\ \|(\widetilde{P}_1,-\widetilde{Q}_1)(s)-(\widehat{P}_1,-\widehat{Q}_1)(s)\|<|r(s)|,\ |r(s)|\|\begin{pmatrix}\widehat{D}_2\\ \widehat{N}_2\end{pmatrix}(s)\|\le1\\&\Longrightarrow\|(\widetilde{P}_1,-\widetilde{Q}_1)(s)\begin{pmatrix}\widehat{D}_2\\ \widehat{N}_2\end{pmatrix}(s)-\mathrm{id}_p\|\\&\quad=\|\left((\widetilde{P}_1,-\widetilde{Q}_1)(s)-(\widehat{P}_1,-\widehat{Q}_1)(s)\right)\begin{pmatrix}\widehat{D}_2\\ \widehat{N}_2\end{pmatrix}(s)\|\end{aligned}$$

$$\leq \|(\widetilde{P}_1, -\widetilde{Q}_1)(s) - (\widehat{P}_1, -\widehat{Q}_1)(s)\| \|\begin{pmatrix} \widehat{D}_2 \\ \widehat{N}_2 \end{pmatrix}(s)\| < |r(s)| \|\begin{pmatrix} \widehat{D}_2 \\ \widehat{N}_2 \end{pmatrix}(s)\| \leq 1$$

$$\underset{\mathcal{K}\text{ compact}}{\Longrightarrow} \|(\widetilde{P}_1, -\widetilde{Q}_1)\begin{pmatrix} \widehat{D}_2 \\ \widehat{N}_2 \end{pmatrix} - \mathrm{id}_p\| < 1 \Longrightarrow \widetilde{U} := (\widetilde{P}_1, -\widetilde{Q}_1)\begin{pmatrix} \widehat{D}_2 \\ \widehat{N}_2 \end{pmatrix} \in \mathrm{Gl}_p(\mathcal{S}). \tag{10.174}$$

Hence Theorem 10.2.33 is applicable. ◊

10.2.4 Operator Norms

In this and the following section we consider the continuous-time standard case with the stability region $\Lambda_1 = \mathbb{C}_- = \{\lambda \in \mathbb{C};\ \Re(\lambda) < 0\}$. *For* $H \in \mathcal{S}^{p\times m} \subseteq \mathrm{C}^0(\mathcal{K})^{p\times m}$ *we have defined the norm* $\|H\|$ *from Lemma* 10.2.22 *and are going to introduce another norm* $\|H\|_1$. *On the other hand Corollary and Definition* 8.2.84 *implies the transfer operators* $H\circ : (\mathrm{L}^q_+)^m \to (\mathrm{L}^q_+)^p$. $1 \leq q \leq \infty$. *We are going to compute the norm of these operators for* $q = 2, \infty$, *compare* [10, Theorem 589, (13.64), p. 518], *and to establish the connection of the norms* $\|H\|$ *and* $\|H\|_1$ on $\mathcal{S}^{p\times m}$ *with these operator norms. The main results are Theorems* 10.2.36, 10.2.46, 10.2.47 *and* 10.2.50. *Several important theorems on the Fourier transform in Sect.* 10.2.5 *are needed for the proofs.*

We need some preparations concerning operator norms. Our approach is more elementary than usual since we neither use the Lebesgue theory, in particular in connection with the convolution product, but only the Riemann integral, nor the theory of temperate distributions. An outstanding presentation of the more advanced material is Hörmander's book [14]. All vector spaces in the sequel use the base field $F = \mathbb{C}$ or $F = \mathbb{R}$.

Consider normed spaces X, Y with norms $\| - \|$ and a linear map (operator) $H : X \to Y$. The map H is continuous in 0 and then indeed everywhere if and only if there is a constant $M \geq 0$ such that $\|Hx\| \leq M\|x\|,\ x \in X$. Then the operator norm

$$\|H\| := \sup_{0\neq x\in X} \|Hx\|\|x\|^{-1} = \sup_{x\in X,\ \|x\|=1} \|Hx\|\|x\|^{-1} \leq M \tag{10.175}$$

is defined and satisfies $\|Hx\| \leq \|H\|\|x\|$ and $\|H_2H_1\| \leq \|H_2\|\|H_1\|$ if the H_i are two composable continuous operators. If $X' \subset X$ is a dense subset then

$$\|H\| = \sup_{0\neq x\in X'} \|Hx\|\|x\|^{-1}. \tag{10.176}$$

If $H_1, H_2 : X \to Y$ are continuous and coincide on X', i.e., $H_1x = H_2x,\ x \in X'$, then $H_1 = H_2$. The space $\mathrm{Hom}_{F,\mathrm{cont}}(X, Y)$ of continuous operators $H : X \to Y$ is again normed with $\|H\|$ from (10.175).

The space X is not necessarily a Banach space or complete, i.e., a Cauchy sequence does not necessarily have a limit in X. But X admits a unique completion $\overline{X}$ that is a Banach space with a unique extended norm such that X is dense in $\overline{X}$ and every Cauchy sequence in X has a limit in $\overline{X}$. One constructs $\overline{X}$ from X like $\mathbb{R}$ from $\mathbb{Q}$ as the space of all Cauchy sequences in X modulo the space of all zero sequences. The continuity of $H : X \to Y$ implies that the image of a Cauchy sequence is again such. Therefore the operator H can be uniquely extended to the continuous operator

$$\overline{H} : \overline{X} \to \overline{Y},\ x = \lim_{n\to\infty} x_n \mapsto \overline{H}x := \lim_{n\to\infty} Hx_n,\ x_n \in X,\ \|\overline{H}\| = \|H\|, \tag{10.177}$$

the last equation following from (10.176). The map

$$\mathrm{Hom}_{F,\mathrm{cont}}(X, Y) \to \mathrm{Hom}_{F,\mathrm{cont}}(\overline{X}, \overline{Y}),\ H \mapsto \overline{H}, \tag{10.178}$$

is injective and therefore we identify $H = \overline{H}$. There follows the isomorphism

$$\begin{aligned} &\mathrm{Hom}_{F,\mathrm{cont}}(X, \overline{Y}) \overset{\cong}{\to} \mathrm{Hom}_{F,\mathrm{cont}}(\overline{X}, \overline{Y}),\ H = \overline{H}|_X \leftrightarrow \overline{H},\ \text{hence} \\ &\mathrm{Hom}_{F,\mathrm{cont}}(X, \overline{Y}) \underset{\text{identification}}{=} \mathrm{Hom}_{F,\mathrm{cont}}(\overline{X}, \overline{Y}). \end{aligned} \tag{10.179}$$

The space $(\mathrm{Hom}_{F,\mathrm{cont}}(X, \overline{Y}), \| - \|)$ is itself a Banach space: If $(H_n)_n$ is a Cauchy sequence in this space then $(\|H_n\|)_n$ and $(H_n x)_n,\ x \in X$, are Cauchy sequences. Therefore the map

$$H : X \to \overline{Y},\ x \mapsto Hx := \lim_{n\to\infty} H_n x$$

is well defined. A simple proof, cf. [15, p. 85], implies the continuity of H and

$$H = \lim_{n\to\infty} H_n,\ \text{i.e.}\ \lim_{n\to\infty} \|H - H_n\| = 0 \Longrightarrow \|H\| = \lim_{n\to\infty} \|H_n\|. \tag{10.180}$$

Analogous results hold for a bilinear map $b : X_1 \times X_2 \to Y$. It is continuous if and only if

$$\|b\| := \sup\left\{\|b(x_1, x_2)\| \|x_1\|^{-1} \|x_2\|^{-1};\ 0 \neq x_i \in X_i\right\} < \infty. \tag{10.181}$$

A continuous b can be uniquely extended to

$$\begin{aligned} &\overline{b} : \overline{X_1} \times \overline{X_2} \to \overline{Y}\ \text{with}\ \|\overline{b}\| = \|b\|\ \text{where} \\ b(x_1, x_2) = &\lim_{n_1\to\infty,\ n_2\to\infty} b(x_{1,n_1}, x_{2,n_2}),\ x_i = \lim_{n_i\to\infty} x_{i,n_i},\ x_i \in \overline{X_i},\ x_{i,n_i} \in X_i. \end{aligned} \tag{10.182}$$

We apply the preceding considerations to the function spaces $\mathrm{L}^q \subseteq \mathrm{C}^{0,\mathrm{pc}},\ 1 \le q \le \infty$, from Sect. 8.2.6 with their norms $\| - \|_q$. By definition, the functions $u \in \mathrm{L}^q$ are piecewise continuous with finite $\|u\|_q = \left(\int_{-\infty}^{\infty} |u(t)|^q \mathrm{d}t\right)^{1/q}$ for $1 \le q < \infty$ and finite $\|f\|_\infty = \sup_{t\in\mathbb{R}} |f(t)|$ for $q = \infty$. An added lower index 0 indicates compact support. The inclusions $\mathrm{C}_0^{0,\mathrm{pc}} \subset \mathrm{L}^q$ are obvious. The completion of $(\mathrm{L}^q, \| - \|_q)$ is denoted by $\mathfrak{L}^q$ in deviation from the standard notation L^q. A concrete representation of $\mathfrak{L}^q$ is the space of Lebesgue measurable functions u with integrable $|u|^q,\ q<\infty$, or essentially bounded $u,\ q=\infty$, modulo the space of functions with support of measure zero. We neither need nor use this result. Notice that in this representation $u\in\mathfrak{L}^q$ is not a function on $\mathbb{R}$, but a residue class, and that $u(t)$ is not defined for $u\in\mathfrak{L}^q$ and a fixed chosen t. A disadvantage of our approach is the lack of a comprehensive space like that of measurable functions modulo almost zero functions that contains all $\mathfrak{L}^q$ and $\mathrm{C}^{0,\mathrm{pc}}$, so an intersection of these spaces is not defined a priori. The space $\mathrm{C}^{0,\infty}:=\mathrm{C}^0\bigcap\mathrm{L}^\infty$ of bounded continuous functions is a closed subspace of $\mathfrak{L}^\infty$, and hence a Banach space. We also consider the space $\mathfrak{St}\subset\mathrm{C}_0^{0,\mathrm{pc}}$ of step functions u with compact support. These are finite sums

$$u=\sum_{i=1}^n a_i\chi_{I_i},\ a_i\in F,\ I_i\subset\mathbb{R},\ \chi_{I_i}(t):=\begin{cases}1 & \text{if } t\in I_i\\ 0 & \text{if } t\notin I_i\end{cases}, \tag{10.183}$$

where the I_i are pairwise disjoint, finite intervals (=convex subsets) of $\mathbb{R}$. The disjointness may be omitted, but can always be satisfied by different I_i. For $q<\infty$ the subspace $\mathrm{C}_0^{0,\mathrm{pc}}$ is obviously dense in L^q and $\mathfrak{St}$ and C_0^∞ are dense in $\mathrm{C}_0^{0,\mathrm{pc}}$. We obtain several dense subspaces of $\mathfrak{L}^q$:

$$\mathrm{C}_0^\infty,\ \mathfrak{St}\subset\mathrm{C}_0^{0,\mathrm{pc}}\subset\mathrm{L}_+^q\subset\mathrm{L}^q\subset\mathfrak{L}^q,\ 1\le q<\infty. \tag{10.184}$$

According to Theorem 8.2.82 and (10.181) the bilinear convolution

$$L_+^1\times\mathrm{L}_+^q\to\mathrm{L}_+^q,\ (u_1,u_2)\mapsto u_1*u_2,\ \|u_1*u_2\|_q\le\|u_1\|_1\|u_2\|_q, \tag{10.185}$$

is continuous. Since L_+^q is dense in $\mathfrak{L}^q$ the convolution can be uniquely extended to the continuous bilinear convolution, cf. (10.182),

$$*:\mathfrak{L}^1\times\mathfrak{L}^q\to\mathfrak{L}^q,\ \|u_1*u_2\|_q\le\|u_1\|_1\|u_2\|_q,\ 1\le q<\infty. \tag{10.186}$$

Since L_+^q is a normed L_+^1-module (with the continuous multiplication $*$) $(\mathfrak{L}^q,*)$ is an $\mathfrak{L}^1$-module and $\mathfrak{L}^1$ is a Banach algebra (without one element).

The case $q=\infty$ requires a different argument since L_+^∞ is not dense in L^∞ or $\mathfrak{L}^\infty$. One defines the convolution

$$*:\mathrm{L}^1\times\mathrm{L}^\infty\to\mathrm{C}^{0,\infty},\ f(t):=(u_1*u_2)(t):=\int_{-\infty}^{\infty}u_1(t-x)u_2(x)dx:=\lim_{n\to\infty}f_n(t),$$
$$\text{where } f_n(t):=\int_{-n}^{n}u_1(t-x)u_2(x)dx\Longrightarrow\|u_1*u_2\|_\infty\le\|u_1\|_1\|u_2\|_\infty. \tag{10.187}$$

The function u_1*u_2 is continuous since the f_n are continuous and the limit $\lim_{n\to\infty}$ is uniform in t on all finite intervals $[-T,T],\ T>0$, due to

$$
\begin{aligned}
&\forall -T \le t \le T:\ |f(t) - f_n(t)| \le \|u_2\|_\infty \int_{|x|\ge n} |u_1(t-x)|dx \\
&= \|u_2\|_\infty \left(\int_n^\infty + \int_{-\infty}^{-n}\right) |u_1(t-x)|dx \underset{y:=t-x}{=} \|u_2\|_\infty \left(\int_{-\infty}^{t-n} + \int_{t+n}^{\infty}\right) |u_1(y)|dy \\
&\le \|u_2\|_\infty \left(\int_{-\infty}^{-(n-T)} + \int_{n-T}^{\infty}\right) |u_1(y)|dy = \|u_2\|_\infty \int_{|y|\ge n-T} |u_1(y)|dy \\
&\underset{u_1\in \mathrm{L}^1}{\Longrightarrow} \sup_{|t|\le T} |f(t) - f_n(t)| \underset{n\to\infty}{\longrightarrow} 0.
\end{aligned}
$$

Since $u_1 * u_2$ is continuous on $\mathrm{L}^1 \times \mathrm{L}^\infty$ and $\mathrm{C}^{0,\infty}$ is complete $*$ can be uniquely extended to

$$
*:\ \mathfrak{L}^1 \times \mathfrak{L}^\infty \to \mathrm{C}^{0,\infty} \text{ with } \|u_1 * u_2\|_\infty \le \|u_1\|_1 \|u_2\|_\infty. \tag{10.188}
$$

Notice that the following are not defined in our approach: $\mathfrak{L}^q \bigcap \mathfrak{L}^\infty$, $q < \infty$, the restrictions of (10.186) and (10.188) to this intersection and their equality. However, as in Theorem 8.2.82 one concludes for $1 \le q < \infty$ that (10.187) induces

$$
\begin{aligned}
&*: \mathrm{L}^1 \times (\mathrm{L}^\infty \textstyle\bigcap \mathrm{L}^q) \to \mathrm{C}^{0,\infty} \textstyle\bigcap \mathrm{L}^q = \left\{v \in \mathrm{C}^0;\ \|v\|_\infty < \infty,\ \|v\|_q < \infty\right\}, \text{ with} \\
&(u_1 * u_2)(t) = \int_{-\infty}^{\infty} u_1(t-x)u_2(x)dx,\ \|u_1 * u_2\|_q \le \|u_1\|_1 \|u_2\|_q.
\end{aligned}
\tag{10.189}
$$

The preceding equations imply

$$
\begin{aligned}
&\forall 1 \le q \le \infty:\ *: \mathfrak{L}^1 \times \mathfrak{L}^q \to \mathfrak{L}^q,\ \|u_1 * u_2\|_q \le \|u_1\|_1 \|u_2\|_q, \text{ where} \\
&\forall u_1 \in \mathrm{L}^1 \forall u_2 \in \mathrm{L}^\infty \textstyle\bigcap \mathrm{L}^q:\ (u_1 * u_2)(t) \text{ from (10.187), (10.189).}
\end{aligned}
\tag{10.190}
$$

Note that for $u_1 \in \mathrm{L}^1$ and $u_2 \in \mathrm{L}^\infty \bigcap \mathrm{L}^q$ the function $(u_1 * u_2)(t)$ is given by the *Riemann* integral (10.187) and that the restriction to functions of left bounded support in Theorem 8.2.82 is not required any more.

In the same fashion, but more simply, the continuous multiplication

$$
\cdot: \mathrm{L}^\infty \times \mathrm{L}^q \to \mathrm{L}^q,\ (u_1 u_2)(t) = u_1(t)u_2(t),\ \|u_1 u_2\|_q \le \|u\|_\infty \|u_2\|_q,\ 1 \le q \le \infty,
$$

is uniquely extended to

$$
\cdot := \cdot_q: \mathfrak{L}^\infty \times \mathfrak{L}^q \to \mathfrak{L}^q,\ \|u_1 \cdot_q u_2\|_q \le \|u_1\|_\infty \|u_2\|_q,\ 1 \le q \le \infty. \tag{10.191}
$$

Remark 10.2.35 We need the following specific norms:

1. A matrix $A = (A_{ij})_{i,j} \in F^{p\times m}$ induces the map $A\cdot : F^m \to F^p$, $x \mapsto Ax$. With the maximum norm $\|x\|_\infty := \max\left\{|x_j|;\ 1 \le j \le m\right\}$ for $x = (x_j)_j \in F^m$ and

likewise on F^p the corresponding matrix norm is

$$\|A\|_\infty = \sup_{0\neq x\in F^m} \|Ax\|_\infty \|x\|_\infty^{-1} = \max_{1\le i\le p} \sum_{j=1}^m |A_{ij}|. \tag{10.192}$$

2. We replace the norm $\|-\|_\infty$ by the 2-norm. The Hermitian (symmetric for $F=\mathbb{R}$) scalar product on F^m is given as $x^*y := \sum_{j=1}^m \overline{x_j} y_j$ where $x^* := (\overline{x_1},\dots,\overline{x_m}) \in F^{1\times m}$ and makes F^m a unitary (Euclidean for $F=\mathbb{R}$) space. The associated 2-norm $\|-\|_2$ is defined by $\|x\|_2^2 := x^*x = \sum_{j=1}^m |x_j|^2$. For $A\in F^{p\times m}$, $x\in F^m$ we conclude

$$\|Ax\|_2^2 = (Ax)^*Ax = x^*(A^*A)x \ge 0,\ A^*\in F^{m\times p},\ (A^*)_{ij} := \overline{A_{ji}}.$$

The matrix A^* is the Hermitian adjoint of A and A^*A is a Hermitian matrix, i.e., $(A^*A)^* = A^*A$, with nonnegative eigenvalues $\lambda_1\ge\lambda_2\ge\dots\ge\lambda_m\ge 0$. There is a unitary (orthogonal for $F=\mathbb{R}$) matrix $U\in F^{m\times m}$ (with $U^*U = \mathrm{id}_m$, $U^{-1}=U^*$ and $\|Ux\|_2^2 = x^*U^*Ux = x^*x = \|x\|_2^2$) such that $U^*(A^*A)U = \mathrm{diag}(\lambda_1,\dots,\lambda_m)$, $\lambda_i\ge 0$. For $x\in F^m$ and $y := U^*x$, $x = Uy$ this implies

$$\begin{aligned}\|Ax\|_2^2 &= \|AUy\|_2^2 = y^*U^*A^*AUy = y^*\,\mathrm{diag}(\lambda_1,\dots,\lambda_m)y = \sum_{j=1}^m \lambda_j|y_j|^2\\ &\le \lambda_1\sum_j |y_j|^2 = \lambda_1\|y\|_2^2 = \lambda_1\|x\|_2^2 \Longrightarrow \|A\|_2 \le \sigma(A) := \lambda_1^{1/2}.\end{aligned} \tag{10.193}$$

If x_A with $\|x_A\|_2 = 1$ is a nonzero eigenvector of A^*A to λ_1, i.e., $A^*Ax_A = \lambda_1 x_A$ then

$$\begin{aligned}&\|Ax_A\|_2^2 = x_A^*A^*Ax_A = x_A^*\lambda_1 x_A = \sigma(A)^2\|x_A\|_2^2 = \sigma(A)^2\\ &\Longrightarrow \|A\|_2 \ge \sigma(A) \underset{(10.193)}{\Longrightarrow} \|A\|_2 = \sigma(A).\end{aligned} \tag{10.194}$$

The norm $\sigma(A) = \|A\|_2$ is called the *first singular value* of A and is a continuous function of A.

The connection of $\|A\|_2$ with $\|A\|_\infty$ is the following:

$$\begin{aligned}&\|x\|_2 = (|x_1|^2| + \dots + |x_m|^2)^{1/2} \le m^{1/2}\max_i |x_i| = m^{1/2}\|x\|_\infty,\ \|x\|_\infty \le \|x\|_2\\ &\Longrightarrow \|Ax\|_2 \le p^{1/2}\|Ax\|_\infty \le p^{1/2}\|A\|_\infty\|x\|_\infty \le p^{1/2}\|A\|_\infty\|x\|_2\\ &\Longrightarrow \|A\|_2 = \sigma(A) \le p^{1/2}\|A\|_\infty = p^{1/2}\max_i\sum_{j=1}^m |A_{ij}| \le p^{1/2} m \max_{i,j}|A_{ij}|.\end{aligned} \tag{10.195}$$

3. Consider a transfer matrix $H \in \mathcal{S}^{p\times m}$, i.e., H is proper and its poles lie in $\mathbb{C}_-$. According to Theorem and Definition 8.2.83 and Corollary and Definition 8.2.84 we get

$$\begin{aligned} &H = H_0 + H_{\mathrm{spr}},\ H\circ\delta = H_0\delta + h,\ H_0 \in \mathbb{C}^{p\times m},\ h := H_{\mathrm{spr}}\circ\delta \in (L^1_+)^{p\times m},\\ &h = \alpha Y,\ \alpha \in (\mathrm{C}^\infty)^{p\times m},\ \alpha \text{ polynomial-exponential}\\ &H\circ : (\mathrm{L}^\infty_+)^m \to (\mathrm{L}^\infty_+)^p,\ u = \delta * u \mapsto y := H\circ u = (H\circ\delta) * u = H_0 u + h * u. \end{aligned} \tag{10.196}$$

The distribution $H\circ\delta$ is the impulse response of H. On $(\mathfrak{L}^\infty)^m$ we use the maximum norm $\|u\|_\infty := \max\left\{\|u_j\|_\infty;\ j = 1,\dots,m\right\}$ and likewise on $(\mathfrak{L}^\infty)^p$. We are going to compute the norm of $H\circ$ for arbitrary $h \in (\mathfrak{L}^1)^{p\times m}$, and hence consider matrices

$$\begin{aligned} &H_0 = (H_{0,ij})_{i,j} \in F^{p\times m},\ h = (h_{ij})_{i,j} \in (\mathfrak{L}^1)^{p\times m} \text{ and the operator}\\ &H\circ : (\mathfrak{L}^\infty)^m \to (\mathfrak{L}^\infty)^p,\ u \mapsto y := H\circ u := H_0 u + h * u. \end{aligned} \tag{10.197}$$

The equations $y_i = \sum_j H_{0,ij}u_j + \sum_j h_{ij} * u_j,\ i = 1,\dots,p$, and (10.188) imply

$$\begin{aligned} &\|y_i\|_\infty \le \sum_j |H_{0,ij}|\|u_j\|_\infty + \sum_j \|h_{ij}\|_1\|u_j\|_\infty,\ i = 1,\dots,p\\ &\Longrightarrow \|y\|_\infty \le \max_i \sum_j \left(|H_{0,ij}| + \|h_{ij}\|_1\right)\|u\|_\infty. \end{aligned}$$

Hence $H\circ :\ (\mathfrak{L}^\infty)^m \to (\mathfrak{L}^\infty)^p,\ u \mapsto H_0 u + h * u$, is continuous and

$$\|H\circ : (\mathfrak{L}^\infty)^m \to (\mathfrak{L}^\infty)^p\| \le N(H_0, h) := \max_i \sum_j \left(|H_{0,ij}| + \|h_{ij}\|_1\right). \tag{10.198}$$

In particular, the operator $H\circ$ depends continuously on $H_0 \in F^{p\times m}$ and $h \in (\mathfrak{L}^1)^{p\times m}$. In Theorem 10.2.36 below we will show equality in (10.198).

4. According to (8.103) the following Hermitian inner product is well defined:

$$\begin{aligned} &\langle u, v\rangle := \int_{-\infty}^{\infty} \overline{u(t)}v(t)dt,\ u, v \in \mathrm{L}^2 \Longrightarrow \|u\|_2^2 = \langle u, u\rangle = \int_{-\infty}^{\infty} |u(t)|^2 dt \text{ and}\\ &|\langle u, v\rangle| \le \int_{-\infty}^{\infty} |u(t)||v(t)|dt \le \|u\|_2\|v\|_2. \end{aligned} \tag{10.199}$$

The completion $\mathfrak{L}^2$ of L^2 or of L^2_+ is a Hilbert space with the unique extended inner product and norm. The inner product is extended componentwise to $(\mathfrak{L}^2)^m$ by

$$\begin{aligned}&\langle u, v\rangle := \sum_{j=1}^{m}\langle u_j, v_j\rangle,\ u=(u_j)_j,\ v\in(\mathfrak{L}^2)^m\\ &\Longrightarrow \|u\|_2^2=\sum_{j=1}^{m}\|u_j\|_2^2 \Longrightarrow \|u\|_2\le\sum_{j=1}^{m}\|u_j\|_2.\end{aligned}\tag{10.200}$$

Consider the data from (10.197), but with $u\in(\mathfrak{L}^2)^m$, and hence $y=H\circ u\in(\mathfrak{L}^2)^p$. The same computation as in (10.198) with (10.188) furnishes

$$\begin{aligned}&y=H_0u+h*u,\ y_i=\sum_j(H_{0,ij}u_j+h_{ij}*u_j)\\ &\Longrightarrow \|y\|_2\le\sum_i\|y_i\|_2\le\sum_{i,j}\left(|H_{0,ij}|+\|h_{ij}\|_1\right)\|u_j\|_2\\ &\le\left(\sum_{i,j}(|H_{0,ij}|+\|h_{ij}\|_1)\right)\|u\|_2\\ &\Longrightarrow \|H\circ:\left(\mathfrak{L}^2\right)^m\to\left(\mathfrak{L}^2\right)^p\|\le\sum_{i,j}\left(|H_{0,ij}|+\|h_{ij}\|_1\right).\end{aligned}\tag{10.201}$$

Hence $H\circ:(\mathfrak{L}^2)^m\to(\mathfrak{L}^2)^p$ is continuous and depends continuously on $H_0\in F^{p\times m}$ and $h\in(\mathfrak{L}^1)^{p\times m}$. The norm $\|H\circ:\left(\mathfrak{L}^2\right)^m\to\left(\mathfrak{L}^2\right)^p\|$ will be computed in Theorem 10.2.46 by means of the Fourier transform.

5. Consider $\widehat{H}\in\left(\mathrm{C}^{0,\infty}\right)^{p\times m}$ and the function $\sigma(\widehat{H}(t))$ from item 2. The entries of $\widehat{H}(t)$ are continuous and bounded, and hence, by 2, so is the function $\sigma(\widehat{H}(t))$. We infer

$$\begin{aligned}&\widehat{H}\cdot:(\mathrm{L}^2)^m\to(\mathrm{L}^2)^p,\ u\mapsto y=\widehat{H}u,\ y(t)=\widehat{H}(t)u(t)\\ &\qquad\ \ \cap\qquad\quad\ \cap\\ &\widehat{H}\cdot:(\mathfrak{L}^2)^m\to(\mathfrak{L}^2)^p\\ &\underset{\text{item 2}}{\Longrightarrow}\ \forall t\in\mathbb{R}:\ \|y(t)\|_2\le\sigma(\widehat{H}(t))\|u(t)\|_2\le\sigma\|u(t)\|_2\ \text{where}\\ &\sigma:=\sup_t\sigma(\widehat{H}(t))=\|\sigma(\widehat{H}(t))\|_\infty\Longrightarrow\|y\|_2:=\left(\int_{-\infty}^{\infty}\|y(t)\|_2^2dt\right)^{1/2}\le\sigma\|u\|_2\\ &\Longrightarrow\|\widehat{H}\cdot\|_2:=\|\widehat{H}\cdot:(\mathfrak{L}^2)^m\to(\mathfrak{L}^2)^p\|\le\sigma=\|\sigma(\widehat{H}(t))\|_\infty.\end{aligned}\tag{10.202}$$

◇

Theorem 10.2.36 (Cf. [16, p. 107], [10, Theorem 589, p. 518]) *Consider*

$$H_0=(H_{0,ij})_{i,j}\in F^{p\times m},\ h=(h_{ij})_{i,j}\in(\mathfrak{L}^1)^{p\times m}\ \textit{and the operator}$$
$$H\circ:(\mathfrak{L}^\infty)^m\to(\mathfrak{L}^\infty)^p,\ u\mapsto H\circ u:=H_0u+h*u.$$

The norm of this operator $H\circ$ is

$$\|H\circ\|_\infty := \|H\circ : (\mathfrak{L}^\infty)^m \to (\mathfrak{L}^\infty)^p\| = N(H_0, h) := \max_{i=1,\dots,p} \sum_{j=1}^m (|H_{0,ij}| + \|h_{ij}\|_1). \tag{10.203}$$

In particular, this applies to

$$H = H_0 + H_{\text{spr}} \in \mathcal{S}^{p\times m},\ h = H_{\text{spr}} \circ \delta \text{ and } \|H\|_1 := N(H_0, h). \tag{10.204}$$

Proof The inequality $\le$ follows from (10.198).
(i) We show the opposite inequality, first for $h \in \mathfrak{St}^{p\times m}$. Assume $N(H_0, h) = \sum_j \big(|(H_{0,ij}| + \|h_{ij}\|_1\big)$.
(a) Define

$$u_0 = (u_{0j})_j \in F^m \text{ by } u_{0j} := \begin{cases} \frac{\overline{H_{0,ij}}}{|H_{0,ij}|} & \text{if } H_{0,ij} \neq 0 \\ 1 & \text{if } H_{0,ij} = 0 \end{cases}$$
$$\Longrightarrow \|u_0\|_\infty = 1 \text{ and } H_{0,i-}u_0 = \sum_j H_{0,ij}u_{0,j} = \sum_j |H_{0,ij}|. \tag{10.205}$$

(b) Recall that the h_{ij} are step functions by assumption in (i). Similarly to (a) we define the step functions

$$f_j(t) := \begin{cases} \frac{\overline{h_{ij}(t)}}{|h_{ij}(t)|} & \text{if } h_{ij}(t) \neq 0 \\ 1 & \text{if } h_{ij}(t) = 0 \end{cases} \in \mathfrak{St} \Longrightarrow h_{ij}f_j = |h_{ij}|. \tag{10.206}$$

Let

$$\epsilon > 0 \text{ and } \mathbb{R} = I(\epsilon) \uplus J(\epsilon),\ I(\epsilon) := [-\epsilon, \epsilon],\ J(\epsilon) = (-\infty, -\epsilon) \uplus (\epsilon, \infty)$$
$$\Longrightarrow \chi_{I(\epsilon)}(x) = \chi_{I(\epsilon)}(-x),\ \chi_{J(\epsilon)}(x) = \chi_{J(\epsilon)}(-x),\ 1 = \chi_{I(\epsilon)}(x) + \chi_{J(\epsilon)}(x).$$

Define

$$u_j(x) := u_{0j}\chi_{I(\epsilon)}(x) + f_j(-x)\chi_{J(\epsilon)}(x),\ u := (u_j)_j \in \mathfrak{St}^m$$
$$\Longrightarrow u_j(-x) := u_{0j}\chi_{I(\epsilon)}(x) + f_j(x)\chi_{J(\epsilon)}(x),$$
$$u_j(0) := u_{0j},\ \|u_j\|_\infty = 1,\ \|u\|_\infty = 1.$$

These data imply

$$
\begin{aligned}
(H\circ u)_i(0) &= \sum_j \left(H_{0,ij}u_j(0) + \int_{\mathbb{R}} h_{ij}(x)u_j(-x)dx \right) \\
&= \sum_j \left(|H_{0,ij}| + u_{0j}\int_{I(\epsilon)} h_{ij}(x)dx + \int_{J(\epsilon)} |h_{ij}(x)|dx \right).
\end{aligned}
\tag{10.207}
$$

By definition of the various norms and since the first and third summand in the last line of (10.207) are nonnegative we conclude

$$
\begin{aligned}
&\| H\circ : (\mathfrak{L}^\infty)^m \to (\mathfrak{L}^\infty)^p \| \underset{\|u\|_\infty=1}{\geq} \|H\circ u\|_\infty \geq \|(H\circ u)_i\|_\infty \geq |(H\circ u)_i(0)| \\
&\underset{|u_{0j}|=1}{\geq} \sum_j \left(|H_{0,ij}| + \int_{J(\epsilon)} |h_{ij}(x)|dx - \int_{I(\epsilon)} |h_{ij}(x)|dx \right) \\
&\underset{\epsilon\to 0}{\Longrightarrow} \| H\circ : (\mathfrak{L}^\infty)^m \to (\mathfrak{L}^\infty)^p \| \geq \sum_j \left(|H_{0,ij}| + \int_{\mathbb{R}} |h_{ij}(x)|dx \right) \\
&= \sum_j \left(|H_{0,ij}| + \|h_{ij}\|_1 \right) = N(H_0, h) \underset{(10.198)}{\Longrightarrow} \|H\circ\|_\infty = N(H_0, h).
\end{aligned}
$$

(ii) Let $h \in (\mathfrak{L}^1)^{p\times m}$ be arbitrary and $H\circ = H_0 + h*$. Since $\mathfrak{St}$ is dense in $\mathfrak{L}^1$ we can and do choose sequences $h^\nu_{ij} \in \mathfrak{St},\ \nu \geq 0$, such that $\lim_{\nu\to\infty} h^\nu_{ij} = h_{ij}$ in $\mathfrak{L}^1$. We define $h^\nu := (h^\nu_{ij})_{ij} \in \mathfrak{St}^{p\times m}$ and $H^\nu\circ := H_0 + h^\nu *$. Then

$$
\begin{aligned}
&\|H\circ - H^\nu\circ\|_\infty \leq N(0, h - h^\nu) = \max_i \sum_j \|h_{ij} - h^\nu_{ij}\|_1 \underset{\nu\to\infty}{\longrightarrow} 0 \\
&\Longrightarrow H\circ = \lim_{\nu\to\infty} H^\nu\circ \underset{(10.180)}{\Longrightarrow} \|H\circ\| = \lim_{\nu\to\infty} \|H^\nu\circ\| \\
&\underset{(i)}{=} \lim_{\nu\to\infty} N(H_0, h^\nu) = N(H_0, \lim_{\nu\to\infty} h^\nu) = N(H_0, h).
\end{aligned}
$$

□

Theorem 10.2.37 *For $\widehat{H} \in (\mathrm{C}^{0,\infty})^{p\times m}$ one gets equality in* (10.202), *i.e.,*

$$
\|\widehat{H}\cdot\|_2 := \|\widehat{H}\cdot : (\mathfrak{L}^2)^m \to (\mathfrak{L}^2)^p\| = \|\sigma(\widehat{H}(t))\|_\infty := \sup_{t\in\mathbb{R}} \sigma(\widehat{H}(t)). \tag{10.208}
$$

Proof (i) Assume $\widehat{H} \neq 0$ and hence $0 < \sigma := \|\sigma(\widehat{H}(t))\|_\infty < \infty$. The inequality $\leq$ in (10.208) follows from (10.202).
(ii) We use the properties of $\sigma(A)$ from Remark 10.2.35, item 2. Choose

$$
\begin{aligned}
&\epsilon_1 > 0,\ t_0 \in \mathbb{R},\ u_0 \in F^m \text{ with} \\
&0 \leq \sigma - \epsilon_1 \leq \sigma(\widehat{H}(t_0)) \leq \sigma,\ \|u_0\|_2 = 1,\ \|\widehat{H}(t_0)u_0\|_2 = \sigma(\widehat{H}(t_0)).
\end{aligned}
$$

Choose $\epsilon_2 > 0$ with $\sigma\left(\widehat{H}(t) - \widehat{H}(t_0)\right) \le \epsilon_1$ for $t_0 \le t \le t_0 + \epsilon_2$ and define

$$u \in \mathfrak{St}^m \subset \left(\mathrm{L}^2\right)^m : \; u(t) := \begin{cases} u_0 & \text{if } t_0 \le t \le t_0 + \epsilon_2 \\ 0 & \text{otherwise} \end{cases}, \; y := \widehat{H}u.$$

These data imply $\|u\|_2^2 = \int_{t_0}^{t_0+\epsilon_2} 1dt = \epsilon_2$ and for $t_0 \le t \le t_0 + \epsilon_2$

$$\begin{aligned}
&\sigma - \epsilon_1 \le \|y(t_0)\|_2 = \sigma(\widehat{H}(t_0)) \le \sigma, \; y(t) = \widehat{H}(t_0)u_0 + (\widehat{H}(t) - \widehat{H}(t_0))u_0 \text{ and} \\
&\|y(t)\|_2^2 = \|y(t_0)\|_2^2 + \|(\widehat{H}(t) - \widehat{H}(t_0))u_0\|_2^2 + 2\langle y(t_0), (\widehat{H}(t) - \widehat{H}(t_0))u_0\rangle \\
&\qquad \ge \|y(t_0)\|_2^2 - \|(\widehat{H}(t) - \widehat{H}(t_0))u_0\|_2^2 - 2|\langle y(t_0), (\widehat{H}(t) - \widehat{H}(t_0))u_0\rangle| \\
&\qquad \ge |y(t_0)\|_2^2 - \sigma\left(\widehat{H}(t) - \widehat{H}(t_0)\right)^2 - 2\|y(t_0)\|_2\sigma\left(\widehat{H}(t) - \widehat{H}(t_0)\right) \\
&\Longrightarrow \|y(t)\|_2^2 \ge (\sigma - \epsilon_1)^2 - \epsilon_1^2 - 2\sigma\epsilon_1 \\
&\Longrightarrow \|\widehat{H}u\|_2^2 = \|y\|_2^2 = \int_{t_0}^{t_0+\epsilon_2} \|y(t)\|_2^2 dt \ge ((\sigma - \epsilon_1)^2 - \epsilon_1^2 - 2\sigma\epsilon_1)\epsilon_2 \\
&\qquad = ((\sigma - \epsilon_1)^2 - \epsilon_1^2 - 2\sigma\epsilon_1)\|u\|_2^2 \\
&\Longrightarrow \|\widehat{H}\cdot\|_2^2 \ge (\sigma - \epsilon_1)^2 - \epsilon_1^2 - 2\sigma\epsilon_1 \underset{\epsilon_1 \to 0}{\Longrightarrow} \|\widehat{H}\cdot\|_2 \ge \sigma.
\end{aligned}$$

□

10.2.5 The Use of the Fourier Transform

In continuous time with $\Lambda_1 = \mathbb{C}_-$ and $H \in S^{p \times m}$ we compute the norm of the operator $H\circ : (\mathfrak{L}^2)^m \to (\mathfrak{L}^2)^p$ and discuss its connection with the norm $\|H\|$ and the robustness of stabilizing compensators. Another norm $\|H\|_1$ on $S^{p\times m}$ is needed to obtain analogous results for $H\circ : (\mathfrak{L}^\infty)^m \to (\mathfrak{L}^\infty)^p$. The proofs require important properties of the Fourier transform on $\mathfrak{L}^1$ and $\mathfrak{L}^2$ that are established first and taken from [14, pp. 159–165]. *Neither the Lebesgue integral nor temperate distributions are needed or used in our derivations.*

The base field for the Fourier transform is $\mathbb{C}$ since the complex exponentials $e^{j\omega t}$ with real time t and real frequency ω play a dominant part. If $\mathcal{F}$ is a function space we use a lower index t or ω ($\mathcal{F}_t$, $\mathcal{F}_\omega$) to indicate the name of the independent variable. The Fourier transform on L^1 is defined as

$$\mathbb{F} : \mathrm{L}_t^1 \to \mathrm{C}_\omega^{0,\infty}, \; u \mapsto \mathbb{F}(u) := \widehat{u}, \; \widehat{u}(\omega) := (u * e^{j\omega t})(0) = \int_{-\infty}^{\infty} u(t)e^{-j\omega t}\mathrm{d}t. \tag{10.209}$$

The function $\widehat{u}$ is well defined due to (10.189). It is the continuous uniform limit of the continuous functions $\widehat{u}_n(\omega) := \int_{-n}^{n} u(t)e^{-j\omega t}dt$, $n \in \mathbb{N}$, since $u \in \mathrm{L}^1$ and

$$\forall \omega \in \mathbb{R} : |\widehat{u}(\omega) - \widehat{u}_n(\omega)| \leq \int_{|t|\geq n} |u(t)|\mathrm{d}t \underset{n\to\infty}{\longrightarrow} 0.$$

Moreover $|\widehat{u}(\omega)| \leq \|u\|_1$ and hence

$$\mathbb{F} : \mathrm{L}_t^1 \to \mathrm{C}^{0,\infty}, \ \|\mathbb{F}(u)\|_\infty \leq \|u\|_1. \tag{10.210}$$

Therefore $\mathbb{F}$ is continuous and can be uniquely extended to

$$\mathbb{F} : \mathfrak{L}_t^1 \to \mathrm{C}_\omega^{0,\infty}, \ \|\mathbb{F}(u)\|_\infty \leq \|u\|_1. \tag{10.211}$$

Since $\mathfrak{S}t$ and C_0^∞ are dense in $\mathfrak{L}^1$ the Fourier transform is uniquely determined by its restriction to these subspaces. The conjugate Fourier transform is

$$\begin{aligned} \overline{\mathbb{F}} : \mathfrak{L}_t^1 \to \mathrm{C}_\omega^{0,\infty}, \ \forall u \in \mathrm{L}_t^1 : \overline{\mathbb{F}}(u)(\omega) &:= \mathbb{F}(u)(-\omega) \\ := (u * e^{-j\omega t})(0) &= \int_{-\infty}^{\infty} u(t)e^{j\omega t}\mathrm{d}t, \ \|\overline{\mathbb{F}}(u)\|_\infty \leq \|u\|_1. \end{aligned} \tag{10.212}$$

By interchanging the roles of t and ω one also obtains

$$\mathbb{F} : \mathfrak{L}_\omega^1 \to \mathrm{C}_t^{0,\infty}, \ \overline{\mathbb{F}} : \mathfrak{L}_\omega^1 \to \mathrm{C}_t^{0,\infty}. \tag{10.213}$$

Under suitable assumptions, discussed below, the *Fourier inversion formula*

$$\overline{\mathbb{F}}\mathbb{F} = 2\pi \,\mathrm{id} \ \text{or} \ u(t) = (2\pi)^{-1} \int_{-\infty}^{\infty} \widehat{u}(\omega)e^{j\omega t}\mathrm{d}\omega, \ \widehat{u}(\omega) := \int_{-\infty}^{\infty} u(t)e^{-j\omega t}\mathrm{d}\omega, \tag{10.214}$$

holds. In this case $u(t)$ is a continuous sum (=integral) of the complex exponentials or sinusoids $e^{j\omega t}$.

Lemma 10.2.38 *Let $H \in \mathbb{C}(s)$ be a strictly proper and stable rational function and $h := H \circ \delta \in \mathrm{L}_+^1$ the corresponding impulse response according to Lemma 8.2.46 and Theorem 8.2.83. Then $\mathbb{F}(h)(\omega) = H(j\omega)$.*

Proof By partial fraction decomposition H is a finite sum of functions

$$a(s-\lambda)^{-k}, \ 0 \neq a \in \mathbb{C}, \ \lambda \in \mathbb{C}, \ \Re(\lambda) < 0, \ k \geq 1.$$

It suffices to show the assertion for $H_k := (s-\lambda)^{-k}$ and $h_k := H_k \circ \delta$. Recall

$$h_k := H_k \circ \delta = e_{\lambda,k-1}(t)Y = \begin{cases} \frac{t^{k-1}}{(k-1)!}e^{\lambda t} & \text{if } t \geq 0 \\ 0 & \text{if } t < 0 \end{cases}$$

from Lemma 8.2.46. By induction we get

$$k=1:\ \mathbb{F}(h_1)(\omega)=\int_0^\infty e^{(\lambda-j\omega)t}\mathrm{d}t=(\lambda-j\omega)^{-1}(0-1)=(j\omega-\lambda)^{-1}=H_1(j\omega)$$

$$k>1:\ \mathbb{F}(h_k)(\omega)=\int_0^\infty \tfrac{t^{k-1}}{(k-1)!}e^{(\lambda-j\omega)t}\mathrm{d}t$$

$$\underset{\text{partial integration}}{=}(j\omega-\lambda)^{-1}\int_0^\infty \tfrac{t^{k-2}}{(k-2)!}e^{(\lambda-j\omega)t}\mathrm{d}t$$

$$=(j\omega-\lambda)^{-1}\mathbb{F}(h_{k-1})(\omega)\underset{\text{ind.}}{=}(j\omega-\lambda)^{-1}(j\omega-\lambda)^{-(k-1)}=(j\omega-\lambda)^{-k}=H_k(j\omega).$$

□

The properties of the Fourier transform are most easily derived on the space $\mathfrak{S}$ of *rapidly decreasing (in ∞)* functions $\varphi\in\mathrm{C}^\infty$. By definition such a φ satisfies

$$\begin{aligned}
&\forall\alpha,\beta\in\mathbb{N}:\ \sup_{t\in\mathbb{R}}|t^\alpha\varphi^{(\beta)}(t)|<\infty,\ \varphi^{(\beta)}:=d^\beta\varphi/dt^\beta,\ \text{or, equivalently,}\\
&\forall n\in\mathbb{N}:\ p_n(\varphi):=\sup_{t\in\mathbb{R}}(1+t^2)^{n/2}\max_{0\le\alpha\le n}|\varphi^{(\alpha)}(t)|<\infty\\
&\Longrightarrow\forall\alpha,\beta\le n\forall t\in\mathbb{R}:\ (1+t^2)|t^\alpha\varphi^{(\beta)}(t)|\le(1+t^2)^{n/2+1}|\varphi^{(\beta)}(t)|\le p_{n+2}(\varphi)\\
&\Longrightarrow|t^\alpha\varphi^{(\beta)}(t)|\le p_{n+2}(\varphi)(1+t^2)^{-1}\\
&\Longrightarrow\|t^\alpha\varphi^{(\beta)}\|_1=\int_{-\infty}^\infty|t^\alpha\varphi^{(\beta)}(t)|\mathrm{d}t\le Mp_{n+2}(\varphi)\ \text{with}\\
&M:=\int_{-\infty}^\infty(1+t^2)^{-1}\mathrm{d}t=\pi<\infty\Longrightarrow\forall\varphi\in\mathfrak{S}\forall\alpha\in\mathbb{N}:\ \varphi^{(\alpha)}\in\mathrm{L}^1.
\end{aligned}\tag{10.215}$$

Obviously

$$\mathrm{C}_0^\infty\subset\mathfrak{S}\ \text{and}\ e^{-t^2}\in\mathfrak{S}\setminus\mathrm{C}_0^\infty.$$

The p_n are an increasing sequence of norms on $\mathfrak{S}$ and make it a (locally convex) topological vector space. The sets $U(n,\epsilon):=\{\varphi\in\mathrm{C}^\infty;\ p_n(\varphi)<\epsilon\}$, $n\in\mathbb{N}$, $\epsilon>0$, are a basis of neighborhoods of 0 with $U(n+1,\epsilon)\subset U(n,\epsilon)$. The space $\mathfrak{S}$ is closed under differentiation and $s\circ:=d/dt:\mathfrak{S}\to\mathfrak{S}$ is continuous since

$$s\circ:\mathfrak{S}\to\mathfrak{S},\ \varphi\mapsto s\circ\varphi=\varphi',\ p_n(\varphi')\le p_{n+1}(\varphi)<\infty.\tag{10.216}$$

Hence $\mathfrak{S}$ is a $\mathbb{C}[s]$-module and also closed under the translations $\varphi(t)\mapsto\varphi(t-a)$, $a\in\mathbb{R}$. The space $\mathfrak{S}$ is not closed under multiplication with functions in C^∞, for instance, $e^{2t^2}e^{-t^2}=e^{t^2}$ is not bounded. However, for $\psi,\varphi\in\mathfrak{S}$ the equation $(\psi\varphi)^{(\alpha)}=\sum_{\beta\le\alpha}\binom{\alpha}{\beta}\psi^{(\alpha-\beta)}\varphi^{(\beta)}$ implies

$$\begin{aligned}
&p_n(\psi\varphi)\le cp_n(\psi)p_n(\varphi),\ c>0,\ \psi,\varphi\in\mathfrak{S}\Longrightarrow\mathfrak{S}\cdot\mathfrak{S}\subseteq\mathfrak{S}\\
&\Longrightarrow\forall 1\le q\le\infty:\ \mathfrak{S}\subset\mathrm{L}^q\bigcap\mathrm{C}^{0,\infty}.
\end{aligned}\tag{10.217}$$

Especially $\mathfrak{S}$ is a subalgebra of C^∞. For $q < \infty$ the space C_0^∞ and therefore the larger space $\mathfrak{S}$ are dense in $L^q \subset \mathfrak{L}^q$. Similarly one shows

$$p_n(\psi\varphi) \le c p_{n+d}(\varphi),\ c > 0,\ \psi \in \mathbb{C}[t],\ d := \deg(\psi),\ \varphi \in \mathfrak{S}, \tag{10.218}$$

i.e., $\mathfrak{S}$ is closed under multiplication with polynomials ψ and thus a $\mathbb{C}[t]$-module and $\psi\cdot : \mathfrak{S} \to \mathfrak{S}$ is continuous.

The proof of the Fourier inversion formula on $\mathfrak{S}$ requires the following additional preparations. By differentiation under the integral one obtains

$$\begin{aligned}
&(s_\omega \circ \mathbb{F}(\varphi))(\omega) = \int_{-\infty}^{\infty} e^{-j\omega t}(-jt)\varphi(t)\mathrm{d}t = -j\mathbb{F}(t\varphi)(\omega) \text{ for}\\
&\varphi, t\varphi \in \mathfrak{S}, \text{cf. (10.218)}, \Longrightarrow s \circ \mathbb{F}(\varphi) = -j\mathbb{F}(t\varphi),\ \mathbb{F}(t\varphi) = js \circ \mathbb{F}(\varphi)\\
&\Longrightarrow \forall \beta \in \mathbb{N}:\ \mathbb{F}(\varphi)^{(\beta)} = (-j)^\beta \mathbb{F}(t^\beta\varphi).
\end{aligned} \tag{10.219}$$

Here s_ω indicates the differentiation with respect ω. Partial integration furnishes

$$\begin{aligned}
&\mathbb{F}(s_t \circ \varphi)(\omega) = \int_{-\infty}^{\infty} e^{-j\omega t}\varphi'(t)\mathrm{d}t = [e^{-j\omega t}\varphi(t)]_{-\infty}^{\infty} + j\omega\int_{-\infty}^{\infty} e^{-j\omega t}\varphi(t)\mathrm{d}t = j\omega\mathbb{F}(\varphi)\\
&\Longrightarrow \mathbb{F}(s \circ \varphi) = j\omega\mathbb{F}(\varphi),\ \omega\mathbb{F}(\varphi) = -j\mathbb{F}(s \circ \varphi)\\
&\Longrightarrow \forall \alpha \in \mathbb{N}:\ \omega^\alpha\mathbb{F}(\varphi) = (-j)^\alpha\mathbb{F}(s^\alpha \circ \varphi) = (-j)^\alpha\mathbb{F}(\varphi^{(\alpha)})\\
&\underset{(10.206)}{\Longrightarrow} \forall \alpha, \beta \le n:\ \omega^\alpha\mathbb{F}(\varphi)^{(\beta)} = (-j)^{\alpha+\beta}\mathbb{F}((t^\beta\varphi)^{(\alpha)})\\
&\underset{(t^\beta\varphi)^{(\alpha)}\in\mathfrak{S}}{\Longrightarrow} |\omega^\alpha\mathbb{F}(\varphi)^{(\beta)}(\omega)| \le \|(t^\beta\varphi)^{(\alpha)}\|_1 < \infty\\
&\underset{(10.215)}{\Longrightarrow} \mathbb{F}(\varphi) \in \mathfrak{S}_\omega,\ p_n(\mathbb{F}(\varphi)) \le c p_{n+2}(\varphi),\ c > 0\\
&\Longrightarrow \mathbb{F} : \mathfrak{S}_t \to \mathfrak{S}_\omega \text{ is well defined and continuous.}
\end{aligned} \tag{10.220}$$

These equations further imply

$$\begin{aligned}
&\overline{\mathbb{F}}(s \circ \varphi)(\omega) = \mathbb{F}(s \circ \varphi)(-\omega) = (j\omega\mathbb{F}(\varphi))(-\omega) = -j\omega\mathbb{F}(\varphi)(-\omega) = -j\omega\overline{\mathbb{F}}(\varphi)(\omega)\\
&\Longrightarrow \overline{\mathbb{F}}(s \circ \varphi) = -j\omega\overline{\mathbb{F}}(\varphi) \text{ and likewise } \overline{\mathbb{F}}(t\varphi) = -js \circ \overline{\mathbb{F}}(\varphi)\\
&\Longrightarrow \overline{\mathbb{F}}\mathbb{F}(s \circ \varphi) = s \circ \overline{\mathbb{F}}\mathbb{F}(\varphi),\ \overline{\mathbb{F}}\mathbb{F}(t \cdot \varphi) = t \cdot \overline{\mathbb{F}}\mathbb{F}(\varphi).
\end{aligned} \tag{10.221}$$

The last line of (10.221) signifies that $\overline{\mathbb{F}}\mathbb{F} : \mathfrak{S}_t \to \mathfrak{S}_\omega \to \mathfrak{S}_t$ is $(\mathbb{C}[s], \circ)$ and $(\mathbb{C}[t], \cdot)$-linear. This property implies the inversion formula on $\mathfrak{S}$.

Theorem 10.2.39 (Cf. [14, Theorem 7.1.5]) *The Fourier transform* $\mathbb{F} : \mathfrak{S}_t \to \mathfrak{S}_\omega$ *is a continuous isomorphism with the continuous inverse* $\mathbb{F}^{-1} = (2\pi)^{-1}\overline{\mathbb{F}}$.

Proof It suffices to show the equations $T := \overline{\mathbb{F}}\mathbb{F} = 2\pi\,\mathrm{id}$ and $\mathbb{F}\overline{\mathbb{F}} = 2\pi\,\mathrm{id}$, the proof of the second following from that of the first.
(i) We show $T = c\,\mathrm{id}$, $T(\varphi) = c\varphi$, with $c \in \mathbb{C}$.
(a) Assume $\varphi(t_0) = 0$ and $n \in \mathbb{N}$. Then

$$\varphi(t) = (t - t_0)\psi,\ \psi \in C^\infty,\ \psi(t) := \int_0^1 \varphi'(t_0 + \tau(t - t_0))d\tau,$$

$$\Longrightarrow \forall t \text{ with } |t - t_0| \geq 1 : |\psi(t)| \leq |\varphi(t)| \Longrightarrow \sup_{t\in\mathbb{R}}(1 + t^2)^{n/2}|\psi(t)|$$

$$\leq \max\left(\max_{t,|t-t_0|\leq 1}(1 + t^2)^{n/2}|\psi(t)|,\ \sup_{t,|t-t_0|\geq 1}(1 + t^2)^{n/2}|\varphi(t)|\right)$$

$$\leq \max\left(\max_{t,|t-t_0|\leq 1}(1 + t^2)^{n/2}|\psi(t)|,\ p_n(\varphi)\right) < \infty.$$

By induction $\varphi = (t - t_0)\psi$ and $0 < \alpha \leq n$ imply

$$\varphi^{(\alpha)} = \alpha\psi^{(\alpha-1)} + (t - t_0)\psi^{(\alpha)} \Longrightarrow \sup_t(1 + t^2)^{n/2}|\psi^{(\alpha)}(t)|$$

$$\leq \max\left(\max_{|t-t_0|\leq 1}(1 + t^2)^{n/2}|\psi^{(\alpha)}(t)|,\ p_n(\varphi) + \alpha\sup_t(1 + t^2)^{n/2}|\psi^{(\alpha-1)}(t)|\right)$$

$$< \infty \Longrightarrow p_n(\psi) < \infty \Longrightarrow \psi \in \mathfrak{S}$$

$$\underset{(10.221)}{\Longrightarrow} T(\varphi) = T((t - t_0)\psi) = (t - t_0)T(\psi) \Longrightarrow T(\varphi)(t_0) = 0.$$

(b) Choose $\varphi \in \mathfrak{S}$ with $\varphi(t) \neq 0$ for all t, for instance, $\varphi := e^{-t^2}$, and let $\psi \in \mathfrak{S}$ be arbitrary. For any t_0 and $a := \psi(t_0)\varphi(t_0)^{-1}$ we get

$$\psi - a\varphi \in \mathfrak{S},\ (\psi - a\varphi)(t_0) = 0 \underset{(a)}{\Longrightarrow} 0 = T(\psi - a\varphi)(t_0) = T(\psi)(t_0) - aT(\varphi)(t_0)$$

$$\underset{t_0 \text{arbitrary}}{\Longrightarrow} \forall\psi \in \mathfrak{S} : T(\psi) = c\psi,\ c \in C^\infty,\ c(t) := T(\varphi)(t)\varphi(t)^{-1}$$

$$\Longrightarrow c\varphi' = T(\varphi') \underset{(10.221)}{=} T(\varphi)' = c\varphi' + c'\varphi \underset{\varphi(t)\neq 0}{\Longrightarrow} c' = 0 \Longrightarrow c \in \mathbb{C}$$

$$\Longrightarrow c \in \mathbb{C},\ \forall\psi \in \mathfrak{S} : T(\psi) = c\psi.$$

(ii) We use the standard result $\int_{-\infty}^{\infty} e^{-t^2/2}dt = c_1 := (2\pi)^{1/2}$, hence, $c_1^2 = 2\pi$. The function $\varphi(t) := e^{-t^2/2} \in \mathfrak{S}$ satisfies the linear differential equation

$$(s_t + t) \circ \varphi = 0 \underset{(10.219),(10.220)}{\Longrightarrow} 0 = \mathbb{F}((s_t + t) \circ \varphi) = j(\omega + s_\omega) \circ \mathbb{F}(\varphi)$$

$$\Longrightarrow \mathbb{F}(\varphi)(\omega) = \mathbb{F}(\varphi)(0)e^{-\omega^2/2},\ \mathbb{F}(\varphi)(0) = \int_{-\infty}^{\infty} e^{-t^2/2}dt = c_1$$

$$c\varphi = \overline{\mathbb{F}}(\mathbb{F}(\varphi)) = c_1\overline{\mathbb{F}}(e^{-\omega^2/2}) = c_1^2 e^{-t^2/2} = c_1^2\varphi \Longrightarrow c = c_1^2 = 2\pi.$$

□

Lemma 10.2.40

$$\int_{-\infty}^{\infty} \mathbb{F}(\varphi)(\omega)\psi(\omega)d\omega = \int_{-\infty}^{\infty} \varphi(t)\mathbb{F}(\psi)(t)dt,\ \varphi \in \mathfrak{S}_t,\ \psi \in \mathfrak{S}_\omega.$$

Proof

$$\int_{-\infty}^{\infty} \mathbb{F}(\varphi)(\omega)\psi(\omega)\mathrm{d}\omega = \int_{-\infty}^{\infty}\left(\int_{-\infty}^{\infty} \varphi(t)e^{-j\omega t}\psi(\omega)\mathrm{d}t\right)\mathrm{d}\omega$$
$$= \int_{-\infty}^{\infty} \varphi(t)\left(\int_{-\infty}^{\infty} e^{-j\omega t}\psi(\omega)\mathrm{d}\omega\right)\mathrm{d}t = \int_{-\infty}^{\infty} \varphi(t)\mathbb{F}(\psi)(t)\mathrm{d}t.$$

□

Recall the inner product on the Hilbert space $\mathfrak{L}^2$ from (10.184):

$$\langle u_1, u_2\rangle,\ u_1, u_2 \in \mathfrak{L}^2,\ \langle u_1, u_2\rangle := \int_{-\infty}^{\infty} \overline{u_1}(t)u_2(t)\mathrm{d}t \text{ for } u_1, u_2 \in \mathrm{L}^2. \quad (10.222)$$

Also recall that C_0^∞ and thus $\mathfrak{S}$ are dense in $\mathfrak{L}^2$ with respect to $\| - \|_2$.

Theorem 10.2.41 *The Fourier transforms* $\mathbb{F}, \overline{\mathbb{F}} : \mathfrak{S} \to \mathfrak{S}$ *can be uniquely extended to continuous isomorphisms* $\mathbb{F} : \mathfrak{L}_t^2 \to \mathfrak{L}_\omega^2$, $\overline{\mathbb{F}} : \mathfrak{L}_\omega^2 \to \mathfrak{L}_t^2$ *and then*

$$(2\pi)^{-1}\overline{\mathbb{F}}\mathbb{F} = \mathrm{id},\ (2\pi)^{-1}\mathbb{F}\overline{\mathbb{F}} = \mathrm{id},\ \textit{and Parzevals's equation holds:}$$
$$\langle\mathbb{F}(u_1), \mathbb{F}(u_2)\rangle = 2\pi\langle u_1, u_2\rangle,\ \|(2\pi)^{-1/2}\mathbb{F}(u)\|_2 = \|u\|_2,\ u_1, u_2, u \in \mathfrak{L}_t. \quad (10.223)$$

Hence $(2\pi)^{-1/2}\mathbb{F} : \mathfrak{L}_t^2 \to \mathfrak{L}_\omega^2$ *is an isometry, i.e., preserves the inner product, with the inverse* $(2\pi)^{-1/2}\overline{\mathbb{F}} : \mathfrak{L}_\omega^2 \to \mathfrak{L}_t^2$.

Proof (i) We show the second line of (10.223) for $u_1, u_2, u \in \mathfrak{S}_t$. Define

$$v_1 := \mathbb{F}(u_1) \in \mathfrak{S}_\omega \Longrightarrow u_1 = (2\pi)^{-1}\overline{\mathbb{F}}(v_1) = (2\pi)^{-1}\int_{-\infty}^{\infty} e^{j\omega t}v_1(\omega)\mathrm{d}\omega$$
$$\Longrightarrow \overline{u_1}(t) = (2\pi)^{-1}\int_{-\infty}^{\infty} e^{-j\omega t}\overline{v_1}(\omega)\mathrm{d}\omega = (2\pi)^{-1}\mathbb{F}(\overline{v_1})(t),$$
$$\Longrightarrow \langle u_1, u_2\rangle = \int_{-\infty}^{\infty} \overline{u_1}(t)u_2(t)\mathrm{d}t = (2\pi)^{-1}\int_{-\infty}^{\infty} \mathbb{F}(\overline{v_1})(t)u_2(t)\mathrm{d}t$$
$$\underset{\text{Lemma 10.2.40}}{=} (2\pi)^{-1}\int_{-\infty}^{\infty} \overline{v_1}(\omega)\mathbb{F}(u_2)(\omega)\mathrm{d}\omega = (2\pi)^{-1}\langle\mathbb{F}(u_1), \mathbb{F}(u_2)\rangle.$$

Hence $\mathbb{F}|_{\mathfrak{S}}$ is a similarity, i.e., preserves the inner product on $\mathfrak{S}_t$ up to the factor 2π, and is thus continuous with respect to $\| - \|_2$. Since $\mathfrak{S}_t$ is dense in $\mathfrak{L}_t^2$ and $\mathfrak{L}_\omega^2$ is complete $\mathbb{F}$ can be uniquely extended to

$$\mathbb{F} : \mathfrak{L}_t^2 \to \mathfrak{L}_\omega^2 \text{ with } \langle\mathbb{F}(u_1), \mathbb{F}(u_2)\rangle = 2\pi\langle u_1, u_2\rangle.$$

(ii) The Fourier transform $\overline{\mathbb{F}} : \mathfrak{L}_\omega^2 \to \mathfrak{L}_t^2$ exists likewise. The composition

$$(2\pi)^{-1}\overline{\mathbb{F}}\mathbb{F} :\ \mathfrak{L}_t^2 \to \mathfrak{L}_\omega^2 \to \mathfrak{L}_t^2 \quad (10.224)$$

is continuous and coincides with id on $\mathfrak{S}_t$ by Theorem 10.2.39. Since $\mathfrak{S}_t$ is dense in $\mathfrak{L}^2_t$ this implies $(2\pi)^{-1}\overline{\mathbb{F}}\mathbb{F} = \mathrm{id}$ on $\mathfrak{L}^2_t$. Likewise one obtains

$$\mathrm{id} = (2\pi)^{-1}\mathbb{F}\overline{\mathbb{F}} : \mathfrak{L}^2_\omega \to \mathfrak{L}^2_t \to \mathfrak{L}^2_\omega. \tag{10.225}$$

The equations (10.224) and (10.225) imply that $(2\pi)^{-1/2}\mathbb{F}$, $(2\pi)^{-1/2}\overline{\mathbb{F}}$ are isometries, especially continuous isomorphisms, and inverse of each other.

□

In the following corollary we distinguish $\mathbb{F}_1 : \mathfrak{L}^1_t \to \mathrm{C}^{0,\infty}_\omega$ from $\mathbb{F}_2 : \mathfrak{L}^2_t \longrightarrow \mathfrak{L}^2_\omega$. By definition these maps coincide on the subspace $\mathfrak{S}_t$, but not on $\mathfrak{L}^1 \bigcap \mathfrak{L}^2$ since the latter intersection is not defined in our approach. However:

Corollary 10.2.42 *(i) If*

$$u \in \mathrm{L}^1_t \bigcap \mathrm{L}^2_t \subset \mathrm{C}^{0,\mathrm{pc}}_t \text{ then } v := \mathbb{F}_2(u) = \mathbb{F}_1(u) \in \mathrm{C}^{0,\infty}_\omega \bigcap \mathrm{L}^2_\omega.$$

(ii) Assume $v \in \mathrm{L}^1_\omega$ *in addition to (i) and hence, by (i),*

$$u_1 := (2\pi)^{-1}\overline{\mathbb{F}}_2(v) = (2\pi)^{-1}\overline{\mathbb{F}}_1(v) = (2\pi)^{-1}\int_{-\infty}^{\infty} e^{j\omega t}v(\omega)\mathrm{d}\omega \in \mathrm{C}^{0,\infty}_t \bigcap \mathrm{L}^2_t.$$

Then $u = u_1$, *in particular* u *is continuous and bounded.*

Proof (i) (a) Let $v_1 := \mathbb{F}_1(u) \in \mathrm{C}^{0,\infty}$ and $v_2 := \mathbb{F}_2(u) \in \mathfrak{L}^2$. We choose functions $u^\nu \in \mathrm{C}^\infty_0$ such that $u = \lim_{\nu\to\infty} u^\nu$ both in L^1_t and in L^2_t, hence

$$v^\nu := \mathbb{F}(u^\nu) \in \mathfrak{S} \subset \mathrm{C}^{0,\infty} \bigcap \mathrm{L}^2,\ \lim_{\nu\to\infty} \|v_1 - v^\nu\|_\infty = 0,\ \lim_{\nu\to\infty} \|v_2 - v^\nu\|_2 = 0$$
$$\Longrightarrow \|v_2\|_2 = \lim_{\nu\to\infty} \|v^\nu\|_2.$$

Let $a > 0$, $\epsilon > 0$. There is ν_0 such that

$$\forall\omega\forall\nu \geq \nu_0 : |v_1(\omega) - v^\nu(\omega)| \leq \|v_1 - v^\nu\|_\infty \leq \epsilon,\ \|v^\nu\|_2 \leq \|v_2\| + \epsilon$$
$$\Longrightarrow \int_{-a}^{a} |(v_1 - v^\nu)(\omega)|^2\mathrm{d}\omega \leq 2a\epsilon^2$$
$$\Longrightarrow \left(\int_{-a}^{a} |(v_1(\omega)|^2\mathrm{d}\omega\right)^{1/2} \leq \left(\int_{-a}^{a} |(v^\nu(\omega)|^2\mathrm{d}\omega\right)^{1/2}$$
$$+ \left(\int_{-a}^{a} |(v_1 - v^\nu)(\omega)|^2\mathrm{d}\omega\right)^{1/2} \leq \|v_2\|_2 + \epsilon + (2a)^{1/2}\epsilon$$
$$\underset{\epsilon\to 0}{\Longrightarrow} \int_{-a}^{a} |(v_1(\omega)|^2\mathrm{d}\omega \leq \|v_2\|_2^2 \underset{a\to\infty}{\Longrightarrow} \|v_1\|_2^2 = \int_{-\infty}^{\infty} |(v_1(\omega)|^2\mathrm{d}\omega \leq \|v_2\|_2^2 < \infty$$
$$\Longrightarrow v_1 \in \mathrm{L}^2 \Longrightarrow v_1, v_2 \in \mathfrak{L}^2.$$

(b) Since $\mathfrak{S}_\omega$ is dense in $\mathfrak{L}^2_\omega$ the equality $v := v_1 = v_2$ follows from $\langle\psi, v_1\rangle = \langle\psi, v_2\rangle$ for all $\psi \in \mathfrak{S}_\omega$. But

$$\begin{aligned}
\langle\psi, v_1 - v^\nu\rangle &= \int_{-\infty}^{\infty} \overline{\psi}(\omega)(v_1 - v^\nu)(\omega)\mathrm{d}\omega \\
&\Longrightarrow |\langle\psi, v_1 - v^\nu\rangle| \le \|\psi\|_1\|v_1 - v^\nu\|_\infty \underset{\nu\to\infty}{\longrightarrow} 0 \\
&\Longrightarrow \langle\psi, v_1\rangle = \lim_{\nu\to\infty}\langle\psi, v^\nu\rangle \underset{v_2=\lim_{\nu\to\infty} v^\nu \text{ in } \mathfrak{L}^2}{=} \langle\psi, v_2\rangle.
\end{aligned}$$

(ii) $u \underset{\text{Theorem 10.2.41}}{=} (2\pi)^{-1}\overline{\mathbb{F}}_2\mathbb{F}_2(u) = (2\pi)^{-1}\overline{\mathbb{F}}_2(v) \underset{(i)}{=} (2\pi)^{-1}\overline{\mathbb{F}}_1(v) =: u_1$. Hence u and u_1 belong to $\mathrm{L}^2 \subset \mathrm{C}^{0,\mathrm{pc}}$ and coincide in $\mathfrak{L}^2$. The piecewise continuity of

$$\begin{aligned}
&u - u_1 \text{ and } 0 = \|u - u_1\|_2^2 = \int_{-\infty}^{\infty} |u(t) - u_1(t)|^2\mathrm{d}t \text{ implies} \\
&\forall t: \; u(t) = u_1(t) = (2\pi)^{-1}\int_{-\infty}^{\infty} e^{j\omega t}v(\omega)\mathrm{d}t.
\end{aligned}$$

□

Employing (10.190) and (10.191) we finally prove the *Exchange Theorem*

$$\mathbb{F}(u_1 * u_q) = \mathbb{F}(u_1)\mathbb{F}(u_q),\; u_1 \in \mathfrak{L}^1,\; u_q \in \mathfrak{L}^q,\; q = 1, 2,\; \mathbb{F}(u_1) \in \mathrm{C}^{0,\infty}. \tag{10.226}$$

For $x, y \in \mathbb{R}$ we use the inequality

$$\begin{aligned}
&1 + (x + y)^2 = 1 + x^2 + y^2 + 2xy \\
&\qquad \underset{(x-y)^2\ge 0}{\le} 1 + x^2 + y^2 + x^2 + y^2 \le 2(1 + x^2)(1 + y^2) \\
&\forall t, x \in \mathbb{R}: \; 1 + t^2 \le 2(1 + (t - x)^2)(1 + x^2) \\
&\qquad \Longrightarrow \forall u_1, u_q \in \mathfrak{S}\forall n \in \mathbb{N}: u := u_1 * u_q \in \mathrm{C}^\infty \text{ and} \\
&(1 + t^2)^{n/2}|u(t)| \le 2\int_{-\infty}^{\infty} (1 + (t - x)^2)^{n/2}|u_1(t - x)|(1 + x^2)^{n/2}|u_q(x)|\mathrm{d}x \\
&\qquad \le 2p_n(u_1)p_{n+2}(u_q)\int_{-\infty}^{\infty} (1 + x^2)^{-1}\mathrm{d}x < \infty.
\end{aligned} \tag{10.227}$$

Lemma 10.2.43 *For all* $u_1, u_2 \in \mathfrak{S}: \; u_1 * u_2 \in \mathfrak{S}$ *and* $\mathbb{F}(u_1 * u_2) = \mathbb{F}(u_1)\mathbb{F}(u_2)$.

Proof The equation $(u_1 * u_2)' = u_1' * u_2$ and thus $(u_1 * u_2)^{(\alpha)} = u_1^{(\alpha)} * u_2,\; 0 \le \alpha \le n$, and the inequality (10.227) imply $u_1 * u_2 \in \mathfrak{S}$ and then

$$\begin{aligned}\mathbb{F}(u_1 * u_2)(\omega) &= \int_{-\infty}^{\infty}\int_{-\infty}^{\infty} e^{-j\omega t} u_1(t-x)u_2(x)\mathrm{d}x\mathrm{d}t \\ &= \int_{-\infty}^{\infty} e^{-j\omega x} u_2(x)\left(\int_{-\infty}^{\infty} e^{-j\omega(t-x)} u_1(t-x)\mathrm{d}t\right)\mathrm{d}x \\ &\underset{t-x=y}{=} \int_{-\infty}^{\infty} e^{-j\omega x} u_2(x)\mathrm{d}x \int_{-\infty}^{\infty} e^{-j\omega y} u_1(y)\mathrm{d}y = \mathbb{F}(u_2)(\omega)\mathbb{F}(u_1)(\omega).\end{aligned}$$

□

Theorem 10.2.44 (Exchange Theorem) *Eq.* (10.226) *holds.*

Proof Since $\mathfrak{S}$ is dense in $\mathfrak{L}^q$, $q = 1, 2$, we can and do choose sequences $u_q^\nu \in \mathfrak{S}$, $\nu \in \mathbb{N}$, such that $u_q = \lim_{\nu\to\infty} u_q^\nu \in \mathfrak{L}^q$. The convolution product and the pointwise product are continuous by (10.190) and by (10.191) and hence $\lim_{\nu\to\infty} u_1^\nu * u_q^\nu = u_1 * u_q$. Since $\mathbb{F} : \mathfrak{L}^1 \to \mathrm{C}^{0,\infty}$ and $\mathbb{F} : \mathfrak{L}^2 \to \mathfrak{L}^2$ are continuous we infer

$$\mathbb{F}(u_1 * u_q) = \lim_{\nu\to\infty} \mathbb{F}(u_1^\nu * u_q^\nu) \underset{\text{Lemma 10.2.43}}{=} \lim_{\nu\to\infty} \mathbb{F}(u_1^\nu)\mathbb{F}(u_q^\nu) = \mathbb{F}(u_1)\mathbb{F}(u_2).$$

□

The Fourier transform is applied to matrices of functions componentwise.

Theorem 10.2.45 *With* $H_0 \in \mathbb{C}^{p\times m}$, $h \in (\mathfrak{L}^1)^{p\times m}$, *and hence* $H_0 + \mathbb{F}(h) \in (\mathrm{C}^{0,\infty})^{p\times m}$ *consider the operators*

$$\begin{aligned}&H\circ : (\mathfrak{L}_t^2)^m \to (\mathfrak{L}_t^2)^p,\ u \mapsto y := H_0 u + h * u,\\ &(H_0 + \mathbb{F}(h))\cdot : (\mathfrak{L}_\omega^2)^m \to (\mathfrak{L}_\omega^2)^p,\ \widehat{u} \mapsto \widehat{y} := H_0\widehat{u} + \mathbb{F}(h)\widehat{u} \text{ and their norms}\\ &\|H\circ\|_2 := \|H\circ : (\mathfrak{L}_t^2)^m \to (\mathfrak{L}_t^2)^p\|,\\ &\|(H_0 + \mathbb{F}(h))\cdot\|_2 := \|(H_0 + \mathbb{F}(h))\cdot : (\mathfrak{L}_\omega^2)^m \to (\mathfrak{L}_\omega^2)^p\|.\end{aligned}$$

Then

$$\begin{aligned}&\mathbb{F}(H\circ u) = (H_0 + \mathbb{F}(h))\mathbb{F}(u) \text{ and}\\ &\|H\circ\|_2 = \|(H_0 + \mathbb{F}(h))\cdot\|_2 = \sup_{\omega\in\mathbb{R}} \sigma(H_0 + \mathbb{F}(h)(\omega)).\end{aligned} \tag{10.228}$$

Proof The equation of the first line of (10.228) follows directly from $\mathbb{F}(h * u) = \mathbb{F}(h)\mathbb{F}(u)$ in Theorem 10.2.44. The first equation in the second line holds since $(2\pi)^{-1/2}\mathbb{F} : (\mathfrak{L}_t^2)^k \to (\mathfrak{L}_t^2)^k$, $k = m, p$, is an isometry. The equation $\|(H_0 + \mathbb{F}(h))\cdot\| = \sup_{\omega\in\mathbb{R}} \sigma(H_0 + \mathbb{F}(h)(\omega))$ follows from (10.208). □

Equation (10.228) and Lemma 10.2.38 immediately imply the following important theorem, announced in Remarks 10.2.35, item 4.

Theorem 10.2.46 (Cf. [16, p. 107], [10, Theorem 589, p. 518]) *Consider the standard stability region* $\Lambda_1 := \mathbb{C}_-$ *and the ring* $\mathcal{S} \subset \mathbb{C}(s)$ *of proper and stable rational* *functions f, i.e., with $\deg_s(f) \le 0$ and $\mathrm{pole}(f) \subset \mathbb{C}_-$. Let $H = H_0 + H_{\mathrm{spr}} \in \mathcal{S}^{p\times m}$, $h := H_{\mathrm{spr}} \circ \delta \in (\mathrm{L}^1_+)^{p\times m}$ and consider the operator, cf. Corollary and Definition 8.2.84,*

$$H\circ : (\mathfrak{L}^2_t)^m \to (\mathfrak{L}^2_t)^p,\ u \mapsto H\circ u = H_0 u + h * u.$$

The norm of this operator is

$$\|H\circ\|_2 = \|H\circ : (\mathfrak{L}^2_t)^m \to (\mathfrak{L}^2_t)^p\| = \sup_{\omega\in\mathbb{R}} \sigma(H(j\omega)) =: \|H\|. \tag{10.229}$$

◊

Theorem 10.2.47 *In the situation of Theorem 10.2.33 and Example 10.2.34 consider the proper and stable closed-loop transfer matrices $H, \widetilde{H}$ of the stabilized interconnected systems. Then $\widetilde{H}\circ$ converges to $H\circ$, i.e.,*

$$\lim_{(\widetilde{P}_1,-\widetilde{Q}_1)\to(\widehat{P}_1,-\widehat{Q}_1)} \|(\widetilde{H}-H)\circ : (\mathfrak{L}^2)^{p+m} \to (\mathfrak{L}^2)^{p+m}\| = 0.$$

Hence the compensator does not only carry out the desired tasks (tracking and disturbance rejection) for all nearby plants, but also leads to nearby closed-loop transfer operators in the norm $\|H\circ : (\mathfrak{L}^2)^m \to (\mathfrak{L}^2)^p\|$.

Proof This follows directly from Theorem 10.2.46 □

The analogue of Theorem 10.2.47 for $H\circ : (\mathfrak{L}^\infty)^m \to (\mathfrak{L}^\infty)^p$ requires some further preparations. A function $u \in \mathrm{C}^{0,\mathrm{pc}}$ has support in $[0,\infty)$ if and only if $u = Yu$ where $Y \in \mathrm{C}^{0,\mathrm{pc}}$ is the Heaviside function with $Y(t) = 1$ for $t \ge 0$ and $Y(t) = 0$ for $t < 0$. We define

$$\mathrm{L}^1_{\ge 0} := \left\{u \in \mathrm{L}^1;\ Yu = u\right\} \subset \mathfrak{L}^1_{\ge 0} := \left\{u \in \mathfrak{L}^1;\ Yu = u\right\}. \tag{10.230}$$

Recall that $\mathrm{L}^1_+ (\supset \mathrm{L}^1_{\ge 0})$ was defined as the space of functions *with left bounded support*. Equation (10.191) implies that $\mathfrak{L}^1_{\ge 0}$ is a closed subspace of $\mathfrak{L}^1$ and therefore a Banach space. From Lemma 8.2.63 we have

$$\mathrm{L}^1_{\ge 0} * \mathrm{L}^1_{\ge 0} \subseteq \mathrm{L}^1_{\ge 0} \Longrightarrow \mathfrak{L}^1_{\ge 0} * \mathfrak{L}^1_{\ge 0} \subseteq \mathfrak{L}^1_{\ge 0}, \tag{10.231}$$

the latter implication following from the continuity of $u \mapsto Yu$ and of $(u_1, u_2) \mapsto u_1 * u_2$ and the density of L^1 in $\mathfrak{L}^1$. Hence $(\mathrm{L}^1_{\ge 0}, *)$ and $(\mathfrak{L}^1_{\ge 0}, *)$ are a normed resp. a Banach algebra without one element. Then

$$\begin{gathered} \mathcal{T} := \mathbb{C} \times \mathrm{L}^1_{\ge 0} \subset \mathfrak{T} := \mathbb{C} \times \mathfrak{L}^1_{\ge 0} \text{ with} \\ (a_1, u_1) * (a_2, u_2) := (a_1 a_2, a_1 u_1 + a_2 u_2 + u_1 * u_2),\ \|(a,u)\|_1 := |a| + \|u\|_1, \end{gathered} \tag{10.232}$$

are commutative normed algebras with the one element $(1, 0)$ and the ideal $\mathrm{L}^1_{\ge 0}$ resp. $\mathfrak{L}^1_{\ge 0}$ and $\mathfrak{T}$ is even a Banach algebra. The latter is important for the applicability of

Lemma 10.2.24. According to Theorems 8.2.68 and 8.2.83 the map

$$\mathcal{S} \to \mathcal{T}\ (\subset \mathfrak{T}),\ H = H_0 + H_{\rm spr} \mapsto (H_0, h),\ H_0 \in \mathbb{C},\ h := H_{\rm spr} \circ \delta \in \mathrm{L}^1_{\geq 0}, \tag{10.233}$$

is an algebra monomorphism. We identify $H = H_0 + H_{\rm spr} \circ \delta = (H_0, h) = (H_0, H_{\rm spr} \circ \delta)$. Lemma 10.2.38 and $\mathbb{F}(h)(\omega) = \int_0^\infty e^{-j\omega t} h(t)dt$ imply

$$\begin{aligned} &H(j\omega) = H_0 + \int_0^\infty e^{-j\omega t} h(t)\mathrm{d}t \\ &\Longrightarrow \|H\| = \sup_{\omega \in \mathbb{R}} |H(j\omega)| \leq |H_0| + \|h\|_1 = \|(H_0, h)\|_1 =: \|H\|_1. \end{aligned} \tag{10.234}$$

Hence $\mathcal{S}$ is a normed subalgebra of $\mathfrak{T}$ with the induced norm $\|H\|_1$. The inequality $\|H\| \leq \|H\|_1$ implies that on $\mathcal{S}$ the $\| - \|_1$-topology is finer than the $\| - \|$-topology. We are going to show that the topology on $\mathcal{S}^{1\times p}$ derived from $\| - \|_1$ has the same properties as that derived from $\| - \|$ in Sect. 10.2.3. The main tool is the relation $\mathrm{U}(\mathcal{S}) = \mathcal{S} \cap \mathrm{U}(\mathfrak{T})$ that replaces $\mathrm{U}(\mathcal{S}) = \mathcal{S} \cap \mathrm{U}(B)$ from Lemma 10.2.26 and implies that $\mathrm{U}(\mathcal{S})$ and $\mathrm{Gl}_p(\mathcal{S})$ are topological groups. To show this we need a variant of the Laplace transform. We define the closed right half-plane

$$\begin{aligned} &\overline{\mathbb{C}_+} := \{s = \sigma + j\omega \in \mathbb{C};\ \Re(s) = \sigma \geq 0\} \Longrightarrow \forall s \in \overline{\mathbb{C}_+} \forall t \geq 0:\ |e^{-st}| = e^{-\sigma t} \leq 1 \\ &\Longrightarrow \forall u (= Yu) \in \mathfrak{L}^1_{\geq 0}:\ e^{-\sigma t} u = e^{-\sigma t} Y u \in \mathfrak{L}^1_{\geq 0}. \end{aligned} \tag{10.235}$$

We define the *Laplace transform* on the Banach algebra $\mathfrak{T}$ by

$$\mathcal{L}(a, u)(s) := a + \mathbb{F}(e^{-\sigma t} u)(\omega),\ (a, u) \in \mathcal{T},\ s = \sigma + j\omega \in \overline{\mathbb{C}_+},\ \mathcal{L}(u) := \mathcal{L}(0, u). \tag{10.236}$$

Due to (10.235) $\mathcal{L}(a, u)(s)$ is well defined. For $H = (s - \lambda)^{-k} \in \mathcal{S},\ \Re(\lambda) < 0$, $k > 0$, and $h := H \circ \delta = \frac{t^{k-1}}{(k-1)!} e^{\lambda t} Y$ this gives

$$\begin{aligned} &\mathcal{L}(h)(\sigma + j\omega) = \mathcal{L}(0, h)(\sigma + j\omega) = \mathbb{F}(e^{-\sigma t} h)(\omega) = \mathbb{F}\left(\tfrac{t^{k-1}}{(k-1)!} e^{(\lambda - \sigma)t}\right)(\omega) \\ &= \mathbb{F}\left((s - (\lambda - \sigma))^{-k} \circ \delta\right)(\omega) \underset{\text{Lemma 10.2.38}}{=} (j\omega - (\lambda - \sigma))^{-k} = H(\sigma + j\omega). \end{aligned}$$

This implies for all

$$\begin{aligned} &H = H_0 + H_{\rm spr} \in \mathcal{S},\ (H_0, h) \in \mathcal{T} \subset \mathfrak{T},\ h := H_{\rm spr} \circ \delta,\ \sigma + j\omega \in \overline{\mathbb{C}_+}: \\ &\mathcal{L}(H_0, h)(\sigma + j\omega) = H_0 + H_{\rm spr}(\sigma + j\omega) = H(\sigma + j\omega) \text{ or} \\ &\forall s \in \overline{\mathbb{C}_+}:\ \mathcal{L}(H_0, h)(s) = H(s). \end{aligned} \tag{10.237}$$

For $u \in \mathrm{L}^1_{\geq 0}$ and $s = \sigma + j\omega \in \overline{\mathbb{C}_+}$ one gets

$$\mathcal{L}(u)(s) = \mathbb{F}(e^{-\sigma t}u(t))(\omega) = \int_0^\infty e^{-j\omega t}e^{-\sigma t}u(t)\mathrm{d}t = \int_0^\infty e^{-st}u(t)\mathrm{d}t = \lim_{n\to\infty} f_n(s)$$

$$\text{where } f_n(s) := \int_0^n e^{-st}u(t)\mathrm{d}t,\ |\mathcal{L}(u)(s) - f_n(s)| \le \int_n^\infty |u(t)|\mathrm{d}t \underset{n\to\infty}{\longrightarrow} 0. \tag{10.238}$$

The functions $f_n(s)$ are continuous for $\Re(s) \ge 0$. As uniform limit of the $f_n(s)$ also $\mathcal{L}(u)(s)$ is continuous. This implies the continuous operator

$$\begin{aligned}
&\mathcal{L} : \mathrm{L}^1_{\ge 0} \to \mathrm{C}^{0,\infty}(\overline{\mathbb{C}_+}),\ |\mathcal{L}(u)(s)| \le \|u\|_1,\ \|\mathcal{L}(u)\|_\infty = \sup_{s\in\overline{\mathbb{C}_+}} |\mathcal{L}(u)(s)| \le \|u\|_1, \\
&\underset{\text{continuous extension}}{\Longrightarrow} \quad \mathcal{L} : \mathfrak{L}^1_{\ge 0} \to \mathrm{C}^{0,\infty},\ \|\mathcal{L}(u)\|_\infty = \sup_{s\in\overline{\mathbb{C}_+}} |\mathcal{L}(u)(s)| \le \|u\|_1, \\
&\Longrightarrow \mathcal{L} : \mathfrak{T} = \mathbb{C} \times \mathfrak{L}^1_{\ge 0} \to \mathrm{C}^{0,\infty}(\overline{\mathbb{C}_+}),\ (a, u) \mapsto \mathcal{L}(a, u) = a + \mathcal{L}(u),\ \text{with} \\
&\|\mathcal{L}(a, u)\|_\infty = \sup_{s\in\overline{\mathbb{C}_+}} |\mathcal{L}(a, u)(s)| \le \|(a, u)\|_1.
\end{aligned} \tag{10.239}$$

The exchange theorem for $\mathcal{L}$ follows from that for $\mathbb{F}$.

Lemma 10.2.48 (Exchange theorem) *The Laplace transform* $\mathcal{L} : \mathfrak{T} \to \mathrm{C}^{0,\infty}(\overline{\mathbb{C}_+})$ *is an algebra homomorphism.*

Proof Since the convolution and $\mathcal{L}$ are continuous it suffices to show $\mathcal{L}(u_1 * u_2) = \mathcal{L}(u_1)\mathcal{L}(u_2)$ for $u_j = Yu_j \in \mathrm{L}^1_{\ge 0}$. But

$$\begin{aligned}
(u_1 * u_2)(t) &= \int_0^t u_1(t - x)u_2(x)\mathrm{d}x \Longrightarrow (e^{-\sigma t}u_1 * e^{-\sigma t}u_2)(t) \\
&= \int_0^t e^{-\sigma(t-x)}u_1(t - x)e^{-\sigma x}u_2(x)\mathrm{d}x = e^{-\sigma t}(u_1 * u_2)(t)
\end{aligned}$$

$$\begin{aligned}
&\Longrightarrow e^{-\sigma t}(u_1 * u_2) = (e^{-\sigma t}u_1) * (e^{-\sigma t}u_2) \Longrightarrow \forall s = \sigma + i\omega,\ \sigma \ge 0 : \\
&\mathcal{L}(u_1 * u_2)(s) = \mathbb{F}\left(e^{-\sigma t}(u_1 * u_2)\right)(\omega) = \mathbb{F}\left(e^{-\sigma t}u_1 * e^{-\sigma t}u_2\right)(\omega) \\
&\underset{\text{Theorem 10.2.44}}{=} \mathbb{F}\left(e^{-\sigma t}u_1\right)(\omega)\mathbb{F}\left(e^{-\sigma t}u_2\right)(\omega) = \mathcal{L}(u_1)(s)\mathcal{L}(u_2)(s).
\end{aligned}$$

□

We are now going to extend the 1-norm on $\mathfrak{T}$ to a new norm $\| - \|_1$ for matrices. Let $H = (A, h) \in \mathfrak{T}^{p\times m} = \mathbb{C}^{p\times m} \times (\mathfrak{L}^1_{\ge 0})^{p\times m}$. From Theorem 10.2.36 we get

$$\|H\circ :\ (\mathfrak{L}^\infty)^m \to (\mathfrak{L}^\infty)^p\| = \max_{i=1,\dots,p} \sum_{j=1}^m \|H_{ij}\|_1,\ \|H_{ij}\|_1 := |A_{ij}| + \|h_{ij}\|_1. \tag{10.240}$$

We therefore define the norm of $H = (A, h) \in \mathfrak{T}^{p\times m}$ as

$$\begin{aligned}
&\|H\|_1 := \max_{i=1,\dots,p} \sum_{j=1}^m \|H_{ij}\|_1 = \|H\circ :\ (\mathfrak{L}^\infty)^m \to (\mathfrak{L}^\infty)^p\| \text{ where} \\
&\|H_{ij}\|_1 := |A_{ij}| + \|h_{ij}\|_1,\ A_{ij} \in \mathbb{C},\ h_{ij} \in \mathfrak{L}^1_{\ge 0}.
\end{aligned} \tag{10.241}$$

We use the index 1 for $\|-\|_1$ since this norm generalizes the 1-norm $\|h\|_1 = \int_0^\infty |h(t)|dt$ for $h \in \mathrm{L}^1_{\geq 0}$. It is easy to check that this is indeed a norm, thus makes $(\mathfrak{T}^{p\times m}, \|-\|_1)$ a Banach space.
Let, more generally, $\mathfrak{A}$ be any Banach algebra, i.e., a $\mathbb{C}$-algebra and a Banach space that satisfies $\|H_2H_1\| \leq \|H_2\|\|H_1\|$ for $H_2, H_1 \in \mathfrak{A}$. Then $\mathfrak{A}^{p\times m}$ is a Banach space with this norm, i.e., with

$$\begin{aligned} &\|H\| := \max_{i=1,\dots,p} \sum_{j=1}^{m} \|H_{ij}\| \text{ for } H \in \mathfrak{A}^{p\times m}, \\ &\|H\| = \max_i \|H_i\| \text{ if } m = 1, \ \|H\| = \sum_j \|H_j\| \text{ if } p = 1. \end{aligned} \tag{10.242}$$

Notice that this norm is not defined as $\|H\| = \sup_{0\neq u\in\mathfrak{A}^m} \|Au\|\|u\|^{-1}$.
This norm satisfies the product rule

$$\|H_2H_1\| \leq \|H_2\|\|H_1\|, \ H_2 \in \mathfrak{A}^{q\times p}, \ H_1 \in \mathfrak{A}^{p\times m}, \ q, p, m \in \mathbb{N}, \tag{10.243}$$

since

$$\begin{aligned} &(H_2H_1)_{ij} = \sum_k H_{2,ik}H_{1,kj} \Longrightarrow \|(H_2H_1)_{ij}\| \leq \sum_k \|H_{2,ik}\|\|H_{1,kj}\| \\ &\Longrightarrow \sum_j \|(H_2H_1)_{ij}\| \leq \sum_j \sum_k \|H_{2,ik}\|\|H_{1,kj}\| = \sum_k \|H_{2,ik}\| \sum_j \|H_{1,kj}\| \\ &\leq \sum_k \|H_{2,ik}\| \left(\max_k \sum_j \|H_{1,kj}\| \right) = \sum_k \|H_{2,ik}\|\|H_1\| \\ &\Longrightarrow \|H_2H_1\| = \max_i \sum_j \|(H_2H_1)_{ij}\| \\ &\leq \left(\max_i \sum_k \|H_{2,ik}\| \right) \|H_1\| = \|H_2\|\|H_1\|. \end{aligned} \tag{10.244}$$

The product rule implies that the multiplication of these matrices is continuous and that $\mathfrak{A}^{p\times p}$ is a Banach algebra. According to Lemma 10.2.24 and Corollary 10.2.25 this, in turn, implies that $\mathrm{U}(\mathfrak{A}^{p\times p}) = \mathrm{Gl}_p(\mathfrak{A})$ is a topological group and open in $\mathfrak{A}^{p\times p}$.
We apply this general result to the commutative algebra $\mathfrak{T}$ with the norm $\|-\|_1$ and obtain, for $H = (A, h) \in \mathfrak{T}^{p\times m} = \mathbb{C}^{p\times m} \times \left(\mathfrak{L}^1_{\geq 0}\right)^{p\times m}$,

$$\|H\|_1 = \max_i \sum_j \|H_{ij}\|_1 = \|H\circ = (\mathfrak{L}^\infty)^m \to (\mathfrak{L}^\infty)^p\| \tag{10.245}$$

where $\|H_{ij}\|_1 = |A_{ij}| + \|h_{ij}\|_1$.

We identify $\mathcal{S} \subset \mathfrak{T}$, $H = H_0 + H_{\rm spr} = (H_0, H_{\rm spr} \circ \delta)$. Hence the norm $\| - \|_1$ on $\mathfrak{T}^{p\times m}$ induces a norm on $\mathcal{S}^{p\times m}$. The $\| - \|_1$-topology on $\mathcal{S}^{p\times m}$ is finer than the $\| - \|$-topology of Sect. 10.2.3 since for

$$
\begin{aligned}
& H = H_0 + H_{\rm spr} \underset{\text{ident.}}{=} (H_0, h) \in \mathcal{S}^{p\times m} \subset \mathfrak{T}^{p\times m},\ h := H_{\rm spr} \circ \delta, \text{ we get}\\
& \|H\| = \sup_{\omega\in\mathbb{R}} \sigma(H(j\omega)) \underset{\text{Thm. 10.2.46}}{=} \|H\circ : (\mathfrak{L}^2)^m \to (\mathfrak{L}^2)^p\|\\
& \underset{(10.201)}{\leq} \sum_{k,l} \left(|H_{0.kl}| + \|h_{k,l}\|_1\right) \leq p \cdot \max_k \sum_l \left(|H_{0.kl}| + \|h_{k,l}\|_1\right) = p\|H\|_1\\
& \Longrightarrow \|H\| \leq p\|H\|_1.
\end{aligned}
\tag{10.246}
$$

These properties especially hold for $p = m = 1$.
The next lemma is the analogue of the important Lemma 10.2.26.

Lemma 10.2.49

$$
\mathrm{U}(\mathcal{S}) = \mathcal{S} \bigcap \mathrm{U}(\mathfrak{T}),\ \mathrm{Gl}_p(\mathcal{S}) = \mathcal{S}^{p\times p} \bigcap \mathrm{Gl}_p(\mathfrak{T}).
\tag{10.247}
$$

Proof As in Lemma 10.2.26 the second equation follows from the first. The inclusion $\mathrm{U}(\mathcal{S}) \subseteq \mathcal{S} \bigcap \mathrm{U}(\mathfrak{T})$ is obvious. For the converse inclusion let $H = ft^{-1} \in \mathcal{S},\ f, t \in \mathbb{C}[s]$, be a proper and stable rational function, i.e., with $t(s) \neq 0$ and bounded $H(s) = f(s)t(s)^{-1}$ for $s \in \overline{\mathbb{C}_+}$. We identify

$$
H = H_0 + H_{\rm spr} = H \circ \delta = (H_0, h) \in \mathfrak{T},\ h := H_{\rm spr} \circ \delta,
$$

and assume $(H_0, h) \in \mathrm{U}(\mathfrak{T})$. Hence there is $v \in \mathfrak{T}$ with

$$
\begin{aligned}
& (H_0, h) * v = (1, 0) \underset{\text{Lemma 10.2.48}}{\Longrightarrow} 1 = \mathcal{L}(1, 0) = \mathcal{L}(H_0, h)\mathcal{L}(v)\\
& \underset{(10.237)}{\Longrightarrow} \forall s \in \overline{\mathbb{C}_+} : 1 = H(s)K(s) = f(s)t(s)^{-1}K(s),\ K(s) := \mathcal{L}(v)(s) \in \mathrm{C}^{0,\infty}\\
& \Longrightarrow \forall s \in \overline{\mathbb{C}_+} :\ f(s) \neq 0,\ t(s)f(s)^{-1} = K(s) \text{ bounded}\\
& \Longrightarrow tf^{-1} \in \mathbb{C}[s]_T \cap \mathbb{C}(s)_{\rm pr} = \mathcal{S} \Longrightarrow ft^{-1} \in \mathrm{U}(\mathcal{S}).
\end{aligned}
$$

□

Notice that for the proof of (10.247) the completeness of $\mathfrak{T}$ and the algebra homomorphism $\mathcal{L} : \mathfrak{T} \to \mathrm{C}^{0,\infty}(\overline{\mathbb{C}_+})$ are essential ingredients. Corollary and Definition 10.2.31 holds with the $\| - \|_1$-topology instead of the $\| - \|$-topology of $\mathfrak{R}$.

Theorem 10.2.50 *(i)* (Robustness) *Theorem 10.2.33 holds with the norm* $\|R\|_1$, $R \in \mathcal{S}^{p\times l}$, *instead of* $\|R\|$. *In words: If* $\mathcal{B}_2$ *is a stabilizing compensator of* $\mathcal{B}_1$ *with the desired properties of tracking and disturbance rejection, then* $\mathcal{B}_2$ *has the same*

properties for all plants $\widetilde{\mathcal{B}}_1$ in some $\|-\|_1$-neighborhood of $\mathcal{B}_1$. This neighborhood is also a $\|-\|$-neighborhood.
(ii) In the situation and with the notations of Theorem 10.2.47 assume that

$$\lim_{(\widetilde{P}_1,-\widetilde{Q}_1)} \|(\widetilde{P}_1,-\widetilde{Q}_1)-(\widehat{P}_1,-\widehat{Q}_1)\|_1 = 0, \text{ hence also}$$
$$\lim_{(\widetilde{P}_1,-\widetilde{Q}_1)} \|(\widetilde{P}_1,-\widetilde{Q}_1)-(\widehat{P}_1,-\widehat{Q}_1)\| = 0.$$

Write $\lim := \lim_{(\widetilde{P}_1,-\widetilde{Q}_1)\to(\widehat{P}_1,-\widehat{Q}_1)\in(\mathcal{S}^{\bullet\times\bullet},\|-\|_1)}$. *Let H resp. $\widetilde{H}$ in $\mathcal{S}^{(p+m)\times(p+m)}$ be the proper and stable closed-loop transfer matrices of $\mathcal{B}_1$ resp. $\widetilde{\mathcal{B}}_1$ interconnected with the compensator $\mathcal{B}_2$. Then*

$$\begin{aligned}
&\lim \|\widetilde{H}-H\|_1 = 0,\ \lim \|\widetilde{H}-H\| = 0\\
&\lim \|(\widetilde{H}-H)\circ\|_\infty = \lim \|(\widetilde{H}-H)\circ : (\mathfrak{L}^\infty)^{p+m} \to (\mathfrak{L}^\infty)^{p+m}\| = 0, \quad (10.248)\\
&\lim \|(\widetilde{H}-H)\circ\|_2 = \lim \|(\widetilde{H}-H)\circ : (\mathfrak{L}^2)^{p+m} \to (\mathfrak{L}^2)^{p+m}\| = 0.
\end{aligned}$$

In words: If the plant $\widetilde{\mathcal{B}}_1$ is sufficiently near the nominal plant $\mathcal{B}_1$ in the $\|-\|_1$-topology of $\mathcal{S}^{\bullet\times\bullet}$ then $\widetilde{H}$ is near H in the $\|-\|_1$- and in the $\|-\|$-topology of $\mathcal{S}^{p\times m}$, $\widetilde{H}\circ : (\mathfrak{L}^\infty)^{p+m} \to (\mathfrak{L}^\infty)^{p+m}$ is near $H\circ : (\mathfrak{L}^\infty)^{p+m} \to (\mathfrak{L}^\infty)^{p+m}$ and $\widetilde{H}\circ : (\mathfrak{L}^2)^{p+m} \to (\mathfrak{L}^2)^{p+m}$ is near $H\circ : (\mathfrak{L}^2)^{p+m} \to (\mathfrak{L}^2)^{p+m}$.

Proof With Lemma 10.2.49 the proof is analogous to that of Theorem 10.2.33 and 10.2.47. The limit $\lim\left((\widetilde{H}-H)\circ : (\mathfrak{L}^\infty)^{p+m} \to (\mathfrak{L}^\infty)^{p+m}\right) = 0$ follows from (10.241). □

According to Theorem 10.2.47

$$\begin{aligned}
&\lim \|(\widetilde{P}_1,-\widetilde{Q}_1)-(\widehat{P}_1,-\widehat{Q}_1)\| = 0 \text{ implies}\\
&\lim \|(\widetilde{H}-H)\circ : (\mathfrak{L}^2)^{p+m} \to (\mathfrak{L}^2)^{p+m}\| = 0, \text{ but not}\\
&\lim \|(\widetilde{H}-H)\circ : (\mathfrak{L}^\infty)^{p+m} \to (\mathfrak{L}^\infty)^{p+m}\| = 0.
\end{aligned}$$

Hence for the robustness of the important BIBO stability the norm $\|H\|_1$ on $\mathcal{S}^{p\times m}$ is more important than $\|H\|$.
In the next theorem we use the results on the Fourier transform to prove the injectivity of the general Laplace transform from Theorem 8.2.89. For the applications in this book this result is not needed.

Theorem 10.2.51 *The Laplace transform*

$$\mathcal{L} : \mathfrak{A}_+ \longrightarrow \{f(s);\ \exists\sigma > 0 \text{ such that } f \text{ is holomorphic for } \Re(s) > \sigma\} \quad (10.249)$$

is injective, i.e., $\mathcal{L}(u) = 0$ implies $u = 0$.
If $u \in \mathfrak{A}_+^0$ is continuous and Laplace transformable and if $\mathcal{L}(u)(\rho + j\omega) \in \mathrm{L}^1_\omega$ for any $\rho > \sigma$, then the following inversion formula holds:

$$u(t) = (2\pi j)^{-1} \int_{\rho-\infty}^{\rho+j\infty} \mathcal{L}(u)(s)e^{st}\mathrm{d}s, \ \rho > \sigma. \tag{10.250}$$

Proof 1. Let $v \in \mathfrak{A}_+^{0,\mathrm{pc}}$ be a Laplace transformable function, i.e., v is piecewise continuous, has left bounded support and

$$\begin{aligned}
&\exists \sigma > 0 \text{ with } v(t)e^{-\sigma t} \in \mathrm{L}_+^\infty \\
&\Longrightarrow \forall s = \Re(s) + j\omega \in \mathbb{C} \text{ with } \Re(s) > \sigma : \ v(t)e^{-\Re(s)t} \in \mathrm{L}_+^1 \cap \mathrm{L}_+^2 \cap \mathrm{L}_+^\infty \\
&\Longrightarrow \mathbb{F}_t\left(v(t)e^{-\Re(s)t}\right)(\omega) = \int_{-\infty}^{\infty} v(t)e^{-\Re(s)t}e^{-j\omega t}\mathrm{d}t = \int_{-\infty}^{\infty} v(t)e^{-st}\mathrm{d}t = \mathcal{L}(v)(s).
\end{aligned} \tag{10.251}$$

By (10.211) and Theorem 10.2.41 we know that $\mathbb{F}_t(ue^{-\Re(s)t})(\omega) = \mathcal{L}(u)(s)$ as function of ω is continuous, bounded, and square-integrable and that the Fourier inversion formula

$$ve^{-\Re(s)t} = (2\pi)^{-1}\overline{\mathbb{F}}_\omega\left(\mathcal{L}(v)(s)\right) \text{ and thus } v = (2\pi)^{-1}e^{\Re(s)t}\overline{\mathbb{F}}_\omega\left(\mathcal{L}(v)(s)\right) \tag{10.252}$$

holds in $\mathfrak{L}^2$. Hence $\mathcal{L}(v) = 0$ implies $v = 0$.

2. Let $u = s^n \circ v \in \mathfrak{A}_+$, $n \geq 0$, $v \in \mathfrak{A}_+^{0,\mathrm{pc}}$, be any Laplace transformable distribution and $\mathcal{L}(u) = 0$. Then

$$\mathcal{L}(u)(s) = s^n\mathcal{L}(v)(s) = 0 \Longrightarrow \mathcal{L}(v)(s) = 0 \underset{1.}{\Longrightarrow} v = 0 \Longrightarrow u = 0.$$

Hence $\mathcal{L}$ is injective on $\mathfrak{A}_+$.

3. In the situation of 1 assume in addition that $\mathcal{L}(v)(\Re(s) + j\omega) \in \mathrm{L}_\omega^1$ is also absolutely integrable. Then Corollary 10.2.42 yields

$$\begin{aligned}
v(t)e^{-\Re(s)t} &= (2\pi)^{-1}\int_{-\infty}^{\infty} \mathcal{L}(v)(\Re(s) + j\omega)e^{j\omega t}\mathrm{d}\omega \\
\Longrightarrow v(t) &= (2\pi)^{-1}\int_{-\infty}^{\infty} \mathcal{L}(v)(\Re(s) + j\omega)e^{(\Re(s)+j\omega)t}\mathrm{d}\omega \\
&\underset{\mathrm{d}s=j\mathrm{d}\omega}{=} : (2\pi j)^{-1}\int_{\rho-j\infty}^{\rho+j\infty} \mathcal{L}(v)(s)e^{st}\mathrm{d}s, \ \rho := \Re(s) > \sigma.
\end{aligned}$$

□

10.2.6 Model Matching

In this section we start with Assumption 10.1.18, i.e., with a T-stabilizable plant $\mathcal{B}_1$ and the derived data from this Assumption. Let F be an infinite base field. The Theorems 10.1.20 and 10.1.34 then furnish all proper T-stabilizing compensators

$\mathcal{B}_2$ with proper feedback behavior $\mathcal{B} = \mathrm{fb}(\mathcal{B}_1, \mathcal{B}_2)$. We use the notations of these theorems. The closed-loop transfer matrix of $\mathcal{B}$, cf. (10.25), is

$$H = \begin{pmatrix} H_{y_1,u_2} & H_{y_1,u_1} \\ H_{y_2,u_2} & H_{y_2,u_1} \end{pmatrix} = \begin{pmatrix} \widehat{N}_1\widehat{Q}_2 & \widehat{D}_2\widehat{Q}_1 \\ \widehat{D}_1\widehat{Q}_2 & \widehat{N}_2\widehat{Q}_1 \end{pmatrix}. \tag{10.253}$$

In addition we consider a *model IO behavior*

$$\begin{aligned} &\mathcal{B}_m := \left\{\left(\begin{smallmatrix} y_m \\ u_m \end{smallmatrix}\right) \in \mathcal{F}^{p+p};\ P_m \circ y_m = Q_m \circ u_m\right\} \text{ with} \\ &(P_m, -Q_m) \in \mathcal{D}^{p\times(p+p)},\ \mathrm{rank}(P_m) = p,\ H_m := P_m^{-1}Q_m. \end{aligned} \tag{10.254}$$

Definition 10.2.52 Consider the just introduced behaviors $\mathcal{B}_1, \mathcal{B}_2 \subset \mathcal{F}^{p+m}$ and $\mathcal{B}_m \subset \mathcal{F}^{p+p}$. The compensator $\mathcal{B}_2$ is called *T-model matching* with respect to $\mathcal{B}_m$ if for any input $u_2 \in \mathcal{F}^p$ and resulting trajectories $(y_1, y_2, u_2, 0)^\top \in \mathcal{B}$ and $(y_m, u_2)^\top \in \mathcal{B}_m$ the output component y_1 of $\mathcal{B}$ T-approximates or matches y_m, i.e.,

$$\begin{aligned} &\left\{y_1 - y_m;\ (y_1, y_2, u_2, 0)^\top \in \mathcal{B},\ (y_m, u_2)^\top \in \mathcal{B}_m\right\}_T = 0, \text{ or, equivalently,} \\ &(y_1, y_2, u_2, 0)^\top \in \mathcal{B}_T,\ (y_m, u_2)^\top \in (\mathcal{B}_m)_T \Longrightarrow y_1 - y_m = 0. \end{aligned}$$

◇

Lemma 10.2.53 *If a T-model matching compensator exists, the model is T-stable, i.e., $P_m \in \mathrm{Gl}_p(\mathcal{D}_T)$.*

Proof We have to show that $\mathcal{B}_m^0$ is T-small (T-autonomous). But let

$$\begin{aligned} y_m \in \mathcal{B}_m^0 &\Longrightarrow \left(\begin{smallmatrix} y_m \\ 0 \end{smallmatrix}\right) \in \mathcal{B}_m,\ (0,0,0,0)^\top \in \mathcal{B} \\ &\underset{\text{model matching}}{\Longrightarrow} y_m - 0\ \ T\text{-small} \Longrightarrow \mathcal{B}_m^0\ \ T\text{-small}. \end{aligned}$$

□

In the remainder of this section we assume the necessary conditions that the plant $\mathcal{B}_1$ is T-stabilizable and that the model IO behavior $\mathcal{B}_m$ is T-stable. We consider T-stabilizing compensators $\mathcal{B}_2$ only with proper $\mathcal{B}_2$ and $\mathcal{B} = \mathrm{fb}(\mathcal{B}_1, \mathcal{B}_2)$.

The T-stability of the closed-loop behavior and of the model implies

$$\begin{aligned} &\widehat{P} \in \mathrm{Gl}_{p+m}(\mathcal{D}_T),\ P_m \in \mathrm{Gl}_p(\mathcal{D}_T) \Longrightarrow \forall \left(\begin{smallmatrix} y \\ u \end{smallmatrix}\right) \in \mathcal{F}_T^{(p+m)+(p+m)}, \forall \left(\begin{smallmatrix} y_m \\ u_m \end{smallmatrix}\right) \in \mathcal{F}_T^{p+p}: \\ &\left(\left(\begin{smallmatrix} y \\ u \end{smallmatrix}\right) \in \mathcal{B}_T \Longleftrightarrow y = H \circ u\right),\ \left(\left(\begin{smallmatrix} y_m \\ u_m \end{smallmatrix}\right) \in (\mathcal{B}_m)_T \Longleftrightarrow y_m = H_m \circ u_m\right) \\ &\Longrightarrow y_1 - y_m = H_{y_1,u_2} \circ u_2 + H_{y_1,u_1} \circ u_1 - H_m \circ u_m \\ &\Longrightarrow \text{for } u_2 \in \mathcal{F}_T^p,\ u_1 := 0,\ u_m := u_2: \quad y_1 - y_m = (H_{y_1,u_2} - H_m) \circ u_2. \end{aligned} \tag{10.255}$$

Corollary 10.2.54 *Assume the T-stable closed-loop behavior $\mathcal{B}$ and the T-stable model $\mathcal{B}_m$. Then the compensator is T-model matching if and only if $H_{y_1,u_2} = H_m$. Since H is T-stable and proper, so is $H_m = H_{y_1,u_2}$.*

Proof According to Definition 10.2.52 the compensator is T-model matching if and only if in the last line of (10.255) the difference $y_1 - y_m$ is zero or, equivalently, $H_{y_1,u_2} = H_m$. □

In the sequel we therefore assume that the model behavior $\mathcal{B}_m$ is T-stable and proper and hence $H_m \in \mathcal{S}^{p\times p}$.

Let $\mathcal{B}_2$ be a compensator. According to Theorems 10.1.20 and 10.1.34 $\mathcal{B}_{2,T}$ is described by matrices

$$\begin{aligned}
&(-\widehat{Q}_2, \widehat{P}_2) = (-\widehat{Q}_2^0, \widehat{P}_2^0) + X(\widehat{P}_1, -\widehat{Q}_1),\ \begin{pmatrix}\widehat{D}_2\\ \widehat{N}_2\end{pmatrix} = \begin{pmatrix}\widehat{D}_2^0\\ \widehat{N}_2^0\end{pmatrix} - \begin{pmatrix}\widehat{N}_1\\ \widehat{D}_1\end{pmatrix} X \text{ with}\\
&X \in \mathcal{S}^{m\times p},\ \det(\widehat{P}_2(0)) = \det\left(\widehat{P}_2^0(0) - X(0)\widehat{Q}_1(0)\right) \neq 0\\
&\Longrightarrow H_{y_1,u_2} \underset{(10.253)}{=} \widehat{N}_1\widehat{Q}_2 = \widehat{N}_1(\widehat{Q}_2^0 - X\widehat{P}_1) \Longrightarrow \widehat{N}_1\widehat{Q}_2^0 - H_{y_1,u_2} = \widehat{N}_1 X\widehat{P}_1.
\end{aligned} \tag{10.256}$$

Recall that the conditions $\det\left(\widehat{P}_2\right) \neq 0$ resp. $X \in \mathcal{S}^{m\times p}$ resp. $\det\left(\widehat{P}_2(0)\right) \neq 0$ ensure that $\mathcal{B}_2$ is an IO behavior resp. that the closed-loop behavior $\mathcal{B}$ is T-stable and proper resp. that $\mathcal{B}_2$ is proper.

Corollary 10.2.55 *A necessary condition for the existence of a T-model matching compensator $\mathcal{B}_2$ is the existence of $X^0 \in \mathcal{S}^{m\times p}$ such that*

$$\widehat{N}_1\widehat{Q}_2^0 - H_m = \widehat{N}_1 X^0\widehat{P}_1. \tag{10.257}$$

◊

Assume that $X \in \mathcal{S}^{m\times p}$ is another solution of (10.257). Then

$$\widehat{N}_1(X - X^0)\widehat{P}_1 = 0 \underset{\widehat{P}_1\in\mathrm{Gl}_p(F(s))}{\Longrightarrow} \widehat{N}_1(X - X^0) = 0. \tag{10.258}$$

Let $r := \operatorname{rank}(H_1) = \operatorname{rank}(\widehat{Q}_1) = \operatorname{rank}(\widehat{N}_1)$ and let $U \in \widehat{\mathcal{D}}^{m\times(m-r)}$ be a universal right annihilator of $\widehat{N}_1$. Since $\mathcal{S} = \widehat{\mathcal{D}}_{\widehat{T}}$ is a quotient ring of $\widehat{\mathcal{D}}$ the matrix U is also a universal right annihilator over $\mathcal{S}$. Therefore the equation $\widehat{N}_1(X - X^0) = 0$ is equivalent with

$$\begin{aligned}
&X - X^0 = UY,\ Y = Y(0) + \widehat{s}Z \in \mathcal{S}^{(m-r)\times p} = F^{(m-r)\times p} \oplus \widehat{s}\mathcal{S}^{(m-r)\times p}\\
&\Longrightarrow X = X^0 + UY,\ \widehat{P}_2 = \widehat{P}_2^0 - (X^0 + UY)\widehat{Q}_1 \in \mathcal{S}^{m\times m}\\
&\Longrightarrow \det\left(\widehat{P}_2(0)\right) = \det\left(\widehat{P}_2^0(0) - X^0(0)\widehat{Q}_1(0) - U(0)Y(0)\widehat{Q}_1(0)\right).
\end{aligned} \tag{10.259}$$

Let Σ be an $(m - r) \times p$ matrix of indeterminates and define the polynomial

$$\begin{aligned}
&g(\Sigma) := \det\left(\widehat{P}_2^0(0) - X^0(0)\widehat{Q}_1(0) - U(0)\Sigma\widehat{Q}_1(0)\right) \in F[\Sigma]\\
&\Longrightarrow \det(\widehat{P}_2(0)) = g(Y(0)).
\end{aligned} \tag{10.260}$$

For an infinite field F we know that $g \neq 0$ implies $g(Y(0)) \neq 0$ for almost all $Y(0) \in F^{(m-r)\times p}$. The simplest way to test $g \neq 0$ is to choose $Y(0)$ by chance and to verify $g(Y(0)) \neq 0$.

Summing up the preceding considerations we obtain the following.

Theorem 10.2.56 *Given are a T-stabilizable plant $\mathcal{B}_1 \subset \mathcal{F}^{p+m}$ and a T-stable and proper model IO behavior $\mathcal{B}_m \subset \mathcal{F}^{p+p}$ with (proper) transfer matrix H_m. Compute the data from Assumption 10.1.18 and Theorems 10.1.20 and 10.1.34.*

1. *There is a T-stabilizing, model matching compensator $\mathcal{B}_2$ with proper* $\mathrm{fb}(\mathcal{B}_1, \mathcal{B}_2)$ *and proper $\mathcal{B}_2$ if and only if there is $X^0 \in \mathcal{S}^{m\times p}$ with $\widehat{N}_1\widehat{Q}_2^0 - H_m = \widehat{N}_1 X^0 \widehat{P}_1$ and nonzero g from* (10.260).
2. *If the condition 1 holds all such compensators are obtained by the following steps:*
   (i) Choose X^0 from 1, $Y(0) \in F^{(m-r)\times p}$ with $g(Y(0)) \neq 0$ and $Z \in \mathcal{S}^{(m-r)\times p}$.
   (ii) According to (10.259) *define*
   $$Y := Y(0) + \widehat{s}Z,\ X := X^0 + UY,\ (-\widehat{Q}_2, \widehat{P}_2) = (-\widehat{Q}_2^0, \widehat{P}_2^0) + X(\widehat{P}_1, -\widehat{Q}_1).$$
   Then the controllable IO realization $\mathcal{B}_2 \subset \mathcal{F}^{p+m}$ of $H_2 := \widehat{P}_2^{-1}\widehat{Q}_2$ is a model matching compensator with the properties from 1., and all such compensators are obtained in this fashion. The additional noncontrollable compensators are obtained as in Theorem 10.1.13.

$\diamondsuit$

Remark 10.2.57 (*The solution of $\widehat{N}_1\widehat{Q}_2^0 - H_m = \widehat{N}_1 X^0 \widehat{P}_1$*): Let $\mathcal{D}$ be any principal ideal domain. The equation (10.257) is a matrix equation of the type

$$A_1 X A_2 = B,\ A_1 \in \mathcal{D}^{I\times K},\ X \in \mathcal{D}^{K\times L},\ A_2 \in \mathcal{D}^{L\times J},\ B \in \mathcal{D}^{I\times J}, \tag{10.261}$$

where I, J, K, L are finite index sets and A_1, A_2, B are given whereas X is unknown and to be determined. The equation $A_1XA_2 = B$ is obviously equivalent with the equations

$$\forall(\alpha,\beta) \in I\times J: \quad \sum_{(\gamma,\delta)\in K\times L} A_{\alpha\beta,\gamma\delta}X_{\gamma\delta} = B_{\alpha\beta},\ A_{\alpha\beta,\gamma\delta} := (A_1)_{\alpha\gamma}(A_2)_{\delta\beta}. \tag{10.262}$$

With $m := \sharp(I\times J) = \sharp(I)\sharp(J)$ and $n := \sharp(K\times L) = \sharp(K)\sharp(L)$ choose bijections

$$\begin{aligned}&\mu : \{1,\dots,m\} \xrightarrow{\cong} I\times J \text{ and } \nu : \{1,\dots,n\} \xrightarrow{\cong} K\times L \text{ and define}\\ &A \in \mathcal{D}^{m\times n},\ x \in \mathcal{D}^n,\ a \in \mathcal{D}^m \text{ by } A_{ij} := A_{\mu(i),\nu(j)},\ x_j := X_{\nu(j)},\ a_i := B_{\mu(i)}.\end{aligned} \tag{10.263}$$

Then the equations $A_1XA_2 = B$ and $Ax = a$ are obviously equivalent. If

$$Ax = a \text{ then } A_1XA_2 = B \text{ with } X_{\gamma\delta} := x_{\nu^{-1}(\gamma,\delta)}. \tag{10.264}$$

Consider the Smith form of A over $\mathcal{D}$, i.e.,

$$\begin{aligned} &UAV = \left(\begin{smallmatrix} E & 0 \\ 0 & 0 \end{smallmatrix}\right),\ r := \operatorname{rank}(A),\ U \in \mathrm{Gl}_m(\mathcal{D}),\ V \in \mathrm{Gl}_n(\mathcal{D}), \\ &E = \operatorname{diag}(e_1, \ldots, e_r),\ 0 \neq e_i \in \mathcal{D},\ e_1 \underset{\mathcal{D}}{|} e_2 \underset{\mathcal{D}}{|} \cdots \underset{\mathcal{D}}{|} e_r \neq 0. \end{aligned} \tag{10.265}$$

Then the following equivalences hold:

$$\begin{aligned} &Ax = a \iff \left(\begin{smallmatrix} E & 0 \\ 0 & 0 \end{smallmatrix}\right) y = UAVy = b,\ y := V^{-1}x,\ b := Ua \iff \\ &\begin{cases} b_i = e_i y_i & \text{if } i = 1, \ldots, r \\ b_i = 0 & \text{if } i = r+1, \ldots, m \end{cases}. \end{aligned} \tag{10.266}$$

These equations have a solution with components in $\mathcal{D}$ if and only if

$$\begin{aligned} &\begin{cases} b_i e_i^{-1} \in \mathcal{D} & \text{if } i = 1, \ldots, r \\ b_i = 0 & \text{if } i = r+1, \ldots, m \end{cases} \Longrightarrow Ax = a,\ A_1 X A_2 = B, \\ &\text{where } x := Vy \in \mathcal{D}^n,\ X_{\gamma\delta} := x_{\nu^{-1}(\gamma,\delta)},\ y_i := \begin{cases} b_i e_i^{-1} & \text{if } i = 1, \ldots, r \\ 0 & \text{if } i = r+1, \ldots, n \end{cases}. \end{aligned} \tag{10.267}$$

In the rewritten equation $\widehat{N}_1 \widehat{Q}_2^0 - H_m = \widehat{N}_1 X^0 \widehat{P}_1$ the matrix A has entries in $\widehat{\mathcal{D}}$ whereas B and a have entries in $\mathcal{S}$. The Smith form of A over $\widehat{\mathcal{D}}$ is also this form over $\mathcal{S}$ and can be used in the preceding algorithm to compute a solution $X^0 \in \mathcal{S}^{m \times p}$ of $\widehat{N}_1 \widehat{Q}_2^0 - H_m = \widehat{N}_1 X^0 \widehat{P}_1$ if one exists. ◊

References

1. W.A. Wolovich, Linear multivariable systems, in *Applied Mathematical Sciences*, vol. 11 (Springer, New York, 1974)
2. F.M. Callier, C.A. Desoer, *Multivariable Feedback Systems* (Springer Texts in Electrical Engineering. Springer, New York, 1982)
3. C.T. Chen, *Linear System Theory and Design* (Harcourt Brace College Publishers, Fort Worth, 1984)
4. M. Vidyasagar, Control system synthesis. A factorization approach, in *MIT Press Series in Signal Processing, Optimization, and Control*, vol. 7 (MIT Press, Cambridge, 1985)
5. A.I.G. Vardulakis, *Linear Multivariable Control. Algebraic Analysis and Synthesis Methods* (Wiley Chichester, 1991)
6. P.J. Antsaklis, A.N. Michel, *Linear Systems*, 2nd edn. (Birkhäuser, Boston, 2006)
7. V. Kučera, *Discrete Linear Control. The Polynomial Equation Approach* (Wiley, Chichester, 1979)
8. D.C. Youla, H.A. Jabr, J.J. Bongiorno, Modern Wiener-Hopf design of optimal controllers. II. The multivariable case. IEEE Trans. Automatic Control, **AC-21**(3), 319–338 (1976)
9. A. Quadrat, On a generalization of the Youla-Kučera parametrization. II. The lattice approach to MIMO systems. Math. Control Signals Syst. **18**(3), 199–235 (2006)
10. H. Bourlès, *Linear systems* (ISTE, London, 2010)

11. D. Hinrichsen, A.J. Pritchard, Mathematical systems theory I. Modelling, state space analysis, stability and robustness, in *Texts in Applied Mathematics*, vol. 48 (Springer, Berlin, 2005)
12. J.W. Polderman, J.C. Willems, Introduction to mathematical systems theory. A behavioral approach, in *Texts in Applied Mathematics*, vol. 26 (Springer, New York, 1998)
13. T. Kailath, *Linear systems* (Prentice-Hall, Englewood Cliffs, 1980)
14. L. Hörmander. The analysis of linear partial differential operators. I. Distribution theory and Fourier analysis, in *Grundlehren der Mathematischen Wissenschaften*, vol. 256 (Springer, Berlin, 1983)
15. H.-J. Schmeißer, *Höhere Analysis* (Universität Jena, Lecture notes, 2005)
16. K. Zhou, J.C. Doyle, K. Glover, *Robust and Optimal Control* (Prentice-Hall, 1996)

Chapter 11
Observers

Assume an injective cogenerator signal module ${}_{\mathcal{D}}\mathcal{F}$ over a principal ideal domain $\mathcal{D}$ where in most cases $\mathcal{D} = F[s]$ is a polynomial algebra over a field F. Also given are a behavior and two operators (functions)

$$\mathcal{B} \subset \mathcal{F}^l,\ P\circ : \mathcal{B} \to \mathcal{F}^m,\ Q\circ : \mathcal{B} \to \mathcal{F}^q, \tag{11.1}$$

where $P \circ w,\ w \in \mathcal{B}$, is measured and thus known, whereas $Q \circ w$ is desired, but unknown. An observer for these data is an IO behavior $\mathcal{B}_{\rm obs}$ that accepts $P \circ w,\ w \in \mathcal{B}$, as input and outputs an estimate $\widehat{w}$ of $Q \circ w$. The interconnection diagram is

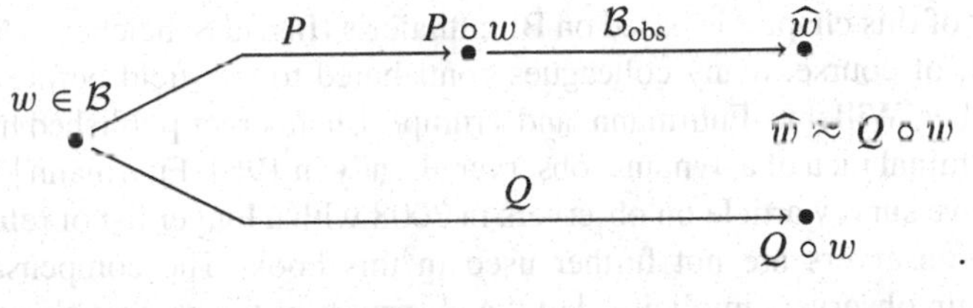

(11.2)

As in Chaps. 9 and 10 the kind of estimation is determined by the choice of a multiplicative submonoid T of $\mathcal{D} \setminus \{0\}$. We assume that $T \subsetneq \mathcal{D} \setminus \{0\}$ or that $\mathcal{F}$ is not a torsion module so that ${}_{\mathcal{D}_T}\mathcal{F}_T$ is also an injective cogenerator by Theorem 9.3.6. For instance, exact, dead-beat, asymptotic, and tracking observers arise from different choices of T. That $\widehat{w}$ T-estimates $Q \circ w$, in signs $\widehat{w} \approx Q \circ w$, means that $\widehat{w} - Q \circ w$ is T-stable. By the appropriate choice of P and Q the theory includes pseudo-state observers for Rosenbrock equations, (Luenberger) state observers and input observers. We focus on observers with proper transfer matrices that can be realized by elementary building blocks.

U. Oberst et al., *Linear Time-Invariant Systems, Behaviors and Modules*, Differential-Algebraic Equations Forum,
https://doi.org/10.1007/978-3-030-43936-1_11

For instance, consider state space equations

$$\begin{aligned}
&s\circ x_1=A_1x_1+B_1u_1,\ y_1=C_1x_1+D_1u_1,\ x_1\in\mathcal{F}^n,\ u_1\in\mathcal{F}^m,\ y_1\in\mathcal{F}^p,\\
&A_1\in F^{n\times n},\cdots,\ \mathcal{B}_1:=\left\{\left(\begin{smallmatrix}x_1\\u_1\end{smallmatrix}\right)\in\mathcal{F}^{n+m};\ s\circ x_1=A_1x_1+B_1u_1\right\},\\
&P\circ:=\left(\begin{smallmatrix}C&D\\0&\mathrm{id}_m\end{smallmatrix}\right)\circ:\ \mathcal{B}_1\to\mathcal{F}^{p+m},\ \left(\begin{smallmatrix}x_1\\u_1\end{smallmatrix}\right)\mapsto\left(\begin{smallmatrix}y_1\\u_1\end{smallmatrix}\right),\ y_1:=C_1x_1+D_1\circ u_1,\\
&Q\circ=(\mathrm{id}_n,0)\circ:\ \mathcal{B}_1\to\mathcal{F}^n,\ \left(\begin{smallmatrix}x_1\\u_1\end{smallmatrix}\right)\mapsto x_1.
\end{aligned}$$

The input u_1 is known and the output y_1 can be measured and is thus known too. In general, this is not the case for all components of the state x_1 of the plant, due to technical restrictions by heat, chemicals, etc. It was Luenberger's seminal idea [1] to estimate, in the continuous-time stability situation, the state x_1 by the state x_2 of an observer described by state space equations

$$s\circ x_2=A_2x_2+B_{2,y}y_1+B_{2,u}u_1,\ x_2\in\mathcal{F}^n,$$

with suitable matrices $A_2,\ B_{2,y},\ B_{2,u}\in F^{\bullet\times\bullet}$. Notice that in contrast to x_1 the complete state x_2 is available, for instance, as output of a computer. We will study constructive Luenberger observers in Sect. 12.5. They are a small, but very important part of the state observers treated in Corollary 11.2.2.

In Theorem 11.1.3 we characterize observers by necessary and sufficient equations, and in Theorem 11.1.5 we characterize the existence of observers and parametrize all of them. Section 11.2 is devoted to observer algorithms, in particular for observers with proper transfer matrices in Theorem 11.2.1.

The theory of this chapter is based on Blumthaler's (Ingrid Scheicher's) Master's thesis [2], but, of course, many colleagues contributed to this field before her, for instance, Valcher, Willems, Fuhrmann, and Trumpf. Luenberger published his paper [1] with the seminal idea of a dynamic observer already in 1964. Fuhrmann [3] wrote a comprehensive survey article on observers in 2008 with a longer list of references.

Functional observers are not further used in this book. The compensators of Chap. 10 contain observers implicitly, but the observers of the present chapter were not needed for the construction of controllers. Therefore we do not present computed examples of the algorithms here and refer to the paper [4] for these.

11.1 Construction and Parametrization of Observers

We assume the behavior and operators from (11.1), i.e.,

$$\mathcal{B}=\left\{w\in\mathcal{F}^l;\ R\circ w=0\right\},\ R\in\mathcal{D}^{k\times l},\ P\in\mathcal{D}^{m\times l},\ Q\in\mathcal{D}^{q\times l}.\tag{11.3}$$

Definition 11.1.1 For $\mathcal{B}$, P, and Q as in (11.3) an IO behavior

$$\begin{aligned}&\mathcal{B}_{\rm obs} := \left\{\begin{pmatrix}\widehat{w}\\ \widehat{u}\end{pmatrix} \in \mathcal{F}^{q+m};\ P_{\rm obs}\circ\widehat{w} = Q_{\rm obs}\circ\widehat{u}\right\} \text{ with}\\ &(P_{\rm obs}, -Q_{\rm obs}) \in \mathcal{D}^{q\times(q+m)},\ \operatorname{rank}(P_{\rm obs}) = \operatorname{rank}(P_{\rm obs}, -Q_{\rm obs}) = q,\\ &H_{\rm obs} = P_{\rm obs}^{-1}Q_{\rm obs} \in K^{q\times m},\end{aligned} \tag{11.4}$$

is called a *(functional) T-observer or -estimator* for the given data if for every $w \in \mathcal{B}$ every output $\widehat{w}$ of $\mathcal{B}_{\rm obs}$ to the input $P \circ w$ gives rise to a T-stable $\widehat{w} - Q \circ w$. In other terms, the associated *error behavior*

$$\mathcal{B}_{\rm err} := \left\{\widehat{w} - Q\circ w \in \mathcal{F}^q;\ w \in \mathcal{B},\ \begin{pmatrix}\widehat{w}\\ P\circ w\end{pmatrix} \in \mathcal{B}_{\rm obs}\right\} \tag{11.5}$$

is T-autonomous, i.e., $\mathcal{B}_{{\rm err},T} = 0$. For $\mathcal{D} = F[s]$ a T-observer $\mathcal{B}_{\rm obs}$ is called *proper* or *strictly proper* if its transfer matrix $H_{\rm obs} = P_{\rm obs}^{-1}Q_{\rm obs} \in K^{q\times m}$ has the respective property. ◊

Lemma 11.1.2 *Every T-observer is T-stable, i.e., its autonomous part $\mathcal{B}_{\rm obs}^0 = \{\widehat{w} \in \mathcal{F}^q;\ P_{\rm obs}\circ\widehat{w}\}$ is T-autonomous or $P_{\rm obs} \in \mathrm{Gl}_q(\mathcal{D}_T)$. In particular, its transfer matrix $H_{\rm obs}$ is T-stable, i.e., $H_{\rm obs} \in \mathcal{D}_T^{q\times m}$.*

Proof Let $\widehat{w} \in \mathcal{B}_{\rm obs}^0$. Then $\begin{pmatrix}\widehat{w}\\ P\circ 0\end{pmatrix} = \begin{pmatrix}\widehat{w}\\ 0\end{pmatrix} \in \mathcal{B}_{\rm obs}$, and since $0 \in \mathcal{B}$, we infer that $\widehat{w} = \widehat{w} - Q\circ 0 \in \mathcal{B}_{\rm err}$. This shows that $\mathcal{B}_{\rm obs}^0 \subseteq \mathcal{B}_{\rm err}$. Since $\mathcal{B}_{\rm obs}$ is a T-observer, the error behavior $\mathcal{B}_{\rm err}$ is T-autonomous, and hence so is its subbehavior $\mathcal{B}_{\rm obs}^0$. □

Theorem 11.1.3 (Characterization of observers) *A T-stable IO behavior $\mathcal{B}_{\rm obs}$ as in* (11.4) *is a T-observer for the data from* (11.3) *if and only if there is a matrix $X \in \mathcal{D}_T^{q\times k}$ such that*

$$Q = -XR + H_{\rm obs}P = (-X, H_{\rm obs})\begin{pmatrix}R\\ P\end{pmatrix}. \tag{11.6}$$

Proof That $\mathcal{B}_{\rm obs}$ is a T-observer signifies $\mathcal{B}_{{\rm err},T}=0$. The assumed T-stability of $\mathcal{B}_{\rm obs}$ implies $P_{\rm obs} \in \mathrm{Gl}_q(\mathcal{D}_T)$ and $H_{\rm obs} \in \mathcal{D}_T^{q\times m}$. For arbitrary $y \in \mathcal{F}_T^q$ and $u \in \mathcal{F}_T^m$ we infer the equivalence

$$\begin{aligned}&y = H_{\rm obs}\circ u \Longleftrightarrow P_{\rm obs}\circ y = P_{\rm obs}\circ(H_{\rm obs}\circ u) = (P_{\rm obs}H_{\rm obs})\circ u = Q_{\rm obs}\circ u\\ &\Longrightarrow \mathcal{B}_{{\rm obs},T} = \left\{\begin{pmatrix}\widehat{w}\\ \widehat{u}\end{pmatrix} \in \mathcal{F}_T^{q+m};\ \widehat{w} = H_{\rm obs}\circ\widehat{u}\right\}.\end{aligned} \tag{11.7}$$

By (11.5) the error behavior is

$$\mathcal{B}_{\rm err} = (\mathrm{id}_q, -Q)\circ\left\{\begin{pmatrix}\widehat{w}\\ w\end{pmatrix} \in \mathcal{F}^{q+l};\ \begin{pmatrix}0 & R\\ P_{\rm obs} & -Q_{\rm obs}P\end{pmatrix}\circ\begin{pmatrix}\widehat{w}\\ w\end{pmatrix} = 0\right\}.$$

The exactness of the quotient functor $(-)_T$ (cf. Theorem 9.3.7, item 4) implies

$$\begin{aligned}\mathcal{B}_{\mathrm{err},T} &= \left\{\widehat{w} - Q \circ w;\ \binom{\widehat{w}}{w} \in \mathcal{F}_T^{q+l},\ \begin{pmatrix} 0 & R \\ P_{\mathrm{obs}} & -Q_{\mathrm{obs}} P\end{pmatrix} \circ \binom{\widehat{w}}{w} = 0\right\}\\ &= \left\{\widehat{w} - Q \circ w;\ w \in \mathcal{B} \cap \mathcal{F}_T^l = \mathcal{B}_T,\ \binom{\widehat{w}}{P \circ w} \in \mathcal{B}_{\mathrm{obs}} \cap \mathcal{F}_T^{q+m} = \mathcal{B}_{\mathrm{obs},T}\right\}\\ &\underset{(11.7)}{=} \left\{\widehat{w} - Q \circ w;\ w \in \mathcal{B}_T,\ \widehat{w} = H_{\mathrm{obs}} P \circ w\right\} = (H_{\mathrm{obs}} P - Q) \circ \mathcal{B}_T.\end{aligned}$$

By means of this representation we obtain the equivalences

$$\begin{aligned}\mathcal{B}_{\mathrm{err},T} = 0 &\Longleftrightarrow \{w \in \mathcal{F}_T^l;\ R \circ w = 0\} \subseteq \{w \in \mathcal{F}_T^l;\ (H_{\mathrm{obs}} P - Q) \circ w = 0\}\\ &\underset{{}_{\mathcal{D}_T}\mathcal{F}_T \text{inj. cogen.}}{\Longleftrightarrow} \exists X \in \mathcal{D}_T^{q \times k}:\ XR = H_{\mathrm{obs}} P - Q,\ \text{i.e., } Q = -XR + H_{\mathrm{obs}} P.\end{aligned}$$

□

T-observability according to the next Lemma 11.1.4 is clearly a necessary condition for the existence of an observer. In item 1 of Theorem 11.1.5 we will show that it is also sufficient.

Lemma and Definition 11.1.4 (T- observability or T-detectability) The following conditions are equivalent.

1. For arbitrary trajectories $w_1, w_2 \in \mathcal{B}$ with the same image $P \circ w_1 = P \circ w_2$ the trajectories $Q \circ w_1$ and $Q \circ w_2$ are T-equivalent, i.e., the behavior

$$\mathcal{C} := Q \circ \ker(P \circ\colon\ \mathcal{B} \to \mathcal{F}^m) = \left\{Q \circ w;\ w \in \mathcal{F}^l,\ \binom{R}{P} \circ w = 0\right\}$$

   is T-autonomous or $\mathcal{C}_T = 0$.
2. There is a matrix $(X_R, X_P) \in \mathcal{D}_T^{q \times (k+m)}$ with

$$X_R R + X_P P = (X_R, X_P) \binom{R}{P} = Q. \tag{11.8}$$

   Then the T-steady state of $Q \circ w$, $w \in \mathcal{B}$, is

$$(Q \circ w)_T = Q \circ w_T = X_P \circ (P \circ w_T) = X_P \circ (P \circ w)_T, \tag{11.9}$$

   so that, up to T-equivalence, the image $Q \circ w$ is determined by $(P \circ w)_T$.

Under these equivalent conditions, the operator $(Q\circ)|_{\mathcal{B}}$ is called *T-observable or -detectable* from $(P\circ)|_{\mathcal{B}}$. Similarly we say that $Q \circ w$, $w \in \mathcal{B}$, is *T-observable* from $P \circ w$ and that $Q \circ \mathcal{B}$ is *T-observable* from $P \circ \mathcal{B}$. Also, we call the triple $(\mathcal{B}, P, Q)$ *T-observable*.

Proof As in the proof of Theorem 11.1.3 the quotient functor is exact and yields

$$\mathcal{C}_T = \left\{Q \circ w;\ w \in \mathcal{F}_T^l,\ \binom{R}{P} \circ w = 0\right\}.$$

Since ${}_{\mathcal{D}_T}\mathcal{F}_T$ is an injective cogenerator, we get

$$\begin{aligned} \mathcal{C}_T = 0 &\iff \left\{w \in \mathcal{F}_T^l;\ \left(\begin{smallmatrix} R \\ P \end{smallmatrix}\right) \circ w = 0\right\} \subseteq \{w \in \mathcal{F}_T^l;\ Q \circ w = 0\} \\ &\iff \exists (X_R, X_P) \in \mathcal{D}_T^{q\times(k+m)} \text{ with } (X_R, X_P) \left(\begin{smallmatrix} R \\ P \end{smallmatrix}\right) = Q. \end{aligned}$$

For $w \in \mathcal{B}$ we conclude

$$\begin{aligned} (Q \circ w)_T &= Q \circ w_T = (X_R R + X_P P) \circ w_T \\ &= X_R \circ (R \circ w)_T + X_P \circ (P \circ w)_T = X_P \circ (P \circ w)_T. \end{aligned}$$ □

Theorem 11.1.5 (Construction and parametrization of T-observers)

1. *The triple $(\mathcal{B}, P, Q)$ from* (11.3) *admits a T-observer if and only if it is T-observable.*
2. *Assume that the conditions of item 1 hold and let $X_P \in \mathcal{D}_T^{q\times m}$ be such that $Q - X_P P \in \mathcal{D}_T^{q\times k} R$, i.e., $Q = X_R R + X_P P$ for some $X_R \in \mathcal{D}_T^{q\times k}$. Let*

$$\begin{aligned} \mathcal{B}_{\mathrm{c}} &:= \left\{\left(\begin{smallmatrix} \widehat{w} \\ \widehat{u} \end{smallmatrix}\right) \in \mathcal{F}^{q+m};\ P_{\mathrm{c}} \circ \widehat{w} = Q_{\mathrm{c}} \circ \widehat{u}\right\} \text{ with} \\ (P_{\mathrm{c}}, -Q_{\mathrm{c}}) &\in \mathcal{D}^{q\times(q+m)},\ \operatorname{rank}(P_{\mathrm{c}}) = q,\ X_P = P_{\mathrm{c}}^{-1} Q_{\mathrm{c}}, \end{aligned}$$

be the unique controllable IO realization of X_P according to Theorem 6.2.4, Then $\mathcal{B}_{\mathrm{c}}$ is the unique controllable T-observer of $(\mathcal{B}, P, Q)$ with transfer matrix X_P.
3. *In the situation of item 2 all T-observers with the same transfer matrix $H_{\mathrm{obs}} = X_P$ are given as*

$$\begin{aligned} \mathcal{B}_{\mathrm{obs}} &:= \left\{\left(\begin{smallmatrix} \widehat{w} \\ \widehat{u} \end{smallmatrix}\right) \in \mathcal{F}^{q+m};\ P_{\mathrm{obs}} \circ \widehat{w} = Q_{\mathrm{obs}} \circ \widehat{u}\right\} \text{ where} \\ (P_{\mathrm{obs}}, -Q_{\mathrm{obs}}) &= Y(P_{\mathrm{c}}, -Q_{\mathrm{c}}) \text{ for some } Y \in \mathcal{D}^{q\times q} \text{ with } \det(Y) \in T. \end{aligned} \tag{11.10}$$

4. *The pairs $(X_P, Y) \in \mathcal{D}^{q\times m} \times \mathcal{D}^{q\times q}$ with*

$$Q - X_P P \in \mathcal{D}_T^{q\times k} R \quad \text{and} \quad \det(Y) \in T$$

parametrize the set of all T-observers $\mathcal{B}_{\mathrm{obs}}$ of $(\mathcal{B}, P, Q)$, where $\mathcal{B}_{\mathrm{obs}}$ is defined as in item 3 and has the transfer matrix $H_{\mathrm{obs}} = X_P$. Two pairs (X_P, Y) and (X'_P, Y') define the same T-observer $\mathcal{B}_{\mathrm{obs}}$ if and only if $X_P = X'_P$ and Y and Y' are row equivalent, i.e., $Y'Y^{-1} \in \mathrm{Gl}_q(\mathcal{D})$.
5. *If F is a field and $\mathcal{D} = F[s]$, the T-observer $\mathcal{B}_{\mathrm{obs}}$ in item 4 is proper if and only if its transfer matrix X_P is proper. Hence a proper T-observer exists if and only if $Q \in \mathcal{D}_T^{q\times k} R + \mathcal{S}^{q\times m} P$, where $\mathcal{S}$ is the ring of proper T-stable rational functions according to Definition 9.5.1.*

Proof 1. (i) Let $\mathcal{B}_{\mathrm{obs}}$ be a T-observer for $(\mathcal{B}, P, Q)$. By Lemma 11.1.2, $\mathcal{B}_{\mathrm{obs}}$ is T-stable; hence, its transfer matrix H_{obs} lies in $\mathcal{D}_T^{q\times m}$ by Theorem and Definition 9.4.2. Moreover, by Theorem 11.1.3, there is a matrix $X \in \mathcal{D}_T^{q\times k}$ with $-XR + H_{\mathrm{obs}} P = Q$. But this is the second condition of Lemma and Definition 11.1.4 with $(X_R, X_P) = (-X, H_{\mathrm{obs}})$. Thus $(\mathcal{B}, P, Q)$ is T-observable.

(ii) Conversely, assume that $(\mathcal{B}, P, Q)$ is T-observable. Let $(X_R, X_P) \in \mathcal{D}_T^{q\times(k+m)}$ be a solution of $X_R R + X_P P = Q$ and $\mathcal{B}_c$ the unique controllable IO realization of X_P from Theorem 6.2.4. Since $\mathcal{B}_c$ is controllable, its system module $M_c := \mathrm{M}(\mathcal{B}_c)$ is torsionfree and so is its quotient module ${}_{\mathcal{D}_T}M_{c,T}$. Hence the ${}_{\mathcal{D}_T}\mathcal{F}_T$-behavior $\mathcal{B}_{c,T}$ is controllable. By construction, its transfer matrix is $X_P \in \mathcal{D}_T^{q\times m}$. Theorem and Definition 9.4.2 implies that $\mathcal{B}_c$ is T-stable. From the condition $Q = X_R R + X_P P = -(-X_R)R + X_P P$ and Theorem 11.1.3 we infer that $\mathcal{B}_c$ is a T-observer.

2. In 1 we have already seen that $\mathcal{B}_c$ is a T-observer. Since it is the unique controllable IO behavior with transfer matrix X_P, it is also the unique T-observer with this property.

3. If $Y \in \mathcal{D}^{q\times q}$ with $\det(Y) \in T$ or, equivalently, $Y \in \mathcal{D}^{q\times q} \cap \mathrm{Gl}_q(\mathcal{D}_T)$, then $\mathcal{B}_{\mathrm{obs}}$ from (11.10) has the same transfer matrix $H_{\mathrm{obs}} = X_P$ as $\mathcal{B}_c$, and it is T-stable since $P_{\mathrm{obs}} = Y P_c \in \mathrm{Gl}_q(\mathcal{D}_T)$, see Theorem and Definition 9.4.2. Theorem 11.1.3 implies that $\mathcal{B}_{\mathrm{obs}}$ is a T-observer.

If, conversely,

$$\mathcal{B}_1 = \left\{\begin{pmatrix}\widehat{w}\\ \widehat{u}\end{pmatrix} \in \mathcal{F}^{q\times m};\ P_1 \circ \widehat{w} = Q_1 \circ \widehat{u}\right\} \text{ with}$$
$$(P_1, -Q_1) \in \mathcal{D}^{q\times(q+m)},\ \mathrm{rank}(P_1) = q,\ X_P = P_1^{-1}Q_1,$$

is an arbitrary T-observer with the same transfer matrix X_P as $\mathcal{B}_c$, then, by Theorem 6.2.4, $\mathcal{B}_c = \mathcal{B}_{1,\mathrm{cont}}$ is the controllable part of $\mathcal{B}_1$, and hence there is a matrix $Y \in \mathcal{D}^{q\times q}$ with $(P_1, -Q_1) = Y(P_c, -Q_c)$. The equation $P_1 = Y P_c$ implies $\det(P_1) = \det(Y)\det(P_c)$. The behavior $\mathcal{B}_1$ is T-stable by Lemma 11.1.2, and hence $\det(P_1) \in T$ and, since T is saturated, $\det(Y) \in T$.

4. By Lemma 11.1.2 and Theorem 11.1.3, the transfer matrix X_P of an arbitrary T-observer satisfies $Q - X_P P \in \mathcal{D}_T^{q\times k} R$. With this, the parametrization of the T-observers follows directly from items 2 and 3.

5. This statement is obvious. □

Example 11.1.6 By using different monoids T, we obtain the following types of observers:

1. *Exact observers.* In the case $T = \{1\}$ or $T = \mathrm{U}(\mathcal{D})$, which is the saturation of $\{1\}$, the equality $\mathcal{D} = \mathcal{D}_T$ holds. Signals or behaviors coincide with their steady states and are T-equivalent if and only if they are equal. A T-observer is called an *exact observer* and Eq. (11.9) gets the form

$$Q \circ w = X_P P \circ w,\ w \in \mathcal{B},\ \text{where } Q = X_R R + X_P P,\ (X_R, X_P) \in \mathcal{D}^{q\times(k+m)}.$$

If $\mathcal{D} = F[s]$ for a field F, notice that a proper rational function is a polynomial if and only if it is a constant in F. Thus there are only trivial proper and exact observers.

If $\mathcal{D}$ is an arbitrary principal ideal domain and $Q = \mathrm{id}_l$, i.e., if one wants to compute the trajectory $w \in \mathcal{B}$ from $P \circ w$, then the triple $(\mathcal{B}, P, \mathrm{id}_l)$ is T-observable according to Corollary and Definition 11.1.4 if and only if $\begin{pmatrix}R\\ P\end{pmatrix}$ is left invertible over $\mathcal{D}$, and this signifies that w is observable or reconstructible from $P \circ w$ in the sense of Definition and Lemma 4.1.1.

2. *Dead-beat observers*, cf. [3, 5]. Assume a field $F, \mathcal{D} = F[s]$, the discrete-time signal space $\mathcal{F} = F^{\mathbb{N}}$, and the submonoid $T = \langle s \rangle = \{1, s, s^2, \dots\}$ or its saturation $T = \{us^n;\ u \in F \setminus \{0\},\ n \in \mathbb{N}\}$. Then the quotient ring $\mathcal{D}_T = \mathcal{D}_s = \bigoplus_{n\in\mathbb{Z}} Fs^n$ is the ring of Laurent polynomials. The T-torsion module of $\mathcal{F}$ is the space $\mathrm{t}_T(F^{\mathbb{N}}) = F^{(\mathbb{N})}$ of sequences with finite support, hence, two signals w_1 and w_2 are T-equivalent if they coincide in almost all time instants $t \in \mathbb{N}$. T-observability is then called *reconstructibilty* and a T-observer is called a *dead-beat observer*.

3. *Asymptotic observers*. This is the standard case. Assume $\mathcal{D} = F[s]$ with $F = \mathbb{R}$ or $F = \mathbb{C}$, the stable complex region

$$\Lambda_1 = \begin{cases} \{\lambda \in \mathbb{C};\ \Re(\lambda) < 0\} & \text{in continuous time} \\ \{\lambda \in \mathbb{C};\ |\lambda| < 1\} & \text{in discrete time} \end{cases}$$

and the associated monoid $T = \{t \in \mathcal{D} \setminus \{0\};\ \mathrm{V}_{\mathbb{C}}(t) \subseteq \Lambda_1\}$. The T-stable signals are the polynomial-exponential signals which converge to zero for $t \to \infty$, T-observability is called *detectability* and a T-observer is called an *asymptotic observer*.

4. *Tracking observers*, compare [3]. If $T = \mathcal{D} \setminus \{0\}$, then $\mathcal{D}_T = \mathrm{quot}(\mathcal{D}) = K$, the T-small behaviors are just the autonomous ones, the module of T-steady states is $\mathcal{F}_T = \mathcal{F}_K$ where $\mathcal{F} = \mathrm{t}(\mathcal{F}) \oplus \mathcal{F}_K$. In this situation, T-observability is called *trackability* and T-observers are called *tracking observers*. ◊

Remark 11.1.7 (*Consistent observers, compare* [3, p. 110]) In the situation of Theorem 11.1.5 we assume without loss of generality that $\mathrm{rank}(R) = k$. Then X_R in (11.8) is uniquely determined by $X_P = H_{\mathrm{obs}}$. The observer $\mathcal{B}_{\mathrm{obs}}$ is called *consistent* if it satisfies the equivalent conditions

$$\begin{aligned}
&\forall w \subset \mathcal{B}:\ \begin{pmatrix} Q\circ w \\ P\circ w \end{pmatrix} \in \mathcal{B}_{\mathrm{obs}},\ \text{i.e., } (P_{\mathrm{obs}}Q - Q_{\mathrm{obs}}P)\circ w = 0 \\
&\iff \mathcal{B} = \{w \in \mathcal{F}^l;\ R\circ w = 0\} \subseteq \{w \in \mathcal{F}^l;\ (P_{\mathrm{obs}}Q - Q_{\mathrm{obs}}P)\circ w = 0\} \\
&\underset{\mathcal{F}\text{ injective cogenerator}}{\iff} \exists Y \in \mathcal{D}^{q\times k}:\ YR = P_{\mathrm{obs}}Q - Q_{\mathrm{obs}}P \\
&\qquad \underset{(11.8)}{=} P_{\mathrm{obs}}(X_R R + H_{\mathrm{obs}}P) - Q_{\mathrm{obs}}P \\
&\qquad = P_{\mathrm{obs}}X_R R + Q_{\mathrm{obs}}P - Q_{\mathrm{obs}}P = P_{\mathrm{obs}}X_R R \\
&\underset{\mathrm{rank}(R)=k}{\iff} \exists Y \in \mathcal{D}^{q\times k}:\ Y = P_{\mathrm{obs}}X_R \iff P_{\mathrm{obs}}X_R \in \mathcal{D}^{q\times k}\ \left(X_R \in \mathcal{D}_T^{q\times k}!\right).
\end{aligned}$$

The preceding argument for $\mathcal{B}_T$ instead of $\mathcal{B}$ shows that $(P_{\mathrm{obs}}Q - Q_{\mathrm{obs}}P)\circ w$ for $w \in \mathcal{B}$ is T-stable, but not zero in general. Consistency of an observer has no practical advantages.

A T-observer is not necessarily consistent. Indeed, consider a T-stable, uncontrollable IO behavior

$$\begin{aligned}
&\mathcal{B} = \left\{\binom{y}{u} \in \mathcal{F}^{p+m};\ (A, -B)\circ \binom{y}{u} = 0\right\},\ \det(A) \in T,\ \text{and} \\
&P := (0, \mathrm{id}_m),\ P\circ \binom{y}{u} = u,\ Q := (\mathrm{id}_p, 0),\ Q\circ \binom{y}{u} = y.
\end{aligned}$$

Let $\mathcal{B}_c = \left\{\binom{y}{u} \in \mathcal{F}^{p+m};\ (A_c, -B_c) \circ \binom{y}{u} = 0\right\} \subsetneq \mathcal{B}$ denote the controllable part of $\mathcal{B}$, which is also T-stable. If $w = \binom{y}{u} \in \mathcal{B}$ and $\binom{\widehat{y}}{P\circ w} = \binom{\widehat{y}}{u} \in \mathcal{B}_c \subset \mathcal{B}$, then $\widehat{y} - Q \circ w = \widehat{y} - y \in \mathcal{B}^0$ is T-stable; hence, $\mathcal{B}_c$ is a T-observer of $(\mathcal{B}, P, Q)$. It is not consistent since $w = \binom{y}{u} \in \mathcal{B}$ does *not* imply $\binom{Q\circ w}{P\circ w} = \binom{y}{u} \in \mathcal{B}_c \subsetneq \mathcal{B}$. ◊

Remark 11.1.8 (Spectral assignment) For $\mathcal{D} = F[s]$, F a field, Theorem 11.1.5 can also be used to construct observers with special spectral properties. For this purpose consider a *finite*, nonempty set $\mathcal{P}_1$ of irreducible polynomials and the associated saturated monoid T of nonzero polynomials whose prime factors belong to $\mathcal{P}_1$, see (9.4). The T-torsion module $\mathrm{t}_T(\mathcal{F})$ of T-stable signals is given by

$$\mathrm{t}_T(\mathcal{F}) = \bigoplus_{q\in\mathcal{P}_1} \mathrm{t}_q(\mathcal{F}) \text{ where } \mathrm{t}_q(\mathcal{F}) = \{y \in \mathcal{F};\ \exists k \in \mathbb{N}:\ q^k \circ y = 0\}.$$

Over $F = \mathbb{C}, \mathbb{R}$ one considers a *finite*, nonempty subset

$$\Lambda_1 = \overline{\Lambda_1} \subset \begin{cases} \mathbb{C}_- = \{\lambda \in \mathbb{C};\ \Re(\lambda) < 0\} & \text{in continuous time} \\ \mathbb{D} = \{\lambda \in;\ |\lambda| < 1\} & \text{in discrete time} \end{cases} \quad \text{and}$$

$$T := \{t \in T;\ \mathrm{V}_{\mathbb{C}}(t) \subset \Lambda_1\}.$$

For the complex base field and the signal space $\mathcal{F} = \mathrm{C}^\infty(\mathbb{R}, \mathbb{C})$ or $\mathcal{F} = \mathrm{C}^{-\infty}(\mathbb{R}, \mathbb{C})$ the space of T-stable signals is $\mathrm{t}_T(\mathcal{F}) = \bigoplus_{\lambda\in\Lambda_1} \mathbb{C}[t]e^{\lambda t}$.

A behavior is T-autonomous if and only if it is autonomous and its characteristic variety is contained in Λ_1. Application of Theorem 11.1.5 furnishes T-observers with the property that both $\mathcal{B}_{\mathrm{err}}$ and $\mathcal{B}^0_{\mathrm{obs}}$ are T-autonomous. Fuhrmann [3, Definition 3.3] talks about *spectral assignability* in this context. The observer algorithms in Sect. 11.2 permit to decide spectral assignability and, if this holds, to construct and parametrize all (proper) T-observers. ◊

Example 11.1.9 1. (Cf. [6, Theorem 7.3.23], [2]) We consider Rosenbrock equations

$$\begin{aligned} &A \circ x = B \circ u,\quad y = C \circ x + D \circ u \text{ with } A \in \mathcal{D}^{n\times n},\ \det(A) \neq 0, \\ &B \in \mathcal{D}^{n\times m},\ C \in \mathcal{D}^{p\times n},\ D \in \mathcal{D}^{p\times m},\ H_1 := A^{-1}B,\ H_2 = D + CA^{-1}B, \end{aligned} \tag{11.11}$$

and an additional operator $K \in \mathcal{D}^{q\times n}$. We want to construct a T-observer of the function $K \circ x$ of the pseudo-state x from the input u and output y. Pseudo-state observers with $K = \mathrm{id}_n$ and state observers of state space behaviors are special cases. To adapt the notations to those of Theorem 11.1.5 we introduce

$$\begin{aligned} &\mathcal{B} := \{w \in \mathcal{F}^{n+m};\ R \circ w = 0\} \text{ with } w := \binom{x}{u} \text{ and } R := (A, -B), \\ &P := \begin{pmatrix} C & D \\ 0 & \mathrm{id}_m \end{pmatrix},\ P \circ w = \binom{y}{u},\ Q := (K, 0),\ Q \circ w = K \circ x. \end{aligned} \tag{11.12}$$

With $(X_y, X_u) := X_P \in \mathcal{D}_T^{q\times(p+m)}$ the equation $Q = X_R R + X_P P$, which characterizes T-observability (see Lemma and Definition 11.1.4), gets the equivalent form

$$\begin{aligned}&(K, 0) = (X_R, X_y, X_u)\begin{pmatrix} A & -B \\ C & D \\ 0 & \mathrm{id}_m\end{pmatrix}\\ &\Longleftrightarrow K = (X_R, X_y)\begin{pmatrix} A \\ C\end{pmatrix} = X_R A + X_y C,\ X_u = X_R B - X_y D \\ &\Longrightarrow X_u = (KA^{-1} - X_y CA^{-1})B - X_y D = KH_1 - X_y H_2.\end{aligned} \tag{11.13}$$

By item 1 of Theorem 11.1.5 a T-observer of $K \circ x$ from $\begin{pmatrix} y \\ u\end{pmatrix}$ exists if and only if K admits a representation

$$K = (X_R, X_y)\begin{pmatrix} A \\ C\end{pmatrix} = X_R A + X_y C \text{ with } (X_R, X_y) \in \mathcal{D}_T^{q\times(n+p)}. \tag{11.14}$$

Then the unique controllable realization of the transfer matrix

$$H_{\mathrm{obs}} := X_P := (X_y, X_u) \in \mathcal{D}_T^{q\times(p+m)} \text{ with } X_u = X_R B - X_y D \in \mathcal{D}_T^{q\times m} \tag{11.15}$$

is the unique controllable T-observer with transfer matrix X_P.

2. [6, Sect. 5.5], [2]. We consider *input T-observers* of an IO behavior

$$\left\{\begin{pmatrix} y \\ u\end{pmatrix} \in \mathcal{F}^{p+m};\ A\circ y = B \circ u\right\} \text{ with } A \in \mathcal{D}^{p\times p},\ \det(A) \neq 0,\ B \in \mathcal{D}^{p\times m} \tag{11.16}$$

and $H := A^{-1}B$. We adapt these data to Theorem 11.1.5 and obtain

$$\begin{aligned}&R := (A, -B),\ w := \begin{pmatrix} y \\ u\end{pmatrix},\ P := (\mathrm{id}_p, 0) \in \mathcal{D}^{p\times(p+m)},\ P\circ w = y,\\ &Q := (0, \mathrm{id}_m) \in \mathcal{D}^{m\times(p+m)},\ Q \circ w = u.\end{aligned} \tag{11.17}$$

The condition (11.8) for T-observability gets the equivalent forms

$$(0, \mathrm{id}_m) = (X_R, X_P)\begin{pmatrix} A & -B \\ \mathrm{id}_p & 0\end{pmatrix} \text{ or } (X_R A + X_P = 0,\ \mathrm{id}_m = -X_R B) \tag{11.18}$$

where $(X_R, X_P) \in \mathcal{D}_T^{m\times(n+p)}$. Hence a T-observer exists if B has a left inverse $-X_R \in \mathcal{D}_T^{m\times n}$. Then an observer with transfer matrix $H_{\mathrm{obs}} = X_P = -X_R A$ can be constructed, and the latter satisfies $H_{\mathrm{obs}} H = -X_R AA^{-1}B = -X_R B = \mathrm{id}_m$. If $y := H \circ u$ is defined, this implies the equation $u = H_{\mathrm{obs}} \circ y$, as required for an observation of the input from the output. ◊

11.2 Observer Algorithms

In this section we derive algorithms to decide the existence of (proper) T-observers and to construct and parametrize all of them if they exist.

Corollary 11.2.1 (Construction and parametrization of T-observers) *We use the same data as in Theorem 11.1.5, namely, a principal ideal domain $\mathcal{D}$, a saturated multiplicative submonoid $T \subseteq \mathcal{D}$, an injective cogenerator signal module ${}_{\mathcal{D}}\mathcal{F}$, and the triple $(\mathcal{B}, P, Q)$ from* (11.3). *By Corollary and Definition 11.1.4, this triple is T-observable if and only if there is*

$$(X_R, X_P) \in \mathcal{D}_T^{q\times(k+m)} \text{ with } (X_R, X_P)\begin{pmatrix} R \\ P \end{pmatrix} = Q. \tag{11.19}$$

To solve this linear matrix equation, we apply Lemma 9.2.6. Let

$$U\begin{pmatrix} R \\ P \end{pmatrix}V = \begin{pmatrix} \operatorname{diag}(e_1,\ldots,e_\rho) & 0 \\ 0 & 0 \end{pmatrix}$$

$$\text{with } U \in \mathrm{Gl}_{k+m}(\mathcal{D}),\ V \in \mathrm{Gl}_l(\mathcal{D}),\ \rho = \operatorname{rank}\begin{pmatrix} R \\ P \end{pmatrix}, \text{ and } e_1 \underset{\mathcal{D}}{\mid} \cdots \underset{\mathcal{D}}{\mid} e_\rho$$

be the Smith form of $\begin{pmatrix} R \\ P \end{pmatrix}$ with respect to $\mathcal{D}$ and to $\mathcal{D}_T$. Then the following holds:

1. *The triple $(\mathcal{B}, P, Q)$ is T-observable, i.e.,* (11.19) *is solvable, if and only if*

$$QV_{-j}\begin{cases} \in \mathcal{D}_T^q e_j & \text{for } j = 1,\ldots,\rho \\ = 0 & \text{for } j = \rho+1,\ldots,l \end{cases}$$

$$\Longrightarrow X' \in \mathcal{D}_T^{q\times(k+m)} \text{ where } X'_{-j} := \begin{cases} QV_{-j}e_j^{-1} & \text{for } j = 1,\ldots,\rho, \\ 0 & \text{for } j = \rho+1,\ldots,k+m. \end{cases} \tag{11.20}$$

   Then, by Lemma 9.2.6,

$$(X_R^0, X_P^0) := X'U \in \mathcal{D}_T^{q\times(k+m)} \text{ and } Q = (X_R^0, X_P^0)\begin{pmatrix} R \\ P \end{pmatrix}. \tag{11.21}$$

   All other solutions of the latter equation with entries in $\mathcal{D}_T$ are given as

$$(X_R, X_P) = (X_R^0, X_P^0) + M(L_R, L_P) \text{ for arbitrary } M \in \mathcal{D}_T^{q\times(k+m-\rho)}, \tag{11.22}$$

   where $L = (L_R, L_P) \in \mathcal{D}^{(k+m-\rho)\times(k+m)}$ is a universal left annihilator of $\begin{pmatrix} R \\ P \end{pmatrix}$, for example, the one given by the last $k+m-\rho$ rows of U.
   The controllable realization of each such $X_P = X_P^0 + ML_P \in \mathcal{D}_T^{q\times p}$ is a controllable T-observer of $(\mathcal{B}, P, Q)$ with transfer matrix $H_{\mathrm{obs}} = X_P$, cf. Theorem 11.1.5.
2. *The Eq.* (11.20) *is satisfied for all Q and in particular for $Q = \mathrm{id}_l$ and 1 is applicable if and only if $\begin{pmatrix} R \\ P \end{pmatrix}$ is left invertible over $\mathcal{D}_T$ or, equivalently, if $\rho = \operatorname{rank}\begin{pmatrix} R \\ P \end{pmatrix} = l$ and if $e_j \in T,\ j = 1,\ldots,l$.* ◊

We finally construct observers with proper transfer matrix.
We assume the data from Sect. 9.5, in particular a field

$$\begin{aligned}
&F,\ \mathcal{D} := F[s],\ \alpha \in F,\ s-\alpha \in T,\ \widehat{s} := (s-\alpha)^{-1},\\
&\widehat{\mathcal{D}} := F[\widehat{s}],\ \widehat{T} = \left\{ \tfrac{t}{(s-\alpha)^{\deg(t)}};\ t \in T \right\} \subseteq \widehat{\mathcal{D}}\\
&\Longrightarrow \widehat{\mathcal{D}} = F[\widehat{s}] \subseteq \mathcal{S} = \widehat{\mathcal{D}}_{\widehat{T}} \subseteq \mathcal{S}_{\widehat{s}} = \mathcal{D}_T \supseteq \mathcal{D}.
\end{aligned}$$

We consider the given matrices R, P, and Q as matrices over $\mathcal{D}_T = \mathcal{S}_{\widehat{s}} = (\widehat{\mathcal{D}}_{\widehat{T}})_{\widehat{s}}$ and apply Corollary 11.2.1 to the polynomial ring $\widehat{\mathcal{D}} = F[\widehat{s}]$, which gives us a parametrization of all T-observers of $(\mathcal{B}, P, Q)$. In particular, we compute the Smith/McMillan form of $\widehat{s}^d \binom{R}{P} \in \widehat{\mathcal{D}}^{(k+m)\times l}$, $d := \deg_s \binom{R}{P}$, over $\widehat{\mathcal{D}}$, viz.,

$$\begin{aligned}
&U\widehat{s}^d \binom{R}{P} V = \begin{pmatrix} \operatorname{diag}(e'_1,\ldots,e'_\rho) & 0\\ 0 & 0\end{pmatrix} \text{ with}\\
&U \in \mathrm{Gl}_{k+m}(\widehat{\mathcal{D}}),\ V \in \mathrm{Gl}_l(\widehat{\mathcal{D}}),\ \rho = \operatorname{rank}\binom{R}{P},\ e'_i \in \widehat{\mathcal{D}},\ e'_1 \underset{\widehat{\mathcal{D}}}{\mid} \cdots \underset{\widehat{\mathcal{D}}}{\mid} e'_\rho \qquad (11.23)\\
&\Longrightarrow U\binom{R}{P} V = \begin{pmatrix} \operatorname{diag}(e_1,\ldots,e_\rho) & 0\\ 0 & 0\end{pmatrix},\ e_i = \widehat{s}^{-d} e'_i = (s-\alpha)^d e'_i,
\end{aligned}$$

is the Smith/McMillan form of $\binom{R}{P}$ over $\widehat{\mathcal{D}}$ and over $\mathcal{D}_T$. Also the last $k+m-\rho$ rows of U form a universal left annihilator

$$L = (L_R, L_P) = (U_{ij})_i \in \widehat{\mathcal{D}}^{(k+m-\rho)\times(k+m)} \subseteq \mathcal{D}_T^{(k+m-\rho)\times(k+m)} \qquad (11.24)$$

of $\binom{R}{P}$ over $\widehat{\mathcal{D}}$ and thus also over $\mathcal{D}_T$, compare Remark 9.2.5.

We assume that the triple $(\mathcal{B}, P, Q)$ is T-observable and that $(X_R^0, X_P^0) \in \mathcal{D}_T^{q+(k+m)}$ solves $Q = (X_R^0, X_P^0)\binom{R}{P} = X_R^0 R + X_P^0 P$. Due to (11.23) Lemma 9.2.6 can be used to test the existence of $(X_R^0, X_P^0) \in \mathcal{D}_T^{q\times(k+m)}$ and to construct it. We are going to determine which of the transfer matrices $X_P = X_P^0 + ML_P$ of Corollary 11.2.1 are proper.

Without loss of generality we also assume $\operatorname{rank}(R) = k$. Therefore there is a right inverse R' of R with entries in $F(s)$. The rank of $L_P \in \widehat{\mathcal{D}}^{(k+m-\rho)\times m}$ follows from the implications:

$$\begin{aligned}
&(L_R, L_P)\binom{R}{P} = 0\\
&\quad\Longrightarrow 0 = (L_R, L_P)\binom{R}{P} R' = L_R R R' + L_P P R' = L_R + L_P P R'\\
&\quad\Longrightarrow L_R = -L_P P R' \Longrightarrow L = L_P(-PR', \mathrm{id}_m) \qquad (11.25)\\
&\quad\Longrightarrow \operatorname{rank}(L_P) = \operatorname{rank}(L) = k+m-\rho.
\end{aligned}$$

Next we compute the Smith form of L_P with respect to $\widehat{\mathcal{D}} = F[\widehat{s}]$, namely,

$$\begin{aligned}
&U^L L_P V^L = (E^L, 0) \in \widehat{\mathcal{D}}^{(k+m-\rho)\times m},\ E^L := \operatorname{diag}(e_1^L, \ldots, e_{k+m-\rho}^L)\\
&\text{with } U^L \in \mathrm{Gl}_{k+m-\rho}(\widehat{\mathcal{D}}),\ V^L \in \mathrm{Gl}_m(\widehat{\mathcal{D}}), \text{ and } e_1^L \underset{\widehat{\mathcal{D}}}{\mid} \cdots \underset{\widehat{\mathcal{D}}}{\mid} e_{k+m-\rho}^L \in \widehat{\mathcal{D}}. \qquad (11.26)
\end{aligned}$$

The relevant part of the parametrization of the T-observer's transfer matrices from (11.22) is the expression $H_{\text{obs}} = X_P = X_P^0 + ML_P \in \mathcal{D}_T^{q\times m}$. We multiply this equation with V^L on the right and obtain

$$X'_P := X_P V^L = X_P^0 V^L + ML_P V^L = X_P^0 V^L + \underbrace{M(U^L)^{-1}}_{=:M'}\underbrace{U^L L_P V^L}_{=(E^L,0)}. \quad (11.27)$$

The columns of this matrix are

$$(X'_P)_{-j} = \begin{cases} X_P^0 V_{-j}^L + M'_{-j} e_j^L & \text{if } j = 1, \ldots, k+m-\rho \\ X_P^0 V_{-j}^L & \text{if } j = k+m-\rho+1, \ldots, m. \end{cases}$$

We solve this equation for $X_P^0 V_{-j}^L$ and obtain, for the individual entries $(X_P^0 V^L)_{ij}$, $i = 1, \ldots, q$, $j = 1, \ldots, m$, the equation

$$\left(X_P^0 V^L\right)_{ij} = \begin{cases} (X'_P)_{ij} - M'_{ij} e_j^L & \text{if } j = 1, \ldots, k+m-\rho \\ (X'_P)_{ij} & \text{if } j = k+m-\rho+1, \ldots, m. \end{cases} \quad (11.28)$$

Since $V^L \in \mathrm{Gl}_m(\widehat{\mathcal{D}}) \subset \mathrm{Gl}_m(\mathcal{S})$, the entries of $X'_P = X_P V^L$ belong to $\mathcal{S}$ if and only if the entries of $X_P = X'_P(V^L)^{-1}$ do.

Theorem 11.2.1 (Construction and parametrization of proper T-observers) *We use the same data as in Theorem 11.1.5 and Corollary 11.2.1, and assume that* $\operatorname{rank}(R) = k$, *that* $(\mathcal{B}, P, Q)$ *is T-observable, and that one solution* $(X_R^0, X_P^0) \in \mathcal{D}_T^{q\times(k+m)}$ *of* $Q = (X_R^0, X_P^0)\binom{R}{P}$ *has been computed. We also need the Smith/McMillan form of* $\binom{R}{P}$ *over* $\widehat{\mathcal{D}}$ *from (11.23), the universal left annihilator* $L = (L_R, L_P)$ *with entries in* $\widehat{\mathcal{D}}$ *of* $\binom{R}{P}$ *from (11.24), and the Smith form of* L_P *over* $\widehat{\mathcal{D}}$ *from (11.26). The following assertions hold:*

1. *If there is a proper T-observer, i.e., with* $H_{\text{obs}} = X_P \in \mathcal{S}^{q\times m}$, *the equation (11.28) implies the necessary condition*

$$\left(X_P^0 V^L\right)_{ij} \in \mathcal{S} \ \text{ for } i = 1, \ldots, q, \ j = k+m-\rho+1, \ldots, m, \ \rho := \operatorname{rank}\binom{R}{P}. \quad (11.29)$$

2. *Assume that 1 holds. We apply Theorem 9.5.6 to obtain* $(X'_P)_{ij} \in \mathcal{S}$ *and* $M'_{ij} \in \mathcal{D}_T$ *such that*

$$\begin{aligned} &\left(X_P^0 V^L\right)_{ij} = -M'_{ij} e_j^L + (X'_P)_{ij} \in \mathcal{D}_T = \mathcal{D}_T e_j^L + \mathcal{S} \text{ for} \\ &i = 1, \ldots, q, \ j = 1, \ldots, k+m-\rho, \ M' := (M'_{ij})_{i,j} \in \mathcal{D}_T^{q\times(m+k-\rho)}. \end{aligned} \quad (11.30)$$

We enlarge the set of entries $(X'_P)_{ij}$ *to a matrix* $X'_P \in \mathcal{S}^{q\times m}$ *by defining*

$$(X'_P)_{ij} := \left(X_P^0 V^L\right)_{ij} \text{ for } i = 1, \ldots, q \text{ and } j = k+m-\rho+1, \ldots, m. \quad (11.31)$$

By the equivalence of (11.27) *and* (11.28), *the matrices*

$$\begin{aligned} &X_P^1 := X'_P (V^L)^{-1} \in \mathcal{S}^{q\times m},\ M^1 := M'U^L \in \mathcal{D}_T^{q\times(k+m-\rho)} \text{ and} \\ &X_R^1 := X_R^0 + M^1 L_R \in \mathcal{D}_T^{q\times k} \text{ satisfy} \\ &(X_R^1, X_P^1) = (X_R^0, X_P^0) + M^1(L_R, L_P) \in \mathcal{D}_T^{q\times k} \times \mathcal{S}^{q\times m}. \end{aligned} \tag{11.32}$$

By Corollary 11.2.1 they give rise to a unique controllable T-observer of $(\mathcal{B}, P, Q)$ with proper transfer matrix X_P. Moreover, all transfer matrices of proper T-observers of $(\mathcal{B}, P, Q)$ are obtained in this fashion.

3. *Let $o_j := \mathrm{ord}_{\widehat{s}}(e_j^L)$, $j = 1, \dots, k+m-\rho$. All other T-observers with proper transfer matrix X_P^2 are given by*

$$\begin{aligned} &(X_R^2, X_P^2) = (X_R^1, X_P^1) + ZU^L(L_R, L_P) \in \mathcal{D}_T^{q\times k} \times \mathcal{S}^{q\times m} \text{ where} \\ &Z \in \mathcal{S}^{q\times(k+m-\rho)} \,\mathrm{diag}(\widehat{s}^{-o_1}, \dots, \widehat{s}^{-o_{k+m-\rho}}) \subseteq \mathcal{D}_T^{q\times(k+m-\rho)}. \end{aligned} \tag{11.33}$$

Hence these matrices Z parametrize the set of all controllable T-observers of $(\mathcal{B}, P, Q)$ with proper transfer matrix.

Proof 1., 2.: See above.
3. Take $(X_R^1, X_P^1) \in \mathcal{D}_T^{q\times k} \times \mathcal{S}^{q\times m}$ from 2. and let $E^L := \mathrm{diag}(e_1^L, \dots, e_{k+m-\rho}^L)$. By Theorem 9.5.6 all other solutions of (11.30) have the form

$$\begin{aligned} &\left(X_P^0 V^L\right)_{ij} = -M''_{ij} e_j^L + (X''_P)_{ij} \in \mathcal{D}_T = \mathcal{D}_T e_j^L + \mathcal{S} \text{ for} \\ &i = 1, \dots, q,\ j = 1, \dots, k+m-\rho \text{ where } M''_{ij} - M'_{ij} \in \mathcal{S}\widehat{s}^{-o_j} \\ &\Longrightarrow X''_P = X_P^0 V^L + M''(E^L, 0) = X_P^0 V^L + M'(E^L, 0) + (M'' - M')(E^L, 0) \\ &\Longrightarrow X_P^2 = X''_P \left(V^L\right)^{-1} = X_P^1 + ZU^L L_P \text{ where} \\ &\quad Z := M'' - M' \in \mathcal{S}^{q\times(k+m-\rho)} \,\mathrm{diag}(\widehat{s}^{-o_1}, \dots, \widehat{s}^{-o_{k+m-\rho}}). \end{aligned} \tag{11.34}$$

□

Corollary 11.2.2 *Assume that the Rosenbrock equations from Example 11.1.9 with $K = \mathrm{id}_n$ are T-observable, i.e., that $\binom{A}{C}$ has a left inverse $(X_R^0, X_y^0) \in \mathcal{D}_T^{n\times(n+p)}$, and that the transfer matrices $H_1 = A^{-1}B$ and $H_2 = D + CA^{-1}B$ are proper. Then the equations admit a T-observer of the pseudo-state x with proper T-stable transfer matrix $(X_y^1, H_1 - X_y^1 H_2) \in \mathcal{S}^{n\times(p+m)}$, cf.* (11.35). *In particular, this is applicable to state space equations and state T-observers with proper transfer matrix. Their state space realizations are generalized* Luenberger state observers *with equations*

$$s \circ x_2 = A_2 x_2 + B_{2,y} y + B_{2,u} u,\quad x_{\mathrm{obs}} = C_2 x_2 + D_{2,y} y + D_{2,u} u.$$

Proof We specialize the considerations of Theorem 11.2.1 to the data from Example 11.1.9. If $(X_R, X_y) \in \mathcal{D}_T^{n\times(n+p)}$ satisfies $(X_R, X_y)\binom{A}{C} = \mathrm{id}_n$ and if $X_y \in \mathcal{S}^{n\times p}$, then there is a pseudo-state T-observer with proper transfer matrix

$$H_{\rm obs} = (X_y, X_u) \in \mathcal{S}^{n\times(p+m)} \text{ where } X_u = X_R B - X_y D = H_1 - X_y H_2.$$

Indeed, the first resp. the second equation for X_u imply that X_u is T-stable resp. proper, and hence contained in $\mathcal{S}^{n\times m}$. Thus it suffices to construct a proper X_y.

Since $p = (n+p) - n = (n+p) - \operatorname{rank}\left(\begin{smallmatrix} A \\ C \end{smallmatrix}\right)$ there is a universal left annihilator

$$(L_A, L_C) \in \widehat{\mathcal{D}}^{p\times(n+p)} \text{ of } \left(\begin{smallmatrix} A \\ C \end{smallmatrix}\right), \text{ hence } \operatorname{rank}(L_C) = p.$$

Then all left inverses of $\left(\begin{smallmatrix} A \\ C \end{smallmatrix}\right)$ in $\mathcal{D}_T^{n\times(n+p)}$ have the form

$$\begin{aligned} &(X_R, X_y) = (X_R^0, X_y^0) + M(L_A, L_C),\ M \in \mathcal{D}_T^{n\times p} \\ &\Longrightarrow X_y = X_y^0 + M L_C,\ \operatorname{rank}(L_C) = p. \end{aligned}$$

The Smith form $U^L L_C V^L = E = \operatorname{diag}(e_1, \dots, e_p) \in \widehat{\mathcal{D}}^{p\times p}$ of L_C over $\widehat{\mathcal{D}}$ implies the equation

$$\begin{aligned} &X'_y := X_y V^L = X_y^0 V^L + M'E,\ M' = M\left(U^L\right)^{-1} \in \mathcal{D}_T^{n\times p} \\ &\Longrightarrow X_y^0 V^L = -M'E + X'_y. \end{aligned}$$

The analogous equation to (11.29) is empty, and hence trivially satisfied. As in (11.30) there is a representation

$$\begin{aligned} &X_y^0 V^L = -M'E + X'_y,\ M' \in \mathcal{D}_T^{n\times p},\ X'_y \in \mathcal{S}^{n\times p} \Longrightarrow M := M'U^L \in \mathcal{D}_T^{n\times p} \text{ and} \\ &X_y^1 := X'_y\left(V^L\right)^{-1} = X_y^0 + M L_C \in \mathcal{S}^{n\times p},\ X_R^1 := X_R^0 + M L_A \in \mathcal{D}_T^{n\times n}, \\ &X_u^1 := H_1 - X_y^1 H_2 \in \mathcal{S}^{n\times m},\ H_{\rm obs} = (X_y^1, X_u^1) \in \mathcal{S}^{n\times(p+m)}. \end{aligned} \tag{11.35}$$

These matrices satisfy $(X_R^1, X_y^1)\left(\begin{smallmatrix} A \\ C \end{smallmatrix}\right) = \mathrm{id}_n$ and give rise to a controllable pseudo-state T-observer with the proper transfer matrix $H_{\rm obs}$. □

Corollary 11.2.3 *In Corollary 11.2.2 all other pseudo-state T-observers with proper transfer matrix $(X_y^2, H_1 - X_y^2 H_2)$ are obtained via*

$$X_y^2 = X_y^1 + ZU^L L_C \text{ with } Z \in \mathcal{S}^{n\times p}\operatorname{diag}(\widehat{s}^{-o_1}, \dots, \widehat{s}^{-o_p}),\ o_j := \operatorname{ord}_{\widehat{s}}(e_j). \tag{11.36}$$

◇

References

1. D.G. Luenberger, Observing the state of a linear system. IEEE Trans. Mil. Electron. **8**(2), 74–80 (1964)
2. I. Blumthaler, Functional T-observers. Linear Algebr. Appl. **432**(6), 1560–1577 (2010)
3. P.A. Fuhrmann, Observer theory. Linear Algebr. Appl. **428**(1), 44–136 (2008)
4. I. Blumthaler, U. Oberst, Design, parametrization, and pole placement of stabilizing output feedback compensators via injective cogenerator quotient signal modules. Linear Algebr. Appl. **436**(5), 963–1000 (2012)
5. M. Bisiacco, M.E. Valcher, J.C. Willems, A behavioral approach to estimation and dead-beat observer design with applications to state-space models. IEEE Trans. Automat. Control **51**(11), 1787–1797 (2006)
6. W.A. Wolovich, *Linear multivariable systems*. Applied Mathematical Sciences, vol. 11 (Springer, New York, 1974)

Chapter 12
Canonical State Space Realizations of IO Systems via Gröbner Bases

In Theorem 6.3.11 we presented Fliess' unconstructive proof that every IO behavior $\mathcal{B}$ over the polynomial algebra $\mathcal{D} = F[s]$ admits an observable state space realization. This is unique up to similarity. In this chapter we use the Gröbner basis theory to *construct*, in Sects. 12.3, 12.4, 12.6 and 12.7, the *canonical observability and observer realizations* of $\mathcal{B}$, and, if $\mathcal{B}$ is controllable, the *canonical controllability and controller realizations*. The canonical realizations depend only on the IO behavior $\mathcal{B}$ and on a chosen term order in the Gröbner basis theory, but not on the special matrices that define the behavior. If the transfer matrix of $\mathcal{B}$ is proper these realizations can be built with elementary building blocks as in Sect. 7.2. In particular, the proper compensators and observers from Chaps. 10 and 11 can be implemented as state space behaviors in this fashion. In Sect. 12.2 we show that every behavior admits a canonical IO structure with *proper* transfer matrix. In Sect. 12.5 we derive *Luenberger's state observer* and *stabilization by state feedback* from these canonical realizations. The realizations of this chapter are related to those of the central Chap. 6 in [1, pp. 345–498], but the Gröbner basis algorithm is much stronger and sharper than the algorithms in [1]. This chapter is based on [2, Example 5.15].

More specifically, we consider a field F, the polynomial algebra $\mathcal{D} := F[s]$, an injective cogenerator $\mathcal{F}$ over $F[s]$, and IO behaviors with trajectories whose components lie in $\mathcal{F}$. For such an IO behavior

$$\begin{aligned} &\mathcal{B} := \left\{ \binom{y}{u} \in \mathcal{F}^{p+m};\ P \circ y = Q \circ u \right\} \text{ with} \\ &(P, -Q) \in \mathcal{D}^{p \times (p+m)},\ \text{rank}(P) = p,\ H := P^{-1}Q, \end{aligned} \tag{12.1}$$

we are going to *construct* observable state space realizations

$$\begin{aligned} &s \circ x = Ax + Bu,\ y = Cx + D \circ u \text{ with} \\ &x \in \mathcal{F}^n,\ u \in \mathcal{F}^m,\ y \in \mathcal{F}^p,\ A, B, C \in F^{\bullet \times \bullet},\ D \in \mathcal{D}^{p \times m}. \end{aligned} \tag{12.2}$$

U. Oberst et al., *Linear Time-Invariant Systems, Behaviors and Modules*, Differential-Algebraic Equations Forum,
https://doi.org/10.1007/978-3-030-43936-1_12

Recall from Theorem 6.3.8 that any two such realizations are similar. The condition for observability is that the map

$$\begin{pmatrix} C & D \\ 0 & \mathrm{id}_m \end{pmatrix} \circ : \quad \mathcal{B}_1 := \left\{ \begin{pmatrix} x \\ u \end{pmatrix} \in \mathcal{F}^{n+m};\ s \circ x = Ax + Bu \right\} \longrightarrow \mathcal{B}, \\ \begin{pmatrix} x \\ u \end{pmatrix} \longmapsto \begin{pmatrix} Cx + D \circ u \\ u \end{pmatrix}, \quad (12.3)$$

is a behavior isomorphism. For the transfer matrices this implies

$$H = D + C(s\,\mathrm{id}_n - A)^{-1}B \in F(s)^{p\times m} = F[s]^{p\times m} \oplus F(s)^{p\times m}_{\mathrm{spr}}. \quad (12.4)$$

Hence D is the polynomial component of H, cf. Lemma 5.5.5. The matrix D has entries in F if and only if the matrix H is proper. For arbitrary IO behaviors state space realizations with polynomial D have to be admitted.

We use the *Gröbner basis theory and algorithms* for the construction of the realizations. We give an introduction to this theory in Sect. 12.1. The Buchberger algorithm for the computation of Gröbner bases is implemented in all computer algebra systems, and therefore we do not describe its details in this book. In Sect. 12.2 we consider any behavior $\mathcal{B} := U^{\perp} \subseteq \mathcal{F}^{l}$, $U \subseteq F[s]^{1\times l}$, and the reduced Gröbner basis of U with respect to a chosen term order on $\{1, \ldots, l\} \times \mathbb{N}$. These data induce a canonical IO structure of $\mathcal{B}$. For a degree-over-position term order the transfer matrix of $\mathcal{B}$ for this IO structure is *proper*, cf. [3, Theorem 3.3.22]. This is very important from an engineering point of view.

Sections 12.3 to 12.7 form the core of this chapter. In order to achieve state space realizations we assume an IO behavior as in (12.1) and a term order on $\{1, \ldots, p\} \times \mathbb{N}$ and apply the reduced Gröbner basis of

$$U^0 := \mathbb{F}[s]^{1\times k}P \text{ to } \mathcal{B}^0 = (U^0)^{\perp} = \left\{ y \in \mathcal{F}^p;\ P \circ y = 0 \right\}.$$

The canonical observability resp. observer realizations are derived in Sects. 12.3 and 12.4, the observer realization requiring the *degree-over-position* term order. For controllable $\mathcal{B}$ the canonical controllability resp. controller realization are derived from the observability resp. observer realizations by duality in Sects. 12.6 resp. 12.7. In Sect. 12.5 the canonical observer realization enables the construction of the Luenberger state observers [4].

12.1 Gröbner Bases

Gröbner bases were developed by Buchberger for computations with *multivariate* polynomial ideals. The theory was later generalized to include polynomial modules. While finding a Gröbner basis of an ideal in the ring $F[s]$ of *univariate* polynomials simplifies to computing greatest common divisors by means of the Euclidean algorithm, the Gröbner basis theory for modules $U \subseteq F[s]^{1\times l}$ over the univariate polynomial ring is nontrivial—although it is easier than in the multivariate case—and relevant, as its application to state space realizations in Sects. 12.3–12.7 shows. In this section we introduce univariate Gröbner bases, which are defined with respect to a term order on $I \times \mathbb{N}$, where I is a finite index set. For Gröbner bases over the multivariate polynomial ring, which are not needed in this book, we refer the reader to the numerous standard books [5], [6], [7, Chaps. 1 and 5], [8, Chap. 15], [9], and [10]. We will only prove the unique existence of a reduced Gröbner basis of a given module U with respect to a term order after Macaulay. The Buchberger algorithm, which is the standard method for computing Gröbner bases, is well documented in all the cited books and is implemented in all computer algebra programs, for instance, in MAPLE, SINGULAR and MATHEMATICA.

We begin by defining term orders, which are necessary for Gröbner bases, and by discussing their basic properties.

Lemma and Definition 12.1.1 *Let $(\Lambda, \leq)$ denote any ordered set.*

1. *The set is* strictly ordered *if any two elements $\lambda, \mu \in \Lambda$ are comparable, i.e., $\lambda \leq \mu$ or $\mu \leq \lambda$. Then, any nonempty finite subset S of Λ has a largest and a least element which we denote by* $\max(S)$ *and* $\min(S)$, *respectively. We introduce another symbol $-\infty$ and define $-\infty < \lambda$ for all $\lambda \in \Lambda$ and* $\max(\emptyset) := -\infty$.
2. *The set Λ is* artinian *if it satisfies the following equivalent conditions:*
   - (a) *Any decreasing sequence $\lambda_1 \geq \lambda_2 \geq \cdots$ in Λ becomes* stationary, *i.e., there is an index n_0 such that $\lambda_n = \lambda_{n_0}$ for all $n \geq n_0$. In other words, every properly decreasing sequence is finite.*
   - (b) *The set satisfies the* minimal condition, *i.e., every nonempty subset Λ_1 of Λ has a minimal element $\lambda_1 \in \Lambda_1$ with the defining property that $\lambda \in \Lambda_1$ and $\lambda \leq \lambda_1$ imply $\lambda = \lambda_1$.*
3. *The set Λ is* well ordered *if it has the following equivalent properties:*
   - (a) *It is nonempty, strictly ordered and artinian.*
   - (b) *It is nonempty and every nonempty subset Λ_1 of Λ has a least element λ_1, i.e., $\lambda_1 \leq \lambda$ for all $\lambda \in \Lambda_1$.*

Proof That the two conditions for *artinian* are equivalent is known from the corresponding result for *Noetherian* where increasing sequences and maximal elements are involved. The proof of the equivalence of the two conditions for *well ordered* is simple and omitted. □

The standard example of a well-ordered set is the set $\mathbb{N}$ of natural numbers with its natural strict order. The induction principle is equivalent to $\mathbb{N}$ being well ordered. The well-ordered sets which arise in the Gröbner theory are natural generalizations of this example. We are first going to show how a well-ordered basis of a vector space permits a generalization of Euclidean division of univariate polynomials and then describe those *term orders* that are relevant for the state space realizations.

We consider a field F, an F-vector space V, and a well-ordered basis $\underline{v} = (v_\lambda)_{\lambda \in \Lambda}$

of V, i.e., a basis indexed by a well-ordered set Λ. For $x = \sum_{\lambda\in\Lambda} a_\lambda v_\lambda \in V$ we consider the finite *support* and *degree*

$$\operatorname{supp}(x) := \{\lambda \in \Lambda;\ a_\lambda \neq 0\} \text{ and } d := \deg(x) := \max(\operatorname{supp}(x)),\ \deg(0) := -\infty. \tag{12.5}$$

A nonzero x can thus be written as

$$x = a_d v_d + \sum_{\lambda<d} a_\lambda v_\lambda = a_d v_d + \cdots,\ a_d \neq 0,\ \operatorname{lc}(x) := a_d,\ \operatorname{lt}(x) := a_d v_d, \tag{12.6}$$

where $\operatorname{lc}(x)$ resp. $\operatorname{lt}(x)$ are the *leading coefficient* resp. the *leading term* of x. For the definition of the degree, it is only necessary that the set Λ is strictly ordered, it need not be well ordered. By this definition, the basis vector v_λ has the degree λ and the leading coefficient 1.

Lemma 12.1.2 *Let $x, y \in V$ and $0 \neq a \in F$.*

1. *If* $\deg(x) \neq \deg(y)$, *then* $\deg(x + y) = \max\big(\deg(x), \deg(y)\big)$. *In general, only the relation* $\deg(x + y) \leq \max\big(\deg(x), \deg(y)\big)$ *holds.*
2. $\deg(ax) = \deg(x)$ *if* $a \neq 0$.

Proof Obvious. □

Example 12.1.3 For the polynomial algebra $\mathcal{D} := F[s]$, the degree with respect to the well-ordered monomial basis $(s^n)_{n\in\mathbb{N}}$ is the usual degree of polynomials. ◇

The following lemma is the main result concerning well-ordered bases.

Lemma 12.1.4 *Consider the vector space V with its well-ordered basis $\underline{v} = (v_\lambda)_{\lambda\in\Lambda}$. Let $\underline{w} = (w_\lambda)_{\lambda\in\Lambda}$ be another family of vectors in V, indexed by the same well-ordered index set, with the property* $\deg(w_\lambda) = \deg(v_\lambda) = \lambda$ *for all* $\lambda \in \Lambda$. *Then, $\underline{w}$ is also a (well ordered) basis of V.*

The proof also furnishes an algorithm to construct, for any vector $x \in V$, the basis representation $x = \sum_{\lambda\in\Lambda} b_\lambda w_\lambda$ from the given representations $x = \sum_{\lambda\in\Lambda} a_\lambda v_\lambda$ and $w_\lambda = \sum_{\lambda'\in\Lambda} c_{\lambda,\lambda'} v_{\lambda'}$.

Proof 1. *Linear independence*: To show that $\underline{w}$ is independent, consider the linear combination $0 = \sum_{\lambda\in\Lambda} b_\lambda w_\lambda$. Assume that $S := \{\lambda \in \Lambda;\ b_\lambda \neq 0\} \neq \emptyset$. Then S is a nonempty finite subset of Λ, and thus has a maximum $\max(S)$ in Λ, in particular, $\max(S) \neq -\infty$. But then

$$-\infty = \deg(0) = \deg\Big(\sum_{\lambda\in\Lambda} b_\lambda w_\lambda\Big) = \deg\Big(\sum_{\lambda\in S} b_\lambda w_\lambda\Big) \underset{\text{Lemma 12.1.2}}{=} \max(S) \in \Lambda,$$

a contradiction. Therefore, $S = \emptyset$, i.e., all b_λ are zero.
2. *Generating system*: We have to show that each vector x is a linear combination of the vectors w_λ. We show first that every nonzero vector x of degree $d := \deg(x)$ has a unique representation

$$x = b_d w_d + y \quad \text{with } b_d = \text{lc}(x)\text{lc}(w_d)^{-1} \neq 0 \text{ and } \deg(y) < \deg(x).$$

(i) *Uniqueness*: Let

$$x = b_d w_d + y = b'_d w_d + y'$$

be two such representations. Then $(b_d - b'_d)w_d = y' - y$. Since $\deg(x) = d$, but $\deg(y)$ and $\deg(y')$ are both strictly smaller than d, the coefficients b_d and b'_d are both nonzero. If $b_d - b'_d \neq 0$, then Lemma 12.1.2 furnishes the contradiction

$$\begin{aligned} d = \deg(w_d) &= \deg\left((b_d - b'_d)w_d\right) = \deg(y' - y) \\ &\leq \max\left(\deg(y), \deg(y')\right) < d. \end{aligned}$$

Hence $b_d - b'_d = 0$, i.e., $b_d = b'_d$, and consequently also $y = y'$.
(ii) *Existence*: With the representations

$$x = \text{lc}(x)v_d + \sum_{\lambda<d} a_\lambda v_\lambda \quad \text{and} \quad w_d = \text{lc}(w_d)v_d + \sum_{\lambda<d} c_{d,\lambda} v_\lambda,$$

we define $b_d := \text{lc}(x)\text{lc}(w_d)^{-1}$ and $y := x - b_d w_d$. Then

$$y = x - b_d w_d = \underbrace{\left(\text{lc}(x) - b_d \text{lc}(w_d)\right)}_{=0} v_d + \sum_{\lambda<d} (a_\lambda - b_d c_{d,\lambda}) v_\lambda$$

has degree $\deg(y) < d = \deg(x)$ as desired.
For $x \neq 0$, we define $x_0 := x$. As long as $x_j \neq 0$ let $d_j := \deg(x_j)$ and let $x_j = b_{d_j} w_{d_j} + x_{j+1}$ be the unique representation from above, i.e., $b_{d_j} := \text{lc}(x_j)\text{lc}(w_{d_j})^{-1}$ and $x_{j+1} := x_j - b_{d_j} w_{d_j}$ with $d_{j+1} := \deg(x_{j+1}) < \deg(x_j) = d_j$. By construction, $d_0 > d_1 > \cdots$ is a properly decreasing sequence in the well-ordered (and hence artinian) set Λ. Thus the sequence is finite. Let d_m be its least and last element with $x_m = b_{d_m} w_{d_m}$. By induction this construction yields the unique representation

$$\begin{aligned} x = x_0 = b_{d_0} w_{d_0} + x_1 &= b_{d_0} w_{d_0} + b_{d_1} w_{d_1} + x_2 = \cdots \\ b_{d_0} w_{d_0} + \cdots + b_{d_m} w_{d_m} + 0 &= \sum_{k=0}^{m} b_{d_k} w_{d_k}. \end{aligned}$$

□

Example 12.1.5 Consider the polynomial algebra $F[s]$ with its well-ordered basis $\underline{v} = (s^k)_{k\in\mathbb{N}}$ with $\deg(s^k) = k$. Let g be a nonzero polynomial of degree $\deg(g) = n$. Then the family

$$w_0 := 1, \ \ldots, \ w_{n-1} := s^{n-1}, \ w_n := g, \ w_{n+1} = sg, \ \ldots, \ w_{n+k} := s^k g, \ \ldots$$

satisfies $\deg(w_k) = k$ for all $k \in \mathbb{N}$, and it is therefore an F-basis of $F[s]$. Every polynomial $f \in F[s]$ has the unique basis representation

$$f = \sum_{k=0}^{n-1} b_k s^k + \sum_{k=0}^{\infty} b_{n+k} s^k g = \sum_{k=0}^{n-1} b_k s^k + \left(\sum_{k=0}^{\infty} b_{n+k} s^k \right) g =: r + hg.$$

This is exactly the unique Euclidean division of f by g with remainder r of degree $\deg(r) < n = \deg(g)$. The algorithm of Lemma 12.1.4 to obtain the b_k is the standard algorithm for polynomial division. ◊

In order to be able to treat modules over the polynomial ring $F[s]$, we use a finite index set I and consider $\Lambda := I \times \mathbb{N}$. This gives rise to the free module

$$F[s]^{1\times I} = \bigoplus_{i\in I} F[s]\delta_i = \bigoplus_{(i,\mu)\in I\times\mathbb{N}} F s^\mu \delta_i, \tag{12.7}$$

where δ_i, $i \in I$, is the standard basis of $F[s]^{1\times I}$. For $\alpha := (i, \mu) \in \Lambda$ we define $v_\alpha := s^\mu \delta_i$ and obtain the F-basis $(v_\alpha)_{\alpha\in\Lambda}$ of $F[s]^{1\times I}$. For $I = \{1, \ldots, l\}$ we obtain the usual row space $F[s]^{1\times I} = F[s]^{1\times l}$, but the slight generalization to arbitrary finite index sets will later be useful.

Due to $s^\lambda(s^\mu\delta_i) = s^{\lambda+\mu}\delta_i$ we define the additive action of $\mathbb{N}$ on $I \times \mathbb{N}$ by

$$+\colon\ \mathbb{N} \times (I \times \mathbb{N}) \longrightarrow I \times \mathbb{N},\quad \big(\lambda, (i, \mu)\big) \longmapsto \lambda + (i, \mu) := (i, \mu) + \lambda := (i, \lambda + \mu). \tag{12.8}$$

Definition 12.1.6 An order $\leq$ on $I \times \mathbb{N}$ is a *term order* if the order is strict, if $(i, 0) \leq (i, \mu)$ for all $\mu \in \mathbb{N}$, and if it is compatible with the action of $\mathbb{N}$, i.e., if $\alpha, \beta \in I \times \mathbb{N}$ with $\alpha \leq \beta$ and $\lambda \in \mathbb{N}$ imply $\alpha + \lambda \leq \beta + \lambda$. For a given term order the degree of a nonzero

$$f = \sum_{(i,\mu)\in I\times\mathbb{N}} a_{i,\mu} s^\mu \delta_i \in F[s]^{1\times I} \text{ is } \deg(f) := \max\big\{(i, \mu) \in I \times \mathbb{N};\ a_{i,\mu} \neq 0\big\}.$$

◊

Lemma 12.1.7 *Every term order on $I \times \mathbb{N}$ is a well order. The family $v_{i,\mu} := s^\mu\delta_i$, $(i, \mu) \in I \times \mathbb{N}$, is a well-ordered basis of $F[s]^{1\times I}$ with the ensuing degree* $\deg(s^\mu\delta_i) = (i, \mu)$.

Proof Let $N \subseteq I \times \mathbb{N}$ be a nonempty subset. We have to show that N has a minimal element. We write

$$N = \biguplus_{i\times I} \{i\} \times N_i \subseteq I \times \mathbb{N} = \biguplus_{i\in I} \{i\} \times \mathbb{N} \quad \text{with } N_i \subseteq \mathbb{N}.$$

If N_i is nonempty, define $d(i) := \min(N_i) \in \mathbb{N}$. For all $(i, \mu) \in N$ the inequality

$$(i,0) \le (i,\mu-d(i)) \text{ implies } (i,d(i)) = (i,0)+d(i) \le (i,\mu-d(i))+d(i) = (i,\mu).$$

This signifies that $\min(N) = \min\left\{(i,d(i));\ i \in I,\ N_i \neq \emptyset\right\}$. The minimum exists because the set on the right-hand side is finite and the order is strict. □

The following orders are the standard examples of term orders for modules.

Example 12.1.8 Let $\le$ be an arbitrary strict order on the finite set I. On $I = \{1,\cdots,l\}$ one generally chooses the order $1 > 2 > \cdots > l$ in order to obtain

$$\deg(\delta_1) = (1,0) > \deg(\delta_2) = (2,0) > \cdots > \deg(\delta_l) = (l,0)$$

for $\delta_j = (0,\ldots,0,\overset{j}{1},0,\ldots,0)$. The lexicographic *position-over-degree* order

$$(i,\mu) = \deg\left(s^\mu\delta_i\right) <_1 (j,\nu) = \deg\left(s^\nu\delta_j\right) :\Longleftrightarrow i < j \text{ or } \left(i = j \text{ and } \mu < \nu\right) \tag{12.9}$$

and the *degree-over-position* order

$$(i,\mu) = \deg\left(s^\mu\delta_i\right) <_2 (j,\nu) = \deg\left(s^\nu\delta_j\right) :\Longleftrightarrow \mu < \nu \text{ or } \left(\mu = \nu \text{ and } i < j\right) \tag{12.10}$$

are obviously term orders. We provide the degree deg and the leading term lt with an index $i = 1,2$ which refers to the term order $\le_i$.

For $l = 3$ and $\xi := (2s+1, 3s^2+2, 4s^3+3) \in \mathbb{Q}[s]^{1\times 3}$, we get

$$\begin{aligned}
&\deg_1(\xi) = (1,1), \quad \mathrm{lt}_1(\xi) = 2s\delta_1 = 2(s,0,0), \quad \text{and}\\
&\deg_2(\xi) = (3,3), \quad \mathrm{lt}_2(\xi) = 4s^3\delta_3 = 4(0,0,s^3).
\end{aligned}$$

As we will see below, the choice of the term order influences the canonical state space representations of IO behaviors. ◊

In the sequel we assume a term order $\le$ on $I \times \mathbb{N}$ and a submodule $U \subseteq F[s]^{1\times I}$. We write the vectors f of $F[s]^{1\times I}$ in the form

$$\begin{aligned}
&f = (f_i)_{i\in I} = \sum_{i\in I} f_i\delta_i = \sum_{(i,\mu)\in I\times\mathbb{N}} a_{i,\mu}s^\mu\delta_i = \sum_{\alpha\in I\times\mathbb{N}} a_\alpha v_\alpha \in F[s]^{1\times I}, \quad \text{where}\\
&f_i = \sum_{\mu\in\mathbb{N}} a_{i,\mu}s^\mu \in F[s],\ a_{i,\mu} = a_{(i,\mu)} = a_\alpha \in F,\ v_{i,\mu} = v_{(i,\mu)} = v_\alpha = s^\mu\delta_i.
\end{aligned} \tag{12.11}$$

Lemma and Definition 12.1.9 *1. If $f \in F[s]^{1\times I}$ is nonzero, then*

$$\deg(s^\lambda f) = \lambda + \deg(f) \ \textit{for all } \lambda \in \mathbb{N}.$$

2. *The set* $\deg(U) := \{\deg(f);\ 0 \neq f \in U\}$ *satisfies* $\deg(U) = \deg(U) + \mathbb{N}$.

Proof 1. With $(j,\nu) := d := \deg(f)$ we get

$$f = a_d v_d + \sum_{\alpha<d} a_\alpha v_\alpha \text{ and } s^\lambda f = a_d s^\lambda v_d + \sum_{\alpha<d} a_\alpha s^\lambda v_\alpha.$$

But $\alpha = (i,\mu) < d = (j,\nu)$ implies

$$\deg(s^\lambda v_\alpha) = \lambda + \alpha < \lambda + d = \deg(s^\lambda v_d),$$

hence, $\deg(s^\lambda f) = \lambda + d = \lambda + \deg(f)$.
2. This follows directly from 1. □

For the submodule $U \subseteq F[s]^{1\times I}$ we define

$$N := \deg(U) =: \biguplus_{i\in I} \{i\} \times N_i. \tag{12.12}$$

Then $N = N + \mathbb{N} \subseteq I \times N$ and $N_i = N_i + \mathbb{N} \subseteq \mathbb{N}$. If $N_i \neq \emptyset$ is nonempty, we write $d(i) := \min(N_i)$ and obtain that $N_i = d(i) + \mathbb{N}$. The complement of the degree set is the set

$$\Gamma := (I \times \mathbb{N}) \setminus N =: \biguplus_{i\in I} \{i\} \times \Gamma_i \text{ with } \Gamma_i = \begin{cases} \mathbb{N} & \text{if } N_i = \emptyset, \\ \{0,\dots,d(i)-1\} & \text{if } N_i \neq \emptyset \end{cases}. \tag{12.13}$$

These data induce the decompositions

$$\begin{aligned}
I &= I^0 \biguplus I^{\text{free}},\ I^0 := \{i \in I;\ N_i \neq \emptyset\},\ I^{\text{free}} := \{i \in I;\ N_i = \emptyset, \text{ or } \Gamma_i = \mathbb{N}\}, \text{ hence} \\
N &= \biguplus_{i\in I^0,\, d(i)>0} \{i\} \times (d(i) + \mathbb{N}) \uplus \biguplus_{i\in I^0,\, d(i)=0} \{i\} \times \mathbb{N}, \\
\Gamma &= \biguplus_{i\in I^0,\, d(i)>0} \{i\} \times \{0,\dots,d(i)-1\} \uplus \biguplus_{i\in I^{\text{free}}} \{i\} \times \mathbb{N}, \\
\Gamma^0 &:= \biguplus_{i\in I^0,\, d(i)>0} \{i\} \times \{0,\dots,d(i)-1\} = \left\{(i,\mu);\ i \in I^0,\ 0 \le \mu \le d(i)-1\right\}.
\end{aligned} \tag{12.14}$$

The indices $i \in I^{\text{free}}$, i.e., with $\Gamma_i = \mathbb{N}$, later exactly identify the *input or free components* w_i of an IO behavior with trajectories $w \in \mathcal{F}^I$. The indices $i \in I^0$ with $d(i) = 0$ ($\Longleftrightarrow N_i = \mathbb{N} \Longleftrightarrow \Gamma_i = \emptyset$) are called *superfluous* or *inessential* since the corresponding w_i can be omitted from the considerations.

In the following theorem we define the normal form of a polynomial vector $f \in F[s]^{1\times I}$, which is a generalization of the remainder after a polynomial division, and we show the unique existence of a *reduced Gröbner basis* for the univariate polynomial module.

Theorem and Definion 12.1.10 (Normal form and reduced Gröbner basis). *We assume a submodule $U \subseteq F[s]^{1\times I}$, a term order on $I \times \mathbb{N}$ and the induced data from* (12.12), (12.13), *and* (12.14).

1. *As an F-vector space $F[s]^{1\times I}$ is the direct sum*

$$F[s]^{1\times I} = \bigoplus_{(j,\mu)\in\Gamma} Fs^{\mu}\delta_j \oplus U \ni f = \mathrm{normf}(f) + (f - \mathrm{normf}(f)). \quad (12.15)$$

The vector normf(f)*, i.e., the component of f in $\oplus_{(j,\mu)\in\Gamma} Fs^{\mu}\delta_j$ with respect to the preceding direct sum decomposition, is called the* normal form *modulo U of the vector f with respect to the chosen term order.*
2. *$M := F[s]^{1\times I}/U = \oplus_{(j,\mu)\in\Gamma} Fs^{\mu}\overline{\delta_j}$, where $\overline{\delta_j} := \delta_j + U$. This means that the elements $s^{\mu}\overline{\delta_j}$, $(j,\mu)\in\Gamma$, form an F-basis of the factor module M.*
3. *The unique elements*

$$g^{(i)} := s^{d(i)}\delta_i - \mathrm{normf}\left(s^{d(i)}\delta_i\right) \in Fs^{d(i)}\delta_i \oplus \bigoplus_{(j,\mu)\in\Gamma} Fs^{\mu}\delta_j,\ i\in I^0, \quad (12.16)$$

are a system of $F[s]$-generators of U with $\deg(g^{(i)}) = (i, d(i))$ and $\mathrm{lt}(g^{(i)}) = s^{d(i)}\delta_i$. Notice that $\Gamma = (I\times\mathbb{N})\setminus\deg(U)$ and the $g^{(i)}$ depend on U and the chosen term order only. This set of generators is called the unique reduced Gröbner basis *of U with respect to the given term order.*
4. *Two submodules U_1 and U_2 of $F[s]^{1\times I}$ coincide if and only if they have the same reduced Gröbner basis (with respect to the same chosen term order). In other words, this finite set of vectors is a complete set of invariants for submodules of $F[s]^{1\times I}$.*

The reduced Gröbner basis of U can be computed from a matrix R with $U = F[s]^{1\times k}R$ with Buchberger's algorithm [7, Theorem 2.11 on p. 216], [9, pp. 140–142], [10, Sect. 2.5 on pp. 121–126].

Proof 1. For every $\alpha\in N = \deg(U)$ we choose any $w_\alpha\in U$ with $\deg(w_\alpha) = \alpha$. For $\alpha = (j,\mu)\in\Gamma = (I\times\mathbb{N})\setminus N$ we define $w_\alpha := v_\alpha = s^{\mu}\delta_j$ of degree α. The family $\underline{w} = (w_\alpha)_{\alpha\in I\times\mathbb{N}^r}$ with $\deg(w_\alpha) = \alpha$ for all $\alpha\in I\times\mathbb{N}$ satisfies the assumptions of Lemma 12.1.4 and therefore it is a well ordered F-basis. Hence

$$\begin{aligned} F[s]^{1\times I} &= \bigoplus_{\alpha\subset I\times\mathbb{N}} Fw_\alpha = \bigoplus_{(j,\mu)\in\Gamma} Fs^{\mu}\delta_j \oplus \bigoplus_{\alpha\in N} Fw_\alpha \\ &\subseteq \bigoplus_{(j,\mu)\in\Gamma} Fs^{\mu}\delta_j + U \subseteq F[s]^{1\times I}. \end{aligned} \quad (12.17)$$

Assume $0\neq f\in\left(\oplus_{(j,\mu)\in\Gamma} Fs^{\mu}\delta_j\right)\cap U$. Since f lies in the first summand, its support is contained in Γ and therefore its degree belongs to Γ. The inclusion $0\neq f\in U$ implies $\deg(f)\in\deg(U) = N$. This contradicts $\Gamma\cap N = \emptyset$, hence,

$$\left(\bigoplus_{(j,\mu)\in\Gamma} Fs^{\mu}\delta_j\right)\cap U = 0 \text{ and } F[s]^{1\times I} = \left(\bigoplus_{(j,\mu)\in\Gamma} Fs^{\mu}\delta_j\right)\oplus U.$$

2. This follows directly from 1.
3. By definition we have the decomposition

$$\begin{aligned} s^{d(i)}\delta_i &= f^{(i)} + g^{(i)} := \operatorname{normf}(s^{d(i)}\delta_i) + \left(s^{d(i)}\delta_i - \operatorname{normf}(s^{d(i)}\delta_i)\right) \\ &\in F[s]^{1\times I} = \bigoplus_{(j,\mu)\in\Gamma} Fs^{\mu}\delta_j \oplus U. \end{aligned}$$

The degree of $f^{(i)}$ is either $-\infty$ or belongs to Γ, whereas

$$\begin{aligned} &\deg(s^{d(i)}\delta_i) = (i, d(i)) \in \deg(U) \\ &\Longrightarrow g^{(i)} = s^{d(i)}\delta_i - f^{(i)} \neq 0 \text{ and } \deg(f^{(i)}) \neq \deg(g^{(i)}) \in \deg(U) \\ &\Longrightarrow (i, d(i)) = \deg\left(f^{(i)} + g^{(i)}\right) = \max\left(\deg(f^{(i)}), \deg(g^{(i)})\right) \in \deg(U) \\ &\underset{\deg(f^{(i)})\notin\deg(U)}{\Longrightarrow} (i, d(i)) = \deg(g^{(i)}),\ \operatorname{lt}(g^{(i)}) = s^{d(i)}\delta_i. \end{aligned}$$

For every $\alpha = (i, \mu) \in \deg(U) = \cup_{i\in I^0}(\{i\} \times (d(i) + \mathbb{N}))$ the inequality $d(i) \leq \mu$ implies that $w_\alpha := s^{\mu-d(i)}g^{(i)} \in U$ is well defined with $\deg(w_\alpha) = \alpha$. Hence we can use these w_α in the proof of item 1 and obtain $U = \oplus_{\alpha\in\deg(U)} F w_\alpha = \sum_{i\in I^0} F[s]g^{(i)}$, i.e., the $g^{(i)}$, $i \in I^0$, generate U.
4. This follows directly from the uniqueness of the reduced Gröbner basis. □

In Theorem 12.1.11, we will show that the $g^{(i)}$, $i \in I^0$, are indeed an $F[s]$-basis of U, but before this, we need to introduce some notation. If $I = I' \uplus I''$ is a disjoint decomposition of a finite index set I and X is an additive monoid, we identify

$$\begin{aligned} X^{I'} &= \{x = (x_i)_{i\in I} \in X^I;\ \forall i \in I'':\ x_i = 0\}, \\ X^{I''} &= \{x = (x_i)_{i\in I} \in X^I;\ \forall i \in I':\ x_i = 0\}, \quad \text{and} \\ X^I &= X^{I'} \times X^{I''} \ni x = \begin{pmatrix} x|_{I'} \\ x|_{I''} \end{pmatrix},\ x|_{I'} := (x_i)_{i\in I'},\ x|_{I''} := (x_i)_{i\in I''}. \end{aligned} \tag{12.18}$$

Let the submodule $U = F[s]^{1\times k}R \subseteq F[s]^{1\times I}$ from (12.12) be generated by the rows of a matrix $R \in F[s]^{k\times I}$ and let $M := F[s]^{1\times I}/U$. Since $I = I^0 \uplus I^{\text{free}}$, we identify $F[s]^{1\times I} = F[s]^{1\times I^0} \times F[s]^{1\times I^{\text{free}}} \ni (\xi, \eta)$, write $R = (P, -Q) \in F[s]^{k\times(I^0\uplus I^{\text{free}})}$, and define $U^0 := F[s]^{1\times k}P \subseteq F[s]^{1\times I^0}$ and $M^0 := F[s]^{1\times I^0}/U^0$.

Theorem 12.1.11 *As F-vector spaces M and M^0 are the direct sums*

$$M = \bigoplus_{i\in I^0} \bigoplus_{\nu=0}^{d(i)-1} Fs^{\nu}\overline{\delta_i} \oplus \bigoplus_{i\in I^{\text{free}}} F[s]\overline{\delta_i} \quad \text{and} \quad M^0 = \bigoplus_{i\in I^0} \bigoplus_{\nu=0}^{d(i)-1} Fs^{\nu}\overline{\delta_i}.$$

The following statements hold:

1. *The elements $\overline{\delta_i} \in M$, $i \in I^{\text{free}}$, are $F[s]$-linearly independent.*

2. *M^0 is a finite dimensional F-vector space of dimension* $\dim_F(M^0) = \sum_{i \in I^0} d(i)$*, and therefore a torsion module.*
3. *U is a free $F[s]$-module of dimension* $\dim_{F[s]}(U) = \sharp(I^0)$*. The reduced Gröbner basis $g^{(i)}$, $i \in I^0$, is linearly independent over $F[s]$, i.e., it is indeed an $F[s]$-basis of U.*

Proof 1, 2.: Since

$$\Gamma = \Gamma^0 \uplus \left\{(i,\mu);\ i \in I^{\text{free}},\ \mu \in \mathbb{N}\right\},\ \Gamma^0 := \left\{(i,\mu);\ i \in I^0,\ 0 \le \mu \le d(i) - 1\right\},$$

the decomposition of M is exactly that of Theorem 12.1.10, item 2. In particular, the $\overline{\delta_i}$, $i \in I^{\text{free}}$, are $F[s]$-linearly independent and the sequence

$$\begin{aligned}
&0 \longrightarrow F[s]^{1\times I^{\text{free}}} \stackrel{\text{inj}}{\longrightarrow} M \stackrel{\text{proj}}{\longrightarrow} M^0 \longrightarrow 0 \text{ with}\\
&\text{inj}(\eta) = \overline{(0,\eta)} = \sum_{i \in I^{\text{free}}} \eta_i \overline{\delta_i} \text{ and proj}\left(\overline{\xi,\eta}\right) = \overline{\xi}, \text{ especially}\\
&\text{proj}(s^\mu \overline{\delta_i}) = \begin{cases} s^\mu \overline{\delta_i} & \text{if } (i,\mu) \in \Gamma^0\\ 0 & \text{if } i \in I^{\text{free}} \end{cases}
\end{aligned} \tag{12.19}$$

from (6.11) is exact. The assertions 1 and 2 follow directly.
3. Recall that the submodule $U \subseteq F[s]^{1\times I}$ is torsionfree and thus free since $F[s]$ is a principal ideal domain. The exact sequence (12.19) and the torsion property of M^0 imply

$$\begin{aligned}
\sharp(I^{\text{free}}) &= \dim_{F[s]}(F[s]^{1\times I^{\text{free}}}) = \text{rank}(M) = \text{rank}\left(F[s]^{1\times I}/U\right)\\
&= \sharp(I) - \dim_{F[s]}(U) \underset{I = I^0 \uplus I^{\text{free}}}{\Longrightarrow} \dim_{F[s]}(U) = \sharp(I) - \sharp(I^{\text{free}}) = \sharp(I^0).
\end{aligned} \tag{12.20}$$

By Theorem 12.1.10,3, the reduced Gröbner basis $g^{(i)}$, $i \in I^0$, of U generates the free module U. Since it has exactly $\dim_{F[s]}(U) = \sharp(I^0)$ elements, it is an $F[s]$-basis of U. □

12.2 The Canonical Proper IO Structure

In this section we apply Theorems 12.1.10 *and* 12.1.11 *to behaviors. Every term order on $I \times \mathbb{N}$ induces a canonical IO structure of $\mathcal{B} := U^\perp$. For the degree-over-position term order the associated transfer matrix is proper, cf. Theorem* 12.2.3, [3, Theorem 3.3.22]. *The notations of Sect.* 12.1 *remain in force.*

For $i \in I^0$ we define coefficients $A_{i,j,\nu} \in F$ for $(j,\nu) \in \Gamma$ by

$$
\begin{aligned}
g^{(i)} &= s^{d(i)}\delta_i - \operatorname{normf}(s^{d(i)}\delta_i) = s^{d(i)}\delta_i - \sum_{(j,\nu)\in\Gamma} A_{i,j,\nu}s^\nu\delta_j \\
&= s^{d(i)}\delta_i - \sum_{j\in I^0}\sum_{\nu=0}^{d(j)-1} A_{i,j,\nu}s^\nu\delta_j - \sum_{j\in I^{\text{free}}}\sum_{\nu=0}^{\infty} A_{i,j,\nu}s^\nu\delta_j,
\end{aligned}
\tag{12.21}
$$

and set $A_{i,j,\nu} := 0$ for $(j,\nu)\notin\Gamma$. From $\deg(g^{(i)}) = (i, d(i))$ we infer that $A_{i,j,\nu} \neq 0$ implies $(j,\nu) < (i, d(i))$. We define the matrix

$$
R^g := \left(g^{(i)}\right)_{i\in I^0} \in F[s]^{I^0\times I} \quad \text{with } R^g_{i-} = g^{(i)} \in F[s]^{1\times I}. \tag{12.22}
$$

This matrix depends on U—or, equivalently, on $\mathcal{B}$—and the chosen term order only, and is called *the Gröbner matrix* of U or $\mathcal{B}$ with respect to this order. We partition R^g as

$$
\begin{aligned}
R^g &= (P^g, -Q^g) \in F[s]^{I^0\times(I^0\uplus I^{\text{free}})} = F[s]^{I^0\times I^0} \times F[s]^{I^0\times I^{\text{free}}}, \quad \text{i.e.,} \\
P^g &= (P^g_{ij})_{i,j\in I^0} := (R^g_{ij})_{i,j\in I^0} \quad \text{and} \\
Q^g &= (Q^g_{ij})_{i\in I^0,\, j\in I^{\text{free}}} := (-R^g_{ij})_{i\in I^0,\, j\in I^{\text{free}}}.
\end{aligned}
\tag{12.23}
$$

By combining this with the last line of (12.21), we obtain

$$
\begin{aligned}
P^g_{ij} &= s^{d(i)}\delta_{ij} - \sum_{\nu=0}^{d(j)-1} A_{i,j,\nu}s^\nu \quad && \text{for } i, j \in I^0, \text{ and} \\
Q^g_{ij} &= \sum_{\nu=0}^{\infty} A_{i,j,\nu}s^\nu && \text{for } i \in I^0,\ j \in I^{\text{free}}.
\end{aligned}
\tag{12.24}
$$

In the same manner, we partition the components of $w \in \mathcal{F}^I$ as

$$
w = \binom{y}{u} \in \mathcal{F}^{I^0\uplus I^{\text{free}}}, \quad \text{where } y := (w_i)_{i\in I^0} \text{ and } u := (w_i)_{i\in I^{\text{free}}}, \tag{12.25}
$$

and conclude that

$$
\begin{aligned}
\mathcal{B} = U^\perp &= \left\{w \in \mathcal{F}^I;\ \forall i \in I^0:\ g^{(i)} \circ w = 0\right\} \\
&= \left\{w \in \mathcal{F}^I;\ R^g \circ w = 0\right\} \\
&= \left\{w = \binom{y}{u} \in \mathcal{F}^{I^0\uplus I^{\text{free}}};\ P^g \circ y = Q^g \circ u\right\} \\
&= \left\{\binom{y}{u} \in \mathcal{F}^{I^0\uplus I^{\text{free}}};\ \forall i \in I^0:\ s^{d(i)} \circ y_i = \sum_{j\in I^0}\sum_{\nu=0}^{d(j)-1} A_{i,j,\nu}s^\nu \circ y_j + Q^g_{i-} \circ u\right\}.
\end{aligned}
\tag{12.26}
$$

This representation of $\mathcal{B}$ is the *Gröbner kernel representation* of $\mathcal{B}$ with respect to the chosen term order.

For $\binom{y}{u} \in \mathcal{B}, i \in I^0$, and $d(i) = 0$, Equation (12.26) furnishes

$$y_i = \sum_{j \in I^0,\ j \neq i} \sum_{\nu=0}^{d(j)-1} A_{i,j,\nu} s^{\nu} \circ y_j + \sum_{j \in I^{\text{free}}} Q^g_{i,j} \circ u_j. \tag{12.27}$$

Such a component y_i of y can be directly computed from the components y_j with $d(j) > 0$ and the u_k, but has no influence on these. Therefore, these components are called *nonessential* [11, p. 142] or *superfluous* with respect to the given term order. We show next that the decompositions $R^g = (P^g, -Q^g)$ from (12.23) and $w = \binom{y}{u} \in \mathcal{F}^{I^0 \uplus I^{\text{free}}}$ from (12.25) are indeed an IO decomposition of $\mathcal{B}$, as our notation already suggests.

Theorem 12.2.1 (The canonical IO structure of a behavior). *The decomposition $I = I^0 \uplus I^{\text{free}}$ is an IO decomposition for $\mathcal{B}$, and it is called the* canonical IO decomposition *with respect to the given term order. The transfer matrix $H^g := (P^g)^{-1} Q^g \in F(s)^{I^0 \times I^{\text{free}}}$ of $\mathcal{B}$ with this IO structure is called* canonical *for the chosen term order.*

Proof According to Theorem and Definition 6.2.2 this follows from the exact sequence (12.19) and the torsion property of M^0. □

In the remainder of this section we prove that for the degree-over-position term order $\leq_2$ from (12.10) the transfer matrix H^g of $\mathcal{B}$ for the canonical IO structure is automatically proper. This is very important from the engineering point of view. The following technical lemma will be used in the proof of Theorem 12.2.3 and for various other results in this section.

Lemma 12.2.2 *Let I be a finite index set with any strict order and let $X \in F(s)^{I \times I}_{\text{spr}}$, where $F(s)_{\text{spr}} = \{h = fg^{-1} \in F(s);\ \deg_s(h) = \deg_s(f) - \deg_s(g) < 0\}$ is the ideal of $F(s)_{\text{pr}}$ of strictly proper rational functions, $\deg_s(f)$ denoting the standard degree of a polynomial $f \in F[s]$. Then $\text{id}_I - X \in \text{Gl}_I(F(s)_{\text{pr}})$.*

Proof From $X \in F(s)^{I \times I}_{\text{spr}}$, we infer that $\text{id}_I - X \in F(s)^{I \times I}_{\text{pr}}$ and $\deg_s(\text{id}_I - X) \in F(s)_{\text{pr}}$ with constant term 1, in other words, $\deg_s(\text{id}_I - X) = 1 + x$ for some $x = \frac{f}{g} \in F(s)_{\text{spr}}$. This means that $\deg_s(g) > \deg_s(f)$ and thus $\deg_s(g + f) = \deg_s(g)$. Consequently,

$$\deg_s(1 + x) = \deg_s\left((g + f)g^{-1}\right) = \deg_s(g + f) - \deg_s(g) = \deg_s(g) - \deg_s(g) = 0.$$

This signifies that $1 + x$ is a unit in $F(s)_{\text{pr}}$, i.e., that also $(1 + x)^{-1} \in F(s)_{\text{pr}}$. By expressing the inverse of $\text{id}_I - X$ via the adjoint matrix, we obtain that

$$(\text{id}_I - X)^{-1} = \underbrace{(1 + x)^{-1}}_{\in F(s)_{\text{pr}}} \underbrace{\text{adj}(\text{id}_I - X)}_{\in F(s)^{I \times I}_{\text{pr}}} \in F(s)^{I \times I}_{\text{pr}}$$

too, and thus $\text{id}_I - X \in \text{Gl}_I(F(s)_{\text{pr}})$. □

Theorem 12.2.3 (Proper canonical transfer matrix, compare [3, Theorem 3.3.22 on p. 90]). *The transfer matrix H^g of the IO behavior $\mathcal{B}$ from Theorem 12.2.1 with respect to the term order $\leq_2$ from* (12.10) *with*

$$(i,\mu) <_2 (j,\nu) :\Longleftrightarrow \mu < \nu \text{ or } \big(\mu = \nu \text{ and } i > j\big)$$

is proper, i.e.,

$$\deg_s(H^g_{i,j}) \leq 0 \text{ for all } i \in I^0 \text{ and } j \in I^{\text{free}}.$$

Proof From (12.21) we recall the representation

$$g^{(i)} = s^{d(i)}\delta_i - \sum_{j\in I^0}\sum_{\nu=0}^{d(j)-1} A_{i,j,\nu}s^\nu\delta_j - \sum_{j\in I^{\text{free}}}\sum_{\nu=0}^{\infty} A_{i,j,\nu}s^\nu\delta_j, \quad i \in I^0. \tag{12.28}$$

Since the leading term of $g^{(i)}$ is $s^{d(i)}\delta_i$, hence $\deg(g^{(i)}) = (i, d(i))$, the monomial $s^\nu\delta_j$ can appear with a nonzero coefficient $A_{i,j,\nu}$ in the second or in the third summand of (12.28) only if $(j,\nu) <_2 (i, d(i))$, and this means that $\nu < d(i)$ or $\big(\nu = d(i)$ and $j > i\big)$. By collecting, in (12.28), all terms $s^{d(i)}\delta_j$ with $j \in I^0$, we obtain

$$\begin{aligned} g^{(i)} = s^{d(i)}\left(\delta_i - \sum_{\substack{j\in I^0,\\ j>i,\ d(i)<d(j)}} A_{i,j,d(i)}\delta_j\right) - \sum_{\substack{j\in I^0,\\ \nu<\min(d(i),d(j))}} A_{i,j,\nu}s^\nu\delta_j \\ - \sum_{j\in I^{\text{free}},\ \nu\leq d(i)} A_{i,j,\nu}s^\nu\delta_j. \end{aligned} \tag{12.29}$$

We use the matrices P^g and Q^g from (12.22)–(12.24), which were defined via $g^{(i)} = (P^g_{i-}, -Q^g_{i-})$, i.e.,

$$\begin{aligned} P^g_{ij} &= s^{d(i)}\delta_{ij} - \sum_{\nu=0}^{d(j)-1} A_{i,j,\nu}s^\nu \quad \text{for } i,j \in I^0, \text{ and} \\ Q^g_{ij} &= \sum_{\nu=0}^{\infty} A_{i,j,\nu}s^\nu \quad \text{for } i \in I^0,\ j \in I^{\text{free}}. \end{aligned}$$

In order to rewrite them according to (12.29), we introduce the matrices

$$\begin{aligned}
&\Delta \in F[s]^{I^0 \times I^0} && \text{by} \quad \Delta := \operatorname{diag}(s^{d(i)};\ i \in I^0),\\
&P^{g,\mathrm{hc}} \in F^{I^0 \times I^0} && \text{by} \quad P^{g,\mathrm{hc}}_{ij} := \begin{cases} 1 & \text{if } i = j,\\ -A_{i,j,d(i)} & \text{if } i < j \text{ and } d(i) < d(j),\\ 0 & \text{otherwise,}\end{cases} \quad \text{and}\\
&P^{g,\mathrm{lc}} \in F[s]^{I^0 \times I^0} && \text{by} \quad P^{g,\mathrm{lc}}_{ij} := \sum_{\nu=0}^{\min(d(i),d(j))-1} A_{i,j,\nu} s^{\nu}.
\end{aligned} \tag{12.30}$$

The indices hc and lc are abbreviations for *highest coefficient* and *lower coefficient* [1, Sect. 6.4]. By definition the matrix $P^{g,\mathrm{hc}}$ is upper triangular with 1 in the main diagonal and thus invertible. We obtain

$$\begin{aligned}
P^g &= \Delta P^{g,\mathrm{hc}} - P^{g,\mathrm{lc}} \quad \text{with } \deg_s(P^{g,\mathrm{lc}}_{ij}) < d(i) \quad \text{and}\\
Q^g_{ij} &= \sum_{\nu=0}^{d(i)} A_{i,j,\nu} s^{\nu} \qquad \text{with } \deg_s(Q^g_{ij}) \le d(i).
\end{aligned} \tag{12.31}$$

The matrix P^g with invertible $P^{g,\mathrm{hc}}$ is called *row-reduced*, cf. [1, Sect. 6.3.2]. Using the inverse $\Delta^{-1} = \operatorname{diag}(s^{-d(i)};\ i \in I^0)$, we get

$$\begin{aligned}
P^g &= \Delta P^{g,\mathrm{hc}} - P^{g,\mathrm{lc}} = \Delta P^{g,\mathrm{hc}} \left(\mathrm{id}_{I^0} - (P^{g,\mathrm{hc}})^{-1} \Delta^{-1} P^{g,\mathrm{lc}}\right) \quad \text{and}\\
H^g &= (P^g)^{-1} Q^g = \left(\mathrm{id}_{I^0} - (P^{g,\mathrm{hc}})^{-1} \Delta^{-1} P^{g,\mathrm{lc}}\right)^{-1} (P^{g,\mathrm{hc}})^{-1} (\Delta^{-1} Q^g).
\end{aligned} \tag{12.32}$$

To show that H^g is proper it suffices to show that the three factors of H^g in (12.32) are proper. For the constant matrix $(P^{g,\mathrm{hc}})^{-1}$ this is obvious. For $\Delta^{-1} Q^g$ this holds since

$$\deg_s\left((\Delta^{-1} Q^g)_{ij}\right) = \deg_s\left(s^{-d(i)} Q^g_{ij}\right) = \deg_s(Q^g_{ij}) - d(i) \overset{(12.31)}{\le} d(i) - d(i) = 0.$$

Consider the matrix

$$\begin{aligned}
X &:= (P^{g,\mathrm{hc}})^{-1} \Delta^{-1} P^{g,\mathrm{lc}} \in F[s]^{I^0 \times I^0}, \quad \text{i.e.,}\\
X_{ij} &= \sum_{k \in I^0} ((P^{g,\mathrm{hc}})^{-1})_{ik} s^{-d(k)} P^{g,\mathrm{lc}}_{kj}.
\end{aligned}$$

The degrees of its entries satisfy

$$\deg(X_{ij}) \le \max_{k \in I^0} \left(\deg\left(s^{-d(k)} P^{g,\mathrm{lc}}_{kj}\right)\right) = \max_{k \in I^0} \left(\underbrace{\deg(P^{g,\mathrm{lc}}_{kj})}_{\overset{(12.31)}{<} d(k)} - d(k)\right) < 0,$$

thus $X \in F(s)_{\mathrm{spr}}^{I^0\times I^0}$. From Lemma 12.2.2, we infer that $\mathrm{id}_{I^0}-X \in \mathrm{Gl}_{I^0}(F(s)_{\mathrm{pr}})$, i.e., $(\mathrm{id}_{I^0}-X)^{-1} \in F(s)_{\mathrm{pr}}^{I^0\times I^0}$. Consequently, $H^g \in F(s)_{\mathrm{pr}}^{I^{\mathrm{free}}\times I^0}$. □

12.3 The Canonical Observability Realization of an IO Behavior

We assume the IO behavior and associated modules from (12.1), i.e.,

$$\begin{aligned}
&\mathcal{B} := \left\{\binom{y}{u} \in \mathcal{F}^{p+m};\ P\circ y = Q\circ u\right\},\ \mathcal{B}^0 := \left\{y\in\mathcal{F}^p;\ P\circ y = 0\right\}\\
&(P,-Q)\in F[s]^{p\times(p+m)},\ \mathrm{rank}(P)=p,\ H:=P^{-1}Q,\\
&U^0 = F[s]^{1\times p}P,\ U = F[s]^{1\times p}(P,-Q) = U^0(\mathrm{id}_p,-H),\\
&M^0 := F[s]^{1\times p}/U^0,\ M := F[s]^{1\times(p+m)}/U.
\end{aligned} \tag{12.33}$$

In contrast to Sect. 12.2 the IO structure of $\mathcal{B}$ is assumed here and not constructed. One can, however, use the canonical IO structures from Theorems 12.2.1 or 12.2.3 and thus ensure by the latter theorem that the transfer matrix is proper.
Let $I := \{1,\dots,p\}$. In contrast to the quoted theorems we choose a term order $\le$ on $I\times\mathbb{N}$ and consider the reduced Gröbner basis of the free submodule $U^0\subseteq F[s]^{1\times p}$ of dimension $\dim_{F[s]}(U^0)=p$ with respect to this term order. Since M^0 is a torsion module, the behavior $\mathcal{B}^0$ is autonomous and has no free components, hence $\{1,\dots,p\}=I=I^0$ in (12.14). According to (12.21)–(12.24) the degree set $\deg(U^0)$ and the reduced Gröbner basis of U^0 have the form

$$\begin{aligned}
&\deg(U^0) = \biguplus_{1\le i\le p}\{i\}\times(d(i)+\mathbb{N}),\ 0\le d(i)\in\mathbb{N},\\
&\Longrightarrow (I\times\mathbb{N})\setminus\deg(U^0) = \{(i,\mu);\ 1\le i\le p,\ 0\le\mu\le d(i)-1\},\\
&g^{(i)} = s^{d(i)}\delta_i - \sum_{j=1}^{p}\sum_{\nu=0}^{d(j)-1} A_{i,j,\nu}s^\nu\delta_j,\quad i=1,\dots,p,\ A_{i,j,\nu}\in F,\ \text{where}\\
&A_{i,j,\nu}=0 \text{ if } (j,\nu)\not< (i,d(i)).
\end{aligned} \tag{12.34}$$

The associated Gröbner matrix is

$$\begin{aligned}
&P^g := \left(g^{(i)}\right)_i \in F[s]^{p\times p} \text{ with } U^0 = F[s]^{1\times p}P = \bigoplus_{i=1}^{p} F[s]g^{(i)} = F[s]^{1\times p}P^g \text{ and}\\
&P^g_{i,j} = s^{d(i)}\delta_{i,j} - \sum_{\nu=0}^{d(j)-1} A_{i,j,\nu}s^\nu \quad \text{for } i,j\in\{1,\dots,p\}.
\end{aligned} \tag{12.35}$$

The $d(i)$, $i=1,\dots,p$, are called the *observability indices* of $\mathcal{B}$. They depend on $\mathcal{B}$, its IO decomposition and the chosen term order only. For degree-over-position term orders they are unique up to a possible permutation, cf. Theorem 12.4.2.

Remark 12.3.1 For the position-over-degree term order

$$(i,\mu) > (j,\nu) :\Longleftrightarrow i < j \text{ or } (i = j \text{ and } \mu > \nu)$$

the Gröbner matrix P^g of P is exactly its *(row) Hermite form*, cf. [1, Sect. 6.7.1]. ◊

Equations (12.34) and (12.35) imply

$$M^0 = \bigoplus_{j=1}^{p}\bigoplus_{\nu=0}^{d(j)-1} Fs^{\nu}\overline{\delta_j},\ \forall i = 1,\dots,p:\ s^{d(i)}\overline{\delta_i} = \sum_{j=1}^{p}\sum_{\nu=0}^{d(j)-1} A_{i,j,\nu}s^{\nu}\overline{\delta_j},$$

$$\mathcal{B}^0 = (U^0)^{\perp} = \left\{y \in \mathcal{F}^p;\ s^{d(i)} \circ y_i = \sum_{j=1}^{p}\sum_{\nu=0}^{d(j)-1} A_{i,j,\nu}s^{\nu} \circ y_j \text{ for } i = 1,\dots,p\right\},$$

$$A_{i,j,\nu} = 0 \text{ if } (j,\nu) \not< (i,d(i)). \tag{12.36}$$

In the next theorem we establish the relation between the elements $s^{\nu}\overline{\delta_j}$ of M^0 and the row vectors $C_{j,-}A^{\nu}$ of the observability matrix of an observable state space representation of the behavior $\mathcal{B}$, cf. [1, Sect. 6.4.6].

Theorem 12.3.2 1. *Assume that the equations*

$$\begin{aligned} &s \circ x = Ax + Bu,\ y = Cx + D \circ u \text{ with}\\ &A \in F^{n\times n},\ B \in F^{n\times m},\ C \in F^{p\times n},\ D \in F[s]^{p\times m} \end{aligned} \tag{12.37}$$

are an observable state space representation for $\mathcal{B}$. Consider $F^{1\times n}$ as $F[s]$-module via $s \circ_A \xi = \xi A$. Then there is the $F[s]$-isomorphism

$$\begin{aligned} &F^{1\times n} \underset{F[s]}{\cong} M^0,\ C_{i,-} \longleftrightarrow \overline{\delta_i},\ C_{i,-}A^{\mu} \longleftrightarrow s^{\mu}\overline{\delta_i}\\ &\Longrightarrow F^{1\times n} = \bigoplus_{j=1}^{p}\bigoplus_{\nu=0}^{d(j)-1} FC_{j,-}A^{\nu}\\ &\forall i = 1,\dots,p:\ C_{i,-}A^{d(i)} = \sum_{j=1}^{p}\sum_{\nu=0}^{d(j)-1} A_{i,j,\nu}C_{j,-}A^{\nu}. \end{aligned} \tag{12.38}$$

In particular, the rows $C_{i,-}A^{\mu}$, $i = 1,\dots,p$, $0 \le \mu \le d(i)-1$, of the observability matrix $\mathfrak{O} = \begin{pmatrix} C\\ \vdots\\ CA^{n-1}\end{pmatrix}$ are an F-basis of $F^{1\times n}$, and hence the matrix $\left(C_{i,-}A^{\mu}\right)_{1\le i\le p,\ 0\le\mu\le d(i)-1}$ is invertible or contained in $\mathrm{Gl}_n(F)$.

2. *In 1 the basis rows $C_{i,-}A^{\mu}$, $i = 1,\dots,p$, $0 \le \mu \le d(i)-1$, consist precisely of those rows of the observability matrix that are F-linearly independent of the rows $C_{j,-}A^{\nu}$, $(j,\nu) < (i,\mu)$, and can be determined as such.*
3. *If also the equations*

$$s \circ x = \widehat{A}x + \widehat{B}u,\ y = \widehat{C}x + \widehat{D} \circ u \tag{12.39}$$

are an observable state representation for $\mathcal{B}$ then there is a unique invertible matrix T with

$$\begin{aligned}
&\widehat{A} = TAT^{-1},\ \widehat{B} = TB,\ \widehat{C} = CT^{-1},\ \widehat{D} = D \Longrightarrow \widehat{C}\widehat{A}^{\mu} = CA^{\mu}T^{-1}\\
&\Longrightarrow \left(\widehat{C}_{i,-}\widehat{A}^{\mu}\right)_{1\le i\le p,\ 0\le\mu\le d(i)-1} = \left(C_{i,-}A^{\mu}\right)_{1\le i\le p,\ 0\le\mu\le d(i)-1} T^{-1}\\
&T = \left(\left(\widehat{C}_{i,-}\widehat{A}^{\mu}\right)_{1\le i\le p,\ 0\le\mu\le d(i)-1}\right)^{-1}\left(C_{i,-}A^{\mu}\right)_{1\le i\le p,\ 0\le\mu\le d(i)-1}.
\end{aligned} \tag{12.40}$$

4. *Items 1 and 2 are applicable, if not $\mathcal{B}$, but the observable equations* (12.37) *are given and $\mathcal{B}$ is* constructed *as*

$$\begin{aligned}
\mathcal{B} &:= \left(\begin{smallmatrix} C & D \\ 0 & \mathrm{id}_m \end{smallmatrix}\right) \circ \left\{\left(\begin{smallmatrix} x \\ u \end{smallmatrix}\right) \in \mathcal{F}^{n+m};\ s \circ x = Ax + Bu\right\}\\
&= \left\{\left(\begin{smallmatrix} y \\ u \end{smallmatrix}\right) \in \mathcal{F}^{p+m};\ P \circ y = Q \circ u\right\},
\end{aligned} \tag{12.41}$$

where according to Theorem 3.2.22, $(-Y, P) \in F[s]^{p\times(n+p)}$ is a universal left annihilator of $\left(\begin{smallmatrix} s\,\mathrm{id}_n - A \\ C \end{smallmatrix}\right)$ and $Q = YB + PD$. Equation (12.37) *are observable if and only $\left(\begin{smallmatrix} s\,\mathrm{id}_n - A \\ C \end{smallmatrix}\right)$ has a left inverse $(X, Z) \in F[s]^{n\times(n+p)}$. Both $(-Y, P)$ and (X, Z) can be easily computed by means of the Smith form of $\left(\begin{smallmatrix} s\,\mathrm{id}_n - A \\ C \end{smallmatrix}\right)$. Notice that P^g and the $d(i)$ depend on (A, C) and the chosen term order only. The $d(i)$, $i = 1, \dots, p$, are then called the* observability indices *of the pair (A, C) for the chosen term order, cf. [1, pp. 413 and 431].*

5. *In item 4 assume the position-over-degree resp. the degree-over-position term orders*

$$\begin{aligned}
&(j, \nu) < (i, \mu) :\Longleftrightarrow j < i \text{ or } j = i \text{ and } \nu < \mu \text{ resp.}\\
&(j, \nu) < (i, \mu) :\Longleftrightarrow \nu < \mu \text{ or } \nu = \mu \text{ and } j < i.
\end{aligned}$$

Then the method from 4 and 2 to compute the basis $C_{i,-}A^{\mu}$, $i = 1, \dots, p$, $0 \le \mu \le d(i) - 1$, of the row space of the observability matrix is called Scheme I *resp.* Scheme II *in [1, p. 426-427].*

Proof 1. From Lemma 4.1.6 recall the $F[s]$-isomorphism

$$F^{1\times n} \underset{F[s]}{\cong} F[s]^{1\times n}/F[s]^{1\times n}(s\,\mathrm{id}_n - A),\ \zeta \mapsto \overline{\zeta} = \zeta + F[s]^{1\times n}(s\,\mathrm{id}_n - A). \tag{12.42}$$

Equation (12.37) imply the dual $F[s]$-isomorphisms

$$\begin{aligned}
&\left\{x \in \mathcal{F}^n;\ (s\,\mathrm{id}_n - A) \circ x = 0\right\} \overset{\cong}{\longrightarrow} \mathcal{B}^0,\ x \longmapsto Cx,\\
&F[s]^{1\times n}/F[s]^{1\times n}(s\,\mathrm{id}_n - A) \overset{\cong}{\longleftarrow} M^0,\ \overline{\xi C} \longleftarrow\!\shortmid \overline{\xi},\ \overline{C_{i,-}} \longleftarrow\!\shortmid \overline{\delta_i}.
\end{aligned} \tag{12.43}$$

Equations (12.42) and (12.43) imply the $F[s]$-isomorphism

$$F^{1\times n} \underset{F[s]}{\cong} M^0,\ C_{i,-} \longleftrightarrow \overline{\delta_i},\ C_{i,-}A^{\mu} \longleftrightarrow s^{\mu}\overline{\delta_i},\ i=1,\cdots,p,\ \mu\in\mathbb{N}. \tag{12.44}$$

This isomorphism and (12.36) imply 1.
2. We study the linear dependencies among the rows $C_{i,-}A^{\mu}$ by means of the isomorphic elements $s^{\mu}\overline{\delta_i}$ of M^0. The linear dependence relation $s^{d(i)}\overline{\delta_i} = \sum_{(j,\nu)} A_{i,j,\nu}s^{\nu}\overline{\delta_j}$, where the summation indices (j,ν) satisfy $1\le j\le p, 0\le \nu\le d(j)-1$ and $(j,\nu) < (i,d(i))$, shows that $s^{d(i)}\overline{\delta_i}$ is F-linear dependent on its preceding elements. For $\mu \ge d(i)$ this implies

$$s^{\mu}\overline{\delta_i} = \sum_{(j,\nu)<(i,d(i))} A_{i,j,\nu}s^{\mu-d(i)+\nu)}\overline{\delta_i} \text{ with}$$
$$(j,\mu-d(i)+\nu) = (j,\nu)+(\mu-d(i)) < (i,d(i))+(\mu-d(i)) = (i,\mu).$$

Hence again $s^{\mu}\overline{\delta_i}$ is F-linear dependent on its preceding elements. In contrast, it is obvious that the basis elements $s^{\mu}\overline{\delta_i},\ 1\le i\le p,\ 0\le\mu\le d(i)-1$, are F-linearly independent of their predecessors.
3. The existence of T was shown in Theorem 6.3.8. The rest follows directly from item 1.
4, 5. Obvious. □

We are now going to construct an observable state space representation of $\mathcal{B}$ by means of P^g. Since U^0 is the row module both of P and of P^g, there is an invertible matrix $X\in \mathrm{Gl}_p(F[s])$ such that $P^g = XP$, i.e., $X = P^gP^{-1}$. We define

$$Q^g := P^gH = XPH = XQ \in F[s]^{p\times m} \tag{12.45}$$

and get

$$\begin{aligned}(P^g,-Q^g) &= X(P,-Q),\ H = P^{-1}Q = (P^g)^{-1}Q^g \text{ and}\\ U &= F[s]^{1\times p}(P,-Q) = F[s]^{1\times p}(P^g,-Q^g),\ X\in \mathrm{Gl}_p(F[s]).\end{aligned}$$

We thus obtain the *canonical Gröbner form* of the IO behavior $\mathcal{B}$ with respect to the chosen term order on $\{1,\dots,p\}\times\mathbb{N}$, namely,

$$\begin{aligned}\mathcal{B} &= \left\{\left(\begin{smallmatrix}y\\u\end{smallmatrix}\right)\in\mathcal{F}^{p+m};\ P^g\circ y = Q^g\circ u\right\}\\ &= \left\{\left(\begin{smallmatrix}y\\u\end{smallmatrix}\right);\ \forall i=1,\dots,p:\ s^{d(i)}\circ y_i = \sum_{j=1}^{p}\sum_{\nu=0}^{d(j)-1} A_{i,j,\nu}s^{\nu}\circ y_j + Q^g_{i-}\circ u\right\}.\end{aligned} \tag{12.46}$$

This representation is similar to that of (12.26), but here the IO structure is given, whereas in (12.26) it was constructed. The defining equations of $\mathcal{B}$ from (12.46) suggest to define

$$\Gamma^{\rm ob} := \{(j,\nu);\ 1\le j\le p,\ 0\le \nu\le d(j)-1\},\ \forall (i,\mu)\in\Gamma^{\rm ob}:\ x_{i,\mu} := s^{\mu}\circ y_i$$
$$\Longrightarrow s\circ x_{i,\mu} = \begin{cases} x_{i,\mu+1} = \sum_{(j,\nu)\in\Gamma^{\rm ob}} \delta_{i,j}\delta_{\mu+1,\nu}x_{j,\nu} & \text{if } \mu < d(i)-1 \\ \sum_{(j,\nu)\in\Gamma^{\rm ob}} A_{i,j,\nu}x_{j,\nu} + \sum_{j=1}^{m} Q^g_{ij}\circ u_j & \text{if } \mu = d(i)-1 \end{cases},$$
$$y_i = \begin{cases} x_{i,0} = \sum_{(j,\nu)\in\Gamma^{\rm ob}} \delta_{i,j}\delta_{0,\nu}x_{j,\nu} & \text{if } d(i)>0 \\ \sum_{(j,\nu)\in\Gamma^{\rm ob}} A_{i,j,\nu}x_{j,\nu} + \sum_{j=1}^{m} Q^g_{ij}\circ u_j & \text{if } d(i)=0. \end{cases} \tag{12.47}$$

Lemma and Definition 12.3.3 (Canonical quasi-state realization of an IO behavior). *Consider the IO behavior from (12.33), the Gröbner matrix P^g from (12.35) and $Q^g := P^g H$ and the data from (12.47). Define*

$$\begin{aligned}
&A^{\rm ob}\in F^{\Gamma^{\rm ob}\times\Gamma^{\rm ob}} && \textit{with}\quad A^{\rm ob}_{(i,\mu),(j,\nu)} = \begin{cases}\delta_{i,j}\delta_{\mu+1,\nu} & \textit{if } \mu<d(i)-1,\\ A_{i,j,\nu} & \textit{if } \mu = d(i)-1,\end{cases}\\
&C^{\rm ob}\in F^{p\times\Gamma^{\rm ob}} && \textit{with}\quad C^{\rm ob}_{i,(j,\nu)} = \begin{cases}\delta_{i,j}\delta_{0,\nu} & \textit{if } d(i)>0,\\ A_{i,j,\nu} & \textit{if } d(i)=0,\end{cases}\\
&Q^{\rm qs}\in F[s]^{\Gamma^{\rm ob}\times m} && \textit{with}\quad Q^{\rm qs}_{(i,\mu),j} = \begin{cases}Q^g_{i,j} & \textit{if } \mu = d(i)-1,\\ 0 & \textit{if } \mu<d(i)-1\end{cases} \quad \textit{and}\\
&Q^{g,0}\in F[s]^{p\times m} && \textit{with}\quad Q^{g,0}_{i,j} = \begin{cases}0 & \textit{if } d(i)>0,\\ Q^g_{i,j} & \textit{if } d(i)=0.\end{cases}
\end{aligned} \tag{12.48}$$

The upper index ob *refers to observability, see below. Let $n := \sharp(\Gamma^{\rm ob}) = \sum_{j=1}^{p} d(j)$. Notice that $A^{\rm ob}$, $C^{\rm ob}$, $Q^{\rm qs}$, and $Q^{g,0}$ are sparse in the sense that most of their entries are zero, and that an index i with $d(i)=0$ complicates $C^{\rm ob}$ and gives possibly rise to a nonzero $Q^{g,0}$. The possibility of $d(i)=0$ is recognized in [11, p. 142], but ignored in [1, p. 414] where [1, (32b)] is false if some $l_i = d(i) = 0$.*

With these matrices we define the IO behavior $\mathcal{B}^{\rm qs}$ with the quasi-state vector *$x = (x_{i,\mu})_{(i,\mu)\in\Gamma^{\rm ob}}$ and the equations*

$$s\circ x_{i,\mu} = \begin{cases} x_{i,\mu+1} & \textit{if } 0\le\mu<d(i)-1,\\ \sum_{(j,\nu)\in\Gamma^{\rm ob}} A_{i,j,\nu}x_{j,\nu} + \sum_{j=1}^{m} Q^g_{ij}\circ u_j & \textit{if } \mu = d(i)-1,\end{cases} \quad \textit{or by}$$
$$\mathcal{B}^{\rm qs} := \left\{ \binom{x}{u} \in \mathcal{F}^{\Gamma^{\rm ob}+m};\ (s\,\mathrm{id}_{\Gamma^{\rm ob}} - A^{\rm ob})\circ x = Q^{\rm qs}\circ u \right\}.$$

Its transfer matrix is

$$H^{\rm qs} := (s\,\mathrm{id}_{\Gamma^{\rm ob}} - A^{\rm ob})^{-1} Q^{\rm qs} \in F(s)^{\Gamma^{\rm ob}\times m}. \tag{12.49}$$

With these data the map

$$\mathcal{B}\longrightarrow\mathcal{B}^{\rm qs},\quad \binom{y}{u}\longmapsto\binom{x}{u},\quad \textit{with}\ \ x_{i,\mu} := s^{\mu}\circ y_i \textit{ for } (i,\mu)\in\Gamma^{\rm ob}$$

is a behavior isomorphism, and its inverse is

$$\begin{pmatrix} C^{\mathrm{ob}} & Q^{g,0} \\ 0 & \mathrm{id}_m \end{pmatrix} : \mathcal{B}^{\mathrm{qs}} \xrightarrow{\cong} \mathcal{B}, \quad \binom{x}{u} \longmapsto \binom{y}{u} = \binom{C^{\mathrm{ob}}x + Q^{g,0} \circ u}{u},$$

where

$$y_i = \begin{cases} x_{i,0} & \text{if } d(i) > 0, \\ \sum_{(j,\nu)\in\Gamma^{\mathrm{ob}}} A_{i,j,\nu} x_{j,\nu} + \sum_{j=1}^{m} Q_{ij}^{g} \circ u_j & \text{if } d(i) = 0. \end{cases}$$

Thus the transfer matrix H of $\mathcal{B}$ is

$$H = P^{-1}Q = Q^{g,0} + C^{\mathrm{ob}} H^{\mathrm{qs}}, \text{ hence}$$

$$H_{ij} = \begin{cases} H^{\mathrm{qs}}_{(i,0)j} & \text{if } d(i) > 0, \\ Q^{g}_{ij} + \sum_{(k,\nu)\in\Gamma^{\mathrm{ob}}} A_{i,k,\nu} H^{\mathrm{qs}}_{(k,\nu)j} & \text{if } d(i) = 0. \end{cases} \tag{12.50}$$

This representation of the IO behavior $\mathcal{B}$ is called its canonical quasi-state realization *with respect to the chosen term order on $\{1, \dots, p\} \times \mathbb{N}$. Since*

$$n = \sharp(\Gamma^{\mathrm{ob}}) = \sum_{i=1}^{p} d(i) = \dim_F\left(\mathcal{B}^{\mathrm{qs},0}\right) = \dim_F(\mathcal{B}^0) = \deg_s(\det(P)), \tag{12.51}$$

it is obvious that, in general, many $d(i)$ will be zero if p is large and $\deg_s(\det(P))$ *is small, as is often the case for electrical networks.*

Proof The proof is contained in Eq. (12.47). □

We are now going to derive the actual state space realization from the quasi-state realization. For this purpose we use the decomposition

$$F(s) = F[s] \oplus F(s)_{\mathrm{spr}} \ni h = h_{\mathrm{pol}} + h_{\mathrm{spr}}$$

of a rational function h into its polynomial part h_{pol} and strictly proper part h_{spr} from Lemma 5.5.5, and also the entrywise extension of this to matrices. We use any strict order of Γ^{ob}, for instance, the term order, in order that the next determinant is uniquely defined, the standard degree $\deg_s$ and define

$$S := s\,\mathrm{id}_{\Gamma^{\mathrm{ob}}} - A^{\mathrm{ob}} \in F[s]^{\Gamma^{\mathrm{ob}} \times \Gamma^{\mathrm{ob}}} \text{ and } f := \det(S) \Longrightarrow \deg_s(f) = \sharp(\Gamma^{\mathrm{ob}}) = n.$$

Let $S_{\mathrm{adj}} \in F[s]^{\Gamma^{\mathrm{ob}} \times \Gamma^{\mathrm{ob}}}$ denote the adjoint matrix of S. Then

$$f\,\mathrm{id}_{\Gamma^{\mathrm{ob}}} = S S_{\mathrm{adj}} = S_{\mathrm{adj}} S, \quad S^{-1} = f^{-1} S_{\mathrm{adj}}.$$

The entries of S have degree ≤ 1, and hence those of S_{adj}, i.e., the $(n-1) \times (n-1)$-minors of S, have degree at most $n-1$. This implies

$\deg_s(S_{\mathrm{adj}}) \le n-1,\ \deg_s(S^{-1}) = \deg_s(S_{\mathrm{adj}}) - \deg_s(f) \le -1 \Longrightarrow S^{-1} \in F(s)_{\mathrm{spr}}^{\Gamma^{\mathrm{ob}}\times\Gamma^{\mathrm{ob}}}.$

Using these matrices we obtain

$$\begin{aligned} H^{\mathrm{qs}} &= S^{-1}Q^{\mathrm{qs}} = f^{-1}S_{\mathrm{adj}}Q^{\mathrm{qs}} = H^{\mathrm{qs}}_{\mathrm{pol}} + H^{\mathrm{qs}}_{\mathrm{spr}} \in F(s)^{\Gamma^{\mathrm{ob}}\times m},\\ Q^{\mathrm{qs}} &= SH^{\mathrm{qs}} = SH^{\mathrm{qs}}_{\mathrm{pol}} + B^{\mathrm{ob}} \text{ with } B^{\mathrm{ob}} := SH^{\mathrm{qs}}_{\mathrm{spr}} = Q^{\mathrm{qs}} - SH^{\mathrm{qs}}_{\mathrm{pol}}\\ S_{\mathrm{adj}}Q^{\mathrm{qs}} &= fH^{\mathrm{qs}} = fH^{\mathrm{qs}}_{\mathrm{pol}} + R \text{ with } R := S_{\mathrm{adj}}B^{\mathrm{ob}} = fH^{\mathrm{qs}}_{\mathrm{spr}}. \end{aligned} \tag{12.52}$$

Lemma 12.3.4 *For the matrices from* (12.52) *the following statements hold:*

1. *The matrix* B^{ob} *is constant, i.e., it is contained in* $F^{\Gamma^{\mathrm{ob}}\times m}$.
2. *The matrices* fH^{qs}, $fH^{\mathrm{qs}}_{\mathrm{pol}}$, *and* R *are polynomial and* $\deg_s(R) < \deg_s(f) = n$. *By the third line of* (12.52) *this signifies that* $H^{\mathrm{qs}}_{\mathrm{pol}}$ *and* R *are obtained as the quotient and the remainder, respectively, of the Euclidean division of* $fH^{\mathrm{qs}} = S_{\mathrm{adj}}Q^{\mathrm{qs}}$ *by* f.

Notice that one needs $H^{\mathrm{qs}}_{\mathrm{pol}} \in F[s]^{\Gamma^{\mathrm{ob}}\times m}$ *only for the computation of*

$$B^{\mathrm{ob}} = Q^{\mathrm{qs}} - SH^{\mathrm{qs}}_{\mathrm{pol}} \quad \textit{and} \quad R = f\left(H^{\mathrm{qs}} - H^{\mathrm{qs}}_{\mathrm{pol}}\right).$$

Proof 1. Since Q^{qs}, S, and $H^{\mathrm{qs}}_{\mathrm{pol}}$ are all polynomial matrices, the second line of (12.52) implies that $B^{\mathrm{ob}} = Q^{\mathrm{qs}} - SH^{\mathrm{qs}}_{\mathrm{pol}}$ is polynomial too. From the second line of (12.52) we infer that

$$\deg_s(B^{\mathrm{ob}}) = \deg_s(SH^{\mathrm{qs}}_{\mathrm{spr}}) \le \deg_s(S) + \deg_s(H^{\mathrm{qs}}_{\mathrm{spr}}) \le 1 - 1 = 0.$$

The matrix B^{ob} is thus polynomial of degree ≤ 0, i.e., it is constant.
2. Since S_{adj} and Q^{qs} are polynomial, also $fH^{\mathrm{qs}} = S_{\mathrm{adj}}Q^{\mathrm{qs}}$ has this property. Since $H^{\mathrm{qs}}_{\mathrm{pol}}$ is polynomial by definition, also $R = fH^{\mathrm{qs}} - fH^{\mathrm{qs}}_{\mathrm{pol}}$ is polynomial. Moreover the last line of (12.52) implies

$$\deg_s(R) = \deg_s(fH^{\mathrm{qs}}_{\mathrm{spr}}) = \deg_s(f) + \deg_s(H^{\mathrm{qs}}_{\mathrm{spr}}) < \deg_s(f) = n.$$

□

We define the matrix

$$D^{\mathrm{ob}} := C^{\mathrm{ob}}H^{\mathrm{qs}}_{\mathrm{pol}} + Q^{g,0} \in F[s]^{p\times m}. \tag{12.53}$$

By (12.48) this matrix satisfies

$$D^{\mathrm{ob}}_{i,j} = \begin{cases} \left(H^{\mathrm{qs}}_{\mathrm{pol}}\right)_{(i,0),j} & \text{if } d(i) > 0,\\ \sum_{(k,\nu)\in\Gamma^{\mathrm{ob}}} A_{i,k,\nu}(H^{\mathrm{qs}}_{\mathrm{pol}})_{(k,\nu),j} + Q^{g}_{i,j} & \text{if } d(i) = 0, \end{cases}$$

hence

$$(D^{\mathrm{ob}} \circ u)_i = \left(H^{\mathrm{qs}}_{\mathrm{pol}} \circ u\right)_{(i,0)} \quad \text{if } u \in \mathcal{F}^m \text{ and } d(i) > 0.$$

This matrix is the last one needed for the state space equations

$$s \circ x = A^{\mathrm{ob}}x + B^{\mathrm{ob}}u, \quad y = C^{\mathrm{ob}}x + D^{\mathrm{ob}} \circ u. \tag{12.54}$$

Theorem and Definion 12.3.5 (The canonical observability realization of an IO behavior, compare [1, p. 422]). *Assume the IO behavior $\mathcal{B}$ with transfer matrix H from* (12.33) *and a term order on* $\{1, \ldots, p\} \times \mathbb{N}$. *Let $g^{(i)}$, $i = 1, \ldots, p$, be the reduced Gröbner basis of U^0, let $P^g := \left(g^{(i)}\right)_i \in F[s]^{p\times p}$ be the associated matrix from* (12.35) *$Q^g := P^g H$ as in* (12.45). *Finally we need the derived matrices A^{ob}, C^{ob}, Q^{qs}, and $Q^{g,0}$ from* (12.48), *H^{qs}, $H^{\mathrm{qs}}_{\mathrm{pol}}$, and B^{ob} from* (12.52), *and D^{ob} from* (12.53).

The following assertions hold for the state behavior

$$\mathcal{B}^{\mathrm{ob}} := \left\{ \begin{pmatrix} x \\ u \end{pmatrix} \in \mathcal{F}^{\Gamma^{\mathrm{ob}}+m};\ s \circ x = A^{\mathrm{ob}}x + B^{\mathrm{ob}}u \right\}: \tag{12.55}$$

1. *The state space equations* (12.54) *are an observable realization of the given IO behavior $\mathcal{B}$, i.e., the map*

$$\begin{pmatrix} C^{\mathrm{ob}} & D^{\mathrm{ob}} \\ 0 & \mathrm{id}_m \end{pmatrix} \circ :\ \mathcal{B}^{\mathrm{ob}} \longrightarrow \mathcal{B},\quad \begin{pmatrix} x \\ u \end{pmatrix} \longmapsto \begin{pmatrix} y \\ u \end{pmatrix} = \begin{pmatrix} C^{\mathrm{ob}}x + D^{\mathrm{ob}} \circ u \\ u \end{pmatrix}, \tag{12.56}$$

is a behavior isomorphism, compare Theorem and Definition 4.1.9.

Recall that the matrices A^{ob} and C^{ob} are sparse while B^{ob} and D^{ob} are not sparse in general, in detail

$$\begin{aligned} A^{\mathrm{ob}}_{(i,\mu),(j,\nu)} &= \delta_{i,j}\delta_{\mu+1,\nu} \text{ and } (A^{\mathrm{ob}}x)_{(i,\mu)} = x_{i,\mu+1} \text{ if } 0 \le \mu < d(i) - 1,\\ C^{\mathrm{ob}}_{i,(j,\nu)} &= \delta_{i,j}\delta_{0,\nu} \text{ and } (C^{\mathrm{ob}}x)_i = x_{i,0} \text{ if } d(i) > 0,\\ y_i &= x_{i,0} + (D^{\mathrm{ob}} \circ u)_i \text{ if } d(i) > 0. \end{aligned} \tag{12.57}$$

The dimension of this state space realization is

$$n := \sharp(\Gamma^{\mathrm{ob}}) = \sum_{i=1}^{p} d(i). \tag{12.58}$$

This realization depends on $\mathcal{B}$, its IO structure and the chosen term order only and is called the canonical observability realization *of $\mathcal{B}$ with respect to that term order, and the numbers $d^{\mathrm{ob}}(i) := d(i)$, $i = 1, \ldots, p$, are the associated* observability indices, *cf. [1, p. 413].*

2. *The inverse of the isomorphism* (12.56) *is given by*

$$\binom{y}{u} \longmapsto \binom{x}{u}, \quad \textit{where } x_{i,\mu} = s^\mu \circ y_i - \left(H^{qs}_{pol}\right)_{(i,\mu)-} \circ u \textit{ for } (i,\mu) \in \Gamma^{ob}. \tag{12.59}$$

This map is called the associated state map *by Willems and Rapisarda.*

3. *The identity* $\left(C^{ob}_{i,-}(A^{ob})^\mu\right)_{(j,\nu)} = \delta_{(i,\mu),(j,\nu)}$ *holds for* $(i,\mu),\ (j,\nu) \in \Gamma^{ob}$. *This identity gives rise to a different proof that the canonical observability realization is indeed observable, and, according to [1, p. 419], this is the origin of the* observability *terminology. Recall from Definition 4.1.7 that the observability matrix* $\mathfrak{O} \in F^{np\times\Gamma^{ob}}$ *of the state equations* (12.54) *consists of the n row blocks* $C^{ob}(A^{ob})^\mu$, $0 \le \mu \le n-1$. *In particular, the row* $C^{ob}_{i-}(A^{ob})^\mu$, $(i,\mu) \in \Gamma^{ob}$, *is a row of* $\mathfrak{O}$. *Hence, the identity matrix* $\mathrm{id}_{\Gamma^{ob}} \in F^{\Gamma^{ob}\times\Gamma^{ob}} = F^{n\times\Gamma^{ob}}$ *is a submatrix of* $\mathfrak{O}$, *and this implies* $\mathrm{rank}(\mathfrak{O}) = n$. *By Theorem and Definition 4.1.9 this is equivalent to the observability of the state space equations.*
4. *The state space equations* (12.54) *are controllable if and only if the IO behavior* $\mathcal{B}$ *is controllable.*
5. *The transfer matrix H of* $\mathcal{B}$ *is*

$$\begin{aligned} H = P^{-1}Q = Q^{g,0} + C^{ob}H^{qs} &= D^{ob} + C^{ob}\left(s\,\mathrm{id}_{\Gamma^{ob}} - A^{ob}\right)^{-1}B^{ob} \\ \in F(s)^{p\times m} &= F[s]^{p\times m} \oplus F(s)^{p\times m}_{spr}, \end{aligned}$$

and therefore, $D^{ob} = H_{pol}$. *In particular, the matrix* D^{ob} *is constant—and thus* $\mathcal{B}^{ob}$ *is a state realization in Kalman's original sense—if and only if the transfer matrix H is proper.*
6. *The rows* $\left(H^{qs}_{pol}\right)_{(i,\mu)-} \in F[s]^{1\times m}$, $(i,\mu) \in \Gamma^{ob}$, *are also given as*

$$\left(H^{qs}_{pol}\right)_{(i,\mu)-} = \sum_{\nu=0}^{\mu-1} B^{ob}_{(i,\nu)-}s^{\mu-1-\nu} + D^{ob}_{i-}s^\mu. \tag{12.60}$$

Notice that, in general, H^{qs}_{pol} *is not constant, even if H is proper and* D^{ob} *is constant.*

Proof 1. From Lemma and Definition 12.3.3, we use the quasi-state behavior

$$\begin{aligned} \mathcal{B}^{qs} &= \left\{\binom{x}{u} \in \mathcal{F}^{\Gamma^{ob}+m};\ (s\,\mathrm{id}_{\Gamma^{ob}} - A^{ob}) \circ x = Q^{qs} \circ u\right\} \\ &\underset{(12.52)}{=} \left\{\binom{x}{u};\ (s\,\mathrm{id}_{\Gamma^{ob}} - A^{ob}) \circ x = B^{ob}u + (s\,\mathrm{id}_{\Gamma^{ob}} - A^{ob})H^{qs}_{pol} \circ u\right\} \end{aligned}$$

and the isomorphism $\begin{pmatrix} C^{ob} & Q^{g,0} \\ 0 & \mathrm{id}_m \end{pmatrix} \circ\colon\ \mathcal{B}^{qs} \xrightarrow{\cong} \mathcal{B}$. These imply the equivalences

$$\begin{aligned}\binom{x}{u} \in \mathcal{B}^{qs} &\Longleftrightarrow (s\,\mathrm{id}_{\Gamma^{ob}} - A^{ob}) \circ x = Q^{qs} \circ u \\ &\Longleftrightarrow (s\,\mathrm{id}_{\Gamma^{ob}} - A^{ob}) \circ x = (s\,\mathrm{id}_{\Gamma^{ob}} - A^{ob}) H^{qs}_{pol} \circ u + B^{ob} u \\ &\Longleftrightarrow (s\,\mathrm{id}_{\Gamma^{ob}} - A^{ob}) \circ (x - H^{qs}_{pol} \circ u) = B^{ob} u \\ &\Longleftrightarrow \begin{pmatrix} x - H^{qs}_{pol} \circ u \\ u \end{pmatrix} \in \mathcal{B}^{ob}.\end{aligned}$$

These, in turn, furnish the isomorphism $\begin{pmatrix} \mathrm{id}_{\Gamma^{ob}} & -H^{qs}_{pol} \\ 0 & \mathrm{id}_m \end{pmatrix} \circ \colon \mathcal{B}^{qs} \xrightarrow{\cong} \mathcal{B}^{ob}$ and its inverse $\begin{pmatrix} \mathrm{id}_{\Gamma^{ob}} & +H^{qs}_{pol} \\ 0 & \mathrm{id}_m \end{pmatrix} \circ \colon \mathcal{B}^{ob} \xrightarrow{\cong} \mathcal{B}^{qs}$. By composing the latter with the isomorphism $\begin{pmatrix} C^{ob} & Q^{g,0} \\ 0 & \mathrm{id}_m \end{pmatrix} \circ \colon \mathcal{B}^{qs} \xrightarrow{\cong} \mathcal{B}$, we obtain the asserted isomorphism

$$\begin{pmatrix} C^{ob} & D^{ob} \\ 0 & \mathrm{id}_m \end{pmatrix} \circ \underset{(12.53)}{=} \begin{pmatrix} C^{ob} & Q^{g,0} \\ 0 & \mathrm{id}_m \end{pmatrix} \begin{pmatrix} \mathrm{id}_{\Gamma^{ob}} & +H^{qs}_{pol} \\ 0 & \mathrm{id}_m \end{pmatrix} \circ \colon \mathcal{B}^{ob} \xrightarrow{\cong} \mathcal{B}.$$

2. The inverse of the preceding isomorphism is

$$\mathcal{B} \xrightarrow{\cong} \mathcal{B}^{qs} \xrightarrow{\begin{pmatrix} \mathrm{id}_{\Gamma^{ob}} & -H^{qs}_{pol} \\ 0 & \mathrm{id}_m \end{pmatrix} \circ} \mathcal{B}^{ob}, \quad \binom{y}{u} \longmapsto \binom{x^{qs}}{u} \longmapsto \binom{x}{u},$$

where $x^{qs}_{i,\mu} = s^\mu \circ y_i$ for $(i,\mu) \in \Gamma^{ob}$ and $x = x^{qs} - H^{qs}_{pol} \circ u$, hence $x_{i,\mu} = s^\mu \circ y_i - (H^{qs}_{pol})_{(i,\mu)-} \circ u$.

3. We define the matrix

$$E \in F[s]^{\Gamma^{ob} \times p} \ \text{ by } \ E_{(i,\mu),j} := s^\mu \delta_{i,j},$$

hence $E_{(i,\mu),-} = s^\mu \delta_i$ and $(E \circ y)_{(i,\mu)} = s^\mu \circ y_i$. The system isomorphism $\mathcal{B}^{ob} \cong \mathcal{B}$ from (12.56) and (12.59) implies the isomorphism of the autonomous parts of these behaviors, namely,

$$\begin{aligned}\{x \in \mathcal{F}^{\Gamma^{ob}};\ s \circ x = A^{ob} x\} &\overset{\cong}{\longleftrightarrow} \{y \in \mathcal{F}^p;\ P \circ y = 0\}, \\ x = E \circ y \quad &\longleftrightarrow \quad y = C^{ob} x = C^{ob} \circ x\end{aligned}$$

By duality this furnishes the $F[s]$-isomorphism

$$\begin{aligned} M^{ob,0} := F[s]^{1\times\Gamma^{ob}} / F[s]^{1\times\Gamma^{ob}} (s\,\mathrm{id}_{\Gamma^{ob}} - A) &\overset{\cong}{\longleftrightarrow} M^0 = F[s]^{1\times p}/U^0 \\ \overline{\eta} = \overline{\xi C^{ob}} \quad &\longleftrightarrow \quad \overline{\eta E} \\ \overline{\delta_{(i,\mu)}} = s^\mu \overline{C^{ob}_{i-}} \quad &\longleftrightarrow \quad s^\mu \overline{\delta_i}. \end{aligned} \tag{12.61}$$

We further consider $F^{1\times\Gamma^{ob}}$ as $F[s]$-module with the scalar multiplication $s \circ_{A^{ob}} \zeta := \zeta A^{ob}$ as in (4.8). Lemma 4.1.6 gives the $F[s]$-isomorphism $F^{1\times\Gamma^{ob}} \xrightarrow{\cong} M^{ob,0}$, $\zeta \longmapsto \overline{\zeta}$, $\delta_{(i,\mu)} \longleftrightarrow \overline{\delta_{(i,\mu)}}$, hence

$$\begin{array}{ccccc} F^{1\times\Gamma^{\mathrm{ob}}} & \xrightarrow{\cong} & M^{\mathrm{ob},0} & \xrightarrow{\cong} & F^{1\times\Gamma^{\mathrm{ob}}} \\ \delta_{(i,\mu)} & \longmapsto \overline{\delta_{(i,\mu)}} \underset{(12.61)}{=} & s^\mu \overline{C^{\mathrm{ob}}_{i-}} & \longmapsto s^\mu \circ_{A^{\mathrm{ob}}} C^{\mathrm{ob}}_{i-} = C^{\mathrm{ob}}_{i-}(A^{\mathrm{ob}})^\mu = \delta_{(i,\mu)} \end{array},$$

hence $\left(C^{\mathrm{ob}}_{i,-}(A^{\mathrm{ob}})^\mu\right)_{(j,\nu)} = \delta_{(i,\mu),(j,\nu)}$ for all $(i,\mu),\ (j,\nu) \in \Gamma^{\mathrm{ob}}$.

4. Controllability is preserved under behavior isomorphisms.
5. The equations $H = D^{\mathrm{ob}} + C^{\mathrm{ob}}(s\,\mathrm{id}_{\Gamma^{\mathrm{ob}}} - A^{\mathrm{ob}})^{-1}B^{\mathrm{ob}}$ and $H_{\mathrm{pol}} = D^{\mathrm{ob}}$ generally hold for state realizations, even nonobservable ones.
6. To compute $s^\mu \circ y_i,\ (i,\mu) \in \Gamma^{\mathrm{ob}},\ d(i) > 0$, we use the output equation $y_i = x_{i,0} + D^{\mathrm{ob}}_{i-} \circ u$, repeatedly insert the state equation $s \circ x_{i,\nu} = x_{i,\nu+1} + B^{\mathrm{ob}}_{(i,\nu)-}u$, $0 \le \nu < d(i) - 1$, cf. (12.57), and obtain

$$s^\mu \circ y_i = x_{i,\mu} + \left(\sum_{\nu=0}^{\mu-1} B^{\mathrm{ob}}_{(i,\nu)-} s^{\mu-1-\nu} + D^{\mathrm{ob}}_{i-} s^\mu\right) \circ u.$$

Together with (12.59), this implies that

$$\left(H^{\mathrm{qs}}_{\mathrm{pol}}\right)_{(i,\mu)-} \circ u = \left(\sum_{\nu=0}^{\mu-1} B^{\mathrm{ob}}_{(i,\nu)-} s^{\mu-1-\nu} + D^{\mathrm{ob}}_{i-} s^\mu\right) \circ u,\ (i,\mu) \in \Gamma^{\mathrm{ob}}.$$

Since this holds for all $u \in \mathcal{F}^m$, the assertion follows. □

Remark 12.3.6 Recall $\Gamma^{\mathrm{ob}} = \left\{(i,\mu);\ 1 \le i \le p,\ 0 \le \mu \le d^{\mathrm{ob}}(i) - 1\right\}$ and $n = \sum_{i=1}^{p} d^{\mathrm{ob}}(i)$. To transform the matrices with indices in Γ^{ob} into standard matrices with indices in $\{1, \cdots, n\}$ one uses the bijection

$$\alpha:\ \Gamma^{\mathrm{ob}} \cong \{1, \ldots, n\},\ (i,\mu) \mapsto \sum_{k=1}^{i-1} d^{\mathrm{ob}}(k) + \mu + 1, \text{ and defines}$$

$$A^{\mathrm{ob}} \in F^{n\times n},\ A^{\mathrm{ob}}_{\alpha(i,\mu),\alpha(j,\nu)} := A^{\mathrm{ob}}_{(i,\mu),(j,\nu)} = \begin{cases} \delta_{i,j}\delta_{\mu+1,\nu} & \text{if } \mu < d^{\mathrm{ob}}(i) - 1 \\ A_{i,j,\nu} & \text{if } \mu = d^{\mathrm{ob}}(i) - 1 \end{cases}.$$

With the bijection α one identifies $F^{\Gamma^{\mathrm{ob}}\times\Gamma^{\mathrm{ob}}} = F^{n\times n}$ and likewise

$$B^{\mathrm{ob}} \in F^{\Gamma^{\mathrm{ob}}\times m} = F^{n\times m},\ C^{\mathrm{ob}} \in F^{p\times\Gamma^{\mathrm{ob}}} = F^{p\times n},\ H^{\mathrm{qs}}_{\mathrm{pol}} \in F[s]^{\Gamma^{\mathrm{ob}}\times m} = F[s]^{n\times m}.$$

This is, for instance, necessary to define $f := \det\left(s\,\mathrm{id}_{\Gamma^{\mathrm{ob}}} - A^{\mathrm{ob}}\right)$ uniquely. ◊

Corollary 12.3.7 *In the situation of Theorem 12.3.5 consider an arbitrary observable state space representation*

$$s \circ x = Ax + Bu,\ y = Cx + D \circ u,\ D = D^{\mathrm{ob}}$$

of $\mathcal{B}$. Theorem 12.3.2 implies that $F^{1\times n}$ has the F-basis

$$C_{i,-}A^{\mu},\ (i,\mu)\in\Gamma^{\rm ob},\ hence\ \left(C_{i,-}A^{\mu}\right)_{(i,\mu)\in\Gamma^{\rm ob}}\in{\rm Gl}_n(F),\ and$$
$$\forall i=1,\dots,p:\ C_{i,-}A^{d^{\rm ob}(i)}=\sum_{(j,\nu)\in\Gamma^{\rm ob}}A_{i,j,\nu}C_{j,-}A^{\nu}\ where \tag{12.62}$$
$$A_{i,j,\nu}=0\ if\ (j,\nu)\not<(i,d^{\rm ob}(i)).$$

Moreover, due to $\left(C^{\rm ob}_{i,-}(A^{\rm ob})^{\mu}\right)_{(i,\mu)\in\Gamma^{\rm ob}}={\rm id}_{\Gamma^{\rm ob}}$, *there is the unique invertible matrix*

$$T=\left(\left(C_{i,-}A^{\mu}\right)_{(i,\mu)\in\Gamma^{\rm ob}}\right)^{-1}\in F^{n\times\Gamma^{\rm ob}}\ such\ that$$
$$A=TA^{\rm ob}T^{-1},\ B=TB^{\rm ob},\ C=C^{\rm ob}T^{-1}. \tag{12.63}$$

◊

Corollary 12.3.8 (Algorithm for the observability realization) *For the computation of the canonical observability realization of* $\mathcal{B}$ *from* (12.33) *the following algorithmic steps have to be taken:*

1. *Use the Buchberger algorithm to compute the reduced Gröbner basis* $g^{(i)}$, $i=1,\dots,p$, *of* $U^0:=F[s]^{1\times p}P$ *and the matrices* $P^g:=\left(g^{(i)}\right)_i\in F[s]^{p\times p}$ *and* $Q^g=P^gH=P^gP^{-1}Q$ *with the equations*

$$U^0:=\oplus_{i=1}^{p}F[s]g^{(i)}=F[s]^{1\times p}P^g,$$
$$g^{(i)}=s^{d^{\rm ob}(i)}\delta_i-\sum_{j=1}^{p}\sum_{\nu=0}^{d^{\rm ob}(j)-1}A_{i,j,\nu}s^{\nu}\delta_j,\ 1\le i\le p,$$
$$P^g_{i,j}=s^{d^{\rm ob}(i)}\delta_{i,j}-\sum_{\nu=0}^{d^{\rm ob}(j)-1}A_{i,j,\nu}s^{\nu},\ 1\le i,j\le p,$$
$$\Longrightarrow\Gamma^{\rm ob}:=\left\{(i,\mu);\ 1\le i\le p,\ 0\le\mu\le d^{\rm ob}(i)-1\right\}.$$

   According to (12.48) *and with the* $A_{i,j,\nu}$ *define the sparse matrices* $A^{\rm ob}$ *and* $C^{\rm ob}$ *and the matrices* $Q^{\rm qs}$ *and* $Q^{g,0}$.
2. *Compute* $H^{\rm qs}=\left(s\,{\rm id}_{\Gamma^{\rm ob}}-A^{\rm ob}\right)^{-1}Q^{\rm qs}$ *and the decomposition* $H^{\rm qs}=H^{\rm qs}_{\rm pol}+H^{\rm qs}_{\rm spr}$ *and define* $B^{\rm ob}:=Q^{\rm qs}-\left(s\,{\rm id}_{\Gamma^{\rm ob}}-A^{\rm ob}\right)H^{\rm qs}_{\rm pol}\in F^{\Gamma^{\rm ob}\times m}$. *Also compute* $D^{\rm ob}=C^{\rm ob}H^{\rm qs}_{\rm pol}+Q^{g,0}\left(=H_{\rm pol}\right)$.

With these data there is the behavior isomorphism

$$\begin{pmatrix}C^{\rm ob}&D^{\rm ob}\\0&{\rm id}_m\end{pmatrix}\circ:\left\{\begin{pmatrix}x\\u\end{pmatrix}\in\mathcal{F}^{n+m};\ s\circ x=A^{\rm ob}x+B^{\rm ob}u\right\}$$
$$\stackrel{\cong}{\longrightarrow}\left\{\begin{pmatrix}y\\u\end{pmatrix}\in\mathcal{F}^{p+m};\ P\circ y=Q\circ u\right\}\ and$$
$$H=P^{-1}Q=D^{\rm ob}+C^{\rm ob}\left(s\,{\rm id}_{\Gamma^{\rm ob}}-A^{\rm ob}\right)^{-1}B^{\rm ob}.$$

◊

Example 12.3.9 With this example, we demonstrate that the observability indices depend on the choice of the term order. On $\{1, 2\} \times \mathbb{N}$, we consider the two term orders

$$(i, \mu) <_1 (j, \nu) :\Longleftrightarrow i > j \text{ or } (i = j \text{ and } \mu < \nu) \quad \text{and}$$
$$(i, \mu) <_2 (j, \nu) :\Longleftrightarrow i < j \text{ or } (i = j \text{ and } \mu < \nu).$$

Consider the matrix P and submodule U^0 defined by

$$P := \begin{pmatrix} s & s^2 \\ s & s^3 \end{pmatrix} \quad \text{and} \quad U^0 := F[s]^{1\times 2} P = F[s](s, s^2) \oplus F[s](s, s^3).$$

The Gröbner matrices $P^{(i)}$ of P with respect to $<_i$ are

$$P^{(1)} = \begin{pmatrix} s & s^2 \\ 0 & s^3 - s^2 \end{pmatrix} \quad \text{and} \quad P^{(2)} = \begin{pmatrix} s & s^2 \\ s - s^2 & 0 \end{pmatrix}.$$

The degree sets $\deg_i(U^0)$ of U^0 with respect to $<_i$ are thus

$$\deg_1(U^0) = \big(\{1\} \times (1 + \mathbb{N})\big) \uplus \big(\{2\} \times (3 + \mathbb{N})\big) \quad \text{and}$$
$$\deg_2(U^0) = \big(\{1\} \times (2 + \mathbb{N})\big) \uplus \big(\{2\} \times (2 + \mathbb{N})\big).$$

Therefore, the observability indices with respect to $<_1$ are $d(1) = 1$ and $d(2) = 3$, while those with respect to $<_2$ are $d(1) = 2$ and $d(2) = 2$. ◊

Example 12.3.10 (*The SISO case*). We demonstrate the algorithm of Theorem and Definition 12.3.5 in the SISO (single input/single output) case, i.e., for $p = m = 1$ and $n := d^{\mathrm{ob}}(1) \geq 1$. The IO behavior is given as

$$\mathcal{B} = \left\{ \begin{pmatrix} y \\ u \end{pmatrix} \in \mathcal{F}^2;\ P \circ y = Q \circ u \right\} \quad \text{with } P = s^n - \sum_{\mu=0}^{n-1} p_\mu s^\mu \text{ and } Q \in F[s] \setminus \{0\}.$$

Its transfer function is $H = P^{-1}Q$. We omit the indices referring to the components of the trajectory, since they are necessarily equal to one. The set Γ^{ob} is given by

$$\Gamma^{\mathrm{ob}} = \big\{(1, 0), \ldots, (1, n-1)\big\} \overset{\text{ident.}}{=\!=} \{0, \ldots, n-1\}.$$

It is clear that P itself is the reduced Gröbner basis of $U^0 = F[s]P$, and hence

$$P^g = P, \qquad Q^g = Q \in F[s] = F[s]^{1\times 1},$$
$$Q^{\mathrm{qs}} = \begin{pmatrix} 0 \\ \vdots \\ 0 \\ Q \end{pmatrix} \in F[s]^n \quad \text{and} \quad Q^{g,0} = 0.$$

Equation (12.48) furnishes

$$A^{\text{ob}} = \begin{pmatrix} 0 & 1 & & 0 \\ \vdots & & \ddots & \\ 0 & 0 & & 1 \\ p_0 & p_1 & \cdots & p_{n-1} \end{pmatrix} \in F^{n\times n}, \qquad C^{\text{ob}} = (1, 0, \ldots, 0) \in F^{1\times n}$$

$$S = s\,\text{id}_n - A^{\text{ob}}, \qquad f = \det(S) = P,$$

$$H = P^{-1}Q, \quad \text{and} \qquad D^{\text{ob}} = H_{\text{pol}}.$$

As described in Lemma 12.3.4 and Eq. (12.52), the column $B^{\text{ob}} \in F^n$ is computed as

$$B^{\text{ob}} = Q^{\text{qs}} - SH^{\text{qs}}_{\text{pol}} \text{ where } Q^{\text{qs}} := \begin{pmatrix} 0 \\ \vdots \\ 0 \\ Q \end{pmatrix} \text{ and } H^{\text{qs}} := S^{-1}Q^{\text{qs}} = H^{\text{qs}}_{\text{pol}} + H^{\text{qs}}_{\text{spr}}$$

is the decomposition into polynomial and strictly proper part. One also obtains $H^{\text{qs}}_{\text{pol}}$ as the quotient of the Euclidean division of $fH^{\text{qs}} = S_{\text{adj}}Q^{\text{qs}} \in F[s]^n$ by f, i.e., $fH^{\text{qs}} = fH^{\text{qs}}_{\text{pol}} + R$, where R is the remainder. The state space equations are

$$s \circ x_\mu = \begin{cases} x_{\mu+1} + B^{\text{ob}}_\mu u & \text{if } 0 \le \mu < n-1, \\ p_0x_0 + \cdots + p_{n-1}x_{n-1} + B^{\text{ob}}_{n-1}u & \text{if } \mu = n-1, \end{cases}$$

$$y = x_0 + D^{\text{ob}} \circ u.$$

and the associated block diagram is

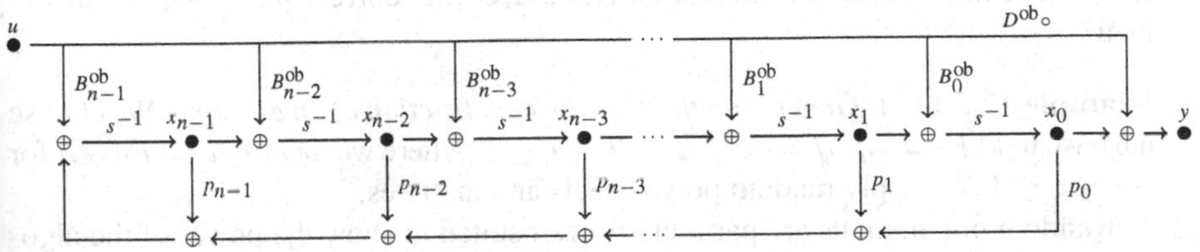

compare [1, Fig. 2.1-7. on p. 42].

We carry out the computation of B^{ob} explicitly in the case $n = 2$ and for proper H, i.e., with

$$P = s^2 - p_1 s - p_0, \; Q = q_2s^2 + q_1s + q_0, \; H = P^{-1}Q, \; H_{\text{pol}} = q_2 = D^{\text{ob}}.$$

We need the matrices

$$A^{\text{ob}} = \begin{pmatrix} 0 & 1 \\ p_0 & p_1 \end{pmatrix}, \; Q^{\text{qs}} = \begin{pmatrix} 0 \\ Q \end{pmatrix}, \; S = s\,\text{id}_2 - A^{\text{ob}} = \begin{pmatrix} s & -1 \\ -p_0 & s-p_1 \end{pmatrix}, \; S_{\text{adj}} = \begin{pmatrix} s-p_1 & 1 \\ p_0 & s \end{pmatrix}.$$

We use $\det(S) = P$ and apply Euclidean division to $PH^{\text{qs}} = S_{\text{adj}}Q^{\text{qs}} = \begin{pmatrix} Q \\ sQ \end{pmatrix} = P\begin{pmatrix} (H^{\text{qs}}_{\text{pol}})_0 \\ (H^{\text{qs}}_{\text{pol}})_1 \end{pmatrix} + \begin{pmatrix} R_0 \\ R_1 \end{pmatrix}$. From the first entry of this row vector, we infer that

$$Q = q_2s^2 + q_1s + q_0 = (s^2 - p_1s - p_0)(H^{\text{qs}}_{\text{pol}})_0 + R_0$$

and thus $(H_{\rm pol}^{\rm qs})_0 = q_2$ and $R_0 = (q_1 + p_1 q_2)s + q_0 + p_0 q_2$. By inserting this in the second entry, we obtain that

$$\begin{aligned} sQ &= P(H_{\rm pol}^{\rm qs})_1 + R_1 \\ &= Psq_2 + sR_0 = P\big(sq_2 + (q_1 + p_1 q_2)\big) + r, \quad \text{with } \deg(r) < 2, \end{aligned}$$

thus $(H_{\rm pol}^{\rm qs})_1 = sq_2 + (q_1 + p_1 q_2)$. Finally, we get

$$\begin{aligned} B^{\rm ob} &= Q^{\rm qs} - SH_{\rm pol}^{\rm qs} = \begin{pmatrix} 0 \\ q_2 s^2 + q_1 s + q_0 \end{pmatrix} - \begin{pmatrix} s & -1 \\ -p_0 & s - p_1 \end{pmatrix} \begin{pmatrix} q_2 \\ sq_2 + (q_1 + p_1 q_2) \end{pmatrix} \\ &= \begin{pmatrix} q_1 + p_1 q_2 \\ q_0 + q_1 p_1 + q_2(p_0 + p_1^2) \end{pmatrix}. \end{aligned}$$

If $H = P^{-1}Q$ is strictly proper or, equivalently, $q_2 = 0$ then $B^{\rm ob} = \begin{pmatrix} q_1 \\ q_0 + q_1 p_1 \end{pmatrix}$.

The computation

$$\begin{aligned} H &= D^{\rm ob} + C^{\rm ob}(s\,{\rm id}_2 - A^{\rm ob})^{-1} B^{\rm ob} = D^{\rm ob} + C^{\rm ob} P^{-1} S_{\rm adj} B^{\rm ob} \\ &= \frac{1}{P}\left(q_2(s^2 - p_1 s - p_0) + (s - p_1, 1)\begin{pmatrix} q_1 + p_1 q_2 \\ q_0 + q_1 p_1 + q_2(p_0 + p_1^2) \end{pmatrix}\right) \\ &= \frac{1}{P}(q_2 s^2 + q_1 s + q_0) = P^{-1}Q \end{aligned}$$

shows that the state space realization has indeed the correct input/output transfer matrix. ◊

Example 12.3.11 1. *General method for the construction of an example*: We choose the base field $F := \mathbb{Z}/\mathbb{Z}7 = \{i \in \mathbb{Z};\ -3 \le i \le 3\}$ where we identify $i := i + \mathbb{Z}7$ for $-3 \le i \le 3$. We apply random polynomials and matrices.

Random polynomials are particularly well suited to show the power of the algorithm and its implementation. It is obvious that the following example cannot be computed by hand. In the example we use the small prime field $\mathbb{Z}/\mathbb{Z}7$ instead of the fields $\mathbb{Q}$, $\mathbb{Q}[j]$, $j := \sqrt{-1}$, $\mathbb{R}$ or $\mathbb{C}$. The many steps of the Buchberger algorithm tend to create large denominators of rational numbers or very long decimal numbers, even if one starts with very simple integer coefficients. This is a problem if the results shall be printed, as in this book, but not otherwise. The stability and robustness of all results, i.e., that small changes in the parameters induce only small changes in the results, is, of course, also an important problem, in particular of numerical analysis. We discussed such problems in Sects. 10.2.3, 10.2.4, and 10.2.5 only.

Random polynomial matrices should be chosen with entries of low degree to save space when they are printed. A random invertible matrix $X \in {\rm Gl}_p(F[s])$ of determinant 1 is most easily created as the product of an upper resp. a lower triangular matrix in $F[s]^{p\times p}$ with 1 in the main diagonal. The analogue applies to ${\rm Gl}_p(F)$. A random matrix $P \in F[s]^{p\times p}$ with ${\rm rank}(P) = p$ is best created in the form $X\,{\rm diag}(f_1, \cdots, f_p)Y$ with nonzero random polynomials $f_1, \cdots, f_p$ and random $X, Y \in {\rm Gl}_p(F[s])$. Then $n := \deg_s(\det(P)) = \deg_s(f_1) + \cdots + \deg_s(f_p)$ is

the state space dimension of the observable realization.
2. The notations are that of Theorem 12.3.5. We use the position-over-degree term order with

$$(j,\nu) < (i,\mu) \Longleftrightarrow i < j \text{ or } i = j \text{ and } \nu < \mu.$$

3. *Construction of the example*: We choose

$$\begin{aligned}
&p := 3,\ m := 2,\ ,\ f_1 = s+3,\ f_2 := s^2 - 3s + 5,\ f_3 := s^5 - s^4 + 2s + 1,\\
&X, Y \in \mathrm{Gl}_3(F[s]),\ Q \in F[s]^{3\times 2},\ P = X\,\mathrm{diag}(f_1, f_2, f_3)Y\\
&\Longrightarrow \det(P) = f_1 f_2 f_3,\ \deg_s(\det(P)) = 8,\ (P, -Q) \in F[s]^{3\times 5},
\end{aligned}$$

where X, Y, Q are chosen randomly. One random choice of $(P, -Q)$ is

$$\begin{aligned}
P &= \begin{pmatrix} -3s^4-3s^3-3s+3 & 3s^6-s^4-2s^2+3s-2 & s^7-2s^6-2s^4+3s^3+s^2-s+3 \\ -3s^5+2s^4-2s^3-3s^2-2s & -2s^7+2s^6+s^4-3s^3+2s^2-s-2 & 3s^8+2s^7-2s^6+2s^5-3s^4-s^2-2s-1 \\ 3s^5+s^4-2s^3-2s^2+3s+1 & s^8-2s^7+2s^6+s^5-2s^4-3s^3-s^2-s & s^9-3s^8-3s^6-2s^5-3s^4-3s^3-2s^2+3s \end{pmatrix}\\
Q &= \begin{pmatrix} 3s^3 & 0 \\ -3s^2+s & -3s^4+2s^3 \\ -s & 0 \end{pmatrix}.
\end{aligned}$$

4. We now apply Corollary 12.3.8 to these data step by step to obtain

$$P^g,\ Q^g,\ d^{\mathrm{ob}}(i),\ \Gamma^{\mathrm{ob}},\ H^{\mathrm{qs}}_{\mathrm{pol}},\ (A^{\mathrm{ob}}, B^{\mathrm{ob}}, C^{\mathrm{ob}}, D^{\mathrm{ob}}),\ A^{\mathrm{ob}} \in F^{\Gamma^{\mathrm{ob}}\times\Gamma^{\mathrm{ob}}} = F^{n\times n} \text{ etc.}$$

The associated Gröbner pair $(P^g, -Q^g)$ is

$$\begin{aligned}
P^g &= \begin{pmatrix} 0 & 0 & s^7+3s^6+s^5+2s^4+2s^3+2s^2+2 \\ 0 & s+3 & 2s^6-2s^5-2s^2+2s-2 \\ 1 & 3 & -3s^6+3s^5+3s^3-2s^2+2s \end{pmatrix}\\
Q^g &= \begin{pmatrix} -2s^{15}-s^{14}+2s^{13}+2s^{12}-2s^{11}+2s^9+s^8-2s^7-s^6+3s^5+s^3-2s^2-s & -2s^{15}+s^{14}-3s^{12}+2s^{11}-s^{10}+2s^9-3s^8+s^7+2s^6-s^5-3s^4+s^3 \\ 3s^{14}+3s^{12}-s^{11}-s^{10}+2s^9-3s^8+3s^7-s^6-3s^5+3s^4+2s^3+3s^2+2s & 3s^{14}-3s^{13}-3s^{12}-2s^{10}-s^9+2s^7+3s^6-2s^5+s^4-3s^3 \\ -s^{14}-s^{12}-2s^{11}-2s^{10}-s^9-s^8-2s^7+3s^6+s^5+s^4+2s^3+3s^2+2s & -s^{14}+s^{13}+s^{12}+3s^{10}+3s^8+s^7+2s^4+3s^3 \end{pmatrix}.
\end{aligned}$$

We have $d(1) = 0$ (corresponding to the third row of P^g), $d(2) = 1$ (corresponding to the second row of P^g), and $d(3) = 7$ (corresponding to the first row of P^g). Hence

$$n = 8 \text{ and } \Gamma^{\mathrm{ob}} = \{(2,0),\ (3,0),\ (3,1),\ (3,2),\ (3,3),\ (3,4),\ (3,5),\ (3,6)\}.$$

The cases of nonproper $H = P^{-1}Q$ and of $d(i) = 0$ are excluded in [1, Sect. 6.4] and often neglected in the literature.
5. The canonical quasi-state realization is given by

$$
A^{\mathrm{ob}} = \begin{array}{c} \\ (2,0) \\ (3,0) \\ (3,1) \\ (3,2) \\ (3,3) \\ (3,4) \\ (3,5) \\ (3,6) \end{array}
\begin{array}{c} \begin{array}{cccccccc} (2,0) & (3,0) & (3,1) & (3,2) & (3,3) & (3,4) & (3,5) & (3,6) \end{array} \\
\begin{pmatrix}
-3 & 2 & -2 & 2 & 0 & 0 & 2 & -2 \\
0 & 0 & 1 & 0 & 0 & 0 & 0 & 0 \\
0 & 0 & 0 & 1 & 0 & 0 & 0 & 0 \\
0 & 0 & 0 & 0 & 1 & 0 & 0 & 0 \\
0 & 0 & 0 & 0 & 0 & 1 & 0 & 0 \\
0 & 0 & 0 & 0 & 0 & 0 & 1 & 0 \\
0 & 0 & 0 & 0 & 0 & 0 & 0 & 1 \\
0 & 2 & 0 & -2 & -2 & -2 & -1 & -3
\end{pmatrix} \end{array}
$$

$$
Q^{\mathrm{qs}} = \begin{array}{c} (2,0) \\ (3,0) \\ (3,1) \\ (3,2) \\ (3,3) \\ (3,4) \\ (3,5) \\ (3,6) \end{array}
\begin{pmatrix}
3s^{14}+3s^{12}-s^{11}-s^{10}+2s^{9}-3s^{8}+3s^{7}-s^{6}-3s^{5}+3s^{4}+2s^{3}+3s^{2}+2s & 3s^{14}-3s^{13}-3s^{12}-2s^{10}-s^{9}+2s^{7}+3s^{6}-2s^{5}+s^{4}-3s^{3} \\
0 & 0 \\
0 & 0 \\
0 & 0 \\
0 & 0 \\
0 & 0 \\
0 & 0 \\
-2s^{15}-s^{14}+2s^{13}+2s^{12}-2s^{11}+2s^{9}+s^{8}-2s^{7}-s^{6}+3s^{5}+s^{3}-2s^{2}-s & -2s^{15}+s^{14}-3s^{12}+2s^{11}-s^{10}+2s^{9}-3s^{8}+s^{7}+2s^{6}-s^{5}-3s^{4}+s^{3}
\end{pmatrix},
$$

$$
C^{\mathrm{ob}} = \begin{array}{c} \begin{array}{cccccccc} (2,0) & (3,0) & (3,1) & (3,2) & (3,3) & (3,4) & (3,5) & (3,6) \end{array} \\
\begin{pmatrix}
-3 & 0 & -2 & 2 & -3 & 0 & -3 & 3 \\
1 & 0 & 0 & 0 & 0 & 0 & 0 & 0 \\
0 & 1 & 0 & 0 & 0 & 0 & 0 & 0
\end{pmatrix} \end{array},
$$

$$
Q^{g,0} = \begin{pmatrix}
-s^{14}-s^{12}-2s^{11}-2s^{10}-s^{9}-s^{8}-2s^{7}+3s^{6}+s^{5}+s^{4}+2s^{3}+3s^{2}+2s & -s^{14}+s^{13}+s^{12}+3s^{10}+3s^{8}+s^{7}+2s^{4}+3s^{3} \\
0 & 0 \\
0 & 0
\end{pmatrix}.
$$

Notice that $d(1) = 0$ implies the more complicated first rows of C^{ob} and $Q^{g,0}$.
6. The canonical observability realization

$$
s \circ x = A^{\mathrm{ob}} x + B^{\mathrm{ob}} u, \quad y = C^{\mathrm{ob}} x + D^{\mathrm{ob}} \circ u
$$

is then given by A^{ob} and C^{ob} as above and the additional matrices

$$
B^{\mathrm{ob}} = \begin{array}{c} (2,0) \\ (3,0) \\ (3,1) \\ (3,2) \\ (3,3) \\ (3,4) \\ (3,5) \\ (3,6) \end{array}
\begin{pmatrix}
-3 & -3 \\
0 & 3 \\
3 & 3 \\
-1 & 0 \\
0 & -2 \\
-3 & -3 \\
2 & 0 \\
1 & 0
\end{pmatrix}
$$

$$
D^{\mathrm{ob}} = \begin{pmatrix}
-s^{11}-2s^{9}+3s^{8}-3s^{7}+s^{6}-3s^{5}+3s^{4}+2s^{3}-3s^{2}-s+3 & -s^{11}+s^{10}-s^{8}-s^{7}+s^{6}-3s^{5}+s^{4}+2s^{3}+3s^{2}-1 \\
2s^{9}+3s^{8}+s^{7}+2s^{6}+3s^{5}+s^{4}-s^{3}-3s^{2}-3s & 2s^{9}+s^{8}+s^{7}+s^{5}+2s^{4}+s^{3}-2s^{2}+3^{s}+1 \\
-2s^{8}-2s^{7}+3s^{6}-s^{5}-s^{4}-s^{3}-s^{2}-s-3 & -2s^{8}+2s^{6}+2s^{5}-2s^{4}+3s^{3}+s^{2}-3^{s}+3
\end{pmatrix}.
$$

7. We check that the state representation (A^{ob}, B^{ob}, C^{ob}, D^{ob}) really solves the realization problem. By construction the state representation is observable. First we verify

$$
\begin{aligned}
&H := P^{-1}Q = D^{\mathrm{ob}} + C^{\mathrm{ob}}(s\,\mathrm{id}_8 - A^{\mathrm{ob}})^{-1}B^{\mathrm{ob}} \text{ and}\\
&\det(s\,\mathrm{id}_8 - A^{\mathrm{ob}}) = f_1 f_2 f_3 = (s+3)(s^2-3s+5)(s^5-s^4+2s+1)\\
&\Longrightarrow \dim_F(\mathcal{B}_s^0) = \dim_F(\mathcal{B}^0) \text{ where}\\
&\quad \mathcal{B}_s^0 = \left\{x \in \mathcal{F}^8;\ s \circ x = A^{\mathrm{ob}}x\right\},\ \mathcal{B}^0 = \left\{y \in \mathcal{F}^3;\ P \circ y = 0\right\}.
\end{aligned}
$$

We compute a universal left annihilator

$$
\begin{aligned}
&(-X, P_1) \in \mathcal{D}^{3\times(8+3)} \text{ of } \begin{pmatrix} s\,\mathrm{id}_8 - A^{\mathrm{ob}} \\ C^{\mathrm{ob}} \end{pmatrix} \text{ and infer}\\
&C^{\mathrm{ob}} : \mathcal{B}_s^0 \cong \mathcal{B}_1^0 := \left\{y \in \mathcal{F}^3;\ P_1 \circ y = 0\right\}.
\end{aligned}
$$

We check

$$
\begin{aligned}
&P_1P^{-1},\ PP_1^{-1} = \left(P_1P^{-1}\right)^{-1} \in F[s]^{3\times 3} \Longrightarrow P_1P^{-1} \in \mathrm{Gl}_3(\mathcal{D})\\
&\Longrightarrow Q_1 := P_1H = (P_1P^{-1})PH = (P_1P^{-1})Q \in \mathcal{D}^{3\times 2},\ \mathcal{B}^0 = \mathcal{B}_1^0, \text{ and}\\
&\begin{pmatrix} C^{\mathrm{ob}} & D^{\mathrm{ob}} \\ 0 & \mathrm{id}_2 \end{pmatrix} :\ \mathcal{B}_s := \left\{\begin{pmatrix} x \\ u \end{pmatrix} \in \mathcal{F}^{8+2};\ s \circ x = A^{\mathrm{ob}}x + B^{\mathrm{ob}}u\right\} \cong\\
&\mathcal{B}_1 := \left\{\begin{pmatrix} y \\ u \end{pmatrix} \in \mathcal{F}^{3+2};\ P_1 \circ y = Q_1 \circ u\right\} = \mathcal{B} := \left\{\begin{pmatrix} y \\ u \end{pmatrix} \in \mathcal{F}^{3+2};\ P \circ y = Q \circ u\right\}.
\end{aligned}
$$

The large degree of the entries of D^{ob} again suggests to use proper matrices $H = P^{-1}Q$ as is also advisable for many technical reasons, as we know. ◊

The following theorem is significant for the simulation of an IO behavior by means of the observability realization. The assumptions and notations are those from Theorem and Definition 12.3.5, in particular, we use

$$\Gamma^{\mathrm{ob}} = \{(i, \mu);\ 1 \le i \le p,\ 0 \le \mu \le d^{\mathrm{ob}}(i) - 1\}.$$

In discrete time over any field F resp. in continuous time over $F = \mathbb{R}, \mathbb{C}$ we use the injective cogenerators $\mathcal{F} = F^{\mathbb{N}}$ resp. $\mathcal{F} := \mathrm{C}^{-\infty}(\mathbb{R}, F)$.

Theorem 12.3.12 (The canonical initial value problem for an IO behavior, cf. [1, Sect. 2.3.1] for SISO behaviors). *Let $y^0 = (y^0_{i,\mu})_{(i,\mu)\in\Gamma^{\mathrm{ob}}} \in F^{\Gamma^{\mathrm{ob}}}$ and let $u \in \mathcal{F}^m$ be an input. The vector y^0 is called an* initial vector.

Discrete time. *There is a unique trajectory $y \in \mathcal{F}^p$ which solves the initial value problem*

$$P \circ y = Q \circ u,\ \text{i.e., } \begin{pmatrix} y \\ u \end{pmatrix} \in \mathcal{B},\ \text{and } y_i(\mu) = y^0_{i,\mu} \text{ for } (i, \mu) \in \Gamma^{\mathrm{ob}}. \tag{12.64}$$

If x is the unique solution of the initial value problem

$$s \circ x = A^{\mathrm{ob}}x + B^{\mathrm{ob}}u,\ x_{i,\mu}(0) = y^0_{i,\mu} - (H^{\mathrm{qs}}_{\mathrm{pol}} \circ u)_{(i,\mu)}(0) \text{ for } (i, \mu) \in \Gamma^{\mathrm{ob}},$$

then $y = C^{\mathrm{ob}}x + D^{\mathrm{ob}} \circ u$ solves (12.64), especially $y_i = x_{i,0} + (D^{\mathrm{ob}} \circ u)_i$ if $d(i) > 0$.

In other words, there is the F-isomorphism

$$\mathcal{B} \overset{\cong}{\longrightarrow} F^{\Gamma^{\text{ob}}} \times \mathcal{F}^m,\ \binom{y}{u} \longmapsto \left(\begin{smallmatrix} (y_i(\mu))_{(i,\mu)\in\Gamma^{\text{ob}}} \\ u \end{smallmatrix} \right).$$

Continuous time. *If u and $H_{\text{pol}}^{\text{qs}} \circ u$ are continuous, there is a unique trajectory $y \in \mathcal{F}^p$ with $s^\mu \circ y_i = \frac{d^\mu y_i}{dt^\mu} = y_i^{(\mu)} \in C^0(\mathbb{R}, F)$ for all $(i, \mu) \in \Gamma^{\text{ob}}$ which solves the initial value problem*

$$P \circ y = Q \circ u,\ \text{i.e.,}\ \binom{y}{u} \in \mathcal{B},\ \text{and}\ y_i^{(\mu)}(0) = y_{i,\mu}^0\ \text{for}\ (i, \mu) \in \Gamma^{\text{ob}}. \tag{12.65}$$

If x denotes the unique C^1-solution of the initial value problem

$$\begin{aligned} s \circ x &= A^{\text{ob}} x + B^{\text{ob}} u \\ \text{with}\ x_{i,\mu}(0) &= y_{i,\mu}^0 - (H_{\text{pol}}^{\text{qs}} \circ u)_{(i,\mu)}(0)\quad \text{for}\ (i, \mu) \in \Gamma^{\text{ob}}, \end{aligned} \tag{12.66}$$

namely, $x(t) = e^{tA^{\text{ob}}} x(0) + \int_0^t e^{(t-\tau)A^{\text{ob}}} B^{\text{ob}} u(\tau)\, d\tau$ (see (3.62) and (3.63)), then the solution y of (12.65) is given by $y = C^{\text{ob}} x + D^{\text{ob}} \circ u$, in particular, $y_i = x_{i,0} + (D^{\text{ob}} \circ u)_i$ if $d(i) > 0$.

Proof We discuss continuous time only, discrete time is easier. We use

$$\begin{aligned} &\mathcal{B}^{\text{ob}} \overset{\cong}{\longleftrightarrow} \mathcal{B},\quad \binom{x}{u} \longleftrightarrow \binom{y}{u},\ \text{with} \\ &y = C^{\text{ob}} x + D^{\text{ob}} \circ u,\quad x_{i,\mu} = s^\mu \circ y_i - (H_{\text{pol}}^{\text{qs}} \circ u)_{(i,\mu)}. \end{aligned} \tag{12.67}$$

Uniqueness. Let y with its differentiability properties solve the initial value problem (12.65). The associated state vector x according to (12.67) is continuous and satisfies the initial value problem (12.66). Since x and u are continuous, so is $s \circ x = A^{\text{ob}} x + B^{\text{ob}} u$, hence $x \in C^1(\mathbb{R}, F)^{\Gamma^{\text{ob}}}$. The initial value problem (12.66) has a unique C^1-solution according to (3.62) and (3.63) and therefore, x and $y = C^{\text{ob}} x + D^{\text{ob}} \circ u$ are indeed unique.
Existence. Let x denote the unique C^1-solution of the initial value problem (12.66) and define $y = C^{\text{ob}} x + D^{\text{ob}} \circ u$. According to isomorphism (12.67), this implies that

$$y_i^{(\mu)} = s^\mu \circ y_i = x_{i,\mu} + (H_{\text{pol}}^{\text{qs}} \circ u)_{(i,\mu)}\quad \text{for}\ (i, \mu) \in \Gamma^{\text{ob}},$$

and these functions are continuous by assumption. Moreover, $\binom{y}{u} \in \mathcal{B}$ and

$$\begin{aligned} y_i^{(\mu)}(0) &= x_{i,\mu}(0) + (H_{\text{pol}}^{\text{qs}} \circ u)_{(i,\mu)}(0) \\ &= \left(y_{i,\mu}^0 - (H_{\text{pol}}^{\text{qs}} \circ u)_{(i,\mu)}(0) \right) + (H_{\text{pol}}^{\text{qs}} \circ u)_{(i,\mu)}(0) \\ &= y_{i,\mu}^0 \quad \text{for}\ (i, \mu) \in \Gamma^{\text{ob}}. \end{aligned}$$

□

Remark 12.3.13 We use the assumptions and notations of Theorem and Definition 12.3.5 and Theorem 12.3.12 in continuous time.

1. The canonical observability realization of an IO behavior with *proper* transfer matrix, i.e., with $H_{\rm pol} = D^{\rm ob} \in F^{p \times m}$, can be technically realized with elementary building blocks, i.e., with integrators, adders, and multipliers. A proper transfer matrix can always be achieved via Theorem 12.2.3 by a suitable choice of inputs and outputs.
2. If the transfer matrix is not proper, the following problems arise:
(i) The behavior may admit impulsive solutions that destroy the system, cf. Sect. 8.2.10.
(ii) The technical realization of $D^{\rm ob}\circ$ requires differentiators (of possibly high order) that have the undesirable property of amplifying noise signals. For instance, the small and bounded signal ϵe^{it^2} with small ϵ gives rise to its unbounded derivative $2\epsilon i t e^{it^2}$. Of course, first and second derivatives can sometimes be measured, as speedometers and accelerometers show. The exact measurement of higher derivatives, however, is generally impossible.
3. Assume that H is proper and that, as generally, $H_{\rm pol}^{\rm qs}$ is not constant. The exact initial value of (12.66) requires $(H_{\rm pol}^{\rm qs} \circ u)(0)$, i.e., higher derivatives of the input u at $t = 0$. In general, these are not available and cannot be measured. Hence, in general and in contrast to [1, Sect. 2.3.1], there is no precise simulation of y by x with given initial conditions $\left(y_i^{(\mu)}(0)\right)_{(i,\mu)\in\Gamma^{\rm ob}}$.
A trivial exception is the case of zero initial conditions for u, i.e., $u^{(\mu)}(0) = 0,\ \mu \in \mathbb{N}$, cf. [1, p. 38, (3)]. Then $x_{i,\mu}(0) = y_i^{(\mu)}(0),\ (i,\mu) \in \Gamma^{\rm ob}$.
4. Assume that H is proper and that $\mathcal{B}$, and hence $\mathcal{B}^{\rm ob}$ are $\mathbb{C}_-$-stable ($\mathbb{C}_- := \{\lambda \in \mathbb{C};\ \Re(\lambda) < 0\}$). The isomorphism $\mathcal{B}^{\rm ob} \cong \mathcal{B}$ from Theorem 12.3.5 induces the isomorphism of the autonomous parts
$$\begin{gathered} C^{\rm ob}\circ:\ \mathcal{B}^{\rm ob,0} = \left\{x \in \mathcal{F}^{\Gamma^{\rm ob}};\ s\circ x = A^{\rm ob}x\right\} \cong \mathcal{B}^0 = \left\{y \in \mathcal{F}^p;\ P\circ y = 0\right\} \\ \Longrightarrow \operatorname{char}(\mathcal{B}^{\rm ob,0}) = \operatorname{spec}(A^{\rm ob}) = \operatorname{char}(\mathcal{B}^0) \subset \mathbb{C}_-. \end{gathered} \tag{12.68}$$
The trajectories w in $\mathcal{B}^0$ and $\mathcal{B}^{\rm ob,0}$ are polynomial-exponential and satisfy $\lim_{t\to\infty} w(t) = 0$.
Let $u \in C^0(\mathbb{R}, F)^m$ be any continuous input, $\widehat{x} \in C^1(\mathbb{R}, F)^{\Gamma^{\rm ob}}$ the unique solution of $s\circ\widehat{x} = A^{\rm ob}\widehat{x} + B^{\rm ob}u$ with arbitrarily chosen $\widehat{x}(0)$ and $\widehat{y} := C^{\rm ob}\widehat{x} + D^{\rm ob}u \in C^0(\mathbb{R}, F)^p$; hence, $\binom{\widehat{y}}{u} \in \mathcal{B}$ and $P\circ\widehat{y} = Q\circ u$. If y is any solution of $P\circ y = Q\circ u$ then $y - \widehat{y} \in \mathcal{B}^0$. Therefore, $y \in C^0(\mathbb{R}, F)^p$ and $\widehat{y}$ is a simulation of $y = \widehat{y} + (y - \widehat{y})$, up to a $\mathbb{C}_-$-small summand.

◊

12.4 The Canonical Observer Realization of an IO Behavior

According to Theorem 6.3.8 all observable state space realizations of $\mathcal{B}$ are similar; hence, Theorem 12.3.5 cannot be essentially extended. However, one can, of course, develop different algorithms for the construction of observable state space realizations. Among the first were those of [12] and of Wolovich [11, Theorem 5.2.14]. In this section we will construct the *canonical observer realization* of the IO behavior

$$\mathcal{B} = \left\{ \left(\begin{smallmatrix} y \\ u \end{smallmatrix}\right) \in \mathcal{F}^{p+m};\ P \circ y = Q \circ u \right\} \text{ with } H = P^{-1}Q \tag{12.69}$$

from (12.33). Its main and important consequence is Theorem 12.4.12.

We use the assumptions and notations of Theorem and Definition 12.3.5, in particular, the rows of $P^g \in F[s]^{p\times p}$ from (12.35), namely,

$$\begin{aligned} P^g_{i-} &= s^{d^\circ(i)}\delta_i - \sum_{(j,\nu)\in\Gamma,\ (j,\nu)<(i,d^\circ(i))} A_{i,j,\nu}s^\nu\delta_j \\ &= s^{d^\circ(i)}\delta_i - \sum_{\nu<d^\circ(j),\ (j,\nu)<(i,d^\circ(i))} A_{i,j,\nu}s^\nu\delta_j, \\ &\text{where } \Gamma = \{(j,\nu);\ 1 \le j \le p,\ 0 \le \nu \le d^\circ(j)-1\}, \end{aligned} \tag{12.70}$$

form the unique reduced Gröbner basis of U^0 with respect to $\le$. Thus the IO behavior $\mathcal{B}$ can be represented as

$$\mathcal{B} = \left\{ \left(\begin{smallmatrix} y \\ u \end{smallmatrix}\right) \in \mathcal{F}^{p+m};\ P^g \circ y = Q^g \circ u \right\},\ Q^g := P^g H.$$

The matrix P^g satisfies

$$P^g_{ij} = s^{d^\circ(i)}\delta_{i,j} - \sum_{\nu=0}^{d^\circ(j)-1} A_{i,j,\nu}s^\nu,\ \text{hence } \deg_s(P^g_{ij}) < d^\circ(j) \text{ for } i \ne j. \tag{12.71}$$

For the canonical observer realization it is decisive to choose a *degree-over-position term order* on $\{1,\ldots,p\}\times\mathbb{N}$ (with dominant second term), i.e., with $\nu < \mu \Longrightarrow (j,\nu) < (i,\mu)$. In the sequel we will use the order

$$(j,\nu) < (i,\mu) :\Longleftrightarrow \nu < \mu \text{ or } \big(\nu = \mu \text{ and } j > i\big). \tag{12.72}$$

In particular, $(j,\nu) < (i,d^\circ(i))$ means that $\nu < d^\circ(i)$ or $\big(\nu = d^\circ(i) \text{ and } j > i\big)$. For the entries of P^g, this implies

$$\forall i:\ P^g_{ii}=s^{d^\circ(i)}-\sum_{\nu=0}^{d^\circ(i)-1}A_{i,i,\nu}s^\nu,\ \forall j\neq i:\ P^g_{ij}=-\sum_{\nu=0}^{d^\circ(j)-1}A_{i,j,\nu}s^\nu$$
$$\forall i,j:\ \deg_s(P^g_{ij})\le d^\circ(i),\ \forall j<i:\ \deg_s(P^g_{ij})<d^\circ(i),$$
$$\forall j>i \text{ with } \deg_s(P^g_{ij})=d^\circ(i):\ d^\circ(i)<d^\circ(j). \tag{12.73}$$

Remark 12.4.1 The conditions of (12.71) and (12.73) are a variant of those for the *Popov echelon form* [1, p. 481]. In contrast to this reference the $d^\circ(i)$ are not increasing and the pivot pairs are (i,i) instead of $(i,p(i))$. The unique coefficients of P^g are called the *Popov parameters*. These depend on the chosen term order. ◊

According to (12.30)–(12.32) we define the matrices $P^{g,\mathrm{hc}}\in \mathrm{Gl}_p(F)$ and $\Delta\in F[s]^{p\times p}$ by

$$P^{g,\mathrm{hc}}_{ij}:=\begin{cases}1 & \text{if } i=j\\ -A_{i,j,d^\circ(i)} & \text{if } i<j,\ d^\circ(i)<d^\circ(j)\\ 0 & \text{otherwise}\end{cases},\quad \Delta=\mathrm{diag}(s^{d^\circ(1)},\dots,s^{d^\circ(p)})$$
$$\Longrightarrow P^g=\Delta P^{g,\mathrm{hc}}-P^{g,\mathrm{lc}},\ P^{g,\mathrm{lc}}:=P^g-\Delta P^{g,\mathrm{hc}}\in F[s]^{p\times p}. \tag{12.74}$$

All entries in the main diagonal of the constant matrix $P^{g,\mathrm{hc}}$ are 1. Moreover, this matrix is upper triangular and, due to our term order, we have that

$$\forall i\neq j:\ \left(P^{g,\mathrm{hc}}_{ij}\neq 0 \Longrightarrow i<j \text{ and } d^\circ(i)<d^\circ(j)\right).$$

As a direct consequence of (12.71) and the second line of (12.45) the matrix $P^{g,\mathrm{lc}}$, which consist of the lower coefficient terms, satisfies

$$\forall i,j:\ \deg_s(P^{g,\mathrm{lc}}_{ij})<\min(d^\circ(i),d^\circ(j)).$$

Thus, P^g can be written as

$$P^g=\Delta Y \quad \text{with} \quad Y:=P^{g,\mathrm{hc}}\left(\mathrm{id}_p-(P^{g,\mathrm{hc}})^{-1}\Delta^{-1}P^{g,\mathrm{lc}}\right)\in \mathrm{Gl}_p(F(s)_{\mathrm{pr}}). \tag{12.75}$$

That Y belongs to $\mathrm{Gl}_p(F(s)_{\mathrm{pr}})$ was shown in the last part of the proof of Theorem 12.2.3.

The matrix $P^g=\Delta P^{g,\mathrm{hc}}-P^{g,\mathrm{lc}}$ from (12.74) is called *row-reduced* in the literature [1, p. 384]. In our derivations it is a consequence of the Gröbner property with respect to degree-over-position term orders.

In the next theorem we admit different degree-over-position term orders and not only the one from (12.72).

Theorem 12.4.2 (cf. [1, Theorem 6.3-13 and Lemma 6.3-14 on pp. 387–388]). *Up to a permutation, the indices $d^\circ(i)$ depend on U^0 only and not on the chosen degree-over-position term order* (12.72).

Proof Assume $U^0 = F[s]^{1\times p}P^{(k)}$ for $k = 1, 2$, as well as representations

$$P^{(k)} = \Delta^{(k)}P^{k,\mathrm{hc}} + P^{k,\mathrm{lc}} = \Delta^{(k)}Y^{(k)}$$
$$\text{with } \Delta^{(k)} = \mathrm{diag}\left(s^{d_k(1)}, \dots, s^{d_k(p)}\right) \ \text{ and } \ Y^{(k)} \in \mathrm{Gl}_p(F(s)_{\mathrm{pr}})$$

as in (12.74) and (12.75). Since $F[s]^{1\times p}P^{(1)} = F[s]^{1\times p}P^{(2)}$ there is

$$\begin{aligned}
&X \in \mathrm{Gl}_p(F[s]) \text{ with } XP^{(1)} = P^{(2)} \Longrightarrow X\Delta^{(1)}Y^{(1)} = \Delta^{(2)}Y^{(2)}\\
&\Longrightarrow (\Delta^{(2)})^{-1}X\Delta^{(1)} = Y^{(2)}(Y^{(1)})^{-1} =: Y \in \mathrm{Gl}_p(F(s)_{\mathrm{pr}})\\
&\Longrightarrow \forall i, j = 1, \dots, p : \ s^{-d_2(i)+d_1(j)}X_{ij} = Y_{ij}, \ X_{ij} \in F[s], \ Y_{ij} \in F(s)_{\mathrm{pr}}\\
&\Longrightarrow \forall i, j : \ -d_2(i) + d_1(j) + \deg_s(X_{ij}) = \deg_s(Y_{ij}) \le 0\\
&\Longrightarrow \forall i, j : \ \left(X_{ij} \ne 0 \Longrightarrow d_1(j) \le d_2(i)\right).
\end{aligned} \tag{12.76}$$

Choose permutations σ and τ such that

$$d_1(\tau(1)) \le d_1(\tau(2)) \le \dots \le d_1(\tau(p)), \quad d_2(\sigma(1)) \le d_2(\sigma(2)) \le \dots \le d_2(\sigma(p))$$

and define

$$\begin{aligned}
&d_1'(j) := d_1(\tau(j)), \ d_2'(i) := d_2(\sigma(i)), \ X' \in \mathrm{Gl}_p(F[s]), \ X'_{ij} := X_{\sigma(i)\tau(j)}\\
&\Longrightarrow d_1'(1) \le d_1'(2) \le \dots \le d_1'(p), \ d_2'(1) \le d_2'(2) \le \dots \le d_2'(p) \text{ and}\\
&\forall i, j : \left(X'_{ij} = X_{\sigma(i)\tau(j)} \ne 0 \underset{(12.76)}{\Longrightarrow} d_1'(j) = d_1(\tau(j)) \le d_2(\sigma(i)) = d_2'(i)\right).
\end{aligned} \tag{12.77}$$

We show $d_1'(i) \le d_2'(i)$ for all i and likewise $d_2'(i) \le d_1'(i)$, and hence $d_1'(i) = d_2'(i)$: If this is false there is a k, $1 \le k \le p$, such that

$$\begin{aligned}
&\forall j \in \{1, \dots, k-1\} : \ d_1'(j) \le d_2'(j), \ \text{ but } d_1'(k) > d_2'(k)\\
&\Longrightarrow d_2'(1) \le \dots \le d_2'(k) < d_1'(k) \le \dots \le d_1'(p)\\
&\underset{(12.77)}{\Longrightarrow} X'_{ij} = 0 \text{ for } i = 1, \cdots, k, \ j = k, \cdots, p,\\
&\Longrightarrow X' = \begin{pmatrix} X'_{11} & \cdots & X'_{1,k-1} & \overset{k}{0} & \cdots & 0\\ \cdots & \cdots & \cdots & \cdots & \cdots & \cdots\\ X'_{k1} & \cdots & X'_{k,k-1} & 0 & \cdots & 0\\ \cdots & \cdots & \cdots & \cdots & \cdots & \cdots\\ \cdots & \cdots & \cdots & \cdots & \cdots & \cdots \end{pmatrix}.
\end{aligned} \tag{12.78}$$

Thus the first k rows of X' are $F(s)$-linearly dependent, and hence $\mathrm{rank}(X') < p$ in contradiction to $X' \in \mathrm{Gl}_p(F[s])$. We conclude

$$d_1(\tau(i)) = d_1'(i) = d_2'(i) = d_2(\sigma(i)) \text{ and hence } d_2(j) = d_1(\tau\sigma^{-1}(j)).$$

This signifies that the vectors $(d_1(1), \dots, d_1(p))$ and $(d_2(1), \dots, d_2(p))$ coincide up to a permutation or the order of the entries. □

Corollary 12.4.3 *For degree-over-position term orders the observability indices $d^{\mathrm{ob}}(i)$ from Theorem and Definition 12.3.5 coincide with the $d^{\mathrm{o}}(i)$ from* (12.74). *Therefore the latter are also called* observability indices. *Example 12.3.9 shows that for other term orders this is not the case.* ◊

Example 12.4.4 The Popov parameters depend on the term order. Indeed , consider two degree-over-position term orders on $\{1, 2\} \times \mathbb{N}$, viz.,

$$(i, \mu) <_1 (j, \nu) :\Longleftrightarrow \mu < \nu \text{ or } (\mu = \nu \text{ and } i > j) \quad \text{and}$$
$$(i, \mu) <_2 (j, \nu) :\Longleftrightarrow \mu < \nu \text{ or } (\mu = \nu \text{ and } i < j).$$

Consider the matrix $P = \begin{pmatrix} s^2 & 2s^2 \\ 3s & 5s^3 \end{pmatrix}$ and the associated module

$$U^0 = F[s]^{1\times 2}P = F[s](s, s^2) \oplus F[s](s, s^3).$$

The Gröbner matrices $P^{(i)}$ of P with respect to $<_i$ and the corresponding degree sets $\deg_i(U^0)$ and observability indices are

$$P^{(1)} = \begin{pmatrix} s^2 & 2s^2 \\ (3/5)s & s^3 \end{pmatrix}, \ \deg_1(U^0) = (\{1\} \times (2 + \mathbb{N})) \uplus (\{2\} \times (3 + \mathbb{N})),$$
$$d(1) = 2, \quad d(2) = 3 \quad \text{and}$$
$$P^{(2)} = \begin{pmatrix} (1/2)s^2 & s^2 \\ -(6/5)s + s^3 & 0 \end{pmatrix}, \ \deg_2(U^0) = (\{1\} \times (3 + \mathbb{N})) \uplus (\{2\} \times (2 + \mathbb{N})),$$
$$d(1) = 3, \quad d(2) = 2.$$

Up to the transposition, the indices $d(i)$ for the two term orders coincide, but the coefficients of the entries, the Popov parameters, differ. ◊

The observer form of $\mathcal{B}$ is constructed in several reduction steps. First we construct two IO behaviors $\mathcal{B}^1$ and $\mathcal{B}^2$ which are isomorphic to $\mathcal{B}$, and then state space equations for $\mathcal{B}^1$. We define the matrices

$$P^{1,\mathrm{lc}} := P^{g,\mathrm{lc}}(P^{g,\mathrm{hc}})^{-1} \text{ and } P^1 := P^g(P^{g,\mathrm{hc}})^{-1} = \Delta - P^{1,\mathrm{lc}}, \tag{12.79}$$

as well as the IO behavior

$$\mathcal{B}^2 := \left\{ \begin{pmatrix} y \\ u \end{pmatrix} \in \mathcal{F}^{p+m};\ P^1 \circ y = Q^g \circ u \right\} \tag{12.80}$$

with the transfer matrix $H^2 = (P^1)^{-1}Q^g = P^{g,\mathrm{hc}}H$, hence

$$\begin{pmatrix} (P^{g,\mathrm{hc}})^{-1} & 0 \\ 0 & \mathrm{id}_m \end{pmatrix} \circ \colon \mathcal{B}^2 \overset{\cong}{\longrightarrow} \mathcal{B}, \quad \begin{pmatrix} y \\ u \end{pmatrix} \longmapsto \begin{pmatrix} (P^{g,\mathrm{hc}})^{-1}y \\ u \end{pmatrix}. \tag{12.81}$$

Moreover

$$
\begin{aligned}
&P_{ij}^{1,\mathrm{lc}} = \sum_k P_{ik}^{g,\mathrm{lc}} (P^{g,\mathrm{hc}})_{kj}^{-1} \Longrightarrow \deg_s(P_{ij}^{1,\mathrm{lc}}) \leq \max_k \deg_s(P_{ik}^{g,\mathrm{lc}}) < d^{\circ}(i)\\
&\Longrightarrow \Delta^{-1} P^{1,\mathrm{lc}} \in F(s)_{\mathrm{spr}}^{p\times p} \underset{\text{Lemma 12.2.2}}{\Longrightarrow} \Delta^{-1}P^1 = \mathrm{id}_p - \Delta^{-1}P^{1,\mathrm{lc}} \in \mathrm{Gl}_p(F(s)_{\mathrm{pr}}).
\end{aligned}
\tag{12.82}
$$

The next reduction step is analogous to (12.52). We decompose $H = H_{\mathrm{pol}} + H_{\mathrm{spr}}$ into its polynomial and strictly proper part and conclude

$$
\begin{aligned}
&H^2 = P^{g,\mathrm{hc}} H = H_{\mathrm{pol}}^2 + H_{\mathrm{spr}}^2 \text{ with } H_{\mathrm{pol}}^2 = P^{g,\mathrm{hc}} H_{\mathrm{pol}},\ H_{\mathrm{spr}}^2 = P^{g,\mathrm{hc}} H_{\mathrm{spr}}\\
&\Longrightarrow Q^g = P^g H = P^g H_{\mathrm{pol}} + Q^1 \text{ with } Q^1 := P^g H_{\mathrm{spr}} = P^1 H_{\mathrm{spr}}^2 \in F[s]^{p\times m}.
\end{aligned}
\tag{12.83}
$$

Moreover

$$
\begin{aligned}
&\Delta^{-1}P^1 \in \mathrm{Gl}_p(F(s)_{\mathrm{pr}}) \Longrightarrow \Delta^{-1}Q^1 = (\Delta^{-1}P^1) H_{\mathrm{spr}}^2 \in F(s)_{\mathrm{spr}}^{p\times m}\\
&\Longrightarrow \forall i = 1, \ldots, p:\ \deg_s(Q_{i,j}^1) < d^{\circ}(i).
\end{aligned}
\tag{12.84}
$$

Therefore, the rows of P^1 and Q^1 have the form

$$
\begin{aligned}
&P_{i-}^1 = s^{d^{\circ}(i)}\delta_i - \sum_{\mu=0}^{d^{\circ}(i)-1} P_{i-}^{1,\mathrm{lc}}(\mu) s^{\mu} \in F[s]^{1\times p}, \quad \text{and}\\
&Q_{i-}^1 = \sum_{\mu=0}^{d^{\circ}(i)-1} Q_{i-}^1(\mu) s^{\mu} \in F[s]^{1\times m}\\
&\text{with constant rows } P_{i-}^{1,\mathrm{lc}}(\mu) \in F^{1\times p} \text{ and } Q_{i-}^1(\mu) \in F^{1\times m}.
\end{aligned}
\tag{12.85}
$$

We define the IO behavior $\mathcal{B}^1$ and its strictly proper transfer matrix

$$
\mathcal{B}^1 := \left\{ \begin{pmatrix} y^1 \\ u \end{pmatrix} \in \mathcal{F}^{p+m};\ P^1 \circ y^1 = Q^1 \circ u \right\},\ H^1 = (P^1)^{-1} Q^1 = H_{\mathrm{spr}}^2.
\tag{12.86}
$$

From (12.85) we infer

$$
\left(d^{\circ}(i) = 0 \text{ and } \begin{pmatrix} y^1 \\ u \end{pmatrix} \in \mathcal{B}^1\right) \Longrightarrow y_i^1 = 0.
\tag{12.87}
$$

Equation (12.85) also implies the equivalences

$$
\begin{aligned}
&\begin{pmatrix} y^1 \\ u \end{pmatrix} \in \mathcal{B}^1 \Longleftrightarrow P^1 \circ y^1 = Q^1 \circ u \Longleftrightarrow \text{for } i = 1, \ldots, p:\\
&s^{d^{\circ}(i)} \circ y_i^1 = \sum_{\mu=0}^{d^{\circ}(i)-1} P_{i-}^{1,\mathrm{lc}}(\mu) s^{\mu} \circ y^1 + \sum_{\mu=0}^{d^{\circ}(i)-1} Q_{i-}^1(\mu) s^{\mu} \circ u\\
&\Longleftrightarrow s^{d^{\circ}(i)} \circ y_i^1 = \sum_{\mu=0}^{d^{\circ}(i)-1} s^{\mu} z_i(\mu),\ z_i(\mu) := P_{i-}^{1,\mathrm{lc}}(\mu) y^1 + Q_{i-}^1(\mu) u.
\end{aligned}
\tag{12.88}
$$

The equivalence

$$P^1 \circ y = Q^g \circ u = P^1 H^2_{\text{pol}} \circ u + Q^1 \circ u \iff P^1 \circ (y - H^2_{\text{pol}} \circ u) = Q^1 \circ u$$

implies the isomorphism of IO behaviors

$$\begin{pmatrix} \text{id}_p & H^2_{\text{pol}} \\ 0 & \text{id}_m \end{pmatrix} \circ : \ \mathcal{B}^1 \xrightarrow{\cong} \mathcal{B}^2. \tag{12.89}$$

Together with (12.81) we obtain the IO behavior isomorphism

$$\begin{pmatrix} (P^{g,\text{hc}})^{-1} & 0 \\ 0 & \text{id}_m \end{pmatrix} \begin{pmatrix} \text{id}_p & H^2_{\text{pol}} \\ 0 & \text{id}_m \end{pmatrix} \circ = \begin{pmatrix} (P^{g,\text{hc}})^{-1} & (P^{g,\text{hc}})^{-1} H^2_{\text{pol}} \\ 0 & \text{id}_m \end{pmatrix} \circ = \begin{pmatrix} (P^{g,\text{hc}})^{-1} & H_{\text{pol}} \\ 0 & \text{id}_m \end{pmatrix} \circ : \quad \mathcal{B}^1 \xrightarrow{\cong} \mathcal{B}^2 \xrightarrow{\cong} \mathcal{B}. \tag{12.90}$$

After these preliminary reduction steps we proceed to the observer canonical form proper by manipulating the behavior $\mathcal{B}^1$. Equation (12.88) suggests to define the state vector

$$x = (x_{i,\mu})_{(i,\mu)\in\Gamma^{\text{o}}} \in \mathcal{F}^{\Gamma^{\text{o}}}, \ \Gamma^{\text{o}} := \left\{(i,\mu);\ 1 \le i \le p,\ 1 \le \mu \le d^{\text{o}}(i)\right\}, \tag{12.91}$$

and the state equations

$$s \circ x_{i,\mu} = \sum_{j;\, d^{\text{o}}(j)>0} P^{1,\text{lc}}_{ij}(\mu-1) x_{j,d^{\text{o}}(j)} + \sum_{j=1}^{m} Q^1_{ij}(\mu-1) u_j + \begin{cases} 0 & \text{if } \mu = 1 \\ x_{i,\mu-1} & \text{if } \mu > 1 \end{cases}. \tag{12.92}$$

We define the *canonical observer matrices* $A^{\text{o}} \in F^{\Gamma^{\text{o}}\times\Gamma^{\text{o}}}$, $B^{\text{o}} \in F^{\Gamma^{\text{o}}\times m}$, $C^{1,\text{o}}$, $C^{\text{o}} \in F^{p\times\Gamma^{\text{o}}}$, and $D^{\text{o}} \in F[s]^{p\times m}$ by

$$\begin{aligned} A^{\text{o}}_{(i,\mu),(j,\nu)} &:= \delta_{i,j}\delta_{\nu,\mu-1} + \delta_{\nu,d^{\text{o}}(j)} P^{1,\text{lc}}_{i,j}(\mu-1), \\ B^{\text{o}}_{(i,\mu),j} &:= Q^1_{i,j}(\mu-1), \ C^{1,\text{o}}_{i,(j,\nu)} := \delta_{i,j}\delta_{\nu,d^{\text{o}}(j)}, \\ C^{\text{o}} &:= (P^{g,hc})^{-1} C^{1,\text{o}}, \text{ i.e., } C^{\text{o}}_{i,(j,\nu)} = (P^{g,\text{hc}})^{-1}_{ij} \delta_{\nu,d^{\text{o}}(j)}, \\ D^{\text{o}} &:= (P^{g,hc})^{-1} H^2_{\text{pol}} = H_{\text{pol}}. \end{aligned} \tag{12.93}$$

With this equation (12.92) gets the standard form

$$s \circ x = A^{\text{o}} x + B^{\text{o}} u. \tag{12.94}$$

Theorem 12.4.5 (Observer canonical form, cf. [11, p. 82], [1, §6.4.1, §6.4.3]). *Assume the behavior $\mathcal{B}$ from* (12.69) *and the derived data from* (12.71)–(12.75) *and* (12.79)–(12.94).

1. *The state space equations*

$$s \circ x = A^{o}x + B^{o}u, \; y = C^{o}x + D^{o} \circ u, \tag{12.95}$$

or, in other form,

$$s \circ x_{i,\mu} = \sum_{j;\, d^{o}(j)>0} P_{ij}^{1,\mathrm{lc}}(\mu-1)x_{j,d^{o}(j)} + \sum_{j=1}^{m} Q_{ij}^{1}(\mu-1)u_j + \begin{cases} 0 & \text{if } \mu = 1 \\ x_{i,\mu-1} & \text{if } \mu > 1 \end{cases} \tag{12.96}$$

$$y = (P^{g,\mathrm{hc}})^{-1}\left(y^1 + H_{\mathrm{pol}}^{2} \circ u\right) = (P^{g,\mathrm{hc}})^{-1}y^1 + H_{\mathrm{pol}} \circ u \text{ where} \tag{12.97}$$

$$y_i^1 = \begin{cases} x_{i,d^{o}(i)} & \text{if } d^{o}(i) > 0 \\ 0 & \text{if } d^{o}(i) = 0 \end{cases} \tag{12.98}$$

are an observable realization of the IO behavior $\mathcal{B}$, i.e., they induce the behavior isomorphism

$$\begin{aligned} &\begin{pmatrix} C^{o} & D^{o} \\ 0 & \mathrm{id}_m \end{pmatrix} \circ : \mathcal{B}^{o} := \left\{ \begin{pmatrix} x \\ u \end{pmatrix} \in \mathcal{F}^{\Gamma^{o}+m};\; s \circ x = A^{o}x + B^{o}u \right\} \\ &\overset{\cong}{\longrightarrow} \mathcal{B},\; \begin{pmatrix} x \\ u \end{pmatrix} \mapsto \begin{pmatrix} y \\ u \end{pmatrix} = \begin{pmatrix} C^{o}x + D^{o} \circ u \\ u \end{pmatrix}. \end{aligned} \tag{12.99}$$

The equations depend on $\mathcal{B}$ and the chosen term order from (12.72) *only and are called the* canonical observer realization *of $\mathcal{B}$. According to Theorem 12.4.2, the observability indices $d^{o}(i)$ do not depend on the choice of the degree-over-position term order, up to a permutation. As can be seen from* (12.93)*, the matrices A^{o} and C^{o} of the observer form are sparse like those of the canonical observability realization from Theorem and Definition 12.3.5.*

2. *The state space dimension of $\mathcal{B}^{o}$ is $n = \sharp(\Gamma^{o}) = \sum_{i=1}^{p} d^{o}(i)$.*
3. *The isomorphism* (12.99) *implies $H = D^{o} + C^{o}(s\, \mathrm{id}_{\Gamma^{o}} - A^{o})^{-1}B^{o}$.*

Proof 1. It suffices to show that the map

$$\begin{pmatrix} C^{1,o} & 0 \\ 0 & \mathrm{id}_m \end{pmatrix} \circ :\; \mathcal{B}^{o} \longrightarrow \mathcal{B}^{1},\; \begin{pmatrix} x \\ u \end{pmatrix} \longmapsto \begin{pmatrix} y^1 \\ u \end{pmatrix},\; y_j^1 = \begin{cases} x_{j,d^{o}(j)} & \text{if } d^{o}(j) > 0, \\ 0 & \text{if } d^{o}(j) = 0, \end{cases} \tag{12.100}$$

is *well defined and an isomorphism*. Composing this map with the isomorphism (12.90) then furnishes the asserted isomorphism

$$\begin{pmatrix} C^{o} & D^{o} \\ 0 & \mathrm{id}_m \end{pmatrix} \circ = \begin{pmatrix} (P^{g,\mathrm{hc}})^{-1} & H_{\mathrm{pol}} \\ 0 & \mathrm{id}_m \end{pmatrix} \begin{pmatrix} C^{1,o} & 0 \\ 0 & \mathrm{id}_m \end{pmatrix} \circ :\; \mathcal{B}^{o} \overset{\cong}{\longrightarrow} \mathcal{B}^{1} \overset{\cong}{\longrightarrow} \mathcal{B}.$$

With

$$z_i(\mu) := \sum_{j,d^{o}(j)>0} P_{ij}^{1}(\mu)x_{j,d^{o}(j)} + \sum_{j=1}^{m} Q_{ij}^{1}(\mu)u_j,\; 1 \le \mu \le d^{o}(i), \tag{12.101}$$

cf. (12.88), the state equations (12.96), (12.98) have the form

$$s \circ x_{i,\mu} = \begin{cases} z_i(0) & \text{if } \mu = 1 \\ x_{i,\mu-1} + z_i(\mu - 1) & \text{if } 1 < \mu \le d^o(i) \end{cases} \quad \text{for } (i, \mu) \in \Gamma^o,$$
$$y_j = \begin{cases} x_{j,d^o(j)} & \text{if } d^o(j) > 0 \\ 0 & \text{if } d^o(j) = 0 \end{cases}. \tag{12.102}$$

Well defined: Let $\binom{x}{u} \in \mathcal{B}^o$, i.e., $\binom{x}{u}$ satisfies (12.96). We have to show $\binom{y^1}{u} \in \mathcal{B}^1$, i.e., $P^1 \circ y^1 = Q^1 \circ u$ with y^1 from (12.100). A simple induction from (12.102) proves $s^\mu \circ x_{i,\mu} = \sum_{\nu=0}^{\mu-1} s^\nu \circ z_i(\nu)$. For $i = 1, \cdots, p$ with $d^o(i) > 0$ this implies

$$s^{d^o(i)} \circ y_i^1 = s^{d^o(i)} \circ x_{i,d^o(i)} = \sum_{\nu=0}^{d^o(i)-1} s^\nu \circ z_i(\nu) \underset{(12.88)}{\Longrightarrow} \binom{y^1}{u} \in \mathcal{B}^1. \tag{12.103}$$

This signifies that the map (12.100) is well defined.
Injective: If $\binom{x}{u}$ maps onto $\binom{y^1}{u} = \binom{0}{0}$ under the map (12.100), we conclude from (12.101) and (12.102) that all $z_i(\mu)$, $(i, \mu) \in \Gamma^o$, are zero. The recursion equations (12.102) starting with $x_{i,d^o(i)} = y_i^1 = 0$ then imply $x_{i,\mu} = 0$ for all $(i, \mu) \in \Gamma^o$.
Surjective: Let $\binom{y^1}{u} \in \mathcal{B}^1$; hence, $y_i = 0$ for $d^o(i) = 0$. With $z_i(\mu) := P_{i-}^{1,\mathrm{lc}}(\mu)y^1 + Q_{i-}^{1,\mathrm{lc}}(\mu)u$ for $(i, \mu) \in \Gamma^o$ we define recursively

$$x_{i,\mu} := \begin{cases} y_i^1 & \text{if } \mu = d^o(i) \\ s \circ x_{i,\mu+1} - z_i(\mu) & \text{if } 1 \le \mu < d^o(i) \end{cases}$$
$$\Longrightarrow s \circ x_{i,\mu} = x_{i,\mu-1} + z_i(\mu - 1) \text{ for } 1 < \mu \le d^o(i),\ \ y_i = x_{i,d^o(i)} \text{ for } d^o(i) > 0. \tag{12.104}$$

For $\binom{x}{u} \in \mathcal{B}^o$ it remains to show $s \circ x_{i,1} = z_i(0)$ according to (12.102). By recursion the equation (12.104) for $x_{i,\mu}$ implies

$$s^{d^o(i)} \circ y_i^1 = s \circ x_{i,1} + \sum_{\nu=1}^{d^o(i)-1} s^\nu \circ z_i(\nu). \tag{12.105}$$

On the other hand $\binom{y^1}{u} \in \mathcal{B}^1$ and (12.88) imply

$$s^{d^o(i)} \circ y_i^1 = \sum_{\nu=0}^{d^o(i)-1} s^\nu z_i(\nu) = z_i(0) + \sum_{\nu=1}^{d^o(i)-1} s^\nu \circ z_i(\nu) \underset{(12.105)}{\Longrightarrow} s \circ x_{i,1} = z_i(0). \tag{12.106}$$

2, 3. Obvious. □

In the literature, mostly either the controller form (Sect. 12.7) or the observer form are derived, and the other form is obtained by duality. The uniqueness of these forms, which we prove here via the uniqueness of the reduced Gröbner basis, is not shown in general.

Remark 12.4.6 Recall $\Gamma^{\mathrm{o}} = \{(i,\mu);\ 1 \le i \le p,\ 1 \le \mu \le d^{\mathrm{o}}(i)\}, n = \sum_{i=1}^{p} d^{\mathrm{o}}(i)$. To transform the matrices with indices in Γ^{o} into standard matrices with indices in $\{1, \cdots, n\}$ one uses the bijection

$$\alpha:\ \Gamma^{\mathrm{o}} \cong \{1, \cdots, n\},\ (i,\mu) \mapsto \alpha(i,\mu) := \sum_{k=1}^{i-1} d^{\mathrm{o}}(k) + \mu, \text{ and defines}$$
$$A^{\mathrm{o}} \in F^{n\times n} \underset{\text{ident.}}{=} F^{\Gamma^{\mathrm{o}}\times\Gamma^{\mathrm{o}}},\ B^{\mathrm{o}} \in F^{n\times m},\ C^{\mathrm{o}} \in F^{p\times n},\ D^{\mathrm{o}} \in F[s]^{p\times m} \text{ by}$$
$$A^{\mathrm{o}}_{\alpha(i,\mu),\alpha(j,\nu)} := A^{\mathrm{ob}}_{(i,\mu),(j,\nu)},\ B^{\mathrm{o}}_{\alpha(i,\mu),j} := B^{\mathrm{o}}_{(i,\mu),j},\ C^{\mathrm{o}}_{i,\alpha(j,\nu)} := C^{\mathrm{o}}_{i,(j,\nu)}.$$

◊

Corollary 12.4.7 (Algorithm for the observer realization) *For the computation of the observer realization of* $\mathcal{B} = \left\{\binom{y}{u} \in \mathcal{F}^{p+m};\ P \circ y = Q \circ u\right\}$ *with the transfer matrix* $H := P^{-1}Q$ *the following algorithmic steps have to be taken:*

1. *Compute the decomposition* $H = H_{\mathrm{pol}} + H_{\mathrm{spr}}$ *into polynomial and strictly proper part by multiple application of the Euclidean algorithm.*
2. *Choose the degree-over-position term order* (12.72), *compute the reduced Gröbner basis* $g^{(i)} \in F[s]^{1\times p}$ *of* $\mathcal{D}^{1\times p}P$ *and define* $P^g := \left(g^{(i)}\right)_i \in F[s]^{p\times p}$ *with*
$$U^0 = F[s]^{1\times p}P = \bigoplus_{i=1}^{p} F[s]P_{i-} = \bigoplus_{i=1}^{p} F[s]g^{(i)} = F[s]^{1\times p}P^g.$$
3. *The entries of* P^g *have the form* $P^g_{ij} = s^{d^{\mathrm{o}}(i)}\delta_{i,j} - \sum_{\nu=0}^{d^{\mathrm{o}}(j)-1} A_{i,j,\nu}s^{\nu}$. *Define*
$$\Gamma^{\mathrm{o}} := \left\{(i,\mu);\ 1 \le i \le p,\ 1 \le \mu \le d^{\mathrm{o}}(i)\right\},$$
$$\Delta := \mathrm{diag}\left(s^{d^{\mathrm{o}}(1)}, \ldots, s^{d^{\mathrm{o}}(p)}\right) \in \mathrm{Gl}_p(F[s]) \textit{ and } P^{g,\mathrm{hc}} \in \mathrm{Gl}_p(F) \textit{ by}$$
$$P^{g,\mathrm{hc}}_{ij} = \begin{cases} 1 & \text{if } i = j \\ -A_{i,j,d^{\mathrm{o}}(i)} & \text{if } i < j \text{ and } d^{\mathrm{o}}(i) < d^{\mathrm{o}}(j) \\ 0 & \text{otherwise} \end{cases}.$$
4. *Compute*
$$\left(P^{g,\mathrm{hc}}\right)^{-1},\ P^1 := P^g(P^{g,\mathrm{hc}})^{-1},\ P^{1,\mathrm{lc}} := \Delta - P^1,\ Q^1 := P^g H_{\mathrm{spr}},$$
$$P^{1,\mathrm{lc}}_{ij} = \sum_{\mu=0}^{d^{\mathrm{o}}(j)-1} P^{1,\mathrm{lc}}_{ij}(\mu)s^{\mu},\ Q^1_{ij} = \sum_{\mu=0}^{d^{\mathrm{o}}(j)-1} Q^1_{ij}(\mu)s^{\mu}.$$
5. *Use* (12.93) *to compute*

$$A^{\mathrm{o}} \in F^{\Gamma^{\mathrm{o}} \times \Gamma^{\mathrm{o}}},\ B^{\mathrm{o}} \in F^{\Gamma^{\mathrm{o}} \times m},$$
$$C^{\mathrm{o}} = \left(P^{g,\mathrm{hc}}\right)^{-1} C^{1,\mathrm{o}} \in F^{p \times \Gamma^{\mathrm{o}}},\ D^{\mathrm{o}} := H_{\mathrm{pol}} \in F[s]^{p \times m}.$$

◊

Example 12.4.8 (The SISO case). We consider the situation of Theorem 12.4.5 in the SISO case with $m = p = 1$ and $\deg(P) = n > 0$. The behavior is

$$\mathcal{B} = \left\{ \binom{y}{u} \in \mathcal{F}^2;\ P \circ y = Q \circ u \right\},$$
$$\text{where } P = s^n - \sum_{\mu=0}^{n-1} p_\mu s^\mu,\quad Q \in F[s],\quad \text{and}\quad H = P^{-1}Q \in F(s).$$

With the notations from Theorem 12.4.5 we easily obtain $P^1 = P^g = P$ and

$$P^{g,\mathrm{hc}} = 1,\ P^{g,\mathrm{lc}} = \sum_{\mu=0}^{n-1} p_\mu s^\mu,\ Q^g = Q,\ H^2 = H,\ d^{\mathrm{o}} = 1,\ \Gamma^{\mathrm{o}} = \{1, \ldots, n\}.$$

By Euclidean division the decomposition $H = H_{\mathrm{pol}} + H_{\mathrm{spr}}$ is computed and furnishes

$$Q = PH = PH_{\mathrm{pol}} + Q^1,\ Q^1 := PH_{\mathrm{spr}},\ \deg_s(Q^1) < n = \deg_s(P)$$
$$\Longrightarrow Q^1 := \sum_{\mu=0}^{n-1} q_\mu s^\mu,\ H_{\mathrm{spr}} = P^{-1}Q^1.$$

The state vector of the canonical observer realization is $x = \begin{pmatrix} x_1 \\ \vdots \\ x_n \end{pmatrix} \in F^{\Gamma^{\mathrm{o}}} = F^n$ by (12.91). According to (12.93) the matrices of this realization are

$$A^{\mathrm{o}} = \begin{pmatrix} 0 & \cdots & 0 & p_0 \\ 1 & & 0 & p_1 \\ & \ddots & & \vdots \\ 0 & & 1 & p_{n-1} \end{pmatrix} \in F^{n \times n},\quad B^{\mathrm{o}} = \begin{pmatrix} q_0 \\ \vdots \\ q_{n-1} \end{pmatrix} \in F^n,$$
$$C^{\mathrm{o}} = (0, \cdots, 0, 1) \in F^{1 \times n},\quad D^{\mathrm{o}} = H_{\mathrm{pol}} \in F[s].$$

The resulting state equations are

$$s \circ x_\mu = \begin{cases} p_0 x_n + q_0 u & \text{if } \mu = 1, \\ x_{\mu-1} + p_{\mu-1} x_n + q_{\mu-1} u & \text{if } 1 < \mu \le n, \end{cases} \qquad y = x_n + H_{\mathrm{pol}} \circ u.$$

The corresponding interconnection diagram is

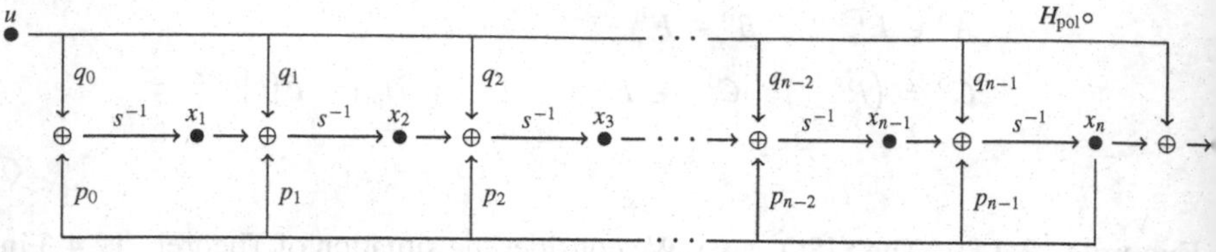

compare [1, Fig. 2.1-9. on p. 43].

If H is proper or, equivalently, $\deg_s(Q) \le n$ then

$$H_{\mathrm{pol}} = D^{\mathrm{o}} = \mathrm{lc}(Q) \text{ and } H_{\mathrm{pol}} \circ u = \mathrm{lt}(Q)u.$$

◊

Example 12.4.9 1. As in Example 12.3.11 we choose the base field $F := \mathbb{Z}/\mathbb{Z}7 = \{-3, \ldots, 3\}$. The notations are that of Theorem 12.4.5. We use the degree-over-position term order with

$$(j, \nu) < (i, \mu) :\Longleftrightarrow \nu < \mu \text{ or } \nu = \mu \text{ and } i < j.$$

We randomly choose $P \in \mathcal{D}^{5\times 5}$ and $Q \in \mathcal{D}^{5\times 2}$ by

$$P = \begin{pmatrix} s^4+s^3-s^2+1 & s^4-2s^3+3s^2 & 2s^5+3s^4-2s^3+2 & s^5+s^4+1 & 2s^2-1 \\ s^2-2 & s^2-s+3 & -s^2+3s & 3s^2-s-3 & 2s^3+3s^2+2s-2 \\ -3s^2+2s+3 & -3s^2-s-2 & s^3-2s^2+2s & -3s^3-3s^2 & 3s^2-2s+1 \\ 2s^3+s^2-2s & 2s^3-s^2-2s+3 & -2s^3-2s^2-3 & -s^3+s^2-2 & -3s^4+s^3-3s^2-3s-2 \\ 3s^4+s^3+3s^2+s-1 & 3s^4-s^3-s^2+s & -s^5-2s^4+2s^3-s^2-2s-2 & 3s^5+s^4-2s^3-1 & 2s^3-2s^2+s+1 \end{pmatrix},$$

$$Q = \begin{pmatrix} -2s^3+s^2-2s+1 & -3s^3+2s^2+3s+3 \\ 2s^3-2s^2-s-2 & -2s^3-3s^2+s+3 \\ 3s^2+2s-1 & -2s^3+3s^2+2s+1 \\ 3s^3+s^2-s+2 & -s^3-3s^2+2s \\ -s^3-3s^2-s+2 & -3s^3-s^2+3s+2 \end{pmatrix}.$$

We now apply the algorithm of Corollary 12.4.7 step by step to compute

$$(A^{\mathrm{o}}, B^{\mathrm{o}}, C^{\mathrm{o}}, D^{\mathrm{o}}),\ A^{\mathrm{o}} \in F^{\Gamma^{\mathrm{o}}\times\Gamma^{\mathrm{o}}} = F^{n\times n} \text{ etc.}$$

2. The reduced Gröbner basis is formed by the rows of

$$P^g = \begin{pmatrix} 1 & 0 & 2 & 0 & 1 \\ 0 & s^2-s+3 & -3s^2+3s-3 & 0 & 1 \\ 0 & 3s & s^3+2s^2-1 & 0 & 1 \\ 0 & 0 & 0 & 1 & -2 \\ 0 & 0 & 0 & 0 & s \end{pmatrix}, \text{ i.e.,}$$

$$P^{g,\mathrm{hc}} = \begin{pmatrix} 1 & 0 & 2 & 0 & 1 \\ 0 & 1 & -3 & 0 & 0 \\ 0 & 0 & 1 & 0 & 0 \\ 0 & 0 & 0 & 1 & -2 \\ 0 & 0 & 0 & 0 & 1 \end{pmatrix}, \quad \left(P^{g,\mathrm{hc}}\right)^{-1} = \begin{pmatrix} 1 & 0 & -2 & 0 & -1 \\ 0 & 1 & 3 & 0 & 0 \\ 0 & 0 & 1 & 0 & 0 \\ 0 & 0 & 0 & 1 & 2 \\ 0 & 0 & 0 & 0 & 1 \end{pmatrix},$$

$$P^1 = \begin{pmatrix} 1 & 0 & 0 & 0 & 0 \\ 0 & s^2-s+3 & -1 & 0 & 1 \\ 0 & 3s & s^3+2s^2+2s-1 & 0 & 1 \\ 0 & 0 & 0 & 1 & 0 \\ 0 & 0 & 0 & 0 & s \end{pmatrix},$$

$$P^{1,\mathrm{lc}} = \begin{pmatrix} 0 & 0 & 0 & 0 & 0 \\ 0 & s-3 & 1 & 0 & -1 \\ 0 & -3s & -2s^2-2s+1 & 0 & -1 \\ 0 & 0 & 0 & 0 & 0 \\ 0 & 0 & 0 & 0 & 0 \end{pmatrix},$$

$$Q^1 = \begin{pmatrix} 0 & 0 \\ 2s & -2s+1 \\ -s^2+s-2 & -2s^2-2s-1 \\ 0 & 0 \\ -3 & 2 \end{pmatrix}.$$

We conclude

$$d^{\mathrm{o}} = (0,\ 2,\ 3,\ 0,\ 1),\ n = 6,\ \{1, 2, 3, 4, 5, 6\}$$
$$\underset{\text{identification}}{=} \Gamma^{\mathrm{o}} = \{(2, 1),\ (2, 2),\ (3, 1),\ (3, 2),\ (3, 3),\ (5, 1)\}.$$

3. The canonical observer realization is

$$A^{\circ} = \begin{array}{c} \\ (2,1) \\ (2,2) \\ (3,1) \\ (3,2) \\ (3,3) \\ (5,1) \end{array} \begin{array}{c} \begin{array}{cccccc} (2,1) & (2,2) & (3,1) & (3,2) & (3,3) & (5,1) \end{array} \\ \begin{pmatrix} 0 & -3 & 0 & 0 & 1 & -1 \\ 1 & 1 & 0 & 0 & 0 & 0 \\ 0 & 0 & 0 & 0 & 1 & -1 \\ 0 & -3 & 1 & 0 & -2 & 0 \\ 0 & 0 & 0 & 1 & -2 & 0 \\ 0 & 0 & 0 & 0 & 0 & 0 \end{pmatrix} \end{array} \quad B^{\circ} = \begin{array}{c} (2,1) \\ (2,2) \\ (3,1) \\ (3,2) \\ (3,3) \\ (5,1) \end{array} \begin{pmatrix} 0 & 1 \\ 2 & -2 \\ -2 & -1 \\ 1 & -2 \\ -1 & -2 \\ -3 & 2 \end{pmatrix}$$

$$C^{\circ} = \begin{array}{c} \begin{array}{cccccc} (2,1) & (2,2) & (3,1) & (3,2) & (3,3) & (5,1) \end{array} \\ \begin{pmatrix} 0 & 0 & 0 & 0 & -2 & -1 \\ 0 & 1 & 0 & 0 & 3 & 0 \\ 0 & 0 & 0 & 0 & 1 & 0 \\ 0 & 0 & 0 & 0 & 0 & 2 \\ 0 & 0 & 0 & 0 & 0 & 1 \end{pmatrix} \end{array}$$

$$D^{\circ} = \begin{pmatrix} 2s^7 - s^6 + 3s^4 + s^3 + 2s^2 & -s^7 - s^6 + s^5 - 2s^4 - 3s + 1 \\ s^6 + s^4 - s^3 - 3s^2 + 1 & 3s^6 + s^5 + 2s^4 - s^3 + 2s^2 + 2 \\ 2s^5 - s^4 + s^3 - 2s^2 + s - 2 & -s^5 - 2s^4 - 3s^3 - 3s^2 + 3 \\ -2s^6 + 3s^5 - 2s^4 + s^3 - 3s + 3 & s^6 - 3s^4 + 3s^3 + 3s^2 + 3 \\ -s^6 + s^5 + s^4 - 3s^3 + 2s^2 + s - 2 & -3s^6 + 2s^5 + 3s^4 - s^3 - 3s^2 + s \end{pmatrix}.$$

4. As in item 7 of Example 12.3.11 we check that $(A^{\circ}, B^{\circ}, C^{\circ}, D^{\circ})$ indeed gives

$$\left(\begin{smallmatrix} C^{\circ} & D^{\circ} \\ 0 & \mathrm{id}_2 \end{smallmatrix}\right) \circ \left\{ \left(\begin{smallmatrix} x \\ u \end{smallmatrix}\right) \in \mathcal{F}^{6+2};\ s \circ x = A^{\circ}x + B^{\circ}u \right\} = \left\{ \left(\begin{smallmatrix} y \\ u \end{smallmatrix}\right) \in \mathcal{F}^{5+2};\ P \circ y = Q \circ u \right\}.$$

Notice again that the randomly chosen IO behavior has a nonproper transfer matrix and $d^{\circ}(1) = d^{\circ}(4) = 0$, an often neglected case in the literature. ◊

Theorems 12.4.5 and 12.3.2 together furnish the canonical observer realization of arbitrary observable state space equations.

Corollary 12.4.10 (Cf. [1, sect. 6.4.6 on pp. 424–439]). *As in item 4 of Theorem 12.3.2 assume observable state equations and the associated behaviors*

$$\begin{aligned} & s \circ x = Ax + Bu,\ y = Cx + D \circ u \text{ with} \\ & \qquad A \in F^{n \times n},\ B \in F^{n \times m},\ C \in F^{p \times n},\ D \in F[s]^{p \times m}, \\ & \mathcal{B} := \left(\begin{smallmatrix} C & D \\ 0 & \mathrm{id}_m \end{smallmatrix}\right) \circ \left\{ \left(\begin{smallmatrix} x \\ u \end{smallmatrix}\right) \in \mathcal{F}^{n+m};\ s \circ x = Ax + Bu \right\} \\ & \quad = \left\{ \left(\begin{smallmatrix} y \\ u \end{smallmatrix}\right) \in \mathcal{F}^{p+m};\ P \circ y = Q \circ u \right\}. \end{aligned} \qquad (12.107)$$

Let $\leq$ be a degree-over-position term order on $\{1, \ldots, p\} \times \mathbb{N}$. Consider the canonical observer realization of $\mathcal{B}$ according to Theorem 12.4.5 and the data from this theorem, i.e.,

$$\Gamma^{\mathrm{o}} = \{(i,\mu);\ 1 \le i \le p,\ 1 \le \mu \le d^{\mathrm{o}}(i)\},\ n = \sharp(\Gamma^{\mathrm{o}}) = \sum_{i=1}^{p} d^{\mathrm{o}}(i),$$
$$s \circ x = A^{\mathrm{o}}x + B^{\mathrm{o}}x,\ y = C^{\mathrm{o}}x + D^{\mathrm{o}} \circ u \text{ with}$$
$$\begin{pmatrix} C^{\mathrm{o}} & D^{\mathrm{o}} \\ 0 & \mathrm{id}_m \end{pmatrix} \circ :\ \mathcal{B}^{\mathrm{o}} = \left\{\begin{pmatrix} x \\ u \end{pmatrix} \in \mathcal{F}^{n+m};\ s \circ x = A^{\mathrm{o}}x + B^{\mathrm{o}}u\right\} \overset{\cong}{\longrightarrow} \mathcal{B}. \tag{12.108}$$

Identify $\Gamma^{\mathrm{o}} = \{1,\dots,n\}$ *via* $(i,\mu) = \alpha(i,\mu) := \sum_{k=1}^{i-1} d^{\mathrm{o}}(k) + \mu$, *cf. Remark 12.4.6, such that* $A^{\mathrm{o}} \in F^{n\times n}$, $C^{\mathrm{o}} \in F^{p\times n}$ *and* $B^{\mathrm{o}} \in F^{n\times m}$.

The state equations (12.107) *and* (12.108) *realize* $\mathcal{B}$ *and are thus similar, i.e., there is a unique invertible matrix* $U \in F^{n\times\Gamma^{\mathrm{o}}}$ *with* $U^{-1} \in F^{\Gamma^{\mathrm{o}}\times n}$ *such that*

$$A = UA^{\mathrm{o}}U^{-1},\ B = UB^{\mathrm{o}},\ C = C^{\mathrm{o}}U^{-1},\ D = D^{\mathrm{o}}. \tag{12.109}$$

According to Theorem 12.3.2 the rows

$$C^{\mathrm{ob}}_{i,-}(A^{\mathrm{ob}})^{\mu} \in F^{1\times\Gamma^{\mathrm{o}}} \text{ resp. } C_{i,-}A^{\mu} \in F^{1\times n},\ 1 \le i \le p,\ 0 \le \mu \le d^{\mathrm{o}}(i) - 1,$$

are an F*-basis of* $F^{1\times\Gamma^{\mathrm{o}}}$, $n = \sharp(\Gamma^{\mathrm{o}})$, *resp. of* $F^{1\times n}$, *and* U *is the unique matrix with*

$$\left(C_{i,-}A^{\mu}\right)_{1\le i\le p,\ 0\le\mu\le d^{\mathrm{o}}(i)-1} U = \left(C^{\mathrm{ob}}_{i,-}(A^{\mathrm{ob}})^{\mu}\right)_{1\le i\le p,\ 0\le\mu\le d^{\mathrm{o}}(i)-1}. \tag{12.110}$$

◊

Example 12.4.11 In Example 12.4.9 let $U \in \mathrm{Gl}_n(F)$ be a random matrix and

$$A := UA^{\mathrm{o}}U^{-1},\ B := UB^{\mathrm{o}},\ C := C^{\mathrm{o}}U^{-1},\ D := D^{\mathrm{o}}. \tag{12.111}$$

Recall $p = 5$, $d^{\mathrm{o}} = (0,2,3,0,1)$, $n = 6$ and that we identify Γ^{o} with $\{1,\dots,6\}$ via $(i,\mu) = \alpha(i,\mu) = d^{\mathrm{o}}(1) + \dots + d^{\mathrm{o}}(i-1) + \mu$. Since the observer realization of (A,B,C,D) for the given term order is unique (canonical), the matrices (A^{o}, B^{o}, C^{o}, D^{o}) are indeed the unique observer realization of the equations $s \circ x = Ax + Bu$, $y = Cx + D \circ u$. Moreover U is the unique invertible matrix with $C_{i-}A^{\mu}U = C^{\mathrm{o}}_{i-}(A^{\mathrm{ob}})^{\mu}$, $1 \le i \le 5$, $0 \le \mu \le d^{\mathrm{o}}(i) - 1$. Thus, in this example, the observer realization of (A,B,C,D) is known and need not be computed.

However, we apply the method of Corollary 12.4.10 to confirm this. We randomly choose

$$U = \begin{pmatrix} -1 & -2 & 3 & -3 & -1 & 2 \\ 1 & -1 & 3 & -2 & 0 & -1 \\ 0 & -3 & -3 & -3 & -3 & 1 \\ 0 & -1 & 1 & 3 & 1 & 3 \\ 0 & -3 & -1 & 1 & 0 & 3 \\ 2 & 1 & 1 & 3 & 2 & -1 \end{pmatrix} \Longrightarrow \det(U) = -3,\quad U^{-1} = \begin{pmatrix} 2 & 3 & -3 & 0 & 3 & 0 \\ -3 & -2 & 2 & -3 & 0 & 3 \\ 1 & 2 & 0 & 2 & -1 & 3 \\ -2 & -3 & 0 & -3 & 0 & -3 \\ 1 & -1 & 3 & 2 & 1 & 1 \\ -2 & 2 & 2 & 1 & 0 & -2 \end{pmatrix}.$$

We obtain

$$A=\begin{pmatrix} 0&-1&3&-1&0&3\\ -1&1&2&2&0&2\\ 3&0&-2&-1&3&0\\ 3&-3&2&-3&1&-1\\ 1&3&-3&1&1&1\\ 0&-3&-2&-3&0&2 \end{pmatrix},\ B=\begin{pmatrix} 3&-2\\ 0&2\\ -3&2\\ 3&-1\\ 2&-3\\ -3&1 \end{pmatrix},\ C=\begin{pmatrix} 0&0&-1&2&-2&0\\ 0&2&-3&3&3&-1\\ 1&-1&3&2&1&1\\ 3&-3&-3&2&0&3\\ -2&2&2&1&0&-2 \end{pmatrix},$$

$$D=\begin{pmatrix} 2s^7-s^6+3s^4+s^3+2s^2 & -s^7-s^6+s^5-2s^4-3s+1\\ s^6+s^4-s^3-3s^2+1 & 3s^6+s^5+2s^4-s^3+2s^2+2\\ 2s^5-s^4+s^3-2s^2+s-2 & -s^5-2s^4-3s^3-3s^2+3\\ -2s^6+3s^5-2s^4+s^3-3s+3 & s^6-3s^4+3s^3+3s^2+3\\ -s^6+s^5+s^4-3s^3+2s^2+s-2 & -3s^6+2s^5+3s^4-s^3-3s^2+s \end{pmatrix}.$$

We compute the associated IO behavior

$$\{\left(\begin{smallmatrix} y\\ u\end{smallmatrix}\right);\ P\circ y=Q\circ u\}=\left(\begin{smallmatrix} C & D\\ 0 & \mathrm{id}_m\end{smallmatrix}\right)\circ\{\left(\begin{smallmatrix} x\\ u\end{smallmatrix}\right);\ s\circ x=Ax+Bu\},\ \text{and obtain}$$

$$P=\begin{pmatrix} 0 & 0 & 0 & 1 & -2\\ 1 & 0 & 2 & 0 & 1\\ 0 & 3s^2+2s+2 & -3s^3-s^2+2s+1 & 0 & 2s^3+2s^2-2s\\ 0 & 3s^2-3s+2 & -2s^2+2s-2 & 0 & -s^2+3s+3\\ 0 & 0 & 0 & 0 & -3s \end{pmatrix},$$

$$Q=\begin{pmatrix} s^5+3s^4+3s^2+2s & 3s^5-2s^4-2s^3+2s^2-2s+3\\ 2s^7-2s^6-2s^5+2s^4+3s+1 & -s^7+3s^6+s^5-3s^4-2s^2-2s\\ -2s^9-3s^8+2s^7+s^6+3s^4+2s^3-3s^2-2 & s^9+3s^8-3s^7+2s^6+3s^5+s^4-s^3-2s^2+2\\ -3s^8+3s^7+3s^{+}2s^5+3s^4+2s^3-3s^2-3s-2 & -2s^8-s^7-2s^6-s^5-2s^4-3s^2+2s\\ 3s^7-3s^6-3s^5+2s^4+s^3-3s^2-s+2 & 2s^7+s^6-2s^5+3s^4+2s^3-3s2+1 \end{pmatrix}.$$

The observer realization of $\{\left(\begin{smallmatrix} y\\ u\end{smallmatrix}\right);\ P\circ y=Q\circ u\}$, i.e., the observer realization of $(A,\ B,\ C,\ D)$, is exactly the same quadruple $(A^{\circ},\ B^{\circ},\ C^{\circ},\ D^{\circ})$ of matrices which we started with. ◇

The following theorem establishes the main consequence of the observer form und its historical and practical significance. It is fundamental for the construction of *Luenberger state observers*, that justify the term *observer form*, and, by duality, for the *stabilization by state feedback*.

Theorem 12.4.12 (Pole shifting theorem, cf. *[11, §6.3], [1, §7.1.1]*).

1. *Consider the observable state space equations and data from Corollary 12.4.10. Choose any polynomial* $f=s^n+\sum_{k=0}^{n-1}a_ks^k\in F[s]$. *Then one can compute a matrix* $L\in F^{n\times p}$ *such that* $\det(s\,\mathrm{id}_n-(A-LC))=f$, *i.e., the characteristic polynomial of* $A-LC$ *can be prescribed arbitrarily.*

2. *Assume, in particular, that* $F=\mathbb{C},\mathbb{R}$ *and that* $\Sigma^{\circ}\subset\mathbb{C}$ *is a nonempty set with at most* n *elements. For* $F=\mathbb{R}$ *assume in addition that* $\Sigma^{\circ}=\overline{\Sigma^{\circ}}$ *(*$\overline{z}$*=complex conjugate) and that* $\mathbb{R}\cap\Sigma^{\circ}\neq\emptyset$ *if* n *is odd. Then there are*

$$f\in F[s],\ L\in F^{n\times p}\ \text{with}\ \deg_s(f)=n\ \text{and}\ \Sigma^{\circ}=\mathrm{V}_{\mathbb{C}}(f)=\mathrm{spec}(A-LC).$$

Proof According to Remark 12.4.6 we use $\alpha:\Gamma^{\circ}\cong\{1,\dots,n\}$ to identify

$$F^{\Gamma^o\times\Gamma^o} = F^{n\times n},\ F^{p\times\Gamma^o} = F^{p\times n},\ F^{\Gamma^o\times p} = F^{n\times p}.$$

1. (i) We construct $L' \in F^{\Gamma^o\times p}$ such that $\det\left(s\,\mathrm{id}_{\Gamma^o} - (A^o - L'C^{1,o})\right) = f$: Let $L' = (L'_{(i,\mu),k})_{(i,\mu),k} \in F^{\Gamma^o\times p}$ and $A' := A^o - L'C^{1,o}$. Recall

$$A^o_{(i,\mu),(j,\nu)} = \delta_{i,j}\delta_{\mu-1,\nu} \text{ for } \nu \neq d^o(j),\ C^{1,o}_{i,(j,\nu)} = \delta_{i,j}\delta_{d^o(j),\nu},$$

$$\Longrightarrow (L'C^{1,o})_{(i,\mu),(j,\nu)} = \begin{cases} 0 & \text{if } \nu \neq d^o(j) \\ \sum_k L'_{(i,\mu),k}\delta_{k,j} = L'_{(i,\mu),j} & \text{if } \nu = d^o(j) \end{cases}, \quad (12.112)$$

$$A'_{(i,\mu),(j,\nu)} = \begin{cases} A^o_{(i,\mu),(j,\nu)} & \text{if } \nu \neq d^o(j) \\ A^o_{(i,\mu),(j,d^o(j))} - L'_{(i,\mu),j} & \text{if } \nu = d^o(j) \end{cases},$$

One can achieve arbitrary columns $A'_{-,(j,d^o(j))}$ by defining

$$L'_{(i,\mu),j} := \begin{cases} A^o_{(i,\mu),(j,d^o(j))} - A'_{(i,\mu),(j,d^o(j))} & \text{if } d^o(j) > 0 \\ 0 & \text{if } d^o(j) = 0 \end{cases}. \quad (12.113)$$

We now choose L' such that $A' = A^o - L'C^{1,o}$ has the form

$$\begin{aligned} A'_{(i,\mu),(j,\nu)} &:= A^o_{(i,\mu),(j,\nu)} = \delta_{i,j}\delta_{\mu-1,\nu} \text{ if } \nu \neq d^o(j) \\ A'_{(i,1),(j,d^o(j))} &:= 1 \text{ if } i > j \text{ and } d^o(j+1) = \cdots = d^o(i-1) = 0 \\ A'_{(i,\mu),(r,d^o(r))} &:= -a_{\alpha(i,\mu)-1} \text{ where } r := \max\left\{j;\ 1 \le j \le p,\ d^o(j) > 0\right\} \text{ and} \\ &\quad \alpha(i,\mu) := d^o(1) + \cdots + d^o(i-1) + \mu - 1 \\ A'_{(i,\mu),(j,d^o(j))} &:= 0 \text{ otherwise.} \end{aligned} \quad (12.114)$$

For A' as matrix in $F^{n\times n}$ the equations from (12.114) furnish

$$A'_{k,l} = \begin{cases} -a_{k-1} & \text{if } l = n \\ 1 & \text{if } 1 \le k-1 = l \le n-1 \\ 0 & \text{otherwise} \end{cases}$$

$$\Longrightarrow \det(s\,\mathrm{id}_n - A') = s^n + \sum_{k=0}^{n} a_k s^k = f$$

since A' is one of the companion matrices for f. The corresponding matrix $L' \in F^{\Gamma^o\times p}$ follows from (12.113).

(ii) From (12.93), (12.109), and (12.110) recall

$$\begin{aligned} &C^{o} = (P^{g,hc})^{-1}C^{1,o} = CU, \ A^{o} = U^{-1}AU \Longrightarrow C^{1,o} = P^{g,hc}CU \\ &\Longrightarrow A' = A^{o} - L'C^{1,o} = U^{-1}\left(A - UL'P^{g,hc}C\right)U \\ &\quad = U^{-1}(A - LC)U, \ L := UL'P^{g,hc} \\ &\Longrightarrow f = \det(s\,\mathrm{id}_n - A') = \det(s\,\mathrm{id}_n - (A - LC)). \end{aligned} \tag{12.115}$$

2. The remaining assertion concerning $\Sigma^{o} \subset \mathbb{C}$ is simple. □

Corollary 12.4.13 (Algorithm for $\det(s\,\mathrm{id}_n - (A - LC)) = f$) *The following constructive steps have to be taken for the given observable equations* (12.107). *With* $n = d^{o}(1) + \cdots + d^{o}(p)$ *recall the identification*

$$F^{\Gamma^{o}\times\Gamma^{o}} = F^{n\times n}, \ M_{(i,\mu),(j,\nu)} = M_{\alpha(i,\mu),\alpha(j,\nu)}, \ \alpha(i,\mu) = d^{o}(1) + \cdots + d^{o}(i-1) + \mu.$$

1. *Compute the matrix P from Corollary 12.4.10 by means of a universal left annihilator* $(-Y, P) \in F[s]^{p\times(n+p)}$ *of* $\binom{s\,\mathrm{id}_n - A}{C}$. *Verify the observability by constructing a left inverse* $(X, Z) \in F[s]^{n\times(n+p)}$ *of* $\binom{s\,\mathrm{id}_n - A}{C}$.
2. *Choose any degree-over-position term order, for instance, from* (12.72). *By means of Theorem 12.4.5 and Algorithm (12.4.7) compute the canonical observer realization of* $\mathcal{B} := \left\{\binom{y}{u} \in \mathcal{F}^{p+m};\ P \circ y = Q \circ u\right\}$. *However, only* A^{o}, C^{o}, *and* $P^{g,hc}$ *have to be determined, so* $D^{o} = H_{\mathrm{pol}}$, H_{spr}, Q^{1}, B^{o} *are not needed. Compute* $U \in \mathrm{Gl}_p(F)$ *by* (12.110).
Notice that P^{g} *and thus* $P^{g,hc}$ *are uniquely determined by the equations*
$$\begin{aligned} &C\left\{x \in \mathcal{F}^{n};\ s \circ x = Ax\right\} = C^{o}\left\{x \in \mathcal{F}^{\Gamma^{o}};\ s \circ x = A^{o}x\right\} \\ &= \left\{y \in \mathcal{F}^{p};\ P \circ y = 0\right\} = \left\{y \in \mathcal{F}^{p};\ P^{g} \circ y = 0\right\} \Longrightarrow \mathcal{D}^{1\times p}P = \mathcal{D}^{1\times p}P^{g}, \end{aligned}$$
since the rows of P^{g} *are the unique reduced Gröbner basis of the equation module of* $C\{x \in \mathcal{F}^{n};\ s \circ x = Ax\}$.
3. *Use* (12.114) *to define* $A' \in F^{\Gamma^{o}\times\Gamma^{o}} = F^{n\times n}$ *and then* (12.113) *and* (12.115) *to compute*
$$L' \in F^{\Gamma^{o}\times p} = F^{n\times p} \text{ and } L = UL'P^{g,hc}.$$

◇

By transposition one obtains the corresponding result for controllable equations $s \circ x = Ax + Bu$ which are needed for stabilization by state feedback.

Corollary 12.4.14 *Cf. [1, pp. 500–505], [3, pp. 318–327]. Let* $(A, B) \in F^{n\times(n+m)}$ *be a controllable pair of matrices, i.e., the matrix* $(s\,\mathrm{id}_n - A, -B)$ *is right invertible, see Theorem and Definition 4.3.8. Let* $d^{c}(1), \ldots, d^{c}(m)$ *be the observability indices of the dual observable pair* $\binom{A^{\top}}{B^{\top}}$. *Let* $f = s^{n} + \sum_{k=0}^{n-1} a_k s^{k}$ *be any monic polynomial of degree n. Use Theorem 12.4.12 and Corollary 12.4.13 to compute the matrix* $L \in F^{n\times m}$ *for the observable pair* $\binom{A^{\top}}{B^{\top}}$ *with* $f = \det(s\,\mathrm{id}_n - (A^{\top} - LB^{\top}))$. *Then*
1. $f = \det(s\,\mathrm{id}_n - (A - BK))$, $K := L^{\top}$, *i.e., the characteristic polynomial of* $A - BK$ *can be prescribed arbitrarily.*

2. Assume $F = \mathbb{C}, \mathbb{R}$ and a nonempty subset $\Sigma^{\mathrm{c}} \subset \mathbb{C}$ with at most n elements. For $F = \mathbb{R}$ assume additionally that $\Sigma^{\mathrm{c}} = \overline{\Sigma^{\mathrm{c}}}$ and $\mathbb{R} \cap \Sigma^{\mathrm{c}} \neq \emptyset$ if n is odd. Then there are

$$f \in F[s] \text{ and } K \in F^{m\times n} \text{ such that } \mathrm{V}_{\mathbb{C}}(f) = \mathrm{spec}(A - BK) = \Sigma^{\mathrm{c}}.$$

◇

Example 12.4.15 We use the state space data (A, B, C, D) from Example 12.4.11 and choose

$$\begin{aligned} f :&= s^6 + 5s^5 + 4s^4 + 3s^3 + 2s^2 + s + 6 \\ &= s^6 - 2s^5 - 3s^4 + 3s^3 + 2s^2 + s - 1 \in (\mathbb{Z}/\mathbb{Z}7)[s]. \end{aligned}$$

We use the algorithm form Corollary 12.4.13 to compute a matrix $L \in F^{6\times 5}$ such that $\det(s\,\mathrm{id}_6 - (A - LC)) = f$. The matrices $A^{\mathrm{o}},\ B^{\mathrm{o}},\ C^{\mathrm{o}},\ D^{\mathrm{o}}$ are those from Example 12.4.11, in particular

$$A^{\mathrm{o}} = \begin{array}{c|cccccc} & (2,1) & (2,2) & (3,1) & (3,2) & (3,3) & (5,1) \\ \hline (2,1) & 0 & -3 & 0 & 0 & 1 & -1 \\ (2,2) & 1 & 1 & 0 & 0 & 0 & 0 \\ (3,1) & 0 & 0 & 0 & 0 & 1 & -1 \\ (3,2) & 0 & -3 & 1 & 0 & -2 & 0 \\ (3,3) & 0 & 0 & 0 & 1 & -2 & 0 \\ (5,1) & 0 & 0 & 0 & 0 & 0 & 0 \end{array},\quad C^{\mathrm{o}} = \begin{pmatrix} \begin{array}{cccccc} (2,1) & (2,2) & (3,1) & (3,2) & (3,3) & (5,1) \\ \hline 0 & 0 & 0 & 0 & -2 & -1 \\ 0 & 1 & 0 & 0 & 3 & 0 \\ 0 & 0 & 0 & 0 & 1 & 0 \\ 0 & 0 & 0 & 0 & 0 & 2 \\ 0 & 0 & 0 & 0 & 0 & 1 \end{array} \end{pmatrix}.$$

With the companion matrix

$$A' = \begin{array}{c|cccccc} & (2,1) & (2,2) & (3,1) & (3,2) & (3,3) & (5,1) \\ \hline (2,1) & 0 & 0 & 0 & 0 & 0 & 1 \\ (2,2) & 1 & 0 & 0 & 0 & 0 & -1 \\ (3,1) & 0 & 1 & 0 & 0 & 0 & -2 \\ (3,2) & 0 & 0 & 1 & 0 & 0 & -3 \\ (3,3) & 0 & 0 & 0 & 1 & 0 & 3 \\ (5,1) & 0 & 0 & 0 & 0 & 1 & 2 \end{array}$$

we obtain

$$L' = \begin{array}{c|ccccc} (2,1) & 0 & -3 & 1 & 0 & -2 \\ (2,2) & 0 & 1 & 0 & 0 & 1 \\ (3,1) & 0 & -1 & 1 & 0 & 1 \\ (3,2) & 0 & -3 & -2 & 0 & 3 \\ (3,3) & 0 & 0 & -2 & 0 & -3 \\ (5,1) & 0 & 0 & -1 & 0 & -2 \end{array}$$

and finally

$$L = UL'P^{g,\mathrm{hc}} = \begin{pmatrix} 0 & 0 & 1 & 0 & 0 \\ 0 & -1 & -2 & 0 & 3 \\ 0 & 2 & 2 & 0 & -1 \\ 0 & 3 & 2 & 0 & 0 \\ 0 & 2 & 2 & 0 & 0 \\ 0 & -1 & -3 & 0 & 3 \end{pmatrix}.$$

The equality $\det\big(s\mathrm{id}_6 - (A - LC)\big) = f$ has been checked. ◊

12.5 Stabilization by State Feedback and the Luenberger State Observer

The construction and parametrization of stabilizing compensators and observers were intensively studied in Chaps. 10 *and* 11. *In the present section we discuss the special case of Corollary* 10.1.40 *where the state dimensions of the plant and the compensator are equal. Applications are the Luenberger state observer* [4] *and stabilization by state feedback with the help of this observer.*

We consider the discrete-time resp. continuous-time stability situation

$$F = \mathbb{R}, \mathbb{C},\ \mathcal{F} := \begin{cases} F^{\mathbb{N}} & \text{(discrete time)} \\ \mathrm{C}^{-\infty}(\mathbb{R}, F) & \text{(continuous time)} \end{cases},$$
$$\mathbb{C} = \Lambda_1 \uplus \Lambda_2,\ \Lambda_1 := \begin{cases} \{\lambda \in \mathbb{C};\ |\lambda| < 1\} & \text{(discrete time)} \\ \mathbb{C}_- = \{\lambda \in \mathbb{C};\ \Re(\lambda) < 0\} & \text{(continuous time)} \end{cases}. \tag{12.116}$$

A controllable and observable state space behavior, the plant, is given with the equations

$$\begin{aligned} &s \circ x_1 = A_1 x_1 + B_1 u_1,\ y_1 = C_1 x_1 + D_1 u_1 \text{ where} \\ &A_1 \in F^{n \times n},\ B_1 \in F^{n \times m},\ C_1 \in F^{p \times n},\ D_1 \in F^{n \times m}. \end{aligned} \tag{12.117}$$

The compensator state equations are chosen as

$$\begin{aligned} &s \circ x_2 = A_2 x_2 + B_2 u_2 + B_3 u_1,\ y_2 = C_2 x_2, \text{ with} \\ &A_2 \in F^{n \times n},\ B_2 \in F^{n \times p},\ B_3 \in F^{n \times m},\ C_2 \in F^{m \times n}. \end{aligned} \tag{12.118}$$

In contrast to Corollary 10.1.40 the input u_1 of the plant is also used as input component of the compensator. The plant and the compensator are interconnected by output feedback with the equations

$$\begin{aligned} &s \circ x_1 = A_1 x_1 + B_1(u_1 + y_2),\ y_1 = C_1 x_1 + D_1(u_1 + y_2), \\ &s \circ x_2 = A_2 x_2 + B_2(u_2 + y_1) + B_3 u_1,\ y_2 = C_2 x_2. \end{aligned} \tag{12.119}$$

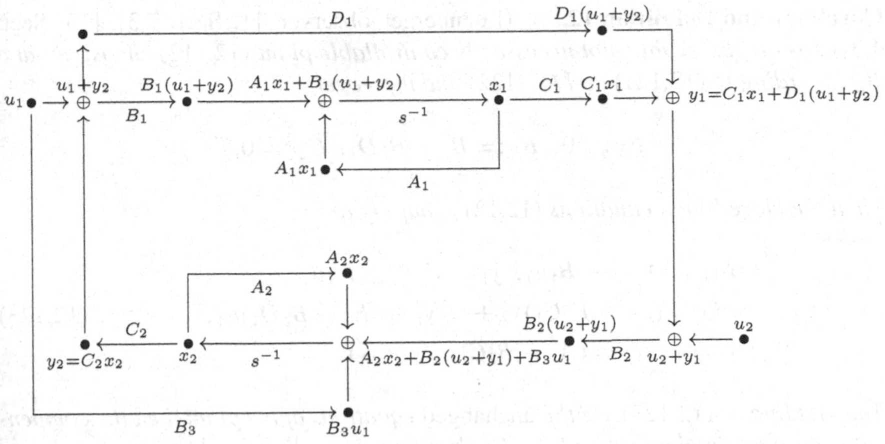

Fig. 12.1 The block diagram of Eq. (12.119)

The inputs u_1 and u_2 are external inputs to the plant and the compensator. The corresponding block diagram is contained in Fig. 12.1.

The closed-loop equations (12.119) have the equivalent form

$$\begin{aligned}
&s\circ x = Ax + Bu,\ y = Cx + Du \text{ where}\\
&x := \begin{pmatrix} x_1\\ x_2\end{pmatrix} \in \mathcal{F}^{n+n},\ u := \begin{pmatrix} u_1\\ u_2\end{pmatrix} \in \mathcal{F}^{m+p},\ y := \begin{pmatrix} y_1\\ y_2\end{pmatrix} \in \mathcal{F}^{p+m},\\
&A := \begin{pmatrix} A_1 & B_1C_2\\ B_2C_1 & A_2+B_2D_1C_2\end{pmatrix},\ B := \begin{pmatrix} B_1 & 0\\ B_2D_1+B_3 & B_2\end{pmatrix},\ C = \begin{pmatrix} C_1 & D_1C_2\\ 0 & C_2\end{pmatrix},\ D := \begin{pmatrix} D_1 & 0\\ 0 & 0\end{pmatrix}\\
&\Longrightarrow s\circ(x_2 - x_1) = (A_2 + B_2D_1C_2 - B_1C_2)x_2 - (A_1 - B_2C_1)x_1\\
&\qquad\qquad + (B_2D_1 + B_3 - B_1)u_1 + B_2u_2.
\end{aligned} \tag{12.120}$$

These equations coincide with (10.77), but with equal state dimension of x_1 and x_2 and additional B_3. The last equation of (12.120) suggests to choose

$$\begin{aligned}
&A_2 + B_2D_1C_2 - B_1C_2 = A_1 - B_2C_1,\ \text{i.e., } A_2 := A_1 - B_2C_1 + B_1C_2 - B_2D_1C_2\\
&\Longrightarrow s\circ x = Ax + Bu,\ y = Cx + Du,\ A = \begin{pmatrix} A_1 & B_1C_2\\ B_2C_1 & A_1 - B_2C_1 + B_1C_2\end{pmatrix}\\
&s\circ(x_2 - x_1) = (A_1 - B_2C_1)(x_2 - x_1) + (B_2D_1 + B_3 - B_1)u_1 + B_2u_2.
\end{aligned} \tag{12.121}$$

Choose a monic polynomial f° of degree n with $\mathrm{V}_{\mathbb{C}}(f^{\circ}) \subset \Lambda_1$. Since the plant is observable Theorem 12.4.12 furnishes a matrix

$$\begin{aligned}
&B_2 \in F^{n\times p} \text{ with } f^{\circ} = \det(s\,\mathrm{id}_n - (A_1 - B_2C_1)) \text{ and thus}\\
&\qquad\qquad \mathrm{V}_{\mathbb{C}}(f^{\circ}) = \mathrm{spec}(A_1 - B_2C_1) \subset \Lambda_1.
\end{aligned} \tag{12.122}$$

In the sequel A_2 and B_2 are chosen according to (12.121) and (12.122).

Corollary and Definition 12.5.1 (Luenberger observer, [1, Sect. 7.3], [13, Sect. 4.3]) *For an observable, not necessarily controllable plant* (12.117) *choose* A_2 *and* B_2 *according to* (12.121) *and* (12.122) *and moreover*

$$u_2 := 0,\ B_3 := B_1 - B_2D_1,\ C_2 := 0.$$

Then the closed-loop equations (12.121) *simplify to*

$$\begin{aligned} s \circ x_1 &= A_1x_1 + B_1u_1,\ y_1 = C_1x_1 + D_1u_1, \\ s \circ x_2 &= (A_1 - B_2C_1)x_2 + B_2y_1 + (B_1 - B_2D_1)u_1, \\ s \circ (x_2 - x_1) &= (A_1 - B_2C_1)(x_2 - x_1). \end{aligned} \tag{12.123}$$

The first line of (12.123) *are the* unchanged *equations of the plant, i.e., the compensator does not influence the plant. The last line shows that* $x_2 - x_1$ *is* Λ_1*- and even* $\mathrm{V}_{\mathbb{C}}(f^{\mathrm{o}})$*-stable and, in particular,* $\lim_{t\to\infty}(x_2 - x_1)(t) = 0$, *i.e.,* x_2 *is an estimate of* x_1 *for all inputs* u_1. *The choice of* f^{o} *and thus of* $\mathrm{V}_{\mathbb{C}}(f^{\mathrm{o}}) = \mathrm{spec}(A_1 - B_2C_1)$ *determines the decay properties of* $x_2 - x_1$. *The compensator is called a* Luenberger state observer. ◊

In the next theorem we choose a suitable C_2 such that the closed-loop equations (12.121) are Λ_1-stable, i.e., $\mathrm{spec}(A) \subset \Lambda_1$

Theorem 12.5.2 (Cf. [3, (10.19), Theorem 10.5.1 on pp. 352, 353], [1, (3a),(3b) on p. 523], [13, (4.7) on p. 364]) *Assume the standard stability situation from* (12.116), *the controllable and observable plant state space equations* (12.117) *and the closed-loop equations* (12.121). *Choose two monic polynomials* $f^{\mathrm{o}}, f^{\mathrm{c}}$ *of degree n with* $\Sigma := \mathrm{V}_{\mathbb{C}}(f^{\mathrm{o}}) \cup \mathrm{V}_{\mathbb{C}}(f^{\mathrm{c}}) \subset \Lambda_1$. *Theorem 12.4.12 and Corollary 12.4.14 furnish matrices* $B_2 \in F^{n\times p}$ *and* $C_2 \in F^{m\times n}$ *such that*

$$f^{\mathrm{o}} = \det\left(s\,\mathrm{id}_n - (A_1 - B_2C_1)\right),\ f^{\mathrm{c}} = \det\left(s\,\mathrm{id}_n - (A_1 + B_1C_2)\right). \tag{12.124}$$

Then the following statements hold:

1. $\det(s\,\mathrm{id}_{2n} - A) = f^{\mathrm{o}}f^{\mathrm{c}}$, *and hence* $\Sigma = \mathrm{V}_{\mathbb{C}}(f^{\mathrm{o}}) \cup \mathrm{V}_{\mathbb{C}}(f^{\mathrm{c}}) = \mathrm{spec}(A) \subset \Lambda_1$, *i.e., for every choice of* B_3 *the closed-loop equations* (12.121) *are* Σ*- and hence* Λ_1*-stable. The choice of* f^{o} *and* f^{c}, *and hence of* Σ *determine the decay properties of the autonomous solutions* $x(t) = e^{tA}x(0)$ *resp.* $x(t) = A^tx(0)$.
2. *In addition to 1 assume* $u_2 := 0$, *i.e., the compensator has no external input* u_2, *and* $B_3 := B_1 - B_2D_1$. *The ensuing simpler closed-loop equations are*

$$\begin{aligned} s \circ x_1 &= A_1x_1 + B_1(u_1 + y_2),\ y_1 = C_1x_1 + D_1(u_1 + y_2), \\ s \circ x_2 &= A_2x_2 + B_2y_1 + B_3u_1,\ y_2 = C_2x_2 \text{ where} \\ &\quad A_2 := A_1 - B_2C_1 + B_1C_2 - B_2D_1C_2,\ B_3 = B_1 - B_2D_1 \\ s \circ (x_2 - x_1) &= (A_1 - B_2C_1)(x_2 - x_1) \end{aligned} \tag{12.125}$$

or, in other form,

$$s\circ x=Ax+B'u_1,\ y=Cx+D'u_1 \text{ where } x:=\begin{pmatrix}x_1\\x_2\end{pmatrix},\ y:=\begin{pmatrix}y_1\\y_2\end{pmatrix},$$
$$A:=\begin{pmatrix}A_1 & B_1C_2\\ B_2C_1 & A_1-B_2C_1+B_1C_2\end{pmatrix},\ B':=\begin{pmatrix}B_1\\B_1\end{pmatrix},\ C:=\begin{pmatrix}C_1 & D_1C_2\\ 0 & C_2\end{pmatrix},\ D':=\begin{pmatrix}D_1\\0\end{pmatrix}. \tag{12.126}$$

The last equation of (12.125) *implies that x_2-x_1 is stable with respect to* $\mathrm{V}_{\mathbb{C}}(f^{\circ})=\operatorname{spec}(A_1-B_2C_1)$ *and, in particular, that* $\lim_{t\to\infty}(x_2-x_1)(t)=0$, *i.e., that x_2 is an estimate of x_1. The input $u_1+y_2=u_1+C_2x_2$ of the plant is then interpreted as* constant state feedback: *The constant multiple C_2x_2 of the estimate x_2 of the plant state x_1 is fed back to the input of the plant. The decay of x_2-x_1 is determined by* $\mathrm{V}_{\mathbb{C}}(f^{\circ})=\operatorname{spec}(A_1-B_2C_1)$. *Notice that in* (12.125), *in contrast to Corollary 12.5.1, x_1 is the state of the plant in the closed-loop behavior and not that of the original plant.*

3. *In case 2 the closed-loop system* (12.125) *is not controllable, and is observable if and only if the equations $s\circ z=(A_1-B_2C_1)z,\ C_2z=0$ are observable, i.e., if $s\circ z=(A_1-B_2C_1)z$ and $C_2z=0$ imply $z=0$.*

Proof 1. Consider the matrices and matrix equations

$$\begin{pmatrix}\mathrm{id}_n & \mathrm{id}_n\\ 0 & \mathrm{id}_n\end{pmatrix},\ \begin{pmatrix}\mathrm{id}_n & -\mathrm{id}_n\\ 0 & \mathrm{id}_n\end{pmatrix}=\begin{pmatrix}\mathrm{id}_n & \mathrm{id}_n\\ 0 & \mathrm{id}_n\end{pmatrix}^{-1},\ \begin{pmatrix}\mathrm{id}_n & \mathrm{id}_n\\ 0 & \mathrm{id}_n\end{pmatrix}\widehat{A}\begin{pmatrix}\mathrm{id}_n & -\mathrm{id}_n\\ 0 & \mathrm{id}_n\end{pmatrix}=A \text{ where}$$
$$\widehat{A}:=\begin{pmatrix}A_1-B_2C_1 & 0\\ B_2C_1 & A_1+B_1C_2\end{pmatrix},\ A=\begin{pmatrix}A_1 & B_1C_2\\ B_2C_1 & A_1+B_1C_2-B_2C_1\end{pmatrix}. \tag{12.127}$$

Hence the matrices A and $\widehat{A}$ are similar and have the same characteristic polynomial and the same set of eigenvalues. Since $\widehat{A}$ is block lower diagonal we infer

$$\det(s\,\mathrm{id}_{2n}-A)=\det\left(s\,\mathrm{id}_{2n}-\widehat{A}\right)=f^{\circ}(s)f^{c}(s) \text{ with}$$
$$f^{\circ}(s)=\det(s\,\mathrm{id}_n-(A_1-B_2C_1)),\ f^{c}(s)=\det(s\,\mathrm{id}_n-(A_1+B_1C_2))\ (\text{cf.}(12.124))$$
$$\Longrightarrow \operatorname{spec}(A)=\operatorname{spec}(\widehat{A})=\operatorname{spec}(A_1-B_2C_1)\cup\operatorname{spec}(A_1+B_1C_2).$$

2. obvious.

3. Recall $x=\begin{pmatrix}x_1\\x_2\end{pmatrix}\in\mathcal{F}^{n+n}$ and consider the behaviors

$$\begin{aligned}\mathcal{B}_1&:=\left\{\begin{pmatrix}x\\u\end{pmatrix}\in\mathcal{F}^{2n+m};\ s\circ x=Ax+B'u_1\right\}\\ &=\left\{\begin{pmatrix}x\\u\end{pmatrix}\in\mathcal{F}^{2n+m};\ (s\,\mathrm{id}_{2n}-A,-B')\circ\begin{pmatrix}x\\u_1\end{pmatrix}=0\right\},\\ \mathcal{C}_1&:=\left\{x\in\mathcal{F}^{2n};\ s\circ x=Ax,\ Cx=0\right\},\\ \mathcal{C}_2&:=\left\{z\in\mathcal{F}^{n};\ s\circ z=(A_1-B_2C_1)z=0\right\}\end{aligned}$$

where $\mathcal{C}_1$ and $\mathcal{C}_2$ are autonomous. By (12.125) there is the system map

$$\phi:=(\mathrm{id}_n,-\mathrm{id}_n,0_m)\circ:\mathcal{B}_1\mapsto\mathcal{C}_2,\ \begin{pmatrix}x\\u\end{pmatrix}\mapsto x_1-x_2, \text{ with the image}$$
$$\mathcal{C}_3:=\phi(\mathcal{B}_1)=\left\{x_1-x_2;\ (s\,\mathrm{id}_{2n}-A,-B')\circ\begin{pmatrix}x_1\\x_2\\u_1\end{pmatrix}=0\right\}\subseteq\mathcal{C}_2.$$

As subbehavior of $\mathcal{C}_2$ also the image behavior $\mathcal{C}_3$ is autonomous.

(a) $\mathcal{C}_3\neq 0$: Assume

$$\mathcal{C}_3 = 0, \text{ i.e., } \left((s\,\mathrm{id}_{2n} - A, -B') \circ \begin{pmatrix} x_1 \\ x_2 \\ u_1 \end{pmatrix} = 0 \Longrightarrow (\mathrm{id}_n, -\mathrm{id}_n, 0) \begin{pmatrix} x_1 \\ x_2 \\ u_1 \end{pmatrix} = 0 \right)$$

$$\underset{\mathcal{F}\text{ cogenerator}}{\Longrightarrow} \exists X \in F[s]^{n \times 2n} \text{ with } (\mathrm{id}_n, -\mathrm{id}_n, 0) = X(s\,\mathrm{id}_{2n} - A, -B')$$

$$\Longrightarrow \exists X \text{ with } (\mathrm{id}_n, -\mathrm{id}_n) = X(s\,\mathrm{id}_{2n} - A)$$

$$\underset{(12.126)}{=} X \begin{pmatrix} \mathrm{id}_n & \mathrm{id}_n \\ 0 & \mathrm{id}_n \end{pmatrix} (s\,\mathrm{id}_{2n} - \widehat{A}) \begin{pmatrix} \mathrm{id}_n & -\mathrm{id}_n \\ 0 & \mathrm{id}_n \end{pmatrix}, \quad \widehat{A} = \begin{pmatrix} A_1 - B_2C_1 & 0 \\ B_2C_1 & A_1 + B_1C_2 \end{pmatrix}.$$

We define $Y := (Y_1, Y_2) := X \begin{pmatrix} \mathrm{id}_n & \mathrm{id}_n \\ 0 & \mathrm{id}_n \end{pmatrix} \in F[s]^{n \times (n+n)}$ and conclude

$$(\mathrm{id}_n, 0) = (\mathrm{id}_n, -\mathrm{id}_n) \begin{pmatrix} \mathrm{id}_n & \mathrm{id}_n \\ 0 & \mathrm{id}_n \end{pmatrix} = X \begin{pmatrix} \mathrm{id}_n & \mathrm{id}_n \\ 0 & \mathrm{id}_n \end{pmatrix} (s\,\mathrm{id}_{2n} - \widehat{A}) = Y(s\,\mathrm{id}_{2n} - \widehat{A})$$

$$= (Y_1, Y_2) \begin{pmatrix} s\,\mathrm{id}_n - (A_1 - B_2C_1) & 0 \\ -B_2C_1 & s\,\mathrm{id}_n - (A_1 + B_1C_2) \end{pmatrix}$$

$$\Longrightarrow \mathrm{id}_n = Y_1(s\,\mathrm{id}_n - (A_1 - B_2C_1)) - Y_2B_2C_1, \; 0 = Y_2(s\,\mathrm{id}_n - (A_1 + B_1C_2))$$

$$\underset{\mathrm{rank}(s\,\mathrm{id}_n - (A_1 + B_1C_2)) = n}{\Longrightarrow} Y_2 = 0 \Longrightarrow \mathrm{id}_n = Y_1(s\,\mathrm{id}_n - (A_1 - B_2C_1))$$

$$\Longrightarrow s\,\mathrm{id}_n - (A_1 - B_2C_1) \in \mathrm{Gl}_n(F[s]).$$

The last line is not true, and hence $\mathcal{C}_3 \neq 0$.
(b) $\mathcal{B}_1$ *is not controllable*: If $\mathcal{B}_1$ was controllable then so would be its image $\mathcal{C}_3$ which, however, is also autonomous and therefore zero. This contradicts (a).
(c) The observability of (12.125) is equivalent with $\mathcal{C}_1 = 0$. Since the plant is observable by assumption, i.e., $s \circ x_1 = A_1x_1, \; C_1x_1 = 0$ imply $x_1 = 0$, a simple computation furnishes the isomorphism

$$\mathcal{C}_4 := \left\{x_2 \in \mathcal{F}^n; \; s \circ x_2 = (A_1 - B_2C_1)x_2, \; C_2x_2 = 0\right\} \cong \mathcal{C}_1, \; x_2 \mapsto \begin{pmatrix} 0 \\ x_2 \end{pmatrix}.$$

This is zero, i.e., (12.125) is observable, if and only if $\mathcal{C}_2$ with C_2 is observable, i.e., if $C_2 \cdot : \mathcal{C}_2 \to \mathcal{F}^m, \; z \mapsto C_2z$, is injective. □

Remark 12.5.3 1. Theorem 12.5.2 can be extended to the case where the plant equations are Λ_1-stabilizable and Λ_1-observable. This means

$$\mathrm{char}\left\{\begin{pmatrix} x \\ u \end{pmatrix} \in \mathcal{F}^{n+m}; \; s \circ x_1 = A_1x_1 + B_1u_1 =\right\} = \mathrm{V}_{\mathbb{C}}(e_n^c) \subset \Lambda_1,$$
$$\mathrm{char}\left\{x \in \mathcal{F}^n; \; s \circ x_1 = A_1x_1, \; C_1x_1 = 0\right\} = \mathrm{V}_{\mathbb{C}}(e_n^o) \subset \Lambda_1,$$

where e_n^c resp. e_n^o are the highest elementary divisors of $(s\,\mathrm{id}_n - A_1, -B_1)$ resp. of $\begin{pmatrix} s\,\mathrm{id}_n - A_1 \\ C_1 \end{pmatrix}$. One has to apply Theorem 12.5.2 to the controllable and observable part of equations (12.117), cf. Theorems 4.1.11, 4.3.11 and Sect. 5.4.1.
2. In contrast to the theory from Chap. 10 the closed-loop behavior here does not allow an input u_2 to the compensator, for instance, of a reference signal that has to be tracked. The asymptotically stable equations (12.125) and (12.126) or similar ones, resulting from Luenberger observers and state feedback, are discussed at length in many of the quoted books, [1, p. 523], [3, Sect. 10.5], [13, Sect. 4.4].

However, the use of these stable closed-loop behaviors with input u_1 for useful tasks is less studied, cf. [14, Chap. 8].

3. If $u_1 \in F^m$ is a constant input of the closed-loop behavior and if $x = \binom{x_1}{x_2}$ is any state to this input, then Corollary 8.2.56 and $(0\ \mathrm{id}_{2n}\ -A)^{-1} = -A^{-1}$ imply

$$x(\infty) := \lim_{t\to\infty} x(t) = -A^{-1}B'u_1 \text{ or } Ax(\infty) = -B'u_1$$
$$y(\infty) = \lim_{t\to\infty} \begin{pmatrix} y_1(t) \\ y_2(t) \end{pmatrix} = (-CA^{-1}B' + D')u_1$$
$$y_1(\infty) = \left(D_1 - (C_1, D_1C_2)A^{-1}\begin{pmatrix} B_1 \\ B_1 \end{pmatrix}\right)u_1.$$

Also here one says that the closed-loop behavior *tracks* the constant signal $y_1(\infty)$ [15, p. 340]. Which constant signals can be tracked in this fashion, depends on the column space $\left(D_1 - (C_1, D_1C_2)A^{-1}\begin{pmatrix} B_1 \\ B_1 \end{pmatrix}\right)F^m$. Section 10.2 contains a much more general theory of tracking, cf. [15, pp. 340–341].

◊

12.6 The Canonical Controllability Realization of a Controllable IO Behavior

In this section, we construct the canonical controllability realization of a controllable behavior

$$\mathcal{B} = \left\{\begin{pmatrix} y \\ u \end{pmatrix} \in \mathcal{F}^{p+m};\ P \circ y = Q \circ u\right\} \tag{12.128}$$

with transfer matrix H by means of the canonical observability realization of the transpose $H^\top$ of its transfer matrix.

Recall from Theorem 6.2.4 that any rational matrix $H \in F(s)^{p\times m}$ admits a unique controllable IO realization

$$\mathcal{B}(H) = \left\{\begin{pmatrix} y \\ u \end{pmatrix} \in \mathcal{F}^{p+m};\ P \circ y = Q \circ u\right\} \tag{12.129}$$

where $P \in F[s]^{p\times p}$, $Q = PH$, and $F[s]^{1\times p}P = \{\xi \in F[s]^{1\times p};\ \xi H \in F[s]^{1\times m}\}$. Equations (6.15)–(6.16) and Lemma 10.1.6 contain algorithms for the computation of $(P, -Q)$ from H. Theorems 12.3.5 and 12.4.5 furnish two canonical observable state space realizations of $\mathcal{B}(H)$ and therefore of H.

Lemma 12.6.1 *Let $H \in F(s)^{p\times m}$ be any matrix and let $\mathcal{B}(H) = \left\{\begin{pmatrix} y \\ u \end{pmatrix} \in \mathcal{F}^{p+m};\ P \circ y = Q \circ u\right\}$ be its unique controllable IO realization. Assume that the state space system*

$$s \circ x = Ax + Bu,\ y = Cx + D \circ u,\ x \in \mathcal{F}^n, \text{ with state behavior}$$
$$\mathcal{B}^s := \left\{\begin{pmatrix} x \\ u \end{pmatrix} \in \mathcal{F}^{n+m};\ s \circ x = Ax + Bu\right\} \tag{12.130}$$

is a controllable state space realization of H*, i.e.,* (A, B) *is controllable and* $H = D + C(s\,\mathrm{id}_n - A)^{-1}B$*. Then the following holds:*

1. $\mathcal{B}(H) = \begin{pmatrix} C & D \\ 0 & \mathrm{id}_m \end{pmatrix} \circ \mathcal{B}^s$*, i.e., (12.130) is a state space realization of* $\mathcal{B}(H)$*. If (12.130) is also observable then* $\begin{pmatrix} C & D \\ 0 & \mathrm{id}_m \end{pmatrix} \circ : \mathcal{B}^s \xrightarrow{\cong} \mathcal{B}(H)$*.*
2. *Let* $\mathcal{B} = \left\{ \binom{y}{u} \in \mathcal{F}^{p+m};\ P^1 \circ y = Q^1 \circ u \right\}$ *be any IO realization of* H*, i.e.,* $H = (P^1)^{-1}Q^1 = P^{-1}Q$*. Then* $\begin{pmatrix} C & D \\ 0 & \mathrm{id}_m \end{pmatrix} \circ \mathcal{B}^s = \mathcal{B}(H) = \mathcal{B}_{\mathrm{cont}}$ *is the controllable part of* $\mathcal{B}$*. Hence, if* $\mathcal{B}$ *is not controllable, there are solutions* $\binom{y}{u}$ *of* $P^1 \circ y = Q^1 \circ u$ *that are not of the form* $y = Cx + D \circ u$*. Thus a representation* $\mathcal{B} = \begin{pmatrix} C & D \\ 0 & \mathrm{id}_m \end{pmatrix} \circ \mathcal{B}^s$ *with controllable* $\mathcal{B}^s$ *is possible only if* $\mathcal{B}$ *is controllable, i.e.,* $\mathcal{B} = \mathcal{B}(H)$*.*

Proof 1. According to Theorem and Definition 6.3.1 and (6.48), the behavior $\begin{pmatrix} C & D \\ 0 & \mathrm{id}_m \end{pmatrix} \circ \mathcal{B}^s$ is an IO behavior with transfer matrix $D + C(s\,\mathrm{id}_n - A)^{-1}B = H$. In particular, it is thus the image of the controllable behavior $\mathcal{B}^s$ under a behavior morphism and thus also controllable, see Lemma 4.2.5. In other words, $\begin{pmatrix} C & D \\ 0 & \mathrm{id}_m \end{pmatrix} \circ \mathcal{B}^s$ is a controllable IO realization of H. By Theorem 6.2.4 the controllable IO realization of H is unique, hence $\begin{pmatrix} C & D \\ 0 & \mathrm{id}_m \end{pmatrix} \circ \mathcal{B}^s = \mathcal{B}(H)$.
2. Obvious. □

Because of item 2 of Lemma 12.6.1, we conclude that controllable state space realizations, in particular, the controllability and the controller realizations, of non-controllable *IO behaviors* do not exist. Therefore, we assume $\mathcal{B} = \mathcal{B}_{\mathrm{cont}} = \mathcal{B}(H)$ in the sequel.

We derive the *canonical controllability form* of H from the observability form of $\widehat{H} := H^\top \in F[s]^{m\times p}$. The unique controllable IO behavior with transfer matrix $\widehat{H}$ is $\mathcal{B}(H^\top) = U^\perp$ where

$$U = \mathcal{D}^{1\times m}(\widetilde{P}, -\widetilde{Q}),\ (\widetilde{P}, -\widetilde{Q}) \in \mathcal{D}^{m\times(m+p)},\ \mathrm{rank}(\widetilde{P}) = m,\ U^0 := \mathcal{D}^{1\times m}\widetilde{P}, \tag{12.131}$$

and, again, $(\widetilde{P}, -\widetilde{Q})$ is computed via (6.15)–(6.16) or via Lemma 10.1.6.

Choose a term order on $\{1, \ldots, m\} \times \mathbb{N}$, let $\widehat{P}$ be the Gröbner matrix of $U^0 = \mathcal{D}^{1\times m}\widetilde{P}$ with respect to this term order, and define $\widehat{Q} := \widehat{P}\widehat{H}$. The degree set and its complement are

$$\begin{aligned} \deg(U^0) &= \biguplus_{i=1}^{m} \{i\} \times (d^{\mathrm{cb}}(i) + \mathbb{N}), \quad \text{and} \\ \Gamma^{\mathrm{cb}} &:= \left\{(i, \mu);\ i = 1, \ldots, m,\ 0 \le \mu \le d^{\mathrm{cb}}(i) - 1\right\}, \end{aligned} \tag{12.132}$$

and we define $n := \sharp(\Gamma^{\mathrm{cb}}) = \sum_{i=1}^m d^{\mathrm{cb}}(i)$. The index "cb" stands for *controllability*. Let

$$\widehat{A} \in F^{\Gamma^{\mathrm{cb}}\times\Gamma^{\mathrm{cb}}},\ \widehat{B} \in F^{\Gamma^{\mathrm{cb}}\times p},\ \widehat{C} \in F^{m\times\Gamma^{\mathrm{cb}}},\ \widehat{D} \in F[s]^{m\times p} \tag{12.133}$$

denote the matrices of the canonical observability form of $\widehat{H}$ or $\mathcal{B}(\widehat{H})$ according to Theorem 12.3.5, hence $\widehat{H} = \widehat{D} + \widehat{C}(s\,\mathrm{id}_{\Gamma^{\mathrm{cb}}} - \widehat{A})^{-1}\widehat{B}$ holds. This state space realization is observable by construction, i.e., $\begin{pmatrix} s\,\mathrm{id}_{\Gamma^{\mathrm{cb}}} - \widehat{A} \\ \widehat{C} \end{pmatrix}$ is left invertible (Theorem and Definition 4.1.9), and it is controllable since $\mathcal{B}(\widehat{H})$ is controllable, i.e., $(s\,\mathrm{id}_{\Gamma^{\mathrm{cb}}} - \widehat{A}, -\widehat{B})$ is right invertible (Theorem and Definition 4.3.8). We define the transposed matrices

$$\begin{aligned} &A^{\mathrm{cb}} := \widehat{A}^{\top} \in F^{\Gamma^{\mathrm{cb}} \times \Gamma^{\mathrm{cb}}}, \quad B^{\mathrm{cb}} := \widehat{C}^{\top} \in F^{\Gamma^{\mathrm{cb}} \times m}, \\ &C^{\mathrm{cb}} := \widehat{B}^{\top} \in F^{p \times \Gamma^{\mathrm{cb}}}, \quad D^{\mathrm{cb}} := \widehat{D}^{\top} \in F[s]^{p \times m} \\ &\text{with } H = \widehat{H}^{\top} = D^{\mathrm{cb}} + C^{\mathrm{cb}}(s\,\mathrm{id}_{\Gamma^{\mathrm{cb}}} - A^{\mathrm{cb}})^{-1} B^{\mathrm{cb}}. \end{aligned} \tag{12.134}$$

Since $\begin{pmatrix} s\,\mathrm{id}_n - A^{\mathrm{cb}} \\ C^{\mathrm{cb}} \end{pmatrix}$ resp. $(s\,\mathrm{id}_n - A^{\mathrm{cb}}, -B^{\mathrm{cb}})$ are left resp. right invertible, the matrices $(A^{\mathrm{cb}}, B^{\mathrm{cb}}, C^{\mathrm{cb}}, D^{\mathrm{cb}})$ define the, essentially unique, controllable and observable state space equations of the transfer matrix H. Summing up, we obtain the following theorem:

Theorem 12.6.2 (The canonical controllability realization of a controllable IO behavior, cf. [1, pp. 417–422]). *Consider a transfer matrix $H \in F(s)^{p \times m}$, its unique controllable IO realization $\mathcal{B}(H)$, a term order on $\{1, \dots, m\} \times \mathbb{N}$ and the derived matrices from* (12.134).

1. *These matrices define a controllable and observable state space realization of $\mathcal{B}(H)$, i.e., $H = D^{\mathrm{cb}} + C^{\mathrm{cb}}(s\,\mathrm{id}_n - A^{\mathrm{cb}})^{-1} B^{\mathrm{cb}}$ and there is the isomorphism*

$$\begin{aligned} &\begin{pmatrix} C^{\mathrm{cb}} & D^{\mathrm{cb}} \\ 0 & \mathrm{id}_m \end{pmatrix} \circ : \mathcal{B}^{\mathrm{cb}} \xrightarrow{\cong} \mathcal{B}(H), \quad \begin{pmatrix} x \\ u \end{pmatrix} \longmapsto \begin{pmatrix} y \\ u \end{pmatrix} = \begin{pmatrix} C^{\mathrm{cb}}x + D^{\mathrm{cb}} \circ u \\ u \end{pmatrix}, \\ &\text{where } \mathcal{B}^{\mathrm{cb}} := \left\{ \begin{pmatrix} x \\ u \end{pmatrix} \in \mathcal{F}^{\Gamma^{\mathrm{cb}} + m};\ s \circ x = A^{\mathrm{cb}} x + B^{\mathrm{cb}} u \right\}. \end{aligned} \tag{12.135}$$

   This realization is called the canonical controllability realization *of H or $\mathcal{B}(H)$. It depends on H and the chosen term order only. The numbers $d^{\mathrm{cb}}(i)$, $i = 1, \dots, m$, are the* controllability indices *of H with respect to the chosen term order.*
2. *The submatrix $\left((A^{\mathrm{cb}})^{\nu} B^{\mathrm{cb}}_{-j}\right)_{(j,\nu) \in \Gamma^{\mathrm{cb}}} \in F^{\Gamma^{\mathrm{cb}} \times \Gamma^{\mathrm{cb}}}$ of the controllability matrix*

$$\mathfrak{C}^{\mathrm{cb}} \in F^{n \times nm} \underset{\text{ident.}}{=} F^{\Gamma^{\mathrm{cb}} \times nm}, \quad \mathfrak{C}^{\mathrm{cb}}_{-(j,\nu)} = \left(A^{\mathrm{cb}}\right)^{\nu} B^{\mathrm{cb}}_{-j}, \quad 1 \le j \le m, \ 0 \le \nu \le n-1,$$

   is the identity matrix $\mathrm{id}_{\Gamma^{\mathrm{cb}}}$.
3. *Algorithmically, the controllability data of $\mathcal{B}(H)$ are obtained from the observability data of $H^{\top}$ via equations* (12.131), (12.133), *and* (12.134).

Proof The proof follows from applying Theorem 12.3.5 to $\mathcal{B}(H^{\top})$ and transposing the results. □

Remark 12.6.3 In the discrete-time case $\mathcal{F} = F^{\mathbb{N}}$ the controllability form can be applied to steer the state from an arbitrary initial state $x(0) \in F^{\Gamma^{\mathrm{cb}}}$ to an arbitrary final state $x(T) = z \in F^{\Gamma^{\mathrm{cb}}}$ if $T \ge \max_i d^{\mathrm{cb}}(i) + 1$. We write $A := A^{\mathrm{cb}}$ and $B := B^{\mathrm{cb}}$. Recall from (3.65) that

$$x(T) = (s^T \circ x)(0) = A^T x(0) + \sum_{\nu=0}^{T-1} A^\nu B u(T-1-\nu).$$

Thus we have for all $(i, \mu) \in \Gamma^{\mathrm{cb}}$ that

$$\begin{aligned} x_{i,\mu}(T) &= \left(A^T x(0)\right)_{(i,\mu)} + \sum_{\nu=0}^{T-1}\sum_{j=1}^{m} (A^\nu B_{-j})_{(i,\mu)} u_j(T-1-\nu) \\ &= \left(A^T x(0)\right)_{(i,\mu)} + \underbrace{\sum_{(j,\nu)\in\Gamma^{\mathrm{cb}}} (A^\nu B_{-j})_{(i,\mu)} u_j(T-1-\nu)}_{\underset{\text{Theorem 12.6.2,2.}}{=} u_i(T-1-\mu)} \\ &\quad + \sum_{\substack{(j,\nu)\notin\Gamma^{\mathrm{cb}} \\ 1\le j\le m,\ 0\le\nu\le T-1}} (A^\nu B_{-j})_{(i,\mu)} u_j(T-1-\nu) \\ &= \left(A^T x(0)\right)_{(i,\mu)} + u_i(T-1-\mu) + \sum_{\substack{(j,\nu)\notin\Gamma^{\mathrm{cb}} \\ 1\le j\le m,\ 0\le\nu\le T-1}} (A^\nu B_{-j})_{(i,\mu)} u_j(T-1-\nu). \end{aligned}$$

If we choose the input components by

$$u_j(T-1-\nu) := \begin{cases} z_{j,\nu} - \left(A^T x(0)\right)_{(j,\nu)} & \text{if } (j,\nu) \in \Gamma^{\mathrm{cb}}, \\ 0 & \text{if } (j,\nu) \notin \Gamma^{\mathrm{cb}}, \end{cases}$$

then we get

$$x_{i,\mu}(T) = \left(A^T x(0)\right)_{(i,\mu)} + \left(z_{i,\mu} - \left(A^T x(0)\right)_{(i,\mu)}\right) + 0 = z_{i,\mu},$$

i.e., $x(T) = z$. According to [1, p. 419], the preceding derivation is the reason for the term *controllability form*. ◊

Example 12.6.4 (The controllability realization in the SISO case). We consider the case $p = m = 1$ with coprime $P = s^n - p_{n-1}s^{n-1} - \cdots - p_0$ and Q and $H = P^{-1}Q = H^\top$. The coprimeness of P and Q ensures that the behavior $(F[s](P, -Q))^\perp$ is controllable and thus equal to $\mathcal{B}(H)$.

The matrices of the state equation are obtained from those of Example 12.3.10 by transposition. We get

$$\Gamma^{\mathrm{cb}} = \{0, \dots, n-1\},$$

$$A^{\mathrm{cb}} = \begin{pmatrix} 0 & \cdots & 0 & p_0 \\ 1 & & 0 & p_1 \\ & \ddots & & \vdots \\ 0 & & 1 & p_{n-1} \end{pmatrix} \in F^{n\times n}, \qquad B^{\mathrm{cb}} = \begin{pmatrix} 1 \\ 0 \\ \vdots \\ 0 \end{pmatrix} \in F^n,$$

$$C^{\mathrm{cb}} = \left(C_0^{\mathrm{cb}} \cdots C_{n-1}^{\mathrm{cb}}\right) \in F^{1\times n}, \quad \text{and} \quad D^{\mathrm{cb}} = H_{\mathrm{pol}} \in F[s]$$

$$\mathfrak{C} = (B^{\rm cb}, A^{\rm cb}B^{\rm cb}, \ldots, (A^{\rm cb})^{n-1}B^{\rm cb}) = \mathrm{id}_n$$

The computation of $C^{\rm cb}$ is described in Example 12.3.10 and in Theorem 12.6.2. The transfer function of this state space system system is

$$H = P^{-1}Q = D^{\rm cb} + C^{\rm cb}(s\,\mathrm{id}_n - A^{\rm cb})^{-1}B^{\rm cb} = D^{\rm cb} + \sum_{\mu=0}^{n-1} C^{\rm cb}_{\mu}\left((s\,\mathrm{id}_n - A^{\rm cb})^{-1}\right)_{\mu 0}.$$

The state space equations for $x = \begin{pmatrix} x_0\\ \vdots \\ x_{n-1}\end{pmatrix} \in \mathcal{F}^n$ are

$$s \circ x_\mu = \begin{cases} p_0 x_{n-1} + u & \text{if } \mu = 0,\\ x_{\mu-1} + p_\mu x_{n-1} & \text{if } 0 < \mu \le n-1, \end{cases}, \quad y = \sum_{\mu=0}^{n-1} C^{\rm cb}_\mu x_\mu + D^{\rm cb} \circ u,$$

and the corresponding block diagram is

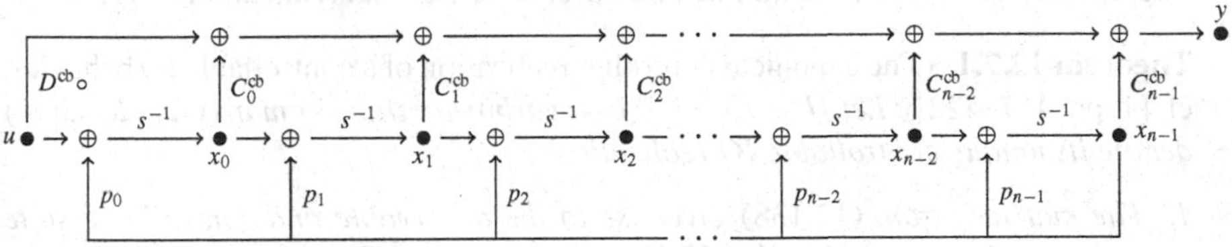

compare [1, Fig. 2.1-10 on p. 44]. By Theorem 12.6.2, the map

$$\begin{pmatrix} C^{\rm cb} & D^{\rm cb}\\ 0 & 1\end{pmatrix} \circ : \left\{ \begin{pmatrix} x\\ u\end{pmatrix} \in \mathcal{F}^{\Gamma^{\rm cb}+1};\ s \circ x = A^{\rm cb}x + B^{\rm cb}u \right\} \xrightarrow{\cong} \mathcal{B}$$

is an isomorphism. ◊

12.7 The Canonical Controller Realization

The canonical controller realization of a controllable IO behavior $\mathcal{B} = \mathcal{B}(H)$ or, equivalently, of its transfer matrix $H \in F(s)^{p\times m}$ is the dual to the canonical observer realization.

We assume a matrix $H \in F(s)^{p\times m}$ and the degree-over-position term order

$$(j, \nu) < (i, \mu) \Longleftrightarrow (\nu < \mu \text{ or } \nu = \mu \text{ and } j > i) \text{ on } \{1, \ldots, m\} \times \mathbb{N}.$$

The module

$$U^0 := \{\xi \in F[s]^{1\times m};\ \xi H^\top \in F[s]^{1\times p}\} \tag{12.136}$$

is the equation module of the autonomous part of the unique controllable IO realization $\mathcal{B}(H^\top)$ of $H^\top$. Let $\widehat{P} \in F[s]^{p\times p}$ be the reduced Gröbner matrix of $U^0 = F[s]^{1\times m}\widehat{P}$ with respect to the term order. The degree set of U^0 is

$$\deg(U^0) = \biguplus_{i=1}^{m} \{i\} \times (d^c(i) + \mathbb{N}) \Longrightarrow \Gamma^c := \{(i,\mu);\ 1 \le i \le m,\ 1 \le \mu \le d^c(i)\}. \tag{12.137}$$

Let $\widehat{A} \in F^{\Gamma^c\times\Gamma^c}$, $\widehat{B} \in F^{\Gamma^c\times p}$, $\widehat{C} \in F^{m\times\Gamma^c}$, $\widehat{D} \in F[s]^{m\times p}$ be the matrices of the canonical observer realization of $\mathcal{B}(H^\top)$ according to Theorem 12.4.5. As in Sect. 12.6 these matrices form a controllable and observable state space realization of $H^\top = \widehat{D} + \widehat{C}(s\,\mathrm{id}_{\Gamma^c} - \widehat{A})\widehat{B}$. By transposition we obtain

$$\begin{array}{ll} A^c := \widehat{A}^\top \in F^{\Gamma^c\times\Gamma^c}, & B^c := \widehat{C}^\top \in F^{\Gamma^c\times m}, \\ C^c := \widehat{B}^\top \in F^{p\times\Gamma^c}, & D^c := \widehat{D}^\top \in F^{p\times m} \\ \multicolumn{2}{c}{\text{with } H = D^c + C^c(s\,\mathrm{id}_{\Gamma^c} - A^c)^{-1}B^c.} \end{array} \tag{12.138}$$

These matrices form a controllable and observable Kalman realization of H.

Theorem 12.7.1 (The canonical controller realization of a controllable IO behavior, cf. [1, pp. 417–422]). *Let $H \in F[s]^{p\times m}$ be an arbitrary transfer matrix and let $\mathcal{B}(H)$ denote its unique controllable IO realization.*

1. *The matrices from* (12.138) *give rise to the observable and controllable state space realization of H and of $\mathcal{B}(H)$, i.e., $H = D^c + C^c(s\,\mathrm{id}_{\Gamma^c} - A^c)^{-1}B^c$ and*

$$\begin{array}{l} \begin{pmatrix} C^c & D^c \\ 0 & \mathrm{id}_m \end{pmatrix} \circ :\ \mathcal{B}^c \xrightarrow{\cong} \mathcal{B}(H),\ \begin{pmatrix} x \\ u \end{pmatrix} \longmapsto \begin{pmatrix} y \\ u \end{pmatrix},\ y = C^c x + D^c \circ u, \\ \text{where } \mathcal{B}^c := \left\{ \begin{pmatrix} x \\ u \end{pmatrix} \in \mathcal{F}^{\Gamma^c + m};\ s \circ x = A^c x + B^c u \right\}. \end{array} \tag{12.139}$$

2. *The matrices A^c, B^c, C^c, and D^c depend on H and the chosen degree-over-position term order only. The controllability indices $d^c(i)$, $i = 1, \dots, m$, depend on H only, up to a permutation. The matrices A^c and B^c are sparse.*
3. *Algorithmically, the controllability data of $\mathcal{B}(H)$ are obtained from the coprime factorization $H^\top = (P^o)^{-1}Q^o$ and from the observer data according to Corollary 12.4.7 of*

$$\begin{array}{l} \mathcal{B}(H^\top) = (U^0)^\perp = \left\{ \begin{pmatrix} \widehat{y} \\ \widehat{u} \end{pmatrix} \in \mathcal{F}^{m+p};\ P^o \circ \widehat{y} = Q^o \circ \widehat{u} \right\} \text{ where} \\ U^0 = F[s]^{1\times m} P^o = F[s]^{1\times m}\widehat{P} \end{array}$$

and $\widehat{P} \in F[s]^{m\times m}$ is the Gröbner matrix of U^0 for the chosen term order.

Proof The proof is analogous to that of Theorem 12.6.2, and uses Theorem 12.4.5 and Lemma 12.6.1. □

Example 12.7.2 We discuss the SISO case $p = m = 1$ with coprime $P = s^n - p_{n-1}s^{n-1} - \cdots - p_0$ and Q and $H = P^{-1}Q = H^\top$. Due to the coprimeness the equation $\mathcal{B}(H) = (F[s](P, -Q))^{\perp}$ holds. With the notations from Corollary 12.4.7 we get

$$\widehat{P}^1 = \widehat{P} = P = s^n - \widehat{P}^{1,\mathrm{lc}},\ \widehat{P}^{1,\mathrm{lc}} = \sum_{\mu=0}^{n-1} p_\mu s^\mu\ \widehat{H}^2 = \widehat{H} = H, \text{ and}$$

$$Q^1 = P^1 \widehat{H}^2_{\mathrm{spr}} = P H_{\mathrm{spr}} = Q - P H_{\mathrm{pol}} = \sum_{\mu=0}^{n-1} q_\mu s^\mu .$$

The matrices of the controller realization are the transposes of those of the observer realization. We get

$$\Gamma^{\mathrm{c}} = \{1, \ldots, n\}$$

$$A^{\mathrm{c}} = \begin{pmatrix} 0 & 1 & & 0 \\ \vdots & & \ddots & \\ 0 & 0 & & 1 \\ p_0 & p_1 & \cdots & p_{n-1} \end{pmatrix} \in F^{n\times n}, \quad B^{\mathrm{c}} = \begin{pmatrix} 0 \\ \vdots \\ 0 \\ 1 \end{pmatrix} \in F^n,$$

$$C^{\mathrm{c}} = \begin{pmatrix} q_0 & \cdots & q_{n-1} \end{pmatrix}, \qquad D^{\mathrm{c}} = H_{\mathrm{pol}}.$$

Notice that in the SISO cases the identities $A^{\mathrm{ob}} = A^{\mathrm{c}} = (A^{\mathrm{o}})^\top = (A^{\mathrm{cb}})^\top$ hold. The state space equations for the canonical controller realization are

$$s \circ x_\mu = \begin{cases} x_{\mu+1} & \text{if } 1 \le \mu < n, \\ \sum_{\nu=1}^{n} p_{\nu-1} x_\nu + u & \text{if } \mu = n, \end{cases}$$

$$y = \sum_{\mu=1}^{n} q_{\mu-1} x_\mu + D^{\mathrm{c}} \circ u,$$

where $x = \begin{pmatrix} x_1 \\ \vdots \\ x_n \end{pmatrix} \in \mathcal{F}^n$. The associated interconnection diagram is

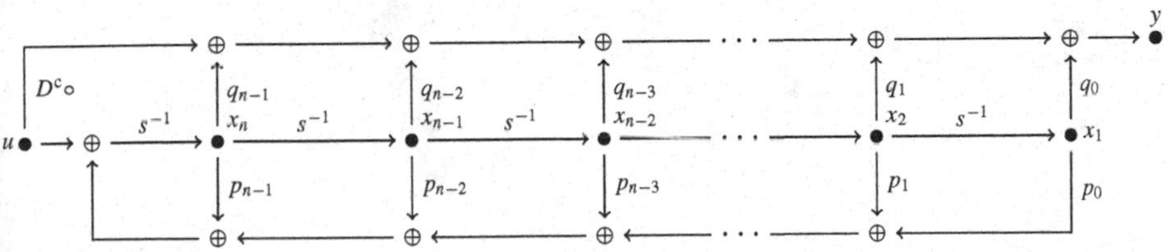

compare [1, Fig. 2.1-4 on p. 39]. ◊

References

1. T. Kailath, *Linear Systems* (Prentice-Hall, Englewood Cliffs, 1980)
2. U. Oberst, Canonical state representations and Hilbert functions of multidimensional systems. Acta Appl. Math. **94**, 83–135 (2006)
3. J.W. Polderman, J.C. Willems, *Introduction to mathematical systems theory. A behavioral approach*, Vol. 26. Texts in Applied Mathematics (Springer, New York, 1998)
4. D.G. Luenberger, Observing the state of a linear system. IEEE Trans. Military Electron. **8**(2), 74–80 (1964)
5. Th. Becker, V. Weispfenning, *Gröbner bases. A computational approach to commutative algebra*, vol. 141. Graduate Texts in Mathematics (Springer, New York, 1993)
6. D.A. Cox, J. Little, D. O'Shea, *Ideals, Varieties, and Algorithms, An Introduction to Computational Algebraic Geometry and Commutative Algebra*, 4th edn. Undergraduate Texts in Mathematics (Springer, Cham, 2015)
7. D.A. Cox, J. Little, D. O'Shea. *Using Algebraic Geometry. Graduate Texts in Mathematics*, vol. 185, 2nd edn. (Springer, New York, 2005)
8. D. Eisenbud, *Commutative Algebra with a View Toward Algebraic Geometry. Graduate Texts in Mathematics*, vol. 150 (Springer, New York, 1995)
9. G.-M. Greuel, G. Pfister, *A Singular Introduction to Commutative Algebra*, 2nd edn (Springer, Berlin, 2008)
10. M. Kreuzer, L. Robbiano, *Computational Commutative Algebra*, vol. 1 (Springer, Berlin, 2008)
11. W.A. Wolovich, *Linear Multivariable Systems. Applied Mathematical Sciences*, vol. 11(Springer, New York, 1974)
12. R.E. Kalman, P.L. Falb, M.A. Arbib, *Topics in Mathematical System Theory* (McGraw-Hill, New York, 1969)
13. P.J. Antsaklis, A.N. Michel, *Linear Systems*, 2nd edn. (Birkhäuser, Boston, 2006)
14. H. Bourlès, *Linear Systems* (ISTE, London, 2010)
15. C.T. Chen, *Linear System Theory and Design* (Harcourt Brace College Publishers, Fort Worth, 1984)

Chapter 13
Generalized Fractional Calculus and Behaviors

In this short chapter we show that the module-behavior duality can also be applied to fractional or symbolic [1, Sect. VI, 5] calculus, to suitably defined fractional behaviors and to the constructive solution of general fractional differential systems, cf. [Wikipedia; https://en.wikipedia.org/wiki/Fractional_calculus calculus, September 4, 2019], [2]. The latter references have long bibliographies. We generalize the standard fractional calculus and differential equations considerably. Schwartz [1, p. 176, below] already suggested *multiple applications, simplifications, and corrections* of the symbolic calculus of the literature by means of the theory of distributions with left bounded support. We do not discuss the applications of the fractional calculus that are comprehensively exposed in [2]. We quote, but do not prove the equations and properties of the Gamma function and the equation that connects the Beta function with the Gamma function.

13.1 Fractional Calculus and Behaviors

The structure of $C_{+}^{-\infty}$ as $\mathbb{C}(s)$-vector space also enables the construction of the generalized Heavyside distributions $Y_m \in C_{+}^{-\infty},\ m \in \mathbb{C}$, according to [1, (II, 2:31), (VI, 5; 6)].

Let $\Gamma(z),\ z \in \mathbb{C}$, denote the meromorphic *Gamma* function, cf. [3, Sect. 3.09], [Wikipedia; https://en.wikipedia.org/wiki/Gamma_function, August 14, 2019], [Wikipedia; https://de.wikipedia.org/wiki/Stirlingformel, July 25, 2019]. It satisfies

$$\begin{aligned}&\Gamma(z+1) = z\Gamma(z),\ z \in \mathbb{C} \setminus (-\mathbb{N}),\ \Gamma(m+1) = m! \quad \text{if } m \in \mathbb{N},\\ &\Gamma(z) = \int_0^\infty t^{z-1}e^{-t}\mathrm{d}t \quad \text{if } \Re(z) > 0,\end{aligned} \tag{13.1}$$

U. Oberst et al., *Linear Time-Invariant Systems, Behaviors and Modules*, Differential-Algebraic Equations Forum,
https://doi.org/10.1007/978-3-030-43936-1_13

has simple poles at $-m,\ m \in \mathbb{N}$, and has no zeros. Moreover, in the real interval $\left(\frac{1}{12}, \infty\right)$ Stirling's approximation formula holds in the form

$$\begin{aligned}&\Gamma(z) = \sqrt{2\pi} z^{z-\frac{1}{2}} e^{-z} e^{\gamma(z)},\ 0 < \gamma(z) < 1\\ &\Longrightarrow \forall \mu > 0 \forall z > \tfrac{1}{12} :\ \tfrac{\Gamma(z+\mu)}{\Gamma(z)} \geq \left(\tfrac{z}{z+\mu}\right)^{1/2} \left(1 + \tfrac{\mu}{z}\right)^z e^{-(\mu+1)} (z+\mu)^\mu.\end{aligned} \tag{13.2}$$

From

$$\lim_{z\to\infty} \tfrac{z}{z+\mu} = 1,\ \lim_{z\to\infty} \left(1 + \tfrac{\mu}{z}\right)^z = e^\mu,\ \lim_{z\to\infty} (z+\mu)^\mu = \infty$$

we conclude that there is $z_0 > 0$ such that $\Gamma(z+\mu) \geq \Gamma(z)$ for $z \geq z_0$. In particular,

$$\forall \mu > 0 \exists N(\mu) \in \mathbb{N} \forall N \geq N(\mu) :\ \Gamma(N\mu) \leq \Gamma((N+1)\mu). \tag{13.3}$$

This will be applied in the proof of Lemma 13.1.2.

If f is a piecewise continuous, complex-valued function on the open interval $(0, \infty)$ we extend this to a function $f_{\mathbb{R}} : \mathbb{R} \to \mathbb{C}$ by

$$\forall t > 0 :\ f_{\mathbb{R}}(t) := f(t),\ \forall t \leq 0 :\ f_{\mathbb{R}}(t) := 0. \tag{13.4}$$

Obviously $f_{\mathbb{R}}$ is piecewise continuous on $\mathbb{R} \setminus \{0\}$. If $f(0+) := \lim_{t\to 0,\, t>0} f(t)$ exists, then $f_{\mathbb{R}}$ is piecewise continuous on $\mathbb{R}$ with the jump $f_{\mathbb{R}}(0+) - f_{\mathbb{R}}(0-) = f(0+)$ at $t = 0$. If for some or all $a > 0$ the Riemann integral

$$\int_{-a}^{a} |f_{\mathbb{R}}(t)| \mathrm{d}t = \int_0^a |f(t)| \mathrm{d}t := \lim_{\epsilon\to 0,\, \epsilon>0} \int_\epsilon^a |f(t)| \mathrm{d}t$$

is finite, then $f_{\mathbb{R}}$ is called *locally integrable*. The corresponding distribution is defined by

$$\begin{aligned}
f_{\mathbb{R}}(\varphi) &= \int_{-\infty}^{\infty} f_{\mathbb{R}}(t)\varphi(t)\mathrm{d}t = \lim_{\epsilon\to 0,\, \epsilon>0} \int_\epsilon^\infty f(t)\varphi(t)\mathrm{d}t,\ \varphi \in C_0^\infty, \Longrightarrow \forall u \in C_+^{0,\mathrm{pc}} :\\
(f_{\mathbb{R}} * u)(t) &= \int_{-\infty}^{\infty} f_{\mathbb{R}}(t-x)u(x)\mathrm{d}x = \int_{-\infty}^{\infty} f_{\mathbb{R}}(x)u(t-x)\mathrm{d}x\\
&= \int_{-\infty}^{t} f(t-x)u(x)\mathrm{d}x = \int_0^\infty f(x)u(t-x)\mathrm{d}x\\
&= \lim_{\epsilon\to 0, \epsilon>0} \int_{-\infty}^{t-\epsilon} f(t-x)u(x)\mathrm{d}x = \lim_{\epsilon\to 0, \epsilon>0} \int_\epsilon^\infty f(x)u(t-x)\mathrm{d}x.
\end{aligned} \tag{13.5}$$

For locally integrable $f_{\mathbb{R}}$ we moreover infer

$$\forall T, T_1 > 0 \forall u \in C^{0,pc} \text{ with } u(x) = 0 \text{ for } x \le -T_1 :$$
$$(f_{\mathbb{R}} * u)(t) = \begin{cases} \int_0^{T+T_1} f(x)u(t-x)dx & \text{if } -T_1 \le t \le T \\ 0 & \text{if } t \le -T_1 \end{cases}. \tag{13.6}$$

This shows that for locally integrable $f_{\mathbb{R}}$ and $u \in C_+^{0,pc}$ the convolution $f_{\mathbb{R}} * u$ is a continuous function. Since $f_{\mathbb{R}} = (s \circ Y) * f_{\mathbb{R}} = s \circ (Y * f_{\mathbb{R}})$, the distribution $f_{\mathbb{R}}$ is the derivative of a continuous function and thus contained in $C_+^{-\infty}$.

For the definition of Y_m, $m \in \mathbb{C}$, we first consider the case $\Re(m) > 1$. Recall

$$\forall m \in \mathbb{N},\ m \ge 1 :\ s^{-m} \circ \delta = (m-1)!^{-1} t^{m-1} Y,\ Y = s^{-1} \circ \delta,$$
$$s^{-m} \circ \delta = \left(s^{-1} \circ \delta\right)^{*m} = Y * Y * \cdots * Y = Y^{*m} \in C_+^{-\infty}.$$

Generalizing this, as apparently already Leibniz considered, one defines

$$\forall m \in \mathbb{C} \text{ with } \Re(m) > 1 :$$
$$Y_m := \left(\Gamma(m)^{-1} t^{m-1}\right)_{\mathbb{R}} = \left(\Gamma(m)^{-1} t^{\Re(m)-1} e^{j\Im(m)\ln(t)}\right)_{\mathbb{R}} \tag{13.7}$$
$$\Longrightarrow Y_m \in C_+^0 \subset C_+^{-\infty},\ Y_m = s^{-m} \circ \delta,\ m \in \mathbb{N},\ m > 1.$$

Due to $\Re(m) > 1$ the function Y_m is indeed continuous, also in $t = 0$.

For $\Re(m) > 1$ the differential relations

$$s \circ Y_{m+1} = Y_m, \quad \text{hence } s^k \circ Y_{m+k} = Y_m,\ k \in \mathbb{N}, \text{ hold since}$$
$$\forall \varphi \in C_0^\infty :\ (s \circ Y_{m+1})(\varphi) = -Y_{m+1}(\varphi') = -\int_0^\infty \Gamma(m+1)^{-1} t^m \varphi'(t)dt$$
$$= -[\Gamma(m+1)^{-1} t^m \varphi(t)]_0^\infty + \int_0^\infty \Gamma(m+1)^{-1} m t^{m-1} \varphi(t)dt = 0 + Y_m(\varphi). \tag{13.8}$$

For arbitrary $m \in \mathbb{C}$ we define Y_m, cf. [1, p. 43, p. 174], by

$$Y_m := s^k \circ Y_{m+k} \in C^{-\infty}[0, \infty) \subset C_+^{-\infty} \text{ where } k \in \mathbb{N},\ \Re(m) + k > 1. \tag{13.9}$$

Due to (13.8) Y_m is well defined, i.e., does not depend on the choice of $k \in \mathbb{N}$. A proof as in (13.8) implies

$$\forall m \in \mathbb{C} \text{ with } \Re(m) > 0 :\ Y_m = s \circ Y_{m+1} = \left(\Gamma(m)^{-1} t^{m-1}\right)_{\mathbb{R}}. \tag{13.10}$$

Notice that for $0 < \Re(m) \le 1$ the function Y_m is not continuous on $\mathbb{R}$, but only locally integrable. An easy consequence of (13.9) and (13.8) is

$$s^k \circ Y_m = Y_{m-k},\ m \in \mathbb{C},\ k \in \mathbb{Z} \Longrightarrow Y_1 = s \circ Y_2 = s \circ s^{-2} \circ \delta = s^{-1} \circ \delta = Y. \tag{13.11}$$

The last equation suggested the notation Y_m. For $l \in \mathbb{N}$ we likewise get

$$\delta^{(l)} = s^l \circ \delta = s^{l+2} \circ (s^{-2} \circ \delta) = s^{l+2} \circ Y_2 = Y_{2-(l+2)} = Y_{-l},\ \delta = Y_0. \tag{13.12}$$

If $m \in \mathbb{C} \setminus (-\mathbb{N}), k \in \mathbb{N}, \Re(m) + k > 1, \varphi \in C_0^\infty$ and $\epsilon > 0$ with $\operatorname{supp}(\varphi) \subseteq [\epsilon, \infty)$, then

$$\begin{aligned} Y_m(\varphi) &= \int_\epsilon^\infty \Gamma(m)^{-1} t^{m-1} \varphi(t) \mathrm{d}t \text{ since} \\ Y_m(\varphi) &= (s^k \circ Y^{m+k})(\varphi) = (-1)^k Y_{m+k}(\varphi^{(k)}) \\ &= (-1)^k \int_0^\infty \Gamma(m+k)^{-1} t^{m+k-1} \varphi^{(k)}(t) \mathrm{d}t \\ &= (-1)^k \int_\epsilon^\infty \Gamma(m+k)^{-1} t^{m+k-1} \varphi^{(k)}(t) \mathrm{d}t \\ &= \cdots \underset{\text{partial integration}}{=} \int_\epsilon^\infty \Gamma(m)^{-1} t^{m-1} \varphi(t) \mathrm{d}t. \end{aligned} \tag{13.13}$$

For $m \notin -\mathbb{N},\ \Re(m) \le 0$, the distribution $Y_m \in C^{-\infty}[0, \infty)$ is called the *finite part* of $\left(\Gamma(m)^{-1} t^{m-1}\right)_{\mathbb{R}}$, cf. [1, (II, 2; 31)]. Notice that this function is not locally integrable on $\mathbb{R}$.

For $\Re(m) > 1,\ \Re(n) > 1$ and $t > 0$ we compute the convolution

$$\begin{aligned} (Y_m * Y_n)(t) &= \int_0^t \Gamma(m)^{-1} \Gamma(n)^{-1} (t-x)^{m-1} x^{n-1} \mathrm{d}x \\ &\underset{x=ty}{=} t^{m+n-1} \Gamma(m)^{-1} \Gamma(n)^{-1} \int_0^1 (1-y)^{m-1} y^{n-1} \mathrm{d}y \\ &= t^{m+n-1} \Gamma(m)^{-1} \Gamma(n)^{-1} B(m, n) \end{aligned} \tag{13.14}$$

where $B(m, n) := \int_0^1 (1-y)^{m-1} y^{n-1} \mathrm{d}y$ is Euler's *Beta function* that satisfies, cf. [3, p. 313]

$$\begin{aligned} & B(m, n) = \Gamma(m)\Gamma(n)\Gamma(m+n)^{-1},\ \Re(m) > 1,\ \Re(n) > 1 \\ & \underset{(13.14)}{\Longrightarrow} Y_m * Y_n = \left(t^{m+n-1} \Gamma(m+n)^{-1}\right)_{\mathbb{R}} = Y_{m+n}. \end{aligned} \tag{13.15}$$

The convolution equation $Y_m * Y_n = Y_{m+n}$ holds generally, cf. [1, (VI, 5; 7)], i.e.,

$$\begin{aligned} & \forall m, n \in \mathbb{C} :\ Y_m * Y_n = Y_{m+n} \text{ since} \\ & \text{with } k, l \in \mathbb{N},\ \Re(m) + k > 1,\ \Re(n) + l > 1 : \\ & Y_m * Y_n = (s^k \circ Y_{m+k}) * (s^l \circ Y_{n+l}) = s^{k+l} \circ (Y_{m+k} * Y_{n+l}) \\ & \quad = s^{k+l} \circ Y_{m+n+k+l} = Y_{m+n}. \end{aligned} \tag{13.16}$$

Due to

$$u^{(k)} = s^k \circ u = (s^k \circ \delta) * u = Y_{-k} * u,\ s^{-k} \circ u = Y_k * u,\ k \in \mathbb{N},\ u \in \mathrm{C}_+^{-\infty}$$

Schwartz [1, (VI, 5; 8)] defines the *fractional integral resp. differential operators*

$$\begin{aligned} &I^m u := Y_m * u \text{ resp. } D^m u := I^{-m} u = Y_{-m} * u,\ m \in \mathbb{C},\ u \in \mathrm{C}_+^{-\infty} \\ &\Longrightarrow D^m I^n = I^n D^m = I^{n-m} = D^{m-n},\ m, n \in \mathbb{C}, \end{aligned} \tag{13.17}$$

where I^m resp. D^m are interpreted as the m-th power with respect to convolution of the integral operator $I := I^1 = s^{-1} \circ = Y*$ with $(Iu)(t) = \int_{-\infty}^{t} u(x)\mathrm{d}x$ for $u \in \mathrm{C}_+^{0,\mathrm{pc}}$ resp. of the differential operator $D := D^1 = s\circ = \delta' * = \mathrm{d}/\mathrm{d}t$ with $Du = u'$. On $\mathrm{C}_+^{0,\mathrm{pc}}$ the operators I^m, and hence $D^m = I^{-m}$ act as

$$I^m u = s^k \circ \Gamma(m+k)^{-1} \int_{-\infty}^{t} (t-x)^{m+k-1} u(x)\mathrm{d}x,\ u \in \mathrm{C}_+^{0,\mathrm{pc}},\ \Re(m) + k > 1. \tag{13.18}$$

Here it suffices to choose $\Re(m) + k > 0$ since then $\left(t^{m+k-1}\right)_{\mathbb{R}}$ is still locally integrable on $\mathbb{R}$. If $u^{(k)} \in \mathrm{C}_+^{0,\mathrm{pc}}$ one additionally gets

$$\begin{aligned} I^m u &= (s^k \circ Y_{m+k}) * u = s^k \circ (Y_{m+k} * u) = Y_{m+k} * (s^k \circ u) = I^{m+k} u^{(k)} \\ &= \Gamma(m+k)^{-1} \int_{-\infty}^{t} (t-x)^{m+k-1} u^{(k)}(x)\mathrm{d}x \\ &= s^k \circ \Gamma(m+k)^{-1} \int_{-\infty}^{t} (t-x)^{m+k-1} u(x)\mathrm{d}x,\ \Re(m) + k > 0. \end{aligned} \tag{13.19}$$

This equation appears in connection with the *Maputo fractional derivative*, cf. [Wikipedia; https://en.wikipedia.org/wiki/Fractional_calculus, September 4, 2019]. Notice that, in general, $I^m u$ and $u^{(k)}$ are distributions for which $\int_{-\infty}^{t} (t-x)^{m+k-1} u^{(k)}(x)\mathrm{d}x$ is not defined. For instance, for $u := Y$ one obtains

$$D^m Y = Y_{-m} * Y_1 = Y_{-m+1} = I^{-m+k} Y^{(k)} = I^{-m+k} \delta^{(k-1)},\ -\Re(m) + k > 0. \tag{13.20}$$

Notice that $Y|_{[0,\infty)} = 1|_{[0,\infty)}$, but $Y \neq 1$ and $Y^{(k)} = \delta^{(k-1)} \neq 1^{(k)} = 0$.

In [Wikipedia; https://en.wikipedia.org/wiki/Fractional_calculus, September 4, 2019] further integral operators are defined:

$$\begin{aligned} &I_a^m u := I^m(uY_a) := s^k \circ \left(\Gamma(m+k)^{-1} Y_a(t) \int_a^t (t-x)^{m+k-1} u(x)\mathrm{d}x \right) \text{ where} \\ &a \in \mathbb{R},\ Y_a(t) := Y(t-a),\ \Re(m) + k > 0,\ u \in \mathrm{C}^{0,\mathrm{pc}}. \end{aligned} \tag{13.21}$$

Equation (13.16) implies the group homomorphism

$$\mathbb{C} \to \mathrm{U}\left(\mathrm{C}_+^{-\infty}, *\right),\ m \mapsto Y_m, \tag{13.22}$$

where $\mathrm{U}\left(\mathrm{C}_{+}^{-\infty},*\right)$ is the group of invertible elements or units of the convolution algebra $(\mathrm{C}_{+}^{-\infty},*)$, i.e., of $\mathrm{C}_{+}^{-\infty}$ with the convolution multiplication.

The integral representation $\Gamma(z)=\int_0^\infty t^{z-1}e^{-t}\mathrm{d}t,\ \Re(z)>0$, and the substitution $t=su,\ s>0$, easily imply

$$\begin{aligned}
&\forall m\in\mathbb{C} \text{ with } \Re(m)>1 \forall s>0:\\
&\mathcal{L}(Y_m)(s)=\Gamma(m)^{-1}\int_0^\infty t^{m-1}e^{-st}\mathrm{d}t=s^{-m}\\
&\Longrightarrow \forall s\in\mathbb{C} \text{ with } \Re(s)>0:\ \mathcal{L}(Y_m)(s)=s^{-m}:=e^{-m\ln(s)} \text{ where}\\
&\quad s=|s|e^{j\alpha},\ |s|>0,\ -\pi/2<\alpha<\pi/2,\ \ln(s)=\ln(|s|)+j\alpha.
\end{aligned}$$

The extension from $s>0$ to $\Re(s)>0$ follows by analytic continuation since the functions $\mathcal{L}(Y_m)(s)$ and s^{-m} are holomorphic for $\Re(s)>0$ and coincide on $(0,\infty)\subset\mathbb{R}$. The same equation holds for general $m\in\mathbb{C}$, i.e.,

$$\begin{aligned}
&\mathcal{L}(Y_m)(s)=s^{-m}=e^{-m\ln(s)},\ m\in\mathbb{C},\ \Re(s)>0,\ \text{since}\\
&\mathcal{L}(Y_m)(s)=\mathcal{L}(s^k\circ Y_{m+k})(s)=s^ks^{-(m+k)}=s^{-m},\ k\in\mathbb{N},\ \Re(m)+k-1>0.
\end{aligned}\tag{13.23}$$

Let $\mu>0$ be a fixed positive real number. The convolution equation $Y_{m\mu}*Y_{n\mu}=Y_{(m+n)\mu}$ implies that $\sum_{m\in\mathbb{Z}}\mathbb{C}Y_{m\mu}$ is a unital subalgebra of $\left(\mathrm{C}_{+}^{-\infty},*\right)$ with one element $\delta_0=Y_0$, and

$$\mathbb{C}[s,s^{-1}]\to\sum_{m\in\mathbb{Z}}\mathbb{C}Y_{m\mu},\ s^m\mapsto Y_{m\mu}=Y_\mu^m,\tag{13.24}$$

is an algebra epimorphism. In particular, $\mathrm{C}_{+}^{-\infty}$ is a $\mathbb{C}[s,s^{-1}]$-module with the scalar multiplication

$$\begin{aligned}
&P\circ_\mu y:=P(Y_\mu)*y=\sum_{m\in\mathbb{Z}}a_mY_{m\mu}*y=\sum_{m\in\mathbb{Z}}a_m\left(I^\mu\right)^m y=\sum_{m\in\mathbb{Z}}a_{-m}(D^\mu)^m y \text{ where}\\
&\quad P=\sum_{m\in\mathbb{Z}}a_ms^m\in\mathbb{C}[s,s^{-1}],\ y\in\mathrm{C}_{+}^{-\infty},\\
&P(Y_\mu)=\sum_{m\in\mathbb{Z}}a_mY_{m\mu}=P\circ_\mu\delta,\ P\circ_\mu y=(P\circ_\mu\delta)*y.
\end{aligned}\tag{13.25}$$

Recall that almost all a_m are zero.

An equation $P\circ_\mu y=u$ is called a *μ-fractional integral/differential equation* in the fractional integral resp. differential operator $I^\mu=Y_\mu*$ resp. $D^\mu=I^{-\mu}=Y_{-\mu}*$. The corresponding *fractional behaviors* or solution $\mathbb{C}[s,s^{-1}]$-submodules are defined as

$$\mathcal{B}:=\left\{w\in\left(\mathrm{C}_{+}^{-\infty}\right)^l,\ R\circ_\mu w=0\right\},\ R\in\mathbb{C}[s,s^{-1}]^{k\times l}.\tag{13.26}$$

Lemma 13.1.1 *The map* (13.24) *is an isomorphism, hence* $\sum_{m\in\mathbb{Z}}\mathbb{C}Y_{m\mu}=\bigoplus_{m\in\mathbb{Z}}\mathbb{C}Y_{m\mu}$.

Proof Assume $\sum_{m\geq m_0} a_m Y_{m\mu}=0$ where, of course, almost all a_m are zero. We have to show $a_m=0$ for all $m\geq m_0$. But

$$\begin{aligned}
&\forall m\geq m_0:\ (-m_0\mu+1)+m\mu=(m-m_0)\mu+1,\\
&\Longrightarrow Y_{-m_0\mu+1}*Y_{m\mu}=Y_{(m-m_0)\mu+1}=\left(\Gamma((m-m_0)\mu+1)^{-1}(t^{\mu})^{m-m_0}\right)_{\mathbb{R}}\\
&0=Y_{-m_0\mu+1}*\sum_{m\geq m_0}a_mY_{m\mu}=\left(\sum_{m\geq m_0}a_m\Gamma((m-m_0)\mu+1)^{-1}(t^{\mu})^{m-m_0}\right)_{\mathbb{R}}.
\end{aligned}$$

But the last sum is a polynomial function in $t^{\mu},\ t\geq 0$. Since it is zero, all coefficients and therefore all $a_m,\ m\geq m_0$, are zero as asserted. □

We are now going to considerably generalize fractional integral/differential equations by replacing the ring $\mathbb{C}[s,s^{-1}]$ by the field of Laurent series with a pole at 0. From Sects. 8.1.1 and 8.1.3 recall the principal ideal domain

$$\mathbb{C}[[s]]=\left(\mathbb{C}^{\mathbb{N}},*\right)\ni(a_n)_{n\in\mathbb{N}}=\sum_{n\in\mathbb{N}}a_ns^n=\sum_{n=0}^{\infty}a_ns^n \tag{13.27}$$

of *formal power series* and its subring of *locally (at 0) convergent* power series

$$\begin{aligned}
\mathbb{C}\langle s\rangle&=\left\{P=\sum_{n\in\mathbb{N}}a_ns^n\in\mathbb{C}[[s]];\ \limsup_{n\in\mathbb{N}}\sqrt[n]{|a_n|}<\infty\right\}=\\
&=\left\{P=\sum_{n\in\mathbb{N}}a_ns^n\in\mathbb{C}[[s]];\ \exists R>0\exists M>0\forall n\in\mathbb{N}:\ |a_n|\leq MR^n\right\}.
\end{aligned}\tag{13.28}$$

Recall that $\rho(P):=\left(\limsup_{n\in\mathbb{N}}\sqrt[n]{|a_n|}\right)^{-1}$ is the convergence radius of P, and that $P(s)$ is a holomorphic function in the disc $\{s\in\mathbb{C};\ |s|<\rho(P)\}$. Thus the coefficients of a locally convergent power series grow at most like a geometric sequence. A convergent power series $P=\sum_{n\in\mathbb{N}}a_ns^n$ is a unit or invertible in $\mathrm{U}\left(\mathbb{C}\langle s\rangle\right)$ if and only if $a_0\neq 0$. The ring $\mathbb{C}\langle s\rangle$ is a discrete valuation ring, i.e., a principal ideal domain with the unique prime s, up to association, cf. (8.3)–(8.9). The unique prime factor decomposition of $0\neq P=\sum_{n\in\mathbb{N}}a_ns^n$ is $P=us^k$ where $k=\mathrm{ord}_s(P):=\min\{n\in\mathbb{N};\ a_n\neq 0\}$ and $u=\sum_{n\in\mathbb{N}}a_{n+k}s^n$. The unique nonzero prime or maximal ideal of $\mathbb{C}\langle s\rangle$ is

$$\mathbb{C}\langle s\rangle_+=\mathbb{C}\langle s\rangle s=\left\{P=\sum_{n=1}^{\infty}a_ns^n\in\mathbb{C}\langle s\rangle\right\}. \tag{13.29}$$

The quotient field of $\mathbb{C}\langle s\rangle$ is the field $\mathbb{C}\langle\langle s\rangle\rangle$ of *Laurent series with at most a pole at 0*. It has the form

$$\mathbb{C}\langle\langle s\rangle\rangle = \left\{\sum_{n=k}^{\infty} a_n s^n;\ k \in \mathbb{Z}\right\} = \mathbb{C}[s^{-1}] \oplus \mathbb{C}\langle s\rangle_+ = \{0\} \uplus \cup_{k\in\mathbb{Z}} \mathrm{U}\left(\mathbb{C}\langle s\rangle\right) s^k. \tag{13.30}$$

In particular, every $H \in \mathbb{C}\langle\langle s\rangle\rangle$ can be written as

$$H = s^{-k} P \in \mathbb{C}\langle\langle s\rangle\rangle,\ k \in \mathbb{N},\ P \in \mathbb{C}\langle s\rangle_+. \tag{13.31}$$

Every nonzero Laurent series $H \in \mathbb{C}\langle\langle s\rangle\rangle$ has an even unique representation

$$H = s^k P \text{ with } k \in \mathbb{Z},\ P \in \mathrm{U}\left(\mathbb{C}\langle s\rangle\right),\ \mathrm{ord}(H) := \mathrm{ord}_s(H) := k. \tag{13.32}$$

The number $\mathrm{ord}_s(H)$ is then called the *order* of H.

We are going to define the expression $H(Y_\mu)$ for all $H \in \mathbb{C}\langle\langle s\rangle\rangle$, first for $H \in \mathbb{C}\langle s\rangle_+$, i.e., we substitute the distribution and locally integrable function $Y_\mu = \left(\Gamma(\mu)^{-1} t^{\mu-1}\right)_{\mathbb{R}}$ for the variable s of a Laurent series to obtain a distribution in $\mathrm{C}_+^{-\infty}$. Let

$$\begin{aligned} &P = \sum_{n=1}^{\infty} a_n s^n \in \mathbb{C}\langle s\rangle_+,\ N \in \mathbb{N},\ N > 0,\ P_N := \sum_{n=1}^{N} a_n s^n \\ &\Longrightarrow P_N(Y_\mu) = \sum_{n=1}^{N} a_n Y_\mu^n = \sum_{n=1}^{N} a_n Y_{n\mu} = \sum_{n=0}^{N-1} a_{n+1} Y_{(n+1)\mu} \\ &\quad = \left(t^{\mu-1} \sum_{n=0}^{N-1} a_{n+1} \Gamma((n+1)\mu)^{-1} (t^\mu)^n\right)_{\mathbb{R}}. \end{aligned} \tag{13.33}$$

The coefficients a_{n+1} grow at most like a geometric sequence or powers, whereas the $\Gamma((n+1)\mu)$ grow very fast according due to Stirling's formula (13.2). As an easy consequence we get

$$\begin{aligned} &\limsup_n \sqrt[n]{|a_{n+1}|\Gamma\left((n+1)\mu\right)^{-1}} = 0 \text{ and hence that} \\ &\widehat{P}(z) := \widehat{P}_\mu(z) := \sum_{n=0}^{\infty} a_{n+1} \Gamma((n+1)\mu)^{-1} z^n,\ z \in \mathbb{C}, \end{aligned} \tag{13.34}$$

is an everywhere convergent power series and an entire holomorphic function on $\mathbb{C}$. For the polynomial P_N from (13.33) this implies

$$P_N(Y_\mu) = \left(t^{\mu-1} \widehat{P_N}_\mu(t^\mu)\right)_{\mathbb{R}}, \tag{13.35}$$

where $\widehat{P_N} = \widehat{P_{N\mu}}$ is the partial sum of $\widehat{P} = \widehat{P}_\mu$ up to the power z^{N-1}. This suggests to *define* the locally integrable function and distribution

$$P(Y_\mu) = \left(t^{\mu-1}\widehat{P}_\mu(t^\mu)\right)_{\mathbb{R}}. \tag{13.36}$$

Since $\widehat{P} = \widehat{P}_\mu$ is entire with partial sums $\widehat{P_N} = \widehat{P_{N\mu}}$ the latter converge uniformly to $\widehat{P}$ on each bounded set of complex numbers, for instance, on discs $\{z \in \mathbb{C};\ |z| \le R\},\ R > 0$. This implies for all $T > 0$ that

$$\lim_{N\to\infty} \epsilon_P(N,T) = 0 \text{ where } \epsilon_P(N,T) := \max\left\{|(\widehat{P} - \widehat{P_N})(t^\mu)|;\ 0 \le t \le T\right\}. \tag{13.37}$$

For $\varphi \in \mathrm{C}_0^\infty$ with $\mathrm{supp}(\varphi) \subseteq [-T,T]$ and $\|\varphi\|_\infty := \max_{t\in\mathbb{R}} |\varphi(t)|$ this implies

$$\begin{aligned}
&P(Y_\mu)(\varphi) - P_N(Y_\mu)(\varphi) = \int_0^T t^{\mu-1}(\widehat{P} - \widehat{P_N})(t^\mu)\varphi(t)\mathrm{d}t \text{ and}\\
&|P(Y_\mu)(\varphi) - P_N(Y_\mu)(\varphi)| \le T^\mu \mu^{-1}\|\varphi\|_\infty \epsilon_P(N,T)\\
&\Longrightarrow \forall \varphi \in \mathrm{C}_0^\infty: \lim_{N\to\infty} P_N(Y_\mu)(\varphi) = P(Y_\mu)(\varphi), \text{ i.e., by definition,}\\
&\quad \lim_{N\to\infty} P_N(Y_\mu) = P(Y_\mu) \text{ in } \mathrm{C}_+^{-\infty}.
\end{aligned} \tag{13.38}$$

Note that the constant polynomial $P = 1 = 1s^0$ gives rise to $P(Y_\mu) = Y_{0\mu} = \delta$ that is not a locally integrable function, and hence the preceding arguments are not applicable. This is the reason for choosing $P \in \mathbb{C}\langle s\rangle_+$. The following lemma is decisive.

Lemma 13.1.2 *If $P, Q \in \mathbb{C}\langle s\rangle_+$ then $(PQ)(Y_\mu) = P(Y_\mu) * Q(Y_\mu)$.*

Proof Let

$$\begin{aligned}
&P = \sum_{n=1}^\infty a_n s^n,\ Q = \sum_{n=1}^\infty b_n s^n,\ PQ = \sum_{n=1}^\infty c_n s^n\\
&\Longrightarrow c_1 = 0,\ \forall n \ge 2:\ c_n = \sum_{i=1}^N a_i b_{N-i}.
\end{aligned} \tag{13.39}$$

Define $P_N, Q_N, \widehat{P}, \widehat{Q}, \widehat{P_N}, \widehat{Q_N}, (PQ)_N = \sum_{n=1}^N c_n s^n$ as in the preceding considerations. We use the ring isomorphism

$$\mathbb{C}[s, s^{-1}] \cong \oplus_{m\in\mathbb{Z}} \mathbb{C}Y_{m\mu},\ s^m \mapsto Y_{m\mu} = Y_\mu^m.$$

We consider the factor ring $\mathbb{C}\langle s\rangle/\mathbb{C}\langle s\rangle s^{N+1} = \oplus_{i=1}^N \mathbb{C}\overline{s}^i$ with its elements $\overline{P} := P + \mathbb{C}\langle s\rangle s^{N+1}$. Then

$$\overline{(PQ)_N} = \overline{PQ} = \overline{P} \cdot \overline{Q} = \overline{P_N} \cdot \overline{Q_N} = \overline{P_N Q_N} \Longrightarrow (PQ)_N - P_N Q_N \in \mathbb{C}\langle s\rangle s^{N+1}$$

$$\Longrightarrow (PQ)_N - P_N Q_N = -R_N,\ R_N := \sum_{k=1}^{N} c_{N,k} s^{N+k},\ c_{N,k} = \sum_{i=k}^{N} a_i b_{N-(i-k)}$$

$$\Longrightarrow (PQ)_N = P_N Q_N - R_N \Longrightarrow (PQ)_N(Y_\mu) = P_N(Y_\mu) * Q_N(Y_\mu) - R_N(Y_\mu).$$

Recall that

$$(PQ)(Y_\mu) = \lim_{N\to\infty} (PQ)_N(Y_\mu),\ P(Y_\mu) = \lim_{N\to\infty} P_N(Y_\mu),\ Q(Y_\mu) = \lim_{N\to\infty} Q_N(Y_\mu).$$

We are going to show

$$P(Y_\mu) * Q(Y_\mu) = \lim_{N\to\infty} P_N(Y_\mu) * Q_N(Y_\mu) \text{ and } \lim_{N\to\infty} R_N(Y_\mu) = 0$$
$$\Longrightarrow (PQ)(Y_\mu) = P(Y_\mu) * Q(Y_\mu).$$

It suffices to consider $N \geq N(\mu)$ with $N(\mu)$ from (13.3).
(i) $\lim_{N\to\infty} R_N(Y_\mu) = 0$: Let $\varphi \in \mathrm{C}_0^\infty$ with $\mathrm{supp}(\varphi) \subseteq [-T, T],\ T \geq 1$. Then

$$Y_{(N+k)\mu}(\varphi) = \Gamma((N+k)\mu)^{-1} \int_0^T t^{(N+k)\mu-1} \varphi(t)\mathrm{d}t$$
$$\Longrightarrow |Y_{(N+k)\mu}(\varphi)| \leq \|\varphi\|_\infty \Gamma((N+k)\mu)^{-1} \int_0^T t^{(N+k)\mu-1}\mathrm{d}t$$
$$= \|\varphi\|_\infty \Gamma((N+k)\mu)^{-1} T^{(N+k)\mu}((N+k)\mu)^{-1}$$
$$\underset{(13.3)}{\leq} \|\varphi\|_\infty \Gamma(N\mu)^{-1}(N\mu)^{-1}(T^{2\mu})^N.$$

Now recall that the a_n and b_n grow at most like powers, hence

$$\exists M_1 > 0, R > 1 \forall n:\ |a_n| \leq M_1 R^n,\ |b_n| \leq M_1 R^n$$
$$\Longrightarrow |c_{N,k}| = |\sum_{i=k}^{N} a_i b_{N-(i-k)}| \leq \sum_{i=k}^{N} M_1^2 R^i R^{N-(i-k)}$$
$$\leq N M_1^2 R^{N+k} \leq N M_1^2 (R^2)^N$$
$$\Longrightarrow |c_{n,k} Y_{((N+k)\mu)}(\varphi)| \leq M_1^2 (R^2)^N \|\varphi\|_\infty \Gamma(N\mu)^{-1} \mu^{-1} \left(T^{2\mu}\right)^N.$$

With $R_N(Y_\mu)(\varphi)| = \sum_{k=1}^N c_{N,k}(Y_\mu)^{n+k}(\varphi) = \sum_{k=1}^N c_{N,k} Y_{(N+k)\mu}(\varphi)$ and $M_2 := M_1^2\mu^{-1}$ we infer

$$|R_N(Y_\mu)(\varphi)| = \sum_{k=1}^{N} |c_{n,k} Y_{(n+k)\mu}(\varphi)| \leq M_2 \|\varphi\|_\infty N \left(R^2 T^{2\mu}\right)^N \Gamma(N\mu)^{-1}.$$

Since the $\Gamma(N\mu)$ grow much faster than polynomials or powers, we conclude

$$\lim_{N\to\infty} R_N(Y_\mu)(\varphi) = 0, \text{ i.e., by definition, } \lim_{N\to\infty} R_N(Y_\mu) = 0.$$

(ii) We have to show $P(Y_\mu) * Q(Y_\mu) = \lim_{N\to\infty} P_N(Y_\mu) * Q_N(Y_\mu)$: We write

$$\begin{aligned}
&P(Y_\mu) * Q(Y_\mu) - P_N(Y_\mu) * Q_N(Y_\mu)\\
&= \big(P(Y_\mu) - P_N(Y_\mu)\big) * Q(Y_\mu) + P(Y_\mu) * \big(Q(Y_\mu) - Q_N(Y_\mu)\big)\\
&- \big(P(Y_\mu) - P_N(Y_\mu)\big) * \big(Q(Y_\mu) - Q_N(Y_\mu)\big).
\end{aligned}$$

We have to show that all three summands of the right side converge to 0. We do this for the first summand only. Recall

$$P(Y_\mu) - P_N(Y_\mu) = \left(t^{\mu-1}(\widehat{P} - \widehat{P_N})(t^\mu)\right)_{\mathbb{R}}, \quad Q(Y_\mu) = \left(t^{\mu-1}\widehat{Q}(t^\mu)\right)_{\mathbb{R}}.$$

Consider $\epsilon(N,T) = \max\left\{|(\widehat{P} - \widehat{P_N})(t^\mu)|;\ 0 \le t \le T\right\}$ with $\lim_{N\to\infty} \epsilon(N,T) = 0$ from (13.37). For all $0 \le t \le T$ this implies

$$\begin{aligned}
&\left((P(Y_\mu) - P_N(Y_\mu)) * Q(Y_\mu)\right)(t)\\
&\qquad = \int_0^t (t-x)^{\mu-1}(\widehat{P} - \widehat{P_N})\left((t-x)^\mu\right) x^{\mu-1}\widehat{Q}(x^\mu)\mathrm{d}x\\
&\Longrightarrow \left|\left((P(Y_\mu) - P_N(Y_\mu)) * Q(Y_\mu)\right)(t)\right|\\
&\qquad \le \epsilon(N,T)\int_0^T (t-x)^{\mu-1}x^{\mu-1}|\widehat{Q}(x^\mu)|\mathrm{d}x.
\end{aligned}$$

Thus $\left((P(Y_\mu) - P_N(Y_\mu)) * Q(Y_\mu)\right)(t)$ converges to 0, uniformly on $[0,T]$, and this implies

$$\begin{aligned}
&\lim_{N\to\infty}\left((P(Y_\mu) - P_N(Y_\mu)) * Q(Y_\mu)\right)(\varphi) = 0\\
&\Longrightarrow \lim_{N\to\infty}(P(Y_\mu) - P_N(Y_\mu)) * Q(Y_\mu) = 0.
\end{aligned}$$

This concludes the proof of $(PQ)(Y_\mu) = P(Y_\mu) * Q(Y_\mu)$ for $P, Q \in \mathbb{C}\langle s\rangle_+$. □

The preceding lemma is the essential ingredient for the following theorem.

Theorem 13.1.3 *Let $\mu > 0$ be any positive real number.*

1. Write any $H \in \mathbb{C}\langle\langle s\rangle\rangle$ in the form

$$\begin{aligned}
&H = s^{-k_1}P_1,\ k_1 \in \mathbb{N},\ P_1 \in \mathbb{C}\langle s\rangle_+ \text{ and define}\\
&H(Y_\mu) := Y_\mu^{-k_1} * P_1(Y_\mu) = Y_{-k_1\mu} * P_1(Y_\mu).
\end{aligned}$$

Then $H(Y_\mu) \in \mathrm{C}_+^{-\infty}$ is well defined and

$$\mathbb{C}\langle\langle s\rangle\rangle \to \left(\mathrm{C}^{-\infty}, *\right), \; H \mapsto H(Y_\mu) = H(Y_\mu) * \delta, \tag{13.40}$$

is a ring monomorphism with the image $\mathcal{F}_{2,\mu} := \left\{H(Y_\mu);\; H \in \mathbb{C}\langle\langle s\rangle\rangle\right\}$. *Hence the latter is a (large) subfield of* $\left(\mathrm{C}_+^{-\infty}, *\right)$. *Note the special cases*

$$H = \sum_{m\in\mathbb{Z}} a_m s^m \in \mathbb{C}[s, s^{-1}] : \; H(Y_\mu) = \sum_{m\in\mathbb{Z}} a_m Y_\mu^m = \sum_{m\in\mathbb{Z}} a_m Y_{m\mu},$$

$$H = \sum_{m=1}^{\infty} a_m s^m \in \mathbb{C}\langle s\rangle_+ : \; H(Y_\mu) = \left(t^{\mu-1}\widehat{H}_\mu(t^\mu)\right)_{\mathbb{R}} \text{ with}$$

$$\widehat{H}_\mu(z) = \sum_{m=0}^{\infty} a_{m+1}\Gamma((m+1)\mu)^{-1} z^m.$$

2. *The space* $\mathrm{C}_+^{-\infty} = \mathrm{C}^{-\infty}(\mathbb{R}, \mathbb{C})_+$ *becomes a* $\mathbb{C}\langle\langle s\rangle\rangle$*-vector space with the scalar multiplication*

$$H \circ_\mu u := H(Y_\mu) * u \Longrightarrow H \circ_\mu \delta = H(Y_\mu). \tag{13.41}$$

For given nonzero $H = \sum_{m\in\mathbb{Z}} a_m s^m \in \mathbb{C}[s, s^{-1}] \subset \mathbb{C}\langle\langle s\rangle\rangle$ *and* $u \in \mathrm{C}_+^{-\infty}$ *the* μ*-fractional integral/differential equation*

$$H \circ_\mu y = \sum_{m\in\mathbb{Z}} a_m Y_{m\mu} * y = \sum_{m\in\mathbb{Z}} a_m (I^\mu)^m y = \sum_{m\in\mathbb{Z}} a_{-m}(D^\mu)^m y = u$$

has the unique solution $y := H^{-1} \circ_\mu u = H^{-1}(Y_\mu) * u$ *in* $\mathrm{C}_+^{-\infty}$.

3. (a) *Let* $0 \neq \lambda \in \mathbb{C}$, $k \in \mathbb{N}$, $k \geq 1$, $H := (1-\lambda s)^{-k} = 1 + H_+$. *Then the binomial series implies*

$$\begin{aligned}
(1-\lambda s)^{-k} &= \sum_{n=0}^{\infty} \tbinom{-k}{n}(-\lambda s)^n = 1 + \sum_{n=1}^{\infty} \tbinom{n+k-1}{k-1}\lambda^n s^n = 1 + H_+ \\
\Longrightarrow \widehat{H_+}(z)_\mu &= \sum_{n=0}^{\infty} \tbinom{n+k}{k-1}\Gamma((n+1)\mu)^{-1}\lambda^{n+1} z^n \\
\Longrightarrow H_+(Y_\mu) &= \left(t^{\mu-1}\widehat{H_+}_\mu(t^\mu)\right)_{\mathbb{R}} =: {}_\lambda Z_\mu^{(k)}, \\
(1-\lambda s)^{-k}(Y_\mu) &= H(Y_\mu) = \left(\delta - \lambda Y_\mu\right)^{-k} = \delta + {}_\lambda Z_\mu^{(k)}.
\end{aligned} \tag{13.42}$$

For $0 < \mu < 1$ *the function* ${}_\lambda Z_\mu^{(k)}$ *is locally integrable since* $\mu - 1 > -1$, *but not piecewise continuous.*

(b) *Consider $H \in \mathbb{C}(s)$ with its unique partial fraction decomposition*

$$\begin{aligned} H &= \sum_{m\in\mathbb{Z}} a_m s^m + \sum_{0\neq\lambda\in\mathrm{pole}(H)} \sum_{k=1}^{m_\lambda} a_{\lambda,k}(s-\lambda)^{-k} \\ &= \sum_{m\in\mathbb{Z}} a_m s^m + \sum_{0\neq\lambda\in\mathrm{pole}(H)} \sum_{k=1}^{m_\lambda} a_{\lambda,k}(-\lambda)^{-k}(1-\lambda^{-1}s)^{-k} \text{ where} \\ &1 \leq m_\lambda \in \mathbb{N},\ a_m, a_{\lambda,k} \in \mathbb{C},\ a_{\lambda m_\lambda} \neq 0,\ a_m = 0 \text{ for almost all } m. \end{aligned} \tag{13.43}$$

We use $s-\lambda = (-\lambda)(1-\lambda^{-1}s)$ here. Then the μ-impulse response $H(Y_\mu) = H \circ_\mu \delta$ of H resp. its action on $u \in C_+^{-\infty}$ are

$$\begin{aligned} H(Y_\mu) &= \sum_{m\in\mathbb{Z}} a_m Y_{m\mu} + \sum_{0\neq\lambda\in\mathrm{pole}(H)} \sum_{k=1}^{m_\lambda} a_{\lambda,k}(-\lambda)^{-k}(\delta-\lambda^{-1}Y_\mu)^{-k} \\ H \circ_\mu u = H(Y_\mu) * u &= \sum_{m\in\mathbb{Z}} a_m Y_{m\mu} * u \\ &+ \sum_{0\neq\lambda\in\mathrm{pole}(H)} \sum_{k=1}^{m_\lambda} a_{\lambda,k}(-\lambda)^{k}(\delta-\lambda^{-1}Y_\mu)^{-k} * u. \end{aligned} \tag{13.44}$$

Recall $(\delta-\lambda^{-1}Y_\mu)^{-1} = \delta + {}_{\lambda^{-1}}Z_\mu^{(k)}$, $Y_{m\mu} * u = (I^\mu)^m u$ and $Y_{-m\mu} * u = (D^\mu)^m u$ for $m \in \mathbb{N}$, and that for $0 < \mu < 1$ the function ${}_{\lambda^{-1}}Z_\mu^{(k)}$ is locally integrable, but not piecewise continuous.

4. *Let $H \in \mathbb{C}\langle\langle s\rangle\rangle$ and, if $H_+ \neq 0$,*

$$\begin{aligned} &N := \mathrm{ord}_s(H_+), \text{ i.e., } N \geq 1,\ a_1 = \cdots = a_{N-1} = 0,\ a_N \neq 0, \text{ and then} \\ &\mathrm{ord}_z(\widehat{H_{+\mu}}) = N-1, \\ &\widehat{H_{+\mu}}(z) = a_N\Gamma(N\mu)^{-1}z^{N-1} + a_{N+1}\Gamma((N+1)\mu)^{-1}z^N + \cdots, \\ &H_+(Y_\mu) = \left(t^{N\mu-1}(a_N\Gamma(N\mu)^{-1} + t^\mu a_{N+1}\Gamma((N+1)\mu)^{-1} + \cdots)\right)_{\mathbb{R}}. \end{aligned} \tag{13.45}$$

(a) *The map $H\circ_\mu$ induces an operator $H\circ_\mu = H(Y_\mu)* : C_+^{0,\mathrm{pc}} \to C_+^{0,\mathrm{pc}}$ if and only if $H \in \mathbb{C}\langle s\rangle$, i.e., $\mathrm{ord}_s(H) \geq 0$.*

(b) *The distribution $H(Y_\mu)$ is a locally integrable function if and only if $H \in \mathbb{C}\langle s\rangle_+$, i.e., $\mathrm{ord}_s(H) \geq 1$ or $H = H_+$.*

(c) *The distribution $H(Y_\mu)$ is a piecewise continuous (at 0) resp. continuous function if and only if $\mathrm{ord}_s(H)\mu \geq 1$ resp. $\mathrm{ord}_s(H)\mu > 1$.*

5. *If H is a rational function, i.e., $H \in \mathbb{C}(s)$, then*

$$H(s)\circ\delta = H(s^{-1})\circ_1\delta,\ H(s)\circ u = H(s^{-1})\circ_1 u,\ \mathrm{ord}_s\left(H(s^{-1})\right) = -\deg_s(H(s)). \tag{13.46}$$

Hence the present theorem can be applied to the vector space $\left({}_{\mathbb{C}(s)}\mathrm{C}_+^{-\infty}, \circ\right)$ *from Theorem 8.2.37, and item 4 implies Theorem 8.2.47,(2).*
If $H \in \mathbb{C}\langle\langle s\rangle\rangle \setminus \mathbb{C}(s)$, *then* $H(s^{-1})$ *is not contained in* $\mathbb{C}\langle\langle s\rangle\rangle$ *and* $H(s^{-1})(Y_1) = H(s^{-1}) \circ_1 \delta$ *is not defined.*

6. *If* H *is a rational function then the Laplace transform of* $H(Y_\mu) = H \circ_\mu \delta$ *is* $\mathcal{L}(H(Y_\mu)) = H(s^{-\mu})$. *This equation is of no significance in the theory of this section.*
7. *Items 1 and 2 are extended to a matrix* $H \in \mathbb{C}\langle\langle s\rangle\rangle^{p\times m}$ *and vectors* $u \in \left(\mathrm{C}_+^{-\infty}\right)^m$ *and* $y := H \circ_\mu u = H(Y_\mu) * u \in \left(\mathrm{C}_+^{-\infty}\right)^p$ *as usual. The equation*

$$y = H \circ_\mu u = H(Y_\mu) * u \text{ means: } \forall \alpha = 1, \ldots, p:\ y_\alpha = \sum_{\beta=1}^{m} H_{\alpha\beta}(Y_\mu) * u_\beta.$$

If, for instance, $(P, -Q) \in \mathbb{C}\langle\langle s\rangle\rangle^{p\times(p+m)}$, $\mathrm{rank}(P) = p$ *or* $\det(P) \neq 0$, *and* $H := P^{-1}Q \in \mathbb{C}\langle\langle s\rangle\rangle^{p\times m}$ *then for given* $u \in \left(\mathrm{C}_+^{-\infty}\right)^m$ *the unique solution of the system*

$$P \circ_\mu y = P(Y_\mu) * y = Q \circ_\mu u = Q(Y_\mu) * u \text{ is } y = H(Y_\mu) * u. \tag{13.47}$$

In other words, the map

$$\left(\mathrm{C}_+^{-\infty}\right)^m \underset{\mathbb{C}\langle\langle s\rangle\rangle}{\cong} \mathcal{B} := \left\{\binom{y}{u} \in \left(\mathrm{C}_+^{-\infty}\right)^{p+m};\ P \circ_\mu y = Q \circ_\mu u\right\},\ u \mapsto \binom{H\circ_\mu u}{u}, \tag{13.48}$$

is a $\mathbb{C}\langle\langle s\rangle\rangle$*-isomorphism. The solution space* $\mathcal{B}$ *is the most general form of a* fractional IO behavior.
If, in particular, $H_2 \in \mathbb{C}\langle\langle s\rangle\rangle^m$ *and* $u := H_2 \circ_\mu \delta = H_2(Y_\mu)$, *then* $y = (HH_2)(Y_\mu)$. *These equations enable the constructive and explicit solution of very general systems and, in particular, of* μ*-fractional integral/differential systems.*
8. *With the definition* $\mathrm{ord}_s(H) := \min\left\{\mathrm{ord}_s(H_{\alpha\beta});\ 1 \le \alpha \le p,\ 1 \le \beta \le m\right\}$ $(\in \mathbb{Z})$ *item 4 implies*

   (a) *The matrix* $H(Y_\mu)$ *induces an operator* $H(Y_\mu)* : \left(\mathrm{C}_+^{0,\mathrm{pc}}\right)^m \to \left(\mathrm{C}_+^{0,\mathrm{pc}}\right)^p$ *if and only if* $\mathrm{ord}_s(H) \ge 0$.
   (b) *The matrix* $H(Y_\mu)$ *of distributions has locally integrable function entries if and only if* $\mathrm{ord}_s(H) \ge 1$ *or* $H \in \mathbb{C}\langle s\rangle_+^{p\times m}$.
   (c) *The matrix* $H(Y_\mu)$ *of distributions has piecewise continuous resp. continuous entries if and only if* $\mathrm{ord}_s(H)\mu \ge 1$ *resp.* $\mathrm{ord}_s(H)\mu > 1$.

This applies in particular to the solution $(HH_2)(Y_\mu)$ *from item 6.*

Proof 1. (a) $H(Y_\mu)$ is well defined: Assume

$$H = s^{-k_1}P_1 = s^{-k_2}P_2,\ k_2 < k_1 \in \mathbb{N},\ P_1, P_2 \in \mathbb{C}\langle s\rangle_+ \Longrightarrow P_1 = s^{k_1-k_2}P_2$$
$$\underset{k_1-k_2>0,\ \text{Lemma (13.1.2)}}{\Longrightarrow} P_1(Y_\mu) = Y_\mu^{k_1} * Y_\mu^{-k_2} * P_2(Y_\mu)$$
$$Y_\mu^{-k_1} * P_1(Y_\mu) = Y_\mu^{-k_2} * P_1(Y_\mu) = H(Y_\mu).$$

(b) $1(Y_\mu) = \delta$, $(H_1H_2)(Y_\mu) = H_1(Y_\mu) * H_2(Y_\mu)$: obvious by Lemma 13.1.2.
(c) $(H_1 + H_2)(Y_\mu) = H_1(Y_\mu) + H_2(Y_\mu)$: Assume

$$H_i = s^{-k_i}P_i,\ i = 1, 2,\ k_i \geq 0,\ P_i \in \mathbb{C}\langle s\rangle_+ \Longrightarrow H_i = s^{-(k_1+k_2)}(s^{k_{3-i}}P_i)$$
$$\Longrightarrow (H_1 + H_2)(Y_\mu) = Y_\mu^{-(k_1+k_2)} * \left(Y_\mu^{k_2} * P_1(Y_\mu) + Y_\mu^{k_1} * P_1(Y_\mu)\right)$$
$$= Y_\mu^{-k_1} * P_1(Y_\mu) + Y_\mu^{-k_2} * P_2(Y_\mu) = H_1(Y_\mu) + H_2(Y_\mu).$$

(d) (i) Assume

$$P = \sum_{n=1}^{\infty} a_n s^n \in \mathbb{C}\langle s\rangle_+ \text{ and } P(Y_\mu) = \left(t^{\mu-1}\widehat{P}_\mu(t^\mu)\right)_{\mathbb{R}} = 0$$

$$\Longrightarrow \forall 0 < t < \infty:\ \widehat{P}_\mu(t^\mu) = 0 \underset{\widehat{P}_\mu \text{ entire}}{\Longrightarrow} \widehat{P}_\mu = \sum_{n=0}^{\infty} a_{n+1}\Gamma((n+1)\mu)^{-1}z^n = 0$$

$$\Longrightarrow \forall n \in \mathbb{N}:\ a_{n+1} = 0 \Longrightarrow P = 0.$$

(ii) Let

$$H = s^{-k}P,\ k \in \mathbb{N},\ P \in \mathbb{C}\langle s\rangle_+,\ H(Y_\mu) = Y_\mu^{-k} * P(Y_\mu) = 0$$
$$\Longrightarrow 0 = H(Y_\mu) * Y_\mu^k = P(Y_\mu) \underset{(i)}{\Longrightarrow} P = 0 \Longrightarrow H = s^{-k}P = 0.$$

2, 3. Obvious.
4. (a) (i) Let $H = a_0 + H_+ \in \mathbb{C}\langle s\rangle$ be a convergent power series. Then

$$H(Y_\mu) = a_0\delta + H_+(Y_\mu) = a_0\delta + H_+(Y_\mu),\ \ H_+(Y_\mu) \text{ locally integrable}$$
$$\underset{(13.6)}{\Longrightarrow} \forall u \in \mathrm{C}_+^{0,\mathrm{pc}}:\ H_+(Y_\mu) * u \in \mathrm{C}_+^0$$
$$\Longrightarrow H \circ_\mu u = a_0 u + H_+(Y_\mu) * u \in \mathrm{C}_+^{0,\mathrm{pc}}.$$

(ii) Assume that $u \in \mathrm{C}_+^{0,\mathrm{pc}}$ implies $H(Y_\mu) * u \in \mathrm{C}_+^{0,\mathrm{pc}}$ and that H is not a power series. Then $H = s^{-k}P$ with $k = -\operatorname{ord}_s(H) > 0$ and $P \in \mathrm{U}(\mathbb{C}\langle s\rangle)$. We infer

$$\forall u \in \mathrm{C}_+^{0,\mathrm{pc}} : H(Y_\mu) * u = Y_{-k\mu} * P(Y_\mu) * u \in \mathrm{C}_+^{0,\mathrm{pc}}$$
$$\underset{(i)}{\Longrightarrow} Y_{-k\mu} * u = P^{-1}(Y_\mu) * H(Y_\mu) * u \in \mathrm{C}_+^{0,\mathrm{pc}}$$
$$\Longrightarrow \text{for } u := Y_{(k-(1/2))\mu} \in \mathrm{C}_+^{0,\mathrm{pc}} : Y_{-k\mu} * Y_{(k-(1/2))\mu} = Y_{-\mu/2} \in \mathrm{C}_+^{0,\mathrm{pc}}.$$

This is a contradiction since $Y_{-\mu/2} \notin \mathrm{C}_+^{0,\mathrm{pc}}$.
(b) Both statements of (b) imply the equivalent conditions of (a), and hence $H = a_0 + H_+ \in \mathbb{C}\langle s\rangle$ and $H(Y_\mu) = a_0\delta + H_+(Y_\mu)$ where $a_0 = H(0)$ and $H_+(Y_\mu)$ is locally integrable. Then $H(Y_\mu)$ is locally integrable too if and only if $a_0\delta = 0$, i.e., $a_0 = 0$ or $H = H_+ \in \mathbb{C}\langle s\rangle_+$.
(c) Both statements of (c) imply the equivalent properties of (b). So assume

$$H = H_+ \neq 0, \text{ hence } H(Y_\mu) = H_+(Y_\mu),\ \mathrm{ord}_s(H) = \mathrm{ord}_s(H_+) = N \geq 1,$$
$$H(Y_\mu) = H_+(Y_\mu) \underset{(13.45)}{=} \left(t^{N\mu-1}(a_N\Gamma(N\mu)^{-1} + t^\mu a_{N+1}\Gamma((N+1)\mu)^{-1} + \cdots)\right)_{\mathbb{R}}.$$

This explicit form of $H(Y_\mu)$ shows that $H(Y_\mu)$ is piecewise continuous (at 0) resp. continuous if and only if

$$N\mu - 1 = \mathrm{ord}_s(H)\mu - 1 \geq 0 \text{ resp. } N\mu - 1 = \mathrm{ord}_s(H)\mu - 1 > 0.$$

5. Assume

$$H(s) = P(s)^{-1}Q(s),\ 0 \neq Q := a_m s^m + \cdots + a_0 \in \mathbb{C}[s],\ a_m \neq 0$$
$$0 \neq P := b_n s^n + \cdots + b_0 \in \mathbb{C}[s],\ b_n \neq 0,$$
$$\Longrightarrow \deg_s(H(s)) = \deg_s(Q) - \deg_s(P) = m - n.$$

Moreover this implies

$$H(s^{-1}) = P(s^{-1})^{-1}Q(s^{-1}),$$
$$Q(s^{-1}) = s^{-m}(a_m + \cdots + a_0 s^m) \in s^{-m}\mathrm{U}(\mathbb{C}\langle s\rangle),$$
$$P(s^{-1}) = s^{-n}(b_n + \cdots + b_0 s^m) \in s^{-n}\mathrm{U}(\mathbb{C}\langle s\rangle),\ H(s^{-1}) \in s^{n-m}\mathrm{U}(\mathbb{C}\langle s\rangle)$$
$$\Longrightarrow \mathrm{ord}_s\left(H(s^{-1})\right) = n - m = -(m-n) = -\deg_s(H(s))$$

and

$$P(s^{-1}) \circ_1 \delta = \sum_{i=0}^n b_i s^{-i} \circ_1 \delta = \sum_{i=0}^n b_i \delta^{(i)} = P(s) \circ \delta$$
$$\Longrightarrow P(s^{-1}) \circ_1 u = P(s) \circ u,\ P(s^{-1})^{-1} \circ_1 \delta = \left(P(s^{-1}) \circ_1 \delta\right)^{-1} = P(s)^{-1} \circ \delta$$
$$\Longrightarrow H(s^{-1})(Y_1) = H(s^{-1}) \circ_1 \delta = Q(s^{-1})P(s^{-1})^{-1} \circ_1 \delta$$
$$= Q(s^{-1}) \circ_1 P(s^{-1})^{-1} \circ_1 \delta = Q(s) \circ P(s)^{-1} \circ \delta = H(s) \circ \delta.$$

6. From (13.23) we know

$$\begin{aligned}
&\mathcal{L}(s^m \circ_\mu \delta) = \mathcal{L}(Y_{m\mu})(s) = s^{-m\mu} = \left(s^{-\mu}\right)^m = s^m(s^{-\mu})\\
&\Longrightarrow \forall Q = \sum_{m\in\mathbb{Z}} b_m s^m \in \mathbb{C}[s, s^{-1}] :\\
&\mathcal{L}(Q \circ_\mu \delta)(s) = \sum_{m\in\mathbb{Z}} b_m \mathcal{L}(s^m \circ_\mu \delta)(s) = \sum_{m\in\mathbb{Z}} b_m \left(s^{-\mu}\right)^m = Q(s^{-\mu}).
\end{aligned}$$

This implies for all

$$\begin{aligned}
H = P^{-1}Q \in \mathbb{C}(s) \text{ with } 0 \neq P \in \mathbb{C}[s, s^{-1}],\ Q \in \mathbb{C}[s, s^{-1}],\ PH = Q :\\
(P \circ_\mu \delta) * (H \circ_\mu \delta) = Q \circ_\mu \delta \Longrightarrow \mathcal{L}(P \circ_\mu \delta)\mathcal{L}(H \circ_\mu \delta) = \mathcal{L}(Q \circ_\mu \delta),\\
\mathcal{L}(H(Y_\mu))(s) = \mathcal{L}(H \circ_\mu \delta)(s) = \mathcal{L}(P \circ_\mu \delta)(s)^{-1}\mathcal{L}(Q \circ_\mu \delta)(s)\\
= P(s^{-\mu})^{-1}Q(s^{-\mu}) = H(s^{-\mu}).
\end{aligned}$$

7, 8. Obvious. □

We have thus shown that for any matrix $R = (P, -Q) \in \mathbb{C}\langle\langle s\rangle\rangle^{p\times(p+m)}$ with $\operatorname{rank}(P) = p$ or $\det(P) \neq 0$ the *μ-IO behavior or $\mathbb{C}\langle\langle s\rangle\rangle$-solution space*

$$\mathcal{B} := \left\{\binom{y}{u} \in \left(\mathrm{C}_+^{-\infty}\right)^{p+m};\ P \circ_\mu y = Q \circ_\mu u\right\} \cong \left(\mathrm{C}_+^{-\infty}\right)^m,\ \binom{H\circ_\mu u}{u} \leftrightarrow u, \quad (13.49)$$

can be explicitly determined. Notice that, by definition, all trajectories of these behaviors have left bounded support like the signals in connection with the Laplace transform of this book or those with left bounded support in electrical engineering. There are no autonomous behaviors since

$$P \in \mathrm{Gl}_p\left(\mathbb{C}\langle\langle s\rangle\rangle\right),\ P \circ_\mu y = 0 \Longrightarrow y = (P^{-1}P) \circ_\mu y = P^{-1} \circ_\mu (P \circ_\mu y) = 0.$$

In practice only rational exponents $\mu \in \mathbb{Q}$ are considered. The preceding theory can then be easily extended to finitely many rational indices $\nu > 0$ and distributions $H(Y_\nu)$, $H \in \mathbb{C}\langle\langle s\rangle\rangle$, and fractional operators $I^\nu = Y_\nu *$ and $D^\nu = Y_{-\nu}$. For this purpose consider positive rational numbers

$$\begin{aligned}
&\nu = r\mu,\ \nu, \mu \in \mathbb{Q},\ \mu > 0,\ r \in \mathbb{N}, r > 0\\
&\Longrightarrow Y_\nu = Y_{r\mu} = Y_\mu^r,\ \forall H \in \mathbb{C}\langle\langle s\rangle\rangle;\ H(Y_\mu),\ H(Y_\nu) \in \mathrm{C}_+^{-\infty}.
\end{aligned} \quad (13.50)$$

Consider the canonical algebra monomorphism

$$\mathbb{C}\langle\langle s\rangle\rangle \to \mathbb{C}\langle\langle s\rangle\rangle,\ H(s) = \sum_{m=m_0}^{\infty} a_m s^m \mapsto H(s^r) = \sum_{m=m_0}^{\infty} a_m s^{rm},\ m_0 \in \mathbb{Z}. \quad (13.51)$$

The following theorem and corollary are important since they enable the simultaneous treatment of finitely many operators I^{ν_i}, $0 < \nu_i \in \mathbb{Q}$.

Theorem 13.1.4 *For all $H \in \mathbb{C}\langle\langle s\rangle\rangle$ and $u \in C_{+}^{-\infty}$ the equations*

$$H(s)(Y_\nu) = H(s^r)(Y_\mu),\ \ H(s) \circ_\nu u = H(s^r) \circ_\mu u \tag{13.52}$$

hold.

Proof (i) Assume $H = \sum_{m\in\mathbb{Z}} a_m s^m \in \mathbb{C}[s, s^{-1}]$ where almost all a_m are zero. Then

$$\begin{aligned} H(s^r) = \sum_{m\in\mathbb{Z}} a_m s^{rm} \Longrightarrow H(s)(Y_\nu) &= \sum_{m\in\mathbb{Z}} a_m Y_{m\nu} = \sum_{m\in\mathbb{Z}} a_m Y_{mr\mu} \\ &= \left(\sum_{m\in\mathbb{Z}} a_m (s^r)^m\right)(Y_\mu) = H(s^r)(Y_\mu). \end{aligned}$$

(ii) Let

$$\begin{aligned} P &= \sum_{m=1}^{\infty} a_m s^m \in \mathbb{C}\langle s\rangle_+, \text{ hence} \\ \widehat{P}_\nu(z) &= \sum_{m=0}^{\infty} a_{m+1}\Gamma((m+1)\nu)^{-1} z^m = \sum_{m=1}^{\infty} a_m\Gamma(m\nu)^{-1} z^{m-1} \\ &= \sum_{m=1}^{\infty} a_m\Gamma(mr\mu)^{-1} z^{m-1}, \\ \widehat{P}_\nu(z^r) &= \sum_{m=1}^{\infty} a_m\Gamma(mr\mu)^{-1} z^{rm-r} = z^{1-r}\sum_{m=1}^{\infty} a_m\Gamma(mr\mu)^{-1} z^{rm-1} \\ P(s)(Y_\nu) &= \left(t^{\nu-1}\widehat{P}_\nu(t^\nu)\right)_{\mathbb{R}} = \left(t^{r\mu-1}\widehat{P}_\nu(t^{r\mu})\right)_{\mathbb{R}}. \end{aligned} \tag{13.53}$$

Notice that we use $\widehat{P}_\nu(z)$ with the index ν to distinguish the two cases with indices ν and μ. Also

$$Q(s) := P(s^r) = \sum_{m=1}^{\infty} a_m s^{rm} = \sum_{n=1}^{\infty} b_n s^n \text{ with } b_n := \begin{cases} a_m & \text{if } n = rm \in \mathbb{N}r \\ 0 & \text{if } n \notin \mathbb{N}r \end{cases}.$$

and then

$$\begin{aligned}\widehat{Q}_\mu(z) &= \sum_{n=0}^{\infty} b_{n+1}\Gamma((n+1)\mu)^{-1}z^n = \sum_{n=1}^{\infty} b_n\Gamma(n\mu)^{-1}z^{n-1}\\ &= \sum_{m=1}^{\infty} a_m\Gamma(mr\mu)^{-1}z^{rm-1} \underset{(13.53)}{=} z^{r-1}\widehat{P}_\nu(z^r),\end{aligned}\tag{13.54}$$

$$\begin{aligned}P(s^r)(Y_\mu) &= Q(Y_\mu) = \left(t^{\mu-1}\widehat{Q}_\mu(t^\mu)\right)_{\mathbb{R}} = \left(t^{\mu-1}t^{\mu(r-1)}\widehat{P}_\nu(t^{r\mu})\right)_{\mathbb{R}}\\ &= \left(t^{\mu r-1}\widehat{P}_\nu(t^{r\mu})\right)_{\mathbb{R}} = P(s)(Y_\nu).\end{aligned}$$

(iii) Let

$$\begin{aligned}&H = s^{-k}P \in \mathbb{C}\langle\langle s\rangle\rangle,\ k \in \mathbb{N},\ P \in \mathbb{C}\langle s\rangle_+ \Longrightarrow H(s^r) = s^{-kr}P(s^r)\\ &\Longrightarrow H(s)(Y_\nu) = Y_{-k\nu}P(s)(Y_\nu) = Y_{-kr\mu}P(s^r)(Y_\mu) = H(s^r)(Y_\mu).\end{aligned}$$

(iv)

$$H(s) \circ_\nu u = H(s)(Y_\nu) * u = H(s^r)(Y_\mu) * u = H(s^r) \circ_\mu u.$$

□

Corollary 13.1.5 *Consider positive rational numbers $\nu_i, i = 1, \dots, k$ and let m be any or the least common denominator of the ν_i such that $\nu_i = r_i m^{-1},\ r_i \in \mathbb{N},\ r_i > 0$. For $H \in \mathbb{C}\langle\langle s\rangle\rangle$ the preceding theorem implies*

$$H(s)(Y_{\nu_i}) = H(s^{r_i})(Y_{1/m}),\ \ H(s) \circ_{\nu_i} u = H(s^{r_i}) \circ_{1/m} u.$$

Therefore the operators $H(s)(Y_{\nu_i})$ and the actions $H(s) \circ_{\nu_i} u = H(s)(Y_{\nu_i}) * u$ can be expressed in the single vector space $\left({}_{\mathbb{C}\langle\langle s\rangle\rangle}C_+^{-\infty}, \circ_{1/m}\right)$ and Theorem 13.1.3 can be applied to this with the k different operators $H(s) \circ_{\nu_i}$.

This applies especially to fractional integral/differential equations in the finitely many operators

$$I^{\nu_i} = Y_{\nu_i} * = Y_{1/m}^{r_i} * = (I^{1/m})^{r_i} \text{ and } D^{\nu_i} = Y_{-\nu_i} * = Y_{1/m}^{-r_i} * = (D^{1/m})^{r_i}.$$

◇

The preceding situation can be reformulated. For any positive rational number μ we consider an indeterminate $s^{(\mu)}$ and the field of Laurent series $\mathbb{C}\langle\langle s^{(\mu)}\rangle\rangle$. We consider a new vector space

$$\begin{aligned}&\left({}_{\mathbb{C}\langle\langle s^{(\mu)}\rangle\rangle}C_+^{-\infty}, \circ_1\right) \text{ with } H(s^{(\mu)}) \circ_1 u := H(s) \circ_\mu u\\ &\Longrightarrow \left({}_{\mathbb{C}\langle\langle s^{(\mu)}\rangle\rangle}C_+^{-\infty}, \circ_1\right) \underset{\text{identification}}{=} \left({}_{\mathbb{C}\langle\langle s\rangle\rangle}C_+^{-\infty}, \circ_\mu\right).\end{aligned}\tag{13.55}$$

The identification is just a renaming of the indeterminates. Now let $\nu = r\mu$ as in Theorem 13.1.4. Then there is the field monomorphism

$$\begin{aligned}&\mathbb{C}\langle\langle s^{(\nu)}\rangle\rangle \to \mathbb{C}\langle\langle s^{(\mu)}\rangle\rangle,\ H(s^{(\nu)}) \mapsto H\left({s^{(\mu)}}^r\right)\\ \Longrightarrow\ &\mathbb{C}\langle\langle s^{(\nu)}\rangle\rangle \underset{\text{identification}}{\subset} \mathbb{C}\langle\langle s^{(\mu)}\rangle\rangle,\ s^{(\mu r)} = s^{(\nu)} = \left(s^{(\mu)}\right)^r \\ &H\left(s^{(\nu)}\right) \circ_1 u = H(s) \circ_\nu u = H\left(s^r\right) \circ_\mu u = H\left({s^{(\mu)}}^r\right) \circ_1 u.\end{aligned} \tag{13.56}$$

This implies that the action $\circ_1$ of $\mathbb{C}\langle\langle s^{(\mu)}\rangle\rangle$ on $\mathrm{C}_+^{-\infty}$ extends that of $\mathbb{C}\langle\langle s^{(\nu)}\rangle\rangle$.
We apply this to

$$\begin{aligned}&m = rn,\ 0 < m, n, r = m/n \in \mathbb{N},\ \nu := 1/n,\ \mu = 1/m\\ &\Longrightarrow \mathbb{C}\langle\langle s^{(1/n)}\rangle\rangle \subset \mathbb{C}\langle\langle s^{(1/m)}\rangle\rangle,\ s^{(1/n)} = \left(s^{(1/m)}\right)^{m/n}\\ &\quad\ \mathbb{C}\langle\langle s^{(1}\rangle\rangle \subset \mathbb{C}\langle\langle s^{(1/m)}\rangle\rangle,\ s^{(1)} = \left(s^{(1/m)}\right)^m\\ &H\left(s^{(1/n)}\right) \circ_1 u = H(s) \circ_{1/n} u = H\left(s^r\right) \circ_{1/m} u = H\left({s^{(1/m)}}^{m/n}\right) \circ_1 u.\end{aligned} \tag{13.57}$$

The equation $s^{(1)} = \left(s^{(1/m)}\right)^m$ shows that $s^{(1/m)}$ is an m-th root of $s^{(1)}$. This suggests to write

$$\begin{aligned}&s := s^{(1)},\ s^{1/m} := s^{(1/m)} \Longrightarrow s^{1/n} = \left(s^{1/m}\right)^{m/n} \text{ if } n|m \text{ or } m/n \in \mathbb{N}\\ &\Longrightarrow \mathbb{C}\langle\langle s^{1/n}\rangle\rangle \subset \mathbb{C}\langle\langle s^{1/m}\rangle\rangle \text{ if } m/n \in \mathbb{N}.\end{aligned} \tag{13.58}$$

Corollary and Definition 13.1.6 *The inclusions from* (13.57) *imply that the union*

$$\mathbb{C}\langle\langle s^{1/\infty}\rangle\rangle := \bigcup_{m=1}^{\infty} \mathbb{C}\langle\langle s^{1/m}\rangle\rangle \text{ with } s^{1/n} = \left(s^{1/m}\right)^{m/n} \text{ if } m/n \in \mathbb{N} \tag{13.59}$$

is a field with the addition and multiplication from the $\mathbb{C}\langle\langle s^{1/m}\rangle\rangle$. *The distribution space* $\mathrm{C}_+^{-\infty}$ *is a* $\mathbb{C}\langle\langle s^{1/\infty}\rangle\rangle$*-vector space with the scalar multiplication*

$$H(s^{1/m}) \circ_1 u := H(s^{(1/m)}) \circ_1 u = H(s) \circ_{1/m} u = H(Y_{1/m}) * u. \tag{13.60}$$

The field $\mathbb{C}\langle\langle s^{1/\infty}\rangle\rangle$ *is called the field of convergent* Puiseux series *and is the algebraic closure of* $\mathbb{C}\langle s\rangle$, *cf. [Wikipedia; https://en.wikipedia.org/wiki/Puiseux_series, September 8, 2019] or [4, Sect. 3.1] for a more detailed discussion. The space* $\left({}_{\mathbb{C}\langle\langle s^{1/\infty}\rangle\rangle}\mathrm{C}_+^{-\infty}, \circ_1\right)$ *is only a different form for the countably many* $\left({}_{\mathbb{C}\langle\langle s\rangle\rangle}\mathrm{C}_+^{-\infty}, \circ_{1/m}\right)$. *Constructive computations always take place in the latter spaces by means of Theorem 13.1.3. In particular, multivariable fractional integral/differential systems with a finite number of differential operators* D^μ *with rational* μ *can thus be solved constructively.* ◊

Example 13.1.7 1.

$$\begin{aligned} &s^{1/2} = \left(s^{1/6}\right)^3, \ s^{1/3} = \left(s^{1/6}\right)^2, \ H \in \mathbb{C}\langle\langle s\rangle\rangle \\ &\Longrightarrow H(s^{1/2}) \circ_1 u = H(s) \circ_{1/2} u = H(s^3) \circ_{1/6} u = H(s^3)(Y_{1/6}) * u, \\ &\quad H(s^{1/3}) \circ_1 u = H(s) \circ_{1/3} u = H(s^2) \circ_{1/6} u = H(s^2)(Y_{1/6}) * u. \end{aligned} \tag{13.61}$$

2. Let $0 \neq m \in \mathbb{N}$, $\lambda \in \mathbb{C}$ and $H \in \mathbb{C}\langle\langle s\rangle\rangle$. The simple differential equation

$$D^{1/m} y - \lambda y = H(Y_{1/m}) \text{ or } (s^{-1} - \lambda) \circ_{1/m} y = H \circ_{1/m} \delta$$

has the solution

$$y = (s^{-1} - \lambda)^{-1} H \circ_{1/m} \delta = \left(s(1 - \lambda s)^{-1} H\right)(Y_{1/m}).$$

Since $\mathrm{ord}_s\left(s(1-\lambda s)^{-1}H\right) = 1 + \mathrm{ord}_s(H)$ the solution y is piecewise continuous if and only if

$$(1 + \mathrm{ord}_s(H))m^{-1} \geq 1 \text{ or } \mathrm{ord}_s(H) \geq m - 1.$$

3. We complete the example from Sect. 2.5 und consider the special case

$$\begin{aligned} &(D^{1/2} - \lambda_1)y = e^{\lambda_2 t} Y, \ \lambda_1 \neq 0, \ \lambda_2 \neq 0, \ \lambda_3^2 = \lambda_2 \neq \lambda_1^2, \ \mu = 1/2, \text{ or} \\ &(s^{-1} - \lambda_1) \circ_{1/2} y = (s - \lambda_2)^{-1} \circ \delta = (s^{-1} - \lambda_2)^{-1} \circ_1 \delta = (s^{-2} - \lambda_2)^{-1} \circ_{1/2} \delta. \end{aligned}$$

of item 2. This has the solution

$$\begin{aligned} &y = (s^{-1} - \lambda_1)^{-1}(s^{-2} - \lambda_2)^{-1} \circ_{1/2} \delta = H(Y_{1/2}) \text{ where} \\ &H := (s^{-1} - \lambda_1)^{-1}(s^{-2} - \lambda_2)^{-1} = s^3(1 - \lambda_1 s)^{-1}(1 - \lambda_3 s)^{-1}(1 + \lambda_3 s)^{-1}. \end{aligned}$$

Since $\mathrm{ord}_s(H) = 3$ and $\mathrm{ord}_s(H)\mu = 3/2 > 1$ the solution y is continuous. The partial fraction decomposition has the form

$$\begin{aligned} &s^3(1 - \lambda_1 s)^{-1}(1 - \lambda_3 s)^{-1}(1 + \lambda_3 s)^{-1} \\ &\quad = a + b(1 - \lambda_1 s)^{-1} + c(1 - \lambda_3 s)^{-1} + d(1 + \lambda_3 s)^{-1} \text{ with} \\ &a, b, c, d, \in \mathbb{C}, \ a + b + c + d \underset{s=0}{=} 0, \ a \underset{s=\infty}{=} (\lambda_1 \lambda_2)^{-1}. \end{aligned}$$

The constants b, c, d are computed by

$$s^3(1-\lambda_3 s)^{-1}(1+\lambda_3 s)^{-1} = b + (1-\lambda_1 s)\cdot\left(a + c(1-\lambda_3 s)^{-1} + d(1+\lambda_3 s)^{-1}\right)$$

$$\begin{aligned}\Longrightarrow \text{ for } s=\lambda_1^{-1} &: b = \lambda_1^{-3}(1-\lambda_1^{-1}\lambda_3)^{-1}(1+\lambda_1^{-1}\lambda_3)^{-1} = \left(\lambda_1(\lambda_1^2-\lambda_2)\right)^{-1}\\ \text{for } s=\lambda_3^{-1} &: c = \lambda_3^{-3}(1-\lambda_1\lambda_3^{-1})^{-1}(1+\lambda_3\lambda_3^{-1})^{-1} = (2\lambda_2(\lambda_3-\lambda_1))^{-1}\\ \text{for } s=-\lambda_3^{-1} &: d = -\lambda_3^{-3}(1+\lambda_1\lambda_3^{-1})^{-1}(1+\lambda_3\lambda_3^{-1})^{-1} = (-2\lambda_2(\lambda_3+\lambda_1))^{-1}.\end{aligned}$$

Then (2.84) furnishes the solution

$$\begin{aligned} y &= a\delta + b\left(\delta + {}_{\lambda_1}Z_{1/2}^{(1)}\right) + c\left(\delta + {}_{\lambda_3}Z_{1/2}^{(1)}\right) + d\left(\delta + {}_{(-\lambda_3)}Z_{1/2}^{(1)}\right)\\ &= b\,{}_{\lambda_1}Z_{1/2}^{(1)} + c\,{}_{\lambda_3}Z_{1/2}^{(1)} + d\,{}_{(-\lambda_3)}Z_{1/2}^{(1)}\end{aligned}$$

with the ${}_{\lambda}Z_{\mu}^{(k)}$ from (13.42). Recall that in standard electrical engineering most input signal components with left bounded support lie in $\left(\oplus_{\lambda\in\mathbb{C}}\mathbb{C}[t]e^{\lambda t}\right)Y$ and are therefore finite $\mathbb{C}$-linear combinations of the

$$\begin{aligned}\tfrac{t^{k-1}}{(k-1)!}e^{\lambda t}Y &= (s-\lambda)^{-k}\circ\delta = (s^{-1}-\lambda)^{-k}\circ_1\delta = s^k(1-\lambda s)^{-k}\circ_1\delta\\ &= Y^{*k} * \left(\delta + {}_{\lambda}Z_1^{(k)}\right) \in \mathcal{F}_{2,1/\infty},\ k\in\mathbb{N},\ k\geq 1,\ \lambda\in\mathbb{C}.\end{aligned}$$

◊

References

1. L. Schwartz, Théorie des distributions. Publications de l'Institut de Mathématique de l'Université de Strasbourg, No. IX-X. Hermann, Paris, nouvelle édition, entiérement corrigée, refondue et augmentée edition, 1966
2. A. Kochubei, Y. Luchko, *Handbook of Fractional Calculus with Applications. Vol. 1: Basic Theory, Vol. 2: Fractional Differential Equations* (De Gruyter, Berlin, 2019)
3. H. Kneser, *Funktionentheorie* (Vandenhoek and Rupprecht, Göttingen, 1958)
4. H. Bourlès, B. Marinescu, U. Oberst, Weak exponential stability of linear time-varying differential behaviors. Linear Algebra Appl. **486**, 523–571 (2015)

Correction to: Linear Time-Invariant Systems, Behaviors and Modules

Correction to:
U. Oberst et al., *Linear Time-Invariant Systems, Behaviors and Modules*, Differential-Algebraic Equations Forum, https://doi.org/10.1007/978-3-030-43936-1

The original version of the book was published with few errors in Chaps. 8 and 10. These have been corrected in the updated version.

The updated versions of these chapters can be found at
https://doi.org/10.1007/978-3-030-43936-1_8
https://doi.org/10.1007/978-3-030-43936-1_10

U. Oberst et al., *Linear Time-Invariant Systems, Behaviors and Modules*, Differential-Algebraic Equations Forum,
https://doi.org/10.1007/978-3-030-43936-1_14

Index

Symbols

U. Oberst et al., *Linear Time-Invariant Systems, Behaviors and Modules*, Differential-Algebraic Equations Forum,
https://doi.org/10.1007/978-3-030-43936-1

A

B

Printed in the United States
by Baker & Taylor Publisher Services

Printed in the United States
by Baker & Taylor Publisher Services